8th International Conference on Physics and its Applications (ICOPIA 2016)

Journal of Physics: Conference Series Volume 776

Denpasar, Indonesia
23 – 24 August 2016

Editor:

Budi Purnama

ISBN: 978-1-5108-3531-3
ISSN: 1742-6588

Printed from e-media with permission by:

Curran Associates, Inc.
57 Morehouse Lane
Red Hook, NY 12571

Some format issues inherent in the e-media version may also appear in this print version.

Printed by Curran Associates, Inc. (2017)

For permission requests, please contact the Institute of Physics
at the address below.

Institute of Physics
Dirac House, Temple Back
Bristol BS1 6BE UK

Phone: 44 1 17 929 7481
Fax: 44 1 17 920 0979

techtracking@iop.org

Additional copies of this publication are available from:

Curran Associates, Inc.
57 Morehouse Lane
Red Hook, NY 12571 USA
Phone: 845-758-0400
Fax: 845-758-2633
Email: curran@proceedings.com
Web: www.proceedings.com

Table of contents

Volume 776

8th International Conference on Physics and its Applications (ICOPIA)
23–24 August 2016, Denpasar, Indonesia

Accepted papers received: 21 November 2016
Published online: 21 December 2016

Preface

Papers

Keynote Speaker

Solar Cell

Material Physics and Characterization

Acoustic

Theoritical Physics

Instrumentation Physics

Geophysics

Journal of Physics: Conference Series **776** (2016) 011001 doi:10.1088/1742-6596/776/1/011001

Proceeding of 8th International Conference of Physics and Its Applications (8th ICOPIA) 2016

Preface

International Conference of Physics and Its Applications (ICOPIA) is a biannual conference, started in 2001, for presenting and discussing current research in Physics and Its Applications. This year of 2016, this conference is the eighth one. Studies about material physics, optics, geophysics, instrumentation, magnetics and theoretical physics are presented either orally or by poster. This conference is mainly supported by Department of Physics, University of Sebelas Maret, Indonesia.

The keynote presentations are provided especially to show the contribution of physicists to medical research. A study of pT resolution magnetic sensor utilizing MI element towards medical use is presented by Prof. Tsuyoshi Uchiyama, from Nagoya University, Japan. Presentation about computational intelligence technique for electroencephalography (EEG) analysis is delivered by Assoc. Prof. Lipo Wang from Nanyang Technological University, Singapore. Keynote presentation about polycapillary optic and its application in medical imaging which is provided by Prof. Carolyn Ann Macdonald, from University At Albany, USA. Understanding the synergic and competitive interaction between Polyvinyl alcohol and plasticizers onto Na-Bentonite Clay is presented by Khairuddin, Ph.D, from University of Sebelas Maret, Indonesia.

This volume contains the papers presented in ICOPIA 2016. The papers are divided into eight sections: magnetism and magnetic material, material physics and characterization, acoustics, theoretical physics, instrumentation physics, optics and geophysics. This structure is made so that readers are easier to find an article in this proceeding.

We would like to thank to all of the participants attending this conference and also to committee for their contribution to this high level conference and its overall success. We also would like to thank to the reviewers for their positive contribution to maintain the quality of the articles presented in this conference. We also appreciate the contribution of our sponsor supporting the conference.

Editor in Chief

Dr. Eng. Budi Purnama

8th International Conference of Physics and Its Applications (8th ICOPIA) 2016

General Chair

Nuryani, M.Si., Ph.D. (Department of Physics, University of Sebelas Maret University, Indonesia)

Organizer

Department of Physics, University of Sebelas Maret, Indonesia

Technical Supports

Kentingan Physics Forum (KPF)
Indonesian Physical Society (HFI)

Advisory Board

Prof. Sutarno, M.Sc., PhD., University of Sebelas Maret
Prof. Muhammad Mad Salleh, University Kebangsaan Malaysia
Prof. Dr. Saiful Amri Mazlan, Universiti Teknologi Malaysia, Malaysia
Prof. Dr. Ir. Benjamin Soenarko, Institut Teknologi Bandung, Indonesia
Prof. Suparmi, M.Sc., Ph.D, University of Sebelas Maret
Prof. Cari, M.Sc. PhD, University of Sebelas Maret
Prof. Ari Handono Ramelan, University of Sebelas Maret
Prof. Warsito, University of Lampung
Dr. Rifai Chai, University of Technology Sydney, Australia
Dr. Mohd Yazed, University of Malaya, Malaysa.

Organizing Committee

Fahru Nurosyid, Sebelas Maret University, Indonesia
Yofentina Iriani, Sebelas Maret University, Indonesia
Risa Suryana, Sebelas Maret University
Agus Supriyanto, Sebelas Maret University, Indonesia
Darsono, Sebelas Maret University, Indonesia
Sorja Koesuma, Sebelas Maret University, Indonesia
Kusumandari, Sebelas Maret University, Indonesia
Harjana, Sebelas Maret University, Indonesia
Fuad Anwar, Sebelas Maret University, Indonesia
Iwan Yahya, Sebelas Maret University, Indonesia
Mohtar Yunianto Sebelas Maret University, Indonesia
Utari, Sebelas Maret University, Indonesia
Artono Dwijo Sutomo, Sebelas Maret University, Indonesia
Hery Purwanto, Sebelas Maret University, Indonesia
Riyatun, Sebelas Maret University, Indonesia
Suharyana, Sebelas Maret University, Indonesia

Khairuddin, Sebelas Maret University, Indonesia
Budi Legowo, Sebelas Maret University, Indonesia
Ahmad Marzuki, Sebelas Maret University
Usman Santosa, Sebelas Maret University, Indonesia
Darmanto, Sebelas Maret University, Indonesia

Reviewer

Prof. Dr. Ir. Akhmad Herman Yuwono, M.Phil. Eng., University of Indonesia (UI)
Dr. Eng. Edi Suharyadi, M.Eng., University of Gadjah Mada (UGM)
Dr. Eng. Rudi W Prastianto, Institut Teknologi Sepuluh Nopember (ITS)
Dr. Eng. Sugeng Wahyudi, M.Eng. Khusyu University Japan
Dr. Eng. Jamhir Syafani, M.Si., University of Halu Oleo (UHO)
Iman Santoso, Ph.D., University of Gadjah Mada (UGM)
Damar Yoga Kusuma, Ph.D. University of Ahmad Dahlan (UAD)
Dede Djuhana, Ph.D dari UI
Nuryani, Ph.D., Sebelas Maret University (UNS)
Dr. Eng. Budi Purnama, Sebelas Maret University (UNS)
Dr. Fahru Nurosyid, Sebelas Maret University (UNS)
Dr. Yofentina Iriani, Sebelas Maret University (UNS)
Dr. Eng. Risa Suryana, Sebelas Maret University (UNS)
Dr. Agus Supriyanto, Sebelas Maret University (UNS)
Sorja Koesuma, M.Si., Sebelas Maret University (UNS)
Dr. Eng. Kusumandari, Sebelas Maret University (UNS)
Dr. Fuad Anwar, Sebelas Maret University (UNS)
Iwan Yahya, M.Si. Sebelas Maret University (UNS)
Mohtar Yunianto, M.Si. Sebelas Maret University (UNS)
Khairuddin, Ph.D. Sebelas Maret University (UNS)
Ahmad Marzuki, Sebelas Maret University (UNS)
Suharyana, Sebelas Maret University, Indonesia (UNS)

Peer review statement

All papers published in this volume of *Journal of Physics: Conference Series* have been peer reviewed through processes administered by the proceedings Editors. Reviews were conducted by expert referees to the professional and scientific standards expected of a proceedings journal published by IOP Publishing.

Journal of Physics: Conference Series **776** (2016) 012001 doi:10.1088/1742-6596/776/1/012001

Polycapillary optics for medical applications

C A MacDonald[1] and J C Petruccelli

Physics Department, University at Albany, SUNY, Albany, NY 12222, USA
[1] Corresponding author: cmacdonald@albany.edu

Abstract. Polycapillary optics can be designed for a wide variety of x-ray applications, including x-ray fluorescence for non-destructive materials analysis and diffraction applications such as protein crystallography. Of particular interest are medical applications, including the removal of Compton scattering with the resultant improvement in contrast and resolution in mammography, the production of monochromatic parallel beams for high-contrast imaging in clinical settings and the localization of radioactive tracers. A recent development is the use of polycapillary optics to improve beam coherence for x-ray phase imaging. Conventional radiographic techniques depend on attenuation, which provides only low contrast between soft tissues. Phase imaging can yield significantly higher contrast but requires spatially coherent beams. Conventional sources small enough to produce high coherence are necessarily low power, requiring long exposures. Polycapillary optics employed to create a small secondary source from high power sources have been shown to yield strong phase effects. Polycapillary optics can also be combined with other x-ray optics. A combination of a toroidally bent crystal with a polycapillary optic has been shown to produce enhanced resolution in a magnified image. Another x-ray imaging technique which does not rely on the small attenuation contrast is the capture of coherently scattered radiation, for which polycapillary optics could be used as angular filters to collect the beam, resulting in a map of tissue or material type.

1. Principles of polycapillary optics

Polycapillary optics are arrays of a large number of small hollow glass tubes. X rays are guided down these curved and tapered tubes by multiple reflections in a manner analogous to the way optical fibers guide light. The x rays can be transmitted down a hollow tube as long as the tube is small enough, and bent gently enough, to keep the angles of incidence less than the critical angle for total reflection, θ_c [1,2,3,4]. The critical angle for borosilicate glass falls with photon energy E as

$$\theta_c \approx \frac{30\ \text{keV}}{E}\ \text{mrad}, \tag{1}$$

and is about 0.1° for 17 keV photons. The angles are somewhat larger for leaded glass [5].

As shown in Figure 1, when a ray enters a curved tube on the side towards the center of curvature, the angle of incidence increases with tube diameter. The requirement that the incident angles remain less than the critical angle necessitates the use of small channel sizes, typically between 2 and 50 μm. Because of the mechanical limitations of producing thin-walled tubes that small, the optics are produced by pulling large diameter glass tubes to create small diameter tubes, stacking and pulling them together, and repeating, as for the example fiber shown in Figure 2.

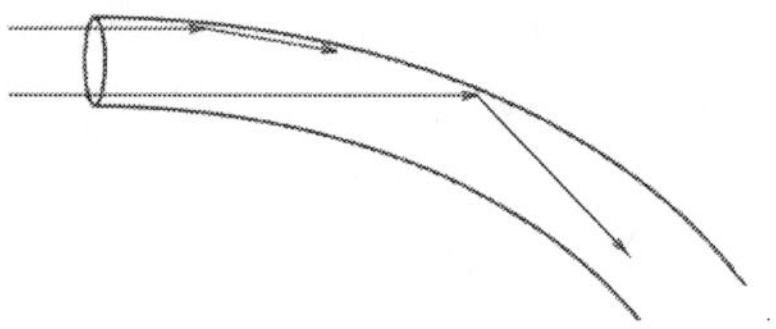

Figure 1. X rays traveling in a bent capillary tube. The trajectory of the ray entering at the top at grazing incidence is projected onto the page, but in three dimensions may "toboggan" in a constant radius spiral. The x ray entering at the bottom (closest to the center of curvature) strikes at a larger angle.

Figure 2. Cross sectional scanning electron micrograph of a polycapillary fiber with 0.55 mm outer diameter and 50 μm diameter channels.

Figure 3. Angular filter composed of multiple polycapillary fibers. The length of the optic is 30 mm.

2. Geometries and Applications

2.1. Angular Filters
Because of the small critical angle for reflection, polycapillary optics provide much higher angular resolution than, for example, anti-scatter grids for radiography, gamma camera collimators for nuclear medicine, or the soller slits for diffraction. A straight angular filter is shown in Figure 3.

2.1.1. *Scatter Reduction.* Compton scattering can result in substantial image degradation in radiography. In a conventional medical imaging system, scatter is partially removed by inserting an anti-scatter grid consisting of a radiotransparent material, such as graphite, embedded with lead ribbons parallel to the incoming beam. Typical anti-scatter grids have primary (unscattered beam) transmissions of 60-70% and scatter transmissions of around 20%. By comparison, the scatter transmission of a polycapillary optic, measured by moving the source off axis to an angle much larger

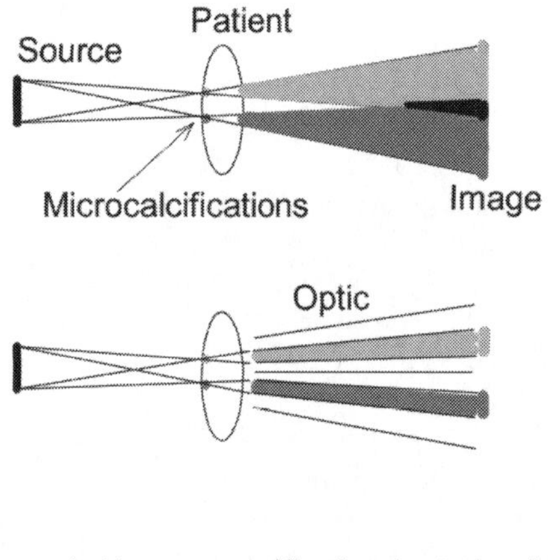

Figure 4 Air gap magnification (top) showing degradation of image due to finite source size. Magnification with a long tapered polycapillary optic (bottom) showing no increased blurring after the exit plane of the patient.

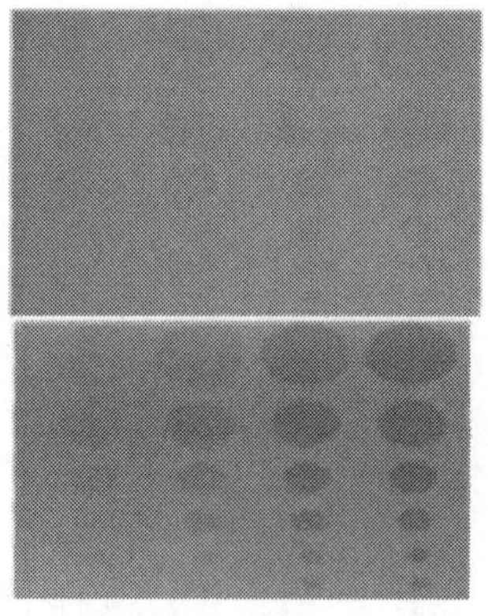

Figure 5. Image without (top) and with (bottom) a polycapillary optic in lieu of an antiscatter grid. Holes in the plastic block range from small to large (bottom to top) and shallow to deep (left to right).

Journal of Physics: Conference Series **776** (2016) 012001 doi:10.1088/1742-6596/776/1/012001

than the critical angle, is less than 1% at 20 keV [6].

Polycapillary optics can also be used to magnify and demagnify images. Magnification can improve image resolution, particularly if the resolution is detector-limited, as is typically the case in digital imaging. Magnification is conventionally performed by increasing the air gap between the patient and the detector, but this will result in a loss of resolution due to the geometrical blurring from the finite source size, as shown in the top part of Figure 4. However, magnification using polycapillary optics, even with a large focal spot, would not be subject to a loss in resolution. Using a magnifying polycapillary optic increased the limiting spatial frequency of the measured modulation transfer functions (MTF) for CR plates by a factor of 1.8. [6] The resolution was not degraded by the capillary structure, which was on a smaller scale (20 μm channel size) than the desired resolution. The optic yields an MTF increase at all spatial frequencies, which may be diagnostically more significant than the increase in limiting MTF. The simultaneous scatter rejection resulted in a contrast enhancement of more than a factor of 2, as demonstrated in Figure 5.[7] Measurements at 40 keV showed a contrast enhancement of more than a factor of 4.[8] Lead glass optics performed well at higher energies.[5] Magnifying capillary optics provide simultaneous contrast enhancement and resolution increase.

2.1.2. Scintigraphy

Collimators are required in nuclear medicine because without them, the rays emitted from the internal radioactive source would create an undifferentiated blur on the detector, as shown in Figure 6(a).[9] Even with the collimator, scatter creates an obfuscating blur, as shown in Figure 6(b), so bulky and relatively low resolution energy sensitive gamma cameras are normally employed to provide discrimination against the Compton scatter. High resolution digital detectors could be employed instead, but only if the signal from the radioactive material is concentrated in small enough areas so that the intensity is larger than the background from the scatter. Scatter discrimination then occurs because the scatter intensity will be spread fairly uniformly across millions of pixels. However, maintaining a high signal requires that the collimator have a resolution on the order of the pixel size of the detector, to avoid dividing the signal over many pixels as well. Typical collimators used in nuclear medicine have resolutions on the order of few millimeters, which is inadequate for this purpose.

Pairing a polycapillary optic such as that shown in Figure 3 with a radiographic detector provided 100-200 μm resolution.[10] Images taken with a parallel 30-mm-thick lead glass polycapillary parallel hole collimator and a computed radiography plate demonstrated the ability to image features of individual ion exchange beads within ^{125}I brachytherapy seeds even in the presence of scattering from 30 mm of tissue-equivalent scatter material, as shown Figure 7.[11] Calculations and measurements

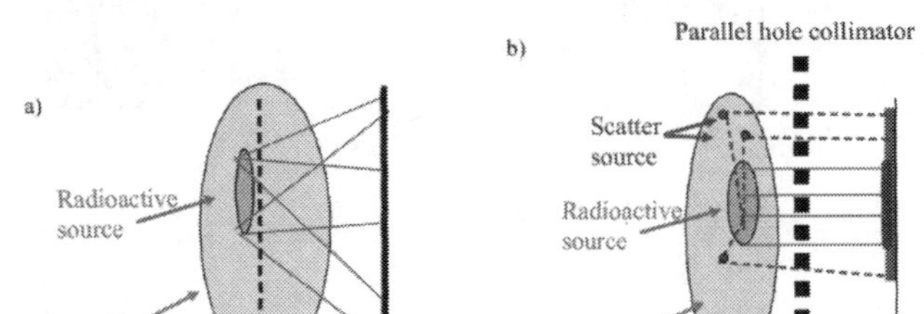

Figure 6. (a) Without a collimator the image is a set of overlapping blurs, with no spatial information. The blur increases with the distance of the detector from the object plane (b) Radiographic detectors are not normally used in nuclear medicine because their failure to discriminate against scatter results in little spatial information in the image.

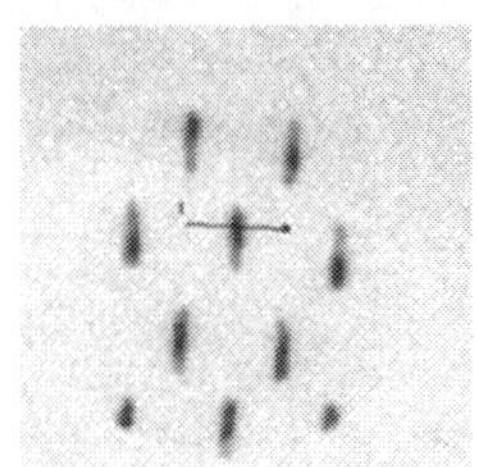

Figure 7. Image of an array brachytherapy seeds with a signal-to-noise ratio of 98.

showed that for highly heterogeneous radiation distributions high signal-to-background ratios could be achieved by using the high resolution to "dilute" the scatter background without energy discrimination. In addition, the high resolution collimator/radiographic detector system is substantially more compact than typical gamma camera systems. [12]

2.1.3. Coherent Scatter

Conventional radiographic techniques depend on attenuation, which provides only low contrast between soft tissues. An alternative x-ray imaging technique which does not rely on the small attenuation contrast is the capture of the coherently scattered radiation, as cancerous and healthy tissues are known to have different coherent scatter signatures. For example, the fat scatter peak is due to the average distance between fatty acid molecules of d = 0.45 nm, which for the Mo Kα characteristic emission at 17.5 keV, corresponds to a peak in the fat scatter at 9°.[13,14,15] The carcinoma scatter peak is at 13°. In addition to peaking at a different scatter angle, the scatter cross section is also higher for carcinoma than for normal tissue.[16,17,18,19] Differences are also seen in collagen alignment, which causes a diminishment of peaks at smaller momentum transfer values.

Rather than a system which requires raster scanning and full analysis of the coherent scatter pattern at each position, a potentially promising design employs the intensity of scatter at a fixed characteristic coherent scatter angle as a measure of the presence of that tissue type. A sketch of the selection of part of the coherent scatter ring with an angular filter (in this case an antiscatter grid) is shown in Figure 8. A schematic of the experimental system is shown in Figure 9. Coherent scatter images were obtained from a system using an antiscatter grid to select the desired peak with an exposure as low as 100-250 mAs. An example image from a checkerboard phantom is shown in Figure 10. Replacing the antiscatter grid, which has an angular acceptance of 6°, with a polycapillary angular filter would improve the angular resolution by an order of magnitude, and allow discrimination between fibroglandular tissue and carcinoma. The reduced intensity due to the higher

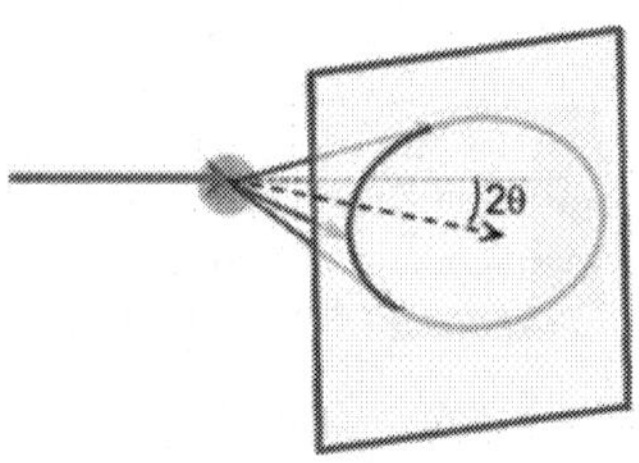

Figure 8. Coherent scatter with an angular filter tilted to the characteristic angle. The dashed black arrow is the grid normal. Only the dark part of the ring passes through the grid.

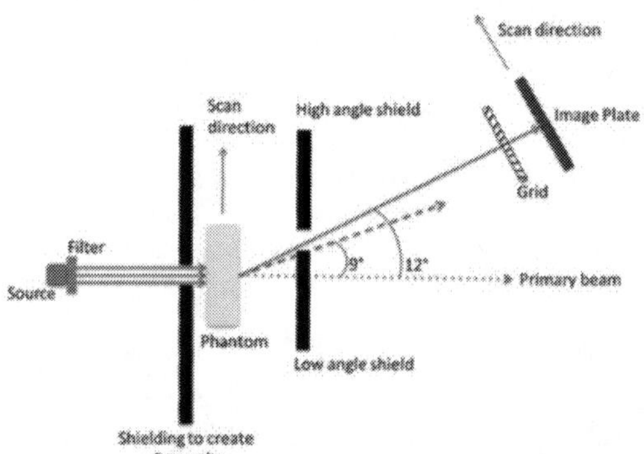

Figure 9. Schematic of the experimental setup. In the final system, a detector would be placed to capture a conventional mammogram from the primary beam. In this test system, only a single detector was employed. In a clinical slot scan system, the patient (phantom) is fixed and the source and slot are scanned past the patient, but the geometry is otherwise very similar. Replacing the grid in the test system with a polycapillary angular filter would greatly improve the angular resolution.

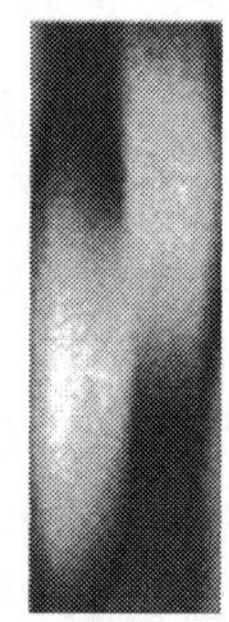

Figure 10. Coherent scatter image of a checkerboard phantom consisting of alternating pieces of polycrystalline graphite and fatty tissue.

angular selectivity would normally require longer exposure times, but the time (and dose) could be reduced by employing a detector with larger pixels. For example, if the coherent scatter is used to highlight regions of the original mammogram, it would be sensible to assume that no region less than 1 mm^2 would be useful. Increasing the pixel size in the coherent scatter image from 200 μm to 1 mm would result in a reduction in required dose of a factor of 25. The pixel size of the final coherent scatter system could be adjusted so that no additional dose is required beyond that required by the simultaneous conventional mammogram. Polycapillary optics could be used as angular filters to collect the beam, resulting in a map of tissue or material type within the image.

2.2. Collimating Optics

Collimating optics are produced by creating longer polycapillary fibers than for angular filters. The final pull is then designed to create a section with the desired shape, from which the ends are cut away, as shown in Figure 11.

2.2.1. Alignment

A typical set up for aligning a polycapillary optic is shown in Figure 12.[20] Either the source or the optic is translated perpendicular to the optic axis in small steps, producing a measurement of intensity versus relative source position, as shown in Figure 13.[21] The plot is symmetric and Gaussian, which indicates good alignment of the source, optic and detector. In order to determine the focal distance of the optic, source scans are performed at different distances from the source. At the focal distance, the ratio of the width of the scan curve to the optic-to-source distance, called the source scan angle, should be the smallest, as shown in Figure 14.[20] Transmission is the ratio of the number of photons passing through the optic to the number incident on the front face of the optic. Transmission with respect to the source-optic distance is also shown in Figure 14. The highest transmission and the lowest source

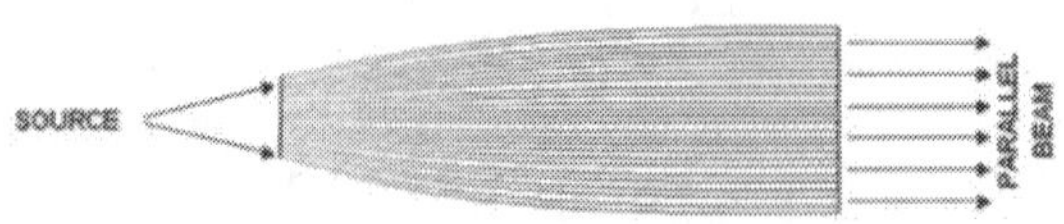

Figure 11. Sketch of the interior channels of a monolithic polycapillary collimating optic. The source is normally placed at the point where lines from the channels converge, and the output is then a nearly parallel.

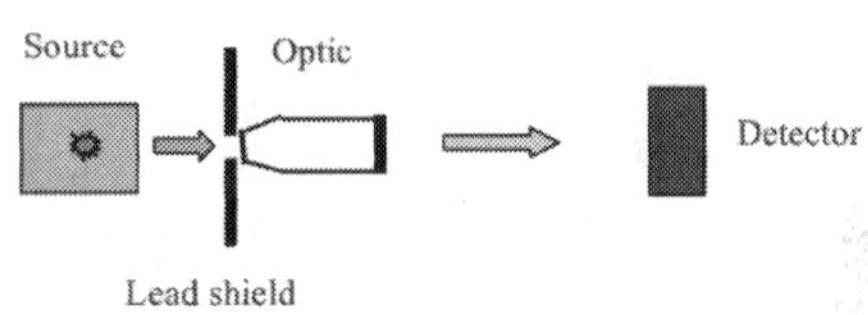

Figure 12. Set up for optics measurement.

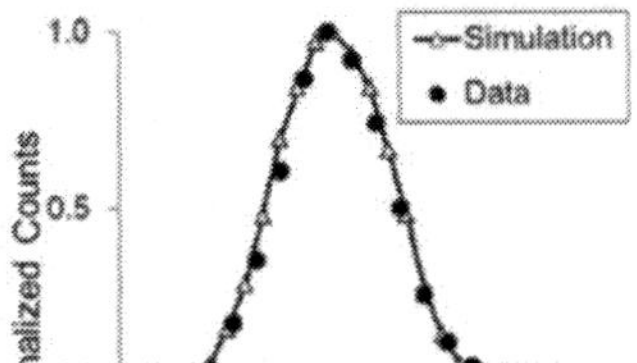

Figure 13. Source scan at 17.5 keV of a collimating optic.

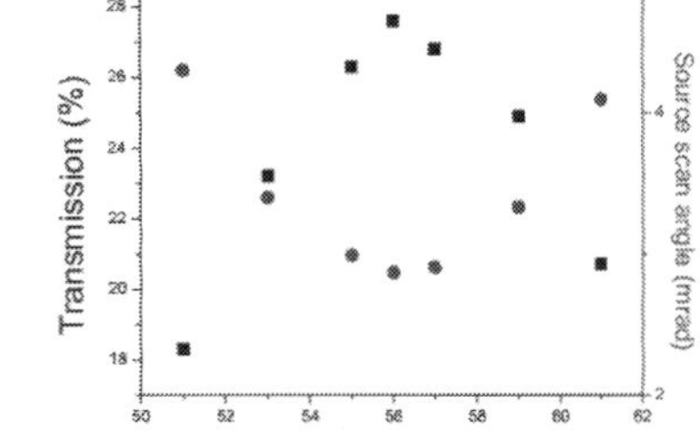

Figure 14 Transmission (■) and source angle (●) with respect to the source optic. The transmission is the highest at the highest at the designed focal distance of 56 mm.

angle occurred at the focal distance.

2.2.2. Divergence

As shown in Figure 15, the output from a multifiber polycapillary collimating optic has both global divergence, α, and local divergence, β. Even if the fibers are parallel (α=0), the output divergence, β, is not zero, but is determined by the critical angle and therefore the x-ray energy. The exit divergence from capillary optics is measured by rotating a high quality crystal in the beam and measuring the angular width of a Bragg peak. Since the Darwin width or crystal mosaicity, and the angular spread due to inherent energy width of the Bragg peak, are both typically much smaller than the exit divergence from the optic, the measurement yields the optic divergence directly.[22] The typical local divergence, β, for collimating optics is ~1.3-1.5 θ_c. The factor is an experimentally determined parameter that arises from the fact that most of the beam has a divergence less than the maximum divergence of $2\theta_c$ produced by reflection at the critical angle. Unlike the case for pinhole collimation, the local divergence of the beam does not depend on the source size. Thus larger, higher power sources may be used without adversely affecting the resolution of the measurement. The maximum useful x-ray source spot size is limited by the acceptance area of the optic.

2.2.3. Simulations

Simulations of polycapillary optics performance are important for understanding defect structure and optimizing and predicting performance for new applications. These are typically based on ray tracing, and must take into account both the complex geometry of the optics, with hundreds of thousands of channels, and the deviation from ideality. Microscopic fluctuations of the channel walls are called roughness, and have the effect of reducing the reflectivity from the value associated with a perfectly flat surface. A typical roughness model assumes a correlation between surface deviations with a correlation length of a few microns, and a roughness height Z. The effect of roughness on photon transmission is more significant when incident angles are high or when the photon bounces many times. Waviness is the midrange slope of the channel walls. The exact geometry of the surface tilt of the polycapillary channel walls is unknown, but the distribution of tilt angles has been found to be Gaussian, characterized with a width σ.[23] The probability distribution must take into account that the photon is more likely to impact a wall tilted toward it than one tilted away. The third length scale usually considered is unintentional bending of straight channels (for example along the central axis of the lens). Results of many different optics give reasonable agreement with an unintentional

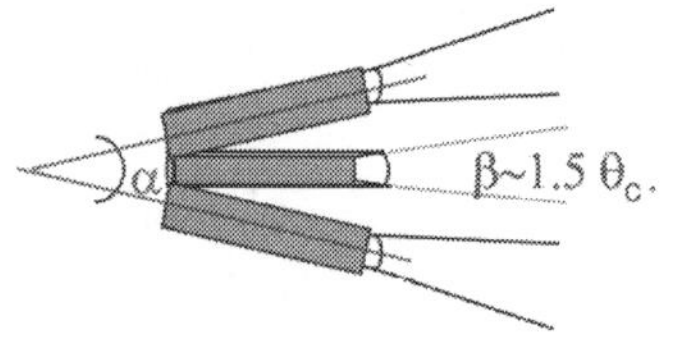

Figure 15. The output divergence from a multifiber collimating optic is characterized by global and local divergence. The minimum global divergence is proportional to the critical angle.

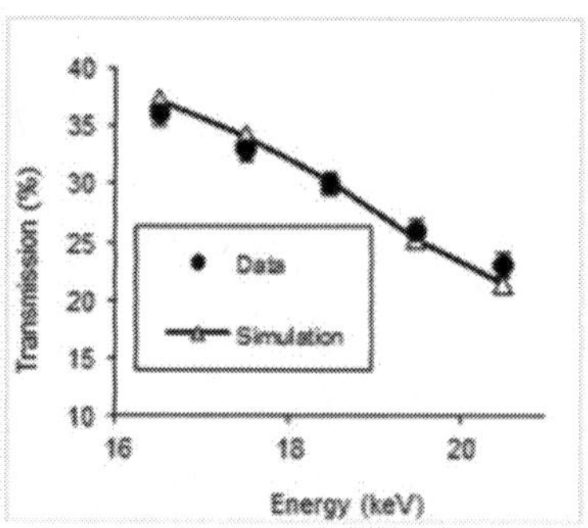

Figure 16: Simulation of the transmission versus energy of a collimating optic compared to experimental data

circular curvature of 225 m, a waviness σ of 0.125 mrad and a roughness height Z of 0.5 nm with a correlation length of 6 μm.

Figure 16 shows the simulated and experimental transmission vs energy for an optic with a focal length of 56 mm. Using the above fixed parameters, the only fitting parameter was the fractional open area (the fraction of the front face which is not glass walls).[21] The best fit fractional open area was 51%, within a typical range of 40-70%. Figure 13 shows a simulated vs experimental source scan at photon energy 17.5 keV for the same optic. For this source scan, the source size was not measured when the experimental data was collected and a best-fit size of 105 μm was used by the simulation. The agreement is quite good for both independent sets of data, with only two adjustable parameters. The simulation tracks the position and velocity of each exiting photon. More recent modeling includes tracking the accumulated phase.[24]

2.2.4. Monochromatic Imaging

As mentioned above, in conventional radiography, subject contrast arises from relatively low differences in absorption coefficients between different tissue types. The already low subject contrast is further reduced in a conventional system by averaging over relatively large energy bandwidths. Synchrotron measurements using monochromatic beams have demonstrated higher contrast, but synchrotrons are not clinically available. Using monochromator crystals with a conventional source without an optic is not practical because the low intensity of the diffracted beam will not allow imaging *in vivo* before motion blur occurs. Polycapillary collimating optics can allow sufficient diffracted beam intensity to make clinical monochromatic imaging possible without a synchrotron.[25] Measurements at 17.5 keV showed subject contrast enhancement of a factor of 2, as illustrated in Figure 17, in agreement with theoretical calculations. This contrast enhancement is in addition to that from the reduction of scattered radiation.

2.2.5. Polarization

The beam from a polycapillary optic is polychromatic and unpolarized. After diffraction from a monochromatizing crystal, the beam can be polarized, as shown in Figure 18, if the diffraction angle is approximately 2θ=90°, so that the new propagation direction is along one of the possible directions of

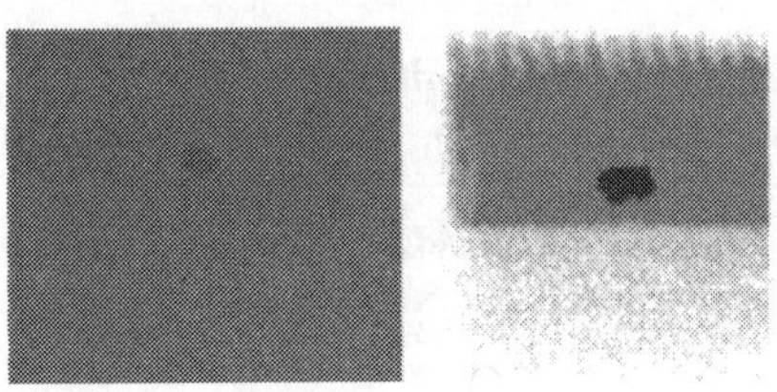

Figure 17. X-ray image of plastic with 6.6 mm step, obtained with a broadband (left) and monochromatic beam (right). The phantom was thin on top and thick on the bottom, with a hole near the middle. The step is readily apparent with the monochromatic beam.

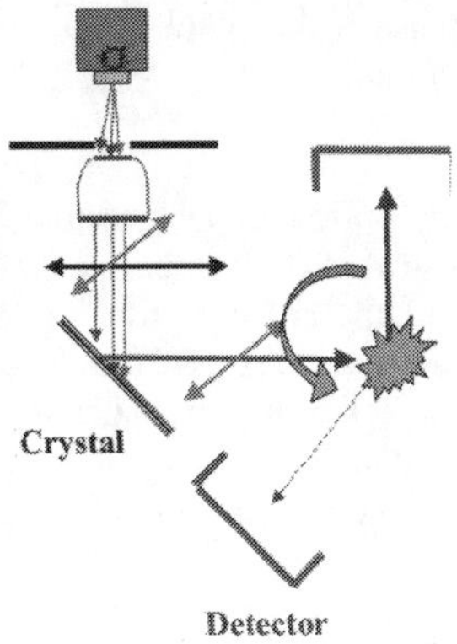

Figure 18. System to measure beam polarization.

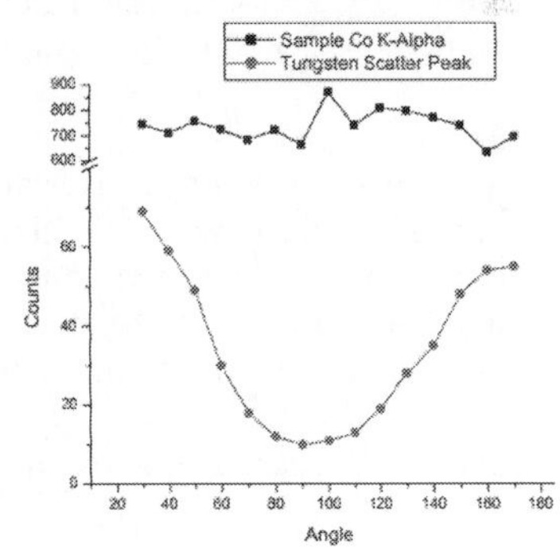

Figure 19. The cobalt fluorescence emission is constant with ϕ angle, indicating good system alignment, while the scatter peak drops near ϕ=90°, indicating a good degree of polarization.

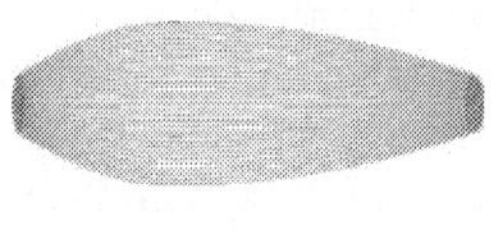

Figure 20. Sketch of the interior channels of a monolithic polycapillary optic with a short input and longer output focal length.

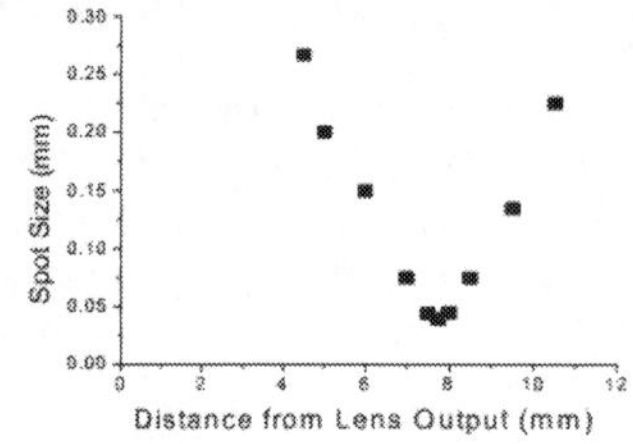

Figure 21. Spot size vs distance from optic.

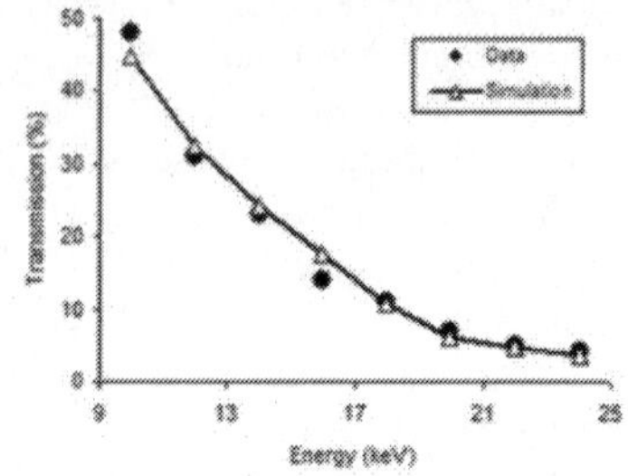

Figure 22: Simulation of the transmission versus energy of a focusing optic compared to experimental data.

initial polarization. Since polarization along the direction of propagation is not possible, the diffracted beam is linearly polarized in the direction perpendicular to both the initial and final directions. In order to verify that the diffracted beam was polarized, a test was performed of the intensity of the fluorescence and scatter detected from a sample. The energy-sensitive detector simultaneously recorded the intensity of the beam scattered off the sample, as well as the fluorescence induced by the incident beam, which occurs at a lower, characteristic energy. Because the fluorescence is emitted isotropically, the fluorescence intensity should remain constant as the polarizing crystal, sample, and detector were rotated together. If the incident beam is polarized, the scattered beam should drop to zero when the ϕ angle nears 90°. The result of the measurement, shown in Figure 19,[26] shows a high degree of polarization.

2.3. Focusing Optics

Polycapillary optics can also be designed to focus the x-ray beam, as shown in Figure 20. The focusing effects come from the overlap of the beams from hundreds of thousands of channels, rather than from the action within a single tube.

2.3.1. Spot Size

The focal spot size of a polycapillary optic can be measured using an imaging detector. The global divergence is then the slope of the size versus distance, as shown in Figure 21.[2] Near the focal point, the spot size can be small compared to the resolution of the imaging detector. The spot can be measured using a knife edge technique as was done in Figure 21, or by scanning a small pinhole across the beam. Assuming perfect overlap, the spot size at the focal point is determined by the spot size from each individual capillary channel, which depends on channel size, c, output focal length, f_{out}, and local divergence, β, as shown in Figure 15, as

$$d_{spot} \approx \sqrt{c^2 + (f_{out} \cdot \beta)^2} \,. \tag{2}$$

The critical angle, θ_c, at 20 keV is 1.5 mrad. Using $\beta \sim 1.3\theta_c$, an optic with c=3.4 μm and f_{out}=9 mm has a predicted spot size of 18 μm. An intensity distribution measurement, by the pinhole technique, gave a FWHM of 21 μm.[27] Because of the divergence from each channel, optics with smaller focal lengths have smaller spot sizes, as do measurements at higher photon energies.

2.3.2. Simulations

Figure 22 shows transmission versus energy results for a focusing optic which is 45.2 mm in length, with input focal length of 57 mm and output focal length of 128 mm. The input radius is 1.63 mm, the output radius is 2.22 mm and the channel diameter has a maximum radius of 5 μm. The transmission

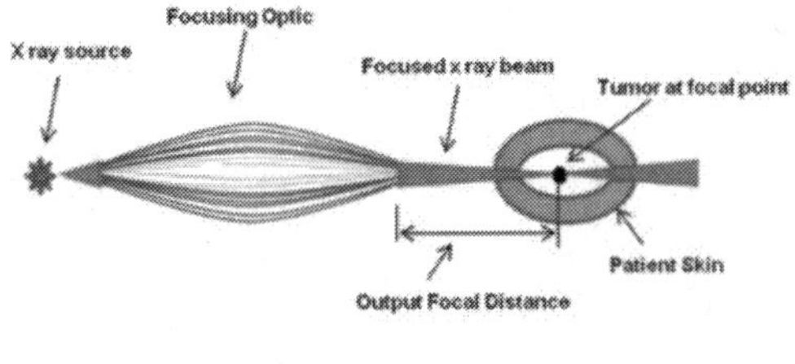

Figure 25. Schematic diagram of a focused x-ray beam using polycapillary optics

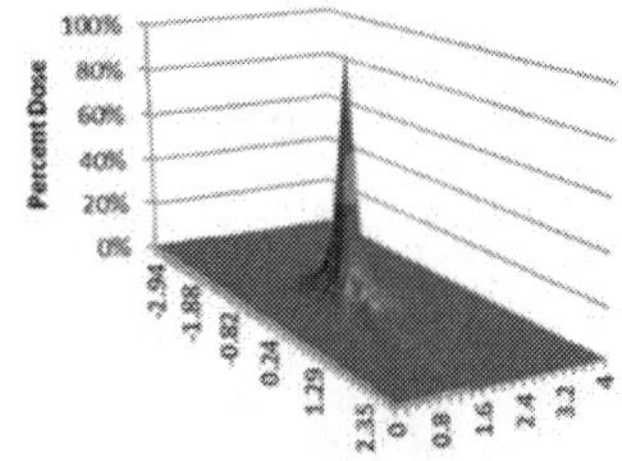

Figure 26. Dose distribution calculated with superimposed mesh tally for a cylindrical tumor at the focal point of the beams which were rotated about the tumor axis.

versus energy experimental results were taken using a 80 μm source. For the simulation the results were fit using a fractional open area (unsupplied by the manufacturer) of 65%, so there was only a single fitting parameter.

2.3.3. Orthovoltage Therapy

Low energy x-ray beams incident on a patient have their highest intensity at the skin surface. In order to reduce the skin dose when treating the deep seated tumors, megavoltage beams were developed for tumor therapy. However, instead of using high energy beams, a focused low energy x-ray beam could spread the beam across a larger area of skin surface while increasing the beam intensity at the tumor. Such a beam can be produced with polycapillary optics,as shown in Figure 25.[28] Experimental measurements of focal spot size with PMMA blocks of varying thickness placed between the optic and focal point showed that the focal point is well maintained at least 4 cm into the plastic. MCNP5 [29] was used to model the patient therapeutic dose from a low-energy focused x-ray beam, using the measured beam parameters. Dose calculations were performed in a target area with a volume of 63 mm^3 inside a spherical breast tissue phantom of radius 30 mm. The optic was rotated in one plane around the sphere. The calculated dose distribution is shown in Figure 26. Skin dose was less than 10% of the tumor dose, demonstrating the potential for skin-sparing radiotherapy with a compact low energy x-ray system employing a focusing optic.

2.3.4. Microscopy

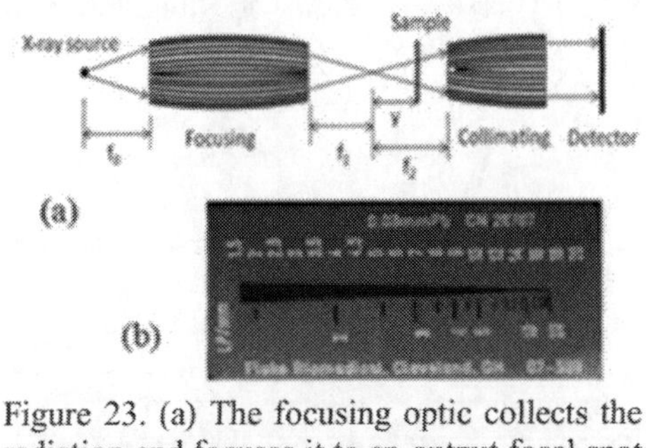

Figure 23. (a) The focusing optic collects the radiation and focuses it to an output focal spot a distance f_1 away. This spot is used as a source for imaging the object placed at a distance y. The collimating lens is placed at its input focal length, f_2, from the focal point of the first optic. (b) Resolution phantom. The images were taken near the right tip.

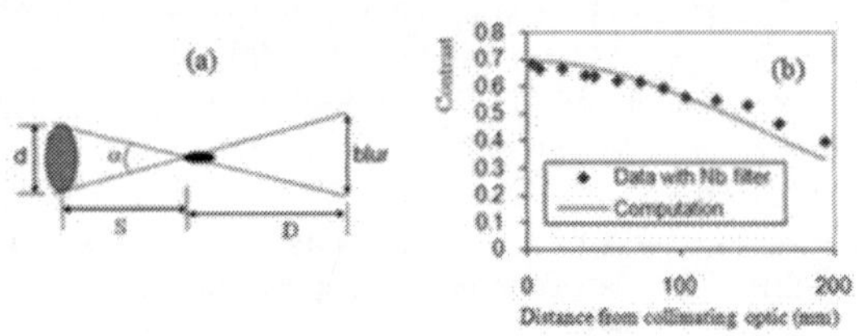

Figure 24. (a) Geometrical blur due to angular divergence of the beam. In this sketch the geometry is for the no optic case, so the divergence is due to the finite source size, and (b) Measured contrast as a function of distance from the collimating lens, and a fit using a post optic divergence of 0.072 mrad.

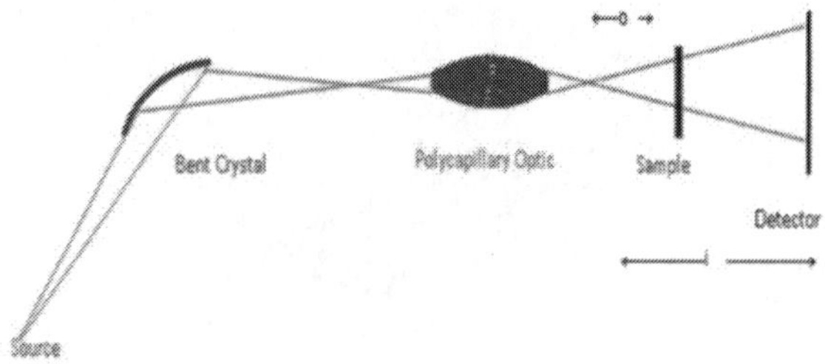

Figure 27. Toroidally bent crystal optic followed by a polycapillary focusing optic.

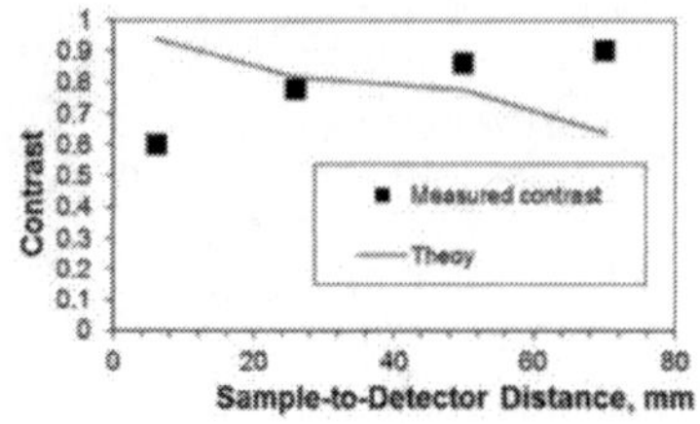

Figure 28. The contrast is not degraded with distance from the object.

The principle of confocal analysis based on polycapillary x-ray optics, in which one optic focuses the incident beam onto the sample, and the second selects only emission from that region, was first proposed in the early 1990s by Gibson and Kumakhov.[30] In recent years, this confocal technology has been widely used in 3D micro x-ray fluorescence technology.[31] For most analytical applications, the optical axes of the focusing and collection optics are typically at right angles, and the output focal spot of the focusing optic and the input focal spot of the collecting optic are adjusted in confocal configuration, so that only the x-rays from the micro-volume defined by the overlap of these foci can be detected. This confocal configuration improves the signal-to-noise ratio of the experiment.

In an experiment to produce magnified imaging, a focusing optic and collimating optic were employed in a parallel confocal geometry to magnify the image of a small sample, as shown in Figure 23.[32] In this application, the collimating polycapillary x-ray optic performs much like a magnifying fiber optic taper. If the detector is placed some distance beyond the optic, the contrast drops due to additional blurring as the image is propagated, as shown in Figure 24. A fit to the contrast drop, using the effective divergence α as a fitting parameter, is shown in Figure 24. The best fit for the effective divergence was 72 μradians; the image is preserved over much larger distances from the optic that might otherwise be expected.

A similar, and even more striking result was obtained if the initial polycapillary optic was replaced with a toroidally bent crystal, as shown in Figure 27.[33] The contrast, shown in Figure 28 remains high as the distance is increased (in fact rises, probably due to a reduction in Compton scatter as the detector is moved farther from the resolution phantom). The result is consistent with simulation modeling.[24] The results imply that the local divergence from the focal spot of some optics is smaller and more complicated than from a point source of similar size.[34]

2.3.5. Phase

As noted, conventional radiography results in poor contrast for soft tissue imaging when the tissues exhibit similar attenuation, *e.g.*, glandular tissue and infiltrating ductal carcinoma in breast imaging. However, x rays can accumulate significant differential phase delay even in weakly absorbing materials. Phase contrast imaging renders the different phase delays as intensity variation in the detector plane. For x-ray energies of 10 to 100 KeV, the variations in phase delay are approximately 100-1000 times greater than the absorption variations. In addition to the higher inherent contrast, because the phase delay falls relatively slowly with photon energy, phase imaging could be performed at higher photon energies, which result in lower patient dose. Propagation-based phase imaging requires significant spatial coherence of the x-ray beam. This requires either synchrotron sources, unsuitable for routine clinical use, or small spot microfocus sources, which are inherently low power and thus require long exposure times. Polycapillary focusing optics can be used to produce an x-ray focal spot of the required size for propagation-based imaging from a rotating anode source.[35] Focusing polycapillary optics can produce a gain of several hundred compared to using a pinhole

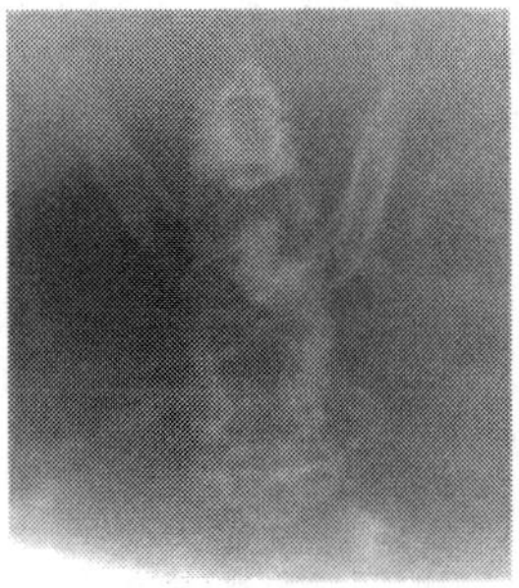

Figure 29. Absorption image of a cricket, taken with the detector close to the insect.

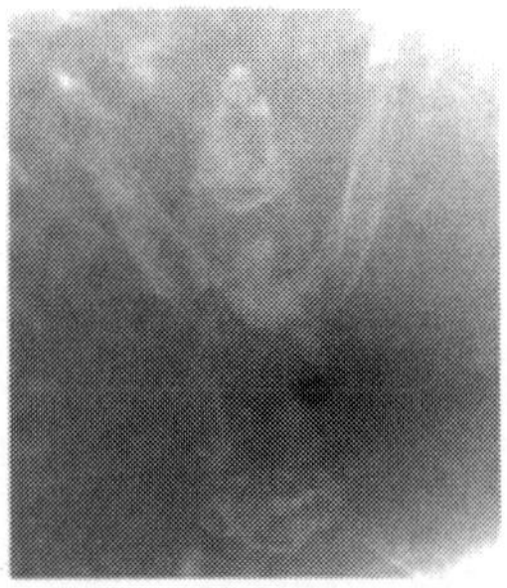

Figure 30. Phase contrast image, taken with the detector displaced. Edge enhancement is visible, as is the background structure of the optic illumination.

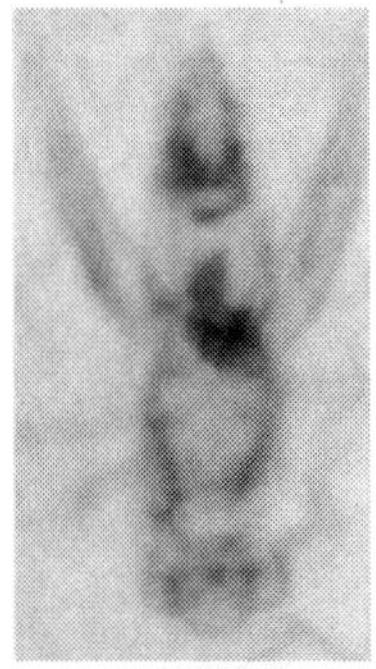

Figure 31. Computed phase delay derived from the phase image, assuming the attenuation is largely due to scatter and hence is proportional to the phase delay.

alone to reduce the spot size. The focal spot of the polycapillary optic was used as a secondary source to illuminate an object. A conventional image of a cricket taken with this illumination is shown in Figure 29. Moving the detector away from contact with the object allows for propagation-based phase effects to appear, seen as enhanced edge contrast in the image of Figure 30. Some of the improvement is due to the smaller effective pixel size due to the magnification; however there is visible edge enhancement. Because the contrast has terms due to both phase delay and attenuation, simplifying assumptions are required to reconstruct the phase delay attributable to the object. The phase-attenuation duality technique assumes that the attenuation is primarily due to Compton scatter, and hence to the electron density of the object, as is the index of refraction decrement and hence the phase delay. Assuming the two terms are proportional to each other allows generation of the image shown in Figure 31.[36]

Acknowledgements

The authors wish to express their great appreciation for data, images, analysis and ideas from many people, including Hassan Abbas, Carmen Abreu Bittel, Sajid Bashir, Henning Von Berlischpe, Ayhan Bingolbali, David Bittel, Cari, , Patrick Conlon, Gianpiero Dellituri, Sonja Dittrich, Sarah Formica, Christi Freinberg-Trufas, Ning Gao, David Gibson, Walter Gibson, Mikhail Gubarev, Laila Hassan Rafik, Danielle Hayden, Porcelain He, Frank Hoffman, Huapeng Huang, Abrar Hussein, Bushra Kanwal, Kardiawarman, Katie Kern, David Kruger, Susanne Lee, Danhong Li, Jan Lohbreier, Dip Mahato, Johanna Mitchell, Charles Mistretta, Robin Moresi, Noor Mail, Scott Owens Rohrbach, Sushil Padiyar, Lubna Peerzada, Wally Peppler, A. Filip Petrescu, Igor Ponomarev, Bimal Rath, Daniel Roemer, Christine Russell, Jared Sachs, Robert Schmitz, Sean Starr-Baier, Francisca Sugiro, Suparmi, Tianxi Sun, Weiyuan Sun, Sajjad Tahir, Christi Trufus-Feinberg, Johannes Ullrich, Mahboob Ur Rehman, Hui Wang, Lei Wang, Russel Youngman, Qi Fan Xiao, and Wei Zhou, and for grant support from the NIH, and the Breast Cancer Research Program.

References

[1] MacDonald C A and Gibson W M 2010 *Handbook of Optics 3rd Ed Vol V: Atmospheric Optics, Modulators, Fiber Optics, X-Ray and Neutron Optics*, ed. Bass M, et al., (McGraw-Hill) p.53.1-53.23.

[2] MacDonald C A, 2010 *XROI*, pp. 867049, doi:10.1155/2010/867049

[3] MacDonald CA 1996, Journal of X-Ray Science and Technology, **6**, pp. 32-47.

[4] MacDonald CA 2017 *An Introduction to X-ray Physics, Optics, and Applications*, to be published, Princeton University Press.

[5] Suparmi, Cari, Wang L, Wang H, Gibson W M and MacDonald CA 2001 Journal of Applied Physics **90** (10), pp. 5363-8.

[6] Abreu C C, Kruger D G, MacDonald C A, Mistretta C A, Peppler W W and Xiao Q F 1995 Medical Physics **22**, pp. 1793-1801.

[7] Kruger D G, Abreu C C, Hendee E G, Kocharian A, Peppler W W, Mistretta C A, MacDonald C A 1996 Medical Physics **23**, pp. 187-196.

[8] Cari, Suparmi, Gibson W M and MacDonald C A 2001 SPIE vol. 4320, pp.163-170.

[9] Mail N, MacDonald C A and Gibson W M 2009 Medical Physics 36 pp. 645-655.

[10] Jorgensen SM, Chmelik MS, Eaker DR, MacDonald CA, Ritman E L **A "**Polycapillary X-ray Optics-based Integrated micro-SPECT/CT Scanner," in **Developments in X-ray Tomography IV**, Ulrich Bonse, ed., SPIE vol. 5535, pp 36-42, 2004.

[11] Patrick Conlon 2013 *Potential For High Resolution Microscintigraphy using Polycapillary Optics,* MS Thesis, University at Albany.

[12] Jorgensen SM, Chmelik MS, Eaker DR, MacDonald CA and Ritman E L A **2014** SPIE vol. 5535, pp 36-42.

[13] Kern K, Peerzada L and MacDonald CA 2016 "Coherent Scatter Imaging Compatible with Screening Mammography," to be published, SPIE Journal of Medical Imaging.

[14] Kern K, Hassan L, Peerzada L, Ur-Rehman M and MacDonald CA, SPIE Vol. 9412, p. 941241.

[15] Hassan L andMacDonald C A, "Coherent Scatter Imaging Monte Carlo Simulation," to be published, SPIE Journal of Medical Imaging.

[16] Fernández M et al. 2004 *Spectroscopy* **18**, pp. 167-176.

[17] Sidhu S et al. 2011 Phys. Med. Biol. **56**, pp. 6779–6791.

[18] Bohndiek S E et al. 2008 Phys. Med. Biol. **53**, pp. 655–672.

[19] Conceição A L C and Polettia M E 2010 *AIP Conference proceedings* Vol. **1266** pp. 72-77.

[20] Li D, Mail N and MacDonald CA, , SPIE vol. 5745, pp.754-763, 2005.

[21] Schmitz R and MacDonald C A 2011 SPIE vol 8141, pp. 81410D-1.

[22] Zhou W, Mahato D, MacDonald C A 2010 Thin Solid Films 518, pp. 5047-5056.

[23] Bilderbeck D, Fontes E. *AIP Conf. Proc.*, 417: 147 (1997)

[24] Kanwal B, 2015 *Phase Imaging Using X-ray Optics*, Ph.D. dissertation University at Albany, SUNY.

[25] Sugiro F R, Li D and MacDonald C A 2004, Med. Phys., **31**, p. 3288.

[26] Robert S, Bingölbali A, Hussain A and MacDonald CA 2008, SPIE 70770Y.

[27] Hoffman F A, Gao N, Owens S M, Gibson W M, MacDonald C A and Lee S M, 1998, *MRS* **502**, pp 133-138.

[28] Abbas H, Mahato D N, Satti S and MacDonald CA 2014 Medical Physics, 41, pp. 041702.

[29] X-5 Monte Carlo Team 2003 *MCNP5 – A General N-Particle Transport Code, Version 5.*

[30] Gibson W M and Kumakhov M A 1992, SPIE Vol. 1736, p. 172.

[31] Tsuji K, Nakano K and Ding X 2007 Spectrochimica Acta **B 62**, pp. 549-553.

[32] Sun T and MacDonald C A 2013 Journal of Applied Physics, 113, pp. 053104-1 to -4.

[33] Sun T and MacDonald C A 2015 Journal of X-ray Science and Technology, 23, pp. 141-146.

[34] Bingolbali A and MacDonald C A 2009 Med. Phys. **36**, pp. 1176-1183.

[35] Bashir S, Tahir S, MacDonald C A and Petruccelli J C, 2016 Optics Communications 369 pp. 28–37.

[36] Petruccelli J C, Bashir S, Tahir S, Kanwal V and MacDonald C A 2015 SPIE Vol. 9412 eds. Hoeschen C and Kontos D, pp. 941254.

Journal of Physics: Conference Series **776** (2016) 012002 doi:10.1088/1742-6596/776/1/012002

The Influence of TiO2 Nanoparticles on LaFeO3/TiO2 Nanocomposites for Reduction of Aqueous Organic Dyes

N. Afifah[1,2] and R. Saleh[1,2,a]
[1]Departemen Fisika, Fakultas MIPA-Universitas Indonesia, 16424 Depok, Indonesia [2]Integrated Laboratory of Energy and Environment, Fakultas MIPA-Universitas Indonesia, 16424 Depok, Indonesia
[a]Corresponding's author: rosari.saleh@gmail.com, rosari.saleh@ui.ac.id

Abstract. A series of Lanthanum ferrite (LaFeO3) nanoparticles over titanium dioxide (TiO2) were synthesized using sol-gel method at room temperature by varying the loading of LaFeO3 on TiO2. The magnetic properties of samples were measured using vibrating sample magnetometer and photosonocatalytic activity towards the degradation of methylene blue under light (UV or visible) and ultrasound irradiation was also evaluated. The morphology and structure of the samples were characterized by field emission scanning electron microscope, energy dispersive analysis and X-ray diffraction. Furthermore the optical properties were also characterized by UV-visible diffuse reflectance. The experimental results showed that the prepared perovskites had sphere-like shape and strong visible light absorption. LaFeO3 demonstrated ferromagnetic properties and the magnetization decreased with the incorporation of TiO2 in the samples. However, the incorporation of TiO2 increased the photosonocatalytic activity and extended the photoresponding to UV light.

1. Introduction

One of methods to degrade organic pollutant in wastewater commonly applied by researches was the photocatalytic process. The photocatalytic process was the process of degrading organic pollutant using light as the irradiation source assisted by catalyst. However, in line with its development, the researchers found that the photocatalytic test greatly depends on the amount of light penetration into the water, so that the light could be absorbed by the catalyst, and then the photodegradation process could take place. Unfortunately, the organic pollutant in environment oftentimes had quite high density and not transparent. Consequently, light only could travel into the water by several millimeters hence inhibiting degradation of organic pollutant [1-2]. Recently researchers found that use of ultrasonic radiation was able to settle the above problem. The ability of ultrasonic irradiation penetration was very high for all water media that could reach depths of up to 15-20 cm [3]. In addition, ultrasonic irradiation was uniformly dispersed in the water medium so that it could reduce the agglomeration of particles that commonly occurred in the catalytic process [4]. The sonolysis process could actually be performed without catalyst, but its low efficiency required semiconductor that was applied as a catalyst [5]. In the sonocatalytic process, semiconductor could also play an active role as catalyst by utilizing sonoluminescene that appeared due to implosion of acoustic bubbles at the high local temperature of 5000 K and pressure of 1000 atm. This caused not only sonocatalytic process but also photocatalytic process occurred simultaneously [1]. Based on the above-mentioned process, the synergy effect might arise if the photocatalytic and sonocatalytic processes were done simultaneously namely the photosonocatalytic process that might draw the attention of researchers. Considering the result to be obtained, the combination of both catalytic processes was predicted to produce better efficiency than if each process was done separately.

One of semiconductors that frequently used as catalyst was TiO2. Its relatively low price, non-toxic and its high catalytic capability drew the attention to use it as a catalyst [6]. This is because solar energy contains around ~46% visible light and only around 4-5% ultraviolet light [7].Therefore, to maximize used of solar energy as an energy source, a new type of catalyst that could play an active role in visible light was interesting to be developed [8-9]. LaFeO3 was one of catalysts that could be active in visible light. In addition, its stability, environmental friendless, and multi functions were the added value of utilization of this material [10]. The photocatalytic process highly depended on the recombination of electron and hole pairs from catalyst materials such as LaFeO3. For that purpose, to prepare LaFeO3 in the form of nanocomposites for example by using TiO2 was known to be able to inhibit the recombination of electron and hole pairs so that its catalytic performance could increase [9, 11]. Several researchers have reported photocatalytic activities using LaFeO3/TiO2 nanocomposites as catalyst [11]. However, photosonocalytic activities performed using the nanocomposites are rarely found.

Hence, in this study several series of LaFeO3/TiO2 nanocomposites with variation of molar ratios have been successfully synthesized using the sol-gel method. The morphology, structure, magnetic properties and optical properties of the samples would be studied. Photosonocatalytic activities would also be performed using methylene blue (MB) as the model of organic pollutant. The influences of TiO2 on the photosonocatalytic activity of LaFeO3/TiO2 were carried out under different light sources (UV or visible).

2. Experimental details

Materials: LaCl3.7H2O, FeCl2.4H2O, titanium dioxide (TiO2) and sodium hydroxide (NaOH) were purchased from Merck. All chemical reagents were analytical grade and used without further purification.

Catalyst preparation: LaFeO3 nanoparticles were synthesized using co precipitation method based on our previous work [12]. Subsequently, several series of LaFeO3/TiO2 nanocomposites with different molar ratio 1:0.1, 1:0.3, and 1:0.5 were prepared using sol-gel method [13]. In brief, 0.00125 mol of TiO2 was dissolved into 30 ml of distilled water under magnetic stirrer. Then, 0.00125 of NaOH that had been dissolved into 65 ml of distilled water was added into the mixture. This mixture was designated as solution A. Afterward solution A was stirred and heated at the temperature of 80°C. Meanwhile, the above synthesized LaFeO3 were dissolved into 30 ml of distilled water and were designated as solution B. Subsequently, solution B was introduced into solution A and the mixtures were continuously stirrer at 80°C for 2h. The obtained mixtures were centrifuged and washed for several times with ethanol and distilled water. Afterward, the final product was allowed to stand overnight at the room temperature and then heated at 100°C for 1h under vacuum condition.

Characterization: The morphology and elemental analyses of samples were studied using LEO 420 field emission scanning electron microscopy (FESEM) and energy dispersive X-ray (EDX) spectroscopy. XRD pattern was obtained using a Rigaku Miniflex 600 with Cu-Kα (λ = 1.54060 Å) as a radiation source operated at 40 kV and 20 mA in the range of 20° up to 90 °. The magnetic properties were measured using the Oxford Type 1.2 T vibrating sample magnetometer (VSM). The measurement was carried out from 0 to ± 1 Tesla. Then, the optical properties were measured using UV-Vis diffuse reflectance spectrophotometer.

Catalytic activity: Methylene blue (MB) was selected as the model of organic pollutant in this study. The photosonocatalytic processes were conducted in a cylindrical glass vessel equipped with a magnetic stirrer under light and ultrasonic irradiation simultaneously. The ultrasonic irradiation was carried out using a commercially ultrasonic bath operated at fixed frequency and power of 40 kHz and 150 W. While, the different light of 40 W- UV lamp and 40 W- Xe lamp were used as the UV and doi:10.1088/1742-6596/776/1/012002

Visible light irradiation source. The solution was analyzed using Hitachi UH5300 UV-visible spectrophotometer. The percentage degradation of MB was recorded as Ct/C0, where Ct is the concentration of organic pollutant in each irradiated time interval and C0 is the initial concentration of organic pollutant after adsorption-desorption equilibrium was reached. In order to study the main active species that play important role in the degradation of MB, different substance such as di-ammonium oxalate, sodium sulfate, and tert-butyl alcohol ware used as holes, electrons and hydroxyl radicals scavengers. The stability of catalyst was also checked by reusing the catalyst four times cycle processes.

3. Results and Discussion

The morphology of the samples was investigated using FESEM image. Figure 1 showed FESEM image from (a) LaFeO3 nanoparticles, (b) TiO2 nanoparticles, and (c) LaFeO3/TiO2 nanocomposites with the molar ratio of 1:0.3. As can be seen from Figure 1a, the synthesized LaFeO3 nanoparticles consist of agglomerated particles, while the individual particles appeared to be spherical [12]. The spherical-like shape of TiO2 nanoparticles was also shown in Figure 1b. Hence in Figure 1c, the morphology of LaFeO3/TiO2 nanocomposites showed spherical-like shape from both nanoparticles and agglomerated particles originating from LaFeO3 nanoparticles. Elemental analysis of the samples has also been studied using EDX measurement. Figure 2a showed the EDX spectra of TiO2 nanoparticles, LaFeO3 nanoparticles, and LaFeO3/TiO2 nanocomposites with the molar ratio of 1:0.3. The presence of La, Fe, Ti, and O in EDX spectrum confirmed the formation of LaFeO3/TiO2 in the form of nanocomposites.

The crystal structure of the samples was studied using XRD spectroscopy. Figure 2b showed the XRD pattern of the LaFeO3/TiO2 sample with the molar variation ratio of 1:0.1, 1:0.3, and 1:0.5. As comparison, the XRD pattern of TiO2 nanoparticles and LaFeO3 nanoparticles were also shown in this figure. The XRD pattern of TiO2 nanoparticles showed tetragonal structure with the anatase phase. While the XRD pattern of LaFeO3 nanoparticles showed orthorhombic structure. No additional phase or impurity found in the diffraction pattern of each nanoparticle. For LaFeO3/TiO2 nanocomposites, the diffraction peaks at 2θ 22.6°, 32.2°, 39.8°, 46.2°, 57.5°, 67.4°, and 76.7° which represent planes (101),(121), (220), (202), (240), (242), and (204) respectively are known to be the diffraction pattern from the orthorhombic structure of LaFeO3 [14]. Meanwhile the diffraction peak at 2θ 25.2° which represent plane (101) constitutes the diffraction pattern of the anatase phase from TiO2 [11]. The dominant XRD pattern of LaFeO3 might be caused by the molar ratio between LaFeO3 and TiO2 which is relatively small. No other segregation of phase was also found in the diffraction pattern of LaFeO3/TiO2 nanocomposites. The lattice parameter obtained by Rietveld refinement method using MAUD program and crystallite size <D> calculated using the Scherrer's formula have been summarized in Table 1.

Figure 1. FESEM image of (a) LaFeO3 nanoparticles; (b) TiO2 nanoparticles; and (c) LaFeO3/TiO2 nanocomposites with molar ratio of 1:0.3. All images are at 20.000× magnification (scale bar = 5μm).

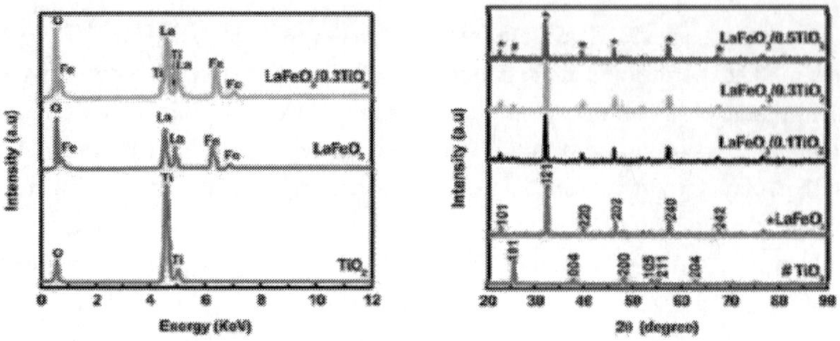

(a) (b)

Figure. 2. (a) EDX spectra; and (b) XRD pattern of LaFeO3/TiO2 nanocomposite.

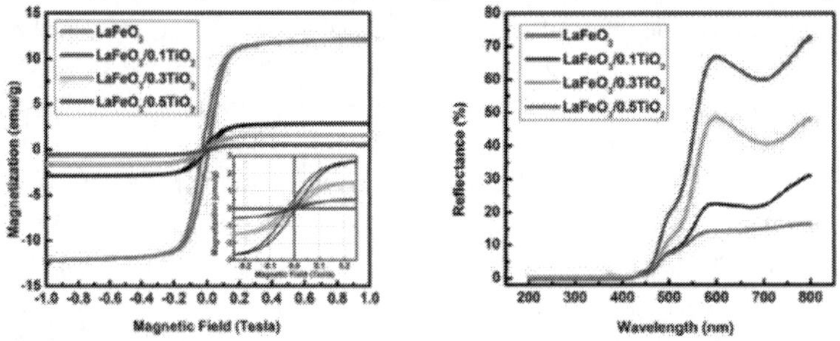

(a) (b)

Figure 3. (a) VSM and (b) UV-vis spectra of LaFeO3/TiO2 nanocomposites.

Table 1. The lattice parameter, grain size <D>, magnetic saturation (M-S) and band gap of LaFeO3/TiO2 nanocomposite.

Sample	Lattice Parameter LaFeO3 a (Å)	b (Å)	c (Å)	Lattice Parameter TiO2 a=b (Å)	c (Å)	<D> LaFeO3 (nm)	<D> TiO2 (nm)	M-S (emu/g)	Band Gap (eV)
LaFeO3	5.5436	7.8456	5.5528	-	-	58	-	12.08	2.45
LaFeO3/0.1TiO2	5.5440	7.7794	5.5500	3.8355	9.6407	28	36	2.91	2.49
LaFeO3/0.3TiO2	5.5590	7.7140	5.6540	3.7822	9.5023	27	30	1.62	2.53
LaFeO3/0.5TiO2	5.5222	7.6865	5.6533	3.7860	9.4950	23	29	0.55	2.58

The influence of TiO2 on the magnetic properties of LaFeO3/TiO2 nanocomposites was also studied using VSM measurement. Figure 3a showed the VSM spectra of LaFeO3 nanoparticles and LaFeO3/TiO2 nanocomposites with different molar ratio. As can be seen in the figure, all samples showed ferromagnetic behaviour under room temperature. With the increasing of TiO2 content the magnetic saturation (Table 1) of LaFeO3/TiO2 nanocomposites decreases respectively. The reduced mass of LaFeO3 in the presence of non-magnetic TiO2 might cause the decrease of magnetic saturation on the samples [13].

The optical property of the samples was performed using UV-Vis diffuse reflectance spectroscopy measured at the room temperature in the range of 200-800 nm. Figure 3b showed UV-Vis reflectance spectra of LaFeO3 nanoparticles and LaFeO3/TiO2 nanocomposites with different molar ratio. By extrapolating the spectra using Kubelka-Munk analysis with following equation: $F(R) = (1-R)^2/2R$ [15], with R as diffuse reflectance, the optical gap of each sample could be known. Based on this function, the band gap was obtained by plotting the value of $F(R)^2$ vs. the energy in electron volts. The linear part of the curve was then extrapolated to become $F(R)^2 = 0$ and subsequently the optical gap could be known. The optical gap of the samples have been summarized in Table 1. The result showed that all samples have optical gap values in the visible light region. However, the optical gap of the nanocomposites increased with the increasing of TiO2 molar ratio. It indicated that the nanocomposites also have good response to UV light. Therefore, in the photosonocatalytic test both UV and Visible light were used as an irradiation sources.

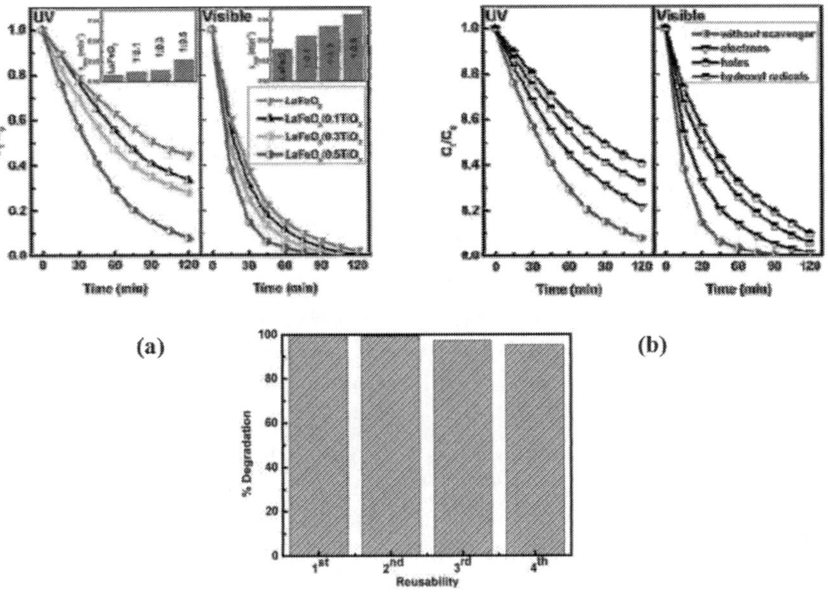

Figure 4. (a) The degradation of methylene blue by using LaFeO3/TiO2 nanocomposite with various molar ratio; (b) The effect of various scavengers on photosonocatalytic degradation of methylene blue by using LaFeO3/TiO2 nanocomposite with molar ratio of 1:0.5; and (c) The reusability on the photosonocatalytic degradation of methylene blue by using LaFeO3/TiO2 nanocomposite with molar ratio of 1:0.5 under visible light irradiation.

Figure 4a showed the photosonocatalytic activity of LaFeO3/TiO2 nanocomposites under UV and visible light irradiation. As a comparison the photosonocatalytic activities of LaFeO3 nanoparticles could also be observed. The degradation of MB increased with the increasing of TiO2 molar ratio and showed better performance than LaFeO3 nanoparticles. LaFeO3/TiO2 nanocomposites with the molar ratio of 1:0.5 showed highest degradation both under UV and visible light irradiation. The MB degradation obtained in the photosonocatalytic follow the equation of pseudo-first order kinetics model: $\ln (Ct/C0) = kapp\ t$, where kapp is the apparent reaction constant for catalysis used as the basic parameter to compare different catalysts. The reaction constant of different catalysts has been examined in this study and the result could be observed in inset Figure 4a. The result indicated that the photosonocatalytic activity using visible light was faster than using UV light irradiation. In addition, the main active species that play the important role in the photosonocatalytic activity was also examined in this study. Degradation of methylene blue with the existence of Na2S2O8, di-ammonium oxalate and tert-butyl alcohol as electrons, holes, and hydroxyl radicals scavenger, could be observed in Figure 4b. The presence of di-ammonium oxalate as holes scavenger inhibited the degradation of MB. This result confirmed that holes play an important role in degrading of MB. Figure 4c showed the stability of catalyst four times cycle processes. The result indicated that LaFeO3/TiO2 nanocomposites showed good stability for degrading MB in the photosonocatalytic activity.

High photosonocatalytic activities of LaFeO3/TiO2 nanocomposites under visible light irradiation compared to using UV light could be explained as follows: In the photosonocatalytic process, the catalyst interacted with light and ultrasonic wave. Interaction between the catalyst and the light produceed the separation process of electron and hole. While interaction between catalyst and ultrasonic wave could increase formation of hydroxyl radicals (•OH) and formation of electron and hole pairs at the same time. Formation of electron, hole and •OH in the photosonocatalytic processes play an important role on the degradation process. LaFeO3 nanoparticles is a catalyst that has optical gap in the range of visible light irradiation, and based on the result of UV-Vis measurement, the optical gap of LaFeO3/TiO2 nanocomposites also showed the absoption in the visible light region. Hence, the possibility of LaFeO3/TiO2 nanocomposites to produce electron and hole pairs was higher if it was conducted using visible light irradiation. In the photosonocatalytic activity using visible light source, TiO2 could be serve as electron transport layer. When visible light was given to the catalyst, the electron that was excitated to the conduction band of LaFeO3 did not directly return to the valence band but first transferred to the conduction band of TiO2. Conversely, when a photosonocatalytic activity was conducted on UV light, the excitated electron in conduction band of TiO2 would first be transferred to the conduction band of LaFeO3. The variation of TiO2 molar ratio in the sample caused the different degradation of MB. Under UV radiation, the higher content of TiO2 was known to be able to increase absorption of UV light, hence LaFeO3/TiO2 nanocomposites with the molar ratio of 1:0.5 indicated a more dominant result compared to with other molar ratio. While in visible irradiation, TiO2 only served as an electron transport layer that could inhibit the recombination of electrons and holes pairs. When an electron was transfred in the conduction band of TiO2 or LaFeO3, at the same time, the hole left in the valence band could move freely to be able to react with OH^-, resulting the total hydroxyl radicals (•OH) obtained in te photosonocatalytic process increase. Hence, this result confirmed that hole play important role in the degradation of MB.

4. Conclusion

Photosonocatalytic activity of LaFeO3/TiO2 nanocomposites for degrading MB was successfully performed under UV or visible light irradiation. The incorporation of TiO2 on LaFeO3/TiO2 nanocomposites showed faster degradation on the photosonocatalytic processes than LaFeO3 nanoparticles both under UV or visible light irradiation. The formed LaFeO3/TiO2 nanocomposites played an important role in inhibiting recombination of electron and hole pairs that further improved the degradation of MB in the photosonocatalytic process.

References

[1] Zhou M, Yang H, Xian T, Li R S, Zhang H M, and Wang X X 2015 *Journal of Hazardous Materials* 289 149
[2] Wang J, Guo B, Zhang X, Zhang Z, Han J and Wu J 2005 *Ultrasonics Sonochemistry* 12 331
[3] Suslick KS, Doktycz SJ and Flint EB 1990 *Ultrasonics* 28 280
[4] Tangestaninejad S, Moghadam M, Mirkhani V, Baltork I M and Salavati H 2008 *Ultrasonics Sonochemistry* 15 815
[5] Hu B, Wu C, Zhang Z and Wang L 2014 *Ceramics International* 40 7015
[6] Li FT, Zhao DS, Luo QZ, Liu RH and Yin R 2008 *Journal of Ceramic Processing Research* 9 398
[7] Niu X, Li H and Liu G 2005 *Journal of Molecular Catalysis A: Chemical* 232 89
[8] Li F-T, Liu Y, Liu RH, Sun Z-M, Zhao D-S and Kou C-G 2010 *Materials Letters* 64 223
[9] Tang P, Tong Y, Chen H, Cao F and Pan G 2013 *Current Applied Physics* 13 340
[10] Thirumalairajan S, Girija K, Ganesh I, Mangalaraj D, Viswanathan C and Balamurugan A 2012 *Chemical Engineering Journal* 209 420
[11] Gao K and Li S 2012 *Applied Surface Science* 285 6460
[12] Afifah N and Saleh R 2016 *Journal of Physics: Conference Series* 710 012030
[13] Arifin S A, Jalaludin S and Saleh R 2015 *Materials Science Forum* 827 49
[14] Li S, Zhao Y, Wang C, Li D and Gao K 2016 *Materials Letters* 170 122
[15] Köferstein R, Jäger L and Ebbinghaus SG 2013 *Solid State Ionics* 249–250 1

Journal of Physics: Conference Series **776** (2016) 012003 doi:10.1088/1742-6596/776/1/012003

Modification of circuit module of dye-sensitized solar cells (DSSC) for solar windows applications

S. D. Hastuti, F. Nurosyid*, A. Supriyanto and R. Suryana
Department of Physics, Faculty of mathematics and Natural Sciences, Sebelas Maret University, Jl. Ir. Sutami 36 A Kentingan Surakarta 57126 Indonesia

*Email: nurosyid@yahoo.com

Abstract This research has been conducted to obtain a modification of circuit producing the best efficiency of solar window modules as an alternative energy for daily usage. Solar window module was constructed by DSSC cells. In the previous research, solar window was created by a single cell of DSSC. Because it had small size, it could not be applied in the manufacture of solar window. Fabrication of solar window required a larger size of DSSC cell. Therefore, in the next research, a module of solar window was fabricated by connecting few cells of DSSC. It was done by using external electrical circuit method which was modified in the formation of series circuit and parallel circuit. Its fabrication used six cells of DSSC with the size of each cell was 1 cm x 9 cm. DSSC cells were sandwich structures constructed by an active layer of TiO_2 as the working electrode, electrolyte solution, dye, and carbon layer. Characterization of module was started one by one, from one cell, two cells, three cells, until six cells of a module. It was conducted to recognize the increasing efficiency value as the larger surface area given. The efficiency of solar window module with series circuit was 0.06%, while using parallel circuit was 0.006%. Module with series circuit generated the higher voltage as the larger surface area. Meanwhile, module through parallel circuit tended to produce the constant voltage as the larger surface area. It was caused by the influence of resistance within the cable in each module. Module with circuit parallel used a longer cable than module with series circuit, so that its resistance increased. Therefore, module with parallel circuit generated voltage that tended to be constant and resulted small efficiency compared to the module with series circuit. It could be concluded that series external circuit was the best modification which could produce the higher efficiency.

1. Introduction

Dye-sensitized Solar Cell (DSSCs) was a solar cell device for converting solar energy into electrical energy. DSSC has advantages in the fabrication process, it fabrication process was easy and was not use high technology. DSSC cells were sandwich structures constructed by an active layer of TiO_2 as the working electrode, electrolyte solution, dye, and carbon layer [1,6]. The working principle of DSSC was when the light absorbed by dye on the working electrode, there will be excitation of electrons in the dye. The excited electrons to be injected into the TiO_2 layer and then flowed through the outer circuit of DSSC towards the counter electrode. The functions of the counter electrode as a catalyst that will help hasten the reaction process between electrons with an electrolyte solution [2]. The results of reaction will charge a hole in the dye process and it will continue to be repeated.

In the previous research, solar window was created by a single cell of DSSC with a small size. Because it had small size, it could not be applied in the manufacture of solar window. Fabrication of solar window required a larger size of DSSC cell. Its can be as an alternative energy for daily usage. The problem right now was the process of making a module of DSSC were difficult. It must has a large size module which then can be applied to solar window. Research of Pranoto (2013) has produced solar module interconnect design type Z series has three pieces of single cells the size of 1 cm x 9.8 cm with a conversion efficiency of 0.77% [3]. Therefore, in the next research, a module of solar window was fabricated by connecting few cells of DSSC. It used external electrical circuit method which was modified in the formation of series circuit and parallel circuit. It's fabrication using six cells of DSSC with the size of each cell was 1 cm x 9 cm. Current-voltages Characteristics of data generated are in the form of voltage, current, and efficiency. This aims for the research is to determine the efficiency and the current-voltages characteristics resulted from modifications of Dye-sensitized Solar Cell (DSSCs). Modification of the circuit in DSSC that good produces current-voltages characteristics can expect to be applied became solar windows DSSC.

2. Methods

Materials are used in this research is FTO glass substrates 7 Ω Dye Sol, TiO_2 nanopowder Aldrich as the working electrode with a size of 21 nm, carbon as the counter electrode, electrolyte solution Dyesol, Dye Ruthenium N719 Dyesol, and silver paste. A module of solar window was fabricated by connecting few cells of DSSC. It using by external electrical circuit method which was modified in the formation of series circuit and parallel circuit. It was fabrication used six cells of DSSC with the size of each cell was 1 cm x 9 cm. FTO substrate was washed using soap water and then rinsed twice using aquades into ultrasonic cleaner, finally is rinsed with 96% alcohol by ultrasonic cleaner. Deposition of TiO_2 paste by screen printing technique [3,7]. TiO_2 paste will be coated on FTO substrate and then be annealed at temperature of 500 ° C in a holding time of 10 minutes. The working electrode is then soaked with dye for 24 hours. Coating the counter electrode used carbon pencil and be annealed with a temperature of 500 ° C. Finally process on the fabrication of DSSC is assembling to arrange the working electrode and the counter electrode with a sandwich structure, then assembled by series circuit and parallel circuit. Silver paste was used to connect between cells in series external circuit and parallel external circuit. DSSC has been formed and then injected with a solution of electrolytes (HSE) and then tested characteristics by using the I-V meter keithley 2602A with intensity of radiation 1000 W / m^2. DSSC modules were connected to the set of tools to probe positive to counter electrode and the negative probe to the working electrode.

3. Result and Discussion

This research resulted DSSC prototype were presented in Figure 1 and the scheme of the series connection and parallel connection were presented in Figure 2a and 2b.

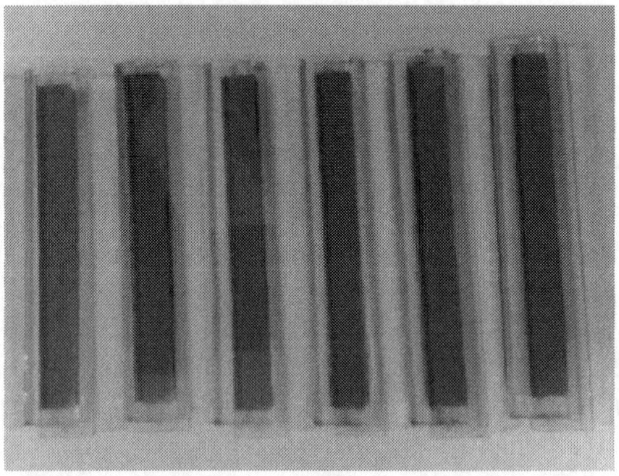

Figure 1. This single cell will be connected in series circuit and parallel circuit

In the series external circuit and parallel external circuit were tested from a single cell, two cells, three cells, four cells, five cells, and six cells. The characteristics produced was voltage, current, and efficiency. The results characteristics of voltage and current were plotted in I-V curve were presented in Figure 3 and Figure 4.

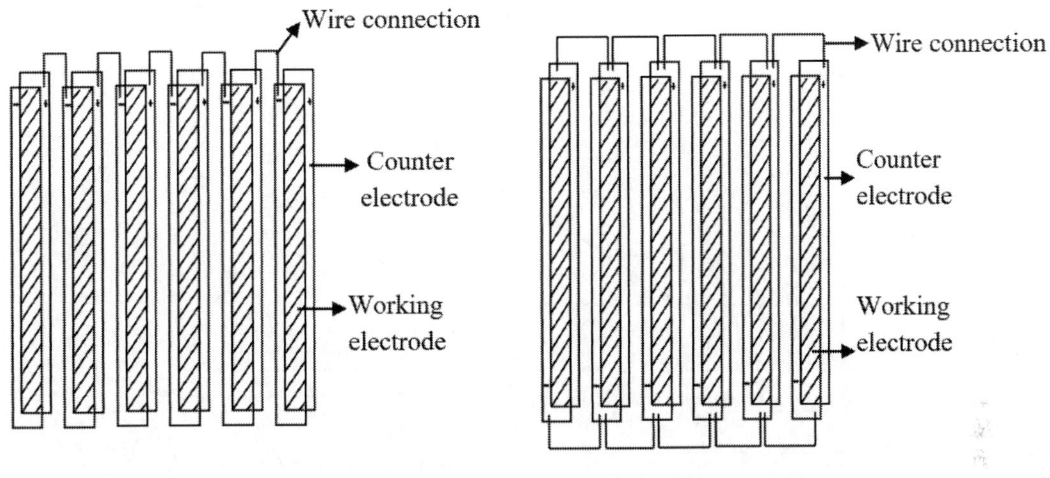

Figure 2. a) Schematics of series external circuit schematic, b) Scematics of parallel external circuit

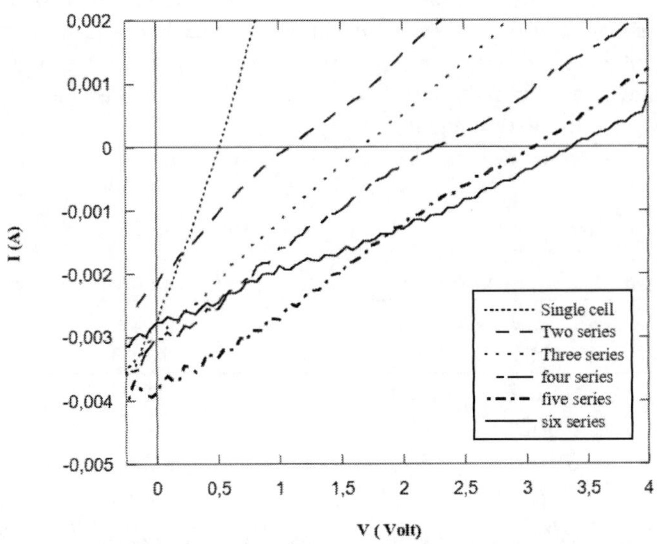

Figure 3. Current-voltage curves for series external circuit

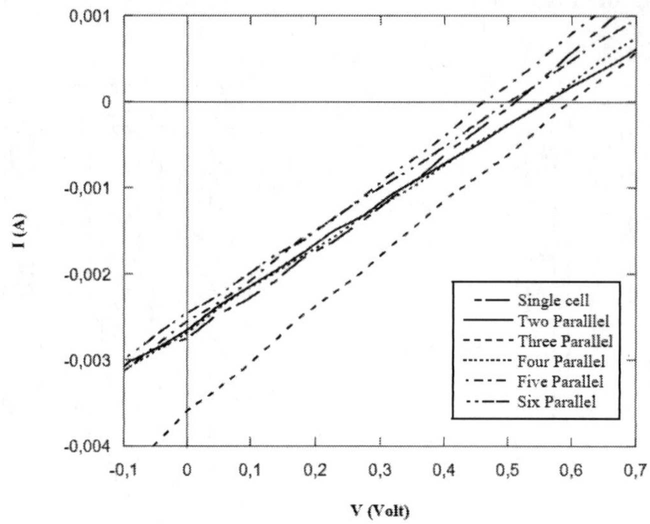

Figure 4. current-voltages curves for parallel external circuit

Figure 3 can be seen that voltages were resulted from series circuit showed shift to the right side which that indicated an increase in voltages, meanwhile the current changes too. Figure 4 showed the changes fluctuated current occur in parallel circuit and it was show resulted voltages were not changed significantly. The current-voltage curve were used to determine of efficiency from each circuit. The efficiency of solar cells depends on multiplication between the maximum voltage and the maximum current it called a maximum power (Pmax) toward the input power (Pin) was resulted from the intensity of radiation and active area of DSSC. The equation was used to find efficiency of a DSSC is presented in equation 1[2].

$$\eta = \frac{Pmax}{Pin} \quad (1)$$

Table 1. current-voltages characteristics of series external circuit

Characteristic	Single cell	Two series	Three series	Four series	Five series	Six series
V oc (volt)	0,5103	0,9903	1,5605	2,1606	2,8908	3,2307
I sc (A)	0,0026	0,0024	0,0036	0,0038	0,0026	0,0033
V max (Volt)	0,3002	0,5302	0,7901	0,9801	1,4304	1,9704
I max (A)	0,0015	0,0011	0,0019	0,0020	0,0025	0,0017
P max (watt)	0,0004	0,0006	0,0015	0,0020	0,0035	0,0033
P in (watt)	0,9000	1,8000	2,7000	3,6000	4,5000	5,4000
eff	**0,0494**	**0,0335**	**0,0549**	**0,0555**	**0,0784**	**0,0610**
FF	0,3380	0,2560	0,2609	0,2438	0,4760	0,3072

Table 2. Current-voltages characteristics of parallel external circuit

Characteristics	Single cell	Two parallel	Three parallel	Four parallel	Five parallel	Six parallel
V oc (volt)	0,5103	0,5614	0,5791	0,5505	0,4405	0,4899
I sc (A)	0,0026	0,0029	0,0037	0,0026	0,0028	0,0026
V max (Volt)	0,3002	0,2412	0,3090	0,2604	0,2299	0,2601
I max (A)	0,0015	0,0016	0,0019	0,0016	0,0016	0,0014
P max (watt)	0,0004	0,0004	0,0006	0,0004	0,0004	0,0004
P in (watt)	0,9000	1,8000	2,7000	3,6000	4,5000	5,4000
eff	**0,0494**	**0,0214**	**0,0217**	**0,0112**	**0,0081**	**0,0065**
FF	0,3380	0,2407	0,2723	0,2810	0,3016	0,2809

Characteristics of series external circuit were presented in Table 1. Characteristics of series circuit generates an increase continuous voltage with an increase in the active area of the DSSC. This is appropriate, if the circuit is arranged in series so it will increase the voltages and it has constant current [4,8,10]. Meanwhile in this research, the current generated in series circuit were not constant. It's also changed with the change of the active cell area. The resulting stream increase to the fourth series and then fell in the five series and start to rise in the six series. Increase and decrease the resulting of current change were not significantly and the range of change was not far, so it can be said that the current relatively constant. It can occur because of the additional resistance of the cable and the silver paste were used in constructed electrical circuits in series so it can affect the value of current. The current flowed in the circuit depends on the amount of load or resistance in the circuit.

Characteristics data on parallel external circuit were presented in Table 2 resulted the relative same voltages. It was a range (0.44 to 0.57) volt which can be said constant value because of the difference was not far too. The current of this circuit generated increased and decrease with an added active area from the cells. The increase occurred in single parallel into three parallel, and began to fall on the four parallel up to six parallel. In the parallel circuit has produced more large current from single cell to addition of the cells [5,8], but this research has not resulted that. This can occurred because of a connection cable that has a longer size and increase the resistance of the silver paste so influence at circuit between cells in parallel DSSC modules.

Efficiency was resulted from series external circuit are presented in Table 1. It was generated increase value in single cell up to five cells, but at six cells the value of efficiency was declined. It can occur because the current was generated from six series down drastically compared with the current of the five series . The amount of the maximum current will affect the maximum power, then it will impact on the resulted efficiency. The addition of the active area caused efficiency of DSSC was changed. Equation 1 showed the efficiency of DSSC was opposite to the active area. It was obtained from the output power, so the larger active area, then was efficiency was resulted will be decreased.

Efficiency at each cell on a parallel external circuit declined in six solar cells. The efficiency of series circuit was a higher than the efficiency of parallel circuits. That situation may occur due to at the series circuit, voltage increases will affect the output power so if voltage were large, output power will be large and then the efficiency were generated also great. Parallel external circuit produced constant voltage at each of the cells, so that the power were not changed significantly the efficiency generated. Power was not increased, then by increased active area of the parallel circuit so it will produce a small efficiency according to equation 1[9].

A module was applied to the solar window has important consider is output current and voltage as well as the best circuit that could make DSSC be durable and can be daily usage. Series circuit has a large voltage and has a greater efficiency but the series circuit has the disadvantage. It when one of the cells in the module damaged or dies so all of the cells in the module will stop, which means not current was flowing and was not reusable. Parallel circuit has a constant voltage and small efficiency, but a parallel circuit becomes durable because it was based on the principle of the parallel circuit, if a single cell was not active then the other cells will remain active so that the module DSSC can still work and produced a current [5] but can not produce a large output power.

4. Conclution

The efficiency of solar window module with series circuit was 0.06%, while using parallel circuit was 0.006%. Module with series circuit generated the higher voltage as the larger surface area. Meanwhile, module through a parallel circuit tended to produce the constant voltage as the larger surface area. It was caused by the influence of resistance within the cable in each module. Module with circuit parallel used a longer cable than a module with series circuit, so that its resistance increased. Therefore, module with parallel circuit generated voltage that tended to be constant and resulted small efficiency compared to the module with series circuit. It could be concluded that series external circuit was the best modification which could produce the higher efficiency.

5. Acknowledgments

This work is supported by Directorate of Research and Community services, Ministry of Research Technology, and Higher Education and Institute for Research and Community services, Sebelas Maret University by PUPT Grant Program (Contacts Number : 353/UN27.21/PN/2016)

6. References

[1] Rosa ES, Lia M, Shobih, Jojo H and Brian Y 2014 *J. Tech.* 18 **2**, 96-100
[2] Irmansyah, Akhiruddin M and Muhammad Z 2008 *J.* Vol. **9** No. 2, 96-103
[3] Pranoto LM 2013 *report.* PPET LIPI
[4] Buwono M C 2010 *Skripsi*.UI
[5] Wijaya A., Rudy S, Slamet W 2011 *J.* UNIBRAW
[6] Alapatt G 2014 *Dissert.*(**2014**)Paper 1311
[7] Ramasamy E, Won J, Dong Y and Jae S 2007 *J.* **165** (2007) 446–449
[8] Chou J, Chin-Hui H,Yi-Hung L,Yu-Jen L,Chia-Ming C and Yu-Hsun N 2016 *J.* Vol **2016**, Article ID 6595639, 8 pages
[9] Stathatos E 2012 *J. Eng. Sci and Tech.* 5 (**4**) (2012) 9 - 13
[10] Wei T, Shien-Ping F, Ya-Huei C, Sheng-Jye C, Yu-Jie L, Chih-Ming C and Han-Hsu C 2012 *J. Elect. Sci.*7 **(2012)** 11904 - 11916

The effect of Cu doping into *Oriza sativa L. indica* dye as photosensitizer for dye sensitized solar cell (DSSC)

U. M. Fadli*, A. B. Prasada, Cari, A. Supriyanto

Physics Graduate Program, Faculty of Mathematics and Science, Sebelas Maret University, St. Ir. Sutami 36A Kentingan Surakarta 57126, Indonesia

E-mail : *oasis.190712@gmail.com

Abstract.The aims of the research to are know the effect of Cu doping into natural dye in increasing the efficiency of DSCC, to determine of the optical and electrical characteristic of natural dye Cu doping. Sandwich structures formed in the sample consisted of working electrode pair Titanium Oxide (TiO_2) and the counter electrode Platina (Pt). Absorbance test is measure by using UV-Visible spectrophotometer Lambda 25, conductivity test by using a two-point probe *El Kahfi/I-V Meter*, and characterization of current and voltage (*I-V*) by using a Keithley 2602A. The Cu doping into dye was increasing the efficiency of 71%.

1. Introduction

DSSC is the third generation of solar cells which can be an alternative to cell conventional solar silicon. DSSC material purchase costs are cheap, easy in production, but the efficiency is still small [1]. There are 4 main components in DSSC, which is a semiconductor as working electrode (TiO_2 and Cu metal), a dye sensitive to the absorption of light, an electrolyte and a counter electrode such as platinum (Pt). TiO_2 nanoparticels in FTO and Cu metal serves as the electron catcher for providing hole. The counter electrode which serves as a catalyst while transferring electrons to the electrolyte. Electrolyte generate electrons for their redox reactions are repeated, which will be forwarded to fill the hole in the dye. The use of glass to work forward light so that the light energy absorbed by electrons in the atoms of dye towards the hole of Cu and TiO_2 layers. Furthermore, electrons move to the FTO layers and to external sircuit, then the electrons towards the platinum layer and electrolyte. Dye used comes from Black Rice. Sianidin 3 glucoside is black rice anthocyanin type of this role to give the red color purple or dark red. Black rice has anthocyanin content of (105.11 ± 7.76) mg / 100 g [2].

To improve the performance of DSSC, using metal inserted into the dye. The synthetic form of Cu metal was used is copper (II) sulphate pentahydrate ($CuSO_4.5H_2O$), because it was easily to found [3]. Titanium dioxide (TiO_2) is material used as the working electrode in DSSC. Energy gap (Eg) from TiO_2 rutile and TiO_2 anatase is 3.0 eV and 3.2 eV. Eg anatase is the highest because the active surface area is larger and more effective photocatalysts [4]. The counter electrode is a double function, the first, it can stream of electrons from the external circuit (lamp) back to the redox system (electrolyte) and the second as a mediator to catalyze the redox reaction [5]. The counter electrode of platinum (Pt) has been the material of choice because it is an excellent catalyst for reducing triiodide [6]. Platina serves to reduce the resistance on the FTO glass. Electrolytes used in this study is a mixture of iodine (I_2) with Potassium Iodide (KI) or Sodium Iodide (NaI) into the polyethylen glycol (PEG). Electrolytes

are used as the electron mediator between the working electrode and the counter electrode. The efficiency value for electrolyte KI is the greatest then NaI, and TEAI each mixed into I_2 [7].
This study aims to determine the absorbance value, conductivity and efficient of DSSC. Determination of absorbance in this study in accordance the law of Lambert Beer, with The Increase value of absorbance of a material which suggested a linear relationship between concentration and absorbance according to the equation 1.

$$A = \varepsilon . b . c \qquad (1)$$

Were A is the absorbance and ε is the molar absorptivity coefficient ($molar^{-1}.L^{-1}$). While b is the thickness of the medium through which the beam (cm), and c is the concentration of the sample solution (ppm) [8]. Determination of the conductivity of a material can be calculated by equations 2.

$$\sigma = d/R.A \qquad (2)$$

With σ is the conductivity (Ωm^{-1}) and R is the resistance of materials (Ω). While d is the distance between the two electrodes (m) and A is the cross-sectional area (m^2).
Performance solar cells is the ability to convert light into electrical energy. Determination of performance solar cells is measured using Keithley 2602A to find the value of short circuit current (I_{sc}) and open circuit voltage (V_{oc}). The value of I_{sc} is determined based on the data in which the voltage value starts from a negative to a positive. Simillary for the value of V_{oc} is determined by the current value starts from a negative to a positive value. Then the value maximum voltage (V_{in}) and maximum current (I_{in}) are obtained by multiplication of voltage and current. Performance solar cells is determined by comparing maximum power from multiplication result V_{oc} and I_{sc}.

2. Experiment

2.1. Preparation of TiO_2 Solution

Titanium (IV) Oxide nano particle is anatase Titanium Dioxide powders, 99.8% trace metals basis. Pasta TiO_2 of 0.5 grams dissolved in 3 ml of ethanol, then stirred of 500 rpm for 30 minutes.

2.2. Preparation of Natural Dye Sensitizers and Doping Copper (Cu)

10 grams of *Oriza sativa L. indica* was crushed using a blender. Then the materials was dissolved in ethanol, acid acetate and distilled water used to have ratio of 25:4:21 so that the value high absorbance of the solution because the best anthocyanin concentrate [9]. Then stirred of 300 rpm at 30°C for 30 minutes. Materialis was allowed to stand for 24 hours and it was filtered with a filter paper quality no.42. Then the dye black rice was mixed with solution of Cu. The solution made of copper sulfate pentahydrate ($CuSO_4.5H_2O$) with mass variation of 0.01 grams, 0.1 grams, 1 grams, 2 grams, 3 grams, and 4 grams. Then was dissolved in ethanol, acid acetate and distilled water then stirred of 300 rpm at 30°C for 60 minutes. Then it was filtered with a filter paper quality no.42. Dye and dye+Cu then tested of absorbance using UV-Vis *Spectrometer Lambda 25* and conductivity using El Kahfi *I-V meter 2602A*. Dye said to be able to work if it has absorption in the ultraviolet and visible, while the conductivity states how large are able to pass an electric current.

2.3. Preparation of Electrolyte

Potassium iodide (KI) solids of 0.8 grams are mixed into 10 ml polyethylen glycol then stirred at 300 rpm for 30 minutes. Then was added iodine (I_2) of 0.127 grams and stirred at 300 rpm for 30 minutes.

2.4. Preparation of Counter Electrode

The counter electrode in the form of platinum (Hexachloroplatinic (IV) acid 10%) and it was mixed with isopropanol (207 ml) was stirred of 300 rpm for 30 minutes. Then the solution was dropped on FTO glass with size 1x1cm and the side masking tape affixed. FTO glass with size 1x1cm is heated at 250^0C for 15 minutes and then the solution of 3 ml was dropped on FTO.

2.5. Preparation of TiO_2

The working electrode is made of glass on which the FTO conductive paste deposited TiO_2 nano with spin coating techniques. At the FTO glass with size 1x1cm and FTO side masking tape affixed as a barrier. Pasta TiO_2 dripped on FTO glass that has been glued in the spinner, then stirring with a speed of 200-300 rpm. FTO glass that has been coated with TiO_2 is heated using a hotplate at 500 ^{0}C for 60 minutes, then cooled at room temperature.

2.6. Assembly of Sandwich DSSC

The composition of the DSSC in the form of glass FTO that has been coated with TiO_2 and has been soaked in a dye or dye+Cu, and then called the working electrode. The working electrode is etched with an electrolyte solution and then covered with glass that has been coated with platinum, and then called the counter electrode. Then the composition of DSSC is clamped on both sides of the right and left. Once formed system Sandwich DSSC then characterized current and voltage (I-V), which indicates how much the DSSC capable of converting light energy.

3. Result and Discussion

3.1. Absorption Spectra of natural dye doped Cu

The data of absorbance using UV-Vis *Spectrometer* lamda 25, and in Figure 1 the maximum absorbance of dye *Oriza sativa L. indica* is 0.1646 a.u at wavelength (λ) 232 nm and 0.018 a.u at λ 531 nm. While maximum absorbance for dye+Cu 4 grams (0.73 M) to shifted between the value of 1.405 a.u (242 nm) and 0.126 a.u (531 nm). Then the maximum absorbance of TiO_2 is 0.069 a.u (341 nm), and the greatest absorbance of TiO_2+dye+Cu 4 grams is 0.33 a.u (324 nm).

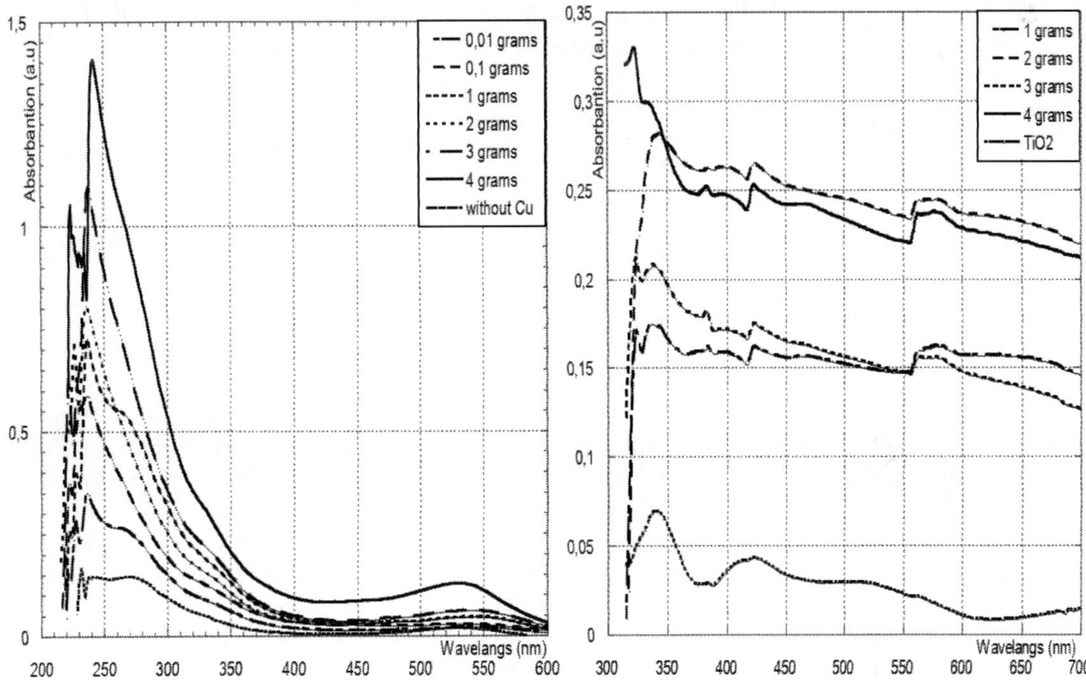

Figure 1. Absorption spectrum of (a) dye+Cu and (b) TiO_2+dye+Cu

Absorbance on figure 1 are able to work on the ultraviolet (UV) and visible (VIS) light, which is the area of energy absorption to excite electrons in the dye molecule. For figure 1(a) and figure 1(b) with the addition of Cu into the dye (dye+Cu), so it will be able to absorb the light energy is greater. The result will be more electrons are released from Higher Occupied Molecular Organic (HOMO), to the bandgap and then headed to Lower Unnocopied Molecular Organic (LUMO) in the dye molecule. The result will be formed holes in the HOMO and free electrons in the LUMO. Free electrons that which will then be forwarded to conducting Cu and TiO_2. But the electrons from the dye can also go directly to the conduction band of TiO_2 because the dye can be attached directly to TiO_2. The conduction band of TiO_2 can be trapping of electrons from the dye or from Cu. So that the number of Cu was affected the electrons can be captured and then channeled to the external circuit. Increase value of absorbance in accordance with the law of Lambert Beer (equation 1). In figure 1(b) the absorbance values of TiO_2 with dye+Cu 3 grams smaller than 2 grams, this is because the sample is less homogeneous when tested, but overall, Figure 1 shows the absorbance value at 4 grams of Cu is the highest, in λ 320 nm.

3.2. Conductivity of dye+Cu

Characterization is electrical conductivity using Elkahfi 100 / IV-Meter with irradiation from halogen lamp of 680.3 W/m^2. The electrical conductivity is the ability of a material to conduct electricity when a potential difference is placed at the ends of a conductor, electrons move through the material.

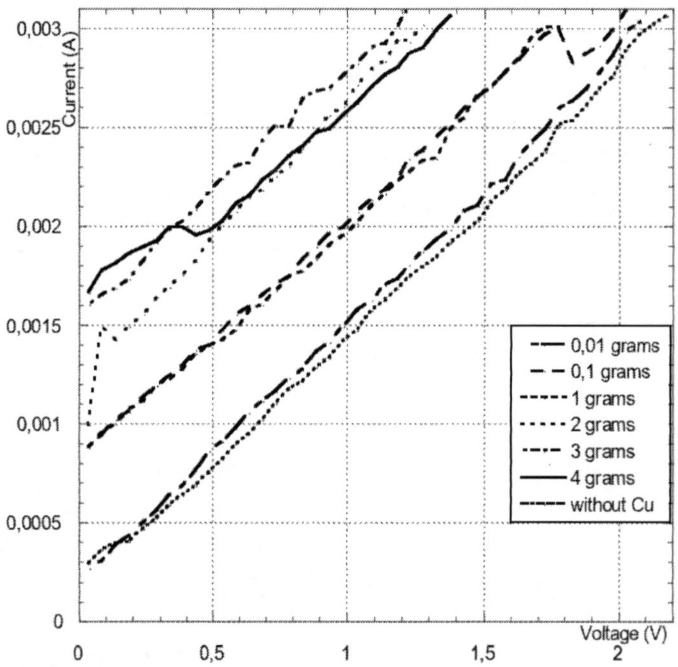

Figure 2. Conductivity Curve for dye with Cu

The Copper is a conductor, which has functions can delivering electrons generated by the dye. Figure 2 shows that the tendency of increase in the value of the conductivity is proportional to the increasing mass of Cu were added in the dye. By using the equation 2, conductivity of dye+Cu 3 grams worth 0.2509 $\Omega^{-1}m^{-1}$, then in the dye+Cu 4 grams only worth 0.2339 $\Omega^{-1}m^{-1}$. Values of 4 grams look smaller then 3 grams, it could be due to a mixture of dye+Cu solution with the solvent is not homogeneous, so the resulting in resistance value solution is still greater then the rate of flow in the solution inhibited. While the value of conductivity for dye is only 0.0669 $\Omega^{-1}m^{-1}$. The results of 3 grams and 4 grams have great conductivity, when compared to the other insertion conductivity values. This suggests that the ability to conduct electrons at the insertion of 4 grams is still the best. The greater of conductivity value is indicates that the resistent it worth getting smaller, so the electrons that are drawn will be more.

3.3. Photovoltaic Properties

Characterization of the current-voltage (I-V) is a method to determine how much ability DSSC can convert light in to electrical energy. Measurements using Keithley 2602A in under irradiation with a halogen light 1000 W/m^2. The test with immersion 24 hours in FTO area of 1 cm^2. The dye+Cu 4 gram in 4 layers the largest value of efficiency, because the value of V_{in} intercept at 3.4 x 10^{-1} mV and the value I_{in} intercept at 2.5 x 10^{-4} mA. (See in Figure 3 and Table 1). The mass of the copper used, it appears that the test absorbance on insertion 4 grams of Cu generate the highest value. Insertion of Cu into a dye, producing the highest absorbance value for the light energy will be absorbed by the dye, and is proportional to the number of excited electrons. Excitation electrons occurs from HOMO to band gap and then passed to LUMO. After that, electron will flow to the Cu, subsequently forwarded to the working electrode. Working electrode layer is TiO_2, that functioning as an electron acceptor from Cu or from the dye. The electrons from the dye can also go directly to the TiO_2, because dye can be attached to TiO_2.

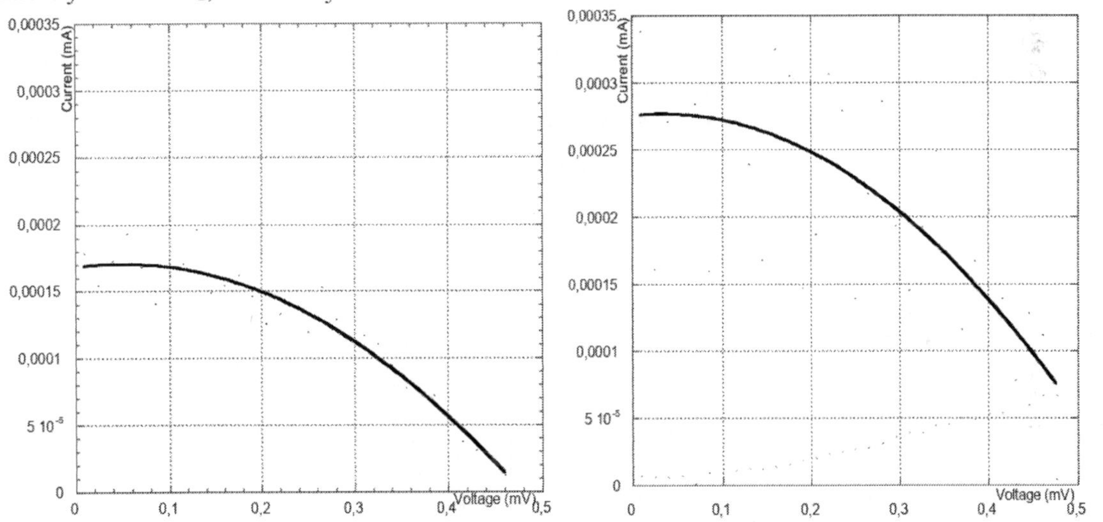

Figure 3. I-V curve with 4 layers of TiO_2 for (a) dye, (b) dye+Cu 4 gram

Besides being able to forward the generated electrons dye, TiO_2 semiconductor can also produces its own electron as a result of absorbing light energy, but relatively small so that it can be ignored. Likewise with dye+Cu conductivity are large, so the ability to transfer electrons from the dye will be faster in proportion to the amount of Cu is inserted in dye. Then the electrons will flow to the Cu and moved toward the working electrode to an external circuit.

The thickness of TiO_2 on FTO glass rather FTO will influence of electrons flow. TiO_2 is a kind of a semiconductor with a band gap value of 3.2 eV which is suitable for use in solar cells. TiO_2 will produce a hole and will be functions as an electron dye trap. Hole will catch or pass on electrons from the dye, or from Cu. If the thickness is too thin, the amount of TiO_2 little hole, then the electrons will be forwarded to an external circuit is not optimal. With proper TiO_2 coating thickness, hence maximizing the trapped electrons with the electrons flowed, so the current flow will also be the maximum, that will produce the highest efficiency. If too thick, the electrons trapped too many, before they could be forwarded to the external circuit, so electrons will be forwarded to an external circuit is not optimal. (See Table 1 and equation 1).

Table 1. Photovoltaic performance with natural dyes from different doping Cu

Dye Source	*Iin (mA)*	*Vin (mV)*	*Isc (mA)*	*Voc (mV)*	*(%)*
Without Cu (4 layers)	0.00014	0.34015	0.00022	0.47521	0.04961
0.01 grams (3 layers)	0.00014	0.32504	0.00013	0.50513	0.04473
0.1 grams (3 layers)	0.00010	0.37007	0.00019	0.50515	0.03797
1 grams (3 layers)	0.00007	0.37008	0.00013	0.43010	0.02574
2 grams (3 layers)	0.00009	0.31006	0.00004	0.44512	0.03051
3 grams (3 layers)	0.00010	0.26508	0.00011	0.47565	0.02774
4 grams (3 layers)	0.00013	0.40003	0.00007	0.46009	0.05169
4 grams (4 layers)	0.00025	0.34026	0.00034	0.53531	0.08460
4 grams (5 layers)	0.00023	0.23507	0.00033	0.46021	0.05419
4 grams (6 layers)	0.00017	0.25002	0.00021	0.44517	0.04167

To improve efficiency need to insert a metal, in this case $CuSO_4.5H_2O$ produce good conductivity, increase absorption, and can reduce the electron hole recombination. The result can be concluded that the DSSC that produces the best performance in 4 layers at the dye+Cu of 4 grams with efficiency is 0.08460%, while the dye only 0.04961%. The value efficiency of the material of organic compounds is still very small (under 1%). For increase the performance DSSC, we can be using different counter electrode with polyaniline (PANI), because PANI can increase efficiency from 6.90% (platinum) to 7.15% using PANI [10], using of electrolyte which has viscosity is smaller than iodine such as PEO polymer gel [11].

4. Summary

The results of the absorbance these material are able to work on the UV-Visible light and the highest value of dye+Cu of 4 grams. The largest conductivity value is dye+Cu of 3 grams. Where as the greatest efficiency in achieved by using dye+Cu of 4 grams (0.73 M) in 4 layers in the amount of 0.0846%, the Cu doping into dye was increasing the efficiency of 71%.

5. Acknowledgement

Thanks to the managers of Physics Materials and Chemistry Laboratory in Faculty of Science UNS Surakarta. This research was supported by LPPM Postgraduate UNS Grant with contract no.343/UN27/HK/2016, April 07th 2016.

6. References

[1] Stathatos, E 2012 *Jurnal of engineering Science and Technology Review 5*. pp 9-13.
[2] Purwandari,FA 2014 Gadjah Mada Univercity: Faculty of Agriculture Technology.
[3] Alloway, BJ 1995 Chapman and Hall India.
[4] Kuznesof PM 2006 *Chemical And Technical Assessment*.Page 1(8)-8(8).
[5] Ting C-C dan Chao W-S 2010 *Applied Energy*, Vol. 87 (2010), hal. 2500–2505.
[6] Calandra P, Calogero G, Sinopoli, A, and Gucciardi, PG 2010 *International Journal of Photoenergy*, Vol. 2010, Article ID 109495, 15 pages, doi:10.1155/2010/109495.
[7] Prasetyo, YH, Wahyuningsih, S, Suryana, R 2014 *Indonesian Physics Journal* No: 53, Vol XVIII, Agusts, ISSN: 1410-2994. Page:47-49.
[8] Dian A, Yanlinastuti, Noviarty, Masrukan 2012 *Urania* Vol. 18 No 2. Juni:59-119.
[9] Misbachudin, MC et al.2013.ISBN 978-602-8047-80-7 LPF 1350: page 1-5.
[10] Li Q, Wu J, Tang Q, Lan Z, Li P, Lin J, Fan L 2008 *Elseiver*.
[11] Duangkaew P, Chindaduang A, Tumcharem 2010 *Science Jurnal Ubonratchathani University*.Vol 1.No 1(January-June).page: 9-14.

Journal of Physics: Conference Series **776** (2016) 012005 doi:10.1088/1742-6596/776/1/012005

Effect of wet etching process on the morphology and transmittance of fluorine doped tin oxide (FTO)

S. L. Triana, Kusumandari*, R. Suryana
Physics Department of Universitas Sebelas Maret,
Jl. Ir. Sutami 36A Kentingan Jebres Surakarta 57126, INDONESIA

*E-mail: kusumandari@staff.uns.ac.id

Abstract. Wet etching process was performed on the surface of FTO. The FTO coated glasses subtrates with size of 2x2 cm covered by screen were patterned using zinc powder and concentrated hydrochloric acid (1 M). The substrates were then cleaned in ultrasonic baths of special detergent(helmanex) diluted in deionized water and isopropanol in sequence. The screens with various of hole size denotes by T32, T49 and T55 were used in order to create a pattern of surface textured. The atomic force microscopy (AFM) image revealed that wet etching process changes the morphology of FTO. It indicates that texturization occured. Moreover, from the UV-Vis Spectrophotometer measurement, the transmittance of FTO increase after wet etching process. The time of etching and pattern of screen were affect to the morphology and the transmittance of FTO.

1. Introduction

Dye sensitized solar cell (DSSC) as a candidate of renewable generation solar cells has advantages relatively to silicon-based solar cells such as it can be fabricated at low cost and it does not require material with high purity. However, it's lower efficiency and stability still being a big problem to overcome. In many recent years, researchers work to improve efficiency of DSSC through modifies some component of DSSC. Kilic et al. was observed that band gap engineering and modifying surface of TiO2 nanostructure by Fe2O3 succesfully increase the solar cell efficiency to 7.27% [1]. Also, various of natural dye as a sensitizer of DSSC have been investigated by Zhou et al.[2].

Other attempt to improve the efficiency of DSSC is by light trapping method. One such method is making pattern at the transparent electrode as surface of solar cell or known as a surface texturing. Some of lights which come on the surface of solar cell will reflect, others will absorb or transmit. With patterning of surface we will minimize reflectance and increasing the transmittance of transparent electrode with obeying the negligible of absorbance. In general, the kind of surface texturing method are acid texturization using chemical [3], reactive ion etching [4], mechanical texturization [5] and laser processing. Each of them has some advantages and drawbacks depend on the application. The application of etches based on acid induces difficult reproducible results due to random etching area and necessity of precise control of composition of etches. Reactive ion etchings such as using plasma need a control of temperature. Mechanical texturing may be effective, but has some limitations since cannot be applied specially for thin and fragile materials [6].

In this research, wet etching method using HCl solution was performed on the surface of FTO. various type of screen were used as pattern mask. UltraViolet-Visible (UV-Vis) spectrophotometer and

Atomic Force Microscope (AFM) were conducted to investigate transmittance and morphology as well as roughness of FTO.

2. Method

The FTO transparent glass substrates with size of 2x2 cm were prepared. Substrates are cleaned with alcohol and aquades by using ultrasonic cleaner for 10 minutes to remove residual contaminant, and then dry. HCl with 1 M of concentartion was used as acid solution. The FTO etching process is more complicated since FTO cannot be reduced to the metal by using hydrogen ions in strong acid solution [7]. It need a catalyst such as zinc particles (Zn). In this research, the paste of Zn powder was applied on the surface of FTO which covered by screen as mask patterning and followed by HCl solution. After etching the samples are washed using a solution of Hellmanex to clean up the rest of the etching chemical and followed by alcohol using a ultrasonic cleaner for 10 minutes. In this research, the screens with various of hole size denotes by T32, T49 and T55 were used in order to create a pattern of surface textured. The T32 means the amount of screen's hole is 32 per inch and so on. The etching times are 1, 2 and 3 minutes for every sample. Characterization of samples is carried out by Atomic Force Microscopy (AFM) to know the morphology and roughness of surface etching. Furthemore, Ultraviolet-Visible (UV-Vis) Spectrophotometer was used to understand the transmitannce of samples. Characterization was also performed on the sample without wet etching as reference.

3. Results and Discussion

The 3D AFM image of FTO with 1 minute of etching using T32, T49 and T55 type of screen in comparison to FTO without etching is depicted in Figure 1. Moreover, picture inset shows the distribution of FTO surface topography.

FTO surface is relatively uniform as shown fig. 1(a) while it is shown several domains after wet etching for 1 min as shown in fig. 1(b)-(d). Also, the distribution of surface show the peak is shifted to the lower topography. This results indicates that wet etching process can modify the FTO surfaces through reaction with the HCl solution. In addition, root mean square roughness (Rq) of FTO surfaces in area of 5 μm × 5 μm shown in fig. 1 are approximately 18 nm for the without etching sample, 14 nm for the etching sample (T32), 13 nm for the etching sample (T49) and 12 nm for the etching sample (T55). Here, the Rq represents the standard deviation of surface heights. The Rq was decrease after wet etching process since some of surfaces were reacted with chemical and leave many spaces.

The optical transmittance of FTO samples of the without etching and with etching using T32, T49 and T55 type of screen for (a) 1 min, (b) 2 min and (c) 3 min are shown in fig. 2. In general, an etching treatment will increase transmittance due to domain formation of FTO will leave many spaces and will be more passed the visible lights of UV-Vis spectrophotometer. However, it is considering that etching time and type of screen were affecting the transmittance. For the shortest etching time, the type of screen does not significantly affect the transmittance as it can be seen in fig. 2 (a). It is because the depth of etch almost same, indicated by Rq. While the transmittance will be seen different for every type of screen when the etching time increase, as shown in fig. 2(c) – 2(d). It is considering to deeper penetration of etching surface will create surface texturing. When light come to the flat surface, it will be directly transmitted or reflected from the surface. However, when light come to surface texturing, some of them will be transmitted and the others will be reflected from one to another side of surface. This process will repeat many time known as light trapping method . And it will reduce or increase the total transmittance depends on the shape of surface texturing and angle of .

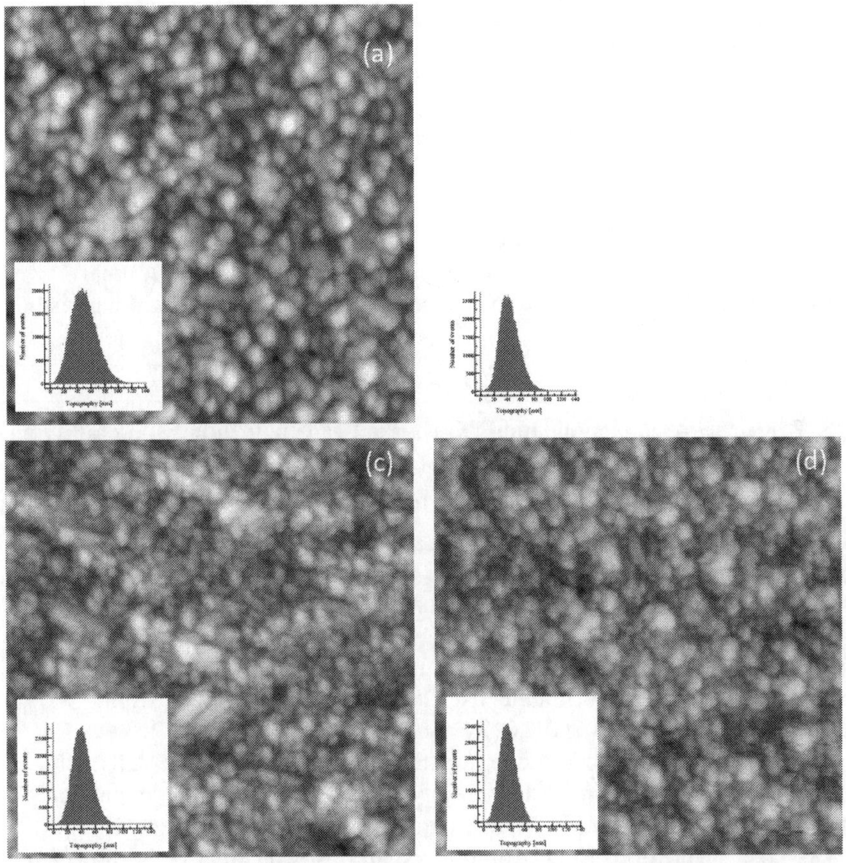

Figure 1. The 3D AFM image of FTO (a) without etching and with 1 minute of etching using (b) T32, (c) T49 and (d)T55 type of screen. Inset : the distribution of surface topography

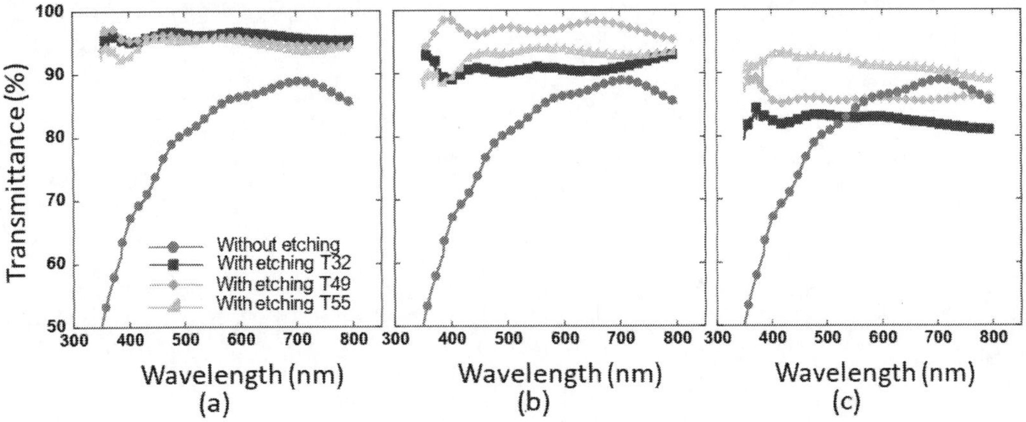

Figure 2. The transmittance of FTO for samples without and with etching using T32, T49 and T55 type of screen. Etching time (a) 1 min, (b) 2 min and (c) 3 min.

Finally, the highest optical transmittance of the FTO films is beneficial for efficient transport of light in DSSC applications. Also, transmittance of FTO can affect the efficiency of DSSC because a light trapping structure contributes to increasing the light path length by scattering the incident light at the different angles.

4. Conclusions

The paper presents results of wet etching process on the surface of FTO. Reaction between surface of FTO with Zn and HCl resulting some domains indicate surface texturing. The time of etching and pattern of screen were affect to the morphology and the transmittance of FTO. In summary, wet etching processing of FTO surface can be used as texturization method to enhance the light harvesting of DSSC.

5. References

[1] B. Kılıc,n, N. Gedik, S. P. Mucur, A. S. Hergul, and E. Gür, Band gap engineering and modifying surface of TiO_2 nanostructures by $Fe_2O_3$for enhanced-performance of dye sensitized solar cell, Materials Science in Semiconductor Processing 31 (2015) 363–371

[2] H. Zhou, L. Wu, Y. Gao, and T. Ma, Dye-sensitized solar cells using 20 natural dyes as sensitizers, Journal of Photochemistry and Photobiology A: Chemistry 219 (2011) 188 - 194

[3] P. Panek, M. Lipiski, and J. Dutkiewicz, Texturization of multicrystalline silicon by wet chemical etching for silicon solar cells, Journal of Materials Science 40/6 (2005) 1459-1463.

[4] L. Jayanti, Kusumandari, T. Sujitno and R. Suryana, The study of FTO surface texturing fabrication using Argon plasma etching technique for DSSC applications, IOP Conf. Series: Materials Science and Engineering 107 (2016) 012028.

[5] P. Fath, C. Marckmann, E. Bucher, and G. Willeke, Multicrystalline silicon solar cells using a new high throughput mechanical texturization technology and a roller printing metallization technique, Proceedings of the 13thEuropean PV Solar Energy Conference, Nice, 1995, 29-32.

[6] L.A. Dobrzanski, A. Drygała, Surface texturing of multicrystalline silicon solar cells, Journal of Achievements in Materials and Manufacturing Engineering, 31 (2008) 77 - 82.

[7] http://informationdisplay.org/IDArchive/2008/April/TransparentConductiveOxidesforDisplayApplicat.aspx

Growth of TiO_2 nanofibers on FTO substrates and their application in dye-sensitized solar cells

R. Suryana[1], L. R. Rahmawati[1] and K. Triyana[2]

[1]Department of Physics, Faculty of Mathematics and Natural Sciences, Universitas Sebelas Maret, Indonesia
[2]Department of Physics, Faculty of Mathematics and Natural Sciences, Universitas Gadjah Mada, Indonesia

Email: rsuryana@staff.uns.ac.id

Abstract. Growth of TiO_2 nanofibers on fluorine-doped tin oxide (FTO) substrates have been performed using electrospinning method. Homogenous TiO_2 solution as nanofibers material was prepared with titanium tetraisopropoxide (TTIP), ethanol, acetic acid and polyvinyl pyrrolidone (PVP) which was stirred for 24 h. TiO_2 solution was loaded into the syringe pump. Electrospun voltage was operated under 15 kV with optimum distance between syringe tip and collector was 15 cm. FTO substrates were attached on the collector surface. Electrospinning coating time was varied at 15 min, 30 min, 45 min, and 60 min. Then TiO_2 nanofibers layer was annealed at temperature of 450° C for 3 h. X-ray diffraction spectrum of TiO_2 nanofibers showed major anatase peaks at 25.3°, 48.0° and 37.8° correlating crystal orientation of (101), (200), and (004), respectively while only one rutile peak at 27.5° (110). TiO_2 nanofibers diameter was measured using atomic force microscopy (AFM). TiO_2 nanofibers have diameter in range of 100–1000 nm. The obtained-TiO_2 nanofibers were applied in dye-sensitized solar cell (DSSC) with beta-carotene as dye, carbon as catalyst, and I^-/I^{3-} redox couple as electrolyte. DSSC performance was analyzed from I-V characterization. Growth of TiO_2 nanofibers at electrospinning time for 45 min has highest efficiency that is 0.016 %. It is considered that TiO_2 nanofibers at electrospinning time for 45 min can produce optimum thickness so that it is speculated many dyes adsorb on the nanofiber surfaces and many electrons diffuse toward the electrodes.

1. Introduction

Nowadays, the solar cell conversion system requires on low cost materials and simple fabrication. Dye-sensitized solar cell (DSSC) as third generation of solar cells have attracted researchers due to its potential to obtain high efficiency and its low cost of component materials such as the natural dyes. Sandwich structures of DSSC consists of natural dyes as photosensitizer, semiconductor oxide layers, electrolytes and carbon counter electrodes which are grown on the fluorine tin oxide (FTO) substrates as photoelectrode of transparent conductive oxide [1-2]. Titanium dioxide (TiO_2) is one of semiconductor oxides which act as the active layer plays role in DSSC due to its performance on the light trapping and dyes absorber and affect the DSSC performance.

Due to the high light trapping and the lifetime of electrons, modification toward one dimensional in the active layer is important because it can improve the DSSC performance [1]. Several methods have been developed to obtain one-dimensional nanostructure. One of the developed and simple method to obtain one-dimensional nanostructure is electrospinning.

Electrospinning works with simple principle by generating high-voltage between the needle tip of the syringe and the collector to produce nanofibers from polymer solution. Lee at al reported that nanofibers layer support to improve absorption in visible wavelength region [2]. In addition, the thickness of nanofibers affected the diffusion and transport electron in a cell [3]. In this research, the TiO_2 nanofibers was grown on fluorine-doped tin oxide (FTO) substrates using the electrospinning method. Thickness of TiO_2 nanofibers was determined by variation of coating time. It is well-known that nanofibers structure of TiO_2 can enhance the surface area so that it could absorb photon at broader wavelength [4]. Then, the obtained-TiO_2 nanofibers was applied in DSSC and it was calculated its performance.

2. Experimental

TiO_2 was synthesized from solutions such as polyvinylpyrrolidone (PVP, Mw= ~29,000), ethanol and acetic acid as catalyst. Titanium (IV) isopropoxide (TTIP) was added to catalyst solution and stirred for 24 h to obtain a homogeneous clear solution of TiO_2. Solution was loaded into a syringe pump. The fluorine-doped tin oxide (FTO) substrate was attached on the collector surface. The electrospinning was operated at 15 kV with optimum distance between the syringe tip and the collector surface was 15 cm. Coating time of electrospun was varied at 15 min, 30 min, 45 min, and 60 min. TiO_2 nanofibers layer formed on FTO substrates. Then TiO_2 nanofibers was annealed at 450° C for 3 h to remove PVP and organic compound and remaining TiO_2 nanofibers layer on FTO substrates [5].

Phase of TiO_2 nanofibers were characterized using the X-ray diffraction (XRD). The diameters of TiO_2 nanofibers were determined using the atomic force microscopy (AFM) in area of 20×20 μm. The performances of DSSC using TiO_2 nanofibers for each coating time were characterized using I-V meter. V_{oc} (open circuit voltage), I_{sc} (short circuit current), V_{max} (maximum voltage) and I_{max} (maximum current) were determined from the I-V curves. These data were used to calculate performance of DSSC. Calculation of DSSC performance has been demonstrated in the previous work [6].

DSSC was formed in a sandwich structure with TiO_2 nanofibers as work electrodes, beta-carotene as dyes, I^-/I^{3-} as electrolytes, and carbon layers as counter electrodes. The layer active was designed in area size of 1.5 × 1 cm^2. Beta-carotene dyes was extracted from the carrot through technique as seen in reference [6]. The beta-carotene dye was confirmed using the ultra-violet visible (UV-Vis) spectroscopy. The TiO_2 nanofibers was immersed into the dye solution for 24 h.

3. Results and Discussion

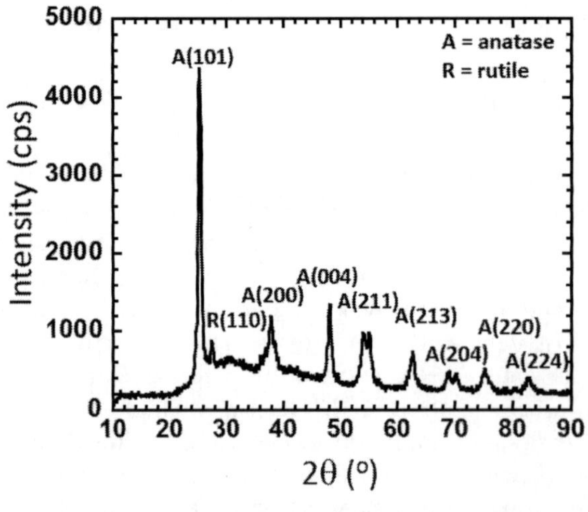

Figure 1. X-ray diffraction peaks of TiO_2 nanofibers after annealed at 450°C for 3 h

Figure 1 shows a diffraction pattern of TiO_2 nanofibers with coating time of 15 min after annealed at 450°C for 3 h. Peaks are dominated by anatase phase at crystal orientation of (101), (200), (004), (211), (213), (204), (220), and (224) while only one rutile phase appears at crystal orientation of (110). These peaks are similar for coating times of 30 min, 45 min, and 60 min (*not shown*).

Absorbance of TiO_2 nanofibers before and after immersed in beta carotene dyes for each coating times of 15 min, 30 min, 45 min, and 60 min are shown in Figure 2. For all coating times, absorbance increases after TiO_2 nanofibers is immersed in dyes. These indicates that dyes can enhance the photon adsorption at ultraviolet and visible wavelength. For coating times of 15 min – 45 min, absorbance of TiO_2 nanofibers increase as coating time increases. This is in agreement with Paez and Matousek [7] who reported that optical absorbance increase with increasing number of coating (thickness). While for coating time of 60 min, absorbance of TiO_2 nanofibers decrease below coating time of 45 min. It is considered that since the FTO substrate is attached on the collector in vertical position, coating time of 60 min is too long so that TiO_2 nanofibers coated on FTO surface will flow down. It is concluded that coating time of 45 min is the optimum time to obtain the optimum thickness of TiO_2 nanofibers layer on the FTO substrates.

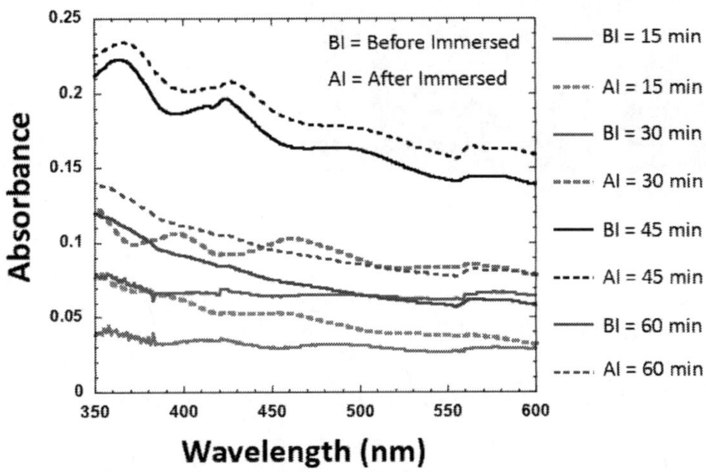

Figure 2. Absorbance of TiO_2 nanofibers before (BI) and after (AI) immersed in beta carotene dyes for 24 h for each coating times of 15 min (red), 30 min (green), 45 min (black), and 60 min (blue)

Ghelich et al [8] evaluated that quality nanofibers are affected by instrument parameters of electrospinning method such as applied voltage, tip-target distance, flow rate, needle diameter and solution parameters. However, in this research, instrument parameters are kept constant such as voltage = 15 kV, distance of syringe tip and collector = 15 cm, and needle diameter = 0.55 mm.

Figure 3 shows morphology of TiO_2 nanofibers in size of 20×20 μm^2 after annealed at 450°C for 3 h. TiO_2 nanofibers have non-uniform diameters in the range between 100 nm – 1000 nm. Line profile (red) shows fiber has diameter of 309 nm and height of 119 nm while line profile (green) shows fiber has diameter of 469 nm and height of 262 nm. Fibers have two types i.e. rounds fiber and ribbon-like fibers. These types always appeared at all coating times after annealed at 450°C. Ghelich et al [8] reported that with increasing the solution concentration, the fibers were tangled together and it could be obtained that after calcination morphology was changed from fibrous-like to flat materials. In this research, it is speculated that the TiO_2 solution was too high but the annealed temperature was still low so that several fibrous-like changed to ribbon-like fibers.

Previous study, we successfully synthesized the TiO_2 nanorods and it was applied in the DSSC [9]. TiO_2 nanorods could enhanced the DSSC performance about 154.20% compare to using TiO_2 nanoparticles only. As TiO_2 nanofibers has function similar to TiO_2 nanorods i.e. to adsorb the dyes and to trap the photon, it is hypothesized that using the TiO_2 nanofibers also will enhance the DSSC performances. However, in this research we just grown TiO_2 nanofibers on FTO substrates using electrospinning method and its application in DSSC. While effect of TiO_2 nanofibers and nanoparticles on the DSSC performance will be reported in the other paper.

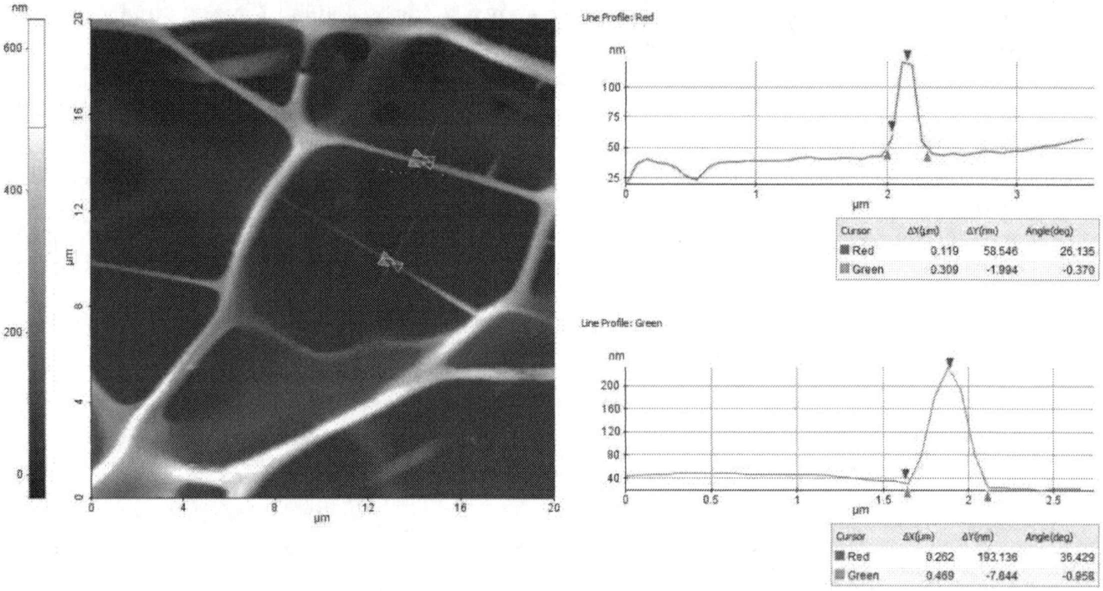

Figure 3. AFM image of TiO_2 nanofibers after annealed at 450°C for 3 h in 20×20 µm² (left side) and line profiles (red and green colors) at certain nanofibers (right side)

Table 1. I-V characteristic of TiO_2 nanofibers for each coating time

I-V Characteristic	Coating time (min) 15	30	45	60
V_{oc} (Volt)	0.08	0.19	0.01	0.16
I_{sc} (A)	6.55×10^{-6}	1.58×10^{-6}	0.11×10^{-3}	3.47×10^{-3}
V_{max} (Volt)	0.13	0.22	0.19	0.01
I_{max} (A)	2.06×10^{-5}	5.25×10^{-5}	0.16×10^{-3}	3.07×10^{-5}
P_{max}(Watt)	2.6×10^{-6}	1.6×10^{-5}	3.0×10^{-5}	5.5×10^{-6}
η (%)	1.3×10^{-3}	8.0×10^{-3}	16×10^{-3}	3.0×10^{-3}

Table 1 is summary of the I-V characteristics for each the coating time i.e. 15 min, 30 min, 45 min, and 60 min. Excluding for coating time of 60 min due to technical problem as explained above, the DSSC performances increase with increasing coating time or thickness. Meanwhile, in the case of TiO_2 nanoparticles, the DSSC performance increases as thickness decreases because of the barrier of the nanoparticle boundaries [10]. It is indicated that nanofiber creates the channeled electron transport through the one-dimensional nanostructures [11]. The nanowire or nanofiber morphology provides direct conduction paths for the electrons from the point of injection to the collection electrode while maintaining high surface area for dye adsorption [4]. Furthermore, nanofibers could solve the trapping/detrapping in the nanoparticles case during electron transport.

Journal of Physics: Conference Series **776** (2016) 012006 doi:10.1088/1742-6596/776/1/012006

4. Conclusion

TiO_2 nanofibers have been successfully grown on the FTO substrates using electrospinning method with instrument parameters such as voltage, distance of needle tip and collector, and TiO_2 solution were kept constant. Fibers have two types i.e. rounds fiber and ribbon-like fibers. Coating times are variated at 15 min, 30 min, 45 min, and 60 min. Excluding coating time of 60 min, the DSSC performance increases with increasing the coating times. It is considered that nanofibers could improve the electron-collection efficiency by promoting faster electron transport and slower recombination while maintaining high surface are for dye adsorption.

Acknowledgments

Authors wish to thank PNBP Universitas Sebelas Maret 2016 for financial support through Hibah Peneliti Utama sebagai Rujukan MRG.

References

[1] Shin I, Seo H, Son M, Kim J, Prabakar K and Kim H 2010 *Curr. Appl. Phys.* **10** S422
[2] Lee J, Ahn K, Kim S H, Kim J M, Jeong S, J. Jin, Jeong E D and Cho C 2014 *Curr. Appl. Phys.* **14** 856
[3] Filipi M, Berginc M, Smole F and Topi M 2012 *Curr. Appl. Phys.* **12** 238
[4] Baxter J B and Aydil E S 2005 *Appl. Phys. Lett.* **97** 043522
[5] Nair S, Jose R, Shengyuan Y and Ramakrishna S 2011 *J. Colloid Inter. Sci.* **353** 39
[6] Suryana R, Khoiruddin and Supriyanto A, "Beta-carotene dye of daucus carota as sensitized on dye-sensitized solar cell", in *Materials Science Forum* **737**, eds Abdullah M *et al.* 20 September 2012 (Switzerland: Trans Tech Publication) pp. 15-19
[7] Paez L R and Matousek J 2005 *Ceramics-Silikaty* **48(2)** 66
[8] Ghelich R, Rad M R and Youzbashi A A 2015 *J. Eng. Fiber Fabr.* **10(1)** 12
[9] Suryana R, Sehati and Kusumandari, "Synthesis of nanorods titanium dioxide via anodic alumina membrane template and their application in dye-sensitized solar cells", in *J. Phys.: Conf. Ser.* **737** 01208, eds Waris A *et al.* 19-20 Agustus 205 (IOP Publishing) pp. 1-4
[10] Ardhani N, Supriyanto A, Yuwono A H and Suryana R, "Influence of mass ratio of aquadest and TTIP on the synthesis of TiO_2 nanoparticles to improve the performance of DSSC with beta-carotene as sensitizer", in *Advanced Material Research* **896**, eds Triyana K *et al.* 17-18 September 2013 (Switzerland: Trans Tech Publications) pp. 481-484
[11] Othman M A, Ahmad B H and Amat F 2013 *J. Semicond. Tech. Sci.* **13(6)** 635

Journal of Physics: Conference Series **776** (2016) 012007 doi:10.1088/1742-6596/776/1/012007

Optical properties of natural dyes on the dye-sensitized solar cells (DSSC) performance

D. D. Pratiwi, F. Nurosyid[1], A. Supriyanto, and R. Suryana
Department of Physics, Faculty of Mathematics and Natural Sciences, Sebelas Maret University, Jl. Ir. Sutami 36A Kentingan Jebres Surakarta 57126, INDONESIA

[1] E-mail: nurosyid@yahoo.com

Abstract. This study reported several natural dyes for application in dye-sensitized solar cells (DSSC). This study aims was to determine the effect of optical absorption properties of natural dyes on efficiency of DSSC. The sandwich structure of DSSC consist of TiO_2 as working electrode, carbon layer as counter electrode, natural dyes as photosensitizer, and electrolyte as electron transfer media. The natural dyes used in this experiment were extracted from dragon fruit anthocyanin, mangosteen peels anthocyanin, and red cabbage anthocyanin. The absorbance of dyes solutions and the adsorption of the dye on the surface of TiO_2 were characterized using UV-Vis spectrophotometer, the quantum efficiency versus wavelength was characterized using incident photon-to-current efficiency (IPCE) measurement system, and the efficiency of DSSC was calculated using *I-V* meter. UV-Vis characteristic curves showed that wavelength absorption of anthocyanin dye of red cabbage was 450 – 580 nm, anthocyanin of mangosteen peels was 400 – 480 nm, and anthocyanin of dragon fruit was 400 – 650 nm. Absorption spectra of the dye adsorption on the surface of TiO_2 which was resulted in the highest absorbance of red cabbage anthocyanin. IPCE characteristic curves with anthocyanin dye of red cabbage, mangosteen peels anthocyanin, and dragon fruit anthocyanin resulted quantum efficiency of 0.058%; 0.047%; and 0.043%, respectively at wavelength maximum about 430 nm. *I-V* characteristic curves with anthocyanin dye of red cabbage, mangosteen peels anthocyanin, and dragon fruit anthocyanin resulted efficiency of 0.054% ; 0.042% ; and 0.024%, respectively.

1. Introduction

Dye-sensitized solar cells (DSSCs) are energy devices for converting light energy into electricity [1,2]. DSSC offered the advantages as new class of low cost and easy to fabricate, and can achieve high solar energy conversion efficiency [3]. Also, they are not involve the complex technology so that can reduce of production cost and the materials are easily obtained from the environment [1]. The standard structure of a DSSC consist of transparent conducting oxide as substrate (e.g. FTO or ITO), semiconductor layer from TiO_2 or ZnO as photo anode, Pt or C as counter electrode, dye molecules as photosensitizer, and electrolyte as electron transfer media [4]. One of the main factors which affect in the DSSC performance is dye.

The dyes have been developed by the researchers, both synthetic and natural dyes. In current studies of DSSCs, synthetic dyes can reach conversion efficiency at 11 – 12% [5,6]. In generally, these kinds of synthetic dyes, such as N3 and N719. However, they have high production costs, containing the heavy metals so that create environmental pollution. Alternatively, natural days can be used as a photosensitizer for the purpose with environment friendliness. The advantages of natural dyes include widely available, cost effective, and environment friendliness. Natural dyes can be extracted from various plants, fruits, flowers, and leaves [7]. Several natural dyes pigments such as flavonoid, betacyanin, anthocyanin, chlorophyll, tannin, and β-carotene have been successfully as sensitizers of DSSCs [8]. Many research have been developed kinds of natural dyes. Chang et al. employed chlorophyll dye from wormwood and anthocyanin dye from red cabbage as sensitizer in DSSC and reported conversion efficiency of 0.9% and 1.47%, respectively [9]. Shalini et al. in the year 2015 have been successfully took extracts from leaves, seeds, flowers, fruits, vegetables, and tree barks as natural dyes. His research results showed that the best of DSSC parameters by pomegranate peel (flavonoid pigment), raspberries (betacyanin pigmen), moses-in-the-cradle (chlorophyll pigmen) [10]. A number studies have been mentioned about dyes have not many studies the optical properties of the dyes adsorption to the working electrode and related the contribution of dyes for energy conversion efficiency and quantum efficiency. The optical properties of the dye adsorption on the surface of TiO_2 is an important parameter to determine the efficiency of DSSC [11]. The dye molecules collected light and produced excitation of electrons [12]. Therefore, in this review article reported the effect of optical properties about dye absorbance for quantum efficiency and energy conversion efficiency of DSSC.

2. Methods

The used materials are natural dyes from red cabbage, mangosteen peels, and dragon fruit as photosensitizer, *Fluorine-doped Tin Oxide* (FTO) conductive glass from Dyesol was cut the size of 2cm x 2.5 cm as substrate, TiO_2 powder 21 nm as working electrode semiconductor (photoanode), carbon as counter electrode (photocathode), electrolyte solution from EL-HPE Dyesol as electron transfer media, different solvents (ethanol, methanol, acetic acid, and distilled water) as extraction solvent of dyes, graphite pencil-2B and candle as carbon for counter electrode, and filter paper to remove solid residues from the anthocyanin extracts.

TiO_2 films were coated on the FTO conductive glass by spin coating using TiO_2 pastes, annealing process of TiO_2 films by furnace, fresh dyes were milled by mortar, homogeneity of dyes solutions by magnetic stirrer hot plate. The absorbance of dyes solutions and the adsorption of the dye on the surface of TiO_2 were characterized using UV-Vis spectrophotometer (Lambda 25), the quantum efficiency versus wavelength was characterized using incident photon-to-current efficiency (IPCE) measurement system (QEX7 Serial #150), and the efficiency of DSSC was calculated using *I-V* meter (Keithley 2602A).

Red cabbage was cut into very small pieces and mashed in a mortar. It was kept in solvents of distilled water, methanol, and acetic acid for 24 hours. Mangosteen peels were milled and dried for 3 – 5 dyes. Sample of mangosteen peels powder was soaked in ethanol with stirring for 12 hours at room temperature. Fresh dragon fruit was mashed in a mortar and added solvents of distilled water, ethanol, and acetic acid. It was kept for 24 hours. The extract of anthocyanin dyes solutions were filtered using filter paper, respectively. The absorption spectra of dyes solutions were characterized by using UV-Vis spectrophotometer.

TiO_2 pastes were prepared using TiO_2 powder in ethanol. TiO_2 solution was stirred by 300 rpm for 30 minutes. TiO_2 pastes were deposited on the FTO conductive glass with an active area of 1cm x 2cm by spin coating technique. TiO_2 films were annealed in a furnace at 400°C for 10 minutes.

Counter electrodes were prepared from carbon catalyst on the FTO conductive glass. Working electrode and counter electrode were assembled into a sandwich type arrangement. Electrolyte solution was filled the space between the two electrodes. The DSSCs were characterized using incident photon-to-current efficiency (IPCE) measurement system and *I-V* Meter.

3. Results and Discussion

3.1. *Absorption spectra of different natural dyes solutions and the dye adsorbed on the TiO_2 surface*

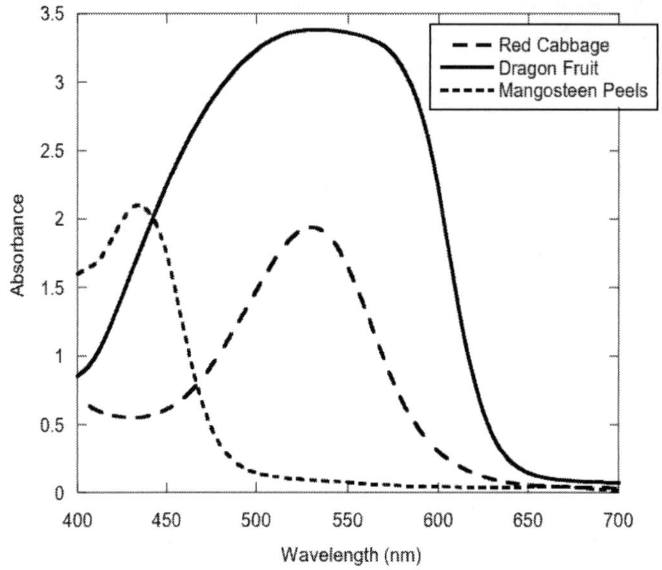

Figure 1. UV-Vis absorption spectrum of different anthocyanin natural dyes solutions

Figure 1 shows the absorption spectra of different anthocyanin dyes solutions. The absorption wavelength of anthocyanin dye of red cabbage was 450 – 580 nm, anthocyanin of mangosteen peels was 400 – 480 nm, and anthocyanin of dragon fruit was 400 – 650 nm. The results of absorption wavelength showed that the dyes solution in Fig.1 can be used as photosensitizer for DSSC because they appeared the absorption of visible region. The graph in Fig.1 showed that the dye solution of dragon fruit anthocyanin was resulted in the widest and highest of absorbance.

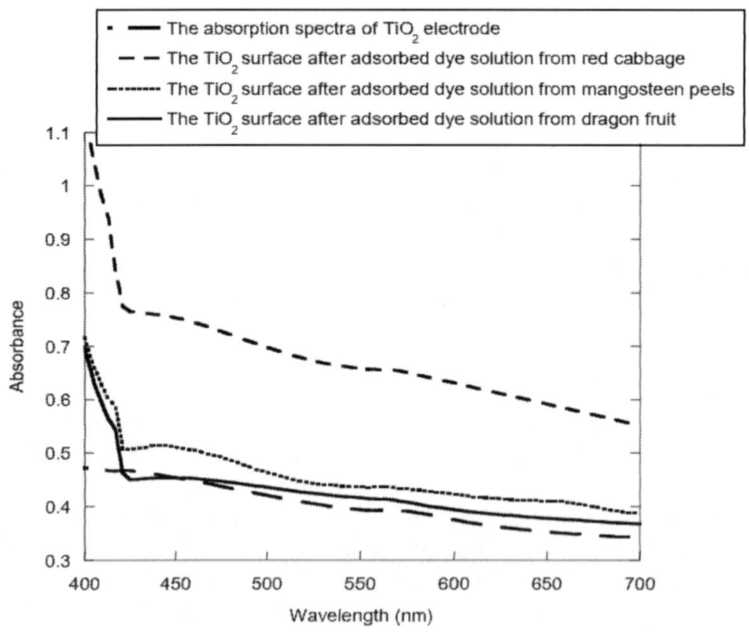

Figure 2. UV-Vis absorption spectrum of TiO_2 electrode and dye after adsorption on the TiO_2 surface

Figure 2 shows the absorption spectra of TiO_2 films can increase after the dye adsorption at wavelength above 400 nm. This result showed that the dye adsorption on the surface of TiO_2 can be used as photosensitizer for DSSCs absorbed of light in the visible region. The dye solution of dragon fruit anthocyanin was resulted in the widest and highest of absorbance, however when the dye adsorption by dragon fruit anthocyanin on the surface of TiO_2 was resulted in the lowest of absorbance. The dye adsorption by red cabbage anthocyanin on the surface TiO_2 was resulted in the highest of absorbance than the others of anthocyanin dyes were presented in Fig.2. This situation may be explained that the dye adsorption by dragon fruit anthocyanin on the surface of TiO_2 less than the dye of red cabbage anthocyanin.

3.2. Incident Photon-to-Current Conversion Efficiency (IPCE) Characterization

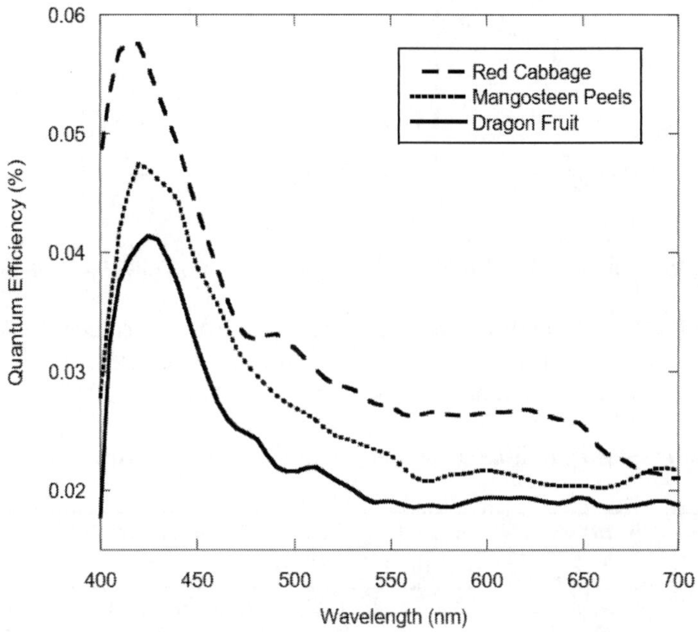

Figure 3. IPCE spectrum of DSSCs with different anthocyanin natural dyes

The IPCE versus wavelength curve obtained is shown in Figure 3. It shows that the dyes has effectively worked the wavelength in the range of 400 nm – 460 nm at wavelength maximum about 430 nm. This result shows the wavelength region contributed in highest current formation. The IPCE spectrum showed the decrease after in wavelength about 460 nm. This situation can be explained that the dye adsorption on the surface of TiO_2 was not work optimally in visible region because the back transfer of electrons excitation on the TiO_2 trapped back into the UV region, so that the photons were converted into the current would be slightly [13,14]. Most of the molecules of dyes that attached on the surface of TiO_2 has not effectively inject the electron to electrode. The result of IPCE curve was comparable with the result of absorbance curve of the dye adsorption on the TiO_2 surface (Figure 2). The dye adsorption on the surface of TiO_2 was higher as the higher contribution of photons was absorbed, so that the result of efficiency quantum was also higher. The DSSC of red cabbage anthocyanin resulted in the highest of quantum efficiency.

3.3. Photocurrent – Photovoltage Characterization

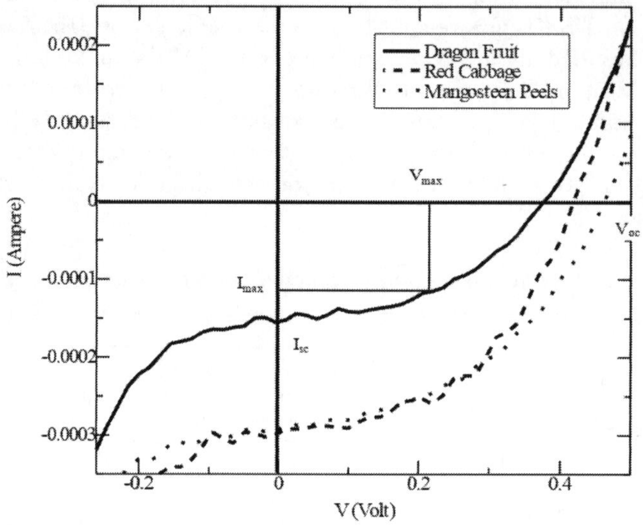

Figure 4. Current-voltage curves for the DSSCs by different anthocyanin natural dyes

For the *I-V* curves by Fig.1, the samples were illuminated with Xe light source using *I-V* Meter. The light intensity was kept constant at 1000 W/m^2 and active area of the cell exposed to light was 2 cm^2 (1 x 2 cm). From *I-V* plot can be calculated parameters of DSSCs as reported in Table 1 :

Table 1. Parameters of DSSCs with different anthocyanin natural dyes extracted from vegetable, fruit, and rind.

Dye solution	Photos	I_{max} (mA)	V_{max} (V)	I_{sc} (mA)	V_{oc} (V)	FF	η (%)
Hylocereus polyrhizus (Dragon Fruits)	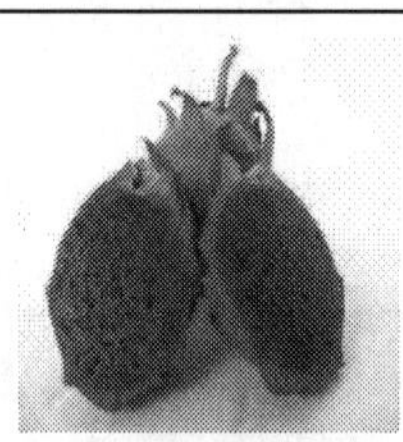	0.17	0.28	0.23	0.34	0.63	0.024
Brassica oleracea var (Red cabbage)		0.35	0.31	0.49	0.43	0.51	0.054
Mangosteen pericarp		0.25	0.34	0.38	0.46	0.48	0.042

The conversion efficiency value of DSSCs were resulted in the highest efficiency of DSSC by red cabbage anthocyanin. This result was comparable with IPCE curve in Fig.3. The quantum efficiency was higher showed as the higher contribution of photons was absorbed, so that the conversion efficiency of DSSC was also higher. It was also supported by UV-Vis characteristic of the dye adsorption on the TiO_2 surface showed that in the highest absorbance of red cabbage anthocyanin. This study reported that the efficiency DSSC was higher as the higher adsorption of the dye on the surface of TiO_2.

Table 1 displayed that the open circuit voltage (V_{oc}) of measured DSSCs were fluctuated, while the measured short circuit current (I_{sc}) increased as the increasing of dye adsorption on TiO_2 surface. This result can be explained that the absorption of photons were converted to electrons have least. The dye adsorption on the TiO_2 surface was not dominant work in the visible region. The efficiency has a low because the output current has least. This situation effected the conversion efficiency is low.

In conclusion, the absorption spectra of dye adsorption on the surface of TiO_2 which was resulted in the highest absorbance of red cabbage anthocyanin. This result was comparable to the results of conversion efficiency and quantum efficiency for DSSCs that solar cells using red cabbage anthocyanin as sensitizers has highest values of efficiency. The efficiency was higher as the higher adsorption of the dye to the surface of TiO_2.

Acknowledgments
This work is supported by Directorate of Research and Community services, Ministry of Research, Technology, and Higher Education and Institute for Research and Community services, Sebelas Maret University by PUPT Grant Program. (Contracts Number : 353/UN27.21/PN/2016)

4. References

[1] O'regan B, and Grätzel M 1991 *Nature.* **353** 737–740
[2] Grätzel M 2001 *Nature.* **414**
[3] Grätzel M 2003 *J.Photochems.Photobio.* **4** 2 145–153
[4] Alhamed M, Issa AS, and Doubal AW 2012 *J.Electron.Devices.* **16** 2012 1370–1383
[5] Chiba Y, Islam A, Watanabe Y, Komiya R, Koide N, and Han LY 2006 *Japanese.J.Appl.Phys.* **45** 24–28 638–640
[6] Buscaino R, Baiocchi C, and Barolo C 2008 *Inorganica.Chimica.Acta.* **361** 3 798–805
[7] Shanmugama V, Manoharanb S, Anandanb S, and Murugan R 2013 *Spectrochimica.Acta.A.* **104** 35–40
[8] Garcia CG, Polo AS, and Iha NYM 2003 *J.Photochem.Photobiol A: Chems.* **160** 87–91
[9] Chang H, Kao M, Chen T, Chen C, Cho K, and Lai X 2013 *J.Photoenergy.* **2013** 8
[10] Shalini S, Prabhu RB, Prasanna S, Mallick TK, and Senthilarasu S 2015 *Renewable and Sustainable Energy Reviews.* **51** 2015 1306–1325
[11] Torchani A, Saadaoui S, Gharbi R, and Fathallah M 2015 *Current. Appl. Phys.* **15** 2015 307–312
[12] Chiba Y, Islam A, Watanab Y, Komiya R, Koide N, and Han L 2006 *Japanese.J.Appl.Phys.* **45** l638–l640
[13] Govorov AO, Zhang H, and Gun'ko YK 2013 *J.Phys.Chems.* **117** 166 16–31
[14] Govorov AO, Zhang H, Demir HV, and Gun'ko YK 2014 *Nano Today.* **9** 85–101

Journal of Physics: Conference Series **776** (2016) 012008 doi:10.1088/1742-6596/776/1/012008

Dye *Oriza sativa glutinosa* doped Fe as a active element of Dye Sensitized Solar Cell (DSSC)

A B Prasada[1], U M Fadli, Cari and A Supriyanto

Physics Department of Post Graduate Program Sebelas Maret University,
Jl. Ir. Sutami 36A Kentingan Jebres Surakarta 57126, INDONESIA

Email: asharibayu01@gmail.com

Abstract. The aims of the research are to determine the effect of doping Fe (III) Sulphate into dye *Oriza sativa glutinosa* on the characteristics parameters of solar cells, to determine the optical characteristic, functional group and electrical characteristic of dye *Oriza sativa glutinosa* doped Fe (III) sulphate. TiO_2 nano size as much as 0.5 gr dissolved in 3 ml ethanol. 100 gr black sticky rice (*Oriza sativa glutinosa*) was immersed in 80 ml ethanol solution (95%) and kept at room temperature without exposing to light. Then it was filtered with a filter paper no.42, and the extracted result was process with chromatography. Furthermore, it was doped with Fe (III) sulphate respectively of 10^{-1} M, 10^{-2} M, 10^{-3} M. The characteristic of dye solution was measured using UV-Visible *Spectrophotometer Lambda 25* for absorbance, *Elkahfi 100/I-V meter* for conductivity amd *Keithey 2602A* for characterization of current and voltage (I-V). The result showed that the area of dye *Oriza sativa glutionosa* doped Fe (III) sulphate with concentration 10^{-1} M the largest, because the value of V_{oc} intercept at 6.40×10^{-1} mV and the value I_{sc} intercept at 1.89×10^{-3} mA, with efficiency value is 0.148%.

1. Introduction

Dye sensitized solar cells (DSSC) is one type of solar cells which is attractive to be studied and development as an alternative way to substitute the conventional solar cells based on silicon semiconductor over the last decade. There are some advantages of DSSC, easy to fabricate, low cost and environmental friendly [1]. A DSSC is composed of a nanocrystalline porous semiconductor electrode-absorbed dye, a counter electrode, and an electrolyte containing iodine and triiodide ions. [2]. The principle of operation of DSSC is based on sensitization of a wide band-gap metal oxide semiconductor to the visible light region by an adsorbed molecular dye. When the metal oxide semiconductor film is immersed in the dye, a monolayer of the dye is anchored onto its surface. Excitation of atoms by sun light occurs in the dye and the photogenerated charges are separated at the interface between the dye and metal oxide [3]. The DSSC sensitized by Ru-containing compounds are reaching, 11-12% the highest efficiency as reported by Chiba and Islam [1]. However, noble metals limited in amount, and costly in production. Therefore, alternative organic dyes such as natural dyes have been studied intensively. Although its absorbance ability is lower than dye synthetic, such as ruthenium complex, but the findings of Hao et al. [4] and Yuliarto et al. [5] showed that extracts of black rice can be used as photosensitizer in DSSC. Furthermore, Hao et al. [4] also report that the dye of black rice extract is the best natural dye be caused, this kind of dye can absorb wavelength in wide spectrum. The anthocyanin pigments contained in black rice was over 80% of cyaniding-3-glucoside, the other minor pigments were peonidin-3-glucoside, malvidin-3-glucoside, and cyanidin-3-ramnoglucoside [6]. Saehana et al. [7] reported that a conversion efficiency of 0.17% was obtained using black rice dye as sensitizer. Until now, variations of the optimization is done by many researchers to improve the efficiency and stability of performance dye sensitized solar cell. Therefore, various engineering has been carried out such as by modifying the TiO_2 electrode, electrolyte and, dye used [8]. This research studied the effect of doping Fe (III) Sulphate into dye *Oryza sativa glutinosa* as photosensitizer for dye sensitized solar cells. Selection of iron complex is based on ferrous metals located in the transition metals in which the electron configuration d^6 as ruthenium and Osmium (used as dye solar cells), have quantum relatively high to produce sensitization of nanocrystalline TiO_2, soluble in polar solvents, and its wavelength at visible region is 551 nm.

2. Experiment Methods

2.1. *Preparation of TiO_2 Solution*

TiO_2 nano powders as much as 0.5 gr dissolved in 3 ml of ethanol, then stirred using a vortex stirrer with a speed of 300 rpm for 30 minutes. TiO_2 paste was stored in closed container lined with aluminium foil.

2.2. *Preparation of Natural Dye Sensitizers and Doping Fe (III) Sulphate*

100 gr Black sticky rice (*Oriza sativa glutinosa*) was immersed in 80 ml ethanol solution (95%) and keep at room temperature without exposing to light [9]. Then it was filtered with a filter paper no.42, and the extracted result was process with chromatography. Furthermore, it was doped with Fe (III) sulphate. $Fe_2(SO_4)_3.7H_2O$ respectively of 10^{-1} M, 10^{-2} M, 10^{-3} M which was dissolved in 40 ml of ethanol and each stirred for 2 hours at 60°C to form a solution of Fe (III) sulphate for each concentration. The solution of Fe (III) sulphate with various concentrations was mixed in 40 ml for each dye extracted from black rice, ratio of Fe (III) sulphate and extracted dye (1:1), while stirred for 2 hours.

2.3. *Preparation of Electrolyte*

Natrium iodid (NaI) of 2 grams are mixed into 3.68 ml acetonitrile added 1 ml propylene carbonate. Further to the solution is added 14.56 ml polyethylene glycol then stirred. Further to the solution is added iodine (I_2) of 0.2 grams was stirred with a vortex stirrer at 300 rpm for 30 minutes. Electrolytes solution was stored in closed containers lined with aluminum foil.

2.4. *Preparation of Counter Electrode*

To prepare platinum catalyst coated on FTO glass substrate, 1 ml hexachloroplatinic acid solution in 207 ml isopropanol was spread on the FTO conducting surface. The Pt-coated FTO counter electrode was then heated in a furnace at 250^0C for 15 minutes.

2.5. *Preparation of the TiO_2/Dye Electrode*

The FTO glass (1×1 cm^2) was cleaned with distilled water and ethanol to remove impurities. The cleaned FTO surface was coated with TiO_2 paste using spin coating techniques. The FTO/TiO_2 electrodes were sintered at 450°C for 30 minutes. The electrode were immersed in the dye solution for 24 hours. The white TiO_2 film will change color when dye is absorbed.

2.6. *Fabrication of Natural Colorant-Sensitized Solar Cells (DSSC)*

Natural colorant-sensitized solar cells with active area about 1 cm^2 were fabricated by sandwiching the electrolyte between TiO_2/dye electrode and Pt counter electrode. The *I-V* characteristics of DSSC were obtained under light illumination (1000 W/m^2) using keithley 2602A source meter.

3. Result and Discussion

3.1. *Absorption Spectra of Black Sticky Rice doped Fe (III) Sulphate*

Before being used as a sensitizer, extract black sticky rice and extract black sticky rice doped Fe (III) sulphate, first characterized using UV-Visible *Spectrophotometer Lambda 25*. Absorbance spectra measured in the wavelength range of 400-800 nm.

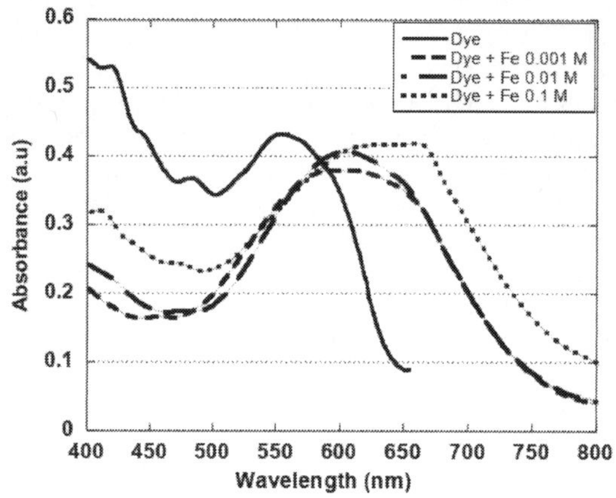

Figure 1 Absorption Spectrum of dye and dye doped Fe (III) sulphate

Figure 1 shows that absorbance maximum extract black sticky rice without doped Fe (III) sulphate at 0.432 a.u with wavelength 553 nm. Extract black sticky rice doped Fe 10^{-3} M at 0.379 a.u with wavelength 595 nm, extract black sticky rice doped Fe 10^{-2} M at 0.406 a.u with wavelength between 601 nm, and extract black sticky rice doped Fe 10^{-1} M at 0.419 a.u with wavelength 660 nm. Absorbance dye of the fourth materials showed that the value of wavelength of these material are able to work on the UV-Visible light. From Figure 1, doping Fe (III) sulphate into the dye causes peak absorbance by dye shifted from wavelength 553 nm becomes 595 nm, 601 nm, and 660 nm. In this case there has been a process of intermolecular copigmentation between anthocyanin extract of black sticky rice with copigmen compound of Fe (III) sulphate which is characterized by a bathochromic shift. Bathochromic shift causes absorption peak shifts towards larger wavelengths.

3.2. Fourier Transform Infrared (FT-IR) Analysis

FTIR spectroscopy is a characterization methods for identification of functional groups by measuring the absorbance at all wavelengths of IR. FTIR spectra of each sample can be shown in the figure 2

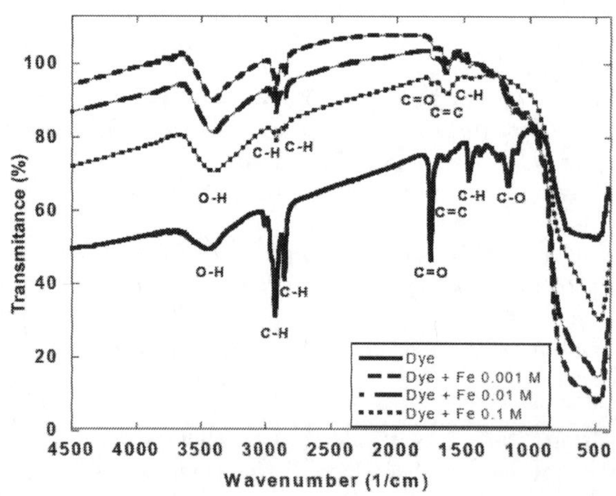

Figure 2 FTIR spectral analysis of dye and dye doped Fe (III) sulphate

Figure 2 shows the different functional groups of extract black sticky rice and black sticky rice doped Fe with concentrations 10^{-1} M, 10^{-2} M, 10^{-3} M. Extract black sticky rice without doped Fe (III) sulphate have the absorbance at wavenumber 3440.19 cm^{-1}, 2925.17 cm^{-1}, 2853.81 cm^{-1}, 1745.65 cm^{-1}, 1632.81 cm^{-1}, 1464.03 cm^{-1}, and 1164.09 cm^{-1} which is a compound of the O-H *stretching*, C-H (alkana), C-H (alkana), C=O *stretching*, C=C *stretching*, C-H (alkana), and C-O *stretching*. In the extract black sticky rice doped Fe (III) sulphate with concentrations 10^{-1} M, 10^{-2} M, 10^{-3} M have a functional groups that is almost the same forms as in extract black sticky rice but different values of the wavenumber and transmitance. In other words, doping Fe (III) sulphate into extract black sticky rice make widen the area of light absorption of dyes to larger wavelengths. In extracts of black sticky rice doped Fe (III) sulphate, O-H bond alcohol *stretching* formed from the value of spectral shifts to shorter wavenumber. This absorption band shift caused by the interaction of the ligand OH on black sticky rice cyanidin molecule with a central atom of Fe (III) sulphate. These interactions multiply the occurrence of hydrogen bonds that make the absorption band widening. In spectra 2925.17 cm^{-1} and 2853.81 cm^{-1} showed the C-H bonds (alkanes) in an area stretching hydrogen, the intensity of absorption that form C-H bonds (alkanes) to be reduced due to vibration in the catchment area occurs with changes in dipole moment that is smaller because no polar OH covalent bonds with Fe (III). In spectra 1745.65 cm^{-1} showed the C=O *stretching* in the double bond with a very high intensity of absorption. Therefore carbonyl group in black sticky rice extract is very polar, so this *stretching* bond is produced change dipole moment is quite large. Then the intensity of absorption form C=O *stretching* bond is reduced because doping Fe (III) sulphate into extract of black sticky rice change molecular structure of the carbonyl group in ketones which causes the frequency of C=O *stretching* vibration become lower.

3.3. Conductivity characteristics

Further characterization is conductivity testing of dye using *Elkahfi 100/I-V meter* under irradiation of the halogen lamp 250 mW/cm^2

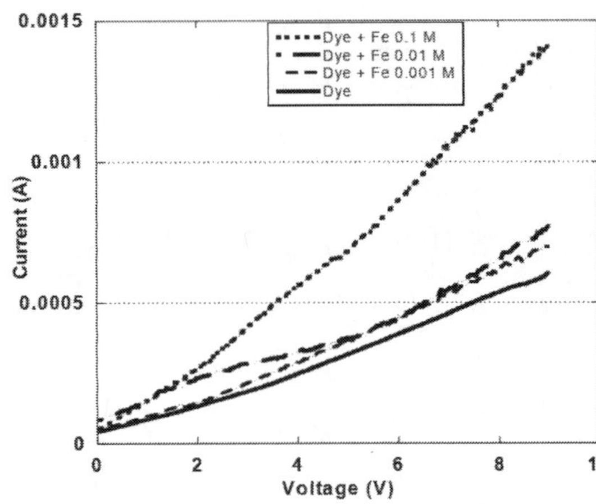

Figure 3 Conductivity Curve for dye and dye doped Fe (III) sulphate

Figure 3 shows the conductivity value of the extract black sticky rice without doped Fe (III) sulphate is 3.20×10^{-3} $ohm^{-1}.m^{-1}$, for black sticky rice doped Fe (III) sulphate with concentrations 10^{-1} M, 10^{-2} M, 10^{-3} M obtained conductivity values of each of 6.27×10^{-3} $ohm^{-1}.m^{-1}$, 4.48×10^{-3} $ohm^{-1}.m^{-1}$ and 3.67×10^{-3} $ohm^{-1}.m^{-1}$, respectively. The result shows that the tendency of increase conductivity value is proportional to the increasing molarity of Fe (III) sulphate were added into the dye.

3.4. Photovoltaic Properties

Characterization of the current-voltage (I-V) is a method to determine how much ability DSSC can convert light into electrical energy. Measurements using *Keithley 2602A* conducted in the dark and the light that is under irradiation with a halogen light intensity of 1000 W/m^2. The test results I-V with 24 hours immersion

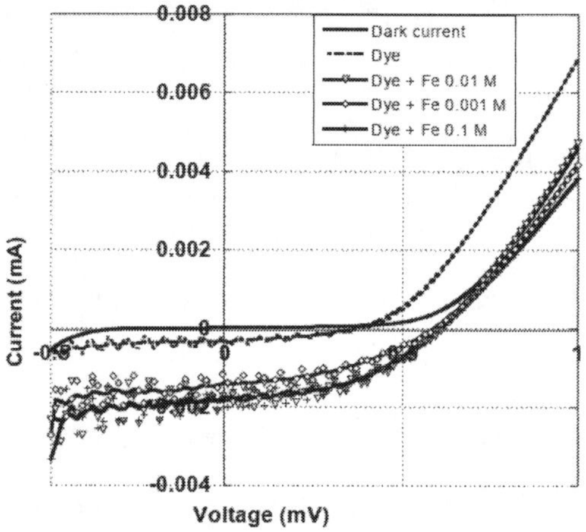

Figure 4 Current-Voltage curve for DSSC extract black sticky rice and black sticky rice doped Fe (III) sulphate

Based on the graph in Figure 4, it can be shown that the dye extract black sticky rice doped Fe 10^{-1} M has the largest area, because the value of V_{oc} intercept at 6.40 × 10^{-1} mV and the value I_{sc} intercept at 1.89 × 10^{-3} mA. Therefore, it has the largest value of efficiency. In generally, the results of the efficiency of the extract black sticky rice without Fe (III) sulphate and black sticky rice doped Fe (III) sulphate with concentrations 10^{-1} M, 10^{-2} M, 10^{-3} M, are shown in Table 1.

Table 1 Photovoltaic performance with dyes extract black sticky rice and black sticky rice doped Fe (III) sulphate concentrations 10^{-1} M, 10^{-2} M, 10^{-3} M

Dye Source	V_{oc} (mV)	I_{sc} (mA)	V_{max} (mV)	I_{max} (mA)	Fill Factor	Efficiency (%)
Black Sticky Rice	0.385	4.03 × 10^{-4}	0.385	4.23 × 10^{-5}	1.71 × 10^{-8}	0.016
Doped Fe (III) sulphate 10^{-1} M	0.640	1.89 × 10^{-3}	0.640	2.31 × 10^{-4}	4.37 × 10^{-7}	0.148
Doped Fe (III) sulphate 10^{-2} M	0.625	1.64 × 10^{-3}	0.625	2.33 × 10^{-4}	3.84 × 10^{-7}	0.145
Doped Fe (III) sulphate 10^{-3} M	0.625	1.55 × 10^{-3}	0.625	1.0 × 10^{-4}	1.55 × 10^{-7}	0.062

From Table 1 and figure 4, we can be concluded that the dye extract of black sticky rice doped Fe (III) sulphate with concentration 10^{-1} M has the largest value of efficiency (0.148%). The result shows that doped Fe (III) sulphate into the dye extract black sticky rice widens absorption spectrum and increase the conductivity of the dye. The presence of a metallic contact layer becomes the path for electrons to flow faster toward the substrate FTO. In general, the value efficiency of these material is still very small (under 1%). There are two kind of efforts to increase the performance DSSC. The first, using different counter electrode with polyaniline (PANI), because PANI can increase efficiency from 6.90% using platinum to 7.15% using PANI [10]. The second, using the kind of electrolyte which has viscosity is smaller than PEG electrolyte such as PEO polymer gel [11].

4. Summary

From the results of this study concluded that the solar cell DSSC has been successfully created by using dye extract black sticky rice and extract black sticky rice doped Fe with various concentrations. The results of the absorbance dye showed that the value of wavelength these material are able to work on the UV-Visible light. The largest conductivity value is dye extract black sticky rice doping Fe (III) sulphate with concentration 10^{-1} M, whereas the greatest efficiency in achieved by using the extract black sticky rice doped Fe (III) sulphate with concentration 10^{-1} M on 24 hour immersion in the amount of 0.148%.

Acknowledgement
This research was partly supported by LPPM Postgraduate UNS Grant with contract no. 343/UN27/HK/2016 April 7th, 2016.

References

[1] Wei YS, Jin QQ and Ren TZ 2011 *Solid-State Electronics* **63** 76
[2] Chiba Y, Islam A, Watanabe Y, Komiya R, Koide N and Han L 2006 *Japanese Journal of Applied Physics* **45** L638
[3] Chen H, Duan Z, Lu YG and Pasquier AD 2009 *Journal of Electronic Materials* **38** 1612
[4] Hao S, Wu J, Huang Y and Lin J 2006 Solar Energy **80** 209
[5] Yuliarto B, Septina W, Fuadi K, Fanani F, Muliani L and Nugraha 2010 *Adv. Mat. Sci. Eng.* **2010** 789541
[6] Lee HL 2010 *Food Sci. Biotechnol.* **19** 391
[7] Saehana S, Yuliza E, Arifin P, Khairurrijal and Mikrajudin A 2013 *Materials Science Forum* **737** 43
[8] Huang ML, Yang HX, Wu JH, Lin JM, Lan Z, Li PJ, Hao SC, Han P and Jiang QW 2007 *J. Sol-Gel Sci. Techn.* **42** 65
[9] Buraidah MH, Teo LP, Yusuf SNF, Noor MM, Kufian MZ, Careem MA, Majid SR, Taha RM and Arof AK 2011 *International Journal of Photoenergy* **2011** 273683
[10] Qinghua L, Jihua W, Qunwei T, Zhang L, Pinjiang L, Jianming L and Leqing F 2008 *Electrochemistry Communications* **10** 1299
[11] Duangkaew P, Chindaduang A and Tumcharem 2010 *Science Jurnal Ubonratchathani University* **1** 9

The influence of material type and composition of TiO_2-ZnO on manufacturing of paste for the application of DSSC

L. Retnaningsih , L. Muliani, P. N. Aggraini and J. Hidayat
Research Center Electronics and Telecomunications – Indonesian Institute of Sciences
Jl. Sangkuriang Bandung 40135, INDONESIA

E-mail: lilisretna@gmail.com

Abstract. Research, fabrication and material selection for the application of Dye-sensitized solar cell (DSSC) has been performed on glass FTO (Flour Tin Oxide). The material is used in the form of TiO_2 paste, TiO_2 powder and ZnO powder. Dye-sensitized solar cell (DSSC), is a fotoelektrokimia-based solar cells where the absorption process light done by the dye molecules and the process of separation of inorganic semiconductor materials by charge of Titanium dioxide (TiO_2) and Zinc oxide (ZnO). The purpose of this research is to know the exact composition of TiO_2 and ZnO materials in order to produce the best efficiency with DSSC. On this research was done making prototype dye-sensitized solar cell using dye Z 907, and semiconductor nanoparticles TiO2 and ZnO powder that is made into a paste by mixing different composition in two variations of samples: A = ZnO (powder) + 40% TiO2 (powder) and B = 60% TiO2 (powder) (40%) + TiO_2 (pasta) 60%. The second variation of this high efficiency is value at sample B i.e. TiO_2 (powder) + 40% TiO_2 (paste) of 60%.

1. Introduction

The main problems facing the developing countries is the energy crisis. Energy is required almost all areas, whereas the reserve source of energy available already thinning. As it is indispensable the existence of alternative energy sources. The use of renewable sources of energy currently has become the alternative energy to replace some of the energy that almost started thinning, this is done to keep the availability of sustainable energy can [1].

The many sources of renewable energy such as wind, biomass and hydro power, the use of energy through solar cell or solar cells is the most potential alternatives. The latest generation of solar cells are dye-sensitized solar cell or better known as DSSC. DSSC is one potential candidate next-generation solar cells, this is because it does not require a material with high purity so the cost of the production process is relatively low and the required materials can be derived from nature.

Solar cells alternative-based dye-sensitized solar cell (DSSC) can use the sorts – range of substrates and environmentally friendly in the fabrication [2]. The dye molecules are already commonly used to dye is a type of DSSC research ruthenium complex. DSSC using layers TiO_2 and ZnO materials as for counter electrodes using a layer of Platinum (Pt), which are mounted on glass with sputtering techniques. TiO_2 and ZnO materials both of which can be either a paste or floured with nano-meter size. Both types of these materials each have advantages and disadvantages depending on the application being performed. One that is done on a nanoparticle TiO_2 and ZnO is heat treatment called Annealing, which is useful for improving maximum DSSC's performance. Improvement of annealing temperature affects the amount of grains, the ability of current density and coloring absorbance [3]. In order to attempt to improve the efficiency of DSSC needed some way to the material selection, fabrication techniques as well as the treatment of the material.

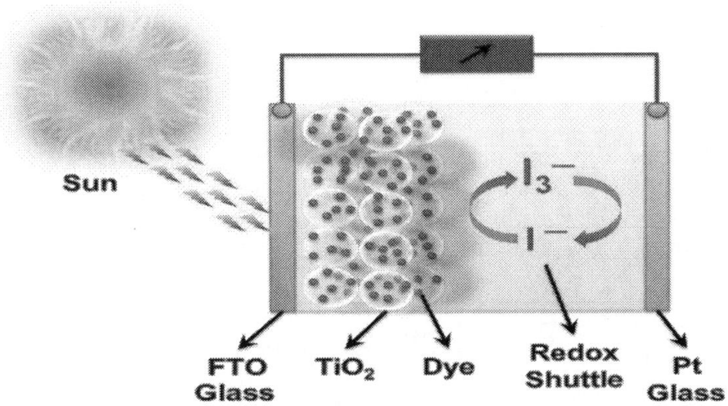

Figure 1: scheme of the DSSC's work

The existence of a layer of oxide semiconductors play an important role as the ' heart ' of a DSSC system to convert light energy into electrical energy. TiO_2 and ZnO material have a large energy gap Ribbon so that it is able to absorb the energy of the photon at the majority of spectrum sun light [4]. For its application on a DSSC, TiO_2 anatase berfasa umunya used because it has a high fotoaktif capability. In addition TiO2 with nanopori structure that is pore size in the nano scale would increase system performance because the structure nanopori has the characteristics of a high surface area, so it will increase the amount of dye absorbed the implication would sharply increase the amount of light absorbed.

The use of substance dye can influence the efficiency of the results of posting is obtained. Types of dye used can be either synthetic or natural dyes [5]. Natural dye that is derived from several types of leaves or fruit that has been extracted and taken the fluid. The use of these natural coloring price is cheaper, because it can be made but the results of their applications have yet to produce high efficiency against values obtained.

This research will be focused on making paste by mixing between the paste that has commercialization and paste that comes from nanometer-sized powder, as well as between the two different pasta. Expected results obtained can increase the value of efficient use of DSSC using pasta from powders.

1. Research Methods

Research DSSC has plenty to do with the use of materials, fabrication method and analysis. Any results obtained in accordance with each characterization.

a. Materials and the Preparation of the Paste

On the research materials used include TiO_2 powder, TiO_2 paste, and ZnO powder , which coupled with Triton X-100, PEG (Polyethylen Glycol) and Ethanol. For the tinting process using Ruthenium Dye powder Z 907, Platinum (Pt) is used as a layer of the counter electrode and electrolyte solution are injected after a merger between the electrode and the counter electrode [6]
.

In this study the paste that is used in the form:

1. the mixture of two kinds of powder particles nano
2. nano powders mixed with nano-paste

The mixing produced two variations of the mixture forms a paste. Mixing the ingredients matched with a mix of comparisons performed. On the mix between A TiO_2 nano powders + ZnO nano-powders , mixture B between TiO_2 nano-paste + TiO_2 nano-powders.
Material in the form of powder, processed into paste before the first at a temperature of sinterring 450° C for 2 hours, then added other chemicals, then processed into paste [7]. On the establishment of a prior solution of paste done stirring using a stirrer for 5 hours.

b. Fabrication of Paste on glass FTO

Mixing paste has been made, then the fabrication is done on top of the FTO glass has been prepared. Paste has been fabrikated on FTO glass which has been cleaned, shapping paste done by doctor blade, active layer with dimensions 1 cm x 1 cm. Some variations of the pasta mixture that has been fabricated, then done drying with temperature 120^0 C for 10 minutes in the oven. Then conducted Firing (burning paste lining in furnaces) and 500^0 C temperature for 30 minutes.
This process aims to form a porous paste lining is formed so that the surface has a large area and form the structure of *anatase* TiO_2.

c. Fabrication of Platinum (pt) Layer

Platinum Plating on glass done by sputtering techniques, this is the Platinum electrode layer opponent serves to accelerate the reaction of reduction of electrolytes on the cathode, and serves as a catalyst in the reaction of formation of triodida becomes iodide.

Figure 2 : Counter Electrode

d. Immersion in aqueous Dye

After firing at a temperature of 500 ^{0}C, the next layer of DSSC done soaking in solution of the Dye. Research on solution of the Dye that is used in the form of ruthenium complex Z 907 dissolved in alcohol. Solution of the dye that has been made, further stirring is done, then the only solution is for 8 hours. Soaking is done for 24 hours, this is done in order to better absorb into ruthenium layer of TiO_2 and ZnO.

e. The Assembling

Process of assembling a DSSC, namely the unification between the electrode and the counter electrode layer by snapping face each other. After both sides poised between an electrode and the counter electrode is plated, and the Samurai affixed with adhesive leaves hole to fill liquid electrolyte.

3. Results and Discussion

The process of the preparation of the paste, fabrication and assembling has been done. Next do some characterization and measurements on the material that has been fabricated. Characterization in the form of SEM, UV-Vis and IPCE has done. Next up is done charging the electrolyte solution between the electrode and the counter electrode layer, once let sit a while, then do the measurement of I-V Sun simulator to find out the efficiency, voltage, current, and the fill factor produced from some fabrication DSSC has been done [8].

Journal of Physics: Conference Series **776** (2016) 012009 doi:10.1088/1742-6596/776/1/012009

a. SEM Characterization

SEM characterization of this research was conducted to find out its homogenity of powder and paste that has been fused, and performed a stirring for a few hours by using a stirrer. After subsequent fabrication done characterization, to find out the porosity of the materials used blend between TiO_2 and ZnO powder or a form of paste.

Figur 3 : ZnO (serbuk) + TiO_2 (serbuk)
40% : 60%

Figur 4 : TiO_2 (serbuk) + TiO_2(pasta)
40% : 60%

From Figure 3 the results of SEM characterization of ZnO powder and mix namely pasta TiO_2, visible grain of the two particles not mixed evenly. From the image still visible two different particle size. The difference in grain and the presence of a difference in space could increase the value of the I-V Sun Simulator generated by a DSSC.

b. UV – Vis Characterization

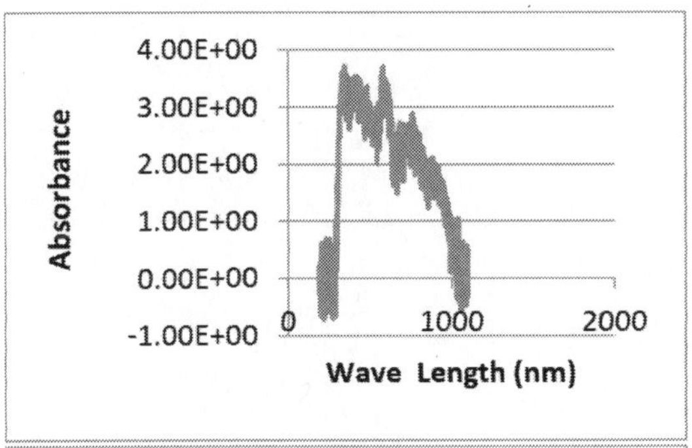

Figure 5 : ZnO (powder) + TiO_2 (powder)
40% : 60%

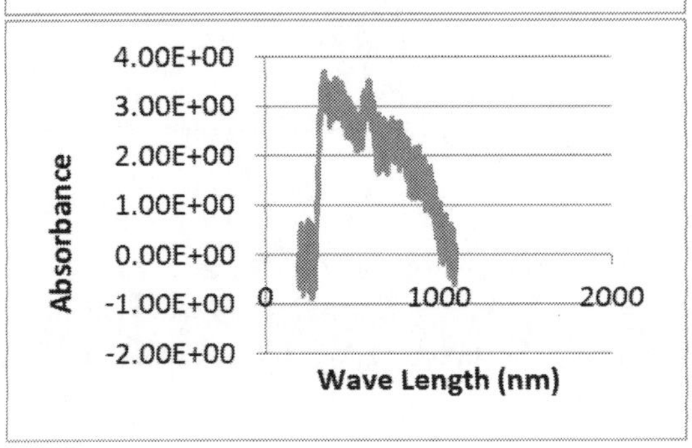

Figure 6 : TiO_2 (powder) + TiO_2 (paste)
40% : 60%

Figure 5 and 6 are the result of UV-Vis characterization of two kinds of mixed-an ZnO powder and pasta TiO2. Figure 5 is a characterization of ZnO (powder) + TiO2 (powder) ---- 40%: 60% and Figure 6 is a characterization: TiO2 (powder) + TiO2 (paste) ------ 40%: 60%.
Characterization of the second image above shows the differences are not that great.

c. *I-V Sun Simulator Characterization*
I-V Sun Simulator characterization is done to find out the capabilities of fabricating the results of DSSC's work. This characterization of the obtained measurement results in the form of efficiency, voltage, current, and fill factor. UV-Vis spectrophotometry is the measurement of the wavelength and intensity of the ultraviolet rays of visible light that is absorbed by the sample. Ultraviolet light and visible light has enough energy to promote-the electrons in the outer shell to a higher energy level. This characterization is obtained from the measurement results in the form of efficiency, voltage, current, and fill factor. Ultraviolet light is at a wavelength of 200-400 nm, while visible light is at a wavelength of 400-800 nm. The results of the characterization work efficiency refers to the ability of the DSSC has been made.
As for the paste made for fabrication above the glass substrate is a mixture between the paste and powder prolific nano particles. Mix first: ZnO (powder) + TiO_2 (powder) with comparison 40:60, the value of the measured efficiency of 0.221. The mix second : TiO_2 (powder) + TiO_2 (paste) with comparison of 40:60, generating efficiency of 0.530. As for the graphics I-V measurement results can be seen in the picture below.

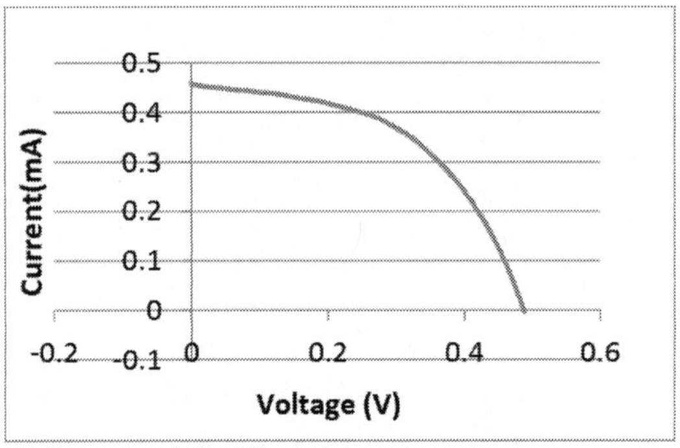

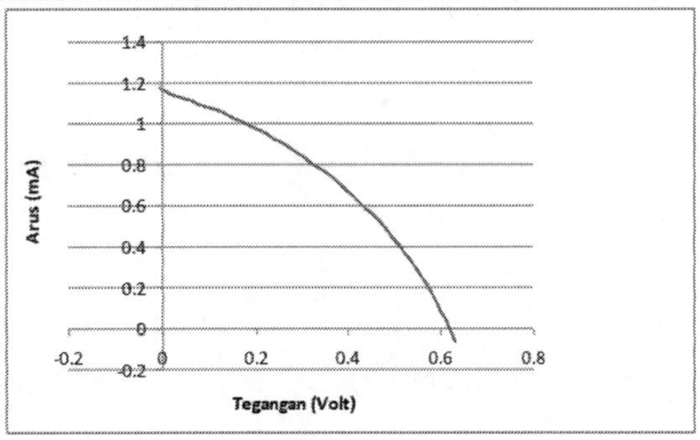

Figure 7 : I-V ZnO (powder) + TiO_2 (Powder)-- 40% : 60%

Figure 8 : I-V TiO_2 (powder) + TiO_2 (paste) -- 40% : 60%

From the graph of figure 7 generated eff = 0.221, Vmax = 0.325 Volt, Isc = 0.450 mA , ff = 0.532 and from the graph of figure 8 above generated eff = 0.530, Vmax =0.366 volt, Isc = 0.144 mA, ff = 0.385.

The results of the measurement of I-V Sun simulator of the second mix, resulting in greater efficiency in the mix : TiO2 (powder) + TiO2 (paste) with 40 : 60, i.e. comparison of 0.530. This is due to paste on the mix contains a TiO_2 paste particles smaller (nano-shaped).

4. Conclusion

From the measurement result and analysis of the characterization , can be conclude :

a. Analysis of SEM obtained that granular particle derived from (TiO_2 powder + TiO_2 Paste) producing granular particle smaller invinite dimensional nano and more evenly.

b. Characterization UV-Vis for ZnO (powder) + TiO_2 (powder)----40% : 60% and TiO_2 (powder) + TiO_2 (paste) ---- 40% : 60% show results characterization UV-Vis TiO_2 (powder) + TiO_2 (paste) ---- 40% : 60% is better.

c. Characterization IV Sun Simulator for ZnO (powder) + TiO_2 (powder) by comparison 40% : 60%, the efficiency of the observable of 0,221. The mixed TiO_2 (powder) + TiO_2 (paste) by comparison 40% : 60% it rewards efficiency of 0,530.

d. Of the three conclusion above that the result of the characterization TiO_2 (powder) + TiO_2 (paste) ---- 40% : 60% is better.

5. Reference

[1] Gratzel M 2003 *Journal of Photochemistry and Photobiology C*, **4**, 145 – 153.

[2] Karasovec UO, Berginc M, Hocevar M, and Topic M 2009 *Solar Energy Materials & Solar Cell*, **93** 379–381.

[3] Janne H 2002 *Master Thesis* Department of Engineering Physics and Mathematics, Helsinki University of Technology, February 12,.

[4] Kay A, Grätzel M 1996 *Solar Energy Materials & Solar Cells*, **44**, 99-117 .

[5] Wang S dkk 2007 *Eletrochimia Acta* **53**

[6] Nygren K 2009

[7] Syrrokostas G, Giannouli M,and Yianoulis P 2009.*Renewable Energy,* **34**(7)

[8] Guillen E, Casanueva F, Anta J, et al. 2008 *Journal of Photochemistry and Photobiology A*, **200**, no.2-3

Journal of Physics: Conference Series **776** (2016) 012010 doi:10.1088/1742-6596/776/1/012010

Perovskite/polymer solar cells prepared using solution process

E. S. Rosa[1], Shobih[1], N. M. Nursam[1] and D. G. Saputri[2]

[1]Research Center for Electronics and Telecommunication Indonesian Institute of Science
Kampus LIPI Gedung 20 Jalan Sangkuriang Bandung 40135 West Java Indonesia
[2]Faculty of Mathematics and Natural Science Sebelas Maret University
Jl. Ir. Sutami No. 36A, Solo 57126 Central Java Indonesia

E-mail: erlytasr@gmail.com

Abstract. We report a simple solution-based process to fabricate a perovskite/polymer tandem solar cell using inorganic $CH_3NH_3PbI_3$ as an absorber and organic PCBM (6,6 phenyl C61-butyric acid methyl ester) as an electron transport layer. The absorber solution was prepared by mixing the CH_3NH_3I (methyl ammonium iodide) with PbI_2 (lead iodide) in DMF (N,N-dimethyl formamide) solvent. The absorber and electron transport layer were deposited by spin coating method. The electrical characteristics generated from the cell under 50 mW/cm^2 at 25 °C comprised of an open circuit voltage of 0.31 V, a short circuit current density of 2.53 mA/cm^2, and a power conversion efficiency of 0.42%.

1. Introduction

In recent years, novel research in the field of halide perovskites nano crystals have drawn much attention as a new class of light absorber layers for solar cells [1]. The perovskites exhibit a semiconductor-to-metal transition with the increase in the dimensionality [2,3]. By increasing the dimension from 2D to 3D, the band gap will be decreased and the electrical properties will change [2]. Solar cells based on perovskite materials were first reported in 2009 by Miyazaka et.al. They applied 3D organometallic halide perovskites in a liquid electrolyte configuration dye-sensitized solar cell (DSSC) with a power conversion efficiency (PCE) of 3.8% [3,4].

The general formula of organometallic halide perovskites is $CH_3NH_3PbX_3$, where X can be Cl, Br, or I. Perovskite is basically a semiconductor that offers a wide range of band gap tuning by modifying the halide content and the organic part [5]. Methyl ammonium lead iodide ($CH_3NH_3PbI_3$) was initially utilized as a sensitizer, in the form of nanoparticles decorating the surface of a mesoporous metal oxide, in a liquid-state dye-sensitized solar cells[6]. $CH_3NH_3PbI_3$ has a band gap of 1.55 eV, close to the theoretical bandgap for an optimum photovoltaic performance (i.e. 1.4 eV) [7]. This coupled with a good extinction coefficient, which is one order of magnitude higher than standard dyes, enables excellent external quantum efficiency spectra (EQE) in the solar cells. This corresponds to an optimum photon absorption in the visible range (400 to 800 nm) and part of the near-infrared of the solar spectrum. Furthermore, when $CH_3NH_3PbI_3$ is heated above 55–60 °C, it undergoes a phase transition from tetragonal to cubic, which is expected to narrow the band gap of the material [7].

The great deal of interest and research activities involving perovskite solar cells have resulted in a tremendous progress in the PCE, with more than 17% of improvement was attained within a very short time scale [8]. However, the breakthroughs have been mostly achieved in a dye-sensitized solar cell architecture that typically requires high-temperature processing due to the use of metal oxides as the charge transporting materials [8]. As an alternative, organic photovoltaic (OPV) structure has gained much consideration as it can be fabricated through simple and low-temperature processing, including being adaptable for flexible photovoltaic applications [4,8]. The combination of a perovskite and OPV concept therefore holds a great potential and deserves further studies.

Herein, we report the fabrication of perovskite/polymer solar cells using $CH_3NH_3PbI_3$ as an absorber and PCBM (6,6 phenyl C61-butyric acid methyl ester) as an electron transport layer. The device comprised of a glass/ITO/PEDOT:PSS/ $CH_3NH_3PbI_3$/PCBM/Ag, wherein the perovskite and polymer layer were deposited by spin coating. Spin-coating is a well-known, rapid technique used for producing uniform, high-quality thin films on various substrates. This coating method has been widely used in the fabrication of organic electronics and optical device coatings, among many others [4]. In this research, we prepared a thin film of the low temperature solution-processed organometallic trihalide perovskite ($CH_3NH_3PbI_3$) absorber by mixing CH_3NH_3I (methyl ammonium iodide) with PbI_2 (lead iodide) in DMF (N,N-dimethyl formamide) solvent. The performance of the solar cell was subsequently evaluated by calculating the power conversion efficiency under solar simulator.

2. Methodology

All chemicals in this work were commercially available and used without further purification. The substrate used was 12.5 mm × 12.5 mm indium tin oxide (ITO) coated glass with a sheet resistance of 12 Ω/square. ITO is a transparent conductive oxide that serves to drain the photogenerated holes. To avoid any short circuit on the cell, it is therefore necessary to form an ITO pattern by partially etching the ITO layer. A black paint was sprayed on the ITO surface that has been covered with an adhesive tape mask with a certain pattern. The substrate was then dried at room temperature overnight, and the mask was then removed once it dried. The next step was the etching process, aiming at eliminating the undesired ITO layer. Etching process was performed by dipping the substrate into 50% HCl solution for 5 minutes, followed by washing in DI water and drying at room temperature. The remaining paint was then removed by consecutively immersing the substrate into tinner solution, isopropyl alcohol, and DI water for 10 min each in an ultrasonic bath.

The next step was the coating of the hole transporting layer, which was made of poly(3,4-ethylenedioxythiophene)-poly(styrenesulfonate) (PEDOT:PSS). PEDOT:PSS coating was carried out using spin coating method at a rotation speed of 1000 rpm for 60 s. The PEDOT:PSS layer was then dried in a vacuum oven at 120 °C for 60 min.

The organometallic halide perovskites were then prepared by mixing CH_3NH_3I with PbI_2 (molar ratio 1:1) in 1 mL DMF. The solution was stirred at room temperature for 24 h. The solution mixture was subsequently spin coated on top of the PEDOT: PSS layer at 2500 rpm for 30 s and then annealed in a vacuum oven at 100 °C for 45 minutes.

The polymer solution was prepared by dissolving 20 mg PCBM in 1 mL chlorobenzene and stirred for 60 s. The PCBM solution was then deposited onto the perovskite layer by spin coating at 1000 rpm for 30 s, and annealed in a vacuum oven at 100 °C for 10 min. Finally, a 60 nm-thick silver electrode was deposited by thermal evaporation onto the polymer layer. The size of the active area of each cell was 5.0 mm × 6.0 mm. Characterization

The ultraviolet–visible (UV-vis) absorption spectra of the $CH_3NH_3PbI_3$ layer were recorded using UV-vis spectrophotometer (Hewlett Packard 8453 Agilent Technologies). The surface morphology of the samples was observed using a scanning electron microscopy (SEM) in an FEI Quanta 250. The photovoltaic performance of solar cells was measured under solar simulator comprising of a Xenon lamp (Oriel, Newport, USA, AM 1.5) and a National Instrument source meter, which was integrated to a Lab-view software. The irradiation intensity was set to 50 mW/cm^2 at 25 °C.

3. Results and Discussion

The basic device configuration of the fabricated perovskite/polymer tandem solar cell is illustrated in Fig. 1, which consists of ITO coated glass substrate deposited consecutively with PEDOT:PSS as the hole transport material, CH3NH3PbI3 as the absorber, PCBM as electron transport material, and finally Ag metal layer as the electrode. Such structure is so called an inverted perovskite architecture, as the electron and hole transporting layer are the opposite of those used in the inorganic-based perovskite solar cells. The crystal structure of the organometal halide CH3NH3PbI3 and the molecular structure of the PCBM is shown in Fig. 2a and 2b, respectively.

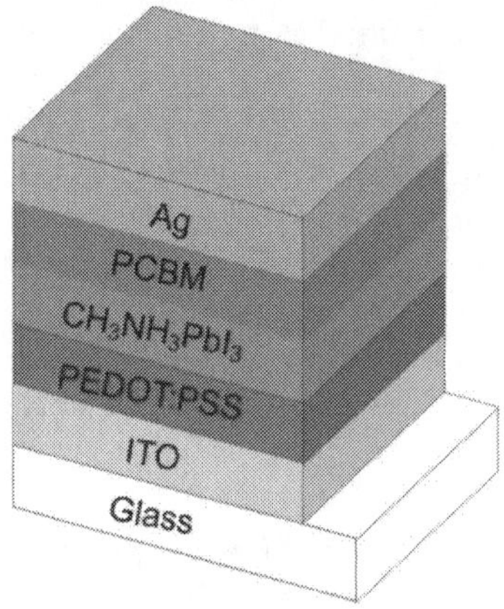

Figure 1. Device configuration of perovskite/polymer solar cells comprising of $CH_3NH_3PbI_3$/PCBM layer.

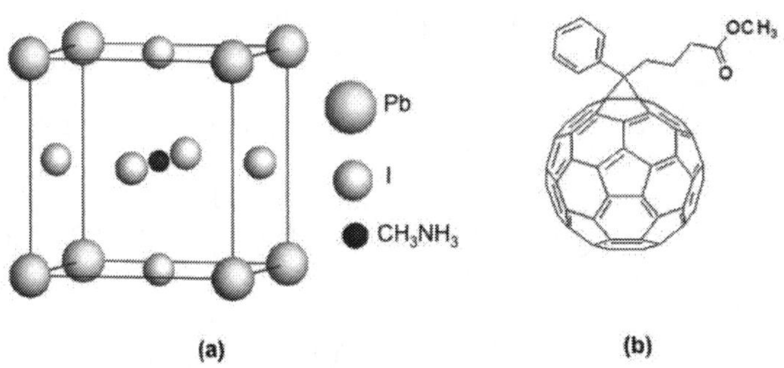

Figure 2. (a) Crystal structure of $CH_3NH_3PbI_3$ and (b) molecular structure of PCBM.

Figure 3. SEM image of the $CH_3NH_3PbI_3$ surface layer at ×256 magnification.

Fig. 3 shows the SEM image of the $CH_3NH_3PbI_3$ surface layer at ×256 magnification. An irregular dendrite-like shape with a length in the range of hundreds of micrometers was formed on top of the substrate. There were some areas thawere not fully covered by the dendrite layers, indicating that the perovskite layer was non-uniform and uneven. The low surface coverage with some uncoated areas could lead to light penetrating through the surface layer and, therefore reducing the light absorption. This phenomenon was similar to the report from Salim et al. [8]. They reported that one of the main challenges to obtain good performance in planar perovskite devices is the fabrication of high quality films with controlled morphology, high surface coverage and minimum pinholes [8].
The UV-vis absorption spectrum of the $CH_3NH_3PbI_3$ layer prepared from the mixture of CH_3NH_3I and PbI_2 precursor is shown in Fig. 4. It can be seen that the perovskite layer exhibited a narrow range of optical absorption from 300 to 440 nm. The first peak absorption with the highest absorption intensity of 0.80 a.u. was located at 320 nm. The second peak was located at ~410 nm with an intensity of 0.65 a.u. This optical profile indicates that the perovskite layer tends to absorb ultraviolet light more than the blue light. Using the transformed Kubelka-Munk function, $(\alpha h\nu)^2$ vs. $h\nu$, for a direct semiconductor [9], the bandgap energy for the obtained $CH_3NH_3PbI_3$ was estimated to be ~2.8 eV.

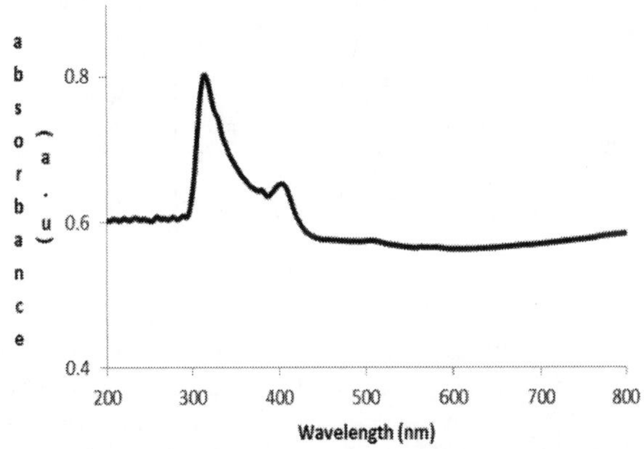

Figure 4. Absorption spectrum of the $CH_3NH_3PbI_3$ layer prepared from the precursor solution of the mixed CH_3NH_3I and PbI_2.

The photovoltaic performance of the perovskite/polymer solar cells based on CH3NH3PbI3/PCBM was characterized under illumination of 50 mW/cm2. The J-V curve and the corresponding electrical parameters are presented in Fig. 5 and Table 1, respectively. As seen in Table 1, an open circuit voltage (Voc) of 0.31 V and a short circuit current density (Jsc) of 2.53 mA/cm2 were obtained, resulting in a power conversion efficiency of 0.42%. The values of these parameters were lower compared to the reported values in the literature [10]. By using same cell configuration with this experiment, Jeng et al. for example, found greater values for all of the electrical characteristic parameters, with Voc, Jsc, FF, PCE, Rs, and Rsh reached 0.60 V, 10.32 mA/cm2, 0.63, 3.90%, 1.82 Ω cm2, and 3930 kΩ, respectively [10].

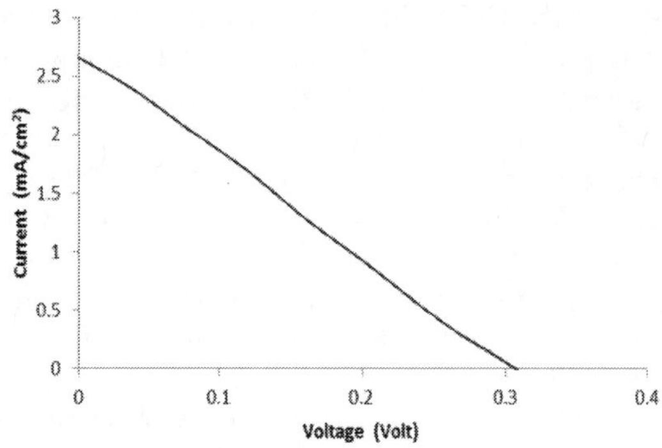

Figure 5. *J*-V curve of the perovskite/polymer solar cells based on $CH_3NH_3PbI_3$/PCBM under solar illumination with an intensity of 50 mW/cm^2.

Table 1. Electrical Characteristics of the perovskite/polymer solar cells based on $CH_3NH_3PbI_3$/PCBM under illumination with an intensity of 50 mW/cm^2.

V_{oc} (volt)	J_{sc} (mA/cm^2)	Fill factor	R_{seri} (Ω cm^2)	R_{shunt} (kΩ)	PCE (%)
0.31	2.53	0.27	132.70	0.47	0.42

There are several factors that could be the reason for the poor performance exhibited by the $CH_3NH_3PbI_3$/PCBM tandem solar cell fabricated in this work. It can be observed in in Fig. 4 that the *J-V* curve formed a relatively linear line, which was far from the ideal square-shape curve. This corresponded to a low fill factor value (FF = 0.27), which was the result of a high series resistance (R_s = 132.7 Ω cm^2) and low parallel or shunt resistance (R_{sh} = 0.47 kΩ). The high R_s could be linked to the thickness of each active layer and the bad contact between the cathode and anode layer in the cell. The high R_s and the low R_{sh} could also contribute to the low J_{sc}. The low J_{sc} might be mainly caused by the thickness of perovskite layer, which was allegedly greater than the charge carrier diffusion length [11]. When the perovskite layer is too thick, the photogenerated charge carriers can recombine before they reach the respective electrode contact. Therefore, the thick perovskite layer can also contribute in the low R_{sh}.

Meanwhile, the low V_{oc} might be attributed to the non-uniform and uneven layers in the cell. This condition could result in interpenetrating layers, so that the energy level of each layer involved in the cell is disturbed [12]. The high resistance was likely to be caused by the humidity during spin coating process of perovskite and PCBM. A recent report by Huang *et al.* shows that a uniform perovskite layer could be formed when the spin coating was performed under controlled atmosphere, *i.e.* with an argon flow in a glove box [13]. Another factor that could contribute to the high resistance was the presence of holes in the surface of the perosvkite layer. The presence of these holes could cause recombination charge at the interface between the perovskite and PCBM [14]. Overall, it was perceived that the poor quality of the perovskite layer has contributed to the low values of the electrical parameters and, thus, low power conversion efficiency was obtained. Optimization of the perovskite layer to obtain a uniform coverage with minimum pinholes will be the focus of our future work based on the findings on the present report.

4. Summary

A perovskite/polymer tandem solar cell was prepared using a solution-based processing, wherein $CH_3NH_3PbI_3$ and PCBM layer was used as an absorber and electron transport layer, respectively. The absorber solution was prepared by mixing CH_3NH_3I with PbI_2 in DMF solvent. The electrical characteristics comprised of an open circuit voltage of 0.31 V, a short circuit current density of 2.53 mA/cm^2, and a power conversion efficiency of 0.42%. Our results in this work demonstrate that perovskite/polymer tandem solar cell could be produced using a simple and facile fabrication process using wet chemical route. Although further improvement is still required, we believe that the fabrication of low-cost perovskite solar cells with excellent performance can be realized in the future.

5. Acknowledgment

This project was financially supported by Research Center for Electronics and Telecommunication Indonesian Institute of Sciences (PPET LIPI) through DIPA Tematik Program 2015.

6. References

[1] Casaluci S, Cin L, Pockett A, Kubiak PS, Niemann RG, Reale A, Carlo AD, Cameron PJ, 2015 *J. Power Sources*, **297** 504-510

[2] Kim HS, Lee CR, Im JH, Lee KB, Moehl T, Marchioro A, Moon SJ, Baker RH, Yum JH, Moser JE, Gratzel M, and Park NG 2015 *Today*, **18** 65-72

[3] Espinosa N, Serrano-Luján L, Urbina A, Krebs FC, 2015 *Sol. En. Mater. & Sol. Cells*. **137** 303–310

[4] Sfyri G, Kumar CV, Raptis D, Dracopoulos V, Lianos P, 2015 *Sol. En. Mater. & Sol. Cells*, **134** 60–63

[5] Salim T, Sun S, Abe Y, Krishna A, Grimsdale AC and Lam YM, 2015 *J. Mater. Chem. A* **3** 943-8969

[6] Boix PP, Nonomura K, Mathews N and Mhaisalkar SG

[7] Yeo JS, Kang R, Lee S, Jeon YJ, Myoung NSS, Lee CL, Kim DY, Yun JM, Seo YH, Kim SS, Na SI, 2015 *Nano Energy* **12** 96–104

[8] Tauc J, Grigorov R and Vancu A, 1966 *Phys. Status Solidi*, **15** 627-637

[9] Jeng JY, Chiang YF, Lee MH, Peng SR, Guo TF, Chen P, and Wen TP, 2013 *Adv. Mater.*, **25** 3727–3732

[10] Susrutha B, Giribabu L, Singh SP, 2015 *Chem. Comm.*, **51** 14696-14707

[11] Chen Q, Zhou H, Hong Z, Luo S, Duan HS, Wang HH, Liu Y, Li G, Yang Y 2014 *J. Am. Chem. Soc.*, **136**, 622-625

[12] Huang F. Dkhissi Y, Huang W, Xiao M, Benesperi L, Rubanov S, Zhu Y, Lin X, Jiang L, Zhou Y, Gray-Weale A, Etheridge J, McNeill CR, Caruso RA, Bach U, Spiccia L, Cheng YB, 2014 *Nano Energy* **10** 10-18.

[13] Zhiang Y, Yu W, Qin W, Yang Z, Yang D, Xing Y, Liu S, Li C, *Nano Energy*, in press.

Effect of screen printing type on transparent TiO_2 layer as the working electrode of dye sensitized solar cell (DSSC) for solar windows applications

F. Nurosyid, L. Furqoni, A. Supriyanto and R. Suryana
Physics Department, Faculty of Mathematics and Natural Sciences
Universitas Sebelas Maret, Surakarta, Indonesia

Email: rsuryana@staff.uns.ac.id

Abstract The working electrode based on semiconductor TiO_2 DSSC has been fabricated by screen printing method. This study aim is to determine the effect of the screen type on TiO_2 layer as the working electrode of DSSC. Screen used for deposition of TiO_2 has the types of; T-49, T-55 and T-61. TiO_2 layer was sintered at temperature of 500°C. DSSC structure was composed of semiconductor TiO_2 adsorbed dye, an electrolyte solution and a platinum counter electrode. TiO_2 layer thickness was characterized by Scanning Electron Microscopy (SEM), while the absorbance was characterized using UV-Vis spectrophotometer and the electrical properties of DSSC were characterized by Keithley I-V measurement. TiO_2 layer fabricated by screen T-49 had the biggest thickness that was 3.2 ± 0.3 μm and the highest UV-Vis absorbance wave at the peak wavelength of 315 nm with the absorbance value was 1.7. The I-V characterization showed that the sample fabricated by screen T-49 obtained the greatest efficiency that was 1.0 x 10^{-1} %

1. Introduction

Solar cell is device which is capable of converting solar energy into electrical energy. The development of solar cell has reached three generation in the world, among them are solar cell made from bulk silicon semiconductor [1], a thin film solar cell [2], and solar cell based on dye-sensitized [3-5]. Dye-sensitized solar cell (DSSC) structures consist of semiconductor layers, dye, an electrolyte solution and a catalyst. Semiconductor layer coated dye acts as the cathode. The catalyst is grown on glass substrates luorine doped in Oxide (FTO) serves as the anode. Dye functions as a light-sensitive substance (photosensitizer). Electrolyte solution performs as substitute of excited electrons from the dye. Catalyst does as the counter electrode [5-6].

Solar window is one of DSSC applications. Solar windows can be installed in the windows of a building. The presence of solar windows is expected to be able to support government programs in electricity savings. Solar windows require large size to be applied in the windows of the building. The suitable method for the fabrication of DSSC in a large scale is screen printing method [7]. The other advantages of screen printing method are investment in equipment that is relative inexpensive and simple fabrication process. The effectiveness of the screen printing method is determined by many factors. Hole size (mesh) and pressure on the screen [8] are important factor in DSSC fabricated by screen printing method. Those factors become the determinant of the amount of TiO_2 particles that can be deposited on substrate. The more deposited particles make TiO_2 layer getting thicker, so that the more dye absorbed. This study is focused on obtaining the effect of screen printing type on each type of TiO_2 pasta to the TiO_2 layer as the working electrode of DSSC. There are three screen types in the screen printing method; those are T-49, T-55 and T-61. Each screen mode represents the number and the mesh diameter of the screen. This research is expected to be able to provide a scientific contribution in the development of DSSC using screen printing methods, as well as application of DSSC as the pioneer of solar windows in Indonesia.

Solar windows require transparent materials in order to be able to pass sunlight into the room. The use of transparent materials is expected to be able to make the absorbed photons in the DSSC become more maximal. The absorbed photons cause electrons excited from the dye molecules. TiO_2 layer is fabricated from transparent TiO_2 pasta with type of 18NRT.

2. Materials and Method

FTO glass substrate (Merck Dyesol) with resistance of 7 Ω/sq was cleaned using ultrasonic cleaner. This study used TiO_2 pasta of 18NR-T which is transparent. TiO_2 coating process on substrate was conducted by screen printing method. Coating of TiO_2 pasta was done by two times in order to obtain two layers of TiO_2 (double layer) without fissure [9]. Types of screen used were; T-49, T-55 and T-61. The code numbers on the screen type indicated the number of mesh/cm. Mesh on the screen became determinant of the amount of TiO_2 particles which can be deposited on the substrate. Screen of T-49 had 49 mesh/cm while T-55 and T-61 had 5 and 61 mesh/cm, respectively.

Active cell area of TiO2 was 1 cm^2. TiO_2 layers deposited on substrate was sintered at temperature of 500°C for 10 minutes. The coloring process was done by immersing the substrate coated TiO_2 into Ruthenium dye N719 for 24 h. The counter electrode was fabricated using FTO glass substrate coated catalyst solution of Counter Electrode Solution/ CELS (Merck Dyesol).

Catalyst deposition process was carried out by droplet method; the catalyst solution was dripped onto the substrate using a syringe. After that, the counter electrode was sintered at temperature of 420°C for 10 minutes. DSSC sandwich structure was composed of dye sensitized TiO_2 working electrode and the counter electrode. The space between the electrodes was injected by electrolyte solution which served as electron recharge through redox reaction. The morphology and the thickness of the deposited TiO_2 layer were characterized by Scanning Electron Microscopy (SEM), while the absorbance was characterized by UV-Vis Spectrophotometer Lambda 25. The electrical properties of DSSC were characterized by using I-V Keithley Measurement.

3. Results and Discussion

Characterization using SEM was conducted to recognize the morphology and the thickness of TIO_2 layers fabricated with different screen. The morphology results of TiO_2 layer are shown by Figure 1.

The shapes of TiO_2 particles are spherical, and formed agglomerations in some parts. Looked there are slits on the surface of TiO_2 layer. Those cracks are useful to adsorb dye molecules[16]. TiO_2 particle size was measured by doing a manual comparison to the scale shown on the results of the SEM image. The results of particle diameter measurement are presented by Figure 2.

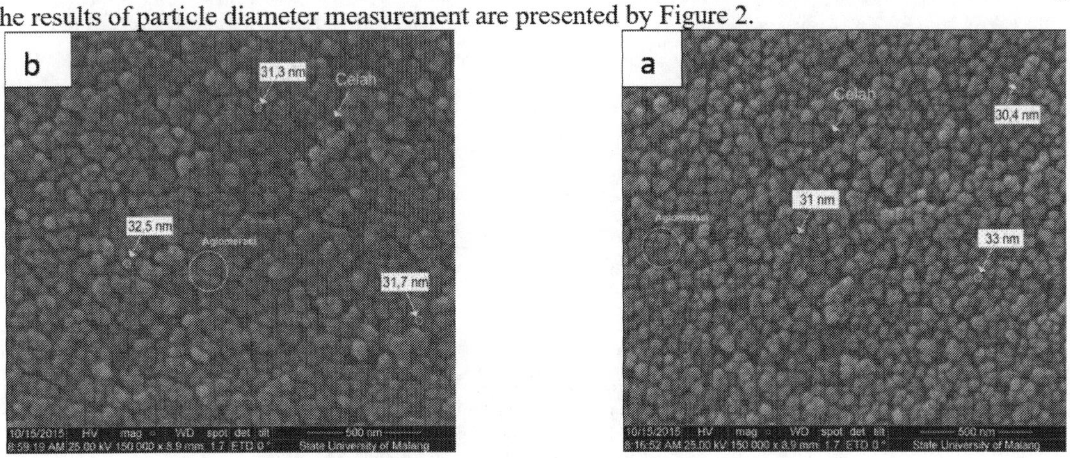

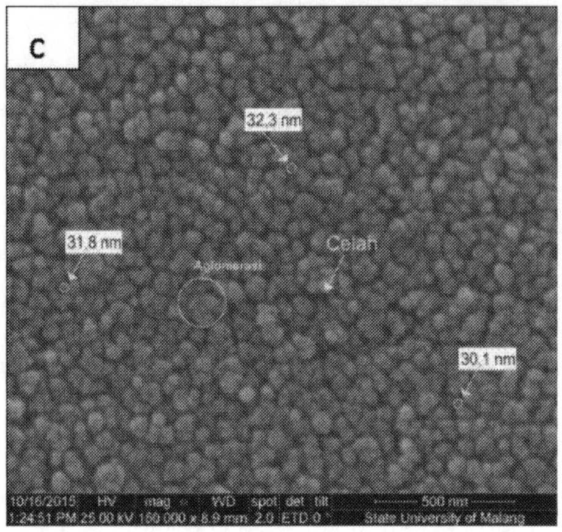

Figure 1. The morphology of TiO_2 layer fabricated by using screen (a) T-49 (b) T-55 (c) T-61

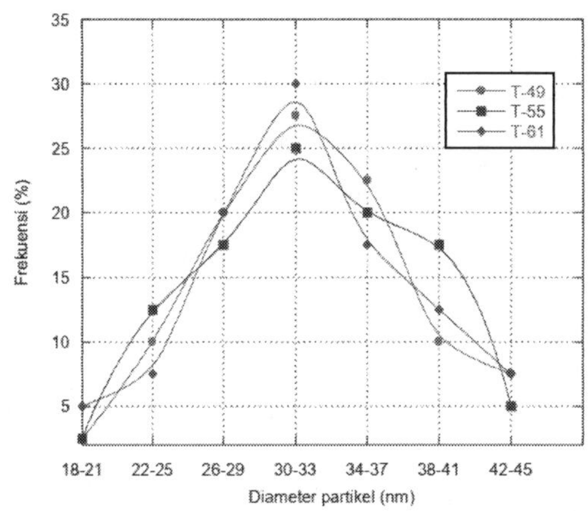

Figure 2. The result of TiO_2 particles size distribution imaged by SEM

Diameter of TiO_2 particles fabricated by screen T-49, T-55 and T-61is nearly same, that is in range of 30-33 nm. Based on this study, the types of screen do not affect the change in diameter of TiO_2 particles. This is due to the size of the particle diameter and TiO_2 crystals can change if they are given different heat treatment [10-11].

Figure 3. The result of SEM cross-section layer deposited by using *screen* T-49

The thickness of TiO_2 layer is shown by Figure 3. It is different for each variation of the screen type. The largest thickness is resulted by TiO_2 layer fabricated using screen T-49. TiO_2 layer deposited by screen T-49 has thickness of (3.2 ± 0.3) μm, while the screen T-55 and T-61 produce thickness of (2.8 ± 0.2) and (1.6 ± 0.2) μm, respectively. The thickness of TiO_2 layer is influenced by the number of mesh on the screen.

The more number of mesh on the screen causing the distance between the mesh becomes closer and the diameter of mesh becomes smaller so that the less of volume of TiO_2 pasta is distributed on the substrate. It causing the TiO_2 layer becomes thinner; otherwise if the more of volume of TiO_2 pasta is distributed, the layer of TiO_2 becomes thicker.

Absorbance of TiO_2 layer before and after immersion in dye was characterized using UV-Vis Lambda Spectrophotometer 25 shown by Figure 4. It is conducted by the absorbance spectrum with wavelengths of 300-700 nm.

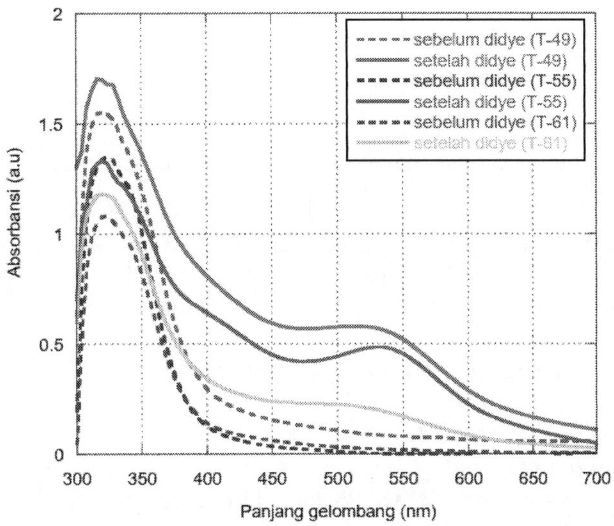

Figure 4. The absorbance curve of TiO_2 layers

Based on the UV-Vis characterization, there is an increasing rate of absorbance after TiO_2 layer soaked in dye. It is due to dye molecules attach to the TiO_2 particles, so it can be concluded that aging dye increases the absorbance of TiO_2 layer. The results of UV-Vis characterization show the absorbance peak of each material. The highest peak of photons absorbance of material TiO_2 is in the wavelength of 200 nm-410 nm [12]. The highest peak of photon absorbance of Ruthenium dye N719 appears in the visible wave range of 500 nm-550 nm [13]. The highest absorbance value is resulted from the TiO_2 layer deposited by screen-type T-49. The highest peak of TiO_2 layer is in wavelength of 315 nm with the absorbance value is 1.7. According to beer lambert law [15], this is due to the TiO_2 layer deposited by screen-type T-49 has the greatest thickness.

Current and voltage DSSC were characterized using I-V Keithley measurement presented by Figure 5. The light source used was a xenon lamp with intensity of 1000 W/m^2. DSSC active cell area is 1.10^{-4} m^2.

I-V curve and datasheet obtained from the characterization can be used to calculate the efficiency of DSSC. The highest efficiency is obtained using screen-type T-49 that is 1.0 x 10^{-1}%. It is occurred because the number of TiO_2 particles deposited using screen T-49 is more than the number of TiO_2 particles deposited using screen T-55 and T-61. The number of TiO_2 molecules is determined by the thickness of TiO2 layer deposited on the FTO glass substrate. The more TiO_2 molecules are deposited on the substrate, the more dye molecules are absorbed. Therefore, when the photons hit the surface of the DSSC, the more electrons are excited from band HOMO (dye) to band LUMO (TiO2 semiconductor) [14]. The number of generated electrons causes the more currents flowing to the external circuit. DSSC efficiency will increase as the increase electric currents flowing in the solar cell system.

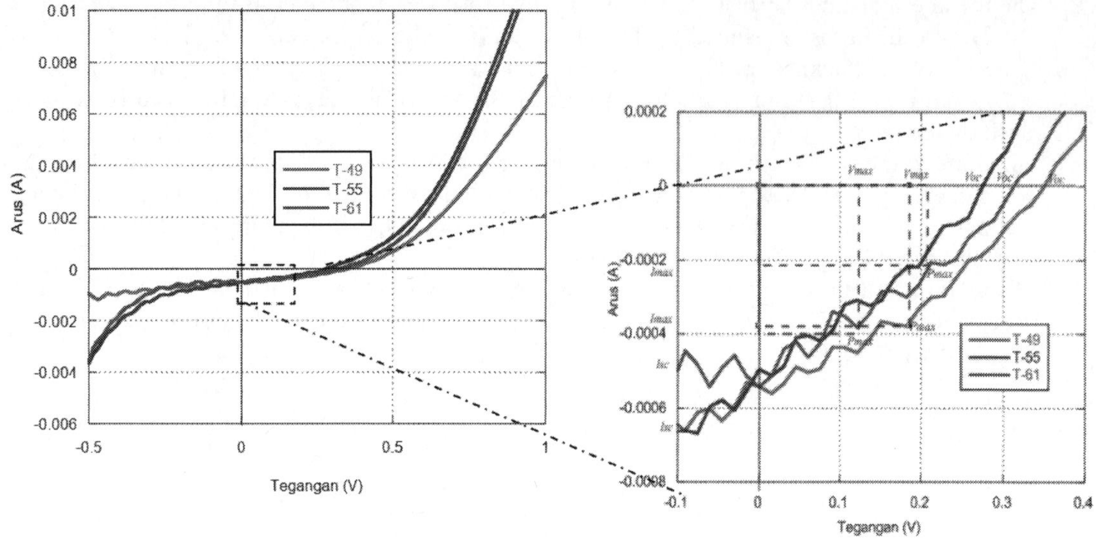

Figure 5. The characteristic curve of DSSC Currents-Voltages (I-V)

4. Conclusion

Based on this study, the screen type affects the thickness of TiO_2 layers, but it does not affect the morphology of the TiO_2 layer. The highest thickness is obtained using screen T-49. The thickness of TiO_2 layer is (3.2 ± 0.3) µm. The result of UV-Vis characterization shows that the thickness of TiO_2 layers affects the absorbance. The highest peak of TiO_2 layer is in wavelength of 315 nm with the absorbance is 1.7. The highest efficiency is obtained from DSSC which its working electrode is deposited by screen T-49. The highest DSSC efficiency is 1.0 x 10^{-1}%.

5. Acknowledgments

This work is supported by DP2M DIKTI and LPPM UNS by PUPT Grant Program (contracts: no 339/UN27.11/PL/2016)

6. References

[1] Kim, DS et al. (18 May 2003). *Proceedings of 3rd World Conference on Photovoltaic Energy Conversion, 2003* **2**: 1293–1296. ISBN 4-9901816-0-3.

[2] Edoff, Marika. (2012). *AMBIO* **41** (2): 112–118.

[3] O'Regan B. and Grätzel, M (1991) *Nature*, 353, 737-739.

[4] Zhou H, Wu L, Gao Y, (2011). *J. Photochem. Photobiol.*, A. 219 188–194

[5] Gratzel M. (2003), *E Journal of Photochemistry and Photobiology C:Photochemistry Review*, 4, 145-153

[6] Nazeeruddin MK, Kay A, Rodicio I, et al. (1993) *Journal of the American Chemical Society*, vol. 115, no.14, pp. 6382–6390,

[7] Muliani, L, Taryana Y, Hidayat J. (2010). *Screen Printing. Jurnal Elektonika.* **10**(1) 126-131

[8] Tsai J, K, Wen, D, H., Tian, C., W Teen, H, M, and Wen, J, C. (2003). *Nanoscale Research Lettrs,* **8**:459

[9] Ito S, Peter C, Pascal, C., Mhammad, K,N., Paul, L, Peter Pand Michael G.(2007) *Prog. Photovolt: Res, Appl. (in press)* Published online in Wiley Interscience (www.interscience.wiley.com) DOI:10.1002/pip.768

[10] Marthua, S, R, N & Diah, S. (2012). *Jurnal Teknik Pomits* **1**, No. 1, hal 1-6

[11] Nadeak, S. M. R. dan D. Susanti (2012). Variasi temperatur dan waktu tahan kalsinasi terhadap unjuk kerja semikonduktor TiO_2 sebagai dye sensitized solar cell (DSSC) dengan dye dari ekstrak buah naga merah, *Tugas Akhir S1*, Jurusan Teknik Material dan Metalurgi, ITS.

[12] Cahyorini K . (2009). Titanium Dioksida Terdoping Nitrogen : Kajian Tentang sintesis, karakterisasi dan aplikasinya. Prosiding seminar nasional penelitian, pendidikan dan penerapan MIPA, K124-K133

[13] Gratzel. (2005). Solar Energy Conversion by Dye Sensitized Photovoltaic Cells, Innorg Chem. 44: 6841-6851

[14] Arman S, Hoda N.M, (2015)"The Effect of Drying and Thickness of TiO2 Electrodes on the Photovoltaic Performance of Dye-Sensitized Solar Cells" *Int. J. Electrochem. Sci.,* 10 ;3354 – 3362

[15] Susanto., Budi, A,S., Khoirun, N., Nita, R., Agus, Y. (2014). Analisis Spektrum Absorbansi Pigmen Flavonoid dari Daun Tanaman Andong (*Cordyline Frutiosa L*) sebagai *Dye Solar* Sel. *Jurnal Fisika,* Vol 4 No 2. (92-95)

[16] Septina, W., D. Fajarisandi, M. Aditia, (2007). "Pembuatan prototip solar cell murah dengan bahan organik-inorganik (dye sensitized solar cell,)" *Laporan Akhir Penelitian Bidang Energi,* Penghargaan PT.RekayasaIndustri

Influences mass concentration of P3HT and PCBM to application of organic solar cells

A. Supriyanto[1], Maya[1], E. S. Rosa[2], Y. Iriani[1], A. H. Ramelan[1], F Nurosyid[1]

[1] Department of Physics, Faculty of Mathematics and Natural Science, Sebelas Maret University, Surakarta

[2] Research Center for Electronics and Telecommunication Indonesian Institute of Science (PPET-LIPI), Bandung

E-mail : agusf22@staff.uns.ac.id

Abstract. Poly (3-hexylthiophene) (P3HT) and [6, 6] -phenyl-C61-butyric acid methyl ester (PCBM) are used for the organic solar cell applications. P3HT and PCBM act as donors and acceptors, respectively. In this study the efficiency of the P3HT: PCBM organic solar cells as function of the mass concentration of the blend P3HT: PCBM with 1, 2, 8, 16 mg/ml. Deposition P3HT:PCBM film using spin coating with a rotary speed of 2500 rpm for 10 seconds. Optical properties of absorption spectra characteristic using a UV-Visible Spectrometer Lambda 25 and electrical properties of I-V characteristic using Keithley 2602 instrument. The results of absoption spectra for P3HT:PCBM within different mass concentration obtained 500-600 nm wavelengths. The Energy-gap obtained about 1.9eV. The organic solar cells device performance were investigated using I-V cahractyeristic. For mass concentration of 1, 2, 8 and 16 mg/ml P3HT:PCBM were obtained 0.5×10^{-3} %, 2.2×10^{-3}%, 5.9×10^{-3} %, and 6.1×10^{-3}% efficiency of organics solar cells respectively.

1. Introduction

Research on solar cells currently widely used organics and conjugated polymer materials[1-2]. Organics Solar cells (OSC) have attracted increasing interest as a lightweight, low-cost, and easy to process replacement for inorganics solar cells. The most efficient organic solar cells produced to date are bulk heterojunction (BHJ) photovoltaic devices based on blends of semiconducting polymers such as poly(3-hexylthiophene-2,5-diyl) (P3HT) and [6,6]-penyl-C61-butyric-acid-methyl-ester (PCBM). Conjugated polymer has a delocalized p-electrons so as to absorb sunlight, forming a carrier, the transport charges and generate electricity[3].

The efficiency of OSC increases significantly with the introduction of the bulk hetero-junction concept consisting of an interpenetrating network of electron donor and acceptor materials. The work done is to maximize the separation exciton. Exciton a pair of attraction between electrons and holes[4]. The separate of exciton the more electrons and holes are obtained. Optimization of the thickness of the active layer is one of the steps that can be taken, it relates to the diffusion length exciton[5]. Other efficiency improvement efforts done by extending the active region including the polymer blend insert[6]. Blending process is blending between donor and acceptor materials using solvents. Blending process can use a variety of polymer materials. The active ingredients are frequently used polymer is poly (3-hexylthiophene) (P3HT) and [6, 6] phenyl C61 butyric acid methyl ester (PCBM) [7-10].

In this article, we investigated the effect of mass concentration P3HT and PCBM in the optical and electrical properties of organics solar cells.

2. Experimental Procedure

In this study, organics solar cells based on poly(3-hexylthiophene-2,5-diyl) P3HT as donor and [6,6]-penyl-C61-butyric-acid-methyl-ester (PCBM) as acceptor were investigated. P3HT:PCBM in accordance with the variation of mass concentration of 1, 2, 8 and 16 mg/ml respectively. P3HT:PCBM were dissolved with blending methods using the ultrasonic cleaner for 30 minutes. P3HT:PCBM blend using a solvent chlorobenzena.

P3HT: PCBM film was deposited on glass substrates. The deposition was performed by spin coater instrument at 2500 rpm of angular velocity for 20 seconds [10]. The optical characterization, which is absorbance properties, of samples were then observed using a UV-Vis Spectrometer Lambda 25 type with a wavelength ranging from 300-800 nm. PEDOT: PSS was spin coated onto FTO at 2500 rpm, 20 seconds. P3HT: PCBM film were dissolved in 1, 2, 8 and 16 mg/ml at different mass concentration. by spin coater as well. The last process was metallization of Al by evaporator. The structure of organic solar cells are FTO/PEDOT: PSS/P3HT:PCBM/Al

The current–voltage characteristic in dark at room temperature was obtained using a 2602A Keithley instrument high voltage source. The performance under dark and illumination was performed to study the photosensitizing effect in the device. The device was illuminated by a visible light from xenon light source. The xenon light source was calibrated by adjusting the light intensity 1000 W/m^2 was obtained. The current-voltage of the device under illumination was recorded using a Keithley high voltage source model 2602A and a personal computer. The intensity was controlled by solar powermeter and temperature was measured by thermocouple. The illuminated cell area was 10 mm^2. The photosensitizing parameters of short circuit current density *(I_{sc})* and open circuit voltage *(V_{oc})* were obtained from the intersection axis of current and voltage from the current-voltage curve under dark and illumination, respectively.

3. Results and Discussion

Fig. 1. Shows UV-Visible absorption spectra of P3HT:PCBM layer spin coated onto glass substrates FTO prepared from different mass concentration. Initial of mass concentration at 1, 2, 8 and 16 mg/ml are WC1, WC2, WC8 and WC16 respectively.

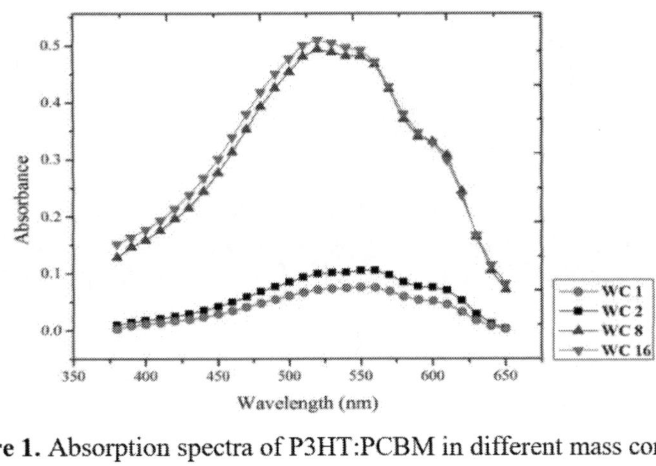

Figure 1. Absorption spectra of P3HT:PCBM in different mass concentration

The results of absoption spectra with different mass concentration have a peak maximum of absorbance 0.72; 0.10; 0.16; 0.49 and 0.51 respectively at 450-600 nm wavelengths. In the first and second peaks had relatively high difference in this case because have mass concentration against mass ratio between the PCBM and P3HT of 1:5, so that comparatively high in the second peak that P3HT.

From absorption spectra graph can determine energy-gap (*Eg*). Absorption spectra curve convert to energy (*hυ*) vs absorption coefficient $(\alpha h\upsilon)^2$ curve. The determination of energy-gap is obtained by Tauc's equation [11] . This method using draw a tangent when the chart has the highest peak and the value $(\alpha h\upsilon)^n$ is zero, Fig.2.

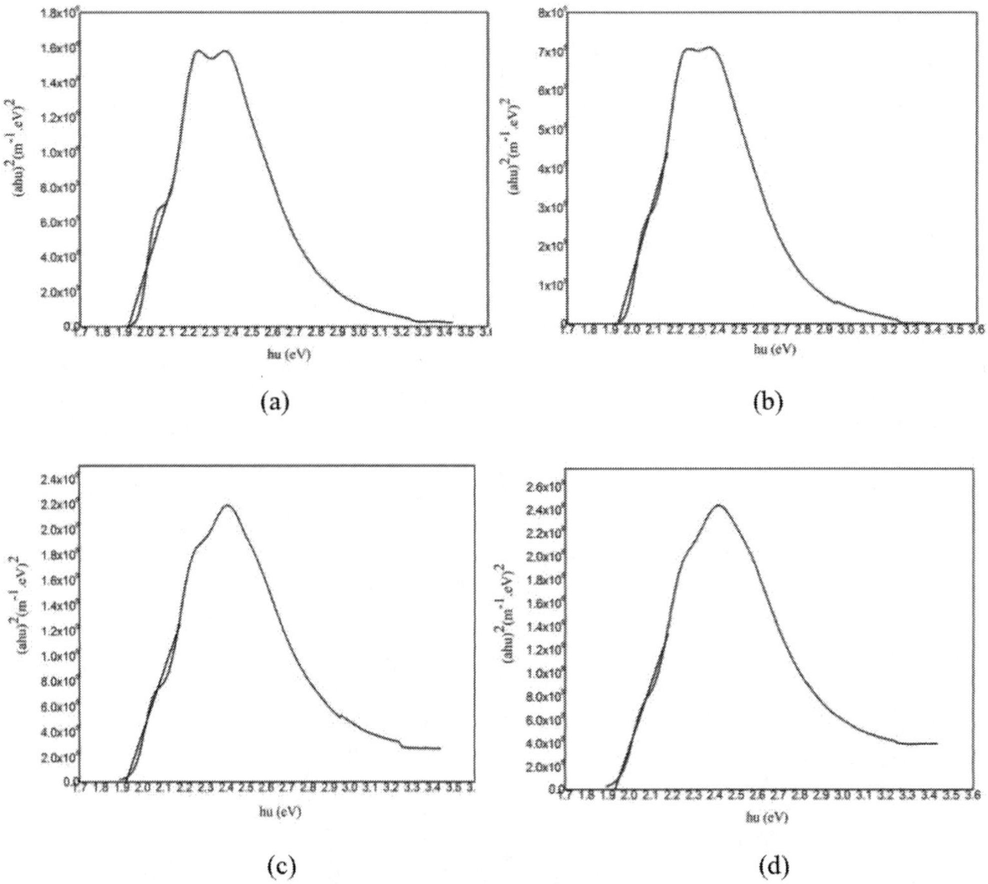

Figure 2. Curves *hυ* vs $(\alpha h\upsilon)^2$ of PCBM: P3HT of the blend mass concentration (a) 1 mg/ml, (b) 2 mg/ml, (c) 8 mg/ml and (d) 16 mg/ml

From the graph on Fig 2, the results of tauc plot methods that energy-gap obtained 1.9 eV. The different mass concentration of P3HT:PCBM have near of energy-gap in P3HT which have 1.8 to 2.0 eV. The result calculation of energy-gap P3HT:PCBM with different mass concentration shows in Table 1.

Journal of Physics: Conference Series **776** (2016) 012012 doi:10.1088/1742-6596/776/1/012012

Table 1. Energy-gap of P3HT: PCBM with different mass concentration

Mass Concenteration (mg/ml)	E_g (eV)
1	1.933 ± 0.037
2	1.957 ± 0.036
8	1.970 ± 0.067
16	1.992 ± 0.036

Electrical properties of I-V characteristic on organic solar cells using Keithley 2602A. The OSC performance was investigated solar cells characteristic on under dark and illumination condition. The device was illuminated by a visible light from xenon light source. The xenon light source was calibrated by adjusting the light intensity 1000 W/m^2 was obtained. Measurements were taken at the dark and the light. The curve in a dark state is above the curve in light of the circumstances. In the light of the circumstances curve is used as a parameter for determine of efficiency. The results of graph I-V can be determine the value of I_{sc}, V_{oc}, FF (fill factor) and the efficiency of organic solar cells. I_{sc} is the current that occurs when a short circuit. V_{oc} is the voltage value when an open relationship.

The performance of organic solar cells in the dark and illumination. In the state of illumination are used as a parameter for determining the value of efficiency. The I-V characteristic results are showed in Fig 3.

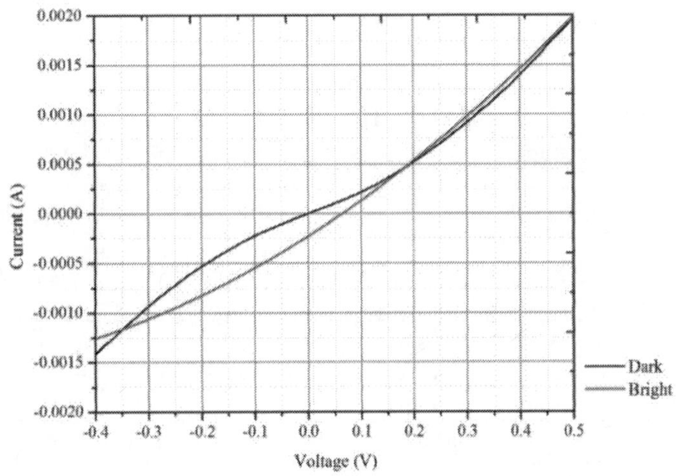

Figure 3. The I-V characteristic on dark and illumination condition of organics solar cells

Dark curve is at the top, while the light curve is on the bottom or fourth quadrant. At the time of dark state no current is flowing so dark curves are not on a four quadrant. For illumination condition of I-V characteristic with the mass concentrations 1, 2, 8 and 16 mg/ml respectively are showed in Fig 4.

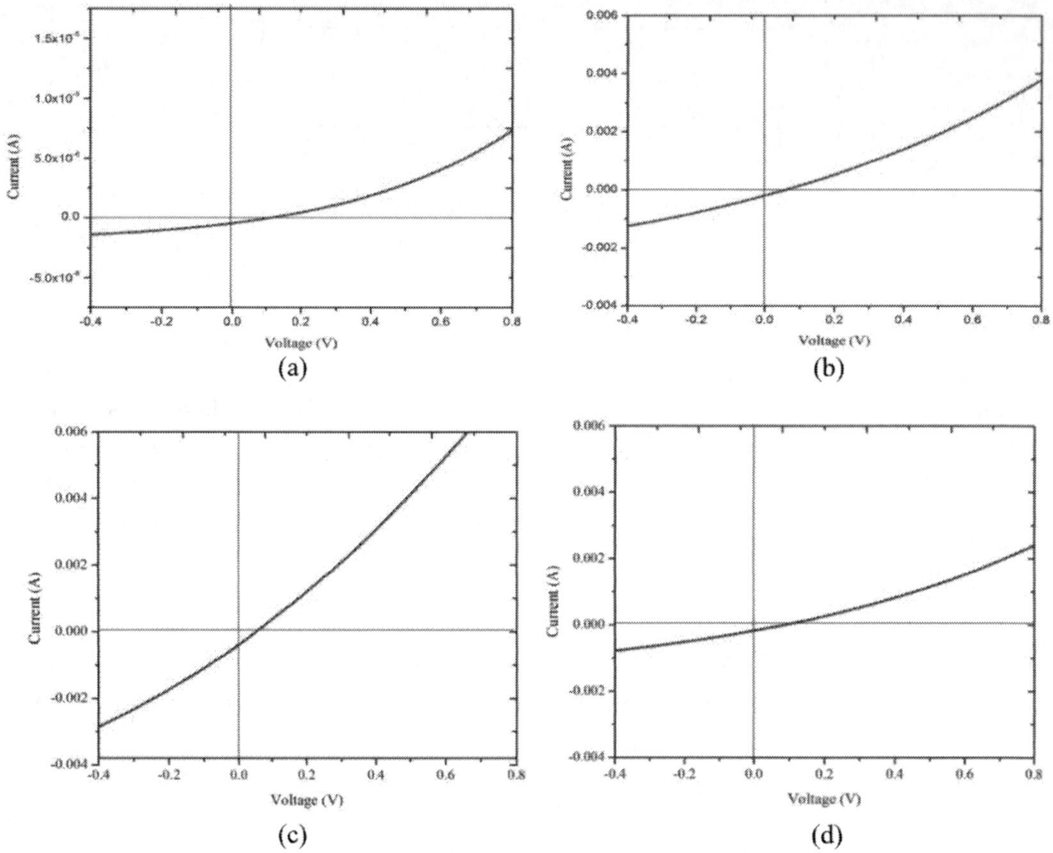

Figure 4. The I-V characteristic of P3HT: PCBM with different mass concentration :
(a) 1 mg/mL (b) 2 mg/mL, (c) 8 mg/mL and (d) 16 mg/mL

The results I-V characteristic of value I_{max}, V_{max}, I_{sc} and V_{oc} are showed in Table 2.

Table 2. The Results of organics solar cells characteristic

Mass concenteration (mg/mL)	I_{max} (10^{-5} A)	V_{max} (V)	I_{sc} (10^{-4} A)	V_{oc} (V)	*FF*	**Efficiency ($x10^{-2}$%)**
1	0.03	0.022	0.004	0.1111	0.15	0.02
2	6.01	0.020	1.631	0.0851	0.09	3.06
8	11.69	0.024	3.136	0.0550	0.21	7.09
16	21.30	0.017	9.438	0.0002	0.18	9.13

From Table 2 shows that, for the mass concentration of 1 mg / ml obtained 0.004 of I_{sc}, 0.1111 of V_{oc}, 0.15 of *FF* and $0.02x10^{-2}$% of efficiency. For the mass concentration of 16 mg/ml was obtained 9.438 of I_{sc}, 0.0002 of V_{oc}, 0.18 of *FF* and $9.13x10^{-2}$% of efficiency. This shows that the higher of the mass concentration will increasing I_{sc} and power conversion efficiency.
Absorbance value affects the value of photons that are absorbed organic solar cells. The more photons are absorbed so the more electrons are obtained. V_{oc} values associated with donor

material HOMO and LUMO acceptor material. The resulting efficiency value also increases with the increase in the value of I_{sc}.

4. Conclusion

P3HT: PCBM blend have 1, 2, 8 and 16 mg/ml at different mass concentration. It is increasing absorption spectra in the range of about 450-600 nm. The results calculation of the energy-gap obtained about 1.9 eV. The characteristic of organics solar cells for efficiency at 1, 2, 8 and 16 mg/ml P3HT:PCBM were obtained 0.5×10^{-3} %, 2.2×10^{-3}%, 5.9×10^{-3} %, and 6.1×10^{-3}% efficiency of organics solar cells respectively. The higher of mass concentration in the active solution of P3HT: PCBM then the value of the efficiency of organic solar cells produced higher as well.

5. Acknowledgment

The authors would like to acknowledge Kemeristek-Dikti Republik Indonesia for financial support in Penelitian Unggulan Perguruan Tinggi (PUPT 2016) of Sebelas Maret University (UNS), Indonesia.

6. References

[1] Dennler, N. S. Sariciftci, and C. J. Brabec.2006. *Conjugated Polymer-Based Organic Solar Cellsin Semiconducting Polymers: Chemistry, Physics and Engineering*, Vol I Second Edition,edited by G. Hadziioannou and G.G. Malliaras, Weinheim : Wiley-VCH Verlag GmbH & Co.KGa. 511-512

[2] Wanzhu Cai, Xiong Gong, and Yong Cao. 2010. *Solar Energy Materials & Solar Cells,* 94(2): 114-127

[3] Brabec,C.J, N.S. Sariciftci, and J.C. Hummelen. 2001. *Adv. Funct. Mater.* 11, 15-26

[4] Kim, J.Y.,K. Lee, N. E. Coates, D. Moses, T-Q. Nguyen, M. Dante, A. J. Heeger. 2007 *Science* 222-225

[5] Nunzi, J.M. 2002 *Journal C.R. Physique*, 03. 523-542

[6] Stubinger, T., & Brutting, W. 2001 *J. of Appl. Phys.* 90, 07. 3632-3641

[7] Deboisblanc, J. 2010. Synthesis & Characterization of P3HT:PCBM Organic Solar Cells. *Thesis*: Department of Physics & Astronomy Pomona College

[8] Gupta, Anubhav; S, Praveen; Kumar, Abhishek; Shree, Priyanka; Mishra, Suchana; Joseph, C M. 2012. *IJEEE 1*(4), 13-15

[9] Yue, S., YanChun, H., & JianGang, L. 2013. *Chinese Science Bulletin*, 2767-2774

[10] Kumar, G. Li, Z. Hong, dan Y. Yang. 2009. *Nanotechnology*, 5202-5205.

[11] Bhatt, R., Bhaumik, I., Ganesamoorthy, S., Karnal, A.K., Swami, M.K., Patel, H.S., Gupta, P.K., 2012. *Physics Status Solidi A*, 209, 1, 176-180.

Crystal structure and magnetic properties of $Nd_2Fe_{14}B$ powder prepared by using high energy milling from elements metal Nd,Fe,B powders

Ramlan[1], Muljadi[2], P. Sardjono[2], F. Gulo[3], D. Setiabudidaya[1]

[1]Physics Department FMIPA Sriwijaya University,Palembang Indonesia
[2]Research Center for Physics, Indonesian Institute of Sciences
[3]S3 MIPA Sriwijaya University, Palembang Indonesia

Email : ramlan041066@gmail.com

Abstract. The $Nd_2Fe_{14}B$ powder has been made by using High Energy Milling (HEM) from mixed metal powders Iron (Fe), Neodymium (Nd) and Boron (B). The Nd, Fe and B powders were mixed according stoichiometric composition (atomic ratio Nd:Fe:B = 2 : 14 : 1) and milled and milling time was varied in 10, 20, and 40 hours by using HEM. Toluene liquid was used as milling media to protect of metal powders from oxygen. The measurement result of x-ray diffraction show that the optimum $Nd_2Fe_{14}B$ phase already formed about 69,46 % after milling 40 hours with crystallite size about 25.64 nm. The magnetic properties of milled powders were measured by using VSM at room temperature. The highest value of magnetic properties are obtained at powder milled in 40 hours, at this condition, it is obtained Ms = 122 emu/g, Mr = 81 emu/g, Hc = 5.54 kOe and BHmax = 11.01 MGOe.

1. Introduction

The demand of NdFeB magnets has increased rapidly in various fields of life since their development in 1984, due to their excellent magnetic performance [1]. The $Nd_2Fe_{14}B$ is one type of permanent magnet based rare earth metals, that the $Nd_2Fe_{14}B$ magnet has high magnetic properties [1]. The $Nd_2Fe_{14}B$ magnets are required by many electric motor industries, automotive industries, industrial electronics, and industrial electricity generator [2,3]. $Nd_2Fe_{14}B$ compound formed from the reaction between powder Neodymium (Nd), iron powder (Fe), and powder Boron (B). There are several technologies to make $Nd_2Fe_{14}B$ powder that the process is generally through the smelting process at high temperature (1700 - 1800 ° C) and the atmosphere of process isolated by inert gas such as nitrogen or argon [4,5] . With this technology, it needed a large electrical energy and high cost for production, and the technical operations are quite difficult. On the basis of the conducted an alternate way, another to make the compound $Nd_2Fe_{14}B$ with costs lighter and easier production, namely with using mechanical alloying techniques where powder Nd, Fe, and B as raw materials are mixed and milled using high energy milling (HEM) with high-speed 4000 - 5000 rpm [5]. Some of Authors have been done mechanical alloying process to produce of $Nd_2Fe_{14}B$ powder with using ball mill in Hydrogen gas atmosphere [6,7]. During milling process has been going on repeating the process of mechanical mixing, cold welding, fracturing, and rewelding powders which can lead to a reduction in particle size, and a milled powder becomes more reactive[6]. The existing atoms (Nd, Fe, B) in starting materials will easily vibrate during milling to facilitation the surface diffusion resulting in a phase transformation into $Nd_2Fe_{14}B$ phase [7,8]. Through the process of mechanical alloying by means of milling to note the atmosphere, because Nd and Fe metals are very easy to interact with oxygen so easily corroded at room temperature. Due to the corrosion problem in the milling process , toluene liquid is used as media so that the powder Nd and Fe can be isolated from oxsigen during the milling process. This research was conducted on the process of synthesis of $Nd_2Fe_{14}B$ powder from mixed of elements metal powders (Nd,Fe,B) by using High Energy MIlling. The purpose of this research to determine the relationship between milling time with percentage of $Nd_2Fe_{14}B$ phase formation and crystal size. Besides, it was also analyzed the magnetic properties by using vibrating Sample Magnetometer (VSM).

2. Experimental

The $Nd_2Fe_{14}B$ was made at stoichiometric composition (atomic ratio of elements Nd: Fe: B = 2: 14: 1) by using mechanical alloying (MA) process. NdFeB powder is made from a mixture of raw material powders Nd with a purity of 99.90% and 200 μm, Fe with a purity of 99.90% and 100 μm and Boron with a purity of 99.90% and 50 μm. Applied MA is through the process of High Energy Milling (HEM) with rotation speed 3000 rpm. Toluene is used as liquid media to protect of corrosion on metals powder Nd,Fe and B. Total weight of raw materials were 15 g consisting of a mixture of Neodymium (Nd), iron (Fe) and Boron (B). Powder Nd, Fe and B were inserted in a vial and toluene is added as much as 15 ml and added steel ball about 120 g, furthermore vial tube was closed and began the milling process. The powder preparation from upload materials to vial tube until closing the vial tube have been done in Glove Box at vacuum condition (P vac = 10 mbar). Subsequently tube vial containing the raw material was placed on a milling machine (HEM), then the HEM engine was operated with a variety of milling time, namely 10, 20 and 40 hrs. The all samples after milling process were analyzed crystal structure by using X-ray diffraction and measured magnetic properties by using Vibration Sample Magnetometer (VSM). The crystallite size of $Nd_2Fe_{14}B$ phase was calculated by using the equation of Scherrer.

3. Results and Discussion

The results of x-ray diffraction measurements of the mixture Nd- Fe-B before milling as shown in figure 1. According the XRD patterns and JCCPDS file as seen at figure 1 show that the mixture does not contain impurities, and only consisted of the phase Nd and Fe. The XRD peaks of Nd and Fe can be obtained by comparing the experiment peaks with reference peaks data (JCCPDS) of Nd, Fe and B.

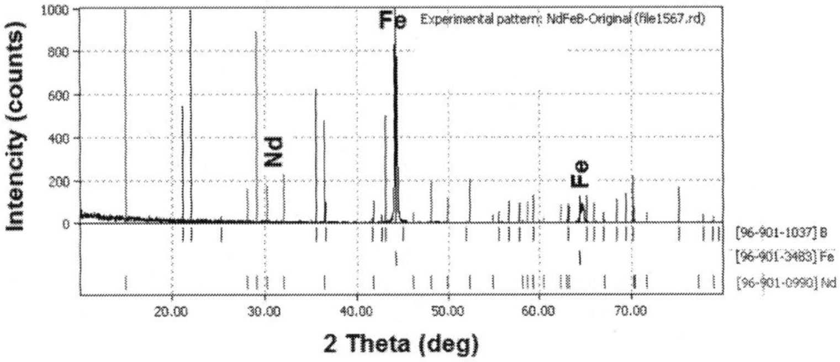

Figure 1. XRD patterns of mixed raw materials before milling.

Boron phase does not appear, it is because the amount of boron to small, so the boron phase was not detected by XRD. Next is the process of milling with milling time variations are: 10, 20 and 40 hours. Figure 2 is shown X-ray diffraction patterns of the sample that has been in the milling for 10 hours.

The refinment and peaks search were done by using software Rietveld and it was found that the peaks around the corner (2θ) : 42.94° and 49.62 which it is an $Nd_2Fe_{14}B$ phase. The XRD curve at figure 2 has fitting value such as : wRp = 25.73, Rp = 20.29 dan χ^2 (*chi-squared*) = 1.377. The result of this refinement produces excellent quality fitting with very small R factor as well. R factor is the criteria of fit and χ^2 factor is the goodness of fit and value of χ^2 (chi-squared) were allowed a maximum of 1.3 [9]. The Results refinement of x-ray diffraction pattern by software Rietveld shows that the composition of $Nd_2Fe_{14}B$ phase is about 13:53%.

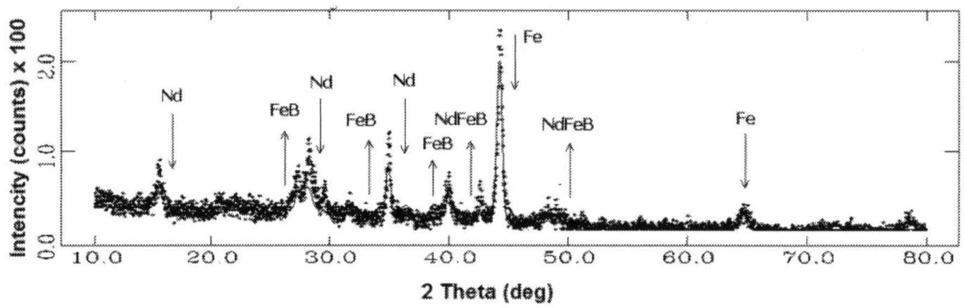

Figure 2. XRD patterns of sample after milling for 10 hrs

It is thought the majority of phases had already begun to react to each other to form alloys. The growth of $Nd_2Fe_{14}B$ phase is followed by growth of FeB phase, so the reaction that occurs after the milling for 10 hours as the equation (1)

$$2Nd + 14Fe + B ==== 0.14\ Nd_2Fe_{14}B + 5.02FeB + 0.38Nd + 9.78Fe \quad (1)$$

XRD analysis of sample that has been in the HEM for 20 hours is shown in figure 3. The refinment and peaks search for sample with 20 hours milling were done by using software Rietveld and the results show that a decline in the peak of Nd and FeB respectively by 3.56% and 4.48%, while the growth of $Nd_2Fe_{14}B$ become more significant about of 41.65%. The XRD curve at figure 3 has fitting value such as : wRp = 34.40, Rp = 25.08 dan χ^2 (*chi-squared*) = 1.345. According to the XRD curve as seen at figure 3 shows that the $Nd_2Fe_{14}B$ phase at sample after 20 hrs milling increases and FeB phase decreases.

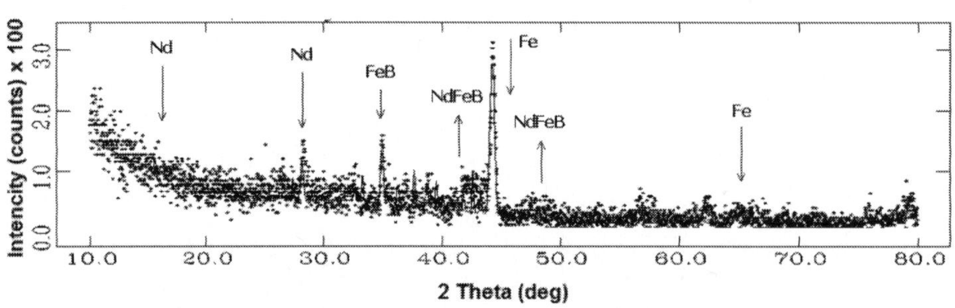

Figure 3. XRD patterns of sample after milling for 20 hrs

Figure 4 shows X-ray diffraction patterns of the sample that has been in the milling for 40 hours. The refinment and peaks search for sample with 40 hours milling were done by using software Rietveld. The XRD curve at figure 4 has fitting value such as : wRp = 33.63, Rp = 25.31 dan χ^2 (*chi-squared*) = 1.365. It has been found that the mostly $Nd_2Fe_{14}B$ phase has been formed, although still leaving phase Nd, Fe, and FeB. From the results of refinement X-ray diffraction pattern shows that the last mmixture is composed of phase $Nd_2Fe_{14}B$, Nd, Fe and FeB according to the following equation (2).

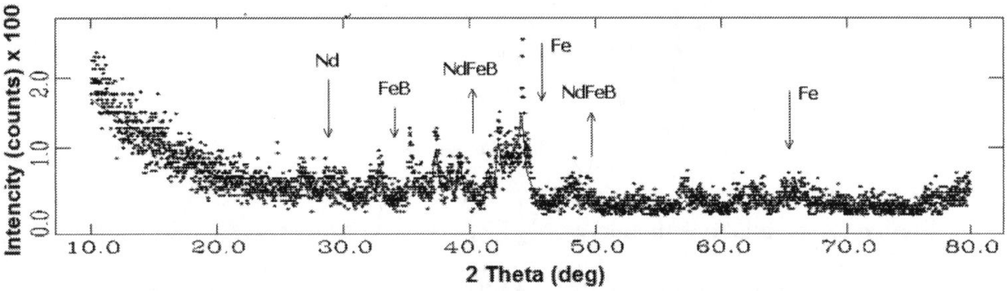

Figure 4. XRD patterns of sample after milling for 40 hrs

$$0.42Nd2Fe14B + 0.73FeB + 0.27Nd + 9.74Fe === 0.69Nd2Fe14B + 0.4FeB + 0.08Nd + 5.22Fe \quad (2)$$

These results show that the longer the process of milling the mass fraction of $Nd_2Fe_{14}B$ phase increases. The $Nd_2Fe_{14}B$ phase grows very well at $2\theta = 42°$. The refinement of the results of x-ray diffraction pattern for a 40-hour milling results showed that a decline peak Nd and FeB respectively by 1.11 % and 2.47%, while the growth of of $Nd_2Fe_{14}B$ phase become more significant about of 69.46%. The combine curve of XRD results from different milling time is shown in figure 5 . The forming of $Nd_2Fe_{14}B$ phase is begun at sample after 10 hrs milling at $2\ \theta = 42.94°$ and $49.62\ °$.

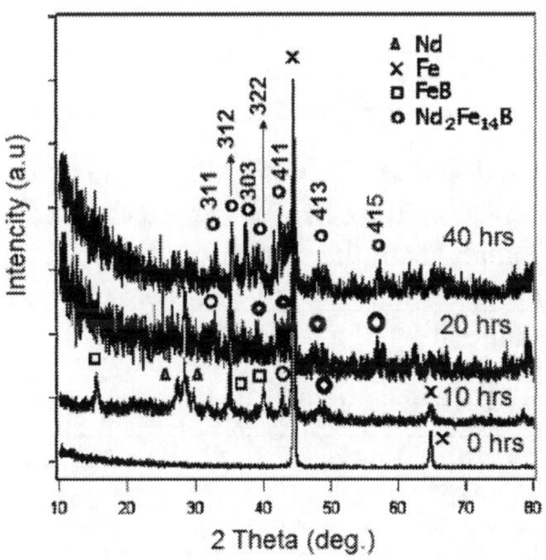

Figure 5. The combine XRD result of samples with different milling time.

The five peaks of $Nd_2Fe_{14}B$ phase look seemed to appear at $2\ \theta = 32.25\ °$, $39.72\ °$, $42.94\ °$, $49.62\ °$ and $57.26\ °$ in sample that had been in the milling 20 hours and seven peaks of $Nd_2Fe_{14}B$ phase are found at $2\ \theta = 32.25\ °$, $35.40\ °$, $37.77\ °$, $39.72\ °$, $42.96\ °$, $49.62\ °$, $57.26\ °$ in sample that had been in the milling 40 hours. Milling time by using HEM can contribute significantly to the formation of $Nd_2Fe_{14}B$ phase. Energy produced during the milling process is sufficient to react the elements Nd, Fe and B to form the $Nd_2Fe_{14}B$ compound according to equation 1 and 2. The $Nd_2Fe_{14}B$ phase growth curve versus milling time obtained through MA process by using HEM is shown in figure 6. The amount of % $Nd_2Fe_{14}B$ phase increases significant with increasing of milling time. The milling process by using HEM is useful for formation of metal alloy $Nd_2Fe_{14}B$. During the milling process, that the mixed powder (metal Fe, Nd, B) became finer and more active and the reaction between metals powder can be easy without addition heat from outside.

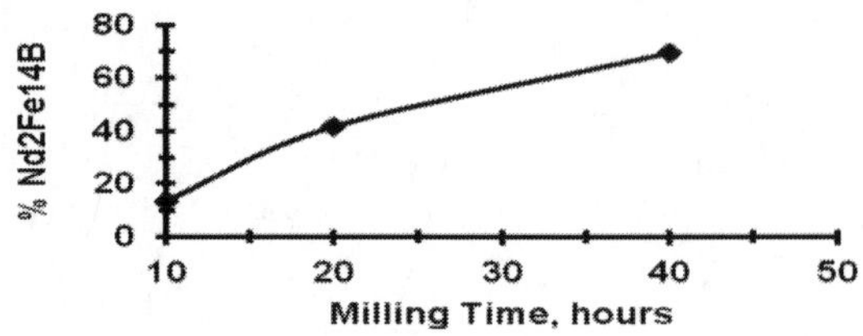

Figure 6. The relationship curve of $Nd_2Fe_{14}B$ phase formation to milling time

These experiment results show that the $Nd_2Fe_{14}B$ alloy can be produce by MA from elements metal powder for minimum 10 hrs milling time. L Sshultz and et all have been doing mechanical milling of mixture Nd, Fe and B for 30 minutes by using planetary ball mill, but $Nd_2Fe_{14}B$ phase can not be formed, and the phase $Nd_2Fe_{14}B$ can only be formed after a heat treatment at a temperature of 600°C [9]. Tetsuji Saito reported that production $Nd_2Fe_{14}B$ alloys by MA, the milling process was used HEM in H_2 atmosfir for 2 hrs and continued with heat treatment at 973K [10]. The authors [3, 9,10] have done some MA process using mechanical miling (HEM or planetary ball mill) in short time milling and limited only to result in a fine powder, and the forming of $Nd_2Fe_{14}B$ alloy is done by heat traetment. That is different with this conducted experiment, where alloying process was done by using HEM for long milling time , it shoud be more than 10 hrs. At this condition, the forming reaction of $Nd_2Fe_{14}B$ alloy can take place easily without additional heat treatment process. The crystallite size (*L* in nm) of $Nd_2Fe_{14}B$ phase were calculated from most intense peaks in XRD patterns using the Scherrer equation 3.

$$L = \frac{k\,\lambda}{B\,\cos\theta} \tag{3}$$

Here, "k" is Scherrer constant and its value is '0.94' for our material, 'B' is full width at half maximum (FWHM) of respective peaks, 'θ' is Bragg's angle and 'λ' is wavelength of Cu-k α radiation (1.5406 Å) used during the XRD analysis of the samples [1]. The calculated value of crystallite size of $Nd_2Fe_{14}B$ samples with different milling time are given in table 1. Calculation of the crystallite size of $Nd_2Fe_{14}B$ each milling time was done at 2 tetha = 42.96°.

Table 1. Value of crystallite size of $Nd_2Fe_{14}B$

Milling time , hours	Crystal Size, nm
10	72.35
20	48.72
40	25.64

According the result of calculated crystallite size, that crystal size is enough small in range 25.64 to 72.35 nm. The crystal size of powder alloy $Nd_2Fe_{14}B$ decreases with increasing of milling time by using HEM. The magnetic properties of sample powder after milling process by using HEM were measured by using VSM. The VSM results are shown in figure 7 and table 2. It can be found that magnetic saturation Ms, moment magnetic (remanence) Mr, coercivity Hc and energy product BHmax strongly depend on milling time. Figure.7 shows that Ms, Mr and Hc increase with increasing milling time. The increasing magnetic properties due to growth of $Nd_2Fe_{14}B$ phase, where the milling time increases so the growth of $Nd_2Fe_{14}B$ increases also. The value of magnetic properties relationship with milling time is presented in table2. Sample powder with milling time 40 hours by using HEM has highest value of magnetic properties, where the amount of $Nd_2Fe_{14}B$ phase is about 69.46 % and it has Mr value = 81 emu/g, Ms =122 emu/g, Hc = 5.54 kOe and BHmax = 11.01 MGOe. The results of the magnetic properties (Mr, Ms, Hc and BHmax) of NdFeB alloy powder obtained from this research through MA were slightly higher than the results achieved by other authors who use the technique microwave assisted combustion process [11].

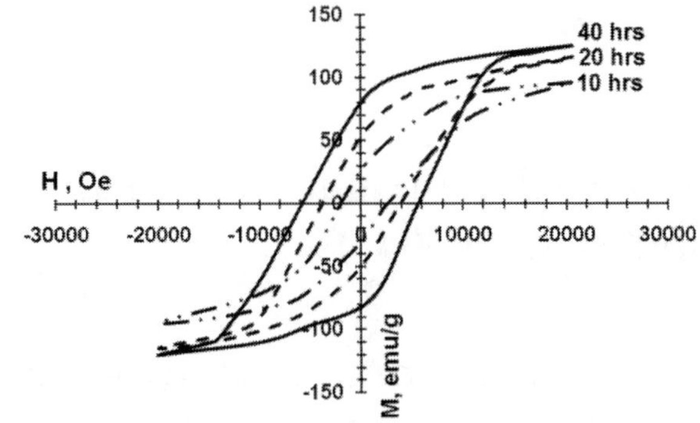

Figure 7 . Magnetic hysteresis loops of the powders after milling by using HEM

Table 2. Value of magnetic properties of powders after milling by using HEM.

Milling Time	Ms, emu/g	Mr, emu/g	Hc, kOe	BHmax, MGOe
10 hours	95	28	2.41	0.55
20 hours	114	53	3.82	5.5
40 hours	122	81	5.54	11.01

4. Conclusion

The $Nd_2Fe_{14}B$ alloy powder can be made through mechanical alloying by using HEM for 40 hours milling, and in these conditions, the $Nd_2Fe_{14}B$ alloy powder was obtained as much as 64.96% with crystallite size about 25.64 mm. The value of magnetic properties tend to increase with increasing milling time, the highest magnetic properties were achieved on the powder in the milling 40 hours namely: Ms = 122 emu/g, Mr = 81 emu/g, Hc = 5.54 kOe and BHmax = 11.01 MGOe.

5. References

[1] Shah IA, Abbas T, Ullah Z, Najam ul Hassan, Rauf A, Ullah K, and Naseem S, 2015, *Armenian Journal of Physics*, **Vol. 8**, issue 4, pp. 185-190.
[2] Liarr H, Gang S, and Erder W, 2003, *Trans. Nonferrous Met. Soc.* Vol **13**(5), pp.1070-1075.
[3] Haavisto M, Tuominen S, Santa-Nokki T, Kankaanpää H, Paju M, and Ruuskanen P, 2014, *Advances in Materials Science and Engineering*, Article ID 760584
[4] Collocott SJ, Dunlop JB, Lovatt HC and. Ramsden VS, 2007 Rare-Earth Permanent Magnets: New Magnet Materials And Applications, CSIRO Division of Telecommunications and Industrial Physics, Lindfield,NSW, Australia.
[5] Brown D, Bao-Min M, and Chen Z, 2002, *J.MMM*, **248**, pp.432–440
[6] Suryanarayana C, 2001, *Progress in Materials Science* **46** 1-184
[7] Liu Xiaoya, Li Yuping, Hu Lianxi, 2013, *J.MMM* **330** pp.25–30.
[8] Chakka VM, Altuncevahir B, Jin ZQ, Li Y, and Liu JP, 2006, *J. of Appl. Phys.* **99**, 08E912-1-08E912-3.
[9] Toby BH, 2006 *Powder Diffr.*, Vol. **21**, No.1 pp.60-67
[10] Saito T, 1998 *Journal of the Japan Society of Powder and Powder Metallurgy* **Vol. 45**, No. 7.
[11] Swaminathan V, Deheri PK, Bhame SD and Ramanujan RV, 2013 *Nanoscale*, **5**, RSC Publishing p.p 2718 – 2725

The effect of mechanical deformation to the magnetic properties of stainless steel 304

N. Mubarok[1], H. A. Notonegoro[2], K. A. Zaini Thosin[3] and A. Manaf[1*]

[1]PPS Materials Science, FMIPA-Universitas Indonesia, Depok 16424, Indonesia.
[2]Mechanical Engineering Dept., FT-Universitas Sultan Ageng Tirtayasa, Cilegon 42435, Indonesia.
[3]Pusat Penelitian Fisika,LIPI, Serpong, Indonesia.

*E-mail: Azwar@sci.ui.ac.id,

Abstract. A study of a non-magnetic 304 austenitic stainless steel alloy through mechanical deformation has been done. These specimens are pipe usually used to deliver gas which contains corrosive oil fields. The metallographic observation of a 20% deformation shows the increase in the value of the magnetization, as compensation for the formation of martensite phase as a result of a mechanical treatment. Martensitic phase formed due to a shift in the structure of the z-axis due to the effects of pressure and shear from the cold rolled. The existence of martensite phase and magnetic properties conducted through x-ray diffraction and permagraf investigation. An identified x-ray diffraction pattern shows the presence of a new peak between 10°-30° angle indicate the mechanical deformation in crystallite structure. Furthermore, at in 20% distortion, the value of magnetization is increased above 0.2 T in small coercivity value and caused decreased the ability of corrosion resistant.

1. Introduction

The AISI 304 stainless steel pipes are important engineering materials mainly used for Oil pipe. These pipes are usually used to deliver gas which contains corrosive and erosion ingredient from oil fields [1,2]. In Several conditions, the pipes undergo the load and stress conduct the deformation which can trigger the formation of the martensite phase [3]. The previous studies found that martensite phase correlates with the appearance of the magnetic properties of the metal structure [4].

There are two types of martensite can induce the formation in austenite stainless steels while deformation process where the transformation sequence $\gamma \rightarrow \varepsilon \rightarrow \alpha'$: ε and α'. The ε martensite is paramagnetic with *hcp* structure, and the α' martensite is ferromagnetic with the *bcc* structure [5,6].

By this paper, we report the experimental process to determine the presence of martensitic phase in AISI 304 stainless steel correlated with magnetic properties at 20% under deformation by mechanical treatment.

2. Experimental procedures

The chemical composition of 304L AISI austenitic stainless steel used in this investigation as is in table 1. The specimen steel was received in pipes form with 11 mm thickness without preheat treatment.

The mechanical treatment by cold rolling was carried out in a pilot rolling machine. Different thickness reductions 0.1 mm in every rolled from 0 to 20% with the *Torsee* 71.8 kg.m were carried out at 27°C. The crystallite structure were examines using X-ray diffraction and PERMAGRAPH® L © MAGNET-PHYSIK Dr. Steingroever GmbH is use to investigate the magnetic properties of the alloy.

Table 1. Chemical composition of stainless steel 304 specimen

	C	Mn	Si	P	S	Ni	Cr	Fe
Mill certificate	0.05	1.08	0.35	0.038	0.001	8.50	18.65	-
AAS	-	1.49	-	-	-	10.6	21.20	66.44

3. Results and discussion

The mechanical treatment illustrated as in Figure 1. The specimen came under pressure from the crush of the two drum suppressor followed by the pull of a result of drum rotation simultaneously. As in Figure 2a, with the normal parallel to the axis, the martensitic plane transformation passes along the horizontal x-axis of the cube (Figure 2b). The macroscopic shear direction is along the horizontal y-axis (Figure 2c). Following the indicated habit plane and shear direction; the shape of the grain will change into the parallelepiped [7].

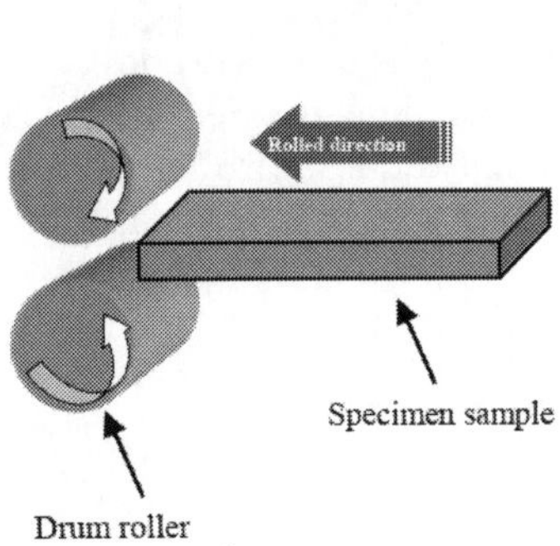

Figure 1. Illustration of mechanic treatment for deformation process

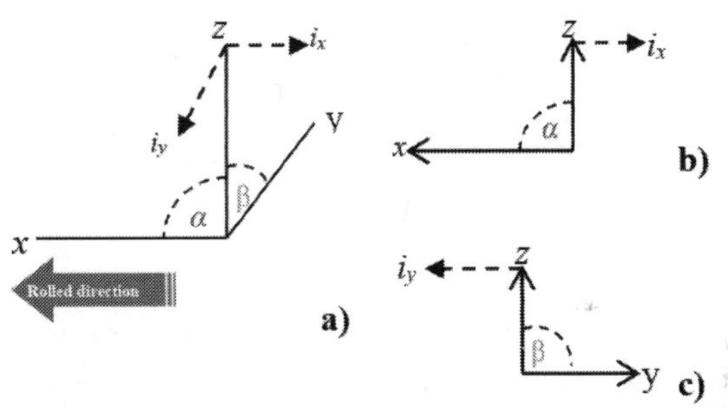

Figure 2. The orientation axis of deformation specimen sample. x-axis perpendicular to y-axis along deformation process. a) along x-axis, the angle of α widen the z-plane, b) along y-axis, the angle of β widen the z-plane.

X-ray pattern by *Cu-Kα* diffraction analysis performed over the 2θ range from 10° to 90°. The results of the XRD analysis done on the surfaces of AISI 304 stainless steel presented in Figure 3. The XRD pattern shows 0% deformation (blue line) and 20% deformation (red line). The symbol of α shows the disappearance of a part of austenite peak and β show of the reveal of martensite phase in deformation sample. As seen, at 20% deformation as unconstrained transformation the peak has consisted the presence of austenite and *α'martensite*, respectively [8]. In fact, the strain energy was mainly consumed for production of *α' martensite* during deformation.

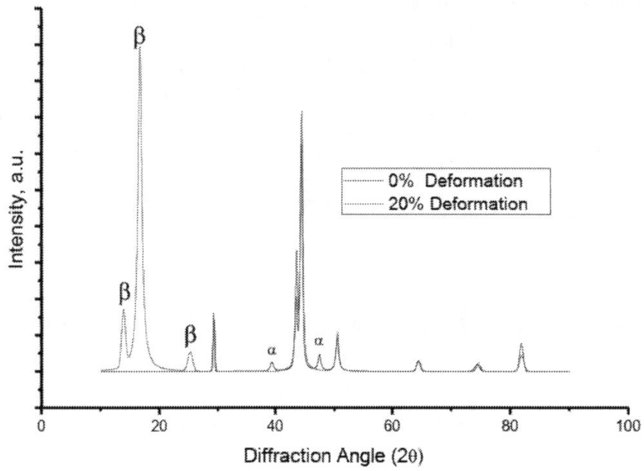

Figure 3. XRD of 0% deformation (blue line) and 20% deformation (red line). symbol of α indent the disappearance a part of austenite peak and β indent the reveal of martensite phase in deformation sample.

The magnetic hysteresis of AISI 304 stainless steel by Permagraf has shown in Figure 4. At in 0% deformation, the magnetic saturation has zero value indicating that specimens are nonmagnetic without deformation. Furthermore, at in 20% distortion, the value of magnetization is increased above 0.2 T and presence their coercivity value on its magnetic properties. This magnetic property indicates the formed of magnetic moments in the sample due to the influence of unconstrained transformations. The impact of the rise of the magnetic properties of the samples of AISI 304 stainless steel deformed caused decreased ability of corrosion resistant [9].

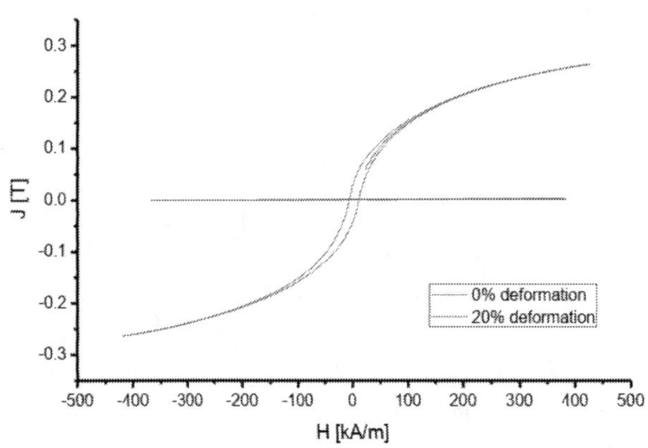

Figure 4. Magnetic hysteresis of 304 stainless steel at 0% deformation (blue line) and 20% deformation (red line) at 22 °C.

4. Conclusion

The emergence of martensitic phase conducted 20% deformed by mechanical treatment identified the presence of new pattern at 2θ = 10° - 30°. The existence of this produce martensite phase changes magnetic properties which have saturation value above 0.2T in small coercivity and caused decreased the ability of corrosion resistant.

5. Acknowledgement

The authors gratefully acknowledge the support of Department of Physics, University of Indonesia for the research facilities.

6. References

[1] Ex N, Electrical NHP, Products E, Nhp NHP. Solutions for the oil and gas industry n.d.

[2] Chawla V, Chawla A, Puri D, Prakash S, Gurbuxani PG. *Corrosion* 2011;**10**:367–85.

[3] Dryzek E, Sarnek M, Wróbel M *J Mater Sci* 2014;**49**:8449–58. doi:10.1007/s10853-014-8555-y.

[4] Surkialiabad R, Hedayati A, Alam AS *Transformation* 2013;**2**:11.

[5] Mubarok N, Notonegoro HA, Thosin KAZ, Manaf A *AIP Conf Proc* 2016;**020022**:020022. doi:10.1063/1.4953947.

[6] Abreu HFG De, Carvalho SS De, Lima Neto P De, Santos RP Dos, Freire VN, Silva PMDO, et al *Mater Res* 2007;**10**:359–66. doi:10.1590/S1516-14392007000400007.

[7] Sade M, de Castro Bubani F, Lovey FC, Torra V *Mater Sci Eng A* 2014;**609**:300–9. doi:10.1016/j.msea.2014.05.018.

[8] Character D. Martensite in Steels 2002:1–12.

[9] Shalash L, Nasher L *ResearchgateNet* 2010;**9**:30–8.

Physical and magnetic properties, microstructure of bonded magnet NdFeB prepared by using synthesis rubber

Suprapedi, P. Sardjono, Muljadi

Research Center for Physics, Indonesian Institute of Sciences

E-mail: suprapu@gmail.com

Abstract. The magnet permanent has been made by using NdFeB (NQP-B) powder with purity 99.90 % and polymer rubber with purity 90%. This research was done to determine the effect of the polymer (rubber) composition as binder in the manufacture of bonded magnet NdFeB on physical properties, microstructure and magnetic properties. Bonded magnets are magnet material made from a mixture of magnetic powder as a filler and the polymer as a matrix material or as binder to bind the magnetic particles. The NdFeB (NQP-B) powder fractions of bonded magnets NdFeB investigated were as follows: a) 97 wt.% of NdFeB (NQP-B) and 3 wt.% of rubber,b) 95wt.% of NdFeB (NQP-B) and 5 wt.% of rubber, c) 93 wt.% of NdFeB (NQP-B) and 7 wt.% of rubber, d) 91 wt.% of NdFeB (NQP-B) and 9 wt.% of rubber. Both raw materials were mixed by using mixer until homogen with total weight about 16 g for each sample, then added 0.3 ml of catalyst and mixed again and put in dies mould and compacted at pressure 30 MPa, then dried for 2 hours at room temperature. The dried samples was characterized such as : bulk density measurement and magnetic properties by using BH-curve permeagraph. The bulk density values of the sample bonded NdFeB magnets using the binder 3% wt. and 5% wt. rubber are respectively 4,70 g/cm^3 and 4.88 g/cm^3. The result from BH-curve shows that the highest value of remanence (Br = 5.12 kGauss) is at sample with 3 % wt. of rubber, but sample with 5 % wt. of rubber has lowest value of remanance (Br = 4.40 kGauss). Based on the observation of the SEM photograph shown that the rubber material has been successfully covered the whole surface of the grain and fill some of the voids that are in the grain boundary.

1. Introduction

Permanent magnet has been developed ; the first of magnetic iron was used then in developing that magnetic iron magnet has been made from some metallic alloy such as : SmCo, AlNCo and rare earth base magnet ($Nd_2Fe_{14}B$) [1]. Also, some of magnetic materials are made from oxide base compounds as like Barium hexa ferrite $BaO6Fe_2O_3$ and it is called hard ferrite [1]. It has limitations of magnetic properties, for example, Ba-hexaferrite has remanence (Br) about 3 – 4 kGauss and energy product [BH]max about 5 MGOe [2]. Compare of Ba-hexa ferrite to rare earth magnet ($Nd_2Fe_{14}B$) that $Nd_2Fe_{14}B$ magnet has higher magnetic performance, it has remanence (Br) about 5 – 13 kGauss and energy product [BH]max 5 - 30 MGOe [2]. Rare earth permanent magnet NdFeB is a new kind of magnetic material developed in the 1980's with excellent magnetic characteristics (high energy product and high coercive force etc.) and relatively low cost. It is getting to replace the traditional magnets of hard ferrite, AlNiCo and SmCo. There are two types of NdFeB magnet namely : (a) sintered magnet and (b) bonded magnet. Sintered NdFeB magnet is made by using sintering process at high temperature in absence of oxygen [2,3]. Bonded NdFeB magnets are generally prepared by blending magnetic powder with binder followed by compacting or molding the material to final shape. The sintered $Nd_2Fe_{14}B$ magnet has higher magnetic properties than bonded $Nd_2Fe_{14}B$ magnet , and both have different application [4]. The magnetic properties of $Nd_2Fe_{14}B$ are shown in Table 1.

Tabel 1. Magnetic Properties of NdFeB magnets [4]

Properties	Bonded NdFeB	Sintered NdFeB
Remanence Br (kG)	6.9	13.1
Energy Product BHÞmax (MGOe)	10	42
Coercivity Hci (kOe)	9	14
Curie TemperatureTc (°C)	200	310

NdFeB magnets have disadvantages such as easy to oxidation or lower corrosion resistance at room temperature. To protection the corrosion or oxidation, the surface of NdFeB magnet should be coated. There are many kinds of coating materials such as : Ni-plated, chrome plated and epoxy or other polymer plated [5,6]. There are some of polymer materials for coating such as : thermosetting polymer epoxy. Bonded NdFeB permanent magnets are mainly composed of NdFeB magnetic powder and polymeric binders such as thermoplastic polyolefin; polyphenylene sulphide (PPS), polyvinyl chloride (PVC), latex or rubber, polypropylene (PP), polyethylene (PE), high density polyethylene (HDPE), polyamide or thermosetting epoxy resin , nitrile rubber or other elastomers [7]. The most important advantages of polymer bonded magnets are low weight, corrosion resistance, easy machining ability, forming and handling, high production rate. Bonded magnets provide an almost infinite variety of combinations of mechanical, physical, chemical, thermal and magnetic properties due to various kinds of polymeric matrices. Several authors have been conducting research manufacture bonded NdFeB magnets using various polymer materials bleak nylon powder, epoxy resin, PEG and so on [8,9,10]. While research manufacture bonded NdFeB magnets made using different binder material which is using polymer rubber . There has been no research manufacture of bonded magnets using rubber polymer, in addition to the price of rubber pasta is much cheaper compared to other polymers and require a low aging temperature is room temperature. The goal of this experiments are to know the influence of rubber composition to density and magnetic properties on preparation of bonded NdFeB magnet.

2. Experimental

Bonded magnets are magnet material made from a mixture of magnetic powder as a filler and the polymer as a matrix material or as binder to bind the magnetic particles. There are four kinds of samples bonded NdFeB magnets that have been made with a variety of polymer adhesive composition (rubber pasta). The raw material used to manufacture bonded NdFeB magnets are NdFeB powder (NQP-B) with a purity of 99.90% and gum paste with a purity of 90%. The NdFeB (NQP-B) powder fractions of bonded magnets NdFeB investigated were as follows: a) 97 wt.% of NdFeB (NQP-B) and 3 wt.% of rubber,b) 95wt.% of NdFeB (NQP-B) and 5 wt.% of rubber, c) 93 wt.% of NdFeB (NQP-B) and 7 wt.% of rubber, d) 91 wt.% of NdFeB (NQP-B) and 9 wt.% of rubber. Both raw materials were mixed by using mixer until homogen with total weight about 16 g for each sample, then added 0.3 ml of catalyst and mixed again and put in dies mould and compacted at pressure 30 MPa, then dried for 2 hours in open room at 30°C. The characterizations were done such as : measurement of bulk density by using Archimedes method. The bulk density value of solid materials (in fragment form) can be calculated by using Archimedes method, mathematically, density is defined as mass divided by volume. A basic formula to calculate the density of materials can be used the following equation [11]:

$$\rho_o = \frac{m\rho_f}{\Delta m}. \qquad (1)$$

Δm is the mass difference value objects hung in the air with the mass of the object in suspension in the liquid , ρf is density of liquid , and m is mass of objects. The observation micro structure was done by using Scanning Electron Microscope (SEM) and magnetic properties measurement by using Gaussmeter and permeagraph (BH-curve analyzer) to obtaine value of magnetic flux density, remanence Br, coercivity Hic and energy product [BH]max.

3. Results and Discussion

The XRD profile of NdFeB (NQP-B) powder is shown at Figure 1. According on analyzing with GSASS is found that major phase is $Nd_2Fe_{14}B$ [ICDD-96-100-8719] about 78.14 % and minor phase is Fe [ICDD-96-901-3473] about 21.86 %. XRD analysis of the raw magnetic powder is important to determine the content of the phase of the material (NQP-B powder), because there are many types of NQP-B powder were sold. With different content of the phases present in the raw magnetic powder will gave different magnetic properties, so the results of this experiment refers to a type of magnetic powder NQP-B.

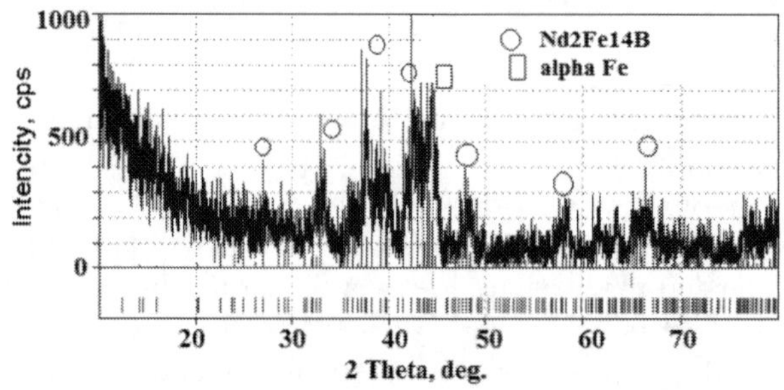

Figure 1. Diffraction patterns of NdFeB powders [NQP-B]

The curve of bulk density versus composition of rubber is shown at Figure 2. The bulk density value has tendency decreasing when rubber composition is higher, because it follows a composite rule that density of rubber is lower then density of NdFeB powder. The highest density value is achieved about 4,88 g/cm^3 at sample with 3 % wt. rubber and lowest density value is 4.20 g/cm^3 at sample with 9 % wt. rubber. Theoritical density of bonded magnet NdFeB is about 5 – 6 g/cm^3 [2]. The densification of the sample bonded NdFeB magnets from this experiment is still low at around 70-80%. It is caused by the pressure on forming process at 30 MPa is still relatively low.

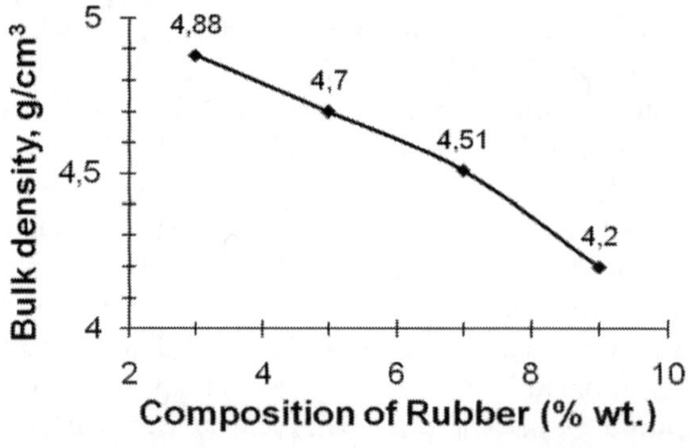

Figure 2. The curve of bulk density $Nd_2Fe_{14}B$ for different rubber compotition

Before magnetic properties measured, all samples are magnetized with field strength about 8 kGauss by using impulse magnetizer. Then the magnetic flux density was measured by using Gaussmeter and the result is shown at Figure 3.

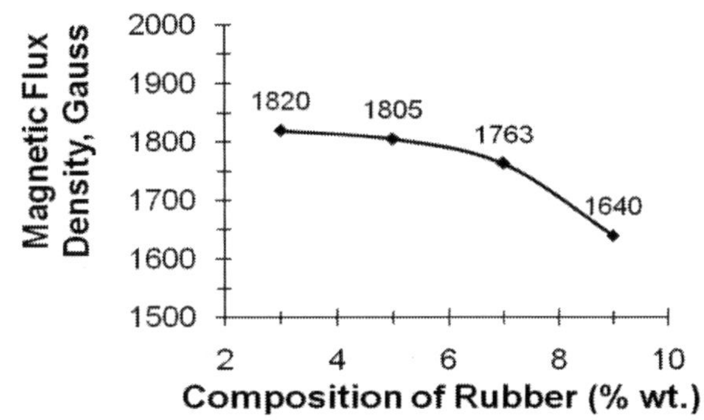

Figure 3. The curve of magnetic flux density as function of composition of rubber

According to Figure 3, The highest value of magnetic flux density is achieved about 1805-1820 Gauss at samples with 3% and 5 % wt. of rubber, but the magnetic flux density tends to decrease again until 1640 Gauss with increasing of rubber composition. Rubber polymer is classified as non-magnetic material, when the rubber composition of the sample bonded NdFeB magnets increases, the magnetic properties of bonded magnets will tend to decline. The BH – curve results for different of rubber composition is shown at Figure 4. The BH-curve shows that when rubber composition increase, the curve hysteresis will be smaller. Rubber polymer is non-magnetic material, when the rubber composition of the sample bonded NdFeB magnets increases, of course, the composition of the magnetic powder (NdFeB) will shrink. This implies that the magnetic properties of bonded magnets will tend to decline. The biggest of BH-curve is at sample with 3 and 5 % wt. of rubber. According to BH-curve result, the magnetic properties such as remanence Br, coecivity Hic, and energy product [BH]max of each sample are presented at Table 2. The rubber composition influenced significantly value of Br, Hc and [BH]max.

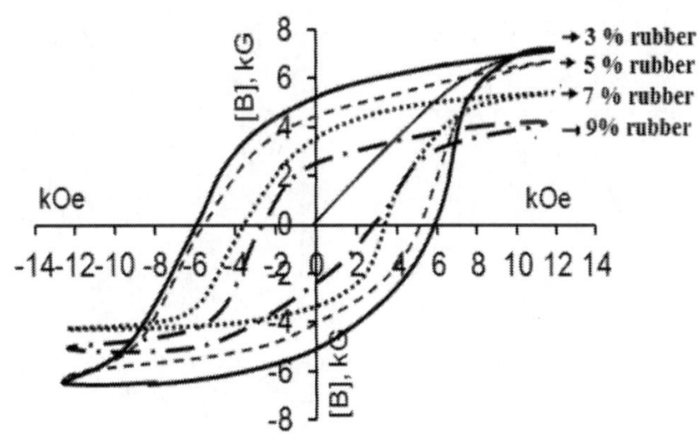

Figure 4. BH-curve of bonded $Nd_2Fe_{14}B$ magnet for different rubber composition

The biggest of BH-curve is at sample bonded $Nd_2Fe_{14}B$ magnet with 3 and 5 % wt. of rubber. According to BH-curve result, the magnetic properties such as remanence Br, coecivity Hic, and energy product [BH]max of each sample are presented at Table 2. The rubber composition influenced significantly value of Br, Hc and [BH]max. Because rubber is non magnetic materials and Polymer that serves as a matrix wrap any $Nd_2Fe_{14}B$ particle that magnetic properties of the particle will be isolated by polymer. Bonded $Nd_2Fe_{14}B$ with composition of 3 and 5 % wt. rubber have the best magnetic properties. They have Br value = 4.40-5.12kGauss, Hc value = 6.02-5.98 kOe and [BH]max = 10.12-8.77 MGOe.

Table 2. The value of Remanence (Br), Coercivity (Hc) and Energy Product (BHmax) of bonded magnet with variation of % of rubber.

% wt.of Rubber	Br,Kgauss	Hc, kOe	BHmax, MGOe
3	5.12	6.02	10.12
5	4.4	5.98	8.77
7	3.85	3.83	5.25
9	2.92	2.94	3.5

Based on the results of measurement of magnetic properties showed that the sample bonded NdFeB magnet with 3% rubber composition has most high value of Br, Hc and BH max respectively, are 5.12 kG, 6.02 kOe and 10.12 MGOe. Magnetic properties of bonded NdFeB magnet from literature [4] and experiments [2,8.9 and 10] that have been done several authors with other polymer binder resulting in a value range of Br = 5-6 KG, Hc = 5-9 KOe and BH Max about 10 MGOe. the results obtained from this study are not much different from the references or the results of other authors.

Figure 5 shows the SEM micro graph of the samples bonded magnet made by using 3 to 9 % wt. of rubber. From SEM images, it can be seen that the rubber binds all surface of grains.The image with grey colour is a grain of magnetic $Nd_2Fe_{14}B$ and the images with white and black colour indicate as rubber polymer. The SEM images for sample with 7% and 9 % of rubber show that the polymer rubber is seen not only bind the grains at the grain boundaries but also cover portion of a grains surface. The polymer rubber on sample with 5 % and 3 % of rubber bind the grain only on grainboundary that are responsible by decreasing density value and magnetic properties for sample with 7 % and 9 % wt.of rubber.

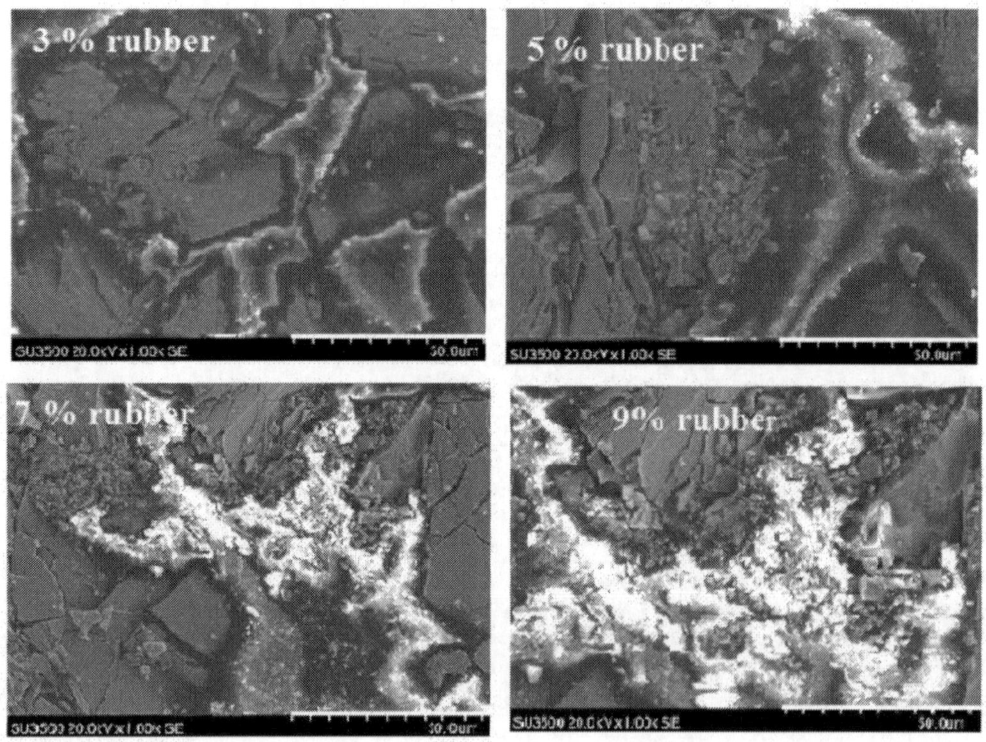

Figure 5. SEM images of samples bonded magnet NdFeB with variation of rubber binder

4. Conclussion

According to XRD result shows that NdFeB powder which used to make bonded magnet has $Nd_2Fe_{14}B$ phase about 78,14 % and Fe phase about 21,86 %. The composition of rubber (3 to 9 % wt.) gives significant effect to bulk density and magnetic properties such as magnetic flux densuty, remanence Br, corcivity Hc and energy product [BH]max. These conditions can be used reference for making of bonded permanent magnet with variation of properties or customer made. Sample bonded NdFeB magnet with 3 % wt. rubber having the best magnetic properties. It has bulk density = 4.88 g/cm^3, Br value = 5.12 kGauss, Hc value = 6.02 kOe, [BH]max = 10.12 MGOe and magnetic flux density = 1820 Gauss. The result of observation using the SEM of sample with 3% wt. rubber showed that almost all of the polymer rubber has been blanketing the surface of the particles and bind the grains with each other.

Acknowledgments

The author would like to thank Program Riset Mandiri LIPI as financial support, and thank you to Mr.Nanang Sudrajat ST PPET-LIPI Bandung for helping in Magnetizing characterization, and thank you for operators XRD, SEM at Research Center for Physics-LIPI.

5. References

[1] L.A. Dobrzański, B. Ziębowicz, M. Drak. 2006 *Journal of Achievements in Materials and Manufacturing Engineering* 18 79-82.

[2] David Brown, Bao-Min Ma, Zhongmin, 2002,*Chen Journal of Magnetism and Magnetic Materials* 248 432–440.

[3] Cristina Stancu, Petru V. Notingher, Valentin Ionita, Virgil Marinescu, Denis Panaitescu,2014, *Electrical Engineering Series*, No 38, 98-106

[4] Mehrdad Kokabi, Fatemeh Arabgol, and Mehrdad Manteghian, 2005 *Iranian Polymer Journal* 14 71-79.

[5] H.R.Kirchmayr, 1996, *J. Phys. D: Appl. Phys.* 29 2763–2778.

[6] B.M. Ma, J.W. Herchenroeder, B. Smith, M. Suda, D.N. Brown, Z. Chen, 2002, *Journal of Magnetism and Magnetic Materials* 239 418–423.

[7] Avtar Singh Dhillon, Lakhvir Singh, Amardeep,2015, *International Journal of Research in Engineering and Technology,* Volume: 04 Issue: 08, 316-320.

[8] Xiuhai Zhang, , Weihao Xiong, 2009 *Rare Metals* Volume 28, Issue 3, pp 248-252.

[9]. Perdamean Sebayang, Muljadi, Priyo Sardjono, Ayu Yuswitasari and Wisnu Ari Adi, 2013 *AIP Conf. Proc.* 1555, 40 - 46

[10] D. Plusa, M. Dospial, B. Slusarek, U. Kotlarczyk and T. Mydlarz, 2008 *Adv.Mater.Sci.* 18(2008) 541-544.

[11] T. Moyo, P.R.J. Silva, H. Saitovitch, 2000 *Bulk Density Measurement of Small Solid Fragments* CBPF-NT-005/03.

Influence of intermetallic Fe and Co on crystal structure disorder and magnetic property of $Ni_{50}Mn_{32}Al_{18}$ Heusler alloy

H. A. Notonegoro[1,2], B. Kurniawan[1*], J. Setiawan[3], D. Nanto[4] and A. Manaf[1*]

[1]PPS Materials Science, FMIPA-Universitas Indonesia, Depok 16424, Indonesia.
[2]Mechanical Engineering Dept., FT-Universitas Sultan Ageng Tirtayasa, Cilegon 42435, Indonesia.
[3]Center for Nuclear Fuel Tecnology-Badan Tenaga Nuklir Nasional, Tangerang Selatan 15310, Indonesia.
[4] Dept. of Physics Education, UIN Syarif Hidayatullah, Jakarta, 15412, Indonesia

*E-mail: Azwar@ ui.ac.id

Abstract. This works reports a study on structure and magnetic properties influenced by both Fe and Co on $Ni_{50}Mn_{32}Al_{18}$ Heusler alloy as a candidate of magnetocaloric effect (MCE) materials. The Ni-Fe-Mn-Co-Al sample was prepared by arc melting furnace (AMF) in high purity argon atmosphere. X-ray diffraction investigation and magnetic hysteresis were conducted to characterize the synthesized sample. X-ray diffraction using Cu-$K\alpha$ pattern shows that both Fe and Co introduce a tungsten type disorder of $Ni_{50}Mn_{32}Al_{18}$ Heusler alloy which partially replace the site position of Ni and Mn respectively. However, in this tungsten type disorder, it is difficult to distinguish the exact position of each constituent atom. Therefore, we believe it may allow any exchange interaction of each electron possessed the atom. Interestingly, it produced a significant increase in the value of the hysteresis magnetic saturation.

1. Introduction

The needs of magnetic refrigerator (MR) operated near room temperature have stimulated much research interest in magnetocaloric effect (MCE) alloy base. To establish main relation between magnetocaloric and magnetic values in MCE mechanisms is studying magnetic properties materials with sharp change of lattice parameters and structure type [1].

Various candidates for magnetocaloric alloys have been studied, including Ni-Mn-Sn-Ga, Ni-Fe-Mn-Ga, Ni-Co-Mn-Al, Ni-Mn-Sn and Ni-Mn-Al system [2–6]. Widespread research has been carried out with the focus on developing Ni-Mn-Al as an MCE alloy base [4]. Liu et al. reported potential application of high-temperature magnetocaloric Heusler alloy (HA) in previous studies [7]. Kainuma et al. found that various compound of Ni-Mn-Al alloy had affected martensitic transformation temperature and magnetic properties [8]. Moreover, Webster et al. found that $Ni_{50}Mn_{25}Al_{25}$ shows antiferromagnetic properties (AFM) in structure *B*2 (Wyckoff) [9,10]. However, Notonegoro et al. reported the similar $Ni_{50}Mn_{25}Al_{25}$ showed diamagnetic (DM) phase in structure *B*2 (Wyckoff). Furthermore, partial replacement of Al with Mn atoms to obtain $Ni_{50}Mn_{32}Al_{18}$ compound shows a ferromagnetic (FM) phase with *B*2 structure [11]. In other hand, Xuan et al. found substitution of Co atom to Ni site in $Ni_{50-x}Co_xMn_{32}Al_{18}$ (x= 3, 4, 5, 6, 7, and 8) had shifts the temperature of martensitic transformation into a lower value and Co concentration play important rule as a tuning of a ferromagnetic austenite-weak-magnetic martensite[12]. Meanwhile, Graf et al. describe the properties of Heusler materials are strongly

dependent on the atomic arrangement of the atoms. Eventually that the size of the band gap decreases with increasing amount of atomic disorder[13].
In this work, we report the study of crystallite structure and hysteresis magnetic properties (Ms) of $Ni_{50}Mn_{32}Al_{18}$ which some Ni atoms replaced by Fe and Mn atoms replaced by Co, respectively to make a $Ni_{44}Fe_6Mn_{24}Co_8Al_{18}$ HA system.

2. Experimental procedures

$Ni_{44}Fe_6Mn_{24}Co_8Al_{18}$ alloy was prepared in an appropriate quantity of high purity Ni, Fe, Mn, and Al metal powder. The mixing compound compacted with 10 tons in a 12 mm cylinder dice for 5 minutes and sealed in quartz cube 1.5 mm diameter. After flushing with high purity argon (HPA) three times, the compacted sample sintered in 900 °C for 1 hour under vacuum condition. This way was necessary to simplify sample preparation and in order to reduce mass lost during smelting process. The arc melting furnace (AMF) methods were smelting on vacuum condition after flushing five times with HPA. These samples were remelted four times to ensure composition homogeneity then followed by the annealing process for 72 hours in 900 °C in vacuum condition. This was a better annealing time then 48 hours in 900 °C as we reported before [14]. Crystallite structure and magnetic properties were examines using XRD with Cu-*Kα* and PERMAGRAPH® L © MAGNET-PHYSIK Dr. Steingroever GmbH.

3. Results and discussion

3.1. Order and Disorder of Heusler Structure compound

Study of crystal structure is essential to understanding properties of Heusler compounds correlated with experimental results. Cu_2MnAl-type structure of Heusler alloy identified by the occurrence of the *fcc*-typical on (111) and (200) plane reflections, and their relation to the (220) plane reflection [13].
XRD pattern in Figure 1 shows structure of $Ni_{44}Fe_6Mn_{24}Co_8Al_{18}$ compound. The type plane reflection shows of Cu_2MnAl-type structure. Thus, as result of refinement base on *JCCPDS - COD 98-063-2933*, the structure of $Ni_{44}Fe_6Mn_{24}Co_8Al_{18}$ determined as tungsten-like structure whereas the *X*, *Y*, and the *Z* atoms become equivalent distributed in all positions. The value $4a$=$4b$=$4c$ shows a *cubic* lattice and a reduced symmetry of $Fm\bar{3}m$ also known as an *A*2-type disorder (Wyckoff) whereas remains only the cubic primary reflection [13].

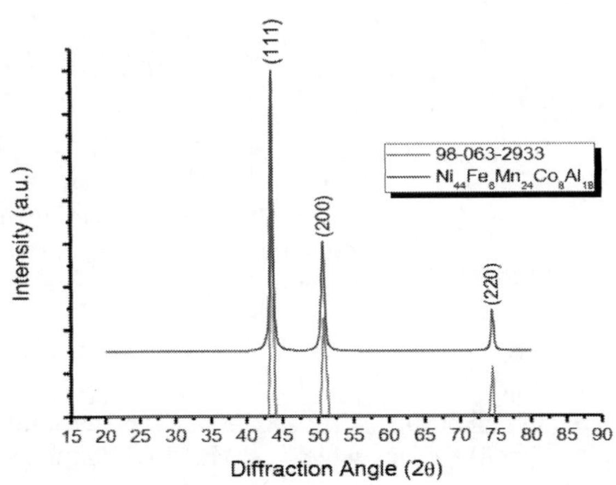

Figure 1. XRD pattern of $Ni_{44}Fe_6Mn_{24}Co_8Al_{18}$ compound at room temperature. $Ni_{44}Fe_6Mn_{24}Co_8Al_{18}$ correspond to tungsten-type disorder of HA compound.

3.2. Microstructure and Morphology

Micrograph from optical microscope results depicted in Figure 2 shows the microstructure of $Ni_{44}Fe_6Mn_{24}Co_8Al_{18}$ HA compound after 72 hours annealed. The general microstructures consist of three type of grain. Grain established in three colors: black, gray and white. On red circle appears the white grain occupy a grain boundary, and grain beside of the white grain.

The microstructures define the effect of tungsten-type disorders onto grain shape. Whereas the atomic structure had initiated the overlapping structure between the grains, generate the proper solubility in $Ni_{44}Fe_6Mn_{24}Co_8Al_{18}$ HA compound.

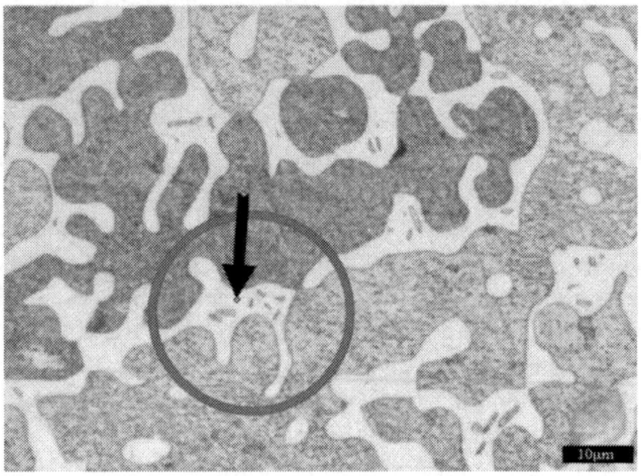

Figure 2. Micrograph from optical microscope results of surface $Ni_{44}Fe_6Mn_{24}Co_8Al_{18}$ HA after 72-hour annealing. Grain established in three colors: black, gray and white. On red circle appears the white grain occupy a grain boundary and grain beside of its grain.

3.3. Magnetic Properties of the compound

The hysteresis magnetic saturation of $Ni_{44}Fe_6Mn_{24}Co_8Al_{18}$ has shown in Figure 3. The magnetic polarization (J) reached 0.3 T when the magnetic field strength (H) below 250 kA/m. We suggest which the tungsten-type disorder had increased hysteresis magnetic saturation value reach more than 0.3 T contrast with magnetic hysteresis curve of $Ni_{50}Mn_{32}Al_{18}$ as a comparison. We believe the overlapping structure inside of the alloy had a correlation with exchange interaction between their electrons whereas aligned the magnetic moment and raised the magnetic saturation [10,15]. Therefore, doping intermetallic Fe and Co on $Ni_{50}Mn_{32}Al_{18}$ positively enhanced magnetic saturation significantly.

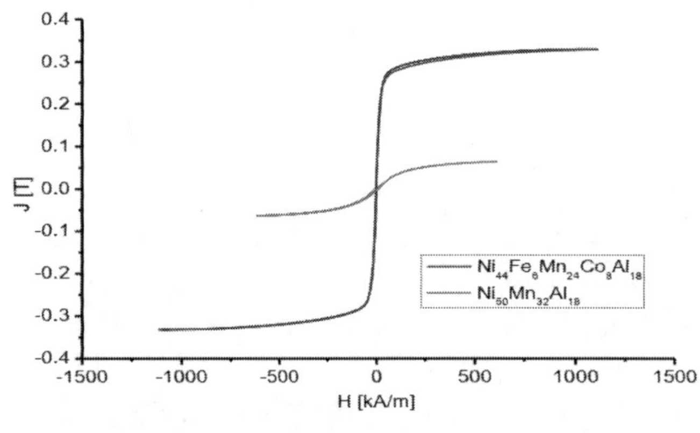

Figure 3. Hysteresis curve of $Ni_{44}Fe_6Mn_{24}Co_8Al_{18}$ had shown. The Magnetic hysteresis curve of $Ni_{50}Mn_{32}Al_{18}$ included as comparison.

4. Conclusion

We successfully synthesized $Ni_{44}Fe_6Mn_{24}Co_8Al_{18}$ alloy by AMF method. The substitution of intermetallic both Fe and Co formed a $Ni_{44}Fe_6Mn_{24}Co_8Al_{18}$ alloy which Cu_2MnAl-type structure, have generated the tungsten-type disorder in Heusler alloy compound. In microstructure view, this type disorder allows the occurrence of the overlapping phases in grain system. These conditions had beneficial in raising significantly the value of magnetic saturation in the alloy.

5. Acknowledgement

The authors gratefully acknowledge the support of Department of Physics, Universitas Indonesia for the research facilities. The authors wish to thank Dr. Vivi Fauzia for support. This work was supported by the Universitas Indonesia under the research grant contract of Hibah Publikasi Internasional Terindeks untuk Tugas Akhir Mahasiswa UI 2016 under the research grant contract no. 1993/UN2.R12/HKP 05.00/2016. References

Reference

[1] Tishin AM, Spichkin YI *Int J Refrig* 2014;**37**:223–9. doi:10.1016/j.ijrefrig.2013.09.012.
[2] Chatterjee S, Giri S, De SK, Majumdar S *J Alloys Compd* 2010;**503**:273–6. doi:10.1016/j.jallcom.2010.05.026.
[3] Sokolovskiy VV, Fayzullin RR, Buchelnikov VD, Drobosyuk MO, Taskaev SV, Khovaylo VV *J Magn Magn Mater* 2013;**343**:6–12. doi:10.1016/j.jmmm.2013.04.069.
[4] Kim Y, Kim EJ, Choi K, Han WB, Kim HS, Shon Y, et al *J Alloys Compd* 2014;**616**:66–70. doi:10.1016/j.jallcom.2014.07.034.
[5] Dan NH, Duc NH, Yen NH, Thanh PT, Bau L V., An NM, et al *J Magn Magn Mater* 2015;**374**:372–5. doi:10.1016/j.jmmm.2014.08.061.
[6] Ghosh A, Mandal K *J Appl Phys* 2015;**117**. doi:10.1063/1.4913951.
[7] Liu C, Zhang W, Qian Z, Hua Z, Zhao Q, Sui Y, et al *J Alloys Compd* 2007;**433**:37–40. doi:10.1016/j.jallcom.2006.06.070.
[8] Kainuma R, Gejima F, Sutou Y, Ohnuma I, Ishida K *Mater Trans - JIM* 2000;**41**:943–9.
[9] Webster JP, Ziebeck KRA Volume 19C: "Alloys and Compounds of d-Elements with Main Group Elements. Part 2" in SpringerMaterials (http://dx.doi.org/10.1007/103532 n.d. doi:10.1007/10353201_31.
[10] Ersoy H, Sandratskii L, Dederichs P, Eriksson O *Exch Organ Behav Teach J* 1975.
[11] Notonegoro HA, Kurniawan B, Setiawan J, Manaf A. Magnetic hysterysis evolution of Ni-Al alloy with Fe and Mn substitution by vacuum arc melting to produce the room temperature magnetocaloric effect material 2016;**020019**:020019. doi:10.1063/1.4953944.
[12] Xuan HC, Chen FH, Han PD, Wang DH, Du YW. Effect of Co addition on the martensitic transformation and magnetocaloric effect of Ni–Mn–Al ferromagnetic shape memory alloys. *Intermetallics* 2014;**47**:31–5. doi:10.1016/j.intermet.2013.12.007.
[13] Graf T, Felser C, Parkin SSP. Simple rules for the understanding of Heusler compounds. *Prog Solid State Chem* 2011;**39**:1–50. doi:10.1016/j.progsolidstchem.2011.02.001.
[14] Notonegoro HA, Ferdian D, Prasetyo Y, Manaf A. Microstructure analysis of synthesized iron substituted Ni44Fe6Mn32Al18 Heusler alloy 2016;**1729**. doi:10.1063/1.4946939.
[15] Mubarok N, Notonegoro HA, Thosin KAZ, Manaf A *AIP Conf Proc* 2016;**020022**:020022. doi:10.1063/1.4953947.

Analysis of the magnetic properties nanoscale barium hexaferrite (BHF) prepared by milling and ultrasonic method

Novizal[1], Sasito Edie[2], Mykel.T.E. Manawan[3]
[1,2] National Institute of Science and Technology, Moh. Kahf 2, South Jakarta, Indonesia
[3]Universitas of Indonesia, Depok, 16424, Indonesia.

Email: novizal@istn.ac.id.

Abstract. Barium hexaferrite (BHF) is well established material which widely used respectively as permanent magnets. In this research, we report our recent investigation on magnetic properties analysis of barium hexaferrite (BHF) compounds with a ratio of Fe/Ba: 11 prepared by a mechanical alloying process and high power ultrasonic destruction to promote the soft magnetic properties. The investigation carried out by Scanning Electron Microscope (SEM) shows the grain size between 500-1500 nm, it indicates that each grain is composed of several crystallites or polycrystalline. By mean of X-ray diffraction revealed the phase composition and the mean crystallite size <70 nm. The Characterization of the magnetic properties of the effects of downsizing the particle size of ~ 200 nm to ~ 50 nm by the ultasonik method provide saturation value of 0.35 T, remanent 0.24 T and the coercivity is 115 kA / m.

1. Introduction

The hexagonal barium hexaferrite ($BaFe_{12}O_{19}$) is known as a hard magnetic material and represents one of the mostly used materials in the area of magnetic recording media, which cannot be easily substituted by other magnets [1–3]. From the scientific point of view, as well as from the standpoint of the technological applications, the barium hexaferrite presents interest due to its special properties, such as: high remanence and coercivity, thermal, electrical and chemical stability [4, 5]. The rapid development of information and telecommunications technology provide the effect of increasing the use of microwave-based electronic instrument (300 MHz - 300 GHz)[6]. In this research we have prepared a barium hexaferrite using the mechanical alloying process and high power ultrasonic destruction then treated at different temperatures. The barium hexaferrite monocrystals resulting after the heat treatments were extracted by solving their structural, and magnetic properties were studied.

2. Numerical Methods

The quantities of powder $BaCO_3$ and Fe_2O_3 with 99% purity (Merck) were weighted according to the stoichiometric formula. The powder were mixed together with a planetary mill Fe_2O_3(99% purity)and $BaCO_3$ (99.8%purity) in proportion of 1:2, 1:1, and 1:0, , reaching a homegenisasi phase, followed by fracturing (embrittlement) which resulted in the particle size reduction of subsequent collisions. The entire amount of the particle size distribution is characterized by a Malvern Zetasizer Ver. 6:20, particle size analyzer (PSA) with the ability to 20 nm. For samples such as BHF magnetic, attractive force between the magnetic particles are very strong need to be separated by the use of ultrasonic and the addition of surfactant to prevent agglomeration so that the data produced is a true particle size. Morphology characterization of samples with a scanning electron microscope (SEM) type FEI-F50 are able to achieve magnification 200.000×. Magnetic characterization of samples was done by using Permagraph EP.3 magnetometer with a maximum induction field in 1700 kA /m (21.5 kOe). Variables were entered in the form of density data acquisition software is done with density meter to generate data that does not depend on the volume or size of the sample. Synthesis samples are taken to form a single phase system BHF by mechanical alloying method with a planetary ball mill. And followed by sonication.

3. Results and Dscussion

The BHF literature review of phase formation does not follow strictly stoichiometric calculations, but follow the ratio of Fe/Ba dependent on the process [7,8]. In the observed to the ratio of Fe/Ba =11 to the mechanical alloying process, with variations of milling time 20, 40, 60 hours are shown in Figure 1. The diffraction pattern BHF which uses the ratio of Fe/ Ba: 11 after the sintering process is then performed manually destruction process by using a mortar and then re-milling for 10 hours, followed by sonication process for 2 hours. BHF seen that the single phase can be generated by using the ratio of Fe /Ba: 11 and the process after forming BHF shows no impurities [9].

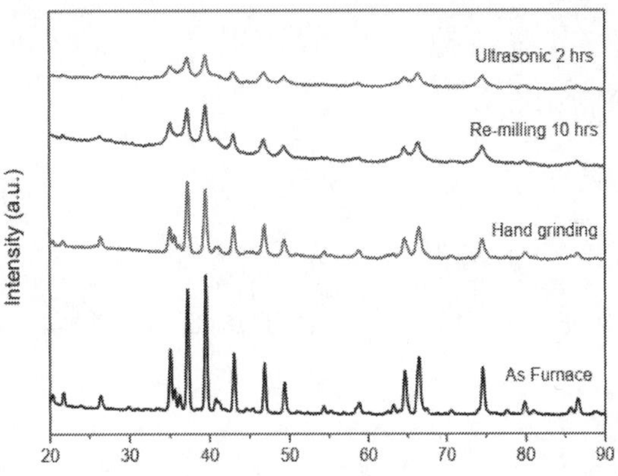

Figure 1. The diffraction pattern in Fe/Ba: 11 with a variety of milling time

This shows that the impurity phase Fe_2O_3 is growing with increasing milling time occurs only at the beginning of the process before the phase BHF formed. It is also seen that a decline in the intensity and widening on the entire diffraction pattern peak as compared with the sample produced after sintering process (as-furnace). The particle size reduction by the process and the size of the smallest one is in the process of sonication.

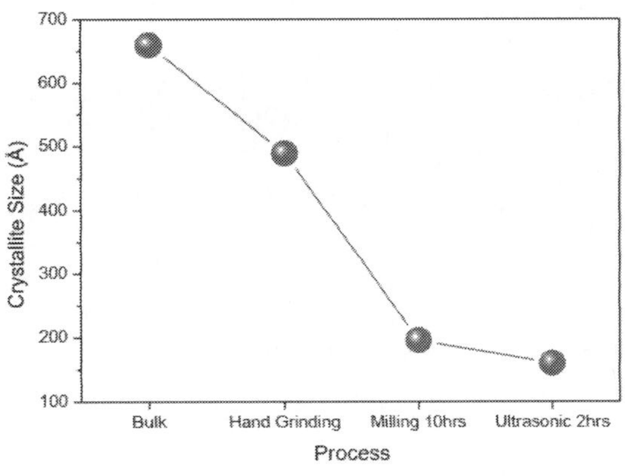

Figure 2. The crystallites of BHF after sintering 1100° C for 2 hours

Therefore, based on a search for the formation of BHF through the process of mechanical alloying with the ratio of Fe/Ba : 11, by milling process 20 hours and sintering 1100° C for 2 hours to produce a single phase BHF.

Figure 2. The size of the crystallites of the processes carried out after sintering. It turned out that the destruction process can manually reduce the size of the crystallites to about 150 nm followed by milling for 10 hours to reduce the size to about 200 nm and a subsequent process which sonication can reduce the crystallite size to about 100 nm. This process is the basis for the production of nano-particles on the method of top-down processing.

Figure 3. SEM photos and topographic after sintering process 1100 °C

Figure 3. SEM morphology of the results of the samples after sintering showed a particle size between 500-1500 nm, while the crystallite size XRD analysis resulted in an average diameter of 65 nm which indicates the sample policrystalline.

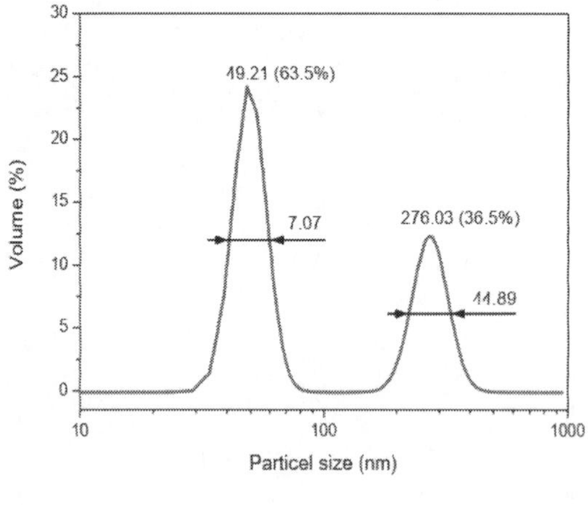

Figure 4. The particle size of 5 hours of sonication with deposition method

Samples that have undergone sonication was deposited for 48 hours, after which the particles do not settle taken and that settles reprocessed. This method provides control with homogeneous results as shown in Figure 4 where the 5 hours of sonication resulting particle diameter of 49.2 nm with a deviation of 7.1 nm and 276 nm with a deviation of 44.9 nm. These results prove that this method is effective and can be used to control the particle size after sonication process. Figure 5 the size of the particles at a magnification of 50000 times (adzan 200000 times (b) produced by the process of sonication for 5 hours. It is seen that the particles produced by the process of ultrasonic spherical with an average diameter ranging between 42-90 nm and a particle size of 276 nm is a measure of the agglomeration of the magnetic particles BHF. To strengthen the results of the PSA and SEM analysis, the crystallite size (crystallite size) on the XRD data by the method of Whole Powder Pattern Modeling (WPPM).

Figure 5. SEM photos after sonication for 5 hours

Input parameters are analyzed by wppm will be directly modeled into forms of diffraction peaks so that the fitting is a real physical properties of materials [10]

Figure 6 shows the results of refinement to the software PM2K[10] that implements the methods wppm. The parameters of the instruments kept constant by using a Si-standard, whereas the parameters direfine is intensity, background, displacement, zero offset and crystallite size distribution

.

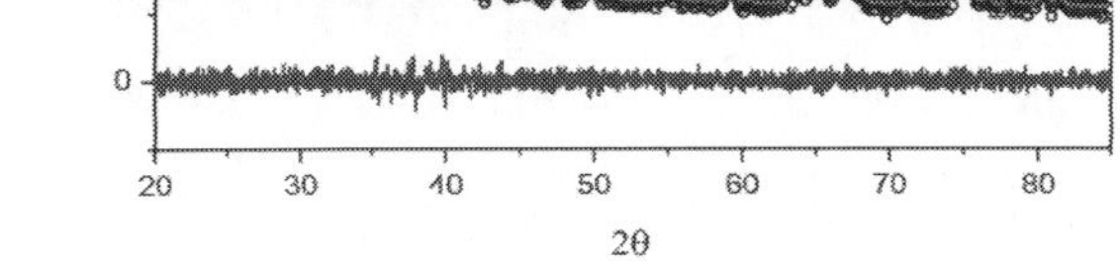

Figure 6. Results of refinement using the WPM

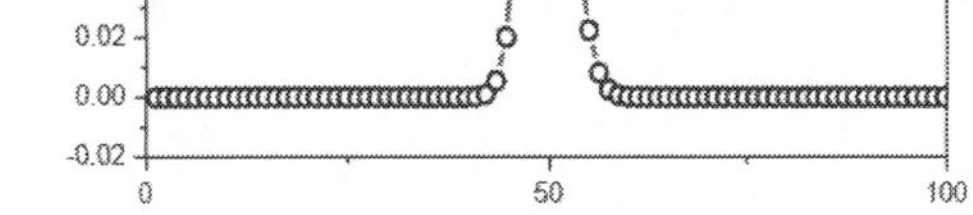

Figure 7. Distribution of particle size analysis results wppm

Figure 7 shows the distribution of crystallite size (crystallite size) at BHF nano particles produced through the process of sonication for 5 hours. Seen that the crystallite sizes were obtained from analysis WPPM (49.79 nm) is almost equal to the PSA (49.28 nm) in figure 6 peaks 2 and smaller than the result of SEM (~ 60 nm) in Figure 5.

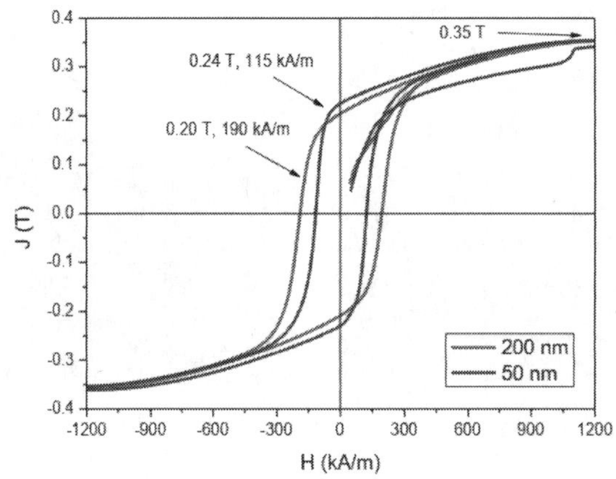

Figure 8. Hysteresis curve with variations in particle size

Figure 8 the sample with a smaller particle size has a higher remanence value but lower coercivity. This is consistent with the theory that particle size reduction over the limit will decrease the single domain limit its coercivity [11]. Saturation value of 0.35 T indicates that the particle size reduction reduces the amount of porosity.

4. Conclusions

Retrieved BHF single phase with a ratio of Fe/Ba = 11 through mechanical alloying process and via sonication can be scaled down to 50 nm particle sizes and narrow distributions (FWHMPSA = 7.07 nm, FWHMWPPM = 5.83 nm). And provide a change of magnetic properties to the value of saturation, remanence and coercivity for a result of the downsizing of the crystals obtained saturation 0:38 T, remanence 0.24 T and coercivity 115 kA/m.

5. References

[1] Nowosielski R, Babilas R, Dercz G, Pajak L and Skowronski W, 2008 *J. Achiev. Mater. Manufact. Eng.* **27**, 51
[2] Brzózka K, Sovák P, Górka B, Szumiata T and Gawroński M 2011 *Acta Phys. Pol. A* **119**, 33

[3] Kouřil K, Chlan V, Štěpánková H, Telfah A, Novák P, Knížek K, Hiraoka Y and Kimura T 2010 *Acta Phys. Pol. A* **118**, 732
[4] Xu P, Han X and Wang M 2007 *J. Phys. Chem. C* **111**, 5866
[5] Dasari MP, Rao KS, Krishna PM and Krishna GG 2011 *Acta Phys. Pol. A* **119**, 378
[6] ITU. (den 20 November 2013). *International Telecommunications Union*. Hämtat från International Telecommunications Union: http://www.itu.int
[7] Topal, Ugur, Ozkan H and Dorosinskii L 2007 *Journal of alloys and compounds* **428** 17-21
[8] Topal U 2008 *Journal of Magnetism and magnetic materials,* **320,** 331-335
[9] Nowosielski R et al. 2007 *Archives of Materials Science and Engineering* **28.12** 735-742
[10] Scardi, Paolo and Leoni M 2006 *Journal of applied crystallography* **39**(1) 24-31
[11] Wohlfart EP 1982 *Handbook of magnetic material Vol.3.* Netherlands: North-Holland
[12] Scard P and Leoni M 2006 ECS Transaction, **3**(9) 125-132

The enhancement of ferromagnetism in Ta-doped anatase TiO_2 system by iron co-doping

S. Muhammady*, **R. Kurniawan, E. Nurfani, I. M. Sutjahja, T. Winata, and Y. Darma**

Department of Physics, Faculty of Mathematics and Natural Sciences, Institut Teknologi Bandung, Ganesha 10, Bandung 40132, INDONESIA

E-mail : *mshibghatullah@gmail.com

Abstract. The ferromagnetic properties of Ta-doped and (Ta,Fe)-doped anatase TiO_2 as diluted ferromagnetic materials has been studied within spin-polarized generalized gradient approximation (GGA) and GGA + U method. We observed a ferromagnetic properties in Ta-doped anatase TiO_2, with a total magnetic moment of 1.00 μ_B/supercell, which mainly arised due to Ti 3d and O 2p states upon E_F. Furthermore, the enhancement of ferromagnetism, mainly arising due to Fe 3d states, in (Ta,Fe)-doped anatase TiO_2 was observed up to 5.00 μ_B/supercell, 5 times larger than that in Ta-doped TiO_2. Our results confirmed the important role of Fe doping for the ferromagnetism enhancement in Ta-doped anatase TiO_2.

1. Introduction

Spintronics have been interesting to be investigated due to the capability of controlling spin currents, in addition to electronic charge utilization [1]. The spintronics development was motivated by the giant magnetoresistance (GMR) discovery in magnetic Fe/Cr magnetic superlattice [2] and current-induced switching due to spin transfer [3]. Hence, the spintronics have provided new fields of material physics and future applications. Regarding semiconductor-based spintronics, diluted magnetic semiconductors (DMS) have potential application in spintronic devices [4].

Because of capability to provide ferromagnetism as well as semiconductor properties, DMS have been developed by doping transition metal in small concentration [5,6]. In early development, DMSs exhibited low-temperature ferromagnetism [7–10] with the highest Curie temperature T_C of 173 K found in GaMnAs [11]. Besides that, room-temperature ferromagnetism in DMS was required to realize applications in real devices [12–14].

Motivated by high-temperature ferromagnetism up to 400 K [15], several works were performed to investigate variated aspects on Co-doped anatase TiO_2 [16–18]. Besides that, Co-doped rutile TiO_2 was also found as ferromagnetic up to 400 K [19]. There was thus several works to investigate both anatase- and rutile-based TiO_2 doped with different transition metals (TM) [20–23].

However, the ferromagnetism in oxide DMS might exist due to the magnetic impurity clustering. The clustering around main DMS material could occur as a consequence of imperfect chemical reaction during the synthesis. Furthermore, the clustering could provide additional magnetism, which made the magnetic properties of the system became more complicated. For avoiding such effect, pentavalent TM was therefore used as the dopant in oxide DMS, such as tantalum (Ta) [24–26].

Additional doping in the pentavalent TM-doped TiO_2 was also performed to obtain co-doping TiO_2 system. Additional Nb, Al, or Cu doping in the V-doped TiO_2 system promoted the lower saturation magnetization of the V-doped TiO_2 and the decreased electrical resistivity [27]. In contrast, higher saturation magnetization was exactly observed in Co-doped and Fe-doped TiO_2 due to additional Nb doping followed by the reduction of their electrical resistivity values [28]. Motivated by those results, we thus investigated magnetic properties of (Ta,Fe)-doped anatase TiO_2 systems, to be compared with Ta-doped anatase TiO_2 within a first-principle calculation.

2. Computational Details

In this work, we used generalized gradient approximation [29] (GGA) + U method within density-functional theory (DFT) [30–31] to calculate electronic properties of Ta-doped anatase TiO_2 using Quantum-ESPRESSO [32]. This software is an open source software which is able to calculate electronic properties of materials within self-consistent (SCF) calculation.

In general, SCF condition is reached when an electron density $n(\mathbf{r})$ in a calculation step approaches next $n(\mathbf{r})$ below a threshold energy. The new $n(\mathbf{r})$ can be obtained by [33]

$$n_{in}^{i+1} = n_{in}^{i} + \beta R\left[n_{in}^{i}\right] \tag{1}$$

with n_{in}^{i} and n_{in}^{i+1} are electron density in ith-step and next step (i + 1), β is a mixing parameter and $R\left[n_{in}^{i}\right]$ is an electron density residual vector. Therefore, in a step of calculation, new optimal electron density input can be obtained as a linear combination of previous electron densities based on Pulay mixing [34]. The new $n(\mathbf{r})$ can be therefore described as

$$n_{in}^{i+1} = \sum_{i} \alpha_i \left(n_{in}^{i} + \beta R\left[n_{in}^{i}\right]\right) \tag{2}$$

with α_i is a coefficient given for each $n(\mathbf{r})$ in ith-step under a condition

$$\sum_{i} \alpha_i = 1 \tag{3}$$

In general, the calculation is initiated by determining ultrasoft (US) pseudopotentials (PP), followed by the definition of $n(\mathbf{r})$. The next step is Hartree potential V_H and exchange-correlation potential V_{XC} calculation forming effective potential [35]

$$V_{eff}(\vec{r}) = V_{nucl}(\vec{r}) + V_{H}(\vec{r}) + V_{XC}(\vec{r}) \tag{4}$$

then, equation (1) can be solved using Schrödinger equation

$$\left(-\frac{1}{2}\nabla^2 + V_{eff}(\vec{r})\right)\psi_i(\vec{r}) = \varepsilon_i \psi_i(\vec{r}) \tag{5}$$

with electron density

$$n = \sum_{i} \left|\psi_i(\vec{r})\right|^2 \tag{6}$$

Inputs included in the calculation are Bravais lattice type, k-point mesh, atomic positions, lattice parameters, kinetic energy cutoff, and Hubbard-U corrections. Our models are 2 × 2 × 2 supercells of $Ti_{1-x}Ta_xO_2$ and $Ti_{1-x-y}Ta_xFe_yO_2$ structures ($x = y = 3.125\%$) in $I4_1/amd$ space group [36]. The Ta and Fe dopants were put at coordinates of (1/2 1/2 1/2) and (1/2 3/4 5/8) to realize two nearest-neighboring O ions sharing. The optimized tetragonal lattice parameters are $a = b = 7.92$ Å and $c = 19.10$ Å. Pseudopotentials (PP) used in this calculation are basically formed by plane wavefunctions (PW). Perdew-Burke-Ernzerhof (PBE) exchange-correlation (XC) functional type was used in the calculation [29]. Besides, we used a Γ-centered k-point mesh of $2 \times 2 \times 2$ and a cut-off kinetic energy of 544 eV. Threshold force in structure optimization was 0.514 eV/Å based on Broyden-Fletcher-Goldfarb-Shanno (BFGS) minimization algorithm [37–40]. We used U_{eff} ($U - J$) potential energy of 5.8 eV, 3.0 eV, and 5.9 eV for Ti 3d, Ta 5d, and Fe 3d orbitals based on previous works [41,42]. In general, electrons in the valence states (weakly-bonded electrons) of each atom are involved, i.e Ti($3p^6 3d^2 4s^2$), Ta($5p^6 5d^3 6s^2$), O($2s^2 2p^4$), and Fe($3p^6 3d^6 4s^2$), while electrons in the remaining inner orbitals were assumed to be strongly bonded as core states.

3. Results and Discussions

Firstly, we investigated magnetic properties of anatase Ta-doped and (Ta,Fe)-doped anatase TiO_2 using calculated total density of states (DOS) as a function of energy relative to Fermi level, as shown in Figure 1.

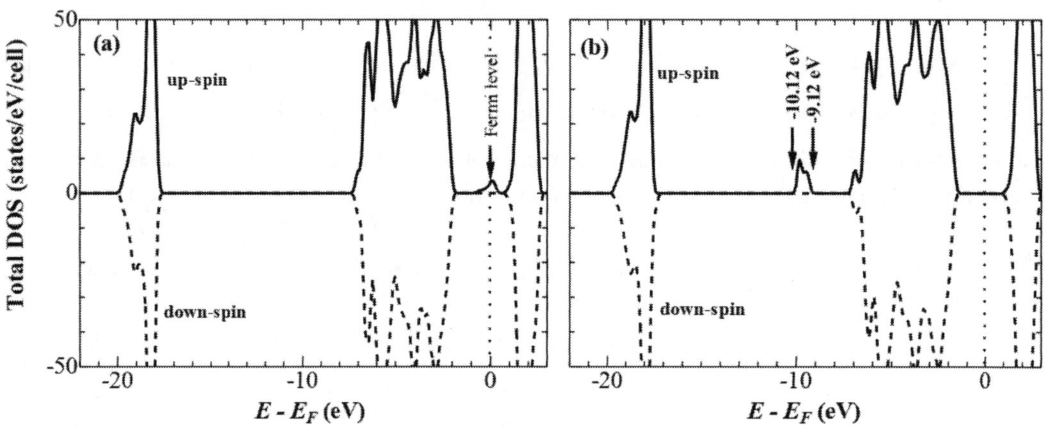

Figure 1. Calculated spin-dependent total DOS of (a) Ta-doped and (b) (Ta,Fe)-doped anatase TiO_2.

In general, both DOS graphs in Figure 1 have a similar symmetrical shape for the spin-up and spin-down electron states. However, Figure 1(a) show a small-range asymmetric DOS curve found around the Fermi level arising ferromagnetism of the Ta-doped anatase TiO_2. We obtain a total magnetic moment of 1.00 μ_B/supercell on this system. The total magnetic moment of the Ta-doped anatase TiO_2 is similar with that was indicated by the previous theoretical study [43]. Besides that, Figure 1(b), a small-range asymmetric DOS curve found in an energy range of –10.12 to –9.12 eV arising ferromagnetism of the (Ta,Fe)-doped anatase TiO_2. We obtain a total magnetic moment of 5.00 μ_B/supercell for this system, namely about 5 times larger than that of Ta-doped anatase TiO_2.

In order to observe a detailed ferromagnetic mechanism of both systems, we plotted projected-DOS (PDOS) as a function of energy relative to the Fermi level, as shown in Figure 2. Based on Figure 2(a), we observed that ferromagnetism in the Ta-doped anatase TiO_2 mainly arises due to observed hybridization between spin-up Ti 3d and O 2p states upon the Fermi level. The hybridization was supported by the previous theoretical result [43] which shows that instead of Ta 3d states, some spin-up Ti 3d states exactly lie upon the Fermi level. The largest local magnetic moment at four Ti ions, where each Ti ion has a local magnetic moment of 0.31 μ_B, as shown in Table 1. Otherwise, Ta 5d states remain to provide a small contribution to ferromagnetism. This small contribution is also indicated by a small local magnetic moment at Ta ion of 0.05 μ_B, as shown in Table 1. The mechanism of magnetism can be described as follows. Since the Ta atom has three electrons at 3d orbital, there is one electron remaining after the Ta atom is ionized (Ta^{5+}). Based on our results, the remaining electron is dominantly shared among four nearest Ti ions. This mechanism has been also previously proposed [43]. Therefore, our Ta-doped anatase TiO_2 is reliable for Fe co-doping.

Table 1. Local magnetic moments of Ta ion and its two nearest Ti ions.

Ions	Coordinate	Site symmetry	Wyckoff letter	Local magnetic moment (μ_B)
Ta	1/2 1/2 1/2	.2/*m*.	*c*	0.05
Ti	0 1/2 1/2	.2/*m*.	*d*	0.31
	1/2 0 1/2	.2/*m*.	*c*	0.31
	1 1/2 1/2	.2/*m*.	*d*	0.31
	1/2 1 1/2	.2/*m*.	*c*	0.31

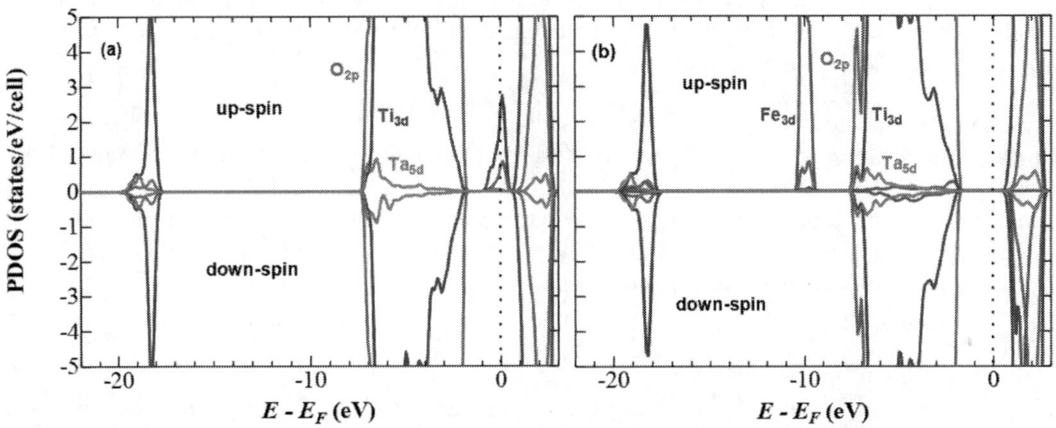

Figure 2. Calculated spin-dependent PDOS (a) of Ta-doped and (b) (Ta,Fe)-doped anatase TiO_2.

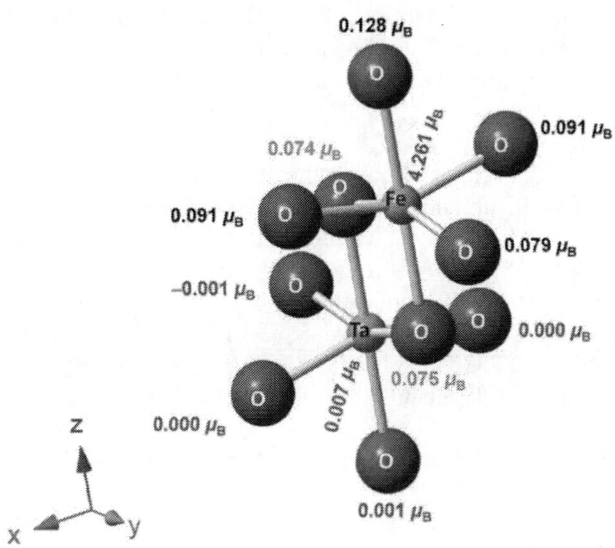

Figure 3. Schematic illustration of local magnetic moments of Ta, Fe, and nearest-neighboring O ions in the (Ta,Fe)-doped anatase TiO_2.

Furthermore, different ferromagnetism mechanism is observed in Figure 2(b). The two peaks in small-range spin-up states are found to be largely dominated by Fe 3*d* states hybridized with O 2*p* states. Moreover, we found a large local magnetic moment at the Fe ion of 4.261 μ_B in the (Ta,Fe)-doped anatase TiO_2 leading to the total magnetic moment of 5.00 μ_B, as shown in Figure 3. Otherwise, a small local magnetic moment of 0.007 μ_B at the Ta ion is observed. Based on Figure 3, there are two O ions which are shared among the Fe and Ta ions. Both shared O ions are polarized up to 0.074 μ_B and 0.075 μ_B (noticed with green letters). The upper O ions have higher local magnetic moments, where the top O ion has a local magnetic moment of 0.128 μ_B. In contrast, the lower O ions, which are also the nearest neighbors of the Ta ion, almost have no local magnetic moments. Based on the analysis, the large ferromagnetism in the (Ta,Fe)-doped anatase TiO_2 is exactly induced by the Fe co-dopant.

4. Conclusions

We observe ferromagnetism in anatase Ta-doped anatase TiO_2, with the total magnetic moment of 1.00 μ_B/supercell, which mainly arise due to Ti 3*d* states hybridized with O 2*p* states upon E_F. The enhancement of ferromagnetism is observed in the (Ta,Fe)-doped anatase TiO_2, which mainly arises due to Fe 3*d* states hybridized with O 2*p* states. The total magnetic moment the (Ta,Fe)-doped anatase TiO_2 is 5.00 μ_B/supercell. Therefore, the high ferromagnetism observed in the (Ta,Fe)-doped anatase TiO_2 provides potential DMS application.

5. Acknowledgements

This work is partly supported by Ministry of Research, Technology, and Higher Education of Republic of Indonesia through 1st Batch Program Pendidikan Magister Menuju Doktor untuk Sarjana Unggul (PMDSU) research scholarship 2013. We also acknowledge QC Cluster of Advanced Computing Laboratory, Department of Physics, Institut Teknologi Bandung, Indonesia for providing calculation facilities and technical support in this work.

References

[1] Dietl T 2010 *Nat. Mat* **9** 965
[2] Baibich M N, Broto J M, Fert A, Nguyen V D F and Petroff F 1988 *Phys. Rev. Lett.* **61** 2472
[3] Barnaś J, Fert A Gmitra M, Weymann I and Dugaev V K 2005 *Phys. Rev. B* **72** 024426
[4] Fert A 2008 *Rev. Mod. Phys.* **80** 1517
[5] Wolf S A, Awschalom D D, Buhrman R A, Daughton J M, von Molnár S, Roukes M L, Chtchelkanova A Y and Treger 2001 *Science* **294** 1488
[6] Ohno H, Shen A, Matsukara F, Oiwa A, Endo A, Katsumoto S and Iye Y 1996 *Appl. Phys. Lett* **69** 363
[7] Furdyna J K 1988 *J. Appl. Phys.* **64** R29
[8] Ferrand D, Cibert J, Wasiela A, Bourgognon C, Tatarenko S, Fishman G, Andrearczyk, T, Jaroszyński J, Koleśnik S and Dietl T 2001 *Phys. Rev. B.* **63** 085201
[9] Karczewski G, Sawicki M, Ivanov V, Ruester C, Gravecki G, Matsukura F, Molenkamp L W and Dietl T 2003 *J. Supercond.: Incorporating Novel Magnetisme* **16** 55
[10] Ohno H, Munekata H, Penney T, von Molnár S and Chang L L 1992 *Phys. Rev. Lett.* **68** 2664
[11] Wang K Y, Campion R P, Edmonds K W, Sawicki M, Dietl T, Foxon C T and Gallagher B L 2005 *AIP Conf. Proc.* **772** 333
[12] Medvedkin G A, Ishibashi T, Nishi T, Hayata K, Hasegawa Y and Sato K 2000 *Jpn. J. Appl. Phys.* **39** L 949
[13] Choi S, Cha G -B., Hong S C, Cho S, Kim Y, Ketterson J B, Jeong S -Y and Yi G –C 2002 *Solid State Commun.* **122** 165
[14] Sharma P, Gupta A, Rao K V, Owens F J, Sharma R, Ahuja R, Guiller J M O, Johansson B and Gehrong G A 2003 *Nat. Mat.* **2** 673

[15] Matsumoto Y, Murakami M, Shono T, Hasegawa T, Fukumura T, Kawasaki M., Ahmet P, Chikyow T, Koshihara S and Koinuma H 2001 *Science* **291** 854
[16] Yang Z, Liu G and Wu R 2003 *Phys. Rev. B* **67** 060402(R)
[17] Kim D H, Yang J S, Lee K W, Bu S D, Noh T W, Oh S –J, Kim Y -W, Chung J -, Tanaka H, Lee H Y and Kawai T 2002 *Appl. Phys. Lett.* **81** 2421
[18] Chambers S A, Thevuthasan S, Farrow R F C, Marks R F, Thiele J U, Folks L, Samant M G, Kellock A J, Ruzycki N, Ederer D L and Diebold U 2001 *Appl. Phys. Lett.***79** 3467
[19] Park W K, Ortega-Hertogs R J and Moodera J S 2002 *J. Appl. Phys.* **91** 8093
[20] Park M S, Kwon S K and Min B I 2002 *Phys. Rev. B* **65** 161201(R)
[21] Droubay T, Heald S M, Shutthanandan V, Thevuthasan S and Chambers S A 2005 *J. Appl. Phys.* **97** 046103
[22] Geng W T and Kim K S 2003 *Phys. Rev. B* **68** 125203
[23] Suryanarayanan R, Naik V M, Kharel P, Talagala P and Naik R 2005 *J.Phys.: Condens. Mat.* **17** 755
[24] Ogale S B 2010 *Adv. Mater.* **22** 3125
[25] Ando K 2006 *Science* **312** 1883
[26] Abraham D W, Frank M M and Guha S 2005 *Appl. Phys. Lett.* **87** 252502
[27] Zhao Q, Wu P, Li B L, and Jiang E Y 2012 *Physica B* **407** 171
[28] Hitosugi T, Kinoda G, Yamamoto Y, Furubayashi Y, Inaba K, Hirose Y, Nakajima K, Chikyow T, Shimada T and Hasegawa T 2006 *J. Appl. Phys.* **99** 08M121
[29] Perdew J P, Burke K and Ernzerhof M 1996 *Phys. Rev. Lett.* **77** 3865
[30] Hohenberg P and Kohn W 1964 *Phys. Rev.* **136** B864
[31] Kohn W and Sham L J 1965 *Phys. Rev.* **140** A1133
[32] Giannozzi P, Baroni S, Bonini N, Calandra M, Car R, Cavazzoni C, Ceresoli D, Chiarotti G L, Cococcioni M, Dabo I, Corso A D, de Gironcoli S, Fabris S, Fratesi G, Gebauer R, Gerstmann U, Gougoussis C, Kokalj A, Lazzeri M, Martin-Samos L, Marzari N, Mauri F, Mazzarello R, Paolini S, Pasquarello A, Paulatto L, Sbraccia C, Scandolo S, Sclauzero G, Seitsonen A P, Smogunov A, Umari P and Wentzcovitch R M 2009 *J. Phys.: Condens. Mat.* **21** 395502
[33] Kresse G and Furthmüller J 1996 *Phys. Rev. B* **54** 11169
[34] Pulay P 1980 *Chem. Phys. Let.* **73** 393
[35] Martin R M 2004 *Electronic Structure: Basic Theory and Practical Methods* (Cambridge: Cambridge University Press)
[36] Mo S -D and Ching W Y 1995 *Phys. Rev. B* **51** 13023
[37] Broyden C G 1970 *J. Inst. Maths. Applics.* **6** 76
[38] Fletcher R 1970 *Computer Journal* **13** 317
[39] Goldfarb D 1970 *Math. Comp.* **24** 23
[40] Shanno D F 1970 *Math. Comp.* **24** 647
[41] Anisimov V I, Zaanen J and Andersen O K 1991 *Phys. Rev. B.* **44** 943
[42] Qi D –C, Barman A R, Debbichi L, Dhar S, Santoso I, Asmara T C, Omer H, Yang K, Krüger P, Wee A T S, Venkatesan T and Rusydi A 2013 *Phys. Rev. B* **87** 245201
[43] Yang K, Dai Y, Haung B and Feng Y P 2014 *J. Phys. D: Appl. Phys.* **47** 275101

Adsorption of methylene blue using Fe_3O_4/CuO/ZnO/ nanographene platelets (NGP) composites with various NGP concentration

H. Tju[1,2], A. Taufik[1,2] and R. Saleh[1,2,*]

[1]Departemen Fisika, Fakultas MIPA-Universitas Indonesia, 16424 Depok, Indonesia

[2]Integrated Laboratory of Energy and Environment, Fakultas MIPA-Universitas Indonesia, 16424 Depok, Indonesia

[*]E-mail: rosari.saleh@ui.ac.id

Abstract. This study will examine the use of Fe_3O_4/CuO/ZnO nanocomposites that have been modified by Nanographene Platelets (NGP) as an adsorbent to degrade organic dye waste Methylene Blue (MB). The nanocomposites were synthesized using the sol-gel method then combined with three variatons of NGP weight percents by simple hydrothermal method. The Fe_3O_4/CuO/ZnO/NGP composites were characterized using the X-Ray Diffraction (XRD) spectroscopy, Fourier Transform Infrared (FTIR), Energy Dispersive X-Ray (EDX), *Thermogravimetric Analysis* (TGA) and Vibrating Sample Magnetometer (VSM). The composites exhibit ferromagnetic behaviour. The presence of hexagonal wurtzite of ZnO, monoclinic of CuO and cubic spinel of Fe_3O_4 were found in the composites. The graphitic-like structure represents the presence of the NGP in the composites. However, the addition of NGP weight percent reduces the thermal stability of the composites. The adsorption capability of the composites are analyzed by observing the degradation of organic dye *Methylene Blue* (MB) under dark condition. The NGP addition of 15 wt% show the best result of the composites to degrade Methylene Blue in alkaline condition. Adsorption mechanism of the composites with NGP addition tend to follow the model Langmuir adsorption kinetic models.

1. Introduction

Each year, the total amount of dye wastes disposed into the environment from the food and textile industry kept increasing which will cause more water pollution [1-2]. To cope with the problem, an effective waste water treatment is required to reduce the content of environmental pollution. There are various methods for dye wastes treatment such as adsorption, coagulation, flocculation, reverse osmosis, biological treatment, and catalytic activity [1-5]. Adsorption is one of reliable methods in overcoming water pollution because of its several advantages such as low cost, simple design, and easy operations [2,6]. However there are constraints applying it since its adsorption capacity is not too high.

In recent years, semiconductor material draws an attention because of its capability in degrading wastes [7-8]. ZnO is a semiconductor that is widely used because of its low cost, high photochemical, biological stability, and low toxicity. However, the use of single material such as ZnO still has shortage in term of efficiency in waste degradation. Combination of two or more semiconductors has been proven to improve the process of waste degradation because of the synergistic effect of each semiconductor [8-10]. CuO is one of semiconductors with relatively small band gap hence it is expected that by combining it with ZnO we could modify the band gap of the combination these two semiconductor to improve the ability of waste degradation. Another problem in using ZnO semiconductor in the degradation process of wastes is the difficulty in separating the ZnO material from the solution after the reaction process. Therefore further modification is required to facilitate the separation process of the adsorbent material from the waste solution. Using magnetic materials such as Fe_3O_4 has been proven to facilitate the separation process of adsorbent materials from the solution [11].

Some researchers have also reported that carbon materials such as *graphene* could be an ideal supporting material for nanoparticles or composites because of its unique properties such as high surface area, high carrier mobility, and high electrical conductivity [6,12]. High surface area of graphene is considered able to increase the adsorption capacity of nanomaterials. Therefore, in this paper, NGP will be combined with Fe_3O_4/CuO/ZnO nanocomposites with several variation of NGP weight percent (wt%). Here the influence of NGP addition into nanocomposites in degrading *Methylene Blue* (MB) as the *model pollutant* was examined under the adsorption method, and the influence of initial pH and MB *concentration* were also investigated.

2. Experiment

Materials: Iron (II) sulfate heptahydrate ($FeSO_4 \cdot 7H_2O$, 99%), copper sulfate pentahydrate ($CuSO_4 \cdot 5H_2O$, 99%), zinc sulfate heptahydrate ($ZnSO_4 \cdot 7H_2O$, 99%), sodium hydroxide (NaOH), ethanol, ethylene glycol (EG), methylene blue, were purchased from Merck. All chemical reagents (Merck) were of analytical grade and used without further purification. Nano graphene platelets (NGP-N006-P) were purchased from Angstron Materials. All solutions were used without further purification.

Nanocomposites Synthesis: Fe_3O_4/CuO/ZnO is synthesized using the method that was applied in the previous study by Taufik, et al. [13-14] without modification. Fe_3O_4/CuO/ZnO/NGP composites were made using simple hydrothermal method [15]. In brief, a certain amount of NGP was first added into *aquades* solution and *ethanol* before being sonicated for 2 hours. After sonication process, Fe_3O_4/CuO/ZnO was poured and stirred for 2 hours to get homogeneous suspension. The resulting solution was further heated at the temperature of 120 °C for 3 hours, and then centrifuged and vacuumed at the temperature of 70 °C for 12 hours. The amount of NGP addition into Fe_3O_4/CuO/ZnO nanocomposites was varied from 5wt%, 10wt%, and 15wt%.

Characterizations: The samples were characterized by *X-Ray Diffraction* (XRD) using Raigaku Miniflex 600 monochromatic Cu-K_α (λ = 1.5 Å) radiation operated at 30 kV and 15 mA at the span angle of 10° to 80° and the absorption spectrum were characterized using Shimadzu *FTIR spectrophotometer* in the range of 400-4000 cm^{-1}. *Thermogravimetric Analysis* (TGA) was measured at the heating rate of 10 °C/min in N_2 atmosphere, and *magnetic* measurement was performed using Oxford type 1.2 T *vibrating sample magnetometer* (VSM).

Adsorption measurement: A certain amount of samples that have been synthesized were dissolved into *Methylene Blue* (MB) solution with the concentration of 20 mg/L at pH 13. The solution was stirred using *magnetic stirrer*. At a determined time, 5 mL of solution was collected and centrifuged to separate it from Fe_3O_4/CuO/ZnO nanocomposites and then the concentration was measured using UV-VIS spectroscopy. The efficiency of the adsorption process can be calculated using the equation of C_t/C_0, where C_t is the MB concentration at a certain interval and C_0 is the initial concentration of MB. Subsequently, several parameter measurements are carried out in the form of initial dye concentration: (20-150 mg/L) and initial pH of the solution: (3-13).

3. Results and discussion

Figure 1 indicates the XRD spectrum of Fe_3O_4/CuO/ZnO/NGP with different variations of NGP wt%. The of XRD diffraction pattern of Fe_3O_4/CuO/ZnO/NGP indicates the existence of cubic spinel phase of Fe_3O_4 detected at the angles of 2θ = 30.14°, 35.49°, 43.28°, 53.76°, 57.20°, and 62.83°. The diffraction pattern at the angle of 2θ =38.81° and 48.7° indicates the existence of monoclinic CuO phase. Several *peaks* at the angle of 2θ =34.47°, 36.26°, 47.50°, 56.70°, 68.06°, and 69.21° conform to the crystal structure of hexagonal wurtzite of ZnO, and at the angle of 2θ =26.40° there is a peak conforming to the graphitic like structure of nanographene platelets. The result of XRD spectra indicates that Fe_3O_4/CuO/ZnO/NGP has been successfully synthesized without any impurity detected. An increased intensity is observed at NGP peak by increasing the wt% of NGP given in composites. This is because of increased amount of NGP in the composites. The average particle size (D) is calculated using *Scherer* formulation $D = 0.9\lambda/(\beta \cos\Theta)$ where λ is the wavelength of incident X-Ray, β is the full width at half maximum and Θ as the Bragg angle. The result of average particle size analysis is tabulated in Table 1.

The FTIR absorption spectrum of Fe_3O_4/CuO/ZnO/NGP is shown in Fig. 2. In the range of wavenumber 1722 cm^{-1} and 1626 cm^{-1}, there are absorptions that correspond with the vibration mode of C=O and C=C stretching while at wavenumber of 3411 cm^{-1}, there is an absorption that corresponds with the vibration mode of O-H stretching. The absorption spectra of Fe_3O_4/CuO/ZnO is presented in Fig. 2b. It can be observed that the absorption spectrum of Fe_3O_4/CuO/ZnO is detected at wavenumber of ~400 cm^{-1}, ~420 cm^{-1} and ~590 cm^{-1} which correspond with the Zn-O [17], Cu-O [18] and Fe-O stretching [19]. The infrared spectra of Fe_3O_4/CuO/ZnO/NGP also consists of the vibration mode of NGP material and the vibration mode from the material of Fe_3O_4/CuO/ZnO nanocomposites. From the FTIR result we could confirm the existence of Fe_3O_4, CuO, ZnO, and NGP that have been mixed in Fe_3O_4/CuO/ZnO/NGP composites.

The thermal stability of Fe_3O_4/CuO/ZnO/NGP composites is identified using *Thermalgravimetry Analysis*. Based on the measurement result (Fig. 3(A)), the Fe_3O_4/CuO/ZnO have quite good thermal stability up to the temperature of 1000°C. When NGP is added by 5, 10 and 15 wt% into Fe_3O_4/CuO/ZnO nanocomposites, there is mass loss by 7.26%, 12.67%, and 18.53%. Such significant

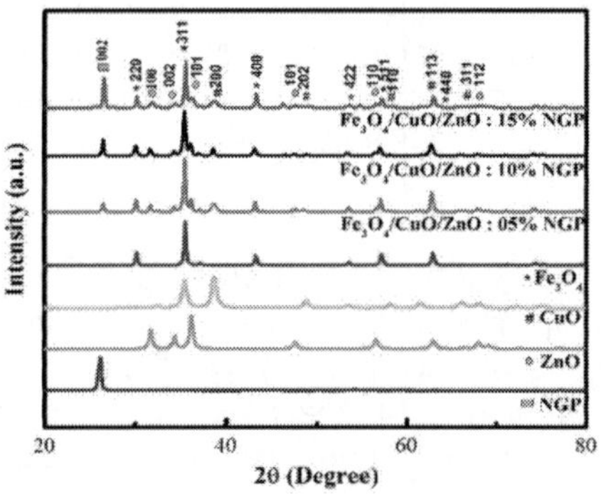

Figure 1. XRD pattern of Fe_3O_4/CuO/ZnO/NGP with variation NGP wt%

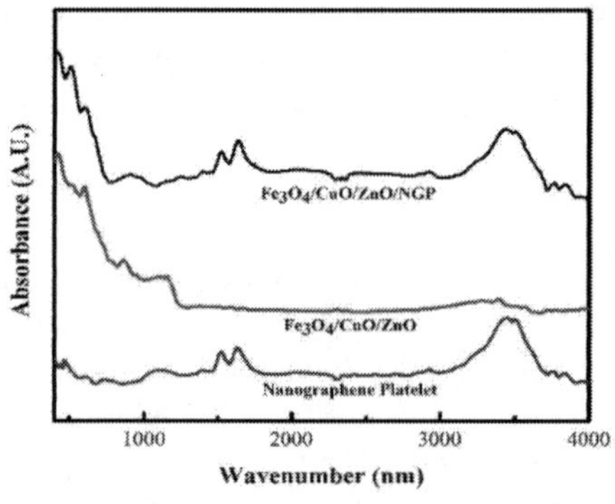

Figure 2. FTIR spectra of NGP, Fe_3O_4/CuO/ZnO, and Fe_3O_4/CuO/ZnO/10wt%-NGP

Table 1. Grain size and magnetic saturation value of Fe_3O_4/CuO/ZnO/NGP nanocomposite

Samples	<D> (nm)			M-S
	ZnO	CuO	Fe_3O_4	
Fe_3O_4	–	–	45	82
CuO	–	15	–	-
ZnO	19	–	–	-

Fe_3O_4/CuO/ZnO	26	15	42	46
Fe_3O_4/CuO/ZnO-05wt% NGP	25	15	41	52
Fe_3O_4/CuO/ZnO-10wt% NGP	22	15	41	57
Fe_3O_4/CuO/ZnO-15wt% NGP	16	15	39	55

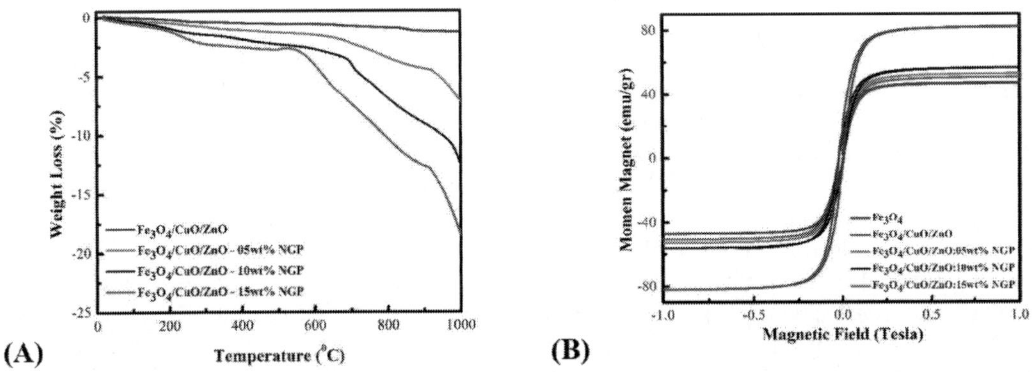

Figure. 3. TGA curve (A) and VSM spectra (B) of Fe_3O_4/CuO/ZnO dan Fe_3O_4/CuO/ZnO/NGP with variation NGP wt%

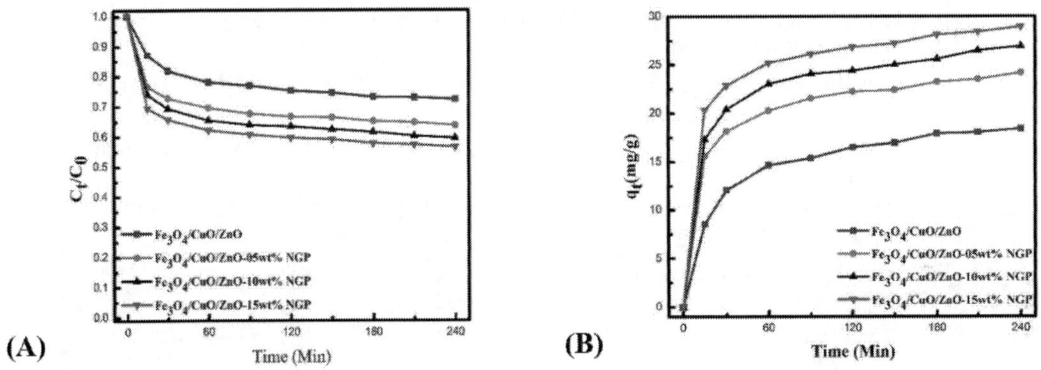

Figure 4. Adsorption (A) and adsorption capacity (B) of methylene blue using Fe_3O_4/CuO/ZnO and Fe_3O_4/CuO/ZnO/NGP with variation NGP wt%

Table 2. Adsorption of Fe_3O_4/CuO/ZnO/NGP nanocomposite

Samples	Degradation (%)	**Adsorption Capacity (mg/g)**
Fe_3O_4/CuO/ZnO	28	18
Fe_3O_4/CuO/ZnO-05wt% NGP	36	24
Fe_3O_4/CuO/ZnO-10wt% NGP	40	27
Fe_3O_4/CuO/ZnO-15wt% NGP	43	29

decreased was caused by combustion from carbon materials in the samples originating from NGP material [20]. The mass lost well correspond to the wt% of NGP in the composites.

The magnetic saturation of Fe_3O_4/CuO/ZnO nanocomposites and Fe_3O_4/CuO/ZnO/NGP composites at room temperature is shown in Fig.3(B). As comparison, the magnetic saturation curve of Fe_3O_4 nanoparticles is also presented in Fig. 3(B). The result obtained indicates that all the samples have ferromagnetic behaviour. The magnetic saturation value of all samples is tabulated in the last column of Table 1. The magnetic saturation value of Fe_3O_4/CuO/ZnO nanocomposites is lower than that of Fe_3O_4 nanoparticles, this may be caused by less amount of Fe_3O_4 when it was formed into Fe_3O_4/CuO/ZnO nanocomposites. However, addition of nanographene platelets by 5wt% and 10wt%

causes an increase of the magnetic saturation value compared to Fe_3O_4/CuO/ZnO nanocomposites and addition of 15wt% NGP lowers the magnetic saturation value but still higher than Fe_3O_4/CuO/ZnO. Increase of magnetic saturation value when NGP material is added can be caused by Fe_3O_4 material sticking on the NGP surface which causes defect and stabilize NGP-based magnetic material [21].

The adsorption capacity measurement of Fe_3O_4/CuO/ZnO and Fe_3O_4/CuO/ZnO/NGP nanocomposites is analyzed by observing the methylene blue degradation represented by C_t/C_0 (C_t =MB concentration at certain interval, C_0=initial concentration) as seen in Fig 4(A). It can be seen in the figure an increase in the adsorption capacity simultaneously with NGP material addition. The qualitative data from the adsorption process for all samples is represented in the form of *adsorption capacity* as indicated in Fig. 4(B). The maximum value of adsorption capacity from all samples is shown in Table 2. The maximum adsorption capacity of Fe_3O_4/CuO/ZnO increases simultaneously with increasing the wt% of NGP.

The adsorption process is also influenced by the acidity of the waste solution being used; therefore in this study we also analyze the influence of methylene blue solution acidity on the adsorption capacity of Fe_3O_4/CuO/ZnO/NGP. The acidity of methylene blue solution is varied from 3, 5, 7, 9, 11, and 13 using NaOH and CH_3COOH. The result obtained can be seen in Fig. 5(A). The adsorption capacity of Fe_3O_4/CuO/ZnO/NGP can reach optimum condition in alkaline condition while in acid condition the adsorption capacity of Fe_3O_4/CuO/ZnO/NGP decreases. This may be caused by the fact that the maximum interaction occurs between negative surface of Fe_3O_4/CuO/ZnO/NGP composites and *positive charge* MB in the alkaline condition [22]. The inset in Fig. 5(A), indicates the adsorption capacity of composites with different pH.

Fig. 5(B) shows the influence of difference in the concentration of methylene blue solution on the adsorption capacity of Fe_3O_4/CuO/ZnO/NGP. In this study the concentration of methylene blue solution is varied from 20 mg/L – 150 mg/L. The result indicates that the adsorption capacity of Fe_3O_4/CuO/ZnO/NGP decreases simultaneously with an increase in methylene blue solution.

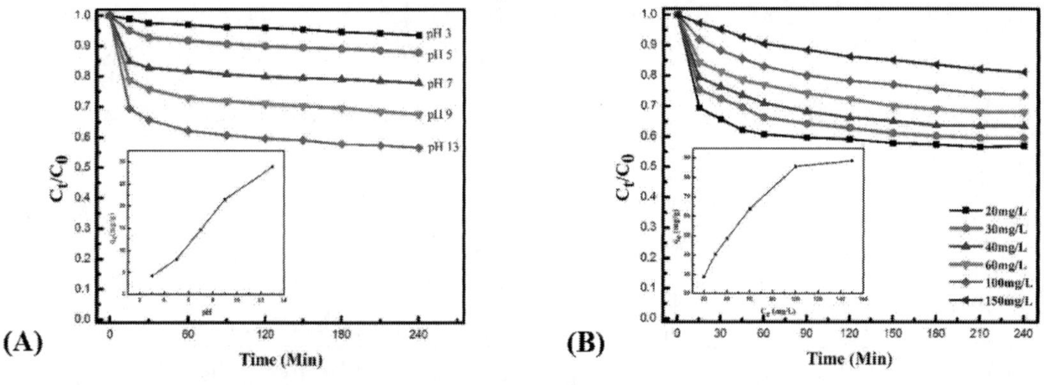

Figure 5. Effect of pH (A) and initial concentration (B) of MB on adsorption capacity

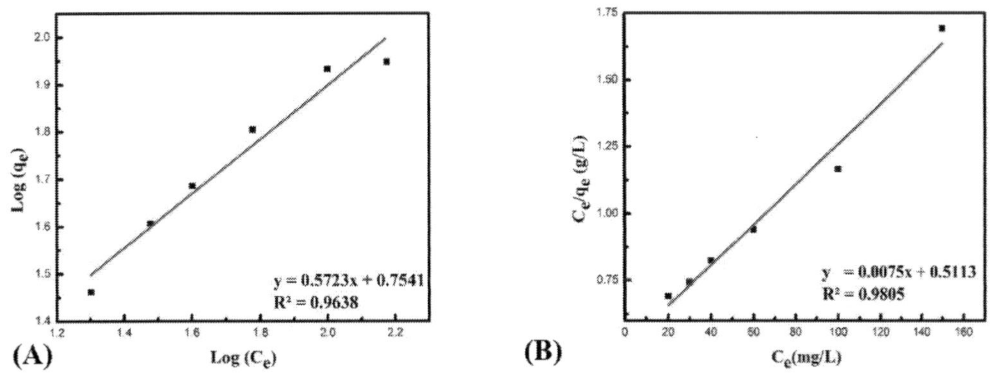

Figure 6. Freudlich adsorption isotherm model (A) and Langmuir adsorption isotherm model (B)

Subsequently, to know the mechanism of adsorption process, Freudlich and Langmuir adsorption kinetic model are applied. The result obtained is indicated in Fig. 6 (A) for Freudlich model and Fig. 6 (B) for Langmuir model. In Fig. 6, we could see the adsorption process of Fe_3O_4/CuO/ZnO/NGP follows the Freudlich and Langmuir adsorption kinetic *model* with the correlation coefficient (R^2) obtained for *Freundlich* and *Langmuir* of 0.9638 and 0.9805 respectively.
Based on the study carried out using several parameters, the optimum condition obtained to degrade MB from *aqueous solution* (*MB concentration of 20* mg/L) using the adsorption method is obtained for Fe_3O_4/CuO/ZnO nanocomposites with 15wt% NGP addition, for the dosage of 0.3 mg/L and pH 13. In this study, maximum condition of the degradation and adsorption capacity is reached 43% and 29 mg/g.

4. Conclusion

In this study, Fe_3O_4/CuO/ZnO/NGP composites with different added amount of NGP has been successfully synthesized using two step method, sol-gel and simple hydrothermal method. The XRD and FTIR characterization result indicate the existence of structure and vibration of Fe_3O_4, CuO, ZnO and NGP in the composites. The result also indicates the amount of NGP addition in accordance with the result of TGA measurement. Addition of NGP may improve efficiency in the adsorption process of methylene blue in aqueous solution, with the addition of 15wt% the degradation efficiency and adsorption capacity increased by 15% and 11 mg/g. The experimental result follows properly the adsorption of Freundlich dan Langmuir kinetic model, where the conformity value of Langmuir is higher than the Freundlich.

Reference

[1] Soltani RDC, et al 2016 *Ultrasonics Sonochemistry* **28** 69-78
[2] Noorimotlagh Z, et al 2014 *Journal of Taiwan Institute of Chemical Engineers* **45** 1783-1791
[3] Eskandarloo H, et al 2015 *Ultrasonics Sonochemistry* **26** 281-292
[4] Soltani T, MH 2013 Entezari *Journal of Molecular Catalysis A:Chemical* **377** 197-203
[5] Chang J, et al 2016 *Applied Clay Science* **119** 132-140
[6] Chang YP, et al 2012 *Applied Surface Science* **261** 504-509
[7] Sanchez D-M, et al 2016 *Ceramics International* **42** 2013-2020
[8] Hou XM, Wang LX, Li F, He GF, Qing L-li 2015 *Materials Letters* **159** 502-505
[9] Akir S, Barras A, Coffinier Y, Bououdina M 2016 *Ceramics International* **42** 10259-10265
[10] Mageshwari K, et al 2015 J. Park *Journal of Alloys and Compounds* **625** 362-370
[11] Bian XF, et al 2013 *Applied Surface Science* **280** 349-353
[12] Ashkarran AA, Mohammadi B 2015 *Applied Surface Science* **342** 112-119
[13] Taufik A, Susanto IK, Saleh R 2015 *Materials Science Forum* **827** 37-42
[14] Susanto IK, Taufik A, Saleh R 2015 *Advance Materials Research* **1123** 227-232

[15] Khalid NR, et al 2013 *Current Applied Physics,* **13** 659-663
[16] Nekahi A, Marashi PH, Haghshenas D 2014 *Applied Surface Science* **295** 59-65
[17] Bandyopadhyay S, et al 2002 *Materials Chemistry and Physics* **74** 83-91
[18] Rytlewski P 2014 *Surface & Coatings Technology* **259** 660-666
[19] Xie XY, et al 2015 *Journal of Chromatography A* **1405** 32-39
[20] Cai D, Song M, 2010 *Journal of Materials Chemistry* **20** 7906-7915
[21] Heidari EK, et al 2015 *Journal of Magnetism and Magnetic Materials,* **379** 95–101
[22] Wu Z, et al 2014 *Water Research* **67** 330-334

Journal of Physics: Conference Series **776** (2016) 012020 doi:10.1088/1742-6596/776/1/012020

The synthesization of Fe_3O_4 magnetic nanoparticles based on natural iron sand by co-precipitation method for the used of the adsorption of Cu and Pb ions

E. A. Setiadi[1], P. Sebayang[1], M. Ginting[1], A. Y. Sari[1], C. Kurniawan[1], C. S. Saragih[2] and P. Simamora[2]

[1]Research Center for Physics, Indonesian Intitute of Sciences (LIPI), Kompleks Puspiptek, Serpong, Tangerang Selatan, Banten 15314 - INDONESIA
[2]Department of Physics, State University of Medan
Jl. Willem Iskandar,Pasar V, Medan, Sumatera Utara 20221 - INDONESIA

E-mail : eko.arief.setiadi@lipi.go.id

Abstract. Magnetic nanoparticles of Fe_3O_4 (magnetite) have been synthesized from natural sand iron by co-precipitation method. The nanoparticles were synthesized using HCl as solvent and NH_3 as co-precipitate. The nanoparticles synthesized at 70°C in two different treatments. Sample without Polyethylene Glycol (PEG) 6000 noted by A and sample with PEG 6000 noted by B symbol. The measurement that have been done for both samples were XRD (X-ray diffraction), FTIR (Fourier Transform Infrared) Spectrometry, SEM (Scanning electron microscopy), VSM (Vibrating sample magnetometer) and SAA (Surface area analyzer). The results showed that both samples were having Fe_3O_4 phases. Particle size, coercivity and magnetic saturation of B samples were smaller than A samples. But the surface area of B sample was larger than A sample. Both samples were then used to adsorb Cu and Pb ions using shaker method. Adsorption analysis from Atomic Adsorption Spectroscopy (AAS) showed that B was more effectivein adsorbing metal ions than A. The adsorption value of Cu and Pb ions were 79 and 91% respectively.

1. Introduction

Magnetic nanoparticles have been widely studied since this material has unique properties. For examples it's super-paramagnetic properties, which is not exist on bulk material. Super-paramagnetic properties of materials have a high saturation magnetization, and the average coercivity equal to zero when those material are not influenced by an external magnetic field [1]. These properties appear only when a ferromagnetic or ferrimagnetic material is in a very small size (in order of nanometer), so this phonomenon had been atrracted to many of researchers in using this material in various applications, such as drugs delivery, hyperthermia treatment, magnetic resonant imaging [2],enhancement agent, manipulating cell membranes, biosensors, bio-labelling [3] and adsorption of metal ions [4]. The materials which have these kind of properties are Fe_3O_4, $CoFe_2O4$, $NiFe_2O_4$ and others. The use of nanoparticles materials as an adsorbent for heavy metal ions should have some properties, such as high saturation value, small coercivity and large surface area per unit mass. Larazza *et al* [4] have been used Fe_3O_4 mixed with polyethylenimine and montmorillonite as magnetic adsorbents for Cr (VI) water treatment. They founded that the surface area of material will be larger if the particle size is smaller.

The natural iron sand contains abundant of ferrite materials. Syntesization of nanoparticle can be used with variuos methods, such as hydrothermal synthesis, thermal decomposition, microwave plasma synthesis [5], sol gel, sono-chemical, thermal hydrolysis, and co-precipitation [1]. The synthesization of nanoparticles based on iron sand has been developed for the last few years. Sunaryono *et al* reported that the average diameter of nanoparticles from iron sand strongly depends on the stirring rate of co-precipitation reaction [6]. Mashuri *et al* substituted Ni and Zn ions in order to improve their sample's magnetic properties [7]. While Bukit *et al* has been used PEG 6000 as nanoparticles template to control the particles size, suface modification and to control the interaction among particles [8]. In this paper will be discussed the synthesization of Fe_3O_4 nanoparticles from nature iron sands by co-precipitation method. This method was chosen because the procedure is quite simple and easy to perform in relatively low temperature [9]. Two different treatments will be applied to the sintesized nanoparticles. Samples A were treated withoutPolyethylene Glycol (PEG 6000 modification) and samples B withPEG 6000 modification. PEG 6000 modification was applied to control the particle size, and improving biocompatibility and chemical stability [1]. Synthesized nanoparticlesthen analized theirability to adsorpCu and Pbions which were classified as heavy metal ions.

2.Experimental Method

Natural sands iron that was used in this research was taken from BuayaRiver in Deli Serdang, North Sumatera. The iron sands was then synthesized by co-precipitation method using400 ml12 M HCl (Merck 37%,) as a solventin, stirred for 90 minutes at 300 rpm using a magnetic stirrer, then filtered using paper filter with 8 μm pore sizes. Two glasses of 50 ml 6.5 M NH_3 solution (Merck 32%) was prepared by stirring with 300 rpm at 70°C. The first glass without any addition and the second glass was added 1 mmol of polyethylene glycol (PEG 6000). Then the filtered iron sands was added into both glasses while stirred for 90 minutes until precipitation appeared at the bottom of both glasses. The precipitate materialthen was separated from the solution, washed with aquades until its PH has reached 7, dried it at 100°C for 5 hours, until Fe_3O_4 nanoparticles powder were obtained.

The nanoparticles powder resulted from the first glass (without PEG 6000) was noted as "A" samples and that from the second glass was noted as "B" samples. Both samples, A and B, were characterized using X-ray Diffraction (XRD - Rigaku SmartLab) at wavelength $\lambda_{CuK\alpha}$ = 1.5418 Å, Scanning Electron Microscopy (SEM - Hitachi SU 3500), vibrating sample magnetometer (VSM - Electromagnetic VSM250), IR Spectroscopy (FTIR - Quanta Crome Intrument) and Surface area analyzer (SAA - Nova 4200e) toanalyzetheir phase, particle size distributions, functional groups, magnetic properties, and surface area.

Analysis of their ability to adsorp heavy metal ions was done by setting up artificial wastewater containing Pb and Cu metals of 706 and 277 ppm, respectively. Nanoparticles powder (50 mg) was poured into the wastewater (25 ml), then mixed using shaker mill for 30 minutes. Atomic Adsorption Microscopy (AAS - Shimadzu AA6800) was used in examining the residual waste. The analysis were performed on both samples, A and B.

3.Results and Discussion

The synthesized nanoparticles from natural iron sand that we obtained are black in color and responsive to external magnetic field. XRD characterization are shown in Figure 1. It shows that both samples have single phase of Fe_3O_4 with cubic spinel structures, and with lattice parameter of 8.393Å [10]. The average particles size were estimated using Scherrer equation,

$$d = \frac{k\lambda}{B\cos\theta} \quad (1)$$

Where d is crystalline size, k is a Scherrer constant, λ is wavelength of the x-ray and B is full width at half maximum (FWHM).

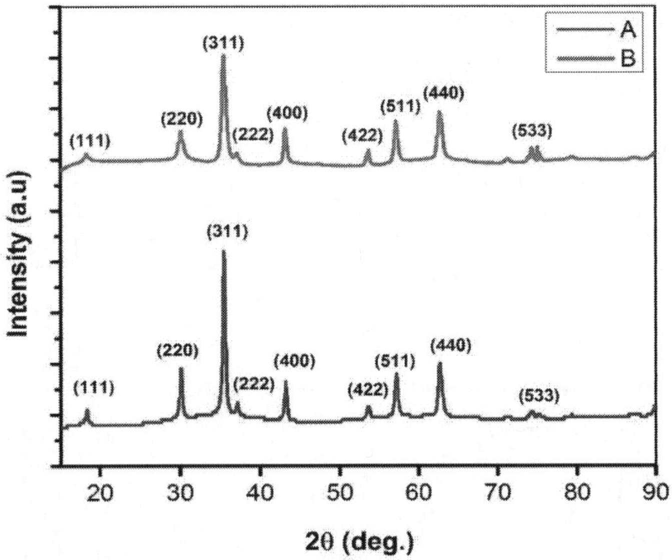

Figure 1.XRD pattern of Fe_3O_4 (a) sample Awithout PEG 6000 and (b) sample B with PEG 6000

The particles size calculated using Scherrer equation were 32.3 nm and 14.4 nm for A and B samples respectively. Based on these results, it is clearly that the nanoparticles had been succesfully synthesized from natural iron sands using this technique.

Figure 2 shows morphologycal properties of nanoparticles for both samples (A and B). The nanoparticles tends to form rough spherical in shape. It has been reported that spherical shape is formed because the nucleation rate per unit area is isotopic at the interface between the Fe_3O_4 magnetic nanoparticles [3]. Particle size distribution of A sample varied around 25-50 nm, while B sample varied around 12-30 nm.

Figure 2.SEM imaging of Fe_3O_4 (a) sample Awithout PEG 6000 and (b) sample B with PEG 6000

These values are consistent with the particle size that we calculated using Scherrer equation. Average particle size of A sample is larger than the B sample. This is due to the addition of PEG 6000 on sample B. The addition of PEG 6000 can contribute as the coating and reducing agents on nanoparticles, thus decrease the agglomeration of nanoparticles. It has been explained also that the use of polymers such as

PEG in nanoparticles materials is to reduce the agglomeration and to prevent further aggregation, so that the size of nanoparticles will be well controlled [2].

From Figure 3, it can been seen that at wave number of 401.19 and 570.93 cm^{-1} for both samples show characteristic of Fe-O stretching, tetrahedral and octahedral of Fe_3O_4. At wavenumber of 1627.9 and 3402.43 cm^{-1} are O-H bending and stretching of water groups. On B sample, there are C-O-C stretching and C-H stretching bond at 1072.42 and 2368.59 cm^{-1} wavenumbers, which belongs to the functional groups of PEG 6000. The presence of those two wavenumbers indicate that the nanoparticles of B sample have coated by PEG 6000 [11]–[13]. The PEG coating improves chemical stability, biocompatibility and particle size control from agglomeration effects [11].

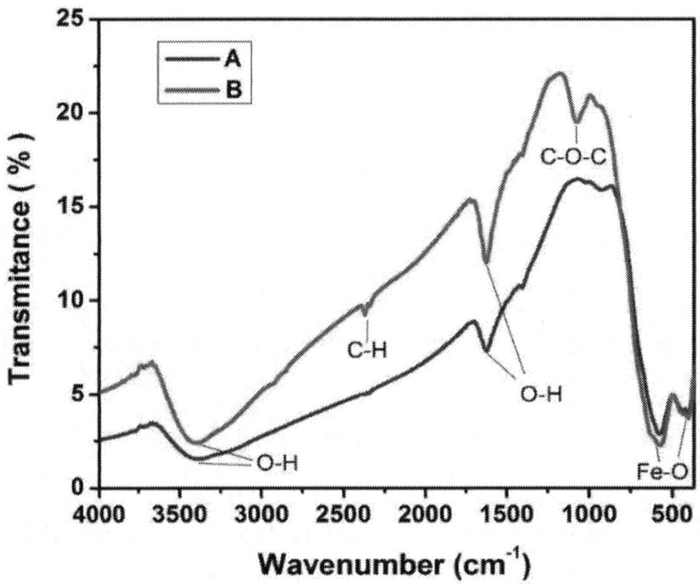

Figure 3. FTIR pattern of Fe_3O_4 (a) sample A without PEG 6000 and (b) sample B with PEG 6000

In Figure 4 is showing the hysteresis curve of the VSM measurement. Based on the hysteresis curve, the coercivity, magnetic saturation, and magnetic remanent of A samples are 92.80 Oe, 1.95 kG and 0.27 kG consecutively. Meanwhile for B sample are 85.65 Oe, 1.41 kG and 0.20 kG, respectively. Based on that value, magnetic saturation and remanence of B sample is smaller than A sample. This is due to the effect of the addition of PEG 6000 which acting as paramagnetic polymer. Hence, when an external magnetic field is applied to the nanoparticles, the magnetic net on B sample will be smaller than on A sample [1]. While the coercivity of B is smaller than A, is because the particle size of B sample is smaller than Asample. When particle size of materials is quite small (nano size), they tend to have a single domain wall, so that the magnetic moments will be easier to switch when an external magnetic field is applied, and also easier to turn back into previous condition when the external magnetic field is removed [1]. The small value of coercivity is one of the characteristic of superparamagnetic material. Then, based on the coersivities values (<132 Oe), both samples are classified as soft magnets [14].

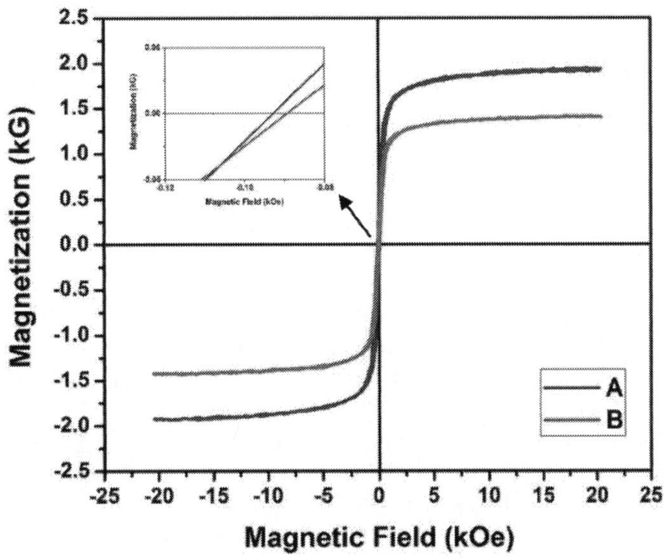

Figure 4. Hysteresis curves of Fe_3O_4 (a) sample A without PEG 6000 and (b) sample B with PEG 6000

In order to know the ability of the obtained sinthesized nanoparticles in adsorbing heavy metal ions, it was conducted an experiment using an artificial wastewater containing Pb and Cu metal ions. The 50 mg Fe_3O_4 nanoparticles was poured into 25 ml wastewater contained 706 and 277 ppm of Pb and Cu ions respectively, then mixed with a shaker mill for 30 minutes. It was observed that the surface area of samples are affecting the adsorption process. The SAA measurement shows the surface area of A sample was 129.82m^2/g while B sample was 140.49 m^2/g. The residual wastewater then was analized using AAS to determine the adsorption value, that is shown in Table 1.

Table 1. The Adsorption of Cu and Pb ions by Fe_3O_4 samples

Indicator	Initial Concentration (ppm)	After Adsorption by			
		Sample A		Sample B	
		Concentration (ppm)	Percentage of Adsorption (%)	Concentration (ppm)	Percentage of Adsorption (%)
Cu	277	46.3	83	24.3	79
Pb	706	509	28	151	91

From Table 1 can bee seen that the B sample is more effective in adsorbing Pb and Cu metal ions, compared with A sample. The concentration of Pb and Cu on wastewater residu on sample A, reduced around 28% and 83% respectively. While the concentration of Pb and Cu wastewater residu on sample B, reduced around 79% and 91% respectively. It is clearly seen that B sample adsorb more metal ions than A sample. This is due to B sample has a smaller particle size than A Sample. Since the smaller the particle size, the larger the surface area is. So the larger the surface area, then more ions will be adsorbed on the surface. Beside, the existence of PEG 6000 coating on nanoparticles surface will also increase its stability and prevent it from oxidation [3].

4. Conclusion
The Fe_3O_4 nanoparticles have been successfully synthesized from natural iron sand of Buaya River in Deli Serdang, North Sumatera by co-precipitation method. Two different treatments were done to get two different samples, such as with and without PEG 6000. XRD results showed that both samples had a single phase Fe_3O_4. Fe_3O_4 with PEG 6000 had smaller particle size and greater surface area than sample without PEG 6000. Fe_3O_4 with PEG 6000 has smaller coercivity and saturation magnetization than Fe_3O_4 without PEG 6000. While the Fe_3O_4 with PEG 6000 was more effective to absorb Cu and Pb metal ions (79% and 91%) compared to Fe_3O_4 without PEG 6000 (28% and 83%).

References
[1] Suharyadi E, Setiadi E A, Shabrina N, Kato T and Iwata S 2014 *Adv. Mater. Res.* **896** 126
[2] Feng B, Hong R Y, Wang L S, Guo L, Li H Z Ding J Zheng Y and Wei D G 2008 *Colloids Surfaces A Physicochem. Eng. Asp.* **328** 52
[3] Anbarasu M, Anandan M, Chinnasamy E, Gopinath V and Balamurugan K 2015 *Spectrochim. ACTA Part A Mol. Biomol. Spectrosc.* **135** 536
[4] Larraza I, López-gónzalez M, Corrales T, and Marcelo G 2012 *J. Colloid Interface Sci.* **385** 24
[5] Liang X, Shi H, Jia X, Yang Y, and Liu X 2011 *Mater. Sci. Appl.* **2** 1644
[6] Sunaryono, Taufiq A, Mashuri, Pratapa S, Zainuri M, Triwikantoro, and Darminto 2015 *Mater. Sci. Forum* **827** 229
[7] Mashuri, Yahya E and Darminto 2011 *J. Mater. Sci. Engeneering A* **1** 182
[8] Bukit N, Frida E, Simamora P, and Sinaga T 2015 *Chem. Mater. Res.* **7** 110
[9] Suharyadi E, Setiadi E A, Riyanto A, Kato T, Iwata S, and Abraha K 2014 *J. Sains Mater. Indones.* **15** 123
[10] Jayanthi S A, Sukanya D, Pragasam A J A, and Sagayaraj P 2013 *Der Pharma Chem.* **5** 90
[11] Junejo Y and Baykal A 2013 *Cent. Eur. J. Chem.* **11** 1527.
[12] Cao X, Zhang B, Zhao F, and Feng L 2012 *J. Nanomater.* **2012** 1
[13] Gao G, Qiu P, Qian Q, Zhou N, Wang K, Song H, Fu H, and Cui D 2013 *J. Alloys Compd.* **574** 340
[14] Coey J M D 2010 *Magnetism and Magnetic Materials* (New York: Cambridge University Press)

Journal of Physics: Conference Series **776** (2016) 012021 doi:10.1088/1742-6596/776/1/012021

The calcination temperature dependence of microstructural, vibrational spectra and magnetic properties of nanocrystalline $Mn_{0.5}Zn_{0.5}Fe_2O_4$

I. P. T. Indrayana[1], N. Siregar[1], E. Suharyadi[1*], T. Kato[2] and S. Iwata[3]
[1] Department of Physics, Gadjah Mada University, Yogyakarta, INDONESIA
[2] Department of Electrical Engineering & Computer Science, Nagoya University
[3] Institutes of Materials and Systems for Sustainability, Nagoya University, JAPAN

*E-mail: esuharyadi@ugm.ac.id

Abstract. Effect of calcination temperature on microstructural, vibrational, and magnetic properties of $Mn_{0.5}Zn_{0.5}Fe_2O_4$ nanoparticles have been successfully investigated. The nanoparticles were synthesized via coprecipitation method and calcined at different temperatures varying from 400, 600, 800, and 1000°C. The X-ray diffraction (XRD) pattern confirmed the formation of cubic spinel structure $Mn_{0.5}Zn_{0.5}Fe_2O_4$ with crystallite size ranging from 18.3 nm to 24.8 nm. The TEM micrograph showed the morphology of nanoparticles change from nearly spherical to cubic form after calcination. The FTIR spectra confirmed the existence of vibrations at 416.6 cm^{-1} - 455.2 cm^{-1} and 555.5 cm^{-1} -578.6 cm^{-1} which corresponds to the intrinsic stretching vibration of metal-oxygen at octahedral and tetrahedral sites, respectively. The maximum specific magnetization and coercivity increase with increasing calcination temperature. The maximum specific magnetization value of 54.7emu/gram was obtained for sample calcined at 1000°C. The results showed that calcination treatment will facilitate the tunability of microstructural and magnetic properties of nanoparticles for expanding the field of application.

1. Introduction

Spinel ferrite nanoparticles with molecular equation MFe_2O_4 (M: Mn^{2+}, Zn^{2+}, Fe^{2+}, Co^{2+}, Ni^{2+}, etc.) are very promising nanomaterial which have many practical applications such as transducers, transformers and electromagnetic interference shielding materials due to their high electrical resistivity and dielectric constant [1]. Meanwhile, because of their high magnetic properties as well as unique crystal and electronic structure allow ferrite nanoparticles to be utilized as high-density information storage media, magnetic printing inks, magnetic drug delivery and magnetic resonance imaging (MRI) enhancement [2,3]. In recent years, spinel ferrite nanoparticles have been also achieved great attention as photocatalyst materials. In this context, ferrite nanoparticles offer several advantages of having narrow band gap for example 1.25 eV for $MnFe_2O_4$ and 1.30 eV for $ZnFe_2O_4$ [4] and high magnetic properties that allow to the development of magnetic separation technology which is very effective, simple, and low cost process compare to centrifugation and filtration [5]. Those advantages can overcome the drawback of TiO_2 and ZnO nanoparticles that are common photocatalyst material have been used. Therefore, in other to obtain high quality photocatalyst materials so engineering ferrite nanoparticles with different metal ion composition, surface modification and synthesis method are the three important efforts that being focussed by researchers up to the present. For example, Ibrahim *et al.* [4] had engineered $MnFe_2O_4$ and $ZnFe_2O_4$ by sol-gel hydrothermal technology to investigate the photocatalytic activity of these ferrite nanoparticles for nitroarenes reduction.

$Mn_{0.5}Zn_{0.5}Fe_2O_4$ ferrite nanoparticles have been long studied for their high magnetic properties. This nanoparticles were formed by introduce nonmagnetic Zn^{2+} ion to $MnFe_2O_4$. Many researchers have varied the molar composition for both Mn^{2+} and Zn^{2+}. The composition of $Mn_{0.5}Zn_{0.5}Fe_2O_4$ gave the most stable magnetic properties [6,7]. Naik *et al.* [6] had found that $Mn_{0.5}Zn_{0.5}Fe_2O_4$ nanoparticles have specific magnetization saturation of 40.65 emu/gram and coercivity of 113.78 Oe. Meanwhile, Baykal *et al.* [7] revealed the band gap of $Mn_{0.5}Zn_{0.5}Fe_2O_4$ nanoparticles in range of 2.60 eV - 2.80 eV. Both magnetic properties and band gap of nanoparticles are influenced by the crystal structure, crystallite size and microstructure of grain nanoparticles.

In regard to obtain good microstructure of nanoparticles, synthesis procedure and heat treatment play important role [8-11]. Therefore, in the present research $Mn_{0.5}Zn_{0.5}Fe_2O_4$ were synthesized via coprecipitation method. Coprecipitation has very simple procedure. The synthesized nanoparticles have small crystallite size [12]. Unfortunately, this method have significant drawback such as the nanoparticles have broad crystallite size distribution and the small size make the nanoparticles tend to agglomerate that causes the weakening of microstructural and magnetic properties [12]. Therefore, controlling the size distribution and microstructure of nanoparticles via heat treatment such as calcination is needed.

This research was a part of project to engineer magnetic ferrite photocatalysts materials and aimed to control the microstructure and vibrational spectra of $Mn_{0.5}Zn_{0.5}Fe_2O_4$ nanoparticles by varying the calcination temperatures. Hopefully, by involving calcination process after synthesis, the crystallinity and microstructure of nanoparticles can be tuned as well as the magnetic properties.

2. Experimental Method

$Mn_{0.5}Zn_{0.5}Fe_2O_4$ nanoparticles were prepared by chemical coprecipitation method. The materials used were $FeCl_3.6H_2O$, $ZnSO_4.7H_2O$, $MnCl_2.H_2O$, HCl and NaOH. All the chemicals were of analytical grade (Merck Emsure) and used without any further purification. The chemical reaction is described by

$$\begin{gathered}4FeCl_{3(aq)} + MnCl_{2(aq)} + ZnSO_{4(aq)} + 16NaOH_{(aq)} + 2H_2O_{(aq)} \\ \downarrow + HCl_{(aq)} \\ 2Mn_{0.5}Zn_{0.5}Fe_2O_{4(s)} + 14NaCl_{(aq)} + Na_2SO_{4(aq)} + 10H_2O_{(aq)}\end{gathered} \qquad 1)$$

First of all, $FeCl_3.6H_2O$ and the combination of $ZnSO_4.7H_2O$ and $MnCl_2.H_2O$ were individually dissolved in 20 ml of distilled water within 2 minutes at room temperature under constant stirring of 500 rpm. Those combined solutions were added of 3.37 ml HCl and mixed within 5 minutes under constant stirring of 500 rpm. The mixture then added dropwise to a boiling of 1.5 M NaOH with pH around 12 throughout burette. Nanoferrites are formed by conversion of metal salts into hydroxides, which takes place immediately and the transformation of hydroxides into ferrites. The solutions were maintained at 120°C under constant stirring of 1000 rpm for 60 minutes. This duration was sufficient for the transformation of hydroxides into spinel ferrite (dehydration and rearrangement involved in the conversion of intermediate hydroxide phase into ferrite). The dark precipitates were obtained and then cooled to a moderate temperature. The precipitates were washed for 6 times using distilled water. The slurries of nanoparticles were obtained through filtration and continued washing by acetone for 2 times. The filtered particles were dried at 90°C in a furnace for 4 hours. Finally, the fine powders were yielded and calcined with four difference temperatures i.e., 400°C, 600°C, 800°C and 1000°C.

The composition and crystallite structure of powder nanoparticles were analysed by X-Ray diffractometer Shimadzu XD using monochromic Cu-Kα radiation (λ = 1.5406 Å). The average crystallite size was calculated by applying peak broadening of (311) using Scherer's equation [13]. The vibrational spectra of nanoparticles were recorded by Fourier transform infrared spectroscope (FTIR) Prestige Shimadzu-21 in range of 4000 – 400 cm^{-1}. The morphology and selected area diffraction (SAED) pattern were performed by Transmission electron microscope (TEM) Jeol Jem-1400. The room temperature magnetic properties measurement of sample nanoparticles was carried out by using Vibrating sample magnetometer (VSM) Riken Denshi Co Ltd.

3. Results and Discussion

3.1 XRD analysis

Figure 1 shows the XRD pattern of calcined and as prepared $Mn_{0.5}Zn_{0.5}Fe_2O_4$ nanoparticles at difference temperatures. The patterns show the characteristic peaks of cubic spinel structures, such as (220); (311); (400); (422); (511); and (440) according to JCPDS (10-0467). The peaks are broad and low intensity that confirm still the low crystallinity of sample nanoparticles. The intensity of nanoparticles increases with an increase of calcination temperature. The peak of (311) shifts to the lower diffraction angle due to increase of calcination temperature that causes strain on crystal lattice of nanoparticles [14]. The lattice exhibits expansion during calcination that makes change ions position in entire crystal lattice, so the interplanar distance of diffraction planes tend to be larger.

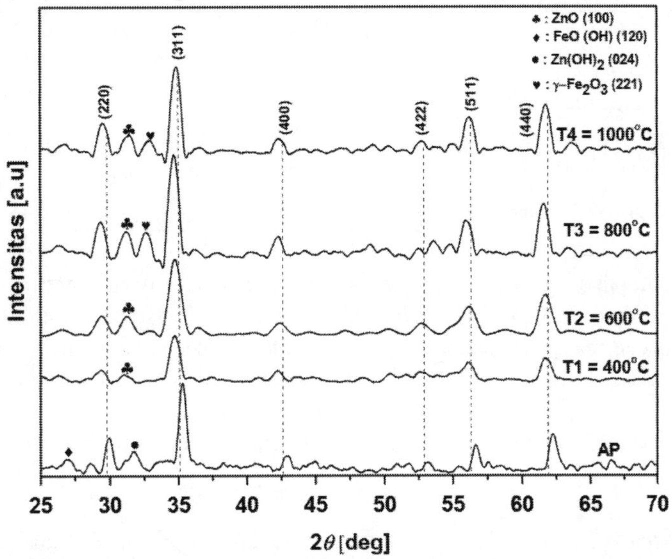

Figure 1. XRD pattern of $Mn_{0.5}Zn_{0.5}Fe_2O_4$ nanoparticles for various temperature: AP (as prepared nanoparticles); T1 (400°C); T2 (600°C); T3 (800°C); and T4 (1000°C)

XRD patterns also confirm the existence of nonspinel $Mn_{0.5}Zn_{0.5}Fe_2O_4$ phase (Table 1), such as $Zn(OH)_2$ (JCPDS no. 01-0360); FeO(OH) (JCDPS no. 17-0536) for as prepared sample and ZnO (JCPDS no. 36-1451) as well as γ-Fe_2O_3 (JCPDS no. 39-1346) for calcined samples. The appearance of $Zn(OH)_2$ and FeO(OH) are considered caused by incomplete ferritization process of precipitated salt precursors to form spinel $Mn_{0.5}Zn_{0.5}Fe_2O_4$ phase. In other hand, the appearance of ZnO and γ-Fe_2O_3 might be caused by thermal decomposition of unreacted $Zn(OH)_2$ and FeO(OH) during calcination. According to thus data can be predicted that the real composition of $Mn_{0.5}Zn_{0.5}Fe_2O_4$ will deviate from the analytical stoichiometry. Therefore, the cationic distribution of nanoparticles will change and influence the crystallite size and lattice parameter [15].

Table 1. Phase compostion of $Mn_{0.5}Zn_{0.5}Fe_2O_4$ nanoparticles

Sample	Spinel phase composition (%)	Nonspinel composition (%)			
		ZnO	$Zn(OH)_2$	FeO(OH)	γ-Fe_2O_3
AP	86.30	-	7.85	5.85	-
T1	94.41	5.59	-	-	-
T2	96.84	3.16	-	-	-
T3	77.88	9.06	-	-	13.06
T4	84.36	6.89	-	-	8.75

The crystallite size and lattice parameter of nanoparticles can be presented in Table 2. The crystallite size tends to increase with an increase of calcination temperature due to grain growth of nanoparticles. The lattice parameter of nanoparticles does not show certain trend due to deviation of cation distribution in both sites of tetrahedral and octahedral caused by thermal fluctuation. The values are higher than that of theoretical calculation, except for as prepared sample. The theoretical calculation based on the following cation distribution and the equation given by [16],

$$(Mn_{0.4}^{2+}Zn_{0.5}^{2+}Fe_{0.1}^{3+})_{tet}\,[Mn_{0.1}^{2+}\,Fe_{1.9}^{3+}]_{oct}O_4^{2-}.$$

Table 2. Structural parameters of $Mn_{0.5}Zn_{0.5}Fe_2O_4$

Structural Parameter / Sample	AP	T1	T2	T3	T4
Crystallite size (nm) (± 0.2)	21.4	21.9	18.3	24.8	22.3
Lattice parameter (Å)	8.424	8.562	8.540	8.565	8.525
Theoretical Lattice parameter (Å)	8.478	8.478	8.478	8.478	8.478
Difference of lattice parameter (%)	0.63	0.99	0.73	1.03	0.55
Lattice strain ($\times 10^{-4}$ /line^2)	16.19	15.85	18.79	13.98	15.52
Specific surface area ($\times 10^{-1}$ m^2/kg)	53.48	54.97	65.30	48.53	53.14

The lattice strain of nanoparticles tends decrease with an increase of calcination temperature. Thus confirm that the crystallinity of nanoparticle increase. Hence, increase the crystallite size makes the specific surface area of nanoparticles decrease. The highest specific surface area was found in sample of calcination temperature 600°C. Technically, for photocatalysis application sample T2 might perform the best catalytic activity because of their crystallinity, the lowest crystallite size and the highest specific surface area for enhancing photoreaction and pollutant degradation.

3.2 TEM analysis

TEM images of nanoparticles can be presented in Figure 2. TEM images show as prepared nanoparticle experience agglomeration due to small crystallite size that produces high surface energy and surface tension of nanoparticles. The grain size calculated from the image was 32.6 ± 0.8 nm which consistent to the XRD result. In other hand, the TEM image of calcined nanoparticles shows the formation of polydisperse grain with morphology changes from nearly spherical into cubic form. The grain of nanoparticles becomes bigger so that the surface energy was reduced as the increase of calcination temperature. The average grain size of nanoparticles was 59.9 ± 0.6 nm. The SAED patterns for both samples confirm nanoparticles were polycrystalline materials.

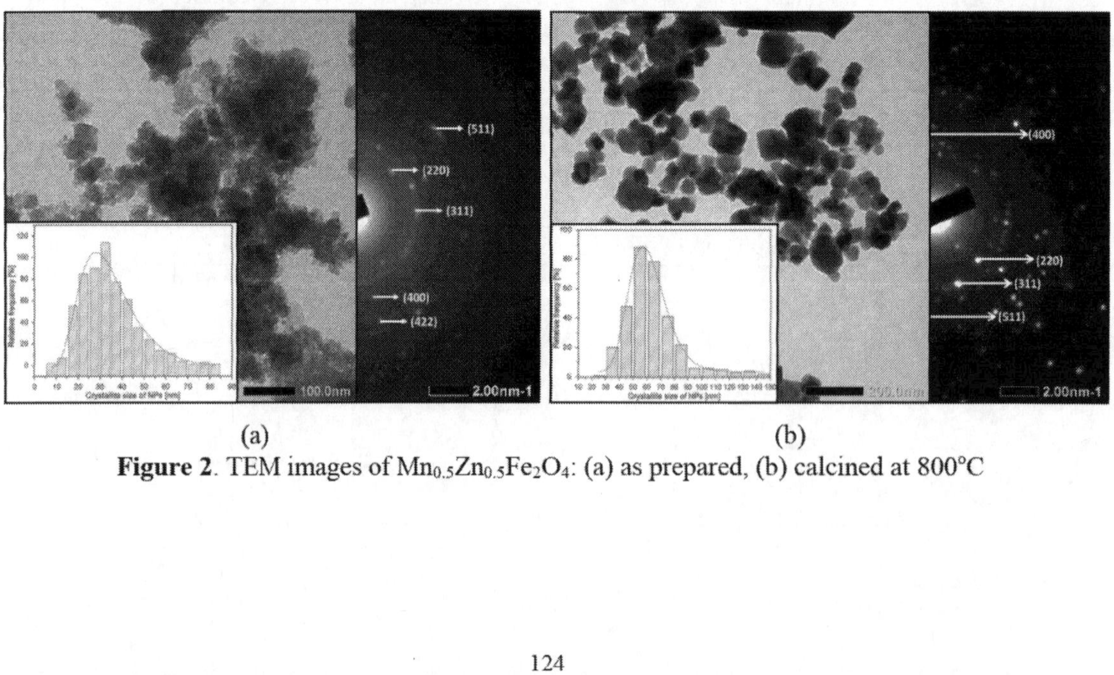

(a) (b)

Figure 2. TEM images of $Mn_{0.5}Zn_{0.5}Fe_2O_4$: (a) as prepared, (b) calcined at 800°C

3.3 FTIR analysis

The FTIR spectra (Figure 3) confirmed the existence of Me_{oct} – O bonds vibration, such as Mn-O and Fe-O stretching bonds in octahedral sites with frequencies of 416.6 cm^{-1} – 455.2 cm^{-1}. The stretching vibration of Me_{tet} – O bonds such as Mn-O; Zn-O and Fe-O were identified in frequencies of 555.5 cm^{-1} – 578.6 cm^{-1}. Increasing the calcination temperature causes shift of frequencies toward smaller values due to the change of cation distribution and bond length between cation- anion in crystal lattice. The Fe-O-H bending vibration occurs in frequencies of 1080.1 cm^{-1} – 1126.6 cm^{-1}. This group might appear from FeO(OH) and hydration of γ-Fe_2O_3 that have been showed in XRD data.

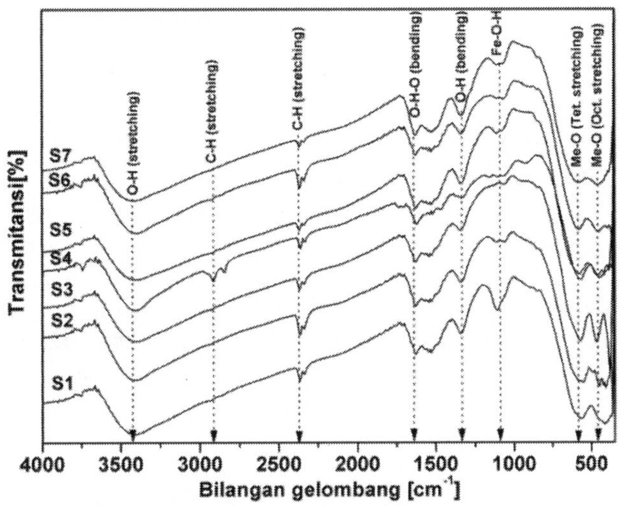

Figure 3. Vibrational spectra of $Mn_{0.5}Zn_{0.5}Fe_2O_4$ nanoparticle for various calcination temperatures

The vibration in frequencies of 1342.5 cm^{-1} and 1627.9 cm^{-1} correspond to O-H and H-O-H stretching bonds which are considered appear from the KBr materials that very hygroscopic to react with oxygen and hydrogen in atmosphere during samples preparation. The appearance of asymmetric stretching vibration of C-H in frequencies of about 2337.7 cm^{-1} and 2916.4 cm^{-1} – 2931.8 cm^{-1} were considered as result of chemical reaction between nanoparticles and CO_2 and H_2 in atmosphere [17].

3.4 VSM analysis

Magnetic parameters of nanoparticles can be presented in Table 3 and Figure 4. The coercivity tends increase with an increase of calcination temperature. The increase in coercivity is a direct result of increase in crystallite size [18]. During calcination nanoparticles experience grain growth and nucleation so that crystal lattice also experience thermal expansion. Thermal energy triggers the driving forces of nanoparticles to decrease their surface energy. The surface spin of nanoparticles will resist their position by enhancing magnetocrystalline anisotropy therefore the coercivity becomes higher. The lower coercivity of sample T3 then T2 might be caused by the existence of ZnO and γ-Fe_2O_3 phases which exhibit diamagnetic and antiferromagnetic properties, respectively.

Table 3. Magnetic properties of prepared and calcined $Mn_{0.5}Zn_{0.5}Fe_2O_4$ nanoparticles

Sample	Crystallite size ($\pm$ 0.2 nm)	*Hc* (Oe)	σ_r (emu/grm)	σ_{max} (15 kOe) (emu/grm)
S4	21.4	47.4	1.0	10.4
T1	21.9	28.2	0.0	24.1
T2	18.3	110.9	0.9	28.7
T3	24.8	104.7	0.5	23.7
T4	22.3	154.5	6.1	54.7

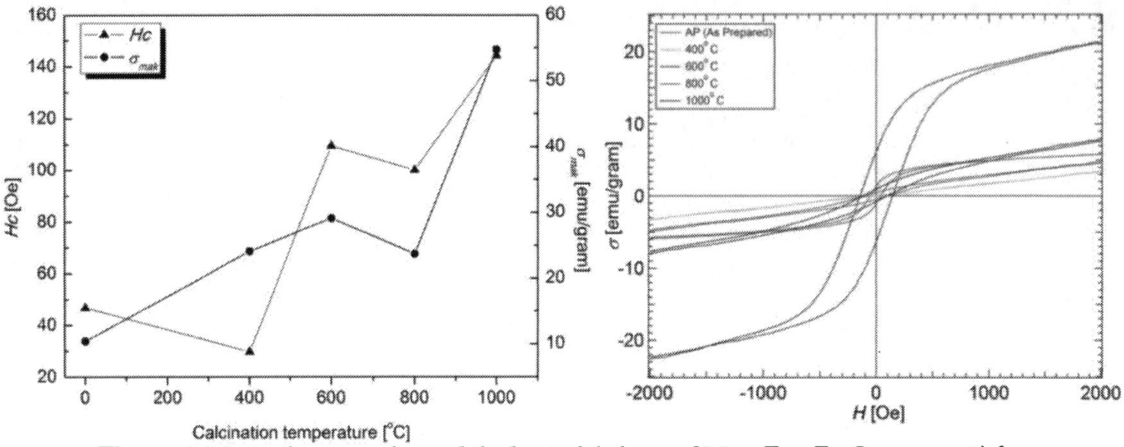

Figure 4. Magnetic properties and the hysterisis loop of $Mn_{0.5}Zn_{0.5}Fe_2O_4$ nanoparticles

Samples of nanoparticles have small specific remanence magnetization due to random canting of particles surface spins [12]. The specific maximum magnetization of nanoparticles tends to increase with an increase of calcination temperature. The values are in range of 10.4emu/gram - 54.7emu/gram. The increase might be attributed by increase of crystallite size that followed by the motion of magnetic domain wall in entire nanoparticles. As the calcination temperature increase, domain wall motion will increase rapidly. The motion of domain wall will contribute to the value of specific maximum magnetisation under external magnetic field of 15 kOe.

4. Conclusion

The $Mn_{0.5}Zn_{0.5}Fe_2O_4$ nanoparticles have been successfully synthesized via coprecipitation. The calcination treatment was proved facilitate the tunability of microstructural, crystal structure, vibrational spectra and magnetic properties of nanoparticles. The calcination treatment of 1000°C results the highest magnetic properties of nanoparticles with the highest quality of microstructure.

5. Acknowledgement

The reported work was financially supported by LPDP Thesis Scholarship 2016 and Nanofabrication Platform Consortium Project of Nagoya University, Minister of Culture, Sports, Science and Technology (MEXT) Nano-Project Platform, Japan of Period 2012 - 2016.

6. References

[1] Choodamani, C., Nagabhushana, G.P and Chandrappa, G.T 2014 *Mater. Lett.* 116 p. 227–230
[2] Gharagozlou, M 2009 *J. Alloys.Compd.* 486 p 660–665
[3] Issa, B., Ihab, M., Obaidat, Albiss, B.A and Haik, Y 2013 *Int. J. Mol. Sci.* 14 p 21266-21305
[4] Ibrahim, I., Ali, I., Salama, T.M., Bahgat and Mohamed, M.M 2016 *App. Catalysis B.* 181 p 389–402
[5] Pang, Y.L., Lim, S., Ong, H.C and Chong, W.T 2006 *Ceramics International.* 42 p 9–34
[6] Naik, P.P., Tangsali, R.B., Sonaye, B and Sugur, S 2015 *J. Magn. Magn. Mater.* 385 p 377–385
[7] Baykal, A., Güner, S and Demir, A 2015 *J. Alloys. Compd.* 619 p 5–11
[8] Gharagozlou, M 2009 *J. Alloys. Compd.* 486 p 660–665
[9] Sharifi, I and Shokrollahi, H 2012 *J. Magn. Magn. Mater.* 324 p 2397–2403
[10] Augustina, M and Balu, T 2015 *Mater. Today: Proc.* 2 p 923 – 927
[11] Reddy, C.V., Vattikuti, S., Ravikumar, Moon, S and Shim, J 2015 *J. Magn. Magn. Mater.* 394 p 70–76
[12] Lan, N.T., Hien, T.D., Duong, N.P and Truong, D.V 2008 J. *the Korean Phys. Soc.* 52 p 1522 – 1525
[13] Amirabadizadeh, A., Farsi, H., Dehgani, M and Arabi, H 2012 *J. Superc. Novel Magn.* 25 p 2763 – 2765
[14] Zak, A.K., Majid, W.H., Abrishami, M.E and Yousefi, R 2011 *Solid State Scie.* 13 p 251-256

[15] Rath, C., Anand, S and Das, R.P 2001 *J. App. Phys.* 91 p 2211 – 2215
[16] Mohammed, K.A., Rawas, A.D., Gismelseed, A.M., and Shongwe, M 2012 *Physica B.* 407 p 795–804
[17] Larkin, P.J 2011 *IR and Raman Spectroscopy: Principles and Spectral Interpretation* (USA: Elsevier) p 38
[18] Vaidyanathan, G., Arulmurugan, R., Likhite, S.D and Anantharaman, M.R 2004 *Ind. J. Eng. Mater. Scie.* 11 p 289 – 294

Influence of CuO nanoparticles and nanographene platelets on the photosonocatalytic performance of Fe_3O_4/TiO_2 nanocomposites

M. Fauzian[1,2], F. F. Harno[1,2], A. Taufik[1,2], R. Saleh[1,2*]
[1]Departemen Fisika, Fakultas MIPA-Universitas Indonesia, 16424 Depok, Indonesia

[2]Integrated Laboratory of Energy and Environment, Fakultas MIPA-Universitas Indonesia, 16424 Depok, Indonesia

E-mail: rosari.saleh@ui.ac.id

Abstract. The effect of adding CuO nanoparticles and Nanographene Platelets (NGP) on Fe_3O_4/TiO_2 nanocomposites to degrade dye waste were examined using photosonocatalytic process. Both nanocomposites Fe_3O_4/TiO_2 with and without the addition of CuO were synthesized using sol-gel method, while the co-precipitation method was used to synthesize those two nanocomposites with NGP. All the samples were analyzed to identify their crystalline phase, magnetic property and thermal stability using X-Ray Diffraction (XRD), Vibrating Sample Magnetometer (VSM) and Thermogravimetric Analysis (TGA) measurements. The photosonocatalytic process of all samples were observed using the ultraviolet and ultrasonic radiation at the same time with the Methylene Blue (MB) as the model of dye waste. The results indicate that the presence of CuO nanoparticles and NGP on Fe_3O_4/TiO_2 nanocomposites could increase its capacibility to degrade MB. Achievement of degradation up to 100 % for a time of 2 hours is obtained by the presence of CuO nanoparticles and NGP simultaneously on Fe_3O_4/TiO_2 nanocomposites.

1. Introduction

Recently, dye waste water is easily found in the textile industry that cause a damage for the environment [1-3]. Therefore it is important to treat dye waste water from textile industry before discharge to the environment. Advance Oxidation Process (AOPs) is very promising in waste water treatment due to its ability to decompose dye waste water into the harmless molecules [4-5]. The AOPs methods that mostly used is photosonocatalytic which combines both photocatalytic and sonocatalytic process at the same time [6-8]. The use of photosonocatalytic process in waste water removal is reported to have better efficiency compared to photocatalytic and sonocatalytic process. This is because of the synergy effect between photocatalytic and sonocatalytic process [7].

Semiconductor materials have been commonly used in the AOPs process such as photocatalytic, sonocatalytic and photosonocatalytic process for waste water removal [8-10]. Titanium Oxide (TiO_2) is very promising materials in the AOPs process because this material is very cheap, environment friendly and active under the exposure of UV (ultraviolet) irradiation [11]. However, studies on the process of waste degradation using the AOPs process such as photosonocatalytic lead to use of material composite as the catalyst. The advantage of this material composite is the synergic effect of each material in the composite that can increase the efficiency of waste degradation through the photosonocatalytic process

[12-13]. The use of Fe_3O_4/TiO_2 nanocomposites have been reported to have better capacibility to degrade waste than the TiO_2 nanoparticle because the recombination rate of electron and hole in TiO_2 is inhibited by the presence of Fe_3O_4 [14].

In addition to Fe_3O_4, other materials which are promising to use with TiO_2 nanocomposites are CuO nanoparticles and carbon nanomaterials such as graphene. The advantage of CuO in the material composite as the catalyst is its ability to inhibit the recombination rate of electron and hole and its ability to extend the absorption toward visible region [15-17]. In addition to having the capacibility to inhibit the recombination rate of electron and hole, graphene also has the ability to expand the surface area of catalyst [18-19]. Therefore, in this study, we investigated the influence of addition of nanoparticles of CuO and Nanographene platelets (NGP) on the ability of Fe_3O_4/TiO_2 nanocomposites to degrade methylene blue (MB) wastes through the photosonocatalytic process.

2. Experimental

The materials used in the synthesis process are among others: Iron (II) sulfate heptahydrate ($FeSO_4$. $7H_2O$, 99%), copper sulfate pentahydrate ($CuSO_4$. $5H_2O$, 99%), titanium dioxide (TiO_2, 99%), sodium hydroxide (NaOH), ethanol, ethylene glycol (EG), and Nanographene Platelets (NGP). All the reagents used do not undergo further purification. Fe_3O_4/TiO_2 and $Fe_3O_4/CuO/TiO_2$ nanocomposites were synthesized under the method already reported in the previous studies [20]. Meanwhile the synthesis process of nanocomposites with NGP addition to Fe_3O_4/TiO_2 and $Fe_3O_4/CuO/TiO_2$ were performed using the co-precipitation method. The NGP was dispersed into the solution of distilled water and ethanol. This solution was afterward put into an ultrasonic bath at the frequency of 40 kHz for 2 hours. Then poured Fe_3O_4/TiO_2 (or $Fe_3O_4/CuO/TiO_2$) nanocomposites into the NGP solution and subsequently stirred the mixture using magnetic stirrer for 1 hour. Afterward, heated the mix at the temperature of 120°C for 3 hours and dry up under vacuum condition for 12 hours.

The crystalline phase and structure of all samples were analyzed using X-Ray Diffraction (XRD) Rigaku Miniflex 600. The magnetic property of the samples were analyzed using Vibrating Sample Magnetometry (VSM) Oxford Type 1.2T. The thermal stability of the samples were analyzed using the Thermogravimetric Analysis (TGA) Rigaku TG/TGA 8121 measurement.

The photosonocatalytic activities of all samples were investigated by observing the degradation of *Methylene Blue* (MB) using ultrasonic bath operated at the frequency of 40 kHz and electrical power of 40 W with ultraviolet (UV) light irradiaton at the same time. First of all, catalyst is dispersed into a beaker containing MB solution with the concentration of 20 mg/L. Prior to the photosonocatalytic process, the solution was stirred with magnetic stirrer under dark condition for 30 minutes to reach the adsorption-desorption equilibrium. The pH of the solution was maintained at pH 13. The photosonocatalytic process were done for 2 hours and with the interval of every 15 minutes the MB solution was analyzed using UV-Vis Spectroscopy to observe the rate of MB degration before and after the process.

3. Results and discussion

Figure 1(a) shows the result of magnetic property measurement using VSM from the samples of Fe_3O_4 nanoparticles, Fe_3O_4/TiO_2, $Fe_3O_4/CuO/TiO_2$ and $Fe_3O_4/CuO/TiO_2/NGP$ nanocomposites. The figure shows that all samples have the ferromagnetic hysteresis curves at room temperature. The magnetic saturation (M-S) of all samples are listed in Table 1. The M-S value of Fe_3O_4 is 82 emu/g, the addition of diamagnetic material of TiO_2 reduces the magnetization value to 65 emu/g. Furthermore, the reduction in magnetization value to 50 emu/g also occurs with the addition of anti-ferromagnetic material of CuO. However the opposite trend happens when NGP is added to $Fe_3O_4/CuO/TiO_2$ nanocomposites, its M-S value increases to 55 emu/g. The risen value in magnetic saturation may due to the stabilize state of Fe_3O_4 in the surface of NGP [21].

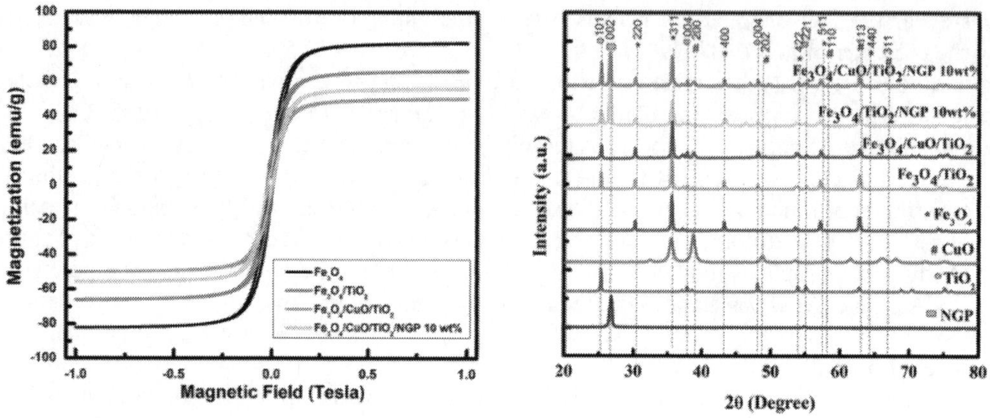

Figure 1. (a) Magnetic curve of pure Fe_3O_4, Fe_3O_4/TiO_2, $Fe_3O_4/CuO/TiO_2$, and $Fe_3O_4/CuO/TiO_2$/NGP 10 wt% nanocomposites, (b) XRD spectra of pure Fe_3O_4, TiO_2, CuO nanoparticles, NGP, Fe_3O_4/TiO_2, $Fe_3O_4/CuO/TiO_2$, Fe_3O_4/TiO_2/NGP and $Fe_3O_4/CuO/TiO_2$/NGP nanocomposites.

Figure 1(b) indicates the result of XRD measurement of the Fe_3O_4/TiO_2, $Fe_3O_4/CuO/TiO_2$, Fe_3O_4/TiO_2/NGP and $Fe_3O_4/CuO/TiO_2$/NGP nanocomposites. As comparison, we include the XRD curves from Fe_3O_4, CuO, and TiO_2 nanoparticles. By comparing the XRD curves, it can be seen that the main peaks in the Fe_3O_4/TiO_2 nanocomposite consists of the cubic spinel phase of Fe_3O_4 and anatase phase of TiO_2. For Fe_3O_4/TiO_2/NGP nanocomposites, there are cubic spinel phase of Fe_3O_4, anatase phase of TiO_2, and graphite-like structure of NGP. For $Fe_3O_4/CuO/TiO_2$ nanocomposite, there are monoclinic phase of CuO present in the XRD curve and also graphite-like structure of NGP for $Fe_3O_4/CuO/TiO_2$/NGP. The XRD curves of all samples confirm that the resulting nanocomposites are synthesized as desired.

TGA measurement from samples of Fe_3O_4/TiO_2, $Fe_3O_4/CuO/TiO_2$, Fe_3O_4/TiO_2/NGP and $Fe_3O_4/CuO/TiO_2$/NGP are shown in Figure 2(a). Both Fe_3O_4/TiO_2 and Fe_3O_4 /CuO/TiO_2 nanocomposites results do not indicate the significant percentage loss. It can be said that both of Fe_3O_4/TiO_2 and $Fe_3O_4/CuO/TiO_2$ nanocomposites are stable until temperature of 1000°C. With the presence of NGP, these nanocomposites have the percentage losses at the temperature of 600-1000°C. Such percentage loss occurs as a result of the combustion process of NGP materials inside these nanocomposites when the temperature is above 600 °C. Together with the XRD curves, TGA results also confirm the presence of NGP in the nanocomposites of Fe_3O_4/TiO_2/NGP and $Fe_3O_4/CuO/TiO_2$/NGP.

Table 1. Magnetic saturation of Fe_3O_4 nanoparticles, nanocomposites of Fe_3O_4/TiO_2, $Fe_3O_4/CuO/TiO_2$, and $Fe_3O_4/CuO/TiO_2$/NGP

Sample	Magnetic saturation
Fe_3O_4	82
Fe_3O_4/ TiO_2	65
$Fe_3O_4/CuO/TiO_2$	50
$Fe_3O_4/CuO/TiO_2$/NGP	55

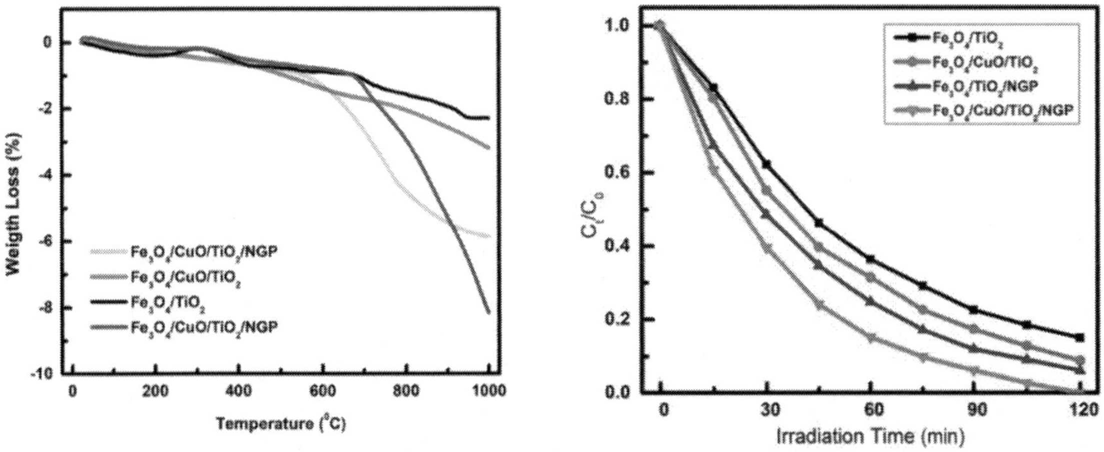

Figure 2. (a) TGA pattern of Fe_3O_4/TiO_2, $Fe_3O_4/CuO/TiO_2$, Fe_3O_4/TiO_2/NGP and $Fe_3O_4/CuO/TiO_2$/NGP nanocomposites, **(b)** Degradation efficiency of Fe_3O_4/TiO_2, $Fe_3O_4/CuO/TiO_2$, Fe_3O_4/TiO_2/NGP and $Fe_3O_4/CuO/TiO_2$/NGP nanocomposites under combination of UV light and ultrasonic irradiation.

Figure 2(b) shows the decolorization process of MB with photosonocatalytic by using Fe_3O_4/TiO_2, Fe_3O_4/TiO_2/NGP, $Fe_3O_4/CuO/TiO_2$, and $Fe_3O_4/CuO/TiO_2$/NGP catalysts with dosage 0.2 g/L. The curves in the figure show that the presence of NGP can improve photosonocatalytic performance of both Fe_3O_4/TiO_2 and $Fe_3O_4/CuO/TiO_2$ nanocomposites. This may be due to the ability of NGP to inhibit the recombination rate of electron and hole since graphene can act as an electron acceptor in the photosonocatalytic reaction [22]. In addition, it can also be seen that the addition of CuO on both Fe_3O_4/TiO_2 nanocomposites with and without NGP improve the performance of both catalyst to degrade MB. Photosonocatalytic performance increase with the presence of CuO due to the formation of heterojunction on TiO_2 coupling with CuO can inhibit recombination of electron and hole [23].

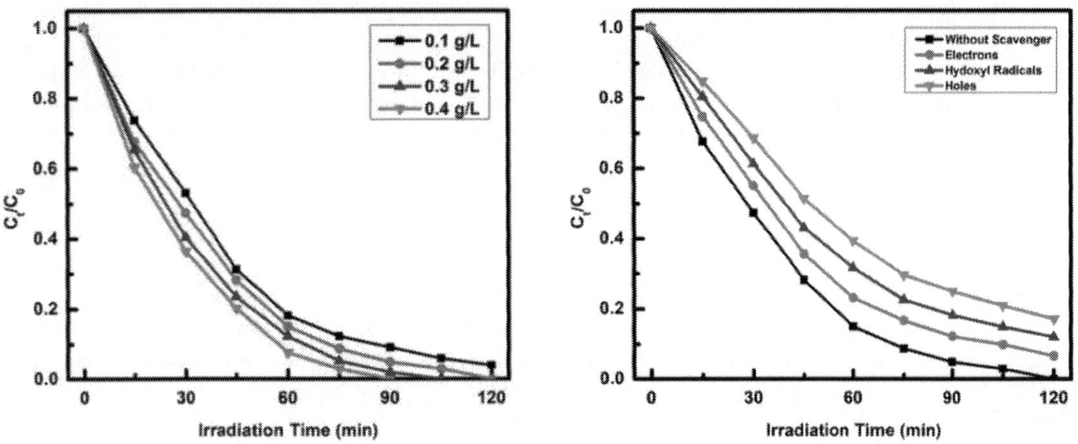

Figure 3. (a) Influence of dosage catalyst and **(b)** Effect of different Scavengers on catalytic activity of $Fe_3O_4/CuO/TiO_2$/NGP.

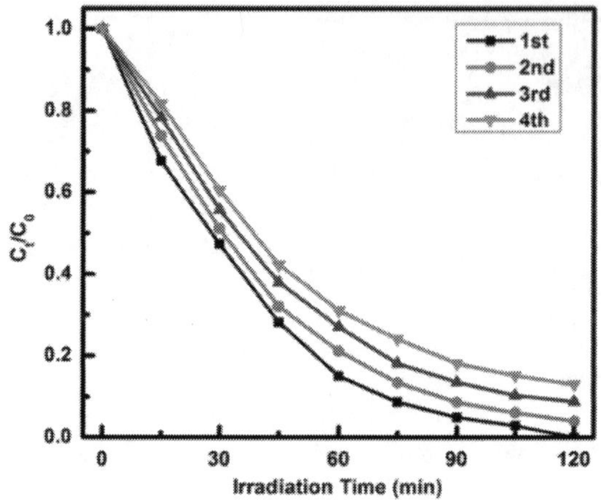

Figure 4. Influence of reusability of catalyst

Photosonocatalytic process in degrading MB for four dosages of the $Fe_3O_4/CuO/TiO_2$/NGP nanocomposites are shown in Figure 3(a). The result shows that the dosage of 0.4 g/L indicates the best photosonocatalytic performance in degrading MB reaches 100% within 90 minutes. Increasing dosage of the catalyst is expected to enlarge the surface area involved in the process [24]. Moreover, the addition of dosage may also increase the formation of electrons and holes that can affect the amount of hydroxyl radicals and superoxide, [24] hence can break the MB molecules.
To examine the species that play an active role in the process of MB degradation, the scavenger materials are added into MB solution to detect the main species that contribute to the photosonocatalytic process. The scavenger materials used are among others tert-butyl alcohol, ammonium oxalate and $Na_2S_2O_8$ each of which plays the role as hydroxyl radicals , holes , and electrons respectively. There are shown in Figure 3(b). The addition of ammonium oxalate scavenger indicates lowest photosonocatalytic activities. It can be seen that ammonium oxalate provoke a large degree of inhibition of MB, which confirmed the indispensable role of holes in the degradation process of MB.
To examine the stability of $Fe_3O_4/CuO/TiO_2$/NGP in photosonocatalytic process, the catalyst is reused four times cycling processes and the results of each performance are shown in Figure 4. $Fe_3O_4/CuO/TiO_2$/NGP catalyst still showed good performance after reuse fourth time and only a slight decline of about 8 % in MB degradation. Decline in catalyst performance is suspected because of its lost when the catalyst is separated from its solution to be reuse. This result proves that the Fe_3O_4/TiO_2 nanocomposites with the presence of CuO and NGP simultaneously has the potential to be used as catalyst to degrade organic dye waste.

4. Conclusion

The presence of CuO and NGP in Fe_3O_4/TiO_2 nanocomposites that have been successfully synthesized using the method of co-precipitation are able to improve the *photosonocatalysis* performance to degrade the organic dye waste methylene blue until 100% by inhibiting recombination rate of electrons and holes in synergy with enlarging their surface areas, and do not reduce much on their potential for reuse.

References

[1] Zhang H,Wei C, Huang Y, Wang J 2016 *Ultrasonics Sonochemistry* **30** 61–69
[2] Salem IA, El-Maazawi MS 2000 *Chemosphere* **41** 1173–118
[3] Fung PC, Huang Q, Tsui SM, Poon CS 1999 *Water Science and Technology* **40** 153-160.
[4] Wang J, Ma T, Zhang ZH, Zhang XD, Jiang YF, Zhang G, Zhao G, Zhao HD, Zhang P 2007 *Ultrasonics Sonochemistry*, **14** 246–252

[5] Suslick KS, Doktycz SJ, Flint EB 1990 *Ultrasonics* **28** 280–290.
[6] Sathishkumar P, Mangalaraja RV, Mansilla HD, Pinilla MAG, Anandan S 2014 *Applied Catalysis B: Environmental* **160–161** 692–700.
[7] Tangestaninejad S, Moghadam M, Mirkhani V, Baltork IM, Salavati H 2008 *Ultrasonics Sonochemistry* **15** 815–822
[8] Khan MAN, Siddique M, Wahid F, Khan R 2015 *Ultrasonics Sonochemistry* **26** 370–377
[9] Reddy DR, Dinesh GK, Anandan S, Sivasankar T, 2016 *Chemical Engineering and Processing* **99** 10–18
[10] Ertugay N, Acar FN 2014 *Applied Surface Science* **318** 121–126
[11] Wang H, Niu J, Long X, He Y, (2008) *Ultrasonics Sonochemistry* **15** 386–392
[12] Li D, Wang J, Li X, Liu H 2012 *Materials Science in Semiconductor Processing* **15** 152–158
[13] Ahmad M, Ahmed E, Hong ZL, Ahmed W, Elhissi A, Khalid NR 2014 *Ultrasonics Sonochemistry* **21** 761–773
[14] Abbas M, Rao BP, Reddy V, Kim C 2014 *Ceramics International* **40** 11177–11186
[15] Samad A, Furukawa M, Katsumata H, Suzuki T, Kaneco S 2016 *Journal of Photochemistry and Photobiology A: Chemistry* **325** 97–103
[16] Liu Z, Bai H, Xu S, Sun DD, (2011) *International journal of hydrogen energy* **36** 13473-13480
[17] Subhan MA, Uddin N, Sarker P, Azad AK, Begum K 2015 *Spectrochimica Acta Part A: Molecular and Biomolecular Spectroscopy* **149** 839–850
[18] Fan C, Liu Q, Ma T, Shen J, Yang Y, Tang H, Wang Y, Yang J 2016 *Ceramics International* **42** 10487–10492
[19] Li Y, Zhang Z, Pei L, Li X, Fan T, Ji J, Shen J, Ye M 2016 *Applied Catalysis B: Environmental* **190** 1–11
[20] Arifin SA, Jalaludin S, and Saleh R 2015 *Materials Science Forum* **827** 49-55
[21] Heidari EK, et al. 2015 *Journal of Magnetism and Magnetic Materials*, **379** 7906-7915
[22] Vinothkannan M, Karthikeyan C, Kumar GG, Kim AR, Yoo DJ 2015 *Spectrochimica Acta Part A: Molecular and Biomolecular Spectroscopy*, **135** 256-264
[23] Qin S, Xin F, Liu Y, Yin X and Ma W 2011 *Journal of Colloid and Interface Science* **357** 257-261,
[24] Saleh R and Djaja N, 2014.*Superlattices and Microstructures* **74** 217-233,

FTIR and structural properties of co-precipitated cobalt ferrite nano particles

E. Hutamaningtyas[a], Utari[a], Suharyana[a], A. T. Wijayanta[b] and B. Purnama[a1]
[a]Department of Physics, Faculty of Mathematics and Natural Sciences,Universitas Sebelas Maret, Jl. Ir. Sutami 36A Kentingan Surakarta 57126, Indonesia
[b]Department of Mechanics, Faculty of Engineering, Universitas Sebelas Maret, Jl. Ir. Sutami 36A Kentingan Surakarta 57126, Indonesia

E-mail: bpurnama@mipa.uns.ac.id

Abstract. The FTIR and structural properties in co-precipitated cobalt ferrite ($CoFe_2O_4$) nanoparticles are discussed in this paper. The synthesis was conducted at temperatures of 75°C and 95°C following post annealing at 1200°C for 5 hours. Other modification samples were synthesis at temperature of 95°C and then annealing at temperature of 1000°C and 1200°C for 5 hours. For both modification of synthesis and annealing temperature, FTIR result showed a metal oxide at a wave number of 590 cm^{-1} which indicated cobalt ferrite nanoparticles. The crystalline structure was confirmed using x-ray diffraction that the high purity of cobalt ferrite was realized. Calculation of the cation distribution by using comparison I_{220}/I_{222} and I_{422}/I_{222} show that the synthesis and annealing temperature succesfully modify cation occupy the site octahedral and tetrahedral.

1. Introduction

Study of ferrite based nano particles show great progress in the past decade due to the interesting techniques and developments to achieve the nano-meter order of magnetic particles. At the nanoscale, there is new behaviors and properties which is different from bulk structures [1-3]. Especially in ferrite which is materials with high chemical stability and low production cost already used for many application field [4]. Ferrite based nanoparticles currently used in application such as optical materials [5], semiconductors [6], nanofluids [7-9], and at the superparamagnetic stage [10] used for magnetic data storage [11], magnetic imaging, and biomedical applications [12].

Ferrite is spinel structured materials which is isostructural with the mineral $MgAl_2O_4$ and have general molecular formula (M^{2+}) $[Fe_2{}^{3+}]$ $O_4{}^{2}$, with M^{2+} is divalent cation and Fe^{3+} is trivalent cation [13]. Its crystallises is face centered cubic containing eight formula unit $[AB_2O_4]_8$ or 56 ions per unit cell. The lattice consist of 32 divalent oxygen arrange with 64 tetrahedral interstitial and 32 octahedral interstitial sites (just 8 tetrahedral and 16 octahedral sites are occupy by divalent and trivalent cations) [14]. In this large empty interstitial sites, there is possibility to cation migration [15]. For normal spinel materials such as $ZnFe_2O_4$, all the divalent cations occupy the tetrahedral sites (A-sites) and all the trivalent cations occupy the octahedral sites (B-sites). In contrast for inverse spinel such as $MgFe_2O_4$, trivalent cations are distributed at the A-sites and B-sites, whereas divalent cations have preference for B-sites. Distribution of cations in each lattice sites form materials as inverse or partially

[1] To whom any correspondence should be addressed.

inverse [16]. The distribution of cations over the two interstitial sites is strongly affected by the variations of lattice constants, size of crystallites, and bond angles-bond length between the cations [17].

One of the most interesting ferrite is cobalt ferrite ($CoFe_2O_4$) due to their strong magnetocrystalline anisotropy, high coercive strength and large magneto-optic coefficient [18,19]. Cobalt ferrite is a spinel structured materials which is known that the Co^{2+} ions have preference for the B-sites [14]. Cobalt ferrite may be complete inverse spinel or partial inverse spinel depend on cation subtitution or heat treatment [16]. The cation distribution between Co^{2+} ions and Fe^{3+} ions over the A-sites and B-sites effect on the physical properties of cobalt ferrite spinel. In cobalt ferrite, the preference of cations is not fixed but cobalt ions show a strong preference for B-sites. The information about cation distribution can be achieved from the peaks of X-rays diagrams [20].

This study focused on the variation of heat treatment (synthesis and high annealing temperature) and a co-precipitation procedure was used to synthesize the cobalt ferrite nanoparticles due to its high simplicity procedure, low cost and good grain size control [21]. The resulting cobalt ferrite nanoparticles were then characterized using FTIR to study their infrared spectra and X-ray diffaction to study their structural properties and cation distribution.

2. Experimental procedure

2.1. Synthesis of $CoFe_2O_4$ nanoparticles

The synthesis of cobalt ferrite nanoparticles was prepared by co-precipitated method like as in the previous study [22]. The synthesis was conducted at temperatures of 75°C and 95°C, where the aqueous solution of 4.8 M NaOH was added dropwise to the mixed solution. The product was then cooled and washed several times with ethanol and deionised water. Precipitated solution were dried overnight at 100°C in order to remove excess water. The following post annealing has been done at 1200°C for 5 hours. Other modification samples were synthesis at temperature of 95°C and then annealing at temperature of 1000°C and 1200°C for 5 hours. The obtained samples were then characterized.

2.2. Characterization

The formation of spinel structure was characteized with Fourier Transform Infra-Red (FTIR). The X-ray diffraction (XRD) patterns of the samples were recorded with Bruker D8 Advanced system X-ray powder diffractometer using Cu Kα source (1.54 Å) radiation. The scans of the selected diffraction peaks were carried out in the step mode 0.064°.

2.3. Calculations

The force constant for the tetrahedral site (k_t) was calculated according to the following relation:

$$k_t = 7{,}62 M_A v_1^2 \times 10^{-7} \frac{N}{m} \tag{1}$$

where M_A is the molecular weights of cations on the A-sites and v_1 is wave number (cm^{-1}).

For the calculation of the crystallite size, Scherrer's method was applied using the following equation:

$$D = \frac{0.9\lambda}{\beta\cos\theta} \tag{2}$$

where D is crystallite size, λ is the wavelength of the X-ray radiation, θ is Bragg's angle, and β is the line width at half maximum height.

The lattice constant 'a' was determined using the following relation:

$$a = \frac{\lambda}{2\sin\theta}\sqrt{h^2 + k^2 + l^2} \tag{3}$$

where (h k l) are the miller indices. The X-ray density (d_x) was calculated according to the following relation:

$$d_x = \frac{Z.M}{N_A \cdot a^3} \tag{4}$$

where Z is number of molecule per unit cell, M is molecular weight, and N_A is Avogadro's number [21].

3. Results and Discussion

3.1 Effect of Synthesis Temperature

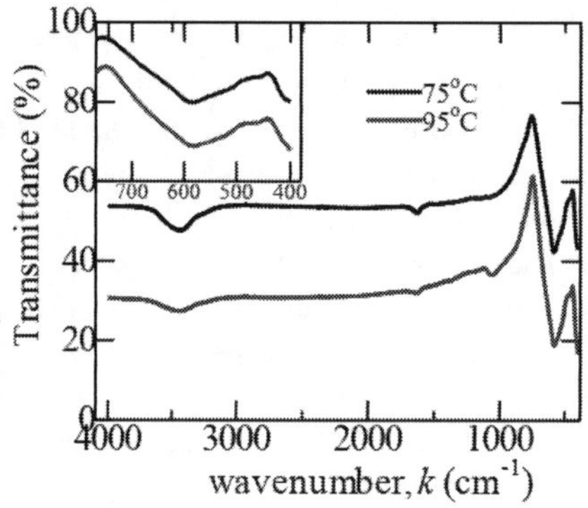

Figure 1. FTIR spectra of the cobalt ferrite nanoparticles synthesized at temperature of 75°C and 95°C and annealed at 1200°C

The formation of the spinel structure of cobalt ferrite and its cation distribution is supported by FTIR analysis. Figure 1 presents the infrared transmittance spectra of the investigated sample cobalt ferrite which vary in the synthesis temperature. The absorption band around 3400 cm^{-1} and 1600 cm^{-1} are ascribed to the stretching mode of O-H and H-O-H bending vibration, respectively [23]. The other absorption band lies at a smaller wavenumber (see insert). The absorption band at this small wavenumber is showed in Table 1 and is the common feature in spinel ferrite oxide. The absorption band are attributed to the stretching vibration of tetrahedral group complexes [23]. This result confirms the appearance of $CoFe_2O_4$ synthesized by co-precipitation method.
The absorption band at wavenumber around 590 cm^{-1} are slighty differ depend on the variations of synthesis temperature. At the variations of synthesis temperature (annealing temperature is constant at 1200°C), from 75°C to 95°C result on the slighty shifted absorption band to a lower frequency side. The boardening of the A-sites and B-sites attribute the different of force constant at tetrahedral sites (k_t). The value of force constants are listed in Table 1. It is observed that the force constant of tetrahedral sites (k_t) is equal 14.65 N/m for synthesis temperature of 75°C and at synthesis temperature of 95°C, the k_t decrease become 14.35 N/m. The change of the k_t is expected to modify the cation-oxygen bond length [21]. Furthermore it will modify the crystalline structure of the obtained cobalt ferrite nanoparticles.

Table 1. FTIR and XRD result of sample with varied synthesis temperature

Synthesis Temperature (°C)	wavenumber (cm^{-1})	k_t (N/m)	2θ (°)	Crystallite size (nm)	a (Å)	d_x (g/cm^3)
75	586	14.65	35.47	43	8.38	5.29
95	580	14.35	35.53	48	8.37	5.31

The X-ray diffraction patterns of the $CoFe_2O_4$ synthesized by co-precipitation vary in synthesis temperatures are shown in Fig. 2. Here, the solid line show the experimental data and the bottom line show the data peaks of the International Centre for Diffraction Data (ICDD) number 221086. This characterization shows a similarity to the data peaks of the ICDD number 221086. These result indicate that the samples are $CoFe_2O_4$ in inverse spinel face centered cubic (*fcc*) structure. The similarity in the data peaks also attribute that there is no other impurities present and the $CoFe_2O_4$ with high purity was successfully produced via co-precipitation method. All the peaks of the XRD pattern were indexed using Bragg's law. The crystallite sizes of cobalt ferrite are becoming larger with the increasing of temperature are showed in Table 1. The crystallite size increase from 43 nm to 95 nm for the change of temperature synthesis 75°C to 95°C. Whereas the lattice parameter a is not significantly different.

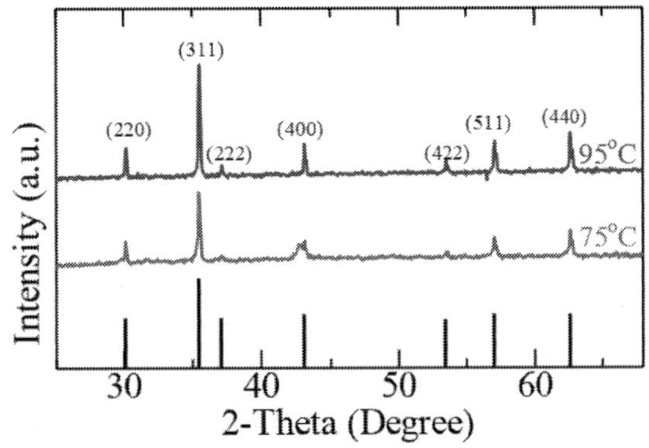

Figure 2. XRD patterns for $CoFe_2O_4$ synthesized at temperature of 75°C and 95°C and annealed at 1200°C

3.2 Effect of Annealing Temperature

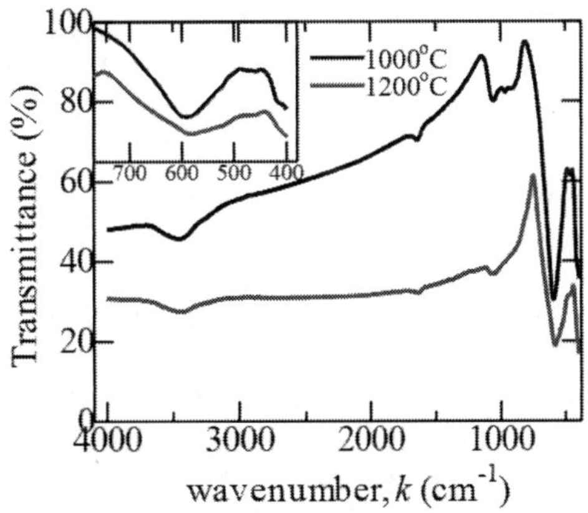

Figure 3. FTIR spectra of the cobalt ferrite nanoparticles synthesized at 95°C and annealed at temperature of 1000°C and 1200°C

Figure 3 presents the infrared transmittance spectra of the cobalt ferrite samples which vary in the annealing temperature. The absorption band around 3400 cm^{-1} and 1600 cm^{-1} are ascribed to the stretching mode of O-H and H-O-H bending vibration, respectively. Similar with the result of synthesis temperature variation, the next absorption band lies at a smaller wavenumber (see insert) is the common feature in spinel ferrite oxide. The absorption band are attributed to the stretching vibration of tetrahedral group complexes [23]. This result confirms the appearance of $CoFe_2O_4$ synthesized by co-precipitation method.

At the variations of annealing temperature (synthesis temperature is constant at 95°C), from 1000°C to 1200°C result on the slighty shifted absorption band to a lower frequency side which is attributed to the increase in the boardening of the A-sites and B-sites. This change show in the Figure 3 where the sample annealed at 1200°C has broad absorption band compared with the sample annealed at 1000°C which has sharper absorption band. The broadening of this band also related with the migration of cations and bond length for A-sites and B-sites complexes. The value of force constants for different annealing temperature are listed in Table 2. It is observed that the force constant of tetrahedral sites (k_t) is at higher level for sample annealed at 1000°C. This result consistent with variation of synthesis temperature. This behaviour show that with the increasing of synthesis or annealing temperature, k_t

can be attributed to the variation in cation-oxygen bond length [21]. To provide more detailed explanation on crystalline structure, samples were characterized by XRD.

Table 2. FTIR and XRD result of sample with varied annealing temperature

Annealing Temperature (°C)	wavenumber (cm^{-1})	k_t (N/m)	2θ (°)	Crystallite size (nm)	a (Å)	d_x (g/cm^3)
1000	592	14.95	35.67	32	8.35	5.36
1200	580	14.35	35.53	48	8.37	5.31

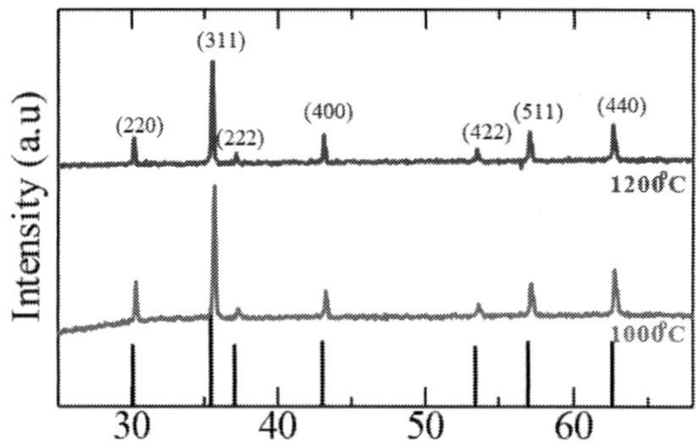

Figure 4. XRD pattern of $CoFe_2O_4$ synthesized at 95°C and annealed at temperature of 1000°C and 1200°C

The X-ray diffraction patterns of the $CoFe_2O_4$ synthesized by co-precipitation which vary in annealing temperatures are shown in Figure 4. This characterization shows a similarity to the data peaks of the ICDD number 221086. These result indicate that the samples are $CoFe_2O_4$ in inverse spinel face centered cubic (*fcc*) structure. The similarity in the data peaks also attribute that there is no other impurities present and the $CoFe_2O_4$ with high purity was successfully produced via co-precipitation method. All the peaks of the XRD pattern were indexed using Bragg's law. The increment of crystallite size also happen at sample which vary in their annealing temperature. The lattice parameter and density of sample have different behaviour to their varied temperature shows in Table 2. For sample that vary in their annealing temperature show the counter behaviour from synthesis temperature due to growth of the crystal with increasing annealing temperature. These two different behaviour reflect the different cation distribution in the sample.

3.3 Cation distribution

From the Figure 2 and Figure 4 the relative intensity of peaks with Miller indices (*h, k, l*) all even such as (220), (222), and (422) may be used to obtain information about sites occupacy. For all even with formula $h+k+l=4n$, like (220) and (422), only tetrahedral sites (8a) contribute to diffraction peaks. In contrast, for all even with formula $h+k+l=4n+2$, like (222), only octahedral sites (16d) contribute to diffraction peaks [20]. The ratio of $I_{220}/I_{222}.I_{422}/I_{222}$ are an indication of the A and B sites occupation is showed in Table 3. For these three results of comparison intensity ratios indicates clearly that in the nanoparticles especially due to the variation of heat treatment, Co^{2+} has a strong preference for tetrahedral sites, contrary with the bulk ferrite.

Table 3. I_{220}/I_{222} and I_{422}/I_{222} ratio of cobalt ferrite vary in synthesis and annealing temperature

Temperature (°C)	I_{220}/I_{222}	I_{422}/I_{222}
75-1200	3.02	1.49
95-1200	2.85	1.52
95-1000	3.48	1.24

For sample which vary in their synthesis temperature, the increasing of temperature effect to the I_{220}/I_{222} is getting smaller (from 3.02 to 2.85 for sample 75 and 95, respectively) hence I_{422}/I_{222} is getting bigger (from 1.49 to 1.52 for sample 75 and 95, respectively). This result can explain that with increasing of synthesis temperature, there is redistribution of Co^{2+} ions from A-sites back to B-sites. Same behaviour occurs at the sample which vary in their annealing temperature. Thus, it can be conclude that the increasing of temperature by heat treatment effect on the redistribution of Co^{2+} ions from A-sites back to B-sites.

4. Conclusion

$CoFe_2O_4$ nanoparticles were successfully synthesized using co-precipitation method. FTIR result indicated the sample was spinel cobalt ferrite at wave number around 590 cm^{-1}. The crystalline structure was confirmed using x-ray diffraction that the high purity of cobalt ferrite was realized. Calculation of the cation distribution by using comparison I_{220}/I_{222} and I_{422}/I_{222} showed that the synthesis and annealing temperature succesfully modify cation occupation on the tetrahedral and octahedral sites.

Acknowledgments
This research was funded by DIPA Universitas Sebelas Maret of Republic of Indonesia (Penelitian Unggulan Perguruan Tinggi Contract No. 353/UN27.21/PN/2016).

References

[1] Morrish AH and Li ZW 2001 *Int. J. Mod. Phys. B* **15** 3312
[2] Velasquez EA, Altbir D, Mazo–Zuluaga J, Duque LF, and Mejia–Lopez J 2013 *JMMM* **348** 154
[3] Vinas SL, Simeonidis K, Li ZA, Ma Z, Myrovali E, Makridis A, Sakellari D, Angelakeris M, Wiedwald U, Spasova M, and Farle M 2016 *JMMM* **doi**: 10.1016/j.jmmm.2016.02.098.
[4] Mathew DS and Juang R-S 2007 *Chem. Eng. J.* **129** 51
[5] Yoon SJ, Murthy A, Johnston KP, Sokolov KV, and Emelianov SY 2012 *Opt. Express* **20** 29479
[6] Chen YF, Spoddig D, and Ziese M 2008 *J. Phys D: Appl. Phys.* **41** 205004
[7] Devendiran DK and Amirtham VA 2016 *Renew. Sust. Energy Rev.* **60** 21
[8] Lu A-H, Salabas EL, and Schuth F 2007 *Angew. Chem. Int. Ed.* **46** 1222
[9] Behrens S, Bonnemann H, Matoussevitch N, Dinjus E, Modrow H, Palina N, Frerichs M,… and Guldbakke M 2006 *Phys. Chem.* **220** 3
[10] Jeppson P, Sailer R, Jarabek E, Sandstrom J, Anderson B, Bremer M, Gner DG, Schulz DL, and Caruso N 2006 *J. Appl. Phys.* **100** 114324
[11] Wu L, Jubert P-O, Berman D 2014 *Nano Lett.* **14** 3395
[12] Jang J-T, Nah H, Lee J-H, Moon SH, Kim MG, and Cheon J 2009 *Angew. Chem. Int. Ed.* **48** 1234
[13] Nlebedim IC, Ranvah N, Williams PI, Melikhov Y, Snyder JE, Moses AJ, and Jiles DC 2010 *JMMM* **322** 1929
[14] B. D. Cullity, *Introduction to magnetic materials,* 2009, Wiley, New Jersey.
[15] Kumar L, Kumar P, Narayan A, and Kar M 2013 Int. *Nano Lett.* **3** 8
[16] Gomes JA, Sousa MH, Tourinho FA, Mestnik-Filho J, Itri R, and Depeyrot J 2005 *JMMM* **289** 184.
[17] Jacob J and Khadar MA 2010 *J. Appl. Phys.* **107** 114310
[18] Wang L, Li J, Liu M, Zhang YM, Lu JB, and Li HB 2012 *JMMM* **324** 4200
[19] Lee SJ, Song SH, Lo CC, Aldini ST, and Jiles DC 2007 *J. Appl. Phys.* **101** 09C502
[20] Ajroudi L, Mliki N, Bessais L, Madigou V, Villain S, and Leroux CH 2014 *Mat. Research Bull.* **59** 49
[21] Amiri S and Shokrollahi H 2013 *JMMM* **345** 18
[22] Purnama B, Rahmawati R, Wijayanta AT, and Suharyana 2015 *J. Magn.* **20** 207
[23] Safi R, Ghasemi A, Shoja–Razavi R, and Tavousi M 2015 *JMMM* **396** 288

Journal of Physics: Conference Series **776** (2016) 012024 doi:10.1088/1742-6596/776/1/012024

Investigation on magnetic field dependent modulus of epoxidized natural rubber based magnetorheological elastomer

N. A. Yunus[1], S. A. Mazlan[1], Ubaidillah[1,2], S. A. A. Aziz[1], M. H. Ahmad Khairi[1], N. A. A. Wahab[1] and S. T. Shilan[1]

[1]Vehicle System Engineering Research Laboratory, Malaysia – Japan International Institute of Technology, Universiti Teknologi Malaysia, Jalan Sultan Yahya Petra, 54100, Kuala Lumpur, Malaysia

[2] Mechanical Engineering Department, Faculty of Engineering, Universitas Sebelas Maret, Jl. Ir. Sutami 36A, Kentingan, Surakarta, 57126, Central Java, Indonessia

E-mail: azhani3510@yahoo.com

Abstract. This paper presents an investigation on the use of epoxidized natural rubber (ENR) as a matrix of magnetorheological elastomers (MREs). Isotropic ENR-based MRE samples were synthesized by homogeneously mixed the ENR compound with carbonyl iron particles (CIPs). The microstructure of the sample was observed, and the magnetic field-dependent moduli were analyzed using rheometer. The influences of excitation frequency, CIPs content and magnetic field on the field-dependent moduli of ENR-based MREs were evaluated through dynamic shear test. The microstructure of MRE samples demonstrated the dispersed CIPs in the ENR matrix. The remarkable increment of storage and loss moduli of the ENR-based MREs has exhibited the magnetically controllable storage and loss moduli of the samples when exposed to the magnetic field. Consequently, the CIPs content, frequency and magnetic field were significantly influenced the dynamic moduli of the ENR-based MREs.

1. Introduction

Magnetorheological elastomers (MREs) have been attracting research interests since rheological properties of MREs corresponding to field-dependent modulus can be controlled by adjustment of magnetic field. The MREs comprise of rubber elastomer and magnetizable particles which contribute to no leakage and settlement of particles as occur in magnetorheological fluids (MRFs) [1]. These benefits provide a wide range of potential applications for MREs, particularly in any applications that use rubber to control vibration and noise such as engine mount, building isolator, prosthetic device, tunable vibration absorber and etc. The magnetizable particles are frequently discussed which refer to the content, shapes and sizes of the magnetic particles. Results showed that all of these parameters have greatly influenced the performance of MREs. Soft spherical magnetizable particles with micronsized are the most common particles that studied by researchers as these particles have low magnetic remnant, high magnetic permeability and magnetic saturation [2-3]. These characteristics contribute to high magneto-induced storage modulus. In terms of magnetic particle dispersion, the MREs can be classified into two categories which are known as anisotropic and isotropic MREs. The anisotropic

MRE is denoted as the one that contains aligned chain-like structure of magnetic particles as the magnetic field is applied during pre-configuration. In case of isotropic MRE, the magnetic particles are uniformly distributed in the matrix. The fabrication of the anisotropic MREs requires modification of the conventional rubber molding equipment's for pre-configuration stage and the chain direction must be taken into consideration for utilizing in any devices [4]. Due to this matter, the isotropic MRE was chosen in this study.

Despite of magnetic particles, previous studies also revealed that the selection of materials for MRE matrix is crucial. Many efforts on synthesizing the MREs using different types of rubber materials such as silicone rubber (SiR), natural rubber (NR), polyurethane (PU), cis-polybutadiene rubber (BR), nitrile rubber (NiR) and thermoplastic elastomer have been performed so as to implement the MREs in those applications [2]. However, fabrication of the MREs that encompasses desirable stiffness and damping properties are the most challenging part in this field. Compared to other types of matrices, more studies have been carried out on using the SiR as MRE matrix due to unnecessary of vulcanization. It is known as a soft elastomer matrix that can reach a high relative MR effect, but the magneto-induced storage modulus is not large enough which confines their practical applications. Other than SiR, the NR-based MREs were also showed interesting outcomes corresponding to higher magneto-induced storage modulus and degradation stability than the SiR [5]. Nevertheless, poor temperature, weather and oil resistance of NR lead to synthesis of the MREs containing dissatisfactory properties [2, 6].

Therefore, the researchers have chemically amended the NR through a process recognized as epoxidation in order to enhance the properties of the NR. The epoxidation process occurs in such a way that NR latex are chemically reacted with hydrogen peroxide or perametic acid, and change the carbon-carbon double bonds into epoxy groups [7]. Epoxidized natural rubber (ENR) is the product of this process, and composed of a combination of natural and synthetic rubber properties. For instance, the ENR has high strength due to their ability to undergo strain crystallization and high degree of damping which are anticipated for synthesizing the MREs [8]. In fact, the epoxidation process improves the compatibility of NR with polar polymers and wears resistance. Similar to the NR, the ENR is also considered as eco-friendly material that derived from renewable resources and lead to less carbon footprint. With growing concerns of global climate change and because of its unique properties, the ENR becomes more prominent recently. Nonetheless, the ENR is not widely explored and there has been a few works focusing on utilizing the ENR as MREs matrix [9]. Therefore, the objective of this study is to investigate and analyze the field-dependent storage and loss moduli of the ENR-based MREs under shear deformation. A total of three isotropic ENR-based MREs were fabricated without application of magnetic field. The effect of magnetizable particles content, excitation frequency and magnetic field will be discussed in details.

2. Experimental Procedure

2.1. Fabrication of MRE and Characterization

Epoxidized natural rubber (ENR) 25 and carbonyl iron particles (CIPs) were used as MRE matrix and magnetizable particles in this study. The ENR 25 has a glass transition temperature of -45 °C, and the density of CIPs is 7.86 g/mL. Generally, the MRE samples were prepared by three stages of mixing process; (a) addition of additives that include ZnO, stearic acid, carbon black and aromatic oil, (b) addition of sulfur and CBS, and finally (c) addition of CIPs. The mixing process was performed via conventional double-roll mill. After mixing, the mixture undergo curing process without magnetic field which was carried out at 150 °C for 15 minutes by compressing the mixture in a mold.

The microstructure observation was carried out using Field Emission Scanning Electron Microscopy (FESEM, Supra 40VP-31-31, Germany). The cross-section of the selected sample was observed at 500x and 2000x magnifications using voltage acceleration of 20 kV. The samples were treated with gold coating prior to the observation.

The rheological measurements were performed experimentally using a Rotational Rheometer (MCR 302, Anton Paar Company, Germany) equipped with MRD 70, which can produce the magnetic field perpendicularly to the shear flow in the test area. A temperature control device (Viscotherm VT2, Anton Paar, Germany) was utilized for controlling the measure ng temperature to be at 25 °C. The samples with diameter of 20 mm were placed between a rotary disk and a parallel base plate. In this study, an oscillatory shear was used for the entire rheological experiments.

The strain sweep test was performed in order to identify the linear viscoelastic region of a sample prior to frequency sweep measurement. The measurement was done by varying the strain amplitude from 0.1 to 10% at a constant frequency of 1 Hz for MRE sample that contained 50 wt% CIPs. The range of frequency at 0.1 to 100 Hz was applied for frequency sweep measurement. Meanwhile, the strain amplitude was kept constant at 0.4% in case of all samples. In order to analyze the effect of magnetic field, the current was increased from 0 to 730 mT by increment of approximately in the range of 150-350 mT at each measurement. The magnetic field values for each current were recorded using teslameter. The rheological measurements were conducted at room temperature.

3. Results and discussion

3.1. Microstructure Observation

Figure 1(a) and (b) illustrate the microstructure of the ENR-based MRE for 50 wt% CIPs at 500x and 2000x magnifications, respectively. Noticeably, the CIPs were available in spherical and white gray color, while the ENR matrix was black in color. From Figure 1(a), it is apparent that the CIPs were scattered uniformly in the ENR matrix, but some agglomerated of CIPs were also observed. In addition, the porosities were also seen in the sample. The porosities were occurred due to a chemical reaction amongst ENR, additives and CIP during vulcanization at high temperature. Based on Figure 1(b), the average diameter of CIPs was determined to be at approximately 3-4 μm. It can be concluded that the CIPs tend to adhere to the ENR matrix and subsequently played a crucial role in inducing the magnetic flux density in the MRE samples. In other words, the magnetic forces of CIPs in the ENRbased MREs were dependent on the adhesion and distance between CIPs which will be discussed in next section.

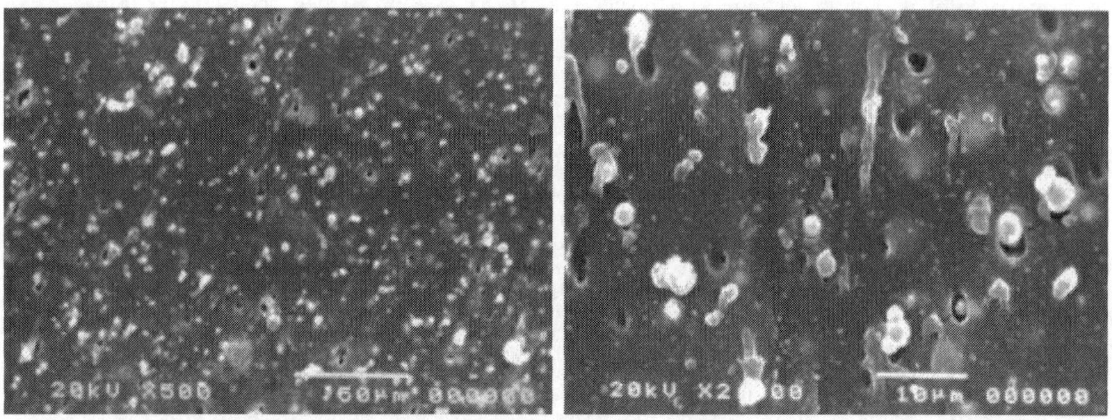

Figure 1. Microstructure observationsat: (a) 500x and (b) 2000x of magnificationfor ENR-based MRE samples containing 50 wt%.

3.2. Rheological properties

3.2.1. Effect of Strain Amplitude

The MREs are classified as viscoelastic materials. In the viscoelastic materials, some part of the shear deformation energy is stored temporarily which could be recovered after each cycle, and the rest of the energy is dissipated as heat. The ability of storing the energy temporarily is referred as stiffness and called as storage modulus (G'). In the meantime, the viscous properties or loss modulus (G") represents the capability of dissipating the deformation energy in the form of heat. Fundamentally, the G' and G" are parameters which describe the rheological properties of MREs. In this paper, selected plots of storage and loss moduli as a function of strain amplitude and oscillation frequency are described without and with the application of magnetic field.

Figure 2 displays the influence of strain amplitude on the storage and loss moduli of ENR-based MRE sample with 50 wt% of CIPs content. Based on this figure, it is apparent that the ENR-based MREs were dependent on strain. Similar to other viscoelastic materials, the storage modulus and loss modulus of the ENR-based MREs was constant up to certain strain amplitude which is signified as linear viscoelastic region. Subsequently, the storage modulus was diminished while the loss modulus was increased as shear strain amplitude decreased. At this point, the structure of polymer network in the samples started to disrupt and this region is called as non-viscoelastic region. Therefore, it is significant to determine the viscoelastic region to avoid from measuring the frequency sweep at region whereby the structure was damaged, and MRE materials become more viscous rather than elastic solid. The viscoelastic region of ENR-based MRE samples was up to approximately 0.4% of strain amplitude.

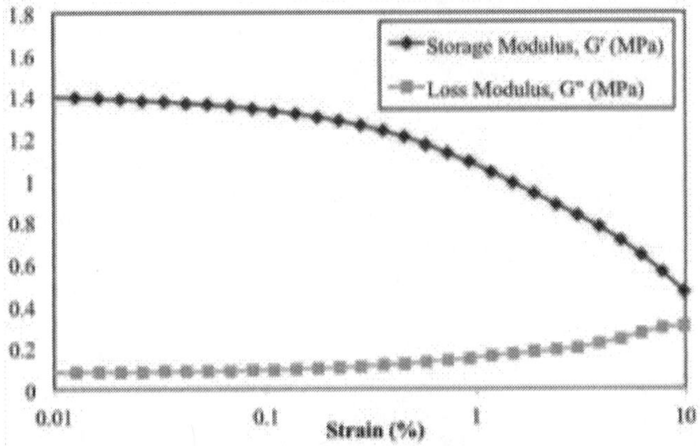

Figure 2. The effect of strain amplitude on the storage and loss modulus of ENR-based MRE for 50 wt% CIPs

3.2.2. *Effect of Frequency*

Figure 3 (a)-(d) demonstrate the effect of frequency on the storage modulus and loss modulus at various CIPs content and magnetic field density. These figures apparently depicted that both moduli were increased with increasing of frequency. As frequency increased from 0.1-100 Hz, the storage modulus was augmented non-linearly from about 0.65 to 0.82 MPa, 0.71 to 0.92 MPa and 0.86 to 1.18 MPa for MRE samples containing 10, 30 and 50 wt% of CIPs, respectively. Similarly, the loss modulus was remarkably increased from approximately 0.025 to 0.12 MPa, 0.026 to 0.14 and 0.05 to 0.17 MPa for 10, 30 and 50 wt% of CIPs, respectively. This is due to the fact that high shear stress was formed at the surface and in the MRE samples during shear deformation particularly at high excitation frequency. Consequently, the ENR-based MRE samples were stiffened, and resulted in increasing of storage modulus. Furthermore, the friction that occurred at the interface between CIPs, and CIPs with the ENR matrix was rose with frequency, leading to increase of loss modulus.

In the meantime, the CIPs content and magnetic field density also significantly augmented both of the moduli. Based on Figure (a) and (c), the small increment of storage and loss moduli were observed as CIPs content rose from 10 to 30 wt%. Meanwhile, both moduli were increased significantly when CIPs content increased to 50 wt%. The same trend of the graph was observed in Figure 3 (b) and (d), in which the storage and loss moduli were enhanced steadily and remarkably as the magnetic field increased from 383 to 483 mT and 483 to 730 mT, accordingly. Besides that, the storage and loss moduli were detected to be the lowest without application of magnetic field. These results indicated that adjusting the applied magnetic field could control the storage and loss moduli. The growth of the moduli with CIPs content and applied magnetic field indicated that the dipole-dipole interparticle magnetic forces among

CIPs in the samples were enlarged. In other words, the magnetic forces were enhanced due to the mitigation of interparticle distance among CIPs as more CIPs were loaded.

Moreover, the increment of the applied magnetic field on the samples resulted in more magnetic field density induced in the samples. Therefore, addition of CIPs and applied magnetic field also contributed to increment of storage and loss moduli as similar to the excitation frequency. The rise of the storage and loss moduli for different wt% of CIPs could be guidance in manufacturing the ENRbased MRE samples with the desirable properties.

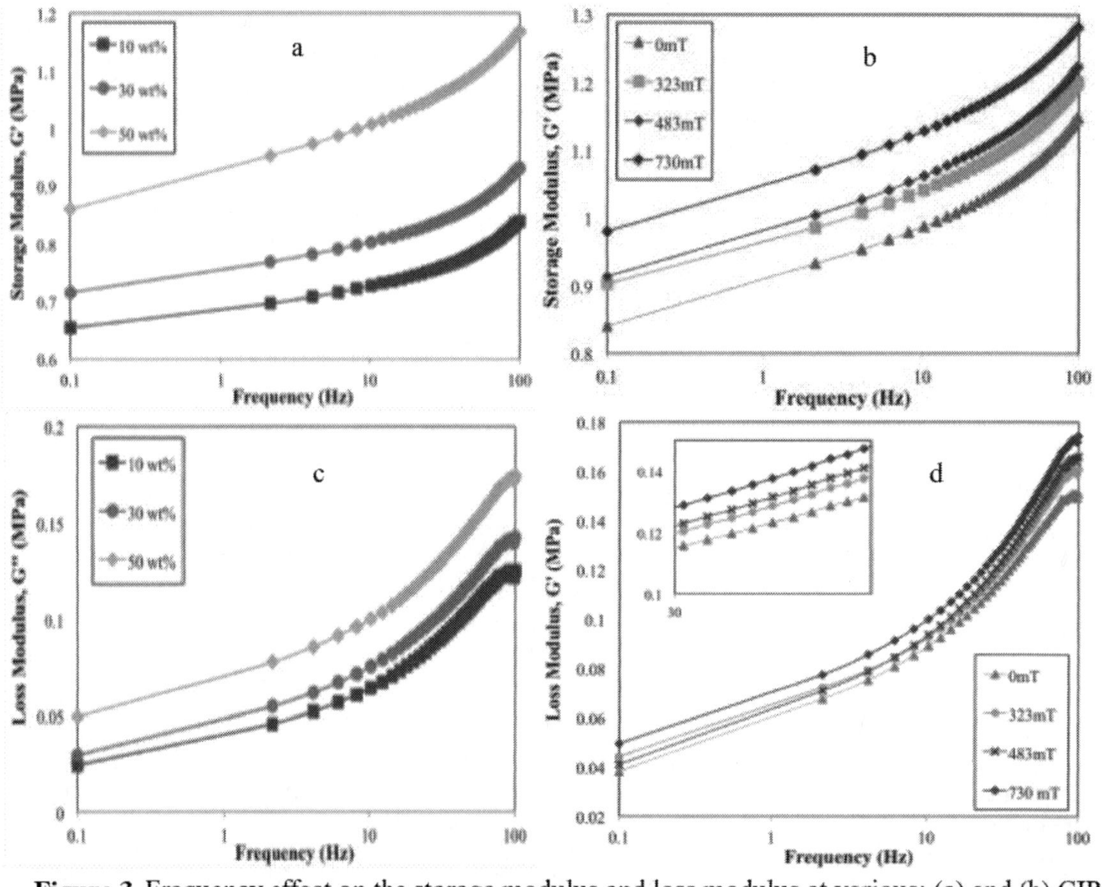

Figure 3. Frequency effect on the storage modulus and loss modulus at various: (a) and (b) CIPs content, and (c) and (d) magnetic field density

4. Conclusion

In this study, the new type of rubber known as epoxidized natural rubber (ENR) was employed as MRE matrix. The microstructure of the ENR-based MREs proved that the CIPs were dispersed uniformly in the ENR matrix which referring to isotropic MRE characteristics. The ENR-based MREs were dependent on strain, and the viscoelastic region of ENR-based MRE sample was 0.4% of strain amplitude. Determination of the viscoelastic region is important to prevent from performing the frequency sweep measurement at the region whereby the network structure in the MRE samples was disrupted. The results of frequency sweep indicated that the storage and loss moduli were dependent on the frequency, magnetic field and essentially relied on the CIPs content. Other than that, both moduli were obviously increased with increasing of these parameters. In other words, the rise of dipole-dipole interparticle magnetic forces and friction between CIPs contributed to the increment of both moduli. The outcomes indicated that adjusting the parameters could control the magnetic fielddependent modulus of

ENR-based MREs. Other than that, the results related to CIPs content could be guidance in manufacturing the ENR-based MRE samples to suit certain requirements for applications.

5. Acknowledgement

The authors gratefully acknowledge the financial funded by the Ministry of Higher Education Malaysia PRGS Grant (Vote No: 4L667) and Universiti Teknologi Malaysia under GUP (Vote No: 13H55).

References

[1] Boczkowska A, Awietjan S F, Wejrzanowski T and Kurzydłowski K J 2009 *J. Mater. Sci.* **44** 3135–40
[2] Ubaidillah, Sutrisno J, Purwanto A and Mazlan S A 2015 *Adv. Eng. Mater.* **17** 563–97
[3] Boczkowska A and Awietjan S 2008 *Polish Soc. Compos. Mater.* **4** 327–31
[4] Gong X L, Zhang X Z and Zhang P Q 2005 *Polym. Test.* **24** 669–76
[5] Xu Y, Gong X, Xuan S, Zhang W and Fan Y 2011 *Soft Matter* **7** 5246
[6] Xu K, He C Z, Wang Y Q, Luo Y Y, Liao S Q and Peng Z 2012 *Adv. Chem. Eng. Pts 1-3* **396398** 478–81
[7] Gong X, Fan Y, Xuan S, Xu Y and Peng C 2012 *Ind. Eng. Chem. Res.* **51** 6395–403
[8] I.R.Gelling 1991 *J. Nat. rubber Res.* **6** 184–205
[9] Wang Y, Zhang X, Oh J and Chung K 2015 *Smart Mater. Struct.* **24** 095006

The epoxy resin variation effect on microstructure and physical properties to improve bonded NdFeB flux magnetic density

N. Rusnaeni[1*], Priyo Sarjono[1], Muljadi[1], Nasrudin Noer[2]
[1)] Research Center for Physics, Indonesian Institute of Sciences,
Kawasan Puspiptek Serpong, Tangerang Selatan, INDONESIA, Tel. 021-7560570
[2)] Department of Physics, University of Sumatera Utara, Indonesia

E-mail: nenenrusnaeni@lipi.go.id

Abstract. NdFeB magnets have been fabricated from a mixture of powder NdFeB (MPQ-B^+) and epoxy resins (ER) with a variation of 0% wt, 2% wt, 4% wt and 6% wt. The pellets samples were made by pressing 4 tons of the mixture powder at room temperature before curing at 100°C for 1 hour. The SEM-EDX results showed the microstructure with ER were evenly smeared the NdFeB magnetic particles due to higher percent C and lower transition metals value. Sample with 2% wt epoxy resin was able to achieve the highest density of 5.35 g/cm^3 and the highest magnetic flux of 2121 Gauss. The magnetic properties characterization using the permagraph indicates that the sample pellets with 2% wt epoxy resin has a value of remanence (Br) = 4.92 kG, coercivity (Hc) = 7.76 kOe, and energy product (Bhmax) = 4.58 MGOe. Despite low remanence value in the pellet samples, the resistance to demagnetization value was still acceptable.
Keywords. Bonded Isotropic NdFeB Magnet, Epoxy Resin, Structure, Magnetic Properties

1. Introduction

Permanent magnet is a very strategic material to be developed in the future. At present, polymer-bonded magnets with different specifications are used in various applications to convert electrical energy to mechanical energy or vice versa. Three important specifications which describe permanent magnet performance are the remanence (Br), intrinsic coercivity (Hci), and maximum energy product (BHmax) [1]. However, the absence of local manufacturers of permanent magnets in the country causing dependency on imported products which mainly come from Japan and China – Tiongkok. Therefore, we need an intensive research to develop a permanent magnet and encourage the growth of local industry. Nonetheless, Indonesian researchers had done research on the manufacture of bonded NdFeB permanent magnets [2, 3,4,5]. However, there are still opportunities to improve the processing technology of the bonded NdFeB magnetic performance. In the present work, the research main topic is to analyze the effect of ER %wt variations on their microstructure and physical properties of a certain particle size bonded NdFeB by using flux magnetic density test and permagraphic characterization. Changes in microstructure and physical properties is caused by ER %wt variations on improve bonded NdFeB flux magnetic density.

2. Experimental

Materials used in this study are magnetic powder with a brand name MQP-B$^+$ (100 mesh), epoxy resin (ER), and hardener. The samples were made by varying the ER %wt; 0 %wt (sample1), 2 %wt (sample2), 4 %wt (sample3), 6 %wt (sample4). The magnetic powder was manually mixed with liquefied epoxy resin for at least 5 min before adding the hardener to achieve a homogeneous mixture. The mixture was molded in a cylindrical cast with a diameter of 12 mm by using hot press method of 4 tonne pressure at 30°C. The sample characteristics were determined from the bulk density, microstructure, and magnetic properties (flux magnetic density and magnetic hysteresis loop). The microstructure of the sample was analyzed using the Scanning Electron Microscope SEM (Secondary electron SE; backscattered electron BSE; energy dispersive X-ray EDX and mapping). Typical remanence (Br) intrinsic coercivity (Hc), and maximum energy product (BHmax) of the samples were obtained from magnetic hysteresis loop derived by permagraph test.

3. Results and Discussion

3.1 The Effect of %wt epoxy resin (ER) to carbon content of compacted bonded $Nd_2Fe_{14}B$

The microstructural profiles of compacted 100 mesh bonded $Nd_2Fe_{14}B$ was observed using SEM-SE (Secondary Electrons) as shown in Figure 1.

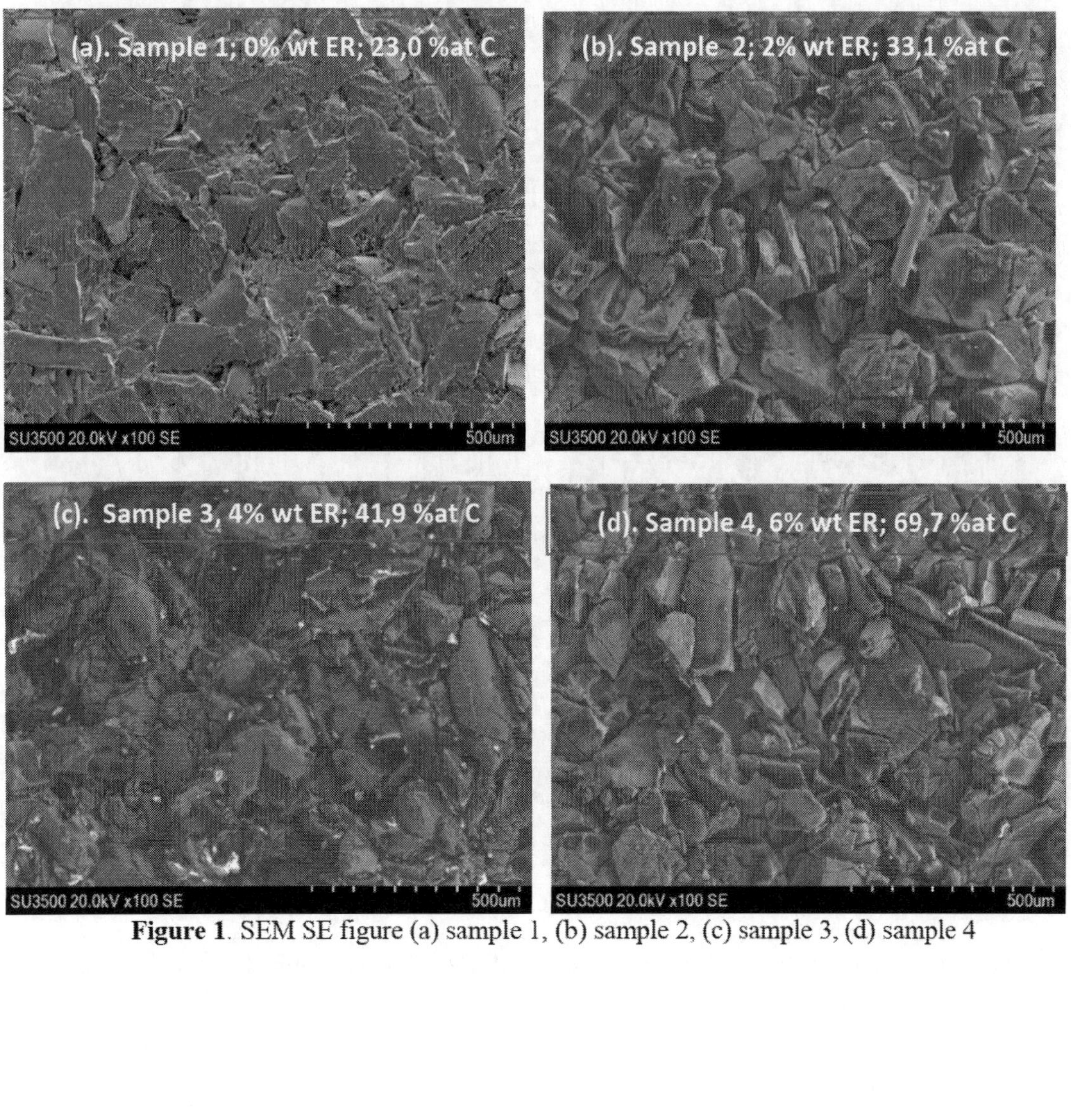

Figure 1. SEM SE figure (a) sample 1, (b) sample 2, (c) sample 3, (d) sample 4

The carbon C content of the samples was determined by using EDX analysis to relate the C content from ER %wt (0 %wt, 2 %wt, 4 %wt, 6 %wt) on the bonded $Nd_2Fe_{14}B$ particles. The bonded $Nd_2Fe_{14}B$ particles have been smeared with resin evenly.The compacted powder NdFeB of all sample (fig.1a to 1d) is homogeneous and dense as the pores between the powders is tight. Meanwhile, the epoxy resin material fill in the space between powders in the second sample based on the SEM-SE microstructural characterization. Higher the epoxy resin percentage resulting in greater epoxy resin material filled the space between powders causing the powders more compact from each other. Based on the analysis, the atomic percentage of the carbon (C) element contained increases as the ER %t increases, i.e: 23.0 %at C (sample 1/fig.1a); 33.1 %at C (sample2/fig.1b); 41.9 %at C (sample 3/fig.1c) and 69.7 %at C (sample 4/fig.1d). The C content in sample 1 is derived from a polymer material that cover the material $Nd_2Fe_{14}B$. However, the addition of C content is related to the increase of epoxy resin content. From the experiment it was found that the atomic percentage of the carbon element content will likely to increase as the epoxy resin %wt increase.

3.2 The Effect of %wt ER to Nd, Pr, and Fe content of compacted bonded $Nd_2Fe_{14}B$
Figure 2 shows the relation and profiles of Nd, Pr, Fe content and ER %wt (0 %wt, 2 %wt, 4 %wt, 6 %wt) of compacted 100 mesh bonded $Nd_2Fe_{14}B$ at 30°C temperature and 4 tonne real pressure.

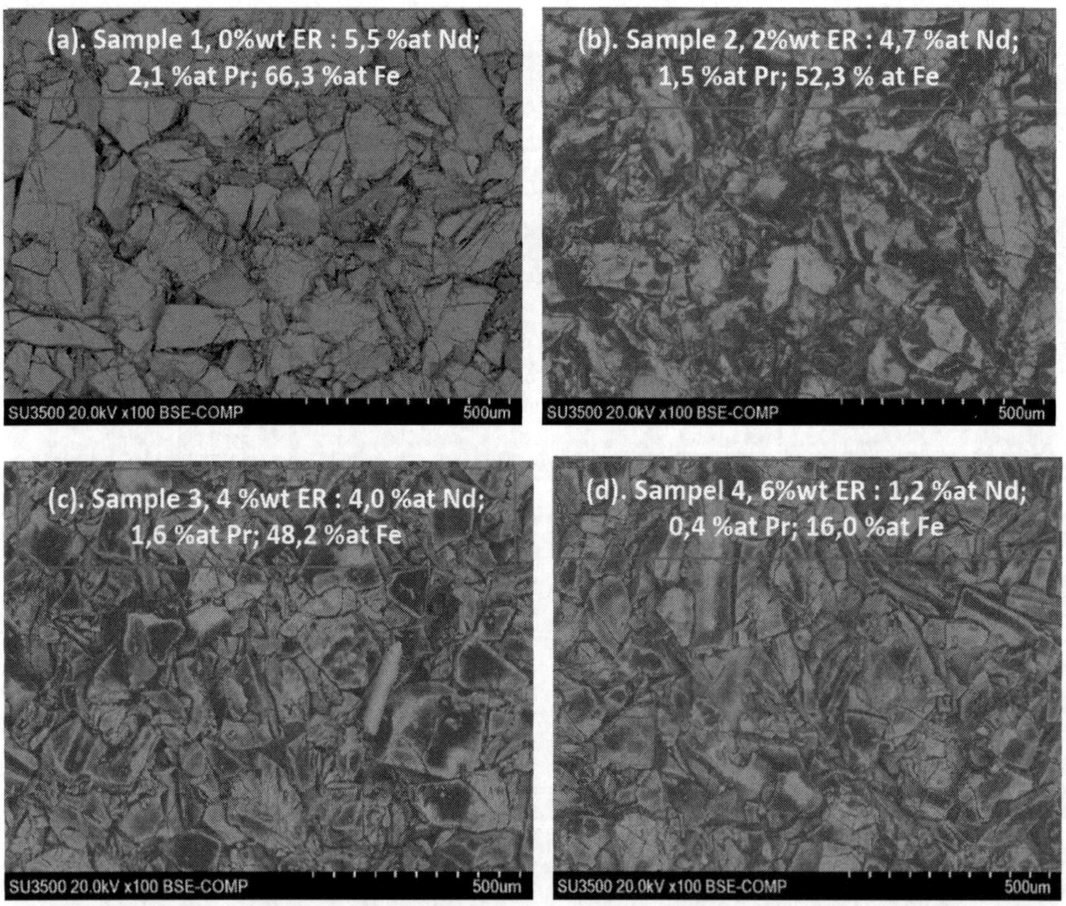

Figure 2. SEM BE figure (a) sample 1, (b) sample 2, (c) sample 3, (d) sample 4

Their microstructural profile was observed using SEM BSE (Back-Scattered Electrons) while the metal Nd, Pr and Fe content was detected using EDX analysis. According to the SEM-BSE profile, the

$Nd_2Fe_{14}B$ particle is quite compact after the compacting process. From the EDX experiment it was found that the atomic percentage of Nd, Pr, Fe element content will likely to decrease as the epoxy resin %wt increase. The results obtained from the EDX analysis shows the atomic percentage of Nd and Pr element decreases as the epoxy resin %wt increases, i.e : 7.6 %at (sample 1/fig. 2a); 6.2 %at (sample 2/fig. 2b); 5.6 %at (sample 3/fig.2c) and 1.6 %at (sample 4/fig.2d). The atomic percentage of Fe element also decreases as epoxy resin %wt increases, i.e: 66.3 %at (sample 1/fig.2a); 52.3 %at (sample2/fig.2b); 48.2 %at (sample 3/fig.2c) and 16.0 %at (sample 4/fig. 2d). From the experiment it was found that the atomic percentage of the sum of (Nd,Pr)Fe element content will likely to decrease as the epoxy resin %wt increase.

3.3 SEM mapping patterns to detected Fe elemen distribution

The previous observation results were strengthened with the pattern characterization results by using SEM mapping as shown in Figure 3. Darker pattern display of the SEM mapping shows less Fe contents on the sample surface as it has been tested on four compacted 100 mesh bonded $Nd_2Fe_{14}B$ under different epoxy resin %wt (0 %wt/sample 1/fig.3a, 2 %wt/sample 2/fig. 3b, 4 %wt/sample 3/fig.3c dan 6 %wt/sample 4/fig. 3d) at 30°C temperature and 4 tonne real pressure.

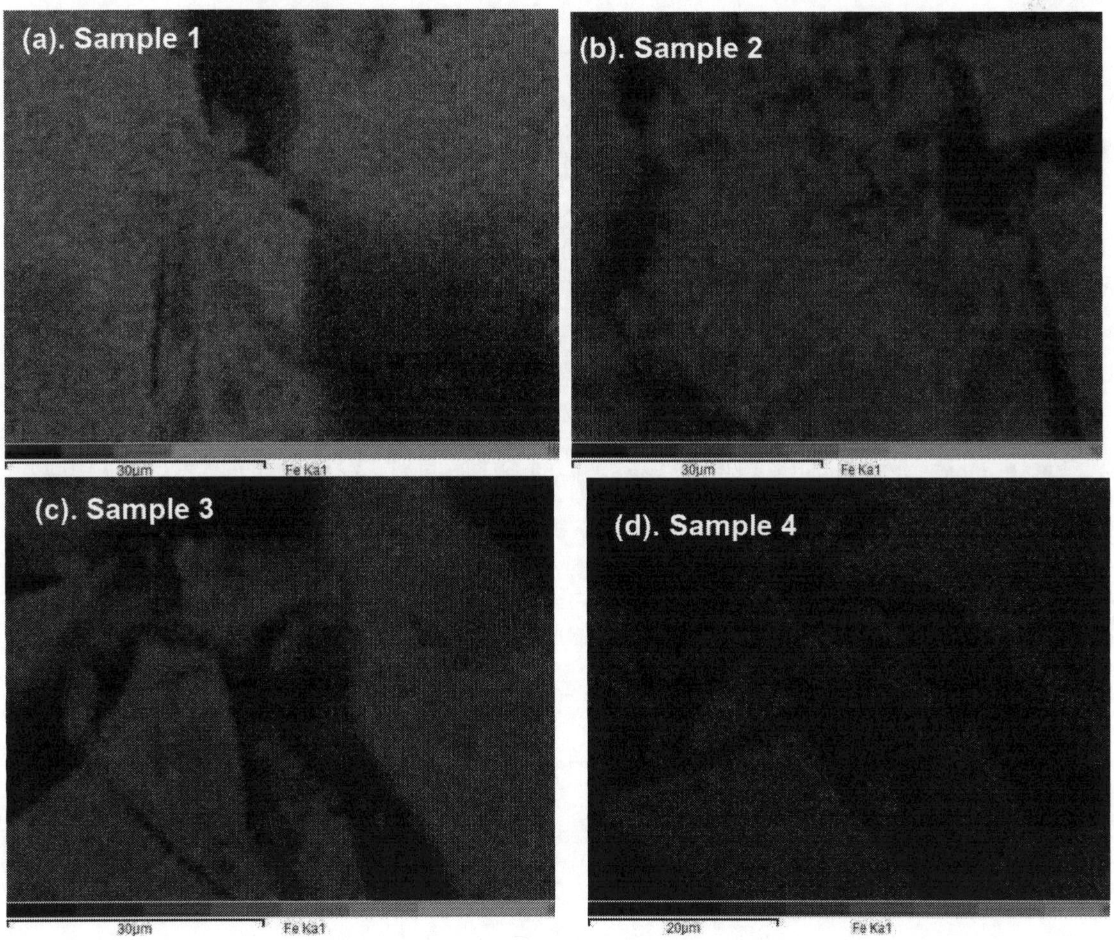

Figure 3. SEM mapping patterns. (a) sample 1, (b) sample 2, (c) sample 3, (d) sample 4

From the experiment it was found that the atomic percentage of the Fe element content in the surface of sample will likely to decrease as the epoxy resin %wt increase.

3.4 The effect of ER %wt to bulk density and flux magnetic density of compacted bonded $Nd_2Fe_{14}B$

Figure 4 below show correlation between bulk density and epoxy resin %wt variations. Four compacted 100 mesh bonded $Nd_2Fe_{14}B$ under different epoxy resin %wt (0 %wt, 2 %wt, 4 %wt dan 6 %wt) at 30°C temperature and 4 tonne real pressure were plotted against the bulk density value, and the result shows that sample 2 (2 %wt ER) has the highest density which equal to 5.35%.

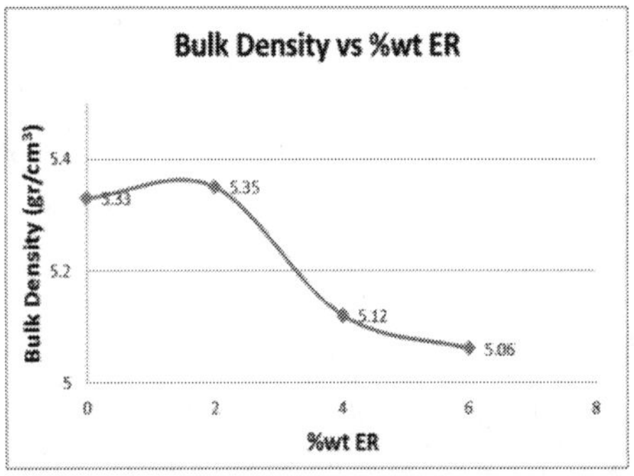

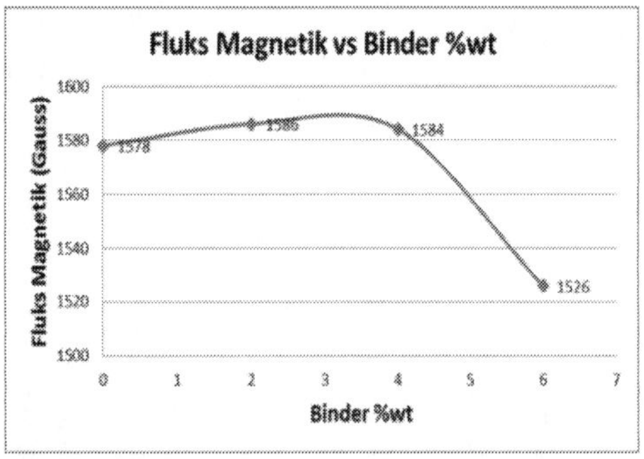

Figure 4. Relation of bulk density and ER %wt **Figure 5.**Relation of Flux Magnetic and ER %wt

Figure 5 shows the correlation between ER %wt and flux magnetic density (magnetization voltage was 1500V). The flux magnetic density value is remain constant. It was found that sample with 2 %wt ER (sample 2) has the slightly higher magnetic flux value of 1586 gauss from the magnetic flux density measurement using gauss meter.

3.5 The Effect of Magnesiatzation Votgage to Flux Magnetic Density and Permagraph Test of Compacted Bonded $Nd_2Fe_{14}B$ 2 ER %wt

The magnetic hysteresis loop of permagraph test shows that the bonded NdFeB magnets with a mixture of epoxy resin is categorized as hard magnet by having 4.92 kG of remanence (Br) value in sample 2 and 4.94 kG in sample 3, 7.76 KOe of coercivity (Hcj) value in sample 2 and 7.34 KOe in sample 3, and 4.58 MGOe of maximum energy product (BHmax) in sample 2 and 4.95 MGOe in sample 3 (Table 1). The coercivity of sample 2 show the the highest result, although the remanence has the slightly lower than the value of sample 3. The coercivity value of the approached data sheet has value of 82.97% [7, 9, 10].

Table 1. Permagraph test of compacted bonded $Nd_2Fe_{14}B$ with 2 %wt and 4 %wt ER and 4 %wt ER

Magnetic Properties	Sample 2	Sample 3
Br (kG)	4,92	4.94
Hcj (kOe)	7,76	7.34
Bhmax (MGOe)	4.58	4.95

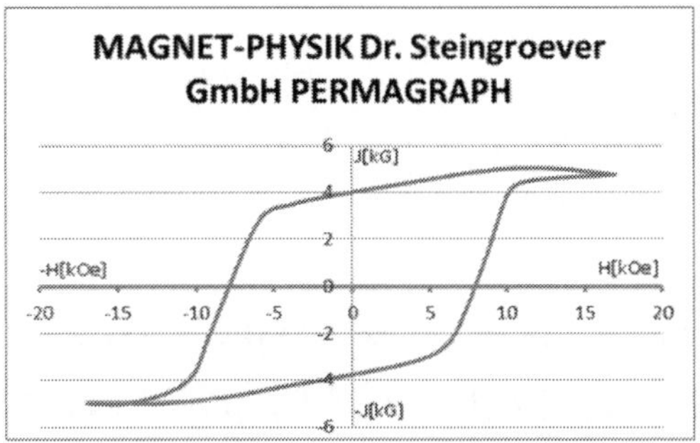

Figure 6. Permagraph Test of compacted bonded Nd2Fe14B with 2%wt ER

The relation between the magnetization voltage variation with flux magnetic is can be seen from sample 2 result where the magnetic flux increases as the applied voltage increases resulting in sample 2 pellets NdFeB, thus increase the magnetic properties performance. Sample with 2% wt epoxy resin was able to achieve the highest density of 5.35 g/cm^3 and the highest magnetic flux of 2121 Gauss (magnetization voltage was 1800V).

4. Conclusions

The effect of different epoxy resin %wt on microstructure and physical properties to improve bonded NdFeB magnetic flux density have been investigated in this research. It was found that the bulk density of the sample are 66-70% of the theoretical mass density (7.63 gr/cm^3). The bonded $Nd_2Fe_{14}B$ magnet made with 2 %wt of epoxy resin has better magnetic performance compared to other epoxy resin variations. The magnetic properties characterization using the permagraph indicates that the sample pellets with 2% wt epoxy resin has a value of remanence (Br) = 4.92 kG, coercivity (Hc) = 7.76 kOe, and energy product (Bhmax) = 4.58 MGOe. Despite low remanence value in the pellet samples, the resistance to demagnetization value was still acceptable.

References

[1] Ma BM, Herchenroeder JW, Smith B, Suda M, Brown DN, Chen Z, 2002, *J. Magnetism and Magnetic Materials* 239, 418-423.

[2] Christina M, Karo-kKaro A, Febri M, Yuwono AH, 2000, *Prosiding Seminar Nasional Bahan Magnet I*, Serpong Indonesia,11 Oktober.

[3] Deswita, Karo Karo A dan Sudirman, 2007, *Jurnal Sains Materi Indonesia*, Edisi khusus Oktober hal 126-131, ISSN 1411-1098.

[4] Aryanto D, Ray Z, Sudiro T, Wismogroho AS and Sudrajat N 2015, *Advanced Materials Research* Vol. 1123, pp. 89 – 91. D10.4028/www.scientific.net/AMR.1123.88

[5] Muljadi, Sardjono P, Suprapedi 2014, *Proceeding of 2nd International Conference on Sustainable Energy Engineering and Application*, Bandung Indonesia, 14 – 15 October.

[6] Siregar DP 2016, Thesis, Pengaruh variasi % berat Epoxy Resin Terhadap Sifat Fisis, Mikrostruktur dan Sifat Bonded Magnet NdFeB, Fakultas Matematika dan Ilmu Pengetahuan Alam Universitas Sumatera Utara.

[7] Kokabi M, Arabgol F, Manteghian M, 2005, *Iranian Polymer Journal* 14(1), pp.71-79.

[8] Sudrajat N and Kristiantoro T, 2013, *Jurnal Fisika dan Aplikasinya*, 9,no 1

[9] Sheth N, 2011, *Magnetics Technology International*

[10] Liu W, Yang Y, Meng Y, Wu FY, 2003, *Materials Transaction*, vol.44, no.6, pp 1159-1162

Preparation of Fe_2O_3-TiO_2 composite from Sukabumi iron sand through magnetic separation, pyrometallurgy, and hydrometallurgy

S. Wahyuningsih[1], A. H. Ramelan[2], H. P. Pranata[1], Q. A. Hanif[1], Y. A. Ismoyo[1], and K. F. Ichsan[1]

[1] Inorganic Materials Research Group, Faculty of Mathematics & Natural Sciences, Sebelas Maret University
[2] Electronic Material and Energy Research Group, Faculty of Mathematics & Natural Sciences, Sebelas Maret University
Jl. Ir. Sutami 36A, Kentingan, Jebres, Surakarta 57126, Indonesia

E-mail: sayekti@mipa.uns.ac.id

Abstract. Preparation of Fe_2O_3/TiO_2 composite from Sukabumi iron sand by magnetic separation, roasting, leaching and precipitation treatment has been carried out. Magnetic separation can separate magnetic particles and non-magnetic particles of iron sand content, while the non-magnetic particles (wustite (FeO), hematite (α-Fe_2O_3), maghemite (γ-Fe_2O_3) and magnetite (Fe_3O_4)) was washing with oxalic acid 1 M. The result product then was roasted at 800 °C treated by sodium carbonate (Na_2CO_3) addition of 1:1; 2:1 and 1:2 (w/w) of iron sand to Na_2CO_3 weight ratio, respectively. The X-Ray Fluorescence (XRF) analysis result shown that Sukabumi iron sand have hematite (Fe_2O_3) and titanium dioxide (TiO_2) content about 72.17% dan 14.42%. XRD analysis of roasted iron sand shown the rutile (TiO_2), Hematite (Fe_2O_3), $NaFeO_2$, FeO, and Na_2TiO_3. Leaching of roasted iron sand using sulphuric acid (H_2SO_4) have influenced by concentrations of the H_2SO_4 solution. The optimum iron sand dissolution occurred in H_2SO_4 9 M, which condensation product of the leachant have a weight ratio of Fe:Ti = 1:1 (w/w). Meanwhile, the settling back-filtrate result of second condensation was obtained a ratio of Fe_2O_3: TiO_2 of 3: 1 (w/w).

1. Introduction

Titanium dioxide (TiO_2) has a great potential application in water purification [1–3]. Because this semiconductor be able to degrade the organic compounds through the photocatalytic activity. This photocatalyst also has high chemical stability and low toxicity. However, the major drawback of TiO_2 is that, TiO_2 alone can only photodegrade organics under ultraviolet light and has large band gap (3.2 eV) that absorbs only 4% of sunlight from solar spectrum [4]. Accordingly, many efforts have been made to extend the absorbance of TiO_2 to visible light by doping the transition metal or composites [5].

Futhermore, Fe_2O_3–TiO_2 composites as a photocatalyst can respond to visible light due to the narrow band-gap of Fe_2O_3. It has also been found that another kind of iron oxide, Fe_3O_4, reacted with TiO_2 powders and produced iron titanium oxide compounds, such as $FeTiO_3$ [6].

Commonly, the Fe_2O_3–TiO_2 composite was prepared from different precursor like Titanyl acetylacetonate (TiAcAc) [7], titanium (IV) isopropoxide and Fe(III) acetylacetonate [8], $Ti(SO_4)_2$ and $Fe_2(SO_4)_3$ [9]. Here we report the preparation of Fe_2O_3–TiO_2 in 1:1 ratio (w/w) from Sukabumi iron sand through magnetic separation, pyrometallurgy, and hydrometallurgy.

2. Experimental section

1.1. Materials and equipment

The materials used was iron sand from Sukabumi (Indonesia), aceton, methanol p.a (Merck), sodium carbonate (Merck), and oxalic acid (Merck). The equipment that require in this research was ultrasonic cleaner (DSA50 GL_1), planetary ball miller, oven (Memmert), sulphuric acid 96% (Merck), thermometer (Futaba, 0 °C-220 °C), and water. The materials characterization both before and after treatment were carried out by X-Ray Diffraction (XRD Bruker D8 Advance), X-Ray fluorescence (XRF Bruker S2 Ranger), and Scanning Electron Microscopy (SEM Quanta 250). Roasting process was carried out by Furnace Thermolyne 4800.

1.2. Preparation of Iron Sand

The Sukabumi iron sand was analyzed by X-Ray Fluorescence (XRF) and Scanning Electron Microscopy (SEM). The magnetic and non magnetic particles was then separated by magnet. Subsequently the separated particles were milled by planetary ball milling (1000 rpm) for 2 hours. Milling process was conducted in w/w ratio, ball milling:material = 10:1 (130 gram:13 gram).

1.3. Fabrication of Fe_2O_3/TiO_2

1.3.1. Washing Non Magnetic Particles under Ultrasonic Irradiation

A 50 gram iron sand powder was washed with 150 mL oxalic acid (1 M) for 2 hours under ultrasonic irradiation. Precipitate and filtrate were separated and subsequently characterized by XRF.

1.3.2. Pyrometallurgy

Sukabumi iron sand was added Na_2CO_3 in ratio 1:2 (w/w) and afterwards roasted 800 °C for 2 hours. The roasted iron sand was analysed by XRF. A 20 gram iron sand was washed by 300 ml water in 90 °C for 2 hours. The precipitation was characterized by XRD and SEM.

1.3.3. Hydrometallurgy using Sulphuric Acid

Sukabumi iron sand was leached under reflux (90 °C) with an addition of 100 mL H_2SO_4 (9 M) for 2 hours. Subsequently the filtrate and precipitate were separated. The precipitate was analyzed by XRF. Whereas, the co-precipitation process from filtrat was washed by water and ethanol. The obtained precipitate was analyzed by XRF.

3. Result and Discussion

Firstly, the pretreatment Sukabumi iron sand was analysed qualitative and quantitative by XRF instrument. The result in Figure 1 and Table 1 shows that Sukabumi iron sand contains Fe 50.48%, Ti 8.65%, and other elements under 5%.

Table 1. Quantitative analysis of pretreatment Sukabumi iron sand (w/w)

Element	Concentration
Fe	50,48%
Ti	8,65%
Si	3,07%
Al	1,16%
Ca	0,78%
Mn	0,57%

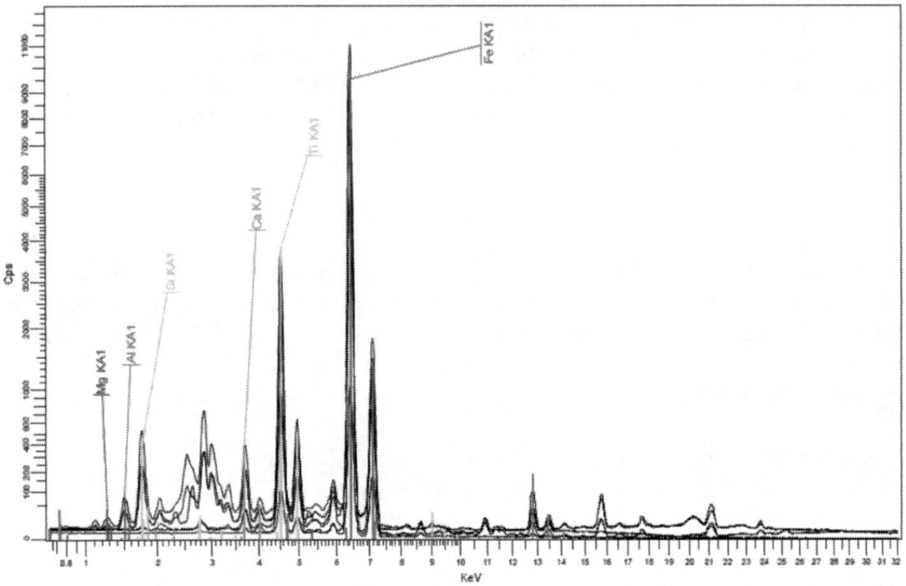

Figure 1. XRF analysis of pretreatment Sukabumi iron sand

Morphology of iron sand in Figure 2, shows that material have a rock shape. While the size particle is about 122,7 μm.

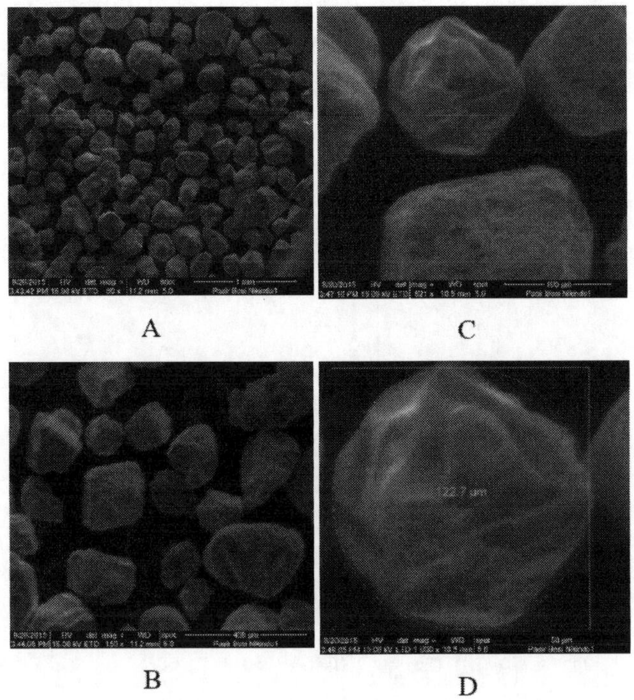

Figure 2. SEM results of Sukabumi iron sand a) 50x, b) 150x, c) 500x, and d) 1000x magnification

Magnetic separation of iron sand be able to reduce the magnetic particles such as wustite (FeO), hematite (α-Fe_2O_3), maghemite (γ-Fe_2O_3), and magnetite (Fe_3O_4). The XRF result (Table 2) have been proof this statement. Fe content in non magnetic particles reduce from 50.48% to 38.81%.

Table 2. XRF result from iron sand after magnetic separation

Element	Concentration
Fe	38,81%
Ti	13,18%
Si	5,47%
Ca	1,55%
Al	1,45%
Mn	0,68%

Washing the iron sand with oxalic acid under ultrasonic radiation has been done for reducing the impurities. The XRF result (Table 3) indicates that Fe and Ti content were decrease. It possibly because of dissolution of such amount Fe and Ti in oxalic acid, while Si content increase due to low solubility of this element in oxalic acid solvent.

The roasting treatment with an addition of sodium carbonate was done for decomposition of materials in iron sand such as Fe_2TiO_5 (pseudobrookite) to form hematite (Fe_2O_3) and TiO_2. This is have a deal with the XRD result in Figure 3. The high peaks were corresponds to JCPDS No. 20-1115 (Na_2CO_3) in $2\theta = 38.177°$ (d_{hkl} = 002) and $2\theta = 43.379°$ (d_{hkl} = 020), JCPDS No. 89-1633 (TiO_2 rutile) in 2θ =

30.46° (hkl = 222), 2θ = 35.316° (hkl = 400) and 2θ = 38.604° (hkl = 331), JCPDS No. 88-2359 (Fe_2O_3) in 2θ = 23.851° (hkl = 012), 2θ = 32.856° (hkl = 104) and 2θ = 35.080° (hkl = 110), FeO also show in 2θ = 35.927° (hkl = 111) and 2θ = 41.725° (hkl =200) according to JCPDS No. 89 – 0687.

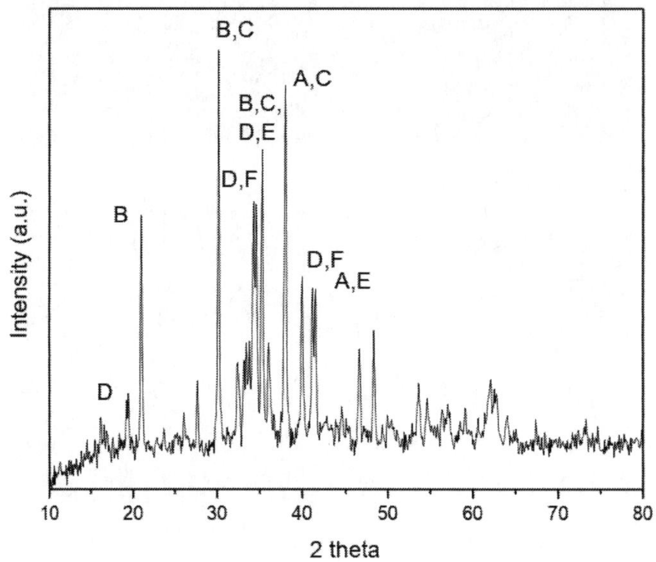

Figure 3. Diffractogram XRD of roasted iron sand (A = sodium carbonate, B = hematite , C = rutile, D= $NaFeO_2$, E = FeO, F = Na_2TiO_3)

The mechanism reaction that may occur during the roasting process is in equation 1.

$$2TiO_{2(s)} + 2Fe_2O_{3(s)} + 2NaCO_{3(s)} \rightarrow Na_2TiO_{3(s)} + NaFeTiO_{4(s)} + NaFeO_{2(s)} + Fe_2O_{3(s)} + 2CO_{2(g)} \quad (1)$$

Iron sand and Na_2CO_3 in roasting process were conducted in several ratio of 1:2 (A) ; 1:1 (B) ; 2:1 (C) (w/w), and the XRF result shows that (Table 3), those different ratio are be able to influence the obtained Fe and Ti.

Table 3. XRF result of roasted iron sand with an addition of Na_2CO_3 in ratio (w/w) 1:2 (A) ; 1:1 (B) ; 2:1 (C)

A		B		C	
Element	**Concentration**	**Element**	**Concentration**	**Element**	**Concentration**
Fe	32.75%	Fe	21.93%	Fe	22.14%
Ti	10.40%	Na	14.59%	Si	11.64%
Na	8.24%	Si	9.68%	Na	10.21%
Si	6.76%	Ti	5.89%	Ti	6.42%
Ca	1.57%	Ca	3.93%	Ca	4.06%
Al	1.24%	Al	2.73%	Al	3.25%
Mn	0.58%	Mn	0.39%	Mn	0.37%

Not only the content of Fe and Ti that changes through the treatment, but also the morphology of iron sand (Figure 4). The materials was leached under reflux with an addition of strong acid H_2SO_4 9 M. This process could dissolve the iron sand reach 86.70%. The possible reaction in leaching process is in equation 2-4. While XRF results can be seen in Table 4.

$$Fe_2TiO_{5(s)} + 2H_2SO_{4(aq)} \rightarrow 2FeSO_{4(aq)} + TiOSO_4 + 2H_{2(g)} \quad (2)$$

$$FeTiO_{3(s)} + 2H_2SO_4 \rightarrow FeSO_{4(aq)} + TiOSO_{4(aq)} + 2H_2O_{(l)} \quad (3)$$

$$TiO_{2(s)} + Fe_2O_{3(s)} + 4H_2SO_{4\,(aq)} \rightarrow TiOSO_{4\,(aq)} + Fe_2(SO_4)_{3(aq)} + 3H_2O_{(aq)} \quad (4)$$

Table 4. XRF result of first precipitate and co-precipitate (w/w)

First precipitate		Co-precipitate	
Element	**Concentrations**	**Element**	**Concentrations**
Si	35.71%	Fe	32.42%
Fe	5.96%	Na	11.69%
Ti	4.23%	Ti	8.09%
S	1.80%	S	4.28%
Ca	1.01%	Al	1.04%
Al	0.38%	Ca	0.83%

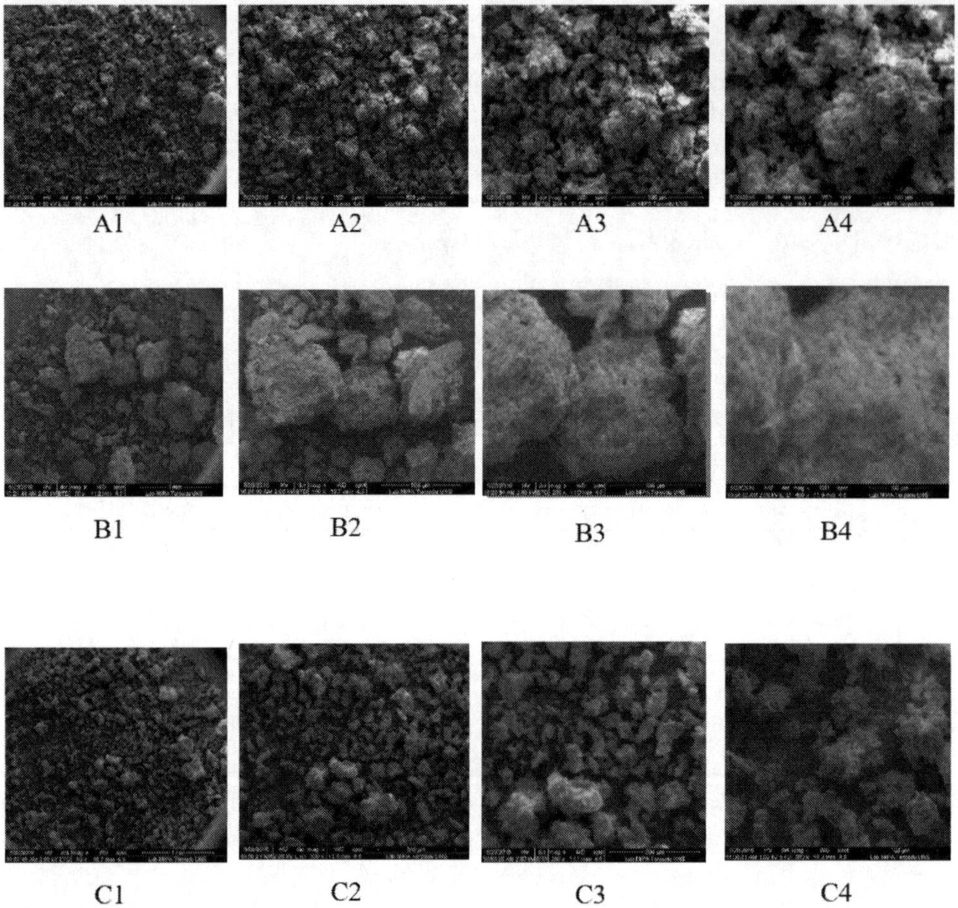

Figure 4. Morphology of roasted iron sand with an addition of Na_2CO_3 in ratio (w/w) 1:2 (A); 1:1 (B); 2:1 (C)

In conclusion, preparation of Fe_2O_3/TiO_2 composite through magnetic separation, pyrrometallurgy, and hydrometallurgy be able to yield Fe 5.96%, Ti 4.23% from precipitate. Then from coprecipitation process produce Fe 32.42% and Ti 8.09%. The more addition of Na_2CO_3 the more decomposition can occurred. While addition of sodium carbonate with ratio iron sand:Na_2CO_3 = 1:2 was the optimum condition for obtained Fe:Ti (1:1).

4. Acknowledgment
The researchers wish to express their gratitude to the Insinas Ristek Dasar, Kemenristek of Indonesia for supporting this research.

5. References
[1] Anpo M, M Takeuchi **2003** *J. Catal.* 216 505–516
[2] Watanabe T, A Nakajima, R Wang, M Minal, S Koizumi, A Fujishima, K Hashimoto **1999** *Thin Solid Films* 351 260–263
[3] Hoffman M R, S T Martin, W Choi **1995** *Chem. Rev.* 95 69–96
[4] Cheng X F, W H Leng, D P Liu, J Q Zhang, C N Cao **2007** *Chemosphere* 68 1976– 1984
[5] Navio J A, G Colon, M Macias, C Real, M I Litter **1999** *J. Appl. Catal. A: Gen.* 177 111–120
[6] Ye F, A Ohmori **2002** *Surf. Coat. Technol.* 160 62–67
[7] Mahadik M A, S S Shind, V S Mohit, S S Kumbhar, A V Moholkar, K Y Rajpur, V Ganesan, J Nayak, S R Barman, C H Bhosal **2014** *J. Photochem. Photobiol. B, Biol.* 133 90–98
[8] Pal B, M Sharon, G Nogami **1999** *Mat.Chem. Phys.*59 254-261
[9] Liu H, H K Shon, X Sun, S Vigneswaran, H Nan **2011** *Appl. Surf. Sci.* 257 5813–5819

Journal of Physics: Conference Series **776** (2016) 012027 doi:10.1088/1742-6596/776/1/012027

Magnetic switching probability of perpendicularly magnetized nano-dot

N. A. Wibowo, and S. Trihandaru

Physics Department, Faculty of Science and Mathematics, Satya Wacana Christian University, Indonesia

E-mail: nurajiwibowo@gmail.com

Abstract. Understanding of magnetization reversal mode plays an important role in spintronic devices performance. Especially in Heat Assisted Magnetic Recording, the contribution of thermal field in the magnetization dynamics becomes a crucial aspect which should be further explored. In this paper, Landau Lifshift-Gilbert equation was employed to investigate the magnetic switching dynamics for nano-dots with different magnetic anisotropies. This micromagnetic study shows that these nano-dots exhibit the different mode of switching. Multi-domain configuration dictates the switching process for nano-dot with strong magnetic anisotropy before magnetized in the driven magnetic field direction. Meanwhile, multi-domain configuration, domain-wall nucleation, and propagation take place in the process to achieve magnetized single-domain for the nano-dot with the lower level of magnetic anisotropy. Furthermore, in the absence of the external magnetic field, magnetic moments of this low magnetic anisotropies nano-dot is more probably to reverse.

1. Introduction

As a new technology, heat-assisted magnetic recording (HAMR) usage will enlarge areal density of magnetic hard disk drive which today's perpendicular magnetic (PMR) product has an areal density in hundreds Gb.in^{-2} (Gigabits/square inch) [1]. As the next major HDD technology, HAMR is projected to achieve areal densities beyond 1,5 Tb.in^{-2} [2].

In realizing large areal density of HDD media, high magnetic anisotropy material is implemented. This usage is purposed to maintain thermal stability of materials when its size is downsized. The uses of temporary heating around Curie temperature well-known can be used to overcome the problem of the usage of high anisotropy magnetic materials so that the conventional write-head-field capable of controlling the magnetic moments direction. However, the various investigation is still attractive to be conducted in this field because of the subtly magnetization reversal dynamics under this thermal stimulation. Magnetization reversal is a process that the direction of magnetic moments of the magnetized materials varies from one to other direction which was driven by induced magnetic field. The reversal mechanism is not only controlled by magnetic properties but also the way the thermal and magnetic field was applied.

Dependable control of the magnetization reversal dynamics is the one of the central subjects in the present magnetic recording devices. Especially for the magnetic recording which thermal field is used to encourage the magnetization reversal process, stochastic magnetic switching dynamics becomes the main aspect of magnetic recording and writing [3]. Magnetic switching probability issues vital role in the spintronic devices performance [4]. In this paper, we investigate numerically the impact of magneto-crystalline anisotropy to reversal probabilities and magnetization mode of nano-dot under heating stimulation near its Currie point.

2. Numerical Method

This micromagnetic study employed Landau-Lifshift Gilbert equation in solving the magnetization dynamics. The effective magnetic field that functioned for driving the magnetic polarization was defined as a partial derivative of energy densities with respect to magnetization. Energy density terms which contribute consist of Zeeman energy, demagnetization energy, magneto-crystalline anisotropy energy as well as exchange energy. In order to activate the thermal aspect in the system, thermal fluctuation field has been added. Thermal field distribution was considered as a Gaussian with the mean value is zero. Theorem of fluctuation-dissipation was adopted in calculating thermal field. Temperature dependence of magnetic properties such as magnetic saturation, magnetic anisotropy, and exchange constant are presented according to the previous simulation of stochastic simulation for thermally assisted magnetization reversal which conducted by Budi Purnama et all [5]. Magnetic bit media was modeled as a nano-dot which its shape is rectangular [6] with the surface area and the thickness are 50×50 nm^2 and 20 nm respectively. It represents particular bit information so that the areal densities can reach hundreds Gb.in^{-2} [3]. A nano-dot is expected as an isolated system which is consists of 225 unit cells that each denotes single magnetic moment. These magnetic moments were initially oriented perpendicular to the surface of nano-dots. The nano-dot was assumed as an isolated system that the interaction effects of the neighboring nano-dots were neglected.

In evaluating the impact of magneto-crystalline, two samples with a different value of magnetic anisotropy were chosen i.e. sample A and B with anisotropy constants as large as 3×10^5 erg/cc and 3×10^6 erg/cc respectively. Other magnetic parameters that used are magnetic saturation 2100 Gauss, exchange constant 1×10^{-7} erg/cm, Gilbert damping 0.3 and Curie temperature 373 K. To simulate HAMR technology, two terms of pulses were applied at this nano-dot i.e. magnetic field and thermal pulses. During the presence of the external magnetic field pulse, nano-dot was temporarily heated for 250 ps near its Curie point i.e. 372.9 K by the thermal pulse and cooled quickly during 750 ps. In accommodating the stochastic effect as a consequence of the thermal presence, 20 random numbers of magnetization were employed and switching probability over these random numbers was evaluated.

3. Result and Discussion

Figure 1 shows the switching probability of nano-dots magnetization over the external magnetic field under the presence of a thermal pulse. As shown that the probability of switching of sample A and B tend to increase for the larger magnitude of magnetic field. The probabilities reach a maximum at the specific value of induced magnetic field which called as threshold field. For Sample A, magnetization probably to reverse even no presence of a driven magnetic field. This magnetic switching probability existence even in the absence of magnetic field implies that writing data could be realized without any writing magnetic field. One factor that can be related to this phenomenon is randomizing of the magnetization during temporary heating process. Through the randomization of magnetic moments mechanism, the thermal field reduces even bridging the gap energy of two stable magnetization states [4], which lead to the unstable state. With the result that the magnetic polarization is enable to be re-orientated without any driven field. The threshold field of this sample is 350 Oe. For Sample B, the probability of switching in the absence of driven magnetic field was not observed. In fact, zero probability was still observed up to ~ 1000 Oe of driven magnetic field before it grows to reach its maximum value at 2550 Oe – the large magnitude of threshold field compared to the first sample.

As observed in Fig. 1, the probability of the zero-field-switching be as one of the key factor that responsible in the reduction of the switching-field (magnetic field required to switch the magnetization of nano-dot). This probabilities is significantly affected by the strength of the magnetic anisotropies of nano-dot. As will be discussed later, the switching mode of nano-dots by which determine the probability and the magnitude of the switching-field is highly impacted by anisotropies strength.

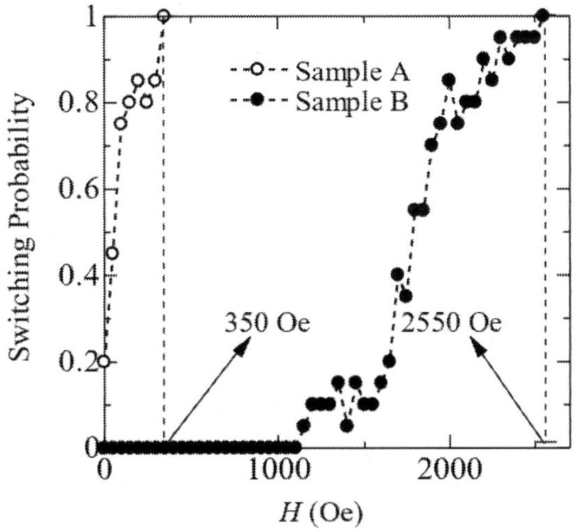

Figure 1 Magnetic switching probability of two different magnetic anisotropy samples.

Figure 2(a) presents the magnetization dynamics of nano-dots which temporary heated near to its Curie temperature during under controlling of the external magnetic field. The samples initially magnetized in $-x$ direction while the driven magnetic field is submitted along the $+x$ direction. While the nano-dot inducted by opposite magnetic field, the magnetization was going to stable on its initial state (normalized magnetization equal to $-$ 1). After the activation of thermal field close to its Curie point, the magnetization was randomized so that its normalized magnetization value increased to zero. It means that no preferences of the magnetic direction even had been supported by the external magnetic field. Subsequently, nano-dot was cooled rapidly. During this cooling phase, magnetic moments of the nano-dot were re-oriented before fully magnetized inline to the external magnetic field direction.

Both samples exhibit a different mode of magnetization dynamics during the time of the cooling process. These magnetization modes can be observed from the micro-graph. Micro-graph of the samples is shown in Figure 2(b). For Sample A, the magnetized state was achieved through two modes. The first mode is a multi-domain configuration which its occurrence started after 1.5 ns, coincide to the rising temperature of the nano-dot from its ambient temperature. These domains were continually randomized during the heating period for 250 ps, whereas at the beginning of the cooling phase, this randomization mode had been ended and replaced by domain-wall nucleation. This domain-wall propagated to the one of nano-dot corner until the single-domain state was realized. At this single-domain state, the magnetization of nano-dot was completely parallelized to the external magnetic field direction. While for Sample B, the single-domain state was achieved through only one mechanism, i.e. multi-domain configuration. This mode was going on since the temperature of the nano-dot was increased up to the single-domain state was realized. The interesting point is that the single-domain state was reached before the temperature of the nano-dot was cooled back to the ambient temperature.

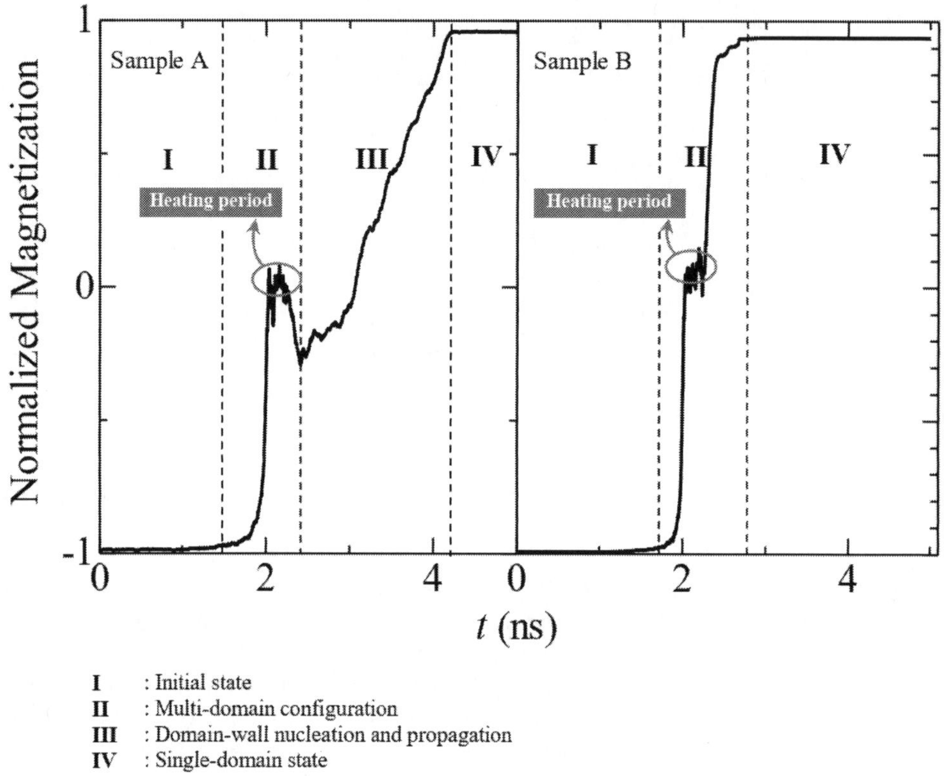

(a)

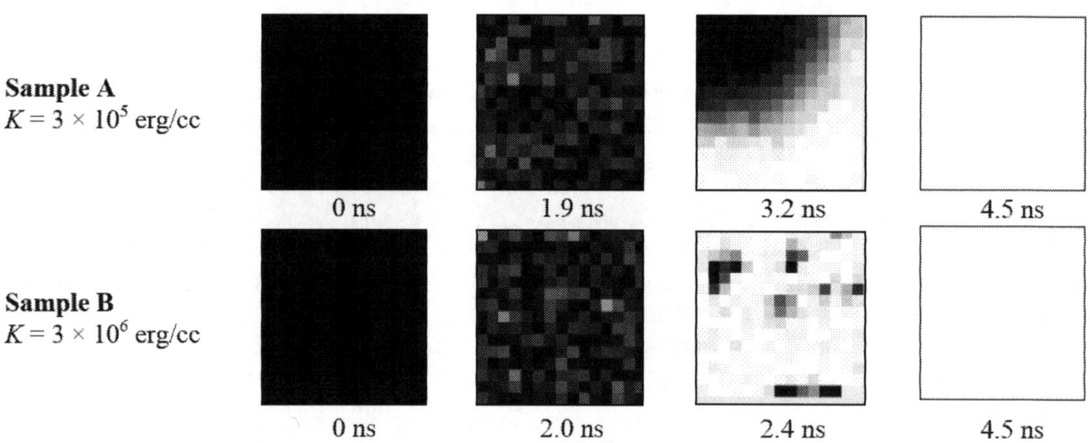

(b)

Figure 2. Magnetic switching during the presence of thermal-magnetic field pulses induction for 5 ns, (a). Normalized magnetization, (b). Micro-graph of nano-dots.

4. Conclusion

Micromagnetic study of switching probability for perpendicular nano-ferromagnetic has been conducted by using Landau Lifshift-Gilbert equation as the main equation to solve. This study results that under thermal stimulation, nano-dot with low magnetic anisotropy is prospective to switch its magnetic orientation when the external magnetic field submitted. In fact, magnetic switching is probably even no driven magnetic field. This zero-field switching occurrence is enabled because of thermal activation through the randomization of magnetic moments. Furthermore, magnetic anisotropy also significantly has an impact on magnetic switching mode. In achieving the completely magnetized nano-dot, domain-wall nucleation and propagation mode was observed after magnetic randomization for low magnetic anisotropy nano-dot. Meanwhile, for nano-dot with strong magnetic anisotropy, the multi-domain configuration had been dominating the magnetic switching mode.

References

[1] Weller D, Mosendz O, Parker G, Pisana S and Santos T S 2013 *Phys. Status Solidi A.* 210 7
[2] Chen Y J, Yang H Z, Leong S H, Cher K M, Hu J F, Sethi P and Lew W S 2015 *J. Appl. Phys.* 117 17
[3] Wang Y and Zhang F C 2013 *EPL Europhys. Lett.* 102 4
[4] Scholz W, Schrefl T and Fidler J 2001 *J. Magn. Magn. Mater.* 233 3
[5] Purnama B, Koga M, Nozaki Y and Matsuyama K 2009 *J. Magn. Magn. Mater.* 321 9
[6] Fidler J, Schrefl T, Tsiantos V D, Scholz W and Suess D 2002 *Comput. Mater. Sci.* 24 1–2

Study of the reduced magnetic field required for thermally assisted magnetization reversal

H. F. Y. Firdausi, Utari, and B. Purnama[a1]
Department of Physics, Faculty of Mathematics and Natural Sciences,Universitas Sebelas Maret, Jl. Ir. Sutami 36A Kentingan Surakarta 57126, Indonesia

E-mail: bpurnama@mipa.uns.ac.id

Abstract. The reduced magnetic field required for thermally magnetization reversal discussed in this paper. Study of thermally assisted magnetization reversal conduct by using micromagnetic simulation. The magnetic dot size of the simulation was 50 nm × 50 nm × 20 nm. The perpendicularly anisotropy constant was 2×10^6 erg/cm^3. Initial condition was set single domain configuration. Then a sufficiently thermal pulse was used to get stochastic effect so that the magnetization along to the induce field direction for pico second duration. The results show that the reduced magnetic field mechanism seem to be temporary antiferromagnetic configuration before single domain configuration in alinging along to field direction. The same mechanims observed for modify of thickness dot particles. The require magnetic field of 145 Oe in thermally assisted magnetization reversal open a posibility for MRAM application.

1. Introduction

Nowadays, the availability of a high density memory is required for realization of a high speed devices electronic in information technology. Memory-based magnetic materials or magnetic random access memory (MRAM) is one of the candidates because they are non-volatile and no-endurance time. In order to realize a high density of MRAM, a magnetic material should be patterned in nano-meter order scale. To overcome the thermal stability of magnetic materials at the nanoscale meter, a high perpendicular anisotropy magnetic to be as one as potential candidate. Whereas thermally assisted of selected storage cell memory during read-write process is one promissing procedure to reduced a magnetic field require for reversal [1-4].

Although the system of termally assited magnetization reversal (TAMR) has been proven to reduce the field required for magnetization reversal, however the reduced field mechanism so far not completely understood [5-7]. In the early development of TAMR, the degree of randomness of the magnetic moment due to heating process is considered as a main factor for the lowering in the magnetic field required for reversal. Here, the energy barrier decrease with increasing temperature within result a few of the magnetic field require for reversal. However, the evaluation TAMR with constant stiffness variations in order to modify the degree of randomness of the magnetic moment on the magnetic material with perpendicular asiotropi showed different results. Furthermore, magnetization reversal also have a chance to occur under zero-field coolling in TAMR scheme.

[1] To whom any correspondence should be addressed.

In this paper, the reduced mechanism of magnetic field require for magnetization reversal is studied by micromagnetic simulation. A several different thickness of nano dot cell memory were modified to obtained the minimal field required for switching. The TAMR of ultra thin of nano dot cell memory become a major atention at this study.

2. Experimental procedure

The reduced mechanism of magnetic field require for aligning to along field in thermally assited magnetization reversal (TAMR) has been studied by micromagnetic simulation [bp, JMMM 2009]. In initial, magnetization set a uniform direction, then magnetic field and heating pulse enter so that randomly magnetized state as a stochastic result of pulse heating is realized. It should be note that for all calculations, a duration pulse of heating is shorter than magnetic field and magnetization reversal process perform in field cooling from near Curie Temperature to room temperature for pico secon order. The thermal fluctuation effect is taken into account in the Landau-Lifshit-Gilbert (LLG) equation by using fluctuation dissipation effect [4]. The probability aligning of magnetic state owing same direction to magnetic field direction at field cooling writing process was performed for 20 different series of randomly magnetic state. The perpendicularly anisotropy magnetic is considered as a storage cell memory with K is 3×10^5 erg/cc and $4\pi M_S$ is 2.10 kG. The moderate T_C of 372.9 K was used the whole simulation. The magnetic dot size is 50 nm × 50 nm × 20 nm. The second-power temperature dependence with the reduced magnetization was assumed for both the exchange stiffness constant, A and the perpendicular crystalline anisotropy, K.

3. Results and Discussion

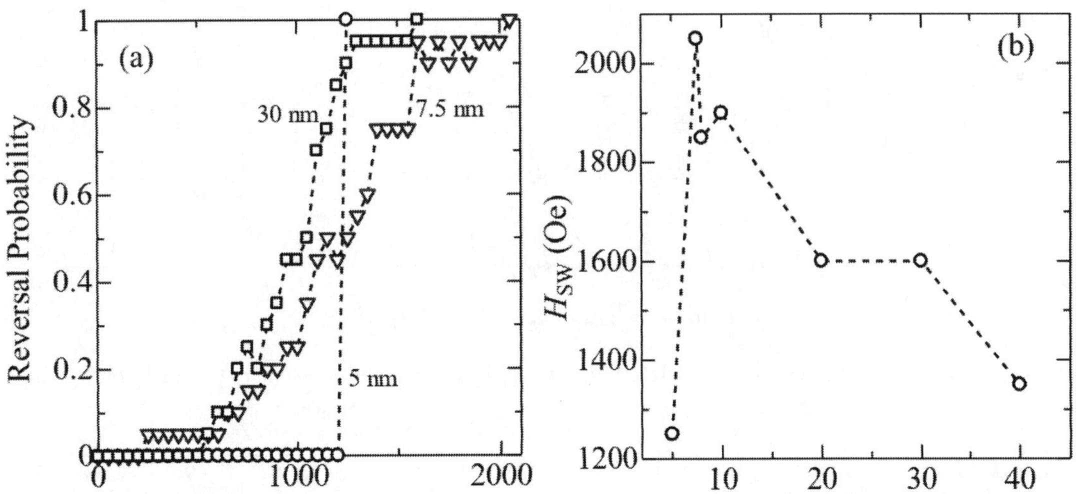

Figure 1. (a) Reversal probability (magnetization aligning along to field direction) versus magnetic field H calculated for 20 randomly magnetized state; (b) the magnetic field require for thermally assisted magnetization reversal (TAMR) H_{SW} as a function of the magnetic dot thickness.

Figure 1a show the reversal probability versus a magnetic field calculated for 20 randomly magnetized state in TAMR scheme. In case ultra thin dot of 5 nm, reversal probability show a step-like curve. Reversal probability calculation is equal zero within the increase of magnetic field untill given the H of 1200 Oe. Then the reversal even is attained for calculated 20 different randomly magnetized state when the H is equal 1250 Oe. It seem that the entire magnetized state configurations of magnetic dot aligning to along field direction do simultaneously. In contrast for both the thickness of 7.5 and 30 nm, the reversal probability increase gradually with increasing the magnetic field H and reversal even attain when the probability is equal 1. The experimental result indicate that the whole magnetized state configurations of magnetic dot do not simultaneously response to magnetic field for aligning along to field direction in TAMR scheme. Calculation for the entire thickness variation of a perpendicular anisotropy magnetic dot films, the magnetic field require for TAMR H_{SW} as a function of the magnetic dot thickness as depicted in Figure 1b. The H_{SW} increase with in decreasing the thickness of the magnetic dot and the highest H_{SW} of 2050 Oe attain at the thickness of 7.5 nm. Then the H_{SW} decrease to 1250 Oe at the thickness of 5 nm. The increase of the H_{SW} with decreasing the thickness of magnetic dot relate to long range interaction from magnetostatic interaction when the thickness of the films reduce to few nanometer order. So that the magnetization reversal realize a single-domain configuration with incoherent rotation along to external field direction. The decrease of the H_{SW} for the dot thickness of 5 nm should attribute from coherent rotation of single domain configuration in later case [8]. Interesting result obtain for the magnetic dot thickness of 8 nm i.e. the magnetic field require for reversal H_{SW} decrease to equal 1850 Oe. In order to understand more detail about the lowering magnetic field, the TAMR perform individual of initial randomly magnetized state as depicted Figure 2.

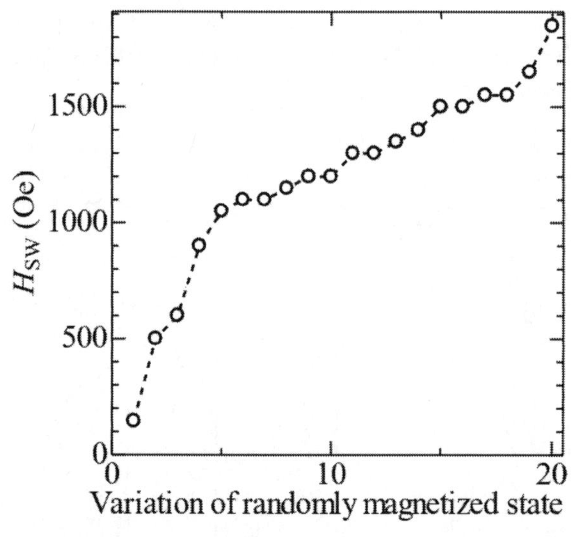

Figure 2. Evaluation of the H_{SW} calculate for 20 randomly magnetized state with the thickness dot magnetif of 8 nm in TAMR scheme.

As seen clearly at the Figure 2 that the magnetic field require for TAMR range from the H_{SW} of 145 untill 1850 Oe. It mean that the individual randomly magnetic state owing typical efficiency of sthochastic effect for reducing the field require H_{SW} for reversal. In high efficience case, the H_{SW} of 145 Oe is enough to realize reversal in TAMR scheme. Contrary for low efficience case, it should exceed of 1850 Oe for thermally assited magnetization reversal realization. Observation of the micromagnetic steps indicate that typical the TAMR have a similar mechanims. The reduced magnetic field mechanism seem to be temporary antiferromagnetic configuration before domain wall nucleation following domain wall propagation realize single domain configuration in alinging along to field direction. Moreover, the similar mechanism also observe for all succefully TAMR despite a different magnetic field required for TAMR. From the view point of application, it is interesting although the reducing mechanism of TAMR is still not clear. Figure 3 show a typical steps micromagnetic graph in TAMR procedure evaluated for the thickness magnetic dot of 8 nm under magnetic field H of 145 Oe. The uniform initial magnetization was set anti-parallel to magnetic field direction. The sufficient stochastic effect of heating observed after t = 200 ps. Randomly pattern of micro-magnetic graphs appear at a time interval of t = 200 ps dan 400 ps. Then the orientation magnetized state of nano dot particles tend to field direction which indicate domain wall nucleation following domain wall propagation and realize single domain configuration.

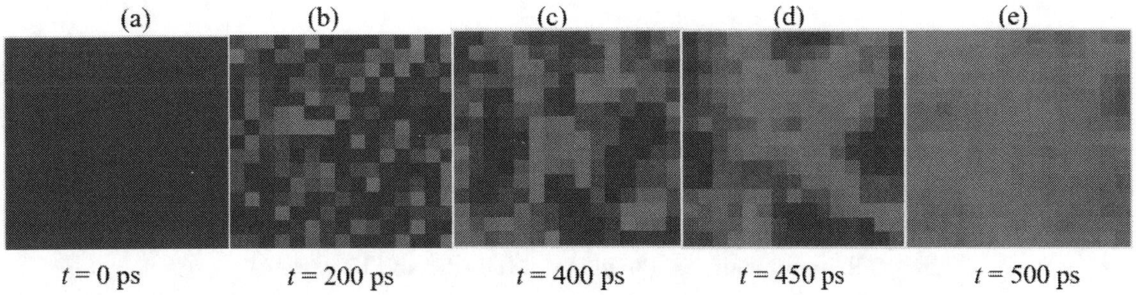

Figure 3. Micromagnetic configuration in the TAMR scheme: (a) initial configuration, (b) randomly magnetized state, (c) domain wall nucleation, (d) domain wall propagation and (e) single domain configuration.

4. Conclusion

The process of the reduced magnetic field required for thermally magnetization reversal discussed in this paper. Study of thermally assisted magnetization reversal conduct by using micromagnetic simulation. The results show that the reduced magnetic field mechanism seem to be temporary antiferromagnetic configuration before single domain configuration in alinging along to field direction. The same mechanims observed for modify of thickness dot particles. The require magnetic field of 145 Oe in thermally assisted magnetization reversal open a posibility for MRAM application.

5. References

[1] Matsumoto K, Inomata A and Hasegawa S, 2006 *Fujitsu Sci. Tech. J.*,**42**,1.p.158-167
[2] Purnama B, Nozaki Y and Matsuyama K 2007 *JMMM* 310 2683
[3] Nozaki Y, Isowaki Y, Hashimoto A, Purnama B and Matsuyama K 2006 *Journal of the Magnetics Society of Japan* **30**, No. 6-2
[4] Purnama B, Koga M, Nozaki Y, Matsuyama K 2009 *Journal of Magnetism and Magnetic Materials* **321** 1325–1330
[5] Sadnawanto W and Purnama B, 2014 *Journal of Physics: Conference Series* **539** 012024
[6] Suess D, Vogler C, Abert C, Bruckner F, Windl R, Breth L, and Fidler J, 2015 *Journal of Applied Physics* **117**, 163913
[7] Iskandarova IM, Ivanov AV, Knizhnik AA, A. F. Popkov AF, Potapkin BV, Skirdkov PN, Zvezdin KA, Stainer Q, Lombard L, dan Mackay K, 2016, *Nanotechnologies in Russia*, **11** pp. 208-214.
[8] Skomski R, Oepen H.-P., and Kirschner J., 1998:*Phys.Rev. B* **56**(6) 3223-3227.

Dependence of electrical property on the applied magnetic fields in spin coated Fe(III)-Phorphyrin films

Utari[1a,c], Kusumandari[a], B. Purnama[a], Mudasir[b] and K. Abraha[c]

[a]Department of Physics, Faculty of Mathematics and Natural Sciences,Universitas Sebelas Maret, Jl. Ir. Sutami 36A Kentingan Surakarta 57126, Indonesia

[b]Department of Chemistry, Faculty of Mathematics and Natural Sciences, Universitas Gadjah Mada, Bulaksumur BLS 21 Yogyakarta 55281, Indonesia

[c]Department of Physics, Faculty of Mathematics and Natural Sciences, Universitas Gadjah Mada, Bulaksumur BLS 21 Yogyakarta 55281, Indonesia

[1]E-mail: utarisby70@yahoo.com

Abstract. We report here on the experimental results of the effect of external magnetic field on the current flow in plane surface of Fe(III)-porphyrin thin layer. The deposition of the Fe(III)-porphyrin thin layer was done by spin coating method. The I-V characteristics of film were measured by means of two point probes. The sample of layer number N = 4 was used to evaluate the magnetic effect on the electrical currents. The ohmic characteristics of the I-V film measurement were obtained. The current decreases when magnetic field is applied to the system and saturated current is obtained at a given magnetic field. Here, the decrease in the current can be attributed to the recombination of carrier charge under the magnetic field. In addition, the magnitude of the saturated current is found to increase with the increase in the voltage used.

I. Introduction

In recent decades, the study of organic materials as electronic devices has grown very rapidly. One of the organic materials that is very attractive for advanced development is porphyrin or metalloporphyrin. Porphyrin is a specific macro cyclic compound with four ring-pyrrole attached to the bridge of methane [1]. A diversity of biochemical functions can be modified by a slight variation of atoms that bind to these macrocyclic compounds. Recently, it has been found that porphyrin layers also show a magnetic field response. Hence, porphyrin- or metalloporphyrin-based layers have a great opportunity to be developed as one of electronic device materials.

For examples, studies on the coupling between porphyrin and a thin layer of ferromagnetic cobalt and nickel substrates have been carried out by Bernien et al, [2], following by the success of the experiment by

Wende et al [3] in preparing the iron molecules octaethylporphyrin (FeEOP) monolayer on a thin layer of nickel and cobalt. This finding suggests that the material is able to act as a new type of molecule magnet. The reported study has indicated that porphyrin coupling with the substrate can create opportunities to accomplish magnetization reversal which is induced by the substrate. This modification, in turn, will be able to alter the electronic transport properties [4]. Other achievements have also confirmed that a large external magnetic field applied to the system is able to reverse spin orientation of the parallel state into perpendicular field of molecular orientation observed as a single molecular switching state [5]. Interestingly, Hu et al. [6] has also reported experimentally that the organic semiconductor of nonmagnetic materials can exhibit a significant response to an applied magnetic field with a low level in the event of electroluminescence, photocurrent and injection-electric current.

So far, only a few numbers of articles have reported the influence of magnetic field on the Fe(III)-porphyrin thin layers. In this experiment, we have systematically studied the effect of external magnetic field on the I-V types of the Fe (III)-porphyrin layers on a Cu-substrate. It should be noted that the I-V types of in-plane surface films used in this study and the magnetic field orientation perform/set to parallel-antiparallel of the current direction.

2. Experimental Methods

Material used in this experiment was Fe(III)–Porphyrin powders (100 mg) purchased from Frontier Scientific- USA (FeM658). Solvent of chloroform was purchased from Merck. Cu-Substrates and ITO-coated glass substrates (Aldrich-578 274) were used separately to deposit Fe(III)-porphyrin layer. Film depositions of Fe(III)-porphyrin layer were performed by using spin coater (Chemat Technology KW-4A) and the electrical property of the obtained films were evaluated by using I-V meter.

The detailed procedure of experiments was conducted in the following steps.
First, Fe((III)–porphyrin powder (100 mg, Frontier Scientific-FeM658) was dissolved in chloroform (10 mL, Merck) and stirred with a magnetic stirrer for 60 min at a rotational speed of 200 rpm. Then, the solution was used to deposit Fe(III)-porphyrin layer by using spin coating method. Second, Fe(III)-porphyrin layers were deposited with rotating speed of 500 rpm for 20 s followed by post-heating at 40 °C for 60 s. This procedure was repeated to obtain four types of layers. Third, the characterization of the layers including electrical property was done by using I-V measurement without and under external magnetic field B.

3. Results and Discussion

Figure 1 shows the I-V graph of voltage versus current for the Fe(III)-porphyrin thin layer with the number of layers N = 4 evaluated without and under magnetic field B = 10 mT. It can clearly be seen that for both samples (with and without magnetic field), the current I increases with the increase in the voltage V. The electric current (I) is observed to decrease for the whole variation of voltage when the magnetic field in plane of the films is applied. This result is consistent with that reported by Hu et al. [6], which states that external magnetic field affects the charge carrier recombination events in the material, resulting in neutral atoms due to combination of charge carriers, therefore some of charge carriers in the material are no longer available. This causes the number of charge carriers in the material reduces significantly, leading to the lower measurable current compared to that of material without the application of magnetic field.

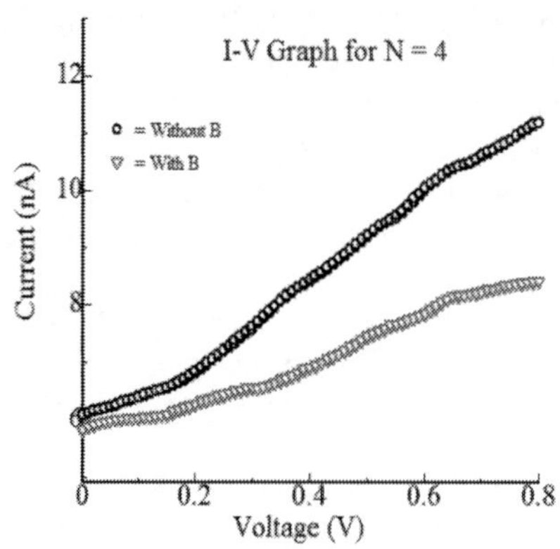

Figure 1. Electrical Characteristic of spin coated Fe(III)-phorpyrin films with four layers, without and with under magnetic field (B).

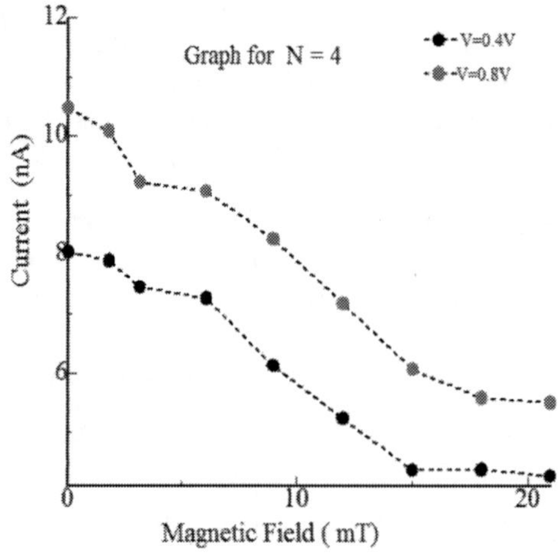

Figure 2. The current profiles of the thin films as a function of the magnetic field in the spin coated Fe(III)-porphyrin layers with the layer number of N = 4 evaluated at a voltage $V = 0.4$ and 0.8 volts.

Figure 2 shows the current profiles of the thin films as a function of the magnetic field in the spin coated Fe(III)-porphyrin layers with the layer number N = 4 evaluated at a voltage $V = 0.4$ and 0.8 volts. For two different voltages, the current I decreases with increasing magnetic field and reaches saturated value at a certain magnetic field. For V = 0.4 volts, current reaches saturation value at $I = 4.3$ nA when the field of 15 mT is applied. And, when $V = 0.8$ volts, saturated current of 5.7 nA is achieved when a magnetic field of 18 mT is used. The result may be explained as follows. When the magnetic field is applied, the charge carriers in the material undergo recombination to yield neutral species so that the number of charge carriers in the system is reduced. With the increase in magnetic field, no recombination occurs because the recombination depends on the applied voltage as a source of carrier charge generation.

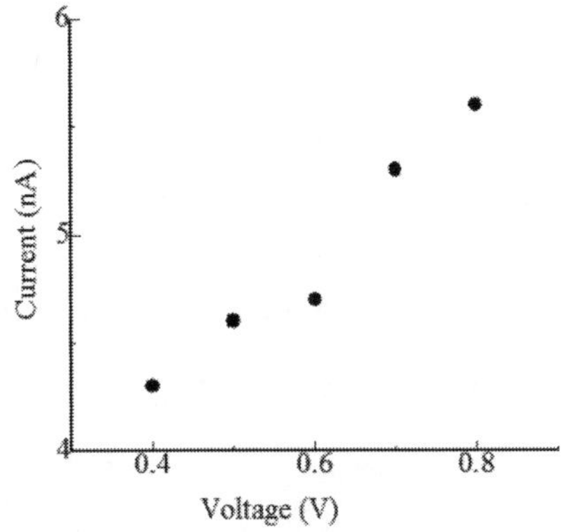

Figure 3. Relationship between saturated current and applied voltage.

Figure 3 shows the saturated current as a function of voltage. As can be seen from the figure, the saturated current increases with the increase of voltage. If the voltage can act as a carrier-charge source, it is easily understood that the increase in the voltage enhances the number of carrier charge. At low value V, i.e. 0.4 volts, the saturated current is 4.3 nA. In contrast, the saturated current of 5.6 nA is obtained for the voltage of 0.8 volts. This indicates that the magnetic field affects the distribution of the whole available carrier charge number.

4. Conclusions

We have reported the experimental results about the effect of external magnetic field on the current flow in plane surface morphology of the Fe(III)-porphyrin thin layer. The deposition of the Fe(III)-porphyrin thin layer was successfully done by spin coating method. The I-V characteristics of the film have been systematically measured by means of two point probe. The sample with layer number N = 4 is used to evaluate the magnetic effect on the electric currents. The ohmic characteristic profiles of the I-V measurement are obtained. The current has been found to decrease when magnetic field is applied and the saturated current is obtained at a given magnetic field. Here, the decrease in the current may be attributed to the recombination process of carrier charge generated by the magnetic field. In addition, it has also been found that the magnitude of saturated current increases with the increase of applied voltage. This indicates that the recombination of carrier charge is a function of used voltage.

5. Acknowledgement

This research was partially funded by the Scholarship for Domestic Postgraduate Program of the Directorate General of Higher Education of the Republic of Indonesia, (BPPDN DIKTI, No.153/DIKTI/KEP/2007) and DIPA Universitas Sebelas Maret of Republic of Indonesia (Penelitian Unggulan Perguruan Tinggi Contract No339/UN27.11/PL/2015).

6. References

[1] Hart, H., Craine, L. E., dan Hart, D. J. 2003 *Organic Chemistry* Houghton Mifflin Company.
[2] Bernien, M., X.Xu, J.Miguel, M.Piantek, Ph.Eckhold, J.Luo, J. Kurde, W. Kuch, and K. Baberschkeet 2007 *Phys.Rev.* **B 76** 214406.
[3] Wende, H.; M.Bernien; J.Luo; C.Sorg; N.Ponpandian; J.Kurde; J.Miguel; M.Piantek; X.Xu; PH.Eckhold; W.Kuch; K.Baberschke; P.M.Panchmatia; B.Sanyal; Oppeneer, P.M. and O.Eriksson 2007 *Nat. Mater.* **6** 516-520.
[4] Gatteschi, D 2007 *Nat.Mater.* **6** 471-472.
[5] Babersckhe, K.2009 *14th International Conference on X-Ray Absorption Fine Structure (XAFS14)* 012012.
[6] Hu.M, Liang Yan and Min Shao 2009 *Adv. Mater.* **21** 1500–1516.

Application of bonded NdFeB magnet for C-Band circulator component

T. Kristiantoro, N. Idayanti, N. Sudrajat, A. Septiani and Dedi
Research Centre for Electronics and Telecommunication – Indonesian Institute of Sciences (LIPI), Jl. Sangkuriang, Kampus LIPI, Gd. 20, Lt. 4, Bandung 40135, INDONESIA

E-mail: tony9910@gmail.com

Abstract. In this paper bonded NdFeB permanent magnets of the crashed-ribbon type were made as an alternative for circulator magnet to improve their magnetic performance. The fabrication process is also easier than the sintered NdFeB because there had no shrinkage of product (such as sintered barium ferrite magnet), with the others advantages as follows; large freeness of product shapes, high precision of dimension and good corrosion resistance. The dimension of the samples was measured to calculate its bulk densities and the magnetic properties were characterized by Permagraph to obtain values such as; Remanence induction (Br) in kG, Coercivity value (Hcj) in kOe, the Maximum energy product (BH max) in MGOe. Whereas the surface magnetic field strength (B) was observed by the Gauss-meter. The bonded NdFeB permanent magnets revealed 6.39 kG of Br, 6.974 kOe of Hcj and 7.13 MGOe of BHmax. The circulator performance was measured using Vector Network Analyzer (VNA). The optimum values of the circulator measurement at a frequency of 5 GHz show a VSWR value of 1.062 and insertion loss of -0.463 dB. The bonded magnet could be used as component of permanent magnets on the circulator that working on C-Band at a frequency range of 4 GHz - 8 GHz.

1. Introduction

A circulator was a passive component which was used for RF communication and microwave system to enhance their stability, performance, and reliability. The demand of permanent magnets for telecommunication devices was high enough [1], one of which was a component for the circulator. The circulator commercial in this research had been working in C band which was produced by *A in Flow* and composed of body, strip line, ferrite disc, and permanent magnets. This research was conducted to replace the permanent magnets on the circulator with the commercial bonded permanent magnets as an alternative component in the circulator. Magnets had become important parts of our daily life and many applications could be applied on devices such as; electrical motor, loud speaker, microwave, telecommunication, and etc. Other applications of magnets were founded many in tools of instrumentations, production and lab research [2]. However the contribution of magnets often overlooked by us due to this component had been mounted in a device and invisible. In fact, these components needed to be very diverse depending on the usability and functionality of the device. In general, demand for magnetic components was distinguished by the shape, dimensions, and magnetic field strength. The NdFeB was known as a rare earth magnet because its compound composed of rare earth elements, which had maximum energy product was higher than ferrite [3]. Fabrication of bonded NdFeB magnets from crashed-ribbon type expected could enhance the performance of permanent magnets in a circulator. Because bonded permanent magnet had the advantage that no shrinkage losses due to sintering effect, such as barium hexaferrite so that it will yield the precise product with the desired dimension.

2. Numerical Methods

In a point of view in making bonded magnets is known two methods of compaction, such hot press [4] and green compact [5]. In this work, we used a green compact technique and magnetic materials have mixed with polymer binder as commercially obtained. The NdFeB magnetic powder was compacted in certain pressure to get an optimum bulk density. When bonded NdFeB permanent magnet have been fabricated, furthermore it applied into a circulator and characterized by Vector Network Analyzer (VNA, Advantest R3770) to know circulator performance.

The fabrication of bonded NdFeB permanent magnets were made using MQEP 16-7 powder. This magnet powder obtained from Magnet quench and have mixed by epoxy resin so that other binders should not be added. The first step is weighing of MQEP 16-7 powder around 10.5 gram for one sample and we prepared ten samples for this experiment. The circulator has a pair of permanent magnets so that 10 samples used for five times to measure the circulator performance. The pressure of compaction was given about 100 kg/cm^2 to sample powder in the mold. Furthermore, the green body was heated to a temperature at 200 °C and held until 60 minutes. This heat treatment was given to the sample to ensure all of the bonding between polymer and MQEP 16-7 magnet powder become optimum. The dimension of samples is measured by Calipers and magnetized using the Impulse magnetizer (Magnet-Physik) with various input voltages to charge the difference of magnetic field strength into samples. Further, the surface magnetic field strength of magnetized samples were measured by Gauss-meter (Yokogawa) as presented in table 1.

Table 1 Various magnetic field strength of samples were resulted from the Impulse Magnetizer (Magnet-Physik)

Voltage (Volt)	Magnetic Field Strength (Gauss)
700	500
800	1000
1700	1600
2000	1800
2500	2000

The intrinsic properties of bonded NdFeB permanent magnets were characterized by Permagraph (Magnet-Physik) with given maximum magnetic energy to rise a saturation magnetization and obtained magnetic properties for that bonded magnets such as Remanence inductance (B_r) in Gauss, Coercivity (H_{cj}) in Oersted and Maximum magnetic energy product (BH_{max}) in MGOe.

3. Results and Discussion

Figure 1 shows a hysteresis curve of a bonded permanent NdFeB magnet which exhibited a saturation magnetization (Ms) at ~ 15 kG. The B_r value of a bonded NdFeB is about 6.26 kG, This result states that after magnetization of the sample until saturated magnetization and followed by demagnetization to zero of magnetic field strength (*H*), however, the magnet still has a remain inductance at 6.26 kG. This value is an intrinsic property which shows the capability of bonded magnet to hold the given energy from outside. In this work, the remanence inductance of bonded NdFeB permanent magnet is higher than the bonded magnet from previously reported by J.S. Trosic *et. all* [6]. For applications in telecommunication fields, we used a magnetic energy on the surface of samples, which is called as a flux magnetic density (*B*). We made samples of bonded NdFeB permanent magnet with several of flux magnetic densities, which were measured by Gauss-meter as follows; 500, 1000, 1600, 1800 and 2000 G. Another measured intrinsic property is Coercivity, which the bonded NdFeB yields a coercivity at 6.974 kOe. This value shows the demagnetization energy level to erase the magnetic field strength from the bonded magnet is 6.974 kOe and Its maximum energy product is 7.13 MGOe, this means that the bonded magnet can store the maximum magnetic energy until 7.13 MGOe after the external magnetic field removed. In this research, all sample have similar treatment so that they have the same intrinsic properties. They have similar hysteresis loop.

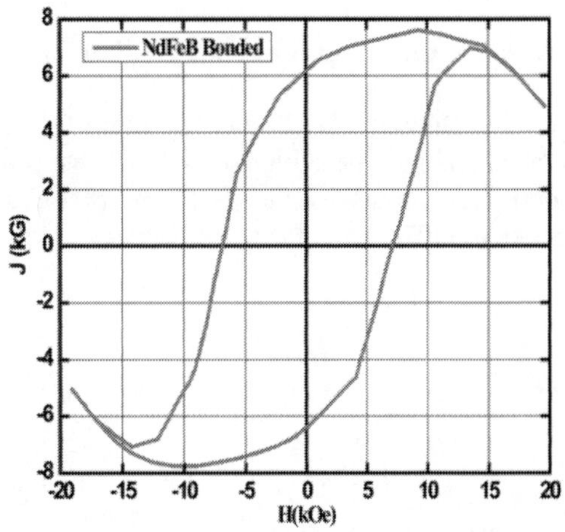

Figure 1 The hysteresis loop of a bonded NdFeB permanent magnet.

After the pair of bonded magnet installed in the circulator, the performance of circulator was measured by VNA. The input signal was given an electrical current at 1 mA on Port 1, the output signal was measured on port 2, and Port 3 was terminated by a terminator of 50 Ohm, in the frequency range 4 to 8 GHz. Figure 2 shows a VSWR curve of bonded NdFeB permanent magnet. VSWR is a signal that reflected back to transmission channel [7], which this value present a performance level of transmission channel system or circulator components. The allowed value of VSWR on the circulator is equal or less than 2[8]. In characterization of circulator for bonded NdFeB permanent magnets have been installed in a circulator, the VSWR value was obtained around 1.062, this for bonded magnet of surface flux magnetic 1800 G. While for bonded permanent magnets of 500, 1000 and 2000 G cannot be applied to the circulator in work frequency range 4 to 5 GHz due to its VSWR is higher than 2. Figure 3 shows the insertion losses of a circulator after installed of our bonded NdFeB permanent magnets in the circulator. Insertion loss is the value of power losses during transmitted signals [9].

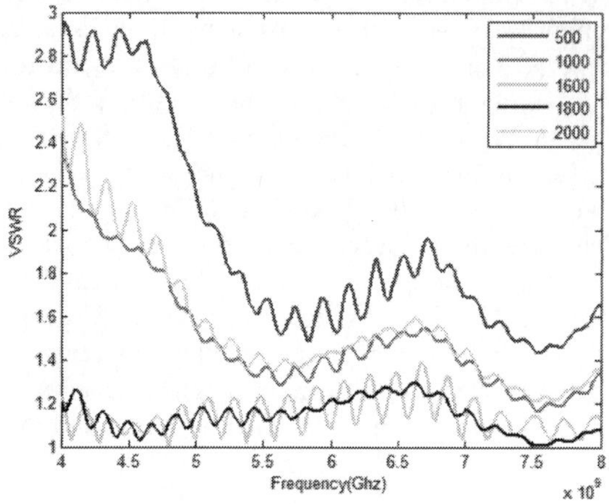

Figure 2 Frequency dependent of VSWR curves for bonded NdFeB permanent magnets.

For Bonded NdFeB permanent magnets with surface field strength levels; 500, 1000 and 2000 G, the circulator shows low performance due to their high power losses. While for bonded permanent magnets with the surface magnetic field strength either 1600 or 1800 G give good insertion losses and their optimum insertion loss is -0.463 dB.

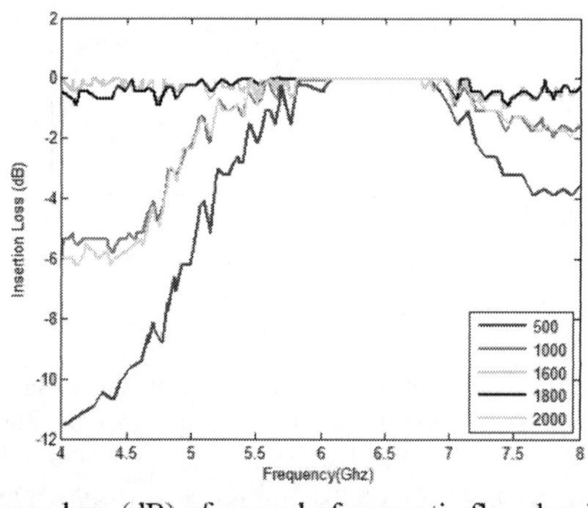

Figure 3 The Insertion loss values (dB) of several of magnetic flux densities versus frequency (GHz) for bonded NdFeB permanent magnets.

Isolation (dB) is the level of reflection power that can be muted by circulator components. Figure 4 show the isolation values of a circulator after installed of bonded NdFeB permanent magnet. The bonded permanent magnets with different of magnetic field strengths exhibit different frequency response. For 500 G of field strength, the circulator has the lowest isolation level; less than 5 dB at frequency ~ 6.3 GHz. While for 1000 G of bonded magnet shows the highest isolation level at 27 dB and the lowest isolation level at 6.5 dB. Whereas the 1800 G of bonded magnet yields optimum isolation with a more linear curve, which the highest and the lowest isolation levels are 18 dB and 10 dB, respectively.

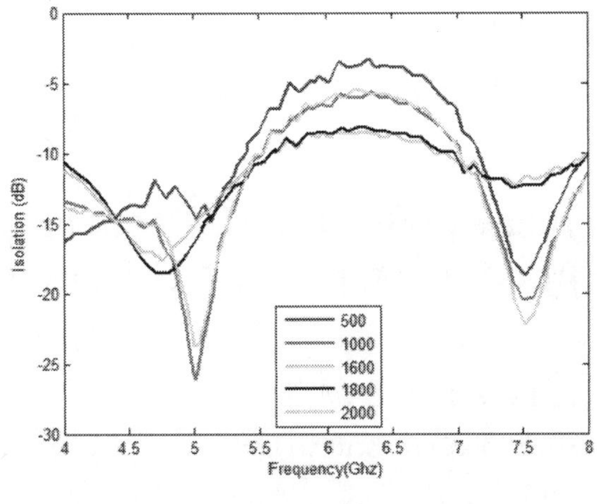

Figure 4 The Isolation curves of bonded NdFeB permanent magnets from a circulator

Figure 4 shows that the bonded magnet with 1800 G has a frequency response that is better to insulation compared with 1600 G of bonded magnets because it can do a reduction to greater isolation. In conclusion, from the characterization results of bonded NdFeB permanent magnets for intrinsic properties, surface magnetic field strength, and circulator performance, such as; VSWR, Insertion loss, and Isolation, which the bonded magnet of 1800 G can be applied as an alternative permanent magnet for A in flow circulator component with work frequency ranging from 4 – 8 GHz in C-Band.

4. References

[1] Rahmouna B, Faozi S. B, D. Vincent and R. Noum 2013 *IOSR-JECE* **8** 31
[2] Li I, Liu Y, Gao SJ, Li M, Wang YQ, and Tu MJ 2006 *JMMM* **195**
[3] Drak M, and Dobrzaanski LA 2007 *AMME* **24** 63
[4] Brown D, Ma BM, and Chen Z 2002 *JMMM* **248** 432
[5] Fidler J, Schreff T, Hoefinger S, and Hajduga M 2004 *Phys.Cond. Matt.* **16** 455
[6] Trosic JS, Stijepovic M, Aleksic R, and Grujic A 2011 *Intech* **25**
[7] Kamo B, Cakaj S, Kolici V, and Mulla E 2012 *JMNSS* **5** 767
[8] Otasowie PO and Ogujor EA 2009 *Phys. Sci.* **11** 651
[9] Chan YK, Koo VC, Chung BK, and Chuah HT 2009 *Proc. Elec. Res. Lett* **7** 139

Journal of Physics: Conference Series **776** (2016) 012031 doi:10.1088/1742-6596/776/1/012031

Effect of nanographene platelets (NGP) surface area on organic dye adsorption using Fe_3O_4-NGP composites

A. Taufik[1,2], and R. Saleh[1,2, a]

[1]Departemen Fisika, Fakultas MIPA-Universitas Indonesia, 16424 Depok, Indonesia

[2]Integrated Laboratory of Energy and Environment, Fakultas MIPA-Universitas Indonesia, 16424 Depok, Indonesia

E-mail: rosari.saleh@ui.ac.id

Abstract. Fe_3O_4-NanoGraphene Platelets (NGP) composites with different surface area were successfully synthesized using sol gel method. The inverse cubic spinel structures as well as graphitic like structure from NGP were detected using X-Ray Diffraction (XRD) Measurement, while the ferromagnetic behavior for all samples were detected using Vibrating Sample Magnetometry (VSM) measurement. The vibrational mode for all samples were characterized using Fourier Transform Infrared Spectroscopy (FT-IR), and thermal stability for all samples were characterized using Thermogravimetric Analysis (TGA). The adsorption process were tested using methylene blue (MB) as a model of organic pollutant. The result showed that the higher NGP surface area could enhance the adsorption capacity of the samples. The kinetic model of adsorption shows that the adsorption process of Fe3O4-NGP materials followed the second order kinetic reaction. The reusability of adsorbent were also performed to analyze the stability of the adsorbent.

1. Introduction

Organic dye is the main organic pollutant compound in wastewater that released from various industries such as leather, textile, paper, and plastic [1-3]. This dye could be toxic and probably mutagenic and carcinogenic. Even at low concentration organic dye is very harmful to human being and microorganism [4]. Therefore it is necessary and important to find an effective method to treat organic dye effluent from industrial process before discharge into water resources. Adsorption process has received special attention in wastewater treatment due to its high efficiency, cost effectiveness, and simple operation [5-6].

Nowadays, there are a lot of adsorbents that have been used in organic dye adsorption such as zeolite, activate carbon, and clay [4,7]. However the low adsorption capacities and low separation efficiencies of these adsorptions limit their application. Recently, magnetic nanomaterials have attracted huge interest due to its high adsorption capacity, high separation efficiency and reusability [8]. Fe_3O_4 has been found as a good adsorption due to high adsorption capacity and unique magnetic properties [9-13]. However Fe_3O_4 nanoparticles are restricted due to its small specific surface area.

Graphene as a new carbon material, has attracted huge attention in adsorption process due to its high surface area [14]. Combination of Fe_3O_4 with graphene materials are expected could enhance the adsorption efficiency. Not only having high adsorption capacity due to its high surface area but also high separation efficiency and reusability of catalyst due to magnetic properties of Fe_3O_4

nanoparticles. So in this research will be reported the adsorption process of Fe_3O_4 with the addition of nanographene platelets with different surface area on the degradation of methylene blue as a model of organic pollutant. The kinetic models of adsorption process are also studied.

2. Experimental

2.1. Chemicals

All chemical reagents are analytical grade and used without further purification. Iron (II) sulfate heptahydrate ($FeSO_4$. $7H_2O$, 99%), sodium hydroxide (NaOH), ethanol and ethylene glycol (EG) were purchased from Merck. Three types of Nanographene Platelets with different surface area were purchased from Angstron Materials, N006-P10 (Surface area 15 m^2/g), N008-100N Surface area (30 m^2/g), and N006-P10 (Surface area 40 m^2/g).

2.2. Preparation Fe_3O_4-NGP composites

Fe_3O_4-NGP were synthesized using the same method in our previous study [15]. Briefly 200 mg of NGP was dissolved in 80 mL water solution and 40 mL ethanol through ultrasonic treatment for 2 h, followed by the addition of 2 g of Fe_3O_4 to the NGP solution and an additional 2 h of stirring to achieve a homogeneous suspension. The suspension was then heated at 120°C for 3 h to affect the deposition on the graphene sheets. The resulting composite was isolated by centrifugation and dried at 70 °C for 12 h. the all samples were denoted as S1, S2, and S3 for Fe_3O_4-NGP composites with surface area 15 m^2/g, 30 m^2/g, and 40 m^2/g respectively.

2.3. Characterizations

The samples were characterized by X-ray diffraction (XRD) measurements using a Rigaku Miniflex 600. The infrared absorption spectra of the samples were recorded using a Shimadzu FT-IR spectrophotometer in the range of 400-4000 cm^{-1}. Thermal stability of the samples were analyze using thermal gravimetric analyzer (Rigaku Thermo Plus EVO2) while the magnetic measurements were performed on Oxford Type 1.2T vibrating sample magnetometer (VSM).

2.4. Adsorption processes

Adsorption experiments were carried out at room temperature for the different time. 0.04 g Fe_3O_4-NGP composites were put into the glass vessel containing 100 mL methylene blue (MB) solution as model of pollutant at concentration of 20 mg/L. The pH solution was adjusted using NaOH or CH_3COOH. After adsorption Fe_3O_4-NGP composites were separated using a permanent magnetic bar from the solution and the dye concentration was analyzed using UV-vis spectrophotometer.

The amount of dye adsorbed at different time intervals was calculated using the following equation:

$$q_t = \frac{(C_0 - C_t)V}{m} \quad (1)$$

where q_t is the amount of dye adsorbed per unit mass of adsorbent (mg/g) at time t and C_t is concentration of dye (mg/L) at time t, m is amount of adsorbent (g), and V is amount of dy solution (L).

The data of adsorption kinetics of MB on adsorbent are fitted by the pseudo-first-order and pseudo-second-order model. The equations of pseudo-first-order and pseudo-second-order model can be described by Eq. (2) and (3) respectively [16].

$$\ln q_e - q_t = \ln q_e - k_1 t \quad (2)$$

$$\frac{t}{q_t} = \frac{1}{k_2 q_e} + \frac{t}{q_e} \quad (3)$$

where q_e and q_t represent the amount of equilibrium adsorption capacity and adsorption capacity at particular times. k_1 is pseoudo first order rate constant while k_2 is pseoudo second order rate constant.

3. Results and discussion

Figure 1(a) shows the XRD spectra from Fe_3O_4-NGP with different NGP surface areas, for comparison the XRD spectra of Fe_3O_4 nanoparticles were also plotted. On Figure 1a all of the composites samples consisted of cubic spinel structure that detected at $2\theta = 30.14°, 35.49°, 43.28°, 53.76°, 57.20°$, and $62.83°$. Thus peak were attributed to (220), (311), (400), (442), (511), and (440) planes respectively. In $2\theta = 26.16°$ the graphitic like structure from NGP were detected for all samples which attributed to (220) plane. The grain size of Fe_3O_4 were estimated using Scheerer equation with analyze peak at (311) plane of Fe_3O_4 and the results were tabulated in Table 1. As can be seen that the grain size value were slightly decrease with the incorporation of NGP materials. The decrease of grain size probably will help in the adsorption process. The lattice parameter of the samples were calculated using Retvield refinement method and the result were also tabulated in Table 1. The lattice parameter of all samples did not change significantly, indicated there were no change in structural properties of Fe_3O_4 nanoparticles and NGP materials.

The molecular vibration of all samples were investigated using FT-IR measurement. The results were plotted in Figure 1(b). In the region of 500 cm^{-1} attributed to binding vibration between Fe and O, while in the region of 1752 cm^{-1} Attributed to C=O binding vibration which attributed to NGP materials [17-18].

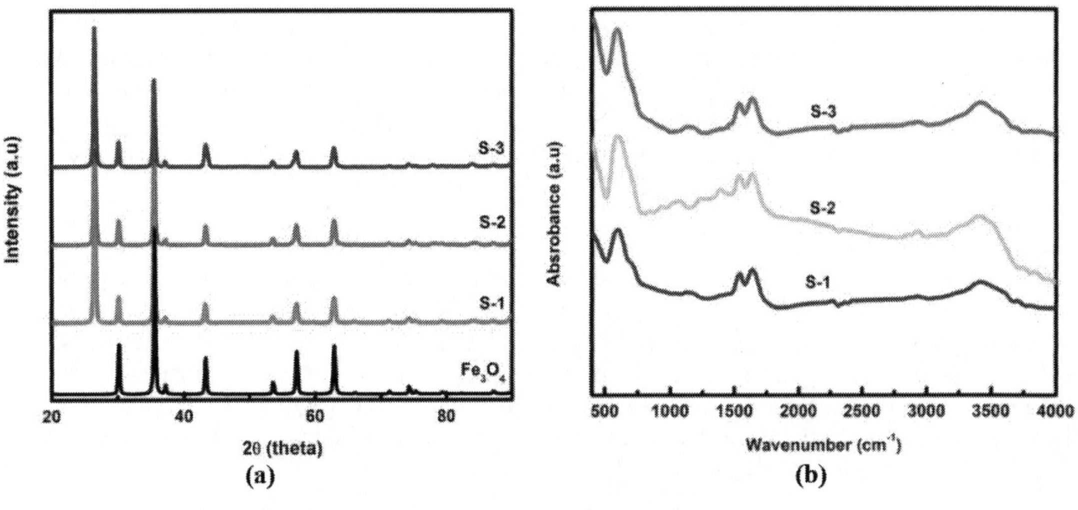

Figure 1. (a) XRD spectra and (b) FT-IR analysis of Fe_3O_4-NGP with different NGP surface area.

Table 1. Lattice parameter, and grain size Fe_3O_4 and Fe_3O_4-NGP composites.

Sample	Lattice Parameter (Å)		<D> (nm)
	Fe_3O_4	NGP	
	a=b=c	a	
Fe_3O_4	8.364	-	45
S-1	8.372	3.632	39
S-2	8.373	3.633	40
S-3	8.375	3.635	38

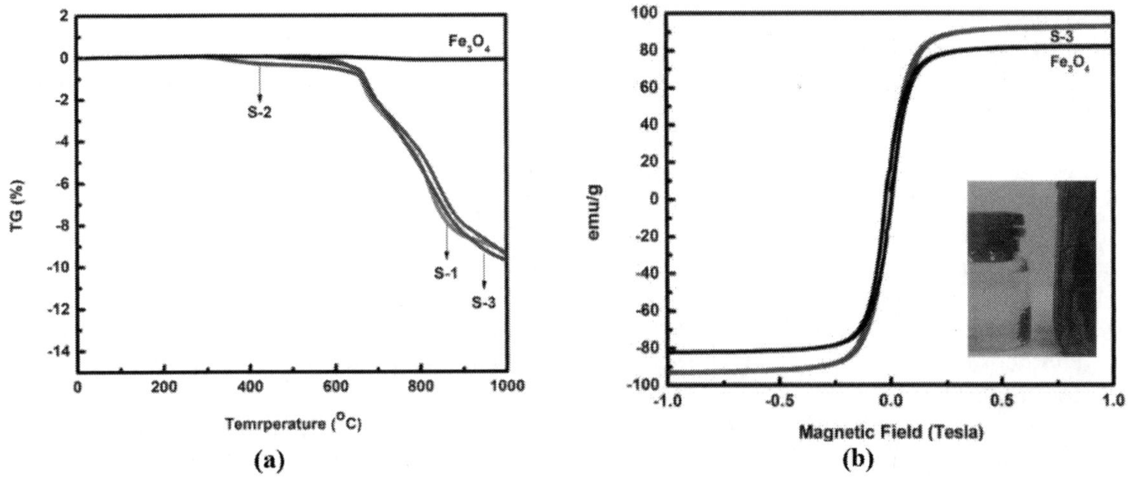

Figure 2. (a) TGA measurement of Fe_3O_4 and Fe_3O_4-NGP composites with different NGP surface area, (b) VSM spectra of Fe_3O_4 nanoparticles and Fe_3O_4-NGP composites.

The TGA result were performed in Figure 2(a). It can be seen from Figure 2a that the samples Fe_3O_4 nanoparticles has good stability until 1000°C. However the incorporation of carbon materials like NGP reduce the stability of the samples under 700°C due to the combustion of carbon materials. Almost 10 wt% were lost after heating treatment 1000°C which attributed to amount of carbon content of the samples.

Magnetic hysteresis curves of Fe_3O_4-NGP composites with the highest surface area (S3) was plotted in Figure 2(b). The magnetic hysteresis from Fe_3O_4 nanoparticle was also plotted as comparison. The both samples exhibited ferromagnetic behavior under room temperature. The saturation magnetization were slightly increase with the incorporation of NGP materials in Fe_3O_4 nanoparticles. These magnetic properties of the samples is important in separation process and also reusability of adsorbent. The inset of Figure 2b is the example of separation process using Fe_3O_4-NGP composites. It is clearly seen that the adsorbent could be easily attracted by external magnetic bar indicated that Fe_3O_4-NGP composites is potential as adsorbent with high efficiency, reusability and stability.

The adsorption ability of the samples were plotted in Figure 3(a) and (b). Figure 3(a) demonstrated the decolorization of MB at particular time. It can be seen from this figure that the incorporation of NGP

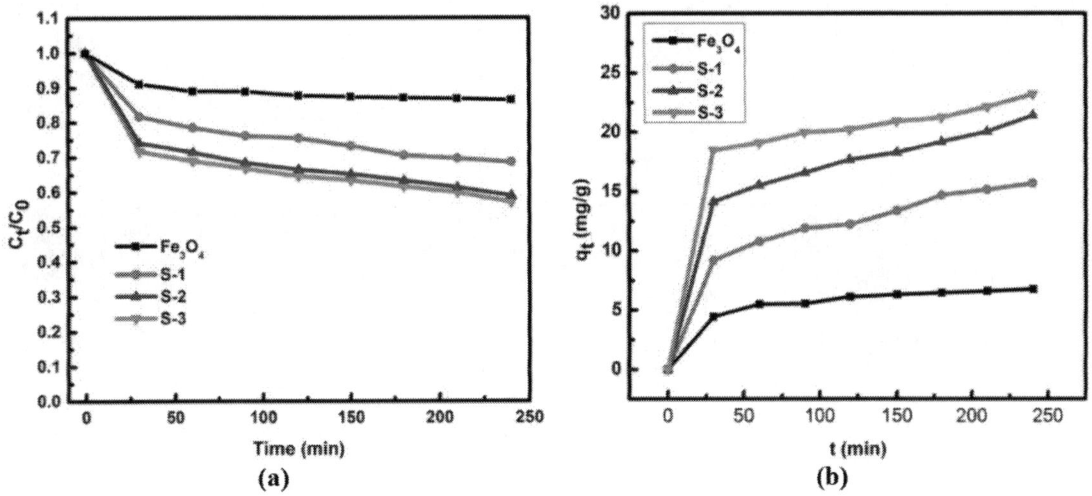

Figure 3. (a) Decolorization and (b) adsorption capacity of Fe_3O_4 nanoparticles, S1, S2, and S3 composites.

materials enhance the decolorization of MB in aqueous solution. The highest decolorization ability occurred using S-3 followed by S-2, S-1 and Fe_3O_4 nanoparticles. Figure 3(b) shows the adsorption capacity for all samples at particular times. As can be seen in Figure 3(b) the highest adsorption capacity was performed using S-3 materials with adsorption capacity approximately 23 mg/g, while the lowest adsorption capacity was performed using Fe_3O_4 nanoparticles with adsorption capacity value 6 mg/g. Based on this result it assumed that specific surface area from NGP is crucial factor in adsorption procces. NGP N008–P10 with highest NGP specific surface area provided more active sites for adsorption process compared to NGP N008–100 N, N006–P10 and Fe_3O_4 nanoparticles.

The fitted data from first and second order were plotted in Figure 4 (a) and (b). As can be seen from Figure 4 (a) and (b) the adsorption process of the samples followed pseudo second order kinetic model. The result of fitted data were tabulated in Table 2. As can be seen in the tables that the equilibrium adsorption capacity value from pseudo second order kinetic reaction almost the same with the experimental value.

The stability of the samples were tested using S-3 materials. Figure 5 demonstrated the percentage degradation of methylene blue using S-3 composites for four repetitive experiment. It can be seen from Figure 5 that the samples has good stability after four repetitive usage. The degradation of methylene blue were decreased only 10% from the initial usage. It means that the magnetic separation process is a powerfull method for separate adsorbent from the solution and produce high stability of catalyst due to it is ability to minimize weight lost of adsorbent.

Table 2. Kinetic adsorption process of Fe_3O_4 nanoparticles and Fe_3O_4-NGP composites.

		Pseoudo First Order			Pseoudo Second Order		
Sample	q_e (mg/g)	q_e cal (mg/g)	k_1	R_2	q_e cal (mg/g)	k_2	R_2
Fe_3O_4	6.75	6.05	0.0172	0.95	7.26	0.005909	0.98
S-1	15.68	14.29	0.0141	0.94	16.36	0.002609	0.97
S-2	23.21	11.03	0.0114	0.81	22.99	0.004576	0.99
S-3	21.40	13.90	0.0108	0.92	21.50	0.002691	0.98

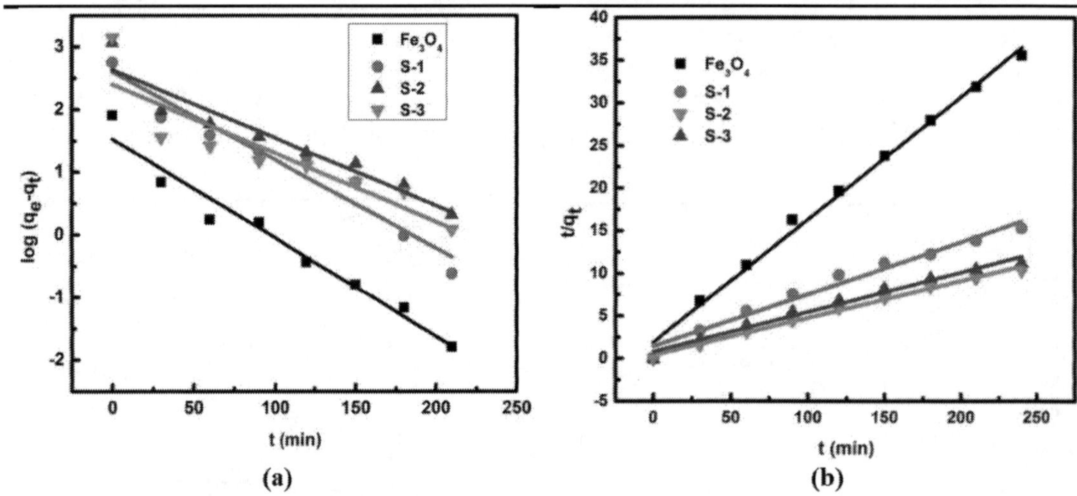

Figure 4. Linear plot of (a) Pseudo First Order Kinetic Models (b) Pseudo Second Order Kinetic Models.

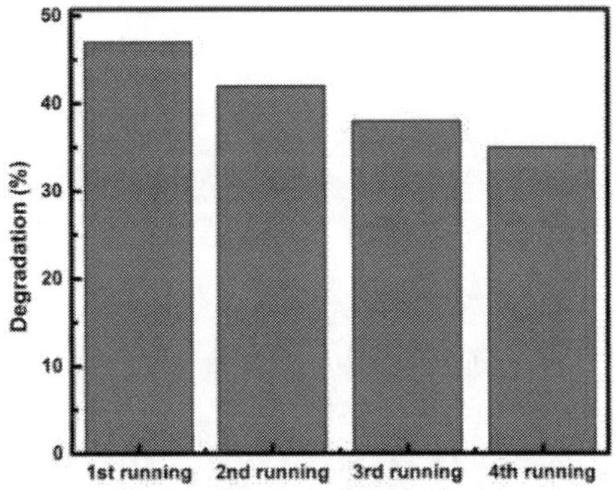

Figure 5. Reusability of Adsorbent.

4. Conclusion
Magnetic Fe_3O_4-NGP composites with different NGP surface area were successfully synthesized using co-precipitation method. The all prepared samples consist of inverse cubic spinel structures and graphitic like structures from Fe_3O_4 and NGP respectively. The adsorption experiment showed that the higher surface area enhance the adsorption efficiency. The adsorption process followed pseudo second order kinetic model and the samples exhibit good stability after four times cycling process.

5. References
[1] Cao C, Xiao L, Chen C, Shi X, Cao Q, Gao L 2014 *Powder Technology* **260** 90–97
[2] Golka K, Kopps S, Myslak ZW 2004 *Toxicology Letters* **151** 203–210
[3] Bao N, Li Y, Wei ZT, Yin GB, Niu JJ 2011 The Journal of Physical Chemistry C **115** 5708–5719
[4] Chang J, Ma J, Ma Q, Zhang D, Qiao N, Hu M, Ma H 2016 *Applied Clay Science* **119** 132–140
[5] Zhang WJ, Zhou CJ, Zhou WC, Lei AH, Zhang QL, Wan Q, Zou BS 2011 *Bulletin of Environmental Contamination and Toxicology* **87** 86–90
[6] Kumar KY, Muralidhara HB, Nayaka YA, Balasubramanyam J, Hanumanthappa H 2013 *Powder Technology* **246** 125–136.
[7] Cheng WP, Gao W, Cui X, Ma JH, Li RF, *Journal of the Taiwan Institute ofChemical Engineers* **62** 192–198
[8] Hoa NV, Khong TT, Quyen TTH, Trung TS 2016 *Journal of Water Process Engineering* **9** 170–178
[9] Lin CC, Lin YS, Ho JM 2016 *Journal of Alloys and Compounds* **666** 153-158
[10] Giri SK, Das NN, Pradhan GC 2011 *Colloids and Surfaces A: Physicochemical and Engineering Aspects* **389** 43-49
[11] Shi HT, Tan LF, Du QJ, Chen X, Li LL, Liu TL, Fu CH, Liu HY, XW Meng 2014 *Dalton Transactions* **43** 12474-12479
[12] Ghaedi M, Hajjati S, Mahmudi Z, Tyagi I, Agarwal S, Maity A, Gupta VK 2015 *Chemical Engineering Journal* **268** 28-37
[13] Saha B, Das S, Saikia J, Das G 2011 *The Journal of Physical Chemistry C* **115** (2011) 8024-8033
[14] Boruah PK, Borah DJ, Handique J, Sharma P, Sengupta P, Das MR 2015 *Journal of Environmental Chemical Engineering* **3** 1974-1985
[15] Taufik A, Saleh R 2016 *Materials Science Forum* **864** 106-111
[16] Anirudhan TS, Deep JR, Binusreejayan 2015 *Chemical Engineering Journal* **273** (2015) 390-400.
[17] Shang Y, Li T, Li H, Dang A, Zhang L, Yin Y, Xiong C, Zhao T 2016 *Composites Part B* **99** (2016) 106-111
[18] Lin CC, Ho JM 2014 *Ceramics International* **40** 10275-10282

Preparation of xerogel SiO_2 from roasted iron sand under various acidic solution

A. H. Ramelan[1], S. Wahyuningsih[2], Y. A. Ismoyo[2], H. P. Pranata[2] & H. Munawaroh[2]

[1]Electronic Materials and Energy Research Group, Faculty of Mathematics and Natural Sciences, Sebelas Maret University
[2]Inorganic Materials Research Group, Faculty of Mathematics and Natural Sciences, Sebelas Maret University
Jl. Ir. Sutami 36A, Kentingan, Jebres, Surakarta 57126, INDONESIA

E-mail: aramelan@mipa.uns.ac.id

Abstract. Xerogel SiO_2 had been prepared from roasted iron sand through variation of Na_2CO_3 addition and sol-gel process under various acidic solution. Roasting treatment was carried out on the compositional variation of iron sand:Na_2CO_3 = 1:2; 1:1 and 2:1 at 1100 °C. While the sol-gel process was conducted at room temperature and neutralized using HCl 0.1 M and 6 M. The color characteristics of roasted iron sand shown light brown, dark brown and dark gray of the compositional variation of iron sand:Na_2CO_3 = 1:2; 1:1 and 2:1, respectively. In addition, the levels of thoughness increased by increasing the ratio of sand in the composition of the mixture. The best composition of roasted treatment was at a variety of iron sand:Na_2SiO_3 = 1:2 with 57.72% had been dissolved in hot water. The addition of Na_2CO_3 will influence the Na_2SiO_3 formation, because of the increase of Na_2CO_3 capable produced the iron sand decomposition product. Na_2SiO_3 gel had been produced after it was neutralized with certain amount of HCl solution. The neutralization was more effective if using high concentration of HCl because of the formation of gel SiO_2 will be easier occurred. The results of SiO_2 had been identified by the FTIR spectra, which an absorption spectra of Si-O-Si asymmetric stretching at 1098.51 cm^{-1}, symmetric stretching of Si-O-Si at 804.35 cm^{-1} and the bending O-Si-O at 469.69 cm^{-1}. The result of SiO_2 content by XRF analysis is about 85.15%.

1. Introduction

Nanomaterials are experiencing a rapid development in recent years due to their existing and/or potential applications in a wide variety of technological areas such as electronics, catalysis, ceramics, magnetic data storage and structural components. Silica as a nanomaterial has been developed due to the nature of nanosilica which has high porosity and surface area so it can be used widely in the field of materials such as filler, pharmaceuticals, catalysis and chromatography [1]. Silica (SiO_2) itself is generally an abundant element and are contained in a variety of natural resources, in an organic source such as rice husk ash which contains SiO_2 reach 86.90 to 97.30% [2, 3] and inorganic sources such as sand.

Indonesia is a country with abundant natural resources which one of them is sand. Silica is usually synthesized from quartz sands because silica content in the quartz sand high enough. Tuban sand contained 81.7% SiO_2 [4] while Bancar sand had a SiO_2 content of 69.3% [5]. Industries usually convert silica (SiO_2) in the sand to form sodium silicate (Na_2SiO_3) from melted quartz sand with sodium carbonate at a temperature of 1300 °C [6]. Sodium silicate is easily soluble in water, so that the metal oxide compounds can be separated by adding water that will dissolve the solid sodium silicate while the metal oxide remains solid form insoluble [7]. Sodium silicate is then made into xerogel through sol-gel method [8]. Sol-gel process was done by adding hydrochloric acid (HCl) to a solution of sodium silicate which was alkaline until the solution has a neutral pH [1].

This paper reported the synthesis of xerogel SiO_2 from iron sand using sol-gel method in order to widen the industrial uses of iron sand and optimized the synthesis of xerogel SiO_2.

2. Experimental methods

2.1. Preparation of iron sand

The composition of the chemical major elements of iron sand from Glagah beach was obtained from the chemical analysis determined by X-ray Fluorescence (XRF). Firstly, iron sand was treated to minimize the content of impurities in the sand then it characterized using XRF to determine the change of the content.

2.2. Pyrometallurgy process

Iron sand was added with sodium carbonate (Na_2CO_3) with the ratio 1:2; 1:1; 2:1 (w/w) and roasted at 1100 °C for 2 hours. Each 10 g roasted iron sand then dissolved in hot water.

2.3. Sol-gel process

Sol-gel process of 50 mL sodium silicate was carried out by neutralization using hydrochloric acid (HCl) in concentration of 0.1 M and 6 M. The results of sol-gel allowed to stand overnight then washed with water and dried. Synthesized xerogel analyzed using XRF and analysis of functional groups using Fourier Transform Infra Red (FTIR).

3. Results and discussions

Visually, Glagah iron sand form relatively uniform smooth black granules with slightly shiny like glass sparkle. Identification of compounds or minerals contained in iron sand was analyzed using X-Ray Fluorescence (XRF). The XRF analysis of iron sand can be seen in Table 1.

The XRF analysis shows that major elements correspond to a high proportion of Fe_2O_3 (about 48%); SiO_2 (about 18%); small content of CaO, Na_2O, TiO_2, Al_2O_3, MgO and traces amount of another elements. After iron sand was treated to minimize the content of impurities, it is known that the content of SiO_2 increased up to 38%.

Table 1. XRF analysis of iron sand

Formula	Content (%)	Formula	Content (%)
Fe_2O_3	48.88	SO_3	0.68
SiO_2	18.97	Cl	0.58
CaO	7.71	Nd_2O_3	0.27
Na_2O	6.62	V_2O_5	0.24
TiO_2	4.72	Pr_6O_{11}	0.15
Al_2O_3	4.35	Cr_2O_3	0.06
MgO	3.45	ZnO	0.06
P_2O_5	1.27	CeO_2	0.05
K_2O	1.01	SrO	0.05
MnO	0.75	SnO_2	0.04

Xerogel SiO_2 can be synthesized either by using organometallic or by the alkaline attack of the sand. In this work, the manufacturing process of silicates by dry way, called furnace process, is used. This corresponds to the fusion of silica and sodium carbonate together at high temperature in a furnace. Iron sand roasting process carried out at a temperature of 1100 °C for 2 hours with the ratio of iron sand:Na_2CO_3 = 1:2; 1:1; 2:1. Roasted iron sands was obtained with the characteristics that can be seen in Figure 1.

Figure 1. Roasted iron sands with the compositional variation of iron sand:Na_2CO_3 = 1:2 (a); 1:1 (b) and 2:1 (c)

In general, roasted iron sands generated molten that hardens like a rock and hard enough to be destroyed, if it compared with one another, the more addition of sand will obtain more harden solid. The color characteristics of roasted iron sands shown light brown, dark brown and dark gray of the compositional variation of iron sand:Na_2CO_3 = 1:2; 1:1 and 2:1, respectively.

At decomposition, Na_2CO_3 presence would disturb the mineral composition which Na ions trapped in the network and reduce the number of bridges/bonds between tetrahedral. Na^+ cations have an effect on the size of the hole/cavity and is expected to the formation of clusters [10]. Therefore, more or less Na_2CO_3 were added to the iron sand affects the decomposition of sand that affected the differences in color characteristics and thoughness as the result of roasting.

Table 2. Results of roasted iron sands dissolution

Compositional ratio iron sand:Na_2CO_3 (w/w)	**(1:2)**	**(1:1)**	**(2:1)**
Rendement mass (g)	4.228	8.127	9.171
Dissolution efficiency (%)	57.72	18.73	8.29

In Table 2 shows that the most soluble roasted iron sands is the variation of iron sand:Na_2CO_3 = 1:2 with dissolution efficiency reach 57% from 10 g roasted iron which has been dissolved. These additional variation is the most effective ratio in decomposing minerals in the sand and generates more sodium silicate seen from the dissolution efficiency.

Sol-gel process in this work has been done by adding a variation of 0.1 M HCl and 6 M to neutralize the alkaline solution of sodium silicate and the results can be seen in Figure 2.

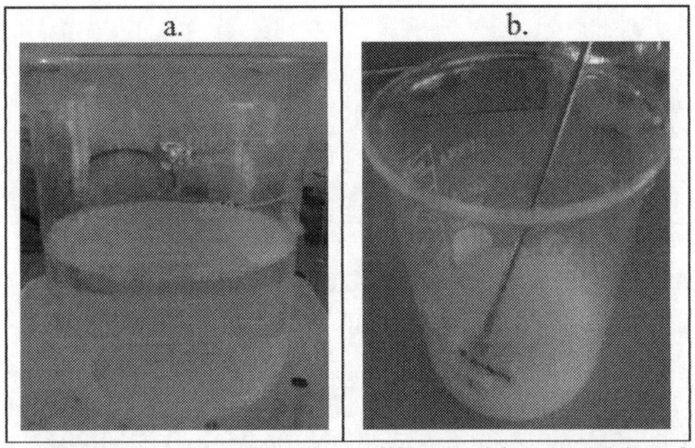

Figure 2. Neutralization results of sodium silicate with 0.1 M (a) and 6 M (b) HCl

The neutralization of sodium silicate solution with 6 M HCl is more effective because at pH 9 xerogel SiO2 is formed and requires only small addition of aqueous HCl. While the use of 0.1 M HCl less effective because at neutral pH only formed very little gel floating in solution and the neutralization required lots aqueous HCl. The use of concentrated HCl produces H^+ ions with large numbers but in a minimal presence of solvent so it would be more effective in catalyzing sodium silicate in the sol-gel process to form xerogel. Content analysis of xerogel is performed by XRF and can be seen in Table 3.

Table 3. XRF analysis of xerogel

Formula	**Content (%)**	**Formula**	**Content (%)**
SiO_2	85.15	V_2O_5	0.16
Al_2O_3	3.41	Fe_2O_3	0.06
K_2O	2.98	ZrO_2	0.02
Cl	2.62	BaO	0.01
P_2O_5	2.33	CuO	0.01
SO_3	1.81	ZnO	0.01
CaO	1.39		

Journal of Physics: Conference Series **776** (2016) 012032 doi:10.1088/1742-6596/776/1/012032

Table 3 shows the final content of SiO_2 reach 85% with other contents as impurity. To clarify the final results, functional groups analysis of xerogel is done using FTIR and can be seen in Figure 3.

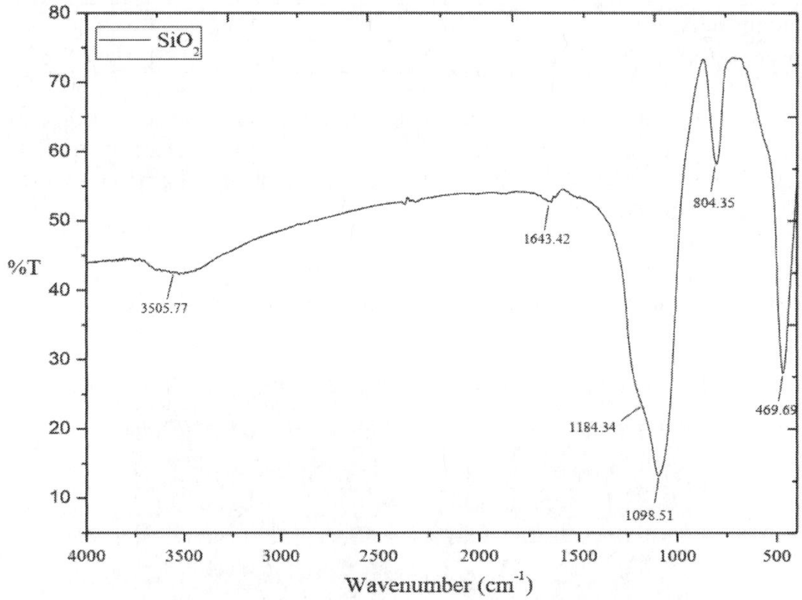

Figure 3. Infra red spectra of xerogel

The results of SiO_2 had been identified by the FTIR spectra, which an absorption spectra of Si-O-Si asymmetric stretching at 1098.51 cm^{-1}, symmetric stretching of Si-O-Si at 804.35 cm^{-1} and the bending O-Si-O at 469.69 cm^{-1}. These absorption characteristics in accordance with the research of Musić et al. (2011), which identifies amorphous SiO_2.

From this evidence, it can be assumed that the xerogel is relatively pure SiO_2 because no other group absorption was analyzed. Impurities were detected on XRF analysis only physically interact and bond does not occur so that does not affect the functionality of the compound of SiO_2. Meanwhile, it appears quite weak absorption at wavenumber 1643.42 and 3505.77 cm^{-1} is the vibration profile of –OH stretching –OH bending which indicates relatively hygroscopic properties of amorphous SiO_2 produced.

In conclusion, xerogel SiO_2 can be synthesized from iron sand with the final result contain 85.15% SiO_2. The more addition of Na_2CO_3 will increase the decomposition of iron sand, with the optimum ratio of sand:Na_2CO_3 = 1:2. While the formation of xerogel is more effectively done with a high concentration of HCl.

Acknowledgment
The researchers wish to express their gratitude to the Insinas Ristek Dasar 2015 Kemenristek Dikti for supporting this research.

Refferences

[1] Liou T-H and Yang C-C **2011** *Materials Science and Engineering B* 521-529.
[2] Trivana L, Sugiarti S and Rohaeti E **2015** *Jurnal Sains dan Teknologi Lingkungan* 66-75.
[3] Sayekti W, Ari HR, Ludfiaastu R, Reva EN, Rizky MIM and Uswatul C **2015** *Journal of Chemical and Pharmaceutical Research* 85-89.
[4] Suparsih THS and Zainuri M **2013** *Jurnal Teknik Pomits* 1-3.
[5] Munasir, Triwikantoro, Zainuri M and Darminto **2013** *Jurnal Penelitian Fisika dan Aplikasinya* 12-17.
[6] Iler RK **1979** *The Chemistry of Silica* John Wiley and Sons New York.
[7] Xu Y, Luo X and Chung DDL **2000** *Journal of Electronic Packaging* 128-131.
[8] Dorcheh AS and Abbasi MH **2008** *Journal of Materials Processing Technology* 10-26.
[9] Smallman RE and Bishop RJ **2000** *Metalurgi Fisik Modern dan Rekayasa Material* Erlangga Jakarta.
[10] Musić S, Filipović-Vinceković S and Sekovanić L **2011** *Brazilian Journal of Chemical Engineering* 89-94.

Properties of plate-like carbonyl iron particle for magnetorheological fluid

S. T. Shilan, S. A. Mazlan, M. H. A. Khairi and Ubaidillah
VSE Research Laboratory, Malaysia-Japan International Institute of Technology (MJIIT) Universiti Teknologi Malaysia, Jalan Sultan Yahya Petra, 54100 Kuala Lumpur, Malaysia.
E-mail: salihahtsl@yahoo.com

Abstract. This work experimentally discussed the characterization, magnetic, and rheological properties of plate-like carbonyl iron particle (CIP) in comparison with conventional spherical CIP. Plate-like CIP was produced by using ball milling method. The effect of plate-like shape on the magnetic behavior of CIP was firstly investigated by vibrating sample magnetometer (VSM). The results indicated that the plate-like CIP obtained higher saturation magnetization (about 8%) than that of the spherical particles. In addition, the field-dependent rheological properties such as yield stress were investigated and the results are compared between two particles as a function of the magnetic field intensity.

1. Introduction

Magnetorheological fluid (MRF) is a magnetic field responsive fluid containing polarizable particles and a liquid carrier. MRF is typically used in automobile application such as dampers to provide varied damping forces, depending on a magnetic field strength. In general, the apparent viscosity of MRF is changeable with the application of magnetic field strength [1]. This happens due to the suspended magnetic particles in non-magnetic medium of MRF which is aligned as a chain-like formation and restricted the fluids flow [2]. This phenomenon developed the yield stress in MRF. The magnetic and rheological properties of MRF are highly dependent on the magnetisable particles features. It has been stated that they play the important roles in achieving high yield stress and low sedimentation of MRF [3]. In recent researches, numerous types of magnetisable particles are widely used as constituents of MRF, such as cobalt alloy, nickel alloy, soft ferrite, manganite and CIP [3, 4]. CIP is the most common magnetic particles used in MRF fabrication due to its promising permeability, relatively low electrical conductivity, high Curie temperature, high saturation magnetization and low cost in comparison with other metallic particles [2, 3, 4]. CIP based MRF is also capable to operate with the yield stress of 100 kPa and has a high saturation magnetization about 2.1 T [5].

The development of MRF with the aim of achieving high yield stress has extensively been studied in order to enhance the maximum output force of the application devices or systems [6]. It is well known that the yield stress can be increased by increasing the particles amount and particle size. However, this may cause high yield stress at the off-state and severe sedimentation problem [7, 8]. The high yield at the off-state restricts controllable range of the field-dependent force or torque in application systems. Therefore, many researchers have focused on improving the rheological properties of MRF through adjusting magnetisable particles behaviours such as coating of magnetic particles, addition of additives and bi-dispersing of nano size particles [9].

Recently, the particle shape has been considered to be the most important factor to overcome the sedimentation problem and enhance the output force [3, 9, 10, 11]. As reported by Upadhyay et al. [9], the flake-based MRF exhibited a higher storage modulus compared to sphere-based MRF. This is due to the field dependent solid friction between the flake shaped particles. In addition, the sedimentation rate decreased about 50% once the flake-like shape particles were used. With this regard, Gomez-Ramırez et al. [10] prepared CoNi-based MRF with different shape of magnetisable particles; CoNi nanofibers (56 nm length, 6.6 nm width) and nano spheres (24 nm diameter). It was found that the MR effect remarkably strengthened once nanofibers were utilized as dispersed phase instead of nano spheres due to the formation of the strong effect of fibre orientation on the demagnetizing field. Comparing the rod-based and sphere-based MRF revealed that the rod based MRF had an enhanced MR performance under oscillatory shear [11].

The main technical contribution of this work is to introduce plate-like CIP based MRF and compare the field-dependent rheological properties with the spherical particles based MRF. The plate-like particles are prepared through the solid state powder processing technique which involves repeated welding, fracturing and re-welding of powder particles in a high-energy ball mill. After testing X-ray diffraction of the particles before and after milling process, the particle shapes are observed via SEM (Scanning Electron Microscopy). Then, the yield stress are evaluated and compared between two different particle shapes based MRF. It is noted here that to the best of our knowledge, no study has been conducted to prepare the plate-like CIP and investigated the effect of the plate-like CIP on the rheological behaviours of MRF.

2. Experimental

In this study, the OM series of CIP manufactured by Basf Company was used with average size of diameter and density are 4 μm and 7.874 g/cm^3 respectively. In order to synthesize plate-like shape of CIP, rotary ball mill method was used. The milling process was operated at 200 rpm for 20 hours in room temperature by using wet milling method. Zirconia ball was used as grinding media with ball to powder weight ratio is 20:1. Pure ethanol was added as process controlling agent (PCA) to overcome adhesion between particles and to improve the efficiency in synthesizing plate-like shape particles.

Polyalphaolefin (PAO) oil produced by Ferrotec Company was used as carrier fluid in MRF preparation, and oleic acid from R&M Chemicals with density of 0.9 g/ml3 is adopted as additive. Two types of MRF are individually prepared using spherical and plate-like CIP. The phase structure analysis of both particles are identified (within 2θ range of 35–90°) using powder X-ray diffractometer (PANalytical XRD) employing Cu-Ka X-radiation of wavelength 1.5418Å. A scanning electron microscopy (JEOL JSM-7001F SEM) was used to analyze the surface morphology and the average size of the particle. Magnetic behavior of the particles were characterized using a vibrating sample magnetometer (Lakeshore 7404 series VSM) in magnetic field range of 10000 Oe. In addition the rotational flow tests are carried out at 25 °C using a parallel-plate configuration in the MCR 302 Anton Paar rheometer. The 20-mm diameter plate and Magneto-Rheological Device (MRD 70 Anton Paar) are attached for on-state measurement. The on-state measurement for the shear stress and viscosity is carried out at conditions; magnetic field from 100 mT until 500 mT and rate sweep until 1000/s.

3. Results and Discussion

3.1. Structural characterization

Figure 1 shows the X-ray diffraction (XRD) pattern for spherical and plate-like carbonyl iron particle (CIP) at 25 ^{0}C. The diffraction peaks for both samples can be indexed to the body centre cubic structure (BCC α Fe) with Im-3m space group. The intensity peak was observed to increase as the CIP undergoes the milling process. A sharper and broaden of diffraction peak was obtained by plate-like CIP due to the decreasing of the crystallite size and the increasing of the internal lattice strain [12]. Furthermore, the peaks widths broaden are also attributed to the accumulation of lattice defects such as grain boundaries and dislocations during the milling process [12]. Herewith, the changes observed in the XRD patterns suggest that the particle shape is changed from the spherical to the plate-like. Although, there is no impurity peak was observed and the structure of CIP was preserved where the diffraction peak for both shapes of CIP was aligned at the same angle.

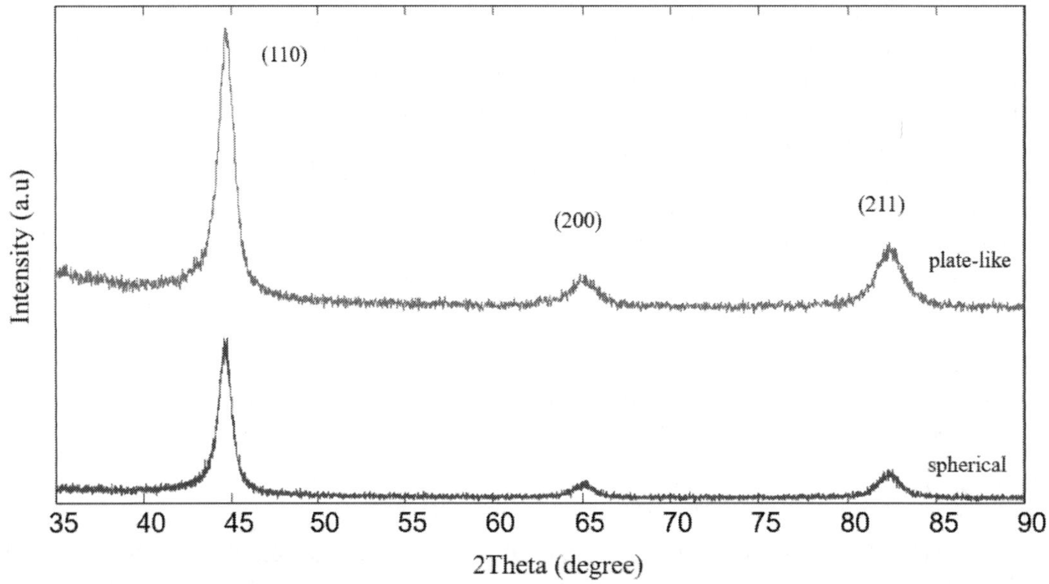

Figure 1 X-ray Diffraction graph for spherical CIP and plate-like CIP at T = 293 K.

The SEM micrographs for both spherical and plate-like CIP are shown in Figure 2. It is evident that the particles are flattened and varied to the plate-like shape with more homogeneous distribution after the ball milling process. It is also obvious that the particles become more dispersed after ball milling (Figure 2b). This is because the collision between balls, vial and powders leads to break the cluster form of particles. From the SEM micrograph, the average particle size is measured and the diameter size for spherical and plate-like CIP is 2 μm and 3.6 μm respectively. The particle length was increased about 80% after the ball milling process with an average thickness about 0.86 μm.

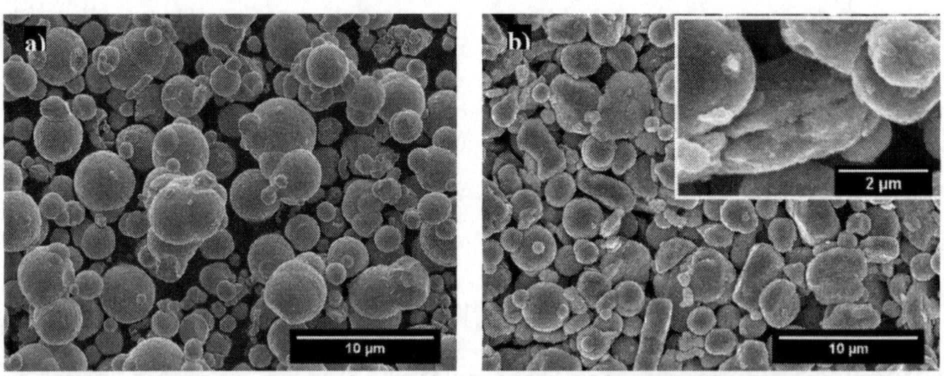

Figure 2 SEM micrograph for a) spherical CIP and b) plate-like CIP at x3000.

3.2. Magnetic characterization

M-H curve for both spherical and plate-like CIP was shown in Figure 3. Magnetic saturation was changed as the shape of CIP changed, which is induced by the milling process. The plate-like CIP exhibits higher saturation magnetization compared to the spherical CIP. Indeed, the saturation magnetization for the plate-like CIP is 211.3 emu/g while the spherical shape CIP is only 195.7 emu/g. This happens due to the shape anisotropy in magnetic anisotropy where it can only be obtained by non-spherical particles. Not perfectly spherical particles lead to the unequal direction of demagnetization field. Hence, one or more easy axis can be created where the magnetic domains tend to align. In addition, the magnetised particles will align with their long axes in the field direction, which results in the polarization charge concentration at the upper and lower edges.

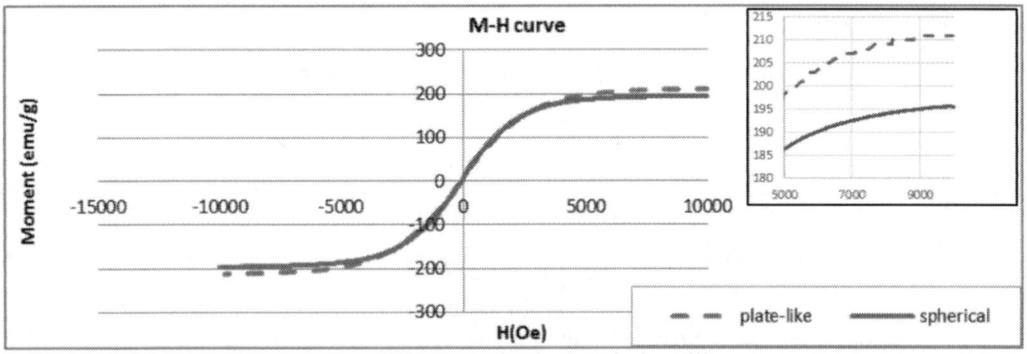

Figure 3 Magnetization curve for spherical and plate-like CIP at T = 293 K.

3.4. Rheological properties

Flow curves of shear stress versus shear rate for both plate-like and spherical CIP based MRF are shown in Figure 4. Different magnetic field strengths are applied to investigate their MR effect. It is found that the shear stress increases much faster at a lower shear rate and then changes slowly at a high shear rate. This is due to the fact that disordered arrangement of particles leads to the enhancement of flow resistance at lower shear rate, while particles begin to arrange their orientation along the direction of shear at higher shear rate, causing to the decrease of viscosity [13]. Further, the flow curves obtained by both samples are best fitted with Herschel-Bulkley (H-B) model to describe the shear yield stress value [6]. In this model, the shear stress is given by

$$\tau = \tau_0 + k\dot{\gamma}^n \quad (1)$$

where τ_0 is the shear yield stress in the state of presence and absence applied magnetic field, τ is the shear stress, k is the consistency index, n is the shear thinning exponent and $\dot{\gamma}$ is the shear rate. The term $k\dot{\gamma}^n$ shows the shear thinning behaviour of the suspensions fluid under the magnetic field. The n describes the degree to which a material is shear thinning ($n < 1$) or shear thickening ($n > 1$. The values of n analysed from this both samples are lower than 1, indicating that the rheological behaviour for both fluids are shear thinning. The shear yield stress values, which are obtained by fitting the H-B model to the experimental curves, are plotted as a function of magnetic field strength as shown in Figure 5.

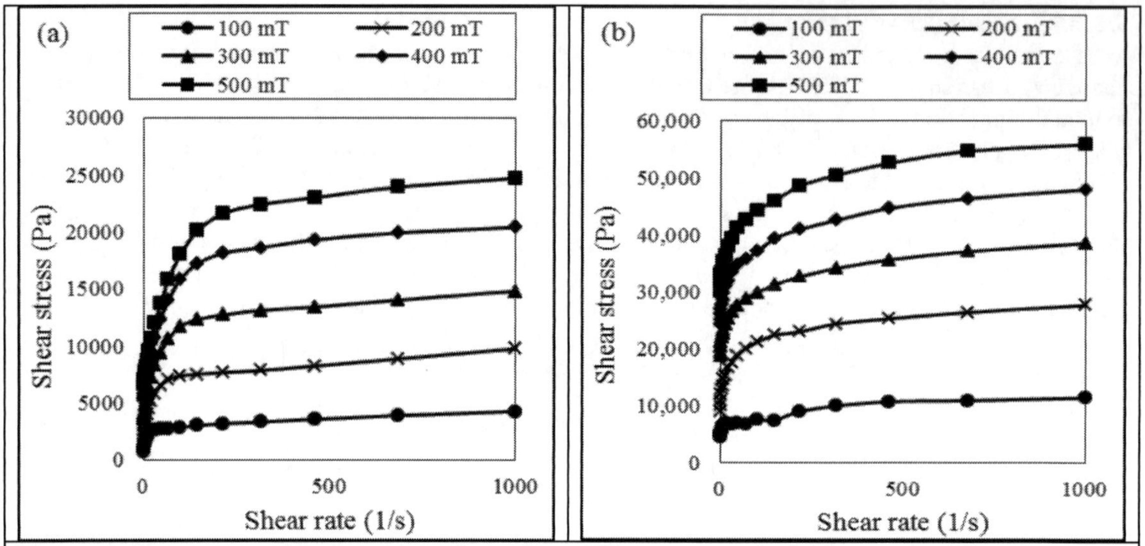

Figure 4. Flow curves of the investigated sample at different magnetic field strengths (H) at T = 293 K (a) spherical CIP and (b) plate-like CIP.

As presented in Figure 5, the shear yield stress for both MRF samples is observed to increase with increase of the applied magnetic field strength. This is more probably due to the stronger pull force among magnetic particles in the presence of applied magnetic field. As the magnetic field intensity is greater, a larger shear stress is needed to break the column-like structures and fluid to start to flow. It is interesting to notice that the plate-like CIP based MRF exhibits a sharp increase of shear yield stress value compared to the spherical CIP based MRF. Hence, a greater yield stress about 81494 Pa obtained by plate-like CIP based MRF compared to the spherical CIP based MRF (about 21979 Pa) at 700mT. Indeed, the yield stress achieved by plate-like based MRF is increased up to 270% compared to its original spherical shape. This is because the plate-like particles have larger contact surface areas which lead to a strong attraction between particles and consequently the clusters become hard to break and flow.

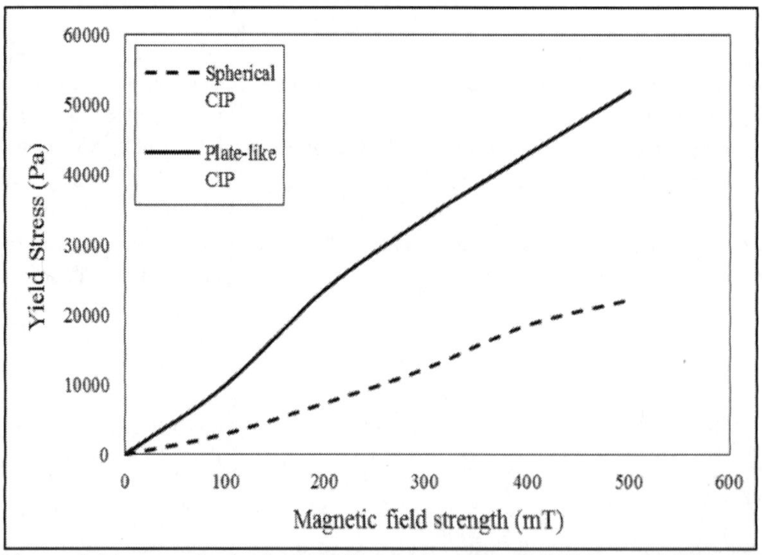

Figure 5 Shear yield stress for both plate-like and spherical CIP based MRF at T = 293 K.

4. Conclusion

In this work, two different shape of CIP for MRF composition were prepared and experimentally investigated the filed-dependent rheological properties. It has been observed from XRD and SEM that no new phase is occurred by milling the spherical shape to the plate-like shape and the plate-like CIP is flattened resulting in 86% of the length enhancement. This resulted in higher saturation magnetization compared to the original un-milled sample. This happen because the plate-like CIP has larger contact surface which formed stronger clusters due to magnetic field. Finally, the field-dependent rheological test has been demonstrated that the yield shear stress of the plate-like CIP based MRF is increased by 270% compared to the spherical particles based MRF.

5. Acknowledgement

This work was funded by the Ministry of Higher Education, Malaysia under PRGS (Vote no. 4L667), Universiti Teknologi Malaysia under GUP (Vote No: 13H55).

6. References

[1] Iglesias GR, Lopez MT-L, Duran JDG, Gonzalez F-C and Delgado AV 2012 *J.Colloid Interface Sci.* **377** 153-159
[2] Yang R-B and Liang W-F 2011 *J.Appl.Phys.* **109** 07A311.
[3] Bell RC, Vavreck AN, and Zimmerman DT 2007 *Int.J.Modern Phys. B* **21** 5018-5025.
[4] Olabi AG and Grunwald A 2007 *Mater.Des.* **28** 2658–64
[5] Goncalves FD 2006 *Shock Vib.Dig* **38** 203–219.
[6] Zhu X, Jing X, and Cheng L. 2012 *J.*Intell.*Mater.Sys.Struct.* **23** 839-873.
[7] Ierardi RF and A. J. F. 2009 *J.Phys.Conf.Ser.* **149** 012037.
[8] Park JH, Chin BD, and Park OO 2001 *J.Colloid Interface Sci.* **240** 349–354.
[9] Upadhyay RV, Laherisheth Z, and Shah K 2014 *Smart Mater.Struct.* **23** 015002.
[10] Gómez A-R, López MT-L, Durán JDG, and González F-C 2009 *Soft Matter* **5** 3888.
[11] de Vicente J, Segovia JP -G, Andablo E-R, Vereda F, and Hidalgo R-Á 2009 *J. Chem. Phys.* **131** 194902.
[12] Hajalilou A, Hashim M, and Mohamed H-K 2014 *J.Mater.Sci.Mater.Electron.* **26** 1709-1718.
[13] Hong R, Ren Z, Han Y, Li H, Zheng Y and Ding J 2007 *Chem.Eng.Sci.* **62** 5912-5924.

Effect of sucrose acetate isobutyrate ester on the epoxidised natural rubber based magnetorheological elastomers

M. H. A Khairi[1], S. A. Mazlan[1], Ubaidillah[2], K. Z. K. Ahmad[3], S. A. A. Aziz[1], N. A. Yunus[1]

[1]Vehicle System Engineering, Malaysian-Japan International Institute of Technology, Universiti Teknologi Malaysia, Jalan Sultan Yahya Petra, 54100 Kuala Lumpur, MALAYSIA

[2]Mechanical Engineering Department, Faculty of Engineering, Universitas Sebelas Maret, Jl. Ir. Sutami 36A Kentingan Jebres Surakarta 57126, INDONESIA

[3]Mechanical Engineering Department, Engineering Faculty, National Defence University of Malaysia, MALAYSIA

E-mail: amri.kl@utm.my

Abstract. In this study, magnetorheological elastomers (MREs) were synthesized by incorporating various amounts of sucrose acetate isobutyrate (SAIB) ester plasticizer dopped in the epoxidized natural rubber (ENR-50) matrix with constant carbonyl iron particles (CIPs) of 60 weight percent. The effect of SAIB insertion on curing characteristics has been observed using a moving die rheometer. The microstructures appearance of the samples were observed using a field emission scanning electron microscope (FESEM). Finally, the magneto-induced storage modulus was measured by using a rheometer 302. The experimental results have demonstrated that the addition of SAIB ester decreased the cure rate index (CRI) by 63% and increased the MR effect by 37%. Observation of microstructure shows that the CIPs embedded well in the matrix and produced the isotropic structure of the MREs. Meanwhile, the fabricated MREs samples were frequencies dependent, where all MREs samples exhibit the incremental trend when increasing the frequencies. The introduction of SAIB has proven to influence the properties of the MREs.

1. Introduction

Magnetorheological elastomers (MREs) belong to the class of the smart materials that have rheological properties alteration under the influence of magnetic field [1]. MREs consist of magnetic particles embedded in a solid analog matrix like elastomers. Particles inside the elastomer can be either isotropic (random distribution) or anisotropic (aligned structure) [2]. To produce anisotropic types, magnetic fields must be applied to the MREs compound during pre-curing so that the magnetic particles locked in their position before vulcanization. In another way, the absent of external magnetic during the curing; makes the structure of MREs becoming isotropic [3]. Viscoelastic and rheological properties such as stiffness and damping properties can be changed rapidly, continuously and reversibly under presentation of the magnetic field [4]. Having magnetic field dependent modulus and damping properties, MREs become a promising material for a wide range of possible applications including vibration absorbers [5], sensing devices [6], and morphing structure [7].

The influence of plasticizer application on MREs performance was seldom reported. Lokander and Stenberg [8] has found that utilization of plasticizer in MREs enhanced the shear storage modulus. The plasticizer is a kind of additions in the rubber technology in which it can be dissolved in the rubber after mixing and vulcanized. Plasticizer acts as the lubricant and lets the molecular chains of rubber glide easily, and then the rubber matrix shows a low adhesiveness. The addition of plasticizers in the MREs is expected to improve MREs performances because plasticizers can change not only the rubber mechanism but also modify magnetic particles properties [9,10]. Therefore, in this study, the sucrose acetate isobutyrate (SAIB) ester is utilized in the MREs preparation process to modify the viscosity of the ENR-50. This ester also can improve the compatibility between the CIPs and the matrix. Four samples containing a different weight percentage of SAIB ester have been fabricated with constant CIPs content (60 wt%). Field emission scanning electron microscopy (FESEM) was used to observe the distribution and the morphology of the CIPs in the matrix. Finally, the shear storage modulus and the MR effect were assessed using rheometer to analyze the effect of SAIB on the magneto-induced rheological properties.

2. Materials and Method

The isotropic MREs samples used in this work were prepared by compounding CIPs having an average diameter of 6µm (BASF, type CN) and epoxidized natural rubber with 50% mole (ENR-50).The formula for the rubber compound is 100 phr ENR 50, 5 phr zinc oxide, 2 phr stearic acid, 19 phr carbon black FEF N550, 5 phr tudalen 65, 2.3 phr sulfur and 0.8 phr CBS. (Rubber and rubber additive supplied by Malaysian Rubber Board). The rubber matrix and rubber additive were mixed homogeneously using two roll mills to produce rubber compound. The compound then divided to four samples to produce MRE with constant CIPs content of 60wt% and various SAIB ester content of 0,5,7.5, and 10 wt%. The MREs sample was put into the mold under the high pressure, and the temperature was fixed at 150° C for 10 minutes. The minimum torque (ML), maximum torque (MH), delta torque (ΔH), cure rate index (CRI), the cure time (t_{90}) and scorch time (ts_2) were determined using a Monsato Moving Die Rheometer (MDR 2000) according to ASTM method D 2084. The cure rate index is a parameter which indicates the speed of curing reaction was determined from rhemometric data. The microstructure image of MRE was observed using the field emission scanning electron microscope (SUPRA 40VP FESEM, Carl Zeiss, AG Germany). The microstructure observation was performed at an accelerating voltage of 1 kV with 1000 times magnification. The shear storage modulus of the MRE samples was measured by the rheometer (Anton Paar, model Physica MCR 302). The samples were tested under different flux densities of 0 and 800 mT and conducted under frequency sweep modes (from 0.1 to 100 Hz).

3. Results and Discussion

Figure 1 shows the curing curve without and with increasing amounts of SAIB ester. The minimum and maximum torques are denoted by ML and MH. The maximum torque depends on curing related factors, including cross-link density and cross-link length. Every new cross-link in rubber mass decreases polymer chain movement and chain flexibility and this cause to more torque for a disk of the rheometer. When all the curing agent is demanded, and all the cross-link is created, cross-link density is at its maximum level, then rheometer curve got plateau and considered it as maximum torque. From Table 1, the difference between MH and ML which is $\Delta M = (MH - ML)$ decreases with increasing the weight percentage of SAIB ester. The ΔM is related to the degree of cross-linking and hence its decreases can be attributed to the decrease in the cross-link density by incorporation of SAIB ester. Subsequently, the scorch properties indicate how long a rubber compound can remain warm for processing before it loses its ability to change its shape. Often, once a rubber compound is scorched, it can no longer be processed.

It is important to measure the scorch time to ensure that no time is wasted processing or molding material that will not create a good product. For example, premature scorch can lead to underfilled mold cavities as the material loses its ability to flow before completely filling the space. From the experiment, the scorch time increases due to the enhancement of matrix mobility with the introduction of SAIB ester. This happening allows ample time for mold filling and produces the perfect shape of MRE. Meanwhile, the cure time is a time required for the samples to reach a desired state of cure, usually to reach 90% of the maximum cure. The cure time increases as the concentration of SAIB ester increased which corresponding with the scorch time. Meanwhile, the cure rate index decreases as a result of the interaction with SAIB ester, which slower the vulcanization reaction. The effects of SAIB ester are shown in Table 1.

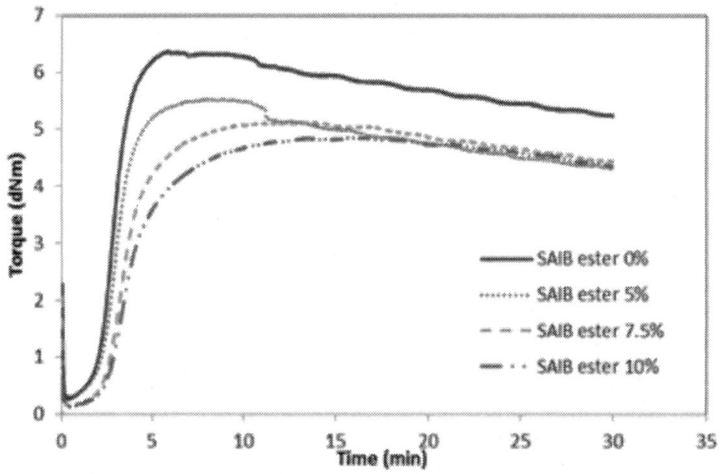

Figure 1. Curing curves of MREs without and with increasing amounts of SAIB ester.

Table 1. The scorch time (ts_2), the optimum curing time (t_{90}), maximum torque, minimum torque and the cure rate index (CRI) of the MRE samples.

SAIB ester content (wt%)	ts_2 (min)	ts_{90} (min)	Max torque	Min torque	CRI
0	2.61	4.13	6.37	0.26	65.8
5	2.87	4.58	5.53	0.29	58.5
7.5	3.30	6.15	5.12	0.13	35.1
10	3.56	7.67	4.86	0.13	24.3

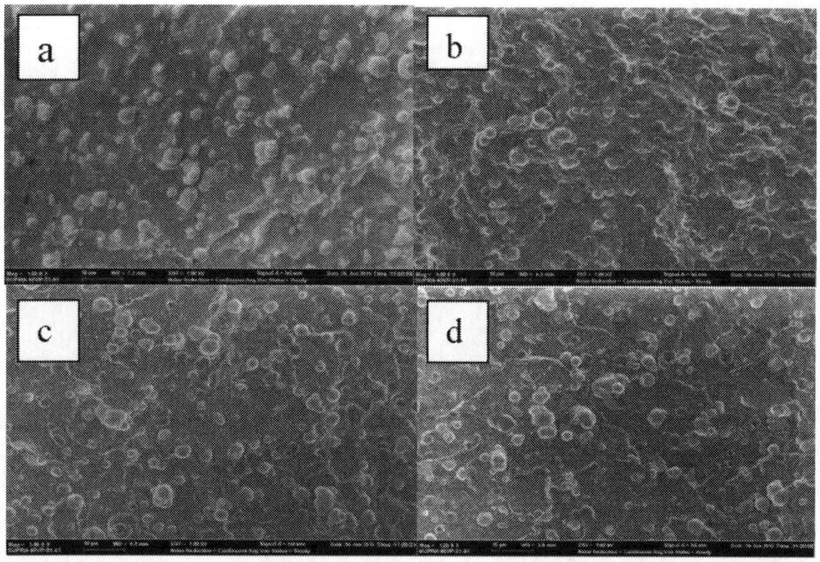

Figure 2. FESEM images of the samples a) SAIB ester 0 wt% b) SAIB ester 5 wt% c) SAIB ester 7.5 wt% and d) SAIB ester 10 wt%.

Figure 2 verifies that the CIPs in the zero-field curing process have been dispersed uniformly and embedded in the elastomer matrix, and MREs has exhibited an isotropic condition. Even though the SAIB ester content in the matrix has increased, the CIPs agglomerates have not been found in the elastomer matrix.The interface between the CIPs and the matrix has been unclear. This fact indicates that the CIPs have good interfacial interaction with the elastomer matrix.

The variation of storage modulus with a frequency of four samples of isotropic MREs is shown in Figure 3. For each test, the strain was set at a constant of 1%, while the frequency varied from 1 to 100 Hz at 0 and 800 mT. The initial value of the storage modulus of the sample without SAIB ester is larger than with SAIB ester. An increasing amount of SAIB ester leads to softer the matrix, therefore, lower the storage modulus. For example, at 800 mT initial storage modulus of MRE with 0 wt% SAIB ester was 0.96 MPa, when increasing from 5 to 10 wt%. The initial storage modulus decreases to 0.88, 0.72, and 0.69 MPa respectively. Introducing plasticizer into MREs promotes the increasing viscosity of the matrix which enhanced the sliding of molecular chains of the matrix that lowers the storage modulus. The input energy due to the stress is lost in such a sliding of the rubber molecular chain instead of being stored. On the other hand, storage modulus monotonically increases with the frequency of all measured samples with and without magnetic field. This increment is an indication that less energy dissipation at higher frequencies. On the other hand, increasing magnetic field from 0 to 800 mT resulted in an increment in the storage modulus of the sample with 0 wt% SAIB ester and the increment in the storage modulus being increased more in sample 2 and even more in sample 3 and 4.

Journal of Physics: Conference Series **776** (2016) 012034 doi:10.1088/1742-6596/776/1/012034

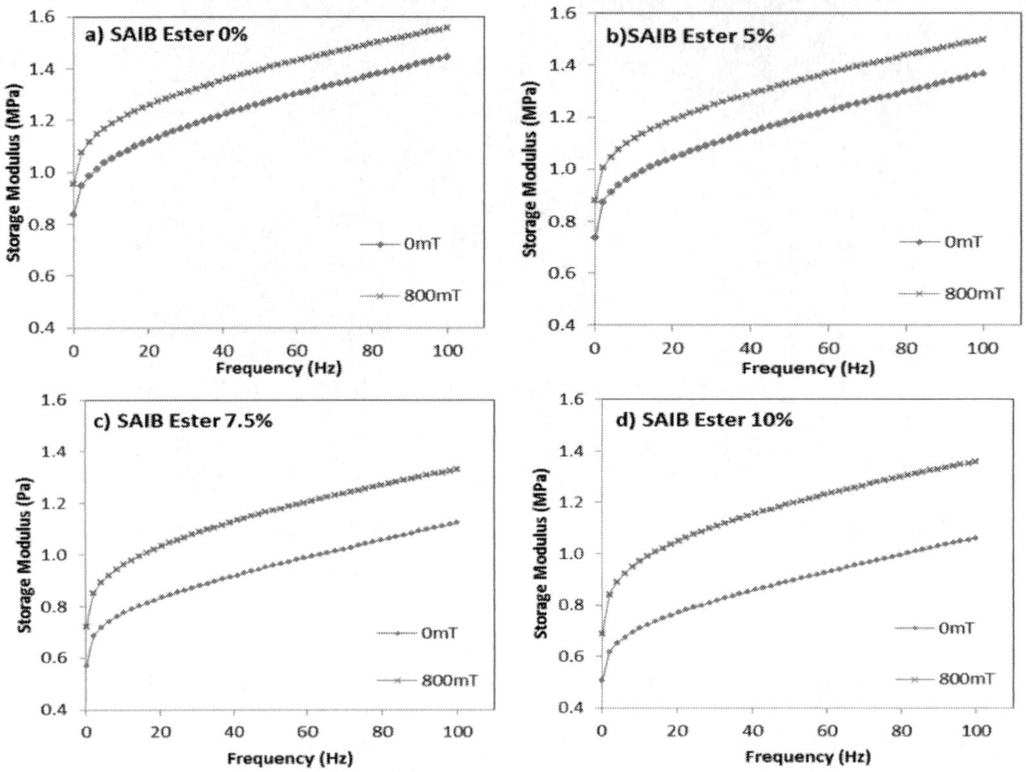

Figure 3. Variation of shear storage modulus with frequency (B=0 mT and B=800 mT) samples a) 0 wt% SAIB ester b) 5 wt% SAIB ester c) 7.5 wt% SAIB ester and d) 10 wt% SAIB ester.

Table 2. Zero-field shear storage modulus (G_0), the magneto-induced modulus (ΔG) and the MR effect of the MREs with different content of SAIB ester.

SAIB ester content (wt%)	G_0 (0 mT)(MPa)	ΔG (800 mT) (MPa)	MR effect (%)
0	0.84	0.09	10
5	0.74	0.14	19
7.5	0.57	0.15	26
10	0.51	0.19	37

These absolute and relative MR effect are clearly shown in Table 2 where all the data were collected from the experimental results (refer to Figure 3). From Table 2, the MR effect of MREs has increased from 10 to 37% under the magnetic field of about 800 mT when SAIB ester content has increased from 0 wt% to 10 wt% (measured at an initial frequency). Therefore, additions of SAIB ester increase the fluidity of the matrix and improved the compatibility between the CIPs and the matrix, and eventually provided the highest MR effect when exposed to the magnetic field.

4. Conclusion

Effect of SAIB on the performance of MREs based on ENR has been experimentally investigated. The addition of the SAIB ester increases the scorch time due to the enhancement of matrix mobility. This condition allows ample time for mold filling and produces the perfect shape of MRE. Meanwhile, the CRI decreases as a result of the interaction with SAIB ester, which slower the vulcanization reaction. Sequentially, SAIB ester also reduced the viscosity of the rubber matrix and CIPs have dispersed well without any agglomerations structures. The MREs with SAIB have had the lower zero-field modulus, and the MR effect of MREs had increased from 10% to 37% when SAIB ester content increased from

0 to 10 wt%. On the other hand, shear storage modulus are frequency dependent, where all MREs samples exhibit the incremental trend when increasing the frequencies.

Acknowledgments
The research described in this paper was supported by FRGS (vote no 4L094) from Ministry of Higher Education, Malaysia. The authors are grateful for the support.

5. References

[1] Gong XL, Chen L, Li JF, 2007, *Int. J. Mod. Physic B.* **21** 4875–4882.
[2] Agirre-Olabide I, Elejabarrieta MJ, Bou-Ali MM, 2015, *J. Intell. Mater. Syst. Struct.* **26** 1880–1886.
[3] Kallio M, Lindroos T, Aalto S, Järvinen E, Kärnä T, Meinander T, 2007, *Smart Mater. Struct.* **16** 506–514.
[4] Fan YC, Gong XL, Jiang WQ, Zhang W, Wei B, Li WH, 2010, *Smart Mater. Struct.* **19** 055015.
[5] Lerner AA, Cunefare KA, 2007, *J. Intell. Mater. Syst. Struct.* **19** 551–563.
[6] Li WH, Tian TF, Du H, 2012, *Recent Adv. Futur. Challenges.* 249–279.
[7] Nguyen VQ, Ahmed AS, Ramanujan RV, 2012, *Adv. Mater.* **24** 4041–4054.
[8] Lokander M, Stenberg B, 2003, *Polym. Test.* **22** 677–680.
[9] Chen L, Gong X, Jiang W, Yao J, Deng H, Li W, 2007, *J. Mater. Sci.* **42** 5483–5489.
[10] Ubaidillah, Sutrisno J, Purwanto A, Mazlan SA, 2015, *Advanced Engineering Materials* **17** 563

Sound absorption enhancement of nonwoven felt by using coupled membrane – sonic crystal inclusion

M.C. Fitriani[1], I. Yahya[1*], Harjana[1], Ubaidillah[1], F. Aditya[2], Y. Siregar[3], M. Moeliono[3], and S. Sulaksono[3]
[1]The Iwany Acoustics Research Group (iARG), Department of Physics, Faculty of Mathematics and Natural Sciences, Universitas Sebelas Maret, Jalan Ir. Sutami 36A, Kentingan, Surakarta 57126, Central Java, Indonesia
[2]PT. Rekadaya Multi Adiprima, Jakarta
[3]Textile Reseach Center (BBT), Bandung

E-mail: iyahya@mipa.uns.ac.id

Abstract. The experimental results from laboratory test on the sound absorption performance of nonwoven felt with an array thin tubes and sonic crystal inclusions reported in this paper. The nonwoven felt sample was produced by a local company with 15 mm in its thickness and 900 gsm. The 6.4 mm diameter plastic straw was used to construct the thin tubes array while the sonic crystal is arranged in a 4 x 4 lattice crystal formation. It made from a PVC cylinder with 17 mm and 50 mm in diameter and length respectively. All cylinders have two holes positioned on 10 mm and 25 mm from the base. The results show that both treatments, array of thin tube and sonic crystal inclusions are effectively increased the sound absorption coefficient of the nonwoven felt significantly especially in the low frequency range starting from 200Hz.

1. Introduction

The patch from textile industries has a potential economic value. Since it made from yarn through an industrial machining process, it can be recycled to produce fibrous sound absorber for various applications including automotive, shipping and building acoustics [1]. Nonwoven felt is a common material that widely uses as a sound absorber in automotive industries such as for doors, roofing, flooring, hood and much more. It made from the mix of the textile patch with a certain commercial binder which is melting on 110°C-200°C [2].

Despite the fibrous structure brings a very good performance on noise reduction for the high frequency band, it needs a more advanced approach to making the nonwoven felt perform the better absorption at the mid and low frequency range. The various approach has been developed by previous researchers accordingly. For example, the inclusion of permeable and perforated membrane have reported by Sakagami et al (2014) and Harjana and Yahya (2015) respectively. It found that the membrane structure affect to the reactance of the cavity backed perforated absorber and produces significant increment to the sound absorption performance on the mid to high frequency range above 500 Hz [3,4].

Another approach by using short thin constrained tube has been reported by Yahya and Harjana (2013) and followed by the compartmented cavities approach by Li et al (2015, 2016). This cavity resonance approaches successfully shifting the sound absorption performance to 200 Hz but it has industrial constraint due to the geometrical design which is impractical for mass production [5-7]. The most attractive approach among researchers recently is sonic crystal assisted sound absorber which is featuring possibility for tuning the local Helmholtz resonant to reduce the noise on certain intended frequency band [8-11]. In this paper, a combined permeable membrane, thin tubes array, and sonic crystal inclusion approach are reported. The proposed combining of three different noise reduction techniques in a single integrated inclusion design is expected to produce a much better sound absorber concept which is less constraint in the industrial perspective.

2. Methods

The transfer function based two microphones technique refer to ASTM E1050 was conducted to measure the sound absorption coefficient of the test sample. Since this work is focused on the low frequency band up to 1.6 kHz, the large tube of B&K 4206 was utilized in the laboratory test. The membrane layer was produced by hand laying up the technique of a commercial leak-proof product on the surface of the nonwoven felt. The layering steps have been repeated for three times followed by an open air drying process. The nonwoven felt has 15 mm in thickness and 900 gsm produced by a local manufacturer while the thin tubes and sonic crystal elements are made from plastic straw and PVC tube of 6.4 mm and 17 mm in diameters respectively. The sonic crystal element has a single geometrical size 50 mm in the height with a 2 mm hole on the wall. It varying in two different positions of 10 mm and 25 mm from the bottom end as illustrated in Figure (1).

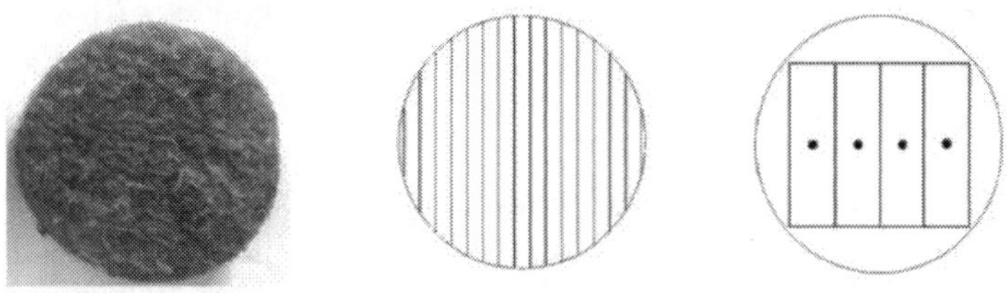

(a) (b) (c)

Figure 1 The test samples: (a). Nonwoven felt, (b). Array of thin tubes, (c). The sonic crystal model

The laboratory setup is presented in Figure (2). B&K 3160-A LAN-Xi type was used in connected with the power amplifier B&K 2716C and impedance tube B&K 4206. Random noise is generated from the loudspeaker and the incident and reflected waves are decomposed from the captured signals by a pair of quarter-inch microphone B&K 4187.

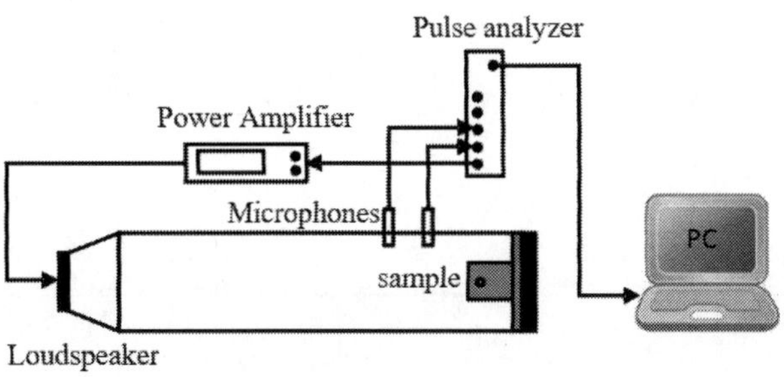

Figure 2 laboratory setup for sound absorption measurement.

There is three frequency response are being calculated in the decomposition process: frequency response H_1, frequency response associated with incident waves H_i, and the frequency response function associated with the reflected wave H_r.

The reflection coefficient is given by,

$$R = \left(\frac{H_1 - H_i}{H_r - H_1}\right) e^{j2k(l+s)} \quad (1)$$

where k is the wave number, l is the distance of the first microphone to the sample surface, and s is the spacing between microphones. The normalized impedance ratio and sound absorption coefficient can be calculated by using the following equation,

$$\frac{z}{\rho c} = \frac{1+R}{1-R} \quad (2)$$

$$\alpha = 1 - |R|^2 \quad (3)$$

3. Results and Discussion

Figure (3) shows that the nonwoven felt does not perform as an effective sound absorber for the low frequency range. It happened since the sound waves passing through the material without energy dissipation. The viscous damping did not occur due to the wavelength are much longer compared to the sample dimension. As the surface of the nonwoven felt covered by the leak-proof silicon layer, the sound absorption coefficient increased significantly. It happened for all frequency starting from 400 HZ and reach the maximum value of 0.75 in the frequency range between 700 Hz to 1.2 kHz.

This phenomenon is in between of the similar results as previously reported by Sakagami et al [3] and Harjana and Yahya [4]. It means that silicon layer brings multiple impacts to the nonwoven felt properties. It does not only affect the surface impedance value but the more significant it is produced membrane type resonance on the surface of the nonwoven felt. As the membrane oscillates on its resonant modes, the fibrous structure of the nonwoven felt below the membrane are also oscillate in the same mode accordingly. It then increases fibrous damping by the nonwoven fibers and porosity and resulting the much higher sound absorption coefficient.

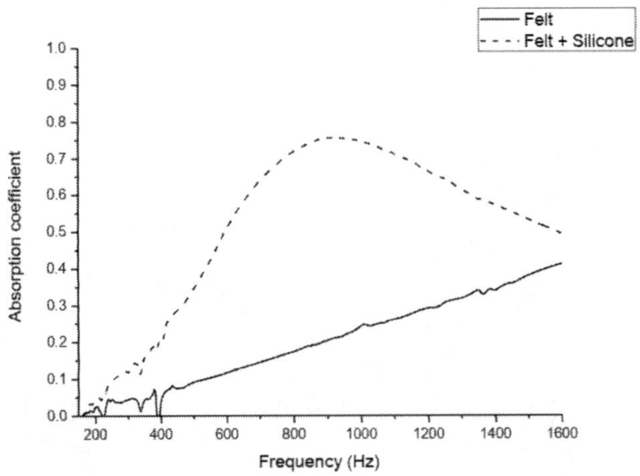

Figure 3 Impact of silicon layering on the sound absorption coefficient of nonwoven felt.

Another result presented in Figure (4). The sound absorption coefficient are slightly increased at two different frequecy range. The first one in 500 Hz to 800 Hz which is associated with coupling effect of the thin tubes array with the surrounding cavity. The second increment on the high frequency band around 1.4 kHz and 1.6 kHz are related to the resonant mode of the half wavelength resonators of the thin tubes array.

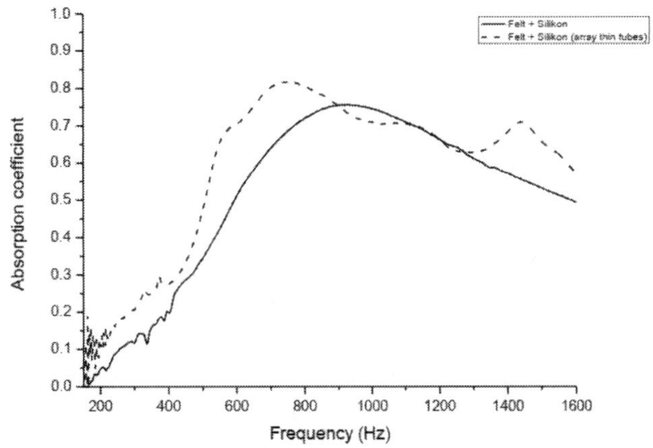

Figure 4. The impact of coupling effect and resonant mode of thin tubes array on the sound absorption.

It is clear from Figure (4) that coupling effect of the thin tubes array which is half wavelength resonators with the surrounding cavity including the membrane layer on the surface of nonwoven felt tends to shift the response. The resonance absorption occurs simultaneously due to the cavity, the thin tubes array, and the membrane. It changes the reactance of the coupled structure and shifting the response to the lower frequency. In the same time, the fibrous and porous layer of nonwoven felt also oscillated and increased the sound energy dissipation into heat. It brings a better understanding of the possibility of manipulating the resonance absorption and coupling effect as the new approach for increasing the sound absorption coefficient of the nonwoven felt.

This new approach was then implemented to create the another test model with the consideration of using the sonic crystal element instead of thin tubes array. It based on a fundamental reason related to the local Helmholtz resonant of the sonic crystal. In addition, it also considering to the scattering and multiple internal reflections which are both not occurred with the thin tubes array.

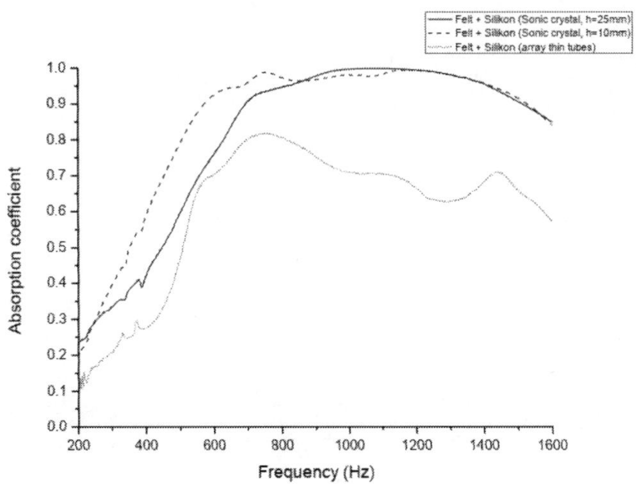

Figure 5 The sound absorption coefficient value on material nonwoven felt with thin tube array and sonic crystal.

The result is presented in Figure (5). The proposed approach with coupled of membrane layer with the sonic crystal inclusion has a very promising performance. The sound absorption coefficient increases at the all frequency range with the smallest absorption coefficient of 0.24 for the frequency of 200 Hz. This result is much higher compared to the initial nonwoven felt which is less than 0.05 for the same frequency. It also clear from Figure (5) that the sound absorption coefficient increase rapidly from 0.24 on 200 Hz to 0.9 on 500 Hz and then featuring flat response on the rest of the frequency range above it.

This evidence emphasizes and concludes that the sound waves energy are effectively dissipated by coupling of the membrane layer, the local resonant, and also by fibrous and porous nonwoven felt. In term of industrial application possibility, it is clear that since the proposed approach consists of a single geometrical dimension sonic crystal element, it much easier to mass production compared to the compartmented cavities and other techniques as proposed by previous researchers.

References

[1] Kucuk M, and Korkmaz, Y, 2015. *Fibers and Polymers* **16**(4) 941-948.
[2] Albrecht W, Fuchs H, and Kittelmann, W, 2003 Nonwoven Fabrics. *Wiley-VCH Weinhelm.*
[3] Sakagami K, Fukuutani, Y., Yairi, M., Morimoto, M, 2014. *Applied Acoustics* **79**. 104-109
[4] Harjana, and Yahya I 2015. 7th ICOPIA, Solo, Atlantis Press. 60-64
[5] Yahya I and Harjana, 2013 ICSV20. Bangkok.
[6] Li D, Chang D, Liu B., and Tian, J, 2015. ICSV22. Florence
[7] Li D, Chang D, and Liu B. 2016. *Applied Acoustics* **102**. 126-132
[8] Redondo J, Pico R, Sanchez-Morcillo, V. J, 2013. *J. Acoust. Soc. Am.* **134**(6). 4412-4417
[9] Morrandi F, Miniaci M, Marzani A, Barbaresi L, Garai M, 2016. *Applied Acoustics* **114**. 294-306
[10] A., Gomez, P. M., Spiousas, I., Eguia, M. C, 2016. *Applied Acoustics* **104**. 1-5
[11] Wang, Z. G., Lee, S. H., Kim, C. K., Park, C. M., Nahm, K, and Nikitov, S. A, 2008, *J. of Appl. Phys.* **116**. 064907 (9 pages)

Journal of Physics: Conference Series **776** (2016) 012036 doi:10.1088/1742-6596/776/1/012036

Steady compression characteristics of laminated MRE isolator

N. A. A. Wahab[1], S. A. Mazlan[1*], Ubaidillah[2], A. H. R. Sharif[1], S. Kamaruddin[3]

[1]VSE Research Laboratory, Malaysia-Japan International Institute of Technology, Universiti Teknologi Malaysia, 54100 Jalan Sultan Yahya Petra, Kuala Lumpur, Malaysia

[2]Mechanical Engineering Department, Faculty of Engineering, Universitas Sebelas Maret, Jl. Ir. Sutami 36A, Kentingan, Surakarta, 57126, Central Java, Surakarta, Indonesia

[3]Engineering Design & Prototype Centre, Centre of Excellence (COE), Malaysian Rubber Board Experimental Station, 47000 Sg. Buloh, Selangor, Malaysia

E-mail: amri.kl@utm.my

Abstract. This paper focused on an experimental setup on laminated magnetorheological elastomer (MRE) isolator under steady state compression test. An isotropic type natural rubber (NR) based MRE were fabricated and layered with a steel plate to form a multilayer sandwich structure adopted from the conventional laminated rubber bearing design. A set of static compression test was conducted to explore the potential of semi-active laminated MRE isolator in field-dependent stiffness properties. Stress versus strain relationship was assessed under different magnetic fields application. Based on the examination, the stress altered as the application of magnetic fields. Consequently, the effective stiffness of isolator also influenced by the magnetic fields induction. The experimental results show that the proposed laminated MRE isolator can effectively alter the compression stiffness up to the 14.56%. The preliminary results have confirmed the tunability of the semi-active laminated MRE isolator in which it would be beneficial for improving building isolator in general.

1. Introduction

Magnetorheological elastomers (MREs) are solid analog of magnetorheological fluids. They consist of iron particles in nonmagnetic matrix materials having viscoelastic characteristics. These materials have rheological properties that can be changed continuously, rapidly, and reversibly by applying magnetic fields [1,2]. Hence, these unique properties of MRE have attracted many researchers to study and develop new vibration isolation system that utilized MRE material for vibration reduction in civil engineering and others application [3–7]. The MRE vibration isolation system with controllable stiffness and damping properties potentially can overcome the major shortcomings of the traditional base isolation system by shifting and avoiding the resonant frequency of the seismic excitations [3,8,9].

Currently, many researchers have proposed their design and tested to study the adaptability of smart MRE vibration isolator with various range of earthquakes wave [5,10–12]. However, the experimental test of MRE isolator is mainly limited to the horizontal movement and only focus on the shear performances of MRE isolator. For example, Usman et al. [5], reported the base isolation system using

MREs have reduced the horizontal structural accelerations and storey drifts compared with the passive base isolation system under different earthquake inputs. Li et al. [11] in their study developed a highly adjustable MRE base isolator and the experimental results on the shear test shows the force and stiffness increase when have applied current. In another study, Eem et al. [12] evaluated the performances of smart base isolation system with different frequencies using shaking table system that gives horizontal movement on the scaled single-story building structure. From the study, the application of smart base isolation system can reduce the responses of the test structure by shifting the natural frequency of the structure. Furthermore, Behrooz et al. [10] examined the adaptability of semi-active variable stiffness and damping isolator (VSDI) on a scaled building system and the results demonstrate a reduction in both acceleration and displacement of the building structure. In theory, a structure should possibly isolate in both horizontal and vertical directions. Nevertheless, the performance of MRE isolator on vertical movement has not been fully explored, and most of the studies only focus on the shear performances of MRE isolator. This is because of the seismic effect was critical in the horizontal direction instead of vertical direction. Unfortunately, some structures that contained sensitive equipment or material such as at nuclear power plant and reactor buildings [13] clearly need vertical isolation system [14,15] to sustain the load from structures and vibration during an earthquake.

Therefore, this paper revealed the performances of laminated MRE isolator in term of stress-strain and stiffness properties during compression loading. The development of a laminated MRE isolator is presented in this paper including design, fabricating and testing of laminated MRE isolator. MRE based NR was utilized in laminated MRE isolator and evaluated using dynamic fatigue machine. The design concept of laminated MRE isolator was adopted from the conventional laminated rubber bearing. The MRE was layered with the steel plate to form the laminated structures that can sustain with large vertical loading from the structures. Magnetic coils are designed to provide sufficient magnetic field across the laminated MRE - steel. Furthermore, after MRE isolator fabrication, an experimental test was conducted to examine the performances of laminated MRE isolator in static compression under various magnetic field intensities.

2. Laminated MRE Isolator

The laminated MRE isolator design was divided into five main parts which are bottom and top plate, laminated structures, electromagnetic coils, housing and coils bobbin. The design concept of laminated MRE isolator has been modified from the conventional laminated rubber bearing by adding more steel plates and adding coil parts to generate magnetic circuits around the laminated MRE isolator. Some parameters need to be considered in designing the laminated MRE isolator to guide the magnetic flux. The selection of materials for each part in laminated MRE isolator, type of coil, thickness of MRE and steel plates are the parameters that must be focused while designing the device. The laminated structures serve as a magnetic core in the isolator because of the properties of MRE can be altered with changing of magnetic strength. The laminated structures consist of two steel blocks with 13 mm thickness, 28 layers of MRE sheets and 27 layers of steel plates. This laminated MRE isolator designed with 1 mm thickness of MRE and steel plate to enlarge distribution of magnetic flux. An electromagnetic coil extending around the coil bobbin with was provided to generate magnetic fields with 1400 number of coil turns. This number of coil turns was calculated based on the area designated coil bobbin, and the total electric resistance was 10.8 Ω.

The working principle of the laminated MRE isolator is as follows: the magnetic field generated by the coil can be controlled by the current from the external direct current power source, its shear modulus is determined by the strength of magnetic field, and the modulus determines the stiffness of the laminated MRE isolator. As the consequence, the natural frequency of the laminated MRE oscillator can be controlled by tuning the current in the coil, and then when the natural frequency of laminated MRE isolator matches the excitation frequency, the vibration can be suppressed significantly.

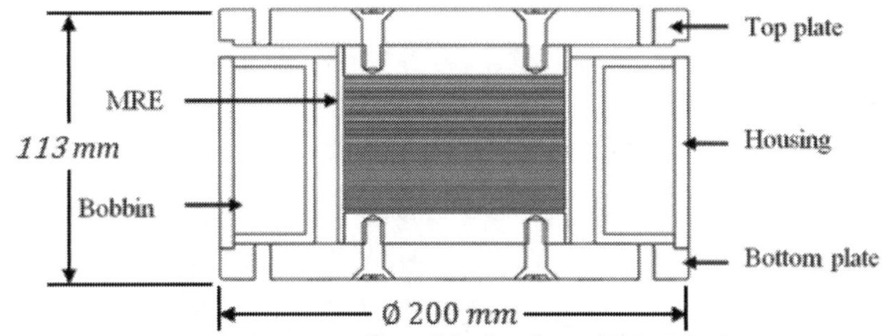

Figure 1. Cross section of the laminated MRE isolator

The materials that involved in fabricating the laminated MRE isolator were 1020 type of mild steel for magnetic materials, aluminium 6061-T6 for non-magnetic materials and AWG 18 coil. Details for each part of MRE isolator are shown in Table 1. The fabrication of laminated structures was done at Malaysian Rubber Board (MRB) experimental station. The others part in the laminated MRE isolators which are bobbin, housing, bottom and top plate as shown in Figure 2 was fabricated at the precision lab (MJIIT) using lathe machine and computer numerical control (CNC) machine.

Table 1. List of parts and materials for laminated MRE isolator

Part no	Part name	Type of material	Material	Type
1	Steel plate	Magnetic	Low carbon steel	1020
2	Top and bottom plate	Magnetic	Low carbon steel	1020
3	Steel block	Magnetic	Low carbon steel	1020
4	Wire coils	Non-magnetic	Copper wire	18SWG
5	Bobbin	Non-magnetic	Aluminium	6061-T6
6	Housing	Magnetic	Low carbon steel	1020

Figure 2. Main parts consist in a laminated MRE isolator

3. Experimental Testing

To observe the performances of laminated MRE isolator, a series of static compression test was conducted. The experiment was carried out using dynamic fatigue machine (Shimadzu Servopulser) equipped with a 20 kN load cell for a compression test. The experimental setup for compression test consisted of four main components; namely Shimadzu Fatigue Dynamic Test Machine, prototype, power supply and computer. The arrangement of the experimental test is shown in Figure 3. The laminated MRE isolator was attached to the testing machine. Then, the testing machine was connected to the data logger to record the result obtain from the compression test. The laminated MRE isolator was connected with a power supply with a capacity of 240 V and 24 A to generate a magnetic field to the magnetic coil of the isolator. The device is mounted on the load cell and compressed until 30% strain from the gap between the top plate and the bobbin. The test was performed without the magnetic field which is in "passive" state, and with magnetic fields exists which is in "active" state varies from

1 to 3 A. A series of tests, with static compression with 30% constant strain and energized with different applied currents from 1 to 3 A. Both stress and strain data can be directly obtained from the experimental result. The stiffness data was developed from the force and displacement data using mathematical equation.

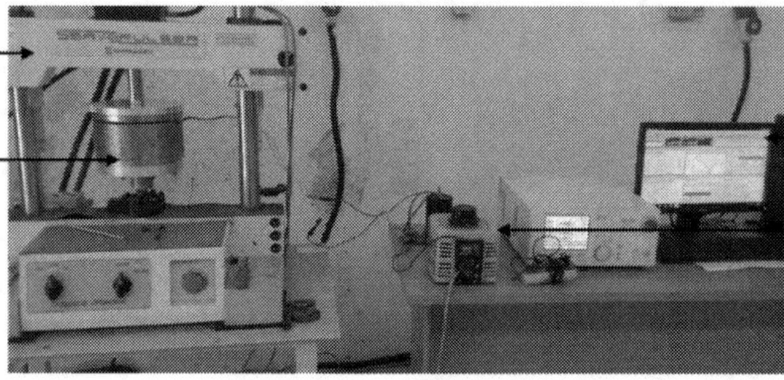

Figure 3. Experimental setups for laminated MRE isolator

4. Result and Discussion

Figure 4 shows the stress measurement of laminated MRE isolator with 30% strain controlled for different applied currents, start with 0 until 3 A. As seen in the figure, the stress increased as applied current increased. The patterns of the graph are similar in such way that the stress was increased with the increased the percentage of strain for four different value of current applied. The stress value has increased from 0.9 to 1. 26 N/mm^2 as the applied current increased from 0 to 3 A. the increment of stress value, however, is not proportional with the increment of applied current. The different in values of stress measurement are becoming lesser as the applied current increases. But then still the stress measurement of laminated MRE isolator are affected by the applied currents, wherein general, the current plays a role in increased the performances of the semi-active device. It is clearly observed that laminated MRE isolator exhibit force increase with the increase of the applied current.

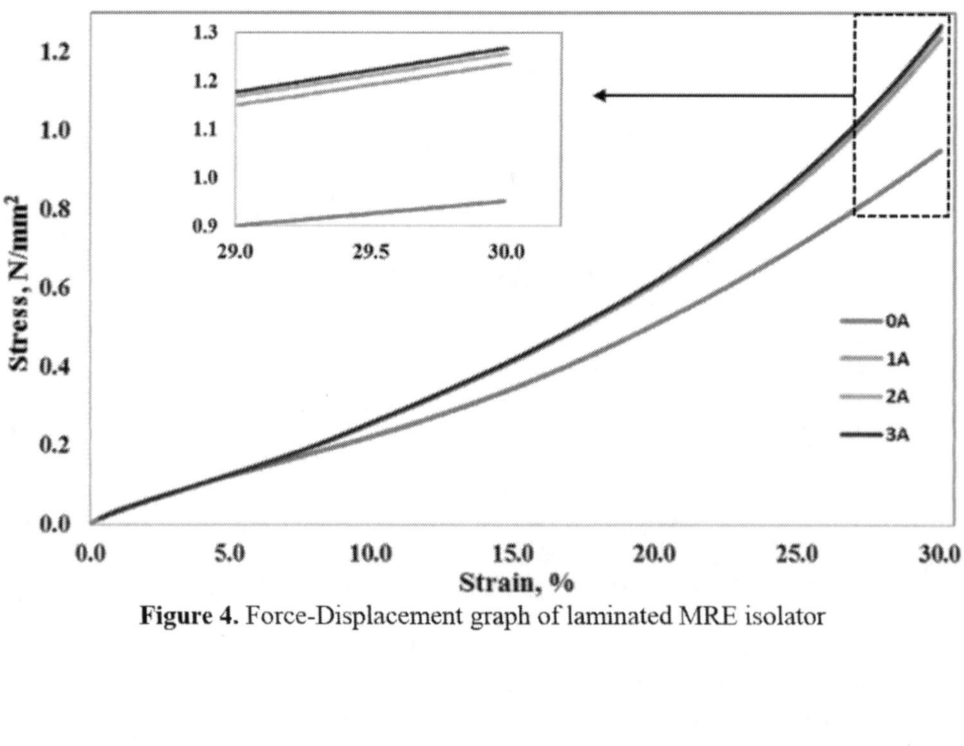

Figure 4. Force-Displacement graph of laminated MRE isolator

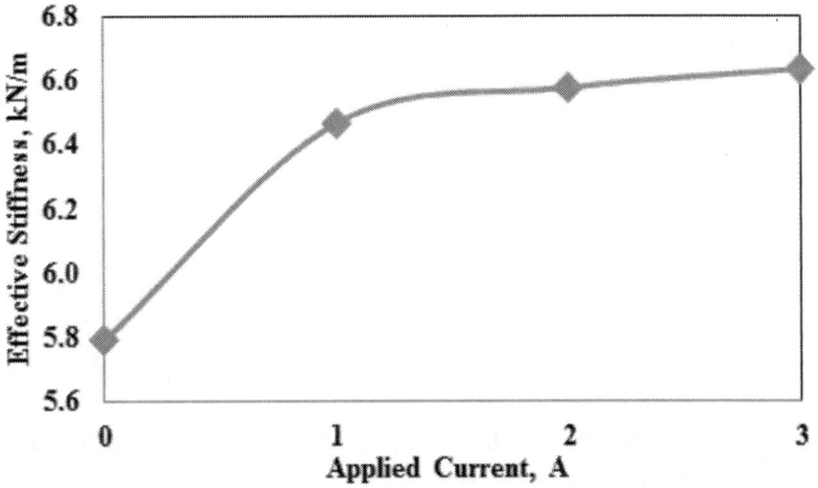

Figure 5. Effective stiffness of the laminated MRE isolator under various applied current

The effective stiffness for laminated MRE isolator during compression loading under applied magnetic field from 0 to 3 A are listed in Table 2. The effective stiffness increase for each applied current varies from 11.7 to 14.6%. From the increasing value of effective stiffness, it shows that the effective stiffness during compression loading was increased with the increased of applied current in the device.

Table 2. Effective stiffness of the laminated MRE isolator under various applied current

Current (A)	0A	1A	2A	3A
Effective stiffness (kN/m)	5.79	6.47	6.58	6.63
Increase (%)	-	11.7	13.6	14.6

5. Conclusion

This paper has delivered development of laminated MRE isolator utilizing MRE based NR material which provides adaptability to a different range of external excitations during a seismic event. The laminated MRE isolator was designed, fabricated and tested. The design of laminated MRE isolator was adapted from conventional laminated rubber bearing to offer high vertical loading capacity and low horizontal stiffness. Experimental investigations were conducted to examine the performances of laminated MRE isolator during compression loading with different magnetic field intensities. During compression loading, the laminated MRE isolator exhibits the field-dependent property where an increase in applied current leads to an increase in stress and effective stiffness. The experimental results demonstrate increasing in both stress and effective stiffness up to 14.56% under increasing the magnetic field intensities from 0 to 3 A. The study on the effect of magnetic field strength during compression loading further improve the adaptability of laminated MRE isolator to sustain the vertical excitation during an earthquake. So that, this laminated MRE isolator have the ability to change targeted isolation excitation range promptly and can sustain high vertical loading. Future work may focus on the experimental test for simultaneous vertical and horizontal and modelling for laminated MRE isolator.

Acknowledgments

The authors would like to express their gratitude to all the technicians in Precision Lab, Malaysia-Japan International Institute of Technology (MJIIT), Universiti Teknologi Malaysia (UTM) for their help and support on handling machine and equipment. The research described in this paper was supported by FRGS (vote no 4L094) from Ministry of Higher Education, Malaysia. The authors are grateful for the support.

References

[1] M. Lokander and B. Stenberg 2003 *Polymer Testing* **22** 677

[2] Ubaidillah, J. Sutrisno, A. Purwanto, and S.A. Mazlan 2015 *Advanced Engineering Materials* **17** 563

[3] Y. Li, J. Li, W. Li, and B. Samali 2013 *Smart Materials and Structures* **22** 035005

[4] S. H. Eem, H. J. Jung, and J. H. Koo 2011 *IEEE Transactions on Magnetics* **47** 2901

[5] M. Usman, S.H. Sung, D.D. Jang, H.J. Jung, and J.H. Koo 2009 *Journal of Physics Conference Series* **149** 012099

[6] S. Opie and W. Yim 2011 *Journal of Intelligent Material Systems and Structures* **22** 113

[7] N.A.A. Wahab, S.A. Mazlan, K. Hairuddin, and H. Zamzuri 2015 *IOP Conference Series Materials Science and Engineering* **100** 012062

[8] C. Yang, J. Fu, M. Yu, X. Zheng, and B. Ju 2015 *Journal of Intelligent Material Systems and Structures* **26** 1290

[9] Ubaidillah, S.A. Mazlan, J. Sutrisno, and H. Zamzuri 2014 *Applied Mechanics and Materials* **663** 695

[10] M. Behrooz, X. Wang, and F. Gordaninejad 2014 *Smart Materials and Structures* **23** 045014

[11] Y. Li, J. Li, T. Tian, and W. Li 2013 *Smart Materials and Structures* **22** 095020

[12] S.H. Eem, H.J. Jung, and J.H. Koo 2013 *Smart Materials and Structures* **22** 055003

[13] J.M. Kelly 1986 *Soil Dynamics and Earthquake Engineering* **5** 202

[14] H. Aslani, C. Cabrera, and M. Rahnama 2012 *Earthquake Engineering and Structural Dynamics* **41** 1549

[15] A.S. Whittaker, M. Kumar, and M. Kumar 2014 *Nuclear Engineering and Technology* **46** 569

Influence of nanographene platelets (NGP) incorporation on Fe_3O_4 nanoparticles as materials additives for enhancement thermal properties stearic acid

M. K. Nuryadin[1,2], R. Andiarto[1,2], A. Taufik[1,2], R. Saleh[1,2,*]

[1]Departemen Fisika, Fakultas MIPA-Universitas Indonesia, 16424 Depok, Indonesia

[2]Integrated Laboratory of Energy and Environment, Fakultas MIPA-Universitas Indonesia, 16424 Depok, Indonesia

[*]E-mail: rosari.saleh@ui.ac.id

Abstract. In this work, Fe_3O_4 nanoparticles, and Fe_3O_4/NGP composite were used as material additive for enhancement thermal properties of stearic acid (SA). The both material additive were synthesized using sol-gel method. Phase change material (PCM) composites SA-Fe_3O_4 and Sa-Fe_3O_4/NGP mixtures were made through the dispersion technique with three different weight % ratio of material additives into stearic acid: 1 wt.%, 3 wt.%, and 5 wt.%. X-Ray Diffractometer (XRD) and Fourier Transform Infrared (FTIR) spectroscopy were used to investigate the structural properties. Magnetic properties also measured by vibrating sample magnetometer (VSM) to see influence of NGP in PCM composites. Differential Scanning Calorimetry (DSC) and Thermogravimetric Analysis (TGA) were used in order to analyse the thermal properties of the samples. The results show an enhancement of the latent heat, thermal stability as well as specific heat by the presence of material additives in SA. Compare to SA-Fe_3O_4, SA-Fe3O4/NGP show better improvement in enhancement of thermal performance of SA. The improvement by about 41.2 % in specific heat and 21.2% in latent heat.

1. Introdcution

Thermal energy storage (TEM) has attracted more attention due to its heating and cooling application for waste heat recovery, solar thermal energy utilization and thermal management of electronics [1-3]. The variety of phase change material (PCM) especially the organic phase change material has been widely studied and has become the key technologies in the thermal energy storage application. Stearic acid which the common preferred organic phase change material is typically used due to their nontoxic, high energy storage density, thermal stability, and inconspicuous volume change [4-6]. Unfortunately, most of the organic phase change material including the stearic acid have a very low thermal conductivity which can limit their application. Therefore, some various methods are proposed in order to overcome this problem and one of the most promising method is by dispersing the nanoparticle into the PCM itself [7-9].

Adapting the idea of nanofluid, the nanomaterial utilization into the PCM is considered as a promising solution to enhance the thermal ability of the PCM in order to absorb and release heat [10]. A recent researchs show that the addition of the high thermal properties metal oxides such as Fe_3O_4, TiO_2 into the PCM has produced an enhancement in both latent heat and thermal conductivity [11-13]. Not only metal oxide, graphene with nano-sized material also shows a tremendous effect to the thermal properties of the PCM. For an example a research by Harish et.al [14] shows that the addition of the single-walled carbon nanohorn into the PCM has produced an increment into the thermal conductivity as well as its thermal latent heat. But unfortunately, there are only a few researches that investigate the addition of metal-oxide and graphene combination into the PCM. Therefore, in this research, Sa/Fe_3O_4 and Sa/Fe_3O_4/NGP are dispersed into the stearic acid and both of structural and thermal properties were observed.

2. Experimental

In this work, Iron (II) sulfate heptahydrate ($FeSO_4$. $7H_2O$, 99), sodium hydroxide (NaOH), nanoraphene platelets (NGP), and Stearic acid (Sa) as a phase change material are used without further purification. Fe_3O_4 nanoparticle prepared by using the same method as our previous study [15] while the Fe_3O_4/NGP with 10% NGP is prepared by sol-gel method. In brief, the NGP is dissolved in 80 ml of distilled water and 40 ml ethanol by using the ultrasonic treatment. The resulted 10% NGP solution is mixed into the 2 gram of Fe_3O_4 and are stirred for 2 h. The homogeneous suspension then is heated in the 120°C for 3h or 3 h to affect the deposition on the graphene sheets. The resulting nanocomposite is isolated by centrifugation and dried in vacuum at 70 °C for 12 h. The PCM mixtures of Sa/Fe_3O_4 and Sa/ Fe_3O_4/NGP composites are prepared by dispersing the additives materials in the Sa with ultrasonic water bath for 2 hours. The ratio of Fe_3O_4 nanoparticle and Fe_3O_4/NGP nanocomposite were varied by 1, 3, and 5 wt% with the stearic acid.

The structural properties and chemical structures of all the samples are measured by the X-Ray Diffraction (XRD) and Fourier Transform Infrared (FTIR) respectively while the magnetic behaviors of the sample are measured by the vibrating sample magnetometer (VSM) measurement. For the thermal properties, Differential Scanning Calorimetry (DSC) is used in order to know the phase change behavior and specific heat capacity of the samples while the Thermo-gravimetric Analysis (TGA) is used to observe the degradation process of the samples.

3. Results and discussion

The XRD spectrum of Sa, Fe_3O_4 and NGP are plotted in Figure 1(a) along with the spectrum of Sa/Fe_3O_4 and Sa/Fe_3O_4/NGP with 5 wt% ratio as a representation of the Sa mixtures. The XRD pattern of the Sa shows the presence of two high peaks below the $2\theta = 25°$. These peaks are due to Sa crystallization peak. Another pattern from the XRD spectra of Fe_3O_4 confirms the inverse cubic spinel phase form at $2\theta = 30.2°$, 35.5°, 53.4°, and 57.0° which are attributed to the (220), (311), (511), and (440) crystalline planes. The strong diffraction pattern is also shown at $2\theta = 26°$ for the (200) crystalline plane of NGP. The peaks of the Fe_3O_4 as well as the peaks of Sa can be obviously seen in the XRD spectra of Sa/Fe_3O_4 and Sa/Fe_3O_4/NGP with the additional NGP peak in the spectra. The XRD results indicate that all of the samples are consisted of the desired materials.

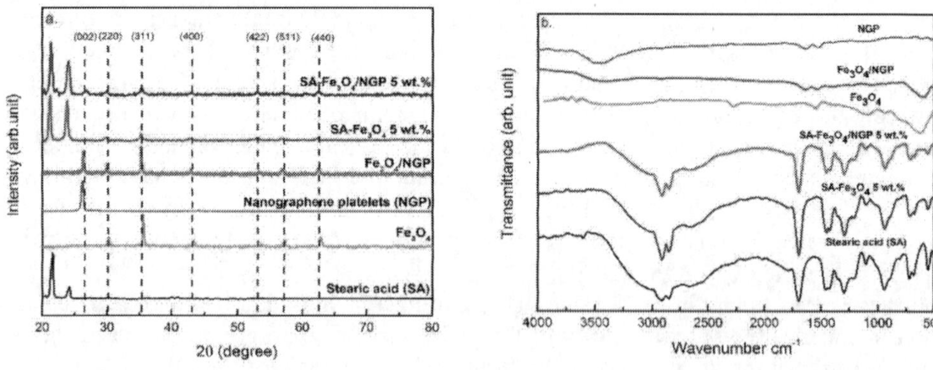

Figure 1. (a) XRD patterns and (b) FTIR spectrum of SA, Fe3O4, NGP, Fe3O4/NGP SA-Fe3O4 wt. % and SA-Fe_3O_4/NGP 5 wt.%.

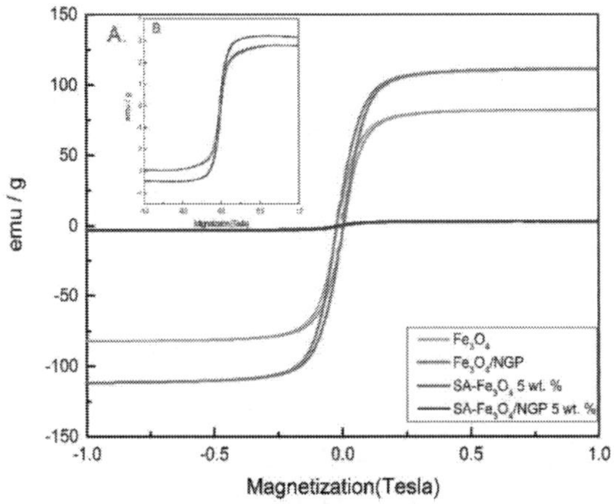

Figure 2. VSM Spectra of (a) Fe_3O_4, Fe_3O_4/NGP, SA-Fe_3O_4 and SA-$Fe_3O_4$4/NGP and (b) SA-Fe_3O_4 and SA-Fe_3O_4/NGP

Table.1 Magnetic Saturation of Fe_3O_4, Fe_3O_4/NGP, SA-Fe_3O_4 and Fe_3O_4/NGP

Sample	M-S (emu/g)
Fe_3O_4	82
Fe_3O_4/NGP	112
SA-Fe_3O_4 5 wt.%	2.76
SA-Fe_3O_4/NGP 5 wt. %	3.16

Figure 1(b) shows the FTIR transmittance pattern of Sa, Sa-Fe_3O_4, and Sa-Fe_3O_4/NGP. The transmittance pattern of Sa shows the strong absorption bands at 2917 cm^{-1}, 2849 cm^{-1} and 1703 cm^{-1} which are attributed to the stretching vibration of $-CH_3$, $-CH_2$ and C=O molecules respectively while another absorption bands at 1464 cm^{-1}, 1296 cm^{-1}, 934 cm^{-1} and 720 cm^{-1} correspond to in plane, out of plane and in plane swinging of OH molecules [16]. There are no transmittance pattern of the Fe_3O_4 and NGP that can be seen in the Sa-Fe_3O_4, and Sa-Fe_3O_4/NGP due to the small amount of the additives materials. However, the similar transmittance pattern of the Sa in Sa/Fe_3O_4 and Sa/Fe_3O_4/NGP spectra indicate that there are no chemical pattern takes place between the PCM and the additives.

The magnetic hysteresis curve of Sa/Fe_3O_4 and Sa/Fe_3O_4/NGP alongside with Fe_3O_4 and Fe_3O_4/NGP as the additive materials are plotted in Figure 2. The magnetic saturation values are also listed in the Table 1. The magnetic behaviour of Fe_3O_4 is increased with the presence of the NGP particles. The same tendency is also observed for Sa/Fe_3O_4 and Sa/Fe_3O_4/NGP with the smaller magnetic saturation due to the small amount of the Fe_3O_4 nanoparticle and Fe_3O_4/NGP nanocomposite in the Sa.

For the thermal degradation process, the TGA measurements results of the Sa/Fe_3O_4 (Figure 3(a)) and Sa/Fe_3O_4/NGP (Figure 3(b)) show that the Sa begins to decompose at about 200°C. However, the slightly change in temperature decomposition are found in the Sa/Fe_3O_4 samples and are more distinguishable in the Sa/Fe_3O_4/NGP samples. Higher temperature decomposition process can provides the higher thermal stability on the PCM which is very important to the thermal energy storage. The TGA measurement also shows the information about the residual weight composition of the sample after the decomposition process. The percentage loss of Sa in Table 2 corresponds to the composition of the samples Sa/Fe_3O_4 and Sa/Fe_3O_4/NGP. DSC curve of the Sa/Fe_3O_4 and Sa/Fe_3O_4/NGP with 1, 3, and 5wt% ratios are shown in Fig. 4a and Fig. 4b respectively in order to know the melting and solidifying process of the Sa composites as the PCM. The melting and solidifying points of the samples from the DSC curves are tabulated in Table 3 along with the latent heat values. The presence of the Fe_3O_4 and Fe_3O_4/NGP do not cause any favourable effect on the melting and solidification point. The Sa completely melt at temperature 70.5 °C while its solidifying temperature is 66 °C. The slightly change occurs in the melting and solidifying temperature for the Sa/Fe_3O_4 and Sa/Fe_3O_4/NGP. Both latent heat of melting and solidifying are higher for Sa/Fe_3O_4 and Sa/Fe_3O_4/NGP than that of Sa. The enhancement of the latent heat can be explained with molecular attraction between Sa and the nanocomposite [17-18]. The influence of the addition of Fe_3O_4 and Fe_3O_4/NGP on enhancing the value of latent heat showed the enhancement ability of the Sa composites to absorb and release heat with greater energy capacity.

The specific heat values of the samples are also calculated from the DSC results and are plotted in

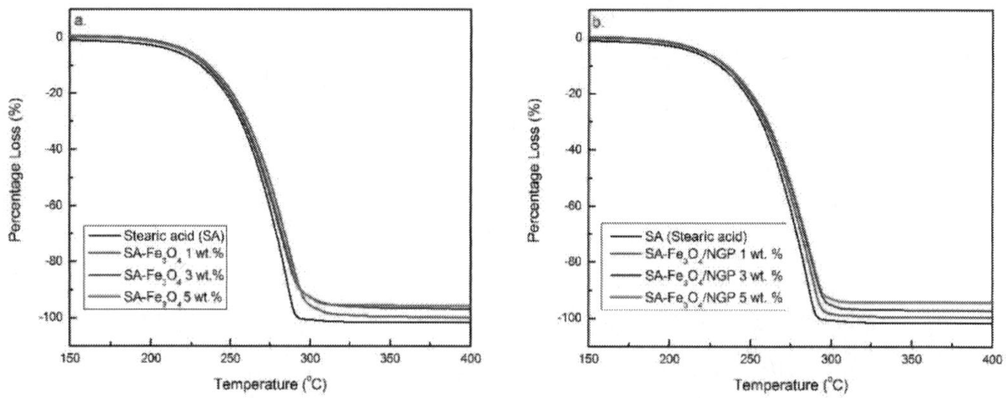

Figure 3. TGA (a) SA-Fe3O4 5 wt. % various wt. % and (b) SA-Fe_3O_4/NGP various wt. %

Table 2. TGA result of SA- Fe_3O_4 5 wt. % various wt. % and SA- Fe_3O_4/NGP various wt. %

Sample	Onset Temp	Percentage Loss
Stearic acid (SA)	251.6°C	100%
SA-Fe_3O_4 1 wt.%	251.4°C	99.89%
SA-Fe_3O_4 3 wt.%	250.2°C	96.41%
SA-Fe_3O_4 5 wt.%	251.6°C	94.49%
SA-Fe_3O_4/NGP 1 wt. %	253.5°C	99.19
SA-Fe_3O_4/NGP 3 wt. %	253.6°C	96.99
SA-Fe_3O_4/NGP 5 wt. %	254.2°C	94.17

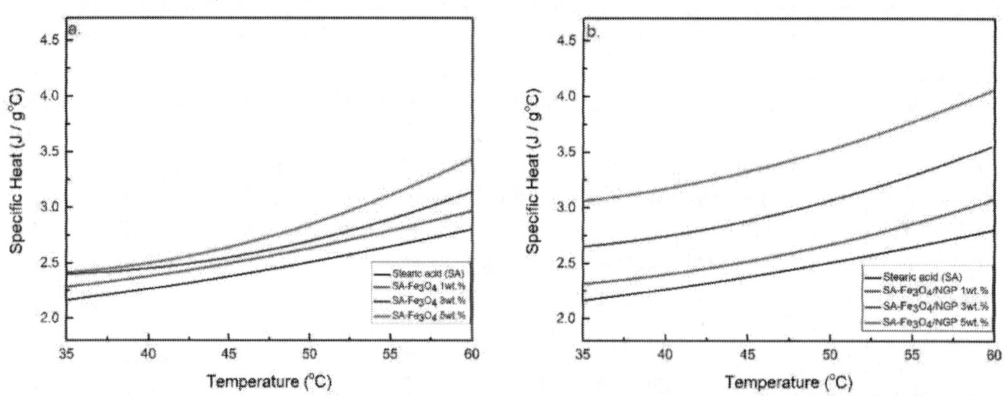

Figure 4. Specific heat of solid polynomial fitting curve (a) SA-Fe3O4 various wt. % and (b) SA-Fe3O4/10% NGP various wt. %

Table 3. DSC result of SA-Fe_3O_4 5 wt. % various wt. % and SA-Fe_3O_4/10% NGP various wt. %

Sample	Solid-liquid			Liquid-solid		
	T_{onset}	T_{peak}	T_{endset}	T_{onset}	T_{peak}	T_{endset}
Stearic acid (SA)	67.5	70.5	76	67.2	66.6	63.1
SA-Fe_3O_4 1 wt.%	67.53	70.48	76.1	66.26	66.72	62.81
SA-Fe_3O_4 3 wt.%	67.52	70.16	75.8	65.88	66.83	62.74
SA-Fe_3O_4 5 wt.%	67.4	70.4	76.1	64.6	66.6	62.2
SA-Fe_3O_4/10% NGP 1 wt. %	67.78	70.57	76.2	67.84	66.73	63.16
SA-Fe_3O_4/10% NGP 3 wt. %	67.61	70.43	76.1	67.89	66.7	63.24
SA-Fe_3O_4/10% NGP 5 wt. %	67.63	70.42	75.9	67.89	66.8	63.45

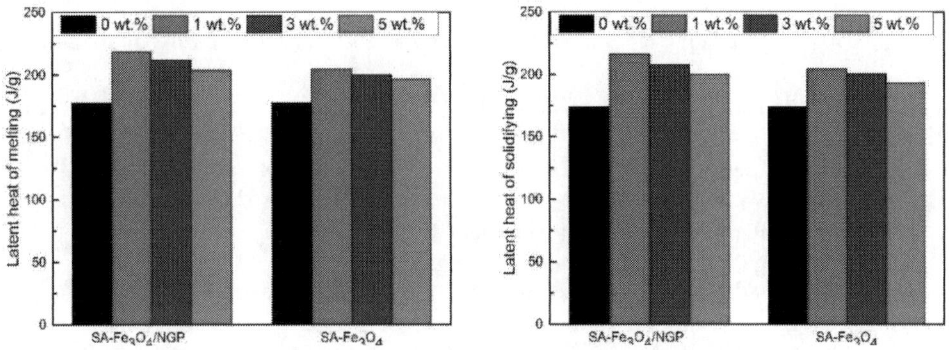

Figure 5. Latent heat melting proses and solidifying proses.

the Figure 4(a) and Figure 4(b) for the Sa/Fe_3O_4 and Sa/Fe_3O_4/NGP respectively. The result showed that Fe_3O_4 nanoparticles increase the specific heat value of Sa and further increase for Fe_3O_4/NGP. The value of specific heat also increase as with the addition of weight percent of the Fe_3O_4 added to either with or without NGP. The enhancement of specific heat indicates that the addition of the Fe_3O_4 and Fe_3O_4/NGP in the Sa also enhance the ability of the Sa in the sensible heat condition.

The comparison of latent heat storage between pure SA and PCM composites and also data of characterization are tabulated in Table 3. In charging process both PCM composites in onset, peak, and endset temperature do not significantly change from pure SA. On the other hand in discharging process, PCM composites shows difference behaviour in onset temperature from pure SA. In SA-Fe_3O_4 discharging curve shows that onset temperature become more smaller with increasing Fe_3O_4 wt.%. While in SA-Fe_3O_4/NGP discharging process curve shows a small increases in onset temperature. It is shows that small amount of NGP in Fe_3O_4/NGP can maintain the onset temperature of pure stearic acid. In Figure 5 plotted latent heat storage charging and discharging process of PCM composites. From Figure 6 shows that latent heat storage of samples increases with content of material additives various wt. %. Maximum enhancement of latent heat PCM composites occurs in content material additives at 1 wt.%. The enhancement of latent heat of SA-Fe_3O_4/NGP shows higher improvement than SA-Fe_3O_4. Latent heat of SA-Fe_3O_4/NGP improved by about 23% while SA-Fe_3O_4 improved by 15%. The enhancement of latent heat storage PCM composite is due to molecular interaction between molecules SA and material additives. Molecular interactions between molecules and material additives is physical interactions that are affected by the large surface area of material additives. The larger surface area of material additives produces greater interaction molecular with molecules SA [18-20]. From the result of XRD spectra Fe_3O_4/NGP have smaller crystallite size than Fe_3O_4 that can be lead to increased surface area. Therefore the enhancement of latent heat SA with addition Fe_3O_4/NGP have better improvement than Fe_3O_4.

4. Conclusion

Fe_3O_4 nanoparticles and Fe_3O_4/NGP nanocomposites were synthesized using sol gel method and used as material additives into PCM composites various wt. % prepared by dispersion using sonication method. Material additives used as enhancement latent heat storage of stearic acid. Based on XRD synthesized PCM composite shows that crystal structure SA and material additives remain unchanged. Thermal stability of SA-Fe_3O_4/NGP slightly increase than SA because the incorporation of NGP in nanocomposites. Enhancement of specific heat and latent heat storage were confirmed via DSC curve show that both of Fe_3O_4 and Fe_3O_4/NGP increased thermal performance of SA. In comparison material additives for percentage of enhancement specific heat and latent heat storage of SA, Fe_3O_4/NGP have higher improvement than Fe_3O_4 because of small content NGP in nanocomposites.

5. References

[1] Sharma A, Tyagi VV, Chen CR, et al. 2009 *Renewable and Sustainable Energy Reviews* **13** 318-345
[2] Farid MM, Khudhair AM, Razack SAK, et.al. **2004** *Energy Conservation and management* **45** 1597-1615
[3] Zalba B, Marin JM, Cabeza LF, et al. 2003 *Applied Thermal Engineering* **23** 251-283
[4] Li B, Liu T, Hu L, Wang Y, and Nie S 2013 *Chemical Engineering* **215** 819-826
[5] Cao L, Tang Y and Fang G 2014 *Energy* **80** 98-103
[6] Chen Z, Cao L, Fang G 2013 *Energy and Buildings* **62** 469-474
[7] Agyenim F, Hewitt N, Eames P, Smyth M. 2010 *Renewable & Sustainable Energy Reviews* **14** 615–628
[8] Sahan N and Paskoy HO 2014 *Solar Energy Materials & Solar Cells* **126** 56-61
[9] Teng TP and Cheng CM 2013 *Applied Thermal Engineering* **50** 637-644
[10] Kim S and Drzal LT 2009 *Solar Energy Materials and Solar Cells* **93** 136–142
[11] Elgafy A and Lafdi K 2005 *Carbon* **43** 3067-3074.
[12] Li Y, Zhu J, Wei S, Ryu J, Sun L, and Macromol ZG 2011 *Chemicals Physics* **212** 1951-1959.
[13] Arifin SA, Jalaludin S, and Saleh R 2015 *Advanced Materials research* **1123** 264-269
[14] Taufik A and Saleh R 2016 *Materials Science Forum* **1725** 020089
[15] Andiarto R, Nuryadin MK,and Saleh R 2016 *Journal of Physics: Conference Series* **710** 012020
[16] Harikrishnan S, Magesh S and Kalaiselvam S 2013 *Thermochimica Acta* **565** 137-145
[17] Cao Y, Feng J and Wu P 2010 *Carbon* **48** 3834–3839
[18] Shaikh S, Lafdi K and Hallinan K 2008 *Journal of Applied Physics* **103** 094302
[19] Şahan, Fois M and Paksoy H 2015 *Solar Energy Materials and Solar Cells* **137** 61–67
[20] Warzoha RJ and Fleischer AS2014 *International Journal of Heat and Mass Transfer* **79** 314-323

Study of non-stoichiometric $BaSrTiFeO_3$ oxide dedicated to semiconductor gas sensors

D. Fasquelle[a], N. Verbrugghe[a], S. Députier[b]
[a]Unité de Dynamique et Structure des Matériaux Moléculaires (UDSMM)
Université du Littoral Côte d'Opale, 50 rue F. Buisson, BP717, 62228 Calais, France
Tel/fax: 00 33 21 46 57 78
[b]Institut des Sciences Chimiques de Rennes(ISCR), UMR 6226 CNRS-Université de Rennes 1
Campus de Beaulieu, 35042 RENNES Cedex, France,

E-mail: didier.fasquelle@univ-littoral.fr

Abstract. Developing instrumentation systems compatible with the European RoHS directive (restriction of hazardous substances) to monitor our environment is of great interest for our society. Our research therefore aims at developing innovating integrated systems of detection dedicated to the characterization of various environmental exposures. These systems, which integrate new gas sensors containing lead-free oxides, are dedicated to the detection of flammable and toxic gases. We have firstly chosen to study semiconductor gas sensors implemented with lead-free oxides in view to develop RoHS devices. Therefore thick films deposited by spin-coating and screen-printing have been chosen for their robustness, ease to realize and ease to finally obtain cost-effective sensors. As crystalline defects and ionic vacancies are of great interest for gas detection, we have decided to study a non-stoichiometric composition of the $BaSrTiFeO_3$ sensible oxide. Non-stoichiometric $BaSrTiFeO_3$ lead-free oxide thick films were deposited by screen-printing on polycrystalline Al_2O_3 substrates covered by a layer of Ag-Pd acting as bottom electrode. The physical characterizations have revealed a crystalline structure mainly composed of $BaTiO_3$ pseudo-cubic phase and $Ba_4Ti_{12}O_{27}$ monoclinic phase for the powder, and a porous microstructure for the thick films. When compared to a BSTF thick film with a stoichiometric composition, a notable increase in the BSTF dielectric constant value was observed when taking into account of a similar microstructure and grain size. The loss tangent mean value varies more softly for the non-stoichiometric $BaSrTiFeO_3$ films than for the perovskite BSTF film as $\tan\delta$ decreases from 0.45 to 0.04 when the frequency increases from 100 Hz to 1 MHz.

1. Introduction

Functional oxides have been intensively studying by the scientific community since the 1990s. They present a wide range of interesting functionalities, such as high dielectric permittivity, piezoelectricity, pyroelectricity and ferroelectricity which have been largely developed in applications such as ceramic capacitors, tunable devices, actuators and sonars, I-R sensors and ferroelectric memories, respectively [1,2]. More recently, the community attention has been attracted by other properties and materials such as high temperature superconductivity, colossal magneto-resistance, multiferroics, gas sensing, half-metallic behaviour, thermoelectrics and oxides for fuel cells. The extended spectrum of involved materials has triggered extensive studies to understand the fundamental nature of existing systems, and then to modify and control novel materials for applications. These materials are characterized by strong interactions between the fundamental degrees of electron freedom (electron spin), charge, orbital and their interplay with the lattice [3]. All oxides exhibit various crystalline structure. One of the best-known family is oxides that exhibit phase transitions in strongly correlated electron systems. A typical example is an oxide with a perovskite-type crystal structure (ABO_3). In these crystalline structures, the grain boundaries, impurities and structural defects strongly influence the dynamics of ferroelectric or ferroic domains, if the material is not in the paraelectric state. And when other phenomena occur like the catalytic activity, often relied on atomic scale defects such as oxygen vacancies, new defects can be generated. So in the same time the dynamics of defects are strongly influenced by local electric fields, domain boundaries, surfaces chemical potentials which are themselves influenced by ferroelectric or ferroic properties and catalytic action.

Today, gas detection on the industrial market mainly uses optical sensors, electro-chemical sensors often containing lead, and catalytic sensors based on platinum. Many commercial gas sensors are composed of electrochemical cells, especially those used in the detection of flammable gas. Industrial firms producing these lead-based sensors can consume more than a ton of lead per year. This is a concern because lead is one of the most hazardous and polluting raw materials; it affects the central nervous system and has the potential to induce infertility and cancer. In 2008, the modified European Restriction of Hazardous Substances (RoHS) directive appeared, prohibiting the use of hazardous substances like lead, cadmium and mercury. As such, there is an urgent need for lead-free components in Europe, and industry is now trying to produce electronic components and devices which comply with RoHS standards. In this frame, developing innovating gas sensors containing lead-free oxides and dedicated to the detection of hazardous substances is highly encouraged by authorities. Among the different technologies that can be used for the detection of dangerous toxic and flammable gases in industrial sites, as well as for monitoring indoor air quality in houses and public buildings, oxides have been implemented to realize gas sensors, and especially semi conductive gas sensors [4, 5]. In this kind of sensor, the material of the oxidation-reduction is not a metal, as is the case for a catalytic detector, but generally a semi-conductor metallic oxide with n or p-type behavior, like the well-known SnO_2 or ZnO, and whose conduction is due to oxygen vacancies (oxide not exactly stoichiometric). The redox reactions, or simply adsorption to the surface, will change the resistivity of the material, by varying the number of charge carriers. The material is heated, as in the case of the catalyst beads, but the measurement is different: it is the variation of the resistance of the material itself which is measured, and not that of the heating element. The measuring range of these devices depends on the implemented signal processing: either the range 0-100% LLE (Low Limit of Explosivity) or the range 0-100% V / V, either the range ppm. Like the catalytic technology, the semiconductor detectors are not selective and their sensitivity to various compounds will depend on the gas calibration. They require the presence of oxygen to function. Indeed, the presence of oxygen in the background determines a higher baseline resistance (formation of ionosorbed oxygen), and generally increases the response and recovery times. The semiconductor detectors for detecting flammable gases are not very used in the industry, as they age (due to heating) and thus derive very rapidly (within weeks for some). In addition, these detectors are influenced by ambient conditions, particularly by moisture. The water vapor adsorbs onto the surface of the semiconductor: this is an interfering problematic. The response time of these devices is very variable, from a few tens of seconds to several minutes. The lifetime of the sensors is of the order of several weeks to several years, depending on the context of use. Therefore, the use of solid-state detectors for the detection of flammable gas is not recommended. In opposition to this point, semiconductor oxides can be used for the detection of toxic gases and vapors.

In fact, gas detectors need to be implemented for monitoring the risk of poisoning caused by the possible presence of toxic gases or vapors, or for the protection of persons by monitoring the risk of suffocation from a lack of oxygen (anoxia) in confined premises and caused by gas leaks such as nitrogen (N2), helium (He). Modern technologies offer high performance in terms of compensation (humidity and / or temperature) and provide long-term stability. To meet different needs, three sensor technologies are currently met in this case: electrochemistry, semiconductor and optical devices. We have firstly chosen to study semiconductor gas sensors implemented with lead-free oxides in view to develop RoHS devices. Therefore thick films deposited by spin-coating and screen-printing have been chosen for their robustness, ease to realize and ease to finally obtain cost-effective sensors. Our previous work enabled us to optimize the conditions of deposition of lead-free thick and thin films ($BaTiO_3$, $BaSrTiO_3$, $BaTiSnO_3$, $BaSrTiFeO_3$, SBN, KTN, ZnO and TiO_2) integrated to semiconductor gas sensors [4,5].

Among these sensors embedding a lead-free film based on a stoichiometric oxide, some materials have ttracted our attention in view to study another possibility: As crystalline defects an ionicvacancies are of great interest for gas detection, i.e. for gas adsorption - desorption, a new emerging question appears: will this phenomenon be emphasized by a non-stoichiometric composition of the sensible oxide? So we have started works on a non-stoichiometric $BaSrTiFeO_3$ material. We present here a complete study of non-stoichiometric $BaSrTiFeO_3$ lead-free oxide thick films deposited on polycrystalline Al_2O_3 substrates, dedicated to the development of semiconductor gas sensors.

2 Experimental procedure

2.1 Preparation of $BaSrTiFeO_3$ powder

The composition $Ba_{0.85}Sr_{0.15}Ti_{0.9}Fe_{0.1}O_3$ was prepared by solid-state reaction, using amounts of BaO (ACROS ORGANICS, 90%), SrO (Alfa Aesar, 97 %), TiO_2 (ACROS ORGANICS, 98 %), and Fe_2O_3 (STREM CHEMICALS, 99.8 %) oxides, with no respect of the required amount of oxygen in aim to make a non-stoichiometry in oxygen. The obtained BSTF precursor was mixed with alcohol in ball mill for 3h. The homogenized powder was dried at 300 °C for 6 h. and then calcined at 1100 °C for 3 h. The samples made with this powder were labeled BSTF-NS1. A second step of milling was done for 3h in order to decrease the grain size. The samples made with this double-milled powder were labeled BSTF-NS2.

2.2 Ink preparation

To formulate the thick-film ink, BSTF powder was combined with an organic vehicle (alpha-terpineol (ACROS ORGANICS, 97 %), polyvinyl butyral (ACROS ORGANICS, Butvar B-98), 2-(2-butoxyethoxy)-ethyl acetate (Alfa Aesar, 98 %) and polyethylene glycol (ACROS ORGANICS, M.W. 400) using a 60:40 ratio. The ink was milled by a roller to obtain a good homogeneity.

2.3 Film preparation and characterizations

High purity alumina substrates were coated by a silver-palladium ink (Ag-Pd, Tanaka Kikinzoku Kogyo K.K.) to realize the bottom electrode and then heated at 850 °C for 1 h to densify the Ag-Pd layer. For the electrical characterizations, BSTF thick films were deposited by screen-printing on Ag-Pd/Al_2O_3 substrates, and then sintered at 1100 °C for 2 h. Gold (Neyco, 99.99 %) / Chromium (CERAC, 99.99 %) (Au/Cr) upper electrodes were deposited by thermal evaporation. The obtained films presented a thickness ranging from 20 to 40 μm. For X-ray diffraction (XRD) analysis, a D8 Advance Bruker AXS (θ-2θ configuration) equipped with a monochromatized $CuK\alpha_1$ radiation source was used. Surface morphology and thickness were determined by field-emission scanning electron microscopy (FE-SEM), using a JEOL JSM 6310F system working at a low accelerating voltage (5 kV). The resistance and capacitance parameters of BSTF films were investigated and measured by a HP4284A impedance analyzer using a metal-insulator-metal (MIM) structure. These parameters were used for calculating the dielectric constant (ε'), loss tangent ($\tan\delta$).

3. Results and discussion

Figure 1 shows the polycrystalline structure of the non-stoichiometric BSTF powder. This structure is mainly a combination of $BaTiO_3$ pseudo-cubic phase and $Ba_4Ti_{12}O_{27}$ monoclinic phase, given as references for identification (JCPDS files).

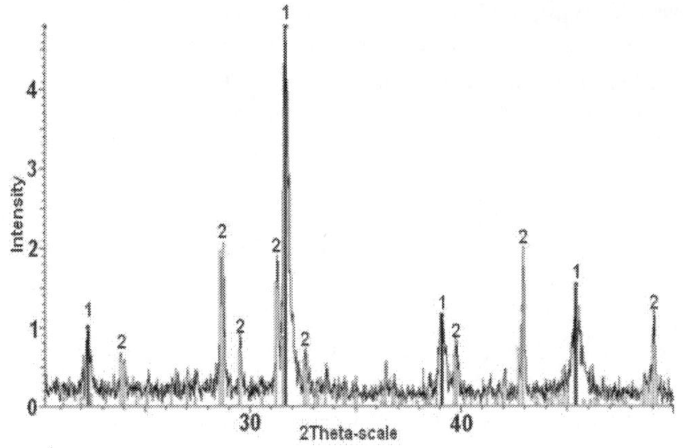

Figure 1. XRD pattern of non-stoichiometric BSTF thick film. (1) identifies the $BaTiO_3$ phase, (2) identifies the $Ba_4Ti_{12}O_{27}$ phase.

Figure 2 shows the organization of the crystalline cells of the main phases: a) the perovskite structure of BT in cubic phase, b) the monoclinic phase of $Ba_4Ti_{12}O_{27}$. Details on the different cells are given in Figure 2.

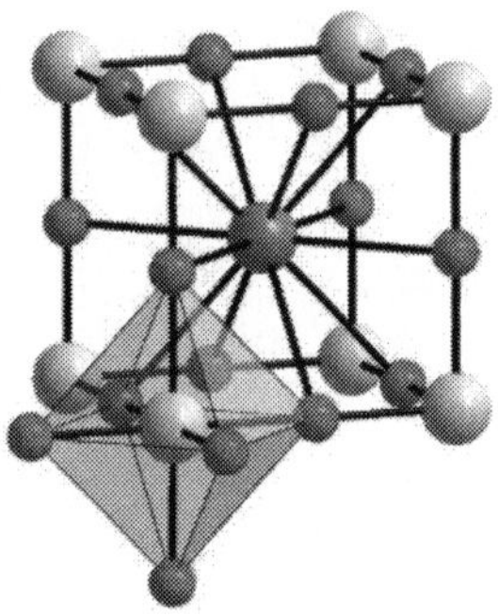

$BaTiO_3$ [a=3.996 Å]

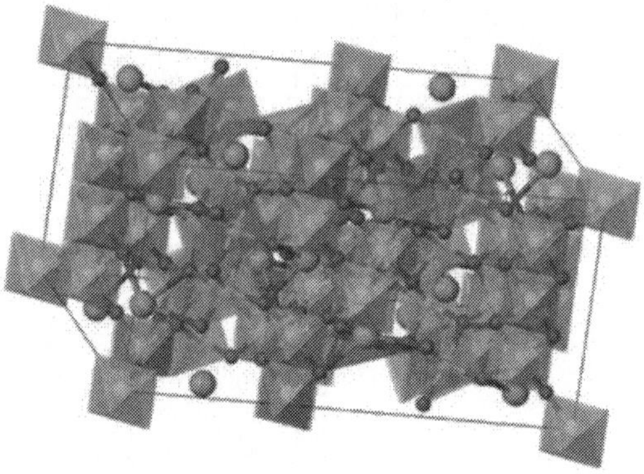

$Ba_4Ti_{12}O_{27}$ [a=19.8165 Å; b=11.4468 Å; c=9.9208 Å; β=109.1847°]

Figure 2. Organization of the crystalline cells.

Figure 3 shows the microstructure of the two thick films deposited by screen printing on Ag-Pd/Al_2O_3 substrates, where the top and cross section views are given from the observation by FE-SEM. The film made with the powder only milled one time was labeled BSTF-NS1, while the powder with 2 milling steps was used to elaborate the film labeled BSTF-NS2. The film thickness, which was measured by observing the cross-section, was of about 30 μm for the BSTF-NS1 sample and of about 20 μm for the BSTF-NS2 sample. The grain size of the BSTF layers is rather regular with agglomerated grains. Although, it can be noticed that the film is porous as a binder was used in the ink preparation. All this binder was burnt during the sintering step and all organics were removed from the ink layer, creating holes in the film. This porous structure is of interest for gas sensing as the detection process can in this case involve the whole volume of the oxide layer, i.e. not only be linked to a surface reaction like in a dense thin film. The grain size ranged from 300 to 400 nm for BSTF-NS1 and from 100 to 150 nm for BSTF-NS2. The observation of the cross-section views clearly identifies the strong adhesion of BSTF/Ag-Pd layers on Al2O3 substrates.

Figure 3 FE-SEM photos of BSTF thick films. Images on left are for BSTF-NS1 and on right for BSTF-NS2. Top and cross-section views are given.

The dielectric characterizations were made on a metal-insulator-metal (MIM) structure. Figure 4 shows the frequency dependence of the dielectric constant (ε') of the BSTF films where the frequency ranges from 100 Hz to 1 MHz at room temperature. As many dots, i.e. upper electrodes, were deposited on the samples, different curves are plotted in Figure 4. The ε' mean values at 1 kHz and 100 kHz for BSTF-NS1 were 125 and 100, respectively; for BSTF-NS2, 68 and 58, respectively. The dielectric constant values of the BSTF-NS2 thick film are lower because of the lower grain size of this film, due to the use of a second step of powder milling. This effect of the grain size on the dielectric properties is well-known for ceramics and films [6,7,8,9]. When compared to a BSTF thick film with a stoichiometric composition [10], a notable increase in the BSTF dielectric constant value was observed when taking into account of a similar microstructure and grain size, i.e. by comparing with BSTF-NS1. A similar value of ε', i.e. lower than 100 at room temperature has been measured on BSTF-NS2. On BSTF ceramics, dielectric constant values lower than 100 are observed as well [11]. All dielectric values are summarized in table 1.

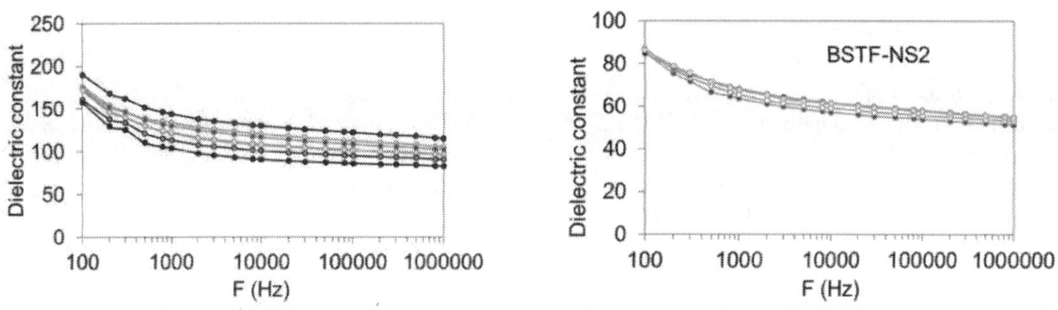

Figure 4 Frequency dependence of the dielectric constant (ε') given for the BSTF-NS1 (on left) and BSTF-NS2 thick films.

Table 1: Dielectric values given for the BSTF-NS1 and BSTF-NS2. Values measured on a stoichiometric composition of BSTF are given as well for comparison [10].

	Dielectric constant (mean value)		**Loss tangent (mean value)**	
Frequency	1 kHz	100 kHz	1 kHz	100 kHz
BSTF-NS1	125	100	0.16	0.04
BSTF-NS2	68	58	0.12	0.04
BSTF [10]	65	52	0.3	0.04

The frequency dependence of the loss tangent (tanδ) at room temperature is shown in Figure 5. In our frequency range, the loss tangent variation is quite similar for the 2 films. As many upper electrodes were deposited on the samples, different curves are plotted in Figure 5. The loss tangent mean value varies more softly for the non-stoichiometric $BaSrTiFeO_3$ films than for the perovskite BSTF film [10]. In fact tanδ decreases here from 0.45 to 0.04 when the frequency increases from 100 Hz to 1 MHz.

About the important decrease of the ε' and tanδ evolutions in function of the frequency, it is interesting to remind that our thick films are very porous due to the relatively low percentage of oxide powder in our inks, i.e. 60% for BSTF films. A high porosity in a film implies that the layer is composed of grains and holes, where the holes are filled with air. This porous structure is therefore in relation with the observed evolutions. But the porosity is not problematic as it is really a desired characteristic of the microstructure because we dedicate these porous thick films to semiconductor gas sensors, and in this case the porosity is of importance for the adsorption-desorption process.

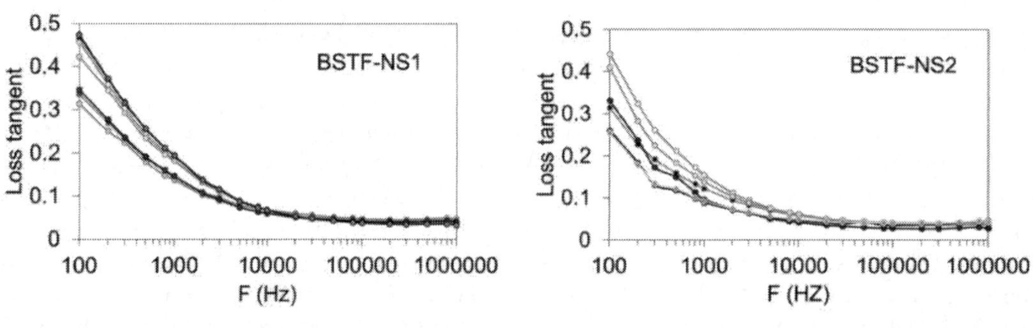

Figure 5 Frequency dependence of the loss tangent (tanδ) for the BSTF-NS1 (on left) and BSTF-NS2 thick films.

4. Conclusion

$Ba_{0.85}Sr_{0.15}Ti_{0.9}Fe_{0.1}O_3$ powder, with a non-stoichiometry in oxygen, has been synthesized by solid-state reaction. BSTF inks were prepared and deposited on Ag-Pd/Al_2O_3 substrates by screen printing to obtain porous thick films dedicated to lead-free gas sensors for RoHS devices. The physical characterizations have revealed a crystalline structure mainly composed of $BaTiO_3$ pseudo-cubic phase and $Ba_4Ti_{12}O_{27}$ monoclinic phase for the powder, and a porous microstructure for the thick films. These porous polycrystalline films have given significant differences in dielectric properties when compared to a perovskite BSTF thick film.

Acknowledgment

The SEM photos were made by Dr Lucie Courcot from the Maison de la Recherche en Environnement Naturel (MREN), Université du Littoral Côte d'Opale.

References

[1] Jaffe B, Cook WR and Jaffe H, 1971, *Piezoelectric Ceramics*, Academic Press London and New York.
[2] Scott J F, 2007, *Science* 315, 954.
[3] Ramesh R, 2013, *Current Science*, Vol. 105, No. 8, pp 1107-1114.
[4] Fasquelle D, Députier S, Mascot M, Uschanoff N, Bouquet V, Demange V, Guilloux-Viry M, Carru J.-C. 2013, *Advanced Materials Research* Vol. 789, pp 105-111.
[5] Fasquelle D, 2014 *International Innovation*, pp 106-108.
[6] Luan W, Gao L, and Guo J, 1999, *Ceramics International* 25 pp.727-729.
[7] Tusseau-Nenez S, Ganne J.-P., Maglione M, Morell A, Niepce J.-C., Paté M., 2004, *Journal of The European Ceramic Society* 24 pp. 3003–3011.
[8] Fasquelle D, Mascot M, and Carru JC, 2011, *Advanced Materials Research* Vol. 277, pp 1-10.
[9] Fasquelle D, Mascot M, and Carru JC, 2012, *Solid-State Electronics* 75 pp. 6–12.
[10] El Romeh MA, Fasquelle D, Députier S, Mascot M, and Guilloux-Viry M, 2016, *Journal of Materials Science: Materials in Electronics*, Vol. 27, Issue 2, pp 2096–2102.
[11] Guo Z, Pan L, Bi C, Qiu H, Zhao X, Yang L, and Rafique MY, 2013, *Journal of Magnetism and Magnetic Materials*, 325, pp. 24-28.

Journal of Physics: Conference Series **776** (2016) 012039 doi:10.1088/1742-6596/776/1/012039

Better adsorption capacity of SnO_2 nanoparticles with different graphene addition

V. Paramarta[1,2], A. Taufik[1,2], and R Saleh[1,2,*]

[1]Departemen Fisika, Fakultas MIPA-Universitas Indonesia, 16424 Depok, Indonesia
[2]Integrated Laboratory of Energy and Environment, Fakultas MIPA-Universitas Indonesia, 16424 Depok, Indonesia

[*]E-mail: rosari.saleh@ui.ac.id

Abstract. The adsorption capacity of SnO_2 nanoparticle has been studied by graphene and nanographene platelets (NGP) additions using co-precipitation method. The crystalline phase, composition, and morphology of the samples are analyzed using X-Ray Diffraction (XRD), Energy Dispersive X-Ray (EDX), Fourier Transform Infrared Spectroscopy (FT-IR), and Transmission Electron Microscope (TEM). Tetragonal structure of SnO_2 is shown for the nanoparticle and its composites. The presence of graphene and NGP is also confirmed. The adsorption capacity of the nanoparticle and its composites are analyzed by observing the degradation of methylene blue (MB) as the organic dye model using UV-Vis Spectroscopy. The result shows that SnO_2 composite with graphene achieves higher adsorption capacity of about 20 % than the composite with NGP. The fitting of equilibrium adsorption capacity result indicates that the adsorption mechanism of SnO_2 composite with graphene tends to follow the Langmuir adsorption-isotherm model.

1. Introduction

Environmental pollution has become a major attention in recent years. One of the major concerns of researchers is water pollution due to industrial dye wastes. A great number of industries use dyes in their products: food, clothing, paper, pharmaceuticals, cosmetics and textiles [1,2]. The existence of dye in waste water, even in low concentration may be very dangerous for human's life and microorganism [3]. Methylene blue (MB) is a type of dye materials which is important and has broad application in chemical analysis for pH indicator, dye for cellulose and bacteria dyes, and culture media for bacteria [4]. Therefore, various researches have been conducted and developed to settle problems being faced due to the use of this dye.

Various methods such as adsorption, coagulation and flocculation, oxidation, and chemical precipitation have been reported as the methods used to overcome issues of water pollution due to dye wastes. Of the various methods, adsorption is considered to be the best method because of its low cost, easily to be applied, and its availability of adsorbents [5]. Nanostructure material has been known to be able to become a good adsorbent because of its large surface area [6]. One of nanomaterials that is widely used in the degradation process of dye wastes is tin oxide (SnO_2). SnO_2 has various advantages such as its sensitivity when exposed to light and thermal energy, its numbers of intrinsic defect in its structure caused by low formation energy and strong attraction between the tin sites and oxygen vacancies [7]. In recent years, the graphene materials draw special attention to scientist in the world because of its unique properties such as its large surface area, its hydrophlicity, its high dispersibility, and its high mobility of charge electron [8]. Several reviews mention that graphene materials can be used as adsorbent in degrading dye wastes [9]. Therefore, considering the advantages of SnO_2 and graphene materials to be applied in adsorption methods for waste water treatment, these two materials are combined to degrade dye wastes through adsorption process which is expected to get a better efficiency in degrading dye wastes than when each applied separately.

In this study, SnO_2 nanoparticle will be combined with two different types of graphene namely Nanographene platelets (NGP) and graphene, SnO_2/NGP and SnO_2/graphene that are applied as adsorbents to degrade dye and the adsorption mechanism of these two composites is also studied.

2. Experimental

2.1. Chemicals

All the reagents used are analytical grade and without going through further purification; tin chloride ($SnCl_2$), sodium hydroxide (NaOH), and ethanol. Nanographene platelets and graphene were bought from Angstron Material.

2.2. Preparation SnO_2 nanoparticles

SnO_2 nanoparticles were synthesized using the modification method reported by Yue Li. et.al [10]. First, $SnCl_2$ is dissolved into ethanol and aquades, then NaOH solution is poured into the solution then stirred magnetically. This mixed solution is then heated at 180 °C for 3 hours, and cooled down at the room temperature. The deposit produced is centrifuged and rinsed using aquades and ethanol several times. The deposited particle is dried out under vacuum condition at 80 °C. SnO_2 nanoparticle is obtained after going through the calcination process for 3 hours at the temperature of 700 °C.

2.3. Preparation SnO_2/NGP and SnO_2/graphene coposites

SnO_2/NGP is synthesized using the co-precipitation method. First, NGP is dissolved into aquades and ethanol through ultrasonic treatment for 2 hours, then SnO_2 nanoparticles is poured into the solution and stirred magnetically. The mixed solution is then heated at 120 °C for 3 hours. The result of solution is then centrifuged and dried at 70 °C under vacuum condition. The same method is also applied to synthesize SnO_2/graphene.

2.4. Characterizations

The samples are characterized using X-Ray Diffraction (XRD), measured using Rigaku Miniflex 600, while the image of transmission electron microscopy (TEM) is recorded using Tecnai G2 Supertwin. The spectrum of infrared absorption from the samples is obtained using Shimadzu FTIR spectrophotometer in the range of 400-4000 cm^{-1}, and the elements of the nanoparticle and composites are revealed using energy dispersive X-Ray (EDX) with the energy range of 0-5 keV.

2.5. Adsorption processes

Adsorption experiment is carried out by mixing SnO_2, SnO_2/NGP, and SnO_2/graphene into methylene blue (MB) solution used as pollutant model with the concentration of 20 mg/L, and the pH solution is adjusted using NaOH. The concentration of dye is analyzed using UV-Vis Spectrometer. Furthermore, adsorption experiment is carried out again with varying MB concentration for sample with best adsorption activity.

3. Results and Discussion

Figure 1 indicates the XRD pattern of SnO_2 nanoparticles, SnO_2/NGP and SnO_2/graphene composites. The diffraction pattern of the samples is observed at 2θ = 26,5°, 33.8°, 38°, 39°, 51.8°, 54.8°, 58°, 62°, 64.7°, 65.8°, 71.2°, 78.2°, 81.2°, 83.7° which indicates the existence (110), (101), (200), (111), (211), (220), (002), (310), (112), (301), (202), (321), (400), (222) planes of the SnO_2 tetragonal structure. The peak at 2θ of 26.0° indicates (002) plane of the NGP structure. For graphene, there is only a broad peak at around 2θ of 25°, which indicates the structure of graphene. The diffraction peaks of SnO_2, NGP and graphene at SnO_2/NGP and SnO_2/graphene composites are overlapped at 2θ of around 26°. The grain

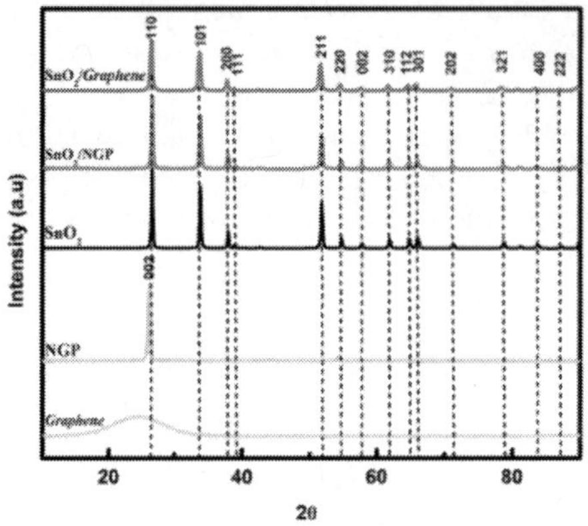

Figure. 1. XRD Spectrum of SnO_2 nanoparticle, SnO_2 /NGP and SnO_2/graphene composites

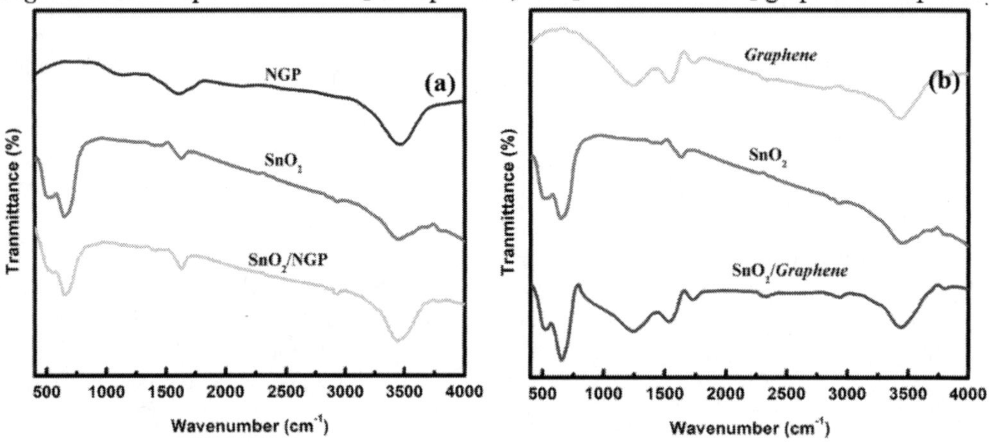

Figure. 2. FTIR Spectrum of SnO_2 nanoparticle, SnO_2/NGP and SnO_2/graphene composites

Table 1. Lattice parameter and grain size of SnO_2, SnO_2/NGP and SnO_2/graphene

Sample	Lattice Parameter (Å)			<D> (nm)
	SnO_2		NGP	
	a=b	c	a=b=c	
SnO_2	4.7384	3.1872	-	44
SnO_2/NGP	4.7442	3.1891	3.6382	33
SnO_2/graphene	4.7586	3.1994	-	21

size is calculated using the Scherrer's formula, and the lattice parameter is calculated using Rietveld refinement method that is used in MAUD software. The calculation results of grain size and lattice parameter is presented in Table 1.

The FTIR spectrum of SnO_2 nanoparticles, SnO_2/NGP and SnO_2/graphene composites are presented in figure 2. Figure 2a shows the FTIR spectra for SnO_2/NGP composites, while the FTIR spectra for SnO_2/graphene composites is shown in figure 2b. The absorption of SnO_2 nanoparticle appears in the range of 540-660 cm^{-1} which indicates O-Sn-O and Sn-O stretching vibration modes. The absorptions in range of 3400 cm^{-1} and 1634 cm^{-1} indicate the O-H group stretching and bending vibration modes originating from water molecule [11] while absorption in the range of 1220 cm^{-1} and 1527 cm^{-1} indicates the C-OH and C-O stretching vibration modes [12,13]. All SnO_2, NGP and graphene characteristics can be revealed with the FTIR spectrum of SnO_2/NGP and SnO_2/graphene.

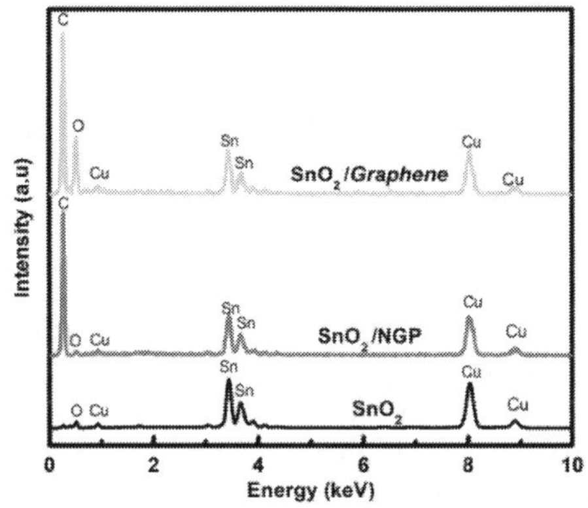

Figure. 3. EDX Spectrum of SnO_2 nanoparticle, SnO_2/NGP and SnO_2/graphene composites

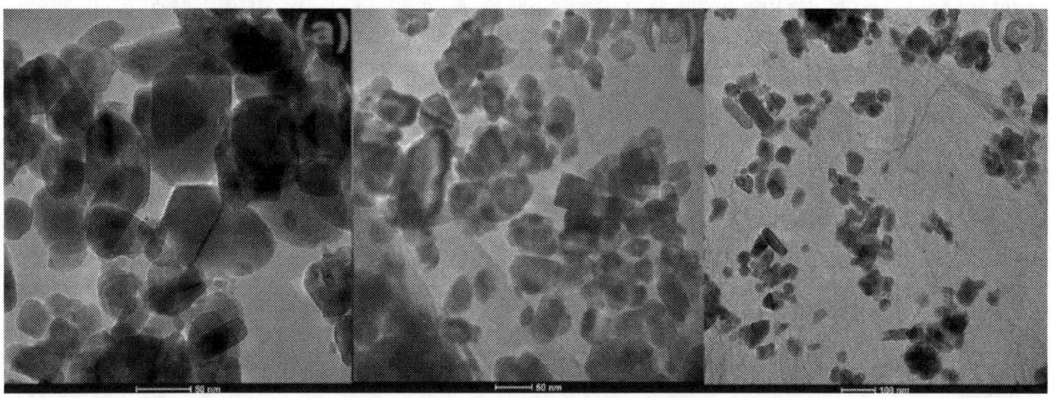

Figure. 4. TEM images from SnO_2 nanoparticle (a), SnO_2/NGP (b) and SnO_2/graphene (c) composites

The Energy Dispersive X-Ray (EDX) spectrum of SnO_2 nanoparticles, SnO_2/NGP and SnO_2/graphene composites are shown in figure 3. Based on the EDX measurement results, the SnO_2 nanoparticle consists of Sn, O and Cu elements. The energy spectra of Cu presents in the results of EDX measurements for the use of copper grid as the sample holder. The existence of NGP and graphene in the SnO_2/NGP and SnO_2/graphene composites are confirmed with the existence of carbon energy spectra in the EDX measurement results. The absence of other elements in the EDX spectrum showed the absence of an impurity in SnO_2 nanoparticle and composites samples.

The morphology of SnO_2 nanoparticles, SnO_2/NGP and SnO_2/graphene composites is analyzed using Transmission Electron Microscope (TEM). The TEM result of these samples is indicated in fiigure 4. Figure 4a is the morphologic image of SnO_2 nanoparticle in irregular shape with the average particle size of 40-50 nm. When NGP and graphene are incorporated in SnO_2 nanoparticle, the morphology of the produced composited is still in irregular shape, but smaller particle size with the average particle size of 30-36 nm for SnO_2/NGP (figure 4b) and 18-25 nm for SnO_2/graphene (figure 4c). Appearence of fine wrinkled structure in these figures indicate the presence of NGP and graphene in SnO_2/NGP and SnO_2/graphene composites samples. The results of TEM image confirm the grain size calculation results from the XRD measurements (Table 1). The clear of TEM images indicates good crystal characteristics [17].

MB degradation process of SnO_2 nanoparticles, SnO_2 /NGP and SnO_2/graphene composites is shown in figure 5. Figure 5a is the proces of decolorization (color change) from MB during the adsorption process using SnO_2 nanoparticles and composites. It can be seen that the decolorization process increases simultaneously with addition of NGP dan graphene from 22% for SnO_2 nanoparticles and 35% and 60%

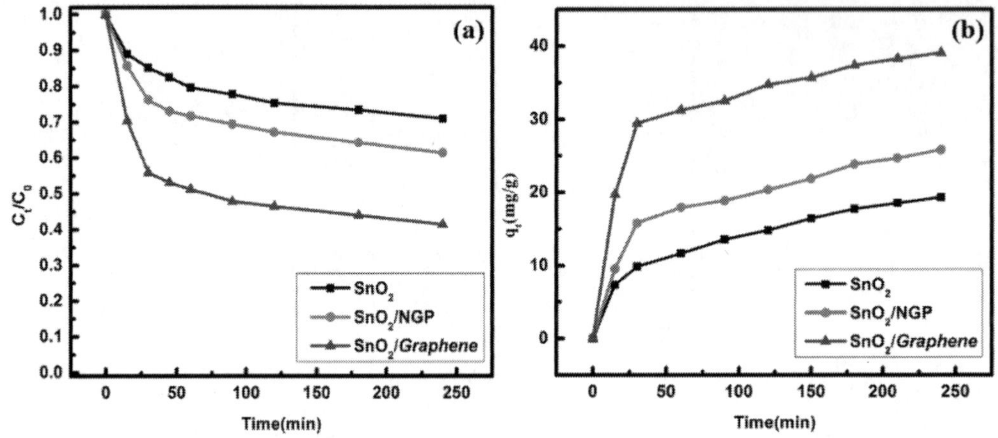

Figure. 5. (a) Decolorization process of MB concentration and (b) adsorption capacity of SnO_2 nanoparticle, SnO_2/NGP and SnO_2/graphene composites

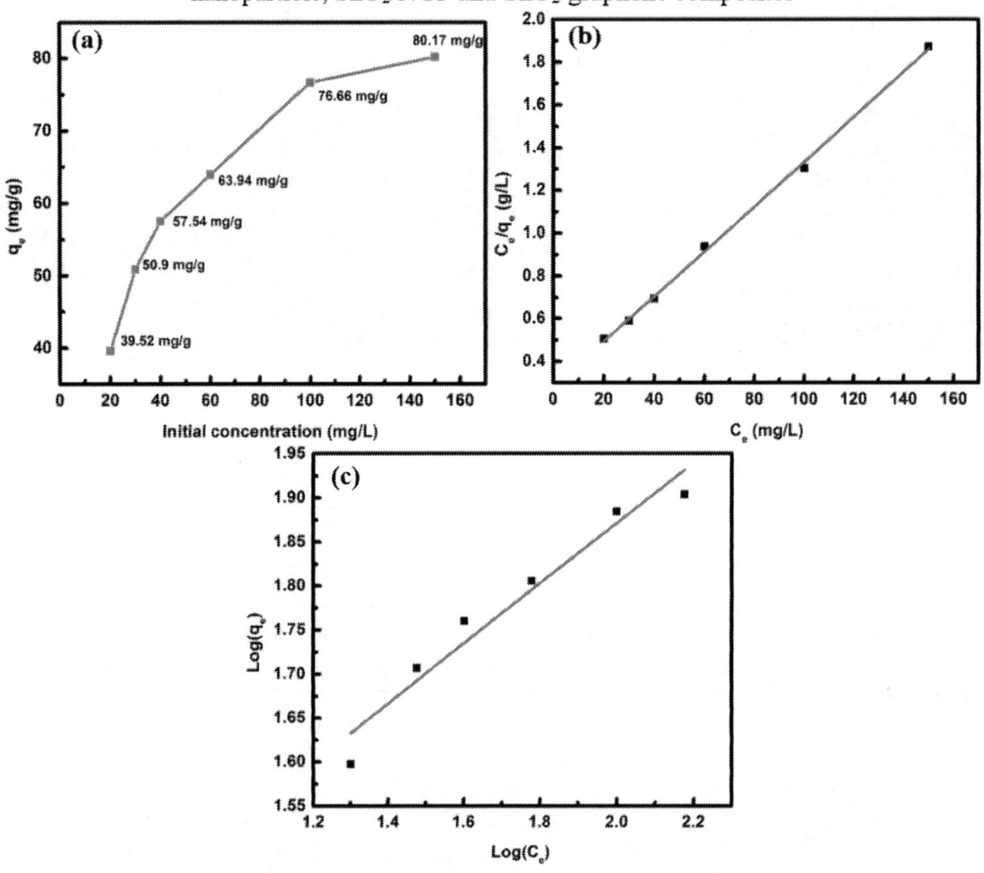

Figure. 6. (a) Equilibrium adsorption capacity of SnO_2/graphene with different MB concentrations, (b) Langmuir adsorption-isotherm model, (c) Freundlich adsorption-isotherm model fitting the experimental data of SnO_2/graphene.

respectively for SnO_2/NGP and SnO_2/graphene composites. To identify the adsorption capability of the adsorben to adsorp the pollutants on its surface per unit of time, there is a certain contact time during an adsorption process [14]. Figure 5b indicates the adsorption capacity against time up to 250 minutes. It can be seen that the adsorption capacity of the SnO_2 nanoparticles and composites nearly reach the equilibrum at 180 minutes referred to equilibrium adsorption capacity. Based on the adsorption result, the highest equilibrium adsorption capacity is reached by SnO_2/graphene at 39.1 mg/g then followed by

SnO_2/NGP at 25.9 mg/g and SnO_2 nanoparticles at 19.2 mg/g. Figure 6a shows the the adsorption capacity of SnO_2/graphene composites to degrade varied MB initial concentrations. In this study, MB concentration is varied from 20 mg/L – 150 mg/L with the adsorbent dosage of 0.3 g/L. The figure shows an increase of total adsorption capacity of SnO_2/graphene as the MB concentration is increased with equilibrium adsorption capacity reach its maximum of 80.17 mg/g. The adsorption mechanism of SnO_2/graphene can be identified using two adsorption models namely Langmuir and Freundlich [12-13]. The Langmuir adsorption-isotherm model indicates monolayer coverage on the surface of adsorbent, while the Freundlich adsorption-isotherm model indicates a heterogeneous system [15-16]. The experimental data of the adsorption capacity are fitted using Langmuir and Freundlich adsorption-isotherm models (figure 6b and 6c). It can be seen that the experimental data are more closely fitting using the Langmuir adsorption-isotherm model (R^2=0.99) than using the Freundlich adsorption-isotherm model (R^2=0.95).

3. Conclusion

The SnO_2 nanoparticles, SnO_2 composites with NGP and graphene addition have been successfully synthesized using co-precipitation method. The XRD spectrum of the SnO_2 nanoparticle and composites show the tetragonal structure of SnO_2. The spectrum of EDX confirm the existence of NGP and graphene in SnO_2 composites and also confirms the absence of impurity in the samples. The equilibrium adsorption capacity of SnO_2 nanoparticles and composites reach the value of 19.2, 25.9 and 39.1. The equilibrium adsorption capacity value inceases with the presence of NGP dan graphene in SnO_2 composites and the best adsorption capacity is obtained by SnO_2/graphene. The adsorption process of SnO_2/graphene follows the Langmuir adsorption-isotherm model.

4. Reference

[1] Wu Z, Zhong H, Yuan X, Wang H, Wang L, Chen X, Zeng G, Wu Y 2014 *Water Research* **67** 330-334

[2] Robati D, Bagheriyan S, Rajabi M, Maradi O, Peyghan AA 2016 *Physica E* **83** 1-6

[3] Chang J, Ma J, Ma Q, Zhang D, Qiao N, Hu M, Ma H 2016 *Applied Clay Science* **119** 132-140

[4] Lin TC, Li YS, Chiang WH, Pei Z, *Biosensors and Bioelectronics* doi:10.1016/j.bios.2016.03.046

[5] Borah L, Goswami M, Phukan P 2015 *Journal of Environmental Chemical Engineering* **3** 1018-1028

[6] Agarwal S, Tyagi I, Gupta VK, Ghasemi N, Shahivand M, Ghasemi M 2016 *Journal of Molecular Liquids* **218** 208-218

[7] Xu H, Qu Z, Zhao S, Yue D, Huang W, Yan N 2016 *Chemical Engineering Journal* **292** 123-129

[8] Zhao H, Qju F, Yan J, Wang J, Li X, Yang D 2016 *Applied Clay Science* **121-122** 137-145

[9] Wu Z, Zhang L, Guan Q, Ning P, Ye D 2014 *Chemical Engineering Journal* **258** 77-84

[10] Li Y, Guo Y, Tan R, Cui P, Li Y, Song W 2009 *Materials Letters* **63** 2085-2088

[11] Luo Q, Wang L, Wang D, Yin R, Li X, An J, Yang X 2015 *Journal of Environmental Chemical Engineering* **3** 622-629

[12] Tan LL, Ong WJ, Chai SP, Mohamed AR 2013 *Nanoscale Research Letters* **8** 465

[13] Peng X, Li Y, Zhang G, Zhang F, Fan X, *Journal of Nanomaterials* http://dx.doi.org/10.1155/2013/841789

[14] Singh D, Verma S, Gautam RK, Krishna V 2015 *Journal of Environmental Chemical Engineering* **3** 2161-2171

[15] Chang YC, Chen DH 2005 *Macromolecular Bioscience* **5** 254-261

[16] Ma W, Ya FQ, Han M, Wang R 2007 *Journal of Hazardous Materials* **143** 296-302

[17] Liu H, Liu T, Dong X, Lv Y, Zhu Z 2014 *Materials Letters* **126** 36-38

Preparation and catalytic performance of ZrO_2 - nanographene platelets composites

Y. Kristianto[1,2,] A. Taufik[1,2] and R. Saleh[1,2,*]

[1]Departemen Fisika, Fakultas MIPA-Universitas Indonesia, 16424 Depok, Indonesia

[2]Integrated Laboratory of Energy and Environment, Fakultas MIPA-Universitas Indonesia, 1624 Depok, Indonesia

[*]E-mail: rosari.saleh@gmail.com

Abstract. The ZrO_2/Nanographene platelets (NGP) composites with different NGP weight percents were successfully synthesized using two-step methods (sol-gel followed by co-precipitation methods). The structural properties of the composites were analyzed using X-Ray Diffraction (XRD), Energy Dispersive X-Ray (EDX), Transmission Electron Microscope (TEM), and Thermalgravimetric Analysis (TGA). The presence of NGP in the composites confirms by the graphitic-like structure in the XRD and layer-like structure in the TEM. Thermal stability of the composites tends to reduce with the incorporation of NGP. The catalytic performances of the composites were evaluated by three catalytic processes. The degradation of methylene blue as a model pollutant by the composites are monitored using UV-Vis Spectroscopy. The result shows that ZrO_2 with 10 wt% of NGP exhibits the highest performance for all of three processes. The scavenger experiment shows that hole acts as an active species that play a role in the catalytic process of the composites.

1. Introduction

The waste water and dyes of the textile industry contain organic dye substance that may cause serious damage to the environment. Moreover, it difficult to be destroyed using traditional waste water treatment [1]. One of commonly-used dye substance is methylene blue (MB), a dangerous cationic with mutagen and carcinogen properties as well as resistant to biological decomposition [2]. There are various techniques to decompose the dye substances, such as electrochemical treatment [3], chemical oxidation [4] and surface adsorption [5]. Other methods to decompose dye substances are using photocatalytic and sonocatalytic. These two techniques are a promising method in decomposing dye substances because of their practical application and do not create any secondary pollutant [6-8].

Use of metal oxide semiconductor (MOS) as a catalyst in catalytic activities has attracted researchers' attention due to it can exhibit good performance in decomposing and destroying organic pollutants to decontaminate water [9]. Among the various of MOS, Zirconium oxide (ZrO_2) is an n type of semiconductor which recently has attracted the attention because of its unique properties such as high photosensivity, high stability, non-toxic, high melting point, and good adsorption [10,11]. Therefore, ZrO_2 has various applications in many field, such as gas sensor, catalyst or supporting catalyst, photocatalyst and waste water treatment [12-14]. In the previous report, ZrO_2 as a catalyst has successfully degrade organic dye substance [15]. It indicates that ZrO_2 has the potency to be applied as a catalyst for catalytic activities. Nevertheless, catalytic performance of ZrO_2 in degrading organic dye substance is relatively low. This is because of fast recombination rate of electron-hole pairs [16]. To deal with this issue, we can combine metal oxide with graphene material. The properties of graphene as electron transport channel could prevent recombination of electron-hole pairs. Moreover, graphene could increase the process of charge separation and improve the adsorptive capacity in catalyst surface to enhance the catalytic performance [17].

The focus of this study is to investigate the catalyst's capacity in degrading MB as the pollutant model and the effect of adding NGP on improvement of catalytic performance in the photocatalytic, sonocatalytic, and photosonocatalytic processes. The scavenger is added to evaluate the behavior of oxidative species that has a role in degrading MB solution.

2. Experimental

Preparation of ZrO_2 nanoparticle: ZrO_2 nanoparticle is synthesized by sol-gel method. The precursor of Zirconium (IV) Chloride ($ZrCl_4$) and Sodium Hydroxide (NaOH) is dissolved into distilled water. Then $ZrCl_4$ solution is poured slowly into NaOH solution under magnetic stirring. The formed gel remains stirred magnetically for 3 hours at the temperature of 80°C. Then, the white colloid obtained is centrifuged and rinsed using distilled water and ethanol several times. After the rinsing process, the samples are dried for 12 hours at 120°C then calcined for 5 hours at the temperature of 550°C. This process will produce ZrO_2 nanoparticles.

Preparation of ZrO_2 - NGP composites: ZrO_2 - NGP composites are synthesized by co-precipitation method. The NGP powder is dissolved into distilled water and ethanol. The NGP solution is put into an ultrasonic bath for sonication process for 2 hours to make the suspense homogenous. Afterward, ZrO_2 nanoparticle is poured into the NGP solution and stirred magnetically for 1 hour. The solution is heated for 3 hours at the temperature of 120°C. The grayish colloid obtained is centrifuged and rinsed using distilled water and ethanol several times. In the final stages, the samples are dried up for 12 hours at the temperature of 70°C under the vacuum condition. This produces grayish ZrO_2 - NGP composite powder.

Characterization: ZrO_2 nanoparticles and ZrO_2 - NGP composites are characterized by X-Ray Diffraction (XRD) Diffractometer Rigaku Miniflex 600 with the radiation source of Cu-K_α, wave length λ=1,5406, Transmission Electron Microscopy (TEM) Tecnai G2 Supertwin with the voltage of 200 kV, Energy Dispersive X-Ray (EDX) Spectroscopy and Thermal Gravimetric Analysis (TGA) Rigaku TG8121 with aluminum oxide (Al_2O_3) as reference to identify the physical and structural properties of the samples.

Catalytic experiments: The photocatalytic, sonocatalytic, and photosonocatalytic activities of the samples are studied in MB degradation. The samples are poured into methylene blue solution and stirred magnetically for 30 minutes under dark condition to reach adsorption/desorption equilibrium between substrate and catalyst. Catalytic activities examination is done for 2 hours after the dark process. At a certain time interval, about 5 mL of suspense is taken from reactor cell and centrifuged to separate the catalyst with MB solution. Then changes in MB concentration are monitored by inspecting the absorption peak at 663 nm using UV-Vis Spectrophotometer. The photocatalytic activity is carried out by irradiating the solution mixture with ultraviolet (UV) light. Two 40-watt UV-C lamps are used as sources of ultraviolet light radiation. Sonocatalytic activity is conducted by giving ultrasonic radiation. Ultrasonic bath with frequency vibration of 40 kHz and power of 150 watts are used as the reactor of ultrasonic radiation. Meanwhile photosonocatalytic activity is performed by combining the photocatalytic process with the sonocatalytic process, namely by giving ultraviolet and ultrasonic radiation at the same time.

Scavenger experiments: To analyze the species involved in the process of degradation, the scavenger is examined by adding certain reagents into the MB solution. As the inhibitor for the species of hole, electron, and hydroxyl radicals ($OH^{\bullet}$), diamonium oxalate ($C_2H_8N_2O_4$), sodium sulphate (Na_2SO_4), and tert-Butyl alcohol ($(CH)_3OH$) are used respectively. The same measurements as described above were used.

3. Result and Discussion

As shown in the Figure 1, ZrO_2 diffraction peak at $2\theta = 30.18°, 35.03°, 50.36°$, and $59.94°$ are associated with ≈ [111], [200], [220], and [311] planes. These planes indicate the tetragonal phase from ZrO_2. Meanwhile, the peak at $2\theta = 26.49°$ is associated with [002] plane of graphitic-like structure of NGP. The diffraction peak owned by ZrO_2 - NGP can be seen as combination from the tetragonal peak of ZrO_2 and NGP structure without any new peak is existed. This confirms that the desired samples are successfully synthesized. Increased intensity of NGP peak at the XRD pattern of ZrO_2 - NGP samples indicates the more content of NGP in the samples. The average of grain size is calculated using the Debye Scherrer's formula for [111] plane and the lattice parameter value is refined using the Rietveld refinement. The resulted calculations are presented in Table 1.

The TEM images of ZrO_2 nanoparticles and ZrO_2 - NGP composites are shown in Figure 2. It is seen that there is nanoparticle agglomeration with nearly spherical topography which indicates the existence of ZrO_2. This nanoparticle agglomeration may occur because of high surface energy of ZrO_2 nanoparticle [18]. The existence of NGP in ZrO_2 - NGP composites is indicated in Figure 2 (b) in the shape similar to layer-like. Based on TEM analysis, the average of grain size is around 13 nm. The result is consistent with the result shown in XRD analysis.

The elements of these samples are analyzed using EDX. The EDX spectrum shows that ZrO_2 sample consists of the elements of Zr, O and Cu. The existence of Cu element in EDX spectrum is derived from grids during measurement. Meanwhile in ZrO_2 - NGP samples there is a carbon (C) element which indicates the existence of NGP in the sample. No other elements indicate that the synthesized samples have good purity.Thermal stability of ZrO_2 and ZrO_2 - NGP samples is represented with the loss of mass

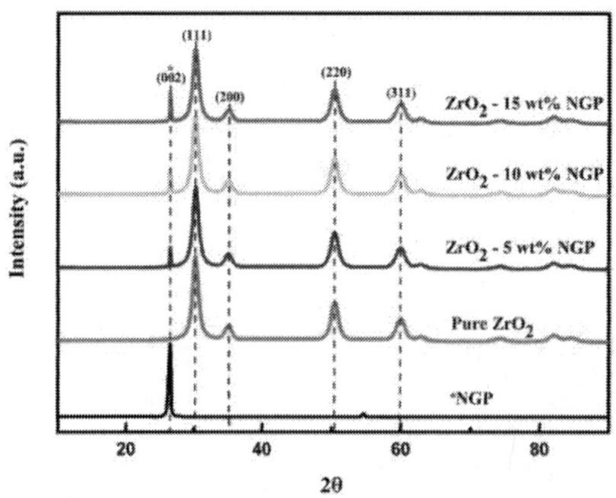

Figure 1. XRD pattern of NGP, ZrO_2, and ZrO_2 - NGP with different NGP weight percents.

Table 1. Lattice parameter, and grain size of ZrO_2 - NGP with different NGP weight percents.

Sample	Lattice Parameter (Å)		<D> (nm)
	ZrO_2 a=b≠c	NGP a=b=c	
ZrO_2	3.605 ; 5.146	-	13.59
ZrO_2 - 5 wt% NGP	3.612 ; 5.153	3.635	12.97
ZrO_2 - 10 wt% NGP	3.614 ; 5.156	3.637	12.41
ZrO_2 - 15 wt% NGP	3.615 ; 5.157	3.638	12.02

in thermal gravimetric analysis indicated in Figure 3b. Both ZrO_2 and ZrO_2 - NGP experience loss of mass by around 100°C. This loss occurs due to possibly desorption (removal) of water molecule which is physically attached to the nanomaterial. At the temperature of around 670°C, ZrO_2 - NGP samples significantly experience loss of mass indicated with the combustion of NGP.

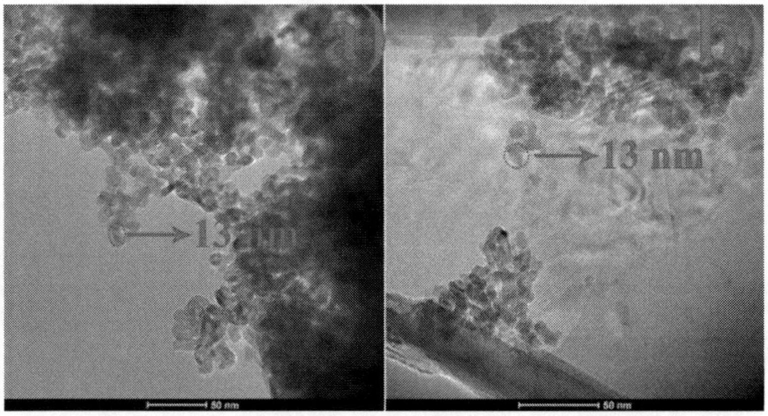

Figure 2. TEM of (a) ZrO_2 and (b) ZrO_2 - NGP.

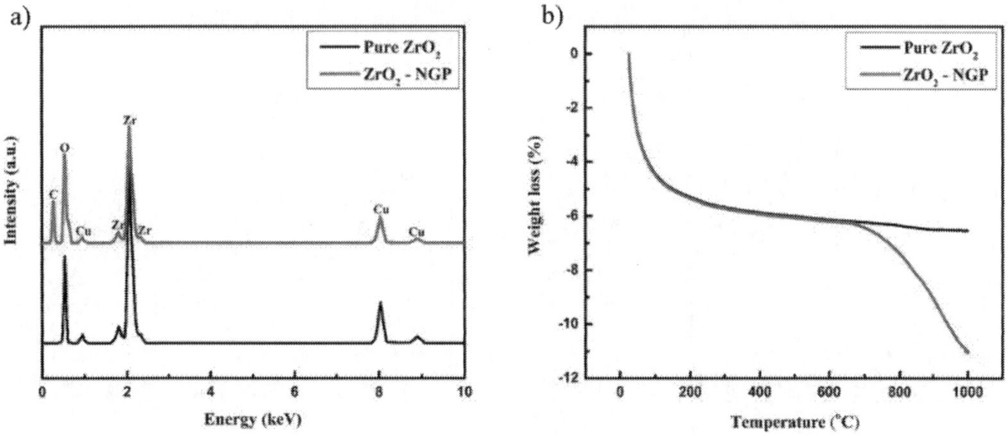

Figure 3. ZrO_2 and ZrO_2 - NGP (a) EDX spectrum and (b) thermal gravimetric plots.

In order to evaluate the ability of catalyst to degrade MB, photocatalytic, sonocatalytic and photosonocatalytic were performed. All catalytic activity measurements are performed under room temperature with the catalyst dosage of 0.3 g/L. In Figure 4 (a), each catalyst in the photosonocatalytic process exhibit the highest degradation efficiency followed by sonocatalytic and photocatalytic processes. Increased degradation efficiency can also be seen for sample with the addition of NGP. The increasing of degradation efficiency occurs due to NGP as a graphene material can reduce the recombination rate of electron-hole pairs, and it also able to increase the charge transfer between metal oxide and dye molecule, as well as improve the adsorptive capacity in catalyst surface [17,19,20]. Moreover, ZrO_2 - 10 wt% NGP is found that exhibit the highest degradation efficiency. Decreased efficiency occurring on ZrO_2 - 15 wt% NGP may be caused by excessive addition of NGP composition that may cover the active sites of ZrO_2 catalyst. In the process of catalytic activities for 120 minutes, ZrO_2 nanoparticle can degrade by 60.6%, 75%, 83.5 % and ZrO_2 - 10wt% NGP composites can degrade by 77.3%, 91.7%, 100% each for the photocatalytic, sonocatalytic, and photosonocatalytic process, respectively. The experiment of MB degradation follows the first-order kinetics [21]. The rate of the constant MB degradation (K_{app}) for all samples and catalytic activities are indicated in Figure 4b. The figure confirms that degradation efficiency of ZrO_2 nanoparticle increases with the addition of NGP and it achieve maximum at 10 wt% of NGP.

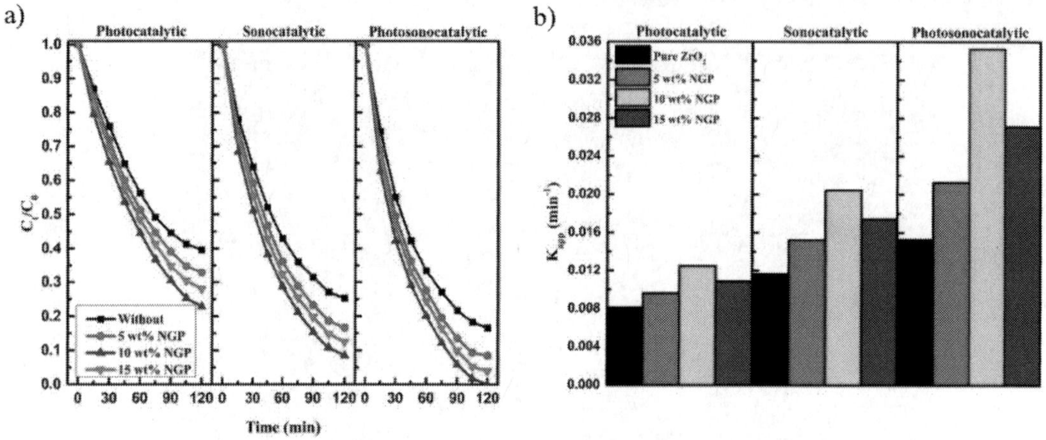

Figure 4. The effect of different NGP weight percents: (a) degradation curve, (b) rate constant diagram

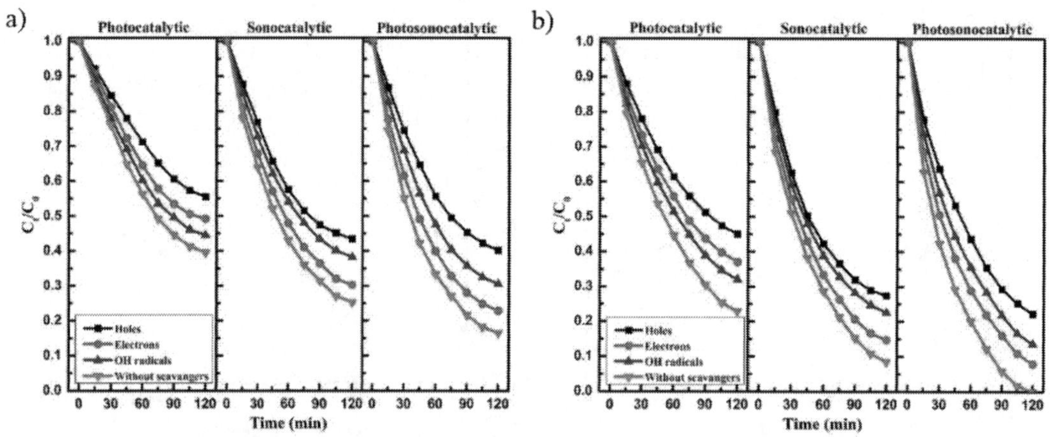

Figure 5. The effect of addition of various scavenger: (a) ZrO_2, (b) ZrO_2 - 10 wt% NGP.

Diamonium oxalate, sodium sulphate, and tert-Butyl alcohol are added to analyze the species that has the role in the process of MB degradation which is useful for inhibiting respectively the species of hole, electron, and hydroxyl radical in photocatalytic, sonocatalytic, and photosonocatalytic as presented in Figure 5. It can be seen that for each catalytic process, the addition of diamonium oxalate can cause highest decrease of degradation efficiency compared to the others. In other words, hole is an oxidative species that plays the most active role in the process of MB degradation. There is a difference in the role of species in the degradation of MB in photocatalytic with sonocatalytic and photosonocatalytic process. In the photocatalytic, the sequence species that play the most role in the process of MB degradation is hole > electron > $OH^{\bullet}$. Meanwhile for sonocatalytic and photosonocatalytic are hole > $OH^{\bullet}$ > electron. This is possible because sonocatalystic involves the hot spot process that can increase the $OH^{\bullet}$ population through pyrolysis of water molecule [22,23].

In evaluating the stability and reusability, the catalyst is used repeatedly for four cycles. At the end of each cycle, the catalyst is separated from MB solution by centrifuged to be re-used under the same MB condition as previously. The result of re-use catalyst in degrading MB are indicated in Figure 6. It can be seen that both ZrO_2 and ZrO_2 - NGP samples do not experience significantly-decreased degradation efficiency for each cycle. It indicates that the catalyst has good stability and reusability capability.

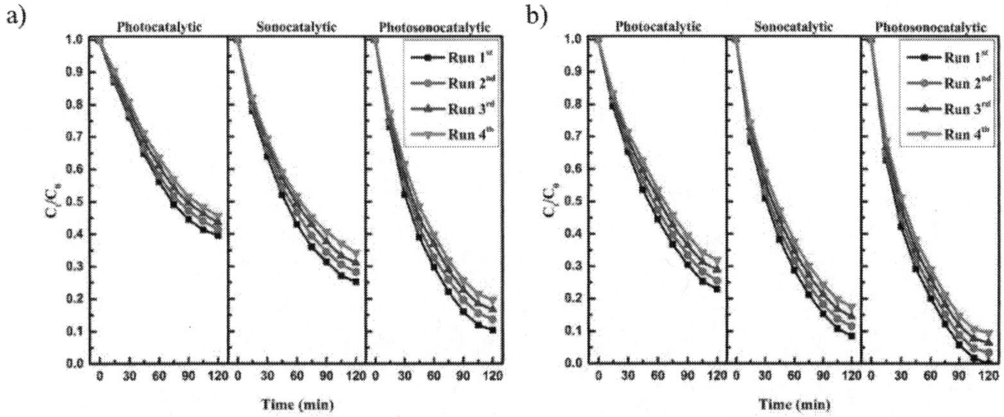

Figure 6. Reusability of (a) ZrO_2 and (b) ZrO_2 - 10 wt% NGP.

4. Conclusion

ZrO_2 - NGP composites have been successfully synthesized by sol-gel method followed by co-precipitation that confirmed with the XRD, TEM, EDX, and TGA characterizations. The degradation efficiency of catalytic activities improves with the NGP addition. The sample with the additional 10 wt% of NGP exhibit the best catalytic performance. ZrO_2 composites - 10 wt% can degrade MB solution by 77.3%, 91.7%, 100% for photocatalytic, sonocatalytic, and photosonocatalytic processes, respectively. In the scavenger experiment, the hole is known as the species that plays the most active role in the process of MB degradation. The ZrO_2 nanoparticle and ZrO_2 - NGP composite have good stability and reusability because the degradation efficiency does not significantly decrease after using for four cycles.

References

[1] P. Pekakis, N.P. Xekoukoulotakis, D. Mantzavinos 2006 *Water Res.* **40** 1276-1286.
[2] S.M. Lee, S.T. Ong 2014 *APCBEE Proc.* **9** 165-169.
[3] L. Fan, Y. Zhou, W. Yang, G. Chen, F. Yang 2008 *Dyes Pigm.* **76** 440-446.
[4] E. Oguz, B. Keskinler 2007 *Dyes Pigm.* **74** 329-334.
[5] S. Wang, Y. Boyjoo, A. Chouebib 2005 *Chemosphere* **60** 1401-1407.
[6] R.A.R. Monteiro, *et.al.* 2015 *J. Photochemistry and Photobiology A: Chemistry* **311** 41-52.
[7] R.A. Torres, *et.al.* 2007 *J. Hazard. Mater.* **146** 546-555.
[8] S.G. Anju, S. Yesodharan, E.P. Yesodharan 2012 *J. Chem. Eng.* **189-190** 84-93.
[9] J.A. Rodriguez, *et.al.* 2000 *J. Phys. Chem. B* **104** 319-328.
[10] G.Y. Guo, Y.L. Chen 2005 *J. Solid State Chem.* **178** 1675-1682.
[11] M. Tahmasebpour, A.A. Babaluo, M.K. Razavi Aghjeh 2008 *J. Eur. Ceram. Soc.* **28** 773-778.
[12] J. Konishi, K. Fujita, S. Oiwa, K. Nakanishi, K. Hirao 2008 *Chem. Mater.* **20** 2165-2173.
[13] R. Malakooti, *et.al.* 2013 *RSC Adv.* **3** 22353-22359.
[14] L. Jing, W. Zhou, G. Tian, H. Fu 2013 *Chem. Soc. Rev.* **42** 9509-9549.
[15] Z.A. Sahar, S.N. Masoud 2016 *J. Moleculer Liquids* **216** 545 – 551.
[16] X. Wang, B. Zhai, M. Yang, W. Han, X. Shao 2013 *Materials Letters* **112** 90-93.
[17] L. Han, P. Wang. S. Dong 2012 *Nanoscale* **4** 5814-5825.
[18] Z.L. Liu, *et.al.* 2004 *J. Magn. Mater.* 283 **2–3** 258–262.
[19] Q. Xiang, J. Yu, M. Jaroniec 2012 *Chem. Soc. Rev.* **41** 782-796.

[20] C.H. Kim, B.-H. Kim, K.S. Yang 2012 *Carbon* **50** 2472-2481.
[21] A. Nezamzadeh, S. Hushmandrad 2010 *Appl. Catal. A* **388** 149–159.
[22] M. Ahmad, *et.al.* 2014 *Ultra. Sono.* **21** 761-773.
[23] M. Zhou, *et.al.* 2015 *J. Hazard. Mater.* **289** 149-157.

Journal of Physics: Conference Series **776** (2016) 012041 doi:10.1088/1742-6596/776/1/012041

Study of $La_{0.67}Sr_{0.33}MnO_3$ structure and resistivity synthesized by using sol gel method

U. Widyaiswari[1], B. Kurniawan[1], A. Imaduddin[2], S. A. Saptari[3]

Department of Physics, Faculty of Mathematics and Natural Sciences, Universitas Indonesia, Depok 16424, Indonesia
Metallurgical Research Center, Lembaga Ilmu Pengetahuan Indonesia, Puspiptek-Serpong 15314, Indonesia
Faculty of Science and Technologyy, Universitas Islam Negeri Syarif Hidayatullah, Jakarta 15412, Indonesia

E-mail: widyaiswari_utami@yahoo.com

Abstract. In this study a colossal magnetoresistance $La_{0.67}Sr_{0.33}MnO_3$ material was synthesized by using a sol gel method with different treatment heating process. First sample calcined at 850°C for 10 hours and pelletized, labelled as LSMO1, while another sample calcined at 850°C for 10 hours, pelletized, and sintered at 1200°C for 2 hours, labelled as LSMO2. The aim of this study was to learn a structure and resistivity of this material. The XRD results showed that both samples were single phase with no peak of impurity and formed a rhombohedral structure with space group R-3c. And the result showed that LSMO1 has smaller crystallite size than LSMO2. Result of resistivity measurement showed that LSMO1 has insulator behavior with resistivity several order of magnitude bigger than LSMO2. While LSMO2 had the peak of resistivity that showed a metal-insulator transition.

1. Introduction

During the last decade, lanthanum manganites $La_{1-x}A_xMnO_3$ (A = Sr, Ba, Ca) have been studied and become subject of research interest because of their magnetic and electrical properties which can be applied in magnetic random access memory, and magnetic sensor [1-2]. Previous study showed that mixed valence manganites have various phase diagram depend on concentration of doping [3]. Doping in La site would change structure and electrical properties of material. Beside the concentration of doping, the difference physical properties caused by different treatment in preparation process that affected the size of particle [3-5].

Yadav et al showed that different annealed temperatur produced the samples with different grain size that caused the different magnetic properties of the samples. The results showed that increased of annealing temperature enhanced the particle size of the samples [5], and increased of particle size leads to increase of magnetic properties indicated by increasing of magnetization saturation of the samples. Another study about the effect of various sintering temperature have been published by Zhang et al and Balcells et al. The results showed that higher sintering temperature increased the grain size and affected the physical properties of samples, such as the resistivity of samples [3], [4]. Smaller particle size gave higher resistivity measurement.

Journal of Physics: Conference Series **776** (2016) 012041 doi:10.1088/1742-6596/776/1/012041

The aim of this study was to learn a structure and resistivity of $La_{0.67}Sr_{0.33}MnO_3$ material synthesized by sol gel method with different treatment heating process.

2. Experimental details

$La_{0.67}Sr_{0.33}MnO_3$ perovskite manganites materials were synthesized using a sol gel method. Suitable proportions of La_2O_3, $Sr(NO_3)_2$, and $Mn(NO_3)_2.4H_2O$ were used as starting materials, and citric acid as a fuel. Each starting material solved in aquabidest. La_2O_3 precursor then solved in diluted HNO_3. The resulting solution of each starting material mixed and stirred at 80°C. Ammonia solution was added to the mixed solution to adjust pH until reached 7 [6]. Then, the solution mixed and stirred at 70-80°C until the water evaporated and dried it at 120°C for 2 h until sol gel formed. The gel then calcined at 500°C for 5 h, grinded, and calcined again at 850°C for 10 h.

The re-calcined samples were divided into two portions. For first portion, sample was pressed into pellet [7] and labelled as LSMO1, while another portion of sample was pressed into pellet, sintered at 1200°C for 2 h, and labelled as LSMO2. Both of sample then characterized using XRD and SEM, while the resistivity were measured using cryogenic magnetometer.

Cryogenic magnetometer using four point probe (FPP) principal include the cryostat system, sehingga resistivity measurement can be done in the variuos temperature with range of temperature 5-300 K.

3. Results and discussion

Samples that have been synthesized by sol gel method then characterized by XRD and cryogenic magnetometer to learn the structure and trtansport properties of the samples.

3.1. Structure characterization

The LSMO1 and LSMO2 samples synthesized using sol gel method were characterized using X-ray Diffraction (XRD). Figure 1 shows XRD pattern of both samples. The result shows that samples have single phase with no impurity peak. The data was fitted with Rietvel refinement. The samples have rhombohedral structure with R-3c space group.

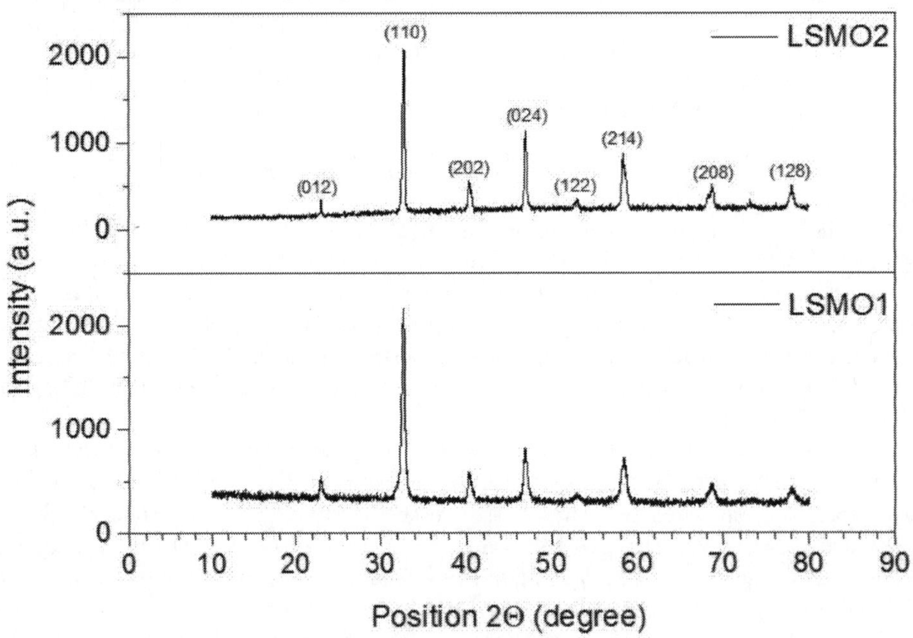

Figure 1. XRD pattern of LSMO1, sintered at 850°C [7], and LSMO2, re-sintered at 1200°C

Rietveld refined parameters were showed in Table 1. The results show that lattice parameter of LSMO1 is smaller than LSMO2. The crystallite size of LSMO2 is bigger than LSMO1. It is caused by crystal growth at higher temperature better than in low temperature.

Table 1. Rietveld refined parameter of LSMO1 [7] and LSMO2

Sample	a = b (Å)	c (Å)	V (Å^3)	Crystallite size (nm)	Structure	Space group
LSMO1	5.4875	13.359	348.38	26.01	rhombohedral	R-3c
LSMO2	5.5005	13.368	350.26	53.56	rhombohedral	R-3c

3.2. Resistivity measurement
Resistivity of samples were measured using cryogenic magnetometer with four point probe work principle. Figure 2 and figure 3 show the results of resistivity measurement as temperature function of LSMO1 and LSMO2 respectively. It is shows that LSMO1 has insulator behavior, it is showed by increased of resistivity with temperature decreased, while LSMO2 has metallic behavior showed by increased of resistivity with temperature increased. In figure 3, resistivity peak of LSMO2 haven't been seen yet. It is predicted happened above room temperature, 300 K, while the maximum working temperature of cryogenic magnetometer used in this study is 300 K.

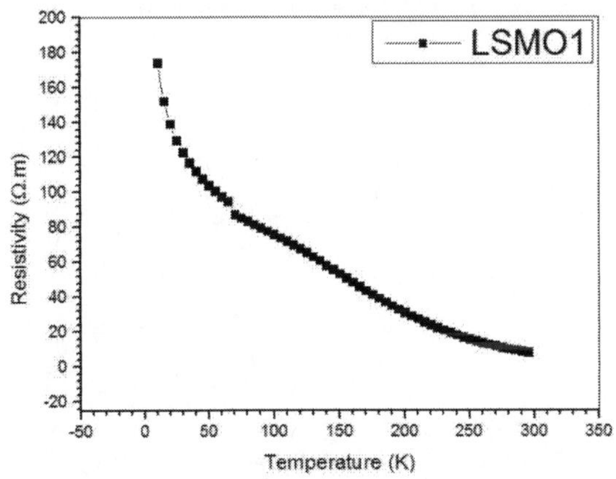

Figure 2. Resistivity as temperature function of LSMO1

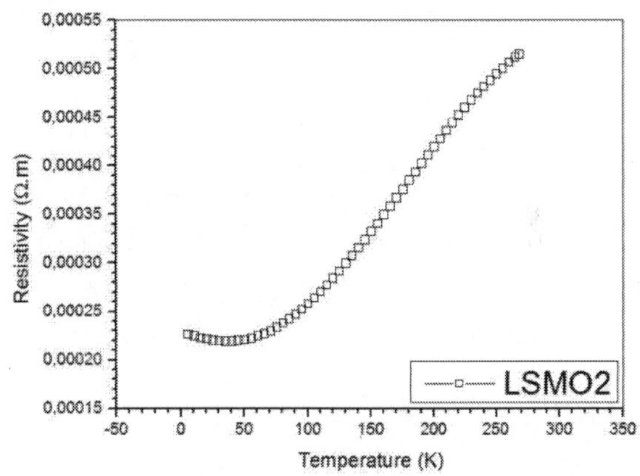

Figure 3. Resistivity as temperature function of LSMO2

Beside the different behavior of LSMO1 and LSMO2, LSMO1 has resistivity several order of magnitude bigger than LSMO2 resistivity. We can stated that sample with smaller crystallite size give higher resistivity than sample with bigger crystallite size. It is related with the content of oxygen. To analyze this phenomena, EDX characterization have been done to observe the element content of LSMO1. In LSMO1, EDX result shows that atomic percent of oxygen element in the amount of 35.15%, half times below the expected value.

The oxygen deficiency indicate the presence of oxygen vacancies [8]. The oxygen vacancies suppresses the Double Exchange (DE) interaction that have a big role in mixed valence manganites transport phenomena. Lakshmi and Reddy showed same result and used same approximation to analyze the decreases of resistivity with increased of grain size. Besides the intrinsic DE interaction between Mn3+ and Mn4+, extrinsic contribution due to spin-polarized tunneling affect the resistivity [8]. In granular sample, smaller grain size give larger number of grain boundaries in sample and enhance the resistivity due to spin-polarized tunneling [8].

The different behavior and resistivity value of both sample can be related to the crystallite size of samples. LSMO1 have smaller crystallite size than LSMO2, so that there will be more local interface between the grain and the more tunneling process.

4. Conclusions

Temperature of sintering process affect the crystallite size of sample. Higher sintered temperature gives bigger crystallite size. It is related with oxygen content in the sample. Lower temperature gives smaller crustallite size and less oxygen content. This oxygen deficiency affected enhancement of resistivity and different electrical behavior. LSMO1 has insulator behavior with high resistivity, while LSMO2 has metallic behavior with smaller resistivity than LSMO1.

Acknowledgement

We sincerely thank to LPDP, Ministry of Finance, Republic of Indonesia and Universitas Indonesia for partially funding this research through Indonesian Education Scholarship program for thesis and PITTA grant.

References

[1] Myron BS and Marcelo J 2001 *Rev. Mod. Phys.* **73** 583
[2] Danyan Cao, Yuanyuan Z, Wenxia D, Jing Y, Wei B, Ying C, Genshui W, Xianlin D and Xiaodong T 2015 *Ceram. Int.* **41** 381
[3] Ll Balcells, J. Fontcuberta, B. Martinez and X. Obradors 1998 *Phys. Rev. B* **58** 697
[4] Ning Zhang, Weiping D, Wei Z, Dingyu X and Youwei D 1997 *Phys. Rev. B* **56** 8138
[5] P A Yadav, A V Deshmukh, K P Adhi, B B Kale, N Basavaih, S I Patil 2013 *J. Magn. Magn. Mater.* **328** 86
[6] D T M Hue, T V Manh, L H Anh, L V Hong, M H Phan, P T Huyen and H D Chinh 2014 *IEEE Trans. Magn.* **50**
[7] Would be published in online proceedings of Seminar Nasional Fisika
[8] Y Kalyana Lakshmi and P Venugopal Reddy 2009 *J. Alloys Comp.* **470** 67

Characteristics of ethylene glycol-Al_2O_3 nanofluids prepared by utilizing Al_2O_3 nanoparticles synthesized from local bauxite

D. G. Syarif
Center for Applied Nuclear Science and Technology (PSTNT)-BATAN
Jl. Tamansari 71, Bandung 40135, Indonesia
Telp. 62-22-2503997, Fax 62-22-2504081,

E-mail: danigus@batan.go.id

Abstract. Nanoparticles of Al_2O_3 have been synthesized from local bauxite mineral, and ethylene glycol (EG)-Al_2O_3 nanofluids have been prepared. Powder Al(OOH) was extracted from local bauxite using bayer process, and heated at 600°C for 3 hours to get Al_2O_3 nanoparticles. XRD analyses showed that the Al_2O_3 nanoparticles crystallizes in γ-Al_2O_3 with crystallite size of 4.12 nm. The specific surface area of the Al_2O_3 nanoparticles was 296.72 m^2/gr. Viscosity of the EG-Al_2O_3 nanofluids was temperature dependent, and decreased with increasing temperature. The viscosity of the nanofluids increased with the concentration of the Al_2O_3 nanoparticles. Meanwhile, Critical Heat Flux (CHF) enhancement of the nanofluids increased with the concentration of the Al_2O_3 nanoparticles. The largest CHF enhancement was 54% at Al_2O_3 concentration of 0.095 vol %.

1. Introduction

Since term "nanofluid" proposed for the first time by Choi in 1995 [1,2], research on nanofluids intensively carried out worldwide. Nanofluid is a dispersion of nanoparticles with 1-100 nm size in a base fluid forming a stable suspension [1-3]. Some researches on nanofluids especially for heat transfer have been being carried out with the goal of increasing economic value and safety of some systems such as nuclear power, automotive, and electronics, through increasing heat transfer capability of the fluid [4]. Based on literature data, the nanofluids can be prepared by utilizing various nanoparticles and base fluids. One of them that is intensively investigated is ethylene glycol-Al_2O_3 (EG-Al_2O_3) nanofluid.

EG-Al_2O_3 nanofluids have been studied by many researchers and some of them are Gallego et al. [5], Hechey et al. [6], Bozorgan [7], Sarafraz et al. [8], Esfe et al. [9], Fal et al. [10], and Yoshino and Torii [11]. In 2013, Yoshino et al. [11] studied thermal conductivity of EG-Al_2O_3 nanofluids by utilizing Al_2O_3 powder with particle size of 31 nm and specific surface area of 55 m^2/g. Other than Yoshino, some researchers prepared the nanofluids with relatively large nanoparticles and/or small specific surface area [5,6,8,10]. The synthesis and characterization of nanofluids with small nanoparticles of 4 nm and large surface area of 296.7 m^2/g has not been reported so far. In addition, although Yoshino and Torii [11] reported the pH of their nanofluids, however, most of researchers did not report the pH of their nanofluids [5,6,7,9,10]. It is known that the pH is a very important characteristic of the nanofluids. It should be known before measuring other characteristics such as thermal conductivity and critical heat flux (CHF).

Based on the above mentioned background, in this work a study of synthesis of Al_2O_3 nanoparticles and characteristics of ethylene glycol-Al_2O_3 nanofluids was carried out. The pH has a strong relation to the application of a nanofluid. If the pH is too low the nanofluid will have problem with corrosion. Since the pH of the nanofluids is about 7 (neutral) the corrosion problem diminishes. Therefore, in this study the nanofluids with pH about 7 were prepared. In addition, in order to step up the added value of the local material, also in this study the Al_2O_3 nanoparticles was synthesized from local bauxite. Characteristics of the Al_2O_3 nanoparticles synthesized from local bauxite and those of the nanofluids with pH about 7 especially the viscosity and critical heat flux (CHF) were discussed.

2. Methodology

2.1. Synthesis and Characterization of Al_2O_3 Nanoparticles

Extraction of $Al(OH)_3$ as precursor from local bauxite was done by using Bayer process. The procedure was used in our previous study [12]. The bauxite was mixed with NaOH with 1:1 weight ratio and heated at 175°C for 3 hours in an autoclave. The heated mixture was placed in a beaker glass. An amount of HCl was added into the mixture until pH = 11. A precipitate formed. The precipitate was washed with aquadest until pH about 7. The washed precipitate was then heated at 600°C for 3 hours to get Al_2O_3 nanoparticles. The Al_2O_3 nanoparticles was subjected to XRD analyses using a diffractometer from Philips PAN Analytical with K_α radiation (λ=1.5406A) in order to investigate the crystal structure and crystallite size, and to TEM analyses in order to know particle size using a TEM from JEOL. Specific surface area of the nanoparticles was measured using a surface area meter from Quanthachrome.

2.2. Preparation and Characterization of Ethylene Glycol- Al_2O_3 nanofluids

Some nanofluids were prepared utilizing ethylene glycol as base fluid and the Al_2O_3 nanoparticles that already synthesized. The nanofluids were prepared with 4 concentrations of the nanoparticles. The nanoparticles were dispersed into 100 ml ethylene glycol and ultrasonicated using an ultrasonic bath from Branson for 2 hours. After 24 hours from ultrasonication step, the upper part of the nanofluids was extracted and separated from the precipitate. The precipitate was dried and weighed for measuring the concentration of the nanofluids. Visual appearance of the nanofluids was taken using a digital camera. Then, pH of the nanofluids was measured using a pH meter from Mettler Toledo. Meanwhile, zeta potential of the nanofluids was measured using a Zetasizer from Malvren. After that, Critical Heat Flux (CHF) was measured using a laboratory made equipment based on measurement principle found in [13]. Cu wires 0.2 mm in diameter were used for the CHF measurement. The CHF was calculated using equation (1) [13].

$$CHF = \frac{I.V}{\pi DL} \quad \text{...(1)}$$

where, I is the imput current (Amp), V is the imput voltage (Volt), D is the wire diameter (m), and L is the wire length (m).

3. Results and Discussions

3.1. Characterization of the Al_2O_3 nanoparticles

Figure 1 reveals the diffraction pattern of the Al_2O_3 nanoparticles synthesized from local bauxite. By comparing to the JCPDS standard no. 29-0063 for γ-Al_2O_3, it is clear that the Al_2O_3 nanoparticles crystallizes in γ-Al_2O_3 with three strong peaks at 2θ = 38, 46, and 67° which are corresponding to peaks of (311), (400), and (440).. From the data of Fig.1, crystallite size was calculated using Debye Scherrer method expressed in equation (2) [14].

$$d = \frac{0.9\lambda}{\beta Cos\theta} \quad \text{...(2)}$$

where, d is crystallite size, λ is wavelength (1.5406A), θ is diffraction angle, and β is FWHM. From the calculation, it is known that the average crystallite size of the Al_2O_3 nanoparticles is 4.12 nm.

Figure 2 reveals the image of Al_2O_3 nanoparticles taken using TEM. According to Fig.2, the Al_2O_3 nanoparticles forms agglomerate. The agglomeration causes the average particle size of the Al_2O_3 nanoparticles much larger than that measured using the XRD.
The specific surface area of the Al_2O_3 nanoparticles was determined using the Brunauer Emmet Teller (BET) method. The specific surface area for the Al_2O_3 nanoparticles was 296.72 m^2/g. This surface area is quite large. It is much larger compared to the surface area of the γ-Al_2O_3 found in work of Yoshino and Torii (55 m^2/g) [4]. If the surface area is converted to particle size by using equation (3) [15] by assuming the nanoparticle is spherical, we get particle size of 5.12 nm. This particle size is comparable to crystallite size calculated from data of XRD (4.12 nm). This data shows that the particles form aglomer consisting of small particles (5.12 nm) as can be seen in Fig.2.

$$d = \frac{6000}{A_s.\rho} \quad \text{...(3)}$$

where, d is the particles size, A_s is the specific surface area (m^2/g), and ρ is the density of particles in gr/cm^3 (In this case is Al_2O_3 (3.9 gr/cm^3)).

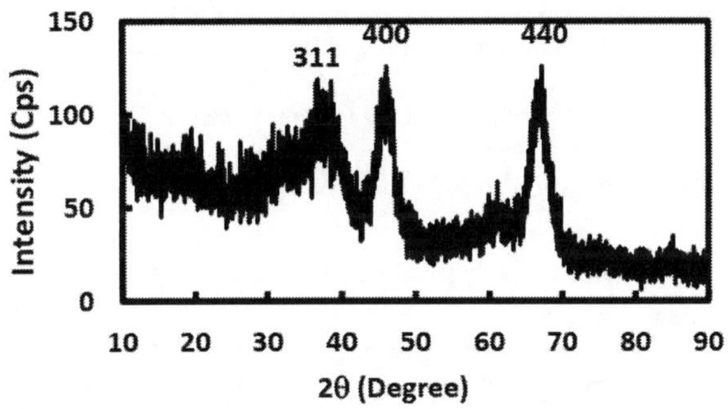

Figure 1. XRD pattern of the Al_2O_3 nanoparticles

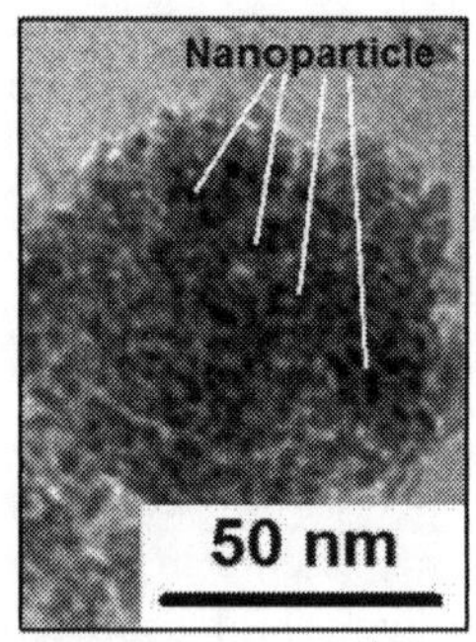

Figure 2. TEM image of the Al_2O_3 nanoparticles forming an aggregate

3.2. Characterization of Nanofluids
3.2.1. Visual Appearance and Zeta Potential

Figure 3 shows the visual appearance of a nanofluid prepared in this work with concentration of Al_2O_3 nanoparticles of 0.058 vol % as an example. The nanofluid is stable. It is shown by a small decrease of the surface height of the nanofluid after 14 days observation. The decrease of the surface height of the nanofluid is about 2 % from the initial height. The EG-Al_2O_3 nanofluid of Sarafraz et al [8] is also stable, however, the pH of their nanofluid is 10.2, larger than that in this study (about 7). The best nanofluid of Sarafraz et al. with pH 10.2 is stable until 45 days.
Zeta potential is a measure of the stability of the nanofluid. When a particle gets into a fluid, it will spontaneously have charge and form an electric double layer at the surface of the particle. Zeta potential of the nanofluids was measured using a Zeta sizer from Malvern. Fig.4 reveals the zeta potential of the nanofluids. As shown in Fig.3, the nanofluids with different concentrations of the Al_2O_3 nanoparticles are stable with the zeta potential is larger than 30 mV. A nanofluid is stable when the zeta potential is larger than 30 mV or smaller than -30 mV. When the zeta potential increases, the repulsion between particles becomes strong and the stability of particles increases. On the contrary, when the zeta potential closes to zero, the nanofluids become easy to precipitate. As shown in Fig. 4, the zeta potential tends to decrease with increasing concentration of the nanoparticles, although the decrease is not significant. It is due to particle aggregation when the concentration of the nanoparticles is relatively high. Compared to the EG-Al_2O_3 nanofluid of Yashino [11], the zeta potential in this work is much larger almost twice. It may be due to small particle size and large specific surface area.

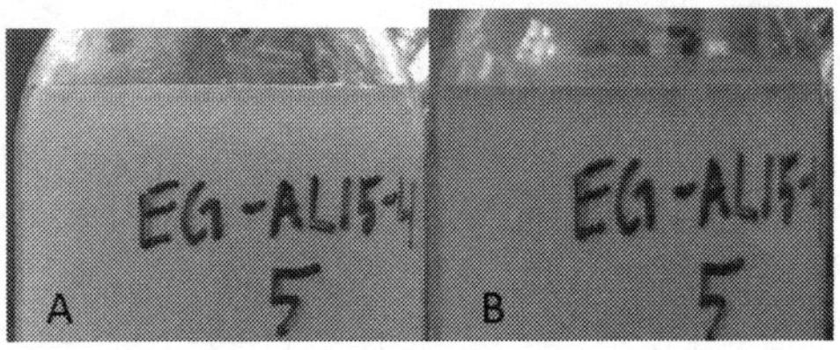

Figure 3. Visual appearance of a nanofluid with Al_2O_3 concentration of 0.058 vol %. A. Initial, and B. After 14 days

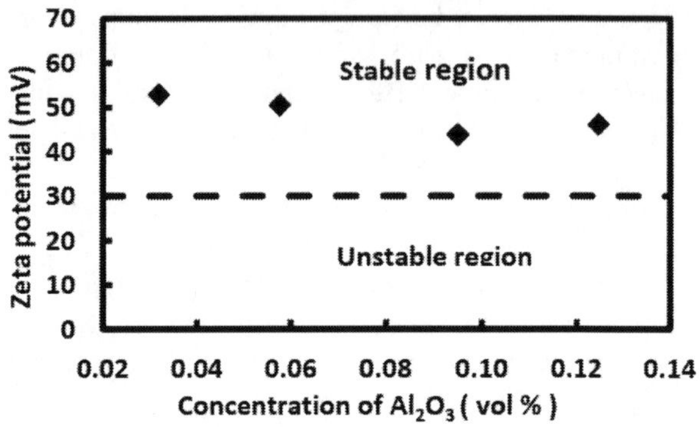

Figure 4. Zeta potential of the EG-Al_2O_3 nanofluids

3.2.2. *Viscosity*

Viscosity is an important characteristic of heat transfer nanofluids. Figure 5 reveals the viscosity of the EG-Al_2O_3 nanofluids as function of temperature. The viscosity of all nanofluids as well as EG decreases with increasing temperature. The relationship between the viscosity and the temperature fits a model of equation (4). B is related to the free activation energy of fluid, and A is fitting parameter [16].

$$\eta = A.Exp\left(\frac{B}{T}\right) \quad (4)$$

where, η is the viscosity, A and B are constant, and T is the temperature in Kelvin.

Figure 6 shows the viscosity of the nanofluids as function of Al_2O_3 concentration at temperature of 27°C. The present data is compared to Einstein model and model of Wang et al [9]. As clearly seen in Fig. 6, neither Einstein model nor model of Wang et al. does not fit the experimental data. This fact shows that the nanofluid is more complicated than the two phase fluid that used for Einstein model.

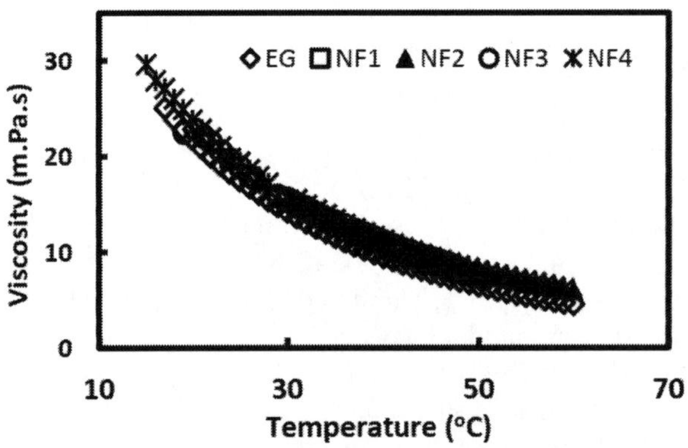

Figure 5. Viscosity of EG-Al_2O_3 nanofluids as function of temperature

3.2.3. *Critical Heat Flux (CHF)*

Figure 7 depicts the data of CHF enhancement as function of concentration of Al_2O_3 nanoparticles. The CHF enhancement is defined as $[(CHF_{NF} - CHF_{EG})/CHF_{EG}] \times 100\%$. where, CHF_{NF} is critical heat flux of the nanofluid, and CHF_{EG} is critical heat flux of the ethylene glycol.

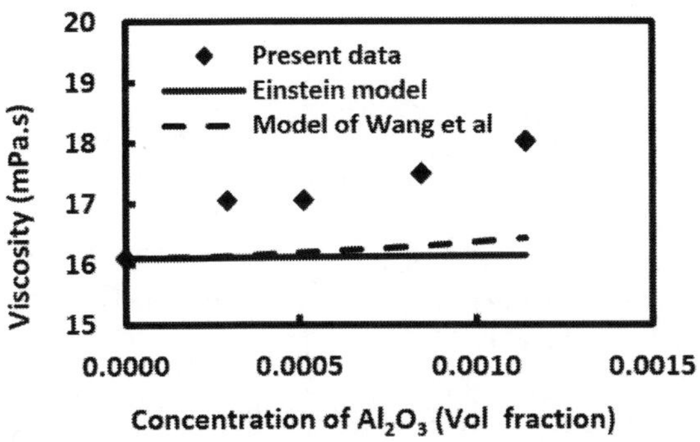

Figure 6. Viscosity of EG-Al_2O_3 nanofluids as function of Al_2O_3 concentration at 27°C

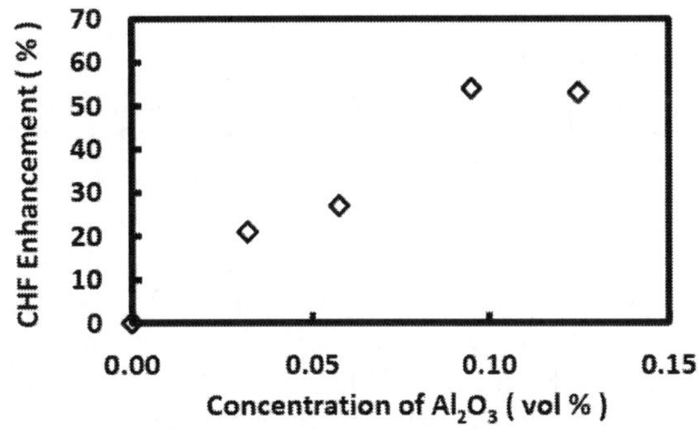

Figure 7. CHF enhancement of EG-Al_2O_3 nanofluids as function of concentration of Al_2O_3 nanoparticles

CHF in pool boiling is defined as the peak heat flux under which a boiling surface can sustain nucleate boiling. CHF is the end of nucleate boiling and means the limit of safe operation of a heat transfer system. The CHF is an important characteristic of the nanofluid other than viscosity and thermal conductivity. The understanding of CHF phenomenon is important for safe and economic design of many heat transfer units. A nanofluid with large CHF is required for the coolant applied in ECCS (Emergency Core Cooling System) dan RVCS (Reactor Vessel Cooling system), electronic coolant, refrigeration and cryogenic, and metal forming.
As shown in Fig.7, the CHF enhancement increases with the concentration of Al_2O_3 nanoparticles. The increase of the CHF enhancement reachs an optimal condition at the concentration of 0.095 vol % where the enhancement of 54% is the largest one. The enhancement of the CHF of the nanofluids is due to increasing surface area of the wire that interfaces with the nanofluid. The nanoparticles in the nanofluid are deposited on the surface of the wire, generating a porous layer with large surface area. Then, when the the surface area of the wire increases, heat transfer from the wire to the nanofluid also increases, resulting in larger CHF. Compared to the CHF enhancement of Water-Al_2O_3 nanofluid with particle size 200-210 nm and concentration of 0.001-0.1 vol % using SS wire in literature [17] of 50%, the enhancement in this study is comparable. However, the enhancement is smaller compared to the CHF enhancement of Water-Al_2O_3 nanofluid with particle size 47 nm and concentration of 0.00001-0.1 vol % using Ni-Cr wire in literature [17] of 100%. Actually, the comparison is not fully true due to different base fluid and wire, and because pH of the nanofluids in literature [17] was not reported as well. Direct comparison to the CHF enhancement of the ethylene glycol-Al_2O_3 nanofluids could not be done due to lack of literature data especially the CHF data.

4. Conclusions

Nanoparticles of Al_2O_3 with crystallite size of 4.12 nm and specific surface area of 296.7 m^2/gr have been successfully synthesized from local bauxite, and nanofluids of ethylene glycol (EG)-Al_2O_3 have been well prepared. The viscosity of the EG-Al_2O_3 nanofluids increases with the concentration of Al_2O_3 nanoparticles. The viscosity is also temperature dependent which decreases with increasing temperature. The CHF of the nanofluids is larger than that of ethylene glycol. The CHF enhancement of the nanofluids increases as the concentration of Al_2O_3 nanoparticles increases, and reaches the optimal value of 54% at the Al_2O_3 concentration of 0.095 vol %.

5. Acknowledgment

This research was funded by the Government of Republic of Indonesia through DIPA Program of BATAN year 2015-2016.

6. References

[1] Sharma P, Baek I H, Cho T, Park S, Lee KB, *Powder Technol.* **208** 7–19.

[2] John T, Krishnakumar TS 2013 International J. Innov. Res. Sci., Eng. Tech. **2** [1] 611-617.

[3] Sridhara V. and L. N. Satapathy 2011 *Nanoscale Res. Lett.* **6**[456] 1-16.

[4] Yu W and Xie H 2012 *J. Nanomater.* **2012** Article ID 435873 1-17.

[5] Gallego MJP, Lugo L, Legido JL, Pineiro MM 2011 *Nanoscale Res. Lett.* 2011 **6** (221) 1-11.

[6] Hechey MA, Nguyen CT, Galanis N, Popa, Catalin V 2012 *9 th International Conference on Heat Transfer, Fluid Mechanics and Thermodynamics*, Malta, July 16-18.

[7] Bozorgan N 2012 *Int. J. Adv. Design and Manuf. Tech.* **5**(2).

[8] Sarafraz MM, Hormozi F 2014 *Int. Commun. Heat Mass Tran.* **58** 96-104.

[9] Esfe MH, Karimipour A, Yan WM, Akbari M, Safei MR, M. Dahari 2015 *Int. J. Heat Mass Tran.* 88 728-734.

[10] Fal J, G. Ya, Gizowska M, Witeka A, Cholewa M 2015 *Acta Phys. Polonica A* 128 (2) 153-155.

[11] Yoshino H and Torii S 2013 *Int. J. Adv. Transp. Phenom.* **02** (01)1-6.

[12] Syarif DG, Prajitno DH, Umar E 2015 *The 5th International Conference on Advanced in Nuclear Sciences and Engineering,* Bandung, Indonesia, November 18th-20th, 2015.

[13] Hiswankar SC, Kashirsagar JM 2013 *Int. J. Eng. Res. Tech. (IJERT)* **2**(7) 2091-2095.

[14] Ravikumar BS, Nagabhushana H, Sharma SC, Vidya YS, Anantharaja KS 2013 *Spectrochimica Acta Part A: Mol. Biomol. Spectrosc.* **136** 1027-1037.

[15] Akbari B, Tavandashti MP, and Zandrahimi M 2011 *Iran. J. Mater. Sci. Eng.* **8** (2) 48-56.

[16] Abareshi M, Goharshadi EK, Zebarjad SM, Fadafan HK, A, Youssefi 2010 *J. Magn. Magn. Mater.* **322** 3895-3901.

[17] Kamatchi R, and Venkatachalapathy S 2015 *Int. J. Therm. Sci.* **87** 228-240.

Journal of Physics: Conference Series **776** (2016) 012043 doi:10.1088/1742-6596/776/1/012043

Influence of annealing treatment on electric polarization behaviour of zinc oxide films grown by low-power dc-unbalanced magnetron sputtering

R. Kurniawan$, E. Nurfani, S. Muhammady, I. M. Sutjahja, T. Winata and Y. Darma*

Department of Physics, Institut Teknologi Bandung, Ganesha 10 Bandung 40132, Indonesia

E-mail: $robi.kurniawan00@yahoo.co.id

Abstract. We study the annealing effect of highly oriented zinc oxide (ZnO) films grown by low-power dc-unbalanced magnetron sputtering (DC-UBMS). In this study, we compare the structural and electrical properties of thermal-annealed ZnO films (ann-ZnO) and as-growth ZnO films (ag-ZnO) by using x-ray diffraction (XRD), scanning electron microscopy (SEM) and RT66A standardized ferroelectric test system. We found that the ag-ZnO films and the ann-ZnO films show a high orientation in (101) plane. SEM images indicate that annealing treatment at 600°C in nitrogen ambient promote the surface atomics arrangement and convert a non-uniform ag-ZnO surface to relatively flat ann-ZnO film surface. Also confirm that the ag-ZnO and the ann-ZnO films have strong ferroelectric characteristics, while the values of remnant polarization and polarization saturation are almost similar. The electric coercivity (E_c) of the ann-ZnO film is larger than ag-ZnO films as an indication of structural defects elimination. Our results are beneficial for high energy electric-based storage devices with less depolarized structural systems.

1. Introduction

Ferroelectric materials were widely used for memory storage application and piezoelectric-based devices because it has a high electric polarization. The high electric polarization can be produced from a material that have non-centrosymmetric structure, such as $Pb(Zr,Ti)O_3$ (PZT) [1, 2], $BiFeO_3$ (BFO) [3, 4] and $(Pb,La)(Zr,Ti)O_3$ (PLZT) [5]. In memory storage application, electric polarization and electric coercivity were the important parameters to measure the quality of the storage devices. Non-volatile memory storage devices developed using material that has high electric coercivity value because it has a high electric energy storage and less depolarized [6]. Among the non-centrosymmetric material, the zinc oxide (ZnO) system was not widely used and still in the development of research for memory storage material systems because ZnO system has lower electric polarization and electric coercivity [7-11]. Many studies have been done to find out how the value of the electric coercivity can be improved, such as material structure modification [12], doping treatment [13], and arrangement of films growth [14].

Sputtering is one of many methods which were widely used for film fabrications. A high power deposition was required to produce high sputter yield and deposition rate [15]. In this work, as an attempt to increase the electric polarization and coercivity, we fabricated ZnO films by using a low-power sputtering deposition to provide a longer time for atomics arrangement on the substrate surface and promote a good orientation on a particular plane. Further studies of the polarization behavior of ZnO material will provide a good understanding of the utilization of this material for non-volatile memory storage devices and ferroelectric-based material application.

2. Experiment methods

Zinc oxide (ZnO) sputtering target was produced from pressed ZnO powder (99.99%, Sigma Aldrich) by using a mold with a diameter of 2.8 cm. Silicon (100) was used as the substrate which has been cleaned by using Radio Corporation of America (RCA) standard cleaning at 80°C. The ZnO films have been grown by using a dc-unbalanced magnetron sputtering (DC-UBMS) with following the parameters:

Table 1. Parameters involved in the ZnO films deposition by low-power DC-UBMS

Parameters	**value**	**unit**
temperature	300	°C
deposition power	4.5	watt
chamber pressure	3×10^{-2}	mbar
gas flow (argon)	100	sccm
growth time	1	hour

To study the annealing effect to the structural and electrical properties of the ZnO films, post-annealing treatment (600°C for 10 minutes) has been performed in a nitrogen ambient. In previous studies have been explained that the annealing temperature of 600°C films showed a high density and uniformity [16]. Annealing treatment was not performed with temperatures above 600°C to avoid the effects of deformation on the Si substrate [17]. The substrate deformation was avoided so that there is no involvement of the influence of the deformation on the properties of the ZnO films. The crystalline phase and the morphology of the ZnO films were characterized using x-ray diffraction (X-RD), analytical brand Bruker, tube anode: Cu, generator tension [kV]: 40, generator current [mA]: 30, alpha1 wavelength [Å]: 1.54060, alpha2 wavelength [Å]: 1.54439, and scanning electron microscopy (SEM), analytical brand JEOL JSM-6510, voltage [kV]: 15, magnification[x]: 20.000, respectively. The determination of the crystalline phase and the lattice parameter were performed with the help of software X'Pert Highscore Plus and PowderCell for Windows (PCW). Moreover, RT66A standardized ferroelectric test system, signal mode: single triangle wave, timing [ms/step]: 1.1, number of steps: 500, V_{max} [V]: 5.00, was used to study the type of hysteresis curve and investigate behaviors of the electric polarization, including remnant polarization, polarization saturation, and electric coercivity of the ZnO films.

3. Results and discussions

As compared by inset in Figure 1, our ZnO films have a highly oriented wurtzite phase in the (101) plane. Another phase was found in the ag-growth ZnO film (ag-ZnO), where diffraction peak O_2 appears in the (012) plane. It showed that our system did not have a pure phase of the ZnO. Another treatment was performed on the ZnO film where the ZnO film was annealed at a temperature of 600°C for 10 minutes (ann-ZnO). The annealing treatment causes the crystallinity of the ZnO film enhanced, as indicated by O_2 peak vanishes and convert the multi-phase ZnO films to the single-phase ZnO films. We note that the crystallinity of the ZnO films increased due to the recrystallization process during annealing, as indicated by intensity of the diffraction peak increases. We report that the annealing treatment promotes a lattice strain in *a*-axis direction. The lattice parameters of the ZnO films was observed by using PowderCell software for Windows (PCW). The lattice parameters of the ag-ZnO film are a = 3.2030Å and c = 5.2510Å, while the lattice parameters of the ag-ZnO film are a = 3.2307Å and c = 5.1010Å. In the previous study, the lattice strain in the ZnO films promoted the polarization effect when the external field applied to the system. This polarization mechanism associated with the reduction of binding energy of the atoms so that the material easier polarized [18].

Journal of Physics: Conference Series **776** (2016) 012043 doi:10.1088/1742-6596/776/1/012043

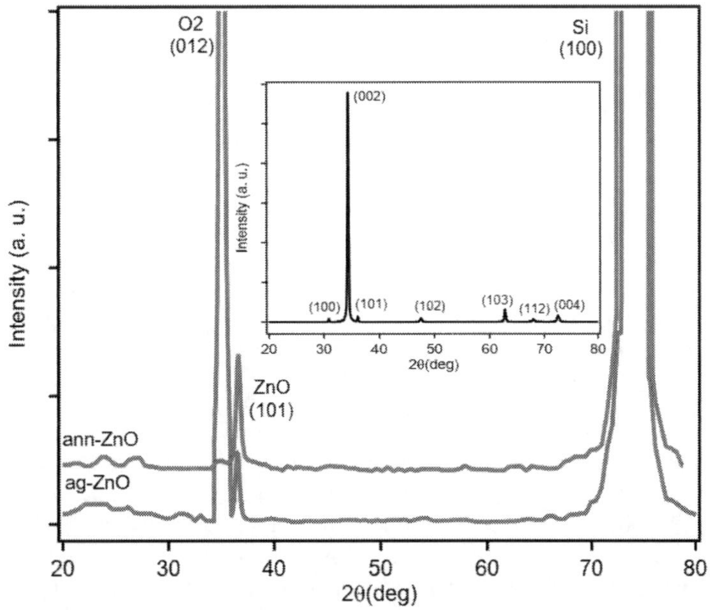

Figure 1. Diffraction pattern of the ZnO films on Si substrate with 1 hour deposition; as growth (red-line) and annealed (green-line). The inset figure shows diffraction pattern on multi-ordered ZnO films grown by dc-unbalanced magnetron sputtering with other parameters (power = 10 W).

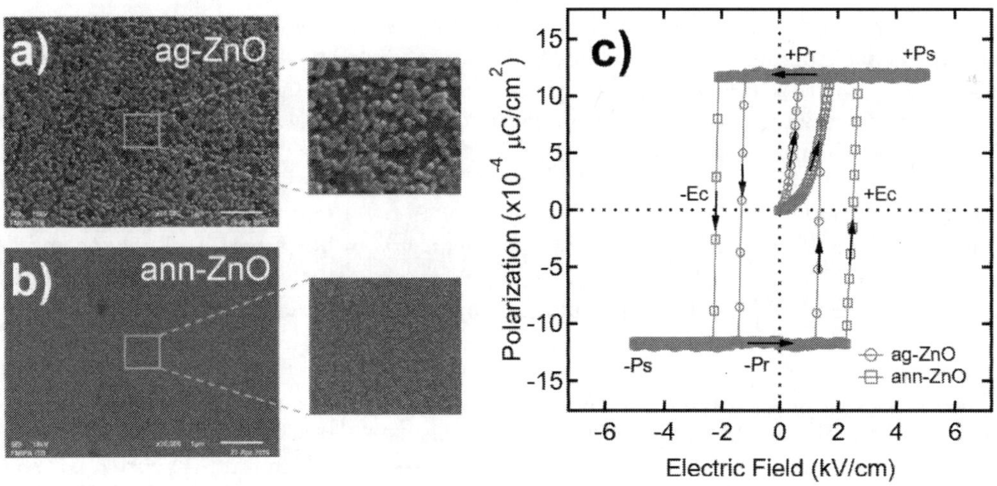

Figure 2. (a)-(b) SEM image of the ag-ZnO and ann-ZnO films (c) the ferroelectric hysteresis curve of the ag-ZnO (red circle-line) and ann-ZnO (green rectangular-line) films.

The applied thermal energy contributed in reducing defect and internal stresses which drive diffusion mechanism and atoms rearrangement so that the ZnO progresses towards its equilibrium state. As shown in Figure 2(a)-(b), the ag-ZnO film has grain boundaries which act as a surface defects. It was showed that the grains were not uniformly distributed on the substrate. Otherwise, the ann-ZnO film has a flat profile with the absence of grain boundaries.

Further analysis was performed to investigated electric properties of the ZnO films by using RT66A standardized ferroelectric test systems that specifically characterize the hysteresis pattern of the materials. Figure 2(c) shows the hysteresis curves of the ag-ZnO and ann-ZnO films. We confirmed the ZnO films have a ferroelectric hysteresis curve profile, which was supported by the saturation curve and large electric coercivity value. The important point in this study is that our systems have a similar value between remnant polarization (P_r) and polarization saturation (P_s). This profile was contributed by high crystal orientation of the films, as reported in previous work. Similar hysteresis patterns are shown by other systems that also has a high orientation [19]. The annealing treatment on the ZnO films induced the increasing of electric coercivity (E_c). We have summarized P_r, P_s, and E_c in Table 2.

Table 2. The parameters involved in the hysteresis curve of the ZnO films by annealing treatment.

Samples	$+P_r$ / $-P_r$ (x10^{-4} μC/cm^2)	$+P_s$ / $-P_s$ (x10^{-4} μC/cm^2)	$+E_c$ / $-E_c$ (kV/cm)
ag-ZnO	11.77 / -11.85	11.78 / -11.85	1.36 / -1.34
ann-ZnO	11.77 / -11.85	11.78 / -11.85	2.50 / -2.20

The annealing treatment contributes to the enhancement of the electric coercivity due to the absence of the grain boundaries, as reported in the previous discussion. The grain boundary is a low permittivity region [20]. The absence of the grain boundaries caused by densification during annealing which drive the interaction between atoms increases. As a result, when the film has polarized it will be more difficult depolarized. The behavior of the electric coercivity which was larger after the annealing showed that our systems have hard-ferroelectric properties. The enhancement of the electric coercivity was associated with an improvement ability of the ferroelectric material to store electrical energy which is required for non-volatile memory storage devices.

4. Conclusions

The annealing treatment causes the crystalline phase of the ZnO films containing O_2 and wurtzite ZnO phase to become single phase wurtzite ZnO. The films are highly oriented in (101) plane which contains the lattice strain in *a*-axis. In addition, the annealing treatment promotes flat profile on film accompanied by the absence of grain boundaries. We report that the absence of grain boundaries causes the ZnO films have a hard ferroelectric property, which indicated by the similar value between the P_r and P_s and enhancement of the E_c.

5. Acknowledgements

The authors acknowledge PMDSU program from Ministry of Research, Technology and Higher Education of the Republic of Indonesia. This work also partly supported by Desentralisasi Kemenristekdikti and RUPT 2016 research program from the Indonesian government.

6. References

[1] Zheng F, Xu J, Fang L, Shen M and Wu X 2008, *Appl.Phys. Lett.* **93** 172101
[2] Cao D, Zhang H, Fang L, Dong W, Zheng F and Shen M 2010, *Appl.Phys. Lett.* **97** 102104
[3] Yang S Y, Martin L W, Byrnes S J, Conry T E, Basu S R, Paran D, Reichertz L, Ihlefeld J, Adamo C, Melville A, Chu Y-H, Yang C-H, Musfeldt J L, Schlom D G, Ager J W and Ramesh R 2009, *Appl. Phys. Lett.* **95** 062909
[4] Yi H T, Choi T, Choi S G, Oh Y S and Cheong S W 2011, *Adv.Mater.* **23** 3403-7
[5] Zhang J, Su X, Shen M, Dai Z, Zhang L, He X, Cheng W, Cao M and Zou G 2013, *Sci. Rep.* **3** 2109
[6] Lines M E and Glass A M 2001 *Principles and Applications of Ferroelectrics and Related Materials* (Oxford: Oxford University Press)
[7] Chung T-K, Wong K, Keller S, Wang K L and Carman G P 2009, *J.of.Appl.Phys.* **106** 103914
[8] Chang H-W, Yuan F-T, Shih C-W, Ku C-S, Chen P-H, Wang C-R, Chang W-C, Jen S-U and Lee H-Y 2012, *Nanoscale Res. Lett.* **7** 435-
[9] Subir R and Sarah P 2007, *J. of Phys. D: Appl. Phys.* **40** 4668

[10] Joshi R, Kumar P, Gaur A and Asokan K 2014, *Appl. Nanoscience* **4** 531-6
[11] Wang Y, Chen W, Wang B and Zheng Y 2014 Ultrathin Ferroelectric Films: Growth, Characterization, Physics and Applications *Materials* **7** 6377-485
[12] Lu C, Dong S, Xia Z, Luo H, Yan Z, Wang H, Tian Z, Yuan S, Wu T and Liu J 2013, *Sci. Rep.* **3** 3374
[13] Costa L V, Rocha L S, Cortés J A, Ramirez M A, Longo E and Simões A Z 2015, *Ceram. Int.* **41** 9265-75
[14] Wu S Z, Miao J, Xu X G, Yan W, Reeve R, Zhang X H and Jiang Y 2015, *Sci. Rep.* **5** 8905
[15] Schepis D J and Seshan K 2012 *Handbook of Thin Film Deposition (Third Edition),* (Oxford: William Andrew Publishing) p 1
[16] Hwang K-S, Jeon Y-S, Kang B-A, Nishio K, Tsuchiya T, An J-H and Kim B-H 2005, *J. of Korean Phys. Soc.* **46** 521-6
[17] Woo Sik Y, Takashi F, Ichiro Y, Kitaek K and Nobuaki T 2002, *Japanese J. of Appl.Phys.* **41** 4442
[18] Kurniawan R, Nurfani E, Muhammady S, Sutjahja I M, Winata T, Rusydi A and Darma Y 2016 Polarity enhancement in high oriented ZnO films on Si (100) substrate *AIP Conf. Proc.* **1725** 020035
[19] Newnham R E 2004 *Properties of Materials : Anisotropy, Symmetry, Structure: Anisotropy, Symmetry, Structure*: OUP Oxford)
[20] Mudinepalli V R, Feng L, Lin W-C and Murty B S 2015, *J. of Adv. Ceram.* **4** 46-53

Synthesis of TiO_2 nanorods from titania and titanyl sulfate produced from ilmenite dissolution by hydrothermal method

S. Wahyuningsih[1], A. H. Ramelan, R. M. I. Munifa, L. N. M. Z. Saputri, and U. Chasanah
Chemistry Department, Faculty of Mathematics and Natural Sciences, Sebelas Maret University
Ir. Sutami 36A Kentingan Jebres Surakarta 57126, Central Java, Indonesia

[1]E-mail: sayekti@mipa.uns.ac.id

Abstract. TiO_2 powder has been synthesized through hydrolysis-condensation of titanyl sulfate solution to a starting material of TiO_2 nanorods formation. This processing was conducted by the solid separation of TiO_2 from ilmenite by roasting ilmenite, acidic leaching (hydrolysis), and co-precipitation (condensation). Roasting of ilmenite was carried out by the addition of Na_2S at a temperature of 800°C. While the acidic leaching process was conducted by sulfuric acid at a various concentrations of 3, 3.5, 4.5, 6, and 9 M. The result shown that the solubility optimum occurs in H_2SO_4 6 M condition. Separation of Fe impurities of TiO_2 gel from titanyl sulfate ($TiOSO_4$) solution was done through complexation using KCNS addition. The characteristic of TiO_2 obtained using X-Ray Fluorescence (XRF) and X-Ray Diffraction (XRD) showed good crystallinity and purity. Further treatment of the TiO_2 is the formation of one-dimensional nano-size (1-D nanorods) through a hydrothermal method under basic condition NaOH 12M solution. TiO_2 nanorods were confirmed by Transmission Electron Microscope (TEM) which indicated that the diameter of TiO_2 nanorods was about 7.02 nm in size.

1. Introduction

The main problems in the world is the energy crisis, it is caused by the impact of the increasing number of population in the world. So, we need alternative energy which has a high potential as a source of renewable energy is the sun. The development of solar cell technology from silicon solar cells to Dye-sensitized Solar Cell (DSSC) was increased recently [1]. In the DSSCs system, absorption of light and charge separation was occurred in a separate process, unlike in the silicon solar cells that the whole process does not separate. Absorption of light is carried by the dye molecules and the charge separation was done by inorganic semiconductor nanocrystal having a wide band gap. Titanium dioxide (TiO_2) has become the preferred semiconductors in a variety of studies because it possesses good photo-sensibility [2] and has a large band gap energy (> 3.00 eV) so as to absorb the energy of photons on most of the spectrum of sunlight [3], TiO_2 is also a material that is inert, harmless and inexpensive, and has good optical characteristics.

The synthesis an characterization of one-dimentional (1-D) nanostructured materials (nanotubes, nanorods, and nanowires) have received considerable attention due to their unique properties an novel applications [4-7]. Much effort has concentraced on the important metal oxide such as TiO_2 and ZnO [4]. Among them, TiO_2 and TiO_2-derived materials are of importance for utilizing solar energy an

[1] Corresponding author : sayekti@uns.ac.id
Chemistry Department, Faculty of Mathematics and Natural Sciences, Sebelas Maret University
Ir. Sutami 36A Kentingan Jebres Surakarta 57126, Central Java, Indonesia

environmental purification. TiO_2 has been widely use for applications such as a semiconductor in dye sensitized solar cell, water treatment materials, catalyst [8]. In our previous works, nanorods TiO_2 were synthesized by hyrothermal method under basic condition using TiO_2 and titanyl sulfate ($TiOSO_4$) source.

2. Experimental Method

Ilmenite was roasted at a temperature of 800°C for 2 hours. Then the roasted ilmenite calcination results was characterized using XRD. About 100 grams of roasted ilmenite was washed with hot water at a temperature of about 90°C for 2 hours. The resulting precipitated was then dried and characterized using XRD. While the filtrate was characterized by ICP-OES. Dissolution of ilmenite performed using H_2SO_4 with concentration variations of 3; 3.5; 4.5; 6; 9 M in a three-neck flask, then was heated at refluxing process under 90°C for 2 hours. The result obtained was precipitated a few minutes in order to separate the residu of the solution. The resulting filtrate was $TiOSO_4$ then characterized using ICP-OES.

About 200 mL of aquades heated to 60°C then added 0.25 mL 1M KCNS. After it was added a solution of 10 mL $TiOSO_4$ then was stirred for one hour. After the hydrolysis process was done a deposition for one hour to obtain a white precipitate at the bottom. After the deposition process, the precipitate be washed with 0.1 M HCl and aquades. The precipitate obtained and then dried at a temperature of 100°C. Titania solids added to a concentration of 12 M NaOH, then stirred for 1 hour at room temperature, then refluxed for 24 hours at a temperature of 120°C. The result refluxed then washed using 0.1 M HCl solution and then dried at a temperature of 60°C for 24 hours. Titania then calcined for 2 hours at of 400°C and characterized using XRD.

3. Results and Discussion

Characterization of ilmenite was characterized by XRF to determine the composition of chemical constituents in the form of metal oxides.

Table 1. Chemical composition of ilmenite by XRF (%)

Content	Level (%)	Content	Level (%)
TiO_2	45,35	Al_2O_3	2,35
Fe_2O_3	31,48	SO_3	2,24
Na_2O	3,86	Cr_2O_3	1,87
ZrO_2	3,57	MnO	1,58
SiO_2	3,29		

Table 1 shows the composition of the which is dominant in TiO_2 ilmenite is as much as 45.35% and 31.48% Fe_2O_3. While the remaining 23.17% is of other metal oxides such as Na_2O, ZrO_2, SiO_2, Al_2O_3, SO_3, Cr_2O_3, MnO.

Roasting of Ilmenite

Roasting of ilmenite was added by Na_2S to simplify the extraction process TiO_2 which will be done in the process of dissolution. The roasting process will produce a surface that is modified for their diffusion processes that occur at high temperatures. Iron moves faster than titanium and they tend to migrate toward the high oxygen potential region [9]. This oxidation processes occur during roasting process shown in equation 1.

$$2FeOTiO_{2(s)} + \frac{1}{2} O_{2(g)} \rightleftharpoons Fe_2O_{3(s)} + 2TiO_{2(s)} \quad (1)$$

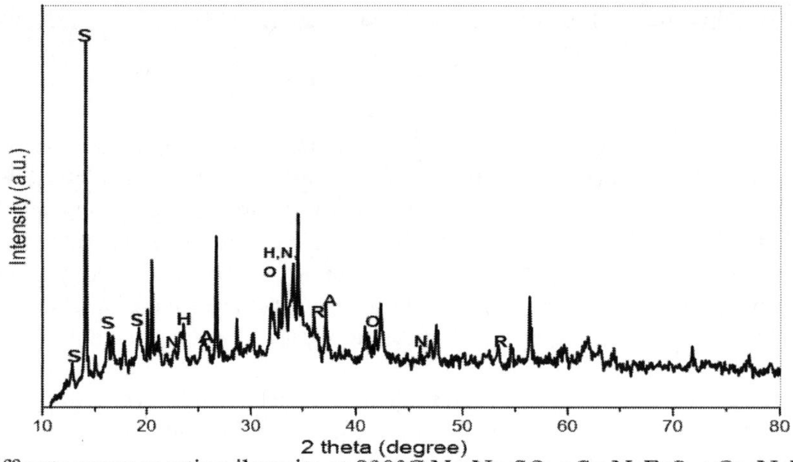

Figure 1. Diffractograms roasting ilmenite at 800°C N= Na_2SO_4 ; S= $NaFeS_2$; O= $NaFeO_2$; H= Fe_2O_3 ; A= TiO_2 anatase ; R= TiO_2 rutile

Diffractogram (Figure 1) showed the results of the roasting ilmenite decomposed into hematite, TiO_2 anatase, TiO_2 rutile and double salts. Peaks at 23,85° (d_{012}); 32,85° (d_{104}) characteristic peaks of hematite in accordance with the standard JCPDS No. 88-2359. The peak of the value 25,53° (d_{101}) and 37,79° (d_{004}) in accordance with the characteristic peaks of anatase TiO_2 standard JCPDS No. 21-1272. Peaks at 36,15° (d_{101}); 54,43° (d_{211}) showed characteristic peaks of rutile TiO_2 in accordance with the standard JCPDS No. 87-0710.

Leaching

Leaching of roasted ilmenite product was carried out by H_2SO_4 at various concentration. In this process, roasted ilmenite dissolved by sulfuric acid, thus obtained titanyl sulfate ($TiOSO_4$) and ferrous sulfate ($FeSO_4$). In this study, the variation in the concentration of sulfuric acid used were 3 M, 3.5 M, 4.5 M, 6 M, and 9 M.

Reactions that may occur in the process of leaching with sulfuric acid is the following equations.

$$FeTiO_{3(s)} + 2H_2SO_4 \rightarrow FeSO_{4(aq)} + TiOSO_{4(aq)} + 2H_2O_{(l)} \quad (2)$$

$$TiO_{2(s)} + Fe_2O_{3(s)} + 4H_2SO_{4(aq)} \rightarrow TiOSO_{4\,(aq)} + Fe_2(SO_4)_{3(aq)} + 3H_2O_{(aq)} \quad (3)$$

$$NaFeO_{2(s)} + NaFeS_{2(s)} + 4H_2SO_{4(aq)} \rightarrow Na_2SO_{4(aq)} + Fe_2(SO_4)_{3(aq)} + 2H_2O_{(aq)} + 2H_2S_{(aaq)} \quad (4)$$

Figure 2. Percentage of ilmenite dissolved in H_2SO_4 solutions

The concentration of sulfuric acid with variations can affect the dissolution process ilmenite. Ilmenite solubility increases with increasing concentration of H_2SO_4 at a concentration of 4.5 M to 6 M in the sample ILM1 with a percentage of 88.4% and 88.7%. Highest solubility occurred at the dissolution of ilmenite at H_2SO_4 6 M. Filtrate dissolution were then characterized using ICP-OES (Table 2) to determine levels of Fe and Ti in the sample. Table 2 shows the content of Ti in the sample about 6 grams / liter and the Fe content in the sample about 4 grams / liter.

Table 2. Result of leachant roasted ilminate by ICP-OES (Sample ILM1)

No	Concentration of H_2SO_4(M)	Analysis result of ICP (g / l)		
		Ti	Fe	Ti : Fe
1	9	6,4425	4,5624	1.41 : 1
2	6	6,2595	4,3373	1.44 : 1
3	4,5	6,6153	4,8408	1.36 : 1
4	3,5	5,8653	4,9271	1.19 : 1
5	3	4,3614	3,8826	1.12 : 1

Preparation TiO_2 from Titanyl Sulfate ($TiOSO_4$)

TiO_2 is synthesized from titanyl sulfate ($TiOSO_4$) with the addition of KCNS. The addition of this KCNS can be separated the iron content with titania to obtain high-purity TiO_2. Reactions hydrolysis that happened shown in equations 5 and 6.

$$TiOSO_{4(aq)} + 2H_2O_{(l)} \rightarrow TiO(OH)_{2(s)} + H_2SO_{4(aq)} \quad (5)$$
$$TiO(OH)_{2(s)} \rightarrow TiO_{2(s)} \quad (6)$$

TiO_2 white solids were then washed using 0.1 M HCl and aquades to reduce impurities that exist in the solids TiO_2.

Table 3. Chemical composition of TiO_2 hydrolysis and complexation

Content	Level (%)
TiO_2	88,03
SO_3	7,94
Fe_2O_3	2,67
SnO_3	0,49
P_2O_5	0,35

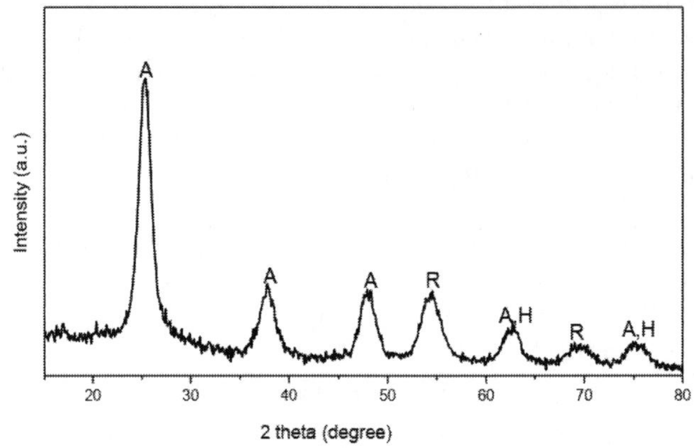

Figure 3. Diffractograms of TiO_2 hydrolysis and complexation
H = Fe_2O_3; A = TiO_2anatase; R = TiO_2rutile

Synthesis of TiO_2 Nanorods

TiO_2 nanorods preparation was done by mixing TiO_2 nanoparticles (4.664 nm) and 12 M NaOH for ± 1 hour to obtain a white solution. Result of the hydrothermal process obtained white material such as mud at the bottom of the autoclave. White solids are then washed using 0.1 M HCl and aquades to reduced impurities.

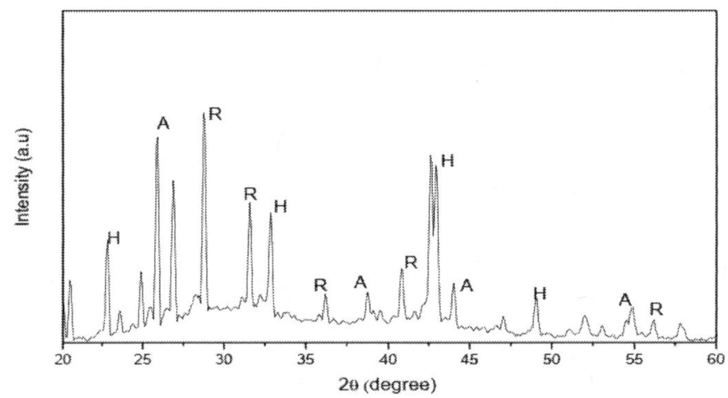

Figure 4. Diffractograms of TiO_2 nanorods H = Fe_2O_3; A = TiO_2 anatase; R = TiO_2 rutile

TiO_2 rutile showed at a peak of 27, 49° (d_{110}), while TiO_2 anatase showed at peak 25,28° (d_{101}). TiO_2 rutile characteristic peak has a higher intensity than TiO_2 anatase that indicated that TiO_2 rutile has high crystallinity. TiO_2 nanorods were characterized by TEM (Transmission Electron Microscope) to determine the morphology of the inside of TiO_2 nanorods (Figure 5) and shows that was formed rod-shaped samples with a diameter of 7.02 nm.

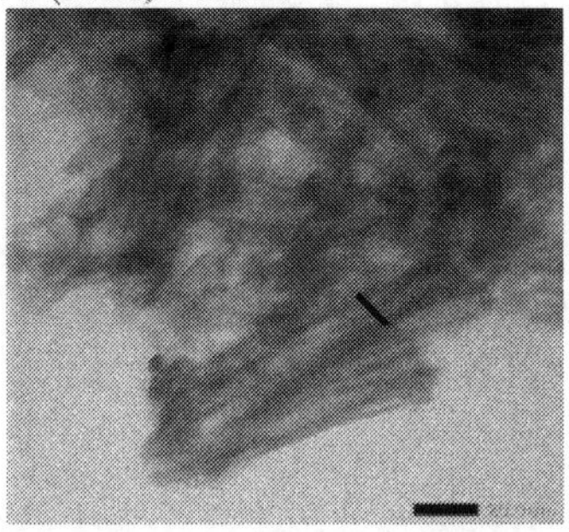

Figure 5. TEM of TiO_2 nanorods

In conclusion, morphology of TiO_2 nanorods from condensation preparation has been formed by hydrolysis and complexation using KCNS, with a hydrothermal process under alkaline conditions. Powder of TiO_2 nanorods has obtained with a diameter of nanorods 7.02 nm.

4. Acknowledgment

The researchers wish to express their gratitude to the Insinas Ristek Dasar, Kemenristek of Indonesia for supporting this research.

5. References

[1] Grätzel M, 2001 *Nature* 414 338
[2] Vasquez R and Molina A, 2008 *Metal* 5 13
[3] Wunderlich W, Oekermann T, Miao L 2004 *Ceramic Processing Research* 5 343
[4] Rao CNR and Nath M 2003 *Dalton Trans* 1 1
[5] Patzke GR, Krumeich F, and Nesper R 2002 *Angew. Chem. Int. Ed.* 41 2446
[6] Huang M, Mao S, Feick H, Yan H, Wu Y, Kind H, Weber E, Russo R, and Yang P *Science* 292 1897
[7] Pan ZW, Dai ZR, and Wang ZL 2001 *Science* 292 1947
[8] Fujishima A, Rao TN and Tryk DA 2000 *J. Photochem. Photobiol. C:Photochem. Rev.* 1 1
[9] Wahyuningsih S, Ari HR, Edi P, Ariantama D 2014 *Sci.Res.Pub.* 4

Journal of Physics: Conference Series **776** (2016) 012045 doi:10.1088/1742-6596/776/1/012045

Effects of electrolytes on ion transport in Chitosan membranes

N. N. Rupiasih
Department of Physics, Faculty of Mathematics and Natural Sciences, Udayana University, Kampus Bukit Jimbaran, Badung, Bali 80362, Indonesia
Group Research Material Sciences and Technology-Polymer and Biomaterial

E-mail: rupiasih@unud.ac.id

Abstract. Recently, charged polymer membranes are widely used for water purification applications involving control of water and ion transport, such as reverse osmosis and electrodialysis. In this study, we have explored the effects of electrolyte solutions on ion transport properties of chitosan synthetic membranes via concentration gradient driven transport. Also, the water uptake of those membranes, before (control) as well used membranes have studied. The membrane used was chitosan membrane 2%. The electrolyte solutions used were HCl, KCl, $CaCl_2$, $MgCl_2$ and $AlCl_3$, with various concentrations of 0.1 mM, 1 mM, 10 mM, 100 mM and 1000 mM. Ion transport experiments were carried out in a cell membrane model which composed of two compartments and the potential difference of membrane was measured using Ag/AgCl calomel electrodes. Those measurements were conducted at ambient temperature 28.8 ^{o}C. The results showed that the current density (J) increased with increased in concentration gradient of solution. The current density was higher in electrolyte solution which has higher molar conductivity than those of a solution with a small molar conductivity. Meanwhile the current density was smaller in electrolyte solution which has larger Stokes radii than those of a solution with small Stokes radii. Except membrane which has been used in HCl solution, the water uptakes of the used membranes were greater than the control membrane. These results can develop and validate a common framework to interpret data of concentration gradient driven transport in chitosan synthetic membranes and to use it to design of membranes with improved performance.

1. Introduction

Chitosan is a biopolymer composed of glucosamine and N-acetyl glucosamine units linked by 1-4 glucosidic bonds, has found wide application in the biomedical field, such as controlled drug release, biomaterial and tissue engineering [1, 2]. The development of synthetic membranes had been inspired by the fact that the selective transport through biological membranes which known as *ion exchange*

---

[3]Author to whom any correspondence should be addresses: rupiasih@gmail.com

membranes (*IEM*) [2]. Ion exchange membranes discriminate cation and anions, thus they should have a high transport number for counter-ions. Such membranes have the potential applications as a new functional material in separation of ionic materials, mostly used in the solutions containing multi components, such as electrodialytic concentration of seawater to produce sodium chloride, demineralization of sugarcane juice, demineralization of saline water, etc [3].

Most ion exchange membranes have made of synthetic polymers; only a few were from natural polymers e.g. chitosan, cellulose and alginate [2, 4]. Chitosan can be used as an *anion exchange membranes* (*AEM*) directly, cellulose derivatives such as cellulose phosphate, cellulose acetate and alginate were used as *cation exchange membranes* (*CEM*) [2].

Since the use of ion exchange membranes has become more diverse, requirements for membranes with unusual properties have increased. These requirements have led to the development of various kinds of newly functionalized ion exchange membranes. Recently the ion exchange membranes has been used not only for the traditional applications, such as electrodialysis concentration or desalting of solutions, diffusion dialysis to recover acids and electrolysis of sodium chloride solution, but also in various fields as a polymeric film having ionic groups [5].

The water uptake in membranes is considered to be a key factor in the ion and water transfer and in polarization processes in electromembrane systems [6]. Also, the water transport mechanism into the polymer is particularly important for assessing the suitability of these materials as drug delivery systems and membranes in separation processes [7]. Concentration polarization is a well-known phenomenon occurring at the interface between an ion exchange membrane and an electrolyte solution when the current passes through it or if there is concentration gradient of electrolyte solution [6].

The objective of this study was to investigate the effects of electrolyte solutions on the ion transport properties of chitosan synthetic membranes. The transport processes were conducted through concentration gradient driven transport and all transport experiments were obtained under the same conditions. Moreover, the water uptake experiments were carried out to investigate the effect of those electrolyte solutions to hydrophilic properties of the membrane.

2. Materials and methods

2.1 Transport in chitosan membranes

HCl, KCl, $CaCl_2$, $MgCl_2$, $AlCl_3$ and other reagents used in this study were obtained from commercial sources. The membrane used was chitosan membrane 2%. The electrolyte solutions used were HCl, KCl, $CaCl_2$, $MgCl_2$ and $AlCl_3$, with various concentrations of 0.1 mM, 1 mM, 10 mM, 100 mM and 1000 mM. Ion transport experiments were carried out in a cell membrane model which composed of two compartments (1 and 2) and the potential difference of membrane was measured using Ag/AgCl calomel electrodes. Those measurements were conducted at ambient temperature 28.8 ^{o}C.

2.2 Water uptake

The membrane was cut into fixed size 1 cm × 1 cm. They were soaked in distilled water for 0.5 – 24 h, then taken out, wiped and then weighed. These samples were then dried in a vacuum oven overnight and the dry weight measured. The water uptake was determined by the following equation [1, 8].

$$Water\ uptake\ (\%) = \frac{w_w - w_d}{w_w} x100$$

Where w_w is the weight of the wet membrane (g) and w_d is the weight of the dry membrane (g).

Table 1. Values of molar conductivity and Stokes radii for selected cations in this study.

Cations	Stokes radii[a] (nm)	Molar conductivity[a] (10^4 S.m^2.mol^{-1})
H^+	0.026	349.8
Al^{3+}	0.125	189
Ca^{2+}	0.308	119
Mg^{2+}	0.346	106.1
K^+	0.437	73.55

[a] From Ref. [9].

3. Results and discussion

3.1 Characteristic of potential difference (V) vs Log (C1/C2)

Figure 1 shows the potential difference (V) of membranes as function of Log (C1/C2) where C1 and C2 is concentration of solution in compartment 1 and compartment 2, when the membranes in contact with electrolyte solutions such as HCl, KCl, $CaCl_2$, $MgCl_2$ and $AlCl_3$. In general, Fig. 1 shows that the potential difference increased as increasing the logarithmic of C1/C2. This result is in accordance to the Nernst equation [10], which illustrates that the potential difference is due to differences in concentration in both compartments e.g. proportional to Log (C_1/C_2). The figure also show that except HCl solution, the increment of potential difference is more in electrolyte with greater Stokes radii than those of an electrolyte with a small Stokes radii and the graph look sharper.

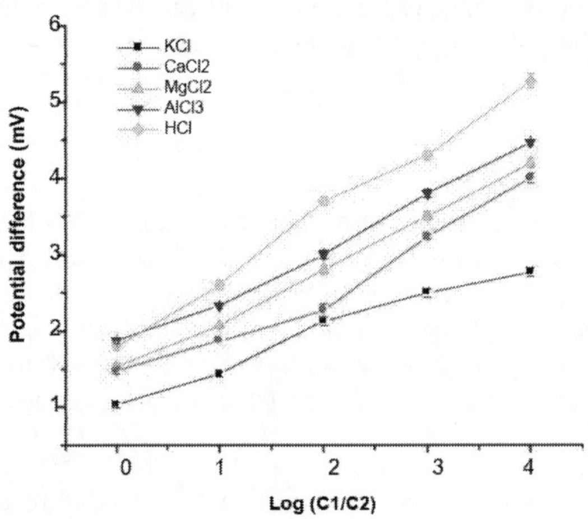

Figure 1. The potential difference (V) as a function of Log (C1/C2) for different electrolyte solutions. HCl, KCl, $CaCl_2$, $MgCl_2$ and $AlCl_3$.

3.2 Characteristic of current density vs. concentration gradient ($\Delta C/\Delta x$)

Figure 2 shows the current density (J) of membranes as function of gradient concentration ($\Delta C/\Delta x$). It shows that the current density is increased as increase in concentration gradient of electrolyte solution. This result is in accordance to Fick law, which states that the current density is proportional to concentration gradient of the electrolyte solution [10]. All curves show that at small concentration gradient ($\Delta C/\Delta x$) e.g. 0 mM/cm, 0.09 mM/cm and 0.99 mM/cm, the current density increased gently and at greater concentration gradient e.g. 9.99 mM/cm the increase was sharp and at the highest concentration gradient 99.99 mM/cm, the increased is very sharp. The same result has been reported at our previous publication [2].

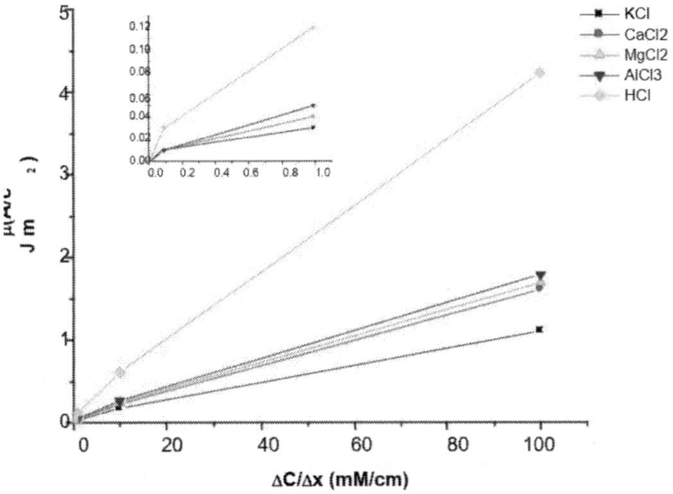

Figure 2. The current density (J) as a function of concentration gradient (ΔC/Δx) for different electrolyte solutions: HCl, KCl, $CaCl_2$, $MgCl_2$ and $AlCl_3$.

3.3 Effect of electrolyte solutions on current density

The relation between current density and electrolyte solutions (HCl, KCl, $CaCl_2$, $MgCl_2$ and $AlCl_3$) in various concentrations gradient 0, 0.09 mM/cm, 0.99 mM/cm, 9.99 mM/cm and 99.99 mM/cm show in Fig. 3a and 3b. Figure 3a, the graphs draw in ascending order of molar conductivity of the electrolyte solutions and Fig. 3b, the graphs draw in ascending order of Stokes radii of the electrolyte solutions. The both figures are shown just to describe them easily. The graphs showed that the current density (J) increased with increased in concentration gradient (ΔC/Δx) of solution. The same result has been reported at our previous publication [2]. The current density is higher in electrolyte solution which has higher molar conductivity than those of a solution with a small molar conductivity (Fig. 3a). This means that the efficiency of ion transport in the membrane is better in the electrolyte solution which has higher molar conductivity. Meanwhile the current density is smaller in electrolyte solution which has larger Stokes radii than those of a solution with small Stokes radii (Fig. 3b). These results explain about the effect of electrolyte solutions to the characteristics of ion transport property of chitosan membrane which expressed as current density via concentrations gradient driven transport.

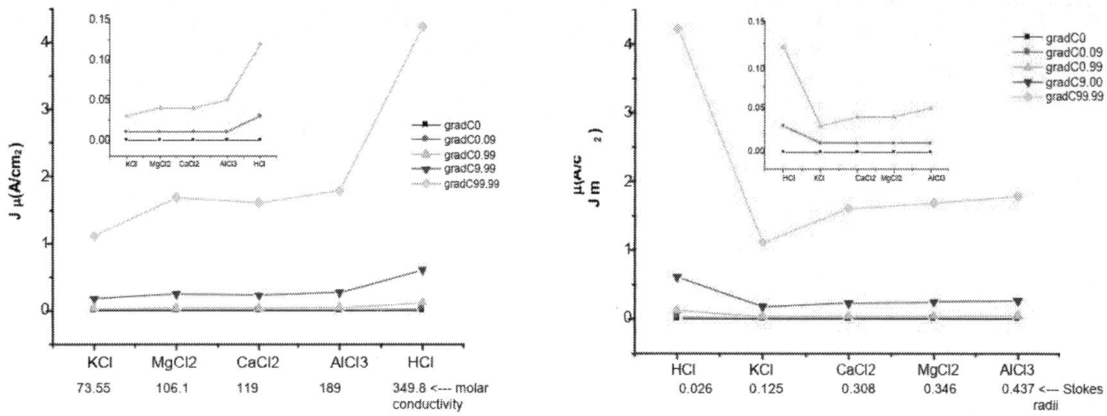

a) **b)**

Figure 3. The relation between current density and electrolyte solutions (HCl, KCl, $CaCl_2$, $MgCl_2$ and $AlCl_3$) in various concentrations gradient 0, 0.09 mM/cm, 0.99 mM/cm, 9.99 mM/cm and 99.99 mM/cm. a) having different molar conductivity and b) having different Stokes radii.

3.4 *Water Uptake*

The water uptake (hydrophilicity) of the membranes, control and used membranes were investigated (Fig. 4a and 4b). Except membrane which has been used in HCl solution, the water uptakes of the used membranes were greater than the control membrane (Fig. 4b). Fig, 4a shows control membrane and the membranes which have been used in HCl, KCl and $MgCl_2$ solutions reached maximum of water uptake after 6 h, membranes which have been used in $AlCl_3$ solutions reached maximum of water uptake after 12 h and membranes which have been used in $CaCl_2$ reached maximum of water uptake very fast that is after 2 h. These means after exposed to those electrolyte solutions some changes have occurred in structure of the membranes which improve their hydrophilic property.

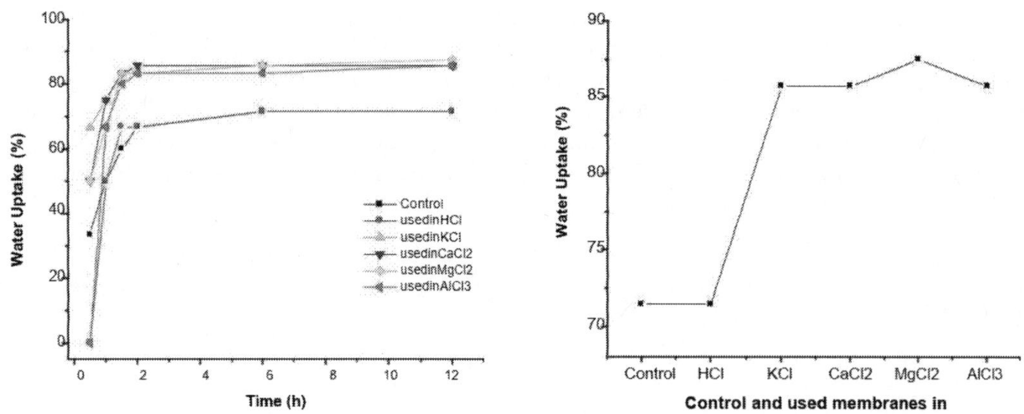

a) **b)**

Figure 4. Water uptake of chitosan membrane, control and used membranes in various electrolyte solutions: HCl, KCl, $CaCl_2$, $MgCl_2$ and $AlCl_3$. a) Water uptake as function of times and b) The percentage maximum of water uptake (%) of chitosan membrane.

4. Conclusion

The study about the effects of electrolyte solutions on ion transport properties of chitosan membranes via concentration gradient driven transport has been described. It was observed that the current density (J) increased with increased in concentration gradient of solution. The current density is higher in electrolyte solution which has higher molar conductivity than those of a solution with a small molar conductivity. Meanwhile the current density is smaller in electrolyte solution which has larger Stokes radii than those of a solution with small Stokes radii. Except membrane which has been used in HCl solution, the water uptakes of the used membranes were greater than the control membrane. These revealed that electrolyte solutions affected the hydrophilic property of the membranes. Those results described that how electrolyte solutions affected ion transport properties of chitosan synthetic membranes via concentration gradient driven transport.

5. Acknowledgments

This work was partially supported by the Ministry of Research, Technology and Higher Education, Republic of Indonesia, under Fundamental Research Grant of Udayana University, Year Budget 2015/2016 are highly acknowledged. Also thanks to my student 'Made Rasmini' for helping in data collection.

6. References

[1] Ricardo MP da Silva, Sofia G-C, Julio S-R'n, Joa˜o F-M and Rui L-R, 2008 *Biomacromolecules* 9, 2132–2138.
[2] Rupiasih NN, Puspita YE and Sumadiyasa M 2015 *Journal of Physics: Conference Series* 622, 012004.
[3] Nagarale RK, Gohil GS and Vinod K-S 2006 *Advances in Colloid and Interface Science* 119, 97 – 130.
[4] Xin C, Jiahao L, Zhicheng F and Zhengzhong S 2005 *Journal of Applied Polymer Science* 96, 1267-1274.
[5] Sata T, 1991 *Reviews of Applied Electrochemistry* 26, *Journal of Applied Electrochemistry* 21, 283-294.
[6] Aguilella VM, Mafe S, Manzanares JA and Pellicer J 1991 *Journal of Membrane Science* 61, 177-190.
[7] Magdalena G-D, Jadwiga O-C 2012 *Progress on Chemistry and Application of Chitin and Its* ...17, 59-66.
[8] Rebecca SLY, Kaisong Z and Bradley PL 2013 *Membranes* 3, 182-195.
[9] Choi J-H, Lee H-J and Moon S-H 2001 *Journal of Colloid and Interface Science* 238, 188–195.
[10] Hobbie, Russel K. 1978, *Intermediate Physic for Medicine and Biology*. Second Edition. John Willey and Son. Singapore, 45-105.

Thermoelectric power of single crystalline lead telluride nanowire

Dedi[1,2], N. Idayanti[1], P-C Lee[2], C-H Lee[3], and Y-Y Chen[2]
[1]Research Center for Electronics and Telecommunication, Indonesian Institute of Sciences, Bandung, Indonesia
[2]Institute of Physics, Academia Sinica, Taipei 11529, Taiwan
[3]Department of Engineering and System Science, National Tsing Hua University, Hsinchu 300, Taiwan

E-mail: dedi030@lipi.go.id

Abstract. In this study, structurally uniform PbTe nanowires (NWs) with high-quality single crystals were synthesized using a stress-induced growth. Selected-area electron diffraction patterns show that the PbTe NWs were grew along the [100] direction. The thermoelectric power *S* of a 142-nm diameter NW plotted for temperature range of 300–350 K exhibited semiconductor *n*-type behavior. The absolute *S* plotted against carrier concentration *n* was size-dependent, compared with our previous work.

1. Introduction

Nanowires (NWs) used for converting waste heat into electricity have become a attractive research topic [1–5]. In general,Thermoelectric can be utilized for power generation and cooling applications [6–11]. PbTe is a semiconductor with an energy band gap of 0.32 eV at 300 K, it is suitable for application in thermoelectricity [10,12–14]. The efficiency of thermoelectric materials is determined according to the dimensionless figure of merit *ZT*, which is defined as $S^2\sigma T/(\kappa_e + \kappa_l)$, where S is the thermoelectric power (or Seebeck coefficient), σ represents the electrical conductivity, κ_e is the electronic thermal conductivity, and κ_l is the lattice thermal conductivity. Numerous studies have reported that nanomaterials enhance the *ZT* as a result of quantum confinement effects and increased surface phonon scattering [11,15–16]. In NWs comprising thermoelectric materials such as PbTe, the nanometer scale is predicted to increase the value of *S* [17–18]. Furthermore, previous studies on the thermoelectric properties of PbTe NWs have reported inconsistent findings, primarily because fabricating electrical Ohmic contacts for PbTe NWs remains extremely difficult. Specifically, the native oxide layer that forms on the outer surface of PbTe NWs can be removed only by using a focus ion beam (FIB). In this study, we report how nanowire diameter influences the electronic contribution to the thermoelectric transport properties in single-crystal PbTe NWs.

2. Experiment Design

PbTe NWs were synthesized using stress-induced growth described in reference [19–20], which is a practical method for growing NWs with a high aspect ratio and high crystallinity. As-received Pb (Alfa Aesar, -200 mesh, 99.9%) and Te (Alfa Aesar, -325 mesh) powders were first mixed at a specific ratio and melted at 1000 °C for 4 h in a vacuumed quartz tube below 5×10^{-6} Torr. The molten compound was slowly cooled to room temperature in the furnace. Subsequently, the compound was cut to form a disk, which served as the target for the pulsed laser deposition. Single-crystal SiO_2/Si (100) wafers (e-light Tech. Inc.; SiO_2 thickness, 413–434 nm; diameter, 100 ± 0.5 mm) with double-side polishing were cut into squares measuring approximately 1.5 cm^2. All substrates were cleaned

[1]**To whom any correspondence should be addressed.**

using acetone, isopropyl alcohol, and deionized water in an ultrasonic bath for 10 min before being dried with N_2 stream. The PbTe films were prepared in an ArF excimer laser (Lamda Physik LPXpro 210) and deposited onto substrates in a vacuum system with a base pressure of 5.0×10^{-7} Torr.
The PbTe thin films were grown with deposition rate at 0.22 Å/s, which the excimer laser was applied at 140 mJ (frequency, 10 Hz) for 15 min at room temperature. The substrate was rotated at approximately 10 rpm, and the thickness of the films was 20 nm. The films were sealed in a vacuumed quartz tube below 5×10^{-6} Torr, annealed at 450 °C for 5 d, and then left in the furnace to cool to room temperature. During the annealing process, the NWs grew from the film and released compressive stress because of the difference between the thermal expansion coefficient of the PbTe film and that of the SiO_2/Si substrate.

3. Results and Discussion

Field emission scanning electron microscopy (FESEM; Hitachi, S-4800) images of the PbTe NWs (Fig. 1a) show that the NWs exhibited diameters ranging from 50 to 300 nm, and lengths up to several micrometers.Figure 1b shows a TEM image and the lattice fringes of the smooth PbTe NW that were separated by a 0.33-nm gap were consistent with a periodicity along the [200] direction with the lattice parameter a of approximately 6.549 Å (Fig. 1b) was 1.87% higher than that of the PbTe annealed thin film (a = 6.43 Å), possibly as a result of antisite defects (Pb_{Te}).

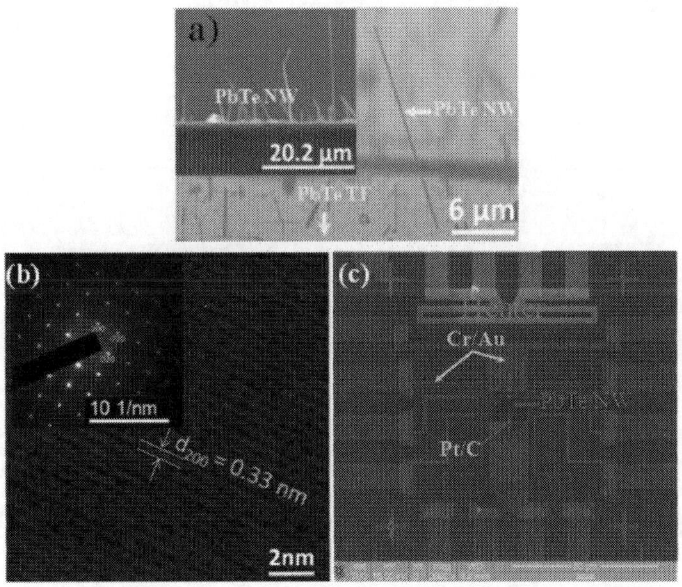

Figure 1. (a) SEM image (top view) showing that all of the PbTe NWs grew several micrometers in length from the surface of the PbTe thin film; (b) The distance between the crystal faces is 0.33 nm, indicating that the lattice fringes grew along the [200] direction; the inset of the Fig. 1b shows the SAED pattern (at the [001] zone axis), confirming the single-crystalline NWs grew in the [100] direction; and (c) SEM images of a single-crystal PbTe NW (diameter d, 142 nm; length l, 18.95 µm) placed on a Si_3N_4 template following electrode formation by using an FIB.

A corresponding selected-area electron diffraction (SAED) pattern show that the PbTe NWs were high-quality single crystals (the inset of Fig. 1b). The chemical composition of the PbTe NWs was examined using energy dispersive X-ray spectroscopy (EDS), and the stoichiometric composition of the individual NW was $Pb_{1.1}Te_{0.9}$. A tungsten needle (d = 100 nm) and a binocular optical microscope were used to extract a single-crystal NW from the PbTe thin film, which was then suspended by electrodes on a Si_3N_4 microchip (Fig. 1c). Subsequently, the microchip was used to obtain the values of S and σ (range, 50–350 K) by using four-point measurements (Fig. 1c).

Figure 2a shows the temperature dependence of σ and S in 50–350 K for a 142-nm NW that exhibited a semiconducting behavior. The σ at near room temperature was 2383 S m^{-1}, which was 96% lower than that of the PbTe bulk (65789.47 S m^{-1}) [14], which was possibly caused by the surface scattering of charge carriers [19,21]. However, the value was higher than those reported in previous studies [19, 22–24].

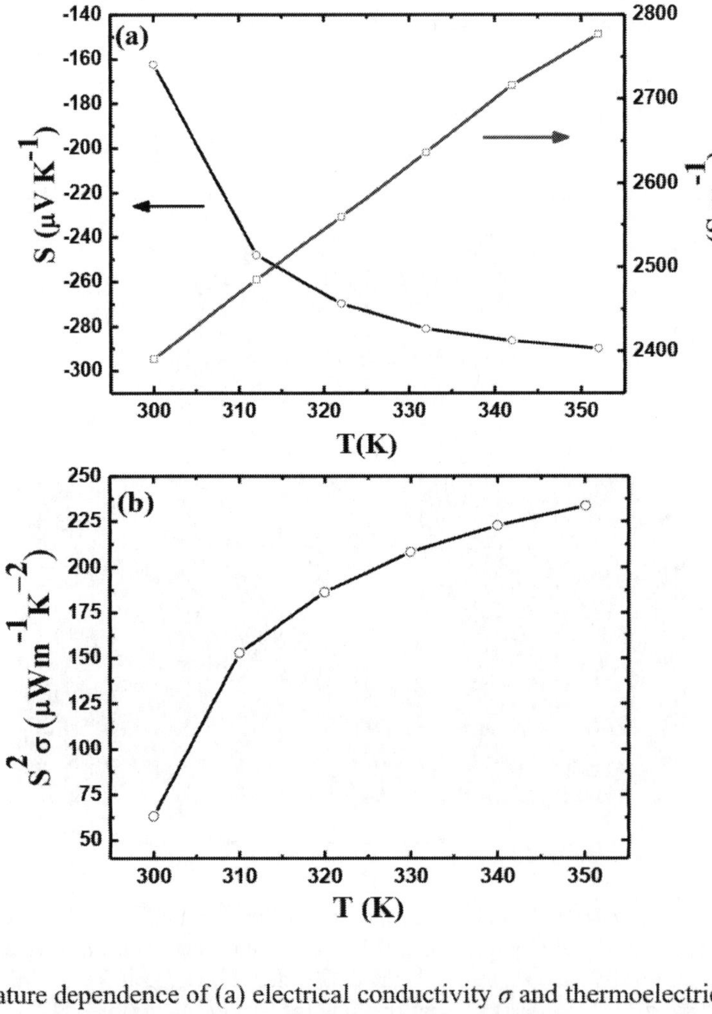

Figure 2. Temperature dependence of (a) electrical conductivity σ and thermoelectric power S; and (b) thermoelectric power factor for a single-crystal PbTe NW (d = 142 nm).

The S value was -162 μV K^{-1} for the 142-nm NW at 300 K, which is approximately 7.43% less than that of the PbTe bulk NW (-175 μV K^{-1}) [14], and might have resulted from structural imperfections

such as antisite defects that were caused by the creation of a vacancy at the Te site, thereby facilitating the enhanced carrier concentration n [14,25]. A negative S value indicates that the PbTe NW is an n-type semiconductor because the electrons exhibited considerably higher carrier mobility μ than the holes did, thereby dominating the electrical transport properties and that the S was controlled using electron diffusion [26–27]. The values of σ correlated closely with S, and the curve of σ was inverse with that of S (Fig. 2a); thus, at 300 K, the thermoelectric power factor ($S^2\sigma = 93.50\ \mu W\ m^{-1}\ K^{-2}$) was greater than those reported in previous studies [22–24,28]. Figure 2b shows the $S^2\sigma$ of the PbTe NW as a function of temperature. The power factor gradually increased in conjunction when the temperature increased, which was primarily attributed to the observed σ trends of the PbTe NW.
According to the Mott relation [6], the carrier concentration n can be estimated from thermoelectric power data as:

$$S = \frac{8\pi^2 k_B^2}{3qh^2} m_d{}^* T \left(\frac{\pi}{3n}\right)^{2/3} \qquad (1)$$

where k_B is Boltzmann's constant, $m_d{}^*$ is the effective mass of the carrier ($m_d{}^* = 0.30\ m_o$ in PbTe) [29], m_o is the electron mass, q represents the electron charge, h is Planck's constant, and T is the temperature. Previous studies have shown that this equation is valid for degenerate semiconductors with an n value ranging from 10^{18} to 10^{20} cm^{-3} [27,30]. The n-value was approximately $3.80–7.14 \times 10^{18}$ cm^{-3} in the 300–350 K range, indicating that the NW exhibited degenerative semiconductor behavior. The value of n decreased as T increased to 320 K, and n remained relatively constant over a wide range of temperatures at 320–350 K [Fig. 3a, marked by the white circles with a black outline], thus indicative of extrinsic doping. The values of μ were calculated using the equation;

$$\mu = \frac{1}{\rho n q} = \frac{\sigma}{nq} \qquad (2)$$

where n is the carrier concentration, σ represents the electrical conductivity, and q is the electron charge. The calculated values of μ ranged from 20.92 to 45.71 $cm^2\ V^{-1}\ s^{-1}$ [Fig. 3a, marked by the white squares with a blue outline], which were considerably lower than those of the PbTe bulk (approximately 209 $cm^2\ V^{-1}\ s^{-1}$) [14], indicating that the carrier mobility of the NW was approximately 90% slower than that of the PbTe bulk at room temperature. The μ values of the 142 nm NW exhibited a continuous increase in conjunction with T from room temperature to 330 K, indicating that the scattering of impurities was dominant throughout the entire temperature range. The effect of ionized impurities in semiconductor on μ is given by the equation $\mu_I \sim (m^*)^{-1/2} N^{-1} T^{3/2}$, where N is the density of the total ionized impurities in the semiconductor [30]; thus, μ remained relatively flat over a wide range above 330 K, thereby indicating that electron scattering had occurred throughout the entire NW (i.e., the scattering was caused by defects or impurities). Figure 3b shows the absolute value of the S for the various PbTe samples as a function of the carrier density at room temperature. In bulk PbTe, in which acoustic-phonon scattering dominates, the $S(n,p)$ curve follows the full line shown [14]. The S value of PbTe NW (d = 142 nm) exist on the full line. The values of S correlated closely with those of n, which was consistent with the samples where a low n coincided with a high S. The absolute values of S plotted against n shows the difference in size (diameter) dependence for the PbTe NWs in comparison with the 75-nm and 217-nm PbTe NWs reported by Chen et al. [19], which may have result from the increase in carrier concentration n coinciding with the decrease in diameter d in that study.

Journal of Physics: Conference Series **776** (2016) 012046 doi:10.1088/1742-6596/776/1/012046

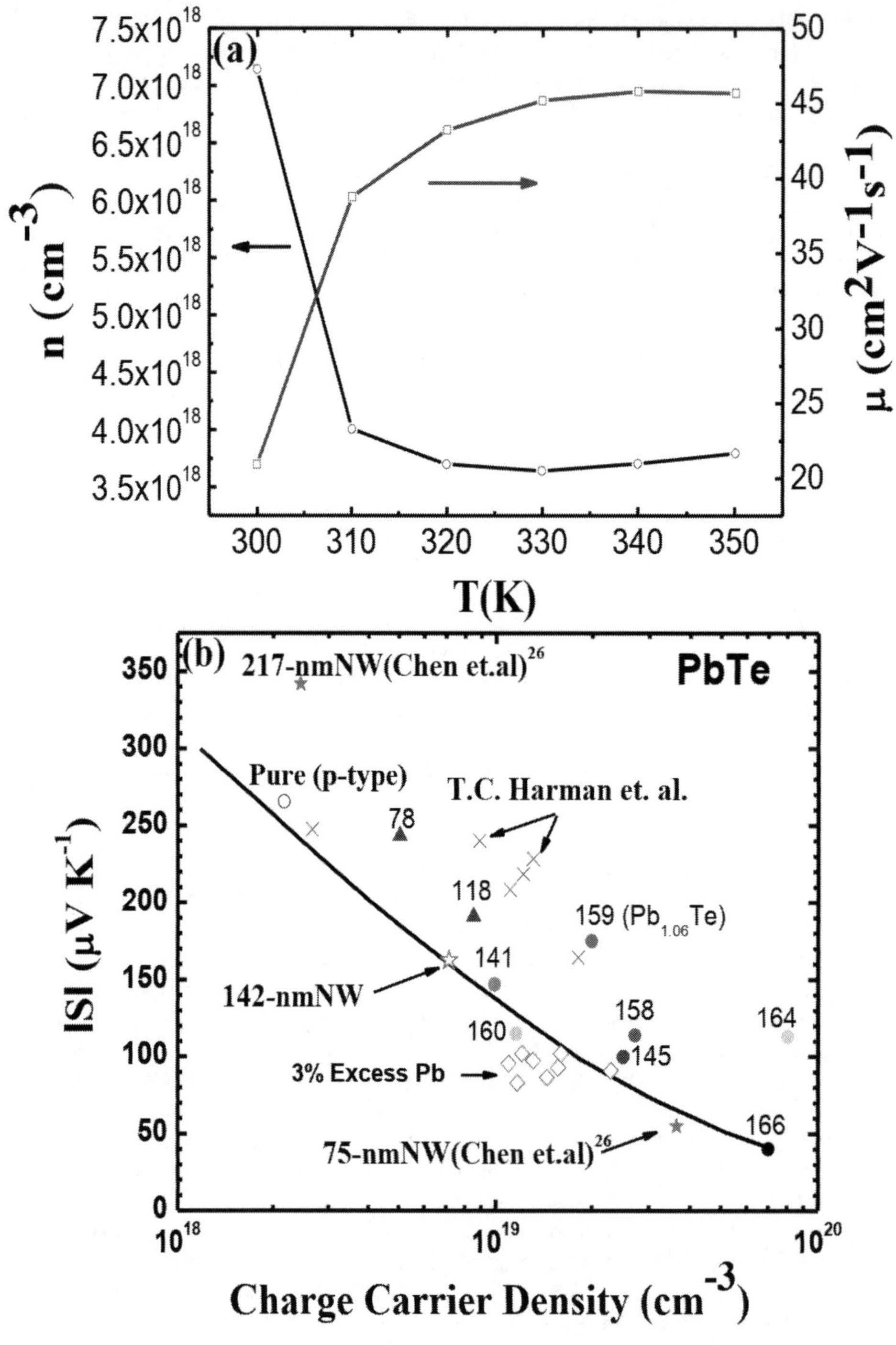

Figure 3. (a) Temperature dependence of carrier concentration and carrier mobility for the a single-crystal PbTe NWs ($d = 142$ nm); (b) Absolute value of the thermoelectric power of various PbTe samples as a function of carrier density; electrons, or holes at room temperature [14]. The solid red stars denote the $Pb_{1.1}Te_{0.9}$ samples with 75- and 217-nm diameter wires [19]. The white star with a red outline represents the $Pb_{1.1}Te_{0.9}$ sample with a 142 nm diameter.

4. Summary

This study reports the electrical conductivity and thermoelectric power were examined as a function of temperature. The NW exhibited *n*-type semiconductor behavior, and the thermoelectric power was size-dependent, compared with our previous work [19]. Specifically, the NWs with a shorter diameter yielded greater carrier concentration and lower thermoelectric power.

Acknowledgment

Technical support was provided by the Core Facilities for Nanoscience and Nanotechnology at the Institute of Physics, Academia Sinica in Taiwan.

References

[1] Boukai AI, Bunimovich Y, Tahir J-K, Yu JK, Goddard WA and Heath JR, 2008 Nature **451**, 168–171.
[2] Bubnova O, Khan ZU, Malti A, Braun S, Fahlman M, Berggren M and Crispin X, 2011 Nat. Mater. **10**, 429–433.
[3] Hochbaum AI, Chen RK, Delgado RD, Liang WJ, Garnett EC, Najarian M, Majumdar A and Yang PD, 2008 Nature **451**, 163–U165.
[4] Minnich AJ, Dresselhaus MS, Ren ZF and Chen G, 2009 Energy Environ. Sci. **2**, 466–479.
[5] Zhao HZ, Pokheral M, Zhu GH, Chen S, Lukas K, Jie Q, Opeil C, Chen G and Ren ZF, 2011 Appl. Phys. Lett. **99**, 163101.
[6] Snyder GJ and Toberer ES, 2008 Nature Materials **7**, 105.
[7] Tritt TM, and Subramanian MA, 2006 MRS Bull **31**, 188.
[8] DiSalvo FJ, 1999 Science **285**, 703.
[9] Tritt TM, 1999 Science **283**, 804.
[10] Rowe DM, 1995 CRC Handbook of Thermoelectrics, (CRC Press, Boca Raton, FL, USA).
[11] Dresselhaus MS, Chen G, Tang MY, Yang RG, Lee H, Wang DZ, Ren ZF, Fleurial JP and Gogna P, 2007 Adv. Mater. **19**, 1043–1053.
[12] Dughaish Z, 2002 Phys. B - Cond. Matt. **322**, 205.
[13] Heremans JP, Thrush C and Morelli DT, 2004 Phys. Rev. B **70**, 115334.
[14] Heremans JP, Thrush C and Morelli DT, 2005 J. Appl. Phys. **98**, 063703.
[15] Bassi ALi, Bailini A, Casari CS, Donati F, Mantegazza A, Passoni M, Russo V and Bottani CE, 2009 J. Appl. Phys. **105**, 124307.
[16] Szczech JR, Higgins JM and Jin S, 2011 Chem MJ. **21**, 4037–4055.
[17] Hicks LD and Dresselhaus MS, 1993 Phys. Rev. B **47**, 12727.
[18] Hicks LD and Dresselhaus MS, 1993 Phys. Rev. B, **47**, 16631.
[19] Dedi, Lee PC, Chien CH, Dong GP, Huang WC, Chen CL, Tseng CM, Harutyunyan SR, Lee CH, and Chen YY, 2013 Appl. Phys. Lett. **103,** 2, p023115.
[20] Dedi, Chien C-H, Te-ChihHsiung, Chen Y-C, Huang Y-C, Lee P-C, Chih-HaoLee, Chen Y-Y, 2014 AIP Advances **4**, 057111.
[21] Wang D, Sheriff BA and Heath JR, 2006 Small **2**, 1153.
[22] S. Y. Jang, H. S. Kim, J. Park, M. Jung, J. Kim, S. H. Lee, J. W. Roh and W. Lee, 2009 Nanotechnology **20**, 415204.
[23] G. Tai, B. Zhou, and W. Guo, 2008 J. Phys. Chem. C, **112,** 11314.
[24] G. Tai, W. Guo, and Z. Zhang, 2008 Cryst. Growth Des. **8**, 2906.
[25] J. Horak, J. Navratil, and Z. Stary, 1992 Phys. Chem. Solids **53**, 1067.
[26] D. L. Partin, J. P. Heremans, D. T. Morelli, C. M. Thrush, C. H. Olk and T. A. Perry, 1988 Phys. Rev. B: Condens. Matter Mater. Phys. **38,** 3818.
[27] A. H. Dekuijper and Bisschop, 1983 J. Thin Solid Films, **110**, 99.

[28] Y. Yang, D. K. Taggart, M. H. Cheng, J. C. Hemminger, and R. M. Penner, 2010 J. Phys. Chem. Lett. **1**, 3004.
[29] H. A. Lyden, 1964 Phys. Rev. A **135**, A514.
[30] S. M. Sze and K. K. Ng, 2007 Physics of Semiconductor Devices, 3rd ed. (Wiley-Interscience, Hoboken, NJ,).

Study of ozone reactor with dielectric barrier discharge plasma (BDBP): variations of inner electrode based on Stainless steel, Galvalume, and Copper

Suraidin[1], M. Nur[2,4*], Gunawan[3,4], Ade Ika Susan[1]
[1)] Magister of Physics, Faculty of Science and Mathematics, Diponegoro University, Jl. Prof. H. Soedarto, SH, Tembalang. Semarang. INDONESIA,
[2)] Department of Physics, Faculty of Science and Mathematics, Diponegoro University,
[3)] Department of Chemistry, Faculty of Science and Mathematics, Diponegoro University,
[4)] Center for Plasma Research, Science and Mathematics Faculty, Diponegoro University, Jl. Prof. H. Soedarto, SH, Tembalang. Semarang. INDONESIA.

E-mail: m.nur@undip.ac.id correspondence author

Abstract: The research of ozone reactor with dielectric barrier discharge plasma (BDBP) with variations of inner electrode based on stainless steel, galvalume, and copper has been done. The ozone was produced with oxygen as inputted gas. The ozone concentration has determined with variations of voltage (from 3 to 5 kV) and flow rate (i.e. 2,3,4,5,6,8 and 10 L/min). Ozone reactor, has a frame with cylinder-cylinder configuration. Inner electrode (cylindrical) made of stainless steel, galvalume and copper. The three of materials was used for evaluates the inner electrode's corrosion resistance characteristics for DBDP reactor. The outer electrodes (cylindrical) made of zinc plate, while Pyrex used as dielectric material. AC High Voltage was used as power supply with frequency of 16 kHz. This research has been conducted with five variations of reactor's length of 5, 10, 15, 20, and 25 cm. The ozone concentration has measured with KI solution and then titrated using sodium thiosulphate. After used as the inner materials of ozone reactor, the corrosion test has been conducted with SEM and EDX analysis. The results showed that ozone concentration was increasing by the increasing of voltage, and the greater gas flow rate gives the less ozone concentration. In addition, the most corrosion electrodes is one made of copper, which supported by SEM EDX result, that has CuO concentration of 60,77%. Otherwise, from the three of materials, stainless steel is obtained as the most stainless material.

1. Introduction

Ozone was first discovered by an European researcher, C.F. Schonbein, in 1839. He identified the smell, that arises from anode during the water electrolysis, as a new compound, named 'ozone'. Ozone was first used commercially in 1907 to provide good water, and then used in 1910 in St. Petersburg [1]
Ozone consists of oxygen atoms with the chemical symbol O_3. Ozone is a relatively unstable molecule when compared with oxygen O_2. Ozone can be generated using ultraviolet radiation, photochemical reaction and dielectric barrier discharge (DBD). Nowadays, DBD known as the most effective method for generating ozone [2]. DBD is a type of non-thermal plasma that generally produced using two electrodes which separated by a gap of several millimetres and covered by a dielectric layer. Both of

electrodes are connected with AC (alternative current) high voltage. Dielectric serves as a current limiter, preventing the formation of spark and distribute the discharge throughout the electrodes area[3]. DBD system is generally consists of a power supply, electrodes and dielectric barrier. According to the operating frequency of power supply, DBD is generated using alternative current (AC), radio frequency (RF), and pulse mode. Power supply and type of reactor provide a great effect on the characteristics of DBD. Reactor types can be classified by the electrodes configuration, electrode material, dielectric barrier material and the presence of catalysts between the electrodes. In addition, dielectric barrier materials used as insulation layer are glass, quartz, ceramic, and polymer coating. The type of material, thickness and surface structure of the dielectric material may affect the plasma discharge [4].

2. Research Methods

Figure 1 shows scheme of experiment set up of this study. The components and size of reactor that used are shown in table 1.

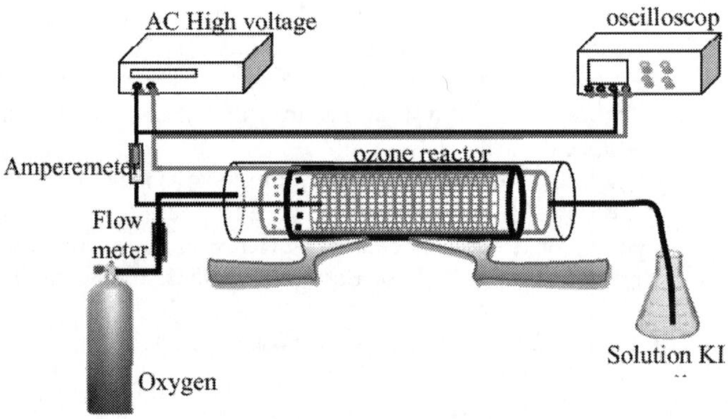

Figure 1. Scheme of experiment set up

Table 1. Component and size of each component of ozone reactor.

No	name of parts	material	length (mm)	Diameter (mm)	Thickness (mm)
1	Dielectric	pyrex glass	225	50	1,1
2	Inner Electrode	stainless staindless, galvalum, and copper,	250	40	2
3	Outter Electrode	zinc plate	250	52	-
4	Discharge gap	-	250	-	2
5	Cover	flexion glas	22	-	3
6	Input line	Cu	20	-	-
7	Output line	Fe	46	-	-
8	Positive terminal	Fe	35	-	1
9	Negative terminal	Cu	-	-	-

Inner electrode in the form of cylindrical framework made of stainless steel ring which connected by the 12 galvalume strings, and the diameter of each cylinder frame was 40 mm. The outer electrode in cylindrical form was made of zinc plate with diameter of 52 mm, and the barrier was made of Pyrex glass of 1.1 mm in thickness and 440 mm in length. AC high voltage used as power supply was varieties of 2, 3, 4 and 5 kV, while oxygen flow rate is variants of 2; 3; 4; 5; 6; 8, and 10 L/minutes.

Ozone's concentration is equal with $Na_2S_2O_3$ solution's volume that used in titration [7][8]. Ozone's concentration can be calculated by the following formula, where M is ozone's mass, V is titration volume, N is sodium Thiosulfat's concentration, v_{O_2} is oxygen volume, and e is constant.

$$C_{O_3} = \frac{MVN}{v_{O_2} e}$$

3. Results and Discussions

Characteristic of Electric Current as Function of Voltage (I-V)

The characteristic of electric current as function of voltage has been conducted five variations of electode's lenght of 5; 10; 15; 20; and 25 cm, as shown in Figure 2. Beside it, applied voltage has been variated of 3; 4; and 5 kV. Figure 2 shows that the increasing of voltage gives the increasing of electric current. It is caused by the charge accumulated in reactor. The electrical charge is formed by the multi-scattering when electric fields accelerate the electrons so that the electrons have enough energy to ionize the oxygen molecules inside of reactor.

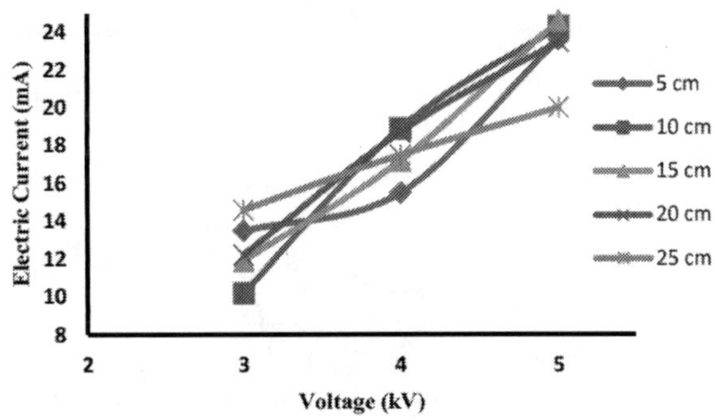

Figure 2. Electric current as function of voltage in some variations of reactor's lenght.

Effect of Voltage to Ozone Concentration.

The effect of voltage to ozone concentration in some variations of electrode's lenght can be seen in Figure 3. The greater the applied voltage provides the higher the potential between the electrodes, and then resulting the higher ozone concentration.

These results are consistent with the results obtained by Syafrudin [9]. The results showed that ozone concentrations increase with smaller flow rate and higher voltage. The higher the voltage, the oxygen molecules inside the reactor will increase, so there are more oxygen molecules collide. These results are consistent with the results obtained by Nur [10].

On the reactor with electrode length of 5 cm, the smallest concentration of 67.2 ppm is obtained at voltage of 3 kV, while the highest ozone concentration of 384 ppm is obtained at voltage of 5 kV. On the reactor with electrode length of 10 cm, the smallest concentration of 76.8 ppm is obtained at voltage of 3 kV, while the highest ozone concentration of 384 ppm is obtained at voltage of 5 kV. On the reactor with an electrode length of 15 cm, the smallest concentration of 76.8 ppm is obtained at voltage of 3 kV, while the highest ozone concentration of 432 ppm is obtained at voltage of 5 kV. On the reactor with an electrode length of 20 cm, the smallest concentration of 96 ppm is obtained at voltage of 3 kV, while the highest ozone concentration of 480 ppm is obtained at voltage of 5 kV. While on the reactor with an electrode length of 25 cm, the smallest concentration of 96 ppm is obtained at voltage of 3 kV, while the highest ozone concentration of 624 ppm is obtained at voltage of 5 kV. Based on the results shown in figure 3, the applied voltage is affects the ozone concentration, the higher voltage provides the higher ozone concentration.

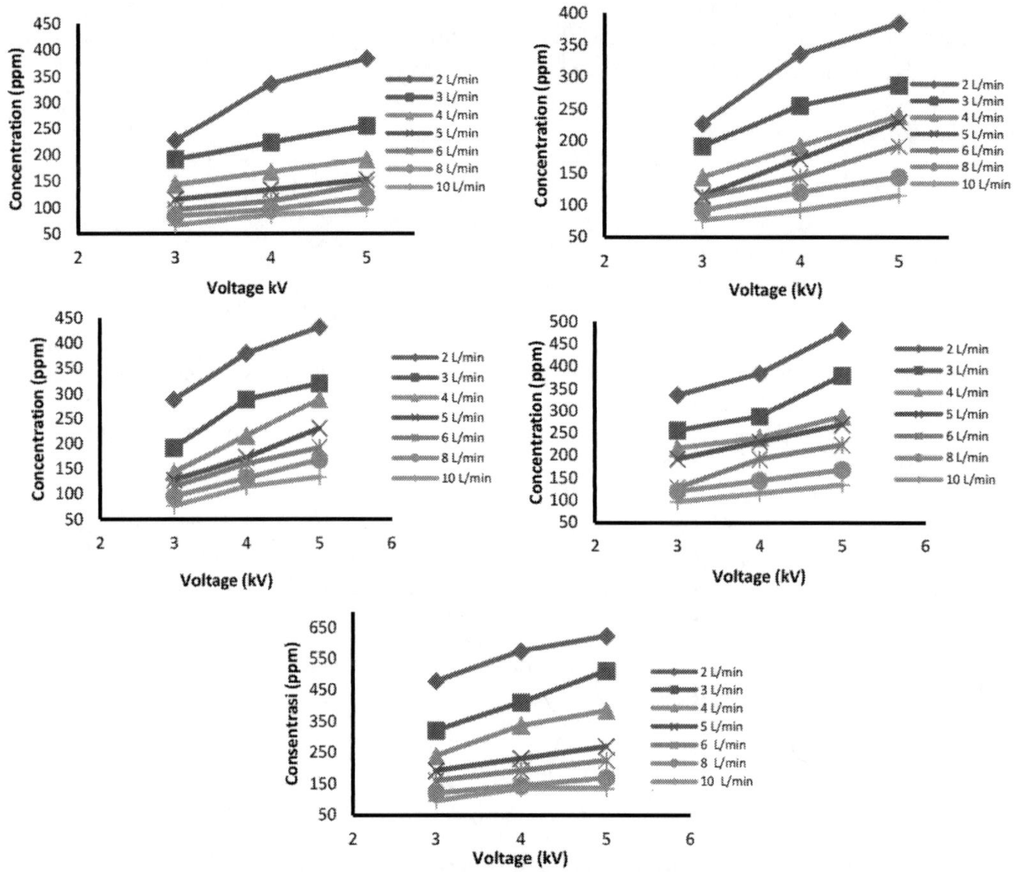

Figure 3. Ozone concentration as function of voltage in a reactor in lenght with (a) 5 (b) 10 cm, (c) 15 cm, (d) 20 cm, and (e) 25 cm.

Morphology Analysis

After being used to generate ozone, the materials used as inner electrode in ozone reactor were evaluated using SEM EDX analysis. The components of inner electrodes is shown in table 2. Morphological analysis aims to determine the surface image on corrosion parts. These parts are magnified by 500×, 1000×, 3000×, and 5000×. The greater magnification of the image shows significant changes on the morphological surface. The increased of surface roughness indicates the defects on sample surface due to corrosion.

The test results of the three electrodes made of stainless steel, galvalume, and copper is shown in Figure 4 for a magnification of 3000x, In the image morphology for galvalume and copper electrodes already seen experience corrosion. The most corrosion electrode is an electrode made of copper. Results the supported by SEM EDX analysis in which electrodes are made of copper with CuO amounted to 60.77%. Figure 4.a, b, and c show the morphological image of stainless steel, galvalume, and copper, respectively, that used as inner electrodes, with magnification of 3000x. The morphology image of galvalume and copper already seen being rusty or corrosion. The most corrosive material is copper. These results are supported by EDX analysis, where the CuO concentration obtained was 60,77 %. The higher magnification provides the better image of morphology.

Staindless Stell

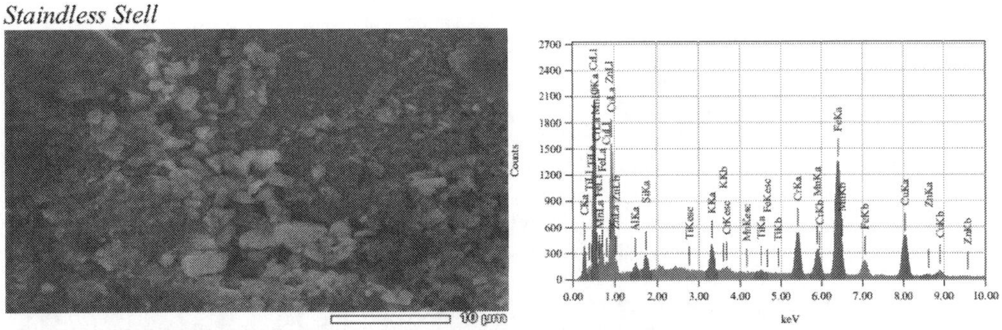

Figure 4a. SEM image of Stainless steel sample 3000x magnification

Galvalum

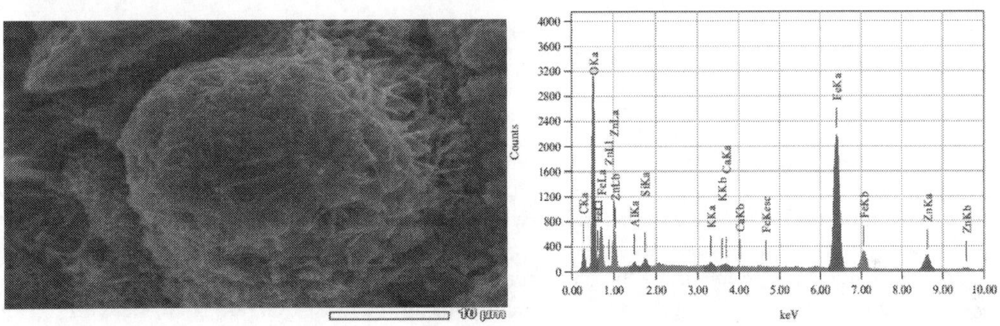

Figure 4b. SEM image of galvalum sampleto 3000x magnification

Copper

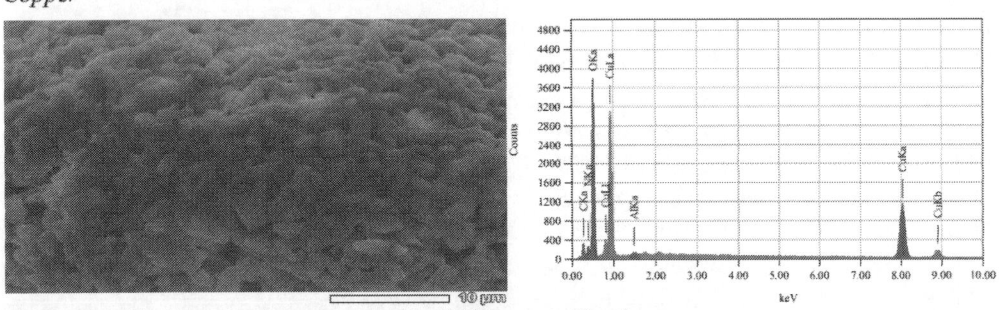

Figure 4c. SEM image of copper sampleto 3000x magnification

For example at a magnification of 5000x seen that copper electrodes in the form of beam, Octoludral, and polypoid (such as flowers). Similar results were also found in galvalume electrode. From the aspect of corrosively, stainless steel was obtained as the most stainless material to used as the inner electrode.

Table 2 Components and composition of inner electrodes in the used ozone reactor.

No	Sample	Component	Composition (% by weight)
1	Stainless steel	Carbon, C	19,39
		Alumina, Al_2O_3	0,90
		silica Dioxide, SiO_2	2,26
		potassium Oxide, K_2O	2,38
		Mercury (II) Chloride, TiO_2	0,57
		chrome oxide, Cr_2O_3	9,57
		manganese Oxide, MnO	5,18
		Iron (II) oxide, FeO	34,70
		Copper (II) oxide, CuO Zink	24,10
		Oxide, ZnO	0,95
2	Galvalume	Carbon, C	19,58
		Alumina, Al_2O_3	0,89
		silica Dioxide, SiO_2	1,62
		potassium Oxide, K_2O	0,60
		calcium Oxide, CaO	0,39
		Iron (II) oxide,FeO	62,06
		Zink Oxide, ZnO	14,85
3	Copper	Carbon, C	20,88
		Nitrogen, N	17,65
		Alumina, Al_2O_3	0,70
		Copper (II) oxide, CuO	60,77

4. Conclusion

The concentration of ozone is always higher with the smaller oxygen flow rate. The highest ozone concentration was obtained with flow rate of 2 L/min. The ozone concentration increases with the increasing of voltage. The length of the electrodes also give influence to the ozone concentration. The longer electrodes provide the higher ozone concentration. In this research, an electrode with the length of 25 cm produces the highest ozone concentration. Copper is the most corrosion materials as inner electrodes than the two others materials used in this research. From the aspect of corrosively, stainless steel was obtained as the best material for utilization as the inner electrode than galvalume and copper.

5. References

[1] Seydim, Z.G., Jr. P.I.B., Greene, A.K., 2004, *Food Microbiology,* 21, pp. 475–479.

[2] Fang Z, Qiu Y, Sun Y, Wang H, Edmund K,2008, *Journal of Electrostatics,* 66, pp. 421-426.

[3] Kim T.H, Kim s.j, 2011, *Sensors and Actuators,* A167, 297-303.

[4] Teke S, dkk., 2014, *Periodic Physics,* Vol 17, 1, pp. 25-31.

[5] Kogelschatz, U., Eliasson, B., Egli, W., 1997, *Journal de Physique* IV, 07 (C4), pp. 47-66.

[6] Nur, M., 2011, *Diponegoro University Press*, Semarang.

[7] Rakness, K., Gordon, G., Langlais, B., Masschelein, W., Matsumoto, N., Richard, Y., Robson, C.M., Somiya, I., 1996, *Ozone Science and Engineering,* Printed in the U.S.A., Vol. 18. pp. 209-229.

[8] Masschelein ,W.J, 1998, *Science and Engineering*, vol. 20, pp. 489-493.

[9] Syafrudin A, Novia., 2013, *Journal of Chemical Engineering,* Vol 19, 2, pp. 9-15.

[10] Nur, M, Restiwijaya, M., and Agustina, T,W., 2014, *IEEE Conference Publications*, pp 129-132

Synthesis and thermal characterization of Al_2O_3 nanoparticles

A. Ismardi[1*], O. M. Rosadi[1], M. R. Kirom[1] and D. G. Syarif[2]
[1]Engineering Physics Department, School of Electrical Engineering, Telkom University, Jl. Telekomunikasi Terusan Buah Batu, Bandung, Jawa Barat, Indonesia
[2]Nuclear Technology Center for Materials and Radiometry BATAN, Jl. Tamansari No. 71 Bandung, Jawa Barat, Indonesia

Email: *abrarselah@telkomuniversity.ac.id

Abstract. Al_2O_3 nanoparticle has been successfully synthesized using sol gel method from AlCl3. The obtained nanoparticles was then characterized for grain size measurement, the size of nanoparticles was 6 nm by using surface area meter (SAM) and Transmission Electron Microscopy (TEM). The crystallinity property of the product was then checked with XRD spectroscopy, the result shows that the diffraction peaks were match with the 10-0425 JCPDS database. Thermal property of the Al2O3 nanoparticles was then studied by mixing it with engine base fluid as nanofluid. The usage of nanofluid was expected to be heat absorber and woulo increase cooling process in cooling machine. The results showed that cooling time increases when the concentration of nanofluid was increased. Finally, it is concluded that thermal property of Al2O3 was studied and applicable to be mixed with engine coolant of cooler machine to reduce cooling time process.

1. Introduction

Recently, many researches about the application of nanosize material to enhance heat transfer process has already been introduced by the researcher, that is popularly known as nanofluid. A nanofluid is a mixture of base fluid with the solid nanoparticle to be used in some applicaton, for example electronic components[1], vehicles[2], nuclear reactor [3-5] and building cooler system[6]. So many metal and metal oxide compound have been synthesized and applied in several kind application. Ravi et.al [7] studied about the application of CuO nanoparticle as nanofluid, they concluded that CuO increase thermal conductivity and sensitivity, they also depend on concentration and different variation of base liquid. Metal oxide nanoparticle is the popular one compare to others since it has high value of thermal conductivity, easy to produce, lighter than dense nanoparticle which is not easier become an agglomeration[8]. Many research has been conducted in order to study the effective thermal conductivity, convective heat transfer and phase change in nanofluid. In this research, nanoparticle powder of Al_2O_3 were synthesized using a very simple, cheap and easier method, sol gel method. Nanoparticles obtained from the synthesis process was applied in a model of coolant engine to characterize their thermal property. As can be seen, the thermal conductivity increases when concentration increases. The effect of different concentration of nanoparticle in the nanofluid was studied, and it is found that higher amount of nanoparticle in a nanofluid absorbed heat more than less concentration.

2. Experimental Method

We divided this research into three sections, the first step was the synthesis of Al_2O_3 nanoparticle using sol-gel method, the second step was the characterization of the powder product, and the last section was the application of nanoparticle as nanofluid in a model of cooling engine system. To synthesis Al_2O_3 nanoparticle, $AlCl_3$ with 95%-98% purity purchased from Yixing Cleanwater Chemicals Co., Ltd. was used as the initial material. $AlCl_3$ powder was then processed by Sol-Gel method, 30 g $AlCl_3$ powder was mixed with 160 g sucrose ($C_{12}H_{22}O_{11}$) as a binder in this method. The mixture of $AlCl_3$ with sucrose were then heating up in the furnace with temperature of 900°C for seven hours. After seven hours heating, there were white powder product found in the ceramic beaker and predicted as the Al_2O_3 nanoparticle. The result product were then characterised for their properties. X-Ray Difraction (XRD) spectroscopy, Surface Area Meter (SAM) and Transmission Electron Microscopy (TEM) were then used for studying the crystalline property, particle size and structure of the nanoparticle obtained, respectively. Al_2O_3 nanoparticle were then applied as nanofluid, it was mixed with base fluid for helping cooling process in a model cooling engine. The mixture of different concentration of Al_2O_3 nanoparticle were then studied and characterised for their thermal conductivity in a cooling engine model.

3. Results and Analysis

Al_2O_3 nanoparticle have been synthesized using sol-gel method. Several characterization have been used to characterize the material obtained from the process. Firstly, to ensure the size of nanoparticle, SAM Quantachrome NovaWin v.11.03 have been used for this purpose by using Brunauer-Emmett-Teller (BET) methods. Surface area for Al_2O_3 nanoparticle was 240.624 m^2/g. The average particle diameter is obtained by assuming the particles is spherical according to the Equation 1 :

$$D = \frac{6}{\rho \times As} \quad (1)$$

Where:

ρ = density of Al_2O_3 (g/cm^3)
A_s = surface area of Al_2O_3 (m^2/g)

By using above equation it can be calculated the size of Al_2O_3 nanoparticle, the result was 6.26 nm. From this result it is confirmed that the result from sol-gel process was in the nanosize. TEM using JEOL JEM 1400 was then used to double confirm the size and structure of the obtained nanoparticle and shown in the Figure 1. From the TEM image it is believed the size of the nanoparticle was less than 20 nm, it is confirmed that sol gel method was successfully done to synthesis the nanoparticle, even though some agglomeration between the particle still occurred.

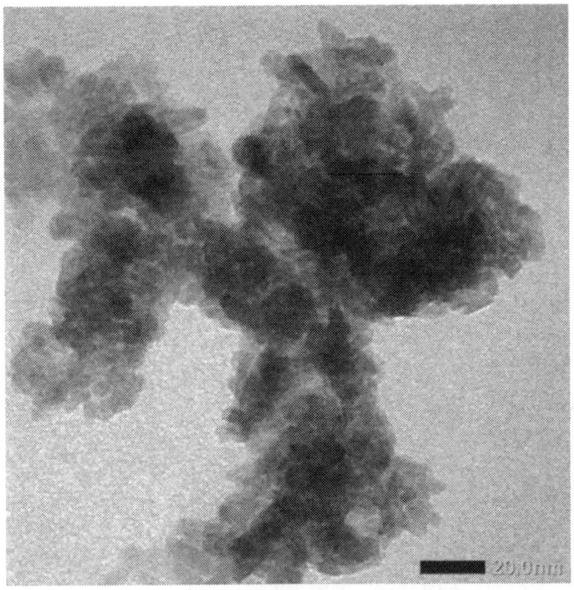

Figure 1. TEM image of Al_2O_3 nanoparticle shown that the size of nanoparticle less than 20 nm and the agglomeration still occurred among them.

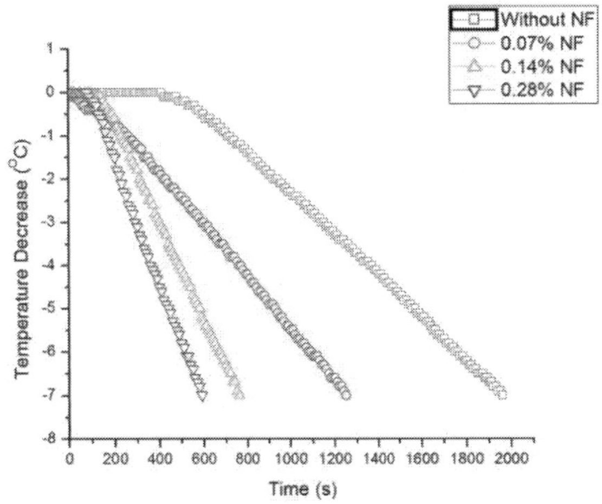

Figure 3. Decreasing time by using different concentration of Al_2O_3 nanoparticle in the nanofluid.

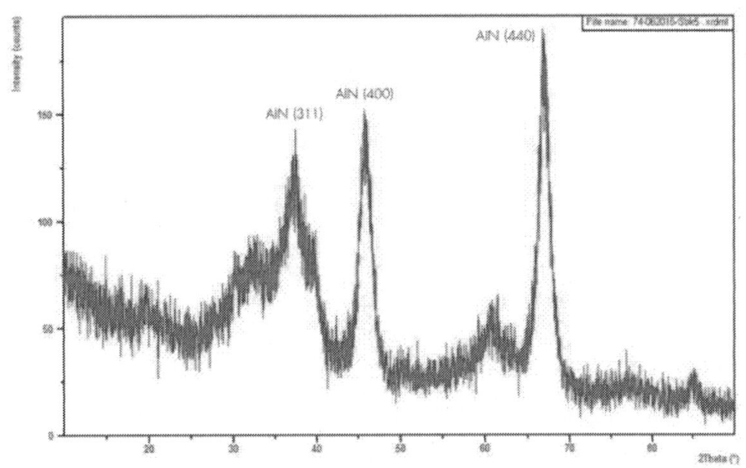

Figure 2. XRD spectrum of the obtained Al_2O_3 nanoparticle using sol gel method match with the 10-0425 JCPDS database.

Crystallinity of the nanoparticle was then characterized with XRD and the result was depicted in Figure 2 above. From the figure it is clearly shown that the obtain particle was a crystal, it is confirmed from the high and narrow peaks of the diffraction spectrum. All the peaks were belong to Al_2O_3 and matched with the 10-0425 JCPDS database.

After doing some characterizations, the nanoparticle was then mixed with base liquid to be used in a model of cooling engine. The performance of cooling engine was then tested by adding different concentration of Al_2O_3 nanoparticle into base fluid of cooling engine and later we can say as nanofluid. Different concentration of Al_2O_3 nanofluid was then applied to decrease 7°C the temperature of cooling engine. The concentration used in this experiment was 0 %, 0.07 %, 0.14 % and 0.28 % and the result shown in Figure 3 below. From the figure it is clearly depicted that the decreasing temperature was depend on the concentration of the nanofluid. When the concentration of Al_2O_3 increased, the decreasing rate will become higher. To decrease 7°C of cooling engine without using nanofluid it is needed around 2000 s, after adding 0.07 % the decreasing time was around 1300 s, and then for 0.14 % and 0.28 % it was around 800 and 600 s respectively.

The shorten time to decrease cooling engine's temperature was predicted by the different physical properties of the nanofluid used comparing to the pure base fluid itself. We believed that the addition of Al_2O_3 nanoparticle would change thermal characteristic of the nanofluid. The usage of nanosize particle in the nanofluid would increase surface area of the Al_2O_3, since the heat absorption property of the Al_2O_3 would take part in this process[9-11]. As the concentration increase, meaning that the available surface area also increase and resulted shorter time to reduce temperature.

Power cooling cycle usage in the system was also would be determined and show in the Table 1. During the experimental process, power can be measured from the system as we compare from input and output power. Power was obtained by measuring power that goes into the system by using a watt meter, since the required power usage is directly proportional to the time. It is shown that more concentration used will applied less power.

Table 1. Electrical power used for the each experimental process based on the concentration of Al_2O_3 nanoparticle

	a	b	c	d
Concentration of Al_2O_3 nanoparticle	0 %	0.07 %	0.14 %	0.21 %
Electric Power usage (kWH)	0.0613	0.032	0.0211	0.0183

The performance of cooling engine was then defined by comparing output and input power. Input and output power defined as the input of electrical power and output thermal power, respectively, and the result shown in Table 4.2 below. It is shown that the performance of the engine was also increase by increasing the concentration of Al_2O_3 nanoparticle.

Table 2. Performance index of the cooling engine directly depend on the concentration of Al_2O_3 nanoparticle

	a	b	c	d
Concentration of Al_2O_3 nanoparticle	0 %	0.07 %	0.14 %	0.21 %
Performance index	0,199	0,383	0,580	0,667

Effective thermal conductivity can be understood by considering all the data obtained, and can be calculated by using the Equation 2. The effective thermal conductivity plotted in the Figure 4 below. From the figure it is clearly understood that the effective thermal conductivity was also increase by increasing the concentration of Al_2O_3 nanoparticle. Actually many parameters will also take part in process, such as viscosity, boiling point of the mixing material, the present of stabilisation agent and so on[12]. In this paper, we just study about the effect of using different concentration of base fluid without considering others parameter.

$$\frac{keff}{kf} = \frac{kp + 2kf - 2\alpha\,(kf - kp)}{kp + 2kf + 2\alpha\,(kf - kp)} \quad (2)$$

Where:

k_{eff} = Effective Thermal Conductivity
k_f = Thermal Conductivity of base fluids (Lubricant)
k_p = Thermal conductivity of Al_2O_3 (90% purity)
$\alpha = k/(1+k)$; k = Concentration ratio between the volume of base fluid and nanoparticles Al_2O_3.

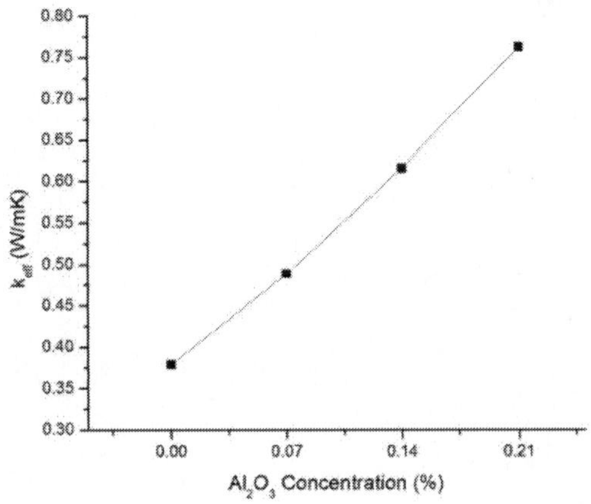

Figure 4. Effective thermal conductivity of the different concentration of Al_2O_3 nanoparticle

4. Conclusion

Al_2O_3 nanoparticle have been successfully synthesized using sol gel method. Some characterization revealed that the size of nanoparticle was around 6 nm, while it has also a high crystallinity structure. Nanofluid, the mixture between Al_2O_3 nanoparticle and base fluid has been applied to reduce the temperature of an engine cooling model system. Thermal properties of nanofluid was proved linearly proportional to the concentration of Al_2O_3 nanoparticle, the higher concentration in nanofluid would have shorter time to reduce temperature. This mechanism was indicated that higher concentration of Al_2O_3 nanoparticle supply higher surface area of the material that resulted higher thermal absorption occurred during the cooling process.

5. Acknowledgement

Authors wishing to acknowledge to Telkom University for the laboratory facilities and for the financial support under the grant of Penelitian Dana Internal Batch 2 2015-2016.

6. References

[1] Vijayakumar, M., P. Navaneethakrishnan, and G. Kumaresan, 2016, *Experimental Thermal and Fluid Science*,. **79**: p. 25-35.
[2] Sidik, N.A.C., M.N.A.W.M. Yazid, and R. Mamat, 2015, *International Communications in Heat and Mass Transfer*, **68**: p. 85-90.
[3] Nematolahi, M., B. Behzadinejad, and A. Golestani, 2015, *International Journal of Hydrogen Energy*,. **40**(44): p. 15192-15197.

[4] Zarifi, E. and G. Jahanfarnia, 2014, *Progress in Nuclear Energy*, **73**: p. 140-152.
[5] Safarzadeh, O., et al., 2014, *Annals of Nuclear Energy*,. **65**: p. 72-77.
[6] Abu-Hamdeh, N.H. and K.H. Almitani,. 2016, *Solar Energy,* **134**: p. 202-210.
[7] Agarwal, R., et al., 2016, *Applied Thermal Engineering*,. **102**: p. 1024-1036.
[8] Ghanbarpour, M., E. Bitaraf Haghigi, and R. Khodabandeh,. 2016, *Experimental Thermal and Fluid Science*, 2014. **53**: p. 227-235.
[9] Bigdeli, M.B., et al.,. *Renewable and Sustainable Energy Reviews*,. **60**: p. 1615-1633.
[10] Sundar, L.S., et al., 2016. *International Communications in Heat and Mass Transfer*, 2016. **76**: p. 245-255.
[11] Azmi, W.H., et al, *International Communications in Heat and Mass Transfer*, **75**: p. 13-23.
[12] Das, S.K., N. Putra, and W. Roetzel,. 2003, *International Journal of Heat and Mass Transfer*,. **46**(5): p. 851-862.

Mid – infrared transmission of polycrystalline (LaSr)(MnNi)O_3

W. D. Laksanawati[1], **B. Kurniawan**[1], **S. A. Saptari**[2]

[1]Department of Physics, Faculty of Mathematics and Natural Sciences, Universitas Indonesia, Depok 16424, Indonesia
[2]Faculty of Science and Technologyy, Universitas Islam Negeri Syarif Hidayatullah, Jakarta 15412, Indonesia

E-mail : winamovic@yahoo.com

Abstract. Polycrystalline (LaSr)(MnNi)O_3 was shintesized using sol gel methods with nitrat precursors $La(NO_3)_3$, $Sr(NO_3)_2$, $Mn(NO_3)_2.4H_2O$, and $Ni(NO_3)_2.6H_2O$ and the different heating process. Sample (LaSr)(MnNi)O_3 with chemical formulation $La_{0,67}Sr_{0,33}Mn_{1-x}Ni_xO_3$ with x = 0,05 and 0,10. We report the crystallite structure of $La_{0,67}Sr_{0,33}Mn_{1-x}Ni_xO_3$ with x= 0,00 and 0,10 are single phase with characterization by X-ray diffraction. Refinement has result that crystallite size of $La_{0,67}Sr_{0,33}Mn_{0,95}Ni_{0,05}O_3$ is 24,67 and $La_{0,67}Sr_{0,33}Mn_{0,9}Ni_{0,1}O_3$ is 21,84 with crystallite system rombohedral, it show us that increasing at Ni composition influence of decreased crystallite size. Sampel (LaSr)(MnNi)O_3 has been characterization with Fourier Transform Infrared with range of wave number from 450 to 4000 cm^{-1} were chategories at mid infrared wave. The FTIR pattern show to us that the Mn-O-Mn bounded has absorp infrared at wave number 605 cm^{-1} and the dominant peak at wave number 3750 cm^{-1} caused the hidroxy compound in sampel $La_{0,67}Sr_{0,33}Mn_{0,95}Ni_{0,05}O_3$.

1. Introduction

The last few years, doped alkaline earth metals such as Sr on the site La in material LaMnO3 make changes in properties of the insulator paramagnetic to the metal magnetic and doped metals transition such as Ni, Fe and Co in the material LaSrMn3 that was previously a metal magnetic turned into is antiferromagnetic with the effect of colossal magneto-resistance (CMR), it indicates a high application potential in the field of electronic magnetic material. In addition, the electromagnetic characteristics of the oxide material make it a good absorber of microwaves. [1] The electromagnetic wave absorbent material has a criterion that high resistivity and magnetic low. This relates to the mechanism Double Exchange (DE) centered on the bond Mn-O-Mn, a third transition metal Mn doped to the site would damage the magnetic properties of materials and the mechanism Double Exchange so that the resistivity of the material will increase [2]. In this study, we want to know whether the functional groups Mn-O-Mn which acts as a central Double Exchange actually formed on the material, therefore the testing Fourier Transform Infrared (FTIR) Spectroscopy with a range of wave number between 450 cm-1 to 4000 cm-1 in the category Mid Infrared.

2. Experimental details

Samples $La_{0,67}Sr_{0,33}Mn_{1-x}Ni_xO_3$ with variations in the value of x = 0.05 and 0.10 is made by sol gel method. Precursor materials used include La(NO3) $_3$, Sr $(NO_3)_2$, $Mn(NO_3)_2.4H_2O$, and Ni $(NO_3)_2.6H_2O$ with scales on each material according to stoichiometric. All the precursor material aquabidest nitrate dissolved, then the solution is mixed and stirred using a magnetic rod with a constant speed 300rpm and the temperature was maintained at 80^0 C. To thicken the solution and make the solution pH around 7 then added a solution of ammonia. When the solution has been stirred for 3 hours, viscous solution will turn into a paste that is increasingly losing the element of water. Once the water content is reduced and the magnetic rod can not rotate anymore, then the samples are in the glass beaker 120^0 put in the oven for 2 hours to remove the water content in the samples. The sample is dried and then calcined at a temperature of 550^0 for 8 hours to remove the element of nitrate and citrate, the sample will be issued black fluffy. Samples black-gray and then crushed using a mortar until it forms a perfect powder, after which the samples through a sintering stage at a temperature of 850^0 for 5 hours, the samples were out of this stage has not changed much from the previous phase. Powder sample is then characterized by XRD and FTIR.

3. Results and discussion

Samples have been synthesized by sol gel method then characterized by XRD and FTIR to learn about percentage transmission toward mid infrared wave number.

3.1. Structure characterization

Refinement of the XRD characterization results showed that the samples are samples of x = 0,05 and x = 0,10 already has a single phase with rhombohedral crystal structure, there is no phase changes that occur in both samples that can be seen in Figure 1.

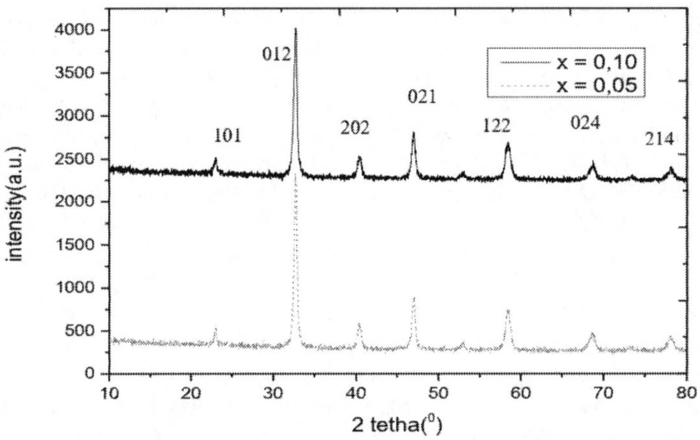

Figure 1. XRD Pattern of $La_{0,67}Sr_{0,33}Mn_{1-x}Ni_xO_3$ with x = 0,05 and 0,10

Refinement result of samples $La_{0,67}Sr_{0,33}Mn_{1-x}Ni_xO_3$ show that these samples are single phase with crystallite structure rombohedral in space group R-3c, The increase in doping Ni on samples $La_{0,67}Sr_{0,33}Mn_{1-x}Ni_xO_3$ resulting lattice parameter decreases and the crystal size also decreased, namely on the sample x = 0,05 at 24,67 and x = 0,10 at 21,84.

3.2. Fourier Transform Infrared (FTIR) measurement

Samples $La_{0,67}Sr_{0,33}Mn_{1-x}Ni_xO_3$ that single phase then characterized by FTIR. This characterization has purpose to know group of Mn-O-Mn bond formed in samples. Mn-O-Mn bond very influence in mechanism double exhange. IR spectra in samples $La_{0,67}Sr_{0,33}Mn_{1-x}Ni_xO_3$ showed at figure 2 below.

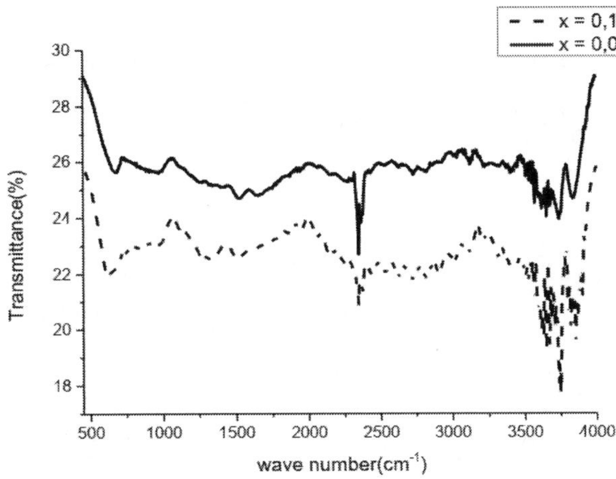

Figure 1. FTIR Spectra of $La_{0,67}Sr_{0,33}Mn_{1-x}Ni_xO_3$ with x = 0,05 and 0,10

Characterization by FTIR returned to see if the functional group Mn-O-Mn has been completely formed. FTIR results reported that the sample $La_{0,67}Sr_{0,33}Mn_{1-x}Ni_xO_3$ have formed the functional groups in the substance. Bond Mn-O-Mn vibrates at a frequency or wave 605 m^{-1} shown in Figure 2. In this discussion area into 3 wave numbers v1, v1 and v3. In v1 is between 550-650 cm^{-1} are formed bounded Mn-O-Mn, at wave number v2 2385-2650 cm^{-1} vibration caused by hydrogen bonds, and the v3 namely in the area wave number 3650-3595 cm^{-1}, in this case the vibrating namely because the hydroxy group [3]. Contradiction in sample x = 0,05 and sample x = 0,10 is in the transmittance percentage, in sample x = 0,05 about 25,68% and sample x = 0,10 is about 21,99%, its mean that if doped Ni increased, so percent transmittance decreased.

4. Conclusions

Samples $La_{0,67}Sr_{0,33}Mn_{1-x}Ni_xO_3$ has been made with sol gel method with the results of refinement in the form of single-phase samples with rhombohedral crystal structure and crystal size decreases with increasing value of x . FTIR results in both samples showed that the functional groups have been formed Mn - O - Mn (metal oxygen Bound) at the wave number of 605cm^{-1}.

5. Acknowledgement
We sincerely thank to University of Indonesia for partially funding this research through PITTA grant.

6. References
[1] Z. Shengming, Zhou; Guien, Zhou; Yuheng, "Infrared Spectra of La0.67Sr x MnO3-δ (x=0.33 and 0.23) around Curie Temperature," *J. Appl. Phys.*, vol. 5075, pp. 5075–5077, 1999
[2] F. Gao, R. A. Lewis, X. L. Wang, and S. X. Dou, "F ar-infrared reflection and transmission of La 1 2," *Elsevier*, vol. 347, no. Alloy and Compound, pp. 314–318, 2002
[3] S. Ravi and A. Karthikeyan, "nanoparticles synthesized with modified sol-gel route,"*Phys. Procedia*, vol. 54, no. 2004, pp. 45–54, 2014

Structure and morphological analysis of various composition of yttrium doped-zirconia prepared from local zircon sand

F. Rahmawati[1*], I. Permadani[1], E. Heraldy[1], D. G. Syarif[2], S. Soepriyanto[3]

[1]Research Group of Solid State Chemistry & Catalysis, Chemistry Department, Sebelas Maret University, Jl. Ir. Sutami 36 A Kentingan Surakarta 57126.
[2]Pusat Sains dan Teknologi Nuklir Terapan (PSTNT) BATAN, Jl. Taman Sari 73, Bandung, Indonesia
[3]Faculty of Mining and Petrochemical, ITB, Jl. Ganesha 10, Bandung

Email: fitria@mipa.uns.ac.id

Abstract. Yttrium ions, Y^{3+} were doped into ZrO_2 that was synthesized from zircon sand. Zircon sand is a side product of tin mining plant in Bangka Island, Indonesia. Solid state reaction was chosen as the method to dope yttrium ions from Y_2O_3 into ZrO_2 at various % mol of 4.5; 8 and 10. This research aims to understand the crystal structure, morphological analysis and particle size analysis. The X-ray diffraction analysis equipped with Le Bail refinement found that the prepared ZrO_2 is in two phases of the monoclinic and tetragonal structure, and the structure changed to cubic after yttrium ions doping. However, the monoclinic and tetragonal still exist. Various yttrium concentrations provide different morphology, in which 4.5 YSZ shows a blocking phase indicated as the presence of impurities. The blocking phase seems to prevent sintering and allows a line crack on the material layer. Meanwhile, 8YSZ and 10YSZ show homogeneous morphology and without provides a line crack. The mean particle size after sintering is in between 1.1 – 1.5 μm.

1. Introduction

Indonesia has tin ore abudant spread over Pulau Karimun, Singkep, in some of the Sumatera lands, and also in Bangka Island, Riau islands, until to the west Kalimantan [1]. Among of those, Bangka island has the largest Tin abundant with a side product is zircon sand ($ZrSiO_4$). Zircon, $ZrSiO_4$, is a stabil molecule due to the strong bond between zirconia and silica. Some methods to extract zirconium from zircon are caustic fusion or with soda ash [2], heat plasma dissociation [3]. Zirconia, ZrO_2, is an oxygen ions conductor with high flexibility and high resistance [4]. It is used as a material for the oxygen sensor, fuel cell material, thermal barrier coating, and some high temperature applications. Tetragonal zirconia has high mechanical streng, meanwhile cubic zirconia has high ionic conductivity [5]. The addition of some metal oxides such as MnO, NiO, Cr_2O_3, Fe_2O_3, Y_2O_3 and Ce_2O_3 could stabilize tetragonally and cubic phase of zirconia at room temperature [6].

A high ionic conductivity of material is required to develop a solid oxide fuel cell, SOFC. Cubic zirconia is known to has highest ionic conductivity. However, the structure is not stable at room temperature. Doping of yttrium ions into zirconia is known to stabilize the cubic structure at room temperature and increase the ionic conductivity due to the formation of oxygen vacancies when Zr^{4+} is replaced by Y^{3+} [7]. Low yttrium concentration might produce tetragonal zirconia that has good mechanical strength, meanwhile high concentration of yttrium dopant might produce cubic structure zirconia that has high ionic conductivity [5]. In this research the zircon sand form Bangka island was used as raw material for zirconia production. Then the prepared zirconia, ZrO_2, was doped with yttrium ions, producing yttria stabilized-zirconia, YSZ, to enhance their electrical conductivity. The yttrium doping was conducted by solid state reaction method, as the most simple method to prepare some inorganic materials [8]. Analysis of structure and electrical conductivity were conducted to identify the prepared materialsat various yttrium ion concentration and to know their functional properties in comparison with the commercial YSZ.

2. Experiment

Zirconia, ZrO_2 was synthesized with caustic fusion method [9] in which the zircon sand concentrate, $ZrSiO_4$, was crushed with NaOH at a ratio of $ZrSiO_4$: NaOH 1:4. The mixture was then heated at 800 °C to produce a greyish white powder. The powder then was leached with distilled water at avolume ratio of 1:10 of powder to water, and followed by filtration. Next step was to leach residue of filtration with 3.5 M of HCl at a volume ratio of 1:10, means that 1 gram of residue was dissolved in 10 mL HCl, at 80 °C and uder stirred condition. The result was a cloudy yellow solution that produced a clear yellow solution after filtration. The yellow solution is $ZrOCl_2.6H_2O$ (ZOC). Zirconia, ZrO_2, was preciptated from ZOC by slowly added 3 M of NH_4OH. A white precipitate was produced. After decantation and heating at 800 °C for 5 hours, a white powder of ZrO_2 wasfounded.

A mixture of Y_2O_3 and ZrO_2 was crushed for 2 hours in a pestle at a stoichiometric ratio to produce a 4.5 %, 8 % and 10 % mol of yttrium. Calcination was at 1000 °C for 2 hours, then followed by sintering of its green pellets at 1500 °C for 5 hours. X-ray diffraction analysis equipped with Le Bail refinement was used to identify the peaks in comparing with YSZ standard diffraction, and also to analyse its crystallinity, crystal structure, and its cell paremeters. The prepared materials also were analyzed by SEM to understand their surface morphology and were analyzed with LCR meter (GW Instek) to study their impedance plot and their electrical conductivity. The impedance was measured at 20 Hz – 5 MHz at 300 – 700 °C.

3. Result and Discussion

Synthesis of zirconia from $ZrSiO_4$ occurs in some steps to produce Na_2ZO_3, ZOC, $Zr(OH)_4$ and then ZrO_2. The reactions can be read in our previous publication [9]. The diffraction pattern of the prepared ZrO_2 in comparison with standard diffraction of tetragonal and monoclinic ZrO_2 is depicted in Figure 1. Compare to the standard diffraction, the diffraction peaks are broader indicating lower crystallinity. The specific peaks of ZrO_2 are identified at 2θ~30.30 °; 34.70 °; 35.24 °; 50.34 ° dan 60.26 °. Meanwhile a peak at 2θ~28.36 ° confirms the present of monoclinic phase ZrO_2 based on ICSD#157403.

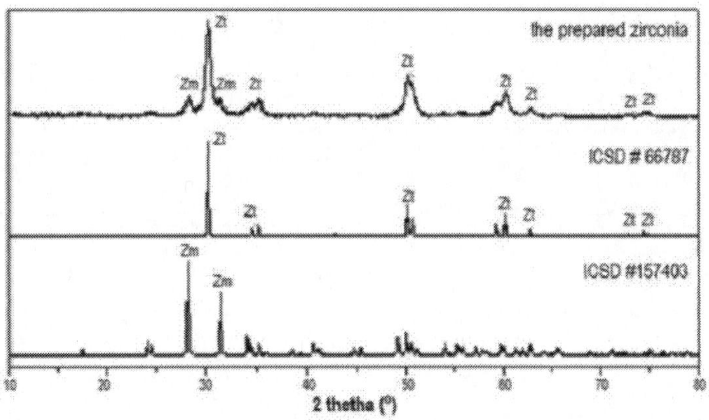

Figure 1. Diffraction pattern of the prepared zirconia compared to the standard diffraction of tetragonal zirconia, ICSD #66787 and monoclinic zirconia, ICSD#157403

A Le Bail refinement that was applied by RIETICA software confirmed the presence of those two phases of tetragonal (space group of *P 42/N M C z*) and monoclinic (space group of *P1 21/C 1*) with cell parameters as listed in Table 1. This result is in agreement with our previous result [9] and it confirms the reproducibility of the synthesis method. Doping of yttrium ions into zirconia was conducted by solid state reaction to produce YSZ at various % mol of yttrium. Two moles of Y^{3+} replaced Zr^{4+} at its crystal site to produce 1 mole vacancy, $V_O^{\bullet\bullet}$, to compensate the lack of positive charge. The stoichiometrical reaction is described in Kroger-Vink Notation (equation (1)).

Tabel 1. The Le Bail refinement result of the diffraction data of the prepared ZrO_2

Cell parameters	Zirconia	
	phase 1 (ZrO_2 m)	phase 2 (ZrO_2 t)
	ICSD#157403	ICSD#66787
	Monoklinik	Tetragonal
	P1 21/C 1	*P 42/ N M C z*
a (Å)	5.233428	3.598713
b (Å)	5.039176	3.598713
c (Å)	5.727037	5.193137
V (Å3)	150.172928	67.254959
α=β=γ	α=γ=90; β=96.12	90
Rp (%)	6.12	
Rwp (%)	6.93	

$$Y_2O_3 \xrightarrow{ZrO_2} 2\,Y'_{Zr} + 3O^x_o + V_o^{\bullet\bullet} \qquad (1)$$

The diffraction patterns of YSZ at the various composition of yttria are depicted in Figure 2.

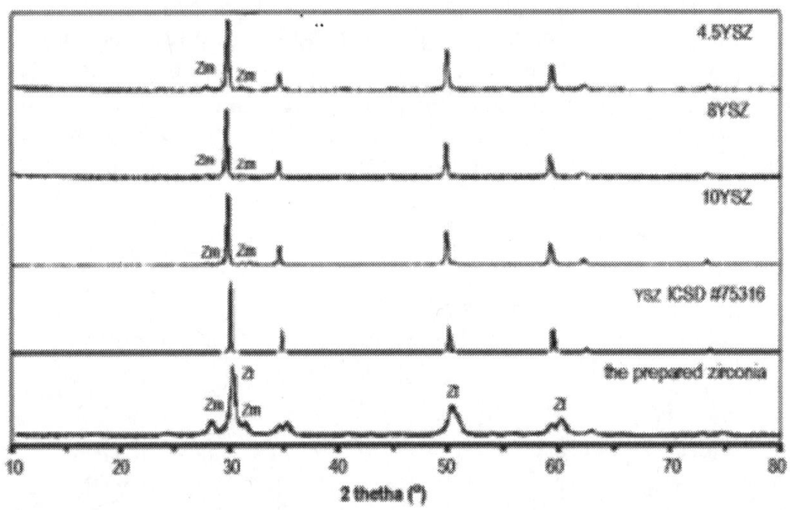

Figure 2.The diffraction patterns of YSZ at various composition.

Figure 2 shows that the materials have spesific peaks as YSZ ICSD#75316 at 2θ~30,8°; 34,92°; 50,15°; dan 59,52°. Meanwhile, small peaks at 2θ~27,91° dan 31,09° confirm the presence of monoclinic phases. It indicates that the yttrium doping did not eliminate the presence of monoclinic zirconia in the prepared zirconia from zircon sand. Le Bail refinement confirms that the materials are in cubic phase with the existence of tetragonal and monoclinic phases. The refinement result is listed in Table 2,3, and 4. Meanwhile, the Le Bail plots are depicted in Figure 3. The refinement shows that yttrium doping into ZrO_2 allows cell parameter to decrease from 3.598713 (Å) into 5.135102 (Å), 5.162044 (Å), and 5.153061(Å) for 4.5YSZ, 8YSZ, and 10YSZ, respectively. It indicates the replacement of Zr^{4+} by Y^{3+}, in which the atomic size is smaller than Zr^{4+}.

Tabel 2. Cell parameters of 4.5 YSZ as resulted from Le Bail refinement by applying monoclinic,tetragonal and cubic phases.

Cell parameters	phase 1 (ZrO2 m)	phase 2 (ZrO2 t)	Phase 3 (4,5YSZ)
	ICSD#157403	ICSD#66787	ICSD#75316
	Monoclinic	Tetragonal	cubic
	P1 21/C 1	*P 42/N M C z*	*F M 3 M*
a (Å)	5.137639	3.631561	5.135102
b (Å)	5.223372	3.631561	5.135102
c (Å)	5.345530	5.164413	5.135102
V (Å3)	141.690506	68.109474	135.408920
α=β=γ	β=98.987 ; α=γ=90	90	90
Rp (%)		6.92	
Rwp (%)		7.47	

Table 3. Cell parameters of 8YSZ as resulted from Le Bail refinement by applying monoclinic, tetragonal and cubic phases.

Cell parameters	phase 1 (ZrO2 m)	phase 2 (ZrO2 t)	phase 3 (8YSZ)
	ICSD#157403	ICSD#66787	ICSD#75316
	Monoclinic	Tetragonal	cubic
	P1 21/C 1	*P 42/N M C z*	*F M 3 M*
a (Å)	5.177034	3.643075	5.162044
b (Å)	5.218873	3.643075	5.162044
c (Å)	5.309533	5.166026	5.162044
V (Å3)	141.691360	68.563469	137.551392
α=β=γ	β=98.99; α=γ=90	90	90
Rp (%)		9.87	
Rwp (%)		14.93	

Table 4. Cell parameters of 10YS as resulted by Le Bail refinement by applying monoclinic, tetragonal and cubic phases.

Cell parameters	phase 1 (ZrO2 m)	phase 2 (ZrO2 t)	phase 3 (10YSZ)
	ICSD#157403	ICSD#66787	ICSD#75316
	Monoclinic	Tetragonal	cubic
	P1 21/C 1	*P 42/N M C z*	*F M 3 M*
a (Å)	5.157311	3.598692	5.153061
b (Å)	5.213131	3.598692	5.153061
c (Å)	5.349006	5.168529	5.153061
V (Å3)	142.194382	66.935455	136.834610
α=β=γ	β=98.60; α=γ=90	90	90
Rp (%)		9.63	
Rwp (%)		15.32	

SEM analysis (Figure 4) on the prepared materials shows a different morphology between that various composition. A Software, MeasureIT (a free edition) that was used to analyse the SEM images found that particle size of 4.5YSZ is 1.58 ± 0.544 μm. Meanwhile, the particle size of 8YSZ and 10YSZ are 2.021 ± 0.716 μm and 1.529 ± 0.498 μm, respectively. A histogram was developed to understand the highest probability particle size, and it was in between 1.1 – 1.5 μm.
Figure 4(a) shows a heteregenous surface morphology with the presence of blocking phases, as shown by a white circle in Figure 4(a) and clearly shown in Figure 5. Those blocking phase may indicate the presence of silica rich surface, as it also found by our previous research [10] in the preparation of doped-ZrO_2 and composite of ZrO_2-LSGM8282 in which the blocking phase presented when the ZrO_2 used was prepared from Indonesian local zircon.

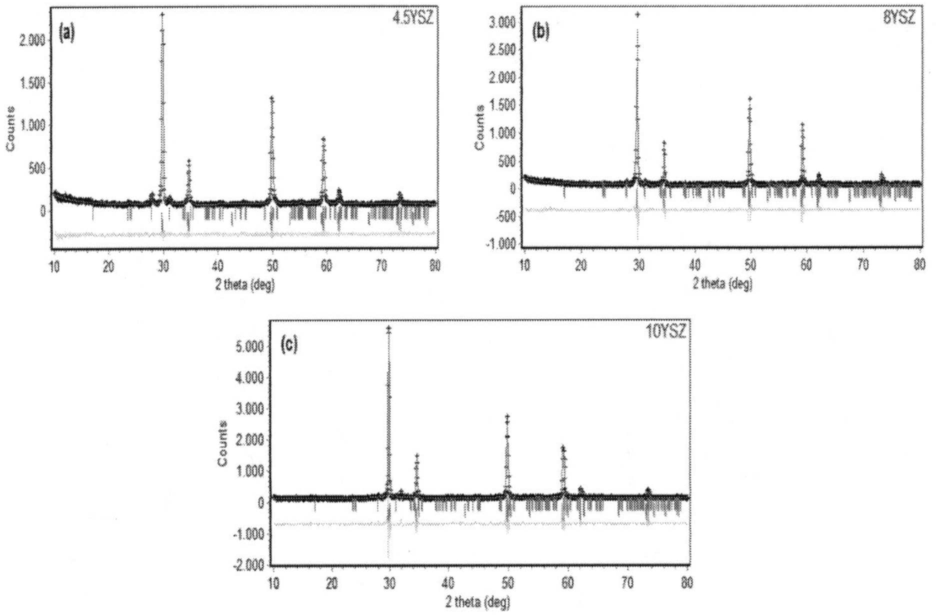

Figure 3. The Le Bail plots of 4.5YSZ(a), 8YSZ(b), dan 10YSZ(c) as resulted from refinement by applying 3 phases of tetragonal, monoclinic and cubic. + isthe experimental data, there line is calculation line, and the green line is different between data and calculation.

Those blocking phase was not found in the composite prepared from commercial ZrO_2. The XRF analysis of ZrO_2 that was synthized from local zircon found the presence of 3.03 % of silica and 14.71 % of sodium oxide. The sodium oxide content might come from NaOH used as destruction agent. It would release during the sintering process at 1500 °C for 5 hours. Liu et al.[11] also found such impurities when producing zirconia from Brazillian zircon. The sodium silicate is formed during mixing process between zircon sand and NaOH, and it hydrolized into H_2SiO_3 during water leaching. Caustic fussion process also results in Na_2ZrSiO_5 which then reacts with HCl during acid leaching to produce $ZrO(OH)_2$-SiO_2 and NaCl.$ZrO(OH)_2$. As the final result the composition of ZrO_2, SiO_2, and Na_2O were 57.169%, 6.77% and 7.049% , respectively[11]. The presence of blocking phase seems to prevent sintering process then it causes line crack as described by Figure 4(a2). Meanwhile, at a high content of yttrium ions of 8YSZ and 10YSZ the blocking phase and a line crack did not exist. It seems that the high content of yttrium ions prevent the impurities to leach out to the surface and inhibite the sintering process. A detailed study is required to understand the mechanism of those such preventing process and to study the material in term of their chemical composition at before and after sintering process, and the detail effect on grain and grain boundary ionic conduction.

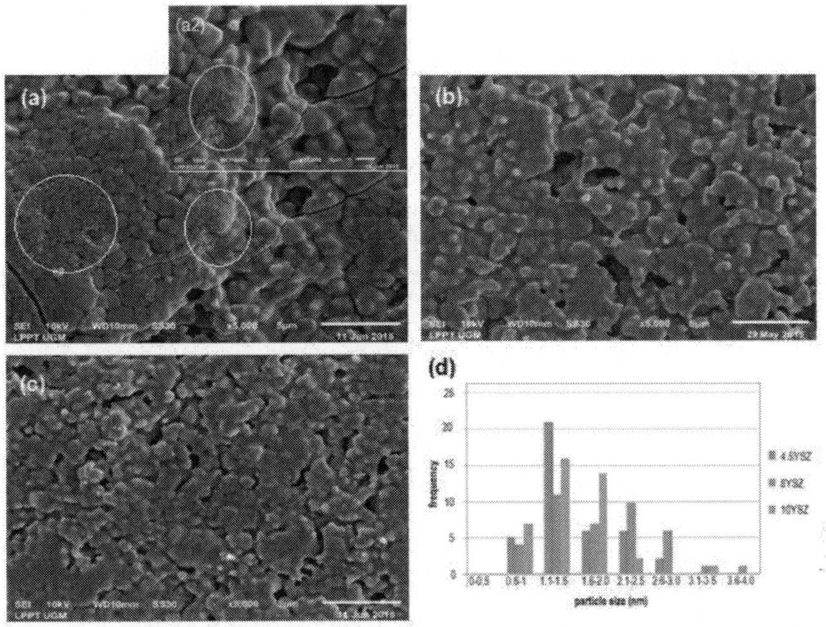

Figure 4. SEM images of 4.5YSZ(a) and 4.5YSZ at high magnification(a2) that showing a blocking phase and line crack, 8YSZ(b), and 10YSZ (c), and the hystogram of particle size distribution (d)

4. Conclusion

Yttrium stabilized zirconia that was prepared from local zircon sand consist of three phases, i.e cubic, monoclinic and tetragonal structure. A high content of yttrium, i.e 8YSZ and 10YSZ show a homogeneus morphology. Meanwhile, the 4.5YSZ shows a blocking phase indicating the presence of silica rich phase. This blocking phase seems to prevent the sintering process that leads to forming a line crack on the material layer. The particle size is ranging from 1.1 – 1.5 μm.

5. References

[1] Sekilas PT Timah Tbk, http://www.timah.com/v3/ina/tentang-kami-sekilas-pt-timah/, access date is July 23, 2016
[2] El-Barawy, K A, El Tawil S Z, Francis A A 2000 *Transaction of the Institute of Mining and Metallurgical Sect. C* **109** 49
[3] Jeong-Gu Y, Sung-Churl C, Jae-Won K, Jae-Ean L, Je-Hyun L, Yeon-Gil, J, 2004 *Mater. Sci. Eng. A* **368** 94
[4] Rivai A K, Takahashi M 2010 *Journal of Nuclear Materials* **398** 160
[5] Shackelford J F, Doremus R H 2008 *Ceramic and Glass Materials Structure, Properties and Processing*, Springer, New York
[6] Munggaran G, Fitriyani D, Rivai A K 2014 *Sintesis Bahan YSZ Dengan Metode Reaksi Padatan dan Karakterisasinya* Jurusan Fisika: Universitas Andalas
[7] Shakthinathan G, Raju S, Chandrasekaren U 2012 *Thermal Science* **16** 481
[8] Ismunandar 2006 *Padatan Oksida Logam, Struktur, Sintesis dan Sifat-sifatnya* Bandung: Institut Teknologi Bandung
[9] Rahmawati F, Permadani I, Soepriyanto S, Syarief D G, Heraldy, E 2014 *Double Step Leaching and Filtration in Caustic Fusion Method To Produce Zirconia From Local Zircon Concentrate*, 7th International Conferences on Physics and its applications (ICOPIA 2014)
[10] Rahmawati F, Prijamboedi B, Soepriyanto S, Ismunandar 2012, International Journal of Minerals, Metallurgy and Materials **19** (9),
[11] Liu R, Qu J, Song J, Qi T, Du A 2014, Ceramics International **40** 1431-1438.

The influence of narrow optical gap silver oxide on zinc oxide nanoparticles produced by microwave-assisted colloidal synthesis: photocatalytic studies

S. P. Prakoso[1], V. Paramarta[1,2], H. Tju[1,2], A. Taufik[1,2] and R. Saleh[1,2*]

[1] Integrated Laboratory of Energy and Environment, Fakultas MIPA-Universitas Indonesia, Kampus UI Depok, Depok 16424, Indonesia
[2] Departemen Fisika, Fakultas MIPA-Universitas Indonesia, Kampus UI Depok, Depok 16424, Indonesia

*E-mail: rosari.saleh@ui.ac.id

Abstract. This paper reports a photocatalytic study on wide band gap zinc oxide (ZnO) incorporated by narrow band gap silver oxide (Ag_2O), namely Ag_2O/ZnO nanocomposites, which were prepared by colloidal synthesis with microwave supports. The Ag_2O/ZnO nanocomposites were prepared with three different molar ratios (MR) of Ag_2O to ZnO (MR: 0.25, 0.5 and 0.75). In order to confirm qualitatively the concentration ratio of Ag_2O in ZnO, crystal phase intensity ratio was executed by peak indexing from x-ray diffraction. The Ag_2O/ZnO nanocomposites properties were further investigated using diffuse reflectance spectroscopy. The nanocomposites were tested for the degradation of organic dyes solutions under visible and UV light irradiations. The photocatalytic activity of Ag_2O/ZnO nanocomposites under visible light increased with increasing molar ratio of Ag_2O to ZnO, while the opposite trends observed under UV light irradiation. The improvement of photo-absorption together with photocatalytic activities might be suspected due to the p-n heterojunction structure in Ag_2O/ZnO nanocomposites. The corresponding mechanism will be discussed in detail.

1. Introduction

In deteriorating water pollutants like organic dyes from waste water of textile industries, researchers have been attempting to find a way by photodegradation process. Photocatalysts in this process are assisting in disintegrating heavy and harmful organic molecules into environment friendly and harmless species. By the aid of oxide semiconductor as a photocatalyst, radical groups (·OH and O_2^-) that are produced by photogenerated electron-hole pairs attack the organic pollutants [1]–[3]. Hence, oxide semiconductors are deemed as a suitable material for photocatalyst owing to their optical gap energies in absorbing photon energy.

Light absorption efficiency rather depends on electronic structure of its semiconductors which is correlated to the band gap energy (Eg). Some interesting facts from oxide semiconductors, its Eg could be tailored in several ways as follows: doping [4], compositing [5], thermal treatment [6], plasma treatment [7], and so on. Composites are the simple one from synthesis point of view among of aforementioned. Moreover, composites are consisted of two or more compounds mixed together in certain comparable amounts for every composer. Thus, normally composites exhibit their combination properties instead of the enhancement from one composer. In regards with the improvement of light absorption for photocatalyst, the unification within wide and narrow band gap oxide semiconductor is a quite strategy. A well-known n-type oxide semiconductor and having wide band gap about 3.3 eV [8] named zinc oxide (ZnO) is widely deployed in the diverse applications. Nevertheless, due to its wide optical gap property, the range of photon energies that could be promoting photogeneration of electron-hole pairs is limited to nearly around UV area. Despite of the broad wavelength that provided from solar spectrum, only small part could be absorbed by ZnO. Therefore, the presence of narrow band gap oxide semiconductors i.e. silver oxide (Ag_2O) might be broadens the light absorption up to the visible region. Ag_2O is possessing an optical gap about 1.4 eV and constitute of hole carrier which is remarkably compatible with ZnO whether as a composite or to form the p-n junction [1], [9]–[11]. In addition, the photogeneration of electron-hole pairs usually come along with annihilation process. This recombination is principally declining photocatalytic activities as well as reducing the production of radical groups. The structure of p-n junction is one of many options to overcome the recombination issue that has been reported recently. Photodegradation rate of ZnO/Ag_2O heterostructures obtained by photochemical synthesis had been showing 27.4 and 15.6 times faster than pure ZnO under UV and visible irradiation, respectively [10]. Whereas, Kadam et al. [11] had been presenting similar results for higher photodegradation rate of ZnO/Ag_2O nanocomposites towards methyl orange as attained by precipitation method. Yet, the photocatalytic studies of Ag_2O/ZnO produced by colloidal synthesis with aid of microwave in one-pot are scantily available and constitute a facile way to obtain nanoparticles. Photocatalytic activities of Ag_2O/ZnO nanocomposites with various ratio of Ag_2O to ZnO were performed in this study. UV and visible light were utilized in photodegradation performance to reduce the concentration of methylene blue in water. Furthermore, characterizations of Ag_2O/ZnO nanocomposites had been done to support the exhibition of photocatalytic improvement.

2. Experimental section

2.1. Synthesis and characterizations of Ag_2O/ZnO nanocomposites
All of precursors were used directly as received without any further purification. In proper chemical reaction, a certain amount of mineral salts ($AgNO_3$ and $ZnSO_4.7H_2O$) and sodium hydroxide (NaOH) were mixed in 200 ml distilled water. Sodium dodecyl sulphate (SDS) was incorporated into the solution as a stabilizing agent in 1:1 ratio to the mineral ions. This mixed-solution then stirred for 10 min to saturate the reaction. Afterward, the mixed-solution was heated-up through microwave heating with irradiation power of 800 W for 90 s. The precipitate then washed three times with distilled water and alcohol to remove the unwanted reaction products. The sample powders were acquired by drying the precipitate at temperature ~ 120°C for 5 hours in vacuum. In order to confirm qualitatively the concentration ratio of Ag_2O in ZnO, crystal phase intensity ratio was executed by peak indexing from x-ray diffraction. The Ag_2O/ZnO nanocomposite properties were further investigated using diffuse reflectance spectroscopy.

2.2. Photocatalytic experiments
Nano-hybrid Ag_2O/Ag catalysts were implemented to degrade methylene blue (MB) which is chosen as a model of organic pollutant in water. Such a 0.03 g of Ag_2O/Ag sample with different silver contents was dispersed in 100 ml MB solution (20 mg-L^{-1}) under constant pH of 13. Prior to the illumination, the suspensions of Ag_2O/Ag photocatalysts in the MB solution were magnetically stirred in the dark chamber for 30 min to assure the steady state of adsorption/desorption MB on the surface of catalyst particles. After that, the mixtures were illuminated by visible-light with power of 40 W for 120 min long. The sampling process was taken for every 15 min and the degraded MB solutions were centrifuged to separate the catalyst. The quantitative analysis of MB in the degraded MB solutions was evaluated by UV-Vis spectrophotometer (Hitachi UH5300) which is recorded the absorption spectrum in the range of 450 – 750 nm.

3. Results and discussion

The transformation of XRD spectra from hexagonal wurtzite-ZnO to cubic-Ag_2O with an addition of MR percentage is obviously observed in Figure 1. At 0%, diffraction pattern can be indexed merely pure ZnO phase which has three main peaks at 2θ of 31.7°, 34.3°, and 36.1° in accordance with hkl indices Miller plane of (100), (002), and (101) wurtzite structure [8], respectively. As the Ag_2O was introduced to the ZnO, intensively crystallographic planes of (111) and (200) cubic-Ag_2O phase begin to appear at 2θ of 32.8° and 38.2° [10], [11], respectively. In addition, either relative intensity or integrated area, they show an opposite trend following the MR percentages of Ag_2O to ZnO (see Table 1 and Table 2). It is indicating quite-well proportionality of molar ratio percentages with the yielded products of Ag_2O/ZnO nanocomposites by one-pot process in this study.

Figure 2 shows the optical reflectance characteristics of Ag_2O/ZnO nanocomposites with various percentages. A contrast trend among pure Ag_2O (x = 100%) and pure ZnO (x = 0%) is shown according to the recorded lowest and highest reflectance of 12.3% and 26% at 800 nm, respectively. The observable abrupt transition ca. 375 nm for Ag_2O/ZnO of 0% to 75% is owing to the excitation in ZnO region as well-known as ZnO has a wide band gap energy [8]. On the higher wavelength range for Ag_2O/ZnO of 25% to 100%, gradual increment was seen. Ag_2O was probably responsible for this transition due to its narrow window gap [9]. Both transition are obviously observed after deploying Kubelka-Munk function [8] as represented on inset Figure 2. Interestingly, those two stages were taken place solely on nanocomposites of 25% to 75%. As can be expected from nanocomposites which the Ag_2O and ZnO are resided as revealed from diffraction patterns (see Figure 1). Moreover, an increase of percentages, lower energy transition was blueshifted (1.34 eV to 1.71 eV) while higher energy transition redshifted (3.24 eV to 3.14 eV). Thus, an emerging of Ag_2O into ZnO gave some affections to their optical properties as demonstrated in Figure 2 and structural properties as listed in Table 1.

Figure 3 displays the photocatalytic activities toward methylene blue for different MRs of photocatalyst nanocomposites during exposure to (a) UV- and (b) visible-light at a given period. The maximum photodegradation rate under UV- and visible-light irradiation could be noticed for MR of 25% and 75%, respectively. Indeed, those photocatalysts showed exceptional performance for each UV- and visible-light irradiation. However, the photocatalyst nanocomposite with MR of 50% exhibits quite-well performance in both UV- and visible-light exposure. Pure ZnO (0%) and Ag_2O (100%) had

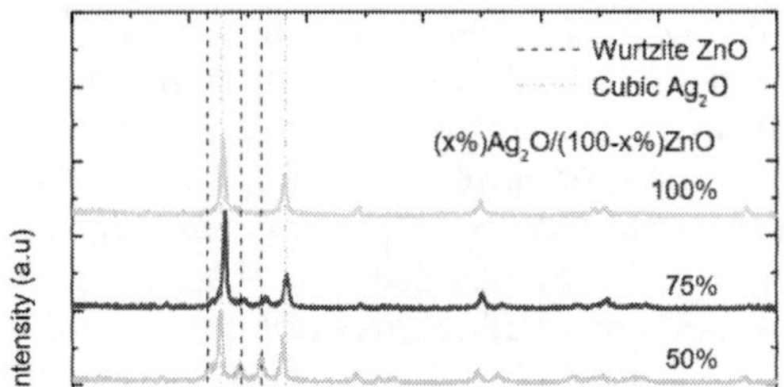

Figure 1. X-ray diffraction patterns from Ag_2O/ZnO nanoparticles with different composition percentages.

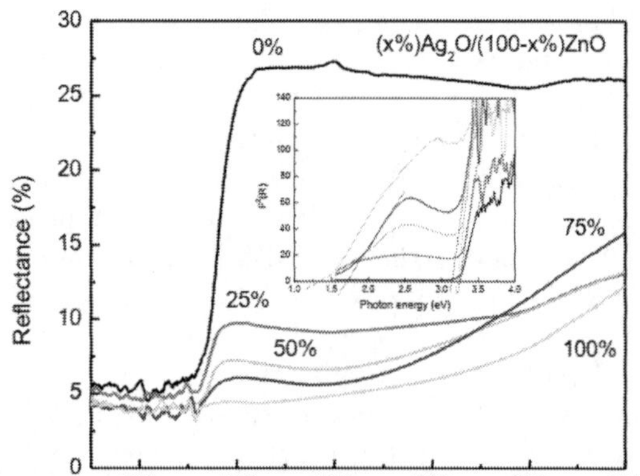

Figure 2. UV-Vis diffuse reflectance of Ag_2O/ZnO nanocomposites with various percentages (inset: Kubelka-Munk function).

Table 1. Crystal properties of ZnO in Ag_2O/ZnO nanocomposites.

Ag_2O (x%) ZnO (100%-x)	Cry. Phase (ZnO)	Crystal Properties				
		<D> (nm)	Rel. Intens (%)	Center (°)	FWHM (°)	Int. Area (%)
	(100)	22.8	27.79	31.65	0.482	28.79
0%	(002)	30.8	26.13	34.3	0.39	21.89
	(101)	22.1	46.08	36.13	0.498	49.31
	(100)	12.3	21.47	32.03	0.792	24.62
25%	(002)	21.5	21.74	34.63	0.507	15.98
	(101)	14.2	35.07	36.47	0.708	35.94
	(100)	8.5	9.57	31.97	1.094	19.38
50%	(002)	13.3	8.69	34.25	0.747	12.01
	(101)	16.1	15.31	36.14	0.638	18.09
	(100)	11.3	4.18	32.02	0.855	8.05
75%	(002)	14.5	5.38	34.65	0.694	8.4
	(101)	15.8	7.29	36.5	0.65	10.66
	(100)	-	-	-	-	-
100%	(002)	-	-	-	-	-
	(101)	-	-	-	-	-

Table 2. Crystal properties of Ag_2O in Ag_2O/ZnO nanocomposites.

Ag_2O (x%) ZnO (100%-x)	Cry. Phase (Ag_2O)	Crystal Properties				
		<D> (nm)	Rel. Intens (%)	Center (°)	FWHM (°)	Int. Area (%)
0%	(111)	-	-	-	-	-
	(200)	-	-	-	-	-
25%	(111)	13.9	6.4	32.83	0.716	6.64
	(200)	13.2	15.32	38.34	0.758	16.82
50%	(111)	31.4	41.03	32.66	0.384	29.17
	(200)	25.2	25.4	37.95	0.454	21.35
75%	(111)	37.9	62.26	33.01	0.339	47.47
	(200)	20	20.9	38.3	0.54	25.41
100%	(111)	25.8	67.85	32.84	0.441	67.04
	(200)	24.9	32.15	38.15	0.458	32.96

slightly lower photocatalytic activities than nanocomposite of 50% under UV- and visible-light, respectively. Nevertheless, they showed the worst photodegradation under reversely light source in accordance with their working optical gap energies. Optical studies in Figure 2 corroborated the efficiency of photocatalytic activities in Figure 3 as well as its photoabsorption. Concisely, wide wavelength optical absorption was achieved by photocatalyst Ag_2O/ZnO with MR of 50% as unveiled in Figure 3.

The photodegradation rate enhancement on Ag_2O/ZnO nanocomposites might be attributed to the p-n heterojunction structure that formed between p-type Ag_2O to n-type ZnO [10], [11]. A simple schematic energy band diagram for isolated p-Ag_2O and n-ZnO is depicted in Figure 4a. The appropriate light source is exciting electrons at valence band (VB) edge to the conduction band (CB) and created electron-hole pairs. And so does to the spontaneous recombination was taken place after illuminating source is shut-off. On the other hand, in case the p-n heterojunction is formed (see Figure 4b), it will cause to the band alignment of Fermi level energy due to the charge balance [12]. The energy band at p-Ag_2O site then bended upwards and at n-ZnO site tends to downward after the equilibrium state accomplished. Upon the photocatalytic activities under UV-light irradiation, ZnO

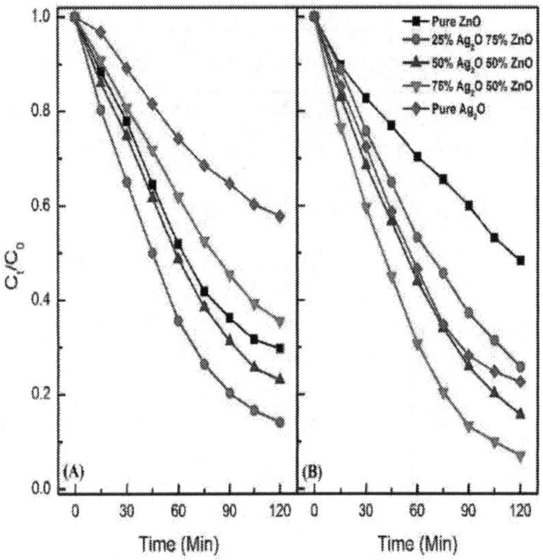

Figure 3. Photocatalysts performance in the reduction of methylene blue during exposure to (a) UV- and (b) visible-light.

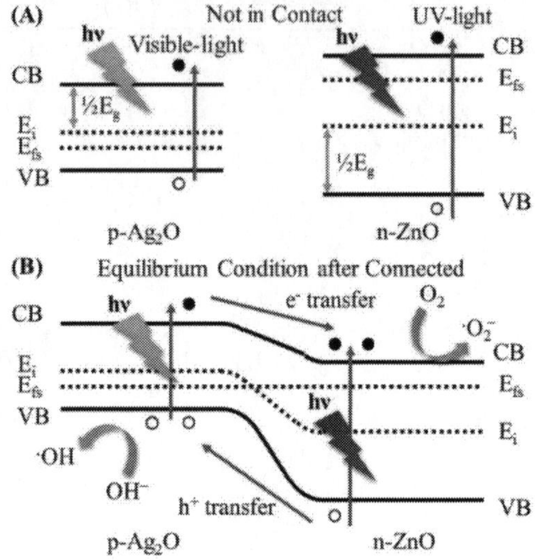

Figure 4. Schematic energy band diagrams of p-Ag_2O/n-ZnO heterojunction in case of (A) isolated p-Ag_2O and n-ZnO and (B) when p-Ag_2O/n-ZnO heterojunction established.

tends to absorb more photon energy rather than Ag_2O. The photogenerated electrons from ZnO site are preserved in CB while the photogenerated holes migrated to the VB of Ag_2O site since lower energy barrier exists within the depletion layer [10], [11], [13]. Therefore, the prolonged charge separation could be effectively improved by p-n heterojunction and lead to the enhancement of photocatalytic activities. Similar photogenerated carrier transport mechanism is happened in case the photocatalytic activities work under visible-light illumination as described in Figure 4b. In this case, the electron-hole pairs are generated dominantly in Ag_2O site. Excited electrons from Ag_2O are transferred to the CB of ZnO and the remained holes are left in VB of Ag_2O that later on can be reacted directly to the organic dyes or water molecules

4. Conclusion

The photocatalyst of Ag_2O/ZnO nanocomposites was successfully prepared by microwave-assisted colloidal synthesis. XRD results described the incorporation of Ag_2O in ZnO as followed by an increase of MR. Optical reflectance studied the influence of Ag_2O incorporation in ZnO. It showed the photo-absorption improved from visible to UV ranges. The photocatalytic activities of Ag_2O/ZnO nanocomposites under visible-light increased with increasing MR of Ag_2O to ZnO, while the opposite trends observed under UV light irradiation. Overall, the photocatalytic efficiency of Ag_2O/ZnO nanocomposites under UV- and visible-light irradiation is better than those of bare-Ag_2O as well as bare-ZnO due to p-n heterojunction structure well-formed in Ag_2O/ZnO nanocomposites.

5. References

[1] Abdullah H, Kuo D-H, Kuo Y-R, Yu F-A, and Cheng K-B *J. Phys. Chem. C*, **vol. 120**, no. 13, pp. 7144–7154, Apr. 2016.
[2] Sinha AK, Manna PK, Pradhan M, Mondal C, Yusuf SM, and Pal T *RSC Adv.*, **vol. 4**, no. 1, pp. 208–211, Nov. 2014.
[3] Wan X, Yuan M, Tie S, and Lan S *Appl. Surf. Sci.*, **vol. 277**, pp. 40–46, 2013.
[4] Saleh R, Prakoso SP, and Adel F *J. Magn. Magn. Mater.*, **vol. 324**, no. 5, pp. 665–670, 2012.
[5] Hendry T, Taufik A, and Saleh R 2016, p. 020027.
[6] Jiang Y-H, Chiu I-C, Kao P-K, He J-C, Wu Y-H, Yang Y-J, Hsu C-C, Cheng I-C, and Chen J-Z

Appl. Surf. Sci., **vol. 327**, pp. 358–363, Feb. 2015.
[7] Choi W-K *J. Vac. Sci. Technol. A Vacuum, Surfaces, Film.*, **vol. 14**, no. 2, p. 359, Mar. 1996.
[8] Prakoso SP and Saleh R *Mater. Sci. Appl.*, **vol. 3**, no. 8, pp. 530–537, 2012.
[9] Ida Y, Watase S, Shinagawa T, Watanabe M, Chigane M, Inaba M, Tasaka A, and Izaki M, *Chem. Mater.*, **vol. 20**, no. 4, pp. 1254–1256, Feb. 2008.
[10] Ma S, Xue J, Zhou Y, and Zhang Z *J. Mater. Chem. A*, **vol. 2**, no. 20, p. 7272, Apr. 2014.
[11] Kadam A, Dhabbe R, Gophane A, Sathe T, and Garadkar K *J. Photochem. Photobiol. B Biol.*, **vol. 154**, pp. 24–33, 2016.
[12] Sze SM and KK. Ng *Physics of Semiconductor Devices*. Wiley, 2006.
[13] Liu S, Wang N, Zhang Y, Li Y, Han Z, and Na P *J. Hazard. Mater.*, **vol. 284**, pp. 171–181, 2015.

Journal of Physics: Conference Series **776** (2016) 012052 doi:10.1088/1742-6596/776/1/012052

Properties of strontium doped barium titanate powder prepared by solid state reaction

A. Jamaluddin[1], Suwarni[2], A. Supriyanto[2], Y. Iriani[2*]

[1]Physics Education Department, Sebelas Maret University, Surakarta, Indonesia
[2]Physics Department, Sebelas Maret University, Surakarta, Indonesia

E-mail: * Yopen_2005@yahoo.com

Abstract. Strontium doping material has been successfully doped to produce Barium Strontium Titanate ($Ba_{1-x}Sr_xTiO_3$) with variation in the percentage of moles of strontium 1%, 2%, 3%, 4% and 5% that processed by the solid state reaction method. Firstly, Ingredients (Barium carbonate, strontium carbonate and titanium oxide) have been mixed, milled during 6 hours to form solid powders. Then, sample pellets were pressed by mold machine which have been sintered in a furnace at 1100 ^{0}C during 2 hours. The properties of sample were observed using X-Ray Diffraction (XRD) for analyzing crystalline structure, Scanning Electron Microscopy (SEM) for showing morphological properties, RCL meter for dielectric constant. The XRD results indicated that the diffraction pattern is Ba_{1-x} Sr_xTiO_3 (BST) pattern. In addition, the results of General Structure Analysis System (GSAS) refinement with the Rietveld method showed that strontium doping has entered into Barium Titanate ($BaTiO_3$). Increasing the number of moles of strontium has changed the lattice parameter and tertagonality of crystal BST. It affected the grain size and dielectric constant of BST too.

1. Introduction

Ferroelectric material has a large contribution on the electronics field such as storage system, Dynamic Random Access Memory (DRAM), phase shifter, sensor and actuator [1-5]. Especially for storage devices (capacitor, multilayer capacitor), ferroelectric material has benefit of storage because it has a higher capacity for storage [1]. The properties of ferroelectric material are shown by the dielectric constant that influenced capacity of storage. Furthermore, researcher improved material properties to achieve the highest dielectric constant. The ferroelectric materials are usually used for storage such as Barium Titanate (BT), Lead Zirconium Titanate (PZT), Bismute Zirconium Titanate (BZT) and Barium Strontium Titanate (BST) [1,4].

This paper presented a process of BT with doping strontium to produce BST using solid state reaction that will be applied as storage. Solid state reaction is a simple method with low cost manufacturing compare to sol-gel, co-preparation and spray phyrolysis [4,5,6,8]. The electronic properties of ferroelectric materials were influenced thermal process, microstructure and doping material [1,5,9]. This paper observed the properties of various moles strontium doping on BT to produce BST on sintering temperature 1100 ^{0}C.

Journal of Physics: Conference Series **776** (2016) 012052 doi:10.1088/1742-6596/776/1/012052

2. Method

The sample was processed by solid state reaction with various percent of strontium (Sr) doping 1%, 2%, 3%, 4% and 5%. Thermal process was done by furnace on temperature 1100°C by holding time during the 2 hours.
The doping process to produce BST was done by mixing powder materials $BaCO_3$, $SrCO_3$ and TiO_2 based on molar ratio which processed by grinding during six hours. The purpose of a grinding process is to reduce particle size and homogenize of BST powders. The bulk sample was produced by hydraulic press and dies with 0.016 m of diameter. Sintering process was done for clustering particle of powders by diffusion phenomena on increasing temperature [10]. Furthermore, the purpose of this process was to produce more solid and strongly particles bound of bulk samples.
International Commission for Data Diffraction (ICDD) database was used for fitting peak diffraction of XRD Result which matched with ICCDD number #741960. It has done for knowing the peak movement between experimental and theoretical data and also lattice parameter of material (a, b and c). Rietveld method using GSAS software was done for refinement to calculate lattice parameter (a, b and c) of doped and un-doped materials. The Surface and Morphological analysis werre investigated by Scanning Electronic Microscopy (SEM) Instrument. Moreover,The electronic properties of sample were observed with RCL meter for measuring capacitance and dissipation lose which measured in Vrms = 1 Volt with range frequency 1 kHz until 100 kHz.

3. Result and Discussion

The results of materials characterization using X-Ray Diffraction (XRD) instrument were peaks diffraction pattern. The relationship between the angle (2θ) and peak intensity of XRD results indicated that crystal planes were growth with a specific orientation. Figure 1 showed a diffraction pattern of Sr dopant on $BaTiO_3$. However, there are shown $BaCO_3$ powders as a precursor (*) on this pattern which indicated that it was not homogenize mixed.

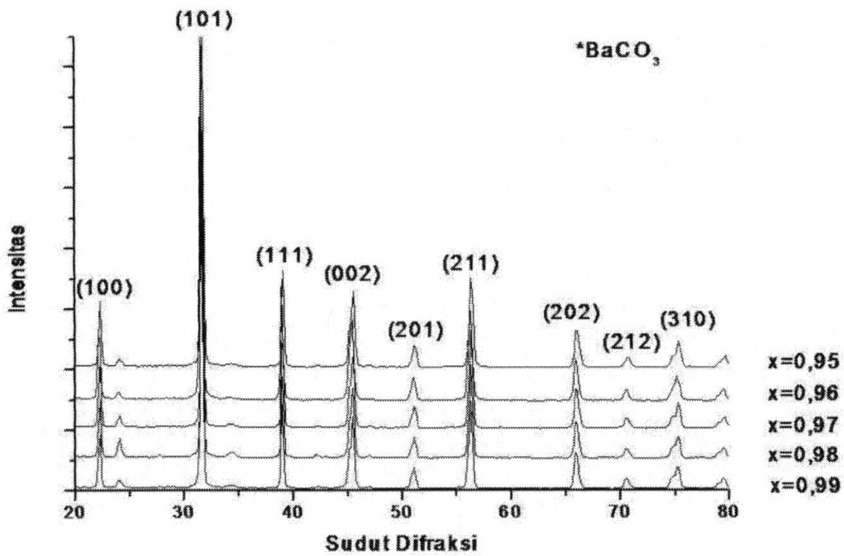

Figure 1. The diffraction pattern of various mol of Sr doped $BaTiO_3$

The influence of a percent of mole Sr doped $BaTiO_3$ effected to shift the diffraction angle (2θ) that is shown on Table 1 which adopted by third peak diffraction of crystal planes (101), (111), 211). The diffraction angle did not change on the percent of mole dopant 1 % to 4%. However, on 5% mole of dopant Sr shifted two peak diffraction of crystal plane (101) and (111).

Table 1. The top three peaks of Sr doped $BaTiO_3$

Crystal Planes	Angle Diffraction(2θ)				
	$Ba_{0.99}Sr_{0.01}TiO_3$	$Ba_{0.98}Sr_{0.02}TiO_3$	$Ba_{0.97}Sr_{0.03}TiO_3$	$Ba_{0.96}Sr_{0.04}TiO_3$	$Ba_{0.95}Sr_{0.05}TiO_3$
101	31.6035	31.6035	31.6035	31.6035	31.7535
111	38.9546	38.9546	38.9546	38.9546	39.1047
211	56.3574	56.3574	56.3574	56.3574	56.3574

The lattice parameters of refinement results were shown in Table 2. It indicated the tetragonal structure of crystalline. Increasing of percent moles of Sr dopant effected lattice parameter raised. Tetragonality value was c/a which showed in table 2. Each sample has tetragonality value more than 1 that indicates that sample have the tetragonal structure of crystalline. The effect of percent moles of Sr dopant on $BaTiO_3$ changed crystal size which shown in Table 2. The increasing of moles of Sr dopant grow up crystal size because the atomic diameters of Sr (1.12 Å) are smaller than a Ba (1.34 Å).

Table 2. Lattice Parameter (Å), tetragonality and Crystal Size of mole Sr dopant variation of $BaTiO_3$

Sampel	Latice parameter (Å)		Tetragonality	Crystal Size
	a = b	c		
$Ba_{0.99}Sr_{0.01}TiO_3$	3.9891	4.0168	1.0069	39
$Ba_{0.98}Sr_{0.02}TiO_3$	3.9888	4.0153	1.0066	59
$Ba_{0.97}Sr_{0.03}TiO_3$	3.9905	4.0187	1.0071	61
$Ba_{0.96}Sr_{0.04}TiO_3$	3.9883	4.0092	1.0052	83
$Ba_{0.95}Sr_{0.05}TiO_3$	3.9905	4.0179	1.0069	86

The Scanning Electron Microscopy (SEM) instrument with 20,000 X magnificent was used for morphological investigation of bulk BaxSr1-xTiO3 which showed on figure 3. The morphological surface of $BaxSr1\text{-}xTiO_3$ with variation of mole dopant Sr influenced grain size that increasing of mole dopant Sr affected grain size to be great and homogenous (Fig 3a). However, a little of mole Sr doped $BaTiO_3$ affected heterogenous grain size and presented holes (Fig 3e) because the mixing process of precursor powders is not consist perfectly yet. The grain size of BaxSr1-xTiO3 with Mole Sr is shown on table 3.

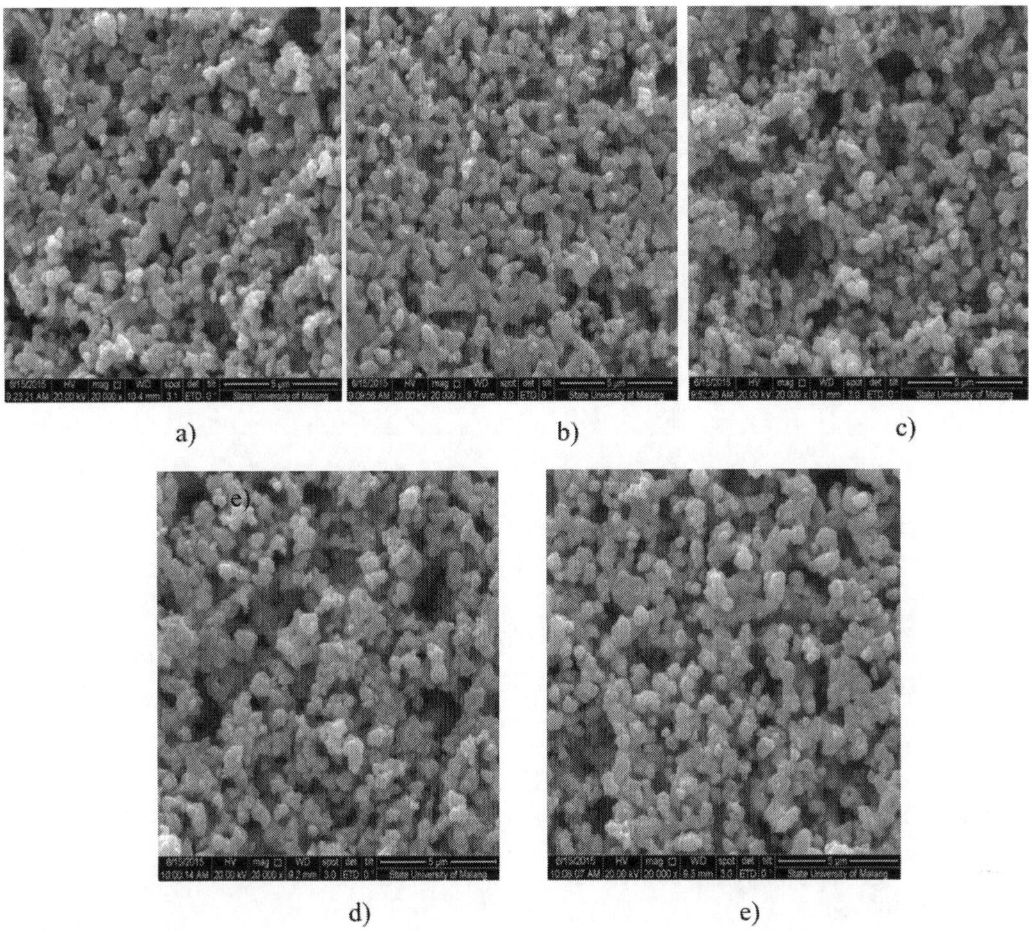

Figure 2. SEM result of various percent moles of Sr doped $BaTiO_3$

a) 1%, b) 2%, c) 3%, d) 4%, e) 5%

Table 3. Grain Size and dielectric constant with variant percent mole of Sr doped $BaTiO_3$

Composition	Grain Size (nm)	Dielectric Constant
$Ba_{0.99}Sr_{0.01}TiO_3$	218	317
$Ba_{0.98}Sr_{0.02}TiO_3$	231	306
$Ba_{0.97}Sr_{0.03}TiO_3$	232	224
$Ba_{0.96}Sr_{0.04}TiO_3$	235	243
$Ba_{0.95}Sr_{0.05}TiO_3$	243	456

The electronic properties $Ba_{1-x}Sr_xTiO_3$ were investigated by the RCL meter to observe the dielectric constant. The dielectric value of $Ba_{1-x}Sr_xTiO_3$ increased because of increasing mole of dopant Sr on $BaTiO_3$ that was showed on table 3. However, the dielectric constant dropped to 224 and 243 for 3% and 4% moles of Sr because there are some holes based on SEM data (fig 2 c and d). So, the density and surface of BST influenced the charge distribution that affected the value of dielectric constant [9]. The addition 5% moles of Sr dopant effected the dielectric constant increase in 456.

4. Conclusion

Increasing the number of moles of strontium has changed the lattice parameter and tertagonality of crystal BST. Furthermore, the grain size and the dielectric constant of BST have been changed because of the addition of moles strontium dopant. The dielectric constant increased to 456 for 5% moles of Sr doped BT for producing BST.

5. Acknowledgement

This project was supported by Hibah Pasca-PNBP project of Universitas Sebelas Maret, Surakarta, Indonesia.

6. References

[1] Vijatovic M M, Bobic J D and Stojanovic B D 2008 *Science of Sintering*. 40 155165
[2] Czekaj A L, Dus B W, Orkisz T, Adamczyk M, Osinska K, Kozielski L and Czekaj D 2007 *Material Science-Poland*. 25 791799.
[3] Scott JF 1998 *Integrated ferroelectrics*. 20 1523
[4] Jamaluddin A, Susilowati E, Budiawanti S and Iriani Y 2014 *Advanced Materials Research* 896 358362
[5] Hu Q, Jin L, Wang T, Li C, Xing Z and We X 2015 *Journal of Alloys and Compounds*. 640: 416420.
[6] Ko YN and KangY C 2012 *Ceramics International*. 38 20712077.
[7] Mahata M K, Kumar K and Rai V K 2014. *Spetrohimica Acta Part A*. 124 285291.
[8] Manzoor U and KimD K 2007 *J. Mater. Sci. Technol*. 23 5.
[9] Sandi K D, Supriyanto A, Jamaluddin A and Iriani Y 2016 *AIP Conf. Proc*. 1710 03000610300064
[10] Callister W D, 1994, *John Wiley & Sons. Inc*. New York

Journal of Physics: Conference Series **776** (2016) 012053 doi:10.1088/1742-6596/776/1/012053

FTIR studies on the effect of concentration of polyethylene glycol on polimerization of Shellac

Khairuddin[1]*, E. Pramono[2], S. B. Utomo[3], V. Wulandari[1], A. Zahrotul W[1], F. Clegg[4]

[1]Department of Physics FMIPA, Universitas Sebelas Maret, Jl. Ir. Sutami 36A, Surakarta 57126, Indonesia.

[2]Department of Chemistry FMIPA, Universitas Sebelas Maret, Jl. Ir. Sutami 36A, Surakarta 57126, Indonesia.

[3]Department of Chemistry FKIP, Universitas Sebelas Maret, Jl. Ir. Sutami 36A, Surakarta 57126, Indonesia.

[4]Materials and Engineering Research Institute, Sheffield Hallam University,Sheffield,S1 1WB, U.K

Email : khairuddin_70@yahoo.com

Abstract In the present paper, it was reported the FTIR studies on the efect of polyethylene glycol on polimerization of shellac. The shellac was shellac waxfree, and the solvent was ethanol 96 %. The shellac films were were prepared by solvent-evaporation method. The concentrations of polyethylene glycol having molecular weight of 400 were 10, 30, 60, and 90 w/w %. Three peak intensity bands of C= O stretching of ester at 1709 cm-1, O-H stretching of hydroxyl group at 3400 cm-1, and C-H stretching vibration at 2942 cm^{-1} were observed and related to polimerization of shellac. It was found that polymerization of shellac was slowed down by polyethyelene glycol, and the degree of polymerization of shellac decreased with increasing the concentration of polyethyelene glycol.

1. Introduction

Shellac is a natural polymer derived from resinous secretion of lac insects found in tropical countries, India and Thailand are major users and exporters of shellac [1]. In recent years, competition from Indonesia as a major shellac cultivating country has emerged [2]. Shellac has been utilised in many areas such as; food, medicine, military, electric appliances, printing ink, leather, plastic, metallurgy, machinery and wood varnishes [3]. Shellac has several natural characteristic advantages for use as packaging material or edible coatings, but also some disadvantages. The disadvantage of shellac is its stability, over time its molecular groups can undergo intra-molecular chemical bonding (polymerisation) which leads to changes in its properties, it becomes more brittle and less soluble. Adding plasticizer into natural polymer was one of the effective method to improve the properties of it. Plasticizer is a relatively low molecular weight molecules that can change the physical properties of polymers (e.g. flexibility) and make them more useful as a film-coating material [4]. The mechanistic action of plasticizers is by interposing themselves between individual polymer strands thus breaking down polymer-

polymer interactions. Its easier for the polymer strands to move past each other and become more pliable. PEG has been used as plasticizer for polymers such as PVOH, starch and shellac. It is non-toxic and used in pharmaceuticals compositions and as a food additive, it is thus suitable for food.applications [5]. The composisition, size and shape of a plasticzer and also its compatibility with polymer could affect the interaction between plasticizer and polymer, including its ability to atrract water [6]. The impact of plasticizers on a polymer system were gennerallay affected by factors such as small size, high polarity, more polar group per molecul, and greater distance betwen polar group within a molecule. In our previous studies, it was found that Polyethylene glycol (PEG) could improve stability of shellac. In the present work, it is reported the effect of concentration of PEG on polymerization of shellac. It was argued that the selection of a plasticizer for a spesific system was normally based on the compatibility and permanence of the plasticizer, the amount of necessity of plasticization, and the desired physical properties of the system [7].

2. Experiments

2.1 Materials

Shellac wax free was supplied by Sigma Aldrich UK. Alcohol 96 % was supplied by Merck Germany. Plasticizer polyethylene glycol was supplied by Sigma Aldrich UK.

2.2. Preparation of shellac and plasticised shellac films

All films presented were prepared by solution-casting. Shellac films were made by dissolving 1.25 gr of shellac in 12.5 ml ethanol (96%), while PEG/shellac films (10 wt%) were prepared by mixing and stirring shellac solution and PEG solution. The shellac solutions were made by dissolving 1.125 gr of shellac in 10 ml ethanol and stirred for 2 hours while PEG solutions were made by dissolving 0.125 gr of PEG in 2.5 ml ethanol and stirred for 2 hours using a magnetic bar/stirrer at room temperature.The solution was then cast into Petri dish, then dried at 50 °C for 7 hours before being analysed.

2.3. Characterization using Fourier Transform Infra Red (FTIR)

A Thermo Nicolet NEXUS spectrometer along with a MCT (mercury cadmium telluride) detector cooled by liquid nitrogen was used for all the FTIR analysis. The spectrometer was coupled with a single reflection diamond ATR cell (GrasebySpecac, U.K). This setup has the trade name "Golden Gate". The film was placed on top of the crystal diamond and the spectra of the films taken. The wavelength of interest ranges from 700 to 4000 cm^{-1} and was collected with resolution of 4 cm^{-1} and 64 scans. Omnic software was used to collect and analyse the spectra.

3. Results and Discusssions

The peak intensity of bands of PEG is assiged as O-H stretching of hyroxyl group at 3400 cm-1, C-H stretching of Alkanes at 2900 cm-1, C-H scissor and bending at 1450-1292 cm-1, C-O stretching of Alcohol at 1250 cm-1, and C-O-C at 1100-1060 of ether. The FTIR spectrum of the PEG shellac (10 wt%) system (Figure 1) was the superimposition of spectral patterns of PEG and shellac.

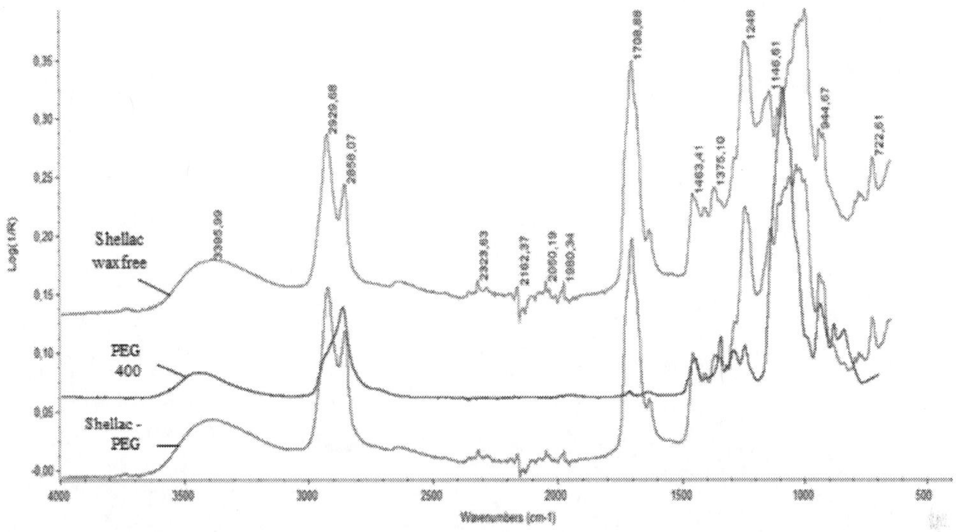

Figure 1. FTIR spectra of shellac, PEG400, and PEG400/shellac system

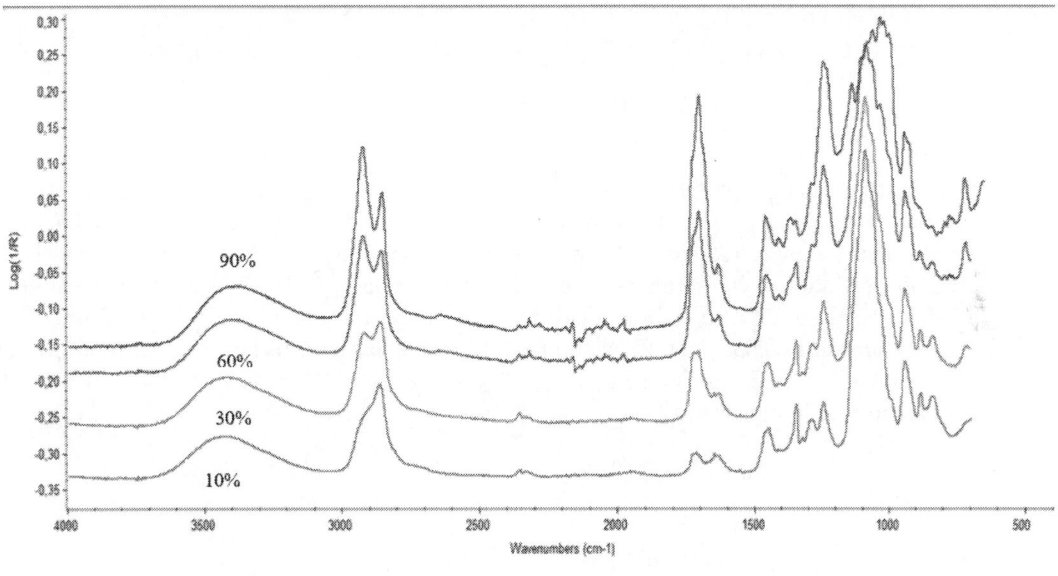

Figure 2. FTIR spectra of PEG400 shellac system with different concentration of PEG400

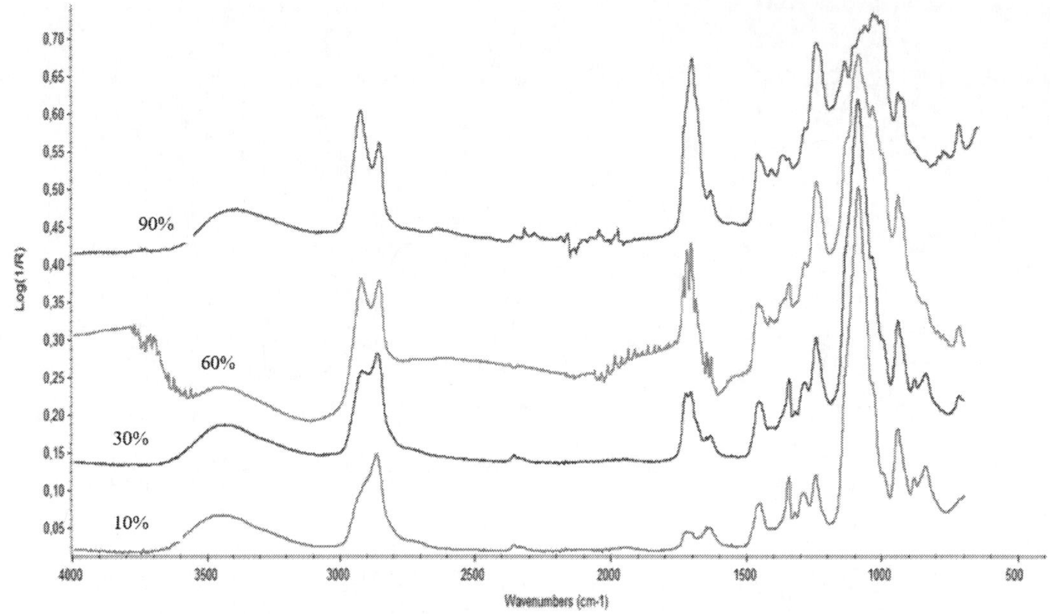

Figure 3. FTIR spectra of PEG400 shellac system with different concentration of PEG400 heated at 125 °C for 30 minutes

FTIR spectrum of PEG400 shellac system with different concentration (i.e, 10, 30, 60, and 90 wt%) with no heating treatment is shown Figure 2, while the FTIR spectrum of PEG400 shellac system when heating at 125 °C for 30 minutes is shown in Figure 3 . Peak intensity of O-H, C=O, and C-H of Figure 2 and Figure 3 are shown in Figure 4. It shows that increasing PEG concentration from 10%, to 90% resulted in the increase of peak intensity of O-H grop in PEG shellac system from 0.052, to 0.0754 (no heat in Figure 4a), but decreasing the peak intensity of C=O rom 0.15 to 0.05 (no heat in Figure 4b), while the peak intensity of C-H tended to be constant (no heat in Figure 4c).

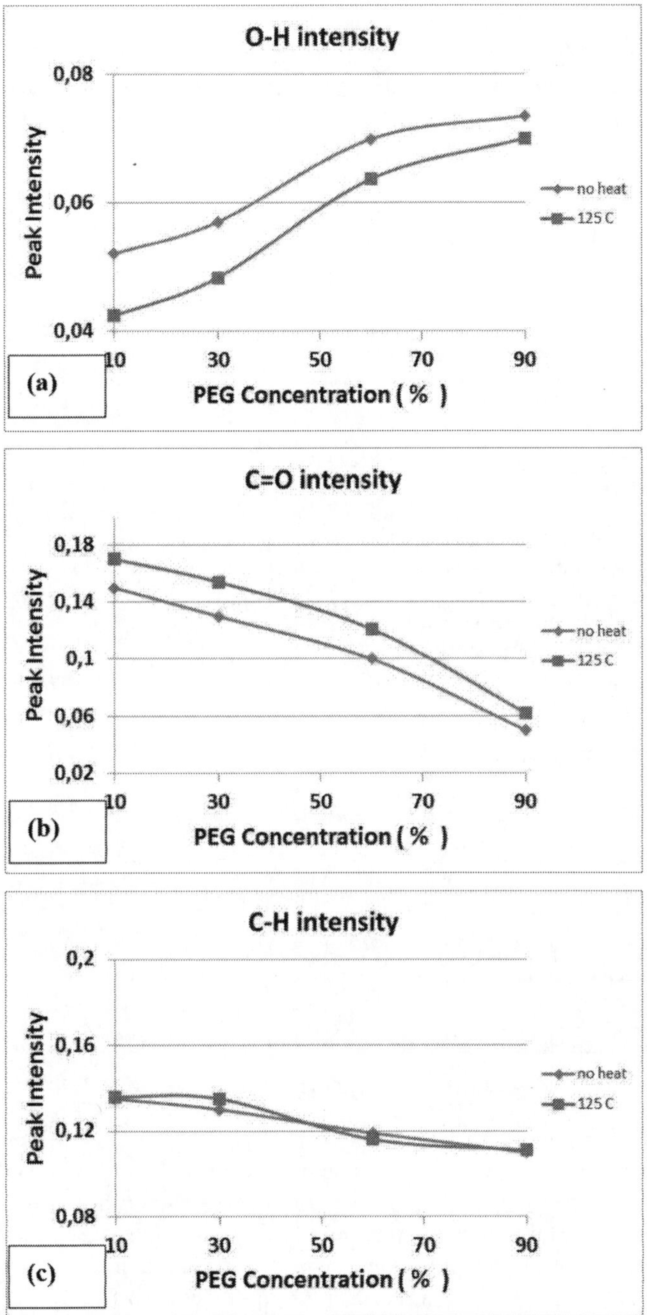

Figure 4. Peak intensity of FTIR spectra of PEG400 shellac system of (a). O-H group, (b). C=O, and (c). C-H group.

One of the main disadvantage of shellac is stability. Over time its molecular groups can undergo intra-molecular chemical bonding called polymerization leading to changes in its properties, it becomes more brittle and less soluble. The polymerization occurs by either heating or exposure to humidity [8]. It is a result of selfesterification of the material [9]. The mechanism of it was proposed by Limmatpavirat [10]. Hence, the stability of shellac can be measured by investigating the hydroxyl and carboxyl group [10,11], and is thus assigned by the increase of peak intensity bands of C= O stretching of ester at 1709 cm-1, and the decrease of as O-H stretching of hydroxyl group at 3400 cm-1. The intensity of the band at 2942 cm^{-1} occurs due to the C-H stretching vibration [12] and does not change with change in degree of shellac polymerization. The intensity ratios of the band of O-H cm^{-1} to C-H, and the intensity ratios of the band of C=O to C-H from all films then correlates to the polymerization in shellac.

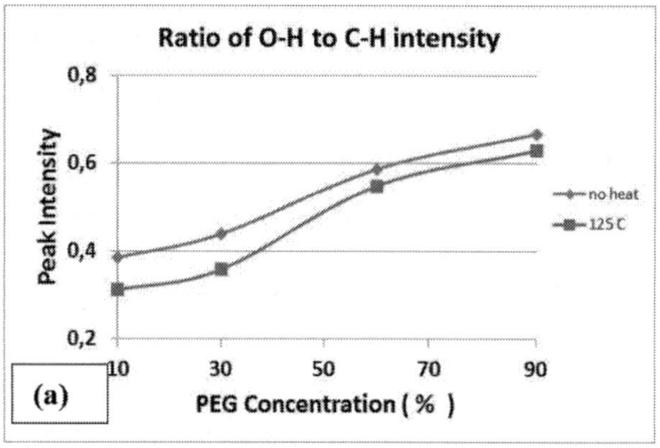

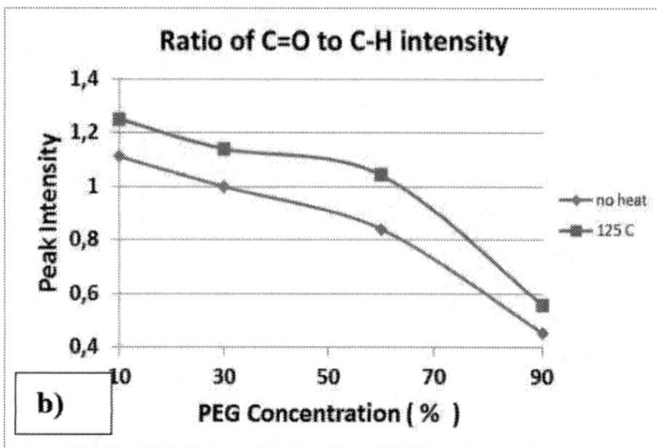

Figure 5. Ratio of peak intensity of FTIR spectra of (a). O-H to C-H (b). C=O, and (b). C=O to C-H

When samples were heated, the energy will break the carboxyl bond to create ester. The peak intensity of O-H, C=O, and C-H of PEG shellac systems heated at 125 °C for 30 minutes is shown in Figure 4, and the ratio of O-H to C-H and C=O to C-H is shown in Figure 5. It shows siilar trend with PEG shellac system with no heating. Heating samples also resulted in the decrease of O-H group but increasing C=O group, while no change in C-H group. These indicates that introduction of PEG into shellac could not stop polymerization, but slow down polymerization, and the effect of it is higher at high concentration than that of low PEG concentration. Further investigation shows that the slow down of polymerization increases with increasing PEG concentration. The amount of carboxyl group that can be protected by PEG may be higher at high concentration than that of low concentration.

4. Conclusion

FTIR stuides showed that polymerization in shellac was assigned by the decrease of peak intensity ratios of O-H to C-H, and the increase of peak intensity ratio of C=O to C-H. The polymerization was slowed down by PEG, and the degree of polymerization of shellac decreased with increasing the concentration of PEG..

Acknowledgement

The authors acknowledge the Indonesian Government for awarding "International colloboration and scientific publication 2014" grants with contract number 6563/UN27.16/PN/2014 dated 10 June 2014.

References

[1] Sengupta,N. *et.al* 2010 Report of the Study on Lac Sub Sector.India.
[2] www.fao.org
[3] Wang,X. *et.al.* 2006 Forestry Studies in China 8, 65-69. References therein.
[4] Anand Shah, N.S. 2009 *Film Coating Technology: An Overview.*
[5] Pillin, I. *et.al.* 2006 *Polymer*, **47**, 4676-4682.
[6] Sothornvit,R,Krochta,J.M, 2001 *J. Food Eng.*, **50**, 149-155.
[7] Cheng,L.H., Karim,A.A., Seow,C.C.,2006 *J.Food Sci.Food Eng.*, **71** (2), 62-67.
[8] Oesman, Z, 2012 Ph.D Thesis, Johannes Guttenberg University, Germany.
[9] Anan, M.L.*et.al.* 2010 *Journal of Agricultural and Food Chemistry*. **58**, 12934-12940.
[10] Limmatvapirat,S. et.al. 2007 Europ. *J. of Pharmac. &Biopharmac.* **67**, 690–698.
[11] Qussi,B & Suess,W.G. 2006 *Drug Development & Industrial Pharmacy* **32** (4), 403-412.
[12] Derry J, 2012 Master Thesis, University of Oslo, Norwegia.

The effect of polyethylene glycol Mw 400 and 600 on stability of Shellac Waxfree

Khairuddin[1]*, E. Pramono[2], S. B. Utomo[3], V. Wulandari[1], A. Zahrotul W[1], F. Clegg[4]
[1]Department of Physics MIPA, Universitas Sebelas Maret, Jl. Ir. Sutami 36A, Surakarta 57126, Indonesia.
[2]Department of Chemistry Kimia FMIPA, Universitas Sebelas Maret, Jl. Ir. Sutami 36A, Surakarta 57126, Indonesia.
[3] Department of Chemistry FKIP, Universitas Sebelas Maret, Jl. Ir. Sutami 36A, Surakarta 57126, Indonesia.
[4]Materials and Engineering Research Institute, Sheffield Hallam University,Sheffield,S1 1WB, U.K

E-mail: khairuddin_70@yahoo.con

Abstract The effect of polyethylene glycol (PEG) having molecular weight of 400 and 600 on stability of shellac waxfree prepared by solvent-evaporation method was reported in the present paper. The stability of shellac was tested by life under heat at 125 °C for 10 , 30 , 90 , and 180 minutes. It was found that that stability of shellac decreased with heating time at 125 °C. PEG400 gave the most stable effect of shellac for 30 minutes of heating, whereas at 90 minutes and above PEG600 gave the most effect of shellac stability as shown by insoluble solid test and FTIR. The WVTR showed that water vapour barrier properties of PEG600/shellac was beter than that of PEG400/shellac system.

1. Introduction

Shellac is a natural polymer secreated by lac insects cultivated in host tree in tropical countries such as India, Thailand and Indonesia, which has been used in food, cosmetic, and pharmaceutical industry [1] .It is a complex mixture of single esters and polyesters containing hydroxyl and carboxyl [2] It is insoluble in water but solublene in alcohol and alkali solution [3]. The application of shellac gradually decreased due to its stability, solubility, and the substitution of semisynthetcs and synthentic polymer. Several efforts have been done to solve these issue including modification into salt formation [4], and shellac derivative [5]. Natural polymer also tended to be brittle, and adding plasticizer may improve the flexibility of polymer by reducing the intermolecular forces, soften the rigidity of the film structure and increase mobility of biopolymer chains, thus affecting mechanical properties such as decreasing tensile strength and increasing percent elongation [6]. Plasticizer castor oil and dibutyl phthalate reduced water favor permeability but Polyethylene glycol (PEG) increased water vapor permeability of hydroxypropyl methylcellulose (HPMC), while in othes case palsticizer did not change the water favor permeability [7] suggesting that the solubility and the affinity of plasticizer to the polymer. Molecular weight of PEG play important role in stability of shellac [8,9]. In our previous studies [8], it showed that PEG1000 gave better effect on shellac stability compared to that of PEG2000. Therefore, the aim of the present study to extend the effect of molecular weight of PEG by using PEG 400 and PEG600. The stability was tested by heating shellac for different times.

2. Experimental

2.1 Sample preparation

The shellac type was wax free supplied by Sigma Aldrich UK. Alcohol 96 % was supplied by Merck Germany. Plasticizer polyethylene glycol was supplied by Sigma Aldrich UK. All films presented were prepared by solution-casting. Shellac films were made by dissolving 1.25 gr of shellac in 12.5 ml ethanol (96%), while PEG/shellac films (10 wt%) were prepared by mixing and stirring shellac solution and PEG solution. The solution was then cast into Petri dish, then dried at 50 °C for 7 hours before being analyzed.

2.2. Characterization

2.2.1. Insoluble solid test

Insoluble solid test is one of the method to quantify polymerization of shellac due to heating or moisture. In this studies, the polymerization was accelerated by heating at 125°C for 10, 30, 90, 180 and 360 minutes. The heated shellac film was dissolved in ethanol (96%) for 3 hours. The solution was then filtered using a metal filter (150 mesh). The insoluble solid was dried at room temperature until a constant weight was obtained and the percentage was calculated.

2.2.2. Fourier Transform Infra Red (FTIR)

FTIR was used to clarify the result of insoluble solid test by monitoring the peak intensities of O-H and C=O group. The intrument was a Thermo Nicolet NEXUS spectrometer along with a MCT (mercury cadmium telluride) detector cooled by liquid nitrogen. The wavelength of interest ranges from 700 to 4000 cm^{-1} and was collected with resolution of 4 cm^{-1} and 64 scans. Omnic software was used to collect and analyse the spectra.

2.2.3. Thermogravimetry analyzer (TGA)

Mettler Toledo TG50 Thermogravimetric with analyser using an alumina crucible was used in TGA measurements at a heating rate of 20 °C/min between 35 and 800 °C after an initial isothermal stage where the sample was kept at 35 °C for 15 minutes. The analysis was performed under a nitrogen atmosphere with a flow rate of 10 ml/min.

2.2.4. Water vapour transmission rate (WVTR).

The water vapor transmission rate (WVTR) which represent the barrier water properties of shellac and plasticized shellac was measured using Payne cups supplied by Sheen Instruments. The cups had an exposed surface area of 10 cm^2 and contained silica gel desiccant (8 g in each cup), which was closed by the material to be tested and dried overnight in an oven at 100 °C. They were placed in a controlled atmosphere (humidity chamber) and were weighed at suitable time intervals. The WVTR result was calculated from the following formula [10]:

$$WVTR = (240 \times m) / (S \times t) \quad (1)$$

where t is the total duration in hours, m is the increase in mass in milligrams of the assembly during the time t, and S is surface area exposed by the Payne cups. The WVTR was measured in tropical packaging standar, i,e, 23 °C and 85 %,. The relative humidities of 85% were achieved by placing saturated potassium chloride salt solutions, in a humidity chamber [11]. Shellac and PEG/shellac films were equilibrated in the humidity oven at 23 °C and 85 % relative humidity (RH) for 24 hours before performing the WVTR measurement.

3. Results and Discussions

Stability of shellac due to polymerization among hydroxyl and carboxyl groups [12] causing the change of properties such as water vapor barrier properties , glass transition temparature, acid value, and mechanical properties. Therefore, some of these properties are investigated in this report. From our previous studies [8], the stability of shellac could be improved when incoporating with PEG1000 and PEG2000, to verify further effect of molecular weight, PEG400 and PEG600 is used in this studies.

The insoluble solid is a direct consequence of shellac polymerization due increased amount of hard resin composed of polyster (Limmatpavirat,2004). Figure 1 shows the insoluble solid of shellac, PEG400-shellac (10 wt/wt %), and PEG600-shellac (10 wt/wt %) as a function time heated at 125 °C. It shows that the insoluble solid is presence not only in native shellac but also in plasticized shellac indicating that PEG could not stop polymerization. The level of polymerization increases significantly after heating for 90 minutes and more. The amount of insoluble solid in both plasticized shellacs is lower that those of native shellac indicating that PEG400 and PEG600 slow down polymerization. Heating time shows different effect of insoluble solid of plasticized shellac. Heating time for 30 minutes and less, the insoluble solid of PEG400-shellac system is lower than that of PEG600-shellac system, whereas heating for 90 minutes and above the insoluble solid of PEG600-shellac system is lower than that of PEG400-shellac system. Mobility of PEG chains in shellac matrix may affect how PEG interacts with shellac. For long heating time, less mobility of PEG600 can keep its interaction with shellac while high mobility of PEG400 distisburbs its contact with shellac resulting less effect of slowing down of shellac polymerization. Our previous studies showed that the trend of insoluble solid in PEG1000-shellac (10 wt/wt %) was lower than that of PEG2000-shellac (10 wt/wt %), and not affected by heating time supporting our present results.

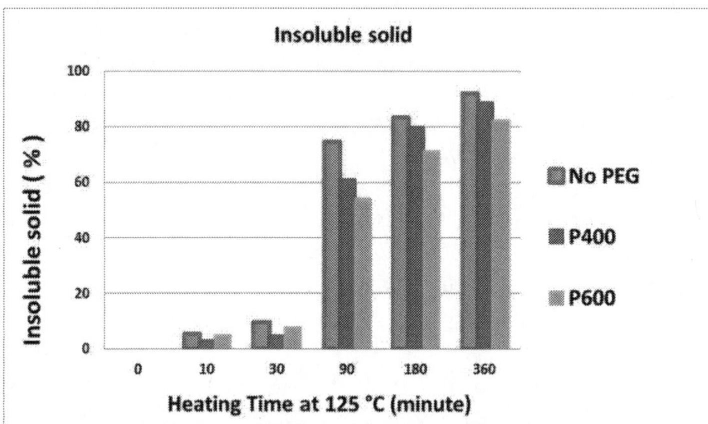

Figure 1. Insoluble solid of shellac, PEG400-shellac, and PEG600-shellac system as a function of time heated at 125 °C

Figure 2 and 3 show the FTIR spectra of PEG400-shellac (10 wt/wt %) and PEG600-shellac (10 wt/wt %) heated at different time at 125 °C (data of native shellac is presented here). There are main peaks that may be affected by polymerization of shellac according to polymerimization mechanisms proposed by Limmatpavirat,*et.al* [13], i.e., O-H group at 3400 cm^{-1} and C=O at 1730 cm^{-1}. Polymerization will be assigned by the decrease of O-H group, and by the increase of C=O group presence in the samples. General trend in both samples show that the peak intensity of O-H decreases but C=O increases with increasing heating time as shown in Figure 4a and b. The peak intensity of O-H in both PEG400-shellac and PEG600-shellac are higher than that of native shellac (Figure 4a), while opposite trend is observed at peak intensity of C=O group (Figure 4b). The peak intensity of O-H group in PEG600-shellac is higher than that of PEG400-shellac, while opposite trend of C=O group peak intensity is observed when samples are heated for 90 minutes and above. The results support the result of insoluble solid (Figure 1).

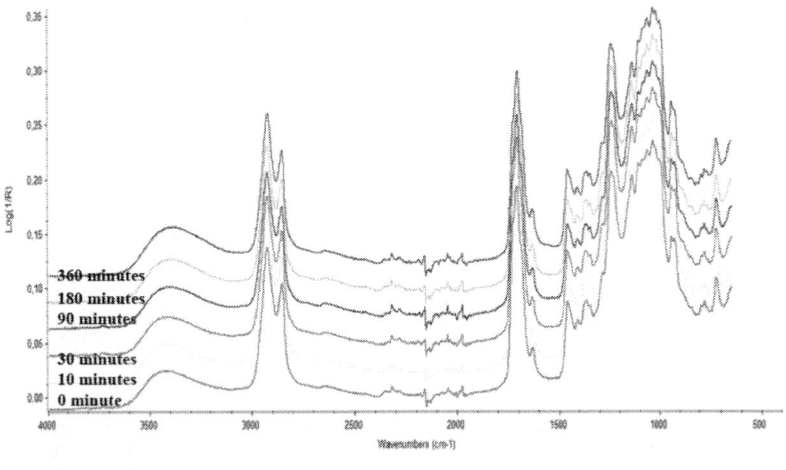

Figure 2. FTIR spectral o Shellac/ PEG400 heated 125 ° with variation of heating time

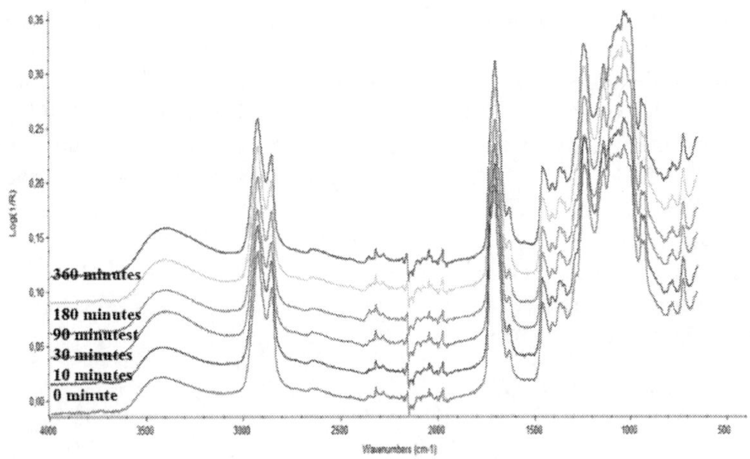

Figure 3. FTIR spectral o Shellac/ PEG600 heated 125 ° with variation of heating time

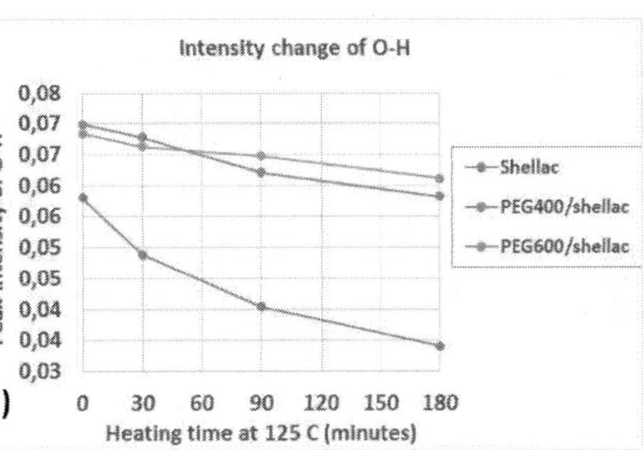

(a)

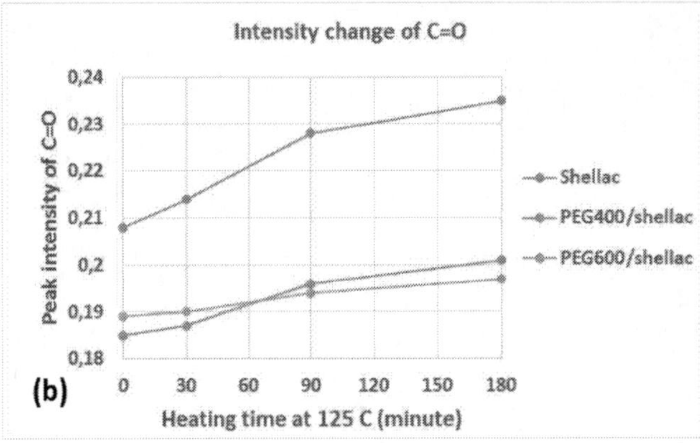

Figure 4. Peak intensity as a funtction heating time at 125 °C of a). O-H, b). C=O group

Thermal stability of samples when they are heated can be given by Thermogravimetric Analysis (TGA). It measures measure the weight change of sample due to the formation of volatile products in relation to change in temperature. Thermogram of shellac, PEG400, and PEG400-shellac is shown in Figure 5, while thermogram of shellac, PEG600, and PEG600-shellac is shown in Figure 6. There are 3 areas of weight loss in all samples,i.e, first: in the range 35-150 °C, second: in the range 250 – 475°C, and third: above 500°C. Each weight lost relates to water or solvent present in the sample, structural decomposition of the shellac, and remained compound or carbon, respectively. No significant different in water/solvent content was observed for all samples. Degradation temperature of PEG-shellac (second weight lost) is higher that those of either PEG or shellac indicating that PEG improved thermal stability of shellac, and vice versa. The weight lost in the range 250 – 475°C of PEG600-shellac was 92,94 wt%, whereas 95,12 wt% for PEG400-shellac.

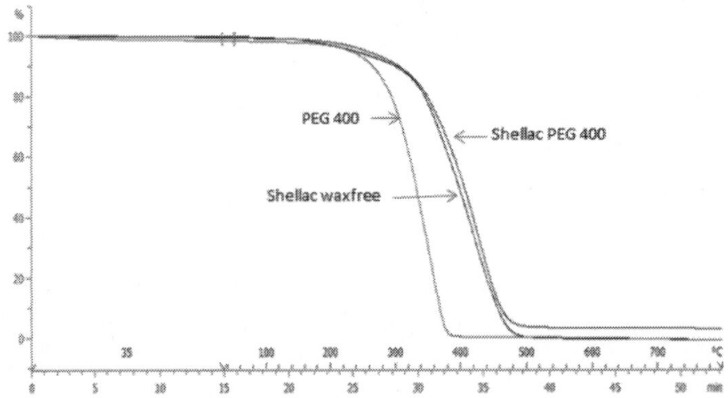

Figure 5. Thermogram of shellac, PEG400, and PEG400-shellac

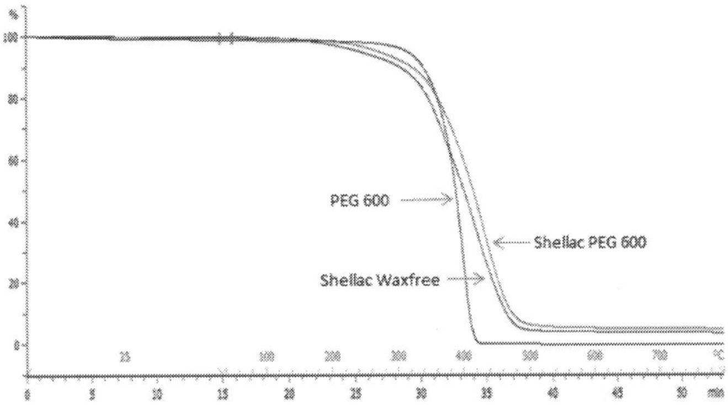

Figure 6. Thermogram of shellac, PEG600, and PEG600-shellac

Water vapor transmission rate (WVTR) of shellac and plasticized shellac is given in Figure 7, which shows that adding either PEG400 or PEG600 reduces water barrier properties of shellac. PEG disturbs the intra molecular bonds of shellac resulting in the reduction of shellac chain packing, in addition to hydrophylic properties of shellac. Similar effect of PEG to polymer was reported by other studies [14,15] PEG600 gave better water barrier properties to shellac compared to given by PEG400. As the WVTR was measured for flms with heating treatment, higher content of hard resin in PEG600-shellac may relate to its water barrier properties.

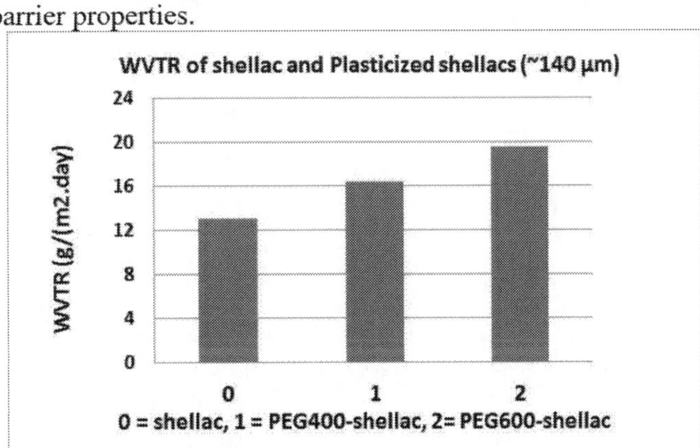

Figure 7. Water vapor transmission rate of shellacm PEG400-shellac, and PEG600-shellac

4. Conclusion

It has been shown that heating of shellac and plasticized shellac at 125 °C caused polymerization, and the level of it increased with increasing heating time. PEG slowed down shellac polymerization, which was affected by molecular weight of PEG. The molecular weight affected the thermal degradation, and the water barrier properties.

Acknowledgement

The authors acknowledge the Indonesian Government for awarding "International colloboration and scientific publication 2014" grants with contract number 6563/UN27.16/PN/2014 dated 10 June 2014.

References

[1] Krause, K.P; Muller, R.H. 2001 Int.J. Pharm, 223 (1-2), 89-92.
[2] Hogan,J.E; Cole; Hogan,J.E ; Aulton,M. 1995 *Pharmaceutical Coating Technology*, Taylor &Francis Books, London pp.6-52.

[3] Banker,G.S; Agyilirah, G.A, 1999 *Polymers for Controlled Drug Delivery*, Tarcha, P.J. Ed,CRS Press, London, pp. 39-66.
[4] Limmatvapirat, S.; Limmatvapirat, C.; Luangtana-anan, M.;Nunthanid, J.; Oguchi, T.; Tozuka, Y.; Yamamoto, K.; Puttipipatkhachorn, S. 2004 *Int. J. Pharm*. 278 (1), 41-49.278.
[5] Sheorey, D. S.; Kshirsagar, M. D.; Dorle, A. K. 1991 *J. Microencapsulation*. 8, 375-80.
[6] Heng,P.W.S; Chan,L.W; Ong,K.T, 2003 *J.Pharm.Pharm.Sci*., 6(3), 334-344.
[7] Saettone, M. F.; Perini, G.; Rijli, P.; Rodriguez, L.; Cini, M..1995 *Int. J. Pharm. 126* (1-2), 83-88.
[8] Khairuddin; Edi P.; Suryadi B.U; Viki W; A'an Z; Clegg,F. 2015 The 10th *International Joint Conference on Chemistry*, Solo.
[9] Anan,L.; Limmatvapirat,L; Nunthanid,J; Wanawonthai,C; Chalongsuk,R; Puttippipatkhachorn, S, 2007 *J.Agr. Food Chem*, 55, 687-692.
[10] British Standard (BS 2782-8:Method 820A). 1996 Method of testing Plastics part 8.
[11] Equilibrium relative humidty at http://www.omega.com/temperature/z/pdf/z103.pdf
[12] Limmatavapirat, S.; Nunthanid, J.; Puttipipatkhachorn, S.; Luangtana-anan, M. 2005 *Pharm. Dev. Technol*.1, 41-46.
[13] Limmatvapirat,S, Limmatvapirat,C, Puttipipatkhachorn,S., Nunthanid,J., Anan, M.L,. 2007 *Europ. J. of Pharmac. &Biopharmac*. 67, 690–698.
[14] Khairuddin, Ph.D Thesis, 2012 Sheffield Hallam University.
[15] Anan,M.L, Nunthanid,J, Limmatpavirat,S. 2010 *J.Agr.Food.Chem*, 58, 12934-12940.

Journal of Physics: Conference Series **776** (2016) 012055 doi:10.1088/1742-6596/776/1/012055

Analysis of thickness dependent on crystallization kinetics in thin isotactic-polysterene films

Khairuddin
Jurusan Fisika FMIPA, Universitas Sebelas Maret, Jl. Ir. Sutami 36A, Surakarta 57126
Indonesia.

E-mail: khairuddin_70@yahoo.com, Tel +62 (0)[illegible]

Abstract Crystalliaztion kinetics of thin film of **Isotactic Polysterene (it-PS) films** has been studied. Thin PET films having thickness of 338, 533, 712, 1096, 1473, and 2185 A° were prepared by using spin-cast technique. The it-PS crystals were grown on Linkam-hostage in the temperature range 130-240°C with an interval of 10°C. The crystal growths are measured by optical microscopy in lateral direction. It was found that a substantial thickness dependence on crystallisation rate. The analysis using fitting technique based on theory crystal growth of Lauritzen-Hoffman showed that the fitting technique could not resolve to predict the mechanism controlling the thickness dependence on the rate of crystallisation. The possible reasons were due to the crystallisation rate varies with the type of crystals (smooth, rough, overgrowth terrace), and the crystallisation rate changes with the time of crystallisation.

1. Introduction

While mechanical applications of polymers took advantage of their bulk properties, there were many technologies in which ultra thin polymer films were used. For instance, they were applied as resist and interlayer dielectrics in microelectronics fabrications, as alignment layers in liquid crystal displays, and as lubricants in magnetic information storage devices [1]. In each of these applications, the polymer chain orientation and state of organization played an important role in determining the final properties [1,2]. There was increasing evidence that the properties of polymer in ultra thin may be different to their bulk properties [3]. This was represented by the recent discovery of substantial thickness dependent changes in mobility. Previous evidence about this was contradictory. Frank et al [1] suggested that in some polymers crystallisation may be completely hidden in ultrathin films i.e with the thickness of order 100 A°. In contrast, Wehrum [4] showed that crystallisation could occur in poly(ethylene terephthalate), albeit with kinetics reduced in speed over values for thicker films. Therefore, studies of polymer crystallization in thin films are of both scientific importance and practical significance.

Below Tg, polymers in the bulk are in the glassy state, where chains are frozen into position. Only recently has increasing experimental evidence been suggesting that near free surface polymer chains may be mobile, while the bulk is in the glassy state. Keddie *et al* [5] suggested the existence of such a mobility from observations of a decrease in Tg in ultra thin polymer films, without a variety of experiments, in effort to recognize the nature of enhanced surface mobility. The aim of this present studies is to extend to which the ordering properties in ultra-thin differ from bulk. Previous evidence about this is contradictory. This preliminary finding is expanded and undertaken a systematic study of the effect of thickness in crystallisation kinetics in ultra thin polymer films.

In order to understand the underlying molecular mechanism of thickness dependence of crystallization, very thin isotactic polystyrene films were chosen where homogeneous nucleation was extremely rare. The growth front nucleation did not represent a significant perturbation due to well-controlled perturbation [6] and large scale single crystals can be grown. Such crystals allow for an clear relation between crystal shape and orientation of all polymer chains within the lamellae. Moreover, crystal morphology can be directly related to the growth kinetics as there is no perturbation of growth front nucleation . As a result, the evolution of the crystal morphology can be followed in detail [3]

2. Experimental

In this study it-PS is supplied in the form of powder by Aldrich Chem. Co. with molecular weight $\overline{M}_w$ of 400000, we use cyclohexanone as its solvent. Because it-PS powder is provided in the form of a crystal (not glassy), it is difficult to dissolved in cyclohexanone.. To overcome this problem, it-PS powder is transformed to glassy state by melting it above its melting point. This is done by placing the it-PS powder onto the glass substrate and then melting it using a Linkam hot stage. Next, the melt it-PS is quenched by removing quickly the glass substrate and puting it in water. The reason for this treatment is to ensure that the it-PS has not had time to form a crystalline state; instead glassy it-PS is obtained. This glassy it-PS can be easily be dissolved in cyclohexanone solvent to make a solution as a raw material to prepare thin films. In order to reduce impurities and filter un-dissolved (crystal) that may exist during preparation of the solution, the solutions are filtered.

The thin it-PS films are prepared by using spin casting. Varying concentration of the solution or speed of the spin controls the thickness of the films., Variation of concentration of the solution produces greater variation of film thickness than variation of speed of the spin. Their thickness is characterised by ellipsometry. Theoretical explanation of the ellipsometry method is given elsewhere [7].
The thermal treatments of thin it-PS films were also similar to the thermal treatment of thin PET films. Under reflected light microscopy, the film was placed onto a Linkam hostage and heated at 250°C (the melting point of it-PS powder is 215°C) for 3 minutes to melt the film and remove any differences in thermal and mechanical history of the film prior crystallisation. The melt film then was under cooled to a constant crystallisation temperature in range 200 – 130°C (with intervals of 10°C). In these studies, the Linkam hot stage was not needed to be purged with liquid nitrogen. This was the case as the thin it-PS film took a long time to crystallise (on average 30 minutes) after it had reached a constant crystallisation temperature.

3. Results and Discussions

The crystals seen in thin it-PS films were different from those of thin PET films [8]. Thin it-PS films formed crystals of hexagonal plates, hexagonal crystal having re-entrant corners to form a star-like shape or sunflower crystal, rounded hexagonal and circular discs rather than the circular discs seen in the thin PET films. The form of crystals produced in thin it-PS films depend on the crystallisation temperature. In other words, the crystals in it-PS films show a morphological transition with crystallisation temperature. Moreover the growth rate of crystals in the same film thickness and crystallisation temperature also shows variation with the type of crystal.

This is illustrated in Figure 1, which shows morphologies of crystals an annealing a 2185A° film for 6 hours at different temperatures. Note the series of qualitatively different morphologies. We also find that the morphologies depend on temperature. It can be seen in the figure that the morphological transition occurs from sunflower-like shape crystal with many re-entrants (190°C), sunflower-like shape crystal with many re-entrants and hexagonal with re-entrants to form star-like shape (180°C), banded circular (170°C), banded and dual circular (160°C), dual circular (150°C) and finally smoothed circular (130°C).

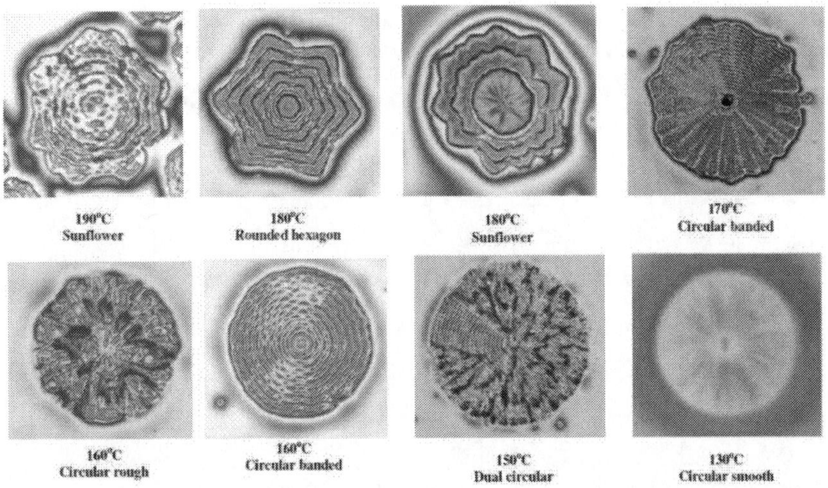

Figure 1. A morphological transition in the 2185A° film it-PS film from high temperature to low temperature

Figure 2 shows the data of crystallisation rate of crystal for all film thickness crystallised in the temperature range of 130-190°C normalised to the thickest film (2185A° in thickness). The growth rate, which was measured, was the growth rate of crystals crystallised for less than 1 hour as displayed in the graph in Figure 3. It can be seen that the rate of crystallisation still depends on film thickness in thin it-PS films, but the thickness dependence on the rate of crystallisation is very complicated. Therefore, it cannot establish a generalised behavioural trend of the thickness dependence on the rate of crystallisation as thin PET films [8].

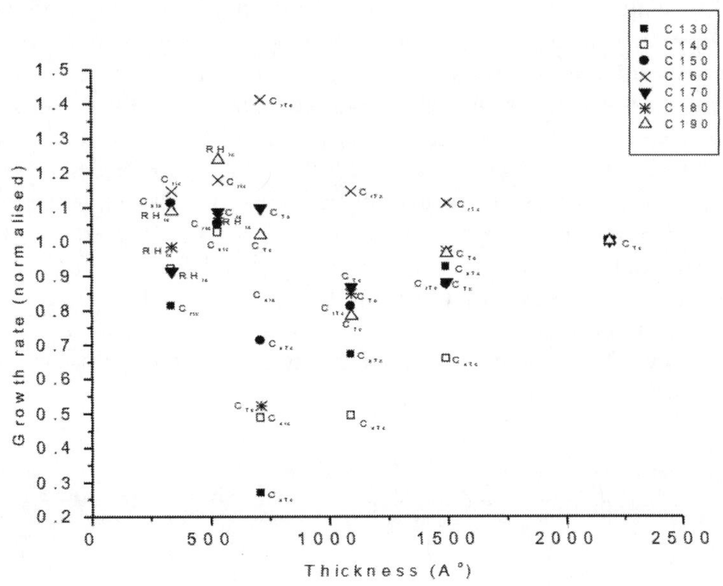

Figure 2. Thickness dependence on crystallisation rate of crystal in thin it-PS films normalised to the thickest one (2185A° in thickness). The data are from measurement of crystal growth for holding less than 1 hour.

PS. R_{tc} =rounded hexagonal thin crystal, C_{stc} = circular smooth thin crystal, C_{rtc} = circular thin crystal, C_{sTc} = circular smooth thick crystal, C_{rTc} =circular thick crystal

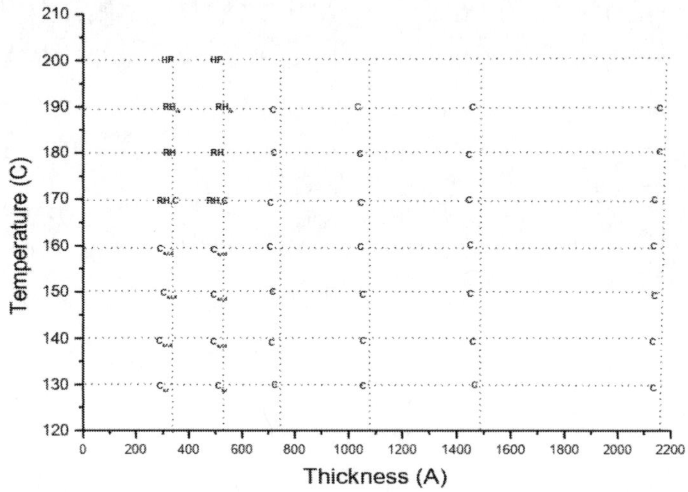

Figure 3. morphological transition as a function of crystallisation temperature and film thickness in it-PS films after holding about hour.PS. HP =hexagonal plate, RH=rounded hexagonal, C =circular, re= re-entrant, s=smooth, r =rough, and d=dual

In further investigate any relation between the thickness dependence and crystallisation rate, it used the equation (1) of spherulite growth rate of Lauritzen-Hofman [9] for fitting to find out the activation temperature (T_A), Vogel temperature (T_V), constant B_o, and equilibrium melting point (T_f^{∞}). The result of spherulite growth of different thickness informs us directly whether the crystallisation rate was a function of film thickness. An analysis of data using the equation (1) for the spherulite growth rate (G), with the dependence of G on the segmental mobility, given by the 'Vogel Fulcher law' represented by the first term and the undercooling 'driving force' represented by the second term of the equation gives us more information which term is affected by thickness of the thin polymer films. An attempt was made to do the non-linear to get T_A, T_V and B_o with T_f^{∞} fixed at 513°K for it-PS thick film, but the software could not resolve this. This was probably because either the crystallisation temperature range of these studies was too short to be fitted by the program or the data was complicated. This procedure was done because the value of parameter B_o, when fitting was done using equation (2), was required. To overcome this problem a trial and error approach was taken. Finally it used a 465 as a value for constant geometry for bulk system.

$$G = \exp(-\frac{T_A}{T - T_V}).\exp(-\frac{B_o}{T_f^{\infty} - T}) \quad (1)$$

$$\left(LnG + \frac{B_o}{T_f^{\infty} - T} \right)^{-1} = -\frac{T}{T_A} + \frac{T_V}{T_A} \quad (2)$$

Journal of Physics: Conference Series **776** (2016) 012055 doi:10.1088/1742-6596/776/1/012055

$$\left[\ln(G)+\left(\frac{T_A}{(T-T_V)}\right)\right]^{-1}=\frac{T}{B_o}-\frac{T_f^{\infty}}{B_o} \tag{3}$$

Table 1

Thickness (A)	T_A (K)	T_V (K)
338	1406.3±	327±8.45
533	1352.7±	330±9.67
712	1511.7±	322±6.87
1095	1661±	311.5±2.98
1473	1468.67±	323.44±6.3
2185	1666±84.2	309±6

The deduced activation temperature T_A and deduced Vogel temperature T_V by the fitting procedure by using equation (2) is shown in table 1. In the fitting, the equilibrium melting point T_f^{∞} and constant geometry B_o were assumed to be 513K and 465, respectively. The equilibrium melting point $T_f^{\infty}$and the constant geometry B_o were obtained using fitting procedure in the opposite way that is by assuming that the activation and Vogel temperature are constant for the bulk system stating equation (1) in the form of equation (3) that was by assuming that the activation and Vogel temperature were constant for the bulk system. Table 2 displays the value of the equilibrium melting point $T_f^{\infty}$and the constant geometry B_o obtained from fitting by assuming the activation temperature T_A and the Vogel temperature T_V of the bulk are constant i.e. 1666 and 309, respectively.

Table 2

Thickness (A)	B_o	T_f^{∞} (K)
338	456.9±26	512.6±5
533	432.6±34	510.2±6
712	480.4±50	505.81±4.4
1095	494.7±32	514.5±6
1473	461.35±36	512±6
2185	410.5±25	508.5±7

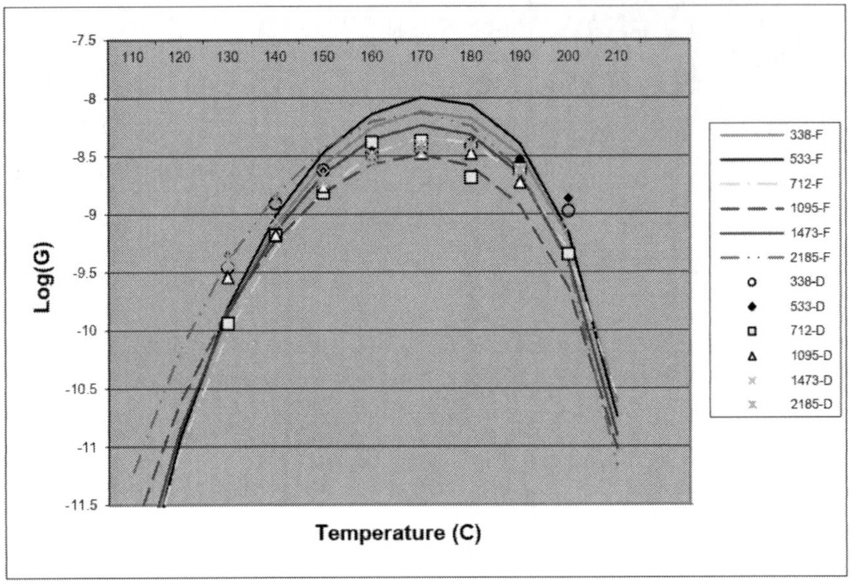

Figure 4. Temperature dependence of crystallisation rate for various it-PS film thickness of experimental data (D) and expected(F).

Figure 4 displays the deducted crystallisation (line) and the experimental crystallisation rate. It can be seen that the fitting technique cannot resolve the thickness dependence on the rate of crystallisation. This was because the deducted crystallisation rate is very spread with the behaviour of the experimental crystallisation rate, especially in the temperature range of 150-180°C and also at low temperature.

4. Conclusion

To conclude therefore, the fitting technique could not resolve to predict the mechanism controlling the thickness dependence on the rate of crystallisation. It was because the rate of crystallisation in it-PS films was complicated due to the following factors; first: the crystallisation rate varied with the type of crystals (smooth, rough, overgrowth terrace). Second: the crystallisation rate changed with the time of crystallisation. It was evidenced by the morphology of crystals dependant on the time of crystallisation, especially in thick films. In other word, the crystallisation rate in it-PS thick films was non-linear.

Acknowledgement
The author would like to thanks Prof R.A.L. Jones, Sheffield University, for the discussion.

References

[1] C.W. Frank, V. Rao, M.M. Despotopoulou, R.F.W. Pease, W.D. Hinsberg, R.D. Miller, J.F. Rabolt, Science 273, 912, 1996.
[2] Y.X. Liu, E.Q. Chen, Coordination Chemistry Reviews 254, 1011, 2010.
[3] Forrest J.A, Jones R.A.L., *Polymer Surfaces, Interfaces and Thin Films*. World Scientific Publishing; Singapore, 1999
[4] Wehrum A. (1999), Ph.D thesis; Cambridge University.
[5] Keddie J.L, Jones R.A.L, and Cory R.A., Europhysics Letters **27**,59, 1994
[6] J.M. Schultz, Polymer Crystallization (Oxford University Press, New York, 2001); Polymer 44, 433, 200); Macromolecules 45, 6299, 2012

[7] Azzam R.M.A, Basharea N.M., Ellipsometry and Polarised Light,North-Holland;Amsterdam, 1997
[8] Khairuddin, Media Fisika,vol 4 (2), 57-62, 2005
[9] Strobl G., *The Physics of Polymers*, Springer; Berlin, 1997

Hidroxyapatite coating on CoCrMo alloy titanium nitride coated using biomimetic method

Charlena, S. G. Sukaryo, and M.Fajar
Department of Chemistry, FMIPA IPB Kampus IPB Darmaga IPB
E-mail: charlena.ipb@gmail.com

Abstract. Bone implants is a way to cure broken bones which is being developed. The implants can be made of metals, ceramics and polymers. Metallic materials commonly used are titanium (Ti), stainless steel, and metal alloys. This study used Co-based alloys, i.e. CoCrMo coated with titanium nitride (TiN) which was then coated on hidroxyapatite (HAp). The HAp coating on the surface of CoCrMo alloy was done by biomimetic methods, first by soaking the metal alloys in simulated body fluid (SBF) solution for 18, 24, and 36 hours. The immersion in the SBF solution produced white coat on the surface of the metal alloy. The layers formed were analyzed by *scanning electron microscope* (SEM) and characterized by *x-ray diffractometer* (XRD). Based on the SEM results of 36 hours treatment, the morphology of apatite crystal formed fine grains. According to XRD result, there were HAp peaks at angles 2θ 31.86, 32.25, dan 39.48. However, there were also $CaCO_3$ peaks at angles 2θ 29.46, 36.04, and 46.79. It indicated the pure HAp is not yet formed.

1. Introduction

Orthopedic field becomes very important because the number of accidents in Indonesia that resulted in fractures is increasing every year. Bone implants is one way of healing a fracture that continues to grow. Implants can be made from metals, ceramics and polymers. Metallic materials used for implants are titanium (Ti), stainless steel and metal alloys. Metal Ti has the highest biocompatibility and has better corrosion resistance, but its price makes this metal began to reduced usage. Similarly, Stainless steel, although the price is cheaper but does not have good corrosion resistance. The use of cobalt-based alloys (Co) as a substrate bone implants continue to be developed in line with human needs. Co-based alloys have a lower level of biocompatibility of Ti, but it is better than stainless steel. Another advantage of the use of Co-based alloys is the price is relatively cheaper and has better mechanical properties compared to Ti alloy [1]. In addition, the Ti metal alloys, namely Ti6Al4V still contains ions of Al and V ions are harmful to the body. NiTi alloys are also often used as an implant material, because it has the corrosion resistance and good biocompatibility. However, the content of Ni is relatively high, reaching 50% make use of metal alloys is reduced because it can cause allergies or even poisoning the human body. Coating of hydroxyapatite (HAp) on metal surfaces can improve the biocompatibility properties of the metal and accelerate bone growth. HAp has been recognized as a bone substitute material and teeth as it has in terms of their biological similarity to human hard bone tissue [2]. This study will use the Co-based alloys, which are coated CoCrMo titanium nitride (TiN). Based on the standard European medical applications, the rate of corrosion for metal implants should be less than 1 mpy. TiN coated CoCrMo alloy has a corrosion rate of 0.0424 mpy. This shows the TiN coated CoCrMo alloys can be used for medical applications. Alloy surfaces TiN coated CoCrMo metal (CoCrMo-TiN) will be coated with HAp, so that the biocompatibility of CoCrMo alloy is getting better and can accelerate bone growth when implanted into the body. CoCrMo alloy coating with HAp-TiN performed using biomimetic methods. The method is a biomimetic coating method that mimics biological systems in the body. The advantages of this method compared with other methods, such as plasma spraying, electrophoretic deposition, and sol-gel is made of apatite crystals have high bioactivity and resorption properties are good, and can accelerate bone growth. Coating by biomimetic method performed by immersion in a solution of simulated body fluid (SBF) [3]. Research Habibovic et al. (2002)[4] reported that HAp can be formed in a solution of 5x SBF with low intensity. After the coating will be performed morphological analysis by scanning electron microscope (SEM) and characterization by x-ray diffractometer (XRD).

2. Experimental Method

Research procedure consists of preparation of metal alloys with a base treatment process, the metal alloy coating CoCrMo-TiN with hydroxyapatite using biomimetic methods, and characterization by SEM and XRD. TiN metal alloys and CoCrMo-CoCrMo soaked in alcohol, and then sonicated for 1 hour and drying. TiN metal alloys and CoCrMo-CoCrMo alloys immersed for 15 minutes in a solution of NaOH 5 M. After that, the metal alloy is dried in an oven at 70°C. CoCrMo alloy and CoCrMo-TiN then coated with HAp. Coating CoCrMo alloy and CoCrMo-TiN done by dipping metal alloy into a beaker containing a solution of 5x SBF placed on a hot plate. Soaking is done for 18, 24, and 36 hours, and the temperature is kept constant at 37C. After 18, 24, and 36 hours of metal removed and dried, then viewed a layer that is formed by optical microscopy and morphological analysis by SEM and XRD characterization.

3. Results and Discussion

After immersion, the surface of the metal alloy CoCrMo-TiN becomes black, because titanium oxide layer on the surface of a metal alloy reacts with ions in the NaOH solution to form a hydrogel titanate, while metal alloy CoCrMo unchanged. Furthermore, the two metal alloys dried in an oven for 30 minutes to dehydrate the hydrogel layer to form crystalline sodium titanate titanate ($Na_2Ti_6O_{13}$) stable. The reaction that occurs in the process of treatment with a base are as follows:

$$Ti + 3\,OH^- \rightarrow Ti(OH)_3^+ + 4e^-$$
$$Ti(OH)_3^+ + e^- \rightarrow TiO_2.H_2O + \frac{1}{2}\,H_2\,(g)$$
$$Ti(OH)_3^+ + OH^- \leftrightarrow Ti(OH)_4$$
$$TiO_2.H_2O + OH^- \leftrightarrow HTiO_3^-.H_2O$$

The method is a biomimetic coating method that mimics biological systems in the body, with this method HAp layer formed naturally on the surface of the sample through immersion in SBF solution. According Colovic et al. (2011)[5], excess biomimetic method compared to other coating methods are that the HAp coating structure and morphology characterization similar to those formed in the body. Coating by biomimetic method performed on CoCrMo alloys with and without treatment with a base first, to see the effect of this treatment on the coating to the surface of a metal alloy. The results of the two alloy coating to the metal (Figure 1) with immersion in SBF solution for 36 hours that preceded the treatment with a white base layer is formed on the surface, whereas the untreated alloy CoCrMo coated with a base yet. The formation of a white layer on the metal alloys given treatment with alkaline ions caused metal alloy constituent, namely Co, Cr, Mo regardless, so it reacts with NaOH and accelerate the coating on the surface of a metal alloy. Widya (2014)[6] reported, on the surface of CoCrMo alloys by immersion in a solution of 5x SBF for 48 h to form a white layer. Based on these results, treatment with a base that is first performed before the coating can accelerate the coating on the metal surface.

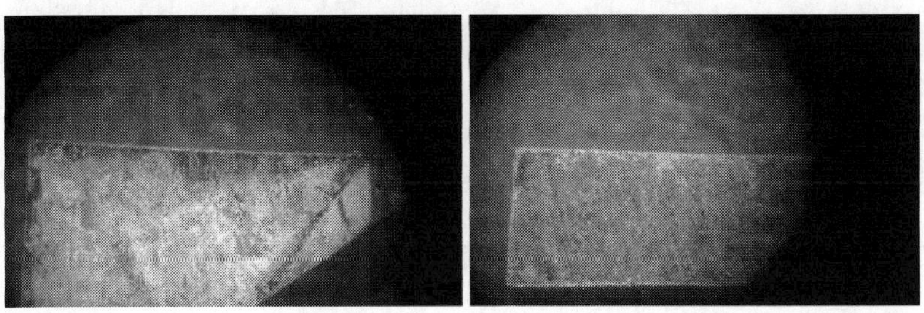

(a) With Alkali Treatment (b) Without Alkali Treatment

Figure 1 Results of the metal alloy coating CoCrMo

The coating also performed on CoCrMo-alloy with TiN treatment process with a base. Metal alloy coating is done with the variation of time 18, 24, and 36 hours. Results CoCrMo-alloy coating of TiN (Figure 2) for 18, 24, and 36 hours by immersion in SBF solution had formed a thin layer of white on the surface. Coating for 36 hours resulted in an increasingly white coating evenly. This suggests immersion in the SBF solution the longer it will create a layer that is formed on the metal surface evenly. Furthermore, the white layer formed on the metal alloy CoCrMo-TiN morphology analysis by SEM.

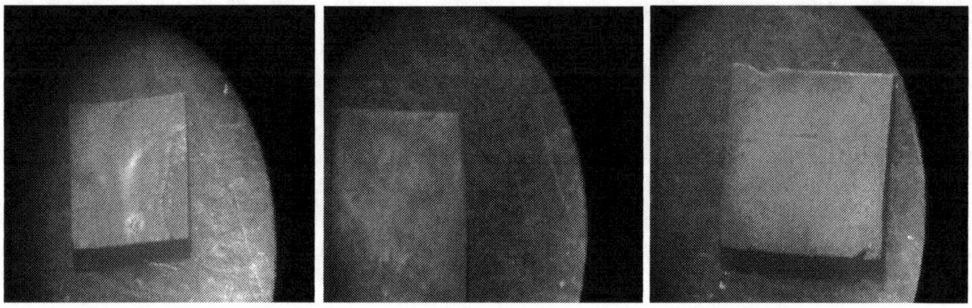

(a) 18 hours (b) 24 hours (c) 36 hours

Figure 2 Results of HAp coating on CoCrMo alloy-TiN

Based on the results of SEM images (Figure 3) is not visible compound apatite crystal morphology of the coating for 18 and 24 hours with a magnification of 1500x. This shows, on immersion for 18 and 24 hours a metal alloy surface is not coated HAp. The results of SEM images for the coating for 36 hours with a magnification of 1500x already appeared several compounds apatite crystal morphology assembled and refined grains are still few. Based on these results, the immersion for 36 hours had already been formed HAp on the surface of a metal alloy CoCrMo-TiN. After that, the characterization by XRD on the white layer formed on the surface of a metal alloy CoCrMo-TiN with immersion in SBF solution for 36 hours.

(a)Coating for 18 hours (b)Coating for 24 hours (c)Coating for 36 hours

Figure 3 SEM photos CoCrMo-alloy coating TiN 1500x magnification

Characterization by XRD performed to identify the crystalline phase layer attached to the metal alloy CoCrMo-TiN by matching the data 2θ diffractogram and the scattering intensity of the JCPDS database. If there are at least 3 highest peaks are identical to the database then allegedly phase equal to the standard. Based on the results diffractogram (Figure 4) and has been matched with JCPDS there HAp peaks at 2θ angle of 31.86, 32.25, and 39.48. In addition, there are also peaks at angles 2θ $CaCO_3$ 29.46, 36.04, and 46.79 with a higher intensity compared with HAp. This indicates that the pure HAp formed yet. Based on the results of characterization by XRD, there are peaks of HAp which indicates that the results are consistent with the SEM images showing the morphology of the apatite compounds in the form of fine grains. $CaCO_3$ sharper peaks of HAp shows the degree of crystallinity owned $CaCO_3$ higher than HAp. HAp does not have a high degree of crystallinity because when its synthesis does not use such a high temperature synthesis of HAp in general. Purnama et al (2006)[7] reported that high temperatures can increase the degree of crystallinity of HAp. These results are consistent with the results reported Habibovic et al. (2002)[4], the HAp formed on the metal surface by immersion in a solution of 5x SBF with low intensity.

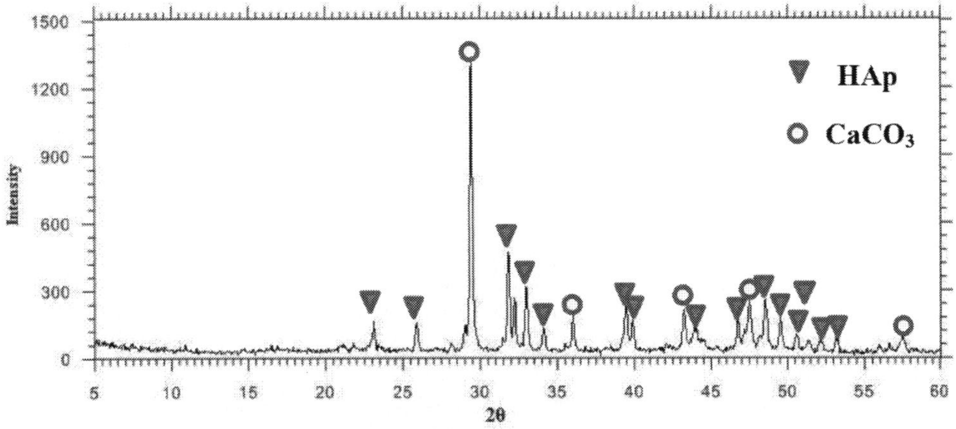

Figure 4 Diffractogram HAp coating on TiN coated CoCrMo alloys

On immersion in SBF solution, sodium titanate formed on the surface of the TiN coated CoCrMo alloy after treatment with a base react with ions in the SBF solution. Sodium ions from the sodium titanate dissolved and replaced by H_3O^+ ions from SBF, then OH^- of interesting metal hydroxy cations Ca^{2+} of the SBF solution and drawn by cation anion PO_4^{3-} previously been deposited [8]. The mechanism that occurs when metal alloy CoCrMo-TiN soaked in SBF solution can be seen in Figure 5 [8]. HAp formation reactions that occur on immersion in SBF solution are as follows:

$$Ca^{2+} + 2OH^- \rightarrow Ca(OH)_2$$

$$10Ca(OH)_2 + 6PO_4^{3-} \rightarrow Ca_{10}(PO_4)_6(OH)_2 + 18OH^-$$

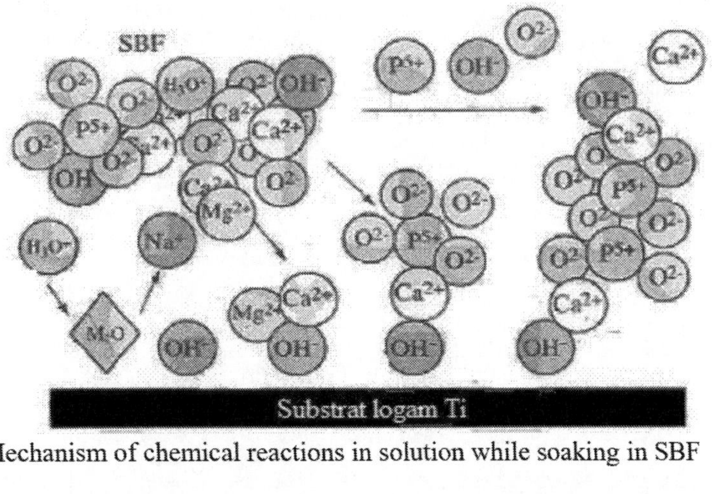

Figure 5 Mechanism of chemical reactions in solution while soaking in SBF

HAp coating on the surface of a metal alloy not only can be done with just a biomimetic method. There are several methods of coating used, the electrophoretic deposition. Electrophoretic deposition method is a method of coating with the aid of an electric field. The advantages of this method is to have high strength coating, coating evenly, and quickly. Prihantoko (2011)[9] reported that HAp-chitosan coating on the surface of a metal alloy CoCrMo-TiN using electrophoretic deposition method produces a thin and fragile layer. . Results coating by biomimetic method in this study is more powerful than the stick EPD method is still fragile. Coating by electrophoretic deposition method is also need to synthesize HAp first. However, the method of biomimetic HAp is formed naturally in the process so that more efisisen HAp coating to the surface of the metal alloy.

HAp has the molecular formula $Ca_{10}(PO_4)_6(OH)_2$ and has the properties of biocompatible, bioactive, and not toxic. Biocompatible is the ability of a material to adjust to the situation in the body. While the bioactive is the ability of a material to stimulate new bone growth around the implant. It can be seen from the phenomenon of interface between HAp with the cell body which shows the surface of HAp dissolution occurred after implant into the body. Dissolution of the HAp surface occurs until equilibrium is reached between the saline condition with HAp surface. After that, the case of adsorption of proteins and bioorganic compounds then occurs until finally the cell adhesion of cells to form new bone [10].

4.Conclusion

Layer of HAp with low intensity managed to form on the surface of the TiN coated CoCrMo alloys by treatment with alkaline process using a biomimetic method by soaking for 36 hours in a solution of 5x SBF. Based on the results of SEM images on the coating for 36 hours, it looks crystal morphology of apatite compounds such as fine grains. Based on the characterization results and there has been matched with JCPDS HAp peaks at 2θ angle of 31.86, 32.25, and 39.48, but other than HAp peaks are also peaks at angles 2θ $CaCO_3$ 29.46, 36.04, and 46.79. This shows pure HAp formed yet.

5.References

[1] Sanchez JF, Balarin A. 2010. *J Maters Sci in Medicine.* **21**(7):2021-2026.
[2] Xiangnan L, Xioming C, Shipu L, Zhiming P. 2010. *Wuhan Univ Technol-Mater.* **25**(2):252-256.
[3] Silva CC, da Silva EC, Marchi J. 2008. *J Mater Research.* **11**(1):47-50.
[4] Habibovic P, Barrere F, Van Blitterswijk CA, de Groot K, Layrolle P. 2002. *J Am Ceram.* **85**(3):517-522.
[5] Colovic B, Markovic D, Jokanovic V. 2011. *J Serbian Dental.* **58**:7-12.
[6] Widya R. 2014. *Pelapisan hidroksiapatit pada logam CoCrMo dengan metode biomimetik* [skripsi]. Bogor (ID): Institut Pertanian Bogor.
[7] Purnama, Nikmatin S, Langenati R. 2006. *J Mater Sci.* Edisi khusus:154-162.
[8] Pratiwi A. 2008. *Pengaruh modifikasi permukaan Ti6Al4V terhadap deposisi apatit melalui metode biomimetik* [skripsi]. Bandung (ID): Institut Teknologi Bandung.
[9] Prihantoko DA. 2011. *Karakterisasi paduan CoCrMo dengan pelapian titanium nitrida dan hidroksiapatit-kitosan* [skripsi]. Bogor (ID): Institut Pertanian Bogor.
[10] Bertazzo S, Zambuzzi WF, Campos DD, Ogeda TL, Ferreira CV, Bertran CA. 2010. *J Biointerface.* **78**:177-184.

Journal of Physics: Conference Series **776** (2016) 012057 doi:10.1088/1742-6596/776/1/012057

Effect of MnO_2 doping and temperature treatment on optical energy band gap properties in Zn-Bi-Ti-O varistor ceramics

M. S. M. Ghazali[1], W. R. W. Abdullah[2], A. Zakaria[3], H. M. Kamari[3] and Z. Rizwan[4]

[1]School of Fundamental Science, [2]School of Ocean Engineering, Universiti Malaysia Terengganu, 21030 Kuala Nerus, Terengganu, MALAYSIA

[3]Department of Physics, Faculty of Science, Universiti Putra Malaysia, 43400 UPM Serdang, Selangor, MALAYSIA

[4]Department of Applied Sciences, Convener Purchase, National Textile University, Faisalabad (37610), PAKISTAN

E-mail: mohdsabri@umt.edu.my

Abstract. In this study, the optical band-gap energy (E_g) was investigated with respect to MnO_2 and sintering temperatures on ZnO based varistor ceramics. E_g of the ceramic (99-x) mol% ZnO + 0.5 mol% Bi_2O_3 + 0.5 mol% TiO_2 + x MnO_2 where x = 0, 0.2, 0.4, 0.6 and 0.8 mol%, were determined using UV-Vis spectrophotometer. The samples was prepared through solid-state route and sintered at the sintering temperature from 1110, 1140 and 1170 °C for 45 and 90 min in open air. At no doping of MnO_2, the values of E_g are 2.991 ± 0.001, 2.989 ± 0.001 eV for 45 and 90 min sintering time; respectively. E_g was decreased to 2.192 ± 0.001 eV at 1140 °C at 45 min sintering time. Similar result of E_g was observed at longer heat treatment. Further addition of dopant causing the E_g decreases rapidly to 2.099 and 2.106 ± 0.001 eV at 45 and 90 min sintering time; respectively. XRD analysis indicates that there is hexagonal ZnO and secondary phases, Zn_2MnO_4, $Bi_4Ti_3O_{12}$ and $Zn_2Ti_3O_8$. The relative density of the sintered ceramics decreased or remain constant with the increase of MnO_2 concentration for 45 min sintering time, however, further prolong sintering time; the relative density decreases form 90.25 to 88.35%. This indicates the pores are increasing with the increase of heat treatment. The variation of sintering temperatures to the optical band gap energy of based ZnO varistor doped with MnO_2 due to the formation of interface states.

1. Introduction

Zinc oxide (ZnO) is used widely in paint, sunblock in cosmetics application and textile industries as well as in electronic industries [1-4]. In electronic industries, ZnO based varistor is enormously used as an electronic component in automobile electronics and also in semiconductor electronic. They are fabricated with different types of dopants such as Bi_2O_3, Co_3O_4, MnO_2, Sb_2O_3, TiO_2 and $CaMnO_3$ [5-11]. Its unique grain boundary feature is responsible for nonlinear current-voltage (*I-V*) characteristics of the device [2, 12] and thus, is used to protect electrical surges. *I-V* studies have been extensively investigated for ZnO based varistor by previous researchers [7, 13] and it is necessary to investigate the electronic states of ceramic ZnO and the effect of doped impurities at different processing conditions. The characterization of the absorption spectrum in semiconductors leads to the determination of the optical band-gap energy [14, 15].

In this study, the investigation regarding the optical band gap energy and relationship with X-ray diffraction (XRD) findings of ZnO-Bi_2O_3-TiO_2 doped MnO_2 with different sintering temperatures of 1110, 1140 and 1170 °C for 45 and 90 min in air.

2. Materials & Method

All oxides precursors' of 99.9% purity (Alfa Aesar) were used. The composition consists of (98-x) mol% ZnO + 0.5 mol% Bi_2O_3 + 0.5 mol% TiO_2 + x MnO_2 where x = 0, 0.2, 0.4, 0.6 and 0.8 mol%. The powder was ball milled for 24 h in deionized water. The slurry was dried at 70 °C using hot plate and continuously magnetically stirred to avoid the sedimentation of the heavy particle and pre-sintered at 800°C for two hours in open air with heating and cooling rate of 6 °Cmin^{-1}. The pre-sintered mixture was pulverized using an agate mortar/pestle and after 1.75 wt.% Polyvinyl Alcohol binder addition, granulated by sieving 75 micron mesh screen. The mixture was then pressed into discs of 10 mm in diameter and 1 mm in thickness, each at a pressure of 2 ton/m^2. Finally, the discs were sintered at 1110, 1140 and 1170 °C in open air for 45 min sintering duration at heating and cooling rate of 2.66°C min^{-1}. The disk from each sample was ground for optical and XRD characterizations. The crystalline phases were identified by an XRD (PANalytical X'Pert Pro PW3040/60, Philips) with CuK_α radiation and the data were analyzed, using X'Pert High Score software. According to Keskenler and co-researcher [16], structural disordering can be obtained by Urbach's Rule for the optical absorption (exponential tail) [17]. The density was measured by the geometrical method taking the average of 10 discs [18, 19]. For the microstructure analysis, each of the disk samples was thermally etched at 150 °C in a tube furnace. The microstructure morphology of sintered discs was examined by Variable Pressure Scanning Electron Microscopy (VPSEM, Leo 1455). The average grain size (d) was determined by lineal intercept method [20], given by Equation 1:

$$d = 1.56\ L/MN \qquad (1)$$

where L is the random line length on the micrograph, M is the magnification of the micrograph and N is the number of the grain boundaries intercepted by lines. The UV-Vis spectrophotometer was used to measure the optical band-gap energy of the ceramics. The transmission signal was measured for the wavelength from 200 to 800 nm and then converted to absorption signal for further evaluation [21]. It was assumed that the fundamental absorption edge of the ceramics is due to the direct allowed transition [22]. The optical band-gap energy is given by Equation 2 [23]:

$$(Ah\upsilon)^2 = C(h\upsilon - E_g) \qquad (2)$$

where A is the optical absorption coefficient, C is the constant independent of photon energy ($h\upsilon$), and E_g is the direct allowed optical band-gap energy. From the plot of $(Ah\upsilon)^2$ versus $h\upsilon$, Figure 1 the value of E_g is obtained by using Origin Pro 8.0 software within the linear fitted regions at $(Ah\upsilon)^2=0$.

3. Results and Discussion

The E_g of the ceramics, Figure 1, decreases from 3.199 eV (pure ZnO) to 2.991 and 2.989 eV for the 0 mol% of MnO_2 at 1140 °C at 45 and 90 min sintering time; respectively. These decrements, 0.208 and 0.210 eV is due to the growth of interface states produced by combined effect of 0.5 Bi_2O_3 and 0.5 TiO_2. As can be seen, prolonged heat treatment make the E_g decreases slightly about 0.02 eV. The slight decrease is due to further growth of the interface states with the longer of the sintering time. When the ceramic ZnO + 0.5 Bi_2O_3 + 0.5 TiO_2 are doped with 0.2 MnO_2 then its E_g suddenly decreases to a value of 2.192 eV at 1140 °C for the 45 min sintering time. Similar result of E_g also was observed at longer sintering time. Further addition of dopant concentration causing the E_g decreases rapidly to 2.099 and 2.106 eV at 45 and 90 min sintering time. A proper interpretation regarding this decrement is due to the growth of the interface states due to the substitution or interstitial of Mn ions in the ZnO lattice as well as in the grain boundaries.

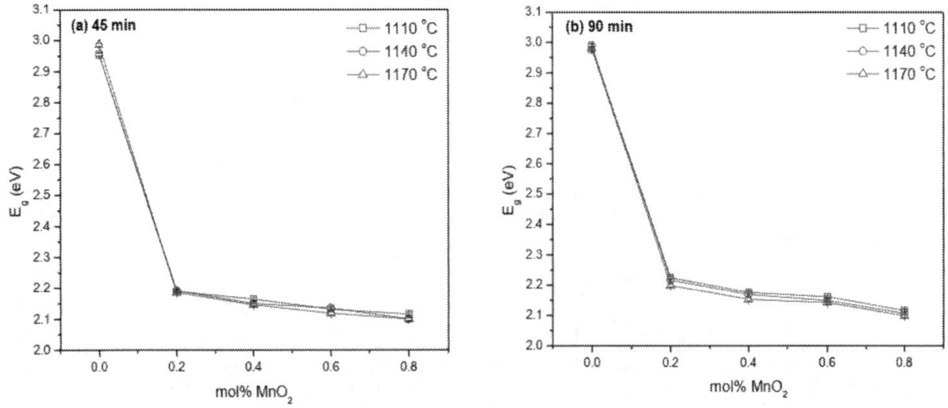

Figure 1 Band gap at (a) 45 min and (b) 90 min sintering time

It is observed that steepness factor σ_A, Figure 2, decreases with the doping level of MnO_2 indicating the increase of structural disordering in the varistor ceramics. Thus, this structural disordering introduces the growth of interface states and as a consequence, the E_g value decreases [23]. The steepness factor σ_B, Figure 2, decreases with the sintering temperatures indicating the increase in thermal energy of displacement [23], hence, increases the structural disordering at all sintering temperatures with slightly decrement of E_g as compared to undoped varistor ceramics.

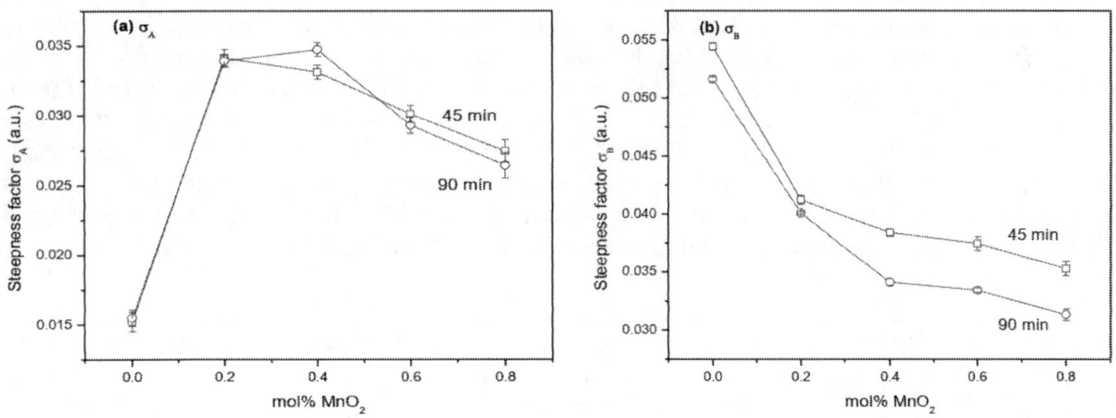

Figure 2 Variation of steepness factor (a) σ_A and (b) σ_B with mol% of MnO_2 at 1140 °C

There are two phases in XRD patterns, Figure 3, which are ZnO (ICSD code: 067454) and secondary phases, Zn_2MnO_4 (ICSD code: 039196), $Bi_4Ti_3O_{12}$ (ref. code: 00-065-2527) and $Zn_2Ti_3O_8$ (ICSD code: 083525). They were observed at all concentration except $Bi_4Ti_3O_{12}$ only seen at 0.2 mol%. Above 0.2 mol%, a few peaks of extra phase Mn_3O_4 (ref. code: 00-001-1127) were observed.

Journal of Physics: Conference Series **776** (2016) 012057 doi:10.1088/1742-6596/776/1/012057

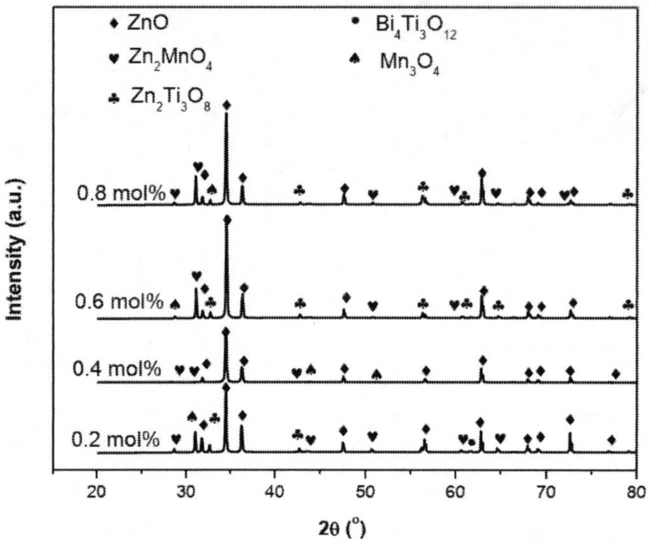

Figure 3 XRD patterns ZnO based varistor at 1140 °C for 45 min sintering time at different MnO_2 dopant concentration

The relative density, Figure 4, of the ceramic without MnO_2 increases from 91.92% up to 93.69% at 1140 °C and slightly decrease or constant at 1170 °C. Relative density decreases at longer sintering time. When dopant of MnO_2 is added in the ceramic the relative density decreases or remains constant with the increase of doping level at 45 min sintering time. However, further prolong sintering time; the relative density decreases from 90.25 to 88.35%. This indicates the pores are increasing with the increase of sintering time. It is also observed that the overall density with the doping of MnO_2 at all sintering condition is low indicating that the MnO_2 lowering the densification manner. It is found that large as well as small grains coexists and can be seen in the SEM micrographs, Figure 5. The pores are clearly observed and many of them have been seen at higher sintering temperature and time, as an additional proof with relative density, Figure 4. Some patches of additives can be seen in the ceramics. As evidence, the EDX results were conducted, Figure 6, show that the Bi is segregated in the grain boundaries at all sintering temperatures. It was found also that Mn and Ti were present near the grain boundaries; hence, this indicates that these elements are substituted in the ZnO lattice [24].

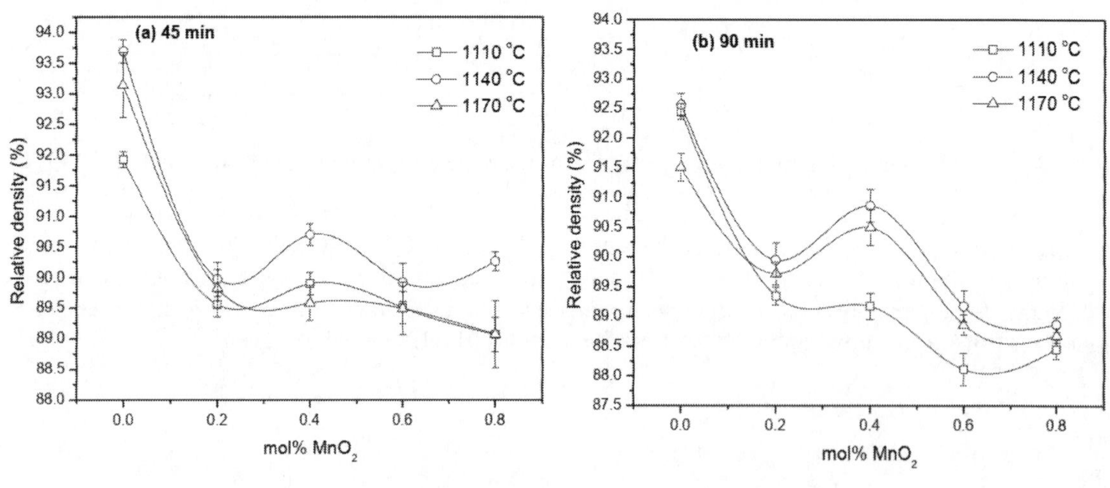

Figure 4 Relative densities of (a) 45 min and (b) 90 min sintering time

Journal of Physics: Conference Series **776** (2016) 012057 doi:10.1088/1742-6596/776/1/012057

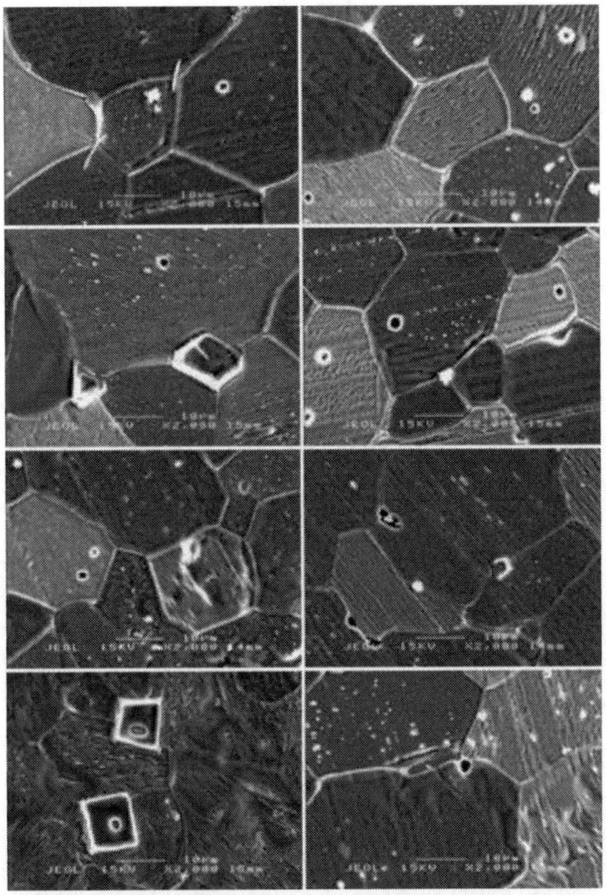

Figure 5 SEM micrographs of sintered varistor ceramics: from left (45 min) and right (90 min) from up to below is 0.2 mol%, 0.4 mol%, 0.6 mol% and 0.8 mol% of MnO_2 dopant at 1140 °C sintering temperature

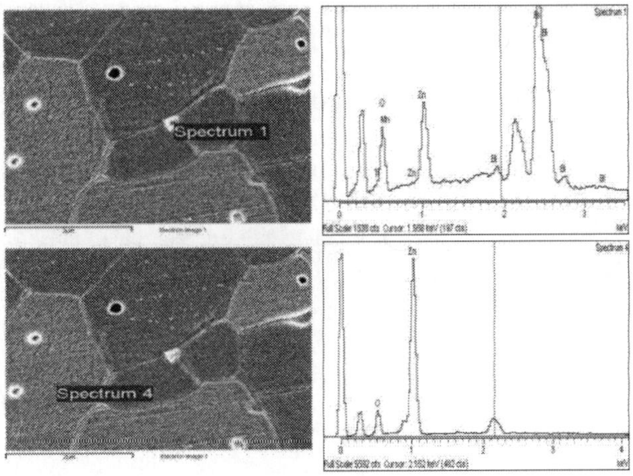

Figure 6 EDX micrographs and spectra of varistor ceramics

4. Conclusion

The lowering optical band gap energy with increment of MnO_2 dopant concentration had clarified through steepness factor σ_A and σ_B. These findings are associated with EDX results that show the segregation Bi_2O_3, TiO_2 and MnO_2 at grain boundaries and possibly substitution of Ti ion with Zn ion, which creates interface states within the forbidden regime.

5. Acknowledgement

The authors gratefully acknowledge the equipment's support for this work from Universiti Putra Malaysia and funding from RAGS (vot no. 57114). First author's is grateful to Universiti Malaysia Terengganu and Ministry of Higher Education for supporting the studies.

6. References

[1] Ammar AH, and Farag AAM 2010. *Phys. B* 405 1518
[2] Clarke DR 1999. *J. Am. Ceram. Soc.* 82(3) 485
[3] Feng H, Peng Z, Fu X, Fu Z, Wang C, Qi L, and Miao H 2010. *J. Alloy. Compd.* 497 304
[4] Look DC 2001. *Mat. Sci. Eng .B-Solid* 80 383.
[5] Matsuoka M 1971. *Jpn. J. Appl. Phys.* 10(6) 736
[6] Snow GS, White SS, Cooper RA, and Armijo JR 1980. *Am. Ceram. Soc. Bull.* 59(6) 617
[7] Eda K 1989. *IEEE Elect. Insul. Mag.* 5 28
[8] Bai SN, Shieh JS, and Tseng Ty 1995. *Mater. Chem. Phys.* 41 104
[9] Toplan O, Gunay V, and Ozkan OT 1997. *Ceram. Int.* 23(5) 251
[10] Fah CP, and Wang J 2000. *Solid State Ionics* 132 107
[11] Sabri MGM, Azmi BZ, Rafizah WAW, Zaid MHM, and Matori KA 2015. *Advanced Materials Research* 1107 20
[12] Souza FL, Gomez JW, Bueno PR, Cassia-Santos MR, Araujo AL, Leiti ER, Longo E, and Varela AJ 2003. *Mater. Chem. Phys.* 80 512
[13] Choon WN, and Byoung-Chil S 2003. *Mater. Lett.* 57 1322
[14] Ghoosh AK, Som KK, Chatterjee S, and Chaudhuri BK 1995. *Phys. Rev. B.* 51(8) 4842
[15] Zelaya-Angel O, Alvarado-Gil JJ, and Lozada-Morales R 1994. *Appl. Phys. Lett.* 64(3) 291
[16] Keskenler EF, Aydin S, Turgut G and Dogan S 2014. *Acta Physica Polonica A* 126(3) 782
[17] Meulenkamp EA 1999. *J. Phys. Chem. B* 103 7831
[18] Wang JF, Wen-Bin Su, Hong-Cun Chen, Wen-Xin Wang, and Guo-Zhong Zang 2005. *J. Am. Ceram. Soc.* 88(2) 331
[19] Sabri MGM, Azmi BZ, Zahid Rizwan, Halimah MK, Hashim M, Zaid, MHM, and Zamiri R 2011. *Int.J. Mol. Sci.*12 1496
[20] Wurst JC, Nelson JA 1972. *J. Am. Ceram. Soc.* 55 109
[21] Zahid Rizwan, Azmi BZ, and Sabri MGM 2011. *Int.J. Mol. Sci.* 12 1625
[22]
[23] Sabri MGM, Azmi BZ, Zahid Rizwan, Halimah MK, Hashim M, and Zaid MHM 2011. Int. J. Phys. Sci. 6(6) 1388
[24] Sabri MGM, Azmi BZ, Zahid Rizwan, Halimah MK, Hashim M, and Sidek HAA 2009. Am. J. App. Sci. 6(8) 1591

Effect of Ca-Doping on the Structure and Morphology of Polycrystalline $La_{0.7}(Ba_{1-x}Ca_x)_{0.3}MnO_3$ (x = 0; 0.03; and 0.05)

S Winarsih[1], B Kurniawan[1,a], A Manaf[1], S A Saptari[2], and D Nanto[3]
[1] Physics Department, Universitas Indonesia, Depok 16424, Indonesia
[2] Faculty of Science and Technology, Universitas Islam Negeri Syarif Hidayatullah, Jakarta 15412, Indonesia
[3] Department of Physics Education, Universitas Islam Negeri Syarif Hidayatullah, Jakarta 15412, Indonesia

E-mail: [a]bkuru07@gmail.com

Abstract. In this paper, we report structure and morphology of polycrystalline $La_{0.7}(Ba_{1-x}Ca_x)_{0.3}MnO_3$ (x = 0; 0.03; 0.05). Basically, these materials are perovskite manganites type with the general structure $AMnO_3$ (A= trivalent rare earth with divalent ion-doped) which have been extensively studied due to their interesting physical properties. It was known that the electron transport in this material influenced by ion doped at A site. Doping with different divalent ions should cause the lattice distortion. Hence, double exchange interaction is enhanced. In this study, we prepared our sample through the sol-gel method. It is show that the method has resulted in powder materials with ultra-fine particle size. The effect of Ca^{+2} and Ba^{+2} doping on the structure did not make any phase changing, but the lattice parameter of $La_{0.7}(Ba_{1-x}Ca_x)_{0.3}MnO_3$ decreased below x = 0.03. Microstructure observed by scanning electron microscope to the sintered samples indicated that the microstructure is homogeneous with fine size of equiaxed grain morphology. Microanalysis by EDS confirmed there is no significant different between designated composition and measured one. It is concluded that effect of Ca^{+2} and Ba^{+2} doped in $LaMnO_3$ has resulted in microstructural and lattice parameter changes. The doped materials are remaining single phase.

1. Introduction

Manganites with perovskite structure has the general formula $AMnO_3$ (A = trivalent rare earth with divalent ion-doped) [1-3]. The material has been extensively studied due to their interested physical properties [1-4]. Electron transport in this material influenced by ion doped at A site [2,3]. Ion doped caused lattice distortion and enhancement double exchange interaction [2,3]. Hence, it would influence transport properties in the material.

Transport properties in polycrystalline material can be explained by a core-shell model [5-7]. Each particle consists of core and shell [5-7]. Double exchange (DE) interaction is very crucial factor that influence the transport properties of the material. DE interaction is electron simultaneously hopping from Mn^{3+} ion to Mn^{4+} ion via O^{2-} ion [1]. The interaction dominant in the core. Hence promotes ferromagnetic behavior in this part. And the other shell is one part where the occurrence insulates region near the grain boundaries. Since this part consist of vacancies, defects, stress, and broken bonds, which causes decrement of DE interaction [5,7]. Intergrain distance plays an important role in the electron transport in a polycrystalline material [6]. When the intergrain distance becomes larger, electron transport becomes more difficult [6]. The consequence is the resistivity of the samples will decrease [6].

Preparation material by sol-gel method has been an interested study of the obtained material which has distribution of homogeneous size and improve physical properties of the material [9]. In this work, we observed the effect Ca-doping on the structure and surface morphology of polycrystalline $La_{0.7}(Ba_{1-x}Ca_x)_{0.3}MnO_3$ (x = 0; 0.03; 0.05). It is expected that Ca-doping would influence structure and morphology of $La_{0.7}Ba_{0.3}MnO_3$ and by using core-shell model we can explain transport properties of $La_{0.7}Ba_{0.3}MnO_3$ due to Ca-doping. Ca-soping was choosen because it can reduce Curie temperature (T_c) of $La_{0.7}Ba_{0.3}MnO_3$. It is important due to $La_{0.7}Ba_{0.3}MnO_3$ is one of candidate for application since the material has high magnetoresistance and magnetocaloric properties [2,3].

2. Experiment Method

Polycrystalline $La_{0.7}$ $(Ba_{1-x}Ca_x)$ $_{0.3}MnO_3$ (x = 0; 0.03; 0.05) were fabricated by the sol-gel method. The precursors were La_2O_3 (Merck, 99.5%), $BaCO_3$(Merck, 99%), $CaCO_3$(Merck, 99.0%), and $Mn(NO_3)_2.4H_2O$ (Merck, 98.5%). Stoichiometries amounts of each precursors were dissolved in nitric acid and then mixed them become one solution. After that, citric acid (Merck, 99%) was added to the solution with a molar ratio of citric acid/total metal ions being 1.0. Ammonia solution was used to adjust the pH value of the solution to 7. Stirred the solution at 353 K - 363 K until gel was formed. The gel was heated at 393 K for 4 hours in the dehydration process until the dried gel was obtained. After dehydration process, dried gel was heated at 873 K for 8 hours and finally the resulting powder was heated at 1123 K for 10 hours. The obtained powder was compacted and then further sintered at 1573 K for 30 hours to further crystal growth and grains become closer packed.

The X-ray diffraction with Cu-Kα radiation (Shimadzu)was used to determine crystal structure and lattice parameters of each sample. Scanning Electron Microscope (SEM) and Energy Dispersive X-ray (EDX) was used to determine the surface morphology and compositional purity of these samples.

3. Results and Discussion

The XRD pattern of polycrystalline $La_{0.7}(Ba_{1-x}Ca_x)_{0.3}MnO_3$ (x = 0; 0.03; 0.05) samples is shown in **Figure 1**. No structural changes occurring in the samples could be seen with increasing x value where the patterns are remaining the same. All diffracted peaks are well indexed correspond to the rhombohedral $LaMnO_3$ structure.

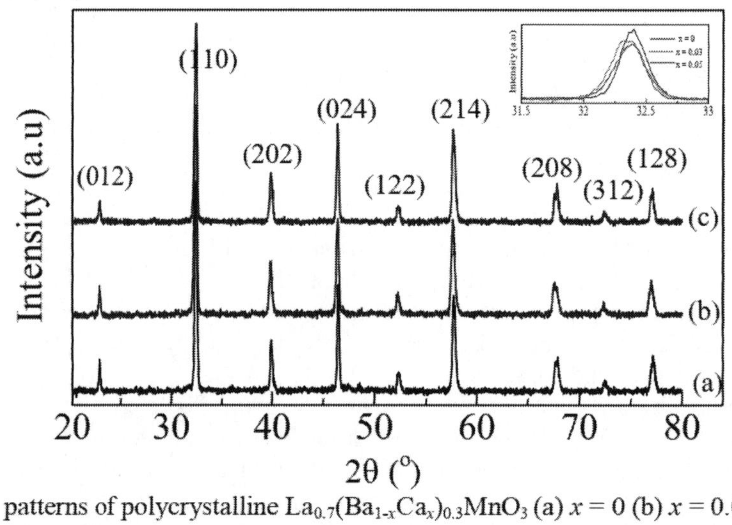

Figure 1. XRD patterns of polycrystalline $La_{0.7}(Ba_{1-x}Ca_x)_{0.3}MnO_3$ (a) x = 0 (b) x = 0.03 (c) x = 0.05

Table 1. Refined structural parameters of polycrystalline $La_{0.7}(Ba_{1-x}Ca_x)_{0.3}MnO_3$

Parameter structure	$x = 0$	$x = 0.03$	$x = 0.05$
Space group	R-3c	R-3c	R-3c
Crystal structure	Rhombohedral	Rhombohedral	Rhombohedral
a (Å)	5.5355	5.5464	5.5426
b (Å)	5.5355	5.5464	5.5426

c (Å)	13.500	13.519	13.522
Volume (Å^3)	358.39	360.79	359.71
<Mn-O> (Å)	1.9590	1.9634	1.9614
<Mn-O-Mn> (°)	172.22	172.22	172.22

However, a small shift in peak position (in set of Fig. 1) as the x value increased can be seen indicating a change in lattice parameter. Results of Rietveld analysis to the XRD data were summarized in **Table 1**. Crystallite size of the material evaluated by XRD measurement using WPPM method (Whole Powder Pattern Modelling). Crystallite size distribution of the material shows in **Figure 2**. The mean crystallite size of the material with $x = 0$, 0.03, and 0.05 were 46 nm, 47.3 nm, and 47.8 nm respectively. It is shown that there is slightly difference in crystallite size distribution due to increasing doping concentration. It is indicated that doping concentration until 5wt% did not affected to the distribution of crystallite size of the material.

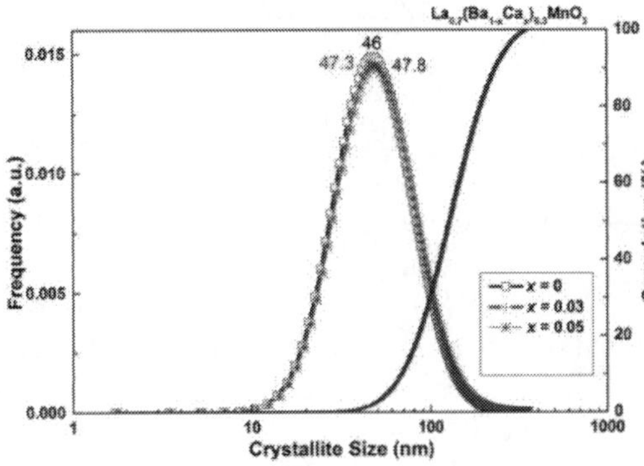

Figure 2. The crystallite size cumulative distribution of polycrystalline $La_{0.7}(Ba_{1-x}Ca_x)_{0.3}MnO_3$

Rietveld refinement analysis confirmed that all samples have rhombohedral structure and R-3c space group. It is evident that the increment of Ca concentration in polycrystalline $La_{0.7}(Ba_{1-x}Ca_x)_{0.3}MnO_3$ caused an increase in lattice parameters compared with un-doped sample. However, the length of the lattice axis for samples x = 0.03 and 0.05 is slightly different. The increase in lattice parameter of $La_{0.7}(Ba_{1-x}Ca_x)_{0.3}MnO_3$ samples is shown in **Figure 3**. These results are in good agreement with literature [2,3]. The increment of lattice parameters and unit cell volume due to Ca doping concentration might occur because the average ionic radius in *A*-site increase with increasing doping concentration [2,3].

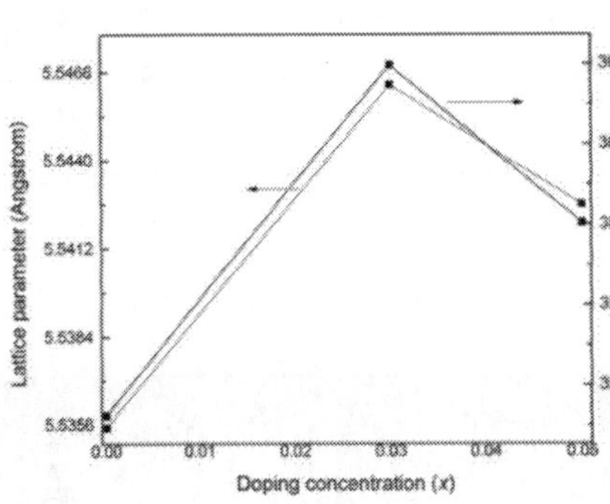

Figure 3. Doping concentration dependent lattice parameter and unit cell volume of $La_{0.7}(Ba_{1-x}Ca_x)_{0.3}MnO_3$

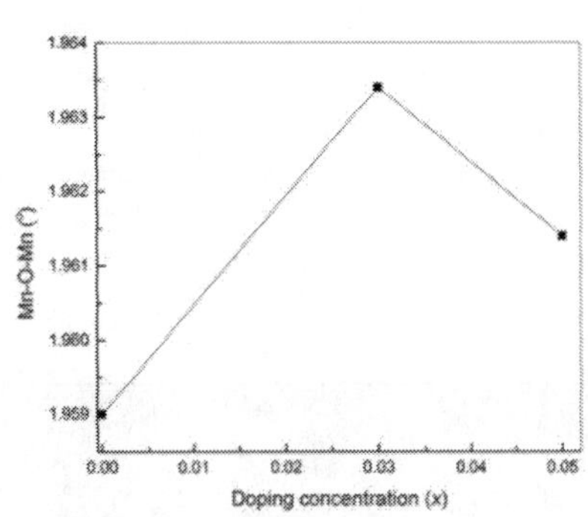

Figure 4. Doping concentration dependent Mn-O-Mn bond length of $La_{0.7}(Ba_{1-x}Ca_x)_{0.3}MnO_3$

Figure 4 demonstrates doping concentration dependent to the Mn-O-Mn bond length. Mn-O bond length and Mn-O-Mn bond angle could get by Rietveld refinement analysis. The change of Mn-O-Mn bond length would impact the transport properties of polycrystalline $La_{0.7}(Ba_{1-x}Ca_x)_{0.3}MnO_3$. Since polycrystalline sample consist of grains with intergrain distance to be about $d/2$, where d is Mn-O-Mn bond length [6]. Core-shell model of perovskite manganites depicts in **Figure 5**. When the intergrain distance becomes larger, electron transport becomes more difficult [6]. The consequence is the resistivity of the material will decrease [6]. It means that there is decrement of DE interaction due to the increment of Mn-O-Mn bond length [6]. The change of Mn-O-Mn bond length occurred because doping in A site of perovskite structure caused lattice distortion [2,3].

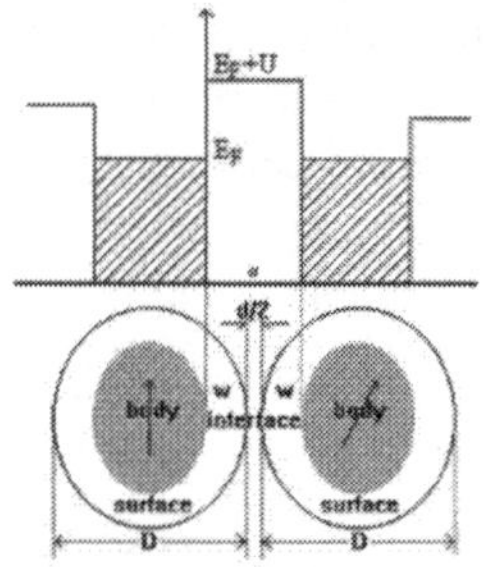

Figure 5. Core-shell model of perovskite manganites [6]

SEM images of $La_{0.7}(Ba_{1-x}Ca_x)_{0.3}MnO_3$ (x = 0; 0.03; 0.05) samples are presented in **Figure 6**. The images showing close pack grains microstructure with equiaxed grain morphology randomly oriented. A change in mean grain size of the samples could be clearly seen in the images. The mean grain size of the samples with x = 0, 0.03, and 0.05 were 274 nm, 229 nm and 1970 nm respectively. A significant increase occurred in sample with x = 0.05 but grains become more packed.

Figure 6. SEM images of $La_{0.7}(Ba_{1-x}Ca_x)_{0.3}MnO_3$ samples (a) $x = 0$ (b) $x = 0.03$ (c) $x = 0.05$

Figure 7 shows Energy Dispersive X-ray (EDX) analysis spectrum of $La_{0.7}(Ba_{1-x}Ca_x)_{0.3}MnO_3$ samples. Identification by EDX analysis in each spectrum indicated that only constituent elements in $La_{0.7}(Ba_{1-x}Ca_x)_{0.3}MnO_3$ were identified. It is concluded that the compositional purity of all samples is confirmed. Moreover, there is no quantifiable loss of each element during the sample preparation process. It was demonstrated in **Table 2** which summarized the compositional analysis of these samples. A slight difference in quantitative value between designated composition and measured one might be due to the EDX method is based on semi quantitative analysis. Hence, the results are not absolute.

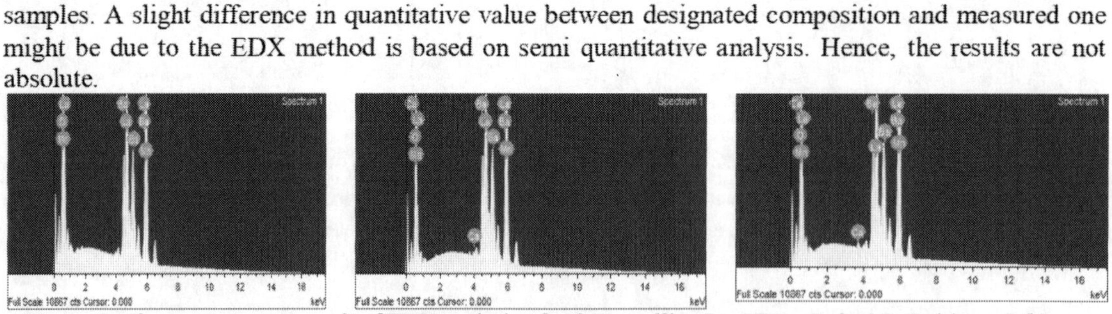

Figure 7. The measurement result of EDX analysis of polycrystalline $La_{0.7}(Ba_{1-x}Ca_x)_{0.3}MnO_3$ (a) $x = 0$ (b) $x = 0.03$ (c) $x = 0.05$.

Table 2. Compositional analysis of polycrystalline $La_{0.7}(Ba_{1-x}Ca_x)_{0.3}MnO_3$

Doping concentration	Element composition	Designated composition (at.%)	Measured composition (at.%)
	La	14.00	13.36
$x = 0$	Ba	6.00	5.70
	Ca	0.00	0.00
	Mn	20.00	17.14
	O	60.00	63.80
$x = 0.03$	La	14.00	13.32
	Ba	5.82	6.00
	Ca	0.18	0.14
	Mn	20.00	17.53
	O	60.00	63.01
$x = 0.05$	La	14.00	11.83
	Ba	5.70	4.80
	Ca	0.30	0.24
	Mn	20.00	15.75
	O	60.00	67.38

4. Summary

The synthesis of polycrystalline $La_{0.7}(Ba_{1-x}Ca_x)_{0.3}MnO_3$ (x = 0; 0.03; 0.05) materials by the sol-gel method were successfully achieved. The effect of Ca^{+2} and Ba^{+2} doping on the structure did not make any phase changing, but the lattice parameter of $La_{0.7}(Ba_{1-x}Ca_x)_{0.3}MnO_3$ decreased below x = 0.03. Ca-doping in synthesized materials leads to the homogeneous microstructure with fine size of equiaxed grain morphology. By using core-shell model, we can conclude that such effects to the structure of the material would give impact to the transport properties of these materials.

5. Acknowledgement

The authors gratefully acknowledge the support of the Physics Department, Universitas Indonesia. This work was partially funded by the LPDP (Lembaga Pengelola Dana Pendidikan) Republic of Indonesia under research grant contract no. PRJ-613/LPDP.3/2016 and by the university under research grant contract no. 1993/UN2.R12/HKP.05.00/2016.

6. References

[1] Salamon M B and Jaime M 2001 *Rev. Mod. Phys.* **73** 583–628.
[2] Manjunatha S O, Rao A, Lin T-Y, Chang C-M and Kuo Y-K 2015 *J. Alloys Compd.* **619** 303–10.
[3] Manjunatha S O, Rao A, Subhashini and Okram G S 2015 *J. Alloys Compd.* **640** 154–61.
[4] Ayadi F, Regaieg Y, Cheikhrouhou-koubaa W, Koubaa M and Cheikhrouhou A 2015 *J. Magn. Magn. Mater.* **381** 215–9.
[5] Hueso L E, Sande P, Miguens D R, Rivas J, Rivadulla F and Lopez-Quintella M . 2002 *J. Appl. Phys.* **91** 9943.
[6] Zhang N, Ding W, Zhong W, Xing D and Du Y 1997 *Phys. Rev. B* **56** 8138–42.
[7] Baaziz H, Maaloul N K, Tozri A, Rahmouni H, Mizouri S, Khirouni K and Dhahri E 2015 *Chem. Phys. Lett.* **640** 77–81.
[8] L. E. Hueso, P. Sande, D. R. Miguens, J. Rivas, F. Rivadulla, and M. Lopez-Quintella 2002 J. Appl. Phys. **91** 9943.
[9] Gaur A and Varma G . D 2006 *J. Phys. Condens. Matter* **18** 8837–46.

The neutron skin thickness in nuclei with clustering at low densities

A. Nooraihan[1], Q. N. Usmani[2] , Z. Sauli[3] and K. Anwar[1]
[1]Institute of Engineering Mathematics, Universiti Malaysia Perlis, MALAYSIA
[2]Centre of Theoretical Physics, Jamia Millia Islamia, New Delhi, INDIA
[3]School of Microelectronics Engineering, Universiti Malaysia Perlis, MALAYSIA

Email: raihan@unimap.edu.my

Abstract. This study concentrates on searching for a dependable, fully microscopic theory to find out new behaviours and understand their consequences for theoretical pictures. The models for nuclear structure are tested, refined and developed by acquiring new data [1][2][3]. This data is useful for astrophysical calculations and predictions. In density functional theories, including the ETF theory, the equation of state (EOS) of symmetric nuclear matter (SNM), is an important measure. Empirically, we receive information about quantities relating to SNM, all these measures are thoroughly tested. In the absence of any unswerving knowledge below this density we shall take that energy still rises up to some density, neglecting possible small fluctuations, as the density is brought down. Our discussion at the moment is without the Coulomb forces applicable only for the hypothetical nuclear matter; they are added finally to correctly portray the actual picture in nuclei. Our approach in this study is macroscopic. This work concludes that the neutron skin thickness in nuclei is found to reduce significantly, for the reason of clustering.

Introduction

We start with the statement that the binding energy of SNM per nucleon in the limit of zero density must get near u_v (≈16 MeV) at the saturation density ρ_0(≈0.16 fm^{-3}), where u_v is the volume term in the Weizsäcker mass formula. The previous statement is based upon the theory that nuclear matter in its ground state (T=0 MeV) at a given density will achieve the lowest possible energy. Our statement can be demonstrated in the following procedure. It was explained in [4] that an idealized α -matter picture of SNM in the neighbourhood of zero density provides energy per nucleon, $E/A \approx -7.3$ MeV, which is the energy of α -particle per nucleon with Coulomb interaction turned off. The previous conclusion is at variance with the mean field theories, results where $E(\rho)\to 0$ as the density $\rho\to 0$, for instance, in the Skyrme-Hartree-Fock calculations. Questions may arise, why α -particles? Why not heavier nuclei that have lower energies per nucleon compared to α -particle? For example, an ideal ^{40}Ca-matter will have lower E/A than an ideal α -matter. This statement of the argument can be expanded further, because, heavier nuclei will lead to lower E/A as the Coulomb interaction is turned off. We might consequently look at SNM as an ideal gas of chunks of NM or very heavy nuclei where we can neglect the surface effects. We consider the densities low enough so that interactions amongst clusters are negligibly small and the idea of the perfect cluster - matter is valid. In that condition, we acquire the exact result $E(\rho)\rightarrow -u_v$ as $\rho \rightarrow 0$. Hence, we get the relation

$$E(\rho \rightarrow 0) \rightarrow E(\rho = \rho_0) = -u_v\,. \tag{1}$$

Extended Thomas-Fermi Theory

Incorporation of (1) into a theory is far from trivial. Clearly Hartree-Fock methods cannot take into account clustering at the nuclear surface since they do not satisfy (1). Although mean-field theories do contain some aspect of clustering, they are not related to surface properties. There are cluster models of nuclei and hypernuclei in terms of specific clusters, mostly α-particle clusters, but these are confined to light nuclei [5]. On the other hand, we have *ab initio* theories which give accurate or nearly exact results, but they are also confined to light nuclei [6]. In the future, it may become possible to extend *ab initio* theories to heavy nuclei with realistic Hamiltonians, but it would be difficult to entangle cluster properties, particularly its universal character, if any, at the nuclear surface. Recently, there has been a spurt of activity in developing microscopically based nuclear density-functional theories [7]–[9] for medium and heavy nuclei with reference to realistic two- and threenucleon or Skyrme interactions. The realistic interactions used are either the Argonne AV18 (two-body) [10] and Urbana IX (three-body) [11] or they are derived from chiral effective field theory. These studies have started with the motivation to search for a universal nuclear energy density functional - a new model for nuclear theory [12], which in principle is supposed to have all the many-body correlations. But none of these theories, because of constraints or choice of density functional, conform to identity (1); on the contrary, they give $E(\rho)\rightarrow 0$ as $\rho\rightarrow 0$. They do not include the clustering aspect in nuclei to the full extent. Besides, on the basis of fits to energy alone it will be difficult to see the impact of (1) unless one probes those properties of nuclei which are sensitive to surface region, such as the neutron skin thickness, symmetry energy at low densities, and the one-neutron separation energies for those pair of nuclei where separation energies are small.

To find a detailed structure of the EOS of SNM with GFMC/DMC methods at low densities is a formidable task and beyond the reach of present-day techniques and computing power. We thus resort to phenomenology and approximate methods. We use an extended version of the Thomas-Fermi model (ETF), which we believe is the only theory at present which can incorporate (1) easily and at the same time describe the properties of a large number of nuclei. The ETF theory provides quick and reliable estimates of energies and other physical quantities of interest with a global insight. This implementation not only explains the experimental large values of symmetry energy [13] at low densities, but also affects the binding energies of nuclei, one- and two-neutron separation, and β-decay energies in the right direction, although by a modest amount. But the influence of clustering on the neutron skin thickness and one-neutron separation energies for those pairs of nuclei with separation energies less than 5 MeV is significant. The latter improvement is crucial for describing the nuclei near the drip line.

Because of cluster formation, the neutron and proton densities at the nuclear surface would tend to equalize each other. This tendency to equalize results in reduced neutron skin thickness as compared to when clustering is absent. We do find considerable reduction in neutron skin thickness. It has bearings on the recently concluded parity-violating electron-scattering experiment on 208Pb and in turn on the neutron star radii [14].

Results and Discussion

We present results for the neutron skin thickness and other relevant quantities. Discussion follows the results as they are described. Finally, we present an overall critique of the results.

A quantity of interest is the neutron skin thickness [15], defined as the difference between the rms radii of neutrons and protons:

$$\delta R = \sqrt{\langle r_n^2 \rangle} - \sqrt{\langle r_p^2 \rangle} \,. \tag{2}$$

In Skyrme-HF theories, δR is sensitive to the slope of the symmetry energy L at the saturation density, defined as

$$L = 3\rho_0 \left. \frac{\partial E_{sym}}{\partial \rho} \right|_{\rho_0} . \tag{3}$$

We expect the clustering to affect δR significantly as it is a direct surface phenomenon. In Table I we give results of our calculations for clustering and no clustering. Results for no-clustering are in reasonable agreement with Skyrme-HF [16], [17] and RMF [18] calculations, and experimental deductions [19], [20] but there is clear discrepancy between the clustering and the other results including the experiment. We find much lower values for δR. Does this imply that experimental deductions are implemented assuming no clustering? They are indeed model dependent. A recently completed parity-violating electron-scattering experiment at the Jefferson Laboratory will greatly help to clarify this inconsistency. Our L value for both cases of clustering and no clustering is the same: $L \approx$ 68 MeV, well within the range of values extracted from isospin diffusion data.

From Table VI [2], we can see that increase in $\sigma(E)$ is not very large compared to the results for 367 spherical nuclei. On average it increases by 0.08 MeV, still within the reasonable limits from the viewpoint of microscopic-macroscopic theories. But $\sigma(R)$ increases by a factor of 2. We do not have any clear explanation for this discrepancy. We may attribute this increase to the absence of self-consistency and particularly, to deformation in our formulation. We also give the *rms* deviations $\sigma(S1)$, $\sigma(S2)$ and $\sigma(Q_\beta)$ for one-($S1$) and two-($S2$) neutrons separation, and β-decay energies, respectively.

Correct prediction of one- and two-neutron separation energies is important from the point of view of neutron drip line. One neutron separation energies govern the asymptotic density of neutrons [21], [22], whereas the two-neutron separation energies reveal the shell structure in an isotopic chain [23]. It is seen in Table VI [2] that these quantities, which are, $\sigma(S1)$, $\sigma(S2)$ and $\sigma(Q_\beta)$, are quite reasonable. These are somewhat larger for the no-clustering case (line 2) implying that clustering does affect the fits though quite modestly; it was more so in case of 367 spherical nuclei the difference is somewhat significant. Line 5 in Table VI [2] gives the results with Wigner term excluded. The value of $\sigma(E)$ increases to 0.12 MeV. This demonstrates the importance of Wigner term. Our preferred results are for $AV8'+V_{2\pi}^{PW}+V_{\mu=300}^{R}$(line 3) and $AV8'$ (line 4), since the *rms* deviations for these are small compared to other results. Last line of the Table VI [2] give results for calculation with $K = 260$ MeV, which demonstrates that dependence on K is weak. It is largely compensated by a change in the parameter M.

Journal of Physics: Conference Series **776** (2016) 012059 doi:10.1088/1742-6596/776/1/012059

Table I Results for neutron skin thickness for a number of nuclei. All items are in fermis. Outcomes for no-clustering correlate with the second row of Table VI [2]. Outcomes for clustering correlate with the averages of the first, third and the fourth rows of Table VI [2]. Observe the significant deduction in δR when clustering is involved.

Nucleus	δR (fm), neutron skin thickness					
	Clustering	No-clustering	HFB-17 [24]	Sky-HF [16], [17], [25], [26]	RMF [18]	Experiment
^{208}Pb	0.10	0.16	0.15	0.22±0.04	0.21	0.16±0.06 [27] 0.18±0.035 [28]
^{132}Sn	0.15	0.23	–	0.29±0.04	0.27	0.24±0.04 [28]
^{124}Sn	0.10	0.16	–	0.22±0.04	0.19	0.185±0.017 [20]
^{90}Zr	0.04	0.07	–	0.09±0.04	–	0.07±0.04[19], [29]
^{48}Ca	0.10	0.16	–	–	–	–

In Table I, we give results of our calculations for clustering and no-clustering. The results for no-clustering are in reasonable agreement with Skyrme-HF [16], [17], [25], [26] and RMF [18] calculations, and experimental deductions [19], [20], [29]. However, there is clear discrepancy between the clustering and the other results including the experiment. We find much lower values for δR. Does this imply that experimental deductions are implemented assuming no-clustering? In fact, they are model dependent. A recently completed parity-violating electron scattering experiment [30] at the Jefferson Laboratory will greatly help to clarify this inconsistency. Our L value for both the cases of clustering and no-clustering is similar; $L \approx 68$ MeV well within the range of values taken from isospin diffusion data.

Clustering at low densities significantly reduces the neutron skin thickness in nuclei. The conventional Skyrme density functionals do not incorporate clustering at low densities; thus their estimates of neutron skin thickness can be considerably off. Also, since they do not contain a quartic isospin term, it would be difficult to explain quantitatively the nuclear binding energies in a fully microscopic density-functional theory and at the same time conform to the realistic EOS of neutron matter, particularly in situations where they are steep, for example with UIX.

References

[1] Sauli Z, Abdullah N, Anwar K, and Usmani QN Sep. 2012 *J. Phys. Conf. Ser.*, vol. 381, p. 012102.
[2] Usmani QN, Abdullah N, Anwar K, and Sauli Z Dec. 2011 *Phys. Rev. C*, vol. 84, no. 6, p. 064313.
[3] Usmani QN, Sauli Z, and Abdullah N Sep. 2011 *J. Phys. Conf. Ser.*, vol. 312, no. 9, p. 092060.
[4] Johnson MT and Clark JW 1980 *Kinam*, vol. 2, pp. 3–33.
[5] Bodmer AR, Usmani QN, and Carlson J 1984 *Phys. Rev. C*, vol. 29, no. 2, pp. 4–7.
[6] Pieper SC and Wiringa RB 2001 *Annu. Rev. Nucl. Part. Sci.*, vol. 51, pp. 53–90.
[7] Stoitsov M, Kortelainen M, Bogner SK, Duguet T, Furnstahl RJ, Gebremariam B, and Schunck N 2010 *Phys. Rev. C*, vol. 82, p. 054307.
[8] Baldo M, Robledo L, Schuck P, and Viñas X 2010 *J. Phys. G Nucl. Part. Phys.*, vol. 37, p. 064015.
[9] Kortelainen M, Lesinski T, Moré J, Nazarewicz W, Sarich J, Schunck N, Stoitsov MV, and Wild S 2010 *Phys. Rev. C*, vol. 82, p. 024313.
[10] Wiringa RB, Stoks VGJ, and Schiavilla R 1995 *Phys. Rev. C*, vol. 51, no. 1.
[11] Pudliner BS, Pandharipande VR, Carlson J, and Wiringa RB 1995 *Phys. Rev. Lett.*, vol. 74, no. 22, pp. 3–6.
[12] Bertsch GF, Dean DJ, and Nazarewicz W 2007 *SciDAC Rev.*, no. 6 Winter, pp. 42–51.
[13] Natowitz JB, Röpke G, Typel S, Blaschke D, Bonasera A, Hagel K, Klähn T, Kowalski S, Qin L, Shlomo S, Wada R, and Wolter HH 2010 *Phys. Rev. Lett.*, vol. 104, no. May, p. 202501.
[14] Horowitz CJ, Pollock S, Souder P, and Michaels R-J 2001 *Phys. Rev. C*, vol. 63, no. 2, p. 025501.
[15] Bodmer AR and Usmani QN-M 2003 *Phys. Rev. C*, vol. 67, no. 3, p. 034305.
[16] Li B-A, Chen L, and Aug CMKo 2008 *Phys. Rep.*, vol. 464, no. 4–6, pp. 113–281.
[17] Chen L-W, Ko CM, and Li B-A Dec. 2005 *Phys. Rev. C*, vol. 72, no. 6, p. 064309.
[18] Yako K, Sagawa H, and Sakai H Nov. 2006 *Phys. Rev. C*, vol. 74, no. 5, p. 051303.
[19] Todd BG-R and Piekarewicz J, Sep. 2005 *Phys. Rev. Lett.*, vol. 95, no. 12, p. 122501.
[20] Terashima S, Sakaguchi H, Takeda H, Ishikawa T, Itoh M, Kawabata T, Murakami T, Uchida M, Yasuda Y, Yosoi M, Zenihiro J, Yoshida H, Noro T, Ishida T, Asaji S, and Yonemura T Feb. 2008 *Phys. Rev. C*, vol. 77, no. 2, p. 024317.
[21] Gambhir YK and Patil SH 1985 *Zeitschrift für Phys. A Atoms Nucl.*, vol. 321, pp. 161–164.
[22] Bhagwat A, Gambhir YK, and Patil SH Sep. 2000 *Eur. Phys. J. A*, vol. 8, no. 4, pp. 511–520.
[23] Bhagwat A, Viñas X, Centelles M, Schuck P, and Wyss R Apr. 2010 *Phys. Rev. C*, vol. 81, no. 4, p. 044321.
[24] Goriely S, Chamel N, and Pearson J Apr. 2009 *Phys. Rev. Lett.*, vol. 102, no. 15, p. 152503.
[25] Steiner AW and Li B-A Oct. 2005 *Phys. Rev. C*, vol. 72, no. 4, p. 041601.
[26] Li B and Steiner AW 2006 *Phys. Lett. B*, vol. 642, pp. 436–440.
[27] Kłos B, Trzcińska A, Jastrzębski J, Czosnyka T, Kisieliński M, Lubiński P, Napiorkowski P, Pieńkowski L, Hartmann FJ, Ketzer B, Ring P, Schmidt R, Von T-E, Smolańczuk R, Wycech S, Gulda K, Kurcewicz W, Widmann E, and Brown BA Jul. 2007 *Phys. Rev. C*, vol. 76, no. 1, p. 014311.
[28] Klimkiewicz A, Paar N, Adrich P, Fallot M, Boretzky K, T Aumann, Cortina D-G, Pramanik UD, Elze TW, Emling H, Geissel H, Hellström M, Jones KL, Kratz JV, Kulessa R, Nociforo C, Palit R, Simon H, Surówka G, Sümmerer K, Vretenar D, and Waluś W Nov. 2007 *Phys. Rev. C*, vol. 76, no. 5, p. 051603.
[29] Piekarewicz J Apr. 2004 *Phys. Rev. C*, vol. 69, no. 4, p. 041301.
[30] Horowitz CJ 2010 *INPC Proceeding*.

Study on the synthesis of Indium Tinoxide (ITO) nanomaterial using sol gel process and its potential for CO gas detection

Goib Wiranto[1], Novrita Idayanti and Lilis Retnaningsih

Research Center for Electronics and Telecommunication, Indonesian Institute of Sciences (LIPI), Jl. Sangkuriang, Bandung 40135, INDONESIA

Email: gwiranto@gmail.com

Abstract. This paper described the synthesis of doped and undoped nanomaterial Indium Tinoxide (ITO) using sol-gel process with the base materials from Indium Nitride (In(NO3)) and Tin Chloride (SnCl4). Doping was done using Palladium Chloride (PdCl), with the molar ratio of 90:10 and 70:30, respectively. Physical characteristics of the nanocrystallites were examined using SEM, EDS, XRD, and FTIR, whereas its resistivity was tested agaist CO gas. The result showed that the ITO powders have a crystal size smaller than 100 nm, with a cubic crystal structure (byxbite type), and having strong molecular bonds for In-O-In and Sn-O-Sn. The response to CO gases showed a decreasing resistivity with increasing CO gas concentration, showing its potential for gas sensor application.

1. Introduction

Indium Tinoxide (ITO) is a semiconducting material that has been widely used in the field of optoelectronics. Many devices such as CCD (Charged Coupled Device), LCD (Liquid Crystal Display), solar cells, LED (Light Emitting Diode), Photo Diode, Photo Transistor, and Laser, have used layers of ITO because of its optical transparancy and high reflectivity in ultraviolet range [1]. In the field of sensor, ITO has been used as an active material to detect gases such as NO_2 [2], methanol [3], H_2 [4], and CO [5]. This widespread application has made ITO an interesting topic to be studied more intensively, especially its synthesis process and characteristics. The process of forming layers of ITO can be conducted using many different techniques, such as sputtering [6], CVD (Chemical Vapor Deposition) [7], thermalevaporation [8], and spray pyrolysis [9], and the resulting layer characteristics are very dependent on the process parameters used during deposition [10].A simple and thus very popular method in synthesising an ITO material is by sol-gel process. Compared with the other techniques, sol-gel process has the advantages in terms of homogeanity, purity, low processing temperature, simplicity in adding dopant elements, and compatibility with microfabrication processes. Previous works in the application of sol-gel process to synthesise ITO nanomaterials have been demonstrated by Li and Ren [11], Hjiri *et al* [12], Silva *et al* [13], and Ibrahem and Moghdad [14]. Therefore this research was conducted to synthesize ITO nanomaterial using sol-gel process. Characterisation was conducted to study the physical characteristics of the resulting ITO powders using SEM, EDS, XRD, and FTIR, and the electrical characteristic was studied using four-point probe in response to CO gas.

[1]To whom any correspondence should be addressed.

2. Experimental

Indium Nitride ($In(NO_3)_3$) and Tin Chloride ($SnCl_4$) have been chosen as the base materials, and $PdCl_2$ was used as the dopant material.Four samples have been prepared in this experiment with each molar composition as seenin Tabel 1. Sample A and B were for the undoped ITO, and sample C and D were for the doped ITO.

Tabel 1. Molar composition of all samples

Sample	$In(NO_3)_3$	$SnCl_4.5H_2O$	$PdCl_2$	% Molar Ratio
A	10 g	1.1 g	-	90 : 10
B	10 g	4.28 g	-	70 : 30
C	10 g	1.1 g	0.28 g	90 : 10
D	10 g	4.28 g	0.36 g	70 : 30

Initially, 10 g of $In(NO_3)_3$was dissolved in 10 ml H_2O. All materials were then mixed for each sample, and added with 15 ml Diethylene Glycol. The resulting mixture was stirred using magnetic stirrer for 2 hours, while maintaining its temperature at 60 °C and pH = 1. A clear solution (sol) was then resulted. The next step was to increase the pH of the sol solution by adding 10 ml of ammonium (NH_3) solution. This has resulted in pH value of the solution = 9. Stirring was then repeated for 2 hours using magnetic stirrer, while maintaining its temperature at 60 °C, and a white gel was formed. The resulting white gel was then filtered and rinsed with DI water, and dried for 24 hours. The final step was to dry the white powder in an oven at 60 °Cfor 3 hours, followed by calcination at 400 °C for 2 hours.

3. Results and discussion

The ITO powders resulted after calcination process at 400 °C have a dark brown color for sample A and B (undoped), and light brown color for sampel C and D (doped). It appeared that the addition of $PdCl_2$ has caused the difference in the resulting ITO powders.

3.1. Morfology

Initial observation showed that the grain size of the resulting ITO powder for sample A was much below 100 nm. As can be seen from the SEM results in Figure 1a, the majority of the grain size measured around 50 nm. For sample B, the average particle size was even smaller than sample A, as seen from Figure 1b. This is in consequence with the increase in Tin (Sn) components of the ITO composition. The atomic radius for ion In^{3+} is 0.81 Angstrom, whereas that of Sn^{4+} is 0.71 Angstrom. In sample B, with the composition of Sn : In = 70 : 30, there are more Sn^{4+} ions replacing the In^{3+} ion position in their lattice structure than in sample A.

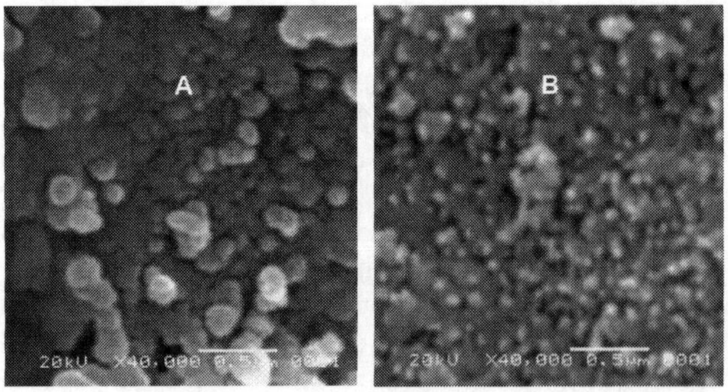

Figure 1. SEM view of: (a). sample A, and (b). Sample B

As can be seen in Figure 2, the addition of Palladium dopant did not have a significant effect to the average particle size of the ITO powder. Although Pd^{2+} ion has an atomic radius of 86 Angstrom, its composition of only 2% from the total ITO composition did not have enough strength to alter the ITO particle size.

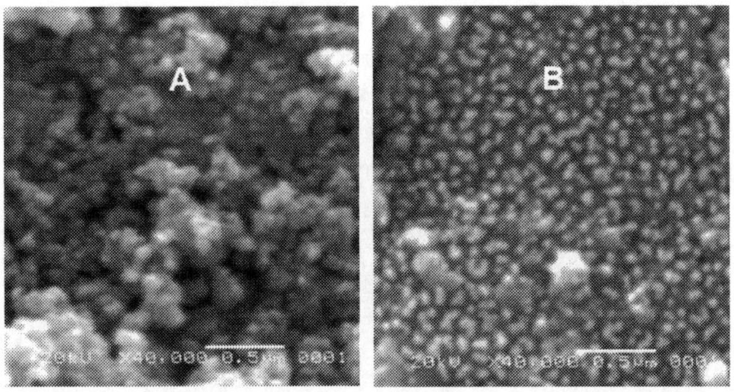

Figure 2. SEM view of: (a). sample C, and (b). Sample D

3.2. Composition

To study the elemental composition of the resulting ITO powders, the EDS spectrum of each sample has been taken. ITO consists of three elements, In, Sn, and O, which are formed by substitution of Sn atom into In^{3+}in In_2O_3compound (through doping substitution mechanism). Therefore the chemical structure of ITO can be expressed in the form of $In_{2-x}Sn_xO_{3-2x}$.

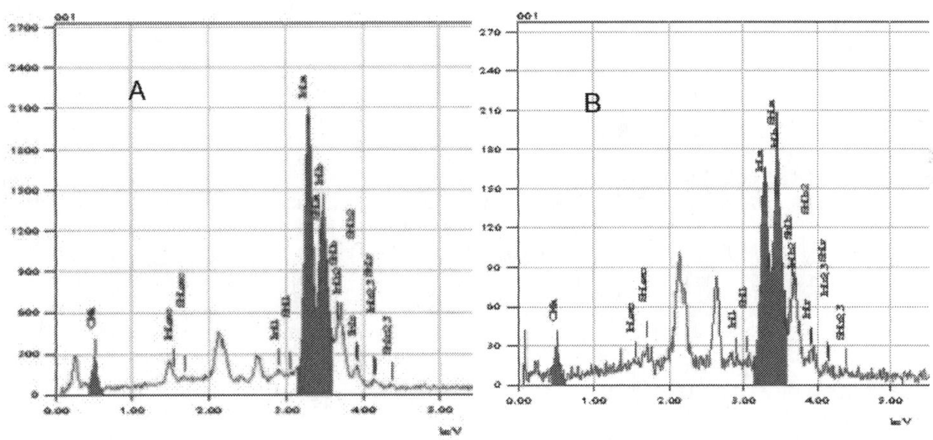

Figure 3. EDS spectra of the ITO powder: (a). sample A, and (b). Sample C

Tabel 2. The content of O, In, and Sn in the ITO compound ekstracted from the EDS spectra, and the corresponding x value from the chemical formula $In_{2-x}Sn_xO_{3-2x}$.

Sampel	% O (8)	% In (49)	% Sn (50)	X value
A	18.87	49.57	31.56	0.75
B	19.67	32.77	47.55	1.27
C	18.47	57.90	23.63	0.53
D	19.07	45.35	35.58	0.90

From the EDS spectra as shown in Figure 3, it was obtained that the percent composition of O, In, and Sn for sample A was 18.87%, 49.57%, and 31.56%, respectively. Since the atomic weight of O, In, and Sn is 8, 49, and 50, therefore the value of X in the chemical formula of ITO was approximately 0.75. By the same technique of calculation, the value of X for sample B, C, and D was obtained, and the results are as seen in Table 2.

3.3. Crystal structure

The XRD results for all samples showed that the ITO compound (In_2SnO_5) has been formed, along with Indium Oxide (In_2O_3). In ITO compound, each Sn^{4+} atom replaces In^{3+} producing one free electron, therefore ITO material will retain its bixbyte type cubic structure from In_2O_3. This can be seen in Figure 4 where the highest peaks of both compounds are located close to one another. For sample A, the highest peak of plane <222> occurs at (2θ angle) 30.44 for In_2O_3, and 32.64 for In_2SnO_5, whereas reflection of planes <400> and <440> occurs at 45.295 and 50.190 (for ITO).

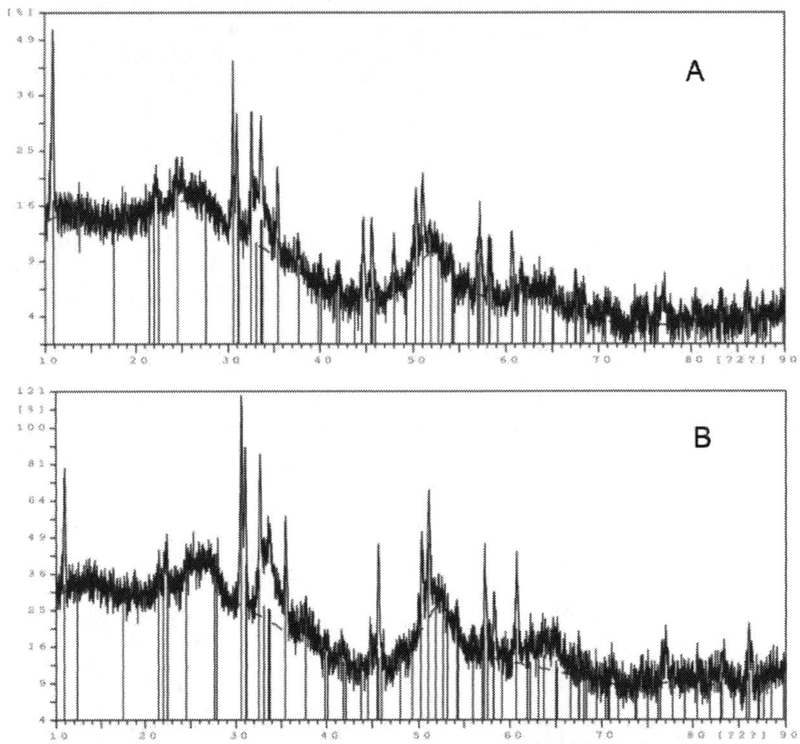

Figure 4. XRD spectra of the ITO powder: (a). sample B, and (b). Sample D

The addition of Pd as dopant material did not shift the location of the spectrum peaks. However, the peak intensity increased even for the same molar concentration. The peak intensity for sample A at 32.64 increased from 151 to 256 (in sample C). This increase showed that the addition of Pd dopant material had a positif effect on the crystal structure of the ITO nanomaterials. In sample B and D, InOCl compound was formed caused by the increase in molar percentage (from 10% to 30% of the $SnCl_4$.

3.4. Molecular bonds

The molecular bonds of the ITO nanomaterial have been analysed using FTIR spectra. Since ITO is composed of In, Sn, and O, therefore the FTIR spectra should provide information on the molecular bond between In-O-In, Sn-O, and Sn-O-Sn. From the spectrum of all samples, it was found that the peaks are located at the the wavelength between ~400 cm^{-1} sampai ~700 cm^{-1}, in which the bonds were dominated by metal and oxygen bonds. Figure 5 shows the FTIR spectra of the ITO powder for all samples, and the wave numbers corresponding to the peaks of the molecular bonds are listed on Tabel 2.

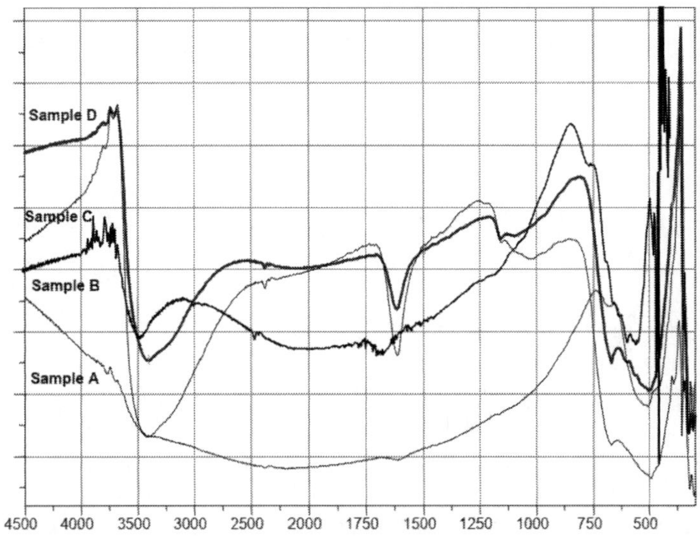

Figure 5. FTIR spectra of the ITO powders for all samples

Tabel 2. Wave numbers (cm^{-1}) for molecular bonds obtained from ITO spectra for all samples

Sample	In-O-In	Sn-O	Sn-O-Sn
A	453.27	503.42	671.23
B	389.62	489.92	667.37
C	443.63	493.78	669.30
D	325.97	499.56	669.30

3.5. Resistivity of the ITO nanomaterials

The study of electrical characteristics of the ITO nanomaterial was aimed to identify its resistivity changes in the presence of CO gas. The measurement of resistivity of the ITO powder was done using four-point-probe. As such, 0.3 g of ITO from each sample was pressed and then put inside a measurement chamber. The ITO sample was then exposed to the CO gas for 5 min, and the amount of CO was determined by the flow rate of the gas (1 ml/s and 5 ml/5). As shown in Figure 6, the results of the measurement showed that for all samples, the resistivity dropped as the amount of CO gas was increased. This is as expected since CO is one of reducing gases for metal oxide materials. In response to increase in molar Sn content, the resistivity decreased for the undoped samples (sample A to B), and increased for the doped samples (sample C to D). The addition of Pd content to the ITO has caused a lower drop in the measured resistivity as the amount of CO was increased.

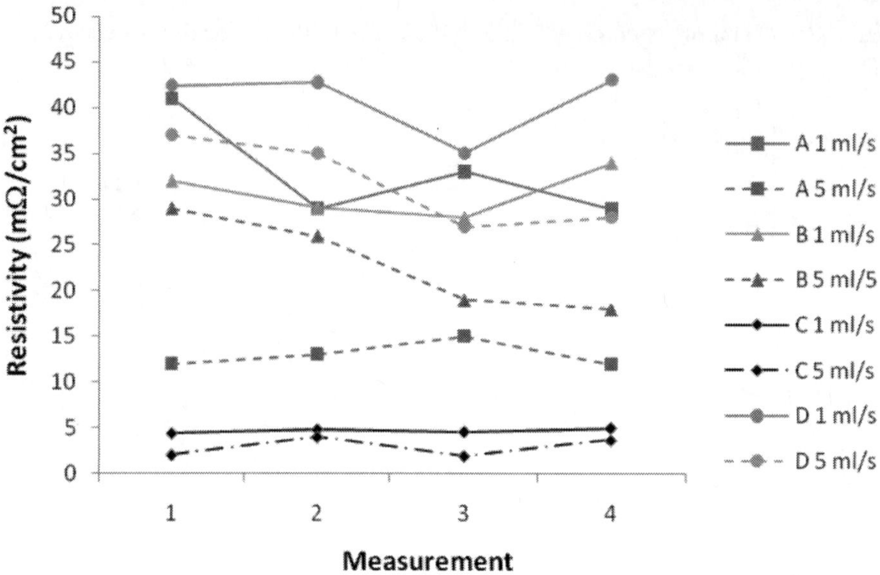

Figure 6. Resistivity of the ITO powders in response to CO gas

4. Conclusion

The synthesis of doped and undoped nanomaterial Indium Tinoxide (ITO) using sol-gel process has been described in this paper. Both physical and electrical characteristics of the resulting ITO powders have been studied. The morfology of the ITO powder showed nanometer grain size and smaller as the Sn content increased. The chemical structure of ITO can be expressed in the form of $In_{2-x}Sn_xO_{3-2x}$, with the value of x as obtained by EDS spectra increased with increasing Sn content. The crystal structure of the ITO nanomaterial as obtained from the XRD peaks is the same as that of In_2O_3. The molecular bonds consiting the ITO are between In-O-In, Sn-O, and Sn-O-Sn. From the FTIR spectra of all samples, it was found that the peaks were located at the the wavelength between ~400 cm^{-1} sampai ~700 cm^{-1}.The response to CO gases showed a decreasing resistivity with increasing CO gas concentration, showing its potential for gas sensor application.

Acknowledgments

The authors greatly acknowledged the financial support provided by Indonesian Institute of Sciences (LIPI) under "Program Unggulan 2016" and "Program Tematik 2016".

References

[1] Singh V, Suman C K, and Kumar S 2006 *Proc. 9th Asia. Symp. on Inf. Dis.* 388.
[2] Vijaylakshmi K, and Ravidhas C 2010 *Mat. Sci. Res. Ind.*7 249.
[3] Jeong C, Shin C, Kim D, Chae J, and Kim Y 2011 *Trans. on Electric. Electro. Mat.* 11 77.
[4] Yoo K S, Park S H, and Kang J H 2005 *Sens. Act. B: Chem.*108 159.
[5] Kim I J, Han S D, Han C H, Gwak J, Lee H D, and Wang J S 2006 *Sens.*6 526.
[6] Tuna O, Selamet Y, Aygun G, and Ozyuzer L 2010 *J. Phys. D: Appl. Phys.* 43 1.
[7] Atabaev I G, Hajiev M U, and Pak V A 2016 *Int. J. Thin. Fil. Sci. Tec.* 5 13.
[8] Dong H, Zhang X, Niu Z, Zhao D, Li J, Cai L, Zhou W, and Xie S 2013 *J Nanosci. Nanotech.* 13 1300.
[9] Gao Y, Zhao G, Duan Z, and Ren Y 2014 *Mat. Sci. Poland* 32 66.
[10] Bhatti M T, Rana A M, Khan A F,and Ansari M I 2002 *J. of Appl. Sci.* 2 570.
[11] Li Z, and Ren D 2006 *Trans. Nonferrous Met. Soc. China* 16 1358.
[12] Hjiri M, Ghribi F, and El Mir L 2014 *Sens. & Trans.* 27 198.
[13] Silva G M, Faria E H, Nassar E J, Ciuffi K J, and Calefi P S 2012 *Quim. Nova* 35 473.
[14] Ibrahem H, and Moghdad M 2013 *IPASJ Int. J. of Electri. Eng.* 1 18.

Fabrication of SnO_2 based CO gas sensor device using thick film technology

I D. P. Hermida[1], G. Wiranto[1], Hiskia[1], R. Nopriyanti[2]
[1]Telecommunication Division, Research Center for Electronics and Telecommunications
Indonesian Institute of Sciences (P2ET-LIPI)
Jl. Sangkuriang, Gedung 20, Lantai 4, Bandung – 40135, INDONESIA
[2]Department of Physics, University of Educations, Jl. Dr.Setiabudhi No.229, Bandung 40154 INDONESIA

E-mail: idewaputu@gmail.com

Abstract-This research was aimed at fabricating a CO gas sensor based on SnO_2 material synthesized using *sol-gel method*. The sensor components include thin film gold electrode and a PdAg heater screen printed on the backside of the alumina substrate. The device was tested to study the effect of temperature variation on the resistivity of the heater component, and the response of the SnO_2 sensitive layer to 10 ppm CO gas. The SnO_2 layer was characterized using SEM and EDS to determine the morphology and composition of the material. It was found that the SnO_2 sensitive layer contained 21.21% C atoms, 22.43% O atoms, 14.98% Si atoms, 0.34% Cr atoms, 1.16% Ag atoms, 1.78% Sn atoms, and 38.11% Au atoms. The sensor sensitivity to CO gas increased with increasing temperature. The Operational temperature of this sensor was 95° C with the highest sensitivity value obtained was 16.59.

1. Introduction

Carbon monoxide (CO) is a gas that is flammable, poisonous, colorless, and tasteless. Because of it is not colored and tasteless carbon monoxide is difficult to detect when without using detection technology [1]. Therefore, carbon monoxide is often dubbed the "silent killer", and needed a technology that can detect the presence of CO gas.

Science and technology of the semiconductor gas sensor were growing rapidly. Device semiconductors gas sensor are generally known as a metal oxide gas sensor because it is made of metal oxides such as TiO_2, ZnO, CeO_2 and SnO_2 [2].

Among all the metal oxide, SnO_2 is one material that is good enough to be used as the active layer of CO gas sensors [3]. Nevertheless, a SnO_2-based gas sensor to date has not made the high sensitivity and selectivity. [4]. According to Pires, one of the factors that can improve the quality of a sensor that is optimizing the quality of the electrode. Electrodes used in a sensor must have high conductivity values. In order for the resulting electrode has a good quality, the electrodes can be fabricated using thin-film technology and made using materials that have good electrical conductivity [5].

Along with the development of microelectronics technology or nanotechnology at this time, electrode fabrication technology that is applied to metal oxide-based gas sensors can be done using the method sputtering. Using Sputtering technology of thin film growth becomes simpler and operational costs become cheaper. In addition, by using the method of Sputtering adhesives between the coating and the substrate surface becomes stronger, coating thickness easily observed and controlled [6].

Journal of Physics: Conference Series **776** (2016) 012061 doi:10.1088/1742-6596/776/1/012061

The purpose of this research is to determine the characterization of SnO_2-based CO gas sensors has been done in terms of the development of electrode components.

2. Experimental

2.1. Design of Gas Sensor

2.1.1. Parts of Sensor

The gas sensor is composed of electrodes that coated by a layer. And the layer is sensitive to the presence of CO gas sensors are printed on a substrate of silicon (Si). Heater sensor printed on the substrate surface alumina (Al_2O_3). Here are specifications of gas sensor CO:

- Dimensions : ≤10 mmx25 mm
- Operating temperature : 25-250^0 C
- Voltage Sensor : 3 Volt dc
- Researching Power of Heater : 1.8 Watt
- Measurement Range : 0 ~ 10 ppm
- Heater Voltage : 3 Volt dc

2.1.1.1 Designing of Electrodes

The electrodes are commonly used in gas sensor components are generally an inter digital electrode structure (inter digital electrodes), which are usually made of materials such as Au or Pt nobel metal. The use of the structure was based on the consideration to minimize space but maximize the sensing area. To calculate the value of the electrode resistance can be seen in the equation and the picture below:

$$R_{el} = \left(\frac{l_1}{w} + \frac{l_2}{w} + 0{,}56\right) R_s \qquad (1)$$

Where: *Rs* is the value of sheet resistance (Ω /sq).

The design layout of electrodes that used can be seen in the following figure:

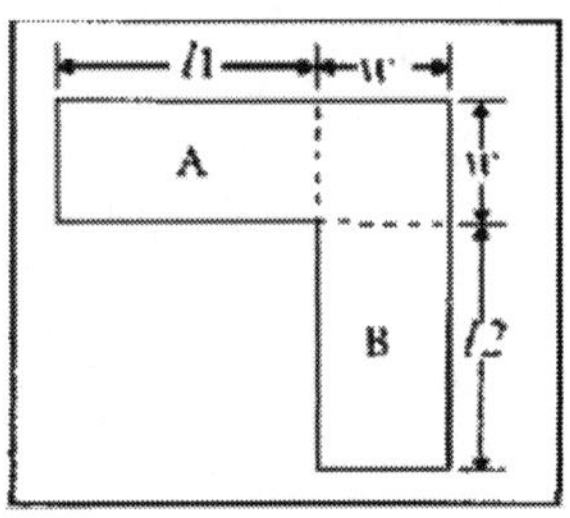

Figure 1 Image Calculation of electrode resistance value.

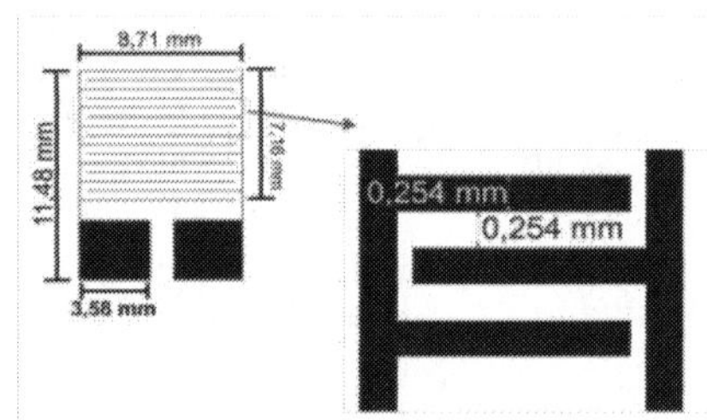

Figure 2 Design of Electrode

2.1.1.2. Design of Heater

One of the factors that determine the success of thin film gas sensors that are temperature factors that determine the success of thin film gas sensors that is temperature. Appropriate temperature distribution will affect the level of selectivity and sensitivity of this sensor element. To determine the characteristics of the heater, the parameters that must be considered are : the desired temperature, the power needed, and the area that you want to heat the area, as well as the character of the material heater itself (TCR, dissipation maximum current that can pass through, etc.). Thus the characteristics of the sensor heater is designed for:

- Operating Temperature : 300°C
- Average power : 1.8 Watts
- Flow : 0.612 A

Based on the calculations that have been carried out, the heater has a length of line conductivity sensor of 94.107 mm and a track width of 0.529 mm. Then the resulting layout design heating gas sensor is as shown below:

Journal of Physics: Conference Series **776** (2016) 012061 doi:10.1088/1742-6596/776/1/012061

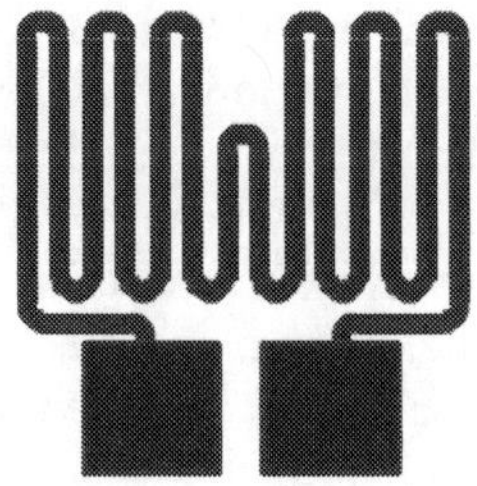

Figure 3 Heater of Gas Sensor Layout Design

3. Sensor Fabrication Processes

SnO_2 gas sensor fabrication process is divided into three phases; there are the electrodes fabrication, heater fabrication, and the active layer fabrication. Here is the exposure steps research that has been conducted by the author in several stages of research as follows:

3.1. *Electrodes fabrication*

The electrodes used in gas sensors, fabricated using sputtering method. Broadly speaking electrode fabrication steps illustrated in the diagram below:

The first steps in the fabrication of electrodes are to do the washing substrate using acetone, 5% HF solution, DI Water, and H_2O_2 solution: HCL:H_2O. After that was done the process of oxidation. This research used dry oxidation process at a temperature of 1100°C for 130 minutes. Then later, the growth of Au thin films on silicon substrates using by DC Sputtering method and the pressure within the chamber to 5.9 x10^{-5} Torr and argon gas flowed at a pressure of about 4mTorr ± 0.1 mTorr. Before coated with gold, the substrate was coated with chromium first. Chromium coating was done for 2 minutes, which is then immediately followed by a coat of gold for 10 minutes.

Then to form the electrode pattern on a silicon substrate performed lithography process consists of several stages.

In the prebake process substrates in an oven at a temperature of 85°C for 3 minutes in the oven. Immediately after completion oven substrate, the substrate is coated resistors ma-P215 positive resist using a spin coater. Then do the soft bake process carried out in an oven with at temperature of 150°C for 15 minutes.

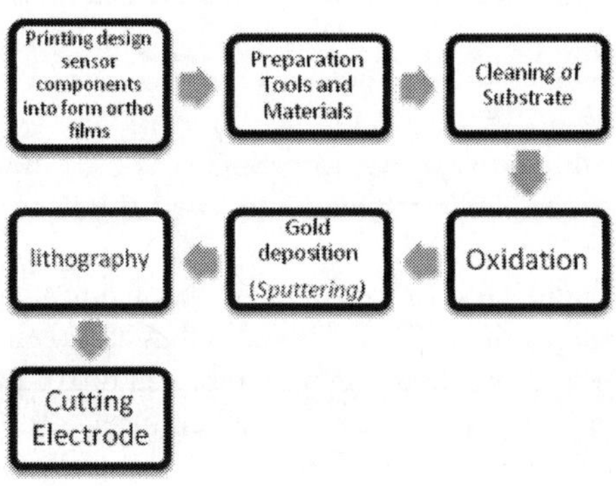

Figure 4 Electrode Fabrication Process

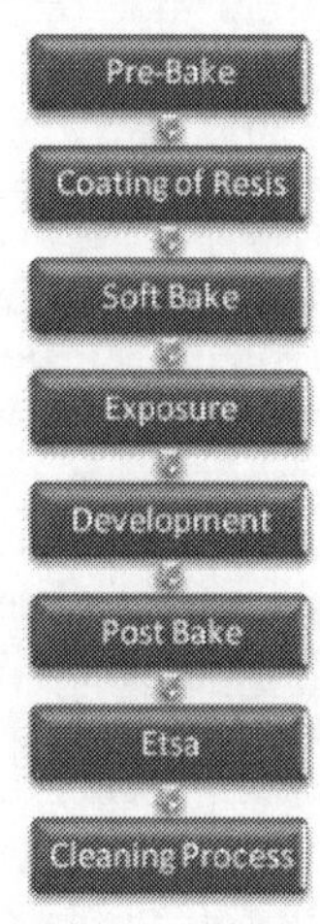

Figure 5 Lithography process.

Then do the process of exposure for 2 minutes in the sun and continued with the process of development by using MF319 developer solution until the pattern emerges and then wash with DI Water. Then, the substrate reheat in the oven for 5 minutes with a temperature of 85°C. The further etching process using a solution of I_2:KI:H_2O with a ratio of 1:4:40 until golden patterns emerge, then wash with DI Water and dry. And then remove a layer of the remaining positive resist using acetone.
And the last thing to do in making the cutting was electrode substrate. The process of cutting the substrates used in this study was done manually using diamond eyes.

3.2. Making Sensor Heater

Heaters were made using thick-film technology and screen printing methods. In the making, to produce heater must pass through several processes including washing the substrate, making the screen, thick film growth, drying, and burning. The substrate used in the fabrication of heater was alumina substrate. Before use the substrate, it must be washed first by soaking in a glass beaker containing water dye, then put in the Ultrasonic cleaner for 5 minutes.
Furthermore, the first step that must be done in this process was the selection screen. In this study, used 325 mesh screen density and size 20 x 20 cm. After determining the screen to be used, the next thing to do is clean the screen by using Ulano 5.
Then after the screen is completely dry, CDF3 film paper that has been cut into a size of 10 x 10 cm is placed in the middle on the front surface of the screen. CDF 3 placed over the screen with the emulsion at the top, and a little duct tape glue on one side of CDF 3 that were not easily moved. Then on the back surface of the screen was coated Ulano 133, just behind the CDF layer 3. Remove the masking tape that attached to one side of CDF 3, then dry it with a hair dryer for 15 minutes.
After Ulano 133 completely dry, a layer of plastic / mylar on CDF 3 removed carefully. Furthermore, ortho-film paste on the upper right that has been released CDF 3 plastic coating. Then the screen was placed in the middle of the field of radiation on the radiation machine, the process was called by photography.
And the last one was on the screen that has not been covered by the CDF 3 coated again with a flattened Ulano 133 using rakel and dried using a hair dryer. After the pattern was formed on the screen, then performed the process of growing a thick film on alumina substrate using screen printing and pastes used the pasta PdAg (Dupont 7484) to the value of the sheet resistance 15-30 mΩ/□. Furthermore, heater dried using an oven for 15 minutes at 195°C temperature used. After heater becoming dry, the heater should be firing in a high temperature furnace. The length of time firing for about 30 minutes.
After the heater is formed, then cut to the size of the alumina substrate heater is formed. Cutting substrate performed using diamond eyes.

3.3 Synthesis Procedures

First, we should prepare materials that will be us is powder SnCl, isopropanol, propanol, and water. Comparison of the composition of the materials to be used is powder SnO_2: Isopropanol: Propanol: Water is 5: 7: 7: 6.
And then, pure SnO_2 powder using a mortar and Pestle for ± 1 hour without stopping. This is done so that the particle size of SnO_2 become nanoscale.Combine SnO2 powder that has been mashed with 2/3 parts of propanol and 1/3 water, then stir for 1 hour while heated at a temperature of 80°C.
Burn the solution that has been made in the oven with a temperature of 300°C. After this stage, the solution will turn into powder.Combine remaining water, isopropanol, and residual of propanol and powder. And then stir for 1 hour without heating, until resulting paste clear and homogeneous. Of the whole process can be generated with sol-gel, then will be used to make a sensitive layer.

Journal of Physics: Conference Series **776** (2016) 012061 doi:10.1088/1742-6596/776/1/012061

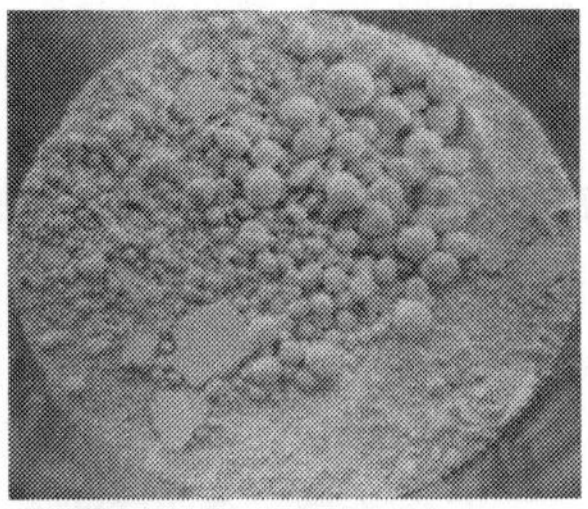

Figure 6 SnCl powder

Figure 7 SnO_2 paste

Furthermore, to form a sensitive layer, SnO_2 paste which was formed dripped right over the electrode which then leveled using a squeegee. In order to coat the pasta in the electrode except for the contact, then before coated the SnO2 contact parts should be covered by using solatip first.After the paste dries then the tape is opened.

4. Results And Discussion

4.1 Micro Structure and Structure Analysis by SEM-EDS

To determine the composition of the sensitive layer formed tested using EDS (Energy Dispersive Spectroscopy) with the type JEOL JSM 6360 LA. Materials used are sensing material SnO2 testing performed by EDS as follows:

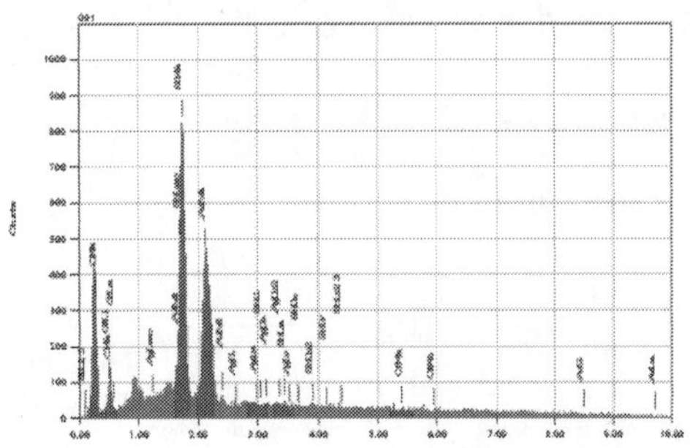

Element		(keV)	mass%	Error%	At%	Compound	mass%	Cation	K
C K		0.277	21.21	0.06	72.98	C	21.21	0.00	29.4350
O			22.43						
Si K		1.739	14.98	0.23	22.04	SiO2	32.04	9.13	58.7864
Cr K	*	5.411	0.34	1.37	0.13	Cr2O3	0.49	0.11	1.1895
Ag L	*	2.983	1.16	0.60	0.22	Ag2O	1.24	0.18	3.1829
Sn L		3.442	1.78	0.95	0.62	SnO2	2.26	0.26	4.7320
Au M		2.121	38.11	0.57	4.00	Au2O3	42.75	3.31	91.8211
Total			100.00		100.00		100.00	12.99	

Figure 8 EDS result from gas sensor

Based on figure 12, it was observed that the elements contained in the SnO_2-based CO gas sensors have the atomic composition of 21.21% C, 22.43% O atom, 14.98% Si, 0.34% Cr atoms, 1.16% atom Ag, 1.78% Sn atoms, and 38.11% Au atoms. The number of C atoms in the sensor was obtained when the combustion process and testing using CO gas sensor, while the O atoms derived from the absorption of free air.

While for the microstructure of the sensitive layer on a silicon substrate to deposition characterized using SEM. The results of the measurements in the form of SEM photographs shown in Figure 8.

From the SEM results above shows that the grain size of the SnO_2 thin film has reached the order of nano, but it can also be seen that the number of pores formed in the sensitive layer. This means, SnO_2 thin layer formed qualify as gas sensors. Because, the gas sensor formed pores that capture of CO gas, which will then lower the resistance value and increases sensor sensitivity.

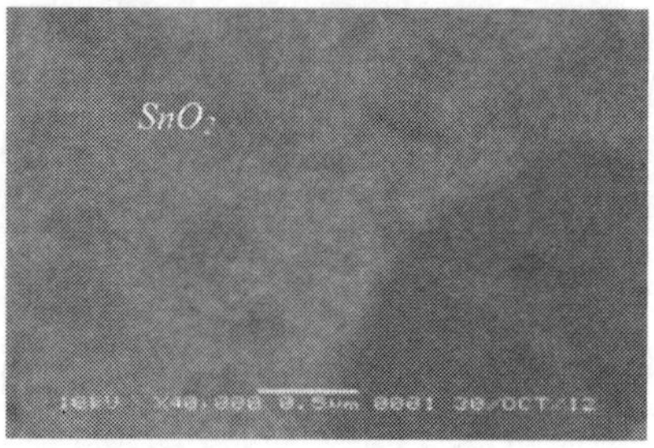

Figure 9 Micro structure of SnO_2 Thin Films with a magnification of 40000 ×

Figure 10 Transverse cross section of SnO_2 Based CO Gas Sensor

Based on figure 11 above, could be seen the thickness of each material coated on a silicon substrate. Figure 11 shows the micro structure of SnO_2 layer with a thickness of 0.118 μm, Au (gold) with a thickness of 0.379 μm, Cr (Chromium) with 0.508 μm thickness, SiO_2 (Silicon Dioxide) with 18.288 μm thick, while the Si (silicon) 39.03 μm.

4.2 Testing of Sensor

Test electrical characteristics initially conducted under 0 ppm CO gas in other words without the granting of CO gas. Testing was done by changing the value of current supplied to the heater, which then change the value of a given temperature gas sensors. Along with the change in the value of the test temperature, then the value will change the resistance.

Then, the test continued with the characteristics of the gas flow rate of 10 ppm CO gas.From the test results obtained can be seen in Figure 11.

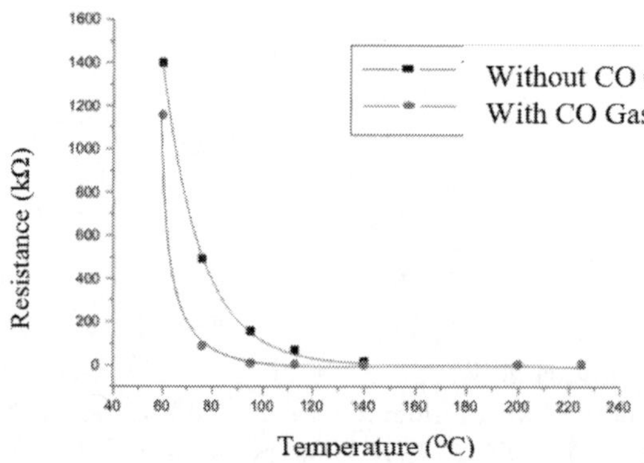

Figure 11 Chart Changes of Resistance Against Temperature Operations Sensor When Given and Without Gas CO

Based on the graph above, it can be seen that the gas sensor resistance decreases exponentially with increasing operating temperature is given, either when given only when given gas CO or CO gas. This is in accordance with the following equation : $R = R_o \exp\left(\frac{E_a}{KT}\right)$

Description:

- R is the resistance at temperature measurement
- R_o is the resistance at the initial temperature measurement
- E_a is the activation energy.

But the decline in the value of the resistance that occurs when no CO gas was not as decrease the resistance value when given gas. According to Andrew et. al. in 1999, it was because the normal air condition and heated sensor, there will be a process adsorbs oxygen from the air around to the surface gas sensors through grain boundaries. The oxygen has been adsorbed on the surface of the gas sensor will attract electrons from the conduction band of an n-type semiconductor gas sensor to the surface, so that the oxygen - oxygen is ionized into ions - oxygen ions O-,.

Thus, it resulted in the potential barrier height (barrier between grains) will be high and hinder the movement of the electrons flowing through the grain boundaries. Then decrease the resistance value is not as big on testing for impairment testing resistance at the same time but fed CO gas.

When compared to the results of the resistance obtained when no CO gas with CO gas is given, can be seen at a temperature of 60-140°C impairment resistance occurs when CO gas was greater than without CO gas. This is due, when the CO gas flow into the heating chamber sensors remain to be done. In this situation, the reaction between CO gas with the gas sensor surface is:

$$2CO + 2O^{-} \rightarrow 2CO_2 + 2e^{-}$$

However, given the testing conditions with gas sensor irregularities. Deviations can be seen when the temperature reaches 200°C and 225°C, the temperature increased resistivity values. According to Khalil et al In 2009, it was caused by the amount of carbon monoxide adsorbed gas molecules and accumulation of gas molecules occurs surface. Thus inhibiting the subsequent adsorption of carbon monoxide gas molecules.

From the test results can also be obtained by the value of the sensor sensitivity. From the data obtained a graph as shown in Figure 12.

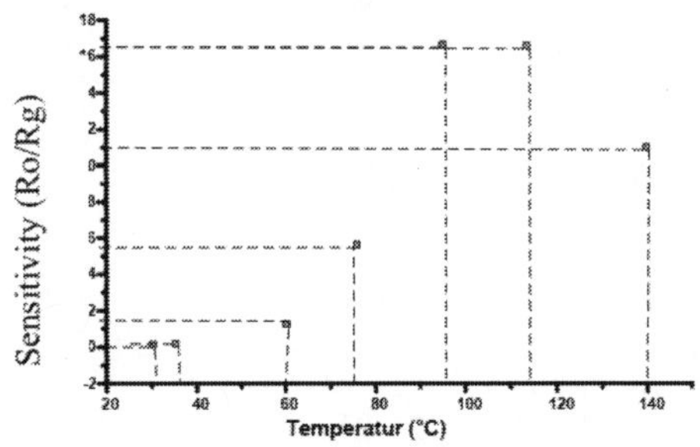

Temperature (°C)

Figure 12 Chart Of Change Sensitivity Sensor Againt Operating Temperature

At temperatures of 60°C to 95°C sensitivity value seen a sharp increase in sensitivity due to the increasing value of the temperature. However, when the temperature is above 95°C decrease with the addition of temperature sensitivity.

5. Conclusion

From the results of testing that has been done, it can be concluded as follows :

1. By using thin-film technology electrodes, gas sensors able to detect the gas at a concentration of 10 ppm with a high value of sensitivity.
2. The gas sensor is able to react to CO gas, a change in the sensor resistance value down when given gas CO.
3. Changes in sensor sensitivity can be affected by temperature resulting heater.
4. The results of the EDS analysis of the elemental composition of the CO gas sensor has a composition of 21.21% C atom, atom O 22.43%, 14.98% Si, 0.34% Cr atom, atom 1.16% Ag, 1.78% Sn atoms, and 38.11% Au atom.
5. On further research, where to get a more sensitive sensor added doping Au/Pt on the surface of the sensor

Acknowledgment
This work was financially supported by LIPI Program.

References

[1] Manjula P.S., Arunkumar, Sunkara V.M. 2011. *Sensor and Actuators B.* pp.168-175.
[2] Kuang Q., Lao C.S., Li Z., Xie Z., Zheng L., and Wang Z. 2008. *Journal Physics Chemistry* C, Vol. 112, No. 30.
[3] Batzill M., Ulrike D. 2005. *"The Surface and M*aterial *Science of Thin Oxide"*. http://www.surface.tulane.edu/pdf/surface science of Tin Oxide.pdf
[4] Hiskia, Hermida I.D.P., 2006. *"Development of Carbon Monoxide Gas Sensor (CO) Based on* SnO_2 *Material"*. http://elib.pdii.lipi.go.id/katalog/index.php/searchkatalog/downloadDatabyId/7039/7039.pdf.
[5] Pires J.M.A. 2003. *Thin Film Gas Sensor.* PhD Dissertation for Universitas Minho, Portugal.
[6] Yunas J., Lia M. 2001. *"Sputtering System Applications for Thin Layer Deposition"*. http://elib.pdii.lipi.go.id/katalog/index.php/searchkatalog/downloadDatabyId/2676/2677.pdf

Application of PVDF composite for lithium-ion battery separator

Q. Sabrina, N. Majid and B. Prihandoko

Research Center for Physics – Indonesian Institute of Science, Kawasan PUSPIPTEK serpong Gd.440-442 Tangerang Selatan

E-mail: qolby89@gmail.com

Abstract. In this study a composite observed in PVDF composite as lithium ion battery separator. Observation of performance cell battery with cyclic voltametry and charge discharge capacity. Surface morphology PVDF separator and commercial separator observed with Scanning electron microscopy (SEM). Cyclic Voltamerty test (CV) and Charge Discharge (CD) showed a capacity value on the coin cell. Coin cell is composed of material $LiFePO_4$ cathode, anode material of lithium metal and varies as graphite, liquid electrolyte varied use LiBOB and $LiPF_6$. While the PVDF as compared to the commercial separator. Coin cell commercial separator has a better high capacity value when compared with Coin cell with the PVDF separator. Life cycle coin cell with the commercial separator material is still longer than coin cell separator with PVDF Copolymer. Development of PVDF as separator remains to be done in order to improve the performance of the battery exceeds the usage of commercial material.

1. Introduction

In recent years, many challenges have been made in battery technology. Specific electrochemical systems still continues improvement, nevertheless there is still no one separator that can be considered "ideal" for all battery chemistries. Separator is a porous membrane placed between electrodes of opposite polarity, permeable to ionic flow but preventing electric contact of the electrodes. Separators play a key role in all batteries to keep the positive and negative electrodes apart to prevent electrical short circuits and at the same time allow rapid transport of ionic charge carriers that are needed to complete the circuit during the passage of current in an electrochemical cell. The capability of conducting ions by either intrinsic ionic conductor or by soaking electrolyte to minimize any processes that adversely affect the electrochemical energy efficiency of the batteries[1].

A separator could be an ionic resistor because lithium ions pass through a separator during charge discharge processes. Electrochemical performance of the lithium-ion battery is closely related to structure of separator, i.e., thickness, porosity, and permeability[2]. Generally, a thin membrane separator can minimize the ionic resistance, enabling high-specific battery power. However, too great of a small membrane thickness can reduce the mechanical strength of the membrane and increase the risk of inner battery electrical shorting. In practice, most separators for liquid electrolyte batteries in use today are 20 to 30 μm thick[3]. The purpose of this paper is to describe the various types of separators based on their applications in batteries and their chemical, mechanical and electrochemical properties, with particular emphasis on separators for lithium-ion batteries. PVDF composite has potential candidate for advanced lithium ion battery separator application, because PVDF copolymer has excellence in thermal stability, chemical resistance and mechanical property[4].

2. Experimental

In this study, performance PVDF as separator compared with commercial separator applied in half cells.

2.1. Material Preparation

2.1.1. Preparation of PVDF Separator. PVDF were dissolved in N-N DMAc at 60°C and continuously stirring for several hours. Entrapped air in polymer solution was removed before using for electrospinning process. Polymer solution was attached to electrospinning machine with applied voltage is 18 kV and flow rate 0.004 ml/h[4]. Measurement thickness of PVDF separator is 15 μm and commercial separator 25 μm. Commercial separator available in Li-ion battery are made out of polyolefins like polyethylene and polypropylene. These materials have been found to be compatible with the cell chemistry and can be cycled for several hundred cycles without significant degradation in chemical or physical properties.

2.1.2. Preparation of half cells. Cathode lithium iron phosphate ($LiFePO_4$) being a safe and low cost material that can offer a relatively high theoretical capacity of 17 0 m Ah/g, it is attracting much attention for applications in lithium batteries. $LiFePO_4$ has good thermal stability and as table charge/discharge property at ~3.4 V, which provides safety margin in terms of electrolytic decomposition ion for many of the organic electrolytes used in batteries. In spite of the several advantages of this material, earlier studies have demonstrated its incapability in achieving high material utilization at lower temperatures and /or higher current densities. The rate capability limitation of $LiFePO_4$ results from the low electronic conductivity of the material (~1 0^{-9} S/c m)[5]. The $LiFePO_4$ cathode appears to be exceptionally stable to long-term cycling[6]. The advantage we use graphite anode to this experiment is low cost, relatively high reversible capacity of 370 mAh g^{-1}, high coulombic efficiency in proper electrolytes[7]. $LiPF_6$ based electrolytes have favourable high conductivity and electrochemical stability, they suffer from a lower than average thermal stability[8]. LiBOB-containing electrolyte has been shown to possess a wide electrochemical stability window on the Pt electrodes, a critical question remaining unanswered is whether it is stable on practical lithium-ion battery electrodes[9]. The $LiFePO_4$ /Graphite half cells were assembled coin cell in the argon-filled glove box using PVDF and commercial celgard 2400 as separator and an appropriate amount of electrolyte $LiPF_6$ and LIBOB.

2.2. Characterization

The morphology of the PVDF separator and commercial separator was characterized by using a scanning electron microscope (SEM) operate at 10-15 kV.

2.3. Electrochemical Evaluation

Electrochemical impedance spectroscopy (EIS) were measured using Hioki LCR. Electrochemical measurements cyclic voltammetry (CV) and Charge discharge (CD) were measured in coin-type cells using WBCS3000. CV is a common technique for studying the properties of an electrochemical system—a cyclic linear potential sweep is imposed onto the electrode and the resulting current is recorded[10]. The cell were initially cycled 3 times between 2 V and 3.7 V at constant current of 0.1 C at room temperature.

3. Results and Discussion

Morphology

Figure 1. Morphology (10 kv x500) of the (a) PVDF separator (b) commercial separator observed by using SEM

Figure 1(a) shows representative morphology of PVDF separator by using SEM. The surface morphology of copolymer separator was smooth without bead. SEM pictures of commercial membrane separator 1(b) shown image that the pores are uniformly distributed. Base on the previous research, the separator with small membrane thickness has better advantage than separator which has big membrane thickness because it can increase the power specific power. Commercial separator has a good mechanical strength so it can decreasing the risk of inner battery electrical shorting. From the SEM data can be conclude that the commercial separator has a better morphology than PVDF separator.

Impedance Analysis

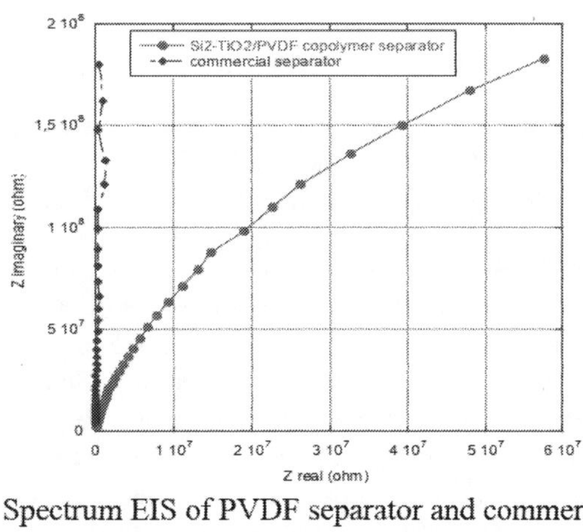

Figure 2. Spectrum EIS of PVDF separator and commercial separator

The function of separator in battery system are to prevent electrodes short circuits and contributing for ionization process between anode and cathode. Electron cannot flow through the separator but it's different of ion that can flow there. One of separator quality pre requirement is its electrical resistance value. A better separator has a small resistance value, because separator resistance affected of internal resistance of the battery so it will be effected of battery performance. Base on EIS spectra analysis performed by the equivalent circuits that simulate the real state of the circuit components for predicting impedance electrode. The commercial separator nyquist graph shows the vertical straight line and PVDF separator nyquist graph shows a half of semicircle. Intersection with x-axis at small frequency value indicate the magnitude of impedance.

Figure 2 shows that impedance component of commercial separator blue line just consist of capacitor and impedance component of PVDF separator red line consist of resistor and capacitor. It's mean that no reduction energy occured at the battery if the separator used for battery. It's different if the PVDF separator used as battery separator that affecting the battery performance is not good. Impedance value of PVDF separator at low frequency was too large than commercial separator because the different thickness, composition, and morphology between PVDF and commercial separator. The conclusion that the commercial separator has a better impedance value than PVDF separator.

Cyclic Voltametry Analysis

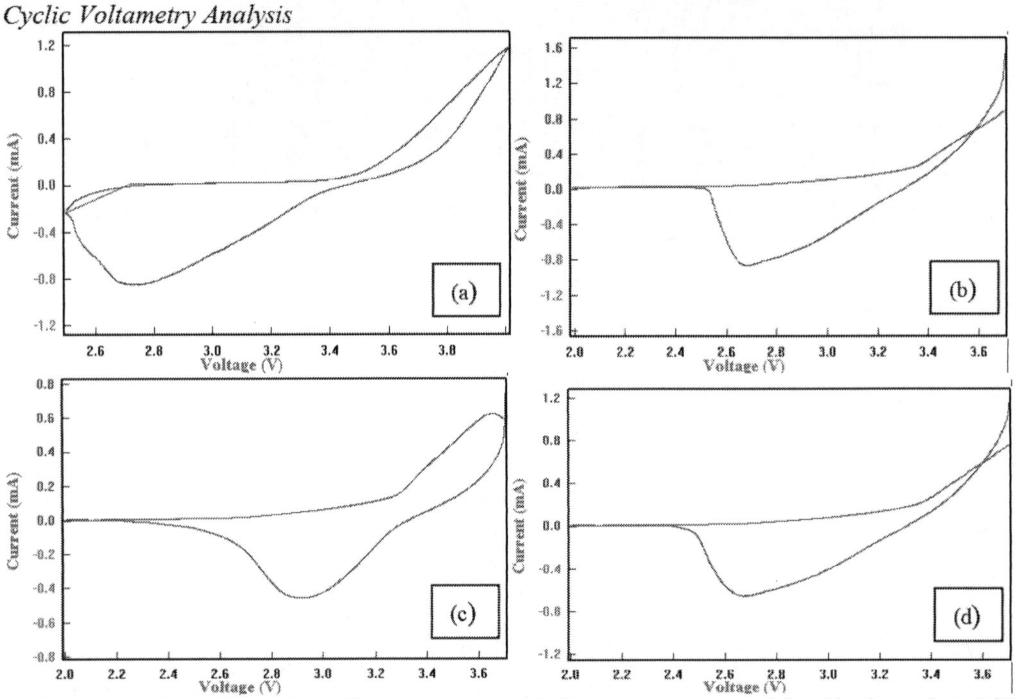

Figure 3. General graph cyclic voltammetry (a) Separator commercial with electrolyte $LiPF_6$ (b) Separator PVDF with electrolyte $LiPF_6$ (c) Separator commercial with electrolyte LiBOB (d) Separator PVDF with electrolyte LiBOB.

Figure 3 shows the cyclic voltammetry characteristics four coin cells. An increasing number of inorganic chemists have been using cyclic voltammetry to evaluate the effects of ligands on the oxidation/reduction potential of the central metal ion in complexes and multinuclear clusters. The current response is plotted as a function of the applied potential. The current depends on two steps in the overall process, the movement of electroactive material to the surface and the electron transfer reaction. Reversible electrochemist mean that the reaction is fast enough to maintain the concentration of the oxidized and reduced forms in equilibrium with each other at the electrode surface. [11]
Figure 3(a) and 3 (c) has the same type that are have the complete redox peak and there are different with figure 3(b) and 3 (d). Figure 3(a) shows that charge peak at 3,8 V and discharge peak at 2,65 V and figure 3(c) shows the charge peak at 3,6 V and discharge peak at 2,85 V. figure 3(b) and 3(d) just has the discharge peak. It was caused the cell which using the commercial separator has better redox process there than the PVDF separator.

Charge/ Discharge Analysis

The capacity retention of the $LiFePO_4$/ Graphite Li-ion cell using separator PVDF with the LiBOB – based electrolyte, was significantly better than for similar separator in coin cells that used the $LiPF_6$ – based electrolyte. This is because the Fe^{2+} ions are still dissolving, migrating to the negative electrode, and being reduced at the electrode surface to cause an impedance increase (similar to what occurs in $LiPF_6$ -based electrolytes, but at a reduced rate).

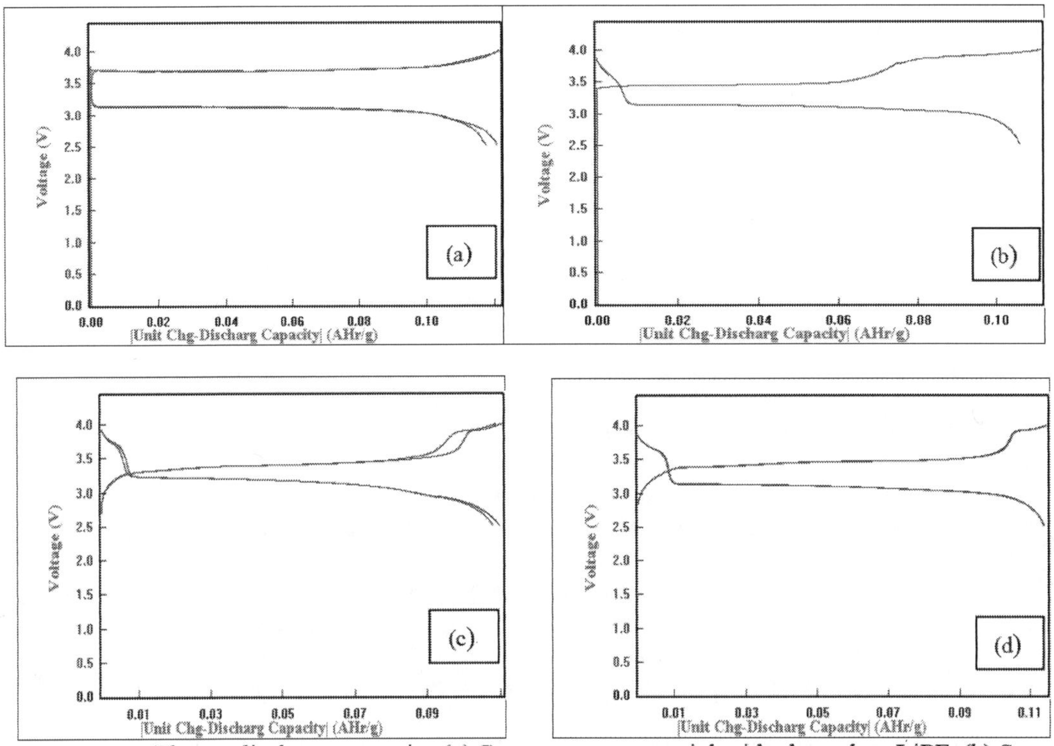

Figure 4. Charge discharge capacity (a) Separator commercial with electrolyte $LiPF_6$ (b) Separator PVDF with electrolyte $LiPF_6$ (c) Separator commercial with electrolyte LiBOB (d) Separator PVDF with electrolyte LiBOB

Table 1. Data report capacity

Half coin cell battery	Data report Capacity
Separator commercial with electrolyte $LiPF_6$	120.39 e-3 (AHr/gr)
Separator PVDF with electrolyte $LiPF_6$	105.50 e-3 (AHr/gr)
Separator commercial with electrolyte LiBOB	100.85 e-3 (AHr/gr)
Separator PVDF with electrolyte LiBOB	113.81 e-3 (AHr/gr)

Figure 4 shows the charge discharge of four cell sample. Sample which using commercial separator and $LiPF_6$ has the best charge discharge figure. That figure shows the standard figure of lithium ion battery charge discharge figure that are contained with the flat zone of charge and discharge voltage. These flat graph zone indicate the open circuit voltage of battery. Different with sample which used PVDF as separator. Figure 4(b) and 4(d) shows the different type if compared with standard lithium ion battery graph. The best battery with commercial separator and $LiPF_6$ as electrolyte has the biggest capacity and the battery with PVDF as separator and LiBOB as electrolyte has the smallest capacity value. It' mean that the redox process and ionization process was occur at commercial separator and $LiPF_6$ electrolyte.

The ideal battery separator would be infinitesimally thin, offer no resistance to ionic transport in electrolytes, provide infinite resistance to electronic conductivity for isolation of electrodes, be highly tortuous to prevent dendritic growths and be inert to chemical reactions. There has been a continued demand for thinner battery separators to increase battery power and capacity. This has been especially true for lithium-ion batteries used in portable electronics. However, it is very important to ensure the continued safety of batteries, and this is where the role of the separator is greatest. Thus, it is essential to optimize all the components of battery to improve the performance while maintaining the safety of these cells. Separator manufacturers should work along with the battery manufacturers to create the next generation of batteries with increased reliability and performance, but always keeping safety.

4. Conclusion

The commercial separator has a better impedance value than PVDF separator. $LiFePO_4$/ Graphite Li-ion cell using separator PVDF with the LiBOB –based electrolyte, was significantly shows capacity better than for similar separator in coin cells that used the $LiPF_6$ –based electrolyte. PVDF separator that has been made thinner than the commercial separator shows increase battery capacity.

References

[1] P. Arora, Z. J. Zhang, S. Lakes, and N. Carolina, "Battery Separators," 2004.
[2] T. H. Cho, T. Sakai, S. Tanase, K. Kimura, Y. Kondo, and T. Tarao, "Electrochemical Performances of Polyacrylonitrile Nanofiber-Based Nonwoven Separator for Lithium-Ion Battery," pp. 159–162, 2007.
[3] X. Huang, "Separator technologies for lithium-ion batteries," pp. 649–662, 2011.
[4] M. Nasir, Juliardi, and P. Bambang, "Available online at www.sciencedirect.com," vol. 16, pp. 184–189, 2015.
[5] J. Kim, G. Cheruvally, and J. Ahn, "Electrochemical properties of LiFePO 4 / C synthesized by mechanical activation using sucrose as carbon source," pp. 799–805, 2008.
[6] K. Striebel, J. Shim, A. Sierra, H. Yang, X. Song, R. Kostecki, and K. Mccarthy, "The development of low cost LiFePO 4 -based high power lithium-ion batteries," vol. 146, pp. 33–38, 2005.
[7] M. Yoshio, H. Wang, K. Fukuda, and T. Umeno, "Improvement of natural graphite as a lithium-ion battery anode material , from raw flake to carbon-coated sphere {," pp. 1754–1758, 2004.
[8] T. For, T. H. E. Degree, O. F. Doctor, I. N. Philosophy, and J. Scheers, *Department of Applied Physics*. 2011.
[9] K. Xu, S. Zhang, T. R. Jow, W. Xu, C. A. Angell, E. S. Lett, and P. A-a, "LiBOB as Salt for Lithium-Ion Batteries : A Possible Solution for High Temperature Operation service LiBOB as Salt for Lithium-Ion Batteries A Possible Solution for High Temperature Operation," vol. 5, no. 1, pp. 2–6, 2002.
[10] D. Y. W. Yu, C. Fietzek, W. Weydanz, K. Donoue, T. Inoue, H. Kurokawa, and S. Fujitani, "Study of LiFePO 4 by Cyclic Voltammetry," pp. 253–257, 2007.
[11] G. A. Mabboil, "An Introduction to Cyclic Voltammetry I ,,,,,," vol. 60, no. 9, pp. 697–702, 1983.

Investigating a cyclotron HM-30 based neutron source for BNCT of deep-seated tumors by using shifting method

Suharyana, Riyatun, E. F. Octaviana

Physics Department, Faculty of Mathematics and Natural Sciences,
Sebelas Maret University

E-mail: suharyana61@staff.uns.ac.id

Abstract We have successfully proposed a simulation of a neutron beam-shaping assembly using MCNPX Code. This simulation study deals with designing a compact, optimized, and geometrically simple beam shaping assembly for a neutron source based on a proton cyclotron for BNCT purpose. Shifting method was applied in order to lower the fast neutron energy to the epithermal energy range by choosing appropriate materials. Based on a set of MCNPX simulations, it has been found that the best materials for beam shaping assembly are 3 cm Ni layered with 7 cm Pb as the reflector and 13 cm AlF_3 the moderator. Our proposed beam shaping assembly configuration satisfies 2 of 5 of the IAEA criteria, namely the epithermal neutron flux 1.25×10^9 n.cm^{-2} s^{-1} and the gamma dose over the epithermal neutron flux is 0.18×10^{-13} Gy.cm^{2} n^{-1}. However, the ratio of the fast neutron dose rate over neutron epithermal flux is still too high. We recommended that the shifting method must be accompanied by the filter method to reduce the fast neutron flux.

1. Introduction

Neutron Capture Therapy (NCT) is not a new technique of neutron cancer therapy. The concept of NCT was first published by Locher in 1936, who proposed the use of slow neutrons with strong neutron absorbers (e.g. boron, gadolinium) injected into tissue for selective destruction of cancerous tissues. To date, only boron has been used in NCT clinical trials and the method is known as Boron Neutron Capture Therapy (BNCT)[1].

BNCT provides a way to selectively destroy malignant cells and spare normal cells. It is based on the nuclear capture reaction that occurs when ^{10}B, which is a nonradioactive constituent of natural elemental boron, is irradiated with low energy thermal neutrons (E < 1eV) to yields high linear energy transfer α particles and recoiling ^{7}Li nuclei. In order for BNCT to be successful, a sufficient amount of ^{10}B must be selectively delivered to the tumor and enough thermal neutrons must be absorbed by them to sustain a $^{10}B(n,\alpha)^{7}Li$ capture reaction. The high Linear Energy Transfer of α particles have limited path lengths in soft tissue (5-9 μm). Since biological cells posses size of 12-13 μm, the destructive effects of these high-energy α particles is limited to boron containing cancer cells [2].

As stated above, the boron capture reaction occurs with thermal neutrons. The treatment of deep-seated tumors, such as glioblastoma multiforme, requires beams of neutrons in the epithermal energy range. Such a beam, if accompany with the least amount of gamma contamination, can be served for treatment of deep tumors. The recommended criteria for the optimal beam for BNCT, including the beam intensity and beam quality, are given in a document published by the IAEA at Table 1.

Journal of Physics: Conference Series **776** (2016) 012063 doi:10.1088/1742-6596/776/1/012063

In order to satisfy the IAEA recommended neutron beam intensity, various neutron sources have been suggested so far. The advantage of epithermal neutrons with respect to thermal ones is widely accepted. At present time, almost neutron sources for BNCT currently are limited to nuclear reactors. However, the availability of neutron flux from the reactors is not high enough. For an example, we have reported that the epithermal neutron flux at the end of the Ring Pierce Beamport of the Kartini reactor at PSTA BATAN Yogyakarta is about 1.06×10^8 n/cm^2/s when operated at 100 kW power [3]. Due to this lack of neutron flux, reactor-based neutron sources practically are no longer available for use. Indonesian BNCT researches are now interested to use accelerator-based neutron sources.

Proton accelerators based neutron sources generate polyenergetic neutrons. The primary neutrons produced from the cyclotron cannot be used directly for BNCT treatment. Suitable beam shaping assembly (BSA) is required to moderate the primary neutrons into the epithermal energy range. While the main purpose of this assembly is to moderate neutrons to epithermal energy range, other considerations are needed to achieve the successful BNCT treatment [4-8].

In this paper we report the feasibility study of designing such a BSA and achieving an optimized beam for BNCT of deep tumors using shifting method. The neutron beams were assumed generated by a HM-30 proton cyclotron, developed by the *Sumitomo Heavy Industries*, Ltd for the *Kyoto University Research Institute* (KURRI).

2. Experiment Method

The specifications of the HM-30 proton cyclotron can be obtained elsewhere [6,7]. In this report, the HM-30 cyclotron was assumed produces 30 MeV monoenergetic proton beams with 1 mA current. The target material was ^{9}Be without cooling system and its temperature was assumed constant at room temperature. The reflector materials were Pb and Ni. The reflector thickness was varied from 0.5 to 4.0 cm with a 0.5 cm step. As moderator materials, we used AlF_3, CaF_2 and MgF_2. A 4 cm thickness disk of uranium nature as a neutron multiplier, attached behind the Be target. The neutronic simulations in this study were carried out with the general purpose particle transport Monte Carlo computer code, MCNPX.

The results reported in this work have been carried out with adequate number of histories so that the relative errors to be less than 1%. The parameters IAEA recommended criteria, were tallied at the BSA beam port. The neutron flux, gamma radiation and neutron current were tallied using F4:n, F4:p and F1:n, respectively. The dose rate was tallied by employing the DE and DF tallies. The normalization factors are respectively 1.4677×10^{14} /s for neutron and 4.8398×10^{13} /s for photon, corresponds to the 1 mA proton current.

3. Results and discussion

Our proposed BSA is presented in Figure 1.

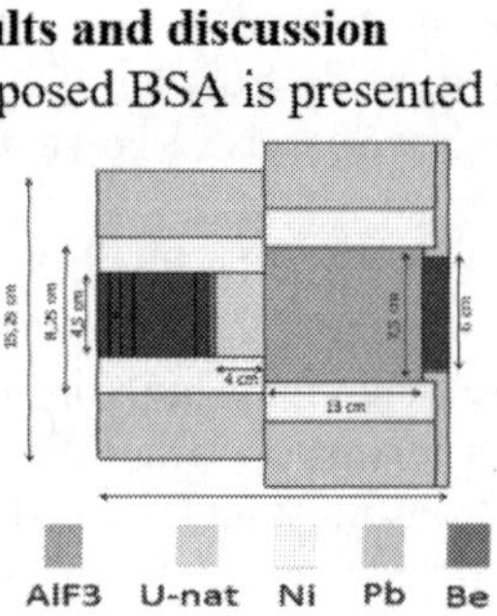

(a)

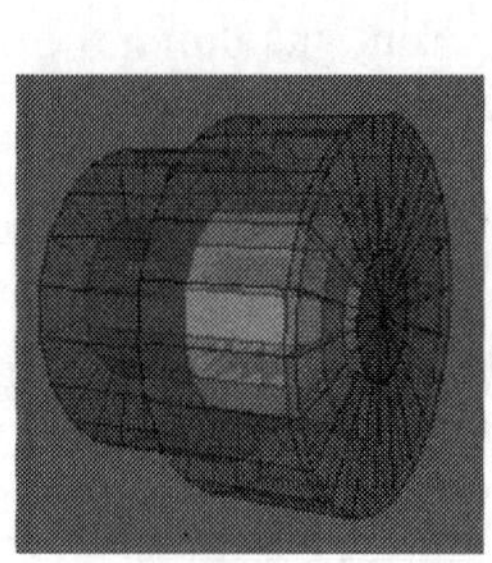

(b)

Figure 1. Our BSA model in (a) two, and (b) three dimensional view.
The symbol U-nat stands for natural uranium.

Figure 2 depicts the simulation results of the neutrons energies spectrum produced by the ^{9}Be(p,n)^{9}B reaction. Is can be seen that the HM-30 produces neutrons almost in fast energy region, with a maximum energy of 28.62 MeV. As a comparison, Tanaka *et al* [5] reported that the same cyclotron yields 28 MeV neutrons maximum energy, in good agreement with us.

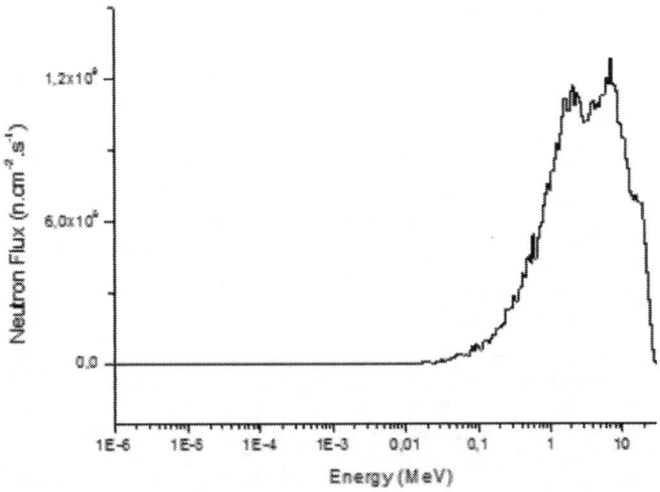

Figure 2. Neutron energy spectrum produced by 1 mA 30 MeV protons through the ^{9}Be(p,n)^{9}B reaction

According to the results presented in Figure 2, almost all the neutrons belong to the fast energy range. Shifting-down the energy of neutrons to the epithermal range is possible by means of neutron interaction with a set of materials mounted in the beam path. Our BSA is responsible for slowing-down the neutrons emitted from the source, and filtering/removing undesired fast and thermal neutrons and gamma contamination as well.

We have calculated the neutron flux of each reflector thickness. In Figure 3 (a) we show the dependence of the neutron flux to the thickness of the Ni reflector. It can be seen clearly the total neutron flux increases significantly. When the reflector thickness is 3.75 cm, the flux of the epithermal neutron raise about 37.08 % when compared to the flux without the Ni reflector. Furthermore, the Ni reflector also raises the fast neutron flux about 4.78 %, compared to that without reflector. We choose the 3.75 cm as the optimized Ni reflector thickness size.

In order to more increase the neutron flux, we layered the 3.75 cm Ni reflector with Pb. We varied the thickness of Pb layer from 1 to 9 cm with 1 cm step. In Figure 3 (b) we show the dependence of the total neutron flux when the Ni reflector is layered with several Pb thicknesses. It can be seen, the neutron flux increase monotonically with the thickness of the Pb layer. When the thickness of Pb layer is 7 cm, the epithermal neutron flux increased by about 21.7%, whereas the flux of the fast neutron decreased although not significantly, just only 0.23%.

In order to maximize neutron yields, we used a natural uranium disk attached behind the Be target. The thickness of the U-nat disk is about 4 cm. We expected that the fast neutron produced by the ^{9}Be(p,n)^{9}B reaction will react with ^{238}U under fission process to yield 2~3 neutrons. In other words, the U-nat acts as a neutron multiplier. In Figure 4 we show the spectrum of neutron, with and without U-nat neutron multiplier, at the BSA beam port. It can be seen, the total neutron flux does not increase. Instead, the fast neutron flux decreased significantly, whereas the flux of the epithermal neutron increases slightly. Most probably, this phenomenon is caused by the amount of the U-nat in the disk is not enough to support fission reaction. The thickness of the U-disk is shorter than the mean free path of neutron in uranium. Consequently, the fast neutrons are moderated by the uranium disk. Therefor the U-nat disk material doesnot act as neutron multiplier but as a moderator from our BSA design.

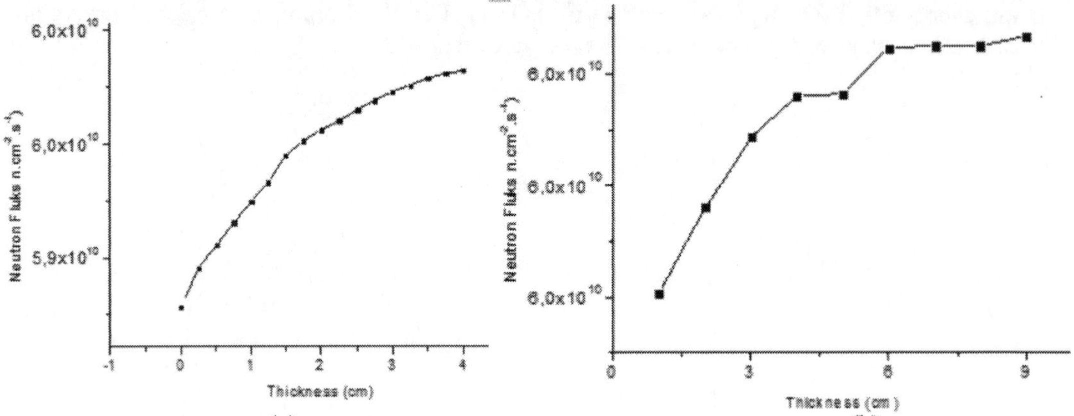

(a) (b)

Figure 3. Total neutron fluxes as a function of the thickness of (a) Ni, (b) Ni (3.75 cm) + Pb

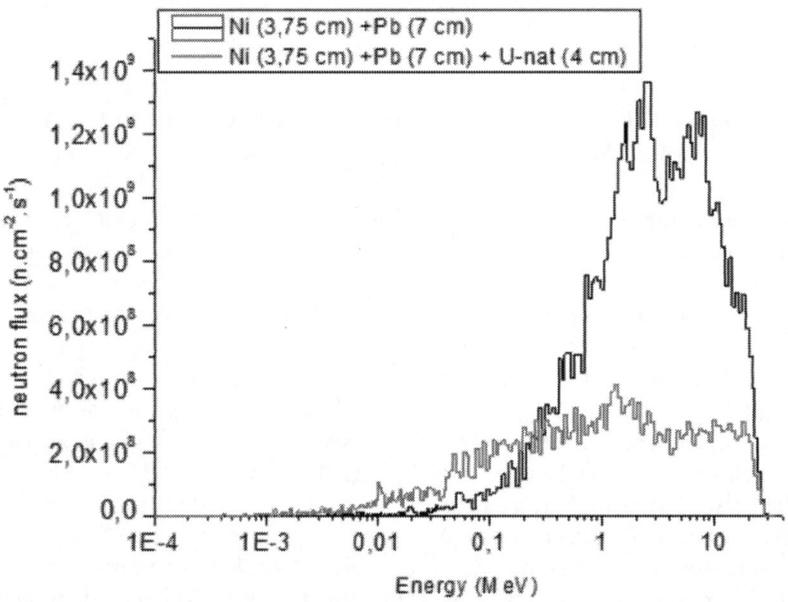

Figure 4. Neutron spectrum at the BSA beam port generated by attaching the U-nat disk and without the addition of the U-nat disk.

In order to find the most appropriate material as moderator, a number of materials have been suggested. We have looked for a moderator which not only increase the number of epithermal neutrons per cm^2, but also remove high-energy neutrons as much as possible. Finally, we decided to use four moderator materials, namely AlF_3, CaF_2, Al_2O_3 and MgF_2 and calculate the neutron flux for each material. Figure 5 depicts the effect of moderator materials to the neutron flux.

The superiority of the AlF_3 moderator among the others is obvious. It can be seen clearly in Figure 5(a), this moderator yields the highest epithermal neutrons flux. In Figure 5(b) we show the ratio of the fast neutron flux over the epithermal neutron flux. Again, the supremacy of the AlF_3 moderator is apparent. It possesses the least ratio number. Although at 13 cm thickness, the advantage of the AlF_3

moderator is still evident, as shown at Figure 5(c). At that thickness or thicker, the fluxes of fast neutron and epithermal are both almost the same.

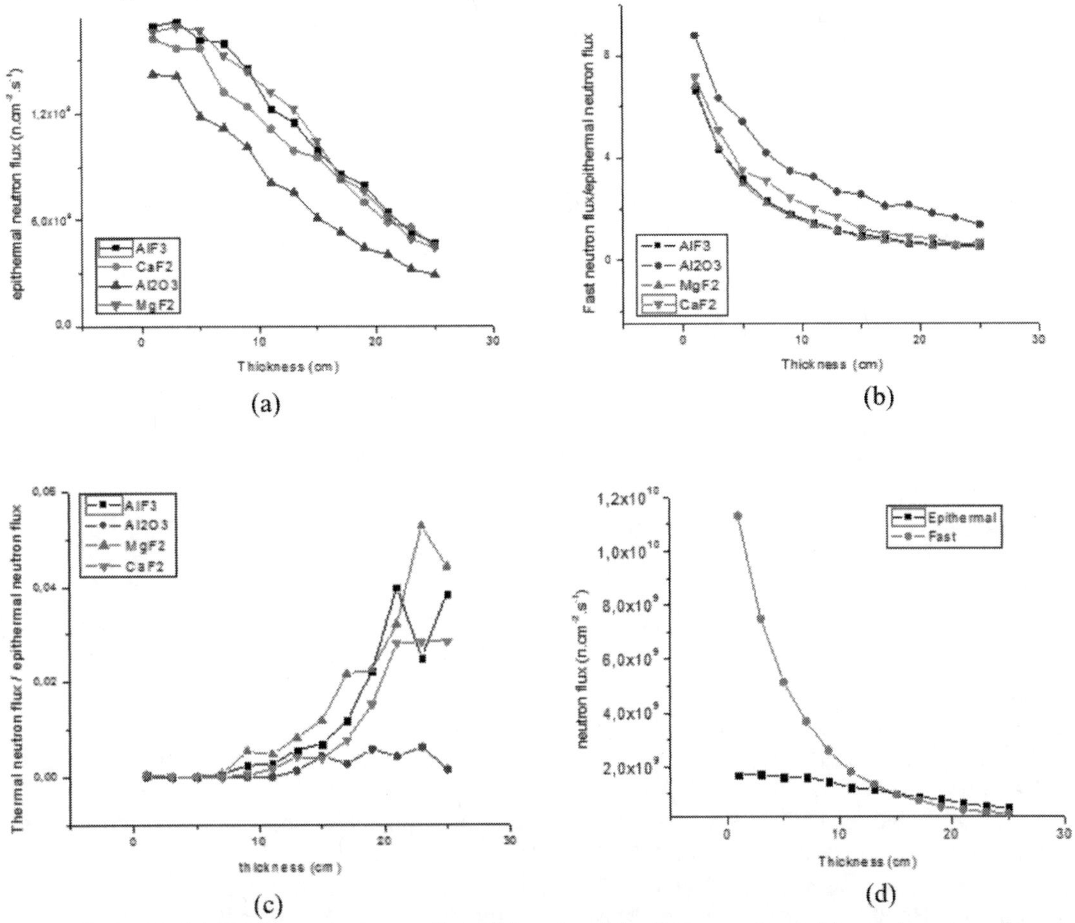

Figure 5. (a) Epithermal neutron flux, (b) ratio of the fast over the epithermal neutron flux, (c) ratio of the thermal over epithermal neutron flux of different moderators, and (d) fast and epithermal neutron fluxes of the AlF_3 moderator.

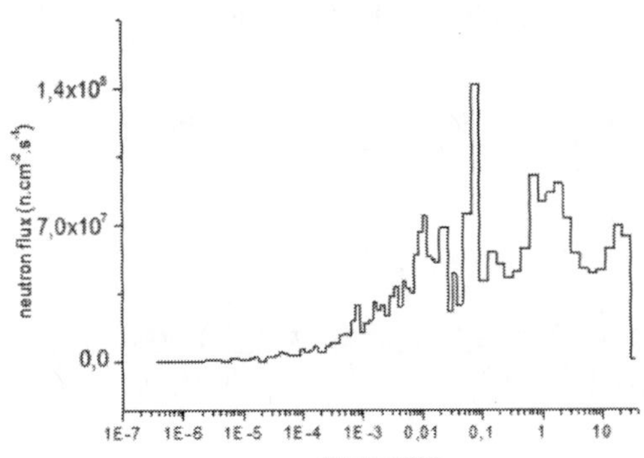

Figure 6. Neutron energy spectrum at our BSA when the AlF_3 moderator is mounted.

Table 1. Characteristics of neutron flux at the end of the BSA

Parameters	unit	IAEA	Result
Φ_{epit}	$n.cm^2s^{-1}$	$>10^9$	1.25×10^9
Φ_{th}	$n.cm^2s^{-1}$	-	5.71×10^6
Φ_{th}/Φ_{epit}	-	<0.05	0.0045
Φ_{fast}	$n.cm^2s^{-1}$	-	1.37×10^9
$\dot{D}_{fast}/\Phi_{epit}$	$Gy.cm^2n^{-1}$	$< 2 \times 10^{-13}$	2.85×10^{-11}
$\dot{D}_{\gamma}/\Phi_{epit}$	$Gy.cm^2n^{-1}$	$< 2 \times 10^{-13}$	0.18×10^{-13}

Based on the characteristic of the moderators, AlF_3 is an interesting material to be used as neutron moderator [9]. Knowing these results was our motivation to study on feasibility of utilizing this material as a neutron moderator produced by accelerator based neutron generator for BNCT purpose. An apperture with a radius of 6 cm is added to the end of beam port. Finally, at Figure 6 we show the neutron energy spectrum at our BSA beam port when the 13 cm AlF_3 moderator is applied. It can be seen, the fast neutron flux reduced significantly and the energy shifts to the epithermal energy range.

4. Conclusion and Suggestions for further work

Critical issues that must be addressed include the use for more selective collimator and moderator materials. Our BSA, composed of 3.75 cm Ni layered Pb with 7 cm reflector and 13 cm moderator AlF_3 only meets 2 of 5 of IAEA criteria. Therefore, our BSA design must be redesign by filtering the fast neutron. Furthermore, the using of U-nat as neutron multiplier have to be reassessed.

5. Acknowledgements

This research was supported by the grant under The Maintenance Research Group scheme of Sebelas Maret University.

6. References

[1] Sauerwein, W., Wittig, A., Moss, R. and Nakagawa, Y., (2012), *Neutron Capture Therapy*, Springer, New York.
[2] Ballarini, F., Bakeine, J., Bortolussi, S., Bruschi, P., Cansolino, L. and Clerici, A. M., (2011). Cell death services as following BNCT: A range approach based on Monte Carlo simulations. *Applied Radiation and Isotopes, 69* , 1745-1747.
[3] Fadli, O. E., Suharyana and Riyatun. (2015). The numerical simulation of the Neutron spectrum in four Beamport Kartini reactor with MCNP5 to its utilization as a pilot Channel Sheaf on the BNCT Neutron. *The Scientific Meeting. XXIX HFI.* Yogyakarta: Proceedings of HFI.
[4] Adib, M., Habib, N., Bashter, L., El-Mesiry, M. and Mansy, M. (2016). The simulation study of accelerator based quasi-mono-energetic epithermal neutron beams for BNCT. *Applied Radiation and Isotopes, 307* , 98-102.
[5] Tanaka, H., Sakurai, Y., Suzuki, M., Takata, T., Masunaga, S. and Kinashi, Y. (2009). Improvement of dose distribution in phantom bank by using heading epithermal neutron source based on the engulfed (p,n) reaction using heading 30 MeV proton cyclotron accelerator. *Applied radiation and Isotopes, 67* , S258-261.
[6] Mitsumoto, T., Yajima, S., Tsutsui, H., Ogasawara, T., Fujita, K. and Tanaka, H. (2015). Source For BNCT neutron Cyclotron-Based. *AIP Publishing*, 319-322.
[7] Mitsumoto, T., Fujita, K., Ogasawara, T., Tsutsui, H., Yajima, S. and Maruhashi, A. (2010). BNCT System Using Heading 30 MeV H- Cyclotron. *Cyclotron Application* , 430-432.
[8] IAEA. (2001). *Current Status of Neutron Capture Therapy.* Viena: IAEA, ISSN 1011-4289.
[9] Monshizadeh, M., Kasesaz, Y., Khalafi, H. and Hamidi, S. (2015). MCNP design of the thermal and epithermal neutron beam for BNCT at the Isfahan MNSR. *Progress in Nuclear Energy* , 427-432.

Deposition barium titanate ($BaTiO_3$) doped lanthanum with chemical solution deposition

Y. Iriani[1*] **, A. Jamaludin**[2] **, N. Nurhadi**[1]
[1]Physics Department, Faculty of Mathematics and Natural Sciences, Sebelas Maret University, Surakarta, Indonesia
[2] Physics Education Program, Sebelas Maret University, Surakarta, Indonesia

*)E-mail: yopen_2005@yahoo.com

Abstract. Deposition of Barium Titanate ($BaTiO_3$) thin films used Chemical Solution Deposition (CSD) method and prepared with spin coater. $BaTiO_3$ is doped with lanthanum, 1%, 2%, and 3%. The thermal process use annealing temperature 900°C and holding time for 3 hours. The result of characterization with x-ray diffraction (XRD) equipment show that the addition of La^{3+} doped on Barium Titanate caused the change of angle diffraction.The result of refine with GSAS software shows that lanthanum have been included in the structure of $BaTiO_3$. Increasing mol dopant La3+ cause lattice parameter and crystal volume become smaller. Characterization result using Scanning Electron Microscopy (SEM) equipment show that grain size (grain size) become smaller with increasing mole dopant (x) La^{3+}. The result of characterization using Sawyer Tower methods show that all the samples (Barium Titanante and Barium Titanate doped lanthanum) are ferroelectric material. Increasing of mole dopant La^{3+} cause smaller coercive field and remanent polarization increases.

1. Introduction

Perovskite ferroelectric is a large group of ferroelectric material, that has a significance and wide application in electronic industry such as Multi Layer Ceramic Capacitor (MLCC), Ferroelectric Random Access Memory (FRAM), thermistor, optoelectronic devices and actuator [1-9]. Barium Titanate is one kind of ferroelectric material which has a perovskite structure (ABO_3). On the perovskite structure ABO_3, A is a atomic position, that has bigger atomic diameter then atomic diameter on B position. It is similiar with structure of $BaTiO_3$, Ba^{2+} (1.35A) occupied A position and Ti^{4+} (0.68 A) occupied B position. Microstructure, dielectric and ferroelectric properties of perovskite structure could be modificated by modifying of Atomic A or B. Ionic doping for subtituting of A atomic position such is such as ionic La^{3+}, Pb^{2+}, Na^{+}, Ga^{2+}, Nd^{3+}, Th $Cr4^{+}$. B atomic position can be substituted by ionic Zr^{4+}, Fe^{3+}, Al^{3+}, In^{3+}, Cr^{3+}, Nb^{5+}, Ta^{5+}, Sb^{5+}, W^{6+}. This paper presented ionic La^{3+} for subtituting Barium Titanate [10].
A thin film was fabricated by many different method such as Pulsed Laser Deposition (PLD), chemical vapour deposition (CVD), CSD (chemical solution deposition) or Sol-Gel spin coating [11-16]. Every method has advantage and disadvantage. A thin film barium titanate ($BaTiO_3$) is doped with (La^{3+})

using metode *Chemical Solution Deposition* (CSD). CSD method is a fabrication method of thin film that combined physic and chemical method. It has step for fabricating thin film : solution development (chemical process), spin coating and annealing process (Physics process). In this research, sol-gel solution material is coated on substrated pt/si using spin coater, then thermal process is done using furnice. Furthermore, the material characterization is done using *X Ray Diffraction* (XRD), *Scanning Electron Microcopy* (SEM-EDX), and electronic characterization is done using Sawyer Tower method for hysteresis properties.

2. Experiment

A Thin film Barium Titanate ($BaTiO_3$) and $BaTiO_3$ with doping lanthanum (BLT) have been fabricated by Chemical Solution Deposition (CSD) method. A thin film Barium Titanate was doped with various molarity of dopant lanthanum (La^{3+}) 1%, 3% and 5%. The precursor material that was used such as Barium acetate ($Ba(CH_3COO)_2$) 99.999%, Strontium acetate ($Sr(CH_3COO)_2$) 99.995%, Titanium (IV) Iso-propoxide ($Ti(C_{12}O_4H_{28})$) 99.000% and Lanthanum acetate ($La(CH_3COOH)_2$). Acetic acid and Methoxy methanol were used as a solvent material. Precursor and solvent material were mixed to be homogeneous solution using magnetic stirrer hotplate. A thin film was deposited $BaTiO_3$ solution (with and without dopant La^{3+}) on Pt/Si Substrate by spin coater instrument with speed 3000 rpm. Furthermore, it was annealed to be crystalline structure using furnace at 900°C with holding time during 3 hours.

Materials properties of sample was investigated by X-Ray Diffraction (XRD) for microstructure and crystallinity analysis. XRD instrument by bruker has a wavelenght λ_{Cu} (1.54056 Å). XRD pattern was analyzed by rietveld method on General Structure Analysis System (GSAS) software that was done for calculation lattice parameter, tetragonality and crystal volume. The morphological investigation included thickness and roughness analysis was done by Scanning Electron Microscopy (SEM). A sawyer tower method was used for investigating of ferroelectric properties.

3. Result and Discussion

The XRD result was shown on figure 1. Based on figure 1, the diffraction pattern was compatible with pattern of Barium Titanate ($BaTiO_3$) on International Center of Diffraction Database (ICDD) with plane orientation (100) and (110). Meanwhile, the other pick diffraction showed plane orientation of Pt/Si substrate.

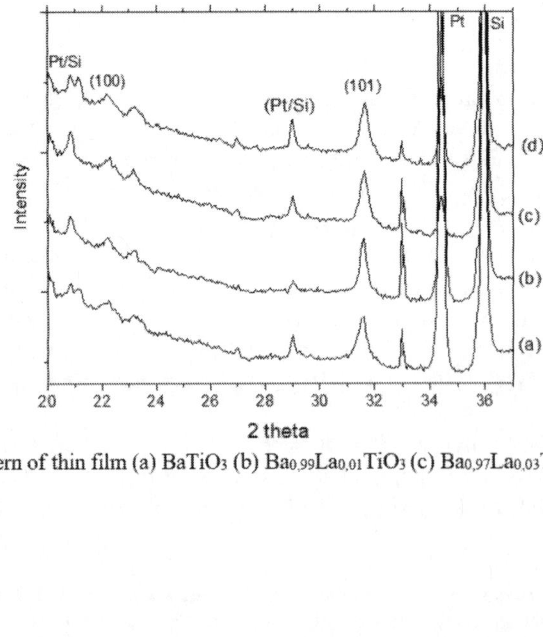

Figure 1. Diffraction pattern of thin film (a) $BaTiO_3$ (b) $Ba_{0,99}La_{0,01}TiO_3$ (c) $Ba_{0,97}La_{0,03}TiO_3$ (d) $Ba_{0,95}La_{0,05}TiO_3$

Based on XRD result, the pick intensity (count) showed crystallinity of thin film Barium Titanate. Table 1 shows pick intensity (count) and shift direction on plane orientation (110). The addition of molarity dopant La^{3+} 1 % will be increase a pick intensity, but the addition molarity dopant La^{3+} (3%, 5%) will be decrease a pick intensity. The dopant La^{3+} was donor dopant with more positive carge (+) than ionic Ba^{2+}. It will substitute ionic Ba^{2+} to be $BaLaTiO_3$. For preserving ionic stability and neutrality, ionic Ti^{4+} released one a positive charge (+), then it effected crystalline deformation especially ionic dislocation of Ti^{4+}. The crystalline deformation changed pattern interference and intensity of XRD.

Tabel 1. Shift direction and Intensity on thin film $BaTiO_3$ and BLT with molarity dopant La^{3+}

Parameters	Plane Orientation (110)			
	BT	BLT 1%	BLT 3%	BLT 5%
Intensity (count)	8206	8807	8635	8450
Shift direction (^{O})	31.55	31.60	31.60	31.65

Table 1 shows a shift direction on plane orientation of $BaTiO_3$ because of increasing dopant La^{3+}. An ionic La^{3+} has smaller diameter than ionic Ba^{2+}. Furthermore, additional of dopant La^{3+} has decreased crystalline volume and gap between other crystal to more shortly (table 2). Then a lattice parameter of crystalline decreased together additional of dopant La^{3+}. Based on brag law, a decreasing of lattice parameter effected shift angle of diffraction (2θ) into large angle diffraction on plannar orientation (110).

Table 2. Lattice parameter, tetragonality and crystal volume of thin film $BaTi0_3$ with various mol of dopant La^{3+}

Sampel	Lattice Parameter (Å)		*c/a* (Tetragonality)	Crystal Volume (Å)^3
	a=b	*C*		
$BaTi0_3$	4.008	4.029	1.005	64.722
$Ba_{0,99}La_{0.01}Ti0_3$	4.003	4.015	1.002	64.336
$Ba_{0,97}La_{0.03}Ti0_3$	3.991	4.014	1.005	63.935
$Ba_{0,95}La_{0.05}Ti0_3$	3.996	4.008	1.003	63.999

A quantitative analysis was done using GSAS software with refinement tool and rietveld method. The result shows that thin film $BaTiO_3$ was successfully growth on Pt/Si substrate with tetragonal structure of perovskite crystalline. A tetragonal structure has value lattice parameter $a=b \neq c$ and value $c/a \geq 1$. The mean of c/a was tetragonality. Table 2 shows alteration of lattice parameter because of dopant La^{3+} into $BaTiO_3$. The effect of increasing dopant La3+ into BaTiO3 will decrease lattice parameter (a,c). Therefore, the volume of crystal will decrease too.

The SEM with magnitude 40.000x was used for observing thin film $BaTiO_3$ and BLT with various dopant La^{3+} whose annealed at 900°C. Figure 2 and 3 show SEM result. Based on figure 2, the morphological of thin film $BaTiO_3$ shows heterogeneous grain size (1137 nm) and grain distance. The increasing of molarity dopant La^{3+} 1%, 3%, and 5% decreased grain size into 389 nm, 364 nm and 336 nm. It was be parallel with decreasing of crystalline volume (table 2). The result, the additional of dopant morality will be decreased crystalline gap and volume, and grain sice. It occurs because ionic diameter of La^{3+} has a smaller diameter than ionic Ba^{2+} which was substituted it.

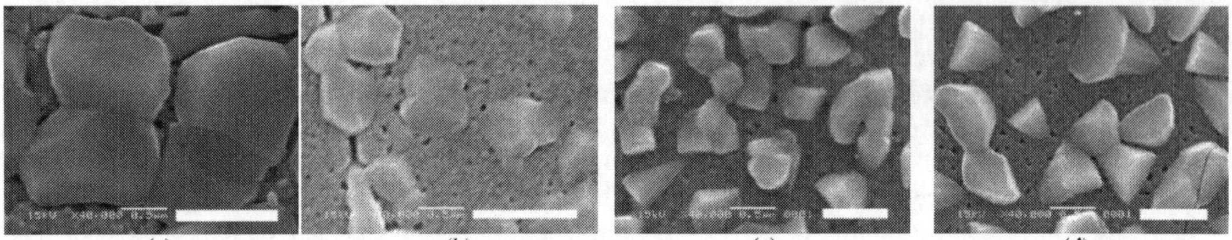

(a) (b) (c) (d)

Figure 2. SEM Thin Film $BaTiO_3$ with various mollarity of dopant La^{3+} (a) $BaTiO_3$ (b) $Ba_{0.99}La_{0.01}TiO_3$ (b) $Ba_{0.97}La_{0.03}TiO_3$ (d) $Ba_{0.95}La_{0.05}TiO_3$

The cross section of thin film was investigated by SEM with 20.000X magnitude. It was done for knowing a thickness of thin film $BaTiO_3$. Figure 3 shows a cross section of thin film $BaTiO_3$ and $BaTiO_3$ with dopant La^{3+} on substrate Pt/Si. Figure 3 (a) shows thin film $BaTiO_3$ without dopant that was deposited on Pt/Si substrate. Thin film $BaTiO_3$ appeared visible roughness film and non-homogeneous thickness. The thickness calculation of $BaTiO_3$ was average of 705 nm. Figure 3 (b), (c) and (d) show cross section of thin film $BaTiO_3$ with various dopant La^{3+} 1%, 2%, 3%. The thickness calculation of it was average 409 nm for dopant La^{3+} 1%, 484 nm for dopant La^{3+} 3% and 646 nm for dopant La^{3+} 5%.

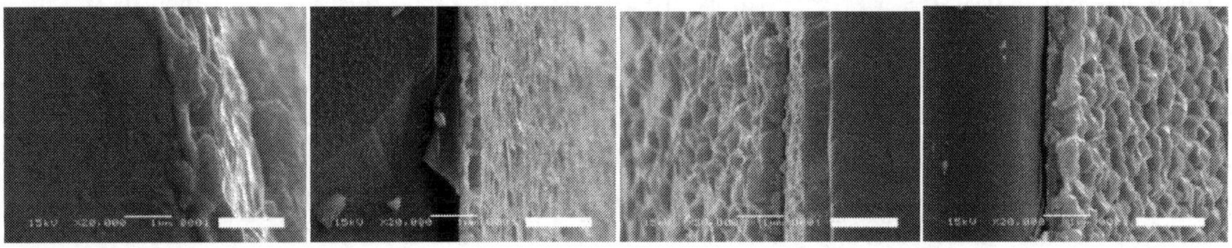

(a) (b) (c) (d)

Figure 3. Cross section of thin film $BaTiO_3$ with various dopant La^{3+} (a) $BaTiO_3$ (b) $Ba_{0.99}La_{0.01}TiO_3$ (b) $Ba_{0.97}La_{0.03}TiO_3$ (d) $Ba_{0.95}La_{0.05}TiO_3$

The Sawyer Tower method was done for investigation of ferroelectric material properties. A ferroelectric properties of material was shown by hysteresis curve while it was given external electric field. Figure 4 shows hysteresis curve of thin film $BaTiO_3$ and with dopant La^{3+} 1%, 3% dan 5%. The increasing doping La^{3+} into $BaTiO_3$ when annealed at temperature 900°C was not significantly influence of hysteresis curve. The coercive field was not change while add molarity of dopant La^{3+}. Furthermore, additional molarity of dopant La^{3+} changed value of remanent polarity. The alteration of remanen polarity was effected by non-uniform grain size. While external electric field offered on sample, domain orientation of sample was not same direction because of non-uniform grain size. When grain size was under critical grain size, domain structure transition happened from multi-domain into mono-domain. A mono-domain structure was a stable domain, so domain orientation transition will be difficulty offered external electric filed.

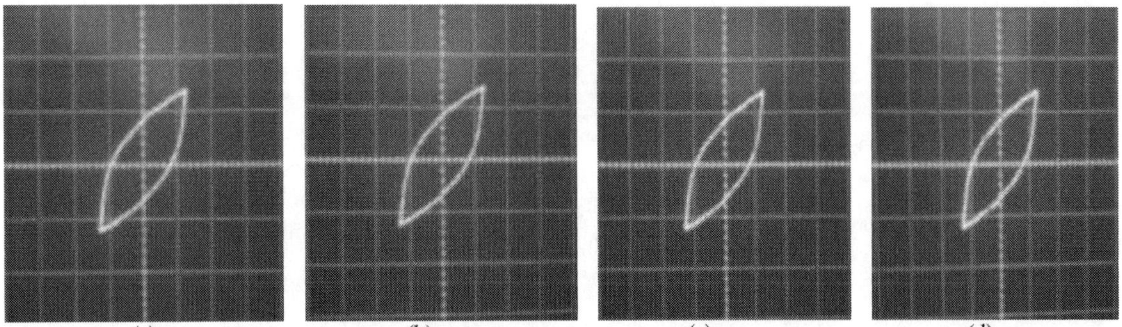

(a) (b) (c) (d)

Figure 4. Hysteresis Curve of Thin film $BaTiO_3$ with various dopant La^{3+} (a) $BaTiO_3$ (b) $Ba_{0.99}La_{0.01}TiO_3$ (c) $Ba_{0.97}La_{0.03}TiO_3$ (d) $Ba_{0.95}La_{0.05}TiO_3$

4. Summary

The addition of La^{3+} doped on Barium Titanate caused the change of angle diffraction.The result of refine with GSAS software shows that lanthanum have been included in the structure of $BaTiO_3$. The addition of molarity dopant La^{3+} 1 % will be increase a pick intensity, but the addition molarity dopant La^{3+} (3%, 5%) will be decrease a pick intensity. Increasing mol dopant La3+ cause lattice parameter and crystal volume become smaller. Grain size become smaller with increasing mole dopant (x) La^{3+}. The result of characterization using Sawyer Tower methods show that all the samples (Barium Titanante and Barium Titanate doped lanthanum) are ferroelectric material. Increasing of mole dopant La^{3+} cause smaller coercive field and remanent polarization increases.

5. Acknowledgments

This work was supported with a grant from Hibah Maintanance Research Grup DIPA BLU Sebelas Maret University, 2015.

6. References

[1] J.B. Xu, J.W. Zhai, X. Yao, Journal Alloys Compound **467**, 567-571 (2009).
[2] A.I.Kingon, J.P.Maria, S.K.Streiffer, Nature **406**, 1032-1038 (2000)
[3] H. Fan and L. Liu, Journal Electroceramics **22**, 291-296 (2009).
[4] F. Wan, J.G. Han, Z.Y. Zhu, Physics Letter A **372**, 2137-2140 (2008).
[5] H. Kishi, Y. Mizuno, H. Chazono, Japanese Journal of Applied Physics 42 (2003) 1–15.
[6] S.B. Narang, D. Kaur, Integreted Ferroelectric **105**, 87-98 (2009).
[7] D.Y. Lu, T. Koda, H. Suzuki, M. Toda, Journal of the Ceramic Society of Japan **113**, 721-7727 (2005).
[8] Y.L. Chen, S.F. Yang, Advanced Applied Ceramic **110**, 257–269 (2011).
[9] W. Cai, C. Fu, J. Gao, Q. Guo, X. Deng, C. Zhang, Physica B **406**, 3583-3587 (2011).
[10] Y. Xu. *Ferroelectric Materials and Their Application* (North-Holland, Amsterdam, 1991), pp. 133-137.
[11] V. Z. Zelezny, D. Chvostova, L. Pajasova, M. Jeli, T. Kocourek, S. Danis, V. Valvoda, Applied Surface Science **255**, 5280-5283 (2009).
[12] L. Goux, M. Gervais, A. Catherinot, C. Champeaux, F. Sabary, Journal of Non-Crystalline Solids **303**, 194-200 (2002).
[13] V. Kumar, I. Packiaselvam, K. Sivanandan , M.A. Vahab, A.K. Sinha, Journal American Ceramic Society **89**, 1136–1139 (2006).
[14] M.M. Vijatovic, J.D. Bobic, T. Ramos, J. Banys, B.D Sjotanovic, Ceramic International **37**, 2669-2677 (2011)
[15] R. Balachandran, B.H.Ong, H.Y.Wong, K.B.Tan, M.M. Rasat, International Journal Electrochemical Science **7**, 11895-11903 (2012).
[16] T.K. Kundu, A. Jana, P. Barik, Doped barium titanate Nanoparticles **31**, 501-505 (2008).

An analytical solution versus half space BEM formulation for acoustic radiation and scattering from a rigid sphere

B. Soenarko and D. Setiadikarunia
Electrical Engineering Department, Universitas Kristen Maranatha,
Jl. Prof.Drg.Surya Sumantri 65, Bandung 40164, INDONESIA

E-mail: bsoenarko@yahoo.com

Abstract. A half space problem in acoustics is described by introducing an infinite plane boundary that reflects the wave coming into the plane. A numerical solution using Boundary Element Method (BEM) has been known which is formulated using a modified Green's function in the Helmholtz Integral Formulation, which eliminates the discretization over the infinite plane. Hence, the discretization are confined to the body or obstacle in question only. This feature constitutes the main advantage of the BEM formulation for half space problems. However, no general analytical solution is available to verify the BEM results for half space problems. This paper is aimed to propose an analytical solution for the BEM to compare with, hence to verify the BEM calculation. This analytical approach is currently developed for a half space problem involving radiation and scattering of acoustic waves from a rigid sphere. The image of sphere as well as the image of the field point are defined with respect to the infinite plane. Then, an ad hoc solution is assumed involving a constant and the distance from the center of the sphere to the field point and the distance from the center of the image of the sphere to the field point. The constant is determined by imposing the boundary conditions. Test cases were run with several configuration involving the location of field points and the sphere. Comparison of the analytical solution with BEM calculations shows a good agreement between the two results..

1.Introduction

The Boundary Element Method (BEM) is one of the numerical methods that has been known to researchers in many area of research. The main advantage offers by the BEM is that it requires only the discretization of the boundary of the body in question. Therefore, in this scheme the dimension of the problem is reduced by one, e.g. a three dimensional problem is solved using a two dimensional calculation in the BEM. In acoustics, the application of the BEM includes many classes of problems involving radiation and scattering of acoustic waves from arbitrary shape bodies. Intensive developments in the BEM have been proposed by quite a number of researchers [1-9]. The acoustic medium may be confined in finite or infinite domain. Finite domain of acoustic media is found in so called interior problems in acoustics, while acoustic problems with infinite domain is known as exterior problems in acoustics. When the infinite domain is bounded by an infinite plane, the problem becomes a half space problem. The BEM formulation for half space problems uses a modified Green's function in the Helmholtz Integral Formulation and includes the image of the vibrating body in the radiation problem or the obstacle in the scattering problem. This formulation avoids the discretization of the infinite plane which simplifies the numerical calculation since it involves only the discretization of the radiating body in the radiation problem or the obstacle in the scattering problem. However, no analytical solution available for half space problems to verify the BEM results. It is the aim of this paper to propose an analytical solution to verify the BEM calculation, specifically for radiation and scattering problems in a half space for a rigid sphere. Test cases were run for several configurations with respect to the field point locations and the distance between the center of the sphere and the infinite plane. Comparison of the results obtained from the analytical solutions with those from the BEM calculations show a very good agreement. Hence the BEM results are well verified.

2.The Half Space BEM Formulation

2.1.Radiation

The BEM formulation for acoustic radiation in a half space is given by the following equation [6,10] :

$$\int_{S_0}\left[\phi(Q)\frac{\partial\psi_H}{\partial v}(P.Q)-\psi_H(P,Q)\frac{\partial\phi}{\partial v}(Q)\right]dS(Q)=C(P)\phi(P)\,, \qquad (1)$$

where S_0 is the surface of the radiating object, ψ_H is the half space Green's function defined by

$$\psi_H(P,Q)=\exp[-ikR(P,Q)]/R(P,Q)+R_p\exp[-ikR'(P',Q)]/R'(P',Q)\,, \qquad (2)$$

where Q is any point on S_0, P may on S_0 or outside of S_0, k is the wave number ω/c where c is the speed of sound and ω is the frequency, and v is the outward unit normal on S_0, R_p is the reflection coefficient of the plane,

$$C(P)=4\pi+\int_{S_0}\frac{\partial}{\partial v}\left[\frac{1}{R(P,Q)}\right]dS(Q)\,, \qquad (3)$$

2.2.Scattering

For scattering problem in a half space, the BEM formulation takes the form [6,10]:

$$\int_{S_0}\left[\phi_t(Q)\frac{\partial\psi_H}{\partial v}(P.Q)-\psi_H(P,Q)\frac{\partial\phi_t}{\partial v}(Q)\right]dS(Q)+4\pi\phi_i'(P)=C(P)\phi_t(P)\,, \qquad (4)$$

where $\phi_t=\phi_s+\phi_i$ is the total velocity potential, ϕ_s is the scattered velocity potential, ϕ_i is the velocity potential of the incoming wave, $\phi_i'=\phi_i+\phi(P)$ is the sum of the velocity potential of the incoming wave and the velocity potential ϕ at P.

3. Analytical solution for radiation from a rigid sphere in a half space

Consider a uniformly pulsating sphere of radius a located with its center O at a distance B from an infinite rigid plane S_H as shown in Figure 1. The origin of the coordinate system is defined at the center of the sphere. Let O_1 be the center of the image of the sphere with respect to the plane. The velocity on the surface of the sphere is $U_a\,e^{i\omega t}$. Now assume an ad hoc solution:

$$\phi(r,t)=A\left[(1/r)e^{i(\omega t-kr)}+(1/r_1)e^{i(\omega t-kr_1)}\right] \qquad (5)$$

where r is the distance from O to any field point P, r_1 is the distance from O_1 to P, and A is a constant.

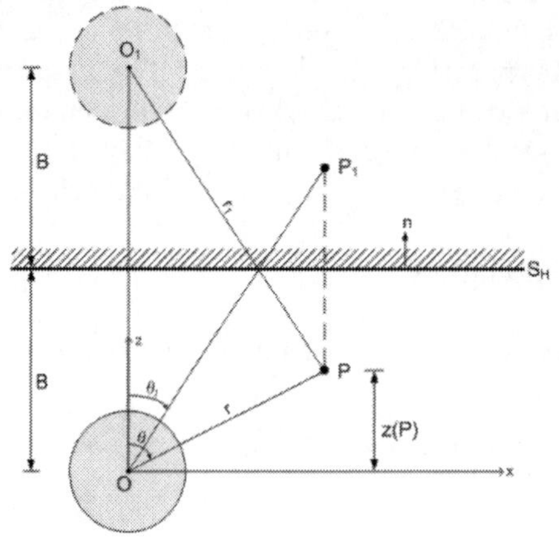

Figure 1 – Radiation from a pulsating sphere in a half space

On S_H :

$$\partial r/\partial n|_{S_H} = \partial r/\partial z = \partial/\partial z(x^2+y^2+z^2)^{1/2} = z/r$$

$$\partial r_1/\partial n|_{S_H} = \partial r_1/\partial z = \partial/\partial z\left[x^2+y^2+(z-2B)^2\right]^{1/2} = (z-2B)/r_1$$

where n is the unit normal on S_H. Furthermore $r=r_1$ and $z=B$ on S_H. Therefore $\partial r/\partial n = B/r$ and $\partial r_1/\partial n = -B/r$ so that

$$\partial r/\partial n|_{S_H} = -\partial r_1/\partial n|_{S_H} \tag{6}$$

Then

$$\partial\phi/\partial n|_{S_H} = A\left[(-1/r^2 - ik/r)e^{i(\omega t-kr)}\partial r/\partial n + \left(-1/r^2 - ik/r\right)e^{i(\omega t-kr)}\partial r_1/\partial n\right]_{S_H} = 0 \tag{7}$$

by virtue of Eq.(6). Thus the ad hoc solution (Eq.(5)) satisfies the boundary condition on S_H. The constant A can be determined from the boundary condition on the sphere, i.e.:

$$\nabla_r\phi|_{r=a} = U_a$$

or

$$-\partial\phi/\partial r|_{r=a} = U_a \tag{8}$$

where the time dependent factor $e^{i\omega t}$ is suppressed for brevity. Express r_1 in terms of r and θ:

$$r_1^2 = r^2 + 4B^2 - 4Br\cos\theta \tag{9}$$

where θ is the angle between r and the z-axis. Then

$$\partial r_1/\partial r = (r - 2B\cos\theta)/\left(r^2 + 4B^2 - 4Br\cos\theta\right)^{1/2} \tag{10}$$

Using Eq.(5) in Eq.(8) in light of Eq.(10) yields:

$$A = U_a a^2 / \left[(1+ika)e^{-ika} + \{(1+ikaQ)/Q^2\}\{(1-2(B/a)\cos\theta)/Q\}e^{-ikaQ} \right] \tag{11}$$

where
$$Q = \{1+4(B/a)^2 - 4(B/a)\cos\theta\}^{1/2} \tag{12}$$

Incorporating Eq.(11) and Eq.(12) in Eq.(5) gives the final solution as follows:

$$\phi(r,t) = (a/r)\left[U_a a Q^3 \{e^{-ikr} + (e^{-ikr\hat{Q}}/\hat{Q})e^{i\omega t}\}\right] / \left[(1+ika)Q^3 e^{-ika} + (1+ikaQ)\{1-2(B/a)\cos\theta\}e^{-ikaQ}\right] \tag{13}$$

where
$$\hat{Q} = \{1+4(B/r)^2 - 4(B/r)\cos\theta\}^{1/2} \tag{14}$$

Eq.(13) and Eq.(14) yield the analytical solution of the radiation of a pulsating rigid sphere in a half space with a rigid infinite plane. It may be noted that A is not a pure constant as assumed in Eq.(5), but $A = A(\cos\theta)$. However, taking into account that $\cos\theta \cong \cos\theta_1$ in the vicinity of the plane, it can be shown that the boundary condition on the rigid plane S_H is satisfied.

4. Analytical solution for scattering from a rigid sphere in a half space

Suppose the sphere described in Fig.1 is not moving and is impinged by a plane wave $\phi_i = \phi_0 e^{i(kx+\omega t)}$ as shown in Fig.2. The incoming wave can be expanded into a series of spherical wave as [10]:

$$\phi_i = \phi_0 e^{i\omega t} \sum_{m=0}^{\infty} i^m (2m+1) P_m(\cos\delta) j_m(kr) \tag{15}$$

where $P_m(\cos\delta)$ are Legendre polynomials of order m, $j_m(kr)$ are spherical Bessel functions of order m and δ is the angle between r and the incoming wave. The image of the sphere has its center at O_1, r_1 is the distance from O_1 to any field point P of interest.

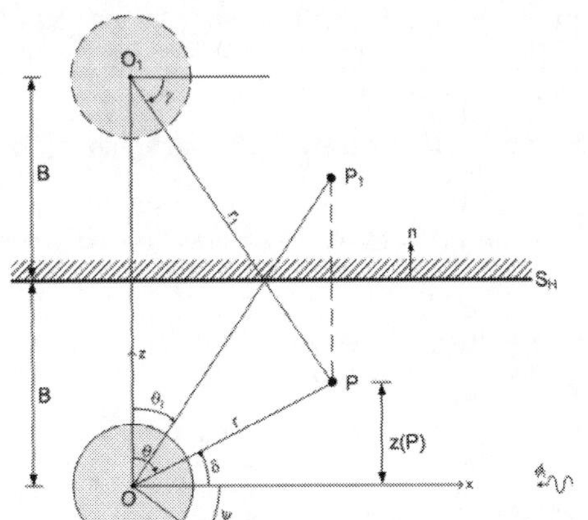

Figure 2 – Scattering from a rigid sphere in a half space

Assume the scattered potential to be given by

$$\phi_s = \sum_{m=0}^{\infty} \left[a_m P_m(\cos\delta) h_m(kr) + b_m P_m(\cos\gamma) h_m(kr_1)\right] e^{i\omega t} \tag{16}$$

where γ is the angle between r_1 and the direction of the incoming wave, $h_m(kr)$ are the Hankel functions of the second kind. Eq.(16) describes the scattered pressure as a series of spherical waves from the origin O and from O_1. Each term in Eq.(16) is associated with the corresponding term in Eq.(15). The total potential ϕ_t is given by

$$\phi_t = \phi_i + \phi_s \tag{17}$$

Since the plane is rigid, the boundary condition on the plane is $-\nabla_z \phi_t|_{S_H} = 0$. Thus,

$$\partial/\partial z(\phi_i + \phi_s)|_{S_H} = 0$$

Drop the time dependent factor $e^{i\omega t}$ for brevity, $\partial\phi_i/\partial z|_{S_H} = \partial/\partial z(\phi_0 e^{ikx})|_{S_H} = 0$, so

$$\partial\phi_s/\partial z|_{S_H} = 0 \tag{18}$$

Using Eq.(16) in Eq.(18) and noting that for field points on the plane: $z=B$, $\delta=\gamma$ and $r=r_1$, it can be shown that $a_m = b_m$. Hence, Eq.(16) can be rewritten:

$$\phi_s = \sum_{m=0}^{\infty} a_m \{P_m(\cos\delta)h_m(kr) + P_m(\cos\gamma)h_m(kr_1)\}e^{i\omega t} \tag{19}$$

The boundary condition on the sphere is (the sphere is rigid) : $-\nabla_r \phi_t|_{r=a} = 0$ or

$$\partial\phi_i/\partial r|_{r=a} + \partial\phi_s/\partial r|_{r=a} = 0 \tag{20}$$

The m-th term of ϕ_I and ϕ_s must satisfy Eq.(20). Substituting Eq.(15) and Eq.(19) in Eq.(20) with little mathematical book keeping gives:

$$a_m = \left[\phi_0 i^m (2m+1)kP_m(\cos\delta)D_m(ka)\sin\delta_m(ka)\right] / \left[Term1 + Term2 + Term3\right] \tag{21}$$

where

$$Term1 = P_m(\cos\delta)kh'_m(ka) = -ikP_m(\cos\delta)D_m(ka)e^{-i\delta_m(ka)} \tag{22}$$

$$Term2 = P_m(\cos\gamma_0)(-ikD_m(kr_{10})e^{-i\delta_m(kr_{10})})\{1 - 2(B/a)\cos\theta\}/Q \tag{23}$$

$$Term3 = \left[m\{\cos\gamma_0 P_m(\cos\gamma_0) - P_{m-1}(\cos\gamma_0)\}/(\cos^2\gamma_0 - 1)\right]\left[\{\cos\delta(4(B/a)^2 - 2(B/a)\cos\theta)\}/(aQ^3)\right]h_m(kr_{10}) \tag{24}$$

$\delta_m(ka)$ is the value of $\delta_m(kr)$ for $r = a$ in which

$\delta_m(kr)$ is defined such that $\frac{\partial h_m(kr)}{\partial(kr)} = D_m(kr)e^{-i\{\delta_m(kr)+\pi/2\}}$

and

$$D_m(kr) = \frac{1}{2m+1}[\{mj_{m-1}(kr) - (m+1)j_{m+1}(kr)\}^2 + \{mn_{m-1}(kr) - (m+1)n_{m+1}(kr)\}^2]^{1/2}$$

$j_m(kr)$ are the spherical Bessel functions with argument kr

$n_m(kr)$ are the spherical Neumann functions with argument kr

$r_{10} = [4B^2+a^2-4Bacos\theta]^{1/2}$

Multiplying the numerator and the denominator by i, Eq.(21) becomes:

$$a_m = \left[k\phi_o i^{m+1}(2m+1)P_m(\cos\delta)D_m(ka)\sin\delta_m(ka)\right]/\left[D_1+D_2+D_3\right] \quad (25)$$

where

$$D_1 = iTerm\ 1,\ D_2 = iTerm\ 2,\ D_3 = iTerm\ 3$$

It may be noted that a_m does not look like a constant as assumed in Eq.(16), but $a_m=a_m(cos\theta)$. However, it can be shown that $\partial(a_m)/\partial z|_{S_H} = 0$ which imply that $\partial\phi_s/\partial z|_{S_H} = 0$. Furthermore, it can also be shown that $\partial a_m/\partial z|_{S_H} = 0$. Thus a_m (Eq.(25)), when used in Eq.(19) will satisfy the boundary condition on S_H , i.e. $\partial\phi_t/\partial z|_{S_H} = 0$. Hence, the solution for the scattered potential can be expressed as:

$$\phi_s = k\phi_0 e^{i\omega t}\sum_{m=0}^{\infty}\left\{i^{m+1}(2m+1)P_m(\cos\delta)D_m(ka)\sin\delta_m(ka)\right\}\left\{P_m(\cos\delta)h_m(kr)+P_m(\cos\gamma)h_m(kr_1)\right\}/\left\{D_1+D_2+D_3\right\} \quad (26)$$

Equation (26) is the analytical solution of the scattering from a rigid sphere in a half space with rigid infinite plane.

5. Results and discussion

A number of test cases were run using the proposed analytical solution involving radiation and scattering of acoustic waves in a half space where comparison with the BEM calculations were examined. The infinite plane is assumed to be rigid. The object involved is a rigid sphere located at a variety of distance from the infinite plane. The first test case is a radiation from a pulsating rigid sphere. The center of the sphere is located at a distance $B=3a$ from the infinite plane. The sphere is vibrating with velocity on the surface equals to U_a. The boundary condition on the sphere is a uniform $U_a = \partial\phi/\partial v$ (pulsating sphere), and the normalized frequency of vibration of the sphere is $ka = 1$.

Figures 3 and 4 show the normalized acoustic pressure ($p=ikz_0\phi$) at distances of $3a$ and $2a$ from the center of the sphere, respectively as a function of polar angle plotted on a plane passing through the center of the sphere and perpendicular to the infinite plane. The plots of the two configurations show a good agreement between the analytical solutions and the BEM results.

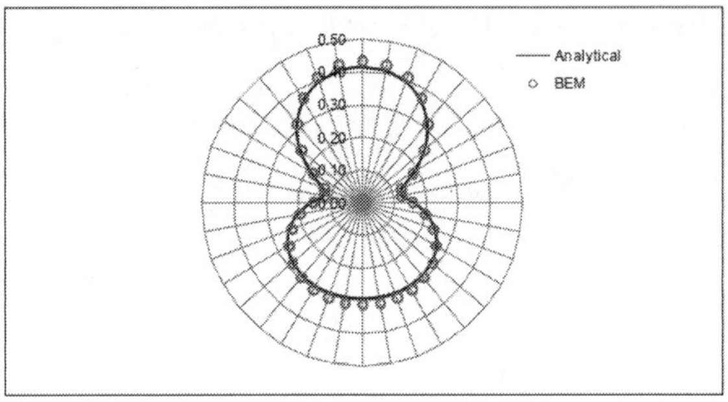

Figure 3 – Normalized radiated pressure at a distance $r = 3a$, when $ka = 1$, $B = 3a$

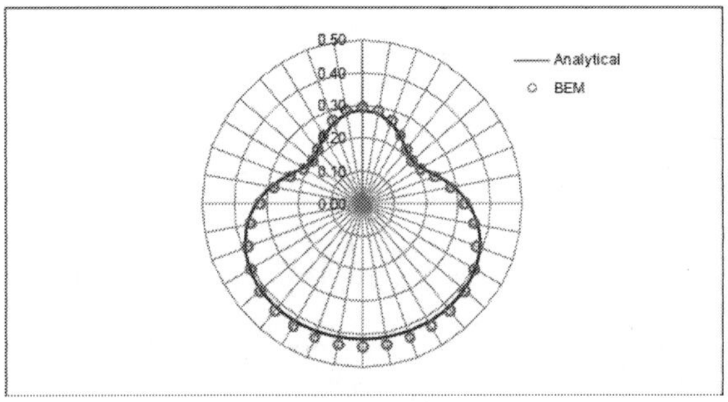

Figure 4 – Normalized radiated pressure at a distance $r = 2a$, when $ka = 1$, $B = 3a$

Next we locate the sphere closer to the plane such that B=2a for the same frequency as the previous case, i.e. ka =1. Figures 5 shows the normalized radiated pressure plotted versus polar angle at distance r=2a. The BEM agrees with the analytical result to within 5% for this configuration.

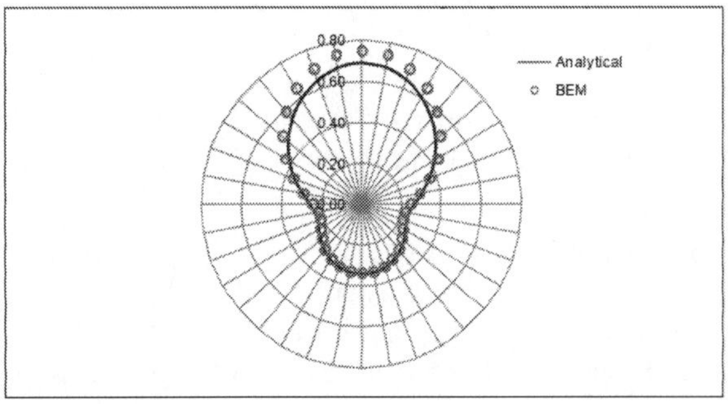

Figure 5 – Normalized radiated pressure at a distance $r = 2a$, when $ka = 1$, $B = 2a$

The next test case for scattering problem considered is the scattering from a rigid sphere of radius a located a distance B=3a from the infinite rigid plane. The incident wave is a plane wave with a frequency of ka=1, propagating parallel to the rigid plane. Figures 6 and 7 are the polar plots of the scattered velocity potential, normalized by the velocity potential of the incident wave, plotted at a distance of 3a and 2a respectively from the center of the sphere. Both Figures 6 and 7 show good agreement between the BEM and the analytical results

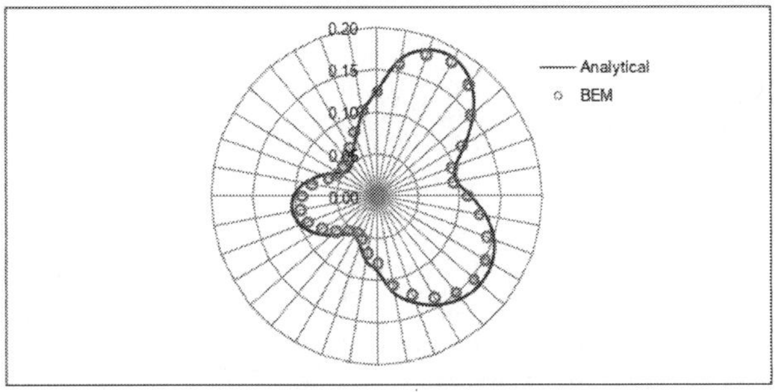

Figure 6 – Scattered velocity potential at a distance $r = 3a$, when $ka = 1$, $B = 3a$

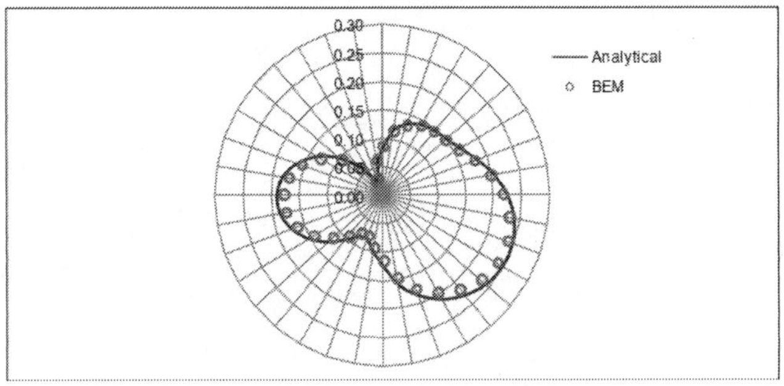

Figure 7 – Scattered velocity potential at a distance $r = 2a$, when $ka = 1$, $B = 3a$

The following test case demonstrates another configuration when the sphere is located closer to the rigid plane, i.e. B=2a. Figure 8 depicts the directivity pattern at a distance r=2a. It can be observed that the analytical solution and the BEM results agree very well.

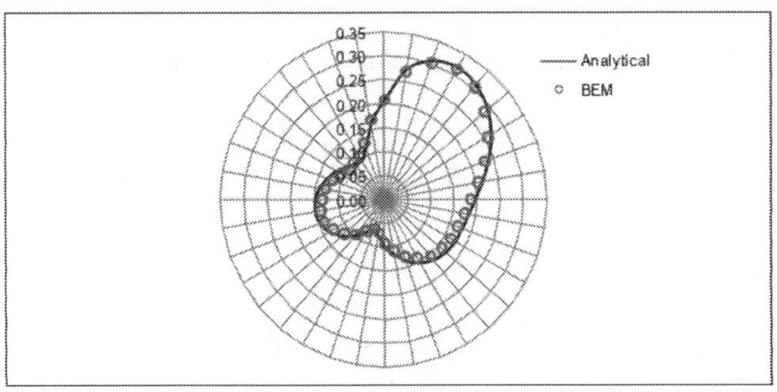

Figure 8 – Scattered velocity potential at a distance $r = 2a$, when $ka = 1$, $B = 2a$

In conclusion, an analytical solution has been derived for half space problems involving radiation and scattering of acoustic waves from a rigid sphere bounded by an infinite rigid plane. Verification has been demonstrated for various configuration taking varieties of the distance of the sphere from the infinite plane and the distance of the field point from the center of the sphere where the acoustic field is anticipated. Comparison with the BEM calculations shows a good agreement between the two results. At the same time the analytical solution may be used as a verification tool for numerical method calculations such as the BEM..

References

[1] Chertock G 1964 *J. Acoust. Soc. Am.* 37 1305-1313
[2] Schenck H.A. 1968 *J.Acoust.Soc.Am.*, 44 41-58
[3] Engblom J.J and Nelson R.B. 1975 *J.Appl.Mech.* 97 295-300
[4] Meyer W.L. Bell W.A., Zinn B.T., and M.P.Stallybrass M.P 1978 *J.Sound and Vib* 59 245-262
[5] Koopman G.H. and Benner H 1982 *J.Acoust.Soc.Am.*71 78-89
[6] Seybert A.F., Soenarko B, Rizzo F.J., and Shippy D.J. 1985 *J.Acoust.Soc.Am*, 77 362-368
[7] Seybert A.F., Soenarko B.,Rizzo F.J., and Shippy D.J. 1986 *J.Acoust.Soc.Am* 80 1241-1247
[8] Latcha M.A. and Akay A. 1986 *J.Vib.Acoust.Stress Reliab Des* 108 447-453
[9] Jiang J.K. and Prasad M.G., 1986 *J.Vib.Acoust.Stress.Reliab Des* 108 454-461
[10] Soenarko B, PhD Thesis, University of Kentucky, 1983

On the role of hanging baffle performance enhancement by using slotted Helmholtz resonator array

R. Hanina, I. Yahya*, **Harjana**
The Iwany Acoustics Research Group (iARG), Physics Department,
Universitas Sebelas Maret, Jl. Ir. Sutami 36A Kentingan, Surakarta

*E-mail:iyahya@mipa.uns.ac.id

Abstract. This paper emphasizes on the experimental investigation performance of hanging baffle with slotted Helmholtz resonator array inclusion. Laboratory test procedure refers to ATMS E-1050 was conducted for measuring the sound absorption coefficient of the Helmholtz resonator, while interrupted noise reverberation time measurement refers to ISO 3382-2 has been done for room acoustic performance test of the proposed design hanging baffle. The result shows that broadband absorption occurs when the Helmholtz resonator inserted to the hanging baffle. Sound energy dissipation increase both in resonance and viscous damping mechanism. It reduces the reverberation time significantly in all frequencies. The proposed design slotted resonator inclusion has brought opportunity for tuning the response and performance of the hanging baffle.

1. Introduction

The needs for high quality and acoustically comforts in a room are increasing rapidly time to time. It leads to the quest for new approach and innovation to gives a better listening and speech experience. One of the most common measures to describe of room acoustics performance is reverberation time [1,2]. Various techniques have been implemented by researchers to get a better room acoustic performance such as variations of wall absorption [3,4], different type resonators [5] and baffles [6,7]. This paper emphasizes on the experimental examination of the performance of hanging baffles with an inclusion of an array of Helmholtz resonators in particular sound diffuser like configuration. The proposing approach has two significant advantages, first for controlling the low frequency modes of the room by using Helmholtz resonators, and the second to stimulate scattering effect caused by its sound diffuser like configuration. These two different noise control mechanism to be expected occurs accumulatively for increasing the energy decay rate inside the room and resulting significant correction to reverberation time. The other advantages of proposed hanging baffle are it could be configured freely to get optimum multiple reflections and also two-sided surface absorption effect which cannot conduct by using wall mounted sound absorber and or diffuser.

2. Methods

The diagram of proposed baffle depicted in Figure 1. It made from commercial sponge with dimension ($l \times w \times h$) are 100cm, 12cm, and 50cm respectively. Each baffle has seven unit of quarter wavelength resonator with internal diameter 8 cm to be slotted with Helmholtz resonators. Each resonator inserted in various depth from baffle surface to form sound diffuser like structure.
The sound absorption coefficient of the Helmholtz resonator was measured using impedance tube after ASTM E-1050. The experiment conducted by using B&K 4206 impedance tube and the whole data analyzed by dedicated B&K material testing software. There are two types of resonators with different

Journal of Physics: Conference Series **776** (2016) 012066 doi:10.1088/1742-6596/776/1/012066

cavity length, h, in three different value of d_2 was measured. The detail of parameters for the test presented in Table 1.

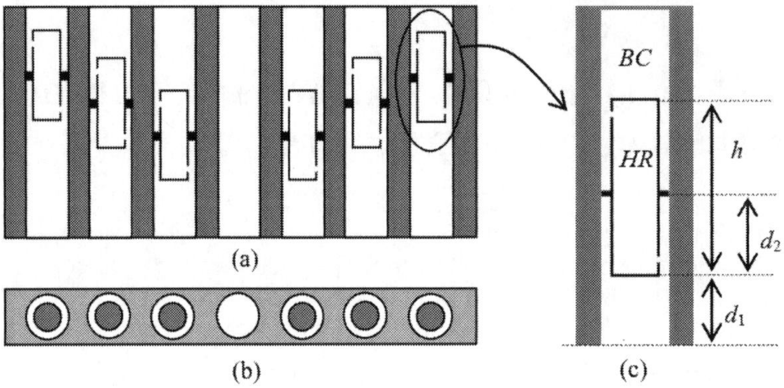

Figure 1. (a). side view of the baffle with seven half wavelength resonators where six of it slotted with Helmholtz resonator fitted inside it. (b). the bottom view, (c). Details of the resonator. *HR* and *BC* is the Helmholtz resonator and back cavity respectively, h is the HR length, d_1 is sound diffuser like well depth while d_2 is slit like cavity depth.

Table 1. Parameters for impedance tube test

No	h (cm)	d_1 (cm)	d_2 (cm)		
			(i)	(ii)	(iii)
1	8	0		Fixed: 4	
2	16	0	4	8	12

A room performance test conducted following the Helmholtz resonator laboratory test. To this purpose, there is six unit of hanging baffle installed inside the room. Reverberation time measurement refers to ISO 3382-2 test procedure was conducted by using B&K Omni-power 4292 for generating the interrupted noise and B&K 2270 handheld analyzer for measuring and analysis the data in one-third octave band. The whole experiment conducted in a fixed value of d_2 for 8 cm and various d1 ranging from 5 cm to 15 cm with 5 cm increment from the edge to center well of the baffle. The center well has no inclusion to keep it work as a quarter wavelength resonator.

3. Results and discussions

The sound absorption coefficient of the two Helmholtz resonator with h value of 8 cm and 16 cm presented in Figure 2. Helmholtz resonator with a longer tube is increasing performance in the low frequency from 150 Hz to 300 Hz. The similar increment also occurred at the frequency range above of 1.4 kHz. These increments are related to the coupling effect of the slotted Helmholtz resonator (HR) with the back cavity (BC) to form a two degree of freedom Helmholtz resonator. When the volume of the first cavity increased, it changes the reactance of the resonator and shifting the resonator performance to the lower frequency band. As a result, an increment at higher frequency band above 1.4 kHz occurred accordingly.

Interesting phenomena occurred with the test for different d_2 values as presented in Figure 3. As the value of d_2 increased, the sound absorption coefficient for frequency range above 800 Hz increased as well. Contrary it reduces the sound absorption for frequency below 800 Hz. The thin tubular shaped empty well area between the slotted cavity and the wall of the quarter wavelength resonator increase sound wave energy dissipation in two different way. First, quarter wavelength resonance effect of the resonator with the depth of d_2. Secondly, this thin area also increases viscous damping according to the porous wall of quarter wavelength resonator. Both improves sound absorption performance at high frequency range. In the same time as the d_2 increased, it slightly reduces the volume of the back cavity of the coupled Helmholtz resonator which takes into account to the changes sound absorption performance in the range between the first and second resonant frequency of the coupled Helmholtz resonator. It gives a possibility for tuning the performance of the proposed slotted resonator design. The combination of coupled or slotted Helmholtz resonators with various d_2 would give an excellent opportunity to get broadband absorption performance in the real room acoustic applications.

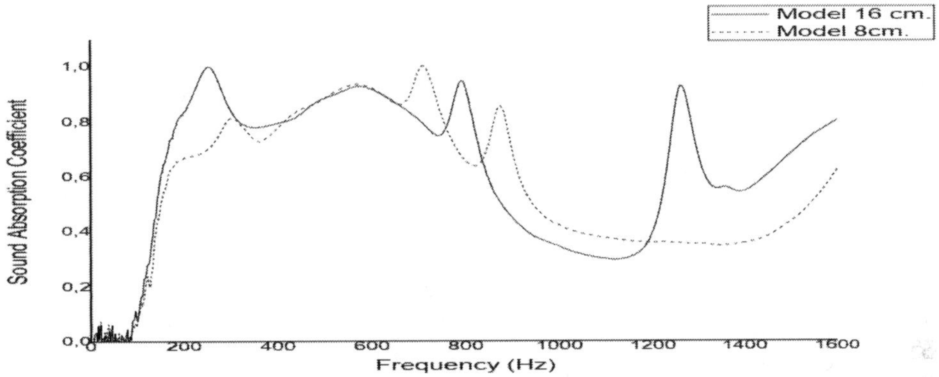

Figure 2. The influence of tube lentgh on the sound absorption cofficient of the slotted Helmholtz resonator..

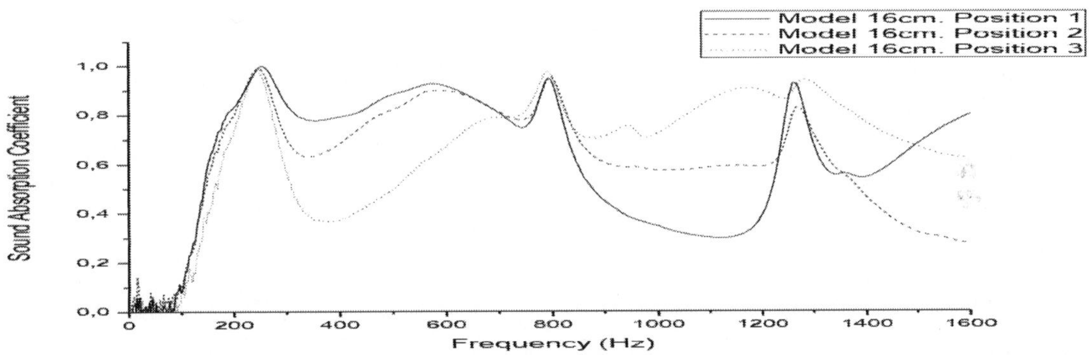

Figure 3. The influence of d_2 depth on the performance of the slotted Helmholtz resonator.

The results of reverberation time measurement presented in Figure 4 and Figure 5 which is the performance of the baffle without and with slotted Helmholtz resonator respectively. As shown in Figure 4, the T_{20} and T_{30} for the whole frequencies are reduced significantly after treatment with the proposed baffle even though without slotted Helmholtz resonator inclusion. The discrepancies values between before and after treatment occurred as the array of seven quarter wavelength resonator on each baffle increase sound waves energy dissipation both in resonance and viscous damping mechanism. For example, T_{20} decreased to 1.74 seconds from 2.72 seconds before treatment, while T_{30} become 1.94 second which is much lower compared to 3.09 second before treatment both for frequency 100 Hz.

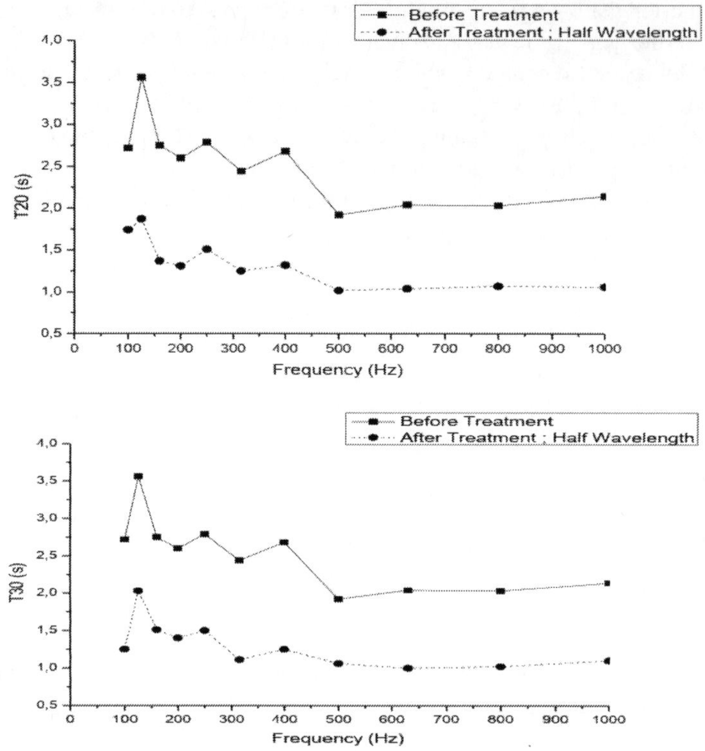

Figure 4. The performance of the proposed baffle without slotted Helmholtz resonator.

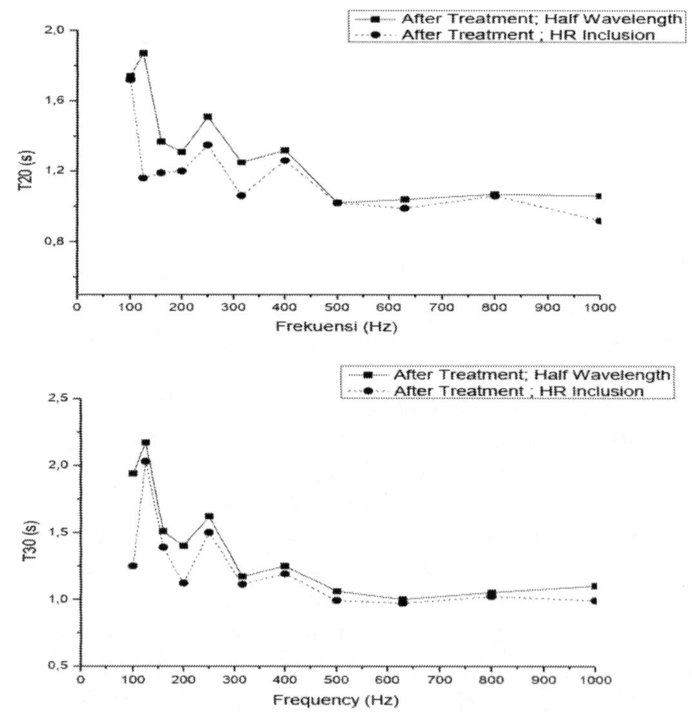

Figure 5. Reverberation time performance of the baffle with slotted Helmholtz resonator inclusion

Journal of Physics: Conference Series **776** (2016) 012066 doi:10.1088/1742-6596/776/1/012066

When the proposed slotted Helmholtz resonator fitted into the hanging baffle, its performance increased as presented in Figure 5. Both T_{20} and T_{30} values decreased for all frequencies. T20 dropped to 1.72 seconds while T_{30} reduced to 1.25 second for 100 Hz. As mentioned above that when the Helmholtz resonator fitted into the baffle, the coupled structure gives a better, and wider sound absorption performance especially in the low frequency band since both of its resonant frequencies are less than the quarter wavelength resonator. Also, the sound diffuser like the configuration of the slotted Helmholtz resonator also plays an important role to produce the scattering effect for controlling the low-frequency noise.

It concludes that inclusion of slotted Helmholtz resonator into hanging baffle giving a significant improvement to the acoustic quality of a room. Also, the proposed design also has brought opportunity for tuning the hanging baffle response with varying front quarter wavelength resonator depth.

4. References

[1] Nowoswiat, A. and Olechowska, M. 2016 *Archives of Acoustics*. **41**(1) 15-26.
[2] Aretz, M., and Orlowski, R., 2009 *Applied Acoustics* **70**, 1099-1110 (2009)
[3] Billon, A., Picaut, J., and Sakaut, A., 2008 *Applied Acoustics* **69**, 68-74
[4] Meissner, M., 2008 *Applied Acoustics* **69**, 583-590
[5] Sohn, H. C., and Park, J. H., 2011 *Aerospace Science and Technology* **15**, 606-614
[6] Cops, A., 1985 *Applied Acoustics* **18**, 435-448
[7] Ohlrich, M., and Hugin, C. T., 2004 *JSV* 277, 405-418

Speech segregation based-on binaural cue: interaural time difference (itd) and interaural level difference (ild)

Mifta Nur Farid, Dhany Arifianto

Dept. of Engineering Physics, Faculty of Industrial Technology, Institut Teknologi Sepuluh Nopember, Kampus ITS Sukolilo, Surabaya 60111, Indonesia

E-mail: miftanurfarid@gmail.com

Abstract. A person who is suffering from hearing loss can be helped by using hearing aids and the most optimal performance of hearing aids are binaural hearing aids because it has similarities to human auditory system. In a conversation at a cocktail party, a person can focus on a single conversation even though the background sound and other people conversation is quite loud. This phenomenon is known as the cocktail party effect. In an early study, has been explained that binaural hearing have an important contribution to the cocktail party effect. So in this study, will be performed separation on the input binaural sound with 2 microphone sensors of two sound sources based on both the binaural cue, interaural time difference (ITD) and interaural level difference (ILD) using binary mask. To estimate value of ITD, is used cross-correlation method which the value of ITD represented as time delay of peak shifting at time-frequency unit. Binary mask is estimated based on pattern of ITD and ILD to relative strength of target that computed statistically using probability density estimation. Results of sound source separation performing well with the value of speech intelligibility using the percent correct word by 86% and 3 dB by SNR.

1. Introduction

Hearing impairment has been an area of research in recent years. Despite the common cause is due to age, excessive noise exposure also reduce hearing ability (hearing loss)[1]. A person who is suffering from hearing loss can be helped by using hearing aids. Kind of hearing aids that provide the most optimal performance if hearing loss in both ears are binaural hearing aids [2] because it has similarities to the human auditory system.

The auditory system can separate many sound source simultaneously. For example, in a conversation at a cocktail party, a person can focus on single conversation even though the background sound and other people conversation is quite loud. This phenomenon is known as the cocktail party effect[3]. The term "cocktail party processing" created in an initial study on the cocktail party effect, this study illustrates that binaural hearing provides an important contribution in auditory analysis that allows us to isolate and localize the sound source[4].

In binaural hearing, if the position of a sound source is not in the median plane hence one ear will be overshadowed by the head while the other ear is fully open to the sound source. The result is a difference in the sound pressure level that is heard in both ears, is called Interaural Level Difference (ILD) and the differences in travel time of the sound source to the ears is called

Interaural Time Difference (ITD). According to Roman[5], changes in the value ITD and ILD, statistically, has the effect to the changes of Relative Strength (RS). So binary mask can be estimated from the value of the RS that obtained by the changes of ITD and ILD in time-frequency domain. So in this study, will be performed separation on the binaural sound input with 2 microphone sensors of two sound sources based on both the binaural cue, interaural time difference (ITD) and interaural level difference (ILD) using binary mask.

To estimate value of ITD, is used cross-correlation method which the value of ITD represented as time delay of peak shifting at time-frequency unit. Then the binary mask is estimated based on pattern of ITD and ILD to relative strength of target that computed statistically using probability density estimation.

The rest of paper is organized as follows. Section 2 describes the process of data training. Section 3 describes the segregation processing. Section 4 present evaluation of segregation results. Section 5 is result and discussion while the last Section 6 is conclusion.

2. Data Training Process

2.1. Binaural Hearing

In this process, spatial information will be given to a target signal and a masker signal by convoluting them with HRTF (head-related transfer function). The HRTF database that used is "HRTF Measurement of Kemar Dummy-Head Microphone" from MIT Media Lab Perceptual Computing. The HRTF database contained head-related impulse response of both ear in anechoic room with distance to sound source is 1.4 m. The location of sound source is azimuth 0° for a target and 5°, 10°, 20°, and 30° for a masker.

2.2. Auditory Periphery

The results of binaural hearing processing only simulate the sound transmission on the outer ear, so it is necessary to simulate the middle and inner ear using auditory periphery processing. Auditory periphery contained 3 stages. The first stage is gammatone-filter, which is the target and masker signal filtered using 4^{th}order gammatone-filter

$$g(t) = at^3 e^{-2\pi bt} \cos(2\pi ft + \phi) \tag{1}$$

with 128 center frequencies distributed in scale ERB 80 Hz to 5000 Hz[5].

$$ERB(f) = 24.7 \left(\frac{4.37f}{1000} + 1 \right) \tag{2}$$

The second stage is middle ear gain that provides gain to simulate the effects of middle-ear function based on equal-loudness of BS3383, Normal Equal-Loudness Level. And the last is the hair cells that model the activity of the auditory-nerve that represented by firing-probability in inner ear[6].

2.3. Binaural Cue Extraction

Interaural time difference (ITD) is estimated as the time delay corresponding to the position of the maximum value of the cross-correlation results of $C(i, j, \tau)$ between the pair HRTF on i-frequency channels, j-time frame, and τ-lag [7].

$$C(i, j, \tau) = \frac{\sum_{k=0}^{K-1} (l_i(j-k) - \bar{l_i})(r_i(j-k-\tau) - \bar{r_i})}{\sqrt{\sum_{k=0}^{K-1} (l_i(j-k) - \bar{l_i})^2} \sqrt{\sum_{k=0}^{K-1} (r_i(j-k) - \bar{r_i})^2}} \tag{3}$$

where l_i, r_i is firing-probabilty of left ear and right ear at i-frequency channel, and $\bar{l_i}$, $\bar{r_i}$ is the average value of firing-probabilty at K-integration window.

Interaural level difference (ILD) is estimated as

$$ILD_i = 20log_{10}\frac{\sum_i l_i^2(t)}{\sum_i r_i^2(t)} \tag{4}$$

where l_i, r_i is firing-probabilty of left ear and right ear at i-frequency channel.

2.4. Relative Strength Estimation

Relative strength is energy ratio between target to mixed signal. Relative strength is estimated as

$$RS_i = \frac{\sum_i s_i^2(t)}{(\sum_i s_i^2(t) + \sum_i n_i^2(t))} \tag{5}$$

where s_i is the target signal and n_i is the masker signal at i-frequency channel.

2.5. Probability Density Estimation

At this stage, the analysis of patterns between both ITD and ILD to the relative strength of a target signal to mixed signal in the time-frequency unit. Pattern analysis conducted by estimating the probability density that represent the statistical relationship between those values (ITD, ILD and relative strength). Estimation of probability density performed at the relative strength of two types, the value of probability density when a target signal is more dominant than a masker signal ($RS > 0.5$) and a masker signal are more dominant or equal than a target signal ($RS \leq 0.5$) in time-frequency domain.

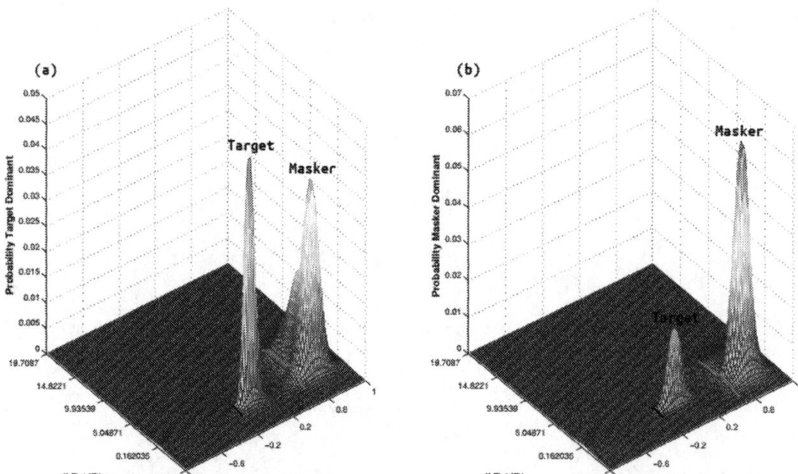

Figure 1. Probability Density Estimation for ITD and ILD to (a) relative strength ($RS \geq 0.5$) and (b) ($RS \leq 0.5$) of target and masker at 0° and 30° with SIR 10 dB and center frequency 500 Hz

The results of probability density estimation (PDE) for both types of relative strength of a single target and masker at 0° and 30° with SIR 10 dB and center frequency 500 Hz shown in figure 1. In figure 1 (a), the highest peak occurs in the value of ITD = 0 ms, which means the relative strength values is greater than 0.5 more distributed on these, whereas in figure 1 (b) highest peak occurs at ITD = 0.8 ms, which means relative strength value is less than or equal to 0.5 more distributed in that ITD. Black line in figure 1 (a) and (b), which represents the distribution of the value of ILD at target, is shorter than the ILD in the masker (red line). The results corresponding to reduction of relative strength value if the position of the sound source away from the median plane. From here it can be concluded, to estimate the binary mask that requires a value relative strength can be estimated with only the known value of the ITD and ILD alone on the mixtured signal.

3. Segregation Process

3.1. Azimuth Estimation

Azimuth of the target signal and masks can be estimated from summing the results of cross-correlation between the mixtured signal on the left ear and right ear in the time domain on all center frequency of 80 Hz to 5000 Hz. The peak value of the sum of cross correlation in ITD function can be represented as azimuth in figure 2. In this figure, the first peak is at a lag of 0 ms is defined as a azimuth of target signal and a second peak at 0.21 ms as azimuth of masker signal. Both of these values as parameters of decision to take probability density estimation database of the results of the training data.

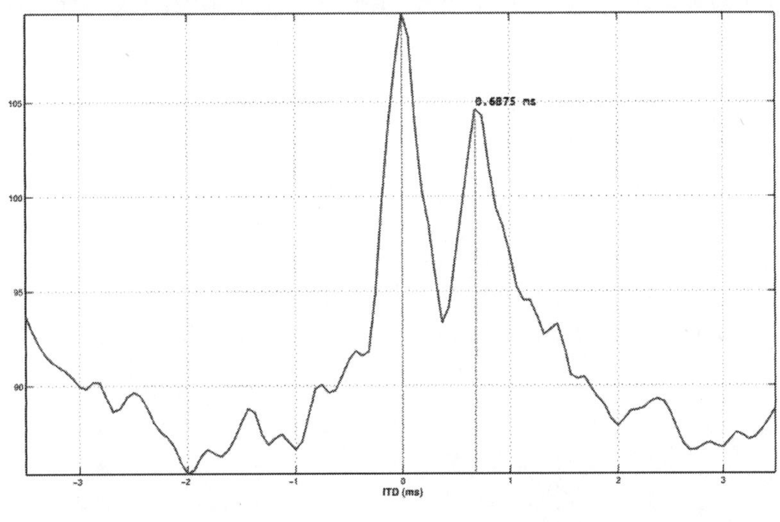

Figure 2. The sum of cross-correlation between mixtured signal at left ear and right ear

3.2. Binary Mask Estimation

Binary mask is estimated at each time-frequency unit. The binary mask value is 1 if the value of PDE with $RS > 0.5$ in figure 1 (a) is greater than PDE with $RS \leq 0.5$ in figure 1 (b) and will be 0 if the value of PDE with $RS > 0.5$ in figure 1 (a) less than or equal to the PDE $RS \leq 0.5$ in figure 1 (b). The value of each PDE obtained from ITD and ILD masker signal in each unit time-frequency of the mixtured signal. In figure 1 is a graph of the PDE of a target at azimuth 0° and a masker at azimuth 30° with the center frequency of 500 Hz and SIR 10 dB. ITD and ILD masker obtained at center frequency of 500 Hz, calculated the value of its binary mask based on the probability density estimation in figure 1. Figure 3 shown spectrogram of estimated target signal compare to clean target signal and mixtured target signal.

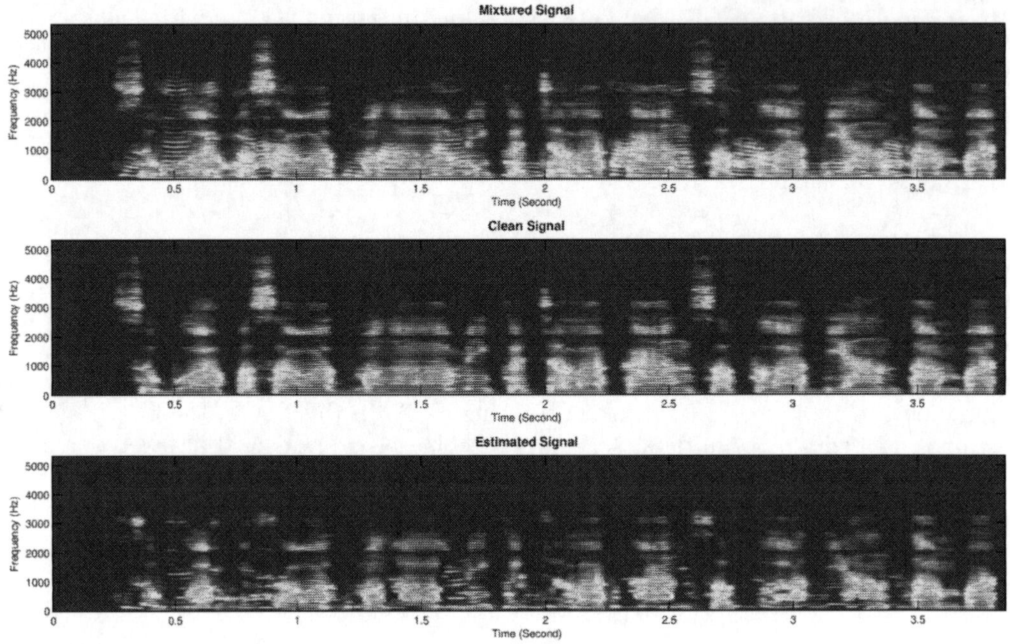

Figure 3. Spectrogram of (a) mixtured signal, (b) clean target signal and (c) estimated target signal

4. Evaluation

Evaluation method that used is percent correct word for subjective evaluation that shown in figure 4 and SNR for objective evaluation that shown in figure 5.

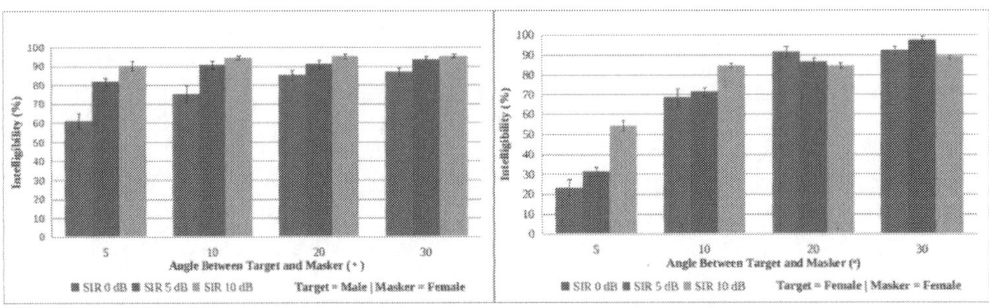

Figure 4. Evaluation using percent correct words with target male speaker and masker female speaker

5. Result and Discussion

Based on the results of subjective evaluation in figure 4, the larger angle and SIR between target and masker signal then speech intelligibility of separation result is increase. For a different speaker between the target and masker signal, has a higher speech intelligibility than the same speaker of target and masker signal. So as for the results of an objective evaluation in figure

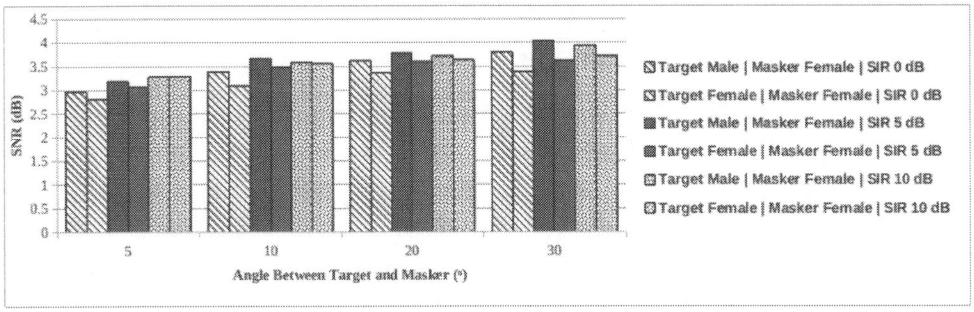

Figure 5. Evaluation using SNR with variation of speaker and angle separator between target and masker

5, the larger angle and SIR between target and masker signal then SNR of separation result is increases. For a different speaker between the target and masker signal, has a higher SNR than the same speaker of target and masker signal.

In figure 4, the increase of speech intelligibility with 0 dB SIR looks significantly by 60% in 5°, 75% in 10°, 85% in 20° and 86% in 30°. Similarly to 5 dB SIR. Meanwhile for 10 dB SIR, speech intelligibility is not very significant increased, which is 90.1% in 5°, 94.4% in 10°, 95.3% in 20° and 95.4% in 30°.

6. Conclusion

The sound quality of separation results for the same speaker between the target and masker signal smaller, that is 72.86% correct words and 3.38 dB SNR, than different speaker that is 86.74% correct words and 3.5 dB SNR.

References

[1] Kates J M 2008 *Digital Hearing Aids* (San Diego: Plural Publishing)
[2] Marin J 2012 Robust Noise Reduction Strategies For Binaural Hearing Aids *PhD thesis* (Georgia Institute of Technology)
[3] Cherry C 1957 *On human communication; a review, a survey, and a criticism* (Cambridge: Technology Press of Massachusetts Institute of Technology)
[4] Hawley M L, Litovsky R Y, and Culling J F 2004 The benefit of binaural hearing in a cocktail party: effect of location and type of interferer *J. Acoust. Soc. Am.* **115** 833.
[5] Roman N, Wang D, Brown G J 2002 Speech segregation based on sound localization *J. Acoust. Soc. Am.* **114** 2236
[6] Meddis R 1986 Simulation of mechanical to neural transduction in the auditory receptor *J. Acoust. Soc. Am.* **79** 702
[7] Jeffress L A 1948 A place theory of sound localization *Journal of comparative and physiological psychology* **41**(1) 35

The effect of boundary shape to acoustic parameters

M. S. Prawirasasra*, R. Sampurna and Suwandi
Engineering Physics Department, School of Electrical Engineering, Telkom University, Jl. Telekomunikasi Terusan Buah Batu, Bandung, Jawa Barat, Indonesia

Email: * bibinprawirasasra@telkomuniversity.ac.id

Abstract. To design a room in term of acoustic, many variables need to be considered such as volume, acoustic characteristics & surface area of material and also *boundary shape*. Modifying each variable possibly change the sound field character. To find impact of *boundary shape*, every needed properties is simulated through acoustic prediction software. The simulation is using three models with different geometry (*asymmetry and symmetry*) to produce certain *objective parameters*. By applying *just noticeable difference* (JND), the effect is considered known. Furthermore, individual perception is needed to gain *subjective parameter*. The test is using recorded speech that is convoluted with room impulse of each model. The result indicates that 84% of participants could not recognize the speech which is emit from different geometry properties. In contrast, JND value of T_{30} is exceed 5%. But for D_{50}, every model has JND below 5%.

1. Introduction

Every room should have unique design for specific purpose in acoustic term. The object of this study is audio-visual room. The daily operation is for distance learning-teaching activities with two-way communication: speech and listening and also supported with visual media tools. It is categorized as room for speech. There are criteria need to be fulfilled and certainly many rooms do not meet the requirements. So, the physical properties of boundary (room) need to be re-designed.

The characteristics of acoustic field in enclosures is very complicated due to multiple reflection from boundary (wall, ceiling and floor). The reflected sound creates two specific features: at receiving point, the inverse square law does not apply and reverberation happened after the sound source has stopped [1]. Those phenomenon creates specific sound field which is function of boundary including shape-dependent [2]. The previous research obtained the prediction of Reverberation Time (RT) for several rooms whose geometry boundaries are different but with slightly the same volume and consist of same scattering coefficient [3].

The purpose of research is to identify the effect of geometry shape to human ear. By applying different shape of room, particular sound field is expected to be noticeable by human ear. There are two assessment methods: objective and subjective acoustic parameters. The selective parameters have been chosen related with speech intelligibility: D_{50}, T_{30} and also sound strength (G). The value of it should no less than certain point regarding to Just Noticeable Difference (JND). Subjective test was applied to random health participants by playing recorded voice which is already convoluted by specific room response of different shape of boundary.

2. Methodology and Definition

During research, there are several systematic steps need to be fulfilled. In general, there are two type of method to produce raw data: 1) field measurement and 2) simulation. Below is the brief description of each stage including the description of objective parameters.

The important parameters (D_{50}, T_{30} and G) is investigated in term of simulation and measurement. The differences are discussed and their significance is concluded within the frame of Just Noticeable Difference (JND). Table 1 shows the parameters used with JND. The varies in geometry shape is expected to perceived differently by human ear. Whenever it has value larger than JND, it means human ear could notice the difference and vice versa.

Table 1. Acoustic Parameters and JND value

Parameter	Symbol	JND
Definition	D_{50} (%)	0.05
Sound strength	G(dB)	1 dB
Reverberation Time (30 dB range)	T_{30} (s)	5%

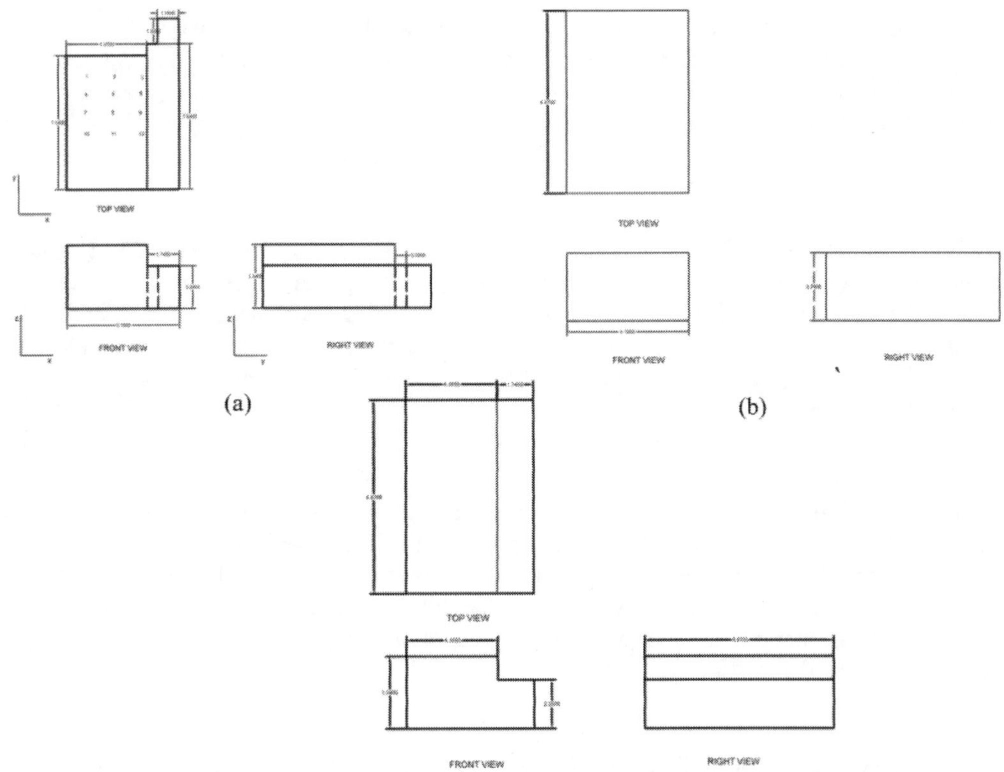

Figure 1 Projection of models (a) Model 1, (b) Model 2 and (c) Model 3.

Measurement is mandatory because the measured variable is used to validate the output of simulation through acoustic prediction software. There are 12 measurement points so simulation requires same point and position to get validated. The simulated model (model 1) is arranged to have identic properties with actual object. Geometry of object has asymmetry cross section refer to xy, xz and yz plane. The projection of model 1 is at Figure 1(a).

To identify the impact of geometry shape to acoustic parameters, it requires evaluation of related objective acoustic parameters represent particular geometry shape with the same acoustic properties. There are two other different shapes of models that will be simulated. Each models have different cross section either asymmetry and symmetry refer to certain planes. Model 2 are considered has symmetry cross section (shoe-box shape) and has the longest length of model 1. Otherwise, model 3 only has symmetry cross section refer to xy and yz plane but asymmetry to xz plane. Similar with model 2, the model 3 has maximum length of model 1. To analyze the impact of geometry shape, the expectation is every models have nearly the same volume. Differently, it is used the comparison of volume and the surface area of models. The projection of last two models and dimension properties are at Figure 1(b) and (c) and Table 2.

Table 2 Comparative of models' dimension

Model	Volume (m^3)	Surface area (m^2)	Volume/Surface area (m)
1	143.306	183.65	0.73
2	212.414	210.1	0.86
3	165.692	196.52	0.84

Procedure of subjective evaluation is different with objective evaluation. However, it requires participants in order to determine the differences of recorded speech. Test voices are convoluted with specific room impulse response of each model. Participants are asked to answer several question related with speech intelligibility and identification whether voices are identic or not. The purpose of assessment is to get individual answer related with ability to differentiate the sound field.

Objective Parameters

D_{50}

At receiving point, the total energy is summation of direct, reflected and noise energy. The reflected energy is potentially either advantage or disadvantage for speech intelligibility. It depends on the amount of energy and receiving time. Useful reflection for speech occurs within early 50 ms after sound source is stopped. The parameters that illustrate the useful energy for speech is definition 50 (D_{50}). In mathematical term it defines as ratio of useful energy (0-50 ms) with total energy (0-∞ ms).

$$D_{50} = \frac{\int_0^{50} p^2(t)dt}{\int_0^{\infty} p^2(t)dt}\ [\%] \tag{1}$$

T_{30}

In physical term, reverberation is repetition of reflected sound by boundary after source is stopped. As the boundary has reflective surface the energy will decay in longer time. The parameter that illustrate the phenomena is defines as reverberant time or RT/T_{60}. The definition of T_{60} is the length of time required for sound to decay 60 dB after source stopped. To get the T_{60} value, it needs to discover the relation between decaying energy and time through linear regression. By implement the extrapolation, time to decay energy of 60, 30, or 10 dB is certain.

Source Strength (G)

This parameter measures of the amplification effect of sound produced by the space enclosure with respect to free space. According to ISO 3882, the definition of G is measured sound level is compared with the sound level in open space (no reflection) at distance 10 m from omnidirectional sound source. Below is the formula to find source strength

$$G = 10log\frac{\int_0^{\infty} p^2(t)dt}{\int_0^{\Delta t} p^2(t)dt}\ [\text{dB}] \tag{2}$$

where Δt is emission duration. Based on definition above, there are two possibilities of G, enhanced or diminish of level pressure.

3. Results

Objective Parameters

The sub subchapter 3.1.1-3.1.3 displayed the calculation result of JND value of each assess parameters. The chart in Figure 1, Figure 2, and Figure 3 has two axis where abscissa is measurement point and difference value is plotted in ordinate. The unit in y-axis is similar with JND unit. For D_{50} and T_{30}, the unit is in percentage and it is calculated by normalized difference value of compared models. Differently, JND of G value is subtraction value between models.

3.1.1. D_{50}

This clarity parameter shows that every measurement point of each model have value no less than 92.3% and the largest value is 99.6%. According to the speech criteria of D_{50} parameter, every shape of geometry has 'very good' quality criteria regarding to speech intelligibility. The effect of geometry shape is trying to find out by compare each value of D_{50} for each models. The difference of D_{50} for one model refer to another is illustrate at Figure 1. Based on JND value, there are no difference value is larger than 5% or below JND. It means human ear could not identify either the sound is produced from model 1, 2 or 3.

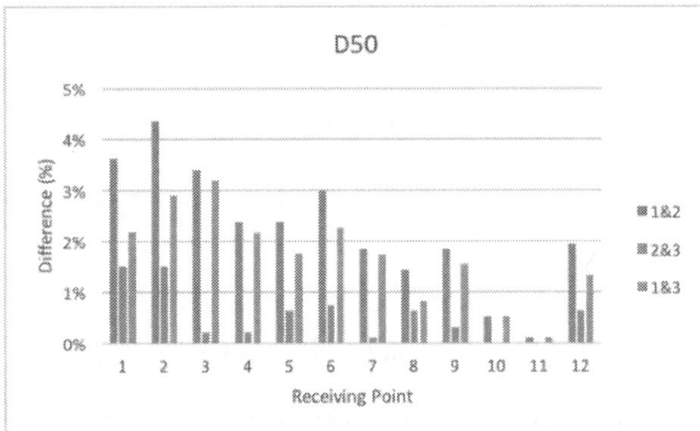

Figure 1 Comparison of D_{50}

3.1.2. T_{30}

In Figure 2, we show that for comparison of every value of T_{30} at measurement point for every models. Almost at every point the JND value fell above the minimum value (5%). Model 1&2 demonstrates that the difference is reach 50%. Only at point 3, the value is below the JND. Similarly, the comparison value of model 2&3 has range of value from 0%-24% and there are 4 measurement point have equal reverberation sound field. Furthermore, the enclosure shape between model 1&3 is clearly has significant effect to the reverberation. Based on the comparison value, there are no point of measurements have difference value less than 5%. The simulation result of this variables are from 11%-25%.

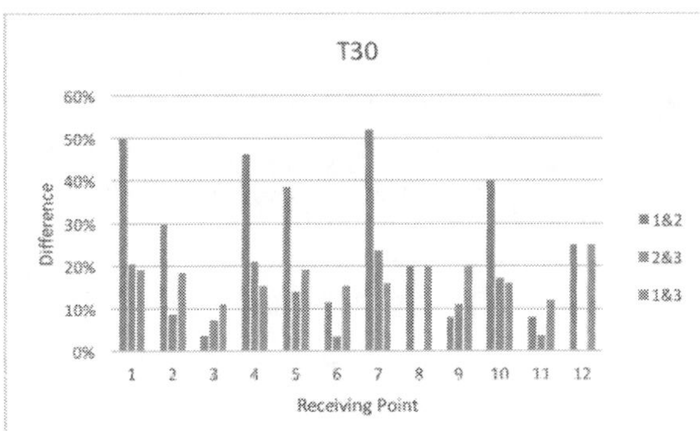

Figure 2 Comparison of T_{30}

3.1.3. Source Strength (G)

In Figure 3, the difference of G value at each measurement point has similar result. The largest difference is located at point 1 for model 1&2. The value is only 0.6 dB while the rest fell between -0.1 - 0.4 dB. Negative mark means that at that point the sound level is decrease and vice versa. According to JND, the source strength at every point does not influenced by geometry shape.

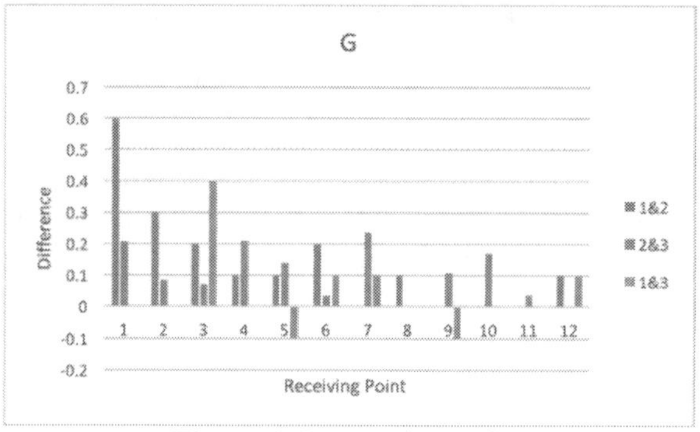

Figure 3 Comparison of G

Subjective parameters

There are 25 participants involved in this questionnaire. Every participant are asked to answer the given questions. The resume of answer related to speech intelligibility are presented in Table 3. The variation of percentage of clarity criteria of each model shows that speech intelligibility in every shape is perceived differently by participants. Only 'fair' criteria has same percentage over the geometry shape.

Table 3 Percentage of chosen criteria

Clarity criteria	Model 1 (%)	Model 2 (%)	Model 3 (%)
Excellent	8	4	8
Good	56	60	20
Fair	36	36	36
Bad	0	0	36

Furthermore, subjective assessment put another question related to hearing ability to differentiate the sound field. After listening the test speech, participants are asked to answer which of sound field were identic or none. The resume of answers is at Table 4. Almost all participants could not recognize the difference of each sound field. According to data, around 84% of participants agreed that the test voice is similar to each other while others say differently.

Table 4 Percentage of sound field recognition

Comparison	Identic (%)	None (%)
Model 1 & 2	72	16
Model 2 & 3	8	
Model 1 & 3	0	
Model 1, 2, 3	4	

4. Conclusion

The conclusion of this research is the geometry shape does not have impact to D_{50} and G because the difference value is lower than JND threshold. There is interesting point related with T_{30} value. The value is equivalent with ratio between volume and surface area. Even though the ratio is alike, it does not mean it has comparable T_{30} and JND. Apparently, the geometry shape has significant to reverberation time.

For subjective assessment, speech intelligibility among models are perceived differently. Not all participants choose the same category over the models. In contrary, participants could not identify the difference of each audio tests. 84 % of participants answered that there are identic to each other. Only 16% of participants could spot the difference among models.

References

[1] Maekawa, Z, and Peter L. 2004 *Environmental and architectural acoustics*. CRC Press, 2004.
[2] Beranek, L, and Daniel W. M.. 1996 *The Journal of the Acoustical Society of America* 99.5 pp. 2637-2637.
[3] Asta Rønning Fjærli, 2015 *Room Geometries with Non-Classical Reverberation Times*, Ms Thesis, NTNU.
[4] ISO 3382-1: 2009 International Standard ISO/DIS 3382-1: Acoustics – Measurement of room acoustic parameters – Part 1: Performance space.
[5] You, J, and Jin Y. J. 2006 *INTER-NOISE and NOISE-CON Congress and Conference Proceedings*. Vol. 2006. No. 1. Institute of Noise Control Engineering.
[6] Pedrielli, F, Eleonora C, and Camilla C. 2008 *Journal of the Acoustical Society of America* 123.5 pp. 3164-3164.
[7] Skålevik, M. 2010 *Proceedings of 20th International Congress on Acoustic, ICA*. Vol. 10.
[8] Haka, C. C. J. M., and Wenmaekersb R. H. C. 2008 *Proceedings of the Internoise.*
[9] Asmoro, W A. 2007. *IPTEK The Journal for Technology and Science*, 18.2.
[10] Araneo, G; Farina, A, Frova, A. 2006 *Saggiatore musicale*, pp 205-220.

Subjective spatiotemporal matrix as a new measure of sound field distribution in a room

E. Wulandari[1], I. Yahya[1], Harjana[1], and Ubaidillah[1,2]
[1]The Iwany Acoustics Research Group (iARG), Physics Department
[2]Mechanical Engineering Department
Universitas Sebelas Maret,Jl. Ir. Sutami 36A Kentingan, Surakarta 57126, INDONESIA

E-mail:iyahya@mipa.uns.ac.id

Abstract.A new approach for representing the acoustic quality of classroom is presented in this paper. The subjective spatiotemporal measure was developed from a 3D scanning plane by using a dedicated scanning gear to measure the subjective volumetric sound field of a listener inside the room. Calculated coherence matrix is derived from spectral comparison on the appropriate scanning planes. It then could be used for indicating the subjective perception of a listener which is may differ to another listener at adifferent position inside the rooms.

1. Introduction

The study of subjective perception to the sound field in a room is one of important field in acoustics nowadays. A room to be said has a high acoustic quality if the sound field from any source are spread and propagate in all direction and exist in any spatial position with the same energy level refer to the particular reference point. In other words, subjective perception is close related with diffuseness of the room. It had been used for various applications including listening condition in church, music signals, and binaural room impulse response [1-3]. Head Related Transform Function (HRTF) is a function that consists of direction, distance, and frequency. It is the most common measure to show the subjectivity, which is known not only varies with the arrival direction of the sound, but with the anthropometric of the subject [4].

In case more than one source are exist in the room, then listener would hear crosstalk and also the reflected sound waves as depicted in Figure 1. In such circumstance, one must consider the binaural room response which is includes of HRTF and the room transfer function. It means that HRTF is could not yet enough to use for describing subjectivity response nor room diffuseness. Subjective perception experience can vary with listener position inside the room and also subject to the room spatial and binaural response [5-7]. This paper proposing an alternative marker for visualization both subjective spatial and temporal sound field based on volumetric scanning measurement approach that can be used in reverberant room condition. The accumulative effect of crosstalk and reflected sound waves heard by the listener are measured in four scanning planeson the front side, rear side, left and right side.

2. Method

The concept of proposing scanning technique for measuring the subjective spatiotemporal matrix can be introduced by using Figure 1. It describes a reverberant room circumstance where more than one source exists in the room in the same times. Both listeners would hear crosstalk and also reflected sound waves from all sources within the room. By using convolution relation, the scanned sound pressure in left and right side scanning plane of listener number one are given by,

$$\begin{Bmatrix} r_{1L} \\ r_{1R} \end{Bmatrix} = \begin{bmatrix} h_{11} & h_{12} \\ h_{21} & h_{22} \end{bmatrix} * \begin{Bmatrix} s_1 \\ s_2 \end{Bmatrix} \quad (1)$$

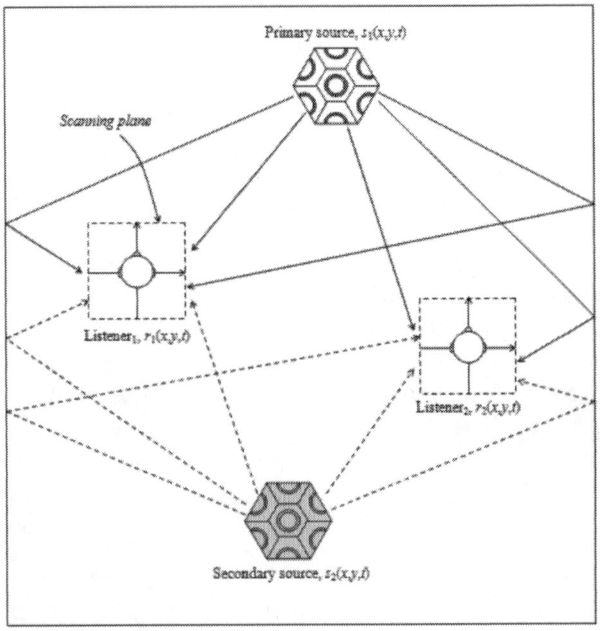

Figure 1. Reverberant room circumstance with more than one sound source.

The first term in equation (1) is binaural room response where the h_{11} representingresponse to primary source s_1, while h_{12} is the response subject to secondary source s_2. According to convolution theorem, the spectra of the scanned sound field the listener number one are given by equation (2). H_{11} and H_{12} are measured frequency response function on listener number one scanning plane subjected to source number one and number two respectively.

$$R_1 = H_{11}S_1 + H_{12}S_2 \quad (2)$$

The spectra for listener number two could be written as,

$$R_2 = H_{21}S_1 + H_{22}S_2 \quad (3)$$

General expression for listenerR_1 within a room with N speakers or sound source given by

$$R_1 = \sum_{i=1}^{N} H_{1i}S_i \quad (4)$$

If there are only one sound source or speaker exist in the room, the equation (2) and (3) becomes to$R_1 = H_1S$ and $R_2 = H_2S$, and the response H_1 and H_2 could be written as,

$$H_1 = \frac{R_1}{S} = \frac{R_1S^*}{SS^*} = \frac{R_1R_1^*}{SR_1^*} = \frac{G_{RS}}{G_{SS}} = \frac{G_{RR}}{G_{SR}} \quad (5)$$

$$H_2 = \frac{R_2}{S} = \frac{R_2 S^*}{SS^*} = \frac{R_2 R_2^*}{SR_2^*} = \frac{G_{RS}}{G_{SS}} = \frac{G_{RR}}{G_{SR}} \quad (6)$$

The R^* and S^* are complex conjugates of the R and S spectra respectively. G_{RR} and G_{SS}areautospectrum and G_{SR} and $G_{RS}$the cross spectrum. Hence from equation (5) and (6) one can derive coherence factor relation $\gamma^2$for sound field onthe scanning plane of listener R_i as,

$$\gamma^2 = \left. \frac{G_{RS} G_{SR}}{G_{SS} G_{RR}} \right|_{R_i} \quad (7)$$

The ratio of $\gamma^2$for the appropriate scanning plane between the two listener R_1 and R_2 can be use for subjective comparison purpose. It is equal to the ratio between H_1 and H_2 for the same appropriate scanning plane.

The data acquisition has been conducting on a 3D scanning gear as depicted in Figure (2). The geometrical dimension of the scanning gear is 2 m in its height and 1 m for the each side. A scanning grid with the element size of 10 cm × 10 cm has been made on the four scanning plane of the scanning gear and denoted as front (F), rear (B), right (R) and left (L). The test room is a real classroom at the Physics department of Sebelas Maret University with the dimensions of 9.3 m, 5.9 m and 4 m in its length, width and ceiling's height respectively.

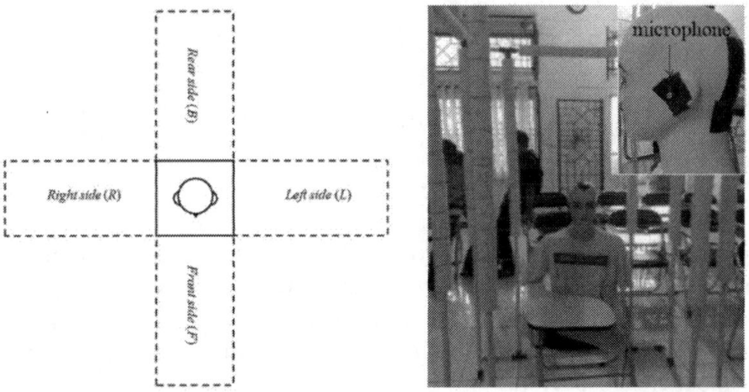

Figure 2. The scanning plane and position of the mannequin model inside the scanning gear.

A mannequin used as listener model. It positioned on a chair inside the scanning gear. An OmniPower B&K 4292 has been used as the sound source for generating random noise. A pair of B&K 4189 microphones attached inside the mannequin ears have been using for capturing the random noise. While another same type microphone was used for scanning purpose. The entire captured data was analysis by using the FFT and CPB analyzer module of the B&K Pulse 3160 LAN-Xi type which is a fully computer controlled system.The intensity spectra were calculated based on the captured data of the scanning microphone in respect to the left and right mannequin ears microphones as the reference.

3. Result and Discussion

Figure (3) shows the time domain sound field distribution on the four scanning plane of thelistener at position number one for t= {0.5; 1.0; 1.4} second. While Figure (4) representing the sound field of the listeners in three different positions for the t=1 second.

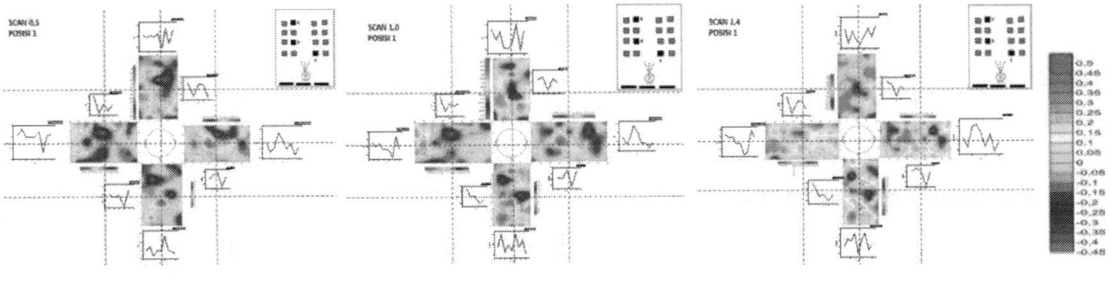

(a) (b) (c)

Figure 3. Sound field distribution on the four scanning planes of the listener on position number one; (a). 0.5 s; (b). 1.0 s, and (c). 1.4 s.

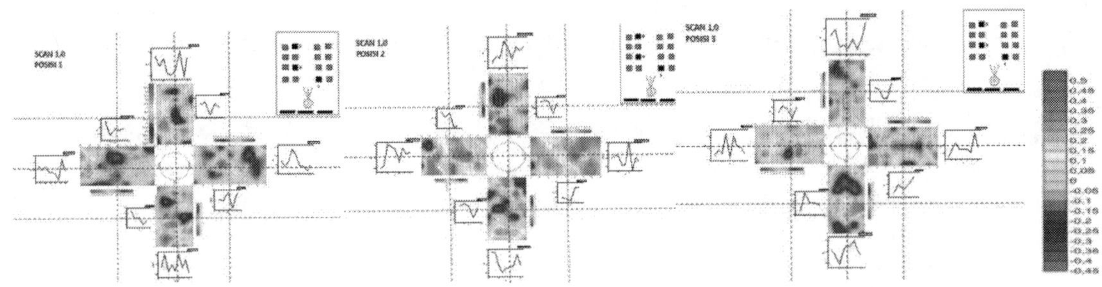

(a) (b) (c)

Figure 4. Sound field distribution of listeners on three different positions on t=1 second; (a). Position number one, (b). Position number two, and (c). Position number three.

It shows from Figure (3) that the sound pressure distribution on the scanning planes is changed subject to the time and its position. Figure (3a) representing the listener at position one which is $t = 0.5$ for the mannequin situated on the chair in the first row, around 2.7 m from the whiteboard on the front wall. It is seen that some yellow to red area spotted on the scanning planes. The fraction area of the yellow to red spot are increased as afunction of time. This phenomenon is associated with multiple reflections of the sound waves inside the classroom. Since the classroom wall constructed by painted coating brick wall without any absorbing layer, it became reverberant inside. Thus, as the sound waves are continuously radiated from the source, the portion of multiple reflectionsincreases as well.

The spotted pattern on Figure (4b) also occurredfor the same reason. It is seen that the left and right scanning plane of the second listener which is the mannequin situated on the chair at therightmid-row position are mostly spotted with the yellow to red colors. As explained above the listener receiving the highly reflected sound waves from the walls on the left and right side. It differs from the listener on position three. It is clearly indicated from Figure (5b) which is a sliced data from the sound field over the left scanning plane on time domain at three different positions on t=1.5 second. Figure (5a) representing sliced data of sound pressure on the front scanning plane for at=1 second.

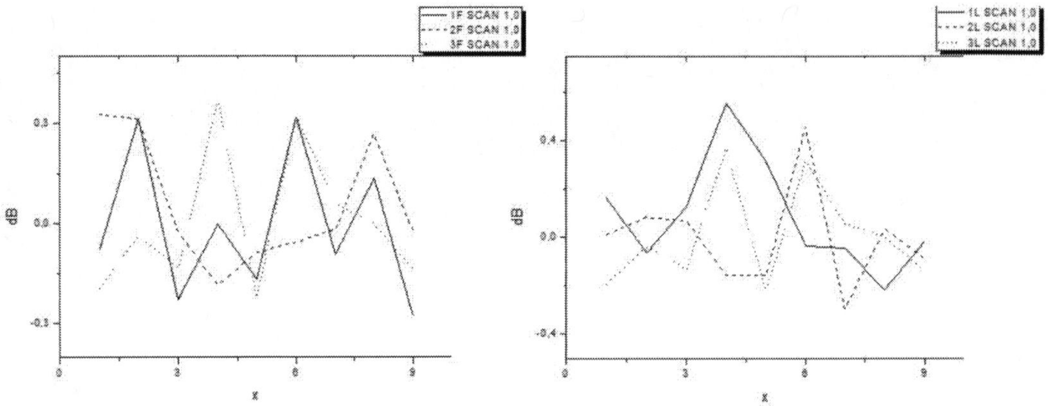

(a) (b)

Figure 5. Slice of sound field distribution on the center of scanning plane on three different positions. (a). Front scanning planes on $t = 1$ second, (b). Left side scanning plane on $t = 1$ second

Further investigation on the differences of the sound field distribution over the scanning planes also investigated on theterm of the coherence matrix. The calculated value for four different frequencies is presented below to show the different sensation received by the listener on three different positions. The coherence matrix for the right scanning plane of position one and two is given by,

$$\left[H_{12(125)}\right] = [|2{,}27 \quad 14{,}93 \quad 2{,}71 \quad 0{,}01 \quad 0{,}002 \quad 0{,}36 \quad 1{,}74 \quad 6{,}34 \quad 0{,}34|]$$

$$\left[H_{12(250)}\right] = [|0{,}46 \quad 0{,}59 \quad 0{,}42 \quad 1{,}00 \quad 2{,}83 \quad 0{,}02 \quad 2{,}68 \quad 25{,}11 \quad 0{,}02|]$$

$$\left[H_{12(500)}\right] = [|1{,}96 \quad 0{,}87 \quad 0{,}20 \quad 0{,}18 \quad 11{,}98 \quad 0{,}54 \quad 0{,}02 \quad 0{,}08 \quad 1{,}74|]$$

$$\left[H_{12(1k)}\right] = [|0{,}29 \quad 12{,}54 \quad 0{,}08 \quad 1{,}93 \quad 0{,}55 \quad 0{,}87 \quad 9{,}08 \quad 0{,}19 \quad 0{,}06|]$$

The results of similar measures for position one and three is given by,

$$\left[H_{13(125)}\right] = [|0{,}34 \quad 0{,}31 \quad 4{,}78 \quad 0{,}71 \quad 0{,}009 \quad 9{,}74 \quad 5{,}05 \quad 99{,}57 \quad 0{,}05|]$$

$$\left[H_{13(250)}\right] = [|0{,}38 \quad 5{,}53 \quad 0{,}95 \quad 2{,}97 \quad 0{,}99 \quad 0{,}01 \quad 0{,}92 \quad 4{,}43 \quad 0{,}98|]$$

$$\left[H_{13(500)}\right] = [|0{,}68 \quad 3{,}46 \quad 0{,}05 \quad 1{,}08 \quad 14{,}15 \quad 0{,}06 \quad 0{,}03 \quad 1{,}63 \quad 0{,}32|]$$

$$\left[H_{13(1k)}\right] = [|0{,}53 \quad 0{,}78 \quad 8{,}65 \quad 1{,}26 \quad 0{,}25 \quad 0{,}47 \quad 5{,}73 \quad 0{,}48 \quad 0{,}09|]$$

The above values strongly indicated the sound field inside the classroom are not distributed diffusively. In the other words, the listener at acertain position is may have the different sensation to the other listeners on the other sitting positions when they listen to the same talk or sound at the same time. Hence, the entire above results conclude that the coherence matrix representing subjective measures for the perception of a listener among the others inside the room. It could be used as anew measure for representing the acoustical quality of a room by using a simple equation as given by equation (7).

References

[1] Martellotta, F., 2008. *JSV* **317**, 378-399
[2] Shimukora R., Tronchin L, Cocchi A, and Soeta Y., 2011. *JSV* **330**, 3526-3573
[3] Wang JH and Pai CS, 2003. *Applied Acoustics* **64**, 1141-1158
[4] Hu H, Zhou L, Ma H, and Wu Z, 2008. *Applied Acoustics* **69**, 163-172
[5] Barbara G and Shinn-Cunningham, 2005. *J. Acoust Soc. Am* **117**, 3100-3115
[6] Hartmann WM, Rakerd B, and Koller A, 2005. *Acust. Acta Acust.* **91**, 451-462
[7] de Vries D, Hulsebos EM, and Baan J, 2001. *J. Acoust Soc. Am* **110**, 947-954

Journal of Physics: Conference Series **776** (2016) 012070 doi:10.1088/1742-6596/776/1/012070

Preference of reverberation time for musicians and audience of the Javanese traditional gamelan music

Suyatno[1]*, H.A. Tjokronegoro[2], I. G. N. Merthayasa[2], R. Supanggah[3]
[1)]Physics Department of Institut Teknologi Sepuluh Nopember Surabaya, Indonesia
[2)] Engineering Physics of Institut Teknologi Bandung, Indonesia
[3)] Karawitan Department of Institut Seni Indonesia, Surakarta, Indonesia

Email: kangyatno@physics.its.ac.is

Abstract-This paper presents results of an investigation of room acoustic parameters those are appropriated particularly to perform a Javanese gamelan. The acoustic parameters were obtained by analysing simulated sounds of performance. Those simulated sounds were obtained by sound convolution technique of a dry sound signal of Javanese gamelan performance with impulse responses rooms whose appropriated reverberation time. The reverberation time were varied between 1.0s to 1.8s, those belong to the characteristic of Pendopo Mangkunegaran Surakarta. In this case, Pendopo Mangkunegaran is assumed as one of the most suitable concert halls for Javanese gamelan performance.

To obtain the acoustic parameters, we used a psycho-acoustics measurement based on paired comparison test that having different of acoustic parameters to determine the most comfortable one to majority of respondents. The respondents who have participated in this research composed of a group of professional musicians of Javanese gamelan and groups of audience who are not musician, nevertheless part of them were familiar with Javanese gamelan music.

The comparison test gave results and showed majority of respondents of group of musicians had a notion sound reverberation time of 1.2s was most comfortable. This corresponds to +6.2dB, clarity and 74% definition. It means the appropriate acoustic condition allows musicians to recognize and distinguish clearly sound of each instrument being played. Meanwhile, group of audience had a notion reverberation time in a range of 1.2s – 1.6s was most comfortable. This range of reverberation time corresponds to +4dB to +6.2dB of clarity, and 66% to 74% of definition.

1. Introduction

The Javanese traditional gamelan is a kind of ensemble of instruments of gamelan called *ricikan*. There are two groups of Javanese gamelan called *pelog* and *slendro*, each has 40-70 types of *ricikan*. Alike a classical orchestra, Javanese gamelan is played by a group of musicians, each musician play a type of *ricikan*. It is very common a performance of the Javanese traditional gamelan is accompanied by vocalist called *sinden*. The existence of *sinden* during a performance is not separated of the Javanese traditional gamelan. During the performance, along with the rhythm of the gamelan music, *sinden* singing a Javanese song, in form of a Javanese poetry, containing a message to audience. The existence of *sinden* together with the gamelan as an ensemble makes a performance of the Javanese traditional gamelan more interesting and traditionally invites much public audience. However, as a kind of music ensemble, there is no a leader who leads the Javanese traditional gamelan.

Since there is no a leader who leads the performance, harmony of performance of the Javanese traditional gamelan then is absolutely depends on ability of each musician (call *pengrawit*) and vocal to recognize and distinguish sound of each instrument of gamelan being played during the performance. Recognizing and distinguishing sound characters is a necessary for musicians to determine tempo, next tones and loudness of each instrument during the show. Therefore, musicians of the Javanese traditional gamelan need a specific acoustic preference in order to be able to take their role in a performance [1,2]. For the same reason, to enjoy a performance of the Javanese traditional gamelan, audience require its own acoustic preference which is different to musicians preferred. Audience need impression of space in addition to the acoustic parameters of the gamelan.

Reverberation time (RT) is an important acoustic parameter that affects to ability of musicians to play their role. The longer reverberation time results less sound clarity in a room [3,4]. In return, musicians would get difficulty to play *ricikan* appropriate to their role in a performance. Using the principle impulse response at a position in space can be determined acoustic parameters associate to that position including of reverberation time, subsequent reverberation time, sound clarity, frequency contents, and loudness [5].

Early in the past, the Javanese traditional gamelan actually was part of traditional and religious ceremony held usually in Kraton. Kraton is the Palace of the Javanese Kingdom in ancient times. Along the performance, *sinden* sing a song enjoy a cultural message to audience. Therefore the performance event invites people to come for the message of goodness. For that reason, in the past, the Javanese traditional gamelan was performed traditionally in an open space or semi-open space in order to accommodate more audience. In the progress, function of performance of the Javanese traditional gamelan has changed gradually becomes a kind of public entertainment. However, in present time, noises that come from environment make open space or semi-open space becomes not very ideal place to conduct the performance. In consequence, performances of the Javanese traditional gamelan move from open or semi-open space to enclosed space, therefore straight changes room acoustics of the performance is unavoidable. To solve the acoustic problem it is very common one installs an electronic system like loudspeaker, which is actually doesn't help but getting worse. The direct affects results clarity of sound in the room is decreasing, then, therefore the musician's loss their precision in the way of playing gamelan. In turn, gradually change of the performance character of the Javanese traditional gamelan has been occurred. Moreover, following the advancement of technology and lifestyle, the presence of the Javanese traditional gamelan in enclosed space whose no appropriated acoustic characteristics finally has shifted number of meanings the authenticity of performances from its originally when it was performed at an open or semi-open space. In order to get the originality of performance, it's necessary to find a design preference of acoustic parameters appropriated to an enclosed space particularly for the Javanese traditional gamelan. That is, an enclosed space which has acoustic parameters equivalently found in a semi-open space where the Javanese traditional gamelan originally was performed. In this case, the acoustic parameters of design must provide acoustic environment desired by musicians and audience as likely it was obtained when the performance was conducted in semi-open space.

Among parameters of a room acoustics, reverberation time is the most important, especially on stage area where musicians are playing gamelan. Long of reverberation time on stage area will influence sound clarity perceived by musicians and vocalist and its can disturb performance harmony and performance communication among the musicians. In addition, the reverberation time of room will present a temporal impression and character of the Javanese traditional gamelan during a performance. For that reason, a study of preference of reverberation time, to find one the most preferred for musicians and audience in an enclosed space, is really important to design a room acoustics for the Javanese traditional gamelan.

Assessments of a quality of performance of the Javanese traditional gamelan usually are based on subjective factors. The results will depend on background and attitude knowledge of who becomes a judge. Because of the attitude of respondents are varying, the results of a subjective assessment are difficult to be realized as standard of acoustic design parameters. In return, because of no standard preference which is appropriated to the acoustic characteristic of the Javanese gamelan makes quality of the performance are inconsistent depends on enclosed room where the performance is. As mentioned previously, it is very often an acoustic modification is made by using a sound system or loudspeaker to adjust sound acoustic parameters of the enclosed room where the performance is conducting. However, loudspeaker may exacerbate other acoustic parameters and even overall of the performance. Therefore, finding objective parameters of the acoustics performance is necessary to have a standard design of enclosed room particularly for Javanese traditional gamelan.

The equivalent objective parameters from the obtained subjective assessment are called as quantified subjective parameters [6]. By quantification of the subjective parameters in order to get the equivalent objective one, the standard of design can be obtained and to be used in designing an enclosed room appropriately for the Javanese traditional gamelan. And then, hypothetically musicians will re-find their ability to keep the quality and authenticity of performances.

It is well known that the Pendopo Mangkunegaran at Surakarta is much closed to history of progress for years of the Javanese traditional gamelan. It is a semi-open space where the Javanese traditional gamelan was played and performed for generations. Therefore, it is a basic reason that acoustic parameters of the Pendopo Mangkunegaran then is referred as standard preference to derive equivalent or appropriate acoustic parameters for enclosed space for the Javanese traditional gamelan performance. Suyatno [7] has reported the room acoustic characteristics of the Pendopo Mangkunegaran at Surakarta. Those parameters are distribution of listening level, reverberation time, clarity (C80), definition (D50) and Inter Aural Cross Correlation (IACC). By referring to the acoustic criteria given by Ribeiro [8], the Pendopo Mangkunegaran actually has acoustics parameters which are appropriately equivalent to prerequisites as a performance venue [8].

Sarwono on his research of an optimum reverberation time appropriate to audience who are unfamiliar to Javanese traditional gamelan is in the range of 500 ms - 650 ms [9]. Those ranges of reverberation time were obtained by psycho-acoustic test using a sound field of performance of the Javanese traditional gamelan that was recorded by mixing, filtering and an artificial effect. While Fariz et al [10] by using the scoring method found the acoustic parameters preference for the Javanese gamelan are 1.4 s of reverberation time, 69 dB - 79 dB of listening level, and 40 ms of initial time delay gap.

In this research, the psycho-acoustic test was conducted to sounds of performance of the Javanese traditional gamelan using a pairwise comparison test method [11]. In this test, each respondent is ordered to listen by using a headset a pair of two sounds which have different acoustic parameters. Then the respondent should select the considered most pleasant one. Using such test method has been obtained acoustic preferable parameters which were chosen by majority of the respondents. Where, hypothetically those acoustic parameters are appropriate to perform the Javanese traditional gamelan.

In section II of this paper will be explained more detail methodology of this research, including creation of the simulated sounds, and method of comparative test used in this research. While in section III will be described results of the comparative test, those are most appropriate acoustic parameters to the musician and audience of the Javanese traditional gamelan. And the conclusion will be given in last section of this paper.

2. Numerical Methods

As stated in previous section, subjective assessment of acoustics quality of performance of the Javanese traditional gamelan depends relatively very much on background and the momently attitude of the "assessors" or respondent, who being a judge. Actually, musicians and audience of the Javanese traditional gamelan have different preference in term of acoustic parameters to enjoy a performance. To conduct their role to a performance, musicians require their own acoustics preference. As has been mentioned previously noted, an important acoustic parameter that influences to acoustics quality is reverberation time. To find the reverberation time which is most pleasant to musicians and audience, it was conducted by a psycho-acoustic test upon appropriated simulated sounds. In this research, simulated sounds were obtained by convolution process of a dry sound signal to impulse response of rooms of differences of reverberation time [12]. Variations of parameters of reverberation time are 1.0 s, 1.2 s, 1.4 s, 1.6 s, and 1.8 s. Those parameters are acoustic parameters of Pendopo Mangkunegaran. Table 1 describes acoustic parameters of impulse responds used in the convolution technique (equation 1).

Table 1. Acustic pameters of responses impulse for synthetic rooms.

Parameters	Room's Parameters				
RT [second]	1.0	1.2	1.4	1.6	1.8
Clarity (C80) [dB]	8.7	6.7	5.2	4.0	1.2
Definition (D50) [%]	82.1	75.4	68.5	66.3	44.8

A convolution technique that was used to create the simulated sound used in this research is represented by the following equation:

$$y_i(t) = \int_{\tau=-\infty}^{+\infty} x(\tau)h_i(t-\tau)d\tau \quad (1)$$

In equation (1), $x(t)$ is dry sound signal, $h_i(t)$ is "synthetic room" impulse response represents "synthetic" acoustic parameters of ith room, while $y_i(t)$ represents a simulated sound that was assumed was conducting in i^{th} room. By using this concept, a simulated sound of performance of the Javanese traditional gamelan can be obtained having the desired acoustic parameters. In this case, the rooms have acoustic parameters belong to Pendopo Mangkunegaran.

Have been stated previously, an objective of this research is to obtain equivalent objective parameters from a subjective assessment. In order to obtain a suitable criterion by musicians and audience, the types of respondents as assessors are the players and audiences. In this research respondents compose of 39 musicians and 94 audiences who were non-musicians or audience. The musicians were professional and expert in playing the Javanese gamelan instruments. The musician respondents came from the laboratory technicians at ISI Surakarta and other professional gamelan artists. Table 2 describes qualification and background of the respondents.

Table 2. Qualification and background of respondents participated in comparison test.

Type of Respondents	Background
Musicians: 39 respondents	30 respondents are expert of the Javanese traditional gamelan 9 respondents are expert and can play all the Javanese gamelan instruments
Audience: 94 respondents	61 respondents are familiar with the Javanese traditional gamelan (audience type A): 37 respondents play the Javanese traditional gamelan 24 respondents don't play the Javanese traditional gamelan 33 respondents are unfamiliar with the Javanese traditional gamelan (audience type B)

To obtain the equivalent objective parameters, it was used Thurstone scaling method. In what follows will be explained steps of quantification in order to obtain the objective parameters from subjective assessment. By Thustone scaling method, each the respondent was ordered to hear a pair of any two sounds. Each respondent has a freedom to choose the most pleasant sound one. In this paper, the type of composition (called *garap*) that was tested by respondents has a moderate-to-fast tempo which called *soran*. During subjective assessment, i^{th} simulated sound will be compared by j^{th} simulated sound, and k^{th} respondent will select which one of the two simulated sounds is preferred. Figure 1 shows example of two simulated sounds of performances of Javanese traditional gamelan should be selected by respondent.

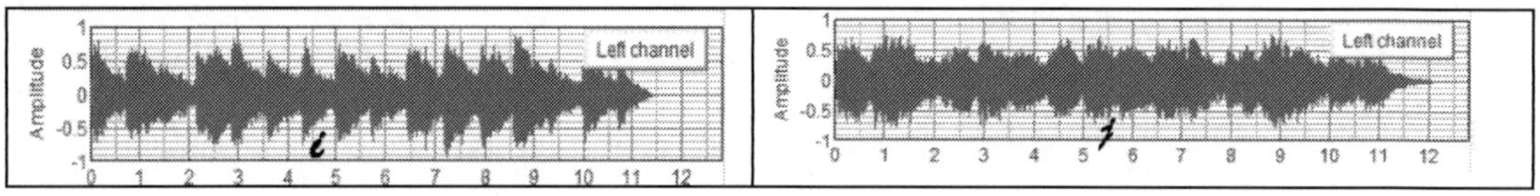

Figure 1. The simulated sounds for pairwise comparison test.

There are 5 simulated sounds represented by Table 1, each represent reverberation time of 1.0s, 1.2s, 1.4s, 1.6s and 1.8s respectively. After heard i^{th} and j^{th} simulated sounds subsequently, respondent should determine only one which is preferred. When i^{th} simulated sound is preferred then the numerical preference scale is given by $Y_{ij}(k) = 1$ and $Y_{ji}(k) = 0$. Otherwise when j^{th} simulated sound is more preferred one, then the numerical preference scale is $Y_{ij}(k) = 0$ and $Y_{ji}(k) = 1$. By this comparative test method numerical value of scale is defined to be $Y_{ij}(k) = 0.5$ when $i=j$. There are 10 pairs of two simulated sounds to be assessed. Then, preference distribution $(S_i(k))$ is calculated using the following equation [11]:

$$S_{i(k)} = \sqrt{2\pi}\left(P_{i(k)} - 0.5\right) \quad i = 1,..,5 \tag{2}$$

where $Pi(k)$ is

$$P_{i(k)} = \frac{1}{M}\sum_{j=1}^{M} Y_{ij(k)} \quad i, j = 1,..,5 \tag{3}$$

And M is number of simulated sounds to be tested, in this experiment, is equal to 5. It is clear every respondent have 10 pairs of two simulated sounds to be assessed to obtained $S_i(k)$, $i=1,..,5$. To avoid a bias that occurs in psychoacoustic tests, pairs of two simulated sounds to be assessed by a respondent was ordered randomly. Table 2 shows an example of results of the psychoacoustics test. According to Table 3, the highest of $S_i(k)$ is equivalent to reverberation time of 1.4 s.

Table 3. Example of numerical values of preference scale calculated by equation (2).

		J					**Si(k)**
	RT[s]	**1**	**1.2**	**1.4**	**1.6**	**1.8**	
	1	0.5	0	0	0	1	-0.501
	1.2	1	0.5	0	1	1	0.501
i	1.4	1	1	0.5	1	1	1.002
	1.6	1	0	0	0.5	1	0.000
	1.8	0	0	0	0	0.5	-1.002

Finally, numerical value of appropriate objective parameter of acoustic preference that which is equivalent to the subjective assessment is obtained by averaging of results given by all respondents:

$$S = \frac{1}{N}\sum_{k=1}^{N} S_{i(k)} \tag{4}$$

The k index in equations (1)-(3) represents respondent was participated, while N in equation (4) is number of respondents were participated, which is 39 musicians and 94 audience (type A and type B all together).

3. Results and Discussion

3.1. Appriopriate Acoustic Parameters for Musicians

Using procedure as explained previously, Figure 2 shows distribution plot of numerical values of reverberation time preferred by respondents of musician.

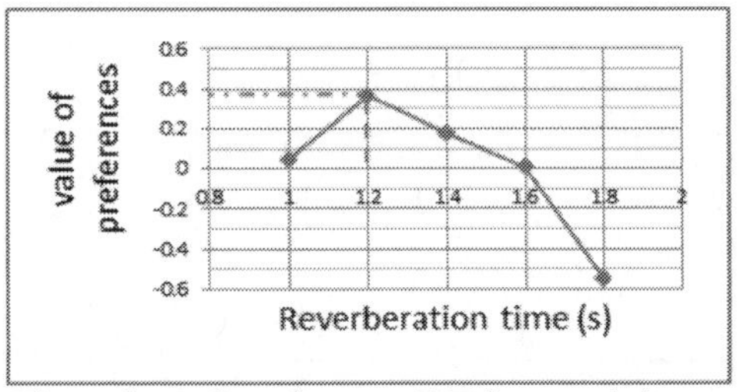

Figure 2. The value of preference RT for musician respondents

As shown by Figure 2, acoustic reverberation time which was most preferred by musicians is 1.2s. Referring Table 1, 1.2 s reverberation time gives good sound clarity as shown by values of C80 is 6.7 dB and D50 is 75.4%. Those values, hypothetically indicate that majority of musicians have an ability to recognize and distinguish the instruments were playing and vocals as well. Such acoustic condition guaranty a good performance communication among musicians and harmony of performance can be obtained.

3.2. Appropriate Acoustics Parameters for Audience

Using the same procedure, assessment by audience respondents (audience of type A and audience type B) have been resulted distribution of acoustic reverberation time shown by Figure 3.

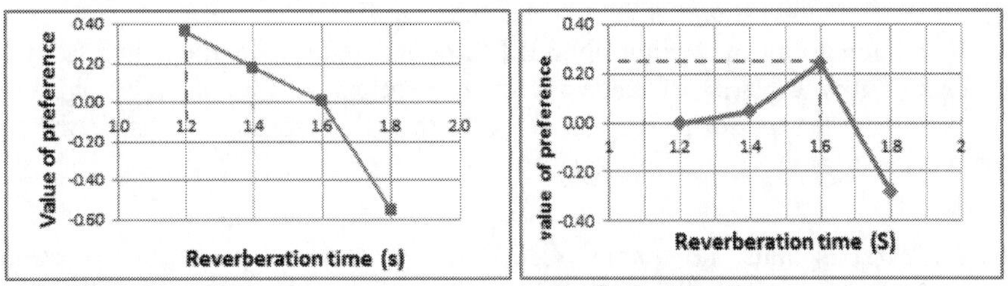

Audience type A Audience type B

Figure 3. The value of preference RT for audience respondents.

A Figure 3 show the highest preference comes from audience of type A is 0.36 which is 1.2 s of reverberation time. Again, referring to Table 1, the equivalent room has acoustic characteristics C80 of +6.7 dB and D50 of 75.4%. While the highest preference comes from audience type B is 0.24 which is 1.6 s of reverberation time. Where, according to Table 1, the equivalent room has acoustic characteristics C80 of +4 dB and D50 of 66%. By those results, in can be concluded the audience feel comfort for the room with reverberation time of 1.4 s and 1.6 s. This may occur because of background of the respondents were familiar and could play the gamelan even they are not expert on it, therefore they require impression of space as well [1]. As noted by Table 2, among audience type A, 55 % are beginner musicians that do Javanese traditional gamelan and the rest are doesn't but just enjoy.

3.3. **Discussion**

The acoustic parameters that suitable to musicians and audience of the Javanese traditional gamelan allow each of them to enjoy and carry out their respective roles in a performance. Refer to Figure 2,

not only was the clarity of sound required by musicians, but also the reverberation of the room still needed. It can be seen from the most respondents' musicians more pleasant with the sound produced by room with RT values of 1.2 s than 1.0 s. The reverberation of room is still needed to emergence the strong impression (called *gagah*) of the Javanese gamelan performances [1,13].
Table 4 shows the acoustics parameter preferred by majority of respondents of musicians and audience obtained by comparison test explained previously.

Table 4. Acoustic parameters preferred by musicians and audience of the Javanese traditional gamelan

Parameters	Musicians	Audience
RT	1.2 s	1.2 s – 1.6 s
C80	6,7 dB	4.0 dB – 8 dB
D50	75.4 %	66 % - 82 %
BR	0.92	0.96 – 1.04
TR	0.99	0.7 – 0.92

According to Table 4, like a classical music, sounds were produced during a performance of the Javanese traditional gamelan chosen by majority of musicians have temporal impressions of warmth (BR=0.92) and brilliance (TR=0.99). Warmth is the subjective impression of the sound at the low frequency (125 Hz and 250 Hz) and the brilliance is high frequency (2000 Hz and 4000 Hz) [3,4]. Like musicians, as shown by Table 3, sounds were chosen by audience have good impression. To be noted, numerical value of TR less than 1 (TR=0.7) is accepted as a brilliant impression, since all instruments *(ricikan)* produce sounds with frequency less than 4000 Hz. Yet, in most of performances, sounds produced by instruments have dominant frequencies less than 3000 Hz [14, 15].

Based on the parameters in Table 4, the player can still identify instrument and vocal, but still have a strong impression. Therefore, in order to maintain communication among the musicians and harmony of performance during the show, acoustics design and material selection for stage must be able to reduce low-frequency and long sound envelope characteristic.
As for the audience, temporal impression obtained through the reverberation time to make the space more real presence, but with clarity of voice and instrument can still be obtained. While, in audience area for Javanese gamelan performance, model of room and ceiling are expected to provide a reflection that is not excessive.

Thus, elements of direct sounds and sounds of reflection in a closed room must be maintained such that to produce reverberation time properly needed by musicians and audience as well. Finally, by the acoustic parameters shown by Table 3, the meaning and originality of performance of the Javanese traditional gamelan conducted in an enclosed space can be obtained as such likely a performance was traditionally conducted in semi-open or open space.

4. Conclusions

To get a performance harmony and performance communication among musicians of the Javanese traditional gamelan, the musicians need acoustic clarity of each instrument of the gamelan during the performance. Sound clarity is influenced by reverberation time of room. In this research has been shown majority of respondents of musicians selected RT of 1.2 s which is equivalent to D50 of 74%, as well as C80 of +6.2 dB. By those acoustic parameters it is guaranteed musicians got an acoustic clarity needed to play well their role. It means musicians can recognize and distinguish well *ricikan* are being played. In return, at a time, each musician knows very well which instrument should be played according to their role of the show. On the other hand, majority of respondents of audience selected reverberation time between 1.2 s - 1.6 s, that is equivalent to C80 between +4 to 6.2 and D50 are 66% - 74%. Through those parameters, the audience can enjoy performances with good acoustic qualities include the impression of room by the RT.

5. References

[1] Supanggah R., 2011, *Bothekan Karawitan*, 2th, ISI Press.
[2] Suyatno, H. A. Tjokronegoro, IGN. Merthayasa, R. Supangah, 2013, *Instrumentation Journal*, vol 37, LIPI, Indonesia.
[3] Beranek L, 2004, *Concert Hall and Opera House*, Springer, Verlag,
[4] M.R. Schroeder, 2007, *Handbook of Acoustics*, Springer,
[5] Beranek L, 2008, *Journal Audio Engineering Society*, vol. 56, no. 7/8,
[6] Marshall L., 2006, *Architectural Acoustics*, Elsevier Academic Press.
[7] Suyatno, H. A. Tjokronegoro, IGN. Merthayasa, R. Supangah, 2015, *Procedia Social and Behavioral Science* 184, 322-327.
[8] Ribeirio, 2008, *Room Acoustic Quality of A Multipurpose Hall*: A Case Study, Acoustics 2008, Paris.
[9] Sarwono J., Y.W. Lam, 2002, Journal of Sound and Vibration,258(30), 451-461 .
[10] Fariz A., Stevanus I, Suyatno, Sarwono J., Merhayasa IGN., 2015, *The Design of Javanese Gamelan Concert Hall Based on Optimum Acoustic Parameters*, SiA.
[11] Ando. Y, 1998, Architecture Acoustics, Springer, Japan,
[12] Farina A, *Convolution of anechoic music with binaural impulse responses*, University of Parma, Italia.
[13] Bernamou M, 2010, *Rasa: Affect and Intuition in Javanese Musical Aesthetics*, Oxford University Press.
[14] Suyatno, H. A. Tjokronegoro, IGN. Merthayasa, R. Supangah, 2013, *The acoustics characteristic of Javanese gamelan instruments, case study of the PSTK ITB Javanese gamelan*, (wrote in Indonesian), Physics and Applications Seminar, Physics ITS.
[15] Sethares W. A., 2005, *Tuning, Timbre, Spectrum, Scale*, 2th edition, Springer.

Preliminary laboratory testing on the sound absorption of coupled cavity sonic crystal

R. Kristiani[1,2], I. Yahya[1*], Harjana[1,2] and Suparmi[2]
[1.] The Iwany Acoustics Research Group (iARG), Department of Physics, Faculty of Mathematics and Natural Sciences.
[2.] Graduate School of Physics.
Sebelas Maret University,
Jl. Ir. Sutami 36A Kentingan, Surakarta 57126, INDONESIA

E-mail: iyahya@mipa.uns.ac.id

Abstract. This paper focuses on the sound absorption performance of coupled cavity sonic crystal. It constructed by a pair of a cylindrical tube with different values in diameters. A laboratory test procedure after ASTM E1050 has been conducted to measure the sound absorption of the sonic crystal elements. The test procedures were implemented to a single coupled scatterer and also to a pair of similar structure. The results showed that using the paired structure bring a better possibility for increase the sound absorption to a wider absorption range. It also bring a practical advantage for setting the local Helmholtz resonant frequency to certain intended frequency.

1. Introduction

Sonic crystals were greatly attracting among researchers in last decade. The local resonant feature brings the possibility for tuning the acoustical properties and response to the propagation of sound waves [1,2]. The various approach has proposed by researchers to create sonic crystal assisted sound absorber and diffusers [3-5]. As the Helmholtz local resonant and the coupling effect is the most important mechanism in this task, some modification, and manipulation on scatterer and the crystal lattice has been reported in the literature. Those modification and manipulation techniques could be divided into two main categories. First is related to the single and multi local resonant scatterer, and the second is associated with crystal defect. Theoretical development [6,7] and laboratory experiment [8] on single and multi local resonant scatterer has been conducted. While crystal defect problems also reported by other researchers [9-11].

That mentioned approach has the limitation according to the scatterer geometrical shapes and dimensions. For example, the scattering effect which is related to the scatterer dimension and configuration was investigated by using the various size of scatterer or by replacing it from the initial position on the lattice. The same issue also happened to the modification of scatterer local resonant which is all approaches are based on various geometrical dimension scatterer.

A preliminary laboratory study on the sound absorption performance of a new coupled scatterer sonic crystal is reported in this work. The basic structure of the proposed scatterer design is constructed by a couple of cylindrical tubes with a different diameter to form multi degree of freedom Helmholtz resonator. it gives the possibility for varying the local Helmholtz resonant by using a single size scatterer.

2. Methods

The proposed coupled scatterer element are illustrated in Figure (1). It is constructed by a pair of the cylindrical tubes with the radius of r_1 and r_2 attached to one another with the axis distance a. The first resonator has a single opening that could be rotated to arbitrary angle β. The second resonator has a bigger diameter. It has two slit-shaped inlets fixed along the connecting wall of the cylindrical resonator. The coupled scatterer could be set to arbitrary skewness angle θ. The inlet, outlet, and first resonator orifice have the same width s. The scatterer was made from PVC tube with the detailed geometrical dimensions provided in Table (1).

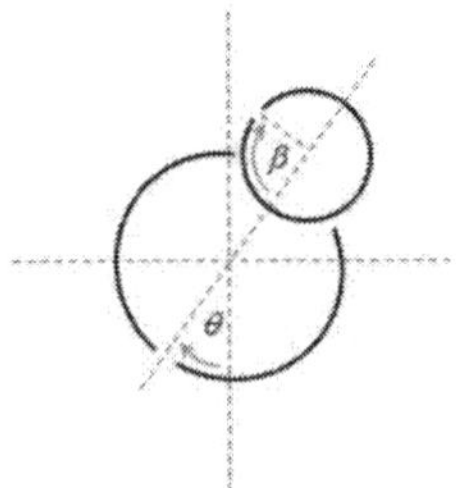

Figure 1. Basic model of the coupled scatterer

Table 1. Scatterer dimension

Parameters	Dimensions
Radius (mm)	r_1: 8.5 r_2: 13
Axis distance (mm)	all fixed: a = 10
Height (mm)	all fixed: 30
θ (degree)	θ: 0, 30, 45
β (degree)	β: 0, 90
Slits (mm)	all fixed: s = 1

The experiment was conducted by using two microphone impedance tube technique refer to ASTM 1050 laboratory test procedure. B&K 4206 impedance tube complete apparatus was utilized for measuring the various samples of single and also the pair of the coupled scatterer. The sample $K_1\_\theta\_\beta$ denotes a single coupled scatterer while the $K_2\_\theta\_\beta$ denotes a pair of coupled scatterer.

3. Results and Discussion

Results from a laboratory test for the single coupled scatterer is presented in Figure (2) while the paired scatterer in Figure (3). Figure (2a-c) shows a similar pattern that when the coupled scatterer has the β=90°, each resonator works separately as a single degree of freedom Helmholtz resonator with a different resonant frequency. It is indicated by the similar two peaks around 300 Hz and 1.5 kHz on the sound absorption coefficient graphs.

It differs to the pattern when the first resonator is set to β=0°. Since the slit of the first resonator is situated inside the second resonator, the coupled cavities become a two degree of freedom Helmholtz resonator. When the sound waves propagate through the two inlets of the secondary resonator, it produces an instability and fluctuated sound pressure circumstance inside the cavity of the secondary resonator. This fluctuated pressure flows through the outlet slit of the secondary resonator and at the same time also stimulating resonant effect on the first resonator. The coupling between the first and secondary resonator affecting and changes the value of the structure reactance, and it changes the sound absorption pattern accordingly. Despite it does not increase the sound absorption performance significantly, varying on values of β and θ shows an interesting result occurred on both low and high frequency. Sound absorption performance is increased at the low frequency band and followed by the shifting absorption range on the high frequency band. The phenomenon indicates the possibility for the proposed coupled scatterer design to be use for tuning the Helmholtz local resonant and in the same time to increase the acoustic performance.

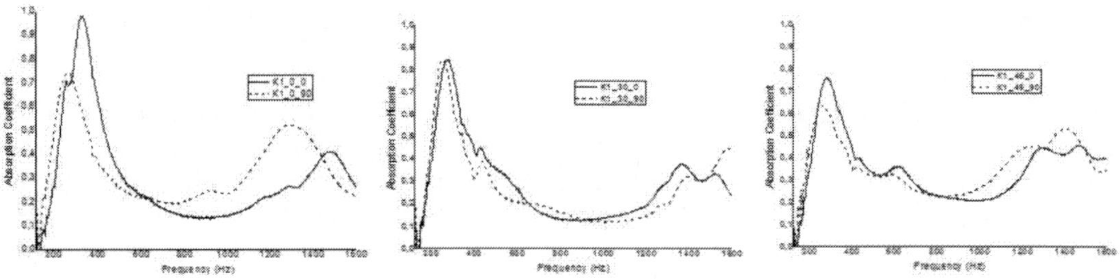

(a) (b) (c)

Figure 2. The sound absorption coefficient of a single coupled scatterer on thre different θ (0, 30, 45) and β (0, 90) configurations.

The almost similar phenomenon also occurs with the paired coupled scatterer models. Two peaks around of 300 Hz and 1.5 kHz and several small peaks indicates that the paired scatterer configuration has a little bit different absorption mechanism compared to the single coupled scatterer model. The two main peaks are associated with the resonant frequency of the two resonators while the other small peaks are related to the coupling of the scatterers with the surrounding air space.

It shows from Figure (3), the paired design brings a better sound absorption performance. The model with skewness $\theta = 30°$ and $\beta = 45°$ has the best performance both for low and high frequency range. The pair of coupled scaterer performance is associated with the response and the sound absorption the four independent single degrees of freedom Helmholtz resonators. Even though cavity of the first scatterer is not coupled to the cavity of the secondary scatterer, it forms coupling effect with empty air space surrounding it. In the same time, the secondary scatterer also forming the similar coupling through the two slits on its cavity walls. Both the first and the secondary cavity construct a higher degree of freedom resonant system with the surrounding empty air space.

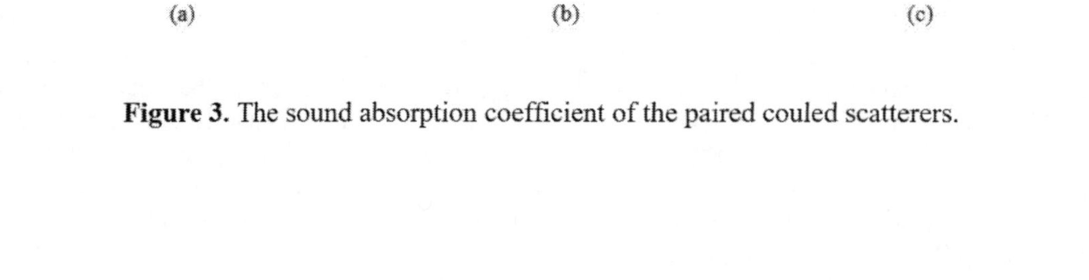

(a) (b) (c)

Figure 3. The sound absorption coefficient of the paired couled scatterers.

It can conclude from the laboratory test results that the using of paired of coupled scatterer shows a better sound absorption compared to the single coupled scatterer. It brings the possibility of improving the sound absorption both through the local Helmholtz resonant and also increases the number of the independent resonators that can perform a coupling effect with its surrounding empty air space.

4. References

[1] Maldovan, M, 2013 *Nature* 503
[2] Khelif, A., Deymier, P. A., Djafari-Rouhani, B., Vasseur, J. O., and Dobrzynski, L, 2003 *J. of Appl. Phys.* 94.3 1308-1311
[3]van der Aa, B., and Forssen, J, 2013 *Applied Acoustics* 74 89-101
[4] Redondo, J., Pico, R., and Sanchez-Morcillo, V. J, 2013 *J. Acoust. Soc. Am.* 134-6 4412-4417
[5] Morrandi, F., Miniaci, M., Marzani, A., Barbaresi, L., and Garai, M, 2016 *Applied Acoustics* 114 294-306
[6] Romero-Garcia, V., Krynkin, A., Garcia-Raffi, L., M, Umnova, O., and Sanchez-Perez, J. V, 2013 *JSV* 332 184-198
[7] Hirsekorn, M., Delsanto, P. P., Batra, N. K., and Matic, P, 2004 *Ultrasonics* 42 231-235
[8] Alberti, A., Gomez, P. M., Spiousas, I., and Eguia, M. C, 2016 *Applied Acoustics* 104 1-5
[9] Wang, X. P., Jiang, P., Chen, T.N., and Yu, K. P, 2016 *Intl Journal of Modern Physics B.* 30 1650025 (17 pages)
[10] Shelke, A., Banerjee, S., Habib, A., Raahani, E. K., Ahmed, R., and Kundu, T, 2014 *Journal of Intelligent Material Systems and Structures*, 25-13 1541-1552
[11] Li, Y., Chen, T., Wang, X., Ma, T., and Jiang, P, 2014, *J. of Appl. Phys.* 116, 024904 (7 pages)

Journal of Physics: Conference Series **776** (2016) 012072 doi:10.1088/1742-6596/776/1/012072

Speech enhancement on smartphone voice recording

Bagus Tris Atmaja, Mifta Nur Farid, Dhany Arifianto

Dept. of Engineering Physics, Faculty of Industrial Technology, Institut Teknologi Sepuluh Nopember, Kampus ITS Sukolilo, Surabaya 60111, Indonesia

E-mail: bagus@ep.its.ac.id

Abstract. Speech enhancement is challenging task in audio signal processing to enhance the quality of targeted speech signal while suppress other noises. In the beginning, the speech enhancement algorithm growth rapidly from spectral subtraction, Wiener filtering, spectral amplitude MMSE estimator to Non-negative Matrix Factorization (NMF). Smartphone as revolutionary device now is being used in all aspect of life including journalism; personally and professionally. Although many smartphones have two microphones (main and rear) the only main microphone is widely used for voice recording. This is why the NMF algorithm widely used for this purpose of speech enhancement. This paper evaluate speech enhancement on smartphone voice recording by using some algorithms mentioned previously. We also extend the NMF algorithm to Kulback-Leibler NMF with supervised separation. The last algorithm shows improved result compared to others by spectrogram and PESQ score evaluation.

1. Introduction

Signal enhancement can be viewed from many perspectives, one of the currently widely used is from the view of source separation perspective. Source separation is the process of decoupling of two or more sources with no or little prior information. From this definition, speech enhancement can be categorized as natural application for source separation because the goal is to enhance target speech from noises.

The use of smartphone as multimedia device has increased exponentially in current era. As we can seen on the media, most of journalist use smartphone devices instead of traditional recorder to record speech sound. In the other speech events, traditional recorder was almost gone replaced by smartphone voice recording feature to record human voice. The voice recording feature on the smartphone also can be used for voice biometrics, voice tapping, video recording, voice commands and voice calling. The use of smartphone as voice recording tools should be evaluated for measurement and it can be enhanced by signal processing technique addressing background noises.

This paper evaluates signal enhancement on smart phone voice recording from traditional method to the advanced one. From the traditional view, spectral subtraction method by Boll [1] was used with three different derivation: power spectral subtraction, magnitude spectral subtraction and over spectral subtraction [2]. The second approach to enhance speech on smartphone voice recording is by wiener filtering to estimate minimum mean square error (MMSE) as proposed by Lim and Oppenheim [3]. On the third approach, estimation of MMSE was done by short-time spectral amplitude (STSA) as suggested by Ephraim and Malah [4].

Moving to current speech enhancement algorithm, Non-Negative Matrix Factorization (NMF) was used as fourth approach based on [5], and finally on the fifth approach, NMF was modified by using supervised separation [6]. All of five approaches was evaluated by spectrogram and PESQ score to evaluate the performance of speech enhancement over smartphone device.

2. Speech Enhancement Algorithms

In the following section we briefly describe some algorithms used on this research: spectral subtraction, Wiener filtering, MMSE STSA and NMF.

2.1. Spectral subtraction

Speech signal recorded by smartphone device can be modelled by addition of clean signal with by noises. This phenomena can be write in frequency domain as follows,

$$Y(\omega) = X(\omega) + D(\omega) \tag{1}$$

where $Y(\omega)$ is recorded noisy signal, $X(\omega)$ is clean target signal and $D(\omega)$ is noise. The goal is to obtain target signal from noisy signal. One of the simple solution is by subtracting $Y(\omega)$ with estimation of $D(\omega)$ to obtain $X(\omega)$,

$$|\hat{X}(\omega)| = |Y(\omega)| - E[|D(\omega)|] \tag{2}$$

By taking real part of that equation, it can be obtained the magnitude of spectral subtraction as follows.

$$|\hat{X}(\omega)| = max\left\{|Y(\omega)| - E[|D(\omega)|], 0\right\} \tag{3}$$

Instead of calculating magnitude of spectral subtraction, the power of each components can be calculated to achieve power of spectral subtraction.

$$|\hat{X}(\omega)|^{\alpha} = max\left\{|Y(\omega)|^{\alpha} - E[|D(\omega)|^{\alpha}], 0\right\} \tag{4}$$

in which $\alpha = 2$. From the power spectral subtraction 4, the gain can be expressed as follows

$$H(\omega) = \frac{|\hat{X}(\omega)|}{|Y(\omega)|} = \sqrt{\frac{|Y(\omega)| - E[|D(\omega)|]}{|\hat{X}(\omega)|^2}} = \sqrt{\frac{\gamma(\omega) - 1}{\gamma(\omega)}} \tag{5}$$

where $\gamma(\omega)$ is called the a-posteriori SNR $\gamma(\omega) = \frac{|Y(\omega)|^2}{E[|D(\omega)|^2]}$. Different gain can be obtained for various α.

$$H(\omega) = \frac{|\hat{X}(\omega)|}{|Y(\omega)|} = \left(\frac{\gamma(\omega)^{\alpha/2} - 1}{\gamma(\omega)^{\alpha/2}}\right)^{1/\alpha} \tag{6}$$

The final method on spectral subtraction is overspectral subtraction which is proposed by Berouti et al. [2] by adding weighting coefficient α and β to improve separation result.

$$|\hat{X}(\omega)|^2 = max\left\{|Y(\omega)|^2 - \alpha E[|D(\omega)|^2], \beta E[|D(\omega)|^2\right\} \tag{7}$$

The weighting coefficient α and β are used for reduce noise peaks which in α should be dependent on the frame segmental SNR (γ), less attenuation (small α) for high SNR and more attenuation (large α) for low SNR.

2.2. Wiener Filtering

Wiener filter is designed to find optimal linear filter that outputs the desired signal i.e. clean signal. It can be obtained by find $h*$ that minimizes $E[e^2(n)]$ by solving $\frac{\partial E[e^2(n)]}{\partial h} = 0$. The gain is defined by,

$$H(\omega) = \frac{\hat{X}(\omega)}{Y(\omega)} \tag{8}$$

Implementing Wiener filter,

$$e(n) = x(n) - \hat{x}(n) \tag{9}$$

In frequency domain, by assuming a non-causal IIR filter and by using the convolution theorem,

$$E(\omega) = X(\omega) - H(\omega)Y(\omega) \tag{10}$$

By minimizing $E[|E(\omega)|^2]$ with respect to $H(\omega)$ we have Wiener filter as follows,

$$H(\omega) = \frac{E[|X(\omega)|^2]}{E[|Y(\omega)|^2]} = \frac{E[|X(\omega)|^2]}{E[|X(\omega)|^2] + E[|D(\omega)^2]} \tag{11}$$

Or more generally,

$$H(\omega) = \left(\frac{E[|X(\omega)|^2]}{E[|X(\omega)|^2] + \alpha E[|D(\omega)|^2]} \right)^{\beta} \tag{12}$$

Since, $E[|X(\omega)|^2]$ is unknown and if $\alpha = 1$, $\beta = 1/2$ then $E[|X(\omega)|^2 = |\hat{X}(\omega)|^2$, with the result

$$|\hat{X}(\omega)| = H(\omega)|Y(\omega)| = \sqrt{\frac{|\hat{X}(\omega)|^2}{|\hat{X}(\omega)|^2 + E[|D(\omega)|^2]}}|Y(\omega)| \tag{13}$$

$$|\hat{X}(\omega)|^2(|\hat{X}(\omega)|^2 + E[|D(\omega)|^2]) = |\hat{X}(\omega)|^2|Y(\omega)|^2 \tag{14}$$

gives two solutions $|\hat{X}(\omega)|^2 = |\hat{Y}(\omega)|^2 - E[|D(\omega)|^2]$ or $|\hat{X}(\omega)|^2$ which is essentially the power spectral subtraction algorithm.

If $E[|X(\omega)|^2]$ is replaced by $|\hat{Y}(\omega)|^2 - E[|D(\omega)|^2]$, then Wiener filter is

$$H(\omega) = \frac{E[|X(\omega)|^2]}{E[|X(\omega)|^2] + E[|D(\omega)|^2]} = \frac{\gamma(\omega) - 1}{\gamma(\omega)} \tag{15}$$

where $\gamma(\omega) = \frac{|\hat{Y}(\omega)|^2}{E[|D(\omega)|^2]}$.

2.3. MMSE-STSA

Minimum mean square error based on short-time spectral amplitude (MMSE STSA) estimator is estimator that minimizes the mean square error of the spectral magnitude. It minimizes,

$$min\, E[(X_k - \hat{X}_k)^2] \tag{16}$$

Defined $\gamma(\omega)$ is called the a-posteriori SNR,

$$\gamma(\omega) = \frac{|Y(\omega)|^2}{E[|D(\omega)|^2]} \tag{17}$$

and

$$\xi_k = \frac{E[|X(\omega_k)|^2]}{E[|D(\omega_k)|^2]} \tag{18}$$

MMSE STSA estimator is function of the gain containing $\xi(\omega)$ and $\gamma(\omega)$. Moreover, the full derivation can be obtained from [4].

$$\hat{X}(\omega) = G(\xi(\omega), \gamma(\omega))Y(\omega) \tag{19}$$

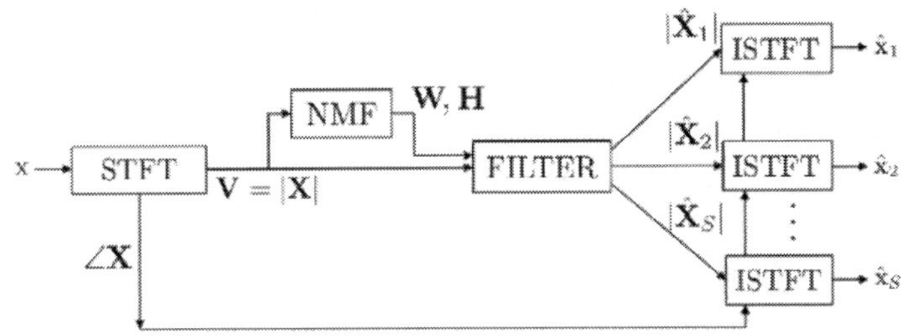

Figure 1. Block diagram of NMF signal enhancement by source separation [8]

2.4. NMF
NMF is matrix Factorization where everything is non-negative. It can be used for signal enhancement by source separation method. Source separation is the core of this work after modeling sound mixture. Separation principle consist of the following steps:

(i) STFT
(ii) NMF (Non-negative Matrix Factorization)
(iii) FILTER (Masking)
(iv) ISTFT

Those steps of signal enhancement by source separation can be organized by the block diagram in figure 1. As seen in the diagram, the source signal can be decomposed into its components, for example $\hat{x}_1$, $\hat{x}_2$ and $\hat{x}_3$. The target signal is the only desired signal while the others can be neglected or assumes as noises.

(i) Unsupervised NMF
In NMF, the matrix output magnitude $|X|$ from STFT process is decomposed into two matrices there are basis vectors W and weights H. Then a subset of basis vectors W_s and activation H_s is chosen to reconstruct source s by estimate the source s magnitude.

$$|\hat{X}_s| = W_s H_s = \sum_{i \in s} (w_i h_i^T) \tag{20}$$

To solve W and H for given a known V, a optimization problem frame is applied as

$$minimize\ D(V||WH) \tag{21}$$

which $W, H \geq 0$ and D is a measure of "divergence" that obtained by Kullback-Leibler, then to minimize

$$D(V||WH) = \sum_i, j \left(v_{ij} log \frac{V_{ij}}{(WH)_{ij}} - V_{ij} + \hat{(WH)}_{ij} \right) \tag{22}$$

Now we just iterate between:

(a) Updating W
(b) Updating H
(c) Checking $D(V||WH)$. If the change since the last iteration is small, then declare the convergence.

(ii) Supervised NMF
In unsupervised NMF, to control which basis vector explain which source is remain difficult, therefore supervised NMF is used. The general idea of supervised NMF is the use isolated

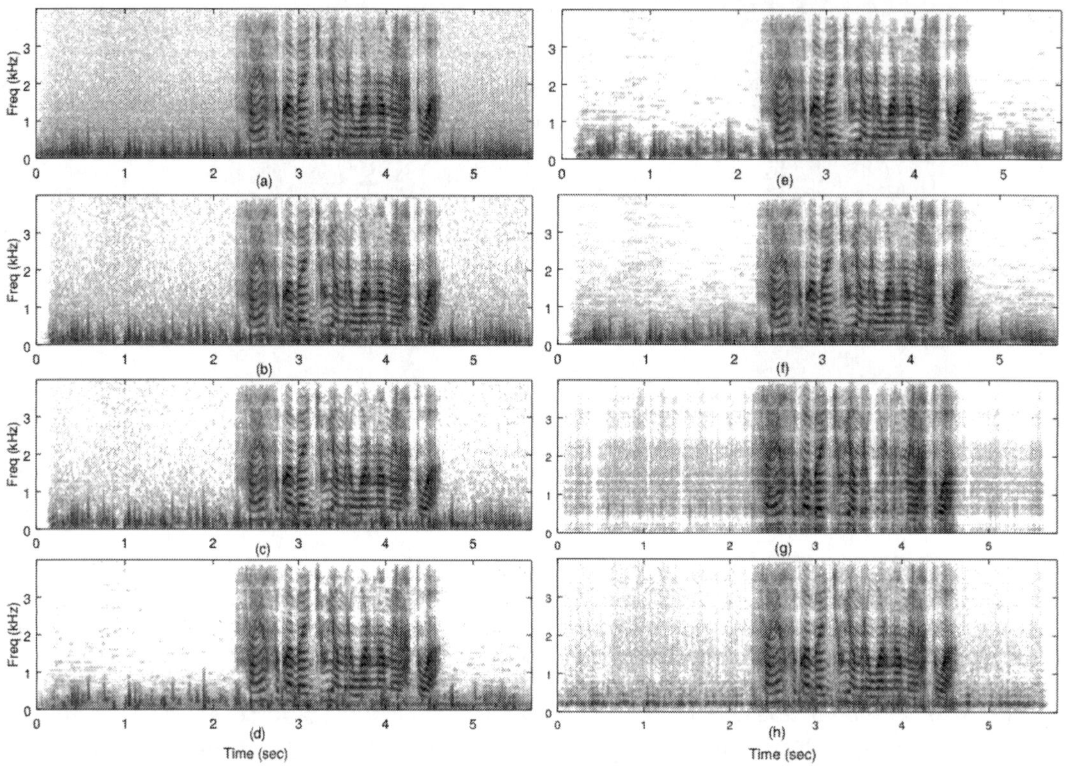

Figure 2. Spectrogram of speech enhancement result: (a) Noisy signal (b) spectral subtraction power (c) spectral subtraction magnitude (d) spectral subtraction over (e) Wiener filtering (f) mmse stsa (g) NMF (h) supervised NMF

training data of each source within a mixture to pre-learn individual models of each source. The process of supervised NMF obtained from [8] consist of the following steps,

(a) Use isolated training data to learn a Factorization (W_sH_s) for each source
(b) Throw away activation H_s for each source s
(c) Concatenate basis vectors of each source (W_1, W_2) for complete dictionary W
(d) Hold W fixed, and factorise unknown mixture of source V (only estimate H)
(e) Once complete, use W and H as before to filter and separate each source

3. Experiment
The experiment is conducted by using smartphone Motorola Razr IS12M to record speech target and fan noise with the voice recording software, Easy Voice Recorder^{TM}. The sampling rate was used at 8000 Hz and recorded in semi anechoic chamber. Sound source used is laptop speaker with Indonesian speech database and background noise from electric fan. This fan noise is made to imitate wind noises when making a call in the road on vehicle. The experiment set-up can be seen in Figure 3. The sound data, Matlab/Octave codes along with .tex files of this paper are openly available at: `http://bitbucket.org/bagustris/icopia2016` in the spirit of open science.

4. Result and Discussion
The recorded sound explained from the previous section analyzed offline by PC computation. Each .wav files were evaluated with seven algorithms explained in section 2 and enhanced

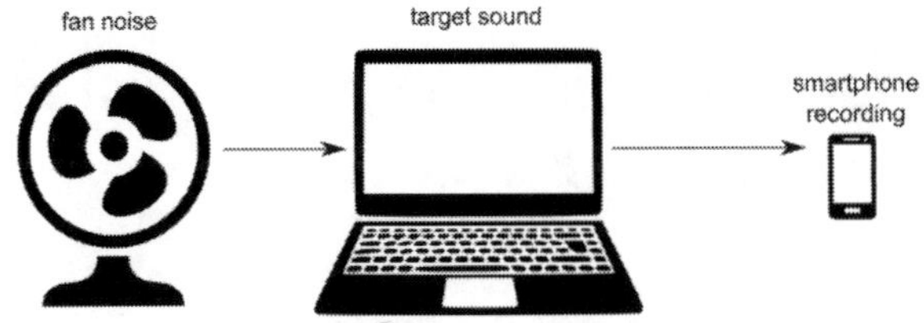

Figure 3. Experiment set-up

Algorithms	PESQ
spec sub pow	2.3982
spec sub mag	2.4493
spec sub over	2.4994
wiener filtering	2.5224
mmse-stsa	2.4922
nmf	1.9156
supervised nmf	2.5843

Table 1. PESQ score

speeches were evaluated by means of spectrogram and PESQ score [7]. Figure 2 shows spectrogram of noisy signal (a) with results from seven different speech enhancement algorithms (b to h).

From the spectrogram, it is shown that NMF algorithms both unsupervised and supervised can suppress noise much more other speech enhancement algorithms. Unsupervised NMF even shows almost no noise in the lower part of spectrogram. However, from the PESQ score, it is clearly shown that supervised has the highest speech intelligibility compared to other algorithms.

The PESQ score also shows that NMF has lowest score although its spectrogram shows it can suppress noise more than other algorithms. When listened, the sound resulted from this unsupervised NMF has degraded voice quality that impact its intelligibility score. However, the spectrogram of enhanced speech from Unsupervised NMF showed filtered noises on lower frequencies more than other methods. We choose PESQ score for objective evaluation because it is the current standard in telecommunication. The listened sound also shows consistency with the PESQ score.

5. Conclusion

This paper review some speech enhancement algorithms from classical to modern method by evaluating its directly in smartphone voice recording. From objective evaluation by using spectrogram and PESQ score, it can be concluded that the modern method, supervised NMF has highest PESQ score and spectrogram similarity compared to the original clean signal. On future research, computation time should be studied and enhanced for real time implementation.

References

[1] Steven Boll 1979 Suppression of acoustic noise in speech using spectral subtraction *Acoustics, Speech and Signal Processing* IEEE Transactions on 27 no. 2, pp. 113—120

[2] M. Berouti, M. Schwartz, and J. Makhoul, 1979 Enhancement of speech corrupted by acoustic noise, *IEEE Int. Conf. Acoust., Speech, Signal Processing (ICASSP âĂŹ79)*, pp. 208—211

[3] Jae S Lim and Alan V Oppenheim 1979 Enhancement and bandwidth compression of noisy speech *Proceedings of the IEEE 67* no. 12, pp. 1586—1604

[4] Y. Ephraim and D. Malah 1984 Speech enhancement using a minimum mean-square error short-time spectral amplitude estimator *IEEE Transactions on Acoustics, Speech and Signal Processing* pp. 1109—1121

[5] P. Smaragdis and J.C. Brown 2003 Non-negative matrix factorization for polyphonic music transcription *IEEE Workshop on Applications of Signal Processing to Audio and Acoustics (WASPAA)* pp. 177—180

[6] P Smaragdis, B Raj, M Shashanka 2007 Supervised and semi-supervised separation of sounds from single-channel mixtures, *International Conference on Independent Component Analysis and Signal Separation* Springer—Verlag pp. 414—421

[7] ITU 2000 Perceptual evaluation of speech quality (PESQ), and objective method for end-to-end speech quality assessment of narrowband telephone networks and speech codecs. *ITU—T Recommendation* P. 862

[8] M. M. Nandakumar and K. E. Bijoy 2014 Performance evaluation of single channel speech separation using non-negative matrix factorization *2014 National Conference on Communication, Signal Processing and Networking (NCCSN)* pp. 1—4

Sound absorption enhancement of nonwoven felt by using coupled membrane – sonic crystal inclusion

M.C. Fitriani[1], I. Yahya[1*], Harjana[1], Ubaidillah[1], F. Aditya[2], Y. Siregar[3], M. Moeliono[3], and S. Sulaksono[3]
[1]The Iwany Acoustics Research Group (iARG), Department of Physics, Faculty of Mathematics and Natural Sciences, Universitas Sebelas Maret, Jalan Ir. Sutami 36A, Kentingan, Surakarta 57126, Central Java, Indonesia
[2]PT. Rekadaya Multi Adiprima, Jakarta
[3]Textile Reseach Center (BBT), Bandung

E-mail: iyahya@mipa.uns.ac.id

Abstract. The experimental results from laboratory test on the sound absorption performance of nonwoven felt with an array thin tubes and sonic crystal inclusions reported in this paper. The nonwoven felt sample was produced by a local company with 15 mm in its thickness and 900 gsm. The 6.4 mm diameter plastic straw was used to construct the thin tubes array while the sonic crystal is arranged in a 4 x 4 lattice crystal formation. It made from a PVC cylinder with 17 mm and 50 mm in diameter and length respectively. All cylinders have two holes positioned on 10 mm and 25 mm from the base. The results show that both treatments, array of thin tube and sonic crystal inclusions are effectively increased the sound absorption coefficient of the nonwoven felt significantly especially in the low frequency range starting from 200Hz.

1. Introduction

The patch from textile industries has a potential economic value. Since it made from yarn through an industrial machining process, it can be recycled to produce fibrous sound absorber for various applications including automotive, shipping and building acoustics [1]. Nonwoven felt is a common material that widely uses as a sound absorber in automotive industries such as for doors, roofing, flooring, hood and much more. It made from the mix of the textile patch with a certain commercial binder which is melting on 110°C-200°C [2].

Despite the fibrous structure brings a very good performance on noise reduction for the high frequency band, it needs a more advanced approach to making the nonwoven felt perform the better absorption at the mid and low frequency range. The various approach has been developed by previous researchers accordingly. For example, the inclusion of permeable and perforated membrane have reported by Sakagami et al (2014) and Harjana and Yahya (2015) respectively. It found that the membrane structure affect to the reactance of the cavity backed perforated absorber and produces significant increment to the sound absorption performance on the mid to high frequency range above 500 Hz [3,4].

Another approach by using short thin constrained tube has been reported by Yahya and Harjana (2013) and followed by the compartmented cavities approach by Li et al (2015, 2016). This cavity resonance approaches successfully shifting the sound absorption performance to 200 Hz but it has industrial constraint due to the geometrical design which is impractical for mass production [5-7]. The most attractive approach among researchers recently is sonic crystal assisted sound absorber which is featuring possibility for tuning the local Helmholtz resonant to reduce the noise on certain intended frequency band [8-11]. In this paper, a combined permeable membrane, thin tubes array, and sonic crystal inclusion approach are reported. The proposed combining of three different noise reduction techniques in a single integrated inclusion design is expected to produce a much better sound absorber concept which is less constraint in the industrial perspective.

2. Methods

The transfer function based two microphones technique refer to ASTM E1050 was conducted to measure the sound absorption coefficient of the test sample. Since this work is focused on the low frequency band up to 1.6 kHz, the large tube of B&K 4206 was utilized in the laboratory test. The membrane layer was produced by hand laying up the technique of a commercial leak-proof product on the surface of the nonwoven felt. The layering steps have been repeated for three times followed by an open air drying process. The nonwoven felt has 15 mm in thickness and 900 gsm produced by a local manufacturer while the thin tubes and sonic crystal elements are made from plastic straw and PVC tube of 6.4 mm and 17 mm in diameters respectively. The sonic crystal element has a single geometrical size 50 mm in the height with a 2 mm hole on the wall. It varying in two different positions of 10 mm and 25 mm from the bottom end as illustrated in Figure (1).

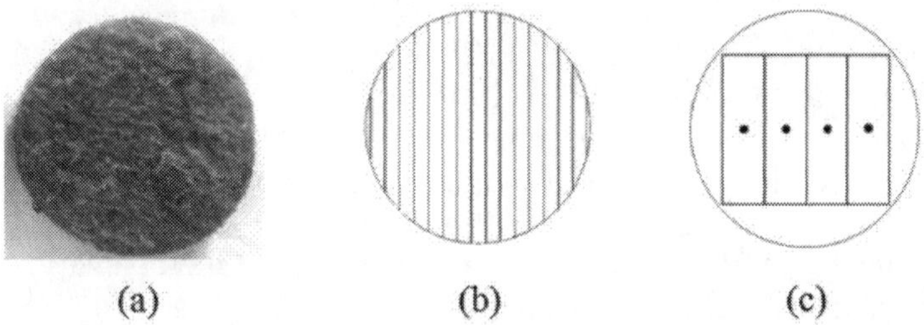

Figure 1 The test samples: (a). Nonwoven felt, (b). Array of thin tubes, (c). The sonic crystal model

The laboratory setup is presented in Figure (2). B&K 3160-A LAN-Xi type was used in connected with the power amplifier B&K 2716C and impedance tube B&K 4206. Random noise is generated from the loudspeaker and the incident and reflected waves are decomposed from the captured signals by a pair of quarter-inch microphone B&K 4187.

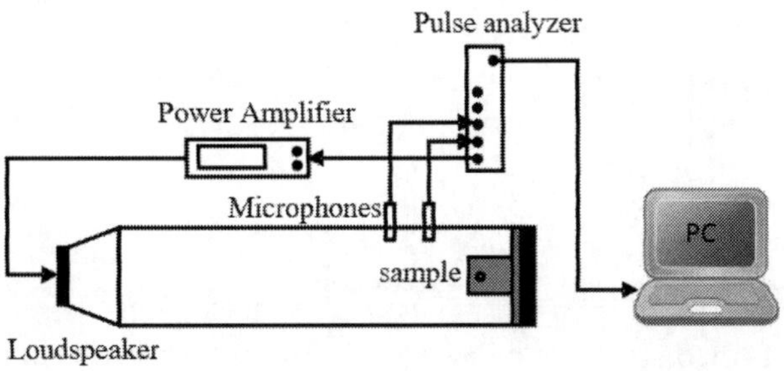

Figure 2 laboratory setup for sound absorption measurement.

There is three frequency response are being calculated in the decomposition process: frequency response H_1, frequency response associated with incident waves H_i, and the frequency response function associated with the reflected wave H_r.

The reflection coefficient is given by,

$$R = \left(\frac{H_1 - H_i}{H_r - H_1}\right) e^{j2k(l+s)} \quad (1)$$

where k is the wave number, l is the distance of the first microphone to the sample surface, and s is the spacing between microphones. The normalized impedance ratio and sound absorption coefficient can be calculated by using the following equation,

$$\frac{z}{\rho c} = \frac{1+R}{1-R} \quad (2)$$

$$\alpha = 1 - |R|^2 \quad (3)$$

3. Results and Discussion

Figure (3) shows that the nonwoven felt does not perform as an effective sound absorber for the low frequency range. It happened since the sound waves passing through the material without energy dissipation. The viscous damping did not occur due to the wavelength are much longer compared to the sample dimension. As the surface of the nonwoven felt covered by the leak-proof silicon layer, the sound absorption coefficient increased significantly. It happened for all frequency starting from 400 HZ and reach the maximum value of 0.75 in the frequency range between 700 Hz to 1.2 kHz.

This phenomenon is in between of the similar results as previously reported by Sakagami et al [3] and Harjana and Yahya [4]. It means that silicon layer brings multiple impacts to the nonwoven felt properties. It does not only affect the surface impedance value but the more significant it is produced membrane type resonance on the surface of the nonwoven felt. As the membrane oscillates on its resonant modes, the fibrous structure of the nonwoven felt below the membrane are also oscillate in the same mode accordingly. It then increases fibrous damping by the nonwoven fibers and porosity and resulting the much higher sound absorption coefficient.

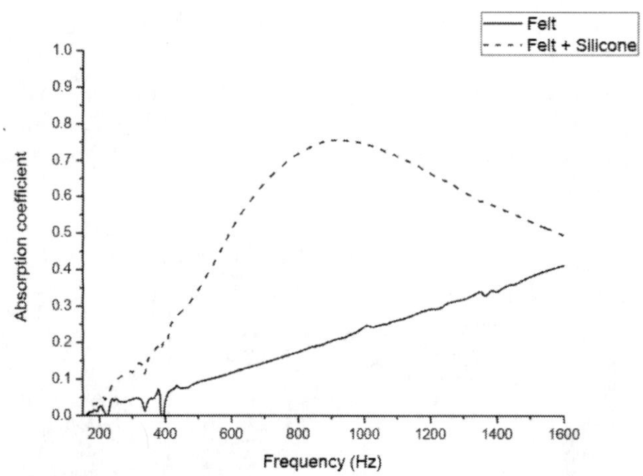

Figure 3 Impact of silicon layering on the sound absorption coefficient of nonwoven felt.

Another result presented in Figure (4). The sound absorption coefficient are slightly increased at two different frequecy range. The first one in 500 Hz to 800 Hz which is associated with coupling effect of the thin tubes array with the surrounding cavity. The second increment on the high frequency band around 1.4 kHz and 1.6 kHz are related to the resonant mode of the half wavelength resonators of the thin tubes array.

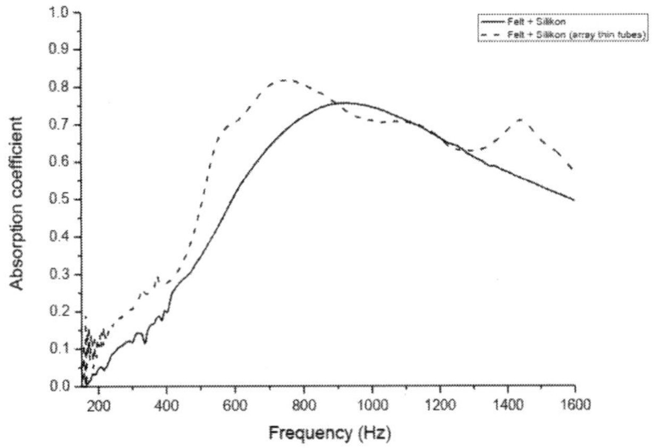

Figure 4. The impact of coupling effect and resonant mode of thin tubes array on the sound absorption.

It is clear from Figure (4) that coupling effect of the thin tubes array which is half wavelength resonators with the surrounding cavity including the membrane layer on the surface of nonwoven felt tends to shift the response. The resonance absorption occurs simultaneously due to the cavity, the thin tubes array, and the membrane. It changes the reactance of the coupled structure and shifting the response to the lower frequency. In the same time, the fibrous and porous layer of nonwoven felt also oscillated and increased the sound energy dissipation into heat. It brings a better understanding of the possibility of manipulating the resonance absorption and coupling effect as the new approach for increasing the sound absorption coefficient of the nonwoven felt.

This new approach was then implemented to create the another test model with the consideration of using the sonic crystal element instead of thin tubes array. It based on a fundamental reason related to the local Helmholtz resonant of the sonic crystal. In addition, it also considering to the scattering and multiple internal reflections which are both not occurred with the thin tubes array.

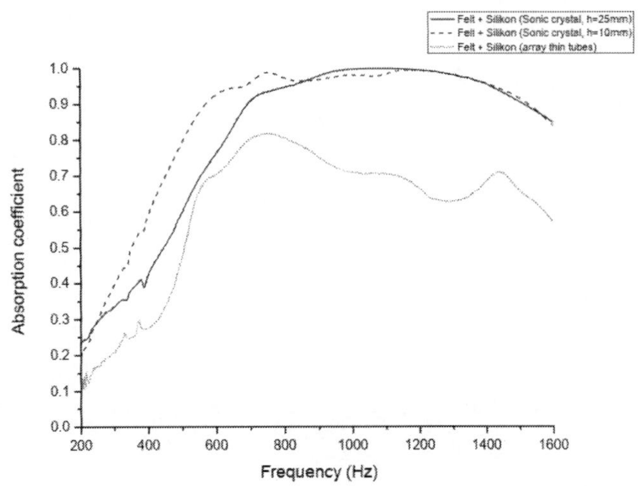

Figure 5 The sound absorption coefficient value on material nonwoven felt with thin tube array and sonic crystal.

The result is presented in Figure (5). The proposed approach with coupled of membrane layer with the sonic crystal inclusion has a very promising performance. The sound absorption coefficient increases at the all frequency range with the smallest absorption coefficient of 0.24 for the frequency of 200 Hz. This result is much higher compared to the initial nonwoven felt which is less than 0.05 for the same frequency. It also clear from Figure (5) that the sound absorption coefficient increase rapidly from 0.24 on 200 Hz to 0.9 on 500 Hz and then featuring flat response on the rest of the frequency range above it.

This evidence emphasizes and concludes that the sound waves energy are effectively dissipated by coupling of the membrane layer, the local resonant, and also by fibrous and porous nonwoven felt. In term of industrial application possibility, it is clear that since the proposed approach consists of a single geometrical dimension sonic crystal element, it much easier to mass production compared to the compartmented cavities and other techniques as proposed by previous researchers.

References

[1] Kucuk M, and Korkmaz, Y, 2015. *Fibers and Polymers* **16**(4) 941-948.
[2] Albrecht W, Fuchs H, and Kittelmann, W, 2003 Nonwoven Fabrics. *Wiley-VCH Weinhelm.*
[3] Sakagami K, Fukuutani, Y., Yairi, M., Morimoto, M, 2014. *Applied Acoustics* **79**. 104-109
[4] Harjana, and Yahya I 2015. 7th ICOPIA, Solo, Atlantis Press. 60-64
[5] Yahya I and Harjana, 2013 ICSV20. Bangkok.
[6] Li D, Chang D, Liu B., and Tian, J, 2015. ICSV22. Florence
[7] Li D, Chang D, and Liu B. 2016. *Applied Acoustics* **102**. 126-132
[8] Redondo J, Pico R, Sanchez-Morcillo, V. J, 2013. *J. Acoust. Soc. Am.* **134**(6). 4412-4417
[9] Morrandi F, Miniaci M, Marzani A, Barbaresi L, Garai M, 2016. *Applied Acoustics* **114**. 294-306
[10] A., Gomez, P. M., Spiousas, I., Eguia, M. C, 2016. *Applied Acoustics* **104**. 1-5
[11] Wang, Z. G., Lee, S. H., Kim, C. K., Park, C. M., Nahm, K, and Nikitov, S. A, 2008, *J. of Appl. Phys.* **116**. 064907 (9 pages)

Beam response analysis of moving vehicle with half car modeling

A.N. Badriyah, D. Arifianto, Y. Susatio
Dept. of Engineering Physics, Faculty of Industrial Technology, Institut Teknologi Sepuluh Nopember, Kampus ITS Sukolilo, Surabaya 60111, Indonesia

E-mail: alinda14@mhs.ep.its.ac.id

Abstract. There were several tragedies concerning damages of bridge which seem to be sooner than the predicted period. One of hypothesis in this situation is an addition of vibration caused by long vehicle such as super long truck which has huge force transferred into the bridge and its long body causes more vibration due to phase difference of front and rear tire. The selected method which is used in this problem is using a simulation for modeling a bridge-vehicle system using half car vehicle model. The simulation is done using ANSYS Workbench 15.0 with some variation such us the thickness of beam and its supports. There are 3 kind of variation used in the thickness variety which are 2 m, 1 m, and 0.5 m. While in supports variation, we have fixed support, knife-edge support, and slider support. The results show that there is addition of vibration caused by long vehicle. It is proved by an oscillation which is showed in every response of beam's total deformation. Highest total deformation is achieved in slider support beam of 0.5 thicknesses, 1.08 mm in 1.12 second. First ripple seen in responses is at 0.84 second. Meanwhile, response of knife-edge and fixed support beam show a similarity. The ripple in this situation is caused by beat modulation from the front and rear tire.

1. Introduction

There are two crucial parts of a bridge, truss and beam. Beam is planar transverse vibrations of one-dimensional elastic continua [1]. In bridges structure, beam is a deck part which accept direct contact with force and vibration from vehicle above it. From elementary theory of elasticity, it is known that when the beam is deflected, certain hypothetical longitudinal lines or fibres are elongated, while others are compressed [2]. From the previous researches, it only used a static mass to see the beam responses. But in reality, vehicles are moving and it can have more or less responses than the static vehicle. Even though there is a suspension system in each vehicle, there are still some vibrations left which is continued into the bridge. This will be discussed in this research as a force that transfer from vehicle's weight or weight force. The differences of mass in order to build a force is directly influence beam responses while in this chapter, the selected vehicle is long truck with huge mass and force. It is a control variable. Variation of beam's thickness and support are done in total 9 models and simulation. Modeling is done by computational simulation based on Finite Element Method which can calculate beam's responses with an output of deformation in time domain. Moving vehicle is the main problem in this research. Beam responses of moving vehicle in bridge-vehicle system are going to be analysed in the further explanation considering each varieties of model.
In previous papers, discussion about beam's deformation is rarely done. Almost all research is talking about beam's natural frequency or modal analysis in bridge's system. So, in this research, I would like

to enhance more about beam's responses in different perspective such as deformation in total direction using analysis method in order to know better about beam's characteristics when there is addition vibration applied at top of the bridge.

2. Modeling Techniques

The general characteristics of the FE modeling techniques used in this research are discussed in this section. Specific information includes geometry of beam for each variation and force setting in simulation.

A. General Specification

The focus of this paper is to discuss beam responses using simulation. Simulation of all models are done by refining 3D finite element modeling that has been performed using package ANSYS [3]. In this particular bridge, modeling was running by half car type as a vehicle because full car analysis can be more complicated and need more consideration from simulation step. Structural behavior is modeled by considering the structure as a whole, with actual boundary and loading conditions included in the finite element [3].

B. Physical Model

Physical model from bridge's beam will be shown below.

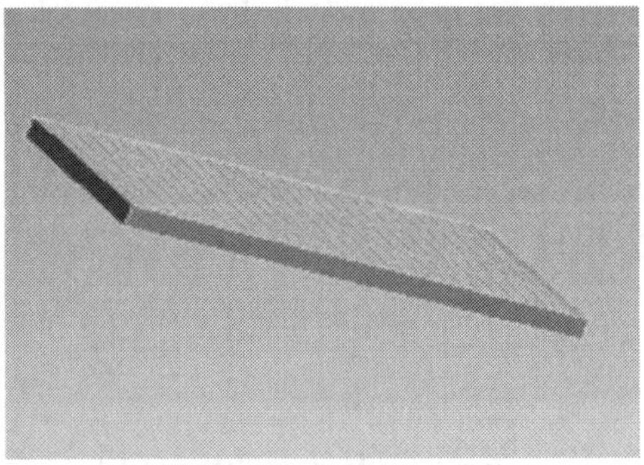

Figure 1. Beam modeling

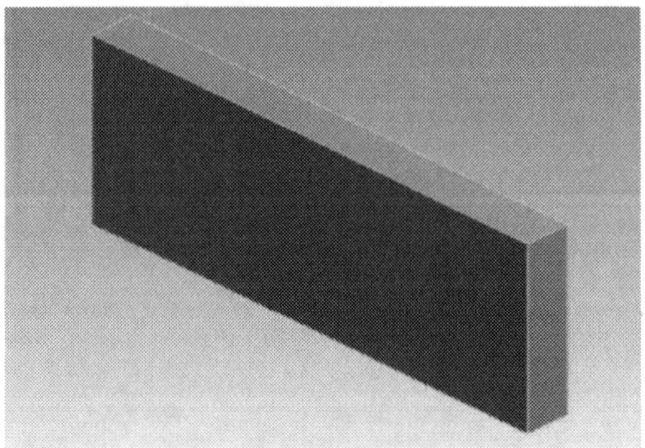

Figure 2. Two meters thickness of beam

The geometry of beam in this research is decided as 22 meters long and 10 meters' width. While thickness of beam will be the variation that is made in order to understand its influence to the response. There are three kinds of thickness which are used, 2 meters, 1 meter, and 0.5 meter.

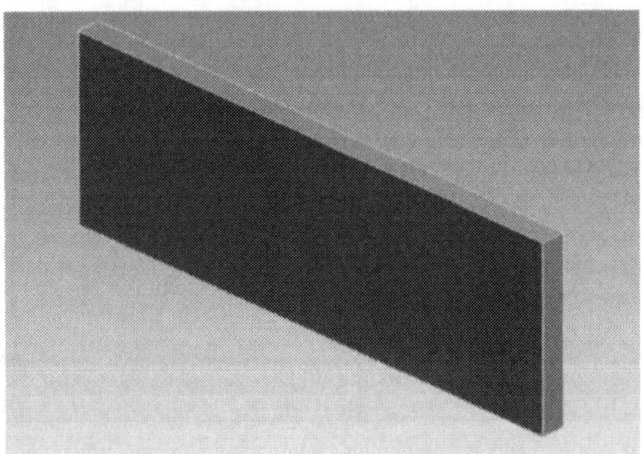

Figure 3. One meter thickness of beam

Figure 4. 0.5 meter thickness of beam

Another variation that will be discussed in this paper is beam support. Initially, there are five kinds of support in beam [4], but when we look up to the reality, there will be mostly three kinds of beam support that is used, fixed support, knife-edge support, and slider support. In this research, there three support

Journal of Physics: Conference Series **776** (2016) 012074 doi:10.1088/1742-6596/776/1/012074

will not stand alone by itself but there will be a combination in one of the model, slider support. For the slider support beam, the slider support will not be set in both edge of beam, but only in one edge. That one is the end of the bridge, the edge when the vehicle is leaving the bridge.

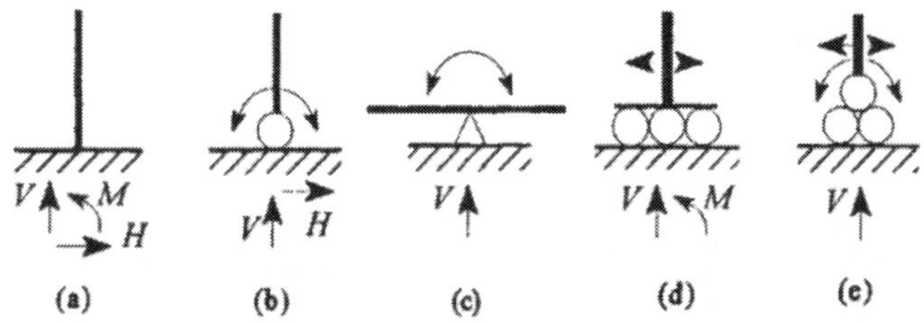

Figure 5. **(a)** *Simple-supported beam* **(b)** *pin support beam* **(c)** *knife-edge support beam* **(d)** *slider support beam* **(e)** *pinned slider support beam* [4]

3. Results and Discussion

The result in this system analysis will be presented by both graph and contour. Deformation in contour result is shown by different color choices as we see below.

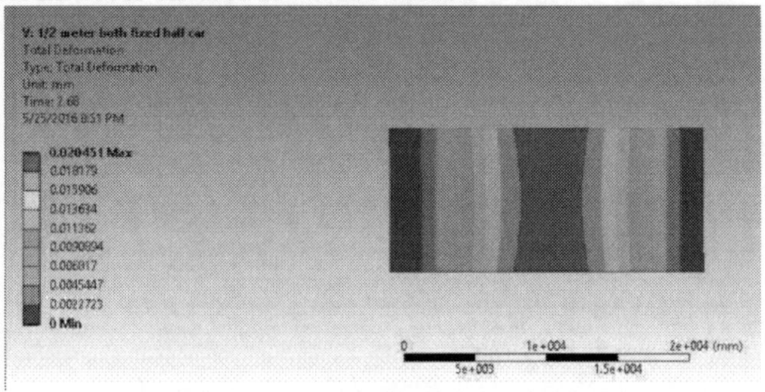

Figure 6. Deformation contour of beam

In order to see the differences from all the models, beam responses will be shown in one graph per variation. So, there are three graphs from thickness variation and another three graphs from support variation.

A. Beam responses of thickness variation

The differences of deformation from each support shown by all kinds of thickness will be shown in graphs below.

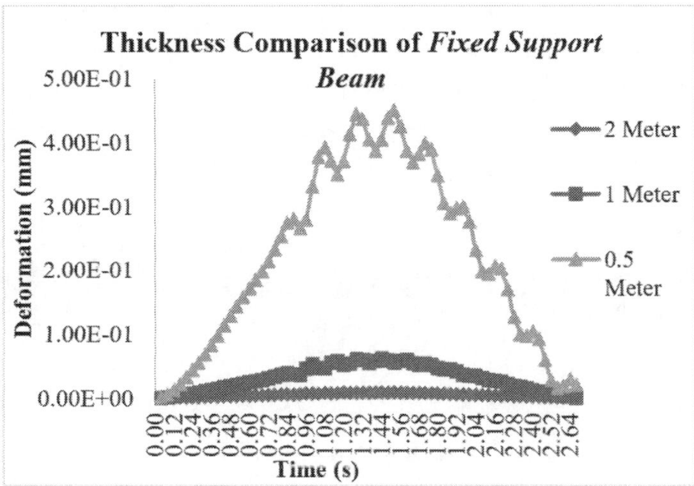

Figure 7. Fixed support beam responses of thickness variation

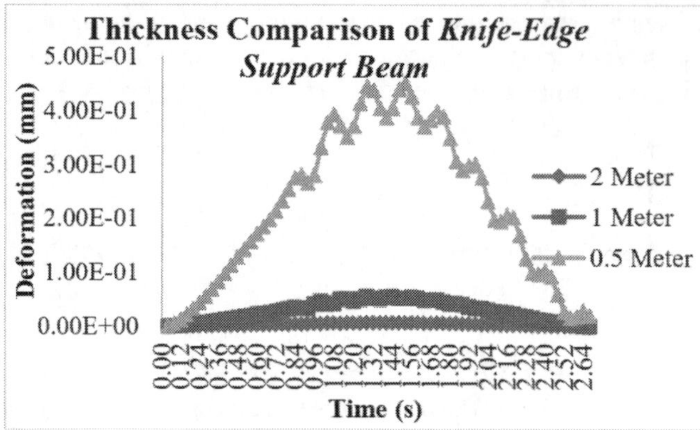

Figure 8. Knife-edge support beam responses of thickness variation

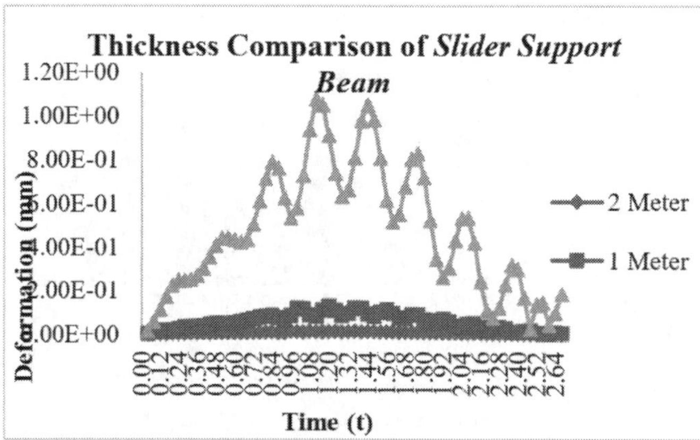

Figure 9. Slider support beam responses of thickness variation

There are significant difference in all three kinds of thickness in each support of beam. The similarity of all graph is shown in 0.5 thickness. It is the model that has the biggest deformation in all thickness variation. Nevertheless, there are different deformation too in 2 meters and 1 meter thickness, but the value is not that significant. There are correlation between thickness of beam and deformation. Increasing of beam thickness caused decreasing of beam deformation by the same condition and force.

B. Beam responses of support variation

The differences of deformation from each thickness shown by all kinds of support will be shown in graphs below.

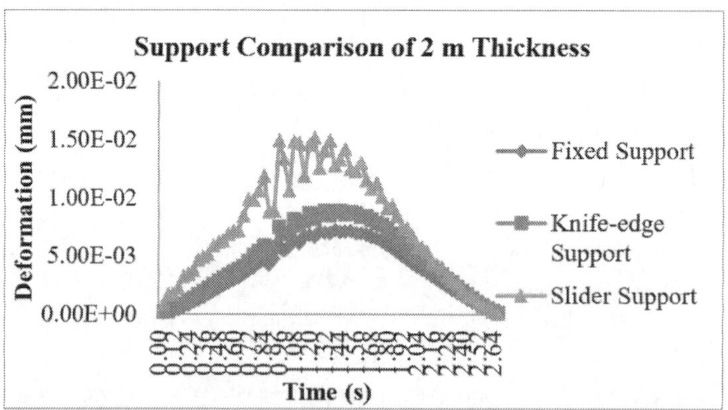

Figure 10. Two meters thickness beam response of support variation

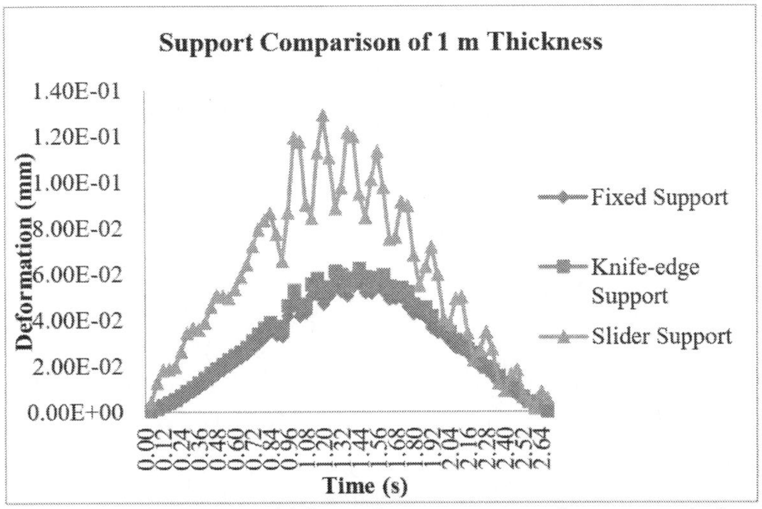

Figure 11. One meters thickness beam response of support variation

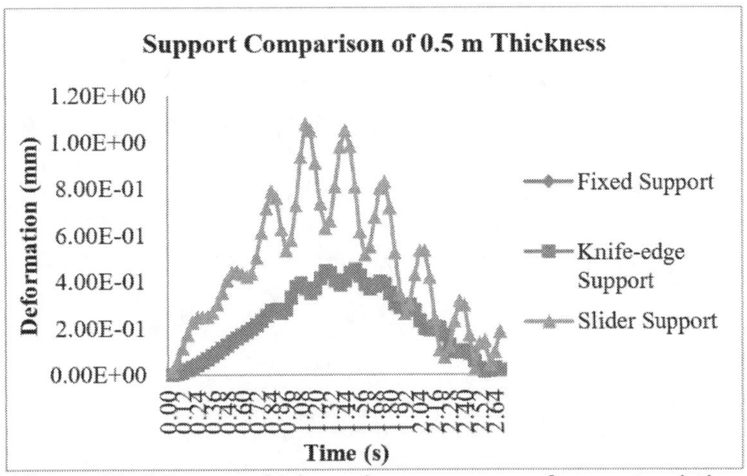

Figure 12. 0.5 meters thickness beam response of support variation

The similarity of responses is found in two kinds of support, fixed and knife-edge. These two beam deformation are almost having the same value in all kinds of thickness. Even in the 0.5-meter thickness beam response, knife-edge and fixed support have exactly the same deformation value so the graph seems to be overlapped. Meanwhile, maximum deformation is shown by slider support beam.

C. Discussion

There is one specific characteristic of beam responses which are analyzed before, oscillation. Oscillation is another name for vibration [1]. This phenomenon has been proved the initial hypothesis about addition of vibration in bridge's beam caused by half car modeling (front and rear tire). Oscillation in this responses is caused by the effect of new waves from rear tire of vehicle which build a new exactly same as the wave from front tire, but has different phase or velocity. This is called beat modulation [4]. Beat modulation is changing first and second wave to a new wave, so it is adding new vibration to the beam.

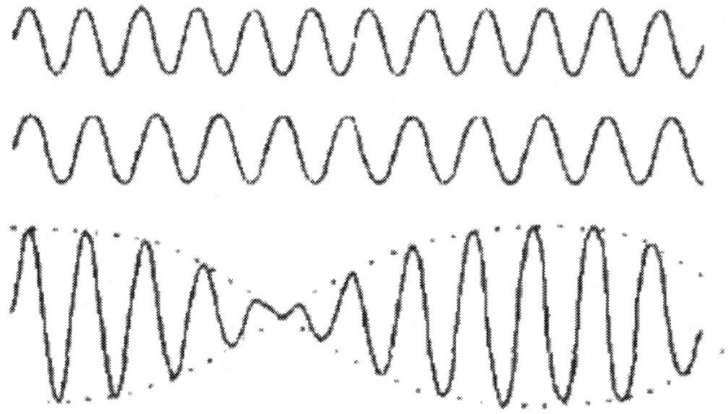

Figure 13. Beat modulation [4]

We can conclude that there are several advantages after I run this research in simulation method. First, we know that there is addition vibration when a half car model going through a beam bridge. It's different with a quarter can responses that has no ripple in it. This phenomenon has added more vibration in beam and influence its total deformation. Second, there is a connection between deflection or deformation in bridges and thickness of beam. When the beam's thickness is increasing, response of beam, which is its total deformation decrease. And when beam's thickness is decreasing, its total deformation will increase. Also we get another conclusion according to the variety of beam's support. Different kind of support determines different responses of beam, in term of total deformation.

References

[1] DaGupta A and Hagedorn P 2007 *Vibration and waves in continuous mechanical system* London Wiley
[2] Kreyszig E., 1969 *Advanced Engineering Mathematics*, 2e, Wiley Eastern Pvt. Ltd., New Delhi.
[3] Yu Z, Des R., Su T, and David C 2009 Predicting Ultimate Punching Failure in Concrete Bridge Deck Slabs through Nonlinear Finite Element Analysis *International Conference on Engineering Computation*
[4] John RD 2002 *Elastic Beams and Frames* Department of Engineering Science University of Oxford. Oxford
[5] Musharraf Z, Taheri MR and Khanna A 1996 *Appl. Math. Model.* Vol. 20(12) pp. 877-889
[6] Chondros TG, Dimarogonas AD, Yao J. 2001 *J Sound Vib* Vol 239(1) pp. 57–67.
[7] Pugno N, Surace C, Ruotolo R. 2000 *J Sound Vib* Vol. 235(5) pp 749–62.

Journal of Physics: Conference Series **776** (2016) 012075 doi:10.1088/1742-6596/776/1/012075

Localization of underwater moving sound source based on time delay estimation using hydrophone array

S. A. Rahman[1], D. Arifianto[1], T. Dhanardono[1], Wirawan[2]
[1] Engineering Physics Department, Faculty of Industrial Technology, Institut Teknologi Sepuluh Nopember
[2] Electrical Engineering Department, Faculty of Industrial Technology, Institut Teknologi Sepuluh Nopember Jl. Arief Rahman Hakim, Surabaya, 0111

E-mail: saptianrahman@gmail.com

Abstract. Signal and noise of an underwater moving sound source is used to track the azimuth of a target. Uniform linear array with four hydrophones is used to detect azimuth of target by obtain the time delay information to get azimuth information. Success rate of time delay estimation influenced by characteristics of sound propagation like reflection, reverberation, etc. Experiment in real environment was done to analyze performance of the cross correlation (CC) and generalized cross correlation with the phase transform (PHAT) weighting to estimate time delay between two signal. The simulation done by convolute two signal that has been given time delay and impulse response of the medium test. Then the time delay of two signal estimated by CC and PHAT algorithm in Matlab in the various SNR. Then the algorithm tested in a pool to detect stationary and moving position of sound source. Result of the simulation and experiment in real environment shown that PHAT better than CC. The best azimuth tracking achieved by using PHAT algorithm with error of 0 - 9.48 degree in stationary position. In moving sound experiments, tracking the bearing and azimuth ofthe mini vessel (sound source) can be done by time delay estimation using PHAT.

1. Introduction

The existence of an object in the water can be detected by its acoustic signal emission. The resulting emission flows in the form of fluid particles pressure on a medium so can be detected by a hydrophone in variable frequency and gain [1]. There are various methods to detect position of the of the sound source, one of them is the estimated time delay. Time delay estimation has advantages compared with other methods to detect the position of the sound source such as Capon 's method or MUSIC. The advantage is more suitable to estimate time delay of broadband signals type, such as vessels signal. Azimuth tracking of underwater sound sources can be obtained based on the estimated time delay using multiple sensors distributed spatially, so that able to capture coming of sound signal [2]. Then value of time delay obtained and converted into azimuth of sound source relative to the hydrophone array. The most basic TDE methods is the cross-correlation (CC). The CC method assumes an ideal sound propagation model which makes it suitable for low-noise and reverberant-free outdoor environments. The improved version of the CC method is the generalized cross correlation (GCC) method [3].

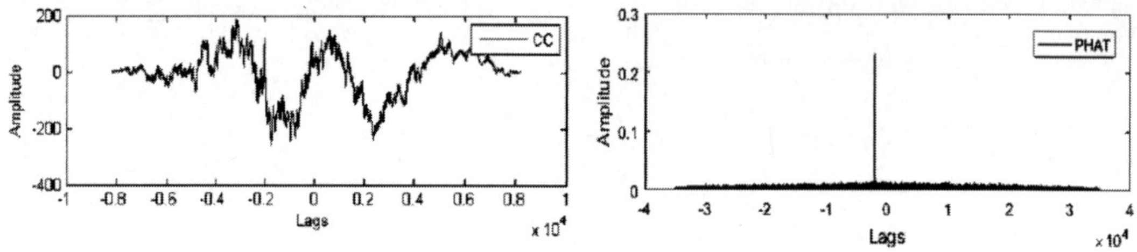

Figure 1. Comaprison peak detection between CC and PHAT algorithm

The GCC is algorithm which applies various weight functions to the received signal (pre-filters) in order to improve time delay estimation in noisy or reverberant environment. But, GCC method more complex computation than CC method [3].

Success rate of time delay estimation influenced by characteristics of sound propagation like reflection, reverberation, etc. By using PHAT method, accuracy of time delay estimation able to improve on all kinds of signals (narrowband and broadband). As the study conducted by [4] that PHAT method can be used to track a moving sound source on the high seas with very well. In that research, the medium used is very broad so that can reduce effect of reflection, reverberation, and changes of medium volume due to the movement of a sound source. Hence, needs to be done research on tracking position of the moving sound source in the pool so that the effect of reflections or noise on the environment to estimate time delay can be known. CC and PHAT method chosen to estimate the delay time in this study.

2. Time Delay Estimation

When a sound source is emitting sound in a field, two signals x1(t) and x2(t), received at two spatially separated hydrophones, 1 and 2, can mathematically be modeled as:

$$x_1(t) = s_1(t) + n_1(t) \tag{1}$$

$$x_2(t) = \alpha s_1(t + d) + n_2(t) \tag{2}$$

where, $x_1(t)$ are the signals picked up by the microphones m_1, m_2 respectively, $s_1(t)$is the sound signal, α is the attenuation factor, s(t) is the signal (direct path) from the source, $n_1(t)$ and $n_2(t)$ are assumed to be mutually uncorrelated noises and d is time delay. This model does not consider multipath or Doppler effects. When there are two microphones are used, it can be estimated based on the delay time difference from propagation of sound waves.

2.1 Cross Correlation

One of the methods commonly used to estimate the time delay d is to calculate the cross-correlation function between the signals received at two microphones. Then look for the maximum peak on the output which is the estimated time delay [5]. Cross power spectrum density function is calculated by the following equation.

$$R_{12}(\tau) = E[x_1(t)x_2(t - \pi)] \tag{3}$$

The time delay can be estimated from $R_{12}(\tau)$:

$$\tau_{12} = \arg max_{\tau ED}\, R_{12}(\tau) \tag{4}$$

2.2 Generalized Cross Correlation Phase Transform

Based on the relationship between the cross correlation and cross power spectrum, can be obtained by the following equation:

$$R_{12}(\tau) = \int_0^{\pi} G_{12}(\omega) e^{j\omega t} d\omega \tag{5}$$

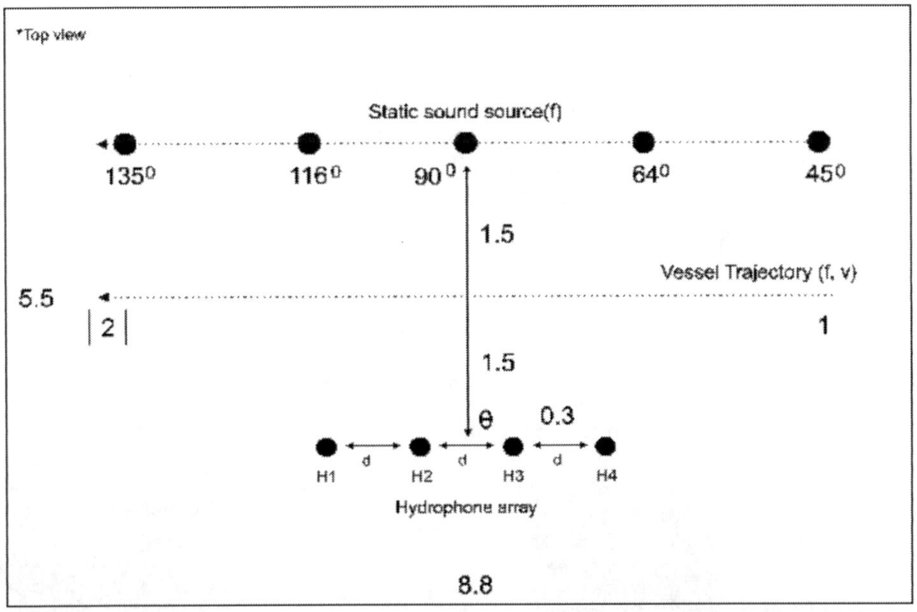

Figure. 2 The experiment setup: stationary and moving sound source straight trajectory

Where G12(ω) is a cross - spectrum x1(t), x2(t) signal captured by the sensor m1 and m2. By looking at the cross- power spectrum between the two signals and then provide weighting corresponding to the frequency domain, the method of whitening the GCC process and array signal thereby increasing the frequency components with higher SNR and inhibiting the effects of the array. Recently performed inverse transform to the time domain to obtain the GCC function of two signals [5]:

$$R_{12}^{g}(\tau) = \int_0^{\pi} \psi_{12}(\omega) G_{12}(\omega) e^{j\omega t} d\omega \qquad (6)$$

Where $\Psi_{12}(\omega)$ is a weighting function GCC. Selection of weighting function depends on two aspects: the array and the condition of the reflectance of the sound source. Shown at Figure 1, the purpose of the use of weighting functions is to make $R_{12}(\tau)$ has a relatively sharp peaks so as to facilitate estimation of the time delay. R_{12} peak (τ) is the time delay between the two sensors. Weighting is very important to obtain a high SNR value it based on field conditions, where the presence of factors such as echo or reverberation time coming from various sides can reduce the value of SNR [5]. Cross correlation function will calculate the peak of the main signal and the reflected signal to estimate the time delay would be difficult to do at a low SNR values. Therefore, needed a weighting Ψ_{12} (ω) to suppress the effects of the array and the reverberation time. GCC PHAT types (Phase Transform) is a type of weighting is most commonly used with the following equation:

$$\Psi_{12}(\omega) = \frac{1}{|G_{12}(\omega)|} \qquad (7)$$

Then do the inverse transform to the time domain to obtain the GCC PHAT function of two signals, the time delay can be estimated from $R_{12}(\tau)$:

$$\tau_{12} = arg\, max_{\tau ED}\, R_{12}(\tau) \qquad (8)$$

3. Experimental Setup

The measurement site done at pool, its bottom in the form of sedimentary soil a depth of 0.5 meters and the wall in the form of concrete.

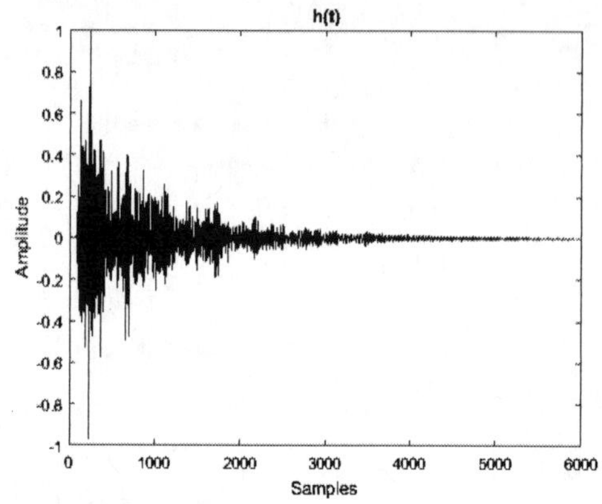

Figure 3. Impulse response of experimental pool

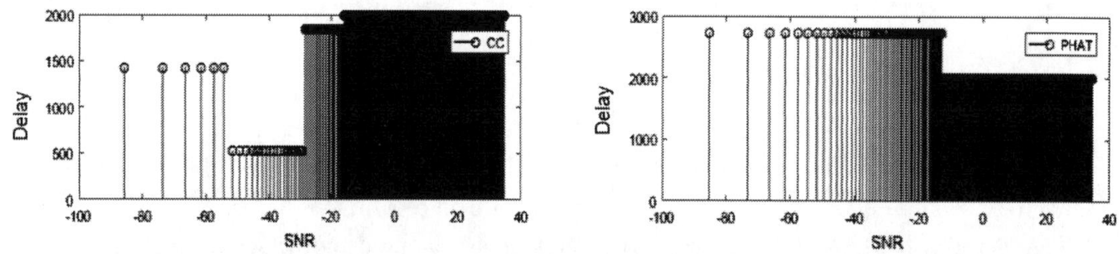

Figure 4. Influence of SNR concerning time delay estimation

Experimental scheme shown at Figure 2. Four hydrophones formed uniform linear array, d = 30 cm. The hydrophones were omnidirectional and placed at 0.25 cm below the water surface. A sampling frequency of 44.1 kHz with a receiver dynamic range of 16 bits used. Increasing the sampling rate improves the TDE resolution which in turn lead to a higher DOA resolution. There are two sound source in this experiment, underwater speaker as stationary and remote control vessel as moving sound source. In the figure, the unit in meters. To represent sound of the vessel that were dominant at low frequencies, generated four frequency variation: 200, 500, 800, 1000 Hz in the stationary experiment. In the moving experiment, used remote control vessel type Sea Wolf Submarine with dimension 35 x 9 x 9 cm and rotational speed of the propeller between 40 – 50 Hz. The vessel move from point 2 to point 1 with angular speed as big as 0.2 m/s and at the time the sound of the vessel recorded.

4. Result and Discussion

4.1 Simulation Results

Two algorithms have been simulated in Matlab, which are Cross Correlation (CC) and Phase Transform (PHAT). In order to simulate the real situation, a pure tone 1000 Hz has been chosen as a sound source signal. Then the sound source convolute by impulse response of the medium so as if generated in the real medium. An additional noise with various signal to noise ratio (SNR) was used to evaluate the methods. A white noise/ Gaussian noise was added to each sound signal. The two noise signals are assumed to be uncorrelated and to have a zero mean value. One of the two sound signals is delayed 2000 samples to simulate the time that the signal needs to reach the further microphone in the real environment. Figure 4 shows the simulation results, it can be seen that all methods was able to estimate the time delay until negative SNR. Figure 4 shows that, as the SNR becomes larger, the localization error becomes smaller. The result of simulation is given in Figure 4.

4.2 Experimental Results

In this subsection, we present the results of time delay based localization experiment. We convert estimated time delay into azimuth information. Consider a setup for DOA estimation consisting of an array of four hydrophones separated by a small distance d and a source that lies in the far field of the array. The azimuth $\boldsymbol{\theta}$ is related to the time delay, which for 1D microphone array is expressed as.

Table 1. Tracking result of stationary sound source

Actual Azimuth	Measured CC	Measured PHAT	Deviation CC	Deviation PHAT	Error CC	Error PHAT
45	70.11	40.14	28.73	0.00	25.11	4.86
64	73.10	61.66	0.00	0.00	9.10	2.34
90	91.50	93.33	17.53	0.00	1.50	3.33
116	136.95	125.48	28.80	1.01	20.95	9.48
135	118.45	135.93	24.90	13.81	16.55	0.93

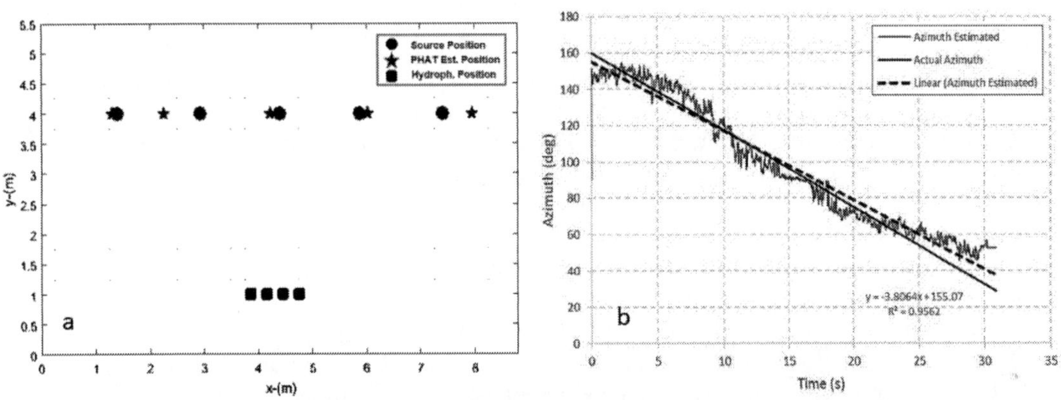

Figure 5. Tracking result of (a) stationary sound source (b) moving sound source

$$\theta = sin^{-1}\frac{R}{D} = cos^{-1}\frac{c.\Delta t}{D} \quad (9)$$

Where c is the sound propagation velocity in the water at 1492.1 m/s. Results of several localization experiments for two different algorithms are presented. In this experiment, was used underwater speaker and remote control vessel as the target. For stationary measurement, underwater speaker placed one by one in line position with angle variation (45, 64, 90, 116, 135 degree). The result of the localizing of azimuth are shown below in the degrees unit.

In this experiment, it was obtained good results of azimuth sound localization. Using four kinds of frequency variation, was obtained azimuth value from each frequency at a certain angle of sound source. The azimuth from each frequency averaged to determine the deviation readings thus the effect of kinds frequency can be known. Shown at Table 1, that PHAT algorithm achieved better accurate reading than CC with smallest error 2.34 degrees and biggest error 9.48 degrees. Meanwhile, CC algorithm achieved smallest error 1.5 degrees and biggest error 25.1 degrees. From the result, also show that PHAT algorithm achieved better precision reading than CC with smallest deviation 0 degrees and biggest error 13.81 degrees. Meanwhile, CC algorithm achieved smallest deviation 0 degrees and biggest error 28.8 degrees. Overall, PHAT has better performance than CC because more accurate and precise in tracking a sound source azimuth. Result of measurement different than simulation, where CC and PHAT cannot obtain actual time delay in SNR 25 dB in the real environment. Inaccurate result be expected because there is some correlation among the signals from the same sound source, we can estimate TDOA value by calculating the correlation function between the signals received by different hydrophones. However, in the real environment, due to the impact of noise and reverberation, the maximum peak of the correlation function will be weakened and sometimes there are multiple peaks, which have caused difficulties in the actual peak detection [5]. The GCC method is to weight signals in the power spectral domain to highlight the relevant signal components and suppress the components disturbed by noise so as to make the peak of relevant function at the time delay more prominent. However, due to the presence of reverberation, there are multiple echo components in the signals, so the cross-correlation function calculated will include the peaks formed by the direct wave and reflected wave that will make the time-delay estimation detection difficult [5].

After that, was done localization of moving sound source by using remote control vessel. From the stationary experiment result, achieved that PHAT algorithm better than CC to localize azimuth of sound source at reverberant, scattering, and reflecting medium. Thus, chosen PHAT algorithm to localize azimuth of moving sound source.

Combination between pair H1 - H2, H2 - H3, H3 - H4 used in this experiment then tracked azimuth in each pair averaged so achieved azimuth value from hydrophone array. Result of the tracking shown in Figure 5(b) where estimated azimuth is in the form of oscillating data. The oscillation occurs because the source of the move resulted in the medium through which it passes also change in volume of the medium. This changes influence direction of propagation so the time delay also changed. But, from the oscillated result can attracted a linear regression so achieved linear result from the tracking. The linear regression is azimuth estimated from hydrophone array. Figure. 5(b) show that PHAT algorithm achieved good tracking, with smallest error 0 degrees and biggest error 6.8 degrees. In this experiment, not occur Doppler effect because ratio between source speed and sound speed in the water is too high so frequency shifting too small.

5. Conclusion

The two algorithms presented in this paper worked well in the simulated environment, where there was added the noise signals were assumed to be uncorrelated and to have a zero mean value. However, PHAT method achieved the best results in the experimental part and CC method had the worst results in experimental part. Our results show also that there are some different between stationary and moving sound source experiment. Where, in stationary experiment achieved that the tracking more accurate and precision than measurement in moving sound condition. The study indicates that hydrophone arrays with CC and PHAT can be used for moving target positioning with satisfying results.

6. References

[1] Urick RJ. 1983. "Principles of Underwater Sound 3rd Edition" New York: Mc Graw Hill.
[2] Khaddour H 2011. "A Comparison of Algorithms of Sound Source Localization Based on Time Delay Estimation" Elektrorevue, ISSN 1213-1539, **2**, p. 1, April
[3] Knapp CH dan Carter GC 1976. *IEEE Transactions on Acoustics* **24**, p 4
[4] Zetterberg V, Pettersson MI and Claesson I 2005. *Proceedings of OCEANS* 2005 MTS/IEEE
[5] Lin C, Yongchun L, Fancheng K. 2011. *Procedia Engineering*, **15**, pp. 4912-4919, 2011.

Airborne sound insulation evaluation and flanking path prediction of coupled room

R. D. Tassia, W. A. Asmoro, and D. Arifianto
Dept. of Engineering Physics, Faculty of Industrial Technology, Institut Teknologi Sepuluh Nopember, Kampus ITS Sukolilo, Surabaya 60111, Indonesia

E-mail: rantidwitassia12@gmail.com

Abstract. One of the parameters to review the acoustic comfort is based on the value of the insulation partition in the classroom. The insulation value can be expressed by the sound transmission loss which converted into a single value as weighted sound reduction index (Rw, DnTw) and also have an additional sound correction factor in low frequency (C, Ctr) .In this study, the measurements were performed in two positions at each point using BSWA microphone and dodecahedron speaker as the sound source. The results of field measurements indicate the acoustic insulation values (DnT w + C) is 19.6 dB. It is noted that the partition wall not according to the standard which the DnTw + C> 51 dB. Hence the partition wall need to be redesign to improve acoustic insulation in the classroom. The design used gypsum board, plasterboard, cement board, and PVC as the replacement material. Based on the results, all the material is simulated in accordance with established standards. Best insulation is cement board with the insulation value is 69dB, the thickness of 12.5 mm on each side and the absorber material is 50 mm. Many factors lead to increase the value of acoustic insulation, such as the thickness of the panel, the addition of absorber material, density, and Poisson's ratio of a material. The prediction of flanking path can be estimated from noise reduction values at each measurement point in the class room. Based on data obtained, there is no significant change in noise reduction from each point so that the pathway of flanking is not affect the sound transmission in the classroom.

1. Introduction

Nowadays, the design of the building for the most part have already consider in some aspects. Acoustic comfort is one that should be considered because basically the building constructed on a specific function. There are some acoustic factors that can be considered in the design of a building, one of which is the transmission loss. Transmission loss declare that the amount of some energy is lost because the sound waves pass through a barrier[1]. The value of transmission loss measured in a certain frequency range and this value can be obtained from the following equation[2]. According to the regulations European standard, ie EN ISO 717-1[3] the value of transmission loss can be calculated per frequency in a certain range and will be transformed into a single value that is weighted sound reduction index (Rw, DnTw) and also have an additional sound correction factor in low frequency (C, Ctr). The higher DnTw + C's value, made the higher ability of a partition to muffle the sound. DnTw value obtained using the following equation[4]. Frequency range used in this measurement is in the range of 100-3150 Hz (1/3 octave frequency band)[5]. The way to measure isolation between the couple room has been standardized in ISO 140-4[6]. The objective of this paper is to measure sound insulation between two class rooms of UPT Bahasa building in Sepuluh

Nopember Institute of Technology. In this paper will provide an evaluation of the type of material used in the partition wall by performing simulations and design calculations based on the single value of sound insulation (DnTw+C) and also predict the flanking path in the classroom based on the noise reduction at each point of observation. The simulated material serves as an insulator and must follow the requirements in terms of weight, density, elasticity, and the principle of isolation [7]

2. Numerical Methods

The study start with determining the existing condition of the partition wall in the classroom which consisting of plywood 12mm on each side and the width of cavity is 66 mm with the size of classrooms 5.97 x 3.97 x 3 meters. Measurements were performed at 4 symmetrical points and 5 points randomly in classrooms.

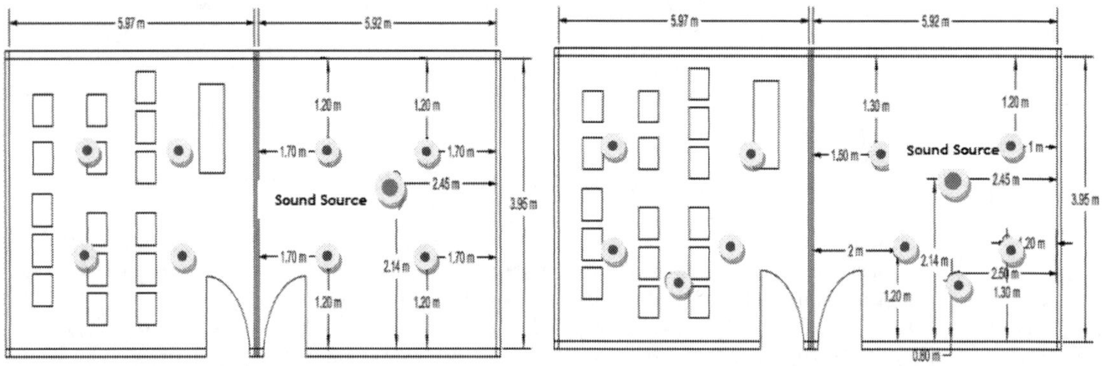

(a) (b)

Figure 1. The sketch of classrooms in two position measurement where the requirements are in the ISO 140-4 standard (at least 0.7 m between the microphone , 0.5 m between the boundary of the room and at least 1 m between the source) (a) The Symmetrical position (b) A random position

Sound source (dodecahedron) raised the high volume and placed in the middle of the room in order to get the diffuse sound area. Sound Pressure Level in the source room is measured with a BSWA microphone connected to the soundcard BSWA where the height of the microphone 1.2 m from the floor as high as the sound source. Measurements performed twice for each predetermined point. Then the parameters will be calculated in a single value (DnTw + C). The calculation is performed on a 1/3 octave band frequency between 100 Hz and 3150 Hz. To obtain a single value DnTw + C, the value of DnT in the field compared to the reference curve. DnT value at a frequency of 500 Hz is used as a reference in the shifting of the value in reference curve. The results were compared with the terms specified in the European standard that is equal DnTw + C> 51 dB. If the insulation value of the existing wall does not qualify the standard, it will be simulated in the selection of materials that include the standards. From the results of the design, the insulation value of the simulation will be compared again with the specified standard. As for predicting the path flanking the classroom, predicted from the amount of noise reduction at observation points in both classroom.

3. Results and Discussion

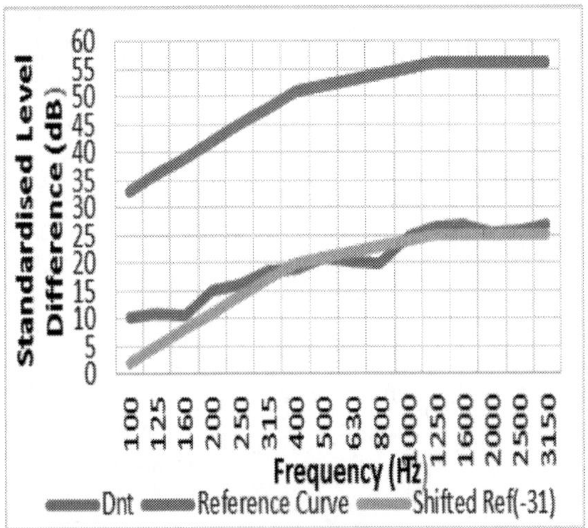

(a)

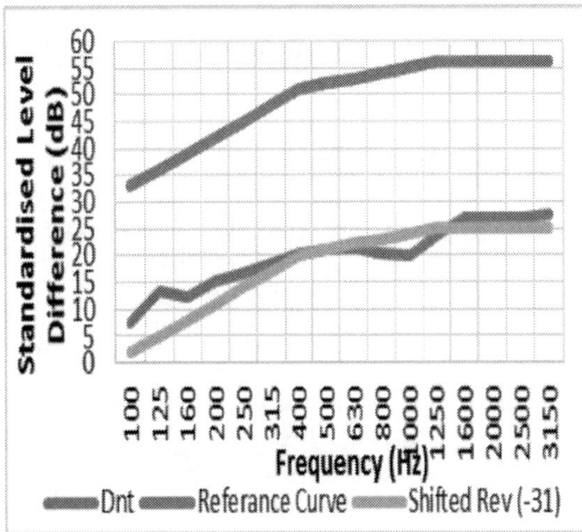

(b)

Figure 2 comparison between DnT curve and reference curve on-site measurement with two position. (a) The symmetrical position (b) A random position

Based on the graph, it can be seen that the performance of a partition wall which represented by the value of DnT is still far below the reference curve so that the reference curve should be shifted down by 31 dB. The final result of both calculation is DnT + C 19.6 dB. This means that the partition wall of classrooms still does not meet the standards because the result is less than 51 dB. Hence, it takes the partition walls with a good sound insulation accordance with the minimum requirements specified standard. Before selecting material, existing partition is simulated according to the actual design of the classroom. This following is the simulation of existing partitions using AFMG SoundFlow software:

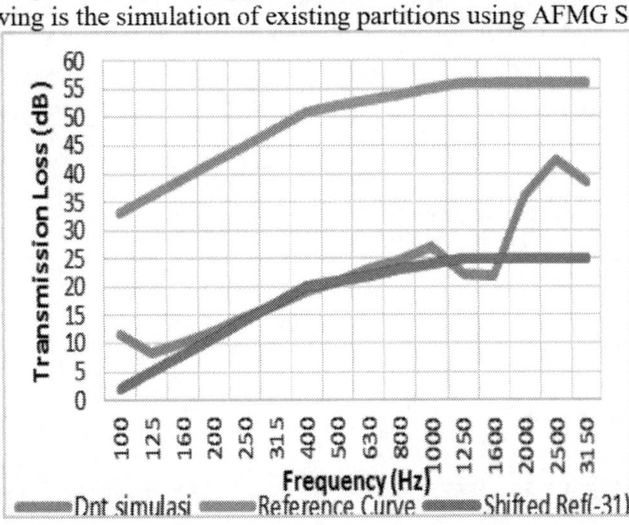

Figure 3. Comparison between DnT curve and reference curve of plywood material based on simulation with material specification; thickness 12 mm ; density 800 Kg/m^3; young's modulus 12 GPa ; poisson ration 0.06; bending loss factor 0.02

Figure 3 show that the insulation value on the simulation (DnT) is approximately 21 dB, while the on-site measurements the insulation value (DnTw + C) is 19.6 dB. The value obtained is still within range of the measurement boundary between measurements in the field (on-site) and measurements in the laboratory. The results of direct measurements is permitted to be worth 5 dB lower than the value of Rw. The design of simulation using gypsum board, plasterboard, cement, and PVC as the material of change for the partition walls. From the result, gypsum board material obtain the insulation value is worth 59 dB. Second, using plasterboard with insulation value is worth 65 dB. Then on the third design by using material cement board the insulation value is worth 69 dB. Lastly, the design obtained by using PVC the insulation value is worth 60 dB. Based on those values, every designs of the materials has already include the standard. The best insulation material owned by cement board. The following is a comparison of DNT curve and reference curve of cement board material:

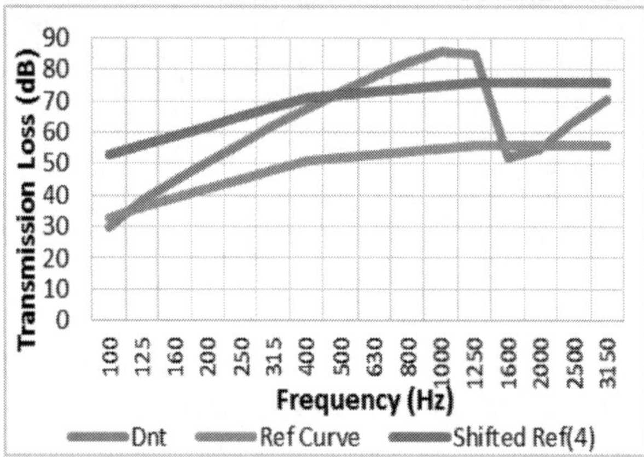

Figure 4. comparison between DnT curve and reference curve of cement board based on simulation with material specification; thickness 12.5 mm ; density 2000 Kg/m^3; young's modulus 20 Gpa ; poisson ration 0.23; bending loss factor 0.01

Considering that the incoming transmission apart from transmission through the partition wall (flanking) can be seen from the value of noise reduction from some of observation points in both classrooms. The following is the graph of noise reduction in the class room of UPT Bahasa:

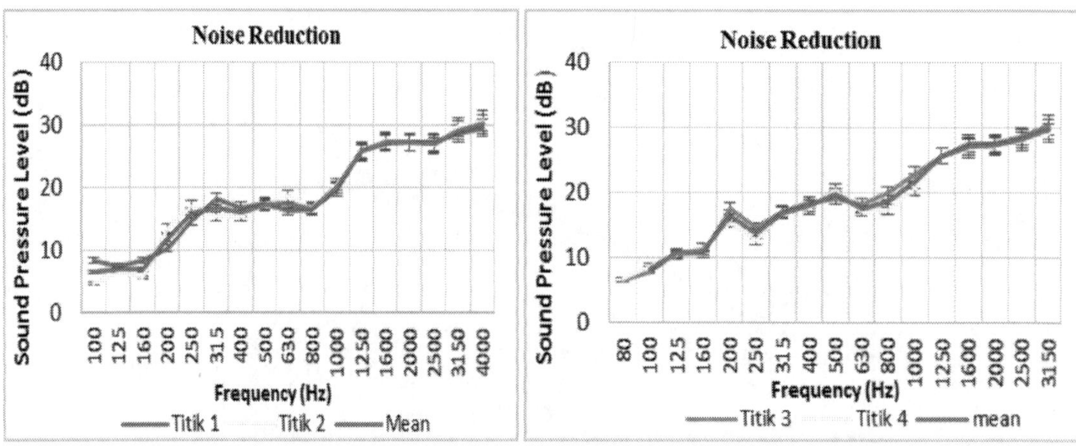

Figure 5. show the graph of noise reduction at two points in symmetrical position, parallel between the source and receiving room.

The comparison of noise reduction at point one and two in both room each worth 23.27 dB and 23.19 dB. While the value of noise reduction at point three and four each worth 24.66 dB and 23.51 dB. Based on the noise reduction of each point, the incoming transmission apart from transmission through the partition wall (flanking) is not significant that the prediction flanking paths do not occur on the side wall in the classroom.

4. Conclusion

The insulation value (DnTw + C) of the partition walls in the classrooms worth 19.6 dB and does not meet the existing standards, so it necessary to redesign in order to improve the performance of acoustic insulation in the classroom. The best insulation value from the design is the third design using cement board material. The insulation value of this material is worth 69 dB. Based on the noise reduction of each point in measurement, the prediction of flanking paths in UPT Language neither too significant nor give an impact in the classroom.

5. References

[1] Hemond, Conrad J. 1983. "*Engineering Acoustic and Noise Control*". New Jersey : Prentice Hall
[2] Long, Marshall. 2006. "*Architectural Acoustic*". USA : Elsevier Inc.
[3] ISO 717-1: 2013 "*Rating of Sound Insulation in Buildings and of Building Elements*".
[4] Harris, D.J. Knight, Steven. 2007. " *Measurement of Airborne Sound Insulation* (Separating Walls) at Gwynne Gardens, East Grinstead. UKAS testing".
[5] Whitefield, W. A. 2005. *Measurement of Airborne & Impact Sound Insulation*. UKAS testing.
[6] ISO 140-4: 1998. " *Field Measurement of Airborne Sound Insulation between Rooms*".
[7] Mediastika, Christina E. 2005. *Akustika Bangunan*. Jakarta : Erlangga.

Energy analysis of four dimensional extended hyperbolic Scarf I plus three dimensional separable trigonometric non-central potentials using SUSY QM approach

A. Suparmi, C. Cari, U. A. Deta, J. Handhika
Physics Department, Graduate Program, Sebelas Maret University, Jl.Ir. Sutami 36A Kentingan Surakarta 57126, Indonesia

Email: soeparmi@staff.uns.ac.id

Abstract. The non-relativistic energies and wave functions of extended hyperbolic Scarf I plus separable non-central shape invariant potential in four dimensions are investigated using Supersymmetric Quantum Mechanics (SUSY QM) Approach. The three dimensional separable non-central shape invariant angular potential consists of trigonometric Scarf II, Manning Rosen and Poschl-Teller potentials. The four dimensional Schrodinger equation with separable shape invariant non-central potential is reduced into four one dimensional Schrodinger equations through variable separation method. By using SUSY QM, the non-relativistic energies and radial wave functions are obtained from radial Schrodinger equation, the orbital quantum numbers and angular wave functions are obtained from angular Schrodinger equations. The extended potential means there is perturbation terms in potential and cause the decrease in energy spectra of Scarf I potential.

1. Introduction

It is well known that the exact solution of non-relativistic Quantum Mechanic for some physical potentials are very important since they provide all necessary information for certain quantum system. Finding an exact solution of Schrodinger equation for some real potentials, that have been applied in molecular physics, solid state and chemistry areas, have been explored intensively by using various methods. Recently, conciderably efforts have been paid to obtain the exact solution of mixture potentials include Killingbeck [1] Sextic and octic [2], extended Cornell [3-4], and Killingbeck in external field [5-7,10] potentials. The bound state energy spectra of these potentials have been investigated by various techniques such as SUSY QM [11-12,16-18], wave function ansatz method [2,6-7,9-10], Nikiforov-Uvarov method [3-4,8,13], and Lie Algebra approach [1]. For l wave, the Schrodinger equation is only solved approximately with suitable approximation scheme for centrifugal term [14].

Furthermore, the extension in higher dimensional spaces for some physical problems is very important in some physics area. The D dimensional system has been constructed to explain the unification of gravitation and electromagnetic fields [9]. It is suspected that the dimensional system is applicable for gravitation field since it is involved in such huge universe. The D-dimensional non-relativistic and relativistic physical systems have been investigated by many authors, Killingbeck potential[1], Poschl-Teller and Manning Rosen non-central potential [12], hyperbolic tangent [13], three anharmonics potential [8].

In this paper we will attempt to solve the four dimensional Schrodinger equation for a charged particle moving in a field governed by an extended radial Scar I potential with simultaneous presence

trigonometric Scarf II, Manning Rosen, and Pöschl-Teller non-central potentials using SUSY QM with the idea of shape invariance. SUSY QM was developed based on Witten's proposal [14], while the idea of shape invariant potentials was proposed by Gendenshtein [15], since then SUSY QM became a powerful tool to determine the energy spectrum and wave function of a class of shape invariant potentials as in Sukumar, Dutt et al., and Gangopadhyaya [16-18]. This potential can be applied to study the non-relativistic effect of the complex vibration-rotation energy structure of multi-electron atoms.
This paper is organized as follows. A brief review of SUSY QM is presented in Section 2. Derivation of hyper-spherical Laplacian is briefly introduced in section 3, Solutions of radial and polar Dirac equations are presented in Section 4. The result and discussion in section 5. The conclusion is presented in Section 6.

2. Review of the SUSY Quantum Mechanics Approach Using Operator and Shape Invariance

According to the definition proposed by Witten [14], in a SUSY quantum system there are super charge operators Q_i that commute with the SUSY Hamiltonian H_{ss} and they also obey to the anti commutation algebra as

$$[Q_i, H_{ss}] = 0 \; i = 1, 2, 3, \ldots N \quad ; \quad \{Q_i Q_j\} = \delta_{ij} H_s \tag{1}$$

SUSY QM is a one-dimensional model of SUSY field theory with N = 2, where

$$Q_1 = \frac{1}{\sqrt{2}}\left(\sigma_1 \frac{p}{\sqrt{2m}} + \sigma_2 \phi(x)\right) \text{and} \quad Q_2 = \frac{1}{\sqrt{2}}\left(\sigma_2 \frac{p}{\sqrt{2m}} + \sigma_1 \phi(x)\right) \tag{2}$$

with σ_i are the usual Pauli spin matrices, $p = -i\hbar \frac{\partial}{\partial x}$ is the usual momentum operator, and $\phi(x)$ is the super-potential. By using equations (3) and (2) we get,

$$H_{ss} = \begin{pmatrix} -\frac{\hbar^2}{2m}\frac{d^2}{dx^2} + \varphi^2(x) + \frac{\hbar}{\sqrt{2m}}\varphi'(x) & 0 \\ 0 & -\frac{\hbar^2}{2m}\frac{d^2}{dx^2} + \varphi^2(x) - \frac{\hbar}{\sqrt{2m}}\varphi'(x) \end{pmatrix} = \begin{pmatrix} H_+ & 0 \\ 0 & H_- \end{pmatrix} \tag{3}$$

with
$$H_- = -\frac{\hbar^2}{2m}\frac{d^2}{dx^2} + V_-(x) = A^+ A \text{ and } H_+ = -\frac{\hbar^2}{2m}\frac{d^2}{dx^2} + V_+(x) = AA^+ \tag{4}$$

where $V_-(x)$ and $V_+(x)$ are the SUSY partner potentials, A^+ and A are raising and lowering operators, and

$$A^+ = -\frac{\hbar}{\sqrt{2m}}\frac{d}{dx} + \phi(x) \text{ and } A = \frac{\hbar}{\sqrt{2m}}\frac{d}{dx} + \phi(x) \tag{5}$$

By constructing super-potential by using the condition of usual Hamiltonian H and SUSY Hamiltinian H_-

$$H = H_- + E_0 = -\frac{\hbar^2}{2m}\frac{d^2}{dx^2} + V_-(x; a_0) + E_0 \tag{6}$$

and by applying the condition of shape invariant potential which is defined as

$$V_+(x; a_j) = V_-(x; a_{j+1}) + R(a_{j+1}) \tag{8}$$

$$V_+(x; a_j) = \phi^2(x; a_j) + \frac{\hbar}{\sqrt{2m}}\phi'(x; a_j) \text{and } V_-(x; a_{j+1}) = \phi^2(x; a_{j+1}) - \frac{\hbar}{\sqrt{2m}}\phi'(x; a_{j+1}) \tag{9}$$

the energy eigen value of H_- [15] and H are obtained from equation (6) given as

$$E_n^{(-)} = \sum_{k=1}^{n} R(a_k) \quad \text{and} \quad E_n = E_n^{(-)} + E_0 \tag{10}$$

with j = 0,1,2,..; a is a mapping parameter. Based on the characteristics of the lowering and raising operators, the ground state and excited state wave functions are obtained from the condition that

$$A\psi_0^{(-)} = 0, \quad \psi_1^{(-)}(x; a_0) \approx A^+(x; a_0)\psi_0^{(-)}(x; a_1), \quad \psi_2^{(-)}(x; a_0) \approx A^+(x; a_0)\psi_1^{(-)}(x; a_1) \tag{11}$$

The energy spectra are obtained from equations. (8-10), while the wave functions is obtained from equation (11).

3. Coordinate System in D Dimension

If x_i s are the coordinate components of Cartesian coordinates then they can be expressed in hyperspherical coordinate components as r, $\theta_1, \theta_2, ..., \theta_{n-1}$ in D dimensional system given as

$$x_1 = r\cos\theta_1 \sin\theta_2 ... \sin\theta_{D-1},\ x_2 = r\sin\theta_1 \sin\theta_2 ... \sin\theta_{D-1}, ..., x_b = r\cos\theta_{b-1}\sin\theta_b ... \sin\theta_{D-1}, ...$$
$$x_{D-1} = r\cos\theta_{D-2}\sin\theta_{D-1}, \text{ and } x_D = r\cos\theta_{D-1} \tag{12}$$

with $b \in [3, D-1]$.The unit vector $\hat{x}$ along x axes is usually expressed as $\hat{x} = x/r$,$r^2 = \sum_{i=1}^{D} x_i^2$ $0 \le \theta_1 \le 2\pi$,$\theta_i \in (0,\pi)$ for $i = 2,3,...,D-1$. The D dimensional Laplacian in polar coordinates is defined as

$$\nabla_D^2 = \frac{1}{h}\sum_{j=0}^{D-1}\frac{\partial}{\partial\theta_j}\left(\frac{h}{h_j^2}\frac{\partial}{\partial\theta_j}\right) \ ; \quad \theta_0 = r \ ; \quad h = \prod_{j=0}^{D-1} h_j \ \text{ and } \quad h_j^2 = \sum_{i=1}^{D}\left(\frac{\partial x_i}{\partial\theta_j}\right)^2 \tag{13}$$

By manipulating equations (12-13) it is obtained D dimensional Laplacian in hyper-spherical coordinates as

$$\nabla_D^2 = \frac{1}{r^{D-1}}\frac{\partial}{\partial r}\left(r^{D-1}\frac{\partial}{\partial r}\right) + \frac{1}{r^2}\sum_{j=1}^{D-2}\frac{1}{\sin^2\theta_{j+1}\sin^2\theta_{j+2}...\sin^2\theta_{D-1}} \times \left\{\frac{1}{\sin^{j-1}\theta_j}\frac{\partial}{\partial\theta_j}\left(\sin^{j-1}\theta_j\frac{\partial}{\partial\theta_j}\right)\right\} \tag{14}$$
$$+\frac{1}{r^2}\left\{\frac{1}{\sin^{D-2}\theta_{D-1}}\frac{\partial}{\partial\theta_{D-1}}\left(\sin^{D-2}\theta_{D-1}\frac{\partial}{\partial\theta_{D-1}}\right)\right\}$$

The four dimensional Schrodinger equation with separable four dimensional non-central potential in hyperspherical coordinates which is obtained from equation (14) is given as

$$-\frac{\hbar^2}{2\mu}\left\{\begin{array}{l}\frac{1}{r^3}\frac{\partial}{\partial r}\left(r^3\frac{\partial}{\partial r}\right) + \frac{1}{r^2\sin^2\theta_2\sin^2\theta_3}\left(\frac{\partial^2}{\partial\theta_1^2}\right) \\ +\frac{1}{r^2}\left\{\frac{1}{\sin^2\theta_3}\left\{\frac{1}{\sin\theta_2}\frac{\partial}{\partial\theta_2}\left(\sin\theta_2\frac{\partial}{\partial\theta_2}\right)\right\} + \left[\frac{1}{\sin^2\theta_3}\frac{\partial}{\partial\theta_3}\left(\sin^2\theta_3\frac{\partial}{\partial\theta_3}\right)\right]\right\}\end{array}\right\}\psi(r,\theta_1,\theta_2,.\theta_3) \tag{15}$$
$$+\left[\frac{\hbar^2}{2\mu}\left\{V(r) + \frac{1}{r^2}\left\{\frac{V_1(\theta_1)}{\sin^2\theta_2\sin^2\theta_3} + \frac{V_2(\theta_2)}{\sin^2\theta_3} + V_3(\theta_3)\right\}\right\} - E\right]\psi(r,\theta_1,\theta_2,\theta_3) = 0$$

where the three dimensional separable trigonometric angular potential is defined as

$$V(\theta_1,\theta_2,\theta_3) = \frac{1}{r^2}\left\{\frac{V_1(\theta_1)}{\sin^2\theta_2\sin^2\theta_3} + \frac{V_2(\theta_2)}{\sin^2\theta_3} + V_3(\theta_3)\right\} \tag{16}$$

with

$$V_1 = \left(\frac{b^2 + c(c-1) + 2b(c-1/2)\sin\theta_1}{\cos^2\theta_1}\right); \quad V_2 = \left(\frac{\upsilon(\upsilon+1)}{\sin^2\theta_2} - 2\zeta\cot\theta_2\right), \quad V_3 = \left(\frac{\rho(\rho-1)}{\sin^2\theta_3} + \frac{\gamma(\gamma-1)}{\cos^2\theta_3}\right) \tag{17}$$

and an extended (perturbed) radial Scarf I potential is

$$V(r) = \left(\alpha^2\left\{\left(\frac{f^2 - g(g+1)}{\cosh^2\alpha r} + \frac{2f(g+1/2)\sinh\alpha r}{\cosh^2\alpha r}\right) + \left(\frac{\delta(\delta-1)}{\sinh^2\alpha r}\right) + 2d\,\mathrm{csch}\,\alpha r\right\}\right)$$

(18)

with α is related to the width of radial potential f, g, δ , d, b, c, υ, ζ , ρ ,γ are parameters related with the depth of potential. By setting $\psi(r,\theta_1,\theta_2,\theta_3) = \mathrm{R(r)\,Y}(\theta_1,.\theta_2,\theta_3) = \mathrm{R(r)\,P_1}(\theta_1)P_2(\theta_2)P_3(\theta_3)$ in equation (15) we get four one dimensional Schrodinger equation given as

$$\left\{r^2\frac{1}{r^3}\frac{\partial}{\partial r}\left(r^3\frac{\partial}{\partial r}\right)-r^2\left(V(r)-\frac{2m}{\hbar^2}E\right)\right\}R=\lambda_3 R \quad , \quad \frac{\partial^2 P_1}{\partial\theta_1^2}-V_1(\theta_1)P_1+\lambda_1 P_1=0 \tag{17}$$

$$\left\{\frac{1}{\sin\theta_2}\frac{\partial}{\partial\theta_2}\left(\sin\theta_2\frac{\partial P_2}{\partial\theta_2}\right)\right\}-V_2(\theta_2)P_2-\frac{\lambda_1 P_2}{\sin^2\theta_2}+\lambda_2 P_2=0 \tag{18}$$

$$\left[\frac{1}{\sin^2\theta_3}\frac{\partial}{\partial\theta_3}\left(\sin^2\theta_3\frac{\partial P_3(\theta_3)}{\partial\theta_3}\right)\right]-V_3(\theta_3)P_3(\theta_3)-\frac{\lambda_2}{\sin^2\theta_3}P_3(\theta_3)+\lambda_3 P_3(\theta_3)=0 \tag{19}$$

with $\lambda_1.\lambda_2$, and λ_3 are variable separation constant.

4. Solution of Schrodinger equation with Extended Hyperbolic Scarf I Radial Potential Plus Trigonometric Non-central Potentials

4.1. Solution of radial part of Schrodinger equation

By using equations (13-14), setting $R=\frac{U}{r^{3/2}}$, and applying centrifugal approximation [19] in equation (17) we get

$$-\frac{\hbar^2}{2\mu}\frac{d^2U}{dr^2}+\left(\frac{\hbar^2\alpha^2}{2\mu}\left\{\left(\frac{f^2-g(g+1)}{\cosh^2\alpha r}+\frac{2f(g+1/2)\sinh\alpha r}{\cosh^2\alpha r}\right)+\left(\frac{\delta(\delta-1)+\lambda_3+3/4}{\sinh^2\alpha r}\right)+2d\operatorname{csc}h\alpha r\right\}\right)U=EU \tag{20}$$

By setting the hypothetical super potential for effective potential expressed in equation (20) as

$$\phi(r)=\alpha\left(B\tanh\alpha r+C\operatorname{sec}h\ \alpha r+D\coth\alpha r\right) \tag{21}$$

and by using equation (6-10) we obtain the superpotential $\phi(r)$, the ground state energy ε_0, the energy spectra E_n and ground state radial wave function R_0 as

$$\phi(r)=\frac{\hbar g\alpha}{\sqrt{2\mu}}\tanh\alpha r+\frac{\hbar f\alpha}{\sqrt{2\mu}}\operatorname{sec}h\ \alpha r+\frac{\hbar(\delta'-1)\alpha}{\sqrt{2\mu}}\coth\alpha r \tag{22}$$

$$\varepsilon_0=-\frac{\hbar^2\alpha^2}{2\mu}(g+\delta'-1)^2,\quad E_n=-\frac{\hbar^2\alpha^2}{2\mu}(g+(\delta'-1)-2n)^2,\quad \delta'=\frac{1}{2}+\sqrt{\delta(\delta-1)+\lambda_3+1} \tag{23}$$

$$R_0=C_0\left[(\cosh\alpha r)^{-\frac{\hbar g\alpha}{\sqrt{2\mu}}}(\sinh\alpha r)^{-\frac{\hbar(\delta'-1)\alpha}{\sqrt{2\mu}}-3/2}\left(-\coth\frac{\alpha r}{2}\right)^{-iC}\right]^{-\frac{\sqrt{2\mu}}{\hbar}} \tag{24}$$

4.2. Solution of Schrodinger for the first angular part θ_1

By using equation (16-17) we get one dimensional Schrodinger equation for θ_1

$$-\frac{\hbar^2}{2\mu}\frac{\partial^2 P_1(\theta_1)}{\partial\theta_1^2}+\frac{\hbar^2}{2\mu}\left(\frac{b^2+c(c-1)+2b(c-1/2)\sin\theta_1}{\cos^2\theta_1}\right)P_1(\theta_1)=\frac{\hbar^2}{2\mu}\lambda_1 P_1(\theta_1) \tag{25}$$

By applying SUSY QM method and the idea of shape invariant given in equations (7-11) and (25) we obtain the superpotential $\phi(\theta_1)$, variable separation constant that related to angular quantum number λ_1 and ground state wave function P_{10} as

$$\phi(\theta_1)=\frac{\hbar}{\sqrt{2\mu}}c\tan\theta_1+\frac{\hbar}{\sqrt{2\mu}}b\sec\theta_1 \tag{26}$$

$$\lambda_1=(c+n_1)^2\quad n_1=0,1,2,\ldots\quad P_{10}=(\cos\theta_1)^{c+b}(1+\sin\theta_1)^{-b} \tag{27}$$

4.3. Solution of Schrodinger for angular part for θ_2 and θ_3

By setting $P_j(\theta_j) = \frac{Q_j(\theta_j)}{\sin^{(j-1)/2}\theta_j}$ in equations(18-19) for $j = 2,3,$ and together with equation (16) we get

$$Q_2''(\theta_j) + \{\lambda_2 + (1/4)\} Q_2(\theta_2) - \left(\frac{\upsilon(\upsilon+1)+\lambda_1-(1/4)}{\sin^2\theta_2} - 2\zeta\cot\theta_2\right) Q_2(\theta_2) = 0 \quad (28)$$

$$Q_3''(\theta_3) + \{\lambda_3 + 1\} Q_3(\theta_3) - \left(\frac{\rho(\rho-1)+\lambda_2}{\sin^2\theta_3} + \frac{\gamma(\gamma-1)}{\cos^2\theta_3}\right) Q_3(\theta_3) = 0 \quad (29)$$

By setting $\upsilon(\upsilon+1)+\lambda_1-(1/4) = \upsilon'(\upsilon'+1)$ and $\frac{\hbar^2}{2M}\left\{\lambda_2 + \frac{1}{4}\right\} = E_2$ in equation (28) and applying equation (7-11) we obtain the superpotential $\varphi(\theta_2)$, the variable separation constant λ_2 which is related to angular quantum number and ground state wave function P_{20} as

$$\varphi(\theta_2) = -\frac{\hbar}{\sqrt{2M}}\left((\upsilon'+1)\cot\theta_2 - \frac{\zeta'}{(\upsilon'+1)}\right) \quad (30)$$

$$\lambda_2 = \left((\upsilon'+(n_2+1))^2 - \frac{\zeta^2}{(\upsilon'+(n_2+1))^2}\right) - 1/4 \quad n_2 = 0,1,2,\ldots \quad \upsilon' = -\frac{1}{2} + \sqrt{\upsilon(\upsilon+1)+\lambda_1} \quad (31)$$

$$P_{20} = (\sin\theta_2)^{\upsilon'+1/2} e^{-\zeta\theta_2/(\upsilon'+1)} \quad (32)$$

By setting $\rho(\rho-1)+\lambda_2 = \rho'(\rho'-1)$ in equation (29) and by manipulating equations (7-11) we obtain the superpotential $\varphi(\theta_3)$, the variable separation constant λ_3 and ground state wave function as

$$\varphi(\theta_3) = \frac{\hbar}{\sqrt{2M}}((-\rho'\cot\theta_3 + \gamma\tan\theta_3) \quad (33)$$

$$\lambda_3 = (2n_3 + \rho' + \gamma)^2 - 1 \quad ; \quad P_{30} = (\sin\theta_3)^{\rho'-1}(\cos\theta_3)^{\gamma} \quad \rho' = 1/2 + \sqrt{(\rho-1/2)^2 + \lambda_2} \quad (34)$$

5. Result and Discussion

From the solutions of four one dimensional Schrodinger equations using SUSY QM and the idea of shape invariant potential are exactly obtained in the scheme of centrifugal approximation. The non-relativistic energy spectra is obtained in the close form given as

$$E_n = -\frac{\hbar^2\alpha^2}{2\mu}(g + (\delta'-1) - 2n)^2 \quad (35)$$

with $\delta' = 1/2 + \sqrt{\delta(\delta-1)+\lambda_3+1}$, $\lambda_3 = (2n_3+\rho'+\gamma)^2 - 1$; $\rho' = 1/2 + \sqrt{(\rho-1/2)^2+\lambda_2}$

(36)

$$\lambda_2 = \left((\upsilon'+(n_2+1))^2 - \frac{\zeta^2}{(\upsilon'+(n_2+1))^2}\right) - 1/4; \ \upsilon' = -1/2 \pm \sqrt{\upsilon(\upsilon+1)+\lambda_1}\ ; \ \lambda_1 = (c+n_1)^2 \quad (37)$$

The extended Scarf I potential is Scarf potential plus Manning Rosen potential. The total ground state wave function is given as

$$\psi_{10} = C_0\left[(\cosh\alpha r)^{-g}(\sinh\alpha r)^{-(\delta'-1)-3/2}\left(-\coth\frac{\alpha r}{2}\right)^{-if}\right]^{-\frac{\alpha\sqrt{2m}}{\hbar}}(\cos\theta_1)^{c+b}(1+\sin\theta_1)^{-b} \quad (38)$$
$$(\sin\theta_2)^{\upsilon'+1/2} e^{-\zeta\theta_2/(\upsilon'+1)}(\sin\theta_3)^{\rho'-1}(\cos\theta_3)^{\gamma}$$

In Table 1 below, are non-relativistic energy spectra for some value of potential constants δ.

From Table 1 is shown that there is perturbation terms in potential and cause the decrease in energy spectra of Scarf I potential.

Table 1. Non-relativistic energy eigenvalues of particle under the influence of Scarf I potential

δ	$E_{nn_1n_2n_3}$					
	E_{0000}	E_{1000}	E_{2000}	E_{3000}	E_{4000}	E_{5000}
0	-2.4351e-37	-1.3568e-37	-5.9155e-38	-1.3947e-38	-4.9379e-41	-1.7464e-38
2	-2.5315e-37	-1.4290e-37	-6.3956e-38	-1.6326e-38	-7.0221e-42	-1.5000e-38
4	-3.0079e-37	-1.7920e-37	-8.8924e-38	-2.9956e-38	-2.2997e-39	-5.9549e-39
6	-3.8472e-37	-2.4516e-37	-1.3691e-37	-5.9972e-38	-1.4345e-38	-2.8667e-41
8	-5.0321e-37	-3.4135e-37	-2.1080e-37	-1.1156e-37	-4.3632e-38	-7.0155e-39

6. Conclusion

The non-relativistic energy spectra for non-central potential which is combination of extended hyperbolic Scarf I potential with three dimensional trigonometric angular potential is obtained in the closed form. The four dimensional energy spectra and wave functions depend on the variable separation constant, the radial and angular quantum numbers. It is worthy to explore more complicated shape invariant potential which is extension of the standard shape invariant potential.

Acknowledgement
This research was partly supported by Higher Education Project Grant with contract no. 632/UN27.21/LT/2016

References

[1] Panahi H, Zarrinkamar S and Baradaran M 2015 *Chinese Physics B* **24** 060301
[2] Kaushal R S 1991 *Modern physics letter* **6** 383
[3] Anake T A and Ita B I 2015 IJTPC **8** 18
[4] Abu-Shady M 2015 *Boson Journal of Modern Physic* **1** 16
[5] Hamzavi M and Ikhdair S M 2013 *Modern Physics Letter B* **27** 1350176-1
[6] Ikhdair S M 2013 *Adv. High En. Phys.* **2013** 491648 1
[7] Sharifi Z, Tajic F, Hamzavi M and Ikhdair S 2015 *Naturforsch* **70** 499
[8] Ikot A N, Antia A D, Akpan I O and Awoga O A 2013 *Revista Mexicana de Fisica* **59** 46
[9] Shi-Hai Dong 2002 *Found Phys. Lett.* **15** 385
[10] Shi-Hai Dong 2011 *Waves Equation in Higher Dimension* (Springer Science: Business Media) pp 41-43, pp 101-102, p 97
[11] Ozer O and Gonul B 2003 arXiv:quant-ph/0308128v1
[12] Suparmi A, Cari C and Pratiwi B N 2016 *J. Phys. Conf. Ser.***710** 012026
[13] Suparmi A, Cari C, Pratiwi B N and Deta U A 2016. *AIP Conf. Proc.* **1710** 030010
[14] Witten E 1981 *Nucl. Phys. B* **188** 513
[15] Gendenshtein L E 1983 *JETP Lett.* **38** 356
[16] Dutt R, Khare A and Sukhatme U P 1988 *Am. J. Phys* **56** 163
[17] Dutt R, Gangopadhyaya A and Sukhatme U P 1997 *Am. J. Phys.* **65** 400
[18] Gangopadhyaya A, Mallow J V, Rasinariu C and Sukhatme U P 2001 *Chin. J. Phys* **39** 101
[19] Greene R I and Aldrich C 1976 *Phys. Rev. A* **14** 2363

Journal of Physics: Conference Series **776** (2016) 012078 doi:10.1088/1742-6596/776/1/012078

Structural dynamic modification using matrix perturbation for vibrations without friction

B. Supriyadi[1] and S. Mungkasi[2]

[1]Department of Civil and Environmental Engineering, Faculty of Engineering, Gadjah Mada University, Yogyakarta, Indonesia
[2]Department of Mathematics, Faculty of Science and Technology, Sanata Dharma University, Yogyakarta, Indonesia

E-mail: bbsupri@tsipil.ugm.ac.id

Abstract. Matrix perturbation is used to determine structural dynamic modification. We want to investigate how much difference of stiffness of a given structure if the involved mass is slightly changed. We start from the scalar frictionless model having one degree of freedom to derive a formula in response to the change of mass to compute the intended change of stiffness. We want to enforce that the new frequency is exactly or approximately the same as the frequency of the original structural setting. The formula is then extended to problems having higher degrees of freedom successfully. Computational experiments confirm the accuracy of our proposed formula.

1. Introduction

Structural dynamic modification has been a popular technique since the late 1970s [1] to determine the effect of structural changes of a system, such as storey buildings, bridges, and other structures (for more examples, see [2-7] and references therein). Finite element method is usually implemented to investigate the effect of such changes. For relatively small scale problems, the effect of structural changes can be computed very fast using computers. For large scale problems, determining the effect of structural changes is computationally expensive [8].

In this paper we consider systems of one degree as well as higher degrees of freedom. The mathematical models governing the systems are differential equations either a single or multiple equations. We limit our work to models without friction. Two parameters, mass and stiffness, are involved. With these two parameters, the frequencies of the systems are determined. Because the extreme cost of computation for large scale problems, it is not a good idea to always run the simulation from the beginning if we have already known the properties of the original structure.

In this paper we assume that properties of the original structure are known. If there is a change in the mass, we want to find how much change of the stiffness, in order that the frequency of the new system is the same or approximately the same as the original structure. Note that we do not want to run another simulation for our new structure, as we assume that running the simulation again is very costly. We shall use only the properties of the original structure and the change of mass to obtain the change of stiffness. With this strategy we can find the answer quickly and cheaply.

This paper is organized simply as follows. First we derive our proposed formulas to answer our research question in Section 2. Computational experiments are presented in Section 3 to confirm the validity of our proposed formulas. Finally concluding remarks are drawn in Section 4.

2. Mathematical models

In this section we present our work for problems having one degree of freedom and those having higher degrees of freedom.

2.1. Problems with one degree of freedom

Let us start by considering the scalar unperturbed frictionless spring-mass model

$$m_0\ddot{x}_0 + k_0 x_0 = 0, \tag{1}$$

where m_0 is a constant representing the mass, k_0 is a constant denoting the stiffness of the structure, $x_0 = x_0(t)$ is the displacement of the mass m_0 at any time t. Its general solution is

$$x_0(t) = c_1 \cos(\omega_0 t) + c_2 \sin(\omega_0 t), \tag{2}$$

where $\omega_0 = \sqrt{k_0 / m_0}$ is the frequency of vibrations, and c_1, c_2 are arbitrary constants. We denote $\lambda_0 = \omega_0^2$. Note that the original problem (1) provides the following generalized scalar eigenvalue problem

$$(k_0 - \lambda_0 m_0)\phi_0 = 0, \tag{3}$$

where ϕ_0 is an eigenvector corresponding to the eigenvalue λ_0.

Assume that there exists a mass modification, usually called perturbation [9], which leads to the change in the frequency. Now the research question is that how much stiffness difference that we must take so that we have the same frequency as the original structural setting. If we denote

$$m = m_0 + \delta m_0, \tag{4}$$

$$k = k_0 + \delta k_0, \tag{5}$$

with a given value of δm_0, $\omega = \sqrt{k / m}$ is the new frequency of vibrations. We also denote $\lambda = \omega^2$. We then have the perturbed system

$$m\ddot{x} + kx = 0, \tag{6}$$

which provides the eigenvalue problem

$$(k - \lambda m)\phi = 0, \tag{7}$$

where ϕ is an eigenvector corresponding to the eigenvalue λ, and $x = x_0 + \delta x_0$ for some δx_0. We assume that the perturbation δm_0 of the mass is given. We want to compute the value of δk_0 such that the modified structure has exactly the same frequency as the original structure.

We derive as follows. We must have

$$\lambda_0 = \lambda = \frac{k}{m} = \frac{k_0 + \delta k_0}{m_0 + \delta m_0}. \tag{8}$$

A straight forward calculation results in

$$\delta k_0 = \lambda_0 \delta m_0. \tag{9}$$

Equation (9) is our proposed formula to response the change of mass in order that we have exactly the same frequency of vibrations as the original unperturbed problem (6). Note that this formula is exact only for the scalar frictionless spring-mass model (1), that is, problems with one degree of freedom. There is no guarantee that it is exact for problems having higher degrees of freedom.

Remark 1: Formula (9) can also be derived using Taylor series of the two variable function

$$\lambda(k,m) = \frac{k}{m} \tag{10}$$

as follows. Taking (4)-(5), then expanding (10) about (k_0, m_0) we obtain

$$\begin{aligned}\lambda(k_0 + \delta k_0, m_0 + \delta m_0) &= \lambda(k_0, m_0) + \delta k_0 \frac{\partial \lambda}{\partial k}(k_0, m_0) + \delta m_0 \frac{\partial \lambda}{\partial m}(k_0, m_0) \\ &+ \frac{1}{2!}\left[(\partial k_0)^2 \frac{\partial^2 \lambda}{\partial k^2}(k_0, m_0) + 2\delta k_0 \delta m_0 \frac{\partial^2 \lambda}{\partial k \partial m}(k_0, m_0) + (\partial m_0)^2 \frac{\partial^2 \lambda}{\partial m^2}(k_0, m_0)\right] + \cdots.\end{aligned} \tag{11}$$

Noting

$$\frac{\partial \lambda}{\partial k}(k_0, m_0) = \frac{1}{m_0}, \qquad \frac{\partial \lambda}{\partial m}(k_0, m_0) = -\frac{k_0}{m_0^2} \tag{12}$$

and enforcing

$$\lambda(k_0 + \delta k_0, m_0 + \delta m_0) = \lambda(k_0, m_0) \tag{13}$$

into equation (11) we have formula (9).

2.2. Problems with multiple degrees of freedom

Even though formula (9) is not exact for problems with higher degrees of freedom, the formula provides a very good approximation for a certain type of problems with higher degrees of freedom. We shall show this using some results of Cha and Solberg [8].

We consider an initial problem, which is an unperturbed frictionless system with N degrees of freedom

$$\mathbf{M}_0 \ddot{\mathbf{x}}_0 + \mathbf{K}_0 \mathbf{x}_0 = \mathbf{0}, \tag{14}$$

where $\mathbf{M}_0$ is the mass matrix and $\mathbf{K}_0$ is the stiffness matrix. The solution to the original problem (14) is not easy to find; this is in contrast with the scalar problem (1).

In this paper, both $\mathbf{M}_0$ and $\mathbf{K}_0$ are assumed to be real and symmetric matrices. Let us assume also that this system has natural frequencies ω_{0j} and mode shapes $\mathbf{x}_{0j}$. Equation (14) is considered as the original unperturbed problem. Following Nad [10] and Sestieri [11], the original problem (14) provides the following generalized eigenvalue problem

$$(\mathbf{K}_0 - \lambda_0 \mathbf{M}_0)\boldsymbol{\varphi}_0 = \mathbf{0}. \tag{15}$$

Now let the original problem (14) be perturbed, as

$$\mathbf{M} = \mathbf{M}_0 + \delta\mathbf{M}_0, \tag{16}$$

$$\mathbf{K} = \mathbf{K}_0 + \delta\mathbf{K}_0. \tag{17}$$

We then have the perturbed system

$$\mathbf{M}\ddot{\mathbf{x}} + \mathbf{K}\mathbf{x} = \mathbf{0}, \tag{18}$$

which provides the eigenvalue problem

$$(\mathbf{K} - \lambda\mathbf{M})\boldsymbol{\varphi} = \mathbf{0}, \tag{19}$$

where $\mathbf{x} = \mathbf{x}_0 + \delta\mathbf{x}_0$ for some $\delta\mathbf{x}_0$. We assume that the perturbation $\delta\mathbf{M}_0$ of the mass is given. We want to find the value of $\delta\mathbf{K}_0$ such that the frequencies of the perturbed problem is approximately the same as those of the original problem (14).
Cha and Solberg [8] proposed the solution to the perturbed eigenvalue problem (19) using a first order approximation of its exact solution, that is,

$$\lambda_j = \lambda_{0j} + \mathbf{x}_{0j}^T(\delta\mathbf{K}_0 - \lambda_{0j}\delta\mathbf{M}_0)\mathbf{x}_{0j}, \tag{20}$$

where $\mathbf{x}_{0j}$ is the mode shape corresponding to the natural frequency ω_{0j}. Here we have the approximation (20) for eigenvalues of the perturbed problem (19). We can compute straight forwardly a good approximation of $\delta\mathbf{K}_0$, so that we have approximately the same frequencies as those of the original problem (14). Enforcing $\lambda_j = \lambda_{0j}$, we obtain

$$\delta\mathbf{K}_0 = \lambda_{0j}\delta\mathbf{M}_0, \tag{21}$$

where $j = 1, 2, \ldots, n$ for matrices $\mathbf{K}$ and $\mathbf{M}$ of the size $n \times n$.

As we have n different matrices $\delta\mathbf{K}_0$ based on (21), we propose to average them. That is, we take

$$\delta\mathbf{K}_0 = \bar{\lambda}_0\delta\mathbf{M}_0, \tag{22}$$

where

$$\bar{\lambda}_0 = \frac{1}{n}\sum_{j=1}^{n}\lambda_{0j}. \tag{23}$$

Equation (22) is our proposed formula to response the change of mass in order that we have approximately the same frequencies of vibrations as the original unperturbed problem (18).

3. Computational experiments

In this section we present our computational examples. All measured quantities are assumed to be with SI units.

3.1. Example on a small scale problem

Consider the scalar model (1) with $m_0 = 1$ and $k_0 = 1$. The frequency of vibrations of this original problem is $\omega_0 = \sqrt{k_0 / m_0} = 1$. If the original mass is perturbed by $\delta m_0 = 0.01$, we want to find the value of δk_0 such that the new frequency is $\omega = \omega_0 = 1$.

Now we compute our stifness response (9) to the change of mass. That is,

$$\delta k_0 = \lambda_0 \delta m_0 = 0.01. \tag{24}$$

With this value of $\delta k_0 = 0.01$, we obviously achieve the intended new frequency

$$\omega = \sqrt{\frac{k}{m}} = \sqrt{\frac{k_0 + \delta k_0}{m_0 + \delta m_0}} = \sqrt{\frac{1.01}{1.01}} = 1, \tag{25}$$

as desired.

3.2. Example on a larger scale problem

Consider the model (14) with [8]

$$\mathbf{K}_0 = k_0 \begin{bmatrix} 14.0 & -4.0 & -2.0 & -2.0 & -3.0 \\ -4.0 & 13.0 & -6.0 & -2.0 & -1.0 \\ -2.0 & -6.0 & 14.0 & -4.0 & -2.0 \\ -2.0 & -2.0 & -4.0 & 13.0 & -5.0 \\ -3.0 & -1.0 & -2.0 & -5.0 & 11.0 \end{bmatrix}, \tag{26}$$

$$\mathbf{M}_0 = m_0 \begin{bmatrix} 4.0 & 0.0 & 0.0 & 0.0 & 0.0 \\ 0.0 & 4.0 & 0.0 & 0.0 & 0.0 \\ 0.0 & 0.0 & 3.0 & 0.0 & 0.0 \\ 0.0 & 0.0 & 0.0 & 5.0 & 0.0 \\ 0.0 & 0.0 & 0.0 & 0.0 & 4.0 \end{bmatrix}, \tag{27}$$

$$\delta\mathbf{M}_0 = m_0 \begin{bmatrix} 0.1 & 0.0 & 0.0 & 0.0 & 0.0 \\ 0.0 & -0.3 & 0.0 & 0.0 & 0.0 \\ 0.0 & 0.0 & 0.2 & 0.0 & 0.0 \\ 0.0 & 0.0 & 0.0 & 0.2 & 0.0 \\ 0.0 & 0.0 & 0.0 & 0.0 & 0.1 \end{bmatrix}. \tag{28}$$

Here m_0 and k_0 are arbitrary positive constants. We want to find the matrix $\delta\mathbf{K}_0$, such that the frequencies of the perturbed problem are approximately the same as those of the original problem. The stiffness response $\delta\mathbf{K}_0$ that we should take to tackle the perturbation $\delta\mathbf{M}_0$ is given by our formula (22).

Table 1. Exact and approximate eigenvalues of the perturbed problem having five degrees of freedom. Here λ_0 represents eigenvalues of the original problem, and λ_p eigenvalues of the perturbed problem.

i	λ_0	λ_p	error
1	0.1265	0.1685	0.3322
2	2.7236	2.7170	0.0024
3	3.5878	3.5806	0.0020
4	4.3637	4.3459	0.0041
5	5.9649	5.9043	0.0102

To be specific we take $m_0 = 1$ and $k_0 = 1$. Using the *eig* command of the matlab software, we can obtain the eigenvalues of the original problem. The natural frequencies of vibrations of this original problem are the square-roots of the eigenvalues. We also compute our stiffness response (22) to the change of mass. The results of eigenvalues are listed in Table 1. The average of eigenvalues of the original problem is $\overline{\lambda}_0 = 3.3533$ The average of eigenvalues of the perturbed problem is $\overline{\lambda}_p = 3.3433$. The average error of the eigenvalues is 0.07, that is, about 2.09%.

Remark 2: With this final example we notice that the norm of $\delta\mathbf{M}_0$ is 0.3 and the norm of $\mathbf{M}_0$ is 5 . Here we have that the perturbation of the mass is about 6% of the norm of the original problem. Using our proposed formula, in order that we maintain the frequency is about the same as the original frequency the norm of $\delta\mathbf{K}_0$ is 0.8697 with a note that the norm of $\mathbf{K}_0$ is 20.3509 . Here we have that the response of the stiffness is about 4.3%. As desired, small change of mass is responded with small change of stiffness in order that the new frequency is about the same as the original frequency.

4. Conclusion

We have proposed a formula to compute a first order approximation of the stiffness difference if the involved mass in the structural dynamics is changed. The formula is limited to the structural mathematical model without friction. The formula is derived from the scalar spring-mass model with its degree of freedom is one. We have shown that an extension of its use for problems of higher degree of freedom has been successful. We project that the formula is applicable for large size problems, as long as the mass perturbation is not so excessive.

Acknowledgments

This work was financially supported by Sanata Dharma University. The support is gratefully acknowledged by both authors.

References

[1] Avitabile P 2003 *Sound Vib.* **January 2003** 14
[2] Kalaycioglu T and Ozguven H N 2014 *Mech. Syst. Sig. Process.* **46** 289
[3] Mazlan A Z A and Ripin Z M 2016 *JSCE* **230** 130
[4] Mottershead J E and Friswell M I 1993 *J. Sound Vib.* **167** 347
[5] Ouyang H and Zhang J 2015 *Mech. Syst. Sig. Process.* **50-51** 214
[6] Skafte A, Aenlle M L and Brincker R 2016 *Smart Mater. Struct.* **25** 025020
[7] Supriyadi B 1993 Recalage des modeles de batiment (Ecole Centrale de Lyon: PhD Thesis)
[8] Cha P D and Solberg K A 2008 *Int. J. Mech. Eng. Educ.* **36** 160
[9] Chen J C and Wada B K 1977 *AIAA J.* **15** 1095
[10] Nad M 2007 *ACM* **1** 203
[11] Sestieri A 2000 *Sadhana* **25** 247

Analytical Solution of Schrodinger Equation for Modified Anisotropic Nonquadratic with Exponential Potential Using Supersymmetric Quantum Mechanics

Miftahul Ma'arif[1], Suparmi[1], Cari[1]

[1]Physics Departement, Faculty of Mathematics and Natural Sciences, Sebelas Maret University, Surakarta, Indonesia
E-mail: miftahulmaarif15@student.uns.ac.id

Abstract. Schrodinger equation for an anisotropic nonquadratic potential that modified by exponential form in axial part is investigated using supersymmetric approach. The three dimensional Schrodinger equation for an anisotropic nonquadratic potential in cylindrical coordinate is separated into three parts that contain one dimensional Schrödinger type equation which are solved using supersymmetric operator and the idea of shape invariant potential. The energy spectrum and the total wave functions are obtained.
Keywords : Schrodinger equation, Anisotropic Nonquadratic, Supersymmetry

1. Introduction

Exact analytical solutions of the Schrodinger equation for some physical potential are important to obtain the wave function and the energy spectrum. More recently, efforts have paid considerably to explore potential solutions right from the central and non-central. There are only few potentials the Schrodinger equation can be solved exactly in the scheme of centrifugal approach.

In recent years, many studies have analysed a bound state of a charged particle moving in a potential vector and scalar potential non-central location, such as an electron moves in the Coulomb field with the same field presence Aharonov-Bohm [1-2], or monopole magnet [3], Makarov potential [4] or potential-shaped ring oscillator [9/5]. Most of these research, the eigen values and eigen functions are determined using variable separation method in the spherical coordinate system. More recently, with the idea of supersymmetric quantum mechanics invariance form [10/9], factorization method [11-12], and Nikiforov-Uvarov method [13-14] are widely used to obtain the energy spectrum and the wave function of the charged particle moved in the non-central potential.

In this paper the anisotropic nonquadratic potential which has been investigated using path integral method [15], the method of algebraic solutions [16], the semmiclassical treatment [17] and now will be investigated by using supersymmetric Quantum Mechanics (SUSY QM) [18] and the idea of the invariance form [19]. SUSY QM is a powerful tool for determining the energy spectrum and the wave functions of the form class invariance potentials [18-22].
Anisotropic nonquadratic potential given as [15-17]

$$V(\vec{r}) = V(r,\theta,z) = -\frac{\alpha}{\sqrt{x^2+y^2}} + \frac{\beta}{y^2} + \frac{\gamma x}{y^2\sqrt{x^2+y^2}} + \alpha' z^2 + \frac{\beta'}{z^2} \tag{1}$$

Modified anisotropic nonquadratic potential with exponential potential in axial part is

$$V(\bar{r}) = V(r,\theta,z) = -\frac{V_0}{\sqrt{x^2+y^2}} + \frac{V_1}{x^2} + \frac{V_2 y}{x(x^2+y^2)} + \frac{V_3 e^{-\alpha z}}{(1-e^{-\alpha z})^2} + V_4 \frac{1+e^{-\alpha z}}{1-e^{-\alpha z}} \tag{2}$$

Schrodinger equation in Eq. (2) is expressed in the three-dimensional type of the Schrodinger equation in cylindrical coordinate and it is solved exactly using variable separation method. Schrodinger equation of radial, angular, and axial parts are solved using SUSY operator and the idea of shape invariant.

2. Schrodinger Equation for Modified Anisotropic Nonquadratic Potential

Schrödinger equation is defined by

$$\frac{-\hbar^2}{2m}\nabla^2\psi(\bar{r}) + V(\bar{r})\psi(\bar{r}) = E\psi(\bar{r}) \tag{3}$$

Modified anisotropic nonquadratic potential in Eq.(2) is changed by applying cylindrical coordinates, and apply the operator ∇^2 for cylindrical coordinate, with the wave function $\psi(\bar{r}) = R(r)P(\theta)\Phi(z)$. The wave function which have variable r, θ, and z can be separated with ordinary algebra method.
So the Schrodinger equation that has been separated may be written by

- Axial part

$$-\frac{\hbar^2}{2m}\frac{\partial^2 Z}{\partial z^2} + \left(\frac{V_3}{4\sinh\frac{\alpha z}{2}} + V_4 \coth\frac{\alpha z}{2}\right)Z = -\lambda_1 Z \tag{4}$$

- Angular part

$$-\frac{\hbar^2}{2m}\frac{\partial^2 P}{\partial\theta^2} + \left(\frac{V_1}{\cos^2\theta} + V_2\tan\theta\right)P = -\lambda_2 P \tag{5}$$

- Radial part

$$-\frac{\hbar^2}{2m}\left[r\frac{\partial}{\partial r}\left(r\frac{\partial R}{\partial r}\right)\right] - \left(E + \lambda_1 + \frac{V_0}{r}\right)r^2 R = \lambda_2 R \tag{6}$$

3. Solution to The Three Dimensional Schrodinger Equations with SUSY QM Method

Supersymmetry quantum mechanics is used to solve one dimensional Schrödinger equation with any shape invariance potentials. After has been separated into three parts, each part of Schrodinger equation would be solved using SUSY QM to find the wave functions and energy spectra .

3.1. Solution of Axial Part

The Schrödinger equation for axial part (Eq. 23) is changed into hyperbolic function. Assume that $\gamma = \frac{\alpha}{2}$, and suppose that $V_3 = \frac{\hbar^2}{2m}\alpha^2 v(v-1)$, $V_4 = -\frac{\hbar^2}{m}q$, and $E' = -\lambda_1$ so Eq.(4) can be resolved into Schrödinger equation with the effective potential is Manning Rosen-like potential

$$-\frac{\hbar^2}{2m}\frac{\partial^2 Z}{\partial z^2} + \frac{\hbar^2}{2m}\left(\gamma^2\frac{v(v-1)}{\sinh^2\gamma z} - 2q\coth\gamma z\right)Z = E'Z \tag{7}$$

We hypothesize that the superpotential is given by $\phi_o(z,a_0) = A\coth\gamma z + \frac{B}{A}$, by applying

$$V(z) = V_-(z;a_0) + E_0 = \phi^2(z;a_0) - \frac{\hbar}{\sqrt{2m}}\phi'(z;a_0) + E_0$$

$$V_-(z,a_0) = \phi^2(z) - \frac{\hbar}{\sqrt{2m}}\phi'(z); \qquad V_+(z,a_0) = \phi^2(z) + \frac{\hbar}{\sqrt{2m}}\phi'(z) \tag{8}$$

Where $V(z)$ is the effective potential. So we get the superpotential, super-partner potentials and ground state energy are

$$\phi_o(\chi,a_0)=\frac{\hbar}{\sqrt{2m}}\gamma(\nu-1)\coth\gamma z-\frac{\hbar}{\sqrt{2m}}\frac{q}{\gamma(\nu-1)} \tag{9}$$

$$V_-(\chi,a_0)=\frac{\hbar^2}{2m}\left(\gamma^2\frac{\nu(\nu-1)}{\sinh^2\gamma z}-2q\coth\gamma z\right)+\frac{\hbar^2}{2m}\left[\gamma^2(\nu-1)^2+\frac{q^2}{\gamma^2(\nu-1)^2}\right] \tag{10}$$

$$V_+(\chi,a_0)=\frac{\hbar^2}{2m}\left(\gamma^2\frac{(\nu-2)(\nu-1)}{\sinh^2\gamma z}-2q\coth\gamma z\right)+\frac{\hbar^2}{2m}\left[\gamma^2(\nu-1)^2+\frac{q^2}{\gamma^2(\nu-1)^2}\right] \tag{11}$$

$$E_0'=-\frac{\hbar^2}{2m}\left[\gamma^2(\nu-1)^2+\frac{q^2}{\gamma^2(\nu-1)^2}\right] \tag{12}$$

A pair of potentials $V_\pm(z)$ in Eqs.(8) are said to be shape invariant if they are similar in shape but different in the parameters and this condition is given as,

$$V_+(z;a_j)=V_-(z;a_{j+1})+R(a_{j+1}) \tag{13}$$

where j = *0,1,2,..*, and a is a parameter in our original potential. The energy eigenvalue of the Hamiltonian H_- is given by

$$E_n^{(-)}=\sum_{j=1}^{n}R(a_j)$$

$$E_n=E_n^{(-)}+E_0 \tag{14}$$

and by using Eqs.(14) and (13) we get the energy spectra of the system given as,

$$E_{n_z}'=-\frac{\hbar^2}{2m}\left[\gamma^2(\nu-n_z+1)^2+\frac{q^2}{\gamma^2(\nu-n_z+1)^2}\right] \tag{15}$$

So we get the first separation variable constant is

$$\lambda_1=\left[\gamma^2(\nu-n_z+1)^2+\frac{q^2}{\gamma^2(\nu-n_z+1)^2}\right] \tag{16}$$

The raising and lowering operator for axial part are given by

$$\mathrm{A}^+=-\frac{\hbar}{\sqrt{2m}}\frac{d}{dz}+\frac{\hbar}{\sqrt{2m}}\gamma(\nu-1)\coth\gamma z-\frac{\hbar}{\sqrt{2m}}\frac{q}{\gamma(\nu-1)} \tag{17}$$

$$\mathrm{A}=\frac{\hbar}{\sqrt{2m}}\frac{d}{dz}+\frac{\hbar}{\sqrt{2m}}\gamma(\nu-1)\coth\gamma z-\frac{\hbar}{\sqrt{2m}}\frac{q}{\gamma(\nu-1)} \tag{18}$$

Using the lowering operator Eq.(18) and applying

$$H_-\psi_0^{(-)}=0\rightarrow A\psi_0^{(-)}=0 \tag{19}$$

So we get the ground state of wave function is

$$Z_0=C'\sinh^{(\nu-1)}\gamma z\, e^{\frac{q}{\gamma(\nu-1)}z} \tag{20}$$

And applying

$$\psi_n^{(-)}(z;a_0)\approx A^+(z;a_0)A^+(z;a_1)....A^+(z;a_{n-1})\psi_0^{(-)}(z;a_n) \tag{21}$$

The first excited wave function is given by

$$Z_1(z,a_0)=C\sinh^{\nu}\gamma z\, e^{\frac{q}{\gamma\nu}z}\left[\coth\gamma z+\frac{q}{\gamma^2\nu(\nu-1)}\right] \tag{22}$$

With $a_0=\nu, a_1=\nu-1,...,a_n=\nu-n$ so we can get the next excited wave functions of axial part by using raising Eq.(17) and lowering Eq.(18) operator and applying Eqs.(14) and (21).

3.2. Solution of Angular Part

The potential of Schrodinger equation for angular part is changed by $V_1 = \frac{\hbar^2}{2m} v(v+1)$ and $V_2 = -\frac{\hbar^2}{m} q$, so the effective potential in angular part becomes Kepler problem in hypersphere potential-like and we assumed that $E'' = -\lambda_2$. Then the angular part becomes,

$$-\frac{\hbar^2}{2m}\frac{\partial^2 P}{\partial\theta^2} + \frac{\hbar^2}{2m}\left(\frac{v(v+1)}{cos^2\theta} - 2qtan\theta\right)P = E''P \tag{23}$$

To find solution of angular part, we put the same method like in the subsection 4.1. We have to define the superpotentials, $\phi_0(\theta) = A\tan\theta + B$, by using Eqs.(8) and (13-14) we get the ground state energy for angular part and the energy spectra are given by

$$E_0'' = -\frac{\hbar^2}{2m}\left((v+1)^2 + \frac{q^2}{(v+1)^2}\right) \tag{24}$$

$$E_{n_\theta}'' = -\frac{\hbar^2}{2m}\left((v+n_\theta+1)^2 + \frac{q^2}{(v+n_\theta+1)^2}\right) \tag{25}$$

So the second variable constant is

$$\lambda_2 = \left(\frac{q^2}{(v+n_\theta+1)^2} + (v+n_\theta+1)^2\right) \tag{26}$$

The raising and lowering operator for angular part are given by

$$A^+(\theta,a_0) = -\frac{\hbar}{\sqrt{2m}}\frac{d}{d\theta} + \frac{\hbar}{\sqrt{2m}}(v+1)\tan\theta - \frac{\hbar}{\sqrt{2m}}\frac{q}{(v+1)} \tag{27}$$

$$A(\theta,a_0) = \frac{\hbar}{\sqrt{2m}}\frac{d}{d\theta} + \frac{\hbar}{\sqrt{2m}}(v+1)\tan\theta - \frac{\hbar}{\sqrt{2m}}\frac{q}{(v+1)} \tag{28}$$

Using the lowering operator Eq.(28) and applying Eq.(19) we get the ground state of wave function is

$$P_0 = D'\left(\cos^{(v+1)}\theta\right)e^{\frac{q\theta}{(v+1)}} \tag{29}$$

And the first excited wave function for angular part is

$$P_1 = D\frac{\hbar}{\sqrt{2m}}\frac{e^{\frac{q\theta}{(v+2)}}}{\sec^{(v+2)}\theta}\left[(v+1)\tan\theta - (v+2)\cos\theta - q\left(\frac{1}{v+1} + \frac{1}{v+2}\right)\right] \tag{30}$$

With $a_0 = v, a_1 = v+1, ..., a_n = v+n$ so we can get the next excited wave functions by using raising Eq.(27) and lowering Eq.(28) operator and applying Eq.(21).

3.3. Solution of Radial Part

Radial part in Eq.(6) must reduce into Schrödinger equation by assumed that $R = \frac{U}{\sqrt{r}}$ and then we change the form of the potentials with $V_0 = e^2$, so the effective potential in radial part becomes Coulomb-like potential and we assumed that $E''' = E - \frac{\hbar^2}{2m}\lambda_1$. Then the radial part has been reduced into Schrödinger equation, given as

$$-\frac{\hbar^2}{2m}\frac{\partial^2 U}{\partial r^2} + \left(\frac{\hbar^2}{2m}\frac{\lambda_2 - \frac{1}{4}}{r^2} - \frac{e^2}{r}\right)U = E'''U \tag{31}$$

We hypothesize that the superpotential for the radial part is $\phi_0(r) = \frac{A}{r} + B$, by using the same method like subsections 1 and 2 so we get the ground state energy and the energy spectra for radial part are given by

Journal of Physics: Conference Series **776** (2016) 012079 doi:10.1088/1742-6596/776/1/012079

$$E_0''' = -\frac{me^4}{2\hbar^2\left(\sqrt{\lambda_2}-\frac{1}{2}\right)^2} \tag{32}$$

$$E_{n_r}''' = -\frac{me^4}{2\hbar^2\left(\sqrt{\lambda_2}-n_r-\frac{1}{2}\right)^2} \tag{33}$$

So we get the energy spectra of the system is

$$E = \frac{\hbar^2}{2m}\left[\gamma^2(v-n_z+1)^2 + \frac{q^2}{\gamma^2(v-n_z+1)^2}\right] - \frac{me^4}{2\hbar^2\left(\sqrt{\frac{q^2}{(v+n_\theta+1)^2}+(v+n_\theta+1)^2}-n_r-\frac{1}{2}\right)^2} \tag{34}$$

The energy spectra of an anisotropic non-quadratic potential expressed in (Eq. 2) is given in (Eq. 57). There are two parts of the energy spectra, the first part is associated with the axial part of the potential, and the second part is associated with radial and polar parts of potential. Without the presence of axial part of potential, the energy spectra reduces to the energy spectra of Coulomb-like potential that is modified by the presence of the angular potential.
The raising and lowering operator for radial part are given by

$$A^+(r,a_0) = -\frac{\hbar}{\sqrt{2m}}\frac{d}{dr} + \frac{\hbar}{\sqrt{2m}}\frac{\sqrt{\lambda_2}-\frac{1}{2}}{r} - \frac{\sqrt{2m}}{2\hbar}\frac{e^2}{\sqrt{\lambda_2}-\frac{1}{2}} \tag{35}$$

$$A(r,a_0) = \frac{\hbar}{\sqrt{2m}}\frac{d}{dr} + \frac{\hbar}{\sqrt{2m}}\frac{\sqrt{\lambda_2}-\frac{1}{2}}{r} - \frac{\sqrt{2m}}{2\hbar}\frac{e^2}{\sqrt{\lambda_2}-\frac{1}{2}} \tag{36}$$

Using the lowering operator Eq.(27) and applying Eq.(19) we get the ground state of wave function is

$$U_0 = F' r^{\sqrt{\lambda_2}-\frac{1}{2}} e^{\frac{m}{\hbar^2}\frac{e^2}{\sqrt{\lambda_2}-\frac{1}{2}}r} \tag{37}$$

And the first excited wave function for radial part is given by

$$U_1 = C\left[\left(\frac{\hbar}{\sqrt{2m}} r^{\sqrt{\lambda_2}-\frac{5}{2}} e^{\frac{m}{\hbar^2}\frac{e^2}{\sqrt{\lambda_2}-\frac{3}{2}}r}\right) - \left(\frac{\sqrt{2m}}{2\hbar}e^2\left(\frac{1}{\sqrt{\lambda_2}-\frac{5}{2}}+\frac{1}{\sqrt{\lambda_2}-\frac{3}{2}}\right)\right) r^{\sqrt{\lambda_2}-\frac{3}{2}} e^{\frac{m}{\hbar^2}\frac{e^2}{\sqrt{\lambda_2}-\frac{3}{2}}r}\right] \tag{38}$$

With $a_0 = \sqrt{\lambda_2}, a_1 = \sqrt{\lambda_2}-1, \ldots, a_n = \sqrt{\lambda_2}-n$ so we can get the next excited wave functions by using raising Eq.(35) and lowering Eq.(36) operator and applying Eq.(19). The total wave functions, the un-normalized ground state and first excited state ones, are obtained from Eqs.(20,22), Eqs.(29-30) and Eqs.(37-38) given as

$$\psi_0(r,\theta,z) = G'\left(\cos^{(v+1)}\theta\right)\left(\sinh^{(v-1)}\gamma z\right) r^{\sqrt{\lambda_2}-\frac{1}{2}} e^{\frac{m}{\hbar^2}\frac{e^2}{\sqrt{\lambda_2}-\frac{1}{2}}r + \frac{q}{(v+1)}\theta + \frac{q}{\gamma(v-1)}z} \tag{58}$$

$$\psi_1(r,\theta,z) \cong \left[\left(\frac{\hbar}{\sqrt{2m}} r^{\sqrt{\lambda_2}-\frac{5}{2}} e^{\frac{m}{\hbar^2}\frac{e^2}{\sqrt{\lambda_2}-\frac{3}{2}}r}\right) - \left(\frac{\sqrt{2m}}{2\hbar}e^2\left(\frac{1}{\sqrt{\lambda_2}-\frac{5}{2}}+\frac{1}{\sqrt{\lambda_2}-\frac{3}{2}}\right)\right) r^{\sqrt{\lambda_2}-\frac{3}{2}} e^{\frac{m}{\hbar^2}\frac{e^2}{\sqrt{\lambda_2}-\frac{3}{2}}r}\right]$$

$$\times \frac{\hbar}{\sqrt{2m}}\frac{e^{\frac{q\theta}{(v+2)}}}{\sec^{(v+2)}\theta}\left[(v+1)\tan\theta-(v+2)\cos\theta-q\left(\frac{1}{v+1}+\frac{1}{v+2}\right)\right] \tag{59}$$

$$\times \sinh^{v}\gamma z\, e^{\frac{q}{\gamma v}z}\left[\coth\gamma z + \frac{q}{\gamma^2 v(v-1)}\right]$$

with λ_2 expressed in (Eq. 26) respectively.

4. Conclusion

By applying a cylindrical coordinate system, the Schrödinger equation for the anisotropic nonquadratic that modified by exponential form in axial part reduces to a perfectly variable separable Schrödinger equation. Three dimensional Schrödinger equation reduces to one radial Schrodinger equation, one angular Schrodinger equation, and one axial Schrodinger equation. These three one dimensional Schrödinger equation are exactly solvable since each of Schrödinger equation with shape invariant potential. By using SUSY Quantum Mechanics method and the idea of shape invariance the energy spectra are calculated and so the total wave functions which are combination of axial, angular and radial parts.

Acknowledgement

This research was partly supported by Higher Education Project Grant with contract number 632/UN27.21/LT/2016.

References

[1] Villalba VM, 1994 *Phys. Lett.* A **193**, pp. 218-222.
[2] Aharonov Y and Bohm D, 1959 *Phys. Rev.* **115**, pp. 485.
[3] Hoang LV, Hai LX, Komarov LI and Romanova TS, 1992 *J. Phys.* a **25**, pp. 6461
[4] Movahedi M and Hamzavi M, 2011, *Int. J. Phys. Sci.* **6** (4) pp.891
[5] Gonul B and Zorba I, 2000 *Phys. Lett. A* 269 pp.83
[6] Ferrer S, Mono. Acad. Ci. Zar, 2006, *Phys. Lett. A* Vol. 30 (2006) 11-21
[7] Chen CY and Dong SH, 2005, *Phys. Lett. A* 335 pp. 374
[8] Quan HX, Guang L, Min WZ, Bin NL, and Yan M, 2010 *Comm.Theor. Phys.* (China) **53** pp. 242
[9] Wu QS, Wen HB, Yun WD, and Yu GZ, Comm.2002, *Theor. Phys.*(China), **38** pp. 139
[10] Sheng JC, Ying Z, Lin ZX, and Tian SL, 2001 *Comm. Theor. Phys.* **36** pp. 641-646
[11] Sadeghi J, 2007, *Act.Phys.Polonica*, A **112** (1) pp. 23
[12] Sadeghi J and Pourhassan B, 2008 *EJTP* **5** (17) pp. 193
[13] Nikiforov AF and Uvarov UB, Special Function in Mathematical Physics, Birkhausa, Basel (1988).
[14] Yasuk F, Berkdemir C, and Onem C, 2005 *Phys. Scr.* **71** pp. 472
[15] Cai, J. M.; Inomata, A., 1989 *Physics Letters A*, Vol **141**, Issue 7, pp. 315-320.
[16] Boschi-Filho H, Vaidya AN, 1990, *Physics Letters A*, Vol **149**, pp. 336-340
[17] Vasan SS, Seetharaman M, and Raghunathan K, 1993, *Journal Pramana* Vol. **40**(1); p. L75-L79
[18] Witten E, 1981, *Nucl. Phys. B* Vol.**188** pp. 513
[19] Gendenshtein LE, 1983 *JETP Lett.*Vol. **38** pp. 356
[20] Dutt R, Khare A, and Sukhatme UP, 1988, *Am. J. Phys.* Vol.**56**, pp.163.
[21] Dutt R, Gangopadhyaya A, and Sukhatme UP, 1997, *Am. J. Phys.* **65**(5).
[22] Gangopadhyaya A, Mallow JV, Rasinariu C, and Sukhatme UP, 2001, *Chin. J. Phys.* Vol. **39**(2)

Journal of Physics: Conference Series **776** (2016) 012080 doi:10.1088/1742-6596/776/1/012080

Steady flow over an arbitrary obstruction based on the gravity wave equations

S. Mungkasi[1], B. Supriyadi[2] and L. H. Wiryanto[3]

[1]Department of Mathematics, Faculty of Science and Technology, Sanata Dharma University, Yogyakarta, Indonesia
[2]Department of Civil and Environmental Engineering, Faculty of Engineering, Gadjah Mada University, Yogyakarta, Indonesia
[3]Department of Mathematics, Faculty of Mathematics and Natural Sciences, Bandung Institute of Technology, Bandung, Indonesia

E-mail: sudi@usd.ac.id

Abstract. We derive an analytical solution to a steady state problem of the gravity wave equations. An arbitrary bottom topography is considered. The problem is assumed to be one dimensional. The depth, discharge and topography elevation at the left-end of the space domain are assumed to have the same values as those at the right-end. We obtain that the fluid surface on the whole interior space domain remains horizontal and is not influenced by the topography shape when we use the gravity wave equations. Furthermore, the analytical solution that we derive is used to test the performance of a finite volume method. We find that the gravity wave equations give some advantages in comparison to the shallow water equations.

1. Introduction

The shallow water equations have been widely accepted as a model for shallow water flows on open channels, such as urban floods, tsunamis, river flows, etc. These equations are often simplified in order to reduce the computational costs and to increase the numerical stability leading to some simpler models, for example, kinematic wave equations, diffusive wave equations and gravity wave equations [1].

Martins et al. [1] presented the analytical solution to the classical dam break problem based on the gravity wave equations. Then they [2] reported the effectiveness of the gravity wave equations for urban flood simulations. Their work [1, 2] was on unsteady state problems. Steady state problems based on the gravity wave equations have, to our knowledge, not been investigated.

Our contributions are as follows. We derive the analytical solution to a steady state problem of the gravity wave equations with an arbitrary bottom topography. In addition, we verify the performance of the finite volume method used by Hidayat et al. [3] in solving steady state problems of the gravity wave equations.

This paper is organized as follows. Mathematical models are recalled in Section 2. We derive the analytical solution to the steady state problem in Section 3. Numerical results for a scenario of the gravity wave equations and the shallow water equations are presented in Section 4. Finally concluding remarks are drawn in Section 5.

2. Mathematical models

The gravity wave equations are [1]

$$\frac{\partial}{\partial t}h+\frac{\partial}{\partial x}(uh)=0\,, \tag{1}$$

$$\frac{\partial}{\partial t}(uh)+\frac{\partial}{\partial x}\left(\frac{1}{2}gh^2\right)=-gh\frac{d}{dx}z\,. \tag{2}$$

Here x is one dimensional space variable, t is time variable, $h=h(x,t)$ represents the fluid depth (height), $u=u(x,t)$ denotes the fluid velocity, $z=z(x)$ is the bed (bottom) topography, and g is the acceleration due to gravity. Note that the gravity wave equations (1)-(2) are simplifications of the standard shallow water equations

$$\frac{\partial}{\partial t}h+\frac{\partial}{\partial x}(uh)=0\,, \tag{3}$$

$$\frac{\partial}{\partial t}(uh)+\frac{\partial}{\partial x}\left(u^2h+\frac{1}{2}gh^2\right)=-gh\frac{d}{dx}z\,, \tag{4}$$

where the convective term $\frac{\partial}{\partial x}(u^2h)$ is neglected [1, 2]. We assume that friction is not involved in the models. The expression uh is the fluid discharge. Equations (1)-(2) and (3)-(4) are one dimensional. Readers interested in two dimensional equations may consult the shallow water literature, such as the work of Aronica et al. [4].

3. Analytical results

Now we suppose that a space domain is given where the fluid depth $h(x,t)$ is positive everywhere on the domain at any time. A fixed bottom obstruction $z(x)$ is also known.

A steady state is achieved when the water depth $h(x,t)$ and the velocity $u(x,t)$ are independent with respect to time t. This leads to

$$\frac{\partial}{\partial t}h=0 \qquad \text{and} \qquad \frac{\partial}{\partial t}(uh)=0\,. \tag{5}$$

As a result we have

$$\frac{\partial}{\partial x}(uh)=0 \qquad \text{and} \qquad \frac{\partial}{\partial x}\left(\frac{1}{2}gh^2\right)=-gh\frac{d}{dx}z\,. \tag{6}$$

That is,

$$uh=q \qquad \text{and} \qquad h+z=c\,, \tag{7}$$

where q and c are constants. Far from the obstacle, the bottom elevation, the water depth and the velocity are assumed to be $z(x)=0$, $h(x,t)=h_0$ and $u(x,t)=u_0$ respectively. Thus for the whole domain, we obtain

$$uh=u_0h_0 \qquad \text{and} \qquad h+z=h_0 . \tag{8}$$

The analytical solution to the steady state problems of the gravity wave equations is as follows. The analytical solution for the fluid depth is

$$h(x,t)=h_0-z(x) , \tag{9}$$

and the analytical solution for the fluid velocity is

$$u(x,t)=\frac{u_0h_0}{h_0-z(x)} , \tag{10}$$

at any space point x and any time $t>0$.

4. Numerical results

In this section we present some numerical results related to the analytical solution that we obtain in the previous section for a scenario of the steady state problem based on the gravity wave equations. All measured quantities are assumed to have SI units. The acceleration due to gravity is 9.81.

Consider the gravity wave equations (1)-(2). We refer to the finite volume method described by Hidayat et al. [5] with first order accuracy. Following Hidayat et al. [5] we take the channel space domain $0 \le x \le 25$. The bottom topography is

$$z(x)=\begin{cases} 0.2-0.05(x-10)^2 & \text{if} \quad 8 \le x \le 12, \\ \\ 0 & \text{if} \quad \text{otherwise.} \end{cases} \tag{11}$$

The boundary condition is

$$q(0,t)=4.42 \qquad \text{and} \qquad u(25,t)=2 , \tag{12}$$

at the ends of the space domain for $t>0$. The initial condition is

$$q(x,0)=4.42 \qquad \text{and} \qquad h(x,0)+z(x)=2.21, \tag{13}$$

for $t>0$.

At time $t=30$ the results of the gravity wave equations are shown in Figure 1. We use 100 cells with the time step is 0.05 times the cell width. The numerical solution matches with the exact analytical solution with the depth error is 7.6983×10^{-15} and the velocity error is 2.9088×10^{-14}. This means that the finite volume method is able to solve the problem very accurately up to the machine precision.

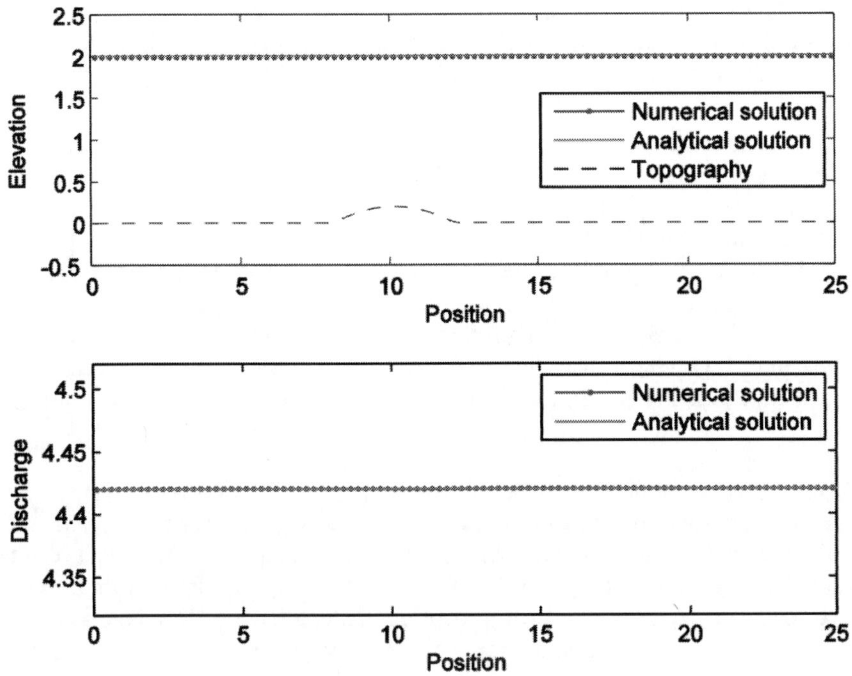

Figure 1. Simulation results of the gravity wave equations. The numerical solution matches the analytical solution up to the machine precision.

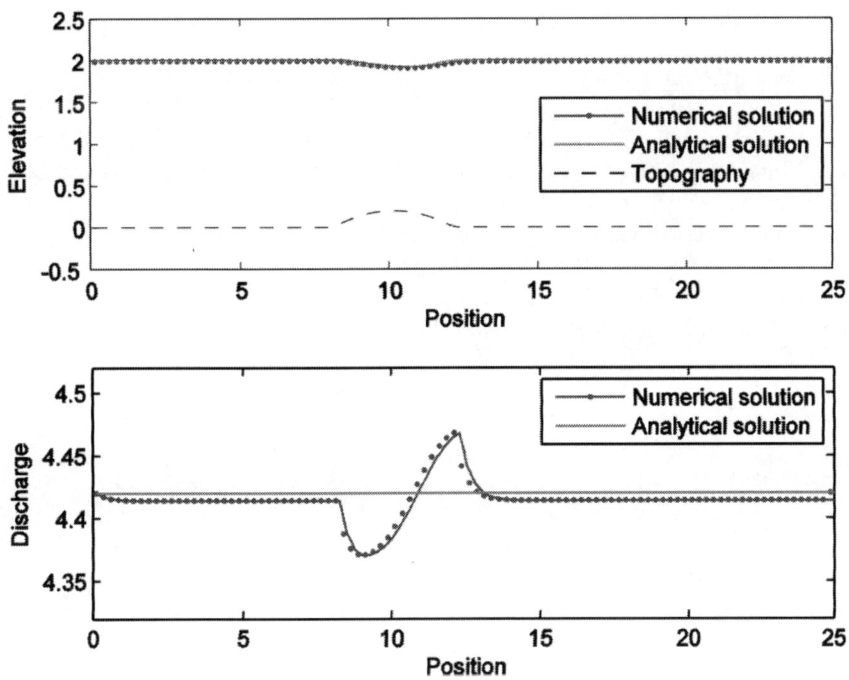

Figure 2. Simulation results of the shallow water equations. The finite volume method solves the shallow water equations not as accurately as the gravity wave equations.

In comparison to the gravity wave equations, we also plot here the results of the same problem, but based on the shallow water equations (3)-(4), as shown in Figure 2. We also take 100 cells with the time step is 0.05 times the cell width. The numerical solution matches with the exact solution with the depth error is 0.0053 and the velocity error is 0.0102. We obtain from Figure 2 that the source of errors is mainly at the region where the topographical obstruction exists. However, these errors can be made small by taking more number of computational cells. More results on numerical solutions to the shallow water equations can be found in the work of Mungkasi et al. [5-8]. Note that the analytical work on the steady state problems of shallow water equations was presented in some literature, for example that done by Houghton and Kasahara [9].

We observe that, for the steady state problem, the finite volume method performs better in solving the gravity wave equations than in solving the shallow water equations. This is another advantage of using the gravity wave equations for water flow simulation in addition to their simpler forms, as long as the gravity effect is much more significant than the convective effect. That is, as long as the convective term of the shallow water equations is negligible for the particular problem under consideration.

5. Conclusion

We have derived an analytical solution to the steady state problem of flows over an arbitrary topography based on the gravity wave equations. The analytical solution can be applied to test the performance of numerical methods used to solve the gravity wave equations. Numerical results show that the finite volume method solves the gravity wave equations more accurately than the shallow water equations. This means that the gravity wave equations indeed have great advantages in saving computational costs and in giving opportunity to be solved more accurately.

Acknowledgments

This work was financially supported by: (1). Sanata Dharma University and (2). a DIKTI research grant of The Ministry of Research, Technology and Higher Education of The Republic of Indonesia year 2016.

References

[1] Martins R, Leandro J and Djordjevic, S 2016 *J. Hydraul. Eng.* **142** 06016003
[2] Martins R, Leandro J and Djordjevic, S 2016 *J. Flood Risk Manag.* DOI: 10.1111/jfr3.12244
[3] Hidayat N, Suhariningsih, Suryanto A and Mungkasi S 2014 *Appl. Math. Sci.* **8** 1411
[4] Aronica G, Tucciarelli T and Nasello C 1998 *J. Water Resour. Plan. Manag.* **124** 210
[5] Mungkasi S 2013 *B. Aust. Math. Soc.* **88** 351
[6] Mungkasi S and Roberts S G 2013 *ANZIAM J.* **54** C18
[7] Mungkasi S and Roberts S G 2015 *ANZIAM J.* **56** C128
[8] Mungkasi S, Li Z and Roberts S G 2014 *Appl. Math. Lett.* **30** 51
[9] Houghton D D and Kasahara A 1968 *Comm. Pure Appl. Math.* **21** 1

Parallel computations using a cluster of workstations to simulate elasticity problems

J. B. B. Darmawan[1] and S. Mungkasi[2]

[1]Department of Informatics, Faculty of Science and Technology, Sanata Dharma University, Yogyakarta, Indonesia
[2]Department of Mathematics, Faculty of Science and Technology, Sanata Dharma University, Yogyakarta, Indonesia

E-mail: b.darmawan@usd.ac.id

Abstract. Computational physics has played important roles in real world problems. This paper is within the applied computational physics area. The aim of this study is to observe the performance of parallel computations using a cluster of workstations (COW) to simulate elasticity problems. Parallel computations with the COW configuration are conducted using the Message Passing Interface (MPI) standard. In parallel computations with COW, we consider five scenarios with twenty simulations. In addition to the execution time, efficiency is used to evaluate programming algorithm scenarios. Sequential and parallel programming performances are evaluated based on their execution time and efficiency. Results show that the one-dimensional elasticity equations are not appropriate to be solved in parallel with MPI_Send and MPI_Recv technique in the MPI standard, because the total amount of time to exchange data is considered more dominant compared with the total amount of time to conduct the basic elasticity computation.

1. Introduction

One of mathematics models describing elastic wave propagation in solid media is the elasticity partial differential equation model. This model can be applied in the real world, such as propagation of earthquakes, acoustic waves and waves on an elastic medium. This paper focuses on wave propagation in solid media.

The elasticity model is solved numerically. It may be solved analytically, but it is very difficult to find an explicit form of its exact solution. Therefore, this paper deals with elasticity problems for their numerical solutions.

Usually it is difficult for a single computer to gain high speed computation. A cluster of workstations (COW) often offers economical and efficient approach to addressing this issue. Parallel programming using COW has been proven to improve the speed of computation to simulate flood flows using ANUGA software [1]. Therefore, a parallel programming might speed up the computation of elasticity simulations, compared to the sequential programming. To analyze the results of the parallel computing, we use speed and efficiency variables. This paper investigates if we obtain high speed and efficiency in parallel computations using a cluster of workstation to solve elasticity problems.

This paper is organized as follows. We recall the mathematical model governing elasticity problems and present the numerical scheme to solve the model in Section 2. In Section 3, the numerical method is implemented in parallel to solve elasticity problems. We provide computational results and discussion in Section 4. Finally, we draw some concluding remarks in Section 5.

2. Mathematical model and numerical scheme

To clarify which mathematical model and numerical scheme used in our work, we recall them in this section.
The mathematical model governing the elasticity problems are [2]

$$\varepsilon_t - u_x = 0, \tag{1}$$

$$u_t - \sigma(\varepsilon)_x = 0. \tag{2}$$

Here t is time variable and x is the space (position) variable, $\varepsilon = \varepsilon(x,t)$ is the strain, $u = u(x,t)$ is the velocity, and $\sigma(\varepsilon)$ is the stress. We consider the space domain $0 \le x \le 100$. We assume that the initial condition is

$$\varepsilon(x,0) = 0, \tag{3}$$

$$u(x,0) = 0, \tag{4}$$

for all x. The boundary condition is

$$\varepsilon(0,t) = 0, \tag{5}$$

$$u(0,t) = \begin{cases} -0.4\left(1+\cos\left(\dfrac{(t-30)\pi}{30}\right)\right) & \text{if} \quad t \le 60, \\ 0 & \text{if} \quad t > 60, \end{cases} \tag{6}$$

$$\varepsilon(100,t) = 0, \tag{7}$$

$$u(100,t) = 0. \tag{8}$$

Note that equations (1)-(2) are conservation laws in the form

$$q_t + f(q)_x = 0, \tag{9}$$

where $q = q(x,t)$ is the conserved quantity and $f(q)$ is the flux function. Conservation laws can be solved using finite volume numerical method.
The fully discrete finite volume scheme for equation (5) is

$$Q_i^{n+1} = Q_i^n - \frac{\Delta t}{\Delta x}\left(F_{i+1/2}^n - F_{i-1/2}^n\right), \tag{10}$$

where Q_i^n is the average approximation of the quantity at the i-th cell at the n-th time step, and $F_{i+1/2}^n$ is the average approximation of the flux function at the $(i+1/2)$ interface from the n-th time step to the $(n+1)$-th time step. All approximations are in the finite volume sense. In this paper we use the Lax-Friedrichs flux formulation. All quantities are assumed to have SI units. Numerical computation is conducted in such a way that the numerical method is stable.

3. Method for parallel computations

Conceptually, distributed processing such as COW is not different from distributed memory MIDP according to Flynn's classification, but the communication between processors is slower as they are connected such as by Ethernet, Fiber Channel or ATM [3, 4]. We implement Message Passing Interface (MPI) on Personal Computer cluster to realize COW for parallel programming with the C programming language.

We apply cluster computations using PC's interconnected by network technology. These computations have great advantages, such as high performance for system scalability, rapid adjustment to new technological and low price [5]. The availability of Gigabit Ethernet, which has speed up the communication performance of PC cluster using MPI than ever before (such as Fast Ethernet, Myrinet and Scalable Coherent Interface), supports those advantages [6].

MPI is one of the most popular communication protocols for PC cluster to support parallel programming. It is a library specification for message-passing in the network [7] and one of its implementations is MPICH which is a high performance and widely portable implementation of the MPI standard [8]. As the implementation of MPI, we use MPICH2 which supports the C programming language.

As execution time varied with problem size, Foster [9] proposed efficiency which is independent of problem size to be a metric to evaluate parallel algorithm performance. However, execution time is also an important issue in computations. Therefore, the execution time as well as the efficiency are considered in this paper to evaluate our parallel algorithm scenarios.

The relative speed up S_{relative} is formulated as

$$S_{\text{relative}} = \frac{T_1}{T_{\text{p}}}, \tag{11}$$

where T_1 is the execution time on one processor and T_{p} is the time on p processors. The relative speed up means the factor by which execution time is reduced on p processors. Relative efficiency E_{relative} is defined as

$$E_{\text{relative}} = \frac{T_1}{p \cdot T_{\text{p}}}. \tag{12}$$

From equations (11)-(12), the relative speed up is clearly related to the relative efficiency.

4. Computational results

In this section we present main results of our research. Computations are conducted in a cluster of workstations in Sanata Dharma University. The cluster consists of 16 PC's, one PC has one i3 processor. Note that an i3 processor has 2 cores and 4 threads. All 16 PCs run on Linux operating system. For network communication we use a one Gigabit Ethernet for each PC connected with switch hub one Gigabit. In addition, we have four gigabytes RAM in each PC. We refer to the work of Foster [9] and Yang *et al.* [10] for concepts of parallel programming.

We have conducted 20 simulations for running program in order to solve the elasticity problem mentioned in Section 2. One simulation is run in sequential and the others are run in parallel. We use scenarios of 1, 2, 3, 4, 8, and 16 PCs for our computations. For each scenario, we consider 1, 2, 3, and 4 processes on each PC.

For a basic elasticity problem we record the total time for each simulation in Table 1. As shown in Figure 1, for the total time we observe that more number of processes leads to slower computation. These anomalies occurred, because the more number of processes needs the more amount of time to distribute data from the master process to several worker processes and to collect data from several worker processes to the master process in the MPI_Send and MPI_Recv techniques in the MPI standard. The total amount of time to exchange data is considered more dominant compared with the total amount of time to conduct elasticity computations with the elasticity model for all processes.

Table 1. Total time for an elasticity simulation scenario (in seconds).

Number of PCs	1 process	2 processes	3 processes	4 processes
1	0.8330	0.8680	1.2655	1.4515
2	20.9475	33.4225	46.0460	58.4080
4	49.5995	86.6760	123.4045	160.9290
8	101.1100	186.9505	273.4295	360.1995
16	200.6775	385.0150	570.1225	755.7885

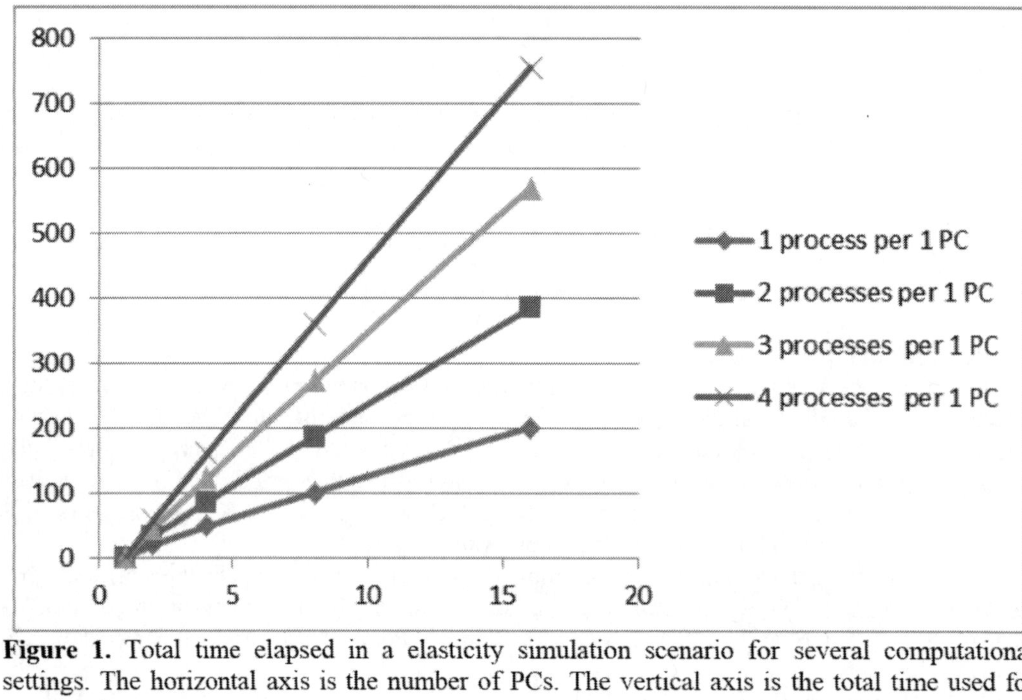

Figure 1. Total time elapsed in a elasticity simulation scenario for several computational settings. The horizontal axis is the number of PCs. The vertical axis is the total time used for computations.

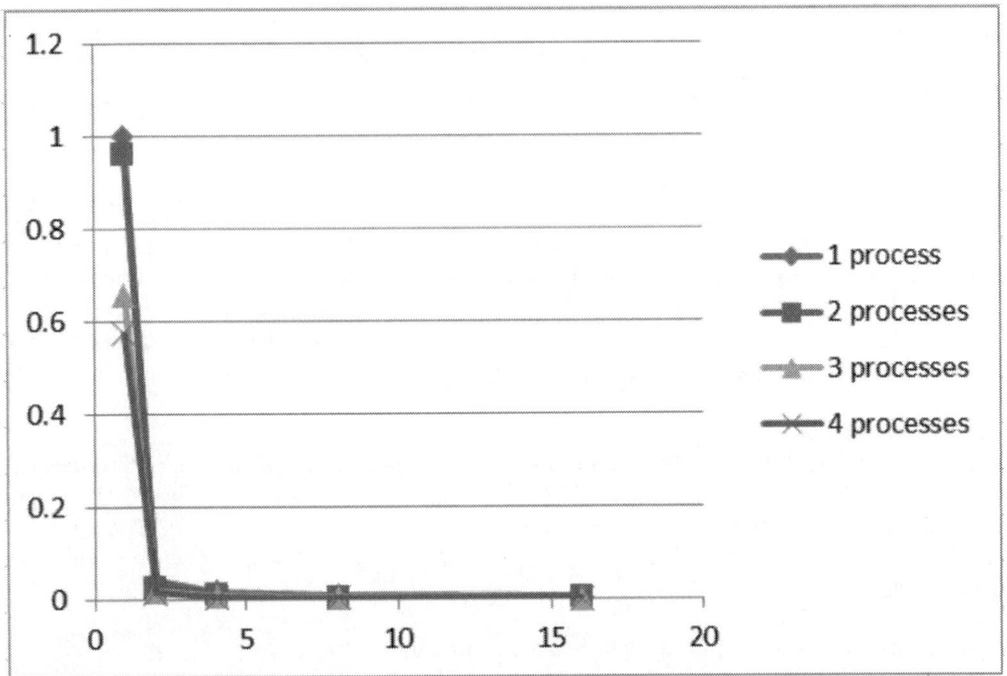

Figure 2. Speed up of parallel computation for an elasticity simulation scenario. The horizontal axis is the number of PCs. The vertical axis is the total time used for computations.

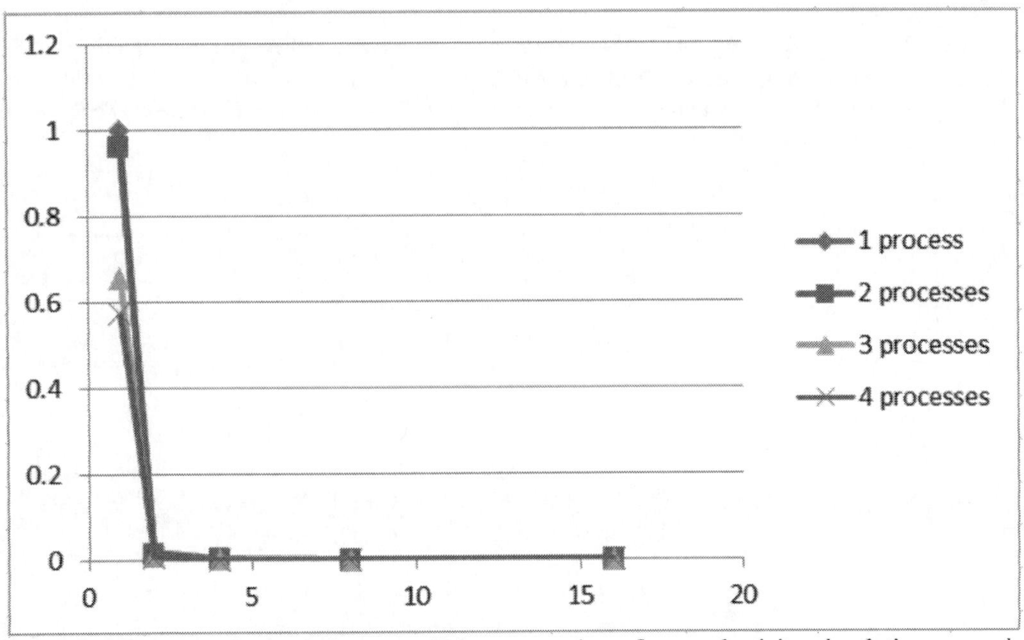

Figure 3. Efficiency obtained from parallel computations for an elasticity simulation scenario. The horizontal axis is the number of PCs. The vertical axis is the total time used for computations.

Figure 2 and Figure 3, for the speed up and the efficiency respectively, show that they get worse, since the number of processes is more than one. This is reasonable because the total time needed for computations also gets worse.

5. Conclusion

We have simulated several scenarios of parallel computations for elasticity problems. We obtain that the one-dimensional elasticity equations are not appropriate to be solved using parallel programming with MPI_Send and MPI_Recv technique in the MPI standard, because the total amount of time to exchange data is considered more dominant compared with the total amount of time to conduct the basic elasticity computation using the finite volume method. Without loss of generality, we recommend that if the total time needed in the data exchange is larger than the total time needed in the basic computations, then the sequential computation may be better to implement than the parallel.

6. Acknowledgments

This work was financially supported by the Institute for Research and Community Services of Sanata Dharma University (LPPM USD). The LPPM USD internal research grant year 2016 is gratefully acknowledged by both authors.

7. References

[1] Mungkasi S and Darmawan J B B 2015 *CCIS* **516** 469
[2] LeVeque R J 2002 *Int. J. Numer. Meth. Fluids* **40** 93
[3] Flynn M J 1972 *IEEE Trans. Comput.* **C-21** 948
[4] van der Steen A J and Dongarra J 2002 *Massive Comp.* **4** 791
[5] Sterling T 2001 *Beowulf Cluster Computing with Linux* (Cambridge: MIT Press)
[6] Mache J 1999 *Proc. 1st IEEE Computer Society Int. Workshop on Cluster Computing* pp 36-42
[7] The Message Passing Interface (MPI) standard http://www.mcs.anl.gov/research/projects/mpi
[8] MPICH http://www.mpich.org
[9] Foster I 1995 *Designing and Building Parallel Programs: Concepts and Tools for Parallel Software Engineering* (Boston: Addison-Wesley)
[10] Yang J, Shen W and Yan G 2009 *Proc. 1st Int. Workshop on Education Technology and Computer Science* pp 895-898

Analysis of D Dimensional Dirac equation for *q*-deformed Posch-Teller combined with *q*-deformed trigonometric Manning Rosen Non-Central potential using Asymptotic Iteration Method (AIM)

Y. Alam, Suparmi, Cari, F. Anwar
Physics Department, Graduate Program, Universitas Sebelas Maret, Jl. Ir. Sutami 36A Kentingan, Surakarta

E-mail: yuniar.alam08@gmail.com

Abstract. In this study, we used asymptotic iteration method (AIM) to obtain the relativistic energy spectra and wavefunctions for D Dimensional Dirac equation. Solution of the D Dimensional Dirac equation using asymptotic iteration method was done by four steps. The first step, we substitutied q deformed Poschl-Teller potential plus q-deformed Manning Rosen Non-Central potential into D dimensional Dirac equation. And then, general term of D dimensioanl Dirac equation for q deformed Poschl-Teller potential plus q-deformed Manning Rosen Non-Central potential was reduced into one dimensioanal Dirac equation, consist of radial part and angular part. The second step, both of one dimensional part must be reduced to hypergeometric type differential equation by suitable parameter change. And then, hypergeometric type differential equation was transformed into AIM type differential equation. For the last step, AIM type differential equation can be solved to obtain the relativistic energy and wavefunctions of Dirac equation. Relativistic energy and wavefunctions were visualized by using Matlab software.

1. Introduction

P. A. M Dirac submit the equation known as the Dirac equation. The Dirac equation has a probability density which is always a positive value, but the solution still provide information of a free particle of negative energy. The Dirac equation describes the basic correspondent for particle spin ½ in electron[1]. Dirac found that a particle that has mass and charge as an electron must have an intrinsic angular momentum and magnetic moment[2]. The Dirac equation is used when a particle is exposed to potential field strong relativistic effects must be considered that emberikan corrections to the nonrelativistic quantum mechanics[3]. In a spin ½ particle, there is the concept of spin symmetry and symmetry pseudospin. Spin symmetry and pseudospin symmetry occurs vector potential $V(r)$ and a scalar potential $S(r)$ is a constant. Spin symmetry occurs when $V(r) = S(r)$ and pseudosopi symmetry occurs when $V(r) = -S(r)$. The concept of spin symmetry has been applied to a spectrum of machinery and antinukleon, and the concept of symmetry pseudospin used to explain quasi-degeneration of double nucleons, superdepormasi in nuclei[6], exotic nuclei[7], and to establish an affective nuclear shell-model scheme[8].

This study presents the completion of D-Dimensional Dirac equation in the case of spin symmetry for q-deformed Posch-Teller potential and q-deformed non-central Manning Rosen using asymptotic iteration method. completion of the D-Dimensional Dirac equation by reduction into hipergeometri equations with variable substitution. Energy spectrum and the wave functions obtained from asymptotic iteration method.

2. Asymptotic Iteration Method (AIM)

This method is used to solve differensial equetion in the following form:

$$y_n''(x)-\lambda_0(x)y_n'(x)-s_o(x)y_n(x)=0 \quad (1)$$

The one-dimensional Dirac equation can be reduced into hypergeometric or confluent hypergeometric type differential equation by suitable changes of variables, and then changes it into the differential equation which has the form in Eq.(1). The solution of Eq.(1) can be obtained by using iteration of λ_i and s_i,

$$\lambda_i(x) = \lambda_{i-1}' + \lambda_{i-1}\lambda_0 + s_{i-1}$$
$$s_i(x) = s_{i-1}' + s_o\lambda_{i-1}$$
$$i = 1,2,3, \dots \quad (2)$$

Eigenvalues can be obtained using equation:[9]

$$\lambda_i(x)s_{i-1}(x) - \lambda_{i-1}(x)s_i(x) = 0 = \Delta_i, i = 1,2,3 \dots . \quad (3)$$

On the other hand, Eq.(1) can be written in term:

$$y''(x) = 2\left(\frac{tx^{N+1}}{1-bx^{N+2}} - \frac{c+1}{x}\right)y'(x) - \frac{Wx^N}{1-bx^{N+2}} \quad (4)$$

Eq.(4) is AIM-type differential equation which is solved by using Eq.(5)[10,11]

$$y_n(x) = (-1)^n C'(N+2)^n {}_2F_1(-n, p+n, \sigma, bx^{N+2}) \quad (5)$$

where

$$(\sigma)_n = \frac{\Gamma(\sigma+n)}{\Gamma(\sigma)}, \qquad \sigma = \frac{2c+N+3}{N+2} \quad (7)$$

$$p = \frac{(2c+1)b+2t}{(N+2)b} \quad (8)$$

C' is normalization constant and ${}_2F_1$ is hypergeometric function.[11]

3. Solution of Dirac Equation in D Dimension

The dirac equation with scalar potential $S(r)$ and vector potential $V(r)$ $(\hbar = 1, c = 1)$[16],

$$\{\vec{\alpha}.\vec{p} + \beta(M + S(\vec{r}))\}\psi(\vec{r}) = \{E - V(\vec{r})\}\psi(\vec{r}) \quad (9)$$

which E is relativistic energy of system and $\vec{p}$ is momentum operator ($\vec{p} = -i\nabla$), while $\vec{\alpha}$ and β is matrix in term:

$$\hat{\alpha}_i = \begin{pmatrix} 0 & \hat{\sigma} \\ \hat{\sigma} & 0 \end{pmatrix}, \quad (10)$$

$$\beta_i = \begin{pmatrix} I & 0 \\ 0 & -I \end{pmatrix} \quad (11)$$

Where $\hat{\sigma}_i$ are Pauli's matrices and **1** is the 2×2 unit matrix. Here we have used the relation for the Pauli's matrices as follow

$$\hat{\sigma}_i\hat{\sigma}_j + \hat{\sigma}_j\hat{\sigma}_i = 2\delta_{ij}\mathbf{1} \quad (12)$$

The wave function of spinor Dirac can be classified in two form, spinor upper $\chi(\vec{r})$ and spinor lower $\varphi(\vec{r})$ as follows:[12,13]

$$\psi(r) = \begin{pmatrix} f_{nk}(r) \\ g_{nk}(r) \end{pmatrix} = \begin{pmatrix} \dfrac{F_{nk}(r)}{r^{\frac{D-1}{2}}} Y^{l}_{l_1 \ldots l_{D-1}}(\hat{x} = \theta_1, \theta_2, \ldots, \theta_{D-1}) \\ i\dfrac{G_{nk}(r)}{r^{\frac{D-1}{2}}} Y^{\bar{l}}_{\bar{l}_1 \ldots \bar{l}_{D-1}}(\hat{x} = \theta_1, \theta_2, \ldots, \theta_{D-1}) \end{pmatrix} \quad (13)$$

we have

$$c\vec{\sigma}.\vec{p}g_{nk}(r) = \left[E - V(\vec{r}) - Mc^2 - S(\vec{r})\right]f_{nk}(r) \quad (14a)$$

$$c\vec{\sigma}.\vec{p}f_{nk}(r) = \left[E - V(\vec{r}) + Mc^2 + S(\vec{r})\right]g_{nk}(r) \quad (14b)$$

For spin symmetry, equation (14a) becomes

$$g_{nk}(r) = \frac{\vec{\sigma}.\vec{p}}{[E + M]} f_{nk}(r) \quad (15a)$$

$$\vec{\sigma}.\vec{p}g_{nk}(r) = [E - 2V(\vec{r}) - M]f_{nk}(r) \quad (15b)$$

Substituting Eq.(15a) into Eq.(15b) yields

$$[\boldsymbol{p}^2 + 2V(r)(E + M)]f_{nk}(\vec{r}) = [E^2 - M^2]f_{nk}(\vec{r}) \quad (16)$$

In spherical coordinates, modified Pöschl-Teller potential combined with trigonometric Manning Rosen non-central potential is defined as

$$V(r,\theta) = \left(\frac{\kappa(\kappa - 1)}{\sinh_q^2 \alpha r} + \frac{\eta(\eta + 1)}{\cosh_q^2 \alpha r}\right) + \frac{1}{r^2}\left(\frac{\nu(\nu - 1)}{\sin_q^2 \theta} - 2q\cot_q \theta\right) \quad (17)$$

Subsitute Eq.(17) into Eq.(16) ans simplify the resuling equation, and let

$$f_{nk} = \frac{F_{nk}(r)}{r}\Theta(\theta)\phi(\varphi) \quad (18)$$

Then we have,

$$\frac{r^2}{F_{nk}(r)}\frac{r^{\frac{D-1}{2}}}{r^{D-1}}\frac{\partial}{\partial r}\left(r^{D-1}\frac{\partial}{\partial r}\frac{F_{nk}(r)}{r^{\frac{D-1}{2}}}\right) + r^2\left(\frac{\kappa(\kappa - 1)}{\sinh_q^2 \alpha r} + \frac{\eta(\eta + 1)}{\cosh_q^2 \alpha r}\right)[E + M] - \left[E^2 + M^2\right]r^2 = l_{D-1}(l_{D-1} + D - 2) \quad (19)$$

Separating the variables in Eq.(20), we obtain

$$\frac{d^2 F_{nk}(r)}{dr^2} - \frac{\left(l_{D-1} + \frac{D-1}{2}\right)\left(l_{D-1} + \frac{D-3}{2}\right)}{r^2}F_{nk}(r) + \left([E + M]\left(\frac{\kappa(\kappa - 1)}{\sinh_q^2 \alpha r} + \frac{\eta(\eta + 1)}{\cosh_q^2 \alpha r}\right) - \left[E^2 - M^2\right]\right)F_{nk}(r) = 0 \quad (20)$$

$$\frac{1}{Y_3(\theta_3)}\frac{1}{\sin^2\theta_3}\left(\frac{\partial}{\partial\theta_3}\sin^2\theta_3\frac{\partial}{\partial\theta_3}\right)Y_3(\theta_3) + \left(\lambda_3 - \frac{\lambda_2}{\sin^2\theta_3} - [E + M]V(\theta_3)\right) = 0 \quad (21)$$

4. Analytical solution of radial and angular parts of the dirac equation

4.1 Solution of the radial part

The radial part of the Dirac equation in Eq. (21) we use the approximation value for the centrifugal term[14,15],

$$\frac{1}{r^2} \approx \frac{\mu^2}{4\sinh^2 \mu r} \tag{22}$$

Substituting Eq.(22) into Eq.(20) and simplifying the equation by substituting variable $\cosh_q^2 \quad = z$, we have

$$z(q-z)\frac{d^2F_{nk}(z)}{dz^2}+\frac{1}{2}(q-2z)\frac{dF_{nk}(z)}{dz}-\left(\frac{A_s}{4(q-z)}-\frac{B_s}{4z}+\frac{E_s'}{4}\right)F_{nk}(z)=0 \tag{23}$$

by substituting,

$$F_{nk}(z)=z^{\delta}(q-z)^{\gamma}f \tag{24}$$

Into Eq(23), we have the second-orde differential

$$z(q-z)f''+\left(\left(2\delta+\frac{1}{2}\right)q+z(2\delta+2\gamma+1)\right)f'-\left((\delta+\gamma)^2-\frac{E_s'}{4}\right)f=0 \tag{25}$$

Where $4\delta^2 q = 2\delta q + B_s$ and $4\gamma^2 q = 2\gamma q + A_s$

Eq.(25) can be transform to differential equation type AIM,

$$f''=\left(\frac{(2\delta+2\gamma+1)z-\left(2\delta+\frac{1}{2}\right)q}{z(q-z)}\right)f'+\left(\frac{(\delta+\gamma)^2-\frac{E_s'}{4}}{z(q-z)}\right)f \tag{26}$$

From Eq.(26), we have

$$\lambda_0 = \frac{\left((2\delta+2\gamma+1)z-\left(2\delta+\frac{1}{2}\right)q\right)}{z(q-z)} \tag{27}$$

$$s_0 = \frac{\left((\delta+\gamma)^2-\frac{E_s'}{4}\right)}{z(q-z)} \tag{28}$$

By using Eq.(3) and using Matlab 2011 software, energy eigenvalue can be obtained, with $\varepsilon = \frac{E_s'}{4}$

$$s_0\lambda_1 - s_1\lambda_0 = 0 \rightarrow \varepsilon_1 = (2\delta+2\gamma+1)+\{(\delta+\gamma)(\delta+\gamma+1)\} = (\delta+\gamma+1)^2$$
$$s_1\lambda_2 - s_2\lambda_1 = 0 \rightarrow \varepsilon_2 = (4\delta+4\gamma+4)+\{(\delta+\gamma)(\delta+\gamma+1)\} = (\delta+\gamma+2)^2$$
$$s_2\lambda_3 - s_3\lambda_2 = 0 \rightarrow \varepsilon_2 = (6\delta+6\gamma+9)+\{(\delta+\gamma)(\delta+\gamma+1)\} = (\delta+\gamma+3)^2$$

can be generalized as follow

$$\varepsilon = (\delta + \gamma + n_r)^2 \tag{29}$$

From Eq. (63) we get relativistic energy Eq. of this system is

$$\frac{\left[E^2-M^2\right]}{\alpha^2}\frac{1}{4}=\left(\frac{1}{2}q\sqrt{B_{s1}+\frac{1}{4}q}\pm\frac{1}{2}q\sqrt{A_s+\frac{1}{4}q}+n_r+\frac{1}{2}q^2\right)^2 \tag{30}$$

where n_r is radial quantum numbers ($n_r = 0,1,2 \ldots$), l is orbital quantum numbers which is obtained from angular part solution. And then, radial wavefunction can be obtain by using Eq.(5), Eq.(6), Eq.(7) and Eq.(8), we have:

$c = \gamma + \frac{1}{4}, N = -1, t = \delta - \frac{3}{2}, b = 1,$

so, $\sigma = \frac{2c+N+3}{N+2} = 2\delta + \frac{1}{2}$ and $p = \frac{(2c+1)b+2t}{(N+2)b} = 2\delta + 2\gamma$

From Eq.(5), we have,

$$f_{nr}(z) = (-1)^{n_r} C_2 (1)^{n_r} \left(2\delta + \frac{1}{2}\right)_{n_r} {}_2F_1\left(-n_r, 2\delta + 2\gamma + n_r, 2\delta + \frac{1}{2}, z\right) \tag{31}$$

By substituting Eq.(31) to Eq.(24), we have radial wavefunction,

$$f_{nr}(z) = z^{\delta}(q-z)^{\gamma}(-1)^{n_r}C_2(1)^{n_r}\left(2\delta+\frac{1}{2}\right)_{n_r}\,{}_2F_1\left(-n_r, 2\delta+2\gamma+n_r, 2\delta+\frac{1}{2}, z\right) \quad (32)$$

which $z = \cosh^2_q$, so

$$f_{nr}(z) = z^{\delta}(q-z)^{\gamma}(-1)^{n_r}C_2(1)^{n_r}\left(2\delta+\frac{1}{2}\right)_{n_r}\,{}_2F_1\left(-n_r, 2\delta+2\gamma+n_r, 2\delta+\frac{1}{2}, z\right) \quad (33)$$

where $C(n_r)$ is radial normalization constant , ${}_2F_1$ is hypergeometric function.

4.2 Solution of the angular part

For angular part in Eq.(21), can be obtain by using AIM to find orbital quantum number.

$$\frac{\partial^2 Y_1(\theta_1)}{\partial\theta_1^2} - [E+M]\left(\frac{v(v-1)}{\sin_q^2\theta_1} - 2q\cot_q\theta_1\right)Y_1(\theta_1) + \lambda_1 Y_1(\theta_1) = 0 \quad (34)$$

Equation (65) can be written as, $cot_q\theta_1 = i(1-2z)$

$$z_1(1-z_1)\frac{d^2Y_1(z_1)}{d\theta_1^2} + (1-2z_1)\frac{dY_1(z_1)}{d\theta_1} + \left([E+M]v(v-1) - \frac{(-2qi[E+M]+\lambda_1)}{4z_1} - \frac{(2qi[E+M]+\lambda_1)}{4(1-z_1)}\right)Y_1(z_1) = 0 \quad (35)$$

by using

$$Y_1(z_1) = z_1^{\alpha}(1-z_1)^{\beta} f_s(z_1) \quad (36)$$

and simplying it, Eq.(35) can be transform to hypergeometric differential equation:

$$z_1(1-z_1)f_s''(z_1) + [(2\alpha+1) - z_1(2\alpha+2\beta+2)]f_s'(z_1) + ([E+M]v(v+1) - (\alpha+\beta)(\alpha+\beta+1))f_s(z_1) = 0 \quad (37)$$

Eq. (37) is hypergeometry type Eq. and we can solve it by AIM as follow

$$\lambda_0 = \frac{(z_1(2\alpha+2\beta+2)) - (2\alpha+1)}{z_1(1-z_1)} \quad (38)$$

$$s_0 = \frac{((\alpha+\beta)(\alpha+\beta+1) - [E+M]v(v+1))}{z_1(1-z_1)} \quad (39)$$

one can generalize Eq. (39)

$$L_1' = \pm\sqrt{\frac{-(qi[E+M])^2}{\left(\sqrt{[E+M]v(v+1)+\frac{1}{4}} - n_l - \frac{1}{2}\right)^4} - \left(\sqrt{[E+M]v(v+1)+\frac{1}{4}} - n_l - \frac{1}{2}\right)^2} \quad (40)$$

Where L is orbital quantum number and n_l is angular quantum number. From Eq.(5), we have

$$Y(z) = (-1)^{n_l}C_2(1)^{n_l}(\sigma)_{n_l}\,{}_2F_1(-n_l, \rho+n_l, \sigma, bz^{N+2}) \quad (41)$$

C_z is angular normalization constant.

Solution for θ ,and θ_3 can be determined by the same way with solution for θ_1, And we get solution for orbital quantum number for θ_2,and θ_3 respectively as follows:

$$L_2' = \pm\sqrt{\frac{-(qi[E+M])^2}{\left(\sqrt{\left(\lambda_1-\frac{1}{4}\right) - \frac{[E+M]}{q}v(v-1)+\frac{1}{4}} - n_l - \frac{1}{2}\right)^4} - \left(\sqrt{\left(\lambda_1-\frac{1}{4}\right) - \frac{[E+M]}{q}v(v-1)+\frac{1}{4}} - n_l - \frac{1}{2}\right)^2 - \frac{1}{2}} \quad (42)$$

$$L_3' = \pm\sqrt{\frac{-(qi[E+M])^2}{\left(\sqrt{\lambda_2 q - [E+M]v(v-1)+\frac{1}{4}} - n_l - \frac{1}{2}\right)^4} - \left(\sqrt{\lambda_2 q - [E+M]v(v-1)+\frac{1}{4}} - n_l - \frac{1}{2}\right)^2 - 1} \quad (43)$$

5. Result and Discussion

In this section, we discuss several results which were obtained in the previous section. From relativistic energy equation in Eq.(30) and orbital quantum number equation in Eq.(41), and by using Matlab software we have numeric solution of relativistic energy are listed in Table 1 with parameters $\kappa = 2, \eta = 2, v = 3, q = 2$ and $M = 5\ fm^{-1}$, the negative value of relativistic energy is taken due to the pseudospin symmetric limit[19]. By inspecting Table 1, show that increase of value α and n_r in the same quantum state causes decrease energy eigenvalue.

Table 1. Relativistic energy corresponding to several sates of a particle under the influence of modified Pöschl-Teller potential and trigonometric Scarf II potential.

n_r	n_l	k	$E_{n_r n_l k}(fm^{-1})$ $\alpha = 0,1fm^{-1}$	$\alpha = 0,2fm^{-1}$	$\alpha = 0,3fm^{-1}$
0	0	0	-5.0805	-5.1083	-5.1411
1	1	0	-5.0715	-5.1181	-5.1351
2	2	0	-5.0798	-5.1289	-5.1738
3	0	0	-5.0887	-5.1358	-5.2678

By varying parameter which corresponding value δ and γ, some of the radial wavefunctions are listed in Table 2. Radial wavefunctions for particle under the influence of modified Pöschl-Teller potential and Manning Rosen potential are affected by potential constants κ, eta, q, v and by α .The parameter α has a dimension inverse of distance in space that describes the reach of Pöschl-Teller potential. If α is enlarged, physically means that the potential reach is smaller in a space. By inspecting Table 2, due to the increase in the value of α causes particles move further away from the nucleus and show that change in radial wavefunctions are affected of potential constants κ, eta, q and v.

Table 2. Energy eigenvalue in fm^{-1} with $n_r = 2$ $n_l = 2, v = 2, q = 4, M = 5, \alpha = 0.05$ for particle under the influence of modified Pöschl-Teller potential and Manning-Rosen non-central potential variation k.

Nr	*Enr*	F_{n_r}
0	-5.14162	$(\cos_q^2 \alpha r)^\delta (\sin_q^2 \alpha r)^\gamma C'\left(2\delta+\frac{1}{2}\right)$
1	-5.09810	$(\cos_q^2 \alpha r)(\sin_q^2 \alpha r)C'\left(-\left(2\delta+\frac{1}{2}\right)\right)\left(1+\frac{(-1)_n(2\delta+2\gamma+')}{\left(2\delta+\frac{1}{2}\right)}(\cos_q^2 \alpha r)\right)$
2	-5.05656	$(\cos_q^2 \alpha r)(\sin_q^2 \alpha r)C'\left(2\delta+\frac{1}{2}\right)\left(1+\frac{(-(4\delta+4\gamma+4))}{\left(2\delta+\frac{1}{2}\right)}\cos_q^2 \alpha r+\frac{(2\delta+2\gamma+2)(2\delta+2\gamma+3)}{\left(2\delta+\frac{1}{2}\right)\left(2\delta+\frac{3}{2}\right)}(\cos_q^2 \alpha r)^2\right)$
3	-5.03558	$(\cos_q^2 \alpha r)(\sin_q^2 \alpha r)C'\left(-\left(2\delta+\frac{1}{2}\right)\right)\left(1-\frac{(6\delta+6\gamma+9)}{\left(2\delta+\frac{1}{2}\right)n_r}\cos_q^2 \alpha r+\frac{(3)(2\delta+2\gamma+3)(2\delta+2\gamma+4)}{\left(2\delta+\frac{1}{2}\right)\left(2\delta+\frac{3}{2}\right)}(\cos_q^2 \alpha r)^2-\frac{(2\delta+2\gamma+3)(2\delta+2\gamma+4)(2\delta+2\gamma+5)}{\left(2\delta+\frac{1}{2}\right)\left(2\delta+\frac{3}{2}\right)\left(2\delta+\frac{5}{2}\right)}(\cos_q^2 \alpha r)^3\right)$

6. Conclusion

The Dirac equation in D dimensions of q-deformed trigonometric Poschl-Teller potential combined with Manning-Rosen non-central potential using Asymptotic Iteration Method (AIM). The radial part of D- dimensions of the Dirac equation reduces to one dimensional Schrodinger type equation in centrifugal approximation scheme. In the exact spin symmetric case, the relativistic energy equation reduces to the non-relativistic energy in the non-relativistic condition. The radial part of the wavefunction is obtained approximately from Eq.(33) and the angular part in Eq.(41). The results show that the disturbance of modified Pöschl-Teller Potential and trigonometric Manning-Rosen non-central potential change in the wave function of the radial part and the angular part. Relativistic energy equatione can be obtained via AIM in Eq.(30) and equation of orbital quantum number l in Eq.(40), Eq.(42) and Eq.(43), where both are interrelated between quantum numbers. Relativistic energy also is solved numerically using Matlab software, where the increase in the radial quantum number n_r causes a decrease in the energy spectrum.

Acknowledgments
This Research was partly supported by Higher Education Project Grant with contract no. 632/UN27.21/LT/2016.

Reference

[1] Greiner W 2000 *Relativistic Quantum Mechanics, Wave Equations* (3rd Edn.) (Berlin:Springer-Verlag)
[2] Beiser A 1987 *Konsep Fisika Modern* (Jakarta: Erlangga)
[3] Ikhdair S M and Hamsavi M 2012. *Nuclear Theory Few-Body Systems* **1** 487.
[4] Ginocchio J.N 2004 *Phys.Rep.C* **69** 034313
[5] Ginocchio J.N 1997 *Phys.Rev.Lett* **78** 436
[6] Meng J, Sugawara-Tanabe K, Yamaji S and Arima A 1999 *Phys.Rev.C* **59** 154
[7] Dudek J, Nazarewicz W, Szymanski Z and Leander G A 1987 *Phys.Rev.Lett* **59** 1405
[8] Troltenier D, Bahri C and Draayer J P 1995 *Nucl.Phys.A* **586** 53
[9] Rostami A and Motavali H 2008 *Progress in Electromagnetic Research B* **4** 171
[10] Soylu A, Bayrak O and Boztosun I 2008 *Journal of Physics A: Mathematical and Theoretical* **41** 065308
[11] Falaye B J, Hamzavi M and Ikhdair S M. 2012 *arXiv*:1207.1218v [nucl-th]
[12] Shi-Hai Dong 2011 *Wave Equations in Higher Dimensions*, Springer Dordrecht Heidelberg.
[13] Tapas Das *arXive*:1409.1457v1
[14] Greene R I and Aldrich C 1976 *Phys. Rev. A* **14** 2363
[15] Ikhdair S M and Sever R 2008 arXiv:0801.4857v3 [quant-ph] 14 Feb 2008
[16] Hu X Q, Luo G, Wu Z M, Niu L B and Ma Y 2010 *Commun. Theor. Phys.* **53** 242

Solution of Dirac equation for modified Poschl Teller plus trigonometric Scarf potential using Romanovsky polynomials method

I. Prastyaningrum[1], C. Cari[2], A. Suparmi[2]
[1]Teaching and Education Institute PGRI Madiun Jl Setia Budi 85 Madiun 63118
[2]Physics Department of Post Graduate Program Sebelas Maret University
Jl. Ir. Sutami 36A Kentingan Jebres Surakarta 57126, INDONESIA

E-mail: arka.prameswara@gmail.com

The approximation analytical solution of Dirac equation for Modified Poschl Teller plus Trigonometric Scarf Potential are investigated numerically in terms of finite Romanovsky Polynomial. The combination of two potentials are substituted into Dirac Equation then the variables are separated into radial and angular parts. The Dirac equation is solved by using Romanovsky Polynomial Method. The equation that can reduce from the second order of differential equation into the differential equation of hypergeometry type by substituted variable method. The energy spectrum is numerically solved using Matlab 2011. Where the increase in the radial quantum number nr and variable of modified Poschl Teller Potential causes the energy to decrease. The radial and the angular part of the wave function also visualized with Matlab 2011. The results show, by the disturbance of a combination between this potential can change the wave function of the radial and angular part.

1. Introduction

Since Dirac equation as relativistic wave Eq. was formulated by P.A.M Dirac in 1928, the exact solution of Dirac equation for some quantum potentials plays a fundamental role in relativistic quantum mechanics [1]. It is well known that the exact energy eigenvalues of the bound state play an important role in quantum mechanics. In particular, the Dirac equation which describes the motion of a spin-1/2 particle has been used in solving many problems of nuclear and high-energy physics [1]. The exact solution of some non-central potentials for l-wave have been investigated intensively by some authors [2].On 2004, Reity, Rubish and Myhalyna has investigated Dirac equation for vector and scalar potential [3], Victor M. Villalba finds the solution in the presence of a gravitational instanton on 2005 [4], and Bakkeshizadeh and Vahidi solving the Dirac equation for Coulomb and NAD potential [5]. In general, the non-central potential is as a function of radial and angular positions simultaneously. These potentials are constructed by the combination of radial shape or nonshape invariance. The potentials that can be used such as hyperbolic Scarf Potential, Manning Rosen, Rosen-Morse, etc. The bound state energy spectra of these potentials have been investigated by Super symmetry method [6], Nikivorof-Uvarof [7], Laplace Transformation Method [8]. Romanovsky polynomials [9], WKB Method [10]. In this paper will be solved the Dirac equation for non-central potential in term of Romanovsky polynomials by numerical method. The Dirac equation of non-central potentials is solved using separation variable method when the non-central potential is separable. We must set the Dirac equation like the hypergeometric term and then make a differential equation in term of Romanovsky polynomials. The last we will get a complicated equation of energy eigenvalue that can't be solved by an analytical method so we must use the numerical method to find the solution. Besides that, we also get angular wave function and that must be solved by Matlab 2011.

2. Review of Formulas for the Polynomial Solution to the Generalized Hypergeometric Equation

The method used to solve the Dirac equations in the limit of spin and pseudospin symmetries is the Romanovski polynomials. The Romanovsky polynomials method were discovered in 1884 by Routh in the form of complexified Jacobi polynomials on the unit circle in the complex plane and were then rediscovered as real polynomials by Romanovsky in a statistics framework [11]. The Romanovsky polynomials are built from the generalized hypergeometric equation is given,

$$\sigma(x)\frac{d^2y_n(x)}{dx^2}+\tau(x)\frac{dy_n(x)}{dx}-\lambda y_n(x)=0 \tag{1}$$

where

$$\sigma(x)=ax^2+bx+c;\quad \tau=dx+e \text{ and } \lambda_n=-\big(n(n-1)+2n(1-p)\big) \tag{2}$$

Eq. (1) is described in the textbook by NikiforovUvarov [12] where it is cast into self-adjoint form and its weight function, *w(x),* satisfies the so-called Pearson differential equation [12]. The corresponding polynomials to the weight function are built up from the Rodrigues representation that is given as,

$$D_n^{(p,q)}(x)=\frac{1}{w(x)}\frac{d^n}{dx^n}((1+x^2)^n w(x)) \tag{3}$$

For Romanovsky polynomial, the values of parameters are,

$$a=1,\ b=0,\ c=1,\ d=2(1-p);\ e=q,\ p>0;\ y_n(x)=D_n^{(p,q)}(x);\ p=-\beta>n \tag{4}$$

And so, the weight function is showed as.

$$w(x)=(1+x^2)^{-p}\,e^{q\,tan^{-1}(x)} \tag{5}$$

By setting Eq. (1) with Romanovski polynomial parameter, we find the new differential equation, called the differential equation of Romanovski polynomial type.

$$(1+x^2)\frac{d^2D_n^{(p,q)}(x)}{dx^2}+(2x(1-p)+q)\frac{dD_n^{(p,q)}(x)}{dx}-\{n(n-1)+2n(1-p)\}D_n^{(p,q)}(x)=0 \tag{6}$$

which is hypergeometric type equation with parameters expressed in equation (4).The Romanovski polynomials obtained from equations (3) and (5) are,

$$D_n^{(p,q)}(x)=\frac{1}{(1+x^2)^{-p}\,e^{q\,tan^{-1}(x)}}\frac{d^n}{dx^n}\left((1+x^2)^n(1+x^2)^{-p}\,e^{q\,tan^{-1}(x)}\right) \tag{7}$$

3. Find The Eigenvalues and Eigenfunctions in terms of Romanovski Polynomial

The Radial part of one dimension Dirac equation with modified Poschl Teller with the centrifugal term plus trigonometric Scarf potential is showed in Eq. (8),

$$\frac{\partial}{\partial r}\left(r^2\frac{\partial R(r)}{\partial r}\right)-r^2\left\{(E+M)t^2\left(\frac{\kappa(\kappa-1)}{sinh^2tr}-\frac{\eta(\eta+1)}{cosh^2tr}+\right)-(E^2-M^2)\right\}R(r)=\lambda R(r) \tag{8}$$

Where $\lambda=l(l+1)$ and by making an approriate change of variable, $cosh\,2tr=ix$ and $R(r)=\frac{\chi(r)}{r}$, so the Eq. (8) can be rewritten as,

$$(1+x^2)\frac{d^2\chi(r)}{dx^2}+x\frac{d\chi(r)}{dx}-\left\{\left(-\frac{\kappa(\kappa-1)(E+M)+l(l+1)+\eta(\eta+1)(E+M)}{2(x^2+1)}\right)+\left(-\frac{\kappa(\kappa-1)(E+M)+l(l+1)-\eta(\eta+1)(E+M)}{2(x^2+1)}\right)ix-\left(\frac{(E^2-M^2)-l(l+1)t^2d_0}{4t^2}\right)\right\}\chi(r)=0 \tag{9}$$

To solve Eq.(9) in terms of Romanovski polynomial, substitude Eq.(5) in equation (9) as,

$$\chi_n = (1+x^2)^{\frac{p}{2}} e^{-\frac{\alpha}{2}\tan^{-1}x} D_n^{(\beta,\alpha)}(x) \tag{10}$$

By inserting Eq.(10) into Eq.(9) and setting the coefficient of $\frac{1}{2(x^2+1)}$ to be zero, we obtain,

$$-\{\kappa(\kappa-1)(E+M)+l(l+1)+\eta(\eta+1)(E+M)\}+2\beta(\beta-1)-\frac{\alpha^2}{2}=0 \tag{11}$$

$$-\alpha+2\alpha\beta-\{\kappa(\kappa-1)(E+M)+l(l+1)-\eta(\eta+1)(E+M)\}i=0 \tag{12}$$

The Eq. (12) can be reduced to the Romanovsky equation like in Eq. (6) and then comparing the parameters we obtain the relation,

$$(2\beta+1)=2(-p+1) \text{ dan } \alpha=-q \tag{13a}$$

$$-\left(\frac{(E^2-M^2)-l(l+1)t^2d_0}{4t^2}\right)-\beta^2=n(n-1)+2n(1-p) \tag{13b}$$

From Eq. (13a), (13b) we have the α and β value,

$$\beta=\frac{\sqrt{\kappa(\kappa-1)(E+M)+\frac{1}{4}+l(l+1)}+\sqrt{\eta(\eta+1)(E+M)+\frac{1}{4}}+1}{2} \tag{14}$$

$$\alpha=i\left(\sqrt{\kappa(\kappa-1)(E+M)+\frac{1}{4}+l(l+1)}-\sqrt{\eta(\eta+1)(E+M)+\frac{1}{4}}\right) \tag{15}$$

And finally, we obtain the energy spectra and the radial wave function by combine Eq. (14), Eq.(15), Eq.(10), Eq.(3), and Eq.(5),

$$(E^2-M^2)=-4t^2\left(\sqrt{\kappa(\kappa-1)(E+M)+\frac{1}{4}+l(l+1)}+\sqrt{\eta(\eta+1)(E+M)+\frac{1}{4}}+1+2n\right)^2+l(l+1)t^2d_0 \tag{16}$$

And the result of the wave function of the nth level is

$$X_{n;l}=\sqrt{(1+x^2)^{\beta}e^{-\alpha\tan^{-1}(x)}}\,\mathcal{R}_n^{(-\beta,-\alpha)}(x) \tag{17}$$

Some the solving of Eq. (20) and Eq. (21) are

$$\mathcal{R}_0^{(-\beta_0,-\alpha_0)}=1 \tag{18a}$$

$$X_{0;l}=\sqrt{(1+x^2)^{\beta}e^{-\alpha\tan^{-1}(x)}} \tag{18b}$$

$$\mathcal{R}_1^{(-\beta_1,-\alpha_1)}=2x(\beta+1)-\alpha \tag{19a}$$

$$X_{1;l}=\sqrt{(1+x^2)^{\beta}e^{-\alpha\tan^{-1}(x)}}\,(2x(\beta+1)-\alpha) \tag{19b}$$

4. Solution of the Angular Part

The angular Dirac equation with non-central potential given as,

$$-\left[\frac{\cot\theta}{P(\theta)}\frac{\partial P(\theta)}{\partial\theta}+\frac{1}{P(\theta)}\frac{\partial^2P(\theta)}{\partial\theta^2}+\frac{1}{\phi(\varphi)}\frac{1}{\sin^2\theta}\frac{\partial^2\phi(\varphi)}{\partial\varphi^2}\right]+(E+M)\left(\frac{b^2+a(a-1)}{\sin^2\theta}-\frac{2b\left(a-\frac{1}{2}\right)\cos\theta}{\sin^2\theta}\right)=\lambda \tag{20}$$

Where,

$$-\frac{1}{\phi(\varphi)}\frac{\partial^2\phi(\varphi)}{\partial\varphi^2}=m^2 \tag{21}$$

By inserting Eq. (21) and Eq.(10) into Eq. (20), with $\lambda=l(l+1)$ and setting $\cos\theta=ix$ we have angular wave function in the Romanovsky Polynomials form,

$$(x^2+1)\frac{\partial^2D_n^{(p,q)}(x)}{\partial x^2}+(2x(\beta+1)-\alpha)\frac{\partial D_n^{(p,q)}(x)}{\partial x}-\left[\frac{\alpha\beta x-\frac{\alpha^2}{4}+\beta^2+(E+M)2b\left(a-\frac{1}{2}\right)ix-(E+M)\left(b^2+a(a-1)\right)-m^2}{(1+x^2)}-\beta^2-\beta+l(l+1)\right]D_n^{(p,q)}(x) \tag{23}$$

By setting the coefficient of $\frac{1}{(1+x^2)}$ to be zero we get the α and β value, that are

$$\beta = \frac{(E+M)2b\left(a-\frac{1}{2}\right)i}{\alpha} \tag{24a}$$

$$\alpha = \frac{(E+M)2b\left(a-\frac{1}{2}\right)i}{\beta} \tag{24b}$$

Because of this condition the solution of Eq. (23) is given as,

$$(x^2+1)\frac{\partial^2 D_n^{(p,q)}(x)}{\partial x^2} + (2x(\beta+1)-\alpha)\frac{\partial D_n^{(p,q)}(x)}{\partial x} - [-\beta^2-\beta+l(l+1)]D_n^{(p,q)}(x) \tag{25}$$

Eq. (25) is a second orde differential equation of Romanovsky Polynomial polar term. The Eq. (25) can be combined with Eq. (9), and then we obtain, the α and β value.

$$\beta = \sqrt{\frac{(E+M)\left(b+\left(a-\frac{1}{2}\right)\right)^2+m^2-\frac{1}{4}(E+M)}{4}} + \sqrt{\frac{(E+M)\left(b-\left(a-\frac{1}{2}\right)\right)^2+m^2-\frac{1}{4}(E+M)}{4}} = \left[\frac{\mathcal{M}_+}{2}+\frac{\mathcal{M}_-}{2}\right] \tag{26}$$

$$\alpha = -i\left[\sqrt{(E+M)\left(b+a-\tfrac{1}{2}\right)^2+m^2-\tfrac{1}{4}(E+M)} - \sqrt{(E+M)\left(b-a+\tfrac{1}{2}\right)^2+m^2-\tfrac{1}{4}(E+M)}\right] = -i[\mathcal{M}_+-\mathcal{M}_-] \tag{27}$$

The quantum orbital number can be found by combine Eq. (28a), Eq. (29b), and Eq. (26),

$$l = \left[\sqrt{\frac{(E+M)\left(b+\left(a-\frac{1}{2}\right)\right)^2+m^2-\frac{1}{4}(E+M)}{4}} + \sqrt{\frac{(E+M)\left(b-\left(a-\frac{1}{2}\right)\right)^2+m^2-\frac{1}{4}(E+M)}{4}}\right] + n_l = \left[\frac{\mathcal{M}_+}{2}+\frac{\mathcal{M}_-}{2}\right]+n_l \tag{28}$$

Eq. (29) is a solving of quantum orbital number, that value used to find angular wave function, radial wave function, and energy eigenvalue. When we were found the quantum orbital number, to determine the angular wavefunction, Eq. (26), (27) and Eq. (28) are inserted into Eq.(3) and Eq.(5) so that we obtain the weight function $w(x)$ and the Romanovsky polynomials $D_n^{(p,q)}(x)$. From the weight function and the Romanovsky polynomials we obtain the angular wave function, are

$$\mathcal{R}^{(-\beta,-\alpha)} = \frac{1}{(1+x^2)^\beta e^{-\alpha\tan^{-1}(x)}}\frac{d^0}{dx^0}\left((1+x^2)^{n_l}(1+x^2)^\beta e^{-\alpha\tan^{-1}(x)}\right) = 1 \tag{29a}$$

$$P_l^m = \sqrt{(1+x^2)^\beta e^{-\alpha\tan^{-1}(x)}}.1 = \sqrt{(1+x^2)^\beta e^{-\alpha\tan^{-1}(x)}} \tag{29b}$$

$$\mathcal{R}^{(-\beta,-\alpha)} = \frac{1}{(1+x^2)^\beta e^{-\alpha\tan^{-1}(x)}}\frac{d}{dx}\left((1+x^2)^{\beta+1}e^{-\alpha\tan^{-1}(x)}\right) = \frac{1}{(1+x^2)^\beta e^{-\alpha\tan^{-1}(x)}}\Bigg[(\beta+1)2x(1+x^2)^\beta e^{-\alpha\tan^{-1}(x)} - \frac{\alpha(1+x^2)^{\beta+1}e^{-\alpha\tan^{-1}(x)}}{(1+x^2)} - \Bigg] = 2x\left(\frac{\mathcal{M}_+}{2}+\frac{\mathcal{M}_-}{2}+1\right)+i[\mathcal{M}_++\mathcal{M}_-] \tag{30a}$$

$$P_l^m = \sqrt{(1+x^2)^\beta e^{-\alpha\tan^{-1}(x)}}\left\{2x\left(\frac{\mathcal{M}_+}{2}+\frac{\mathcal{M}_-}{2}+1\right)+i[\mathcal{M}_++\mathcal{M}_-]\right\} \tag{30b}$$

5. Result and Discussion

In this section, we discuss several results obtained in the previous section. From the energy eigenvalue in Eq. (16) and the orbital quantum number in Eq. (28), by using Matlab, we have the numerical solution of the energy eigenvalue, with parameters n, M, κ, η, l, and t. Where there is two variation κ and η, we get the energy eigenvalue that be showed at the table below

Table 1. Energy eigenvalue with κ variation

n	M	κ	η	l	t	Energy (eV)
1	1	1	2	2	2	-0.3265
1	1	2	2	2	2	-0.6162
1	1	3	2	2	2	-0.7499
1	1	4	2	2	2	-0.8241
1	1	5	2	2	2	-0.8697

Table 2. Energy eigenvalue with η variation

n	M	κ	η	l	t	Energy (eV)
1	1	2	1	2	3	-0.0298
1	1	2	2	2	3	-0.5491
1	1	2	3	2	3	-0.7792
1	1	2	4	2	3	-0.8724
1	1	2	5	2	3	-0.9223

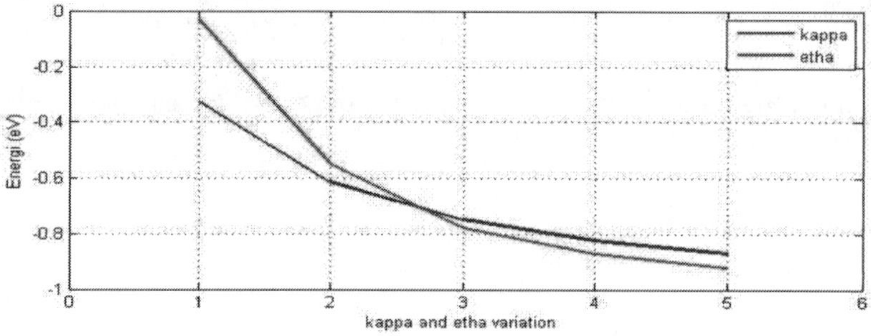

Figure 1. Graph of energy eigenvalue versus κ and η

From the Figure 1 can be showed that if the κ and η value is increase so the energy eigenvalue is decrease. With little increasing of κ and η can make little change of energy eigenvalue. By varying parameters corresponding to value of κ and η, some of the radial wave function are showed in Eq. (18) and Eq. (19). Radial wave function under the influences of modified Poschl Teller potential and trigonometric Scarf potential. Poschl Teller potential are affected by potential constant κ and η. The eigen function from ground state condition can be plot from the solution of $X_{0;l}$, the eigen function can be seen at Figure 2

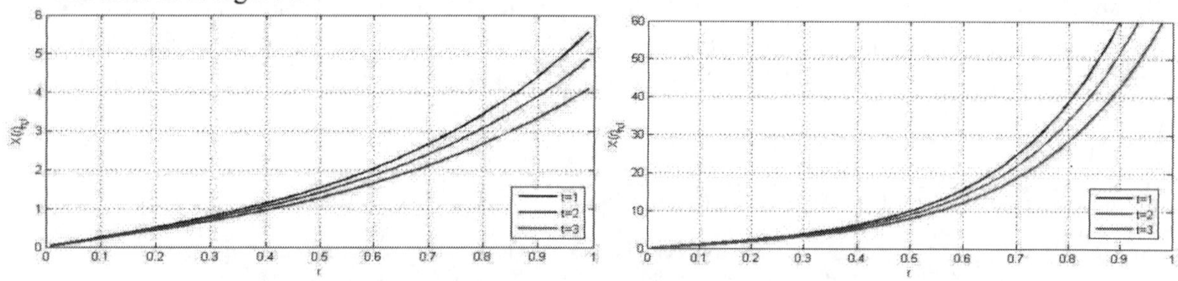

Figure 2. Eigenfunction of ground state condition and first energy with t variation

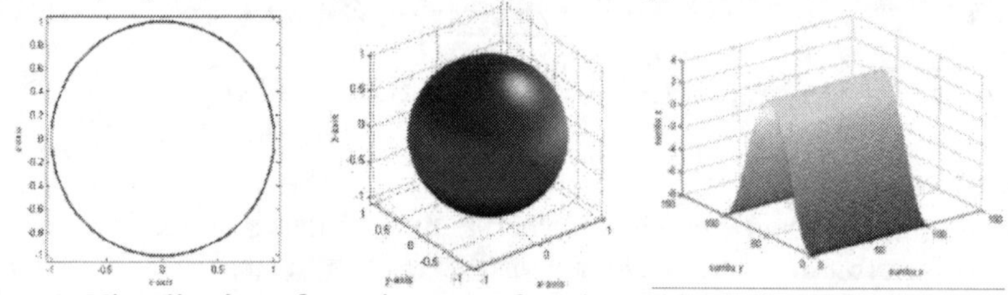

Figure 3. Visualisation of angular wave function without disturbance factor

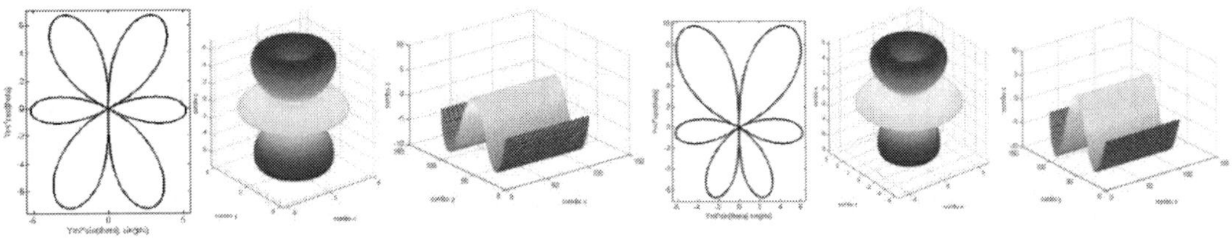

Figure 4. Visualisation of angular wave function for modified Poschl Teller potential with centrifugal term plus trigonometric Scarf

Wavefunction explains about probability density to find of the electron because statistical interpretation about wave function basically shows that the result of measuring in the quantum system unpredictable. From the Figure 2 left, that explain about ground state condition, we can see that with the increase of t constant, the amplitude of wave function decreased, and with the increase of the r value, the amplitude of wave function also increased. That means, by the increasing of amplitude, the probability density to find of the electron is bigger. The fact also happens when the first condition (Figure 2 right). But the differences from Figure 2 left and Figure 2 right, in Figure 2 left, by the small increase of r value, causes a small increase in amplitude, but from Figure 2 right, by the small increase of r value causes a bigger increase in amplitude. From the solved special function and by using the variation of b and a value, we can get some angular wave function. The visualisation of the angular wave function result be showed by Figure 3 and Figure 4.

The form of angular wave function related to the direction of angular momentum and describing depending on probability density to the angular. The general definition of angular wave function same with the radial wave function, but both have differences. Radial wave function about far or near electron from the nucleus, if angular wave function correlation with around of electron.

For the value of $m_l = 0$, electron can be found on the z axis. But for the value of $m_l = \pm 1$, electron can be found on the x and y shape. If $m_l = 1$, the electron counter clock wise, but if $m_l = -1$ the electron opposite.

The form of angular wave function it influenced by the value of v and q that give different influence to form of angular wave function. Increasing q value but the is constant, the noise that be given by Rosen Morse potential can make smaller wave function on the x,y, and z direction. But if the increasing v value with the q constant, the noise that be given by Rosen Morse potential can make bigger wave function. If v and q value not constant, all the wave function can be smaller. The changes of angular wave function to the v and q value explained that the noise from Rosen Morse potential make change to probability on the angular function, but not influence to the direction of electron circle.

6. Conclusion

In this paper, we study the Dirac equation for modified Poschl Teller plus trigonometric Scarf potential. The radial wave function part is obtained approximately from Eq.(19) and the angular part from Eq.(30). The result under influences by modified Poschl Teller plus trigonometric Scarf potentials changes the angular and radial part of the wavefunction. The energy eigenvalue can be obtained via Romanovsky Polynomials method in Eq. (16) and the equation of orbital quantum number l as expressed in Eq. (28). Both correlated between quantum number. The energy spectrum is also numerically solved using Matlab. From the value of energy eigenvalue, we found that by a little increased of γ can make large change of energy eigenvalue. Than from wave function we also know that the probability density to find of electron under influences the amplitude of wave function.

References

[1] M. Azizi, N. Salehi and A. Akbar 2013 *High Energy Physics*, vol **2013** ISRN Article ID 310392, 6 pages.

[2] Cari and Suparmi 2012 *International Journal of Applied Physics and Mathematics* **2**159-161.

[3] Reity, Rubish and Myhalyna 2014 *Proceedings of Institute of Mathematics of NAS of Ukraine*, **50** 1429–1434.

[4] Victor M. Villalba 2005 *Journal of Physics Conference Series* **24** 136-140.

[5] S. Bakkeshizadeh and V. Vahidi 2012 *Exact Solution of the Dirac Equation for the Coulomb Potential Plus NAD Potential by Using the Nikiforov-Uvarov Method* vol 6 Adv. Studies Theor. Phys, p 733 – 742.

[6] Cari, Suparmi and H.Marini 2012 *Penentuan Spektrum Energi dan Fungsi Gelombang Potensial Morse dengan Koreksi Sentrifugal Menggunakan Metode SWKB dan Operator SUSY* vol 2 Indonesian Journal of Applied Physics, p 112.

[7] Ikot, A. N., and Akpabio 2010 *Approximate Solution of the Schrödinger Equation with Rosen-Morse Potential Including the Centrifugal Term* vol 2 Applied Physics Research, pp. 202-208.

[8] Eshghi, Hamzavi, and S. M. Ikhdair 2012 *Exact Solutions of a Spatially Dependent Mass Dirac Equation for Coulomb Field plus Tensor Interaction via Laplace Transformation Method* vol 2012 Advances in High Energy Physics Article ID 873619.

[9] Castillo, David 2009 *Exactly Solvable Potentials and Romanovski Polynomials in Quantum Mechanics*. Instituto de F´ısica Universidad Aut´onoma de San Luis Potos. arXiv:0808.1642v2 [math-ph].

[10] Sadeghi, Pahlavani, Naderi and Banijamali 2005 *Solution of the Relativistic Dirac Equation for Woods-Saxon potential* Proceedings QPF Sept. 22.

[11] H.J. Weber 2008 *Connections between Romanovski and other polynomials* arXiv:0706.3153v1[math.CA].

[12] A. F. Nikiforov and U. B. Uvarov, Special Function in Mathematical Physics, Birkhausa, Basel 1988

Journal of Physics: Conference Series **776** (2016) 012084 doi:10.1088/1742-6596/776/1/012084

Solution of five dimensional Dirac equation with Asymptotic Iteration in the case of spin symmetry

Y.A. Kurniawan, A. Suparmi, C. Cari and F. Anwar
Physics Department , Sebelas Maret University,
Jl. Ir. Sutami 36A Kentingan Jebres Surakarta 57126, Indonesia
E-mail: yosuaardi@student.uns.ac.id

Abstract The relativistic energies and wave function of particle s wave spin 1/2 which is governed by separable non central potential in five dimensions are obtained determined using AIM in the case of spin symmetry. The separable five dimensional shape invariant potentials consist of radial potentials of Hulthen and Manning Rosen angular potentials of θ_1, θ_2, θ_3, and θ_4. The relativistic energies are numerically calculated from relativistic energy equation using MatLab software.

1. Introduction

Exact solution for relativistic energy can be obtained from Dirac equation. Dirac equation works in half spin particle such as electron and positron. Like Klein-Gordon equation [1] Dirac equation [2] can always be reduced to a Schrodinger type equation [3]. Many types of potential also been done for specific system, Ring-Shaped like potential [4], Hulthen potential [5], Eckart potential [6] and so on. Here an electron works on Manning-Rosen [7] and Hulthen potential. Manning-Rosen used to explain system of diatomic molecules and Hulthen potential is one of the most important short-range potentials which behaves like a Coloumb potential for small values of r and decreases exponentially for large values of r. The higher dimensional field theory have been studied of many authors. Variation of dimensional case have been obtained in D-dimensional[8] for Dirac equation.

Nowadays AIM used for many research to find exact solution[9-10]. AIM can be used in many potential Here Hulthen potential works on radial, and Manning-Rosen potential works on angular called θ_1, θ_2, θ_3 and θ_4. Hulthen potential and Manning rosen potentials used in this research are :

$$V(r)=\frac{-V_0 e^{-\alpha r}}{1-e^{-\alpha r}} \quad (1)$$

$$V(\theta_1)=\frac{V_1}{\sin^2\theta_1}-V_2\cot\theta_1 \quad (2)$$

$$V(\theta_2)=\frac{V_1'}{\sin^2\theta_2} \quad (3)$$

$$V(\theta_3)=\frac{V_1''}{\sin^2\theta_2}-V_2''\cot\theta_3 \quad (4)$$

$$V(\theta_4)=-V_3'''\cot\theta_4 \quad (5)$$

2. Asymptotic Iteration Method

AIM solve homogeneous linear second order differential equation to get exact solution of the form

$$y_n''(x) = k_0(x) y_n'(x) + s_0(x) y_n(x) \tag{5}$$

Where $k_0 \neq 0$, $s_0(x)$ is coefficient differential equation and n clarifying quantum number. We differentiate Eq (5) with respect to x :

$$y_n'''(x) = k_1(x) y_n'(x) + s_1(x) y_n(x) \tag{6}$$

Where

$$\begin{aligned} k_1(x) &= k_0'(x) + k_0^2(x) + s_0(x) \\ s_1(x) &= s_0'(x) + s_0 k_0 \end{aligned} \tag{7}$$

And the second derivative of Eq (5) gives :

$$y'''' (x) = k_2(x) y_n' + s_2(x) y_n(x) \tag{8}$$

Where

$$\begin{aligned} k_2(x) &= k_1'(x) + k_1(x) k_0(x) + s_1(x) \\ s_1(x) &= s_1'(x) + s_0(x) k_1(x) \end{aligned} \tag{9}$$

It can iterate up to $(i+1)th$ and $(i+2)th$ derivatives, with $i = 1,2,3...$
So we have

$$\begin{aligned} y_n^{(i+1)}(x) &= k_{i-1}(x) y_n'(x) + s_{i-1}(x) y_n(x) \\ y_n^{(i+2)}(x) &= k_i(x) y_n'(x) + s_i(x) y_n(x) \end{aligned} \tag{10}$$

Where

$$\begin{aligned} k_i(x) &= k_{i-1}'(x) + k_{i-1}(x) k_0(x) + s_{i-1}(x) \\ s_i(x) &= s_{i-1}'(x) + s_0(x) k_{i-1}(x) \end{aligned} \tag{11}$$

From Eq (13) we got relation :

$$\frac{y_n^{(i+2)}(x)}{y_n^{(i+1)}(x)} = \frac{k_i\left[y_n'(x) + \frac{s_i}{k_i} y_n(x) \right]}{k_{i-1}\left[y_n'(x) + \frac{s_{i-1}}{k_{i-1}} y_n(x) \right]} \tag{12}$$

Asymptotic aspect of the method makes large $i(>0)$, Eq (14) becomes

$$\frac{s_i}{k_i} = \frac{s_{i-1}}{k_{i-1}} \equiv \Delta \tag{13}$$

Eq (14) can reduced to simpler equation

$$\frac{y_n^{(i+2)}(x)}{y_n^{(i+1)}(x)} = \frac{k_i}{k_{i-1}} \tag{14}$$

The integration of Eq (16) is :

$$y_n^{(i+2)}(x) = Ce^{\int \frac{k_i}{k_{i-1}} dx} \tag{15}$$

Where C is constant outcome from integration. Inserting Eq (11) and Eq (13) to Eq (15) become

$$y_n^{(i+1)}(x) = C\lambda_{i-1} e^{\int [\alpha(x)+\lambda_0(x)]dx} \tag{16}$$

Inserting Eq (16) to Eq (10), the first order differential equation is obtained as :

$$y_n' + \alpha y_n(x) - Ce^{\int [\alpha(x)+k_0(x)]dx} = 0 \tag{17}$$

This first order differential equation general solution can be write as :

$$y_n(x) = e^{-\int \alpha(x)dx}\left[C' + C\int e^{\int [k_0(x)+2\alpha(x)]dx} dx\right] \tag{18}$$

For given potential, we convert five dimensional Dirac equation to the form of Eq (5). Than determine k_0 and s_0. Using Eq (11) $s_i(x)$ and $k_i(x)$ are obtained. Energy eigenvalue we get from

$$k_i(x)s_{i-1}(x) - k_{i-1}(x)s_i = \Delta_i = 0,\ i = 1,2,3... \tag{19}$$

Where k is the iteration number, and radial quantum number n is equal with iteration number k for this case

3. Bound State Solution

In natural unit $(\hbar = c = 1)$ the Dirac equation in case of symetry spin for half spin particle of rest mass M is written as

$$\left[-\nabla^2 + 2V(E+M)\right]\psi = \left(E^2 - M^2\right)\psi \tag{20}$$

Where E represent relativistic energy of half spin particle. In case of spin symmetry vector potential $V(x)$ assuming equal with scalar potential $S(x)$. We propose a solvable potential model which is expressed as

$$V(r,\theta_1,\theta_2,\theta_3,...,\theta_{D-1}) = \frac{h^2}{2m}\left[V(\bar{r}) + \frac{1}{r^2}\left(\frac{V(\theta_1)}{\sin^2\theta_2 \sin^2\theta_3 \sin^2\theta_4 ... \sin^2\theta_{D-1}} + \frac{V(\theta_2)}{\sin^2\theta_3 \sin^2\theta_4 ... \sin^2\theta_{D-1}} + \frac{V(\theta_3)}{\sin^2\theta_4 ... \sin^2\theta_{D-1}} + ... + V(\theta_{D-1})\right)\right] \tag{21}$$

And we use D dimensional Laplacian in hyper-spherical coordinates given as

$$\nabla^2 = \frac{1}{r^{D-1}}\frac{\partial}{\partial r}\left(r^{D-1}\frac{\partial}{\partial r}\right) + \frac{1}{r^2}\left[\begin{array}{l} \frac{1}{\sin^2\theta_2 \sin^2\theta_3 \sin^2\theta_4 ... \sin^2\theta_{D-1}}\left(\frac{\partial^2}{\partial\theta_1^2}\right) \\ + \frac{1}{\sin^2\theta_3 \sin^2\theta_4 ... \sin^2\theta_{D-1}}\frac{\partial}{\partial\theta_2}\left\{\frac{1}{\sin\theta_2}\left(\sin\theta_2\frac{\partial}{\partial\theta_2}\right)\right\} \\ + \frac{1}{\sin^2\theta_4 ... \sin^2\theta_{D-1}}\frac{\partial}{\partial\theta_3}\left\{\frac{1}{\sin^2\theta_3}\left(\sin^2\theta_3\frac{\partial}{\partial\theta_3}\right)\right\} \\ + \frac{1}{\sin^2\theta_{D-1}}\frac{\partial}{\partial\theta_{D-1}}\left\{\frac{1}{\sin^{D-2}\theta_{D-1}}\left(\sin^{D-2}\theta_{D-1}\frac{\partial}{\partial\theta_{D-1}}\right)\right\} \end{array}\right] \tag{22}$$

With $\psi(r,\theta_m)=R(r)Y(\theta_m)$, and $Y(\theta_m)=P_1(\theta_1)P_2(\theta_2)P_3(\theta_3)P_4(\theta_4)$ by inserting Eq (21) and (22) to Eq (20) become

$$\left[-\nabla_r^{\ 2}-\frac{\nabla_\theta^{\ 2}}{r^2}+V(r)+\frac{V(\theta)}{r^2}(E+M)\right]RY=\left(E^2-M^2\right)RY \tag{23}$$

We separated Eq (23) to one dimensional equation of radial, θ_1, θ_2, θ_3, and θ_4.

3.1 Radial Energy Equation

From the separation of Eq (23) one dimensional radial equation written as

$$\frac{1}{R}\left(-r^2\nabla_r^{\ 2}R\right)+r^2V(r)(E+M)-r^2(E^2-M^2)=\lambda_4 \tag{24}$$

Where λ_4 is constant and r is distance a half particle with source of potential field. ∇_r represent of radial laplacian D dimensional written as

$$\nabla_r^{\ 2}=\frac{1}{r^{D-1}}\frac{\partial}{\partial r}\left(r^{D-1}\frac{\partial}{\partial r}\right) \tag{25}$$

By using $R(r)=\frac{\chi(r)}{r^{(D-1)/2}}$, $\coth\left(\frac{\alpha r}{2}\right)=1-2z$ and $\chi(z)=(z)^{\delta}(1-z)^{\gamma}f_n(z)$ and inserting Eq (1) and Eq (25) to Eq (24) we have

$$z(1-z)f_n''(z)+\left[1+2\delta-(2\delta+2\gamma+2)z\right]f_n'(z)+\left[-(\delta+\gamma)^2-\delta-\gamma+A_s\right]f_n(z)=0 \tag{26}$$

Where

$$\delta=\sqrt{E_s-B_s} \qquad \gamma=\sqrt{B_s+E_s} \tag{27}$$

$$A_s=2-\lambda_4 \qquad B_s=\frac{V_0(E+M)}{2\alpha^2} \qquad E_s=\frac{V_0}{2\alpha^2}(E+M)+\frac{\left(E^2-M^2\right)}{\alpha^2} \tag{28}$$

Eq (26) can reduce to AIM type homogeneous linear second order differential equation. s_0 and k_0 are obtained from Eq (26), so we can easy to find Δ_n form Eq (11) and (19) represent as

$$(\delta+\gamma+n)(\delta+\gamma+(n+1))-A_{sn}=\Delta_0=0 \tag{29}$$

Eq (29) expand mathematically. By inserting Eq (27) and (28) to equation (29) we have

$$\left[\sqrt{2-\lambda_4+\frac{1}{4}}-\left(n+\frac{1}{2}\right)\right]^2+\frac{4\left[\frac{V_0(E+M)}{2\alpha^2}\right]^2}{\left[\sqrt{2-\lambda_4+\frac{1}{4}}-\left(n+\frac{1}{2}\right)\right]^2}=4\left[\frac{V_0}{2\alpha^2}(E+M)+\frac{\left(E^2-M^2\right)}{\alpha^2}\right] \tag{30}$$

3.2 θ_1 Equation

The result of separation Eq (22) in variable θ_1 written as

$$\frac{1}{P_1}\left(\frac{d^2P_1}{d\theta_1^{\ 2}}\right)-V(\theta_1)(E+M)=\lambda_1 \tag{31}$$

Where λ_1 is constant and P_1 is an equation respect to θ_1. Eq (31) can reduce to homogeneous linear second order differential equation by inserting Eq (2) and consider $\cot\theta_1=(1-2z_1)i$, $P_1=z_1^{\ \delta}(1-z_1)^{\gamma}f_n(z_1)$, than solve second differential of P_1 and we have

$$z_1(1-z_1)f_n^{\ ''}+[1+2\delta-z_1(2\delta+2\gamma+2)]f_n^{\ '}-[(\delta+\gamma)^2+\delta+\gamma+V_1(E+M)]f_n=0 \tag{32}$$

We can determine s_0 and k_0 from Eq (32). With Eq (19) we have

$$(\delta+\gamma+n)(\delta+\gamma+n+1)+V_1(E+M)=\Delta_3=0 \tag{33}$$

Can be rewritten as

$$(\delta+\gamma+n+1)^2-(\delta+\gamma+n+1)+V_1(E+M)=\Delta_n=0 \tag{34}$$

Where

$$\delta=\sqrt{\frac{V_2i(E+M)+\lambda_1}{4}} \qquad \gamma=\sqrt{\frac{-V_2i(E+M)+\lambda_1}{4}} \tag{35}$$

By inserting Eq (35) to Eq (34) we have

$$\lambda_1=\frac{[V_2i(E+M)]^2}{4\left[\sqrt{-V_1(E+M)+\frac{1}{4}}-\left(n+\frac{1}{2}\right)\right]^2}+\left[\sqrt{-V_1(E+M)+\frac{1}{4}}-\left(n+\frac{1}{2}\right)\right]^2 \tag{36}$$

3.3 θ_2 Equation

Variable θ_2 from separation of Eq (23) written as

$$\frac{\lambda_1}{\sin^2\theta_2}+\frac{1}{P_2}\left(\frac{1}{\sin\theta_2}\frac{d}{d\theta_2}\left(\sin\theta_2\frac{dP_2}{d\theta_2}\right)\right)-V(\theta_2)(E+M)=\lambda_2 \tag{37}$$

By using $P_2=\frac{Q_2(\theta_2)}{\sin^{1/2}\theta_2}$, $\cot\theta_2=i(1-2z_2)$, and $Q_2(z_2)=z_2^{\ \alpha}(1-z_2)^{\beta}f_n(z_2)$ Eq (45) become

$$z_2(1-z_2)f_s^{\ ''}(z_2)=-[1+2\alpha-z_2(2+2\alpha+2\beta)]f_s^{\ '}(z_2)+\left\{(\alpha+\beta)^2+\alpha+\beta+\lambda_1-V'(E+M)+\frac{1}{4}\right\}f_s(z_2) \tag{38}$$

Where

$$\alpha=\sqrt{\left(\frac{1}{4}-\lambda_2\right)/4} \qquad \beta=\sqrt{\left(\frac{1}{4}-\lambda_2\right)/4} \tag{39}$$

We can find Δ_n from Eq (11), and Eq (19):

$$\Delta_n=(\alpha+\beta+n+1)^2-(\alpha+\beta+n+1)+\lambda_1-V'(E+M)+\frac{1}{4} \tag{40}$$

So we have eigenvalue from θ_2 equation

$$\lambda_2 = \frac{1}{4} - \left[\sqrt{-\lambda_1 + V'(E+M)} - \left(n + \frac{1}{2} \right) \right]^2 \tag{41}$$

3.4 θ_3 Equation

Separation variable of Eq (25) for θ_3 is Eq (49)

$$\frac{\lambda_2}{\sin^2\theta_3} + \frac{1}{P_3}\left(\frac{1}{\sin^2\theta_3}\frac{d}{d\theta_3}\left(\sin^2\theta_3 \frac{dP_3}{d\theta_3} \right) \right) - V(\theta_3)(E+M) = \lambda_3 \tag{42}$$

With the same way, consider $P_3 = \dfrac{Q_3(\theta_3)}{\sin^{2/2}\theta_3}$, $\cot\theta_3 = i(1-2z_3)$, $Q_3 = z_3^{\alpha}(1-z_3)^{\beta} f_n(z_3)$ we reduce Eq (42) to be homogeneous linear second order differential equation

$$z_3(1-z_3)f_n'' + [2\alpha+1-z_3(2+2\alpha+2\beta)]f_n' - \{(\alpha+\beta)^2+\alpha+\beta-4V_1''(E+M)+4\lambda_2\}f_n = 0 \tag{43}$$

Using Eq (11) and Eq (19) we have

$$\Delta_n = (\alpha+\beta+n+1)^2 - (\alpha+\beta+n+1) - 4V_1''(E+M) + 4\lambda_2 \tag{44}$$

Where

$$\alpha = \sqrt{V_2''i(E+M)-(\lambda_3+1)} \qquad \beta = \sqrt{-V_2''i(E+M)-(\lambda_3+1)} \tag{45}$$

By inserting Eq (45) to Eq (44) we obtain

$$\lambda_3 = -\frac{[V_2''i(E+M)]^2}{\left[\sqrt{4V_1''(E+M)-4\lambda_2+\frac{1}{4}} - \left(n+\frac{1}{2}\right)\right]^2} - \frac{\left[\sqrt{4V_1''(E+M)-4\lambda_2+\frac{1}{4}} - \left(n+\frac{1}{2}\right)\right]^2}{4} - 1 \tag{46}$$

3.5 θ_4 Equation

The last equation is for variable θ_4

$$\frac{1}{\sin^3\theta_4}\frac{d}{d\theta_4}\left(\sin^3\theta_4 \frac{dP_4}{d\theta_4} \right) - \left(V(\theta_4)(E+M) - \lambda_4 + \frac{\lambda_3}{\sin^2\theta_4} \right) P_4 = 0 \tag{47}$$

With $P_4 = \dfrac{Q_4}{\sin^{3/2}\theta_4}$, $\cot\theta_4 = i - 2iz_4$, and $Q_4 = z_4^{\delta}(1-z_4)^{\gamma} f_n(z_4)$ we have

$$z_4(1-z_4)f_n''(z_4) + [1+2\delta-z_4(2+2\delta+2\gamma)]f_n'(z_4) + \left\{-(\delta+\gamma)^2-\gamma-\delta+\lambda_3+\frac{3}{4}\right\}f_n(z_4) = 0 \tag{48}$$

Than we find Δ_n

$$(\delta+\gamma+n+1)^2 - (\delta+\gamma+n+1) - \lambda_3 - \frac{3}{4} = \Delta_n = 0 \tag{49}$$

Where

$$\delta=\sqrt{\frac{V_3\eta(E+M)+\left(\frac{9}{4}+\lambda_4\right)}{4}} \qquad \gamma=\sqrt{\frac{-V_3\eta(E+M)+\left(\frac{9}{4}+\lambda_4\right)}{4}} \tag{50}$$

By inserting Eq (50) to Eq (49) the eigenvalue represent as

$$\left[\sqrt{\lambda_3+1}-\left(n+\frac{1}{2}\right)\right]^2+\frac{\frac{1}{4}\left(V_3\eta(E+M)\right)^2}{\left[\sqrt{\lambda_3+1}-\left(n+\frac{1}{2}\right)\right]^2}-\frac{9}{4}=\lambda_4 \tag{51}$$

4. Result and Conclusion

By inserting Eq (36) to Eq (41), than Eq (41) to Eq(46), Eq (46) to (51), and Eq (51) to Eq (30), we have as shown in Table 1

V1	V2	Vak	Vak1	Vtak3	V0	N1	N2	N3	N4	alpha	M	Energy
1	0	0	0	0	0	1	1	1	1	0.5	5	-4.70554
2	0	0	0	0	0	1	1	1	1	0.5	5	-4.71991
3	0	0	0	0	0	1	1	1	1	0.5	5	-4.73098
0	1	0	0	0	0	1	1	1	1	0.5	5	-4.68491
0	2	0	0	0	0	1	1	1	1	0.5	5	-4.68838
0	3	0	0	0	0	1	1	1	1	0.5	5	-4.69407
0	0	1	0	0	0	1	1	1	1	0.5	5	-4.84598
0	0	2	0	0	0	1	1	1	1	0.5	5	-4.85014
0	0	3	0	0	0	1	1	1	1	0.5	5	-4.85383
0	0	0	3	0	0	1	1	1	1	0.5	5	-4.90994
0	0	0	4	0	0	1	1	1	1	0.5	5	-4.91249
0	0	0	5	0	0	1	1	1	1	0.5	5	-4.91474
0	0	0	0	3	0	1	1	1	1	0.5	5	-4.97549
0	0	0	0	4	0	1	1	1	1	0.5	5	-4.9759
0	0	0	0	5	0	1	1	1	1	0.5	5	-4.9764
0	0	0	0	0	1	1	1	1	1	0.5	5	-4.99404
0	0	0	0	0	2	1	1	1	1	0.5	5	-4.9943
0	0	0	0	0	3	1	1	1	1	0.5	5	-4.99454

The amount of energy must be not too far from M. Five-dimensional with AIM can use to solve eigenvalue of Dirac equation. Nevertheless five-dimensional equation is interesting for the next study.

5. Acknowledgment

This Research was partly supported by Higher Education Project Grant with contract no. 632/UN27.21/LT/2016

6. References

[1] Isonguyo C N Okon I B Ikot A N and Hassanabadi H 2014 *Bull. Korean Chem. Soc. 2014, Vol. 35, No. 12.*
[2] Arda A Sever R and Teczan C 2013 *arXiv:0909.0821v1 [quant-ph].*
[3] Saregar A A Suparmi and C Cari, Yuliani H 2013 *International Journal Of Engineering And Science ISSN: 2278-4721, Vol. 2, Issue 3.*
[4] Ahmadov A I Aydin C and Uzun O 2014 *arXiv:1401.8163v1 [math-ph].*
[5] Roy A K 2013 *arXiv:1312.5900v1 [quant-ph].*
[6] Ikot A N Maghsoodi E and Antia A D 2015 *Arab J Sci Eng DOI 10.1007/s13369-015-1602-4.*
[7] Roy A K 2014 *arXiv:1409.2405v1 [physics.atom-ph].*
[8] Hall R L and Zorin P 2015 *Ann. Phys. (Berlin) 572, No. 5-6, 408-422 (2015)/DOI 10.1002/andp.201500123.*
[9] Bayarak O Boztosun I and Ciftci H 2006 *Wiley InterScience DOI 10.1002/qua.21141.*
[10] Ateser E Cifti H and Ugurlu M 2011 *Chinese Journal Of Physics Vol. 45, No. 3.*

Solution of three dimensional Schrodinger equation for Eckart and Manning-Rosen non-central potential using asymptotic iteration method

M. N. Farizky, A. Suparmi , C. Cari and M. Yunianto
Physics Department, Undergraduate Program, Sebelas Maret University, Jl. Ir. Sutami 36A Kentingan Jebres Surakarta 57126, INDONESIA

E-mail: mnf53@student.uns.ac.id

Abstract. Solution of Schrodinger equation in three dimensions for Eckart and Manning-Rosen potential has been obtained by using the asymptotic iteration method. Energy spectrum and wave function for these potentials was obtained. It is known that the wave function for the corresponding potentials contains hypergeometric series due to the type of Schrodinger equation. However, the wave function for radial part is not normalizable, due to its equation that reaches to infinity when r equals to zero. The energy spectrum and wave function for corresponding potentials had also been analyzed with the help of Matlab R2013a software.

1. Introduction

Schrodinger equation is one of the most powerful tool to describe the phenomenon in quantum physics. One of its application is Coulomb potential, which is used to determine the probability of electron can be found in hydrogen atom. Nowadays, researchers had found many potentials that can describe the particular phenomena in the development of theoretical physics. These potentials are known as Poschl-Teller [1], Gendenshtein [2], Rosen-Morse [3], Eckart [4], Manning-Rosen potential [5], and so on.

Many researchers had studied the Schrodinger equation for these potentials. However, these potentials cannot be able to be solved exactly, so researchers studied these equations by using different approximation and methods. Here, we attempt to study the Schrodinger equation which is influenced by Eckart and Manning-Rosen potential, where Eckart potential as a radial function and Manning-Rosen as angular function. This is known as non-central potential.

Eckart potential is used to study the electron tunneling correction. Dong *et al.* [6], Falaye [7], and Resita [8] had studied Schrodinger equation for this potential in a different way, in particular, different variable substitution. In our work, we attempt to solve the Schrodinger equation with different variable substitution. Thus, in our work, there is Manning-Rosen potential which will determine the azimuthal and magnetic quantum number.

Our purpose of this work is to determine the energy spectrum and wave function of three dimensional Schrodinger equation for these potentials. To find the energy spectrum, we're using the *Asymptotic*

Iteration Method (AIM). Researchers who had studied the Schrodinger equation used a various methods like Nikiforov-Uvarof (NU) [9], Romanovski polynomials [10], AIM [11], and so on. However, AIM is one of the most practical methods and often used by many researchers.

The approachment of our work in order to solve the Schrodinger equation lies on variable substitution. Determined variable substitution may lead the equation to the type of hypergeometric equation. Our first "goal" is to find the hypergeometric type equation of corresponding Schrodinger equation before we treat the equation into AIM. Our hypergeometric type of equation is Gauss hypergeometric type equation. The other obtained Schrodinger equation will be treated similarly. The computer software that will be used to support this work will be MATLAB R2013a.

2. Overview of Asymptotic Iteration Method

Suppose we have the second order differential equation expressed as :

$$y_n'' = \lambda_0(x)y_n' + s_0(x)y_n \tag{1}$$

If we differentiate Eq. (1) in respect to x in k times, Eq. (1) will become :

$$y_n^{(k+2)} = \lambda_k(x)y_n' + s_k(x)y_n \tag{2}$$

Where

$$\lambda_k(x) = \lambda'_{k-1}(x) + s_{k-1}(x) + \lambda_0(x)\lambda_{k-1}(x) \tag{3}$$

$$s_k(x) = s'_{k-1}(x) + s_0(x)\lambda_{k-1}(x) \tag{4}$$

Next we're going to examine the ratio between $\lambda_k(x)$ and $s_k(x)$. For a higher k, it is found that

$$\frac{\lambda_k(x)}{s_k(x)} = \frac{\lambda_{k-1}(x)}{s_{k-1}(x)} = \alpha \tag{5}$$

So that we'll know the termination condition as

$$\Delta = \lambda_k(x)s_{k-1}(x) - \lambda_{k-1}(x)s_k(x) = 0 \tag{6}$$

This Eq. (6) will later be used to determine the energy spectrum and quantum numbers of the corresponding Schrodinger equation. For a complicated equation of $\lambda_k(x)$ and $s_k(x)$, calculation of Eq. (6) can be done easier with the help of computer.

3. Obtaining the Wave Function

In order to gain the wave function of the corresponding Schrodinger equation, we can use this second-order differential equation as reference.

$$y_n'' = 2\left(\frac{az^{N+1}}{1-bz^{N+2}} - \frac{t}{z}\right)y_n' - \frac{wz^N}{1-bz^{N+2}}y_n \tag{7}$$

By transforming the Schrodinger equation to Eq. (7) form, we will be able to determine the parameters contained in Eq. (7) such as a, b, N, and t. The solution of Eq. (7) is expressed as

$$y_n(x) = (-1)^n C_2(\sigma)_n (N+2)^n \; {}_2F_1(-n, \rho+n; \sigma; bz^{N+2}) \tag{8}$$

Where

$$\rho = \frac{(2t+1)b+2a}{(N+2)b} \tag{9}$$

$$\sigma = \frac{2t+N+3}{N+2} \tag{10}$$

$$(\sigma)_n = \sigma(\sigma+1)(\sigma+2)\ldots(\sigma+n-1) \tag{11}$$

$${}_2F_1(-n, \rho+n; \sigma; bz^{N+2}) = \sum_{j=0}^{\infty} \frac{(-n)_j(\rho+n)_j}{(\sigma)_j} \frac{\left(bz^{N+2}\right)^j}{j!} \tag{12}$$

And C_2 is normalization constant. Parameter $(\sigma)_n$ in Eq. (11) is known as Pochammer symbol, and have a similar expression as Pochammer symbols contained in Eq. (12), which is known as Gauss hypergeometric series.

4. Three-Dimensional Schrodinger Equation for Eckart and Manning-Rosen Non-Central Potential

It is well-known that time-independent Schrodinger equation is expressed as follows

$$\frac{\hbar^2}{2m}\frac{\partial^2\Psi}{\partial x^2} + (E-V)\Psi(x) = 0 \tag{13}$$

Where the potential we're studying is non-central potential, expressed as

$$V(r,\theta,\varphi) = V(r) + \frac{V(\theta)}{r^2} + \frac{V(\varphi)}{r^2 sin^2\theta} \quad (14)$$

Here, the radial part of the non-central potential is Eckart potential, which is used to describe the electron penetration in potential barrier, or to describe the electron tunneling correction. The general Eckart potential is

$$V(r) = -V_0 \frac{e^{-r/a}}{1-e^{-r/a}} + V_1 \frac{e^{-r/a}}{\left(1-e^{-r/a}\right)^2} \quad (15)$$

Where V_0 and V_1 are the depth of potential well and a is the length of potential. And trigonometric Manning-Rosen potential can be expressed as

$$V(\theta) = \frac{V_2}{sin^2\,\theta} - V_3 \cot\theta \quad (16)$$

And

$$V(\varphi) = \frac{V_4}{cos^2\,\varphi} - V_5 \tan\varphi \quad (17)$$

Where the indexed variable V represents the depth of potential well, similar to Eq. (15). By substituting these potentials into Eq. (14) and (13), we obtain the Schrodinger equation for these potentials as

$$\frac{\hbar^2}{2m}\frac{\partial^2\Psi}{\partial x^2} + \left(E + V_0 \frac{e^{-r/a}}{1-e^{-r/a}} - V_1 \frac{e^{-r/a}}{\left(1-e^{-r/a}\right)^2} - \frac{1}{r^2}\left(\frac{V_2}{sin^2\,\theta} - V_3 \cot\theta\right) - \frac{1}{r^2 sin^2\theta}\left(\frac{V_4}{cos^2\,\varphi} - V_5 \tan\varphi\right)\right)\Psi = 0 \quad (18)$$

Using the parameter separation of $\Psi(x) = R(r)r^{-1}P(\theta)H(\varphi)$ and transforming the Schrodinger Equation into three dimension in spherical coordinate, we'll be able to obtain three Schrodinger equation in radial, angular, and azimuthal part.

$$\frac{\partial^2 R(r)}{\partial r^2} - \left[\frac{2m}{\hbar^2}\left(-V_0 \frac{e^{-r/a}}{1-e^{-r/a}} + V_1 \frac{e^{-r/a}}{\left(1-e^{-r/a}\right)^2} - E\right) - \frac{l'(l'+1)}{r^2}\right] R(r) = 0 \quad (19a)$$

$$\frac{\partial^2 P(\theta)}{\partial\theta^2} + \frac{\cos\theta}{\sin\theta}\frac{\partial P(\theta)}{\partial\theta} - \left[\frac{2m}{\hbar^2}\left(\frac{V_2}{sin^2\,\theta} - V_3 \cot\theta\right) - \frac{m'^2}{sin^2\,\theta} - l'(l'+1)\right] P(\theta) = 0 \quad (19b)$$

$$\frac{\partial H(\varphi)}{\partial\varphi} - \left[\frac{2m}{\hbar^2}\left(\frac{V_4}{cos^2\,\varphi} - V_5 \tan\varphi\right) - m'^2\right] H(\varphi) = 0 \quad (19c)$$

And the next step is to find the hypergeometric type of the corresponding Schrodinger equations, by using the variable and parameter substitution.

5. Solving the Radial Part

In Eq. (19a), if we substitute

$$\frac{e^{-r/a}}{1-e^{-r/a}} = -x \quad (20)$$

Then using the approximation $\frac{1}{r^2} \approx \frac{1}{a^2}\frac{e^{-r/a}}{\left(1-e^{-r/a}\right)^2}$ and substituting equations into Eq. (19a), we get

$$\frac{\partial^2 R}{\partial x^2} + \frac{1-2x}{x(1-x)}\frac{\partial R}{\partial x} - \left[\frac{(\ell'(\ell'+1) - V_1 a^2)}{x(1-x)} - \frac{Ea^2}{x^2(1-x)} - \frac{Ea^2 - V_0 a^2}{x(1-x)^2}\right] R = 0 \quad (21)$$

Equation (21) is the radial part of Schrodinger equation, which has a singular point at $x = 0$ and $x = 1$. By using series solution of R

$$R = z^s \sum a_n z^n \quad (22)$$

and substituting it into Eq. (21), we're able to get the s parameter, and get the proportionality relation

$$R \propto z^s \quad (23)$$

based on Eq. (22). When $x \to 0$, the third term becomes a single term, instead of three terms as in Eq. (21), as its second term will become larger than the other term. Then Eq. (21) will become

$$\frac{\partial^2 R}{\partial x^2} + \frac{1-2x}{x(1-x)}\frac{\partial R}{\partial x} + \frac{Ea^2}{x^2(1-x)} R = 0 \quad (24)$$

Substituting R in Eq. (24) with Eq. (22), we'll get the relation

$$s^2 = \alpha^2 = -Ea^2 \quad (25)$$

We can use the similar way for $x \to 1$ to obtain the different parameter s. For this condition,

$$s^2 = \beta^2 = -(Ea^2 - V_0 a^2) \quad (26)$$

And from the proportionality relation in Eq. (23) for $x \to 0$ and $x \to 1$, we can obtain the wave equation in series function

$$R = z^{\alpha}(1-z)^{\beta}u(x) \tag{27}$$

Where $u(x)$ is a function. Then we can substitute Eq. (25), (26), and (27) into (21) in order to obtain the hypergeometric type Schrodinger equation. For the radial part, the hypergeometric type Schrodinger equation is written as

$$x(1-x)\frac{d^2u}{dx^2} + [(2\alpha+1) - (2\alpha+2\beta+2)]\frac{du}{dx} - [(\alpha+\beta)(\alpha+\beta+1) - (\ell'(\ell'+1) - V_1 a^2)]u = 0 \tag{28}$$

Where u is the function which contains hypergeometric series function. This function is similar to Eq. (8), and the parameters can be obtained by comparing the new Schrodinger equation with Eq. (7).

6. Solving the Angular and Azimuthal Part

We can use the same method as radial part to obtain the hypergeometric type Schrodinger equation for the angular and azimuthal part. For the angular part, we use the variable substitution as

$$\cot\theta = i(1-2y) \tag{29}$$

While for the azimuthal part is

$$\tan\varphi = i(1-2z) \tag{30}$$

So we're able to obtain the hypergeometric type Schrodinger equation for angular and azimuthal part

$$y(1-y)\frac{d^2v}{dy^2} + \left[\left(2\gamma+\frac{1}{2}\right) - (2\gamma+2\delta+1)\right]\frac{dv}{dy} - \left[(\gamma+\delta)^2 + ({m'}^2 - V_2)\right]v = 0 \tag{31}$$

$$z(1-z)\frac{d^2w}{dz^2} + [(2\xi+1) - (2\xi+2\eta+2)]\frac{dw}{dz} - [(\xi+\eta)(\xi+\eta+1) - V_4]w = 0 \tag{32}$$

Eq. (31) is for angular part, and (32) is azimuthal part, where the parameter substitution for angular part is written as

$$-2\gamma(2\gamma-1) = \ell'(\ell'+1) - V_3 i \tag{33a}$$

$$-2\delta(2\delta-1) = \ell'(\ell'+1) + V_3 i \tag{33b}$$

And for azimuthal part is

$$-4\xi^2 = m'^2 - V_5 i \tag{34a}$$

$$-4\eta^2 = {m'}^2 + V_5 i \tag{34b}$$

7. The Energy Spectrum

By using AIM we're able to obtain the energy spectrum of Schrodinger equation for Eckart and Manning-Rosen Potential. From the three Schrodinger equations before, we know that the energy parameter is located in the radial part of Schrodinger equation. So, we'll calculate the radial part of the hypergeometric type Schrodinger equation first using AIM. And the result of iteration is as follows :

$$k = 0 \to c = (\alpha+\beta)(\alpha+\beta+1)$$

$$k = 1 \to c = (\alpha+\beta+1)(\alpha+\beta+2)$$

$$k = 2 \to c = (\alpha+\beta+2)(\alpha+\beta+3)$$

$$k = 3 \to c = (\alpha+\beta+3)(\alpha+\beta+4)$$

And so on, where $c = (\ell'(\ell'+1) - V_1 a^2)$. so we're able to obtain the pattern for this calculation

$$c = (\alpha+\beta+n_r)(\alpha+\beta+n_r+1) \tag{35}$$

By substituting c, α, and β from equations before, we're able to obtain the energy spectrum for non-central Eckart and Manning-Rosen potential.

$$E_{nl} = \frac{V_0}{2} - \frac{1}{a^2}\left\{\frac{V_0^2 a^4}{\left[\sqrt{1+4\ell'(\ell'+1)-4V_1 a^2} - (2n_r+1)\right]^2} + \frac{\left[\sqrt{1+4\ell'(\ell'+1)-4V_1 a^2} - (2n_r+1)\right]^2}{4}\right\} \tag{36}$$

Where ℓ' is the orbital quantum number which depends on the Manning-Rosen potential. Table 1 shows the energy spectrum with the variation of n_r and ℓ' when the particle isn't influenced by Manning-Rosen potential, where $a = {V_0}^{-1}$, $V_0 = 0.005$, and $V_1 = 0.00005$.

Table 1. The energy spectrum of a particle under the influence of Eckart potential with $a = 1 / V_0$, $V_0 = 0.005$, $V_1 = 0.00005$.

No.	n_r	l	E_{nl}
1	1	0	-0.00532
2	2	0	-0.01519
3	2	1	-0.06010
4	3	0	-0.01116
5	3	1	-0.02550

If the particle is influenced by Manning-Rosen potential, the ℓ' parameter will depend on the potential depth of Manning-Rosen potential and quantum number n_ℓ. Note that ℓ' will also depend on the magnetic quantum number m', while m' also depends on the potential depth of Manning-Rosen potential and quantum number n_m. Both ℓ' and m' can be determined by using AIM to the remaining Schrodinger equations, which are the angular and azimuthal part. Table 2 shows the energy spectrum of a particle under the influence of Eckart and Manning-Rosen potential.

Table 1. The energy spectrum of a particle under the influence of Eckart and Manning-Rosen potential with $a = 1 / V_0$, $V_0 = 0.005$, $V_1 = 0.00005$. $V_2 = 10$, $V_3 = 5$, $V_4 = 80$, $V_5 = 150$.

No.	n_r	n_l	n_m	E
1	1	0	0	-0.12099
2	2	0	0	-0.03609
3	2	1	-1	-0.04676
4	2	1	0	-0.02406
5	2	1	1	-0.01076

8. Wave Function

By comparing the hypergeometric type Schrodinger equation into Eq. (7), we're able to obtain the wave function for corresponding Schrodinger equation. However, the obtained wave equations are not normalizable, due to the variable substitution, that cotangent and exponential terms diverge to infinity at $\theta = 0$ and $r = 0$, and the tangent term reaches infinity at $\theta = \frac{\pi}{2}$. Below is the graph of unnormalized wave function of radial part of Schrodinger equations for $n_r = 2$ and parameter values listed below.

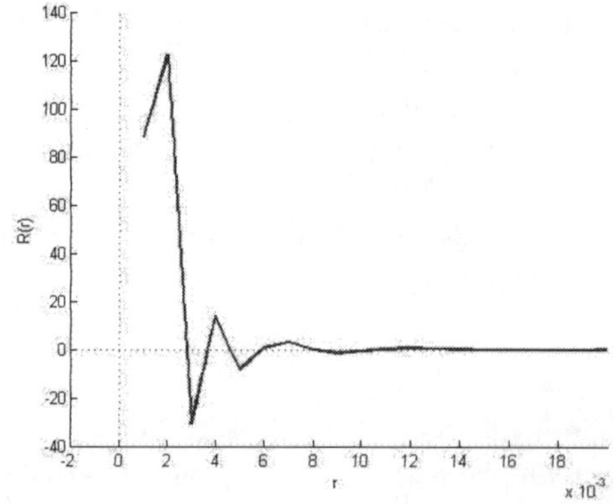

Figure 1. Unnormalized radial part wave function of a particle influenced by Eckart and Manning-Rosen potential, where $n_r = 2$, $n_l = 1$, $n_m = 0$, $a = 73.2$, $V_0 = 0.021$, $V_1 = 0.0021$, $V_2 = 0.5$, $V_3 = 0.1$, $V_4 = 1$, $V_5 = 0.1$.

The wave function of radial part has decreasing amplitude, as r goes higher. The curve is consistent for other value of n_r. However, the amplitude of wave function increases greatly in powers of ten as n_r goes higher. The dead end at r less than 0.002 denotes that as r approach zero, the amplitude becomes negatively higher, and reaches minus infinity. For the angular and azimuthal part is shown in Figure 2.

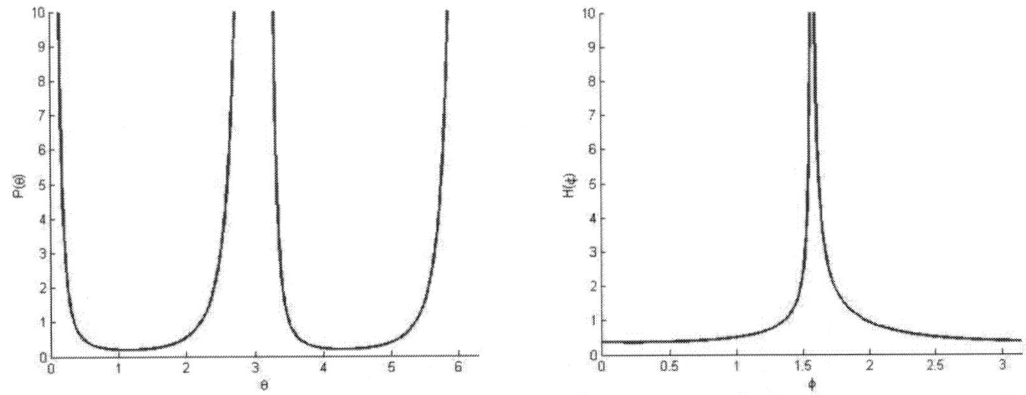

(a) (b)

Figure 2. (a) Unnormalized angular part wave function of a particle influenced by Eckart and Manning-Rosen potential, where n_l = 1, n_m = 0, V_2 = 10, V_3 = 5, V_4 = 5, V_5 = 0.3. (b) Unnormalized azimuthal part of wave function of a particle influenced by Eckart and Manning-Rosen potential, where $n_m = 1$, $V_4 = 5$, $V_5 = 0.3$.

However, these wave functions are solvable only for certain number of n_l and n_m. The azimuthal part one doesn't have a solution at n_m higher than one.

9. Conclusions

In this paper, we've solved the three dimensional Schrodinger equation for Eckart and Manning-Rosen non-central potential using AIM. We've obtained the energy spectrum for particle which influenced by Eckart and Manning-Rosen potential for the certain value of quantum numbers. Also, we've obtained the unnormalized wave function of the particle which influenced by Eckart and Manning-Rosen potential, and visualize it into two-dimensonal plot.

10. Acknowledgement

This research was partly supported by Higher Education Project Grant with contract no. 698/UN27/PN/2015.

11. References

[1] Poschl G, and Teller E 1933 *Z. Physik.* **83** 143
[2] Gendenshtein LE 1983 *Zh. Eksp. Teor. Fiz.* **38** 299
[3] Rosen N, and Morse PM 1932 *Phys. Rev.* **42** 210
[4] Eckart C 1930 *Phys. Rev.* **35** 1303
[5] Manning MF, and Rosen N 1933 *Phys. Rev.* **44** 953
[6] Dong SH, Qiang WC, Sun GH, and Bezerra VB 2007 *J. Phys. A: Math. Theor.* **40** 10535
[7] Falaye BJ 2012 *Cent. Eur. J. Phys.* **10** 960
[8] Resita AS 2016 *Chin. Phys. B* **25** 010301
[9] Antia AD, Ikot AN, Ituen EE, and Akpabbio LE 2012 *Palestine Journal of Mathematics* **1** 104
[10] Cari C, and Suparmi A 2012 *Int. J. Appl. Phys. Math.* **2** 159
[11] Hall RL, and Saad N 2015 *De Gruyter Open* **13** 83

Journal of Physics: Conference Series **776** (2016) 012086 doi:10.1088/1742-6596/776/1/012086

A model of a nonlinear DNA-protein interaction system with Killingbeck potential and its stability

E. Syahroni, A. Suparmi, C. Cari and A. Fuad
Physics Department, Graduate Program, Sebelas Maret University,
Jl. Ir. Sutami 36A Kentingan Jebres Surakarta 57126, INDONESIA

E-mail: edy.syahroni1988@gmail.com

Abstract. In this paper, we presented a model of a nonlinear DNA-protein interaction system. The interaction system consisted of a molecule of protein bound with a single chain of DNA. The interaction between DNA chain, especially adenine and thymine, and DNA-protein bound to glutamine and adenine. The forms of these bonds are adapted from the hydrogen bonds. The Killingbeack potential were used to describe both of the interactions. We proposed the Hamiltonian equation to describe the general model of interaction. The interaction model is satisfied when a protein molecule triggers pulses on a DNA chain. An initial shift in position of protein x_m should trigger the shift in position of DNA y_m, or alter the state. However, an initial shift in DNA, y_n, should not alter the state of a rest protein (i.e. $x_m = 0$), otherwise, the protein would not steadily bind. We also investigated the stability of the model from the DNA-protein interaction with Lyapunov function. The stability of system can be determined when we obtained the equilibrium point.

1. Introduction

Deoxyribonucleic acid is a bimolecular which becomes the main constituent of an organism. It is well known that DNA molecules are double helix. This means that it consists of two complementary polymeric chains twisted around each other [1]. The specific structure was described in 1953 in a paper of historical one page [2]. There have been many efforts to describe its complex dynamics. Some of them are both linear and nonlinear as described by Yakushevich [3]. The nonlinear effect might play an important role in DNA dynamic that was suggested in 1980 by Englander *et al.* [4]. The study of nonlinear physics of DNA published the article: 'nature of the open state in long polynucleotide double helices: possibility of soliton excitation'. This paper first determined the concept of nonlinear conformational excitation or DNA solitons with imitating the opening local of base pairs that were introduced. A powerful impulse for theoretical investigation was given by the first nonlinear Hamiltonian of DNA article.

Dynamic system of DNA cannot be separated from how interaction process of the DNA molecules with molecule around them, either DNA molecule links directly or the others like protein molecule. Observed from the chemical bond, there are three kinds of interaction that occurred among molecules, they are

hydrogen, van der Waals, and electric bond. Hydrogen interaction among the three bonds interaction is the easiest one to observe.
The molecular interaction between DNA and protein is a phenomenon of utmost importance for all of biological regulations. Almost all aspects of cellular function, such as transcriptional regulation, chromosome maintenance, replication, and DNA repair depend on the interaction of proteins with DNA. DNA together with histone and non-histone proteins is the main constituent of the gene which forms the threads of chromatin then forming a chromosome [5]. At some cases, DNA-binding protein has diverse roles and functions such as making up the nucleosome, enzyme modulation chromatin structure to control the gene expression, becoming transcription factors, and as well as being cofactors. Before the DNA either replicates or transcribes, there is a certain protein which triggers the signals so that the DNA can replicate and/or transcribe, and this protein is called as transcription factor [6]. The signal comes from regulatory protein and is transmitted at high efficiency through DNA-protein binding process. One of the wide studious examples of DNA-binding is the transcription factor. Several techniques are available to aptly determine these interactions and their understanding is imperative. For example, the *in vitro* level, molecular biology is based on techniques such as foot printing assay, EMSA, southwestern blotting, Y1 phage display and proximity ligation assay (PLA) screen DNA-protein reliable interaction. The high dynamics in vivo tools of chromatin immune precipitation and its variants, DNA Adenine Methyl Transferase Identification (DAMID) and ChIP-loop assay are robust techniques to characterize several DNA–protein interactions in cells. *In silico* approaches have also evolved drastically over the years to supplement the available information to researchers. Various recent biophysical techniques including fluorescence-based techniques, CD, NMR, AFM, SPR and microcalorimetry have a great potential for the detection of protein-based interactions [7].
Hamiltonian is one of the methods to describe the interaction of any system (8). In our model, to describe the interaction of DNA chain and interaction between molecules of protein and DNA, we propose a Hamiltonian equation consisting of Hamiltonian of DNA, Hamiltonian of Protein, and Hamiltonian Interaction.

2. MODEL DESCRIPTION

The model interaction between DNA and Protein can be observed with the focal point of Hydrogen's bond atom of the both elements. In this paper, we present the model of interaction about molecules of proteins that bind to one of the DNA molecules. Our models only review the interaction of these two molecules on the x-axis and y-axis. It means that our model has two degrees of freedom such as on x_m, the axis describing the interaction of hydrogen atoms between molecules of proteins and DNA, and y_m, the axis which portray the interaction of the hydrogen atom of the DNA chains. The number of DNA molecules we use only two molecules of DNA from their chains and it is illustrated in Figure 1. Both of those molecules are consisted of adenine and thymine. We use the glutamine molecule for example of the side chain of interaction. The indices *m* and *n* is notation of the position of the chain molecules as DNA and protein. The potential of the hydrogen bonds is estimated by the Killingbeck potential. The Killingbeck potential is one of these central potential models. It consists of harmonic oscillator plus Cornel potential, I *i.e.*, ar^2+br-c/r, which is found in application particle physics [9,10]. To distinguish the interaction between the chains of DNA and DNA-protein binding we propose a general Hamiltonian equation represented by the following equation.

$$H = H_{DNA} + H_{prot} + H_{int} \tag{1}$$

where the first part is DNA chain interactions containing the Killingbeck potential, Kinetic Energy and coupling interaction with the following equation.

$$\mathrm{H_{DNA}} = \sum_n \frac{p_{yn}^2}{2m} + \left[-\frac{a}{y_n} + by_n + \lambda y_n^2 \right] + \frac{k}{2}(y_n - y_{n-1})^2 \quad (2)$$

We assume that the mass m of nucleotide bases and momentum p_y is homogenous. The second term of the equation (2) is general equation of Killingbeck potential. The parameters for Killingbeck potential consist of a, b and λ parameters. The third term, the harmonic coupling between two neighboring base pair and constant k is coupling constant. Whereas for Hamiltonian equation of protein is represented by

$$\mathrm{H_{prot}} = \frac{p_{xm}^2}{2M} + \left[-\frac{c}{x_m} + dx_m + \beta x_m^2 \right] \quad (3)$$

As seem as with the equation (2), the second term is the Hamiltonian equation of protein interaction with DNA. Parameter c, d and β are the Killingbeck potential parameter for DNA and protein interaction. The mass of protein is M and momentum of protein is p_{xm}. The last part of Hamiltonian interaction about those two bonds is represented with general form

$$\mathrm{H_{int}} = \sum_n \chi x_m^s y_n^t \quad (4)$$

The coupling constant χ determines the sensitive and strong interaction. The value is not specified yet, but it could be set to obtain a good value for the interaction.

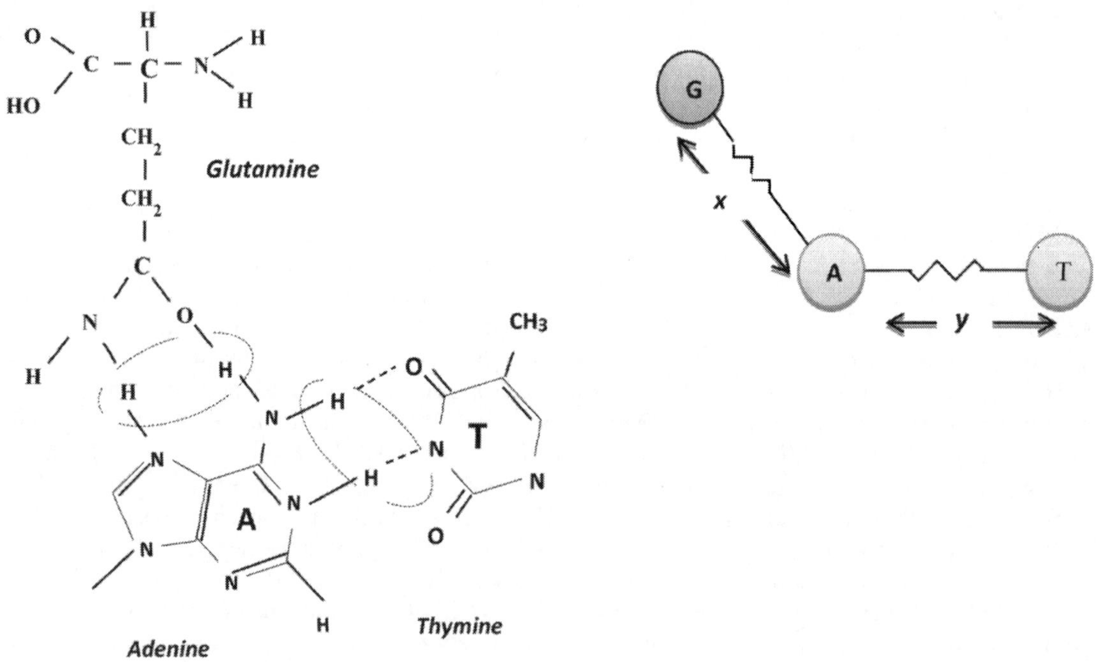

Figure 1. (Left) Chemical bonds among the molecules of DNA (adenine-thymine bonding) and protein (glutamine), and (Right) the model of interaction.

To determine the model of dynamic interaction between DNA-protein, we need to retract and set the value of a and b in (4). The value of the parameter depends on biological requirement influenced by oscillator's initial conditions [11]. In our model, we expected that a protein molecule will trigger pulses on a DNA chain. It means that this requires an initial shift in x_m to trigger a shift in y_n, or to alter the state. However, an initial shift in DNA, y_n, should not alter the state of a rest protein (i.e. $x_m = 0$), as if there is no protein. Otherwise, the protein would not steadily bind. These specific requirements trigger the result of special condition for integers: t=1 and s > 1. This can be understood by considering the equation of motion derived from Hamiltonian,

$$m\ddot{y}_m = \left[-\left\{\frac{a}{y_n^2} + 2\lambda y_n + b\right\} + k\left(y_{m+1} - 2y_m + y_{m-1}\right) - b\chi x_m^s y_m^{t-1}\right]$$

$$m\ddot{x}_m = \left[-\left\{\frac{c}{x_m^2} + 2\beta x_m + c\right\} - a\chi x_m^{s-1} y_m^t\right] \tag{5}$$

We use the interaction Hamiltonian equation with the simplest form $H_{\text{int}} = \frac{1}{2}x_m^2 y_n$ to get a good phase portraits.

3. STABILITY ANALYSIS

To discuss the dynamics of non-linear system, especially on the stability of a nonlinear system, we need a method to discover their stability. In this section, Lyapunov function will be helpful to determine the stability of nonlinear systems [12]. In addition, we can also determine the equilibrium point of the equation. First of all, the Hamiltonian equation will be treated as a lyapunov function. To obtain the stability, based on [13], we are only required to consider the Killingbeck potentials and the interaction term. In the case of coupling interaction, if we approach the equation of harmonic $\frac{k}{2}(y_n - y_{n-1})^2$ by assuming that y_{n-1} comparable to uy_n, with u is parameter, the equation can be written in the simpler form. Consider the potential function

$$U(x,y) = \left[-\frac{a}{y_n} + by_n + (\lambda + \frac{k}{2}(1-u)^2)y_n^2\right] + \left[-\frac{c}{x_m} + dx_m + \beta x_m^2\right] + \frac{1}{2}\chi x^2 y \tag{6}$$

The local minima can be determined with $\nabla U(x_0, y_0) = 0$, we obtain

$$\left\{\frac{a}{y_n^2} + (2\lambda + k(1-u)^2)y_n + b\right\} - \frac{1}{2}\chi x^2 = 0,$$

$$\left\{\frac{c}{x_m^2} + 2\beta x_m + d\right\} - \chi xy = 0. \tag{7}$$

The equilibrium and stability can be determined after we set the value of each parameter. Now we will determine the system in stable or unstable. It can be done by determining the determination of Hamiltonian (det H) using hessian matrix.

$$\textbf{Det H} = \left[2\lambda + k(1-u)^2 - 2\frac{a}{y^3})\right]\left[2\beta - 2\frac{c}{x^3} + \chi y\right] - \left[\chi x\right]^2 \qquad (8)$$

$\det(H)_{(x0,y0)}>0$ with the requirement $x \neq 0$ $y \neq 0$, indicates the system is stable, while the $\det(H)_{(x0,y0)}$ <0 with the same requirement indicates that the system is unstable. In this stage, we can determine the system stability in general form, but we cannot claim that the system is already fully stable. To determine the stability, we pick a, b, c, d, u, k, λ, β and χ as an example when the hydrogen bonds between the base pair are stronger than those between the DNA-protein. The values of parameter on the hydrogen bond of the DNA chain are k=0.1 a = 0.02, b = 0.2, λ =3, and on DNA-protein bond, we set c = 0.01, d = 0.1, β =2and χ = 1.

After we insert the parameter, we obtain the phase portraits of the system as illustrated by Figure 2 and 3,

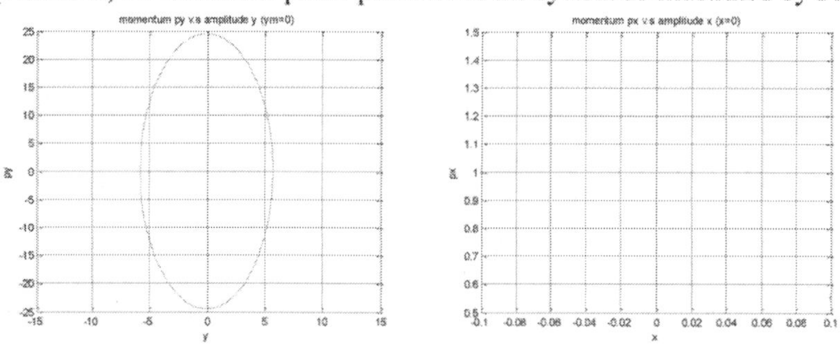

Figure 2. Phase portraits of the systems with no initial amplitude y_m=0 and x_m=0

Based on the Figure 2, it can be inferred that the form of the trajectory of p_{yn} with y_n with no initial amplitude in x_m ,it is the simple trajectory for DNA dynamic (left) and there is no trajectory at p_{xn} with x_n because the interaction of DNA and protein has not happened(right).

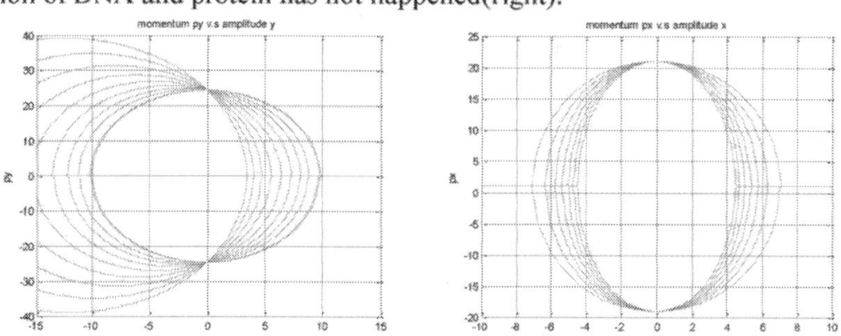

Figure 3. Phase portraits of the systems (left) changes in phase space trajectory due to the presence of protein and (right) the shape of the trajectory on the interaction of DNA with proteins to change by the interaction of the DNA chain.

Figure 3 (left) shows trajectory changes as a result of the relationship between momentums p_y with y because of the existing protein on x axis. The value of x not only affected the value of amplitude range in y

axis but also caused shift of centre-line as well as affected the value of momentum. Figure 3 (right) shows trajectory changes as a result of the relationship between momentums p_x with x because of the existing DNA change on y axis. However, the value only affected the value of amplitude range without shifting the centre-line and affecting the momentum. The value of parameter affects the trajectory especially for momentum which α and β causing the momentum = 0 at the fairly large amplitude. The value of α and β caused the momentum of each interaction rapidly turn to zero. The value of the momentum tends to be stagnant (for y interaction is constant =0 for x interaction = constant) until it reaches maximum and minimum values of amplitude. It is accordance with the expected condition that initial shift in x_m will trigger a shift in yn, or to alter the state. However, an initial shift in DNA, y_n, should not alter the state of a resisting protein (i.e. x_m=0) as if there is no protein.
From equation (7) we obtain the equilibrium points: $\approx (0,0)$ and $\approx (-6.96,3.98)$. After we insert the value of the equilibrium point, we obtain a stability point with $\approx (0,0)$ equilibrium is stable and $\approx (-6.96,3.98)$ is unstable.

4. Conclusion
We have developed a simple model of molecule DNA-protein interaction. This model is only an early step towards how to understand the interaction. The model described short range interaction. In our model, proper trajectories are obtained only with a high energy. For use low energy, the shape of the trajectory will not be as expected. Another problem in this model is the result of velocity constant or zero when the displacement amplitude reaches the maximum or minimum values.

5. Acknowledgments
This Research was partly supported by Higher Education Project Grant with contract no. 632/UN27.21/LT/2016

6. References
[1] Watson J D 1976 *Molecular Biology of the Gene* (W. A. Benjamin, Menlo Park, California).
[2] Watson J D and Crick F H C, *Nature* **171** :737–738
[3] Yakushevich L V 1998 *Nonlinear Physics of DNA, Wiley Series in Nonlinear Science* (John Wiley, Chichester).
[4] Englander S W, Kallenbach N R, Heeger A J, Krumhansl J A and Litwin S 1980, *Proc. Natl. Acad. Sci.* (USA) 777 :7222–7226.
[5] Bulyk M L, Gentalen E, Lockhart D J and Church G M, *Nat Biotechnol* **17**:573–577.
[6] Brivanlou A H and Darnell Jr J E 2002 *Signal Transduction and Control of Gene Expression science's Compass Review* Vol **295** :813-818
[7] Dey B, Thukral S, Krishnan S, Chakrobarty M, Gupta S, Manghani C, Rani V 2015 *DNA–protein interactions: methods for detection and analysis*, 2015, ISSN 0300-8177
[8] Greiner W 2000 *Relativistic Quantum Mechanics (*Springer Verlag Berlin Heidelberg, New York).
[9] Plante G and Antippa A F 2005 *J. Math. Phys.* **46**:062108.
[10] Killingbeck J, *Phys. Lett. A* **65**:87 (1978).
[11] Dwiputra D *et al* 2016 *J. Phys.: Conf. Series* **694** :012076
[12] Rink B W and Tuwankotta J M 2000 *Stability in Hamiltonian Systems* (Universiteit Utrecht. Department of Mathematics).
[13] Verhulst F 1996 *Nonlinear differential equations and dynamical systems* (Springer Science & Business Media)

8th International Conference on Physics and its Applications (ICOPIA) IOP Publishing
Journal of Physics: Conference Series **776** (2016) 012087 doi:10.1088/1742-6596/776/1/012087

Theoretical analysis of Sloshing effect on Pitch Angel to optimize quick dive on litoral submarine 22 M

L. T. P. Sinaga
Senior Researcher at Indonesian Hydrodinamic Laborator, BPPT
Jl. Hidrodinamika, Komplek ITS Sukolilo, Surabaya

Email: luhuttps@yahoo.com.sg

Abstract
This study considers the analytic theoretical model. The Submarine was considered to be rigid body are free sailing model with various angle of attack to be quick dive as pitching motion. By using Floating Body Mechanism supported by analytic model to describe the theoretical model analisys test. For the case of fluid level on 30% of the front balast tank and various angle of pitch. The paper describes a study on Analytic theoretical and modeling in CFD (Computational Fluid Dynamics). For Analyzing at special care of sloshing on free surce ballast tank after peak and fore peak were taken into consideration. In general, both methods (analytic model and CFD model) demonstrated such a good agreement, particularly in the consistent trend of RAO.

1. Introduction

The problem of water sloshing in closed forepeak ballast of . This phenomenon can be described as a free surface movement of the contained fluid due to sudden loads [1]. Sloshing is a liquid vibration phenomenon caused by the movement of the ballast. When the ballast is in transit, the sloshing would affect stability of the system severely caused by pitch of angle when submarine in quick dive cause couple motion [2]. So it is necessary to minimize the impact of sloshing and avoid large amplitude resonance. Sloshing has been studied for many years by analytical and numerical methods[3].
Originally the normal way of diving a submersible was by adjusting the amount of water ballast carried. Towards the end of the nineteenth century the most promising way was to get the boat to an 'awash' condition with a small amount of positive buoyancy and then drive it underwater. The main controversy was whether the boat should be dived horizontally or at an angle and this also applied to changing depth when submerged.

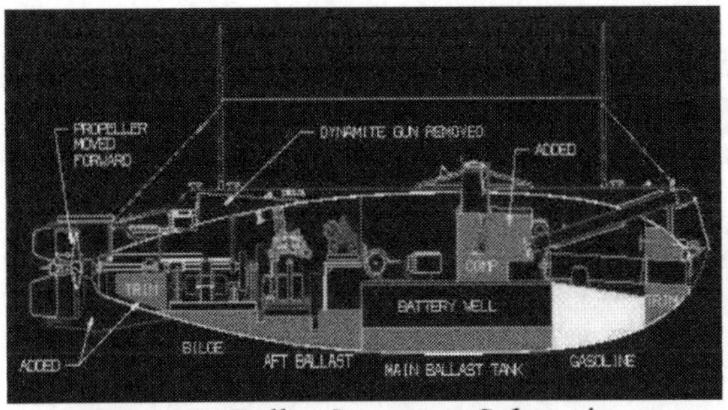

Figure 1. Ballas System at Submarine

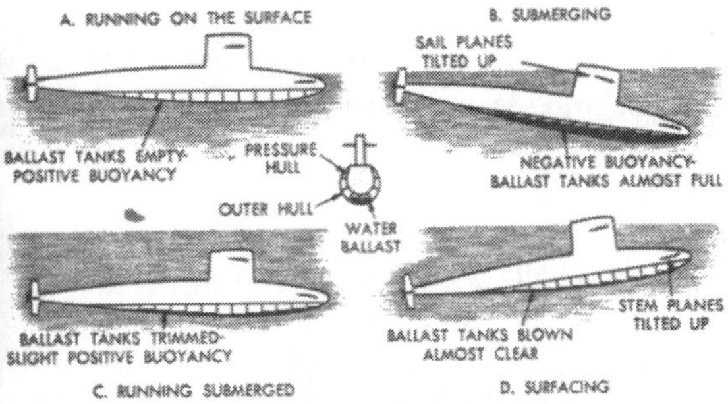

Figure 2. Submersion Proccess

Submarine is designed with the main ballast tanks fulls, the weight of water displaced will be as close as possible to the weight of the boat, see Fig.1. That is, there should be a balance between displacement and buoyancy, referred to as neutral buoyancy. If true neutral buoyancy is achieved the boat will float at whatever depth it presently is, unless something acts on it to make it rise or sink, see Figure 2. In practice, submariners prefer to maintain very slight positive buoyancy, so that if power if lost the boat may be expected to slowly rise to the surface.

In a quick dive, the negative tank is important. By flooding this tank the boat is made negatively buoyant, which gets the boat under water quicker. Once submerged, though, the negative tank is blown "to the mark." the submarine to be heavier than the water it displaced while diving, because you want to get under as quickly as possible.

2. Methods

2.1 Floating Body Mechanism

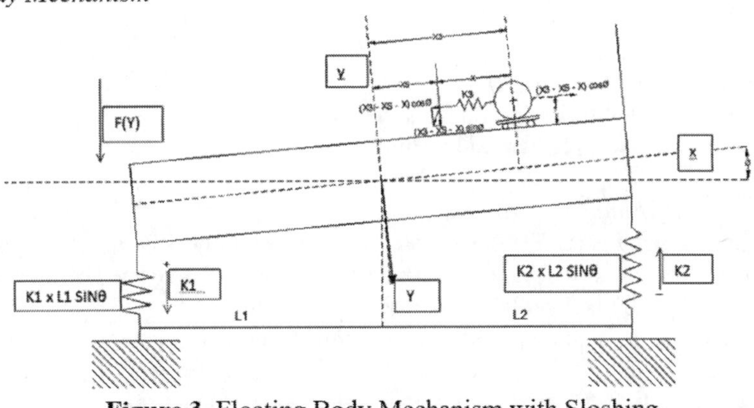

Figure 3. Floating Body Mechanism with Sloshing

In the pitching model system using 2 degrees of freedom, namely the mass beam or submarine described as the mass (M) ship and the mass of fluid in the ballast and described as two (2) spring of stiffness in the form of k_1 and k_2 for pitching motions, see Figure 3.

Mass of fluid in the tank load space is like a mass (m) in a moving train unhindered translational and 1 (one) spring of stiffness k_3 it also causes movement of the coupling to the model. If in the model gets a load (Fy) because of the influence of energy waves received by the model then the model will perform vertical and horizontal translation.

With the outside force that coordinates spring k_1 to the balance point (center of gravity) and is indicated for the spring k_2 toward the center of gravity is indicated l_2. With equation: $\sum F = m \times a$, and force coupling can be summarized as follows;

$k_1 l_1 - k_2 l_2$ So, $k_1 l_1 = k_2 l_2$

For fluid motion in the tank above the carriage modeled as a pendulum having stiffness without hindrance caused by the moving mass is equal to k_3 translansi as $(x_3 - x_s - x)sin\emptyset = (x_3 - x_s - x)cos\emptyset$.

Translation motion of fluid mass (m) will also affect the coupling pitch motion. The amount received by the vertical force spiral at an angle $\emptyset$ is;

$k_1\ l_1\ \emptyset =$ F_1 has a positive direction (compressed) (1)

$k_2\ l_2\ \emptyset =$ F_2 has a negative direction (pulled) (2)

So that the beam free motion coupled equations and moment caused due to a force from the outside of beam or submarine (Fy) and fluid in the ballast tank can be summarized as follows [4];

$$(M + m)\ddot{Y} = -2y(k_1 + k_2) - 2\emptyset(k_1.l_1 - k_2.l_2) - 2k_3(x_3 - x_s - x).(sin\emptyset + cos\emptyset) \quad (3)$$

$$(J + j)\ddot{O} = -2y(k_1.l_1 - k_2.l_2) - 2\emptyset(k_1.l_1^2 + k_2.l_2^2) - k_3(x_3 - x_s - x) [(sin\emptyset.d) - \{(x_3 - x_s - x)cos\emptyset\}] \quad (4)$$

Where:
$k_1.l_1 - k_2.l_2$ is static Couple
$y = A\sin(\omega t + \varphi)$ serta $\emptyset = Csin(\omega t + \varphi)$ (5)
M is displasment ship
m is the mass of water in the fluid tank
$M\ddot{Y}$ is Fy is an external force due to wave energy
$m\ddot{Y}$ is the force due to sloshing
$\emptyset$ is the angle of the incident wave direction
J is moment due to an external force caused by wave
j is moment due to an internal force caused by fluids

2.2. CFD (Computational Fluids Dynamic)

Computational Fluid Dynamics (CFD) is the process of numerically solving fluid dynamics equations to predict resultant flow fields. It is used to model a rich variety of flow phenomena, as well as to define thermal and structural loads on bodies immersed in fluids[5].

The submarine is assumed to be traveling through an infinite domain of stagnant water. The motion of the submarine is controlled by a three-bladed propeller, rudder and stern planes. The entire computational mesh including the submarine body is assumed to be moving with the body without any deformation. The flow field computations were performed in the inertial frame of reference, which makes the specification of boundary conditions easier. Since the body moves through infinite volume of stagnant water, the velocity specified at the far field boundaries of the computational domain is zero.

For the case of horizontal overshoot maneuvering, held in this position until the body reached various pitch angle 6-14 degrees with speed 6 knots. Figure 7 to Figure 15 shows predicted pressure distribution on walls and streamlines around body for pitch motion.

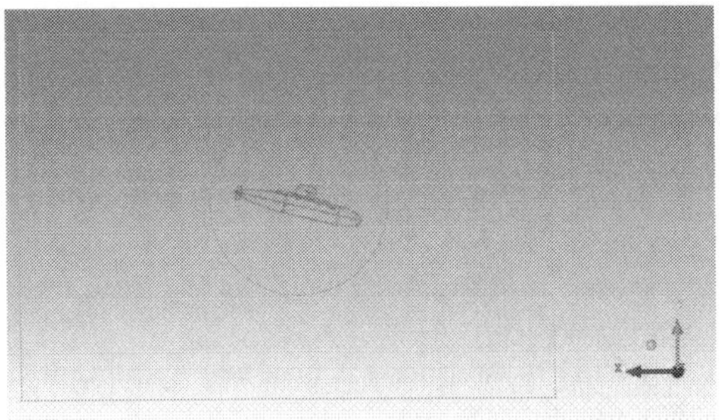

Figure 4 Design model at ICEM CFD

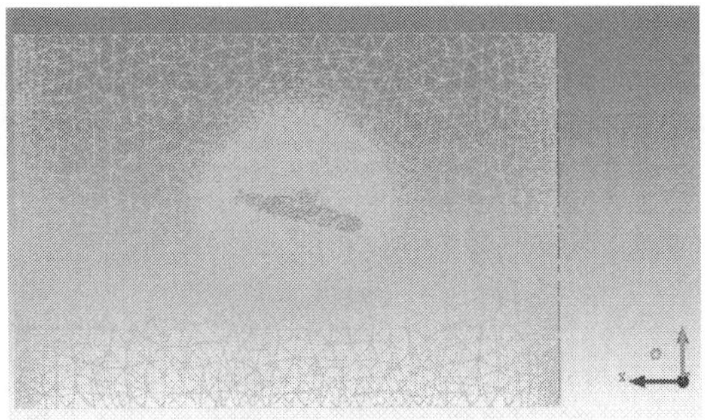

Figure 5 Meshing Model

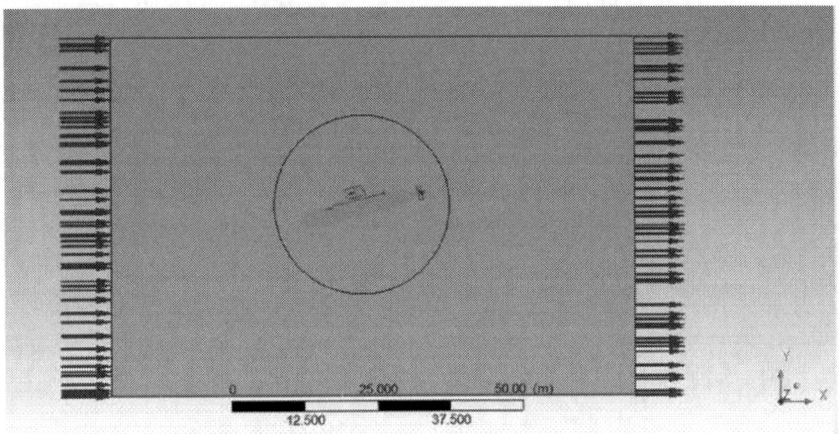

Figure 6 Setting boundary for CFD Pre

As described above (Figure 3), it is assumed that the global fluid motion plays the most important role in the sloshing-induced impact occurrence. The incompressible Euler equation and continuity equation are the governing equations of the present method. To solve these equations, let us consider the discretization of the tank volume into finite meshes (Table 1) and as shown in Figure 3 for meshing model. Adopting the concept of the Cartesian staggered grid, the velocity components are defined on the cell boundaries, while the pressure is computed at the center of each cell. The simulation of a motion requires the coupled solution of rigid body motion equation (in six degrees of freedom) with unsteady Reynolds-averaged Navier-Stokes equations (URANS).

The integration and rigid body mesh motion are performed automatically using CD-adapco's Dynamic Fluid-Body Interaction (DFBI). In this case submarine move forward and acquire force of the front at angle 180^0 (head seas), see figure 6.

Table 1. Generated Mesh Information

Description	Meshing Specification
Object Name	Mesh
State	Meshed
Number of Nodes	1358
Number of element	1314
Number of Nodes (Diffracting Bodies)	952
Number of Elements (Diffracting Bodies)	904

3. Result

In order to exercise the method developed above, submarine considered. A sketch of the submarine is shown in Fig. 7 to 15. The Submarine has the following geometric and mass properties: length = 22 m, reference diameter = 3 m, mass = 113.9 ton

As part of a validation of the coupled Navier-Stokes and six-degree-of-freedom method, time-accurate unsteady numerical computations were performed to predict the flow field, hydrodynamic coefficients, and the quick dive paths of submarine at an initial speed = 6 knots. Full three-dimensional computations were performed and no symmetry wall was used.

Here, the primary interest is in the validation of coupled CFD techniques for accurate simulation of free sailing and quick dive of submarine. Numerical computations were made for the submarine configuration at an initial velocity of 6 knots. The initial angle of attack (trim) was, $\alpha = 10^0$ Figure 9 shows the computed pressure contours at a given effective motion. It clearly shows the orientation of the body at that instant in time and the resulting flow field due to the body at angle of attack.

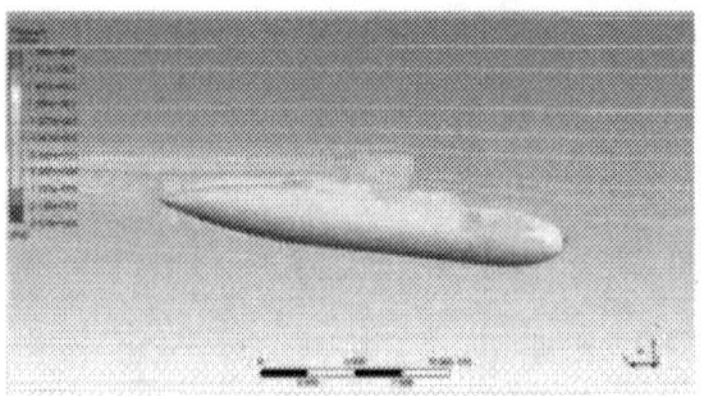

Figure 7 Trim angle at 6^0

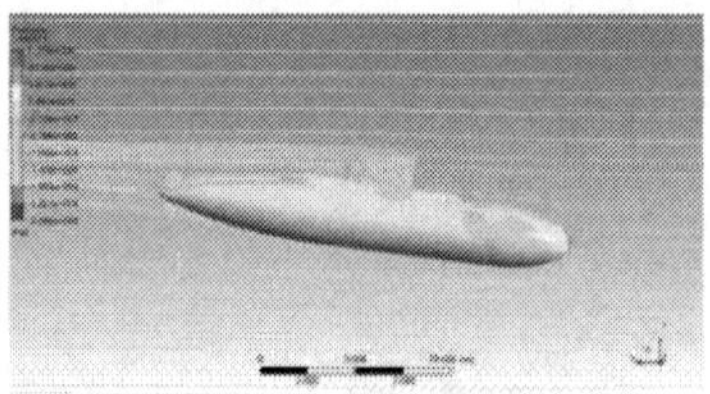

Figure 8 Trim angle at 7^0

Figure 9 Trim angle at 8^0

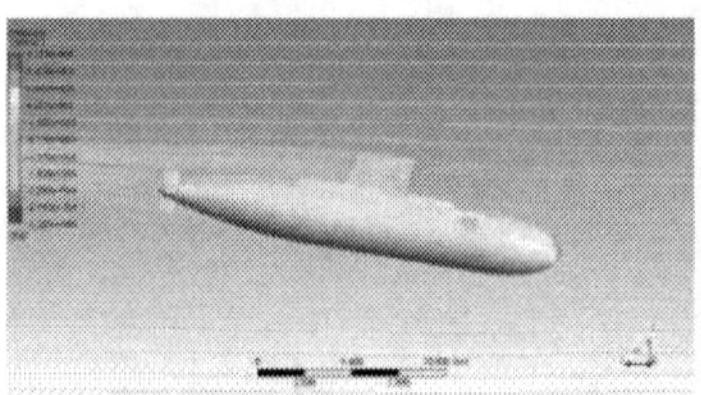

Figure 10 Trim angle at 9^0

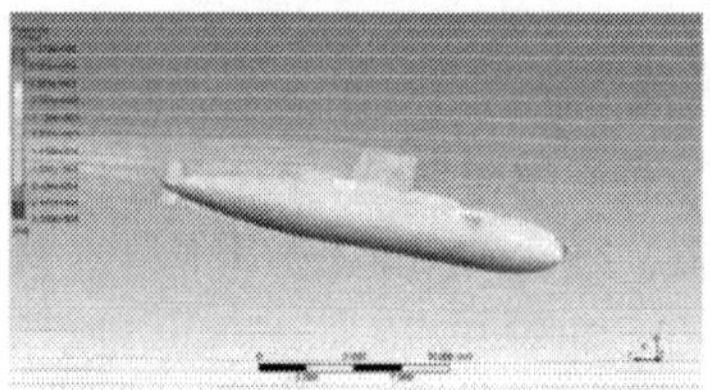

Figure 11 Trim angle at 10^0

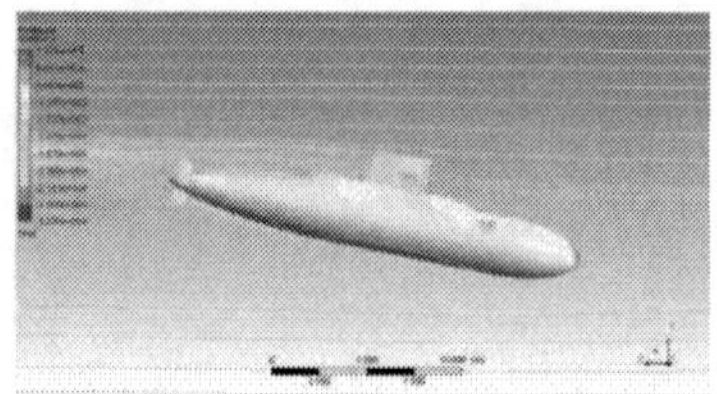

Figure 12 Trim angle at 11^0

Figure 13 Trim angle at 12^0

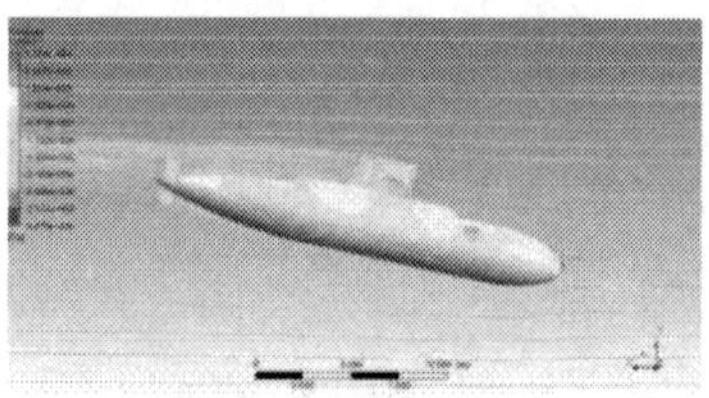

Figure 14 Trim angle at 13^0

Figure 15 Trim angle at 14^0

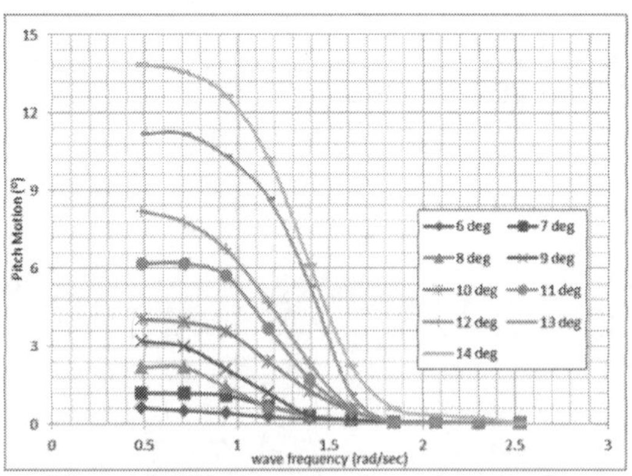

Figure 16 Pitch motion in various angle

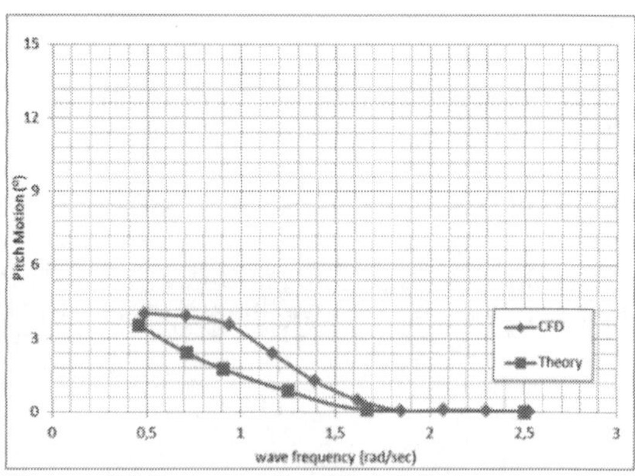

Figure 17 Comparation Pitch Motion at angle 10^0

Journal of Physics: Conference Series **776** (2016) 012087 doi:10.1088/1742-6596/776/1/012087

4. Conclusion

In the present study, Theoretical and CFD of sloshing flows in ballast of submarine are described. Numerical methods, Free Floating Body Equation running by MAT-LAB version 6.5 methods, are applied to solve violent sloshing. Based on the present study, the following conclusions are made:

- The result of the CFD and Mathematical modeling is similar but RAO for CFD the numerical due to some impact like damping force and added mass. CFD and Theoretical analysis shown same trend both on 180^0 of pitch.
- At wave heading 180^0 which cause pitching significant consequences sloshing of 50% in mass (m).
- The sloshing-induced force and moment are not linearly proportional to excitation amplitude. Therefore, the submarine motion with sloshing does not vary in a linear manner with respect to wave amplitude.

The vessel had effective angle of inclination when diving was 10° when submerged the speed of 6 knots after the required depth had been obtained this angle became so small as to be inappreciable and had to be maintained to keep effective depth control.

5. References

[1] Journée, J M 1997 *Six International Conference on Stability of Ships and Ocean Structures* p 137-150. Varna-Bulgaria, September 22-27
[2] Rognebakke OF and Faltinsen, OM 2003 *Jour Ship Research* **47**, 208-221.
[3] Sinaga LTP, Utama IKAP and Sulisetyono A 2014 *Journal IPTEK ITS*. Surabaya
[4] Sinaga LTP, Utama IKAP and Sulisetyono A 2014 *Euro Int J Sci Tech* (EIJST)-UK, **3** No.5. June 30
[5] Godderidge B, Tan M, Turnock S, and Earl C 2006 Multiphase CFD modelling of a lateral sloshing tank. In *9th Numerical Towing Tank Symposium* 2006 (Nutts'06)
[6] Sinaga, LTP, Utama IKAP and Sulisetyono A 2014 *Applied Mechanics and Materials* **664** p 153-157

Journal of Physics: Conference Series **776** (2016) 012088 doi:10.1088/1742-6596/776/1/012088

Estimation of weekly ^{99}Mo production by AHR 200 kW

I. H. Siregar[1], Suharyana[2], A. Khakim[3], D. Siregar[4], A. R Frida[1]

[1]Faculty of Science and Technology, Islamic State University of Sunan Kalijaga
[2]Physics Department, Faculty of Mathematics and Nature Science, Sebelas Maret University
[3]BAPETEN, Jl. Gajah Mada No.8 Jakarta
[4]DISPENDIKBUD Supervisor of North Labuhanbatu, Aekkanopan

E-mail: suharyana61@staff.uns.ac.id

Abstract. The estimation of weekly ^{99}Mo production by AHR 200 kW fueled with Low Enriched Uranium Uranyl Nitrate solution has been simulated by using MCNPX computer code. We have employed the AHR design of Babcock & Wilcox Medical Isotope Production System with ^{9}Be Reflector and Stainless steel vessel. We found that when the concentration of uranium in the fresh fuel was 108 gr U/L of $UO_2(NO_3)_2$ fuel solution, the multiplication factor was 1.0517. The ^{99}Mo concentration reached saturated at tenth day operation. The AHR can produce approximately 1.96×10^3 6-day-Ci weekly.

1. Introduction

Isotope ^{99m}Tc is produced from ^{99}Mo undergoing beta decay having half-life of about 66 hours with approximately 88% of probability. Isotope ^{99m}Tc is the most widely used radionuclide worldwide for nuclear medicine tomographic imaging technique using single photon emission computed tomography (SPECT). It's used approximately 85% of nuclear medicine or about 12,000 six-day curies per week. The global demand for ^{99m}Tc grows at an average annual rate 3%-8% in the new market, especialy in Asia [1,2]. According to Pasqualini's report on 2011 in "33rd International Meeting on Reduced Enrichment for Research and Test Reactor", the major production of ^{99}Mo around the world is from ten heterogeneous nuclear reactors. However, majority of these reactors have been working for more than 40 years. These reactors are listed in Table 1 [3].

Actually there are several method to produce ^{99}Mo. According to Wolterbeek et al. [4], at least there are eight possible production routes. However, only the neutron-fission route is the most effective [4]. Comparing to heterogen reactor, Aqueous Homogeneous Reactor (AHR) offers a better method with several advantages, namely low cost, small critical mass, inherent passive safety, simply fuel handling and also simple processing and purification characteristic [5]. In this paper we reported a simulation of estimation of weekly ^{99}Mo production by AHR. The reactor was assumed operated at 200 kW and the solution was 19.75% uranil nitrate enriched.

2. Experimental Method

In this research we used MCNPX computer code to simulate AHR design and find out the interval time of saturation of ^{99}Mo production. MCNPX is a general purpose Monte Carlo radiation transport code that design to track many particle types in the large range energies. This version is the next generation of Monte Carlo series that began in Los Alamos National Laboratory since 60 years ago

[6]. General equation base on Monte Carlo calculation in this software have been constructed as shown in equation (1) [7].

Table 1. Major current ^{99}Mo production reactor was working more than 40 years [3]

Reactor	Country	Annual Operating days	Production 6- day curies/week	Weekly % of world demand	Fuel/Target	Commis-sioning year
BR-2	Belgium	140	5200	25-65	HEU/HEU	1961
HFR	Netherlands	300	4680	35-70	LEU/HEU	1961
LVR-15	Czech Republic	Just Started	>600	No data	HEU/HEU	1957
MARIA	Poland	Just Started	700-1500	No data	HEU/HEU	1974
NRU	Canada	300	4680	35-70	LEU/HEU	1957
OPAL	Australia	290	1000-1500	No data	LEU/LEU	2007
OSIRIS	France	180	1200	10-20	LEU/HEU	1966
SAFARI-1	South Africa	305	2500	10-30	LEU/HEU	1965
RA-3	Argentina	230	200	< 2	LEU/LEU	1967
RSG-GAS	Indonesia	147	150	< 2	HEU/LEU	1967

$$\frac{dX_i}{dt} = \sum_{j=1}^{N} l_{ij}\lambda_j X_j + \emptyset \sum_{k=1}^{N} f_{ik}\sigma_k X_k - (\lambda_i + \emptyset\sigma_i + r_i)X_i + F_i, \qquad i = 1, \dots, N \tag{1}$$

where:
X_i = atom density of nuclide i;
N = number of nuclides;
l_{ij} = fraction of radioactive disintegration by nuclide j which leads to formation of nuclide i;
λ_j = radioactive decay constant;
$\emptyset$ = position and energy averaged neutron flux;
f_{ik} = fraction of neutron absorption by nuclide k which leads to formation of nuclide i;
σ_k = spectrum averaged neutron absorption cross section of nuclide k;
r_i = continuous removal rate of nuclide i from the system;
F_i = continuous feed rate of nuclide i.

$$\phi = \frac{6.242 \times 10^{18}.P}{\sum_{i=1}^{N} X_i^f \ \sigma_i^f \ R_i} \tag{2}$$

where
$\emptyset$ = instantaneous neutron flux (n.cm^{-2}. s^{-1})
P = power (MW)
X_i^f = amount of fissile nuclide i in fuel (g. atom)
σ_i^f = microscopic fission cross section for nuclide i (barn)
R_i = recoverable energy per fission for nuclide i (MeV/fission)

The input of MCNPX consists of a reactor core geometry specification, a source definition, and some tallies. The reactor core parameters are shown in Table 2. The fuel solution consists of 19.75% enrichment uranyl nitrate dissolved in water. The concentrations of the solution are varied from 75 up to 120 gr-U/L. The critycallity of the AHR were calculated for each concentration solutions.

Table 2. Reactor core parameters

Parameter	Value
Reactor type	AHR
Reactor power	200 kW (thermal)
Enrichment	19.75 wt%
Chemical	$UO_2(NO_3)_2$ liquid
Temperature	80°C
Reactor height (cm)	122
Core diameter (cm)	63.4
Reactor vessel	Stainless steel - 304
Vessel thickness (cm)	3.0
Reflector	Beryllium (radial)
Reflector thickness (cm)	3.0

Since there are many kind of stainless steel, in this report we used stainless steel 304 for the reactor vessel material. The characteristics of the 304 SS are shown in Table 3.

Table 3. Reactor vessel densities [8]

Stainless Steel Nuclide	Atom/barn.cm x 10^{-2}
Cromium (Cr)	1.74
Mangan (Mn)	1.52
Besi (Fe)	5.81
Nikel (Ni)	0.851

The mathematical model of the AHR core is shown in Figure 1.

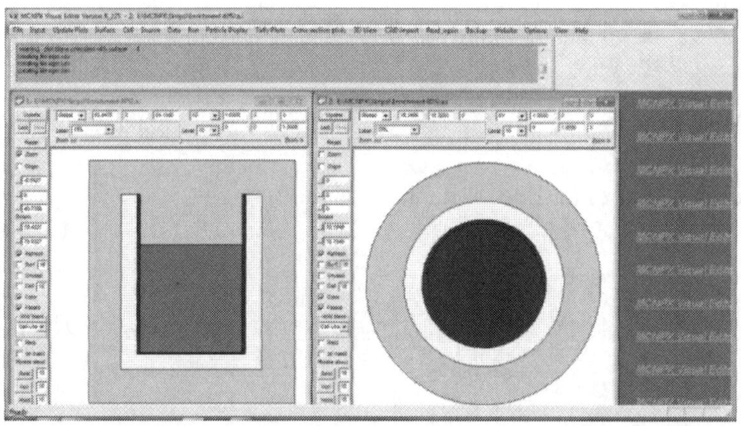

Figure 1. Geometrical model of AHR. Left: Side view. Right: Top view

The colors in the Figure 1 are code for materials:

- Air surounding the AHR core
- Stainless steel: the AHR core vessel
- Fuel $UO_2(NO_3)_2$ solution
- Reflector Be

3. Results and Discussion

In Table 4 we show the values of k_{eff} for each fuel densities. It can be seen clearly that the criticality of the AHR depends on the uranium concentration presents in the fuel solution. The k_{eff} increases monotonically with the uranium concentration, ranging from 0.90685 up to 1.05885. It is generally accepted that the k_{eff} value of any reactor should be around 1.052. Accordingly, we decided to use the concentration of uranium was 108 gr-U/L as the best choice. We apply this value to calculate the production of ^{99}Mo in various interval time from 1 up to 40 days. The results are presented in Figure 2.

Table 4. The k_{eff} values of fresh fuel densities

Enrichment 19,75%		
No	gr-U/L	k_{eff}
1	75	0.90685
2	80	0.93446
3	85	0.95826
4	90	0.98107
5	95	1.00473
6	100	1.0232
7	105	1.04189
8	106	1.04614
9	107	1.05003
10	108	1.0517
11	109	1.05552
12	110	1.05885
13	115	1.07653
14	120	1.09139

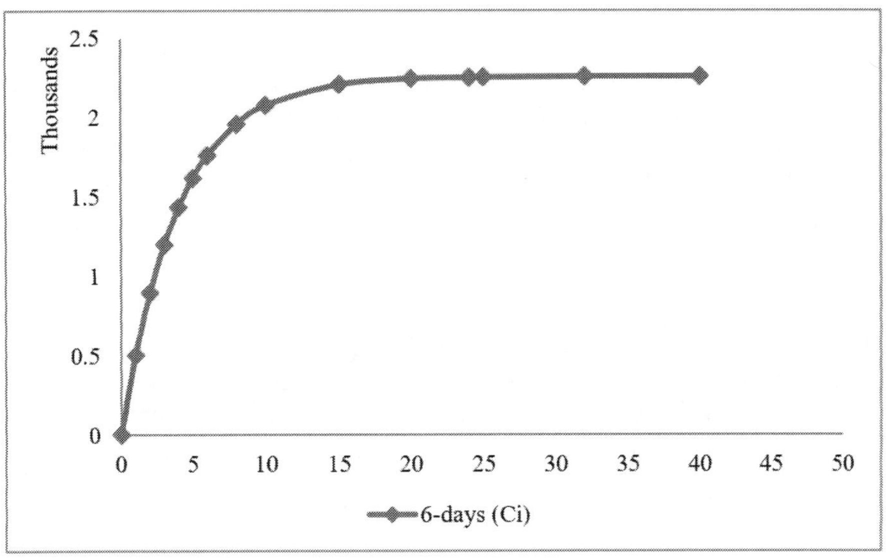

Figure 2. ^{99}Mo production in interval various time

It can be seen from the figure 2, the ^{99}Mo production increases over the operation time and reached saturated after ten days operation. In other words, if the reactor is operated up to 40 days, the production of ^{99}Mo does not grow any more. The rate of ^{99}Mo production is in balance with it's decays. Thus the ^{99}Mo should be extracted from the fuel solution before day-10th. More importantly, this simulation showed that the 200 kW AHR could produce ^{99}Mo approximately 1.96×10^3 Ci 6-days weekly.

After the reactor has been operated over periode time, the concentration of the fissile nuclide ^{235}U decreases. It means the criticallity of the reactor became lower than the original value. The k_{eff} of the AHR should not be lower than 1 at any time. Therefore, it is instructive to calculate the k_{eff} values over periode time of operation. The calculated k_{eff} is presented on Figure 3. It can be seen, althought the k_{eff} decreases however its value is still higher than 1. It means after the reactor has been operated for 40 days, the operator should not add the fuel.

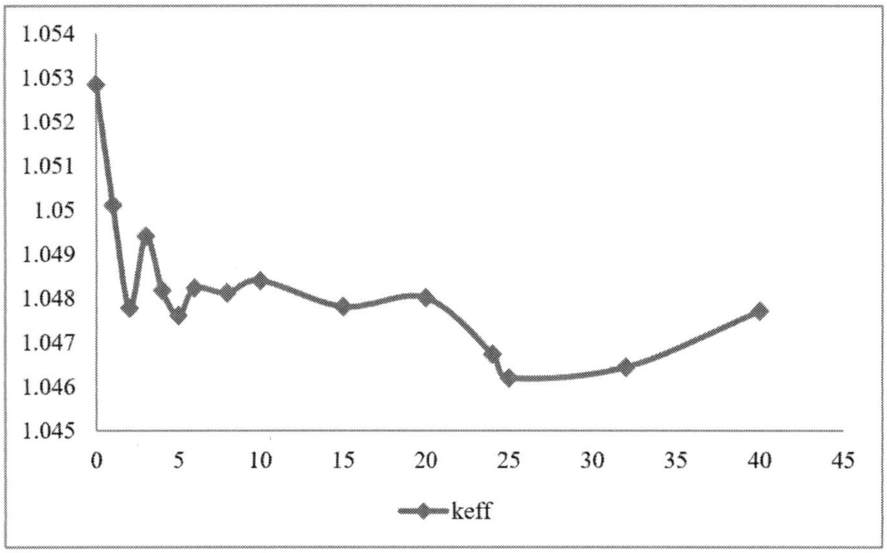

Figure 3. Multiplication factor during operation

4. Conclusion

The simulation of AHR by MCNPX code shows the reactor was critical during operation. The concentration of uranium in fuel solution shows 108 gr U/L impact to critical of reactor, with the multiplication factor about 1.0517. It implies that we have successfully simulated the weekly production ^{99}Mo. Our AHR model when fueled with 108 gr U/L $UO_2(NO_3)_2$ solution and operated at 200 kW yields about 1.96×10^3 6-days Ci.

5. References

[1] IAEA 2013 *Non-HEU Production Technologies for Molybdenum-99 and Technetium-99m*. IAEA NUCLEAR ENERGY SERIES NO.NF-T-5.4. (Austria: IAEA) p 3.

[2] Verbeek P, 2008 *Report on Molybdenum-99 Production for Nuclear Medicine 2010-2020. Association of Imaging Producers & Equipment Suppliers*. e.g *Association of Imaging Producers and Equioment Suppliers Report*

[3] Pasquailini EE 2011 *33rd Int. Meeting on Reduced Enrichment for Research and Test Reactor (Santiago)*. Semi-Homogeneous Reactor fo Mo-99 Production: Conceptual Design. (Chile) October 23-27, 2011

[4] Wolterbeek B, Kloosterman JL, Lathouwers D, Rohde M, Winkelman A, Frima L and Wols F 2014 *Journal Radioanal Nuclear Chemistry (2014)* **302**: 773-779.

[5] IAEA 2008 *Homogeneous Aqueous Solution Nuclear Reactors for the Production of Mo-99 and*

other Short Lived Radioisotopes. IAEA-TECDOC-1601. (Austria: IAEA) p 1

[6] Pelowitz DB 2008 *MCNPX—User's Manual version 2.6.0:* A technical document, LA-CP-07-1473, (New Mexico: Los Alamos National Laboratory) p 1-1 – p 1-3.

[7] OAK RIDGE NATIONAL LABORATORY 2002 *RSICC COMPUTER CODE COLLECTION. ORIGEN 2.1.* (Tennessee: Oak Ridge) p 2-8 – p 2-10.

[8] Harmon, CD, Busch, RD, Briesmeister, JF and Forster, RA 1994 *Criticality Calculation with MCNP^{TM}: A Primer. Los Alamos National Laboratory.* (New Mexico: Los Alamos) p C-10.

Steganography on multiple MP3 files using spread spectrum and Shamir's secret sharing

N. M. Yoeseph, F. A. Purnomo, B. K. Riasti, M. A. Safiie, T. N. Hidayat
Informatics Engineering Program Sebelas Maret University,
Jl. Ir. Sutami 36A Kentingan Jebres Surakarta 57126, INDONESIA

Email: nanangmy@mipa.uns.ac.id

Abstract. The purpose of steganography is how to hide data into another media. In order to increase security of data, steganography technique is often combined with cryptography. The weakness of this combination technique is the data was centralized. Therefore, a steganography technique is develop by using combination of spread spectrum and secret sharing technique. In steganography with secret sharing, shares of data is created and hidden in several medium. Medium used to concealed shares were MP3 files. Hiding technique used was Spread Spectrum. Secret sharing scheme used was Shamir's Secret Sharing. The result showed that steganography with spread spectrum combined with Shamir's Secret Share using MP3 files as medium produce a technique that could hid data into several cover. To extract and reconstruct the data hidden in stego object, it is needed the amount of stego object which more or equal to its threshold. Furthermore, stego objects were imperceptible and robust.

1. Introduction

Steganography is art and science to hide a secret message into another media. The hiding of message is done so the third person is not aware of the existence of such message.To improve data security, steganography usually combined with cryptography [1, 2, 3, 4]. According to Tiend and Lin as quoted by [5], the weakness of combination technique of steganography and cryptography is that the data was centralized.

Solution to overcome the weaknesses is by using secret sharing. In this method, secret message processed to produce shares which will be distributed to a number of parties. One of secret sharing technique is *(k,n)* threshold secret share [6] (Shamir, 1979), or known as Shamir's Secret Share.

Spread Spectrum is a transmission technique using a pseudo-noise code which is independent of the data to modulate a signal with a greater bandwidth than the bandwidth required to transmit the data signals. One of the advantages of Spread Spectrum is its resistance to jamming and interference.

Therefore, a steganography technique is required to hide secret data into several cover object to increases security of the data. In this research, cover object used was MP3 files. It was chosen because of its popularity. To hide data into its cover, the combination of spread spectrum and secret sharing was used. Thus, it can increase the security of data hidden, and simultaneously produce a robust stego object.

2. Analysis
System steganography in the MP3 file created consists of two processes, namely the process to cover data hiding and extraction process and the reconstruction of data from the stego object. In the process of data hiding, secret data is splitted into a number of shares. Splitting is using Shamir's Secret Sharing algorithms. Hiding data using Direct Sequence Spread Spectrum (DSSS). The results of the process of data hiding is several stego objects.
In the extraction process and the reconstruction of data from the stego objects, several stego data are processed to be taken the data which had been hidden. Then stego data is reconstructed to obtain its secret data. The reconstruction process using the inverse of the algorithm Shamir's Secret Sharing.

3. Basic Concepts
In this section, basic theories used in hiding scheme and also in data extraction and reconstruction is presented. In principle, the steganography process is done by adding an addition signal to the original signal. Figure 1 ilustrate the steganography process, where *y(n)* is a stego object, *x(n)* is cover signal, and *c (n)* is the signal hidden.

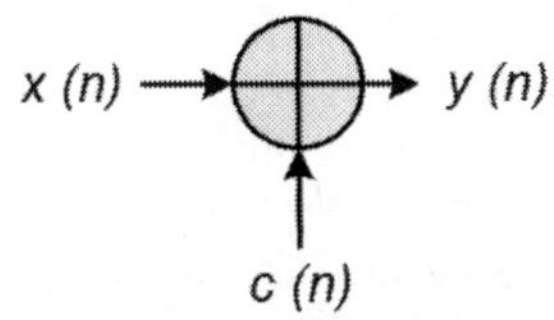

Figure 1. Steganography Process

3.1. Direct Sequence Spread Spectrum
Direct Sequence Spread Spectrum (DSSS) used a pseudorandom sequence and a signal modulator to modulating bit stream data. DSSS technique that uses redundancy and scrambles the data before modulated bit stream and reverse the process on the receiving side of recovery is called Direct Sequence [7].
Because a pseudorandom sequence (PN) is used at the transmitter to modulate the signal, the first requirement at the receiver is to have a local copy of this PN sequence. The copy is needed to de-spread the incoming signal. This is done by multiplying the incoming signal by the local PN sequence copy. To accomplish a good de-spreading, the local copy has to be synchronized with the incoming signal and the PN sequence that was used in the spreading process [7].
DSSS signal is indicated by the equation (1).

$$x(t) = \sqrt{2S}\, sin\left[\omega_o t + \frac{d_n c_{nN+k}\pi}{2}\right] = d_n c_{nN+k}\sqrt{2S}\, cos(\omega_o t) \quad (1)$$

3.2. Psychoacoustic Auditory Model
Psychoacoustic Auditory Model (PAM) is an algorithm that tries to model human auditory systems. PAM is used to process information from the audio signal and determining its masking threshold. Masking threshold is used to form the stego signal that will be hidden inside the media cover. Stego signal formed from this process will not be detected by the human sense of hearing [7].
The psychoacoustic model used by calculate the doorstep of masking on the critical strip scale to respect to best the physiology of the auditory perception. The DSP of the audio signal is transposed in the scale of Bark by integration of the power in every coins strip. The unique display function is applied for the modeling of the effects of masking of the membrane basilar [8]. Algorithm uses by [8] presented in Figure 2.

Journal of Physics: Conference Series **776** (2016) 012089 doi:10.1088/1742-6596/776/1/012089

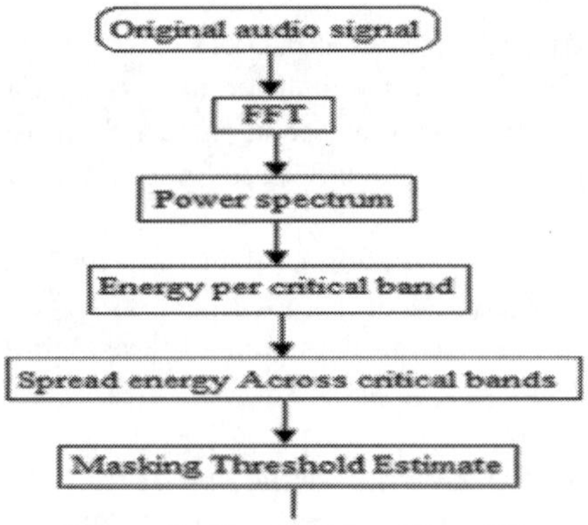

Figure 2. Garcia Algorithm [8].

3.3. Secret Sharing
The idea of secret sharing is to divide the secret key into several parts, called shares, and share it with some people. Only a subset of those people who can not/are allowed to establish initial key back.
The fundamental approach of secret sharing is done by Shamir and Blakley in 1979 with the name (k, n)-threshold scheme. Threshold secret sharing scheme is a protocol between 1+n player (including a dealer and n participant). Dealer distributes partial information (share) on a secret into n participan (distribution) so that (a) each group with a number k participant can know the secret data stored (recovery) in time polynomial, (b) the scheme is declared perfect if the group with the number of members smaller than k will not be able to obtain any information about the secret data stored, (c) the scheme is declared ideal if every size bits of the same shares with confidential data that is hidden [9].
If a secret data, *S*, will be devided, then *S* should be changed to polynomial. If *S* will be divided into *n* sections (S_1, S_2, ..., S_n), then a random coefficient c_{n-1} is generated and formed a polynomial equation of degree (*n-1*), as shown in equation (2).

$$f(x) = c_0 + c_1x + c_2x^2 + \cdots + c_{(n-1)}x^{(n-1)} \tag{2}$$

Then, shares is calculated according to equation (3)

$$S_1 = f(1), S_2 = f(2), \ldots, S_n = f(n) \tag{3}$$

4. Hiding Scheme
In this research, a system which can do a steganographic process will be built.The medium used as cover is several audio with MP3 filetype. Secret data to be hide is text. The result of the process are multiple audio files which hid the secret data, called stego object. Data hidden in stego object called stego data.
In the Data Hiding, data splitted before it hide into several cover. Splitting is using Shamir's Secret Sharing algorithm. Hiding process using Direct Sequence Spread Spectrum (DSSS). The result of Data Hiding process are stego objects. Phases in the Data Hiding process are preprocessing data and data hiding. In preprocessing data, secret data is splitted into shares using Shamir's Secret Share method. Meanwhile, MP3 files which will be uses as cover object is being decoded. Data is hide using Direct Sequence Spread Spectrum (DSSS). Data hiding scheme is shown in Figure 3.

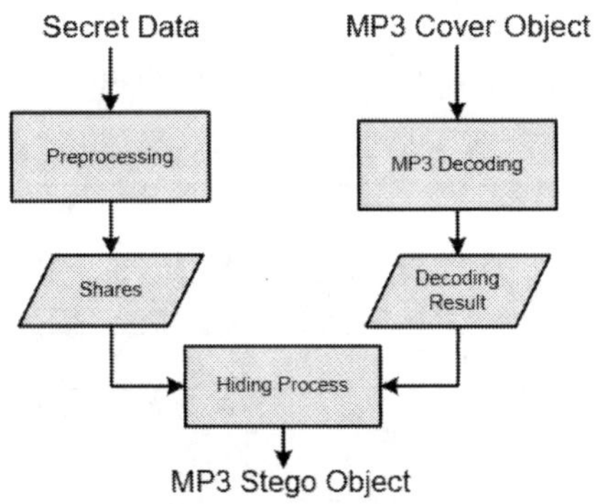

Figure 3. Data Hiding Scheme

In hiding process, data hiding is done in a segments of signal called frames. Psychoacoustic models used to analysis and calculate masking threshold. PN sequence is generated and modulated with shares. Using masking threshold and modulated signal, new signal with secret data is created in frequency domain. New audio signal is created by adding secret data signal and MP3 cover object. Hiding process scheme is presented in Figure 4.

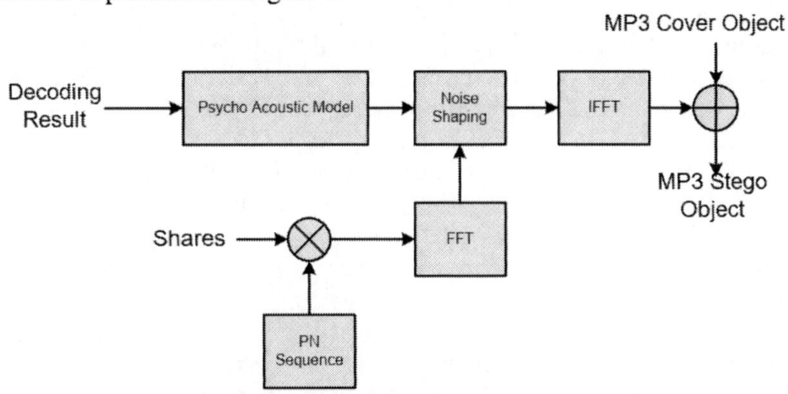

Figure 4. Hiding Process Scheme

5. Experiments

The steganography system using MP3 files as cover. It has ability to hide secret data into multiple cover. Besides that, the system can extract and reconstruct hidden secret data from several MP3 files.
To reconstruct data correctly, it requires a minimum number of shares at least equal to its threshold. In extraction data, not all stego data could be recovered with 100% accuracy. It is because of multiple factors that affect quality of the embedded stego data, such as the number of audio components replaced, and the gain affect. Data reconstruction done to get secret data from its shares. The result was secret data hidden in the cover.

5.1. Functionality test

The testing was done by hiding secret data into several MP3 files, then extraction and reconstuction process was done to retrieve hidden data. Meanwhile, threshold value was also varied to know the affect to the retrieved data. The data hidden into the cover is 1080 bits. Variations made to the amount of cover and its threshold value. In this test, Bit Error Rate (BER) is calculated from recovered data. Functionality test result is shown in Table 1.
Secret data hide in its cover is 1080 bits. Threshold shows the minimun number of shares required to sucessfully construct secret data. Stego object shows the number of stego object extracted to get its stego data (share). Bit Error Ratio (BER) shows error rate between the secret data which reconstruct

from stego object and secret data which is hidden in cover. The result shows that to reconstruct data correctly, the minimum number of shares required is at least equal to its threshold.

Table 1. Functionality Test Result

No	Data (bits)	Cover	Threshold	Stego Object	BER (%)
1	1080	5	5	4	37.5
2	1080	5	5	5	0
3	1080	6	5	4	54.7
4	1080	6	5	5	0
5	1080	6	5	6	0
6	1080	6	6	4	62.5
7	1080	6	6	5	62.5
8	1080	6	6	6	0

5.2. Audio Quality Test

Audio quality testing was done by measuring audio quality of cover before and after hiding process. Parameter measured was Mean Squared Error (MSE) and Peak Sognal-to-Noise Ratio (PSNR). In experiment, audio quality tested by comparing MSE and PSNR of a pair of cover-stego object. Variation done in the value of bits which to be hide, i.e. 360, 420, 600, and 1080 bits.

MSE value of the average for the fourth sample was 0,023. MSE is small enough shows that the audio quality before and after data hiding has not changed much. While the average PSNR for the fourth sample is 129.426 dB. This suggests that the process of data hiding have an effect that is small enough to change the audio quality of the cover is used.

5.3. Robustness Test

Robustness test was done by applying modification to stego object using StarMark for Audio (SMFA). With the change of signal of stego object, the data hidden is expected to change so it could not be reconstruct. The robustness of data from modification shows robustness of algorithm from modification type that applied. Modification on stego object called attack. The type of attack applied on stego object was single attack.

After being attack, extraction and reconstruction process will be doing on stego object. If secret data in stego object was still could be retrieve, it means that the data in stego object is robust against that type of modification applied.

The result of robustness test is presented in Table 2. Data in stego object before the attack and data after the attack applied is compared. If data is changed, showed by the increase of BER, the the attack had change the data hidden in cover object. On another word, stego object is not robust to this type of attack. Meanwhile, if BER is 0, then the attack has no effect to the data hidden.

Table 2. Robustness Test Result

Type of Attack	Effect on Data	Type of Attack	Effect on Data	Type of Attack	Effect on Data
AddNoise	BER increase when Strength>700	Amplify	BER increase when Factor>300	Invert	BER increase 100%
AddDynNoise	BER increase when Strength>20	Normalizer1	Effects	FFT_Invert	BER increase 100%

AddFFTNoise	No Effects	Normalizer2	BER increase	CopySample	BER increase
NoiseMax	No Effects	Compressor	No Effects	FlippSample	No Effects
AddBrumm	No Effects	BassBoost	No Effects	Exchange	No Effects
Echo	No Effects	RC_HighPass	No Effects	CutSample	BER increase
LSBZero	No Effects	RC_LowPass	No Effects	ZeroCross	No Effects
		Stat1	No Effects	ZeroLength1	No Effects
		Stat2	No Effects	ZeroLength2	No Effects
		Smooth1	No Effects	ZeroRemove	No Effects
		Smooth2	No Effects		

6. Conclusion

Steganographic system is constructed by combining a spread spectrum technique with Shamir's Secret Sharing can perform data hiding inside some of the audio files to MP3 format as the media cover.
Based on the functionality testing, steganography system can extract and reconstruct the secret data in stego stego object only when the object is used as an input amount is greater or equal to the threshold value.
The audio quality of media cover before and after hiding the data has not changed much, judging from the average value of MSE, ie 0,023, and amounted to 129.426 dB PSNR. Of the 29 types of attacks are simulated using StirMark Bechmark for audio, data stored in the object Stego has a resistance of 72.41%, or able to survive from the 21 of 29 types of attacks.

References

[1] Rao BR, Kumar PA, Rao KRM, Nagu M 2010 *Indian Journal of Computer Science and Engineering,* **1**(4), pp.327–332
[2] Atoum MS, Al Rababaa MS, Ibrahim S, Ahmed OA 2011 *International Journal of Computer Science and Network Security,* **11** (5) p 184-188
[3] Khan MdS, Bhasker VV, Nagaraju VS 2011 *International Journal of Computer Application* (0975-8887) **33**(4)
[4] Negrat AM and Kumar A 2010 In *Proceedings of the 10th WSEAS International Conference on Signal Processing, Computational Geometry and Artificial Vision* p 13–16
[5] Chan KFP 2011 *Proceedings of the Information Security for South Africa* Johannesburg
[6] Shamir A 1979 *Communications of the ACM* **22**(11), p 612–613. Available at: http://portal.acm.org/citation.cfm?doid=359168.359176
[7] Garcia RA 1999 *Digital Watermarking of Audio Signal Using A Psychoacoustic Auditory Model and Spread Spectrum Theory,* Master Thesis, School of Music University of Miami, Florida.
[8] Barkouti W, Salhi L, Cherif A 2011 *An International Journal (SIPIJ)* **2**(2)
[9] Lv C, Jia X, Tian L, Jing J, Sun M 2010 *Fourth International Conference on Network and System Security,* doi:10.1109/NSS.2010.82

Journal of Physics: Conference Series **776** (2016) 012090 doi:10.1088/1742-6596/776/1/012090

Asymptotic iteration method for the eigenfunctions and eigenvalue analysis in Schrodinger equation with modified anisotropic nonquadratic potential

B. N. Pratiwi, A. Suparmi , C. Cari and F. Anwar
Physics Department, Graduate Program, Sebelas Maret University, Jl.Ir. Sutami 36A Kentingan Surakarta 57126, Indonesia

Email: pratiwi.bn@gmail.com

Abstract. Analysis eigenfunctions and eigenvalue for the modified anisotropic nonquadratic potential in Schrodinger equation is solved using asymptotic iteration method (AIM). Cylindrical coordinates is used to separate three dimensional Schrodinger equation into three one dimensional Schrodinger type equation and then it is solved using AIM to obtain the eigenfunction and eigenvalue. The nonrelativistic energy equation are calculated, while the wavefunctions are discussed.

1. Introduction

The wavefunctions and energy spectra are important fundamental in quantum mechanics to describe behaviour of microscopic particles. Schrodinger equation for the central and non-central potentials have been studied in many research. Numerous studies have analyzed the bound state of charged particle moving in vector potential and non-central scalar potential, such as an electron moving in a Coulomb field with simultaneously precence of Aharonov-Bohm field,[1-2] Hulthen potential,[3] Hylleraas potential,[4-6] or ring-shaped-oscillator potential.[7-9] Very recently, some methods are used to derive the energy spectrum and the wavefunction of a charge particle moving in non-central potential, such as supersymmetric quantum mechanics with the idea of shape invariance, [9-16] factorization method, [17-18] Nikiforov-Uvarov method [19-20] and asymptotic iteration method (AIM)[21-24]. In this paper, will be investigated Schrodinger equation with a modified anisotropic nonquadratic potential using asymptotic iteration method (AIM) [21-24]. In previous research, the anisotropic nonquadratic potential has been investigated using Path integral method, [25] algebraic solution method, [26] and semmi classical treatment [27]. AIM is a powerful tool to determine eigenvalue and eigenfunction of second order differential equation. The modified anisotropic non-quadratic potential is given as [25-27]

$$V(\vec{r})=V(r,\theta,z)=\frac{1}{\alpha^2}\left(\frac{e^{\sqrt{^2+y^2}/\alpha}}{\left(1-e^{-\sqrt{x^2+y^2}/\alpha}\right)^2}\right)+\frac{\beta}{y^2}+\frac{\gamma x}{y^2\sqrt{x^2+y^2}}+\frac{\alpha'+\beta'\sinh z}{\cosh^2 z} \quad (1)$$

α, β, γ, α', β' are positive constant. By applying the cylindrical coordinate, the potential which is expressed in equation (1) is reduced into a simple and separable variable potential. For cylindrical coordinate we have

$$\nabla^2=\frac{1}{r}\frac{\partial}{\partial r}\left(r\frac{\partial}{\partial r}\right)+\frac{1}{r^2}\frac{\partial^2}{\partial\theta^2}+\frac{\partial^2}{\partial z^2};x=r\cos\theta;y=r\sin\theta;z \quad (2)$$

with $0<r\leq\infty, 0\leq\theta\leq 2\pi, 0<z<\infty$ and therefore the potential energy in equation (1) becomes

1 pratiwi.bn@gmail.com

$$V(r,\theta,z)=\frac{1}{\alpha^2}\left(\frac{e^{-r/\alpha}}{\left(1-e^{-r/\alpha}\right)^2}\right)+\frac{\beta+\gamma\cos\theta}{r^2\sin^2\theta}+\frac{\alpha'+\beta'\sinh z}{\cosh^2 z} \tag{3}$$

In cylindrical coordinate by using equation (2), the three dimensional Schrodinger equation for an anisotropic non-quadratic in equation (3) is given by

$$-\frac{\hbar^2}{2m}\left[\frac{1}{r}\frac{\partial}{\partial r}\left(r\frac{\partial}{\partial r}\right)+\frac{1}{r^2}\frac{\partial^2}{\partial\theta^2}+\frac{\partial^2}{\partial z^2}\right]\psi(r,\theta,z)+\left[\frac{1}{\alpha^2}\left(\frac{e^{-r/\alpha}}{\left(1-e^{-r/\alpha}\right)^2}\right)+\frac{\beta+\gamma\cos\theta}{r^2\sin^2\theta}+\frac{\alpha'+\beta'\sinh z}{\cosh^2 z}\right]\psi(r,\theta,z)=E\psi(r,\theta,z)$$

(4)

2. Asymptotic Iteration Method

Asymptotic Iteration Method (AIM) is used to solve second order differential equation in terms:

$$y_n''(k)-\lambda_0(k)y_n'(k)-s_o(k)y_n(k)=0 \tag{5}$$

where $\lambda_0(k)\neq 0$ and $s_o(k)$ are coefficients of the differential equation and are well defined functions as well as sufficiently differentiable. On the other hand, equation (5) can be written in term:

$$y''(x)=2\left(\frac{tx^{N+1}}{1-bx^{N+2}}-\frac{c+1}{x}\right)y'(x)-\frac{Wx^N}{1-bx^{N+2}} \tag{6}$$

The one-dimensional Schrodinger equation can be reduced into hypergeometric or confluent hypergeometric type differential equation by suitable changes of variables, and then changes it into the differential equation which has the form in equation (5). The solution of equation (5) can be obtained by using iteration of λ_i and s_i:

$$\lambda_i(k)=\lambda_{i-1}{}'+\lambda_{i-1}\lambda_0+s_{i-1},\ s_i(k)=s_{i-1}{}'+s_0\lambda_{i-1},\ i=1,2,3,\ldots \tag{7}$$

Eigenvalues can be obtained using equation (8): [23]

$$\lambda_i(k)s_{i-1}(k)-\lambda_{i-1}(k)s_i(k)=0=\Delta_i,\ i=1,2,3\ldots. \tag{8}$$

Equation (8) is AIM-type differential equation which eigenfunctions can be obtained by using equation (9) below [24,28]

$$y_n(k)=(-1)^nC'(N+2)^n\,{}_2F_1\left(-n,p+n,\sigma,bk^{N+2}\right)$$

(9)

where, $$(\sigma)_n=\frac{\Gamma(\sigma+n)}{\Gamma(\sigma)};\ \sigma=\frac{2c+N+3}{N+2};\ p=\frac{(2c+1)b+2t}{(N+2)b} \tag{10}$$

Here C' is normalization constant and ${}_2F_1$ is hypergeometric function. Equations (9-10) is used to obtain wavefunctions of the Schrodinger equation. The non-relativistic energy equation can be formulated by equating eigenvalue by using equations (7-8).

3. Schrodinger Equation for Modified Anisotropic Non-quadratic Potential

The three dimensional Schrodinger equation with an anisotropic non-quadratic potential which is expressed in equation (4) is solved by using variable separation method by setting the total wave function as $\psi(r,\theta,z)=R(r)P(\theta)Z(z)$, so we have the axial part, angular part and radial part, given as

$$\frac{1}{Z}\frac{\partial^2 Z(z)}{\partial z^2}-\frac{2m}{\hbar^2}\left[\frac{\alpha'+\beta'\sinh z}{\cosh^2 z}\right]=\lambda_1 \tag{11}$$

$$\frac{1}{P}\frac{\partial^2 P(\theta)}{\partial\theta^2}-\frac{2m}{\hbar^2}\left[\frac{\beta+\gamma\cos\theta}{\sin^2\theta}\right]=\lambda_2 \tag{12}$$

$$\frac{r}{R}\frac{\partial}{\partial r}\left(r\frac{\partial R)}{\partial r}\right)-\frac{2m}{\hbar^2}\left(\frac{r^2}{\alpha^2}\frac{e^{-r/a}}{\left(1-e^{-r/a}\right)^2}\right)+\frac{2m}{\hbar^2}Er^2+\lambda_1 r^2=-\lambda_2$$
(13)

with λ_1, λ_2 is variable separation constant.

3.1. Solution of Z part Schrodinger equation

The Schrodinger equation of axial (z) part in equation (11) is rewritten as

$$\frac{\partial^2 Z(z)}{\partial z^2}-\frac{2m}{\hbar^2}\left[\frac{\alpha'+\beta'\sinh z}{\cosh^2 z}+\lambda_1\frac{\hbar^2}{2m}\right]Z=0 \quad (14)$$

Equation (14) is solved by AIM with change of variabel which, $\sinh z = i(1-2u)$ (15)

so, $$\cosh^2 z = 4u(1-u); \frac{d^2}{dz^2}=-u(1-u)\frac{d^2}{du^2}-\frac{1-2u}{2}\frac{d}{du} \quad (16)$$

By using equations (15-16) in equation (14), we get

$$u(1-u)\frac{d^2 Z}{du^2}+\left(\frac{1}{2}-u\right)\frac{dZ}{du}+\frac{2m}{\hbar^2}\left[\frac{\alpha'+\beta' i(1-2u)}{4u(1-u)}+\lambda_1\frac{\hbar^2}{2m}\right]Z=0 \quad (17)$$

Equation (17) is reduced to the hypergeometry type differential equation using: $Z=u^{\chi}(1-u)^{\varepsilon}f(u)$ (18)
and then equation (17) become,

$$u(1-u)f''+\left[\left(2\chi+\frac{1}{2}\right)-(2\chi+2\varepsilon+1)u\right]f'+\left[\lambda_1-(\chi+\varepsilon)^2\right]f=0 \quad (19)$$

where $$\chi=\frac{1}{2}\sqrt{-\frac{2m}{\hbar^2}(\alpha'+\beta' i)+\frac{1}{4}}+\frac{1}{4}; \varepsilon=\frac{1}{2}\sqrt{-\frac{2m}{\hbar^2}(\alpha'-\beta' i)+\frac{1}{4}}+\frac{1}{4} \quad (20)$$

Equation (19) must be transformed to AIM type, so we obtain eigenvalues and eigenfunction by using AIM. By divide equation (19) with $u(1-u)$, yields the second order differential equation,

$$f''+\frac{\left[\left(2\chi+\frac{1}{2}\right)-(2\chi+2\varepsilon+1)u\right]}{u(1-u)}f'+\frac{\left[\lambda_1-(\chi+\varepsilon)^2\right]}{u(1-u)}f=0 \quad (21)$$

Equation (21) is AIM type differential equation. By using equation (5) and from equation (21), we get,

$$\lambda_0=\frac{(2\chi+2\varepsilon+1)u-\left(2\chi+\frac{1}{2}\right)}{u(1-u)}=\frac{-\left(2\chi+\frac{1}{2}\right)}{u}+\frac{\left(2\varepsilon+\frac{1}{2}\right)}{1-u}; s_0=\frac{\left[(\chi+\varepsilon)^2-\lambda_1\right]}{u(1-u)} \quad (22)$$

Eigenvalue is obtained from equation (7-8) and equation (21), is given as

$$\Delta_i=s_i\lambda_{i+1}-s_{i+1}\lambda_i \rightarrow \lambda_{1_{n_z}}=(\chi+\varepsilon+n_z)^2 \quad (23)$$

Equation (23) is eigenvalue from axial part of Schrodinger equation, which is generalized by using Matlab software. Equation (23), where χ and ε in equation (20), n_z is axial (z) quantum number, will be used to find energy spectra. Then, we can obtain eigenfunction or wavefunction using equation (6,9 -10). By compare equation (6) and equation (21) and then using equation (10), we get:

$$f(u)=(-1)^{n_z}C_1(1)^{n_z}\left(2\chi+\frac{1}{2}\right)_{n_z}{}_2F_1\left(-n_z, 2\chi+2\varepsilon+n_z, 2\alpha+\frac{1}{2}, u\right) \quad (24)$$

By substitute equation (24) in equation (18), and by using: $\sinh z = i(1-2u)\rightarrow u=\frac{i\sinh z+1}{2}$,we get wavefunction of the axial part for groundstate and the first ecxitation condition,

$$n_z=0\rightarrow Z_0(z)=\left(\frac{i\sinh z+1}{2}\right)^{\chi}\left(\frac{1-i\sinh z}{2}\right)^{\varepsilon}C_1; n_z=1\rightarrow Z_1(z)=-\left(\frac{i\sinh z+1}{2}\right)^{\chi}\left(\frac{1-i\sinh z}{2}\right)^{\varepsilon}C_1\left(2\chi+\frac{1}{2}\right)\left(1-\frac{2\chi+2\varepsilon-1}{2\chi+\frac{1}{2}}\left(\frac{i\sinh z+1}{2}\right)\right) \quad (25)$$

The solution of Schrodinger equation for axial part provides the value of variable separation constant, λ_1, and the wave functions of axial part.

3.2. The solution of angular Schrodinger equation

The Schrodinger equation for angular part which is expressed in equation (12) can be rewritten as

$$\frac{d^2P(\theta)}{d\theta^2}-\frac{2m}{\hbar^2}\left[\frac{\beta+\gamma\cos\theta}{\sin^2\theta}+\frac{\hbar^2}{2m}\lambda_2\right]P=0 \tag{26}$$

Equation (26) can be transformed to hypergeometry type differential equation by using

$$\cos\theta=1-2x\rightarrow\sin^2\theta=4x(1-x);\ \tfrac{d^2}{d\theta^2}=x(1-x)\tfrac{d^2}{dx^2}+\left(1/2-x\right)\tfrac{d}{dx} \tag{27}$$

so equation (27) become,

$$x(1-x)\frac{d^2P}{dx^2}+\left(\frac{1}{2}-x\right)\frac{dP}{dx}-\frac{2m}{\hbar^2}\left[\frac{\beta+\gamma(1-2x)}{4x(1-x)}+\frac{\hbar^2}{2m}\lambda_2\right]P=0 \tag{28}$$

and then by parameter $\qquad P=x^{\eta}(1-x)^{\kappa}g(x) \qquad (29)$

we get hypergeometry type differential equation for angular part of Schrodinger equation:

$$x(1-x)g''+\left[(2\eta+1/2)-(2\eta+2\kappa+1)x\right]g'+\left[-\lambda_2-(\eta+\kappa)^2\right]g=0 \tag{30}$$

Equation (30) is devided with $x(1-x)$, yields

$$g''+\left(\left[(2\eta+1/2)-(2\eta+2\kappa+1)x\right]/x(1-x)\right)g'+\left(\left[-\lambda_2-(\eta+\kappa)^2\right]/x(1-x)\right)g=0 \tag{31}$$

where, $$\eta=\frac{1}{2}\sqrt{\frac{2m}{\hbar^2}(\beta+\gamma)+\frac{1}{4}}+\frac{1}{4};\kappa=\frac{1}{2}\sqrt{\frac{2m}{\hbar^2}(\beta-\gamma)+\frac{1}{4}}+\frac{1}{4} \tag{32}$$

Equation (31) is AIM type differential equation. By compare equation (5) and equation (31), we obtain λ_0 and s_0 and by using equation (7-8), we obtain the generalized eigenvalue,

$$\Delta_n=s_n\lambda_{n+1}-s_{n+1}\lambda_n\ \rightarrow-\lambda_{2_{n_\theta}}=(\eta+\kappa+n_\theta)^2 \tag{33}$$

where $\lambda_{2_{n_\theta}}$ is separation variable constant in angular part, n_θ is angular quantum number, η and κ are constants which is expressed in equation (32). Then for angular wavefunction, we compare equation (31) with equation (6), and by using equation (9-10), we have

$$g(x)=(-1)^{n_\theta}C_2(1)^{n_\theta}(2\eta+1/2)_{n_\theta}\,{}_2F_1(-n_\theta,2\eta+2\kappa+n_\theta,2\eta+1/2,x) \tag{34}$$

From equation (31) and equation (34), and by using: $\cos\theta=1-2x\rightarrow x=\frac{1-\cos\theta}{2}$, we obtain the angular wavefunction, for groundstate and the first ecxited,

$$n_\theta=0\rightarrow P_0(\theta)=\left(\frac{1-\cos\theta}{2}\right)^{\eta}\left(\frac{1+\cos\theta}{2}\right)^{\kappa}C_2;n_\theta=1\rightarrow P_1(\theta)=-\left(\frac{1-\cos\theta}{2}\right)^{\eta}\left(\frac{1+\cos\theta}{2}\right)^{\kappa}C_2\left(2\eta+\frac{1}{2}\right)\left(1-\frac{2\eta+2\kappa-1}{2\eta+\frac{1}{2}}\left(\frac{1-\cos\theta}{2}\right)\right) \tag{35}$$

3.3 Solution of radial Schrodinger equation

From radial part, equation (13) is multiplied with (R/r^2) so we have

$$\frac{1}{r}\frac{d}{dr}\left(r\frac{dR)}{dr}\right)-\left[\frac{2m}{\hbar^2}\left(\frac{1}{\alpha^2}\frac{e^{-r/\alpha}}{\left(1-e^{-r/\alpha}\right)^2}\right)-\frac{2m}{\hbar^2}E-\lambda_1-\frac{\lambda_2}{r^2}\right]R=0 \tag{36}$$

By setting $R(r)=\left(U(r)/\sqrt{r}\right)$ in equation (36) we obtain

$$\frac{\partial^2U}{\partial r^2}-\left[\frac{2m}{\hbar^2}\left(\left(\frac{1}{\alpha^2}\frac{e^{-r/\alpha}}{\left(1-e^{-r/\alpha}\right)^2}\right)-E\right)-\lambda_1-\frac{\left(\lambda_2+\frac{1}{4}\right)}{r^2}\right]U=0 \tag{37}$$

Equation (37) is solved by using approximation of the centrifugal term [29-30] for $\frac{r}{\alpha}<<<1$ is given as

$$\left(1/r^2\right) \cong \left(1/4\alpha^2\right)\left(d_0 + \left(1/\sinh^2\left(r/2\alpha\right)\right)\right) \tag{38}$$

with $d_0 = 1/12$. In term of hyperbolic functions, $\left(\frac{1}{\alpha^2}\frac{e^{-r/\alpha}}{\left(1-e^{-r/\alpha}\right)^2}\right) = \left(\frac{1}{\alpha^2}\frac{1}{4\sinh^2\left(r/2\alpha\right)}\right)$ (39)

and then by using equation (38-39), equation (37) is written as

$$U'' - \left[\frac{\left(2m/\hbar^2\right) - \left(\lambda_2 + \frac{1}{4}\right)}{4\alpha^2 \sinh^2\left(r/2\alpha\right)} - \frac{2m}{\hbar^2}E - \lambda_1 - \frac{\left(\lambda_2 + \frac{1}{4}\right)d_0}{4\alpha^2}\right]U = 0 \tag{40}$$

and by setting change of variable:

$$\coth\left(r/2\alpha\right) = 1 - 2w; \frac{d^2}{dr^2} = \frac{1}{\alpha^2}\left(w(w-1)\right)^2\frac{d^2}{dw^2} + \frac{1}{\alpha^2}w(w-1)(2w-1)\frac{d}{dw} \tag{41}$$

in equation (40), we get

$$w(1-w)\frac{d^2U}{dw^2} + \left(1-2w\right)\frac{dU}{dw} - \left[\left(\lambda_2 + \frac{1}{4}\right) - \frac{2m}{\hbar^2} - \frac{2m\alpha^2}{\hbar^2}\frac{E}{w(1-w)} - \frac{\lambda_1\alpha^2}{w(1-w)} - \left(\lambda_2 + \frac{1}{4}\right)\frac{d_0}{4w(1-w)}\right]U = 0 \tag{42}$$

To reduce equation (42) in terms of hypergeometry type, we set $U = w^{\varsigma}(1-w)^{\tau}h(w)$ (43)

By inserting equation (43) into equation (42), we obtain

$$w(1-w)h'' + \left[(2\varsigma+1) - (2\varsigma+2\tau+2)w\right]h' + \left[-\left(\left(\lambda_2 + 1/4\right) - \left(2m/\hbar^2\right)\right) - \left(\varsigma+\tau\right)^2 - \varsigma - \tau\right]h = 0 \tag{44}$$

where, $\varsigma = \sqrt{-\frac{2m}{\hbar^2}\alpha^2E - \lambda_1\alpha^2 + \frac{1}{4}d_0\left(\lambda_2 + \frac{1}{4}\right)}; \tau = \sqrt{-\frac{2m}{\hbar^2}\alpha^2E - \lambda_1\alpha^2 + \frac{1}{4}d_0\left(\lambda_2 + \frac{1}{4}\right)}$ (45)

Equation (45) is reduced to AIM type by divide it with $w(1-w)$, so equation (45) become

$$h'' + \left(\left[(2\varsigma+1) - (2\varsigma+2\tau+2)w\right]/w(1-w)\right)h' + \left(\left[-\left(\left(\lambda_2 + 1/4\right) - \left(2m/\hbar^2\right)\right) - \left(\varsigma+\tau\right)^2 - \varsigma - \tau\right]/w(1-w)\right)h = 0 \tag{46}$$

and then we compare equation (46) with equation (5) to obtain λ_0, s_0 and by using equation (7-8), can be obtained the generalized eigenvalue in following relation:

$$\Delta_n = s_n\lambda_{n+1} - s_{n+1}\lambda_n \rightarrow \left(2m/\hbar^2\right) - \left(\lambda_{2_{n\theta}} + 1/4\right) = \left(\varsigma + \tau + n_r\right)\left(\varsigma + \tau + n_r + 1\right) \tag{47}$$

From generalized eigenvalue in equation (47), we obtain the non-relativistic energy equation, by inserting ς and τ which is expressed inn equation (45) to equation (47). The non-relativistic energy equation of Schrodinger equation for modified anisotropic nonquadratic potential is,

$$E = \frac{\hbar^2}{8m\alpha^2}\left(\left(\sqrt{\frac{2m}{\hbar^2} - \lambda_{2_{n\theta}}} - \frac{1}{2} - n_r\right)^2 + 4\alpha^2\lambda_{1_{n_z}} + d_0\left(\lambda_{2_{n\theta}} + \frac{1}{4}\right)\right) \tag{48}$$

where $\lambda_{2_{n\theta}}$ is separation variable constant of angular part in equation (33), $\lambda_{1_{n_z}}$ is separation variable constant of axial part in equation (23), n_r is radial quantum number. Table 1 are non-relativistic energy with variation potential constants α, β, γ, α', β' in $n_z = 1, n_\theta = 0, n_r = 1$ In Table 1, $E(\alpha)$ is energy as a function of α, $E(\alpha')$ is energy as a function of α' and so on.

Table 1. Non-relativistic energy corresponding to $n_z = 1, n_\theta = 0, n_r = 1, d_0 = 1/12$ of particle under the influence of an anisotropic non-quadratic potential

value of constants	Non-relativistic energy (Joule)				
	$E(\alpha)$	$E(\alpha')$	$E(\beta)$	$E(\beta')$	$E(\gamma)$
1	0.3425	0.3425	0.3425	0.3425	0.3425
2	0.2410	0.2535	0.0295	0.7535	0.3959
3	0.2222	0.2166	0.2107	1.2166	0.4715
4	0.2156	0.1970	0.4450	1.6970	0.5606
5	0.2125	0.1849	0.6772	2.1849	0.6576

Table 1 show that the increasing radial constant α, and axial constant α', cause the value of energies is smaller, while the energies value increase due to the value of angular constants γ, β is larger.

Then for radial wavefunction, we compare equation (6) with equation (46) and by using equation (9-10), we get

$$h(w) = (-1)^{n_r} C_3 (1)^{n_r} (2\varsigma+1)_{n_r} \, {}_2F_1(-n_r, 2\varsigma+2\tau-1+n_r, 2\varsigma+1, w) \tag{49}$$

By inserting equation (49) into equation (43), and by change of variable:

$$\coth(r/2\alpha) = 1-2w \rightarrow w = (1-\coth(r/2\alpha)/2) \tag{50}$$

we obtain radial wavefunction, for groundstate and the first excited, are given as

$$n_r = 0 \rightarrow \left(\frac{1-\coth(r/2\alpha)}{2}\right)^{\varsigma}\left(\frac{1+\coth(r/2\alpha)}{2}\right)^{\tau} C_3; n_r = 1 \rightarrow P_1(\theta) = -\left(\frac{1-\coth(r/2\alpha)}{2}\right)^{\varsigma}\left(\frac{1+\coth(r/2\alpha)}{2}\right)^{\tau} C_3(2\varsigma+1)\left(1-\frac{2\varsigma+2\tau}{2\varsigma+1}\left(\frac{1-\coth(r/2\alpha)}{2}\right)\right) \tag{51}$$

4. Conclusion

Schrodinger equation for the modified anisotropic non-quadratic potential is investigated by asymptotic iteration method. By using cylindrical coordinate, three dimensional Schrodinger equation is reduced to one radial Schrodinger equation, one angular Schrodinger equation, and one axial Schrodinger equation. By using suitable variable substitution and asymptotic uteration method, the eigenvalue and eigenfunction are obtained. The non-relativistic energy are calculated in variation value of potentials constants and the wavefunctions are expressed in hypergeometry term.

5. Acknowledgement

This research was partly supported by Higher Education Project Grant with contract no. 632/UN27.21/LT/2016

6. References

[1] Villalba V M 1994 *Phys. Lett. A* **193** 218-222
[2] Aharonov Y and Bohm D 1959 *Phys. Rev.* **115** 485
[3] Ikot A N, Akpabio L E and Uwah E J 2011 *EJTP* **8** 225
[4] Ikot A N, Antia A D, Akpan I O and Awoga O A 2013 *Revista Mexicana de Fisica* **59** 46
[5] Ikot A N, Maghsoodi E, Awoga O A, Zarrinkamar S and Hassanabadi H 2014 *Quant. Phys. Lett.* **3** 7
[6] Ikot A N 2013 *Commun. Theor. Phys* **59** 268
[7] Chen C Y and Dong S H 2005 *Phys. Lett. A* **335** 374
[8] Quan H X, Guang L, Min W Z, Bin N L and Yan M 2010 *Com.Theor. Phys.* (China) **53** 242
[9] Wu Q S, Wen H B, Yun W D and Yu G Z 2002 *Comm. Theor. Phys.*(China) **38** 139
[10] Sheng J C, Ying Z, Lin Z X and Tian S L 2001 *Commun .Theor. Phys.* **36** 641-646
[11] Suparmi and Cari 2014 *J. Math. Fund. Sci.* **46** 3 205-223

[12] Cari C and Suparmi A 2014 *AIP Conference Proceedings* **1615** 101
[13] Alemgadmi K I K, Suparmi, Cari, and Deta U A 2015 *AIP Conference Proceedings* **1677** 030004
[14] Dutt R, Khare A and Sukhatme U P 1988 *Am. J. Phys.* **56** 163
[15] Suparmi A, Cari C and Pratiwi B N 2016 *Journal of Physics: Conference Series* **710** 012026
[16] Gangopadhyaya A, Mallow J V, Rasinariu C and Sukhatme U P 2001 *Chin. J. Phys.* **39**
[17] Sadeghi J 2007 *Act.Phys.Polonica A* **112** (1) 23
[18] Sadeghi J and Pourhassan B 2008 EJTP **5** (17) 193
[19] Nikiforov A E and Uvarov V B 1988 *Special Function in Mathematical Physics* (Birkhausa: Basel)
[20] Falaye B J, Oyewumi K J and Abbas M 2013 arXiv:1109.3894v4 [quant-ph]
[21] Ikhdair S M, Falaye B J and Adepoju A G 2013 arXiv:1308.0155v1 [quant-ph]
[22] Oyewumi K J, Falaye B J, Onate C A, Oluwadare O J and Yahya W A 2014 arXiv:1402.2181v1 [quant-ph]
[23] Rostami A and Motavali H 2008 *Progress in Electromagnetic Research B* 4 171
[24] Soylu A, Bayrak O and Boztosun I 2008 *Journal of Physics A: Mathematical and Theoretical* **41** 065308
[25] Cai J M and Inomata A. *Physics Letters A* **141** 7 315-320.
[26] Boschi-Filho H and Vaidya A N 1990 *Physics Letters A* , **149** 336-340
[27] Vasan S S, Seetharaman M and Raghunathan K *Journal Pramana*, **40** (1) L75-L79
[28] Falaye B J, Hamzavi M and Ikhdair S M 2012 *arXiv*:1207.1218v [nucl-th]
[29] Greene R I and Aldrich C 1976 *Phys. Rev. A* **14** 2363
[30] Suparmi A, Cari C and Handhika J 2013 *Journal of Physics: Conference Series* **423** 012039

Student's Preconception and Anxiety When They Solve Multi-Representation Concepts in Newton Laws, And It's Application

C. Cari[1], A. Suparmi[2], J. Handhika[3]
Physics Graduate, Sebelas Maret University[12],
IKIP PGRI Madiun, Graduate Student's, Sebelas Maret University[3]
Jl. Ir. SUtami 36A Kentingan Jebres Surakarta 57126, INDONESIA
E-mail: cari@staff.uns.ac.id

Abstract. The purpose of this study was to describe of preconceptions and anxieties students in solving the representation concepts in newton laws, and it's application. This research was conducted for junior undergraduate student's in physics department (36 Students) and physics education (31 Students). The method of this study is a qualitative descriptive. The data was collection using test for multi representation concept, questionnaires for anxiety, and interviews. Based on the analysis it can be concluded that (1) the higher is anxiety, the higher is inconsistency (67,16%), (2) the higher is anxiety, the higher is consistency but wrong answer (29,85%), (3) the lower is anxiety, the higher is consistency of right answer (2,98%). Mostly students have understood fewer physics concept in newtons laws.

1. Introduction

Conceptions, misconceptions have become an important part of the study of physics. Understanding of both to be important in learning physics, because it affects the overall understanding of the physics [1]. Problem solving is important to be studied, but should not memorize how to solve it, but how to find a solution. Problems in physics can be presented in various presentations, which have the same concept of meaning. Presentations were presented in the form of a problem, will be represented again by students in the form of conception. The conception of student representation is the result of information processing in itself and can be used as a benchmark of the concepts of physics understanding [2]. Misconceptions experienced by students need to be clarified [3], so it can be the cause of misconceptions. Clarifying misconceptions can be done through interviews and provide arguments. Source misconceptions can be derived from student experience, books, language, and teachers [4]. The use of language becomes a source of misconceptions because they could create representations. It can be consistent or not. So that the test of consistency can also be used as a tool for identifying the cause of misconceptions.

In this study, it is assumed that students understand the concept if students are correct and consistent in representing a concept. Multi-representation test questions adapted from the Force Concept Inventory (FCI) [5] and Representational Force Concept Inventory (RFCI) [6]., the instrument has been used for identification understanding of the concept [1,2,7]. In solving problem, the students experienced the anxiety. Anxiety is a factor that also affects the students in solving the problems [8,9]. The initial assumption that students understand the concept well, if they solve the problems will be solved with the low anxiety level (confidence) and the answer is correct.

2. Method

The method used in this study is a qualitative descriptive. The data was collection using test for multi representation concept, questionnaires for anxiety, and interviews. Multi-representation of test consisted of six items that represent Newton's laws I, kinematics, and Newton III. Each material is represented by two questions with a different representation (visual and language representation). Students categorized as inconsistent if more than two pairs of the test are inconsistent, then the student is classified consistent with the wrong answer if there are more than two pairs of test that are consistent, students categorized correctly consistent if all test questions are given consistently and correctly. Indicators of anxiety [10] The questionnaire was developed to create an indicator of anxiety. Interviews are used to strengthen the argument conclusions and clarify students answer. This research was conducted for junior undergraduate student's in physics department (36 Students) and physics education (Students).

3. Result and Discussion

The description of the consistency of the student's answers and the level of anxiety (Table 1).

Table 1 Percentage of the level of consistency and anxiety

consistently Level	Percentage (%)	anxiety Level
inconsistent	67.16	High
correctly consistent	2.98	High
consistently with the wrong answer	29.85	Low

Based on (Table 1) concluded that 67.16% of the students are not consistent with the answer. Based on interviews, the change is due to answer the student does not understand and confused by the graphic representation, verbal, or do not understand the concept of kinematics and dynamics. Comparison of "wrong answers" between visual and verbal representations (Figure 1)

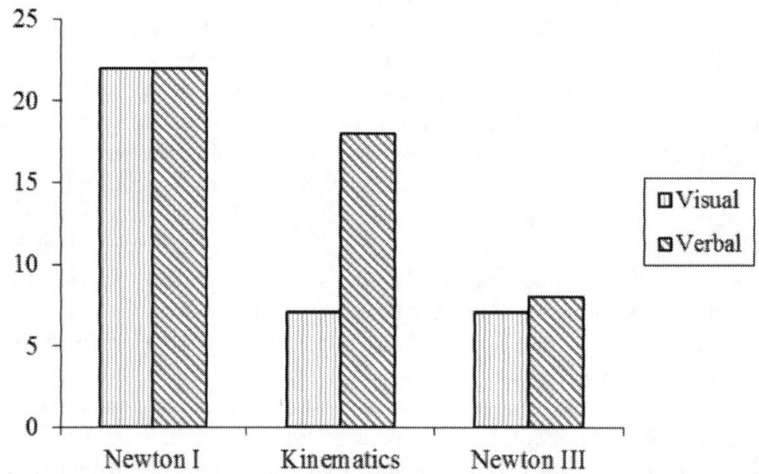

Figure 1 Comparison of "wrong answers" between visual and verbal representations

Figure 1 shows that based on Newton's laws I found a lot of answers consistent but wrong, in kinematics, many of which are not consistent, verbal representations are found errors, whereas the laws of Newton III, better than the verbal representation of a visual. Questions used is a modification of test FCI number 25. Students who answered correctly, can explain and describe uniformly accelerated motion well, for those students who answered incorrectly, could not explain and describe uniformly accelerated motion, even if only partially explains. Results anxiety questionnaire and interview process gave the information that the students are anxious when faced with two problems of the different types of answers, many who understand verbally about the change of pace but misrepresent the form of a graph (Figure 2).

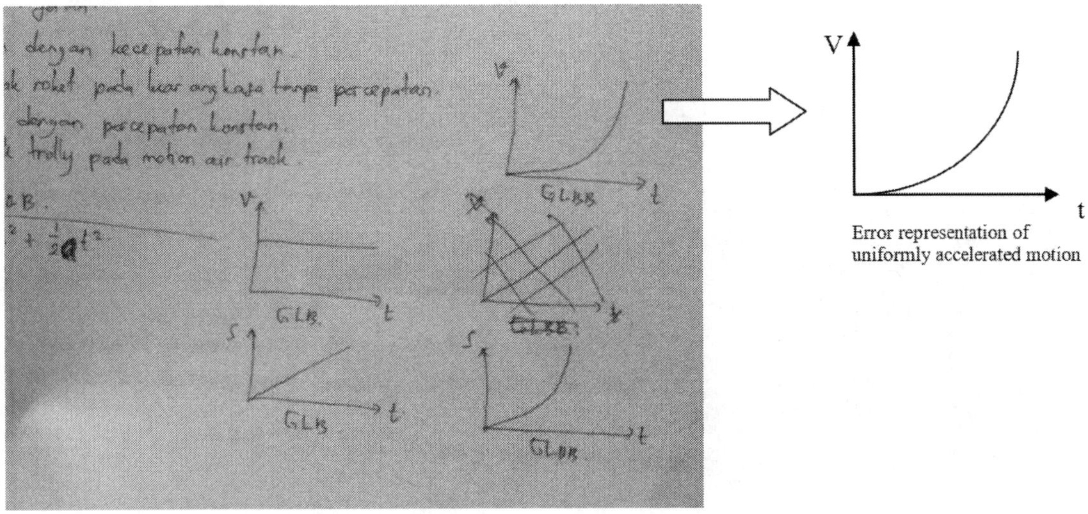

Figure 2 Error representation of the relationship changes in speed and travel time

Based (table 1) it can be concluded that 2.98% of the students answered consistently and correctly. The initial assumption that students can be said to understand the concept, but after the triangulation with interviews obtained information that in high school, it turns out the student had practiced doing about FCI and RFCI. based on the debriefing also obtained information that the students do not mention the word net force in interpreting the sigma symbol of force, but a force in the verbal pronunciation of Newton's laws. based on the results of the description of Newton's laws of understanding obtained information that the student is only a partial understanding of Newton's laws, by writing newton first law $\sum \mathbf{F} = 0$ and $\sum \mathbf{F} = constant$ has the same meaning (Figure 3).

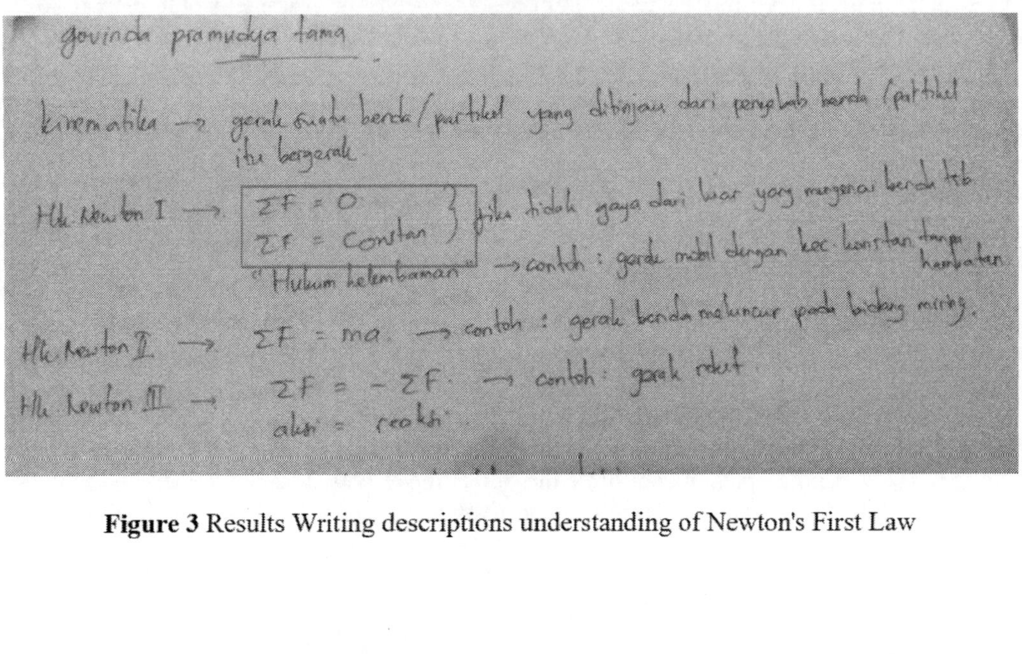

Figure 3 Results Writing descriptions understanding of Newton's First Law

There is a difference in meaning between "0" and "constant." Making of the physical is not yet understood by the students. Constant has a fixed value and magnitude, certainly different from the number "0". Mathematical language learning should be emphasized in physics so that students are able to understand well the laws of physics that is written in the form of an equation.

Next results, based on (Table 1) was 29.85% of the students are the consistent answer, but it is wrong. Newton's first law is most prevalent consistent answers wrong, even 88.89% of students are categorized as inconsistent (table 1), the answer is wrong consistently in this case. Visual and verbal representation present many mistakes. For representation of verbal and visual, the term constant speed on the motion of objects, (synonymous with about FCI number 18) has not been understood by the students, so that students choose answer vector direction y positive have a greater magnitude than the vector direction y negative (Figure 4).

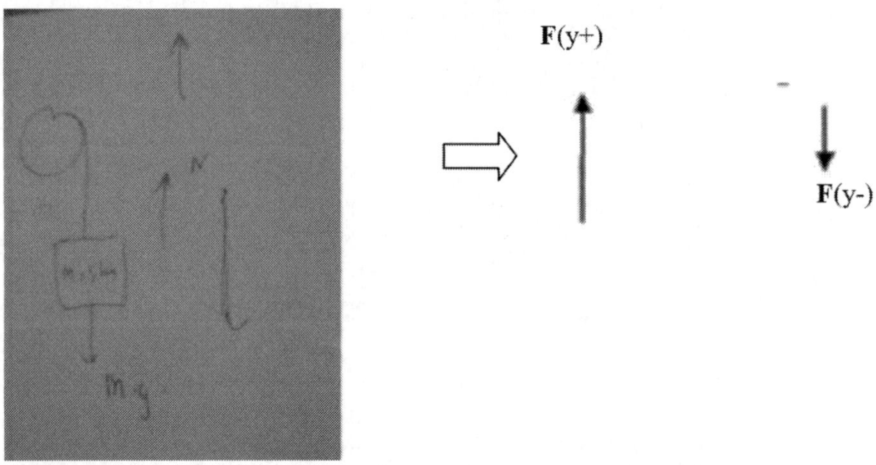

Figure 4 Results descriptions understanding of Newton's second law, in the case of moving objects upwards at a constant speed

These results (Figure 3), providing information that the representation of visual and verbal does not affect the student's understanding, understanding students more influenced by intuition and experience that already exists, that the direction of motion of the object produces a force, so the object is to move in the direction of the force, has a larger force than others. The concept of physics is not all easily be found in everyday life. The concept of constant speed, for example, would be hard to find in everyday life. The use of the laboratory is very important to clarify this concept.

In the third law of Newton the inconsistent arise because of anxiety, students feel the weight influence the magnitude of the force in action and reaction (FCI No. 11). On the other without involving the representation of weight, the student answered correctly.

4. Conclusion

From the analysis and discussion, it can be concluded that the students only a partial understanding of concepts, even though there are two students (2.98%) answered correctly and consistently, based on interviews and descriptions of the concept is obtained that the student does not understand it correctly. Test multiple representations of concepts consistency has not been able to measure student understanding of concepts, but if equipped concept description written by the students, and supported by interviews, will obtain more accurate data than just testing the consistency of representation. The visual representation (graphs) and important mathematical introduced to the students [1], [2], to overcome the inconsistent concept (Figure 2) and the problems that arise in Figure 3 and 4. The use of laboratory to demonstrate the object move with constant velocity (newton's laws I). The emphasis of the concept of kinematics and dynamics need to be reviewed again at the end of the course so that students can clearly distinguish and determine the relationship between these two concepts.

Acknowledgement
This research was partly supported by Higher Education Project Grant with contract no. 632/UN27.21//LT/2016

References

[1] Handhika J, et al 2016 AIP Conf. Proc. **1708**, 070005 dx.doi: 10.1063/1.4941178
[2] Handhika, J, et al 2015 ICMSE 2015 PE 34 37
[3] Yadav MK 2014 Physics Education. 523
[4] Kaltakçi D, Didiş N 2007 AIP Conf. Proc 899 499
[5] Hestenes D, Wells M, and Swackhamer G 1992. Phys. Teach. 30 141
[6] Nieminen P, Savinainen A, and Viiri J 2010 Phys. Rev. St Phys. Educ. Res. 6 020109
[7] Savinainen, Viiri J 2008 IJSME 6 719
[8] Larue A, D'elia LF 1985 Experimental Aging Research 215
[9] Karasel N, Ayda O, Tezer M 2010 Procedia Social and Behavioral Sciences. 2 5804
[10] BaoL, Kim Y, Raplinger A, Han J, Koenig K 2013 Research in Education Assessment and Learning

Supersymmetric approach for Killingbeck radial potential plus noncentral potential in Schrodinger equation

C. Cari, A. Suparmi, M. Yunianto and B. N. Pratiwi
Physics Department, Graduate Program, Sebelas Maret University, Jl.Ir. Sutami 36A Kentingan Surakarta 57126, Indonesia

Email: cari@staff.uns.ac.id

Abstract. Killingbeck radial potential, which consists of harmonic oscillator, linier and Coulomb potentials, is combined with non-central potential. The solution of three dimensional Schrodinger equation for Killingbeck potential is combined with Poschl-Teller potential and Symmetrical Top non-central potentials are investigated using supersymmetry (SUSY) operator. The non-relativistic energy is obtained which is infuenced by potentials and the wave functions are produced by using SUSY operator.

1. Introduction

The exact solutions of Schrodinger equations for some physical potentials are very important since they provided the essential information of the system. Recently, conciderably efforts have been paid to obtain the exact solution of the non-central potentials. These potentials include ring-shaped oscillator, modified Kratzer, Eckart, Poschl-Teller, Scarf, Manning Rosen, Rosen Morse, Hulthen, and Wood Saxon potentials. The bound state energy spectra of these potentials have been investigated by various techniques such as factorization methods[1-3], supersymmetric quantum mechanics[4-7], asymptotic iteration method [8-10], Nikiforov-Uvarov method[11-12] etc.

This paper is organized as follows. In section 2, we reviewed the supersymmetric quantum mechanics method briefly. In section 3, we found the energy solution in term of new orbital quantum number and radial and also angular wave functions. A brief results and discussion is presented in section four and conclusion in section five.

Killingbeck potential has potential application in particle physics, quantum field theory, molecular and solid state physics.The Killingbeck potential is a model of central potential consisting of radial oscillator harmonics potential and Cornell potential, which consists of linier and Coulomb potentials; given as[13-15]

$$V(r)=\frac{\hbar^2}{2m}\left(ar^2+br-\frac{c}{r}\right) \tag{1}$$

and for non-central potential, angular part and azimuth part, we take Poschl-Teller potential and Symmetrical Top potential are given as[16]:

$$V(\theta)=\frac{\hbar^2}{2m}\left(\frac{\alpha(\alpha-1)}{\sin^2\theta}+\frac{\beta(\beta-1)}{\cos^2\theta}\right) \tag{2}$$

$$V(\varphi)=\frac{\hbar^2}{2m}\left(\frac{(\delta^2+\gamma^2-\frac{1}{4})}{\sin^2\varphi}-\frac{2\delta\gamma\cos\varphi}{\sin^2\varphi}\right) \tag{3}$$

so we obtained potentials combination for Schrodinger equation in term,

[1]cari@staff.uns.ac.id

$$V(\vec{r}) = V(r,\theta,\varphi) = V(r) + \frac{V(\theta)}{r^2} + \frac{V(\varphi)}{r^2 \sin^2\theta} \tag{4}$$

2. Supersymmetry Quantum Mechanics (SUSY QM)

According to the definition proposed by Witten, the Hamiltonian in a SUSY QM is expressed as [4]

$$H_{ss} = \begin{pmatrix} -\frac{\hbar^2}{2m}\frac{d^2}{dx^2} + \frac{\hbar}{\sqrt{2m}}\frac{d\varphi(x)}{dx} + \varphi^2(x) & 0 \\ 0 & -\frac{\hbar^2}{2m}\frac{d^2}{dx^2} - \frac{\hbar}{\sqrt{2m}}\frac{d\varphi(x)}{dx} + \varphi^2(x) \end{pmatrix}; H_{ss} = \begin{pmatrix} H_+ & 0 \\ 0 & H_- \end{pmatrix} \tag{5}$$

with

$$V_-(x,a_0) = \varphi^2(x) - \frac{\hbar}{\sqrt{2m}}\varphi'(x); \qquad V_+(x,a_0) = \varphi^2(x) + \frac{\hbar}{\sqrt{2m}}\varphi'(x) \tag{6}$$

Here $H_-$and H_+, are supersymmetry partner of the Hamiltonian, $V_-(x)$ and $V_+(x)$ are the supersymmetry partner potential. To simplify the determination of the energy spectrum and the wave functions, the new operators, raising and lowering operators, are introduced as

$$A^+ = -\frac{\hbar}{\sqrt{2m}}\frac{d}{dx} + \phi(x) \text{ and } A = \frac{\hbar}{\sqrt{2m}}\frac{d}{dx} + \phi(x) \tag{7}$$

By inserting equation (7) into equation (6) we get the SUSY Hamiltonian as

$$H_-(x) = A^+A \text{ and } H_+(x) = AA^+ \tag{8}$$

It is always possible to factorize the usual Hamiltonian as

$$H = H_- + E_0 = -\frac{\hbar^2}{2m}\frac{d^2}{dx^2} + V_-(x;a_0) + E_0 \tag{9}$$

By using equations (6) and (9) it is obtained that

$$V(x) = V_-(x;a_0) + E_0 = \phi^2(x;a_0) - \frac{\hbar}{\sqrt{2m}}\phi'(x;a_0) + E_0 \tag{10}$$

where $V(x)$ is the effective potential, while $\phi(x)$ is determined hypothetically from equation (9) based on the shape of effective potential from the associated system. In addition to the supersymmetry that gives the relationshipbetween the eigenvalues and eigenfunctions between the two Hamiltonian partners but does not provide the actual spectrum . A pair of potentials $V_\pm(x)$ in equation (6) are said to be shape invariant if they are similar in shape but different in the parameters and this condition is given as,

$$V_+(x;a_j) = V_-(x;a_{j+1}) + R(a_{j+1}) \tag{11}$$

where $j = 0,1,2,..$, and a is a parameter in our original potential, V-, whose ground state energy is zero, $a_j = f_j(a_0)$ where f $j$is a function applied j times, the remainder $R(a_j)$ is a's dependence but is independent of x, then $V_\pm(x,a_0)$ is said to be shaped invariant. The energy eigenvalue of the Hamiltonian H_- is given by

$$E_n^{(-)} = \sum_{k=1}^{n} R(a_k) \tag{12}$$

and by using equations (11) and (12) we get the energy spectra of the system given as,

$$E_n = E_n^{(-)} + E_0 \tag{13}$$

Based on the characteristics of lowering operator, the ground state wave function of H-, whose ground state energy is zero, is obtained from condition that,

$$H_-\psi_0^{(-)} = 0 \rightarrow A\psi_0^{(-)} = 0 \tag{14}$$

Subsequently, the excited wave function, $\psi_1^{(-)}(x,a_0),..\psi_n^{(-)}(x,a_0)$ of H_- are obtained by using raising operator operated to the lower wave function, given as

$$\psi_n^{(-)}(x;a_0) \approx A^+(x;a_0)A^+(x.a_1)....A^+(x,a_{n-1})\psi_0^{(-)}(x;a_n) \tag{15}$$

[1]cari@staff.uns.ac.id

The SUSY QM and the idea of shape invariance potential are suitable to be used in solving one dimensional Schrodinger equation problems. D dimensional Schrodinger equation reduces to one dimensional Schrodinger type equation by suitable change in parameters. By obtaining the super-potential, the potential partners, $V_+(x,a_0)$ and $V_-(x,a_0)$, and the SUSY operators, A^+ and A are obtained and so the energy spectra and the wave functions.

3. Solution Schrodinger equation for Killingbeck radial potential is combined by noncentral potential

Schrodinger equation in Killingbeck potential is combined by noncentral potential, given as

$$-\frac{\hbar^2}{2m}\frac{1}{r^2}\left(\frac{\partial}{\partial r}\left(r^2\frac{\partial}{\partial r}\right)+\frac{1}{\sin\theta}\frac{\partial}{\partial\theta}\left(\sin\theta\frac{\partial}{\partial\theta}\right)+\frac{1}{\sin^2\theta}\frac{\partial^2}{\partial\varphi^2}\right)\psi(\vec{r})+\left(\begin{array}{l}\frac{\hbar^2}{2m}\left(ar^2+br-\frac{c}{r}\right)+\frac{1}{r^2}\frac{\hbar^2}{2m}\left(\frac{\alpha(\alpha-1)}{\sin^2\theta}+\frac{\beta(\beta-1)}{\cos^2\theta}\right)+\\ \frac{1}{r^2\sin^2\theta}\frac{\hbar^2}{2m}\left(\left(\frac{(\delta^2+\gamma^2-\frac{1}{4})}{\sin^2\varphi}-\frac{2\delta\gamma\cos\varphi}{\sin^2\varphi}\right)\right)\end{array}\right)\psi(\vec{r})=E\psi(\vec{r}) \quad (16)$$

and by substitution wavefunction $\psi(\vec{r})=R(r)P(\theta)Q(\varphi)$, we obtained three parts of Schrodinger equation, consist of radial, angular and azimuth part are given as below,

$$-\frac{\hbar^2}{2m}\frac{d^2U}{dr^2}+\frac{\hbar^2}{2m}\left(ar^2+br-\frac{c}{r}+\frac{\lambda_1}{r^2}\right)U=EU \quad (17)$$

where $U(r)=Rr$, and for angular part,

$$-\frac{\hbar^2}{2m}\frac{d^2G}{d\theta^2}-\frac{\hbar^2}{2m}\left(\frac{(1/4)-\lambda_2}{\sin^2\theta}-\left(\frac{\alpha(\alpha-1)}{\sin^2\theta}+\frac{\beta(\beta-1)}{\cos^2\theta}\right)\right)G=\frac{\hbar^2}{2m}\left(\frac{1}{4}+\lambda_1\right)G \quad (18)$$

with $P=\frac{G}{\sqrt{\sin\theta}}$ and $\lambda_1=l(l+1),\lambda_2=k^2$ then for azimuth part is given as

$$-\frac{\hbar^2}{2m}\frac{d^2Q}{d\varphi^2}+\frac{\hbar^2}{2m}\left(\frac{(\delta^2+\gamma^2-\frac{1}{4})}{\sin^2\varphi}-\frac{2\delta\gamma\cos\varphi}{\sin^2\varphi}\right)Q=\frac{\hbar^2}{2m}\lambda_2 Q \quad (19)$$

3.1. Solution of radial part

Schrodinger equation for Killingbeck potential is given in equation (17). By using SUSY QM, we can get,

$$V_{ef}=\frac{\hbar^2}{2m}\left\{ar^2+br-\frac{c}{r}+\frac{\lambda_1}{r^2}\right\} \quad (20)$$

We can take the superpotential in term:

$$\phi(r)=Ar+\frac{B}{r}+C \quad (21)$$

and from equations (6, 20-21), we get

$$V_-(r)=A^2r^2+\frac{B^2}{r^2}+C^2+2AB+2ACr+2\frac{BC}{r}-\frac{\hbar}{\sqrt{2m}}\left(A-\frac{B}{r^2}\right) \quad (22)$$

where

$$V_{ef}=V_-+\varepsilon_0 \quad (23)$$

By inserting equation (20) and (22) into equation (23) and equalizing between coefficients of the similar terms we get

$$A=\hbar\sqrt{\frac{a}{2m}}\ ;\ B=-\frac{\hbar}{\sqrt{2m}}(l+1) \quad (24)$$

[1]cari@staff.uns.ac.id

$$2AB-\frac{\hbar A}{\sqrt{2m}}+C^2+\varepsilon_0=0;2AC=\frac{\hbar^2}{2m}b;2BC=-\frac{\hbar^2}{2m}c \quad (25)$$

and then from equations (6,22-25), we got

$$V_-(r)=A^2r^2+\frac{B\left(B+\frac{\hbar}{\sqrt{2m}}\right)}{r^2}+\frac{\hbar^2}{2m}r-\frac{\hbar^2}{2m}\frac{c}{r}+2AB-\frac{\hbar A}{\sqrt{2m}}+\left(\frac{\hbar^2 c}{4mB}\right)^2$$

$$V_+(r)=A^2r^2+\frac{B\left(B-\frac{\hbar}{\sqrt{2m}}\right)}{r^2}++\frac{\hbar^2}{2m}r-\frac{\hbar^2}{2m}\frac{c}{r}+2AB+\frac{\hbar A}{\sqrt{2m}}+\left(\frac{\hbar^2 c}{4mB}\right)^2 \quad (26)$$

By using equations (11-13, 26), we obtained parameter

$$R(a_n)=2A\left(B-(n-1)\frac{\hbar}{\sqrt{2m}}\right)+\frac{2\hbar A}{\sqrt{2m}}+\left(\frac{\hbar^2 c}{4m\left(B-(n-1)\frac{\hbar}{\sqrt{2m}}\right)}\right)^2-2A\left(B-n\frac{\hbar}{\sqrt{2m}}\right)-\left(\frac{\hbar^2 c}{4M\left(B-n\frac{\hbar}{\sqrt{2m}}\right)}\right)^2 \quad (27)$$

so we can obtain the nonrelativistic energy equation using equations (13,27), is given as

$$E_n=2A\left(\frac{\hbar}{\sqrt{2m}}(2\mathrm{n}+l+3/2)\right)-\left(\frac{\hbar^2 c}{4m\left(\frac{\hbar}{\sqrt{2m}}(\mathrm{n}+l+1)\right)}\right)^2 \quad (28)$$

where n is radial quantum numbers and l is orbital quantum numbers. And for wavefunctions, by using equations (7-8,14,21-25) which applied uppering and lowering operator, we obtained the groundstate wavefunctions, and the first exited wavefunctions, are given as

$$U_0=\exp\left(-\frac{\sqrt{2M}}{\hbar}Ar^2-\left(\frac{\sqrt{2M}}{\hbar}Cr\right)\right)r^{-\frac{\sqrt{2M}}{\hbar}B} \quad (29)$$

$$U_1=\left(3Ar+C_0+C_1+\left(\frac{2B-\frac{\hbar}{\sqrt{2M}}}{r}\right)\right)r^{-\frac{\sqrt{2M}}{\hbar}\left(B-\frac{\hbar}{\sqrt{2M}}\right)}\exp\left(-\frac{\sqrt{2M}}{\hbar}Ar^2-\left(\frac{\sqrt{2M}}{\hbar}C_1r\right)\right) \quad (30)$$

where

$$C=C_0=\frac{\hbar^2 c}{4mB}\ ;\ C_1=\frac{\hbar^2 c}{4m\left(B-\frac{\hbar}{\sqrt{2m}}\right)} \quad (31)$$

3.2. Solution of angular part

For angular part from equation (18), we solved using SUSY QM. Equation (18) can be written

$$-\frac{\hbar^2}{2m}\frac{d^2G}{d\theta^2}+\frac{\hbar^2}{2m}\left(\frac{(\alpha+\frac{1}{2})^2+k^2}{\sin^2\theta}+\frac{\beta(\beta-1)}{\cos^2\theta}\right)G=\frac{\hbar^2}{2m}(l+\tfrac{1}{2})^2G \quad (32)$$

where

$$V_{ef}=\frac{\hbar^2}{2m}\left\{\frac{(\alpha+\frac{1}{2})^2+k^2}{\sin^2\theta}+\frac{\beta(\beta-1)}{\cos^2\theta}\right\} \quad (33)$$

We can take the superpotential in term:

$$\phi(\theta)=D\cot\theta+F\tan\theta \quad (34)$$

and from equations (6, 33-34), we get

$$V_-(\theta)=-D^2-F^2+DF+\frac{F}{\cos^2\theta}\left(F-\frac{\hbar}{\sqrt{2m}}\right)+\frac{D}{\sin^2\theta}\left(D+\frac{\hbar}{\sqrt{2m}}\right) \quad (35)$$

[1]cari@staff.uns.ac.id

By inserting equation (33) and (35) into equation (23) and equalizing between coefficients of the similar terms we get

$$D=\pm\frac{\hbar}{\sqrt{2m}}\sqrt{\left(\alpha+\tfrac{1}{2}\right)^2+k^2+\tfrac{1}{4}}-\frac{\hbar}{2\sqrt{2m}};F=\pm\frac{\hbar}{\sqrt{2m}}\left(\beta-\tfrac{1}{2}\right)+\frac{\hbar}{\sqrt{2m}};\varepsilon_0=D^2+F^2-DF \quad (36)$$

and then from equations (6) we get

$$V_+(\theta)=-D^2-F^2+2DF+\frac{F}{\cos^2\theta}\left(F+\frac{\hbar}{\sqrt{2m}}\right)+\frac{D}{\sin^2\theta}\left(D-\frac{\hbar}{\sqrt{2m}}\right) \quad (37)$$

By using equations (11-13, 35-37), we obtain parameter

$$R(a_n)=\frac{\hbar}{\sqrt{2m}}2n_\theta(F-D)+2n_\theta{}^2\frac{\hbar^2}{m} \quad (38)$$

so we can obtain the orbital quantum number l equation using equations (13,38), is given as

$$l=\pm\sqrt{\frac{\hbar}{\sqrt{2m}}2n_\theta(F-D)+2n_\theta{}^2\frac{\hbar^2}{m}+D^2+F^2-2DF-\frac{1}{4}} \quad (39)$$

where n_θ is angular quantum numbers and l is orbital quantum numbers. And for wavefunctions, by using equations (7-8,14,34) which applied lowering operator, we can obtain the groundstate for angular wavefunctions is given as:

$$G_0=Y\left(\sin\theta\right)^{\left(-\frac{\sqrt{2m}}{\hbar}A\right)}\left(\sec\theta\right)^{\left(-\frac{\sqrt{2m}}{\hbar}B\right)} \quad (40)$$

Y is integration constant.

3.2. Solution of azimuth part

Equation (19) can be written in term:

$$-\frac{\hbar^2}{2m}\frac{d^2Q}{d\varphi^2}+\frac{\hbar^2}{2m}\left(\frac{(\delta^2+\gamma^2-\frac{1}{4})}{\sin^2\varphi}-\frac{2\delta\gamma\cos\varphi}{\sin^2\varphi}\right)Q=\frac{\hbar^2}{2m}k^2Q \quad (41)$$

where

$$V_{ef}=\frac{\hbar^2}{2m}\left(\frac{(\delta^2+\gamma^2-\frac{1}{4})}{\sin^2\varphi}-\frac{2\delta\gamma\cos\varphi}{\sin^2\varphi}\right) \quad (42)$$

We can take the superpotential in term:

$$\phi(\varphi)=H\cot\varphi+J\csc\varphi \quad (43)$$

and from equations (6, 42-43), we get

$$V_-(\theta)=H^2\csc^2\varphi-H^2+J\csc^2\varphi+2HJ\cot\varphi\csc\varphi-\frac{\hbar}{\sqrt{2m}}\left(-H\csc^2\varphi-J\cot\varphi\csc\varphi\right) \quad (44)$$

By inserting equation (42) and (44) into equation (23) and equalizing between coefficients of the similar terms we get

$$H=-\frac{\hbar}{\sqrt{2m}}\left(\delta+\tfrac{1}{2}\right);J=\frac{\hbar}{\sqrt{2m}}\gamma;\varepsilon_0=H^2 \quad (45)$$

And then from equations (6) we get

$$V_+(\theta)=H^2\csc^2\varphi-H^2+J\csc^2\varphi+2HJ\cot\varphi\csc\varphi+\frac{\hbar}{\sqrt{2m}}\left(-H\csc^2\varphi-J\cot\varphi\csc\varphi\right) \quad (46)$$

By using equations (11-13, 44-46), we obtain parameter

$$R(a_n)=\frac{\hbar^2}{2m}\left\{a_n^2-a_{n-1}^2\right\};a_n=(\delta+\frac{1}{2}+n_\varphi) \quad (47)$$

so we can obtain the magnetic quantum number k equation using equations (13,47), is given as

$$k=\pm(\delta+\frac{1}{2}+n_\varphi) \quad (48)$$

[1]cari@staff.uns.ac.id

where n_φ is azimuth quantum numbers and . And for wavefunctions, by using equations (7-8,14,43) which applied lowering operator, we can obtain the groundstate for azimuth wavefunctions is given as:

$$Q_0 = T(\sin\varphi)^{\left(\delta-\gamma+\frac{1}{2}\right)}(1+\cos\varphi)^{\gamma} \tag{40}$$

T is integration constant.

4. Result and Discussion

The non-relativistic energy in equation (28) dependent of Killingbeck potential, Poschl-Teller potential and Symmetrical Top potential. Orbital quantum number l in energy equation is determined from solution angular part, while magnetic quantum number k from solution azimuth part. Without presence the non-central potentials, the non-relativistic energy equation in equation (28) depending only from radial Killingbeck potential. And for total wavefunctions of groundstate condition can be written as:

$$\psi(\vec{r}) = L\exp\left(-\frac{\sqrt{2M}}{\hbar}Ar^2 - \left(\frac{\sqrt{2M}}{\hbar}Cr\right)\right) r^{-\frac{\sqrt{2M}}{\hbar}B-1}\frac{1}{\sqrt{\sin\theta}}(\sin\theta)^{\left(-\frac{\sqrt{2m}}{\hbar}A\right)}(\sec\theta)^{\left(-\frac{\sqrt{2m}}{\hbar}B\right)}(\sin\varphi)^{\left(\delta-\gamma+\frac{1}{2}\right)}(1+\cos\varphi)^{\gamma} \tag{41}$$

L is integration constant.

5. Conclusion

The solution of three dimensional Schrodinger equation for Killingbeck potential is combined with Poschl-Teller and Symmetrical Top non-central potential are investigated using supersymmetry operator. The non-relativistic energy is obtained exactly in the closed form and the wave functions are produced by using SUSY operator.

6. Acknowledgement

This research was partly supported by Higher Education Project Grant with contract no. 632/UN27.21/LT/2016

7. References

[1] Infeld L and Hull T D 1951 *Rev. Mod. Phys.* **23** 21
[2] Sadeghi J 2007 *Act.Phys.Polonica A* **112** (1) 23
[3] Sadeghi J and Pourhassan B 2008 EJTP **5** (17) 193
[4] Gendenhtein L E 1983 *JETP Lett.*, **38** 356
[5] Cooper F, Ginocchio J N and Khare A 1987 *Phys. Rev. D* **36** 2458
[6] Witten E 1981 *Nucl. Phys. B* **188** 513
[7] Dutt R, Khare A and Sukhatme U P 1988 *Am. J. Phys* **56** 163
[8] Oyewumi K J, Falaye B J, Onate C A, Oluwadare O J and Yahya W A 2014 arXiv:1402.2181v1 [quant-ph]
[9] Rostami A and Motavali H 2008 *Progress in Electromagnetic Research B* 4 171
[10] Soylu A, Bayrak O and Boztosun I 2008 *Journal of Physics A: Mathematical and Theoretical* **41** 065308
[11] Nikiforov A E and Uvarov V B 1988 Special Function in Mathematical Physics (Birkhausa: Basel)
[12] Falaye B J, Oyewumi K J and Abbas M 2013 arXiv:1109.3894v4 [quant-ph]
[13] Hamzavi M , Ikhdair S M and Thylwe K E 2012 *Z.Naturforsch.* **67a** 567
[14] Neyazi H, Rajabi A A and Hassanabadi H 2016 *Nuclear Physiscs A* **945** 80
[15] Hamzavi M and Rajabi A A 2011 *Commun. Theor. Phys.* **55** 35
[16] Suparmi 2011 *Mekanika Kuantum II* Surakarta: Universitas Sebelas Maret

[1]cari@staff.uns.ac.id

Relativistic energy analysis for D-Dimensional Dirac equation with Eckart plus Hulthen central potential coupled by modified Yukawa tensor potential using Romanovski polynomial method

D. A. Dianawati, A. Suparmi, C. Cari, and Y. Mohtar
Physics Departement Sebelas Maret University,
Jl. Ir. Sutami 36A Kentingan Jebres Surakarta 57126, INDONESIA

E-mail: dyahayu@student.uns.ac.id

Abstract. Romanovski Polynomial method was used to analyze D-dimensional Dirac equation with Eckart plus Hulthen central potential coupled by modified Yukawa tensor potential in the case of spin symmetry and pseudospin symmetry. By using parameter and variable substitution, the Dirac equation was reduced into one dimensional Schrodinger like equation with centrifugal approximation that can be solved using Romanovski polynomial. The D-dimensional relativistic energy are obtained from D-dimensional relativistic energy equation by using Matlab R2008b software. The relativistic energy spectra with dimensional variation were obtained.

1. Introduction

In quantum mechanics, Dirac equation is used to explain dynamic particle with spin ½ [1]. Shape invariance central potential or non-central potential are used to discribe relativistic effect on vibration energy which its complex stucture of molecule [2], By assumption that scalar potential equal to vector potential, then Dirac equation was reduced to be resemble equation with Schrodinger equation and one dimension Dirac equation can be solved with method used to solved Scrodinger equation [3].
In Dirac equation, there are two cases with depend on the relation between scalar and vector potential. Spin symmetry occurs when $S(r) = V(r)$ and poseudopin symmetry (p-spin) occurs when $S(r) = -V(r)$[4]. The concepts of spin symmetry and pseudospin symmetry are used to study the change of core in nuclear physics. Spin symmetry is applicable at meson and antinucleon spectrum [5], while the concept of pseudospin symmetry used to explain degeneration-quation from nucleon doublet [6].
Romanovski polynomial method can be used for some central potentials, especially the spin symmetry and pseudospin symmetry [7]. This method is traditional method consisting differentials Schrodinger equation by variable change from general hypergeometric equation [8-9].
The aim of the research is to calculate relativistic energy spectra as a function of dimension of system with Matlab R2008b Software from relativistic energy equation.
In this paper, there are several sections are in the basic theory of Dirac equation in this section, Romanovski polynomial method, result and discussion, and the last section is conclusion.

Basic Theory Dirac Equation

The motion of a particle with mass M in a repulsive vector potential $V(r)$ and an attractive scalar potential $S(r)$ and coupled by a tensor potential $U(r)$ is discribed by the Dirac equation is [10] :

$$[\vec{\alpha}.\vec{p}+\beta(M+S(r))-i\beta\vec{\alpha}.\hat{r}U(r)]\psi(\vec{r})=[E-V(r)]\psi(\vec{r}) \tag{1}$$

where E is relativistic energy system and $\vec{p}$ is the three dimensional momentum operator,$-i\nabla$

$$\vec{\alpha}=\begin{pmatrix}0 & \sigma\\ \sigma & 0\end{pmatrix} \quad \text{and} \quad \beta=\begin{pmatrix}I & 0\\ 0 & -I\end{pmatrix} \tag{2}$$

The D-dimensional Dirac spinor equation is [11]:

$$\psi_{n\kappa}(r)=\begin{pmatrix}\zeta_{n\kappa}(\boldsymbol{r})\\ \Omega_{n\kappa}(\boldsymbol{r})\end{pmatrix}=\frac{1}{r^{\frac{D-1}{2}}}\begin{pmatrix}F_{n\kappa}(r)Y_{jm}^{l}(\theta,\varphi)\\ iG_{n\kappa}(r)Y_{jm}^{\bar{l}}(\theta,\varphi)\end{pmatrix} \tag{3}$$

Inserting equation (3) and (2) into equation (1), we obtain :

$$[\boldsymbol{\sigma}.\vec{\boldsymbol{p}}-i\hat{\boldsymbol{r}}\boldsymbol{\sigma}\,U(r)]\;\Omega_{n\kappa}(\boldsymbol{r})=[E-M-V(r)-S(r)]\;\zeta_{n\kappa}(\boldsymbol{r}) \tag{4}$$

$$[\boldsymbol{\sigma}.\vec{\boldsymbol{p}}-i\hat{\boldsymbol{r}}\boldsymbol{\sigma}\,U(r)]\;\zeta_{n\kappa}(\boldsymbol{r})=[E+M-V(r)+S(r)]\;\Omega_{n\kappa}(\boldsymbol{r}) \tag{5}$$

For exact spin symmetry case, happen when the scalar potential is equal to vector potential $S(r)=V(r)$, and for exact pseudospin symmetry case, happen when $S(r)=-V(r)$, then the upper Dirac spinor obtained from equation (4) and (5), and $(\boldsymbol{\sigma}.\vec{\boldsymbol{p}})(\boldsymbol{\sigma}.\vec{\boldsymbol{p}})=\boldsymbol{p}^2$, we have :

$$\left[\boldsymbol{p}^2+2\mathrm{V}[M+E]-(U(r))^2\right]\;\zeta_{n\kappa}(\boldsymbol{r})=[E^2-M^2]\;\zeta_{n\kappa}(\boldsymbol{r}) \tag{6}$$

$$\left[\boldsymbol{p}^2-2\mathrm{V}[M-E]-(U(r))^2\right]\;\Omega_{n\kappa}(\boldsymbol{r})=[E^2-M^2]\;\Omega_{n\kappa}(\boldsymbol{r}) \tag{7}$$

when $\boldsymbol{p}^2=-\nabla_{\mathrm{D}}{}^2=\frac{\partial^2}{\partial r^2}+\left(\frac{(D-1)}{r}\right)\left(\frac{\partial}{\partial r}\right)+\left(\frac{L_{D-1}{}^2}{r^2}\right)$ with the Laplacian, eigen value of $L_{D-1}{}^2$ is $l(l+D-2)$ and the angular momentum operator is $\left(\frac{L_{D-1}{}^2}{r^2}\right)=\frac{1}{sin^{D-2}\theta_{D-1}}\frac{d}{d\theta_{D-1}}\left(sin^{D-2}\theta_{D-1}\frac{d}{d\theta_{D-1}}\right)-\frac{1}{sin^2\theta_{D-2}}\left(L_{D-2}{}^2\right)$ [11-12]. So that, from equation (6) and (7), we obtain :

$$\left[\frac{\partial^2}{\partial r^2}+\left(\frac{(D-1)}{r}\right)\left(\frac{\partial}{\partial r}\right)+\left(\frac{L_{D-1}{}^2}{r^2}\right)-2\mathrm{V(r)}[M+E]+(U(r))^2\right]\zeta_{n\kappa}(\boldsymbol{r})=-[E^2-M^2]\;\zeta_{n\kappa}(\boldsymbol{r}) \tag{8}$$

$$\left[\frac{\partial^2}{\partial r^2}+\left(\frac{(D-1)}{r}\right)\left(\frac{\partial}{\partial r}\right)+\left(\frac{L_{D-1}{}^2}{r^2}\right)+2\mathrm{V(r)}[M-E]+(U(r))^2\right]\Omega_{n\kappa}(\boldsymbol{r})=-[E^2-M^2]\;\Omega_{n\kappa}(\boldsymbol{r}) \tag{9}$$

In this paper, V(r) is Eckart plus Hulthen central potential and $U(r)$ is modified Yukawa tensor potential.

2. Method

This researcher used finite Romanovski polynomials method used to solve the Dirac equation in the spin symmetry and pseudospin symmetry cases. The Dirac equation for spin symmetry and pseudospin symmetry was reduced into one dimensional Schrodinger like equation. One dimensional Schrodinger equation for the potential of interest was reduced into differential equation of Romanovski polynomial by approximate variable and wave function substitution.

Generalized hypergeometric equation is [9] :

$$\sigma(s)\frac{d^2y_n}{ds^2}+\tau(s)\frac{dy_n(s)}{ds}+\lambda_n y_n(s)=0 \quad (10)$$

where :

$$\sigma(s)=as^2+bs+c,\ \tau(s)=ds+e,\text{ dan } \lambda_n=-\{n(n-1)+2n(1-p)\} \quad (11)$$

The Pearson differential equation is :

$$\frac{d(\sigma(s)w(s))}{ds}=\sigma(s)w(s) \quad (12)$$

Weight function (w(s)) can be determained from the solution Person differential equation :

$$w(s)=\exp(\int\frac{(d-2a)s+(e-b)}{as^2+bs+c}ds\ ) \quad (13)$$

The appropiate polinomial for weight function equation (13) from Rodrigues representation :

$$y_n=\frac{1}{w(s)}\frac{d^n}{ds^n}\left((as^2+bs+c)w(s)\right) \quad (14)$$

For Romanovski polynomials, the value from parameter in equation (13) are :

$$a=1, b=0,\ c=1, d=2(1-p),\text{ and } e=q \text{ with } p>0 \quad (15)$$

Substitution of equation (15) into (13) it can be obtained weight function :

$$w(s)=(1+s^2)^{-p}e^{q\tan^{-1}(s)} \quad (16)$$

The required condition that the differential Romanovski equation is convergen when $m'+m<2p-1$ and obeys to the integral equation :

$$\int_{-\infty}^{\infty}w^{(p,q)}R_m{}^{(p,q)}(s)R_{m'}{}^{(p,q)}(s)ds \quad (17)$$

The differential Romanovski polynomials equation determined with substitution equation (11) and (15) into (10) obtained as:

$$(1+s^2)\frac{\partial^2 R_n{}^{(p,q)}(s)}{\partial s^2}+\{2s(-p+1)+q\}\frac{\partial R_n{}^{(p,q)}(s)}{\partial s}-\{n(n-1)+2n(1-p)\}R_n{}^{(p,q)}(s) \quad (18)$$

In equation (18), $y_n=R_n{}^{(p,q)}(s)$. When $r=f(s)$, eigen function in Romanovski polynomials is :

$$\Psi(r)=g_n(s)=(1+s^2)^{\frac{\beta}{2}}e^{\frac{-\alpha}{2}\tan^{-1}(s)}D_n{}^{(\beta,\alpha)}(s) \quad (19)$$

where :

$$D_n{}^{(\beta,\alpha)}(s)=R_n{}^{(p,q)}(s) \quad (20)$$

Romanovski polynomial from Rodrigues equation with weight function in equation (14) is :

$$D_n{}^{(\beta,\alpha)}(s)=R_n{}^{(p,q)}(s)=\frac{1}{(1+s^2)^{-p}e^{q\tan^{-1}(s)}}\frac{d^n}{ds^n}(1+s^2)^n\,(1+s^2)^{-p}e^{q\tan^{-1}(s)} \quad (21)$$

Wave function in equation (19) to be :

$$\Psi_n(r)=\frac{1}{\sqrt{\frac{df(s)}{ds}}}(1+s^2)^{\frac{-p}{2}}e^{\frac{q}{2}\tan^{-1}(s)}R_n{}^{(p,q)}(s) \quad (22)$$

Orthogonal integral from Romanovski polynomial is :

$$\int_0^{\infty}\Psi_n(r)\Psi_{n'}(r)dr=\int_{-\infty}^{\infty}w^{(p,q)}R_n{}^{(p,q)}(s)R_{n'}{}^{(p,q)}(s)ds \quad (23)$$

In this case, the value of p and q are not n dependent where n is the degree of the poynomials.

3. Result and Discussion

Equation for Eckart plus Hulthen central potential and equation modified Yukawa like tensor potential can be written [13] :

$$V(r) = -V_0 \frac{\cosh(\alpha r)-\sinh(\alpha r)}{2\sinh(\alpha r)} + 4V_1 cosech^2(\alpha r) - V_2 \coth(\alpha r) \quad (24)$$

$$U(r) = -V_3 \frac{(1+e^{-\alpha r})}{r} \quad (25)$$

3.1.Spin symmetry case

The D-dimensional Dirac equation for spin case in equation (8), can be written :

$$\left[\frac{\partial^2}{\partial r^2} - \left(\frac{\left(L_{D-1}+\frac{(D-1)}{2}\right)\left(L_{D-1}+\frac{(D-3)}{2}\right)}{r^2}\right) + \frac{V_0}{2}\coth(\alpha r)[M+E] - \frac{V_0}{2}[M+E] - 4V_1 cosech^2(\alpha r)[M+E] + V_2\coth(\alpha r)[M+E] + \frac{{V_3}^2}{r^2}(1+2(\cosh(\alpha r)-\sinh(\alpha r)) + (\cosh(\alpha r)-\sinh(\alpha r))^2)\right] F_{n\kappa}(r) = -[E^2-M^2]\ F_{n\kappa}(r) \quad (26)$$

From equation (26), use centrifugal appoximation to the $\frac{1}{r^2}$ term. For $\alpha r \ll 1$ is :

$$\frac{1}{r^2} \cong \frac{a^2}{\sinh^2(\alpha r)} \quad (27)$$

Substitution equation (27) into equation (26), we have :

$$\left[\frac{\partial^2}{\partial r^2} - \frac{a^2}{\sinh^2(\alpha r)}\left(\left(L_{D-1}+\frac{(D-1)}{2}\right)\left(L_{D-1}+\frac{(D-3)}{2}\right) + 4V_1[M+E] - {V_3}^2 + 2{V_3}^2 sinh(\alpha r)\right) + \left(\frac{V_0}{2}[M+E] + V_2[M+E] + 2a^2{V_3}^2 cosech(\alpha r) + a^2{V_3}^2\coth(\alpha r) - a^2{V_3}^2\right)\coth(\alpha r)\right] F(\mathbf{r}) = \left[\frac{V_0}{2}[M+E] - [E^2-M^2] - a^2{V_3}^2\right] F(\mathbf{r}) \quad (28)$$

We can setting equation (28) become :

$$A_s = \left(\left(L_{D-1}+\frac{(D-1)}{2}\right)\left(L_{D-1}+\frac{(D-3)}{2}\right) + 4V_1[M+E] - {V_3}^2 + 2{V_3}^2 sinh(\alpha r)\right) \quad (29)$$

$$B_s = -\frac{\left(\frac{V_0}{2}[M+E]+V_2[M+E]+2a^2{V_3}^2 cosech(\alpha r)+a^2{V_3}^2\coth(\alpha r)-2a^2{V_3}^2\right)}{a^2} \quad (30)$$

$$E_s' = -\frac{\left(\frac{V_0}{2}[M+E]-[E^2-M^2]-a^2{V_3}^2\right)}{a^2} \quad (31)$$

In equation (27), we can reduce to one dimensional Schrodinger type equation :

$$\left(\frac{d^2}{dr^2} - \frac{a^2 A_s}{\sinh^2(\alpha r)} - a^2 B_s\coth(\alpha r)\right) F_{n\kappa}(r) = -a^2 E_s'\, F_{n\kappa}(r) \quad (32)$$

So that, substitution $\coth(\alpha r) = ix$ in equation (32), we obtain :

$$\left((x^2+1)\frac{d^2}{dx^2}F_{n\kappa}(r) + 2x\frac{d}{dx}F_{n\kappa}(r)\right) - \left(A_s - B_s\frac{\coth(\alpha r)}{(x^2+1)} + \frac{E_s'}{(x^2+1)}\right) F_{n\kappa}(r) = 0 \quad (33)$$

From equation (33) and wave equation in equation (19), we have new wave function, then we get :

$$(x^2+1)\frac{d^2}{dx^2}(1+x^2)^{\frac{\beta}{2}}e^{\frac{-\alpha}{2}tan^{-1}(x)}D_n{}^{(\beta,\alpha)}(x)+2x\frac{d}{dx}(1+x^2)^{\frac{\beta}{2}}e^{\frac{-\alpha}{2}tan^{-1}(x)}D_n{}^{(\beta,\alpha)}(x)-\left[A_s-B_s\frac{\coth(\alpha r)}{(x^2+1)}+\frac{E_s'}{(x^2+1)}\right](1+x^2)^{\frac{\beta}{2}}e^{\frac{-\alpha}{2}tan^{-1}(x)}D_n{}^{(\beta,\alpha)}(x)=0 \quad (34)$$

If the coeficient $\frac{1}{(1+x^2)}$ term to zero, we have :

$$-\frac{\alpha^2}{4}+\beta^2-E_s'=0 \qquad \text{and} \qquad \alpha\beta+iB_s=0 \quad (35)$$

So that, equation (34) become :

$$(1+x^2)\frac{d^2D}{dx^2}+[2x(\beta+1)-\alpha]\frac{dD}{dx}-[A_s-\beta^2-\beta]D=0 \quad (36)$$

And the compairing the parameter between equation (18) and (36), we get relations $\alpha=-q$ and $p=-\beta$. From equation (35), we get new equation is :

$$\beta=\sqrt{A_s+\frac{1}{4}}-n-\frac{1}{2} \qquad \text{and} \qquad \alpha=\frac{iB_s}{\left(-\sqrt{A_s+\frac{1}{4}}+n+\frac{1}{2}\right)} \quad (37)$$

From equatin (37), we can obtain the value of β and α, there is :

$$\beta=\sqrt{\left(\left(L_{D-1}+\frac{(D-1)}{2}\right)\left(L_{D-1}+\frac{(D-3)}{2}\right)+4V_1[M+E]-V_3{}^2+2V_3{}^2sinh(\alpha r)+\frac{1}{4}\right)}-n-\frac{1}{2} \quad (38)$$

$$\alpha=\frac{iB_s}{\sqrt{-\left(\left(L_{D-1}+\frac{(D-1)}{2}\right)\left(L_{D-1}+\frac{(D-3)}{2}\right)+4V_1[M+E]-V_3{}^2+2V_3{}^2sinh(\alpha r)+\frac{1}{4}\right)}+n+\frac{1}{2}} \quad (39)$$

Equation (35), we can setting :

$$\beta^4-E_s'\beta^2+\frac{B_s{}^2}{4}=0 \quad (40)$$

Using quadratic equation, energy eigenvalues from equation (40) and (37), we obtain :

$$\sqrt{A_s+\frac{1}{4}}-n-\frac{1}{2}=\frac{E_s'\pm\sqrt{E_s'^2-B_s{}^2}}{2} \quad (41)$$

Substitution E_s from equation (31), we get energy eigenvalue is :

$$(M^2-E^2)=\frac{a^2\left(\left(4\sqrt{A_s+\frac{1}{4}}\right)(2n-1)-(2n+1)^2-B_s{}^2-4\left(A_s+\frac{1}{4}\right)\right)}{\left(\frac{4}{a^2}\sqrt{A_s+\frac{1}{4}}-\frac{4n}{a^2}-\frac{2}{a^2}\right)}+a^2V_3{}^2-\frac{V_0}{2}[M+E] \quad (42)$$

The relativistic energy for spin symmetry case is given in equation (42).

3.2. Pseudospin symmetry case

The D-dimensional Dirac equation for pseudospin case in equation (9), can be written :

$$\left[\frac{\partial^2}{\partial r^2} - \left(\frac{\left(L_{D-1}-\frac{(D-1)}{2}\right)\left(L_{D-1}-\frac{(D-3)}{2}\right)}{r^2}\right) - \frac{V_0}{2}\coth(\alpha r)[M-E] + \frac{V_0}{2}[M-E] + 4V_1 cosech^2(\alpha r)[M-E] - V_2\coth(\alpha r)[M-E] + \frac{V_3{}^2}{r^2}(1+2(\cosh(\alpha r)-\sinh(\alpha r)) + (\cosh(\alpha r)-\sinh(\alpha r))^2)\right] G_{n\kappa}(r) = -[E^2-M^2]\ G_{n\kappa}(r) \quad (43)$$

The solution for pseudospin symmetry case in equation (43) is similar method for spin symmetry case. Therefore, by repeating the steps in equation (27)-(41), we have :

$$\left[\frac{\partial^2}{\partial r^2} - \frac{a^2}{\sinh^2(\alpha r)}\left(\left(L_{D-1}-\frac{(D-1)}{2}\right)\left(L_{D-1}-\frac{(D-3)}{2}\right) - 4V_1[M-E] - V_3{}^2 + 2V_3{}^2 sinh(\alpha r)\right) + \left(-\frac{V_0}{2}[M-E] - V_2[M-E] + 2a^2V_3{}^2 cosech(\alpha r) + a^2V_3{}^2\coth(\alpha r) - a^2V_3{}^2\right)\coth(\alpha r) + \frac{V_0}{2}[M-E] + a^2V_3{}^2\right] G(\mathbf{r}) = \left[-V_0[M-E] - [E^2-M^2] - a^2V_3{}^2\right] G(\mathbf{r}) \quad (44)$$

We can setting equation (44) become :

$$A_{ps} = \left(\left(L_{D-1}-\frac{(D-1)}{2}\right)\left(L_{D-1}-\frac{(D-3)}{2}\right) - 4V_1[M-E] - V_3{}^2 + 2V_3{}^2 sinh(\alpha r)\right) \quad (45)$$

$$B_{ps} = -\frac{\left(-\frac{V_0}{2}[M-E]-V_2[M-E]+2a^2V_3{}^2 cosech(\alpha r)+a^2V_3{}^2\coth(\alpha r)-a^2V_3{}^2\right)}{a^2} \quad (46)$$

$$E'_{ps} = -\frac{\left(-\frac{V_0}{2}[M-E]-[E^2-M^2]-a^2V_3{}^2\right)}{a^2} \quad (47)$$

we can obtain the value of β and α, there is :

$$\beta = \sqrt{\left(\left(L_{D-1}-\frac{(D-1)}{2}\right)\left(L_{D-1}-\frac{(D-3)}{2}\right) - 4V_1[M-E] - V_3{}^2 + 2V_3{}^2 sinh(\alpha r) + \frac{1}{4}\right)} - n - \frac{1}{2} \quad (48)$$

$$\alpha = \frac{iB_{ps}}{\sqrt{-\left(\left(L_{D-1}-\frac{(D-1)}{2}\right)\left(L_{D-1}-\frac{(D-3)}{2}\right)-4V_1[M-E]-V_3{}^2+2V_3{}^2 sinh(\alpha r)+\frac{1}{4}\right)+n+\frac{1}{2}}} \quad (49)$$

Equation (35), we can setting :

$$\beta^4 - E'_{ps}\beta^2 + \frac{B_{ps}{}^2}{4} = 0 \quad (50)$$

Using quadratic equation, energy eigenvalues from equation (50) and (37), we obtain :

$$\sqrt{A_{ps}+\frac{1}{4}} - n - \frac{1}{2} = \frac{E'_{ps} \pm \sqrt{E'_{ps}{}^2 - B_{ps}{}^2}}{2} \quad (51)$$

Substitution E_{ps} from equation (47), we get energy eigenvalue is :

$$(M^2-E^2) = \frac{a^2\left(\left(4\sqrt{A_{ps}+\frac{1}{4}}\right)(2n-1)-(2n+1)^2-B_{ps}{}^2-4\left(A_{ps}+\frac{1}{4}\right)\right)}{\left(\frac{4}{a^2}\sqrt{A_{ps}+\frac{1}{4}}-\frac{4n}{a^2}-\frac{2}{a^2}\right)} + a^2V_3{}^2 + \frac{V_0}{2}[M-E] \quad (52)$$

The relativistic energy equation for pseudospin symmetry case is given in equation (52).

The result from this research is relativistic energy was numerically calculated from relativistic energy equation with Matlab R2008b software. The result of relativistic energy spectra is shown in Table 1 for spin symmetry case (E_s) and pseudospin symmetry case (E_{ps}) with variation of D value.

Table 1. The Relativistic Energy for Spin and pseudospin Symmetry Cases with $M = 5fm^{-1}$, $a = 0.5\ fm^{-1}, V_0 = 2fm^{-1}, V_1 = 0.6fm^{-1}, V_2 = 0.5fm^{-1}, V_3 = 0.4fm^{-1}$

D	L_{D-1}	E_s	E_{ps}
3	2	-6,76693	4,41158
4	3	-8.35021	4,53653
5	4	-10,34434	4,65566
6	5	-12.71695	4,73661

The relativistic energy spectra for Eckart plus Hulthen central potential coupled by modified Yukawa tensor potential for each parameter D and L_{D-1} for spin symmetry (E_s) and pseudospin symmetry (E_{ps}) cases. For spin symmetry case, the increase value of D causes the decrease of relativistic energy value, while for the pseudospin symmetry case is otherwise.

4. Conclusion

The D-dimensional Dirac equation for Eckart plus Hulthen central potential is coupled by modified Yukawa like tensor potential can be solved by using Romanovski polynomial method for spin symmetry and pseudospin symmetry cases. The numerical relativistic energy for D-dimensional is obtained from the relativistic energy equation with Matlab R2008 software. The result is relativistic energy data with D-dimensional variation from 3 to 6, is for spin symmetry case, the increase value of D causes the decrease of relativistic energy value, while for the pseudospin symmetry case is otherwise.

Acknowledgement

This research was partly supported by progress Higher Education Project Grant with contract no. 632/UN27.21/LT/2016.

References

[1] A.Suparmi and C.Cari 2013 *Atom Indonesia.* **39** 3 112-123
[2] A.Ikhair and R.Sever 2010 *Appl.Math.Comput.* 216 911
[3] N.Ikot and L.E.Akpabio 2010 *Applied Physics Research.* **2** 202-208
[4] Ginocchio, J.N, Leviatan, A, Meng J & Zhou, S.G 2004 *Physical Review. C.* **69** 3
[5] Giocchio, J.N 2005 *Physics Report.* 414 **4-5** 165-261
[6] Ginocchio, J. N 1997 *Physical Review Letters.* 78 **3** 436-439
[7] A.Suparmi 2011 *Mekanika Kuantum II*:FMIPA UNS
[8] W.Koepf and M.Masjed-jamei 2006 **17** 559
[9] A.F.Nikiforov and U.B.Uvarov 1988 *Special Function in Mathematical Physics.*
[10] Akpan N.Ikot 2014 *Quantum Physics Letters.***1** 7-14
[11] S.M.Ikhdair and R.Sever 2008 *Quantum Physics.* arXiv 0801.4857 **3**
[12] Falaye B.J and Oyewumi K.j 2011 Af.rev.Phys. **6** 25 211
[13] D.Anita A, N.Ikot A, E.Ituen Eno, E.Akpabbio Louis 2012 *Palestine Journal of Mathematics.* **1** 104-109

An experimental work on wireless structural health monitoring system applying on a submarine model scale

W. H. Nugroho[1] , N. J. H. Purnomo[1] , T. Soedarto[2]

[1]Pusat Teknologi Rekayasa Industri Maritim, BPPT, Komp. ITS Sukolilo, Surabaya, Indonesia
[2]Balai Teknologi Hidrodinamika, BPPT, Komp. ITS Sukolilo, Surabaya, Indonesia

E-mail : wibowo.harso@bppt.go.id

Abstract. This paper presents an experimental work to monitor the health of submarine hull structures using strain sensors and wireless communication technology. The monitored - submarine hull was built in a hydro elastic model scale 1: 30 with a steel bar backbone and tested on water tank of Indonesian Hydrodynamic Laboratory (IHL). Specifically, this health monitoring system for the submarine model was developed using wireless modems, data communication software and conventional strain sensors. This system was used to monitor the loads on a steel bar backbone of the running submarine model from the edge of the water tank. Commands were issued from a notebook to instruct the health monitoring system to acquire data from sensors mounted externally to the steel bar. Data from measurements made on the structure are then transmitted wirelessly back to a notebook computer for processing and analysis. The results of the tank test have been validated and showed no loss of communication signal over an area of the tank. This work also presents a potential use of involving complete automation of this system with an in-service structure coupled with an on-line warning/damage detection capability.

1. Introduction

Pressure hulls which is schematically shown on Figure 1 are the main load bearing structures of naval submarines, commercial and research submersibles, and autonomous underwater vehicles (AUVs) whose primary load-bearing responsibility is to withstand hydrostatic pressure associated with diving. The most efficient pressure hull geometries are circular thin-walled cross- sections that transfer the normal pressure load to in-plane compressive forces. The structural integrity of the pressure hull is very important because pressure hulls are subject to several different load types such as those from weapons (underwater explosions), wave slap on superstructure and other sea loads, and the predominant hydrostatic pressure. Non quantified loading on submarines further extends the uncertainty of predicting future performance. With all of these uncertainties, as well as the possibility of manufacturing defects and errors or material flaws, the possibility exists for a submarine to experience damage (cracking, corrosion, etc.) long before any predictive technique would indicate. Methods of detecting submarine damage prior to being found by routine inspections could therefore provide a great benefit by both reducing maintenance costs and extending ship service life. Structural health monitoring is one of the techniques proposed to bridge the gap between extensive testing and frequent inspections and maintenance. Damage detection systems deployed in difficult to access areas could greatly improve the time and cost required for routine maintenance. Hull monitoring systems also offer opportunities to better understand the performance of hulls during extreme sea states.

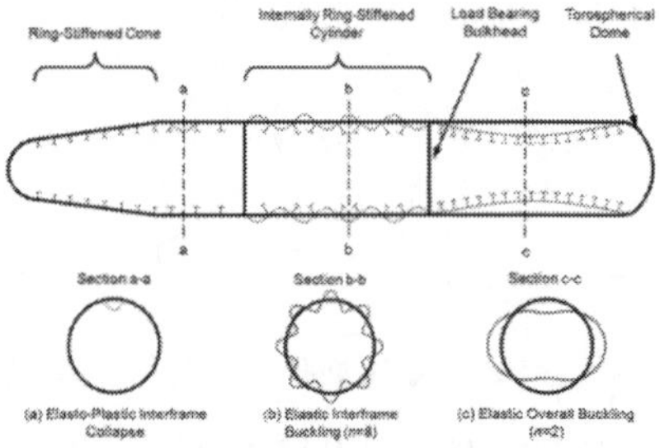

Figure 1. Typical pressure hull structure and buckling modes [3]

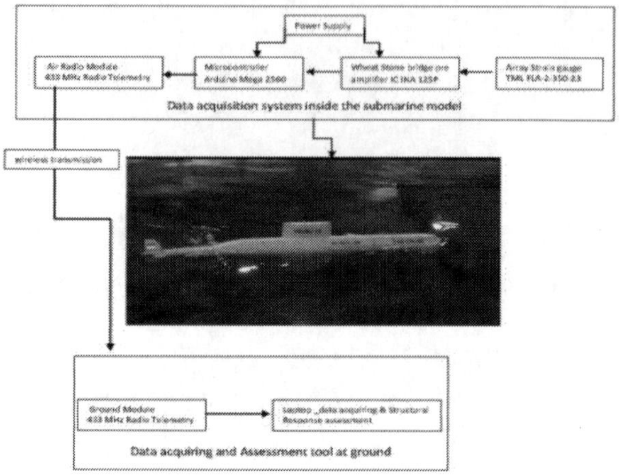

Figure 2. Remote SHM System on submarine model scale

To begin this experimental study there was very little information available about a structural monitoring specifically in submarine hull, since it belongs to military domain. However, the study of structural health monitoring has been motivated for the naval application such as reported on the ISSC report no 468 [9], since the US Navy has recently been incorporating non-traditional construction materials, namely aluminium and composites, into ship structures with greater frequency. This experimental study was supported by the previous works from fields such as wireless technologies, structural health monitoring (SHM) and ship model testing. A conceptual approach using smart sensors and wireless communication technology to monitor the health of large civil structures remotely was introduced by Pinesy [5]. In this paper the concept has successfully applied on a laboratory test specimen with a bolted lap joint from as far away as one mile. A good work presented by Bielen [2] which described the development of a wireless measurement system for multiple strain measurements on rotational structures. Their multi channel systems could measure wirelessly strains in up to 16 points with low cost per channel. Arms [1] has summarized the development of wireless strain sensing systems for a variety of customer driven structural health monitoring (SHM) applications where in the future development these systems could provide a long term, maintenance free, wireless strain monitoring solution. Swartz [8] has shown on his paper a hull monitoring system assembled from wireless sensors that consists of foil strain gauges, accelerometers, gyroscopes installed throughout a vessel to measure hull strain and ship motion. The study presented very encouraging evidence that wireless networks can function within the marine environment for reliable data collection. Drummen [6] also presented a useful material about the setup and results of the full and model scale tests conducted for the US Coast Guard's fatigue life assessment project. In this experimental setup he used a self propelled and self steering model of the NSC USCG BERTHOLF that was built to a scale of 1:25 and tested in MARIN's Seakeeping and Manoeuvring basin.

This paper presents an experimental work to monitor the health of submarine hull structures using strain sensors and wireless communication technology. The monitored - submarine hull was built in a hydro-elastic model scale 1: 30 with a steel bar backbone and tested on water tank of Indonesian Hydrodynamic Laboratory (IHL). Specifically, this health monitoring system for the submarine model was developed using wireless modems, data communication software and conventional strain sensors. The measuring structural response from array strain sensors was processed based on work of Wibowo [11] to calculate severity of the structure. Some of the results of this study was also based on work on slamming loads of frigate hull by Nugroho [7]. This health monitoring application hopefully will be useful since measurements systems in the naval structure domain are mainly strain gauge to avoid electromagnetic interference usually encountered in military environments.

2. Model Scale of Wireless Submarine Structural Health Monitoring (SHM) System

The advantages of spread spectrum technology used in radio telemetry include system flexibility, interference immunity, error free communication and real-time data throughput. However, the most unique feature about the radio telemetry is to be operated over greater distances with minor modifications in the transmitting antenna. The radio is a 2-way half-duplex wireless communication system with a standard TTL UART interface, based on Hope RF's HM-TRP data link modules, and custom firmware that improves upon the modules' original features and performance. Figure 2 displays a schematic of a simple and inexpensive remote health monitoring system for retrieving data from submarine model scale structures before, during and after an adverse loading condition.
The system consisted of train gauge sensor type TML FLA 2-350 – 23, a data acquisition system (preamp INA 125P and Arduino Mega 2560), air and ground module of 433 MHz radio telemetry, a notebook/ personal computer installed communication software for up - linking commands and downloading sensor data. Sensors, mounted externally to the backbone structure, were monitored by a ruggedized data acquisition system in close proximity to submarine model structure of interest. The system was powered by battery. Commands from a notebook computer were uplinked via the radio telemetry to instruct the data acquisition system to record sensor measurements. Upon retrieving these data, additional commands were issued to download the data back to a remote notebook/personal computer for analysis. Damage assessment performed when the data were downloaded to a remote notebook / personal computer. The use of the notebook computer permitted the operator to be mobile or stationary. To test and verify the approach described above which is displayed in Figure 2, a hydro - elastic model scale 1: 30 with a steel bar backbone located in the keel which has flexural stiffness similar to the full scale submarine (see Figure 3 and Table 1). Strain sensor arrays were placed on top and sides of the steel back bone to detect changes in strain due to changes of load. The sensor positions were at a quarter length of LOA from the bow (ST 16), mid-ship (ST 10) and at a quarter of LOA from the stern ship (ST 04) where presumably the slamming and highest normal stress loads may occur. During the experiment a dynamic load was simulated by a falling object to the submarine model which excites the backbone structure in bending. Measurements were then recorded from all 12 strain sensors. Commands to take data were uplinked from a notebook / personal computer through ground module located approximately in range of 50 to 100 m away from the submarine model.
Commands from the remote notebook computer were received through air module connected to the Arduino board (microcontroller) to drive the data acquisition system inside of the submarine model. The 3DR Radio telemetry system had configuration of serial board (for the air) and USB (for the ground) at rates up to 250 kbps in the frequency 433 MHz. Response of the backbone structure due to the dynamic loads was then acquired by microcontroller board and transmitted to notebook or personal computer on the ground. Once the sensor data were acquired to the notebook computer, they could be retrieved wireless for assessment of structural response in the backbone steel. Communication between the two computers and the two radio modules was established using a 3DR Radio open source X bee communication software. This communication software enabled the user to make the remote notebook emulate the data from microcontroller. Thus, commands issued in the Windows environment of the remote notebook computer become the commands issued on the microcontroller board. However, the communication software was more than just an emulation tool; it also enables the operator to view commands being carried out on the microcontroller board. The functional test of the wireless system with configuration for the experiment is presented on Figure 4. Table 2 above summarizes components used to create this experiment demonstration of a remote wireless health monitoring system. The system was validated at ranges up to 100 m for approximately 1 hour without loss of communication link. Using the communication software, a communication link was established via the 3DR Radio telemetry system to make the remote notebook/PC to read the data from the acquisition system. A micro - controller software was applied on the data acquisition board to have the strain sensor response data. Then a download command was then issued from the communication software to transfer data to the remote notebook

computer. Upon receiving the downloaded strain sensor response data, the in – house software was initiated on the remote notebook computer to run assessment of structural response in the backbone steel. Once the communication link is established during the setup mode, sensor data can be retrieved within seconds.

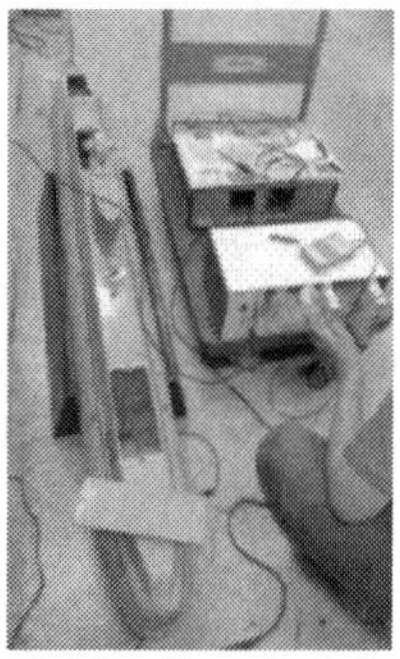

Figure 3. Strain Sensor placement on Submarine Model

Figure 4. Functional test of wireless monitoring system

Table 1. Properties of Submarine hydro- elastic model

Physical property	Units
EA/L (stiffness of backbone steel)	108.9 MN/m
LOA (length over all)	2.04 m
Breadth	0.206 m
Draught	0.190 m
Displacement	55.45 kg

Table 2. Components of wireless SHM of submarine model

Item	Manufacturer
Strain Sensor	TML FLA – 2- 350 – 23
Data Acquisition System	Preamp INA 125P and Arduino Mega R3 2560
DAAS Software	Arduino
Radio Telemetry (air and Ground module)	433Mhz Single TTL 3D Robotics 3DR Radio Telemetry Kit, Type cp2102Silicon Lab
Radio Telemetry Software	Open source 3 DR
Notebook/ Personal Computer	Sony/Any
Damage Detection Software	IN HOUSE

3. Results obtained on experimental test specimen

The submarine model in the water tank is shown on the Figure 5. Using this remote structural health monitoring system, the back bone of the submarine model was excited by a falling object to obtain bending moment response data to strain sensor. These data were retrieved across the wireless link for various dynamic loads. These loads was in form of hydro-dynamics impact load where in this study obtained by measuring the stresses that occurs on the backbone of model submarine when it hits the surface of the water tank . Results for the measurement from the sensors were plotted to the screen of the remote notebook computer and later stored as image files. Figure 6 presents the results in operation during a typical data run from the twelve strain sensors. Notice that in Figure 6 the screen of the remote notebook computer displays the data across the wireless link that were being taken from backbone steel of submarine model.

Figure 5. Submarine Model scale in the water tank

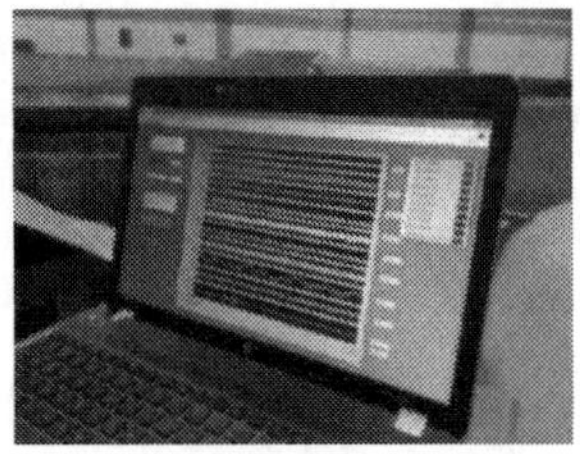

Figure 6. A notebook for recording data from sensor measurements

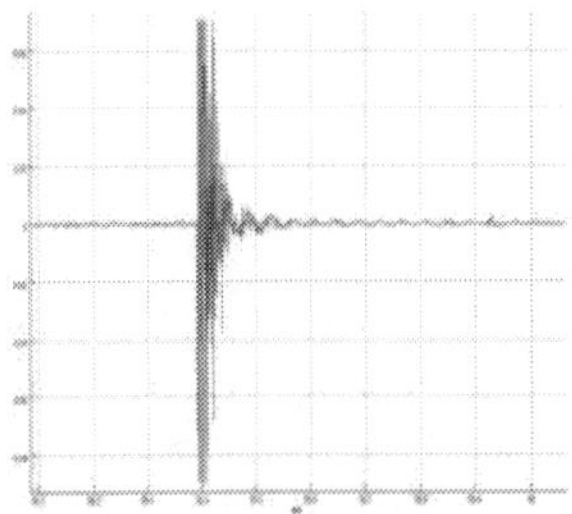

Figure 7. Stress data due to falling object at ST04 of the submarine model

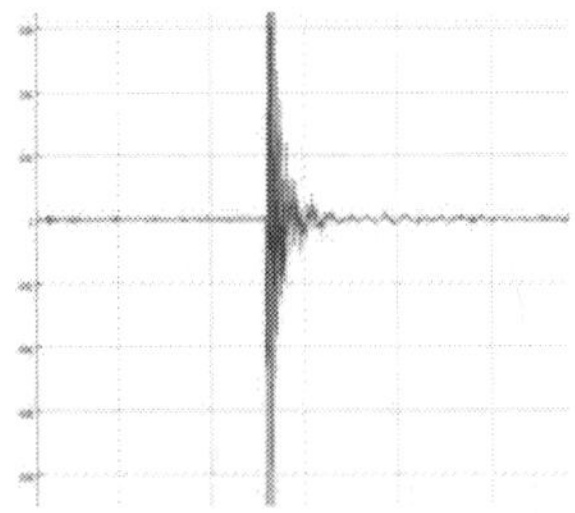

Figure 8. Stress data due to falling object at ST10 of the submarine model

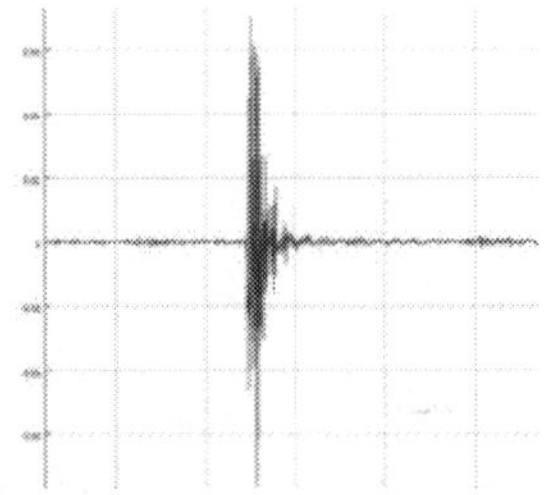

Figure 9. Stress data due to falling object at ST16 of the submarine model

The excitation by the falling object was based on the results of the numerical calculations. The numerical calculation used stripe theory to calculate relative vertical motion and velocity of the submarine during operation on the sea. The relative motion represented the height of the falling object. To determine the height from relative velocity the free - fall physics formula was applied. There were five variations of the height of the falling object were used on this experiment. Examples of data reading of the stresses from strain sensors of the ST 04, ST 10 and ST 16 are presented on Figure 7 to 9 respectively.

The data stresses were collected with sampling rate of 1000Hz, to have the transient response of the load. To validate the results the stress measurements of this experiment were compared with stresses that predicted by make use of slamming pressure values from Stavovy and Chuang in report of ISSC-385 [4]. Their methods provide a procedure to compute slamming impact pressure of high performance vehicles (HPV). Their empirical methods based on two- and three-dimensional data. The method determines the impact pressure in an infinitesimal area of the hull bottom. In that area the dead rise, buttock trim and heel angles are determined from ship lines, body plan, ship motions and wave profile. The empirical formula for impact pressure can be shown as follows [4]:

$$p_i = k_1 V_n^2 \qquad (1)$$

Where,

Vn = relative normal velocity of impact body to wave surface on plane normal to wave surface. And values for k1 are shown in Figure 10 as a dashed line.

By multiplying the impact pressure with the water plane area of the submarine in a particular draught the excitation force F can be found. This force is a useful quantity to calculate a bending moment at a particular station of the submarine. Hence, the normal stress on pressure hull of the submarine (full scale) by following formula [10];

$$\sigma_{ni} = \frac{M_i Y_i}{I_i} \qquad (2)$$

where;

M_i = bending moment at station i of submarine hull; Y_i = distance of the neutral axis to surface of the backbone structure at station i; I = sectional moment inertia of the backbone structure at station i

Figure 11 shows the comparison of maximum normal stress between measurement and prediction. From the picture above it shows that the patterns produced by these two methods are in a good agreement. However, the stress values from the experiment seem to have a higher value than the predicted one. The difference value of stress is approximately 30%. This is probably because the value of k_1 in Statovy and Chuang are more suitable for surface ship hull forms and numerical approximations. This actually supported by the Figure 10 where the value k_1 of the wedge test tend to be larger than other numerical approximation.

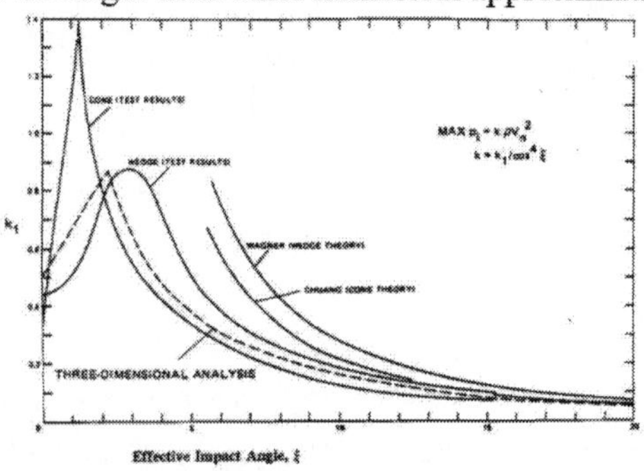

Figure 10. Values for k1 of Stavovy and Chuang method [4]

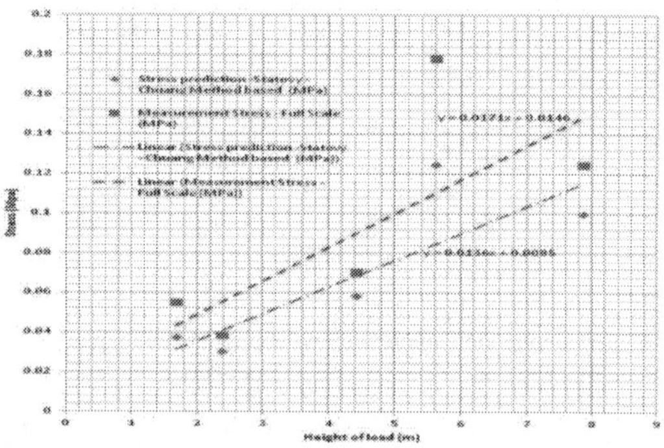

Figure 11. A comparison between stress measurement and prediction

4. Conclusions

This paper has presented a simple model and inexpensive approach for remote health monitoring for submarine structures involving spread spectrum wireless modems, a note book based data acquisition system, off-the-shelf communication software, conventional sensors and custom structural response software. The wireless system can work with a submarine hydro-elastic model over ranges of up to approximately 100 m with no loss of communication signal. The results of test measurements from the tank test have a good agreement with a modified well known impact pressure method. Data rates achieved on transferring data files are of the order of 250 kbps which are suitable enough for this particular monitoring application. Future work will involve complete automation of this system with an in-service structure coupled with an on-line warning/damage detection capability

Acknowledgments

This work was funded by the Ministry of Research Technology and Higher Education, Republic Indonesia through INSINAS RT-2016 -314 research program

References

[1] Arms SW and Townsend CP 2003 *Wireless Strain Measurement Systems – Applications & Solutions*, NSF-ESF Joint Conference on Structural Health Monitoring Strasbourg, France

[2] Bielen P, Lossi, and Vandepitte M 2002 *Proceedings Of Isma* **II**, p 663 – 670

[3] Fricke W, Bronsart R (Eds.) 2012 *18th International Ship And Offshore Structures Congress, Rostock Germany,* Vol **2**

[4] Daidola JC and Victor M 1995 *Hydrodynamic impact on displacement ship hulls, an assessment of the state of the art*, Ship Structure Committee (SSC – 385)

[5] Pinesy DJ and Lovellz PA 1998 *Journal of Smart Material and Structure* **7**, 627–636.

[6] Drummen I, SchiereM, Dallinga R and Stambaugh K 2014 *Ship Structure Symposium, Linthicum Heights*, MD

[7] Wibowo NH and Mujahid AS 2015 *World Maritime Technology Conference, Providence*, USA 03 – 07

[8] Swartz, Zimmerman RA, Lynch AT, Brady JP, Thomas F, Salvino JR, Law LW and Kincho H, 2009 *Wireless Hull Monitoring Systems for Modal Analysis of Operational Naval Vessels*, Proceedings of the IMAC-XXVII, Orlando, Florida USA, Society for Experimental Mechanics Inc. 9-12

[9] Ship Structure Committee 2013 *Development Of A Structural Health Monitoring Prototype For Ship Structures,* Report no 468

[10] Vidosic JP 1987 *Mechanics of Material*, Sec 5.2 of Marks Standard Handbook for Mechanical Engineers, 9th Ed Mc Graw – Hill Book Company

[11] Wibowo HN and Chiu WK 1998 *COMADEM Proceedings, Launceston, Tasmania, Australia* p 905-917

Identification resistivity value of soil: alternative substitute resistance material on resistor

I. Kusumawati
Physics Education Department of Undergraduate Program,
College Teaching and Education Trainning Singkawang (STKIP Singkawang),
Jl. STKIP-Naram, Singkawang Utara, Singkawang 79251, Indonesia

E-mail: intankusumawati10@gmail.com

Abstract. This study utilized soil as a raw material for making resistor. The type of soil use is latosol soil, organosol soil, and clay. The method used is an experiment. These soils are very densely compressed into a tube with a surface area (A) = 6.3 x 10^{-4} m^2. Each of the tube have a length (l) = 1.5 x 10^{-2} m and diameter (d) = 1 x 10^{-2} m. The results of resistance value can be observed through measurement using a multi-meter on the artificial resistor. Found the resistance value from latosol soil, organosol soil, and clay in a row is at 66.3 kΩ; 58.6 kΩ; 47.1 kΩ. The results are used to determine the value of resistivity through a formula. The results showed that latosol soil, organosol soil, and clay respectively have a value of resistivity of 2.78 kΩm; 2.46 kΩm; 1.98 kΩm. This study shows that soil can be used as an alternative to resistance on the resistor material. These results also indicate that resistivity value of latosol soil higher than organosol soil and clay.

1. Introduction

The soil has resistivity and conductivity values [1-6], while resistor containing carbon as constituent components also have two values [7]. This equation reveals that the soil can be used as an alternative material on resistor. Resistivity is an important parameter to identify characteristic of soil ant its content [8]. Soil resistance values can be measured [9-11] through a variety of experiments. It was found that the soil resistance values vary depending on the type or texture of the soil [9,12]. The soil used in this study is latosol soil, organosol soil, and clay. This soil selected as the study sample because these soils are found in Kalimantan, Indonesia. It is expalined [13] Latosol soil is formed from the weathering of sedimentary and metamorphic rocks, while organosol soil formed from the weathering of organic matter such as plants, peat and swamp. There are in an area that has a wet climate and high rainfall. As for the clay is the type of soil consist of a mixture of aluminum and silicate which has a diameter of not more than four micrometers. Clay formed from the weathering of silica process conducted by carbonic acid and some of them produced from geothermal activity. Latosol soil, organosol soil, and clay in this study are packaged in a specific container to determine the resistivity of the material. Thus, if the resistivity is known, can facilitate and be a consideration for further research to produce a resistor made from latosol soil, organosol soil, and clay having regard to the standard specification resistance material on resistor to safety when assembled with other electrical components. Resistors are used to limit or impede the flow of electricity through a series resistor [14-15]. This is made clear by [16] the resistor in a circuit has a function as a flow divider, voltage divider, lowering the voltage, overcurrent safety, inhibiting the flow of electricity, and so on (depending on the design of components of different types of resistor). Resistors have diverse types, one of which is resistors carbon film. Resistor carbon film made of carbon materials that consist of a tubular resistive element with a wire or metal lids on both ends [17]. In this soils are an option to be change component substitution on resistor.

2. Numerical Methods

This research method using an experimental method, while the method of measurement in this study using a length measuring tools (ruler), and measuring barriers (multi-meter). Resistivity soil of value determined by calculations using the formula $R = \rho\frac{l}{A}$. The ruler is used to set the control variables, namely the size of the radius and height of the pipette used. Multi meter is used to measure the resistance value which becomes the dependent variable. The soil used as independent variables in this study.

In this study, begins with soil to compressed slowly using blunt spikes and rubber mallet into a tube. The tube is made from a pipette with both ends covered with aluminum foil adhesive. Length (l) or high (t) pipette of 1.5 x 10^{-2} m and a diameter of 1 x 10^{-2} m, so that the surface area (A) of the tube amounted to 6.3 x 10^{-4} m^2. The calculation is as follows:

$$\begin{aligned} A &= 2\,\pi\, r\,(r + t) \qquad (1) \\ &= 2\left(\frac{22}{7}\right)(0.5 \times 10^{-2})\,[(0.5 \times 10^{-2}) + (1.5 \times 10^{-2})] \\ &= 6.285 \times 10^{-4}\ m^2 \\ &\approx 6.3 \times 10^{-4}\ m^2 \end{aligned}$$

The last, giving the wire on each side of the tube. Wire each side should not be a fused, and the aluminum foil should not exceed big tear wire that is attached to it. Then resistance has measured using a multi meter and a calculation of the resistivity soil. To determine the value of resistivity, it is necessary to first know the value of resistance. The relationship between resistance (R), the cross-sectional area (A), length (l), and resistivity of materials (ρ) shown in the equation:

$$R = \rho\frac{l}{A} \qquad (2)$$

The equation for calculating the value of the resistivity material is as follows:

$$\rho = \frac{RA}{l} \qquad (3)$$

As for the procedure scheme of making resistor from soil is as follows.

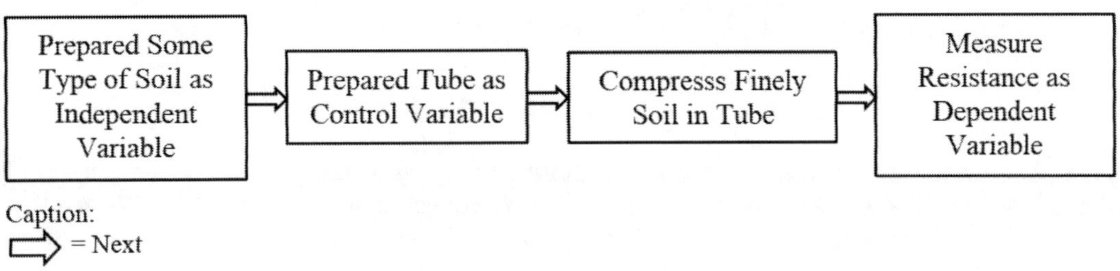

Figure 1. Procedure scheme of making resistor from soil materials

As for the schematic diagram illustrating the measurements resistance and resistivity shown is as follows. The same step to some different soil type.

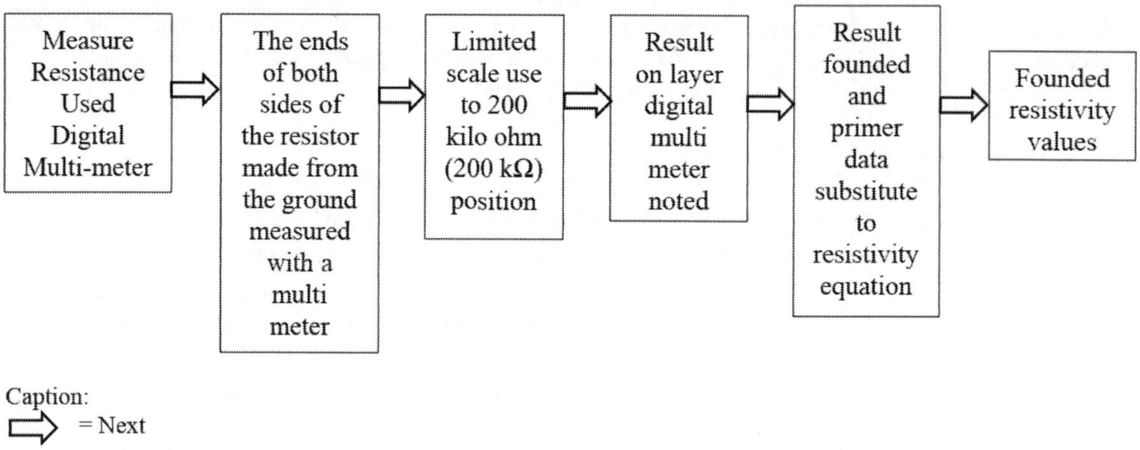

Figure 2. Procedure scheme of measurement resistor from soil materials

3. Results and Discussion

This study aims to determine the resistivity value latosol soil, organosol soil, and clay as alternative substance resistance material on resistor. It is expected to make a significant contribution in the world of technology through the utilization of soil. Components of the resistor in the form of resistance material which has been put into focus as a core material in the manufacture of carbon film resistor can be changed. In this research material soil must be packed solid very tightly into the container. The density of the material is very important, because the resistance value can only be read when the material is really tight. The amount of container or surface area also determines the value of resistivity material. The container used in this study is a tubular pipette with a height (t) = 1.5×10^{-2} m, and a diameter of $1\text{x}10^{-2}$ m. By the equation: $A = 2\pi r(r + t)$, it was found that the surface area was $6.3\text{x}10^{-4}$ m^2. The surface area of a cylindrical tube made of materials similar to latosol soil, organosol soil, and clay intended to do the comparison value of resistance and resistivity between the three soil materials. It value constraints can only be read when the player scale digital multi meter showed limit the scale of 200 kilo ohm (200 kΩ), so that the values read on-screen digital multi meter also has the same unit that kilo ohm (kΩ or 10^3 Ω). This shows that the resistor is made using these three ingredients can only be used to produce a resistor with a large resistance value. The value of constraints for latosol soil, organosol soil, and clay, shown in Table 1 as follows.

Table 1. Soil resistance value

Type Material	Resistance Value (Ω)
Latosol Soil	66.3×10^3
Organosol Soil	58.6×10^3
Clay	$47,1 \times 10^3$

It was found that the value of resistance can determined by using a digital multi meter. The third ingredient is soil can be used as an alternative substitute resistance material in resistor. It found that latosol soil at higher than organosol soil and clay. For simple illustration, the chart resistance values organosol soil, latosol soil, and clay in this study are shown in Figure 2 as follows.

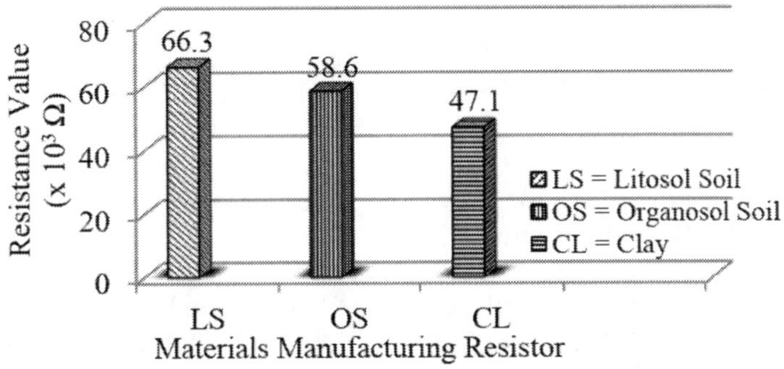

Figure 2. Graphs the resistance value of latosol soil, organosol soil, and clay

After being found by measuring the resistance value, the value of resistivity material searchable through the calculation of the data obtained. The calculation is performed using equation (3). The resistivity value of latosol soil, organosol soil, and clay, shown in Table 2 below.

Table 2. Resistivity value of soil

Type Soil	Resistivity (Ωm)
Latosol Soil	2.78×10^3
Organosol Soil	2.46×10^3
Clay	1.98×10^3

Calculation of resistivity material on latosol soil, organosil oil, and clay in this study showed that the latosol soil has a higher resistivity than organosol soil and clay. This is because latosol soil is formed from the weathering of sedimentary and metamorphic rocks [13], it is contains of iron and aluminum. This metal material has high resistivity values [17]. Resistivity comparison latosol soil, organosol soil, and clay, shown in Figure 3 as follow.

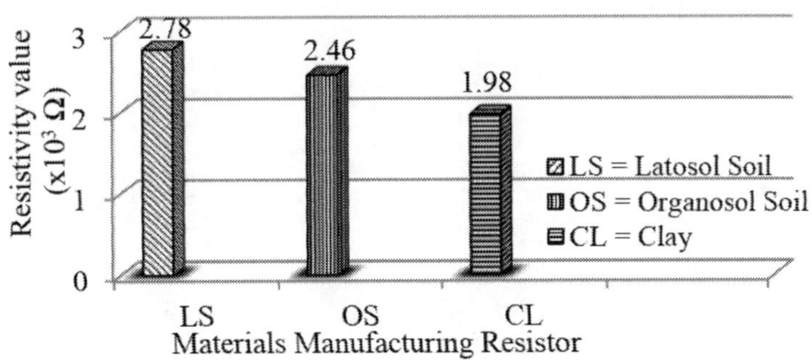

Figure 3. Graph resistivity value of latosol soil, organosol soil, and clay

In conclusion, soil in this study can be used as an alternative substitute resistance material on resistor. Found resistance values with measurements on latosol soil at 66.3 kΩ, organosol soil at 58.6 kΩ, and clay at 47.1 kΩ. Cross-sectional area of each of the materials used at 6.3×10^{-4} m^2 and a length of 1.5×10^{-2} m. Based on calculations using the equations $\rho = \frac{RA}{l}$ can be identified that the resistivity for latosol soil by 2.78 kΩm, organosol soil 2.46 kΩm, and clay 1.98 kΩm. Resistance value and resistivity materials have a proportional relationship. The greater the value of its resistance, the greater the value of resistivity, otherwise the smaller the resistance value is smaller the resistivity.

4. References

[1] Bozkurt A, Cengiz K, and Hasan E 2009 *Journal of Physical Sciences* vol 4 (12) pp 784-795
[2] S Seladji, P Cosenza, A Tabbagh, J Ranger and G Richard 2010 .*European Journal of Soil Science* vol. 61 Issue 6 pages 1043–1055 DOI: 10.1111/j.1365-2389.2010.01309.x.
[3] Z Chik and T. Islam 2011 .*International Journal of Electrochemical Science*) pages 6733-6740.
[4] Bai W, Lingwei K, and Aiguo G 2013. *Journal of Rock Mechanics and Geotechnical Engineering* Vol 5 pp 406–411 DOI:10.1016/j.jrmge.2013.07.003
[5] Liu S Y, Yan J D, Lei C, Zhi B L, and Fei J 2009.*Proceedings of The International Symposium on Geoenvironmental Engineering in Hangzhou China* September 8–10. Part 2 PP 259-264 DOI 10.1007/978-3-642-04460-1_18 Print ISBN 978-3-642-04459-5 Online ISBN 978-3-642-04460-1 Publisher Springer Berlin Heidelberg
[6] G H Cai, Y J Du, S Y Liu, and D N Singh 2015 .*Canadian Geotechnical Journal*) p 52(11): 1699-1713 DOI: 10.1139/cgj-2015-0053
[7] Irianto E A dan Endah R 2014.*Prosiding Pertemuan Ilmiah XXVIII HFI Janteng dan DIY*) pp 322-325. ISSN: 0853-0823
[8] Roy E 1984 *Geotechnical Engineering Investigation Manual* (New York: Mc Graw Hill)
[9] Ishaq Z M 2008 *Studi Resistivitas dan Gravitasi Untuk Investigasi Akuifer Air Bawah Tanah di Kampus UI Depok* Tesis (Jakarta : FMIPA-UI)
[10] Zulaikah S, Era B P, Nugroho A P, Yoyok A L, Ninik M, Rizka A Hdan Husni CK 2015 *Spektra: Jurnal Fisika dan Aplikasinya* vol 16
[11] Loke M H 1999 *Electrical Imaging Surveys for Environmental and Engineering Studies* (Malaysia : Penang)
[12] Yulia 2015 *18 Jenis Jenis Tanah di Indonesia: Manfaat, Persebaran, Gambarnya* [http://ilmugeografi.com/ilmu-bumi/tanah/jenis-jenis-tanah] accessed July 3 2016
[13] Aiyub S 2013 .*Jurnal Energi Elektrik V*ol 2.
[14] Fidianti N 2014 *Resistor Papers*
[15] Susanto B 2015 *Definitions Resistor and Complete Functions* Available at: [http://www.seputarpengetahuan.com/2015/02/pengertian-resistor-dan-fungsinya-lengkap.html] accessed February 8 2016
[16] Hariyanto D 2009 *Telkomnika* vol 7 pp 13 – 22
[17] Sugeng 2015 *Hambatan Kawat Logam*

Journal of Physics: Conference Series **776** (2016) 012096 doi:10.1088/1742-6596/776/1/012096

The design of an embedded system for controlling humidity and temperature room

Dwi Teguh R[1], Didik Eko S[1], Pringgo D Laksono[2], Anif Jamaluddin[1]*

[1]Department of Physic Education, Teaching and Training Faculty, Sebelas Maret
[2]University, Surakarta, Indonesia
Industrial Engineering Department, Engineering Faculty, Sebelas Maret University, Surakarta, Indonesia

*elhanif@uns.ac.id

Abstract. The aim of the system is to design an embedded system for maintenance confortable room. The confortable room was design by controlling temperature (on range 18 – 34 ^{0}C) and humidity (on range 40% - 70%.) of room condition. Temperature and humidity of room were maintained using four variable such as lamp for warm, water pump for distributing water vapour, a fan for air circullation and an exhaust-fan for air cleaner. The system was constucted both hardware (humidity sensor, microcontroller, pump, lamp, fan) and software (arduino IDE). The result shows that the system was perfectly performed to control room condition.

1. Introduction

Temperature and humidity relativity of rooms are an essential factor in an industrial process, farming, research activity, meeting room, etc that must be controlled [1] . There are two factors to keep pleasure room by controlling temperature [2] and humidity [3]. The integration of ventilation, heating and cooling system is needed to maintenance temperature and humidity in room as the result inhabitants feel pleasure [4].

Temperature and humidity controller system that implementation on fungus cultivation almost use two key component for input variable include water pumping (for controlling humidity) and lamp (for controlling temperature) [5]. It is common used to maintenance temperature and humidity. However, this system has problems to distribute temperature or humidity room.

This paper present the control system using 4 degree variable included water pumping, lamp, fan and exhaust fan for distributing temperature and humidity. This system proposed to maintains comfortable room on range 18-34 C for temperature and 40%-70% for humidity. Moreover, designing of temperature and humidity control system consider the competitive aspects of component prices, easily available, and easy to use. The hardware has been used included ATmega328 (Arduino UNO) as the main control and sensor DHT11 as temperature and humidity sensors. DHT11 has the advantage that the calibration has been equipped with digital signals, serial communication interface using one data point, has a temperature measurement range of 0 ° C to 50 ° C with errata + -2oC and humidity measurement range from 20% to 95% with error about + -5% [6].

2. Method

The embedded system for controlling temperature and humidity was design include hardware and software. The room design and hardware for controlling temperature and humidity is showed on figure 1.

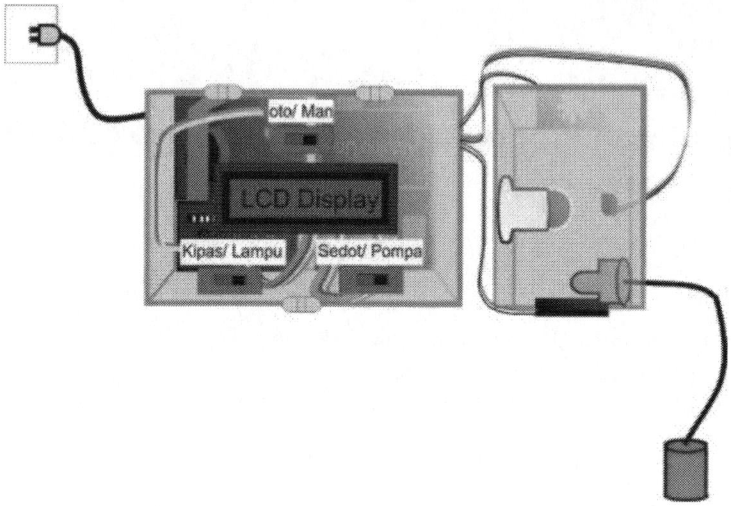

Figure 1. Minature of design of experiment

Other paragraphs are indented (BodytextIndented style).

2.1. Software

2.2. Hardware

DHT11 Sensor, .

2.2.1. DHT11. DHT11 sensor temperature and humidity sensor is used for this system

2.2.2. Microcontroller

2.2.3. Fan/Exhoust Fan

2.2.4. Lamp

2.2.5. Spray/Pump for Humidifier

2.2.6. Relay

3. Result

The system was design with two model control include both automatic system and manual system. The automatic control system has been done for switching lamp,

The system was callibrated by comparationg actual measurement of sensor with data acquisition of system. table 1 showed the comparasion of temperature. The humidity data was showed on table 2.

Average ± Standar Deviasi (°C)	Actual (°C)	Error (%)
18,8 ± 0,40	18	2,13%

21,8 ± 0,75	22	3,43%
28,2 ± 0,75	28	2,65%
32,8 ± 0,40	32	1,22%
33,8 ± 1,33	34	3,92%

Average ± Standar Deviasi (%)	**Aktual (%)**	**Error (%)**
40,6 ± 1,02	40	2,51%
43,4 ± 0,80	43	1,84%
50,2 ± 1,33	50	2,64%
60,2 ± 1,33	60	2,20%
70,4 ± 1,62	70	2,31%

Controlled Test

No	Temperature	Humidity	Lamp	Fan	Pump	Exhoust Fan	compatibility
1	>30	>65	OFF	ON	OFF	ON	Yes
2	>30	45≤H≤65	OFF	ON	OFF	OFF	Yes
3	>30	<45	OFF	ON	ON	OFF	Yes
4	20≤T≤30	>65	OFF	OFF	OFF	ON	Yes
5	20≤T≤30	45≤H≤65	OFF	OFF	OFF	OFF	Yes
6	20≤T≤30	<45	OFF	OFF	ON	OFF	Yes
7	<20	>65	ON	OFF	OFF	ON	Yes
8	<20	45≤H≤65	ON	OFF	OFF	OFF	Yes
9	<20	<45	ON	OFF	ON	OFF	Yes

4. Conclussion

The embedded system was successfully implemented for controlling the humidity and temperature room as confortable zone of temperture (range 18-34 C) and humidity(40%-70%) . The forth degree variable of controle was proposed to design the system include lamp, pump, exhoust fan and fan. There are two mode control (automatic and manual).

References

[1] Ashrae. 1999 Ashrae *Handbook: Heating, Ventilating and Air-conditioning Applications, A: American Society of Heating, refrigerating and Air Conditioning Engineers*, Atlanta.
[2] Chua KJ, Ho JC, and Chou SK 2007 *Energy and Buildings Journals*, **39**, 537–545.
[3] Orosa JA, Baalina A 2008 *Building and Environment Journal*. **43** (12) 2005–2012.
[4] Soyguder S, & Alli H 2009 *Energy and Buildings Journals*, **41**, 814-812.
[5] Laksono PD, Jauhari WA, Iftadi I, Christina AK, Ibnu PBP, Jamaluddin A, Saputro DE, Harjunowibowo D 2015 *Join International Conferences on Electric Vehicular Technology and Industrial, and Chemical Engineering (ICEVT&IMECE)*, Surakarta.
[6] Maidah EN, Putra AE dan Pulungan RMI 2012 *Perancangan Perangkat Keras Pengendali Fuzzy Berbasis Mikrokontroler ATmega32 sebagai Pengendali Suhu dan Kelembaban*. Skripsi Tidak Dipublikasikan, Universitas Gadjah Mada, Yogyakarta.

Slip control design of electric vehicle using indirect Dahlin Adaptive Pid

I. R. Fauzi*, F. Koko, M. R. Kirom
Department of Engineering Physics, School of Electrical Engineering, Telkom University.Jln. Telekomunikasi No 1, 40257, Bandung, Indonesia

E-mail: *1rezafauzii@telkomuniversity.ac.id

Abstract In this paper the problem to be solved is to build a slip control on a wheel that may occur in an electric car wheel. Slip is the difference in vehicle velocity and wheel tangential velocity and to be enlarged when the torque given growing. Slip can be reduced by controlling the torque of the wheel so that the wheel tangential speed does not exceed the vehicle speed. The experiment in this paper is a simulation using MATLAB Simulink and using Adaptive control. The response adaptive PID control more quickly 1.5 s than PID control and can controlled wheel tangential speed close to the vehicle velocity on a dry asphalt, wet asphalt, snow and ice surface sequent at time 2s, 4s, 10s, and 50s. The maximum acceleration of the vehicle (V) on the surface of the dry asphalt, wet asphalt, snow, and ice surface sequent at 8.9 m/s^2, 6.2 m/s^2, 2.75 m/s^2, and 0.34 m/s^2.

1. Introduction

Several control methods to control the slip in the various types of vehicles have developed enough such as Anti-lock Braking System (ABS) and Traction Control System (TCS). ABS and TCS are braking systems on the car in order to avoid locking the wheels when braking the vehicle suddenly. Likewise with slip control Model Following Control (MFC) which not need information of vehicle body velocity or acceleration sensor equipment.

Dejun Yin and Hori [1] conduct research to obtain a method control which is based on the maximum torque which allowed that slip can be limited without consider vehicle velocity. Maximum torque determined to ignore the existence of some motion resistance, such as motion wheel resistance and airflow on the vehicle.

In this paper the problem can be solved is to build slip control by reducing the value of the slip that may occur on a wheel. By building a system that has the ability to organize themselves according to the environmental conditions or adaptability. Then determine the dynamic response of electric cars with testing in various types of tracks. This experiment is simulated by using MATLAB Simulink.

2. Commercial Electric Vehicle (COMS)

This experiment simulated by using MATLAB Simulink from Dejun and Yin's electric car research [1]. The electric car is Commercial Electric Vehicle (COMS) created by TOYOTA AUTO BODY Co. Ltd., The electric car have modified and suitable for research needs. Each drive wheel is equipped by Interior Permanent Magnet Synchronous Motor (IPMSM) so that can be controlled freely.

Figure 1. Dejun Yin & Hori's electric car research [1]

Table 1. COMS Characteristic

Symbol	Definition
M	Vehicle mass (kg) (360 kg)
J_w	Wheel Inertia (kg m/s^2) (0.5 kgm^2)
r	Wheel radius (m) (0.2 m)
T	Driving torque (Nm) (100 Nm)
F_{dr}	Driving resistance (N) (230 N)
F	Driving force (N)
λ	Slip ratio
v	Vehicle velocity (m/s)
v_w	Wheel tangential velocity (m/s)
ω	Wheel rotation (rad/s)
N	Vehicle Weight (N)
μ	Friction coefficient

Characteristic of electric car which used for research refer to characteristic of Dejun Yin & Hori's electric car research [1]. See Table 1.

3. Control Design

3.1 Vehicle and Wheel Dynamic Model

Defining an equation from the motion of one wheel vehicle can be derived from Newton's second law (See Figure 2). Figure 2. show the physical quantities contained in the longitudinal motion of electric car.

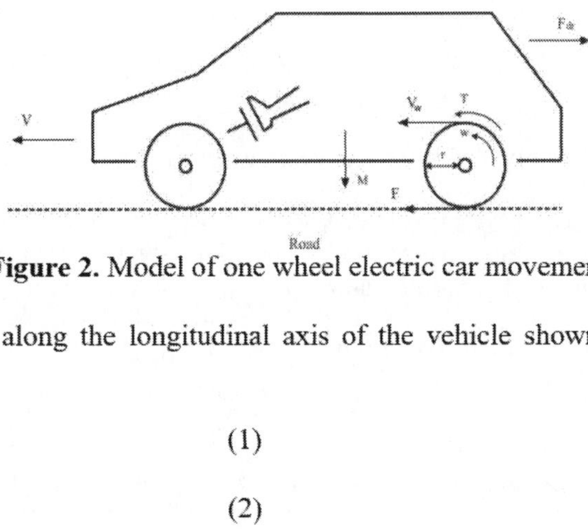

Figure 2. Model of one wheel electric car movement

The equation of motion along the longitudinal axis of the vehicle shown in equation (1) and (2) below:

$$M\dot{v} = F - F_{dr} \quad (1)$$

$$\dot{v} = \frac{F - F_{dr}}{M} \quad (2)$$

Then the linear relationship tangential velocity (v_w) with wheel rotation (ω) on a wheel model shown in equation (3) below:

$$\dot{v_w} = r\dot{\omega} \quad (3)$$

For one wheel model, the physical quantities contained in the longitudinal motion shown in Figure 3.

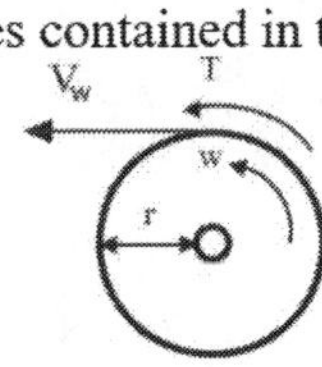

Figure 3. One wheel model

However, when the wheel touching the surface of the track with the input torque, the friction force will occur so that the vehicle body can be drove. See equation (4), (5), and (6).

$$J_W\dot{\omega} = T - Fr \quad (4)$$

$$F = \mu N \quad (5)$$

$$\dot{v_W} = \frac{r(T-rF)}{J_W} \quad (6)$$

Slip ratio (λ) is percentage of the wheel tangential velocity (v_W) with the vehicle velocity (v). Calculation of slip ratio (λ) is shown in equation (7) below:

$$\lambda = \frac{v_W - v}{v_W} \quad (7)$$

In this paper, when the vehicle slip, slip ratio can be detected friction coefficient (μ) and described into a vehicle model (See figure 4).

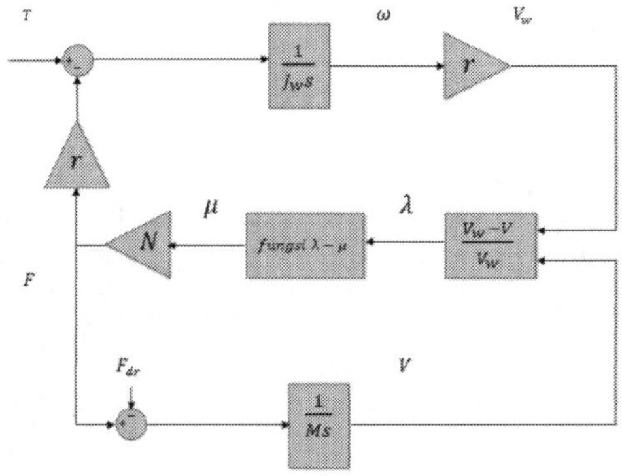

Figure 4. One wheel vehicle dynamic model [1]

The method used to detect the vehicle slip ratio becomes the value of the friction coeficient is using Magic Formula or pacejka formula discovered by Hans B. Pacejka based on experimental data [13]. The equation of *Magic formula* shown in equation (8).

$$\mu = D\ Sin\ [C\ arctan\ \{B\lambda - E(B\lambda - arctan(B\lambda))\}] \quad (8)$$

wich:

B, C, D, E = Coefficient of the tires movement to the tracks (See Table 2).

Table 2. *Magic formula* characteristic in various types of tracks [14]

Coefficient	Name	Dry asphalt	Wet asphalt	snow	ice
B	Stiffness	10	12	5	4
C	Shape	1.9	2.3	2	2
D	Peak	1	0.82	0.3	0.1
E	Curvature	0.97	1	1	1

3.2 Slip Control Design

Control design to be proposed in this paper are PID control and Adaptive PID control and then comparing the outputs. PID diagram block shown in Figure 5, then Adaptive PID shown in Figure 6.

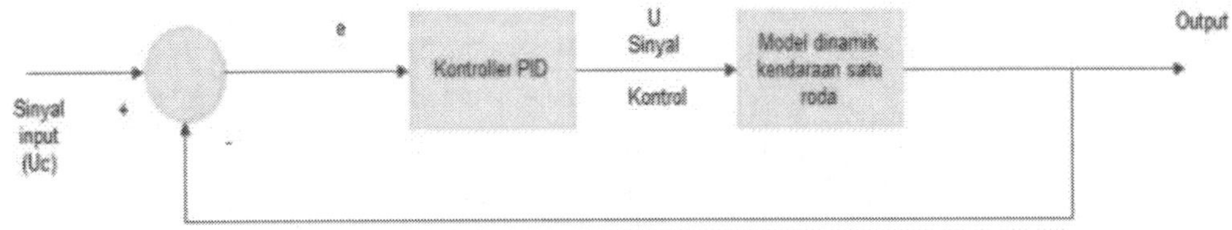

Figure 5. PID control diagram block

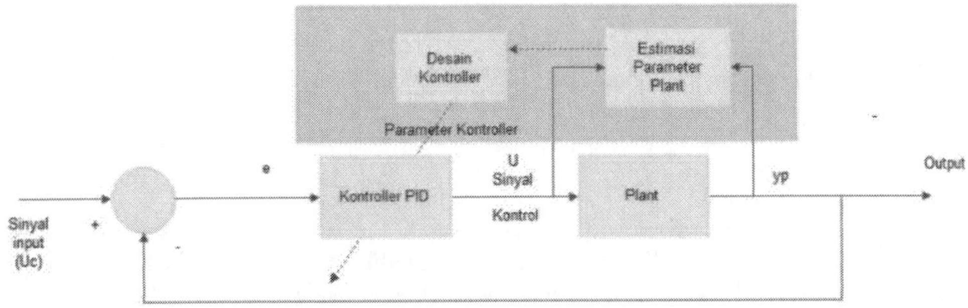

Figure 6. Adaptive PID control diagram block

In PID control design, determining the Kp, Ki, Kd parameters are using Ziegler-Nichols metode first type. Parameters obtained Kp = 198,8, Ki = 82,3, Kd = 0,3. The difference of Adaptive PID control is performed with the addition of Adaptive block from the PID control design. Adaptive block consists of block Parameter Estimation Plant and block Design Controller that serve to adjust PID controller block to generating a new K_p parameter. Generating a new Kp parameter can use the Dahlin PID Controller [12] in equation (9).

$$K_P = -\frac{(a_1+2a_2)Q}{b_1} \qquad (9)$$

Variable Q in equation (9) is defined by equation (10).

$$Q = 1 - e^{-\frac{T_0}{B}} \quad (10)$$

where B is known as the adjustment factor which characterizes the dominant time constant of the transfer function according to changes made to the process output of a closed control loop. The smaller the value of B, the faster the response of the closed control loop [12]. Thereafter, T_0 use the settling time of the output process in a closed loop before using this type of control. In this paper, value of B is 400 and value of T_0 is 17. Parameters of a_1, a_2, and b_1 can be searched with ARMAX block in MATLAB simulink and will automatically generate parameters that adjust from environmental conditions. So we get the parameters:

a_1 = - 0,9874; a_2 = 1; b_1 = 0,006;

The next step makes MATLAB Simulink diagram block with PID control (Figure 7) and adaptive PID control (Figure 8). The tests conducted in various types of tracks. The tracks are on the surface of dry asphalt, wet asphalt, snow, and ice.

Journal of Physics: Conference Series **776** (2016) 012097 doi:10.1088/1742-6596/776/1/012097

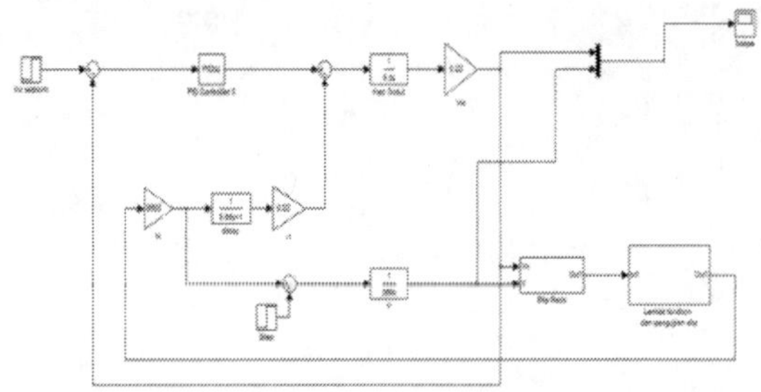

Figure 7. One wheel vehicle dynamic model with PID control diagram block in Simulink

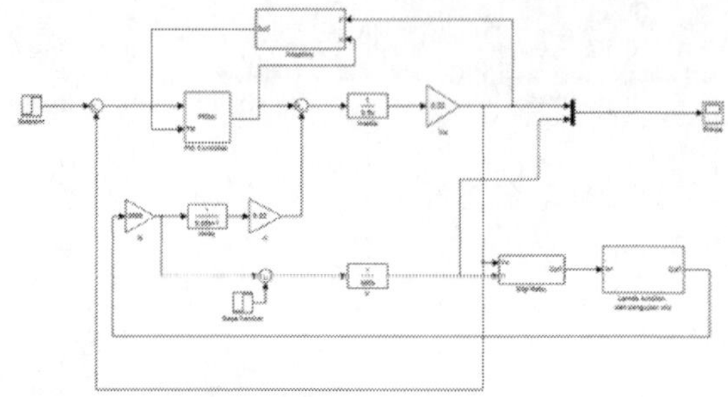

Figure 8. One wheel vehicle dynamic model with Adaptive PID control diagram block in Simulink

4. Test Results and Analysis

The next step is testing the dynamic model of the vehicle by reading the signal V (vehicle velocity) and the signal Vw (wheel tangential velocity). Setpoint using Vw by using step signal. Setpoint value in this paper is used 25 m/s by testing during 100 s. PID control test results shown in Figure 9 and Adaptive PID control testing in Figure 10. Figure 9 shows the dynamic response of vehicle speed PID control with constant input (step) experienced a constant velocity (settling time) on the surface of the dry asphalt, wet asphalt, snow and ice in a row at a time 10 s, 12 s, 20 s, and 80 s. The maximum acceleration of the vehicle (V) on the surface of the dry asphalt, wet asphalt, snow, and ice is 8.9 m/s^2, 6.2 m/s^2, 2.75 m/s^2, and 0.34 m/s^2.

Wheel tangential velocity (Vw) adaptive PID ; ———	Vehicle velocity (V) PID ; ———
Vehicle velocity (V) adaptive PID ; ———	*Setpoint* ; ....
Wheel tangential velocity (Vw) PID ; ———	

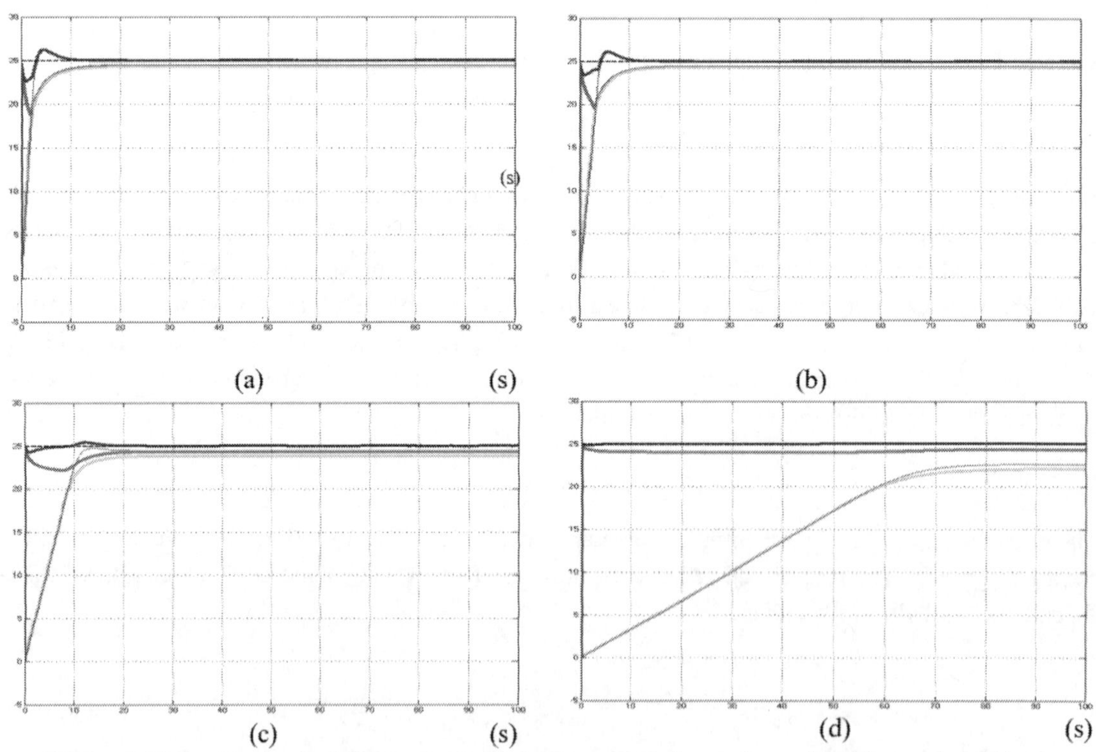

Figure 9. PID and adaptive PID testing with input step on the surface of (a) dry asphalt (b) wet asphalt (c) snow (d) ice.

On the dry asphalt surface (Figure 9(a)) at time 2s, the wheel tangential velocity (Vw) adaptive PID have experienced reduction of 6 m/s from setpoint and towards vehicle velocity value at 19 m/s. Then, on the wet asphalt surface (Figure 9(b)) at time 4 s, the wheel tangential velocity (Vw) adaptive PID have experienced reduction of 5 m/s from setpoint and towards vehicle velocity value at 20 m/s. This proves have adaptation process from adaptive PID control. Dynamic response on adaptive PID control and has no overshoot. On dry asphalt surface have experienced settling time at time 10 s and on wet asphalt surface have experienced settling time at time 12 s.

On the snow surface (Figure 9(c)), the wheel tangential velocity (Vw) adaptive PID have experienced reduction of 2.5 m/s from setpoint and towards vehicle velocity. Then, reduction of the wheel tangential velocity (Vw) adaptive PID on the ice surface (Figure 9 (d)) quite small around 1 m/s from setpoint, because a large setpoint (25 m/s) for slippery surfaces. Settling time from vehicle velocity (V) on the snow and ice surface is slower than dry

Asphalt surface (settling time for snow = 20 s and settling time for ice = 80 s). This proves in adaptive PID control, settling time value of vehicle velocity (V) inversely proportional to the wheel tangential velocity reduction. The acceleration of the vehicle on ice surface have smaller acceleration than on dry asphalt surface. Likewise with vehicle velocity (V) does not reach the full setpoint. This proves have a slip, because wheel tangential velocity not fully modified into vehicle velocity (V). Another reason is the frictional force on the ice surface is very small nearly zero. Measurement of slip ratio λ (t) performed during 20 s in four surfaces that are dry asphalt, wet asphalt, snow, and ice. Slip ratio (λ) range shows the number 0 to 1. Slip ratio (λ) = 0, it shows the wheel tangential velocity (Vw) is equal to vehicle velocity (V). While slip ratio (λ) = 1, it shows vehicle velocity = 0 m/s or vehicle body in stationary condition. In figure 10. The value of slip ratio (λ) with input step at time 0 s shows the slip ratio value (λ) = 1. This signify the wheel tangential velocity (Vw) that directly responds to reach setpoint value with input step. However, at the next time interval, slip ratio (λ) value close to the value 0.

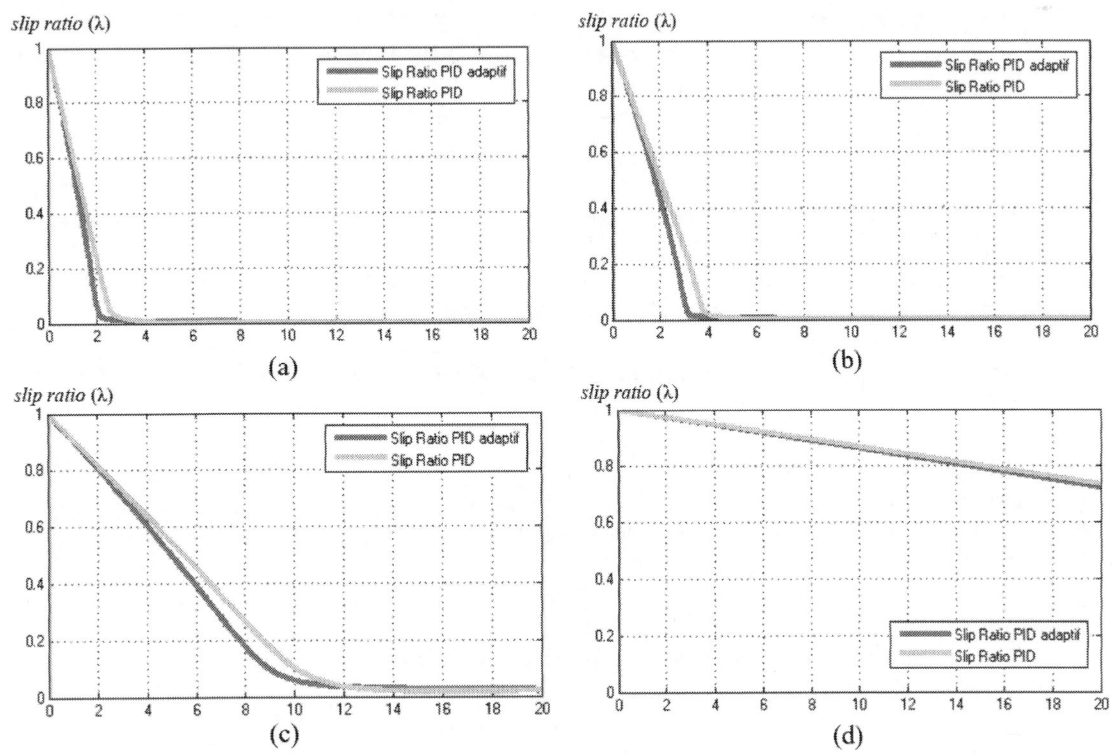

Figure 10. Responds of slip ratio with PID and adaptive PID control with input step on the surface of (a) dry asphalt (b) wet asphalt (c) snow, and (d) ice

Testing on the dry asphalt surface (figure 10(a)) at time 2 s, slip ratio (λ) in PID control shows number at 0.2. But, slip ratio (λ) in adaptive PID control shows number at 0.01. Thus, there is a reduction (reduction) slip ratio (λ) on dry asphalt surface by 0.19. On the other of tracks, there is a reduction (reduction) slip ratio (λ) on wet asphalt, snow, and ice surface sequent at number 0.1, 0.05, and 0.005. On the wet surface (figure 10(b)), respond of slip ratio (λ) value close to the value 0 at time 4 s slower than respond of slip ratio (λ) on dry asphalt surface. On the snow surface (figure 10(c)), respond of slip ratio (λ) value close to the value 0 at time 10 s and on ice surface (figure 10(d)) at 50 s.

The respond of the adaptive PID control close to the value 0 more quickly than PID control. On the dry asphalt surface adaptive PID 0.5 s more quickly than PID control. On the wet asphalt surface 1 s more quickly and on the snow surface 4 s. So the role of adaptation (adaptive PID) works on the slip control.

5. Conclusion

From the results of this simulation, it could be conclude as follows:

1. Dynamic response of vehicle speed control PID and Adaptive PID with step input of 25 m / s experiencing constant velocity (settling time) on the dry asphalt, wet asphalt, snow, and ice surface sequent at time 10 s, 12 s, 20 s, and 80 s.
2. The simulation results prove the PID control vehicle speed maximum acceleration of vehicles on dry asphalt surface, wet asphalt, snow, and ice sequent at 8,9 m/s^2, 6,2 m/s^2, 2.75 m/s^2, and 0.34 m/s^2.
3. Wheel tangential velocity (Vw) on the adaptive PID control is reduced to the velocity of setpoint and the value of the vehicle velocity (V). This proves have adaptation process from adaptive PID control and has no overshoot. Velocity reduction from setpoint and towards vehicle velocity value (V) on the dry asphalt, wet asphalt, snow, and ice surface sequent at time 2 s, 4 s, 10, and 50 s.
4. The role of adaptation (adaptive PID) works on the slip control with the reduced of slip ratio (λ) value. Reduction of slip ratio on dry asphalt surface at 0.19. Then on the wet asphalt, snow, and ice surface sequent at 0.1, 0.05, and 0.005

6. Reference

[1] Yin D and Yoichi H 2009 *A Novel Traction Control without Chassis Velocity for Electric Vehicles.* Norway, EVS24, Online Jurnal.

[2] Hori Y dkk *Traction Control of Electric Vehicle.* Tokyo, University of Tokyo, Online Jurnal.

[3] Shang C and. Peng H 1996 *Road Friction Coefficient Estimation For Vehicle Path Prediction,* **25** Suppl.

[4] Pratikto. dkk 2010 *Pengembangan Sistem Kontrol Traksi Mobil Elektrik Berbasis Rekontruksi Keadaan Kecepatan Model Roda,* Bandung, Institute Teknologi Bandung, Online Jurnal.

[5] *Indonesia's Oil Production and Consumption 1992-2005*, [Internet, WWW]. Address http://www.eia.gov

[6] Andrada P dkk. *Power Losses in Outside-Spin Brushless D.C. Motors*, Spanyol, Universitat Politècnica de Catalunya, Online Jurnal.

[7] Saifizi M. dkk. 2013, *Comparison Of ARX and ARMAX Model For Thermoelectric Refrigerator,* Malaysia, ICMER2013, Online Jurnal.

[8] Burhaumudin and Safwan M dkk 2012 *Modeling and Validation of Magic Formula Tire Road* Malaysia ICAMME Online Jurnal.

[9] Brown W 2002 *Brushless DC Motor Control Made Easy*, Microchip Technology Inc.

[10] Irsyadi, Fakih. dkk. 2011, *Perancangan dan Implementasi Manajemen Kecepatan Mobil Listrik Berbasis Energi dan Target Jarak Tempuh* , Bandung, Institut Teknologi Telkom, Tugas Akhir.

[11] Handri, Bima. dkk. 2011, *Perancangan dan Implementasi Sistem Kontrol Kecepatan Pada Mobil Listrik dengan Penggerak Motor Brushless*, Bandung, Institut Teknologi Telkom, Tugas Akhir.

[12] Bobal, V. dkk. 2005, *Digital Self-tuning Controllers,* Czech Republic, University of Pardubice, Penerbit Springer.

[13] Erdogan, Gurkan. 2009, *Tire Modeling Lateral and Longtudinal Tire Forces.*

[14] *Pacejka '94 Parameter Explained – A Comprehensive Guide*, [Internet, WWW]. Address http://www.edy.es/d

Journal of Physics: Conference Series **776** (2016) 012098 doi:10.1088/1742-6596/776/1/012098

Thermal conductivity measurement using thermoelectric module

T. A. Ajiwiguna[1], S. Y. Kim[2]
[1]Engineering Physics Department, School of Electrical Engineering, Telkom University, Bandung 40252, Indonesia
[2]Center for Urban Energy Systems, Korea Institute of Science and Technology, Seoul 136-79, South Korea

E-mail: triayodha@telkomuniversity.ac.id

Abstract. In this study, thermoelectric module is used as a heater for thermal conductivity measurement of solid materials. Principle of temperature gradient is adopted in this measurement. Stainless steel is utilized as a reference material. Each temperatures are measured by t-type thermocouple and the thermal conductivity is then calculated. The result shows that the thermal conductivity of tested materials is 0.303 W/mK with maximum uncertainty of 3.79 %. To assure the result of this measurement, the thermal conductivity of tested material is also measured by laser flash method. The difference result of both measurement is below 5%..

1. Introduction

Thermal conductivity which is defined as a measure of the ability of substance to conduct heat is one of the important thermal properties. In engineering field, the information of this property is needed to select material so that it meets the requirements of applications [1]. Moreover, thermal conductivity is enhanced by using nano-material technology [2-3]. This causes higher performance at the applied technology. Several methods and improvements have been introduced to measure thermal conductivity of materials [4-10]. Two of them are temperature gradient and laser flash method. In the first method, the substance is heated by specific amount of heat from heating element so that temperature gradient occurs and the thermal conductivity can be calculated [4]. While in the second method, the substance is shot by laser radiation which causes the increasing of temperature. The time dependent of temperature is observed and thermal conductivity is then determined. [5].

Thermoelectric module is a device that can act as a heat pump. Figure 1 shows the sample of the module. When DC electric current is supplied to the module, the temperature difference between the two sides occurs. The temperature of one side is lower than ambient so it absorbs heat [11]. Contrarily, the temperature of another side is higher than ambient so it emits heat. This device has many advantages which is the dimension is relatively small, no moving parts, long life, easy to control, and use direct current electricity. This module can be used as heater element when the hot side is utilized.

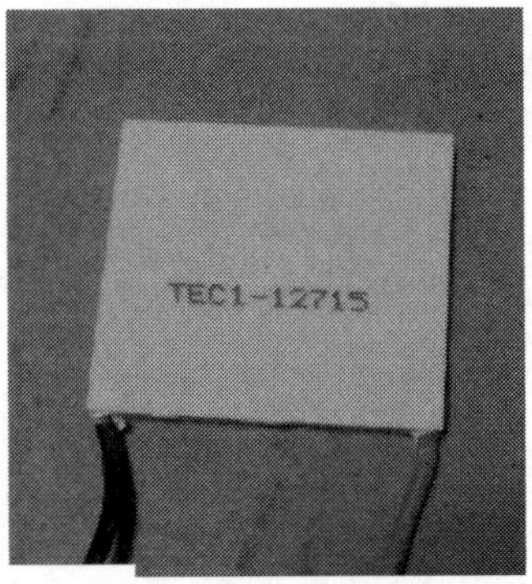

Figure 1. Thermoelectric module

In this study, thermoelectric module is experimentally used as a heater for thermal conductivity measurement by adopting temperature gradient method. The aim of the study is to build measurement system with reasonable accuracy with less than 5% error. The calibrated laser flash apparatus is used for the standard so that the result of experimental thermoelectric module is compared.

2. Experimental Setup

Figure 2 (a) shows the components arrangement. In this measurement, thermoelectric module with 4 x 4 cm surface area was used as a heater. Stainless steel block with the same surface area and 1 cm thickness was utilized as a reference material since its thermal conductivity was already known. The tested sample material was a 2 mm thickness of plastic material and its surface area was conditioned so that the same with the reference material. Heat sink and fan were used to enhance heat dissipation. Insulation materials were used to avoid heat loss from the vertical surface therefore heat loss from vertical surfaces were negligible.

Figure 2 (b) shows the scheme of heat flows. Heat transfer by conduction occur from the hot side thermoelectric module to the heat sink-fan. When the thermocelectric module was being turned on, temperature gradient occurred and temperatures at point T_1 to T_3 were measured by T-type thermocouple. Since heat conduction at the reference material was the same with at the tested sample, therefore the thermal conductivity was calculated by:

$$k_s = \frac{k_{ref} L_s (T_1 - T_2)}{L_{ref} (T_2 - T_3)} \quad (1)$$

where k_s, k_{ref}, L_s, L_{ref} were the thermal conductivity of tested sample material, thermal conuctivity of reference material, thickness of tested sample, and thickness of reference material respectively.

The experiment was performed ten times to analyze the uncertainty of measurement at different average temperatures between the two sides of tested sample material. Increasing the current flow was done to get higher average temperature. The obtained data were then plotted to the graph and analyzed. The thermal conductivity of tested sample material was also measured by calibrated laser flash apparatus. The result of this measurement was used as a reference to analyze the accuracy of the experiment.

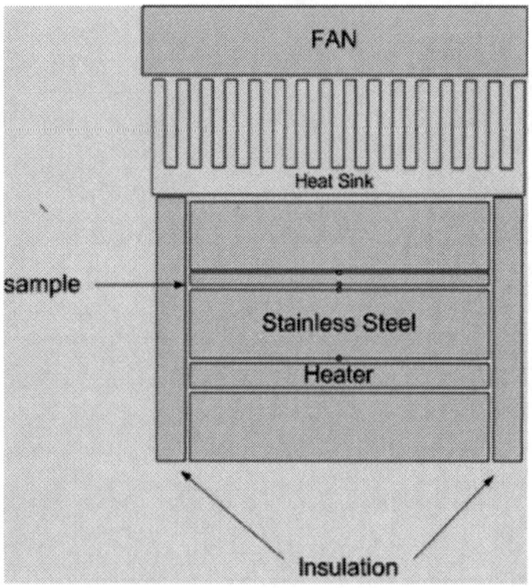

(a)

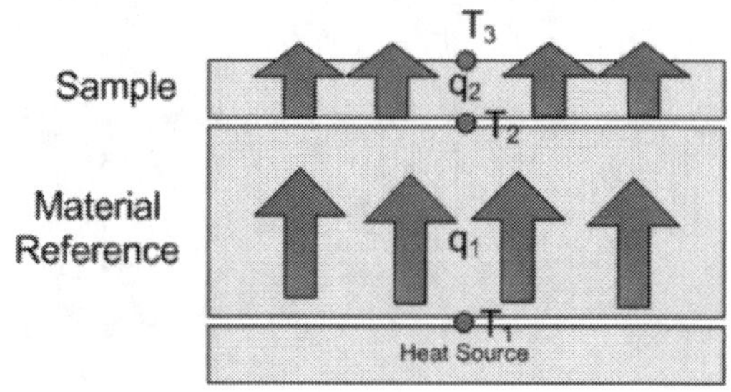

(b)

Figure 2. . (a) Measurement setup, (b) Scheme of heat flows

3. Results and Discussion

The result of measurement using thermoelectric heater is shown in figure 3. The horizontal axis shows the average temperature of tested materials while the vertical axis is the result of thermal conductivity measurement. The result shows the thermal conductivity of tested sample material at 25 °C to 45 tends to decrease linearly if the average temperatures are increased. By using linear regression, the temperature dependence of thermal conductivity is expressed by:

$$k = -0.0027\, T_{avg} + 0.3709 \quad (2)$$

Where k is thermal conductivity of tested sampe and T_{avg} is average temperature. The uncertainty of this measurement is 3.79 % which is obtained from maximum error between experimental result and calculation result by using equation 2. The uncertainty is relatively good since t-type thermocouple for measuring temperature has 0.1 °C uncertainty.

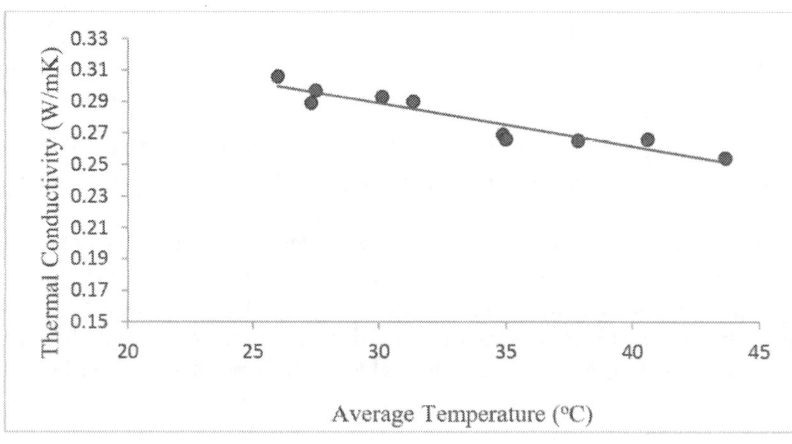

Figure 2. Measurement result using thermoelectric heater

Figure 3.

The thermal conductivity measurement by using laser flash method is shown in table 1. The value of thermal conductivity is obtained 0.295 W/mK at ambient temperature of 25 °C. The precision of this measurement is very high with zero standard deviation based on three time repeated measurement. It also means that the uncertainty is very low. This result is used as a reference since the apparatus has been calibrated.

Table 1.

Table 1. Result of Laser Flash Method

No	T (°C)	k (W/mK)
1	25	0.294
2	25	0.294
3	25	0.294

The result of the thermal conductivity measurement by using thermoelectric heater at that is calculated by equation 2 and obtained 0.303 W/mK. The comparison shows that the difference with laser flash method is 0.0094 W/mK. It is concluded that the measurement result by using thermoelectric heater is reasonable since the error is 3.2%. The experiment result is relatively good since t-type thermocouple has very small uncertainty to measure temperature.

4. Conclusions

Thermal conductivity measurement using thermoelectric module as a heater has been performed in this study. Temperature sensor selections is one of the important step to get accurate result. T-type thermocouple is the best for this measurement system since its dimension is very small and it has very small uncertainty. The result shows that higher temperature causes the lower thermal conductivity of tested sample material. The uncertainty of measurement is 3.79 %. The thermal conductivity of tested sample material at 25 °C is 0.303 W/mK dan 0.294 W/mK by using thermeoelctric module and laser flash method respectively. The error of measurement using thermoelectric is below 5%.

5. References

[1] Yunus A Cengel 2002 Heat Transfer A Practical Approach 2nd Edition, Mc Graw Hill. 40
[2] K Karunamurthy, K Murugumohankumar, S Suresh 2012 Digest Journal of Nanomaterials and Biostructures Vol. 7, No. 4 pp 1833-1841
[3] Miqdam T Chaichan, Shaimaa H Kamel, Abdul-Mohsen N M Al-Ajeely 2015 SAUSSUREA Vol. 5(6): pp 48-55
[4] Robert A. Miller and Maria A. Kuczmarski 2009 NASA/TM 215460
[5] M A Sheikh, S C Taylor, D R Hayhurst and R Taylor 2000 J. Phys. D: Appl. Phys pp 1536–1550
[6] Michael Smith, Bruce Jamieson 2014 Proceedings International Snow Science Workshop pp 507-510
[7] Robert Powell, William M. Rogers, and Don O .Coffin 1957 Journal of Research of the National Bureau of Standards Vol. 59, No.5. pp 349-355
[8] C. Buratti, E. Belloni, L. Lunghi, A. Borri, G. Castori, M. Corradi 2016 Journal of Building Engineering Volume 7 pp 63–70
[9] Cong Zhang, Yong Du, Shuhong Liua, Yuling Liu, Bo. Sundman 2016 Thermochimica Acta Volume 635 pp 8–16
[10] Peter Teertstra 2007 Proceedings of IPACK 33026 pp 1-8
[11] Simon Lineykin and Shmuel Ben-Yaakov 2007 IEEE TRANSACTIONS ON INDUSTRY APPLICATIONS, VOL. 43, NO. 2 pp 505-512

Characterization of copper and nichrome wires for safety fuse

E. Murdani
Study Program of Physics Education, STKIP Singkawang (College of Teacher Training and Education-Singkawang),
Jalan STKIP, Kota Singkawang-79251, INDONESIA

E-mail: ekamurdani@gmail.com

Abstract. Fuse is an important component of an electrical circuit to limiting the current through the electrical circuit for electrical equipment safety. Safety fuses are made of a conductor such as copper and nichrome wires. The aim of this research was to determine the maximum current that can flow in the conductor wires (copper and nichrome). In the experiment used copper and nichrome wires by varying the length of wires (0.2 cm to 20 cm) and diameter of wires (0.1, 0.2, 0.3, 0.4 and 0.5) mm until maximum current reached that marked by melted or broken wire. From this experiment, it will be obtained the dependences data of maximum current to the length and diameter of wires. All data are plotted and it's known as a standard curve. The standard curve will provide an alternative choice of replacing fuse wire according to the maximum current requirement, including the wire type (copper and nichrome wires) and wire dimensions (length and diameter of wire).

1. Introduction

We often hear the fuse in everyday life. Electrical installation uses as a safety fuse for limiting the electric current flowing in the electrical circuit. The use of a safety fuse is essential to limit the electric current flowing in the electronic equipment so that electronic equipment be durable and not easily damaged by high currents. If the electronic device is written 2 A, the maximum current in the electronic device is 2 Ampere [1-3]. So we have to choose a wire that is able to flow through the maximum current 2 A. If the current flows more than 2 A (beyond the ability of fuse wire), the wire will break. The mechanism of breaking the fuse wire is the maximum current in the wire produces high heat, because the heat in the wire exceeds the melting point of the wire, the wire will melt and eventually the wire would break up [4-8]. Therefore, the importance of knowing the dependence of wire resistance to temperature (melting temperature and ambient temperature), the type of wire and wire dimensions to be used as a fuse wire.

Fuse wire is a conductor. Conductor is a material that is easy to conduct electric current [1-8]. Some wire conductors being sold in the market are copper and nichrome [9-10]. Based on the above background, the researchers are interested to characterize the copper and nichrome wire to a safety fuse.

When an electric current flowing in the conductor, the electrical energy is converted into heat continuously. Heat is generated from collision events charge carriers that move with the metal atoms. Heat conductor will be red because the electrons in the skin conductor moved from the tracks to the track lower while emitting a photon with a wavelength that is emitted in the region of the color spectrum of red (λ = 650 nm - 700 nm) [3, 8, 11]. If this heat causes the conductor is at its melting point, the wire will melt and eventually the wire will break. The electric current is decided wire called breaking current (maximum current that can flow in the wire) [4-12].

Dependence resistance to the type of wire and wire dimensions (length and diameter) is defined as:

$$R_o = \frac{\rho\ell}{A} = \frac{4\rho\ell}{\pi d^2} \tag{1}$$

With Ro is the initial resistance of wire, ρ is the resistivity, A is the cross-sectional area of the wire (area of the circle), ℓ is the length of the wire and d is the diameter of the wire [1-10].

The dependences of the resistance on the temperature to be formulated as:

$$R_{T_{melt}} = R_o\left[1+\alpha\left(T_{melt}-T_o\right)\right] = \frac{4\rho\ell}{\pi d^2}\left[1+\alpha\left(T_{melt}-T_o\right)\right] \tag{2}$$

R_T is the resistance wire on the melting temperature (resistance on maximum current), T_0 is the initial temperature (20°C), temperature coefficient constants at 20°C [1-10].

According to Ohm's law formula:

$$I_{max} = \frac{V_{in}}{R_{T_{melt}}} \tag{3}$$

Dependence maximum current to the wire type and dimensions (length and diameter) is defined as:

$$I_{max} = \frac{V_{in}\pi d^2}{4\rho\ell\left[1+\alpha\left(T_{melt}-T_o\right)\right]} \tag{4}$$

With Vin is the input voltage of the power supply (current source). Resistivity, temperature coefficient and the melting point of the wire can be seen in Table 1 [4-12].

Table 1. Value of resistivity (ρ), temperature coefficient (α) at 20°C, and melting point (T_{melt})

materials	ρ (10^{-8} Ωm)	α $(°C)^{-1}$	T_{melt} (°C)
Copper	1.68	0.0068	1083
Nichrome	100	0.0004	1672

2. Experiment

Tools and materials used in this study is a multimeter, power supply (current source), micrometer screw and two types of wire conductor (copper wire and nichrome). Experiments carried out by flowing or varying the electric current until it reaches the maximum value on the copper wire and nichrome indicated with melted wire. Wire length varied from 0.2 cm to 20 cm while the variation of the wire diameter is 0.1 mm; 0.2 mm; 0.3 mm; 0.4 mm and 0.5 mm respectively. The series of these experiments are shown in Figure 1.

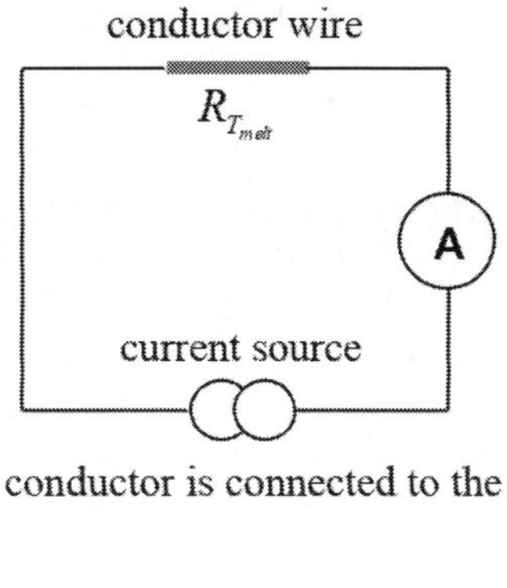

Figure 1. The series of wire conductor is connected to the current source [12].

Journal of Physics: Conference Series **776** (2016) 012099 doi:10.1088/1742-6596/776/1/012099

Wire conductors (copper and nichrome) is connected to the current source and the ammeter in series to the wire. A symbol in the circuit of Figure 1 is ammeters. Volume on the current source is increased step by step until the wire broke. Flows were recorded on ammeter shortly before dropping out is the maximum current that can flow in the wire. Meanwhile, when the wire breaking ammeter will show zeros because the breakdown of the circuit in the measurement system.

3. Results and Discussion

Experiments carried out by flowing or varying the electric current until it reaches the maximum value on the copper wire and nichrome indicated with melted wire. The dependence of the maximum current on the length and diameter of the wire, expressed in a graph called a standard curve. Standard curves will provide instructions to select the fuse wire as needed. Fuse wire options include the type of wire (copper wire and nichrome) and wire dimensions (length and diameter of the wire). The dependence of the maximum current on the length of wire for copper and nichrome wire is shown in Figure 2.

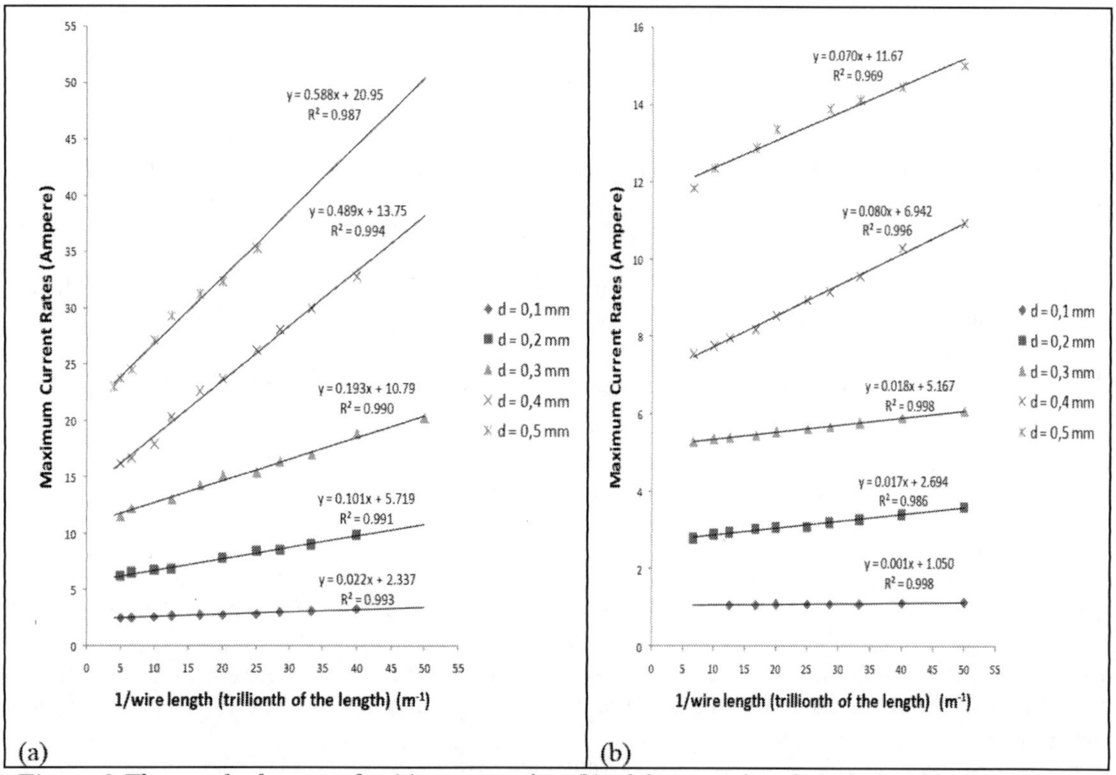

Figure 2 The standard curves for (a) copper wire, (b) nichrome wire, the relationship between the maximum current to trillionth of the length of wire for each wire diameter.

From Figure 2 (a), on the same type of wire (copper wire), and on the the same diameter, maximum current is inversely proportional to the length of the wire. These results are in accordance with the equation (4). Based on the equation (1), if the length of the wire is great, the wire will have greater initial resistance. Because the wire resistance is great, electric current flowing in the wire will be small.

At different wire types (Figure 2), the length and diameter of the wire is the same, it appears that the maximum current of copper wires is greater than nichrome wire. This difference is caused by the resistivity and the melting point of the wire. The maximum current is inversely proportional to the wire resistivity and melting point (according to the equation (4)). In Table 1, the resistivity and melting point for a nichrome wire is greater than copper wire, so that the resistance at nichrome wire will be greater than copper wire. Therefore, the electric current can flow (maximum flow) on the nichrome wire will be smaller than the copper wire.

The dependence of the maximum current on the wire diameter for the copper and nichrome wire are shown in Figure 3.

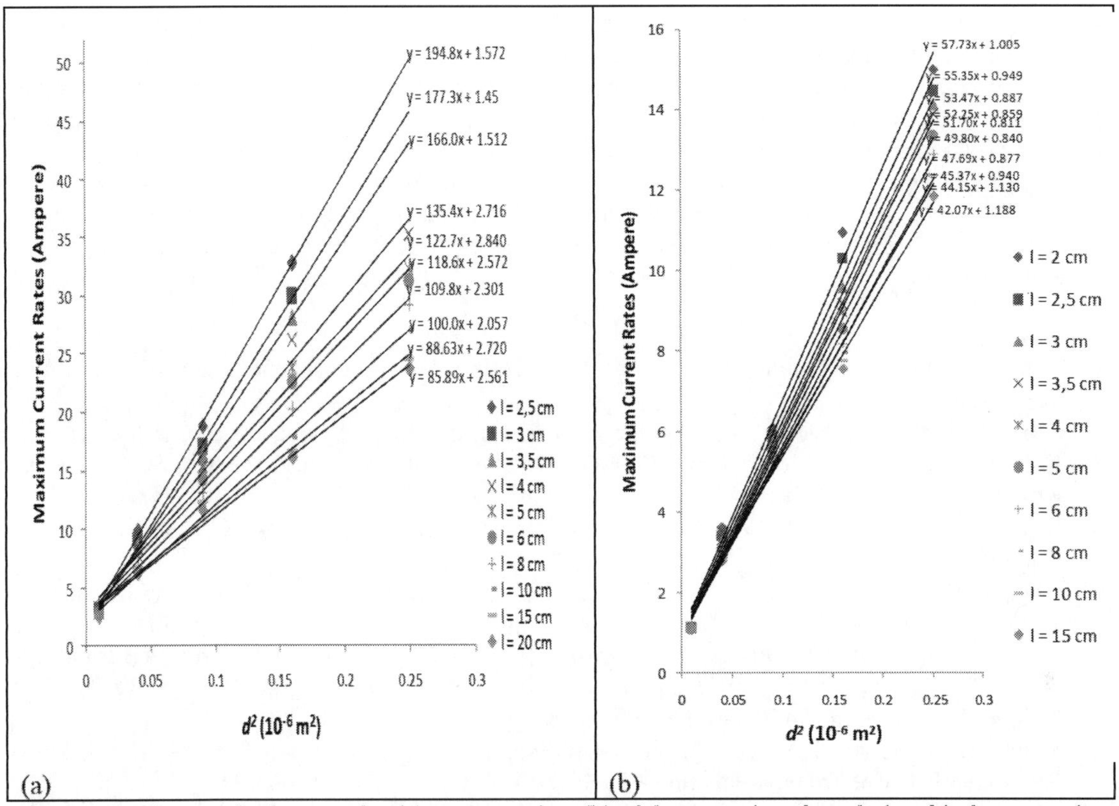

Figure 3. The standard curve for (a) copper wire, (b) nichrome wire, the relationship between the maximum current to the square of the diameter for each wire length.

From Figure 3 on the same wire length, maximum current is proportional to the square of the diameter. These results are in accordance with the equation (4). If the wire diameter is great, on the wire would have a small initial resistance. Because the wire resistance is small, the electric current flowing in the wire would be great.

Fuse is an important component of an electrical circuit to limiting the current through the electrical circuit for electrical equipment safety. Safety fuses are made of a conductor such as copper and nichrome wires. The aim of this research was to determine the maximum current that can flow in the conductor wires (copper and nichrome). In the experiment used copper and nichrome wires by varying the length of wires (0.2 cm to 20 cm) and diameter of wires (0.1, 0.2, 0.3, 0.4 and 0.5) mm until maximum current reached that marked by melted or broken wire. From this experiment, it will be obtained the dependences data of maximum current to the length and diameter of wires. All data are plotted and it's known as a standard curve. The standard curve will provide an alternative choice of replacing fuse wire according to the maximum current requirement, including the wire type (copper and nichrome wires) and wire dimensions (length and diameter of wire).

When an electric current flowing in the conductor, the electrical energy is converted into heat continuously. Heat is generated from collision events charge carriers that move with the metal atoms. Heat conductor will be red because the electrons in the skin conductor moved from the tracks to the track lower while emitting a photon with a wavelength that is emitted in the region of the color spectrum of red (λ = 650 nm - 700 nm) [3,8,11]. If this heat causes the conductor is at its melting point, the wire will melt and eventually the wire will break. The electric current is decided wire called breaking current (maximum current that can flow in the wire) [4-12].

3. Consclusion

By knowing the maximum current dependence on the type of wire and wire dimensions, will allow one to determine or replace the fuse wire as needed. The maximum current depends on the type of wire that is on the resistivity and melting point of the wire. The electrical current is inversely proportional to the resistivity and melting point of the wire. In the same type of wire, the maximum current depends on the dimensions of the wire (proportional to the square of the diameter and inversely proportional to the length of the wire). Standard curves of the experimental results provide guidance for selecting the fuse wire in accordance with current needs. Selection of fuse wire on the standard curve covering the type of wire (copper wire and nichrome) and wire dimensions (length and diameter of the wire).

4. References

[1] Tipler PA 1996 *Fisika untuk Sains dan Teknik* (Terjemahan Bambang Soegijono) Edisi Ketiga Penerbit Erlangga Jakarta
[2] Soeharto 1992 *Fisika Dasar II Listrik-Magnet* Penerbit PT Gramedia Pustaka Utama Jakarta
[3] Halliday D Resnick R 1984 *Fisika Jilid 2* (Terjemahan Pantur Silaban dan Erwin Sucipto) Edisi Ke-3 Penerbit Erlangga Jakarta
[4] Sears FW Zemansky MW 1994 *Fisika untuk Universitas 2 Listrik Magnet* (Terjemahan Nabris Chatib) Penerbit Bina Cipta
[5] Giancoli DC 1998 *Fisika Jilid 2* (Terjemahan Yuhilza Hanum) Edisi Kelima Penerbit Erlangga Jakarta
[6] Soetrisno Tan Ik Gie 1979 *Fisika Dasar: Listrik, Magnet dan Termofisika* Penerbit ITB Bandung
[7] Sears FW Zemansky MW 1994 *Fisika untuk Universitas 3 Optika-Fisika Modern* (Terjemahan Nabris Chatib dan Amir Achmad) Penerbit Bina Cipta
[8] Young HD Fredman RA 2004 *Fisika Universitas Jilid 2* (Terjemahan Pantur Silaban) Edisi Kesepuluh Penerbit Erlangga Jakarta
[9] Isliyanti A Murdani E et al 2010 *Pembelajaran Praktikum Fisika Berbasis Penelitian: Karakteristik Kawat Pada Sekering Pengaman* Prosiding Seminar Nasional Fisika 2010 Bandung
[10] Murdani E Sutarno D 2011 *Karakterisasi Kawat Untuk Sekering Pengaman* Prosiding Simposium Nasional Inovasi Pembelajaran dan Sains (SNIPS) 2011 Bandung
[11] Beiser A 1992 *Konsep Fisika Modern* (Terjemahan The Houw Liong) Edisi Keempat Penerbit Erlangga Jakarta
[12] Ismail B 1995 *Rangkaian Listrik Jilid 1* Penerbit ITB Bandung

Study of EHD flow generator's efficiencies utilizing pin to single ring and multi-concentric rings electrodes

Sumariyah[1,3 a],Kusminarto[3,b] ,A. Hermanto[3] and P. Nuswantoro[3]
[1]Physics Department, Faculty of Science and Mathematics, Diponegoro University, Indonesia
[2]Center for Plasma Research, Faculty of Science and Mathematics, Diponegoro University, Indonesia
[3]Physics Department, Faculty of Mathematics and Natural Sciences, Gadjah Mada University, Indonesia

E-mail: asumariyah@undip.ac.id

Abstract. EHD flow or ionic wind yield corona discharge is a stream coming from the ionized gas. EHD is generated by a strong electric field and its direction follows the electric field lines. In this study, the efficiency of the EHD flow generators utilizing pin-multi concentric rings electrodes (P-MRE) and the EHD pin-single ring electrode (P-SRE) have been measured. The comparison of efficiencies two types of the generator has been done. EHD flow was generated by using a high-voltage DC 0-10 KV on the electrode pin with a positive polarity and electrode ring/ multi-concentric rings of negative polarity. The efficiency was calculated by comparison between the mechanical power of flow to the electrical power that consumed. We obtained that the maximum efficiency of EHD flow generator utilizing pin-multi concentric rings electrodes was 0.54% and the maximum efficiency of EHD flow generator utilizing a pin-single ring electrode was 0.23%. Efficiency of EHD with P-MRE 2.34 times Efficiency of EHD with P-SRE

1. Introduction

Electrohydrodynamic (EHD) flow and is also called ion wind or the electricity wind is a stream coming from the ionized fluid generated by a strong electric field and the direction of flow follows the direction of the field lines. EHD flow prospects in the industry are quite large, as be applied for printing patterns of micro/nano size [1]. The study of EHD flow used as the coolant carried out by Alamgholilou[2] and Shin[3]. The applications of EHD flow for the textile finishing is done by Joshi [4], Guljarani[5] and Zhao[6] and EHD as a drying apparatus is conducted by Singh[7] and Dinani[8]. Devices that use EHD flow has many advantages such as power required is quite low.

Various attempts have been made to improve the efficiency of energy EHD flow generators such as by modifying the electrode configuration, for example electrode configured needle-arrayed electrodes combined with a plate fin heat sink by Huang[9] and six stages in series. As for increasing the number of pin electrodes for EHD pumps has been conducted by Zhang[10]. Configuring wire electrodes and a semi-cylindrical contour shaped electrodes of the cooling system has been done by Kim[11]. Electrode Configuration of the needle to parallel plates for cooling the device by Shin[3].

Research methods in this paper are the same as methods of our previous research.They are determining the maximum angle of EHD flow [12], modeling of electric potential distribution in the zone of EHD flow [13], Measurement of the velocity of EHD flow [14] and determination of Ion Mobility in EHD Flow Zone [15].

Theoretical examination of the EHD Flow

In the experimen the value of the velocity of EHD flow **u** which is the output of generating system of EHD is proportional to the voltage between electrodes V and is comparable to the root of the current I (Robinson, 1967) as indicated by the equation:

$$`\mathbf{u} \propto V \propto I^{1/2} \tag{1}$$

According to the law of conservation of energy:

$$\frac{\rho_g S d\mathbf{u}^2}{2} = \int_0^d q\mathbf{E}dz \tag{2}$$

With ρ_g is mass density of air (kg m^{-3}), S is a ring electrode surface area and q is the density of the charge (C m^{-3}). Substitution Gauss law to equation (1) and integrate it to find:

$$\mathbf{u} = \sqrt{\frac{\varepsilon_0}{\rho_g}}\mathbf{E} = \sqrt{\frac{\varepsilon_0}{\rho_g}}\frac{V}{d} \tag{3}$$

Where ε_0 is the dielectric permittivity of vacuum. If the threshold voltage of corona discharge V_0 considered, relationship of velocity of EHD flow **u** with voltage shown by the equation:

$$\mathbf{u} = \sqrt{\frac{\varepsilon_0}{\rho_g}\frac{V - V_0}{d}} \tag{4}$$

Energy efficiency is defined as:

$$\eta = \frac{\frac{d}{dt}\left(\frac{1}{2}m\mathbf{u}^2\right)}{P} \tag{5}$$

where m is the mass of air (kg), **u** is the velocity of EHD flow and P is electric power (W). Meanwhile, according to Timothy [16], relationship of velocity of EHD flow and electrical power shown by the equation:

$$\mathbf{u} = \frac{2^{1/3}\varepsilon_0^{1/6}}{(S\mu)^{1/3}\rho^{1/2}}P^{1/3} \tag{6}$$

where μ is the ion mobility and ρ is a density of the air. Substitution of equation (6) into the equation (5) resulting the equation:

$$\eta = \frac{\frac{d\left(\frac{1}{2}mu^2\right)}{dt}}{u^3}\frac{2\varepsilon_0{}^{1/2}}{S\mu\rho^{3/2}} \tag{7}$$

With derivating the air mass (m) with respect to time in equation (7), equation (8) is found

$$\eta = \frac{\rho S u^3}{2u^3} \frac{2\varepsilon_0{}^{1/2}}{S\mu\rho^{3/2}} \tag{8}$$

2. Research Methods

Two configurations of the EHD flow generator are studied, they are a pin-multi concentric ring (P-MRE) and pin- single ring electrode (P-SRE) which is as the control of measurement. Pin electrodes made of stainless steel sewing needle with a diameter of pointed end 0.09 mm. Multi concentric ring electrodes consist of 3 concentric ring electrodes are arranged concentrically interconnected. The third ring of the electrode constituent having diameters in each of 8 mm, 16 mm and 24 mm. In a single ring electrode, the diameter is 24 mm. All electrode metal rings with the same width and thickness, respectively 2 mm.The composition of the generating system of EHD flow in this study includes P-MRE, P-SRE, (high voltage power, set voltage and current measurement tools and a set of measuring tools EHD flow velocity). The air can come from the top and out of the bottom, as shown in Figure.1

The EHD flow generated by installing a high voltage DC 10 KV on the electrodes of the corona discharge system with a positive polarity on the pin electrodes and a negative polarity on the ring/ the multi-concentric ring. A high-voltage probe 1000: 1 was used to measure the high voltage, which is connected to the CRO and the ion current was measured using a multimeter. Measure the velocity of EHD flow, Hot-Wire Anemometer was placed perpendicular the pin electrode direction with distance was 15 mm.

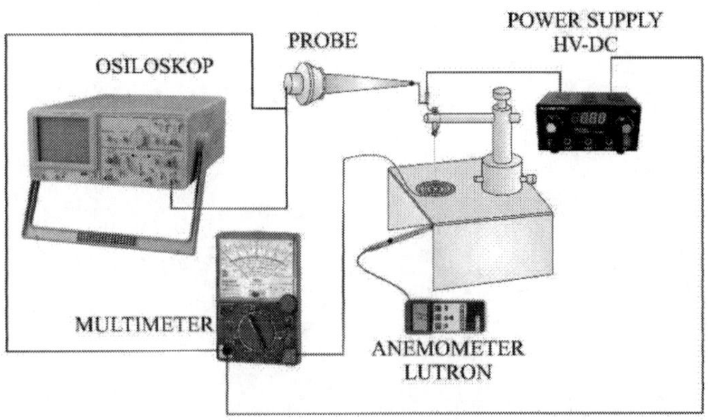

Figure 1. The Composition of a Generating System of the EHD Flow

3. Results and Discussion

Velocity of EHD Flow

Graph of velocity of EHD flow as a function of voltage (characteristics **u**-*V*) with some distance between the electrodes (*d*) fixed for the electrode configuration of pin-multi concentric rings (**uT**) and pin-single ring (**uS**) is shown in Figure 2.

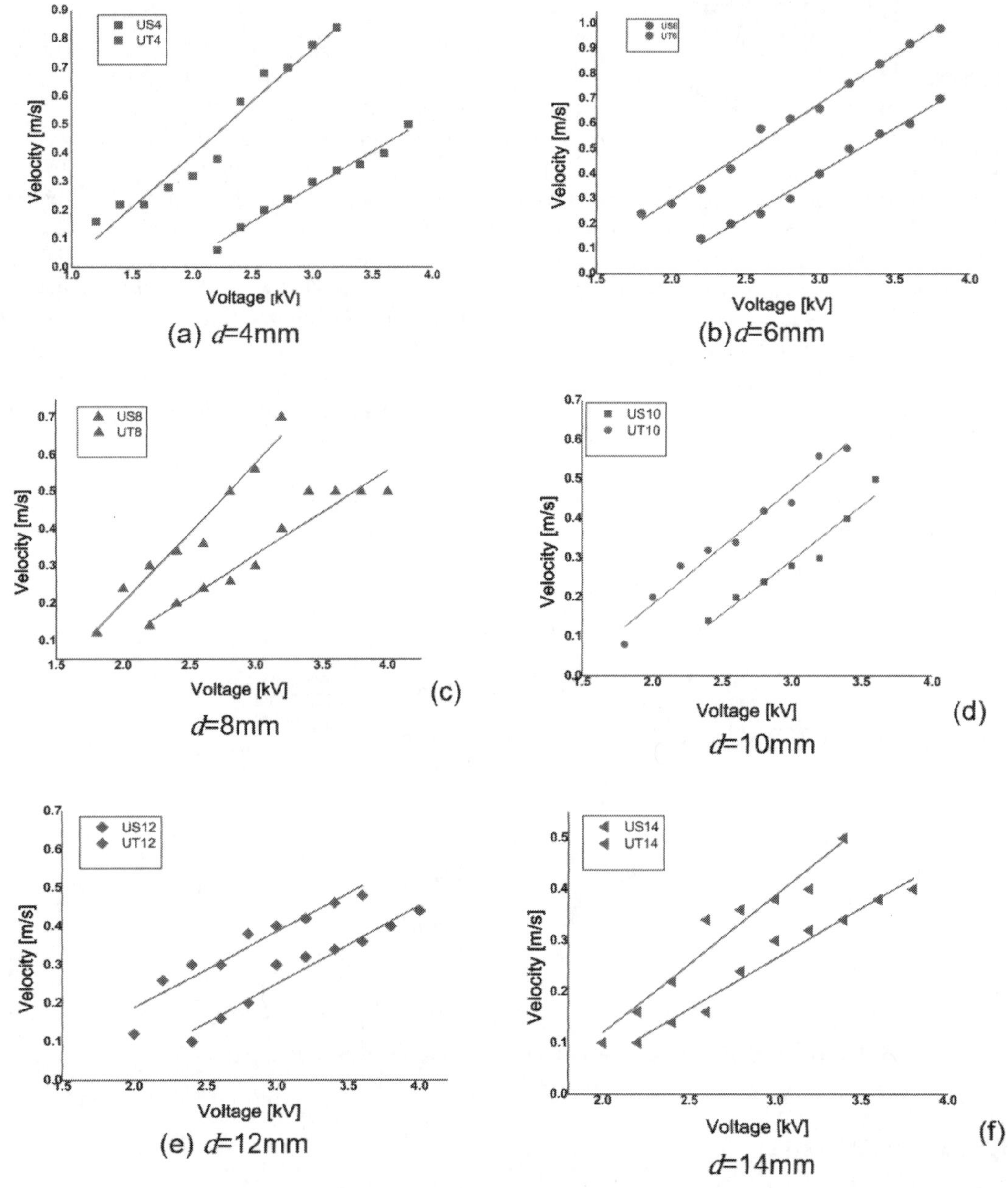

Figure 2. Characteristics u-*V* of (a) pin-multi concentric rings electrodes (*ET*) and (b) pin-single ring electrode(*ES*).

In Figure 2, It can be seen that in all the distance between the electrodes (*d*), the velocity of EHD flow increase with increasing the applied voltage. Our results are in agreement with a theory where a relationship between the applied voltage and the velocity of EHD flow is a linear function. The theoretical formulation is shown in the equation (3). This equation indicates that $\mathbf{u} \propto V$

Furthermore, in Figure 2. with the same distance inter-electrodes, we found that the velocity of EHD flow for **uT** is always greater than the velocity of EHD for **uS**. This is due to the electric field density of corona discharge with pin-multi concentric rings electrodes is greater than the electric field density of corona discharge with pin-single ring electrode. Electric field flux density led to increasing space charge of ions. The ions in its movement in drift region collide with neutral air molecules. It is very high collisions frequency. There is great momentum transfer from positive ions to the molecules of gas. In consequence, in the drift region, EHD flow will be produced with high velocity.

Energy Efficiency of generator system of EHD flow

Graph of maximum energy efficiency of the EHD flow generator (η) for various distance between the electrodes (d) respectively 4 mm, 6 mm, 8 mm, 10 mm, 12 mm, 14 mm as function of voltage (V) by using the pin-multi concentric rings electrodes and pin-single ring electrode as shown in Figure 3(a) and Figure 3(b).

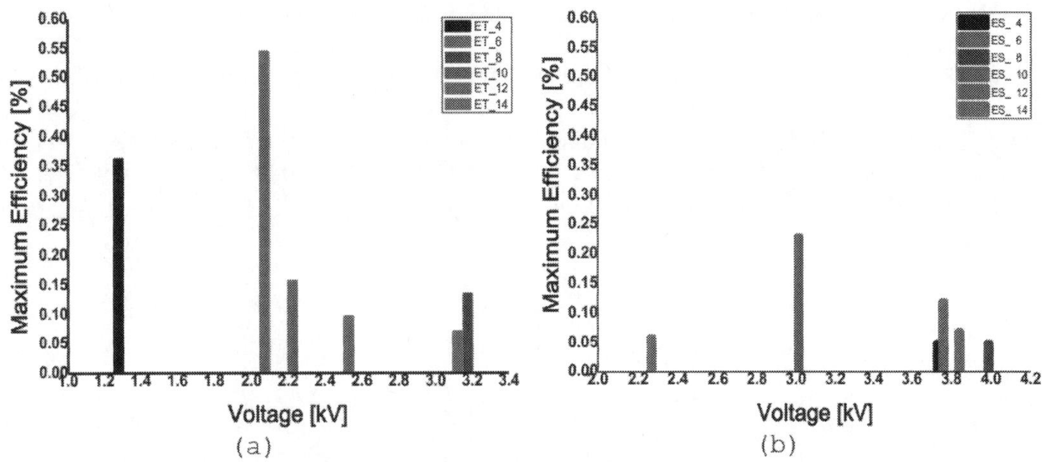

Figure 3. Chart of energy efficiency of EHD Flow Generator utilizing (a) pin-multi concentric rings electrodes (*ET*) and (b) pin-single ring electrode(*ES*).

Figure 3 (a) shows that the greatest value of the energy efficiency (0.54%) of the system generator EHD flow with pin-multi concentric rings electrodes on the voltage $V = 2.0$ kV, with a distance between the electrodes $d = 12$ mm. In Figure 3.(b), we can see that the greatest value of the energy efficiency (0.23%.) of the system generator EHD flow generator system with pin-single ring electrode on the voltage $V = 3.0$ kV, with a distance between the electrodes $d = 10$ mm.

4. Conclusion

At the same distance, we found that velocity of EHD flow with pin-multi concentric rings electrode configuration **uT** is always obtained greater than the flow of the electrode configuration EHD-pin single ring **uS**. The efficiency of the system generator of EHD flow with a pin-multi concentric rings electrodes is greater than the efficiency of system generator of the EHD flow with a pin-single ring electrode.

References

[1] Tae Heon Hwang T. H., Kim Y.J., Haseung Chung H., Ryu W.H., 2016, Motionless Electrohydrodynamic (EHD) Printing of Biodegradable Polymer Micro Patterns, *Microelectronic Engineering* 161 pages 43–51

[2] Alamgholilou A and Esmaeilzadeh E 2012 *Experimental Thermal and Fluid Science*, **38** p 61-73
[3] Shin DH, Yoon JS and Ko HS 2015 *International Journal of Heat and Mass Transfer*, **84** p 35-45
[4] Joshi M and Butola BS 2013 *Advanced in the Dyeing and Finishing of Technical Textiles,* p 355-411
[5] Guljarani ML 2013 *Advanced in the Dyeing and Finishing of Technical Textiles,* p 280-308
[6] Zhao T and Chen Q 2016 *Antimicrobial Textile*, p 141-153
[7] Singh A, Vanga SK, Raveendran NGR, Gariepy Y, Valerie OV and Raghavan V 2015 *LWT-Food Science and Technology protein conformation*, **64** p 750-758
[8] Dinani ST and Havet M 2015 *Journal of Cleaner Production* **95** 203-2011
[9] Huang RT, Sheu WJ and Wang CC 2009 Heat transfer enhancement by needle-arrayed electrodes – An EHD integrated, cooling system, *Energy Conversion and Management*
[10] Zhang JI and Lai FC 2011 *Journal of Electrostatics* **69**, 486-493
[11] Kim C, Park D, Noh KC dan Hwang J 2010 *Journal of Electrostatics* **68** 36–41
[12] Nur M, Azzulkha AH, Restiwijaya M, Muchlisin Z and Sumariyah 2014 *Advances in Physics Theories and Applications* ISSN 2224-719X (Paper) ISSN 2225-0638 (Online) **30**
[13] Sumariyah, Kusminarto, Arief H, Pekik N, Zainul M and Evi S 2015 *Procedia Environmental Sciences* **23** 260 – 265
[14] Sumariyah, Kusminarto, Arief H and Pekik N 2015 *Applied Mechanics and Materials* **271** p 227-231
[15] Sumariyah, Kusminarto, Arief H and Pekik N 2015 *International Conference on Chemical and Material Engineering (ICCME) 2015 AIP Conf. Proc.* **1699** , 040019-1–040019-5
[16] Goodenough TIJ, Goodenough PW and Goodenough SM 2007 *Journal of Food Engineering* **80**, 1233–123

Power consumption analysis DBD plasma ozone generator

M. Nur[1,2*], M. Restiwijaya[1*], Z. Muchlisin[1], I. A. Susan[2], F. Arianto[1,2], S. A. Widyanto[3]
[1]Physics Department, Faculty of Science and Mathematics, Diponegoro University
Jl. Prof. Soedarto, SH, Tembalang, Semarang. Central Java, Indonesia. 50275,
[2]Center for Plasma Research, Faculty of Science and Mathematics, Diponegoro University, Jl. Prof. Soedarto, SH, Tembalang, Semarang. Central Java, Indonesia. 50275,
[3]Faculty of Engineering, Diponegoro University. Jl. Prof. Soedarto, SH, Tembalang, Semarang. Central Java, Indonesia. 50275, INDONESIA

*E-mail: m.nur@undip.ac.id

Abstract.
Studies on the consumption of energy by an ozone generator with various constructions electrodes of dielectric barrier discharge plasma (DBDP) reactor has been carried out. This research was done to get the configuration of the reactor, that is capable to produce high ozone concentrations with low energy consumption. BDBP reactors were constructed by spiral-cylindrical configuration, plasma ozone was generated by high voltage AC voltage up to 25 kV and maximum frequency of 23 kHz. The reactor consists of an active electrode in the form of a spiral-shaped with variation diameter Dc, and it was made by using copper wire with diameter Dw. In this research, we variated number of loops coil windings N as well as Dc and Dw. Ozone concentrations greater when the wire's diameter Dw and the diameter of the coil windings applied was greater. We found that impedance greater will minimize the concentration of ozone, in contrary to the greater capacitance will increase the concentration of ozone. The ozone concentrations increase with augmenting of power. Maximum power is effective at DBD reactor spiral-cylinder is on the Dc = 20 mm, Dw = 1.2 mm, and the number of coil windings N = 10 loops with the resulting concentration is greater than 20 ppm and it consumes energy of 177.60 watts

1. Introduction

A dielectric barrier discharge (DBD) will appear in a gas gap when an ac or rf voltage is applied to an electrode system with one or both electrodes covered by a dielectric layer. The gas-filled gap is small. Typical materials used for the dielectric layer are glass, quartz,ceramics, and polymer coatings [1-4]. Ozone can be produced by using electrical discharge in pure oxygen gas or air that was in the between two electrodes in the DBD plasma system. The factors that can affect ozone formation in general are voltage, dielectric materials, the pressure, the system configuration of the plasma reactor and a gas insert in the plasma reactor [3-6]. The main mechanism of ozone formation are ionization, recombination, dissociation and association. The reaction of ozone formation can occur in a plasma reactor has been

explained for example by Fridman in his book [7]. Nur et al. produced ozone by using DBD plasma and it has been implemented in the rice storage for maintaining the quality of rice [8]. Experimental study on the current-voltage characteristics for atmospheric pressure DBD has been done by Pongsathon [9]. DBDP can be approximated by an equivalent electronic circuit. The equivalent circuit is based on the model proposed by F. Massines et al [10].
This paper tries to discuss the energy consumption of ozone generators and production capabilities. Several studies that have been done related to energy consumption and the concentration of ozone generated. The ozone chamber was developed using the RLC circuit base on RLC resonance tank circuit. This model was used to determine the resonance frequency for inverter as a power supply to ozone generator [11]. Research on quantification of capacitance and power consumption [12]. Influence of electrode configuration on ozone synthesis and micro-discharge property in dielectric barrier discharge reactor has been studied [13].

2. Research Methods.

Figure 1 shows a series of experiments in this study. Ozone has been produced by using Dielectric Barrier Discharge Plasma (DBDP) with spiral-cylindrical reactor configured. DBDP generated by high voltage AC voltage up to 25 kV and maximum frequency of 23 kHz. Electrical parameters of DBPP determined through a voltage divider (HV Probe DC Voltage DC max 40 kV; 28 kV AC EC code number 1010, En G1010). Electrical signal from the probe detected by an Oscilloscope GOS-653, 50 MHz. The electric current, that was generated in the reactor was measured by using a multimeter (Sunwa TRXn 360). Photo taken from experiments using a CCD camera (Creative, DV Cam).

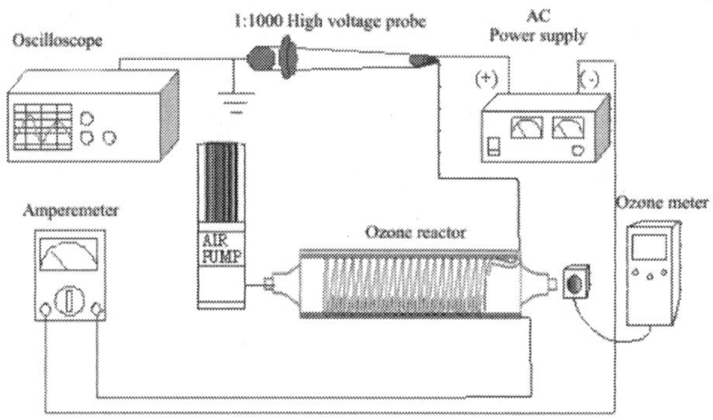

Figure 1. Experimetal set up

The concentration of ozone generated DBDP reactor with spiral-cylindrical configuration measured using an ozone monitor (Quant Ozone "2"). Measurements were taken at applied voltage variations and multiple electrode diameter cylinders/Pyrex tube.

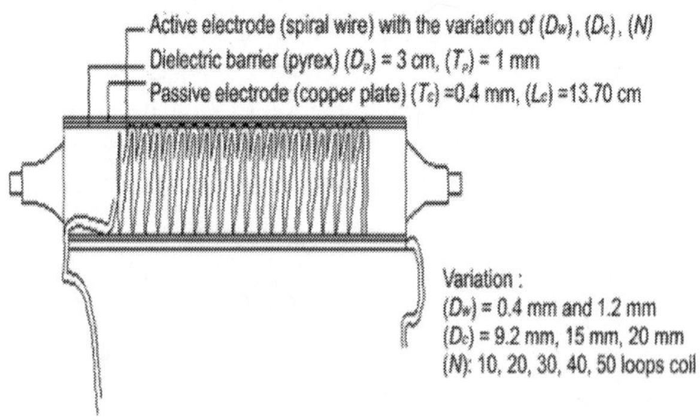

Figure 2. Dielectric Barrier Discharge Plasma Reaktor

The reactor used in this study was Dielectric Barrier Discharge Plasma reaktor (DBDP) reactor, with wire spiral - cylinder configuration (Figure 2). The reactor consists of an active electrode in the form of a spiral-shaped copper wire (wire diameters variation (D_W) are 0.4 mm and 1.2 mm), the diameters of the wire coil windings (D_C) are 9.2 mm, 15 mm, 20 mm, number of loops coil windings (N) are 10, 20, 30, 40, 50 loops. Wire spiral was inserted in a dielectric barrier was shaped pyrex tube pyrex diemeter (D_P = 30 mm) and pyrex thickness (T_P = 1 mm) and pyrexlength (L_P = 15 cm). the outer part of the dielectric barrier will be veiled with a copper plate that serves as a passive electrode, with copper thickness (T_C = 0.4 mm) and copper length (L_C = 13.70 cm).

3. Results and Discussion

We used air as source of ozone generator. The air most constituent is nitrogen and oxygen. The air introduced into the reactor with a pump and with certain flow rate. According to Lofthus and Krupenie [14], O_2 molecules have ionization and dissociation energies lower than N_2 molecules. So the O_2 molecule more susceptible to ionize and dissociate of than nitrogen molecules. Energy levels of molecules O_2 and N_2 molecules that can be found in [15] and the ionization energy of O_2 molecule is 12 eV. If the O_2 molecules gain energy 12 eV can ionize the O_2. Because the conditions of ionized molecules in an unstable state and tends to return to its normal state (recombination), O_2^+ ions can capture electrons and emit an energy of 12 eV. How the same is true of N_2 molecules having ionization energy of 15 eV can ionize and go back to the normal state (recombination) to emit an energy of 15 kV.

Effect of Impedance and Capacitance Against Ozone Concentration

The influence of impedance to ozone production was investigated to get the best value of the parameters for an ozone generator that constructed base on DBDP. In this study, the impedance value was obtained by calculation using equation (9).

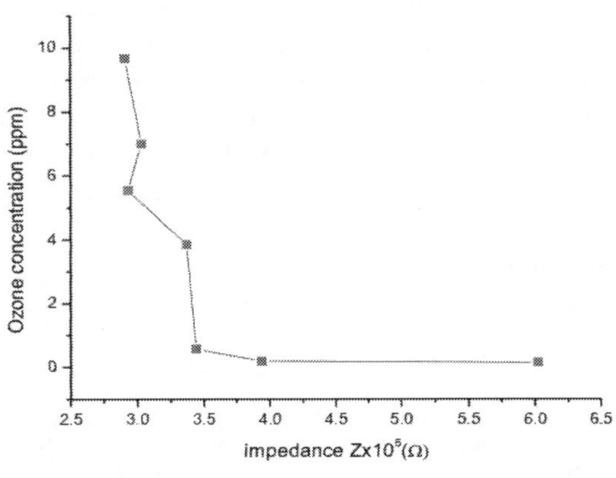

Figure 3. Ozone concentration as function of impedance for N= 50 coils D_W =0.4 mm, D_C= 20 mm, and D_P = 30 mm

The impedance value depends on voltage variation (V), the wire diameter (D_W), the diameter of the wire windings (D_C) and the number of wire windings (N). Graph ozone concentrations as a function of the impedance shown in Figure 3 for the number of wire windings (N = 50 coils), the diameter of the wire windings (D_C = 20 mm) and wire diameter (D_W = 0.4 mm). We found that the greater the value of the impedance then the concentration of ozone generated is getting smaller. This is because the electric field near the electrode active decline. So the process of ionization and recombination also decreased.

Capacitance influence on ozone production also investigated in this study. In the DBD reactor, the electrical equivalent circuit consists of two capacitances, the capacitance of gas (C_g) and dielectric capacitance (C_d). Total capacitance can be obtained with equation $\frac{1}{C} = \frac{1}{C_d} + \frac{1}{Cg}$. The reactance value was determined by the equation $Xc = 1/2\pi fC$. While the dielectric capacitance obtained from measurements using capacitance LCR meter. Graph of ozone concentrations as a function of gas capacitance shown in Figure 4.

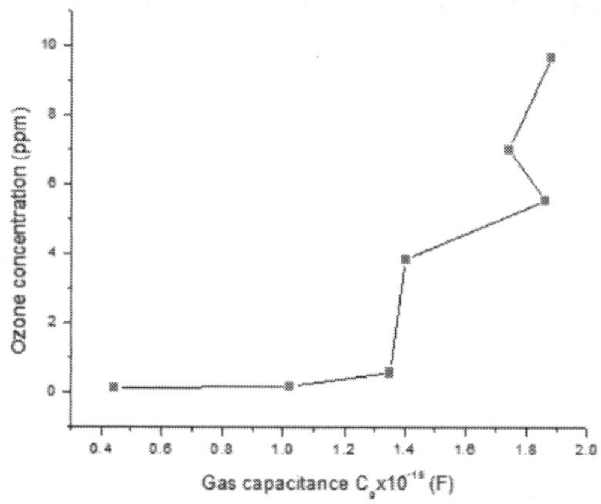

Figure 4. Ozone concentration as function of voltage with several number of winding coil N= 50 coil spools , for D_W =0.4 mm, D_C = 15 mm, and D_P = 30 mm

From this figure, it can be seen that the concentration of ozone enlarges with the enlargement capacitance of gas discharge. It can be understood that the electric current had been enlarged with increasing capacitance of gas discharge. Already mentioned earlier that in the gas discharge, electric currents are capacitive current. That's mean, the greater the capacitance, the greater the charge in the discharge. The total electrical charge determined transformation to the electric current. Moreover, it was followed by greater of the electrical charge changes in change of time (dQ/dt). Electric current shows the occurrence of ionization, and other processes such as dissociation. Energetic electrons produced in the plasma that will be able to break up oxygen molecule (O_2) becomes oxygen atomic (O). These process are the main condition occurrence of tree reaction bodies in the formation of ozone reaction such as: $O_2 + O + M \rightarrow O_3 + M$. In this reaction M is other molecules inside reactor such as, N_2, O_2^-, etc.

The effect of power consumption to produce ozone

The influence of power by generator to the ozone concentration produced is shown in Figure 5.

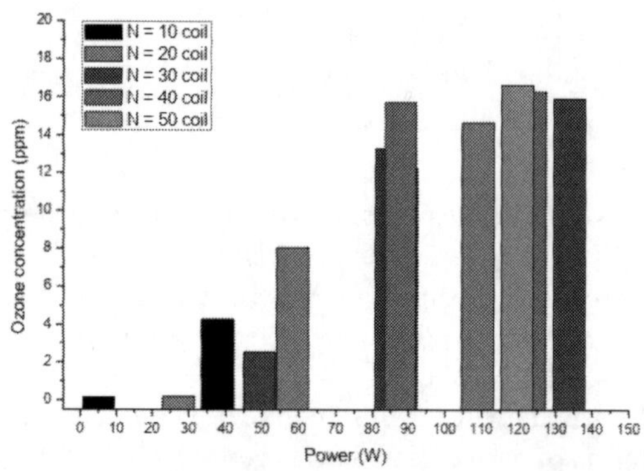

Figure 5. Ozone concentration as function of power for several number of coil spools (N) , for D_w= 1.2 mm, Dc = 20 mm and D_P = 30 mm

From all the reactors that tested, we select on the DBD reactor with a construction which was capable to produce large ozone concentration. Construction selected is reactor that has components such as: D_W = 1.2 mm, D_C = 20 mm and D_P = 30 mm, L_P= 150 mm. Figure 6 shows that the change in the number of windings coil (N) affect the value concentration of ozone generated. It is closely related to the electric field distribution and decomposition of ozone.

When the number of windings coil on the more active electrode electric field between the electrodes is much more uniform and almost complete in the area so that the ionization electrode is also much more intense and uniform. The concentration of ozone produced is also higher. In the figure 6, we can see that windings coil N = 50 coil spools produce ozone concentration is more high than N= 30 and 40 coil spools, with less power consumption.

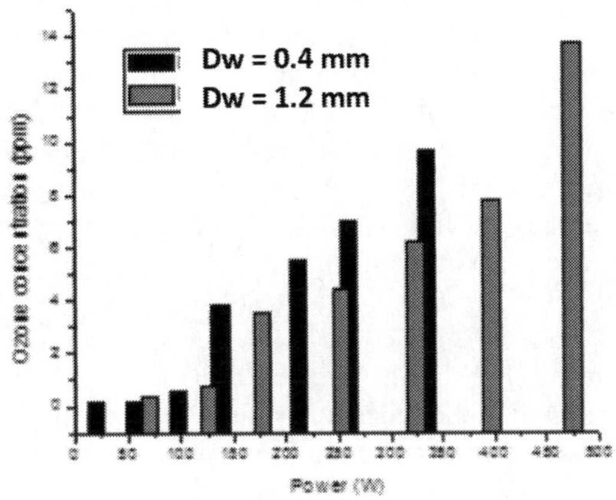

Figure 6. Ozone concentration as function of power for coil spools N = 50, for D_W =0.4 mm, Dw= 1.2 mm, Dc = 15 mm, and DP = 30 mm

The graph shown in Figure 7 is ozone concentration as a function of power. From the figure it can be seen that no ozone is generated for all input power. It is very influential here is the distance between the active electrode with Pyrex tube. This graph is for diameter spiral made of copper wire of 9.2 mm or 4.8 mm from the axis (radius).

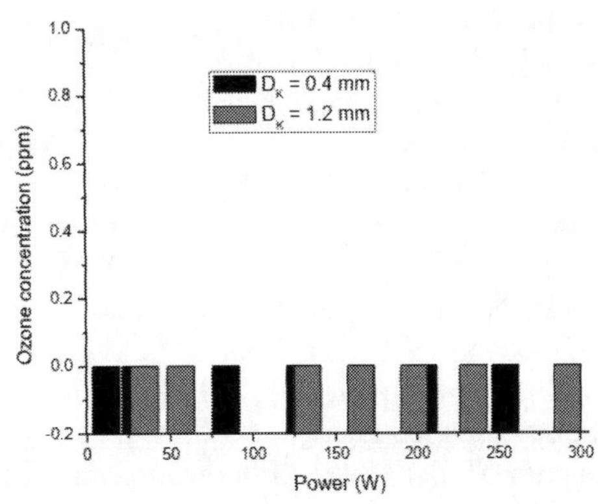

Figure 7. Ozone concentration as function of power for D_W = 0.4 mm, D_w= 1.2 mm, Dc = 9.2 mm, N = 50 coil spools and D_P = 30 mm

So that the edge of the active electrode to a distance of 11.2 mm tube (tube radius = 15 mm). The distance is so large that the electron energy is not enough to dissociate the oxygen molecules. Furthermore, tree bodies reaction does not take place so that ozone is not formed.

Results are shown in Figure 7 corresponds with previous research that has been reported [16]. Differences in diameter spiral-shaped of the active electrode give effect to the concentration of ozone generated. The larger the diameter spiral-shaped of the active electrode closer to the Pyrex, it can generate the greater the concentration of ozone.

4. Conclusions
Ozone concentrations greater when the wire's diameter D_W and the diameter of the coil windings applied was greater. We found that impedance greater will minimize the concentration of ozone, in contrary to the greater capacitance will increase the concentration of ozone. The ozone concentrations increase with augmenting of power. Maximum power is effective at DBD reactor spiral-cylinder is on the diameter of coil winding (D_C = 20 mm), diameter of wire (D_W = 1.2 mm), and the number of coil windings (N = 10 loops) with the ozone concentration produced is greater than 20 ppm and it consumes energy of 177.60 watts. Differences in diameter spiral-shaped of the active electrode give effect to the concentration of ozone generated. The larger the diameter spiral-shaped of the active electrode closer to the Pyrex, it can generate the greater the ozone concentration.

5. Acknowledgment
All members of group of research thanks to Ministry of Research, Technology and High Education for financial support of this research in the schema Flagship Research for High Education (PUPT Program) of the year 2016 with contract number 176-09/UN7.5.1/PG/2016.

6. References
[1] Changquan Wang, Guixin Zhang, and Xinxin Wang 2012 *Journal fo Vacuum* 86 960-964.
[2] S. Jodpimai, S. Boonduang, and P. Limsuwan, 2015 *Journal of Electrostatics* 74, 108-114
[3] Bogaerts, A., Neyts, E., Gijbels, R., and van der Mullen, J 2002 *Spectrochimica Acta Part B: Atomic Spectroscopy* 57 609-658.
[4] Nur, M. Supriati, A. Setyaningrum, D.H. Gunawan, Munir, M., and Sumariyah 2009 *Journal of Berkala Fisika* 12 69-76.
[5] R.G. Rice, 1999 *Ozone. Sci.Eng.* 21 99-118.
[6] A. Bohme 1999, *Ozone. Sci. Eng.* 21 163-176.
[7] Fridman, A. 2012 *Cambridge University Press,* Cambridge
[8] Muhammad Nur, Aribat Solichin, Endang Kusdiayantini, Tri A. Winarni, Susilo, Dian Arif Rahman, Resti Maryam, Sosiowati Teke, Wuryanti, and Harjum Muharam 2013 *The IEEE, Conference Publications, ICICI-BME,* 221-225.
[9] Pongsathon Jitsomboonmit, Mudtorleb Nisoa, and Somsak Dangtip 2012 *Journal of Physics Procedia* 32 723-731.
[10] F. Massines, N. Gherardi, N. Naudé, and P. Ségur 2005 *IOP Science, Plasma Physics and Controlled Fusion* 47 577-588.
[11] Mochammad Facta, Zainal bin Salam and Zolkafle Bin Buntat, 2009 *IEEE Conference Publications,* CITISIA 2009 pp 440 - 445
[12] Jochen Kriegseis, Benjamin Möller, Sven Grundmann, and Cameron Tropea 2011 *Journal of Electrostatics* 69 302-312.
[13] Jae-Duk Moon and Jae-Seung Jung 2007 *Journal of Electrostatics* 65 660-666.
[14] A. Lolthus and P. H. Krupenie 1977 *Journal of Physical and Chemical Reference Data* 6.
[15] Terrell, C.A., Hansen, D.L., Ajello, J.M., 2004, *Journal Of Physics B: Atomic, Molecular And Optical Physics* 37 1931–1949.
[16] Muhammad Nur, Maryam Restiwijaya, Tri Agustina Winarni 2014 *IEEE Conference Publications,* ISTMET 2014, pp. 129 - 132

Study of an atmospheric pressure plasma jet of Argon generated by column dielectric barrier discharge

M. Nur[1,2,*], A. W. Kinandana[1], P. Winarto[1], Z. Muhlisin[1,2] and Nasrudin[3]

[1]Physics Department, Faculty of Science and Mathematics, Diponegoro University Jl. Prof. Soedarto, SH, Tembalang, Semarang. Central Java, Indonesia. 50275,

[2]Center for Plasma Research, Faculty of Science and Mathematics, Diponegoro University, Jl. Prof. Soedarto, SH, Tembalang, Semarang. Central Java, Indonesia. 50275,

[3]Faculty of Health Science, Universitas Muhammadiyah Magelang, Magelang, Central Java, Indonesia

E-mail: *m.nur@undip.ac.id

Abstract. An atmospheric of argon plasma jet was generated by using column dielectric barrier discharge has been investigated. In this study, argon gas was passed through the capillary column by regulating the flow rate of gas. This atmospheric pressure plasma jet (APPJ) was generated by a sinusoidal AC high voltage in the range of 0.4 kV to 10 kV and at frequencies of 15 kHz and 26 kHz. APPJ has been produced with flow rate of argon gas from 1 litter/min - 10 litters/min. The electric current has been taken with variation of voltage and each interval argon gas flow rate of 1 litter/min. The results show that electric current increase linearly and then it trends to saturation condition by the increasing of applied voltage. We found also that the length of the plasma jet increase by augmenting of applied voltage both for frequencies of 15 kHz and 26 kHz. Furthermore, our results show that length of plasma jet optimum for flow rate of argon gas of 2 litters/minute. In addition, we obtained that the larger applied voltage, the greater the temperature of the plasma jet.

1. Introduction

In the last decades, cold atmospheric pressure plasma jets (APPJ) have attracted much attention due to their versatility, low-cost operation and also ability to produce reactive chemistry at room temperature [1]. Many applications have been developed and studied using the discharge plasma based on different geometries using various materials for electrodes. This great flexibility in producing plasma causes the diversity of the use of plasma in a wide range of applications in industrial, energy, biomedical, analytical and quality of life [1,2,3]. Plasma Jet is the result of ionized gas flow from plasma discharge column [1,2]. Plasma jet shaped like flares. The phenomenon of atmospheric pressure plasma jet (APPJ) can date back to 1960s when it was first generated as a local thermodynamic equilibrium plasma jet. Cold plasma jet in atmospheric pressure can be applied to material processing and biomedical applications such as disinfectant and decontamination [4] and for medicine such as wound healing [2,5,6]. In the atmospheric pressure plasma jet has various dimensions of coverage, ranging from the micrometre scale suitable for local treatment and accurate, up to large scale suitable for large scale sample treatment. So many different configurations and applications of non-thermal plasma [7]. Plasma jet can be generated with a length up to several centimetres and can be touched by hands. Argon plasma jet in atmospheric pressure at room temperature can be produced from a stainless steel electrode in dielectric barrier discharge system [7]. Dielectric barrier discharge is also called as a silent discharge plasma, the discharge can be used to produce non-equilibrium atmospheric pressure plasma. Discharge between the electrodes, one or both are coated by dielectric layer [8]. In the produce plasma jet, argon gas was flowed in the reactor plasma reactor. Argon gas was included in the group of noble gases which are unreactive element [9,10]. Noble gas is added so that the process of ionization, excitation and dissociation in the plasma takes place optimally. Working gas is used in the range of atmospheric pressure, or a device to generate plasma in the open air [3]. Therefore, the need for research to determine the characteristics of the plasma jet using dielectric barrier discharge column by the addition of argon gas. In this paper, we presented our results on characteristic of APPJ argon gas such as, I-V characteristic, length and temperature of jet influence of argon gas flow rate and influence of applied voltage.

2. Research Methods.

This research was conducted by characterizing of argon plasma jet using a dielectric barrier discharge column at atmospheric pressure. This atmospheric pressure plasma jet (APPJ) was generated by AC high voltage. A sinusoidal voltage at 15 kHz and 26 kHz are applied for the excitation and sustaining of the discharges. AC high voltage source connected with a resistor of 100 kΩ and HV probe, of the resistor is then connected to the upper electrode.

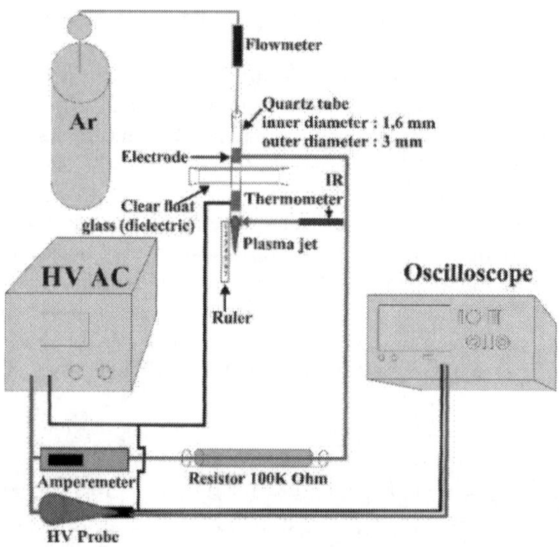

Figure 1. Schematic of research equipment set-up

Output of AC high voltage source was connected to the bottom electrode and the HV probe. HV probe connected to an oscilloscope (GW, GOS 563, 50 MHz). Argon plasma jet was generated by using capillary column as tube of jet. Capillary column made of quartz glass with an inner diameter of 1.6 mm and an outer diameter of 3 mm with a length of 100 mm. At the center of the length capillary column, we added a clear glass with a length of 100 mm, width of 74 mm and a thickness of 3 mm. The clear glass serves as a dielectric barrier (dielectric barrier discharge) to separate the two electrodes. The electrodes used in this study is made of a thin layer of aluminium (aluminium foil) encircling the quartz capillary column. We used two type of reactor. Reactor "A" with the top electrode length of 20 mm and 30 mm of lower electrode, the distance between the electrodes of 10 mm and the spacing between lower electrode to the tip of the capillary column is 5 mm and HV-AC with frequency at 26 kHZ. Other, reactor B with the top electrode length of 30 mm and 20 mm of lower electrode, the distance between the

electrodes of 10 mm and the spacing between lower electrode to the tip of the capillary column is 5 mm, and HV-AC with frequency at 15 kHZ.
Argon gas as the working gas (Ar, 99.95%) is inserted into the capillary column by regulating the gas flow rate in the range of 1-10 L/min with intervals of 1 L/min. Voltage and electronic signals are generated can be observed using an oscilloscope. Electrical current was measured using ammeters (Kyoritsu, AC/DC digital Clamp meter). Gas flow rate was measured using a flow meter (Koploc Kojima Model RK 1600 R). The jet length is measured using a ruler, long jet results are documented using a digital camera. The temperature of plasma jet was measured by IR thermometer (Krisbow, KW06-280). The research equipment scheme can be seen in figure 1. The signal frequency was fixed at 26.0 kHz and 15 kHz the voltage amplitude was varied in the range of 0.4 - 10.0 kVpp. The exit of the DBD reactor, located on its bottoms side, is connected to a commercial flexible plastic tube (3.2 mm outer diameter, 2 mm inner diameter), which guides the plasma species downstream. At certain conditions (especially if a conductive object is closed to the tube end) a plasma jet can be extracted from the plastic tube.

3. Results and Discussion
Characteristics of Electrical Current (I) - Applied Voltage (V)
In this study, we carried out experiment of variation in voltage from 0.4 kV to 10 kV. Average of electrical current as function of voltage for argon gas flow rate of 1-10 liters/min, can be seen in graph shown in Figure 2.

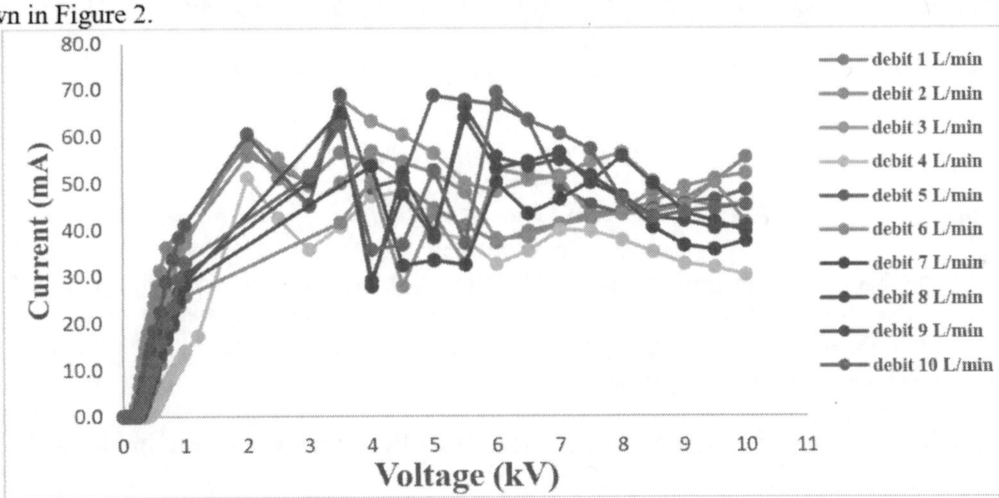

Figure 2. Characteristic of electric current - applied voltage, with power frequency of 26 kHz

Figure 2 shows that there are two areas of current as a function of voltage. The first regions (with applied voltage of 0.4 kV – 2 kV) indicates that the discharge current rises linearly with increasing voltage. In this area can be approached as if the laws apply ohm (called "ohmic" region). After this region, we found that electric current tends to volatile saturate with high range diversity (from 28 mA to 68 mA). Characteristic of I-V in the figure 2, has been taken from reactor A plasma jet with frequency of power supply at 26 kHz. Electric current will increase along with the increase in voltage up to 1 kV, then there will be saturation current at a voltage of 1 kV to 10 kV. The peak of maximum current occurs at a voltage of 3.5- 4 kV and the maximum current of 68.9 mA current occurs in argon gas flow rate of 5 L/min.
Generally, the characteristic of electric current as function of applied voltage on argon plasma jet using a dielectric barrier discharge column at atmospheric pressure, after the electric current start to appear, the electric current will increase proportional to the voltage, this region is categorized as ohmic region. Ohmic region is the enactment area of law Ohm.

The total current in the plasma jet is a combination of capacitive currents to currents. High-energy electrons striking atoms or molecules will transform kinetic energy to produce ionization energy of excited species. The absorption of collision energy by electrons resulting in the displacement of electrons of lower energy levels to higher energy levels. Excitation will take place temporary and the electrons will return to its original energy level accompanied by emission of photons. Interactions that occur repeatedly and continuously, which results generated currents going up and down.

Characteristic of Jet Length as Function of Voltage (V)

Based on the results, the plasma jet can be seen inside and outside the capillary column. The plasma jet emerging and the voltage is increased, the plasma jet will be pushed out through the end of capillary column. Length of the plasma jet was measured from the end of the capillary column to the end of the plasma jet. Plasma jet length will increase with the increase of applied voltage and reaches the maximum length in highest voltage. The measurement results can be seen in the Figure 3.

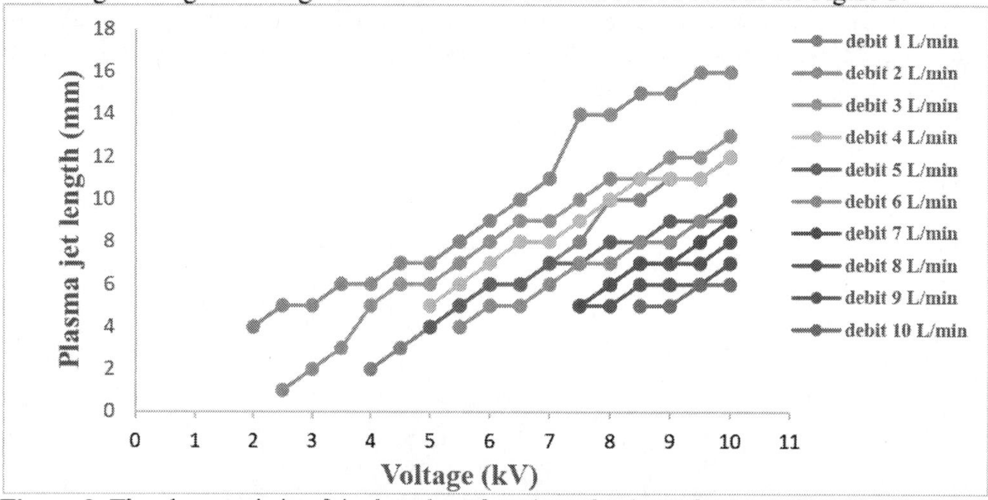

Figure 3. The characteristic of jet length as function of voltage for several gas flow rate and frequency of power supply at 26 kHz.

In general, the characteristic of electric current as function of voltage in argon plasma jet using dielectric barrier discharge in atmospheric pressure, jet length will increase by the increase of voltage. It is occurring due to the role of electrons in plasma jet generation, when the voltage increased, electric field will increase and there are more electron striking the argon gas atoms, so that the plasma jet will become longer due to the encouragement plasma species of argon gas comes out of the capillary column. Photographs of plasma jet on argon gas flow rate of 2 litters/min can be seen in Figure 4.

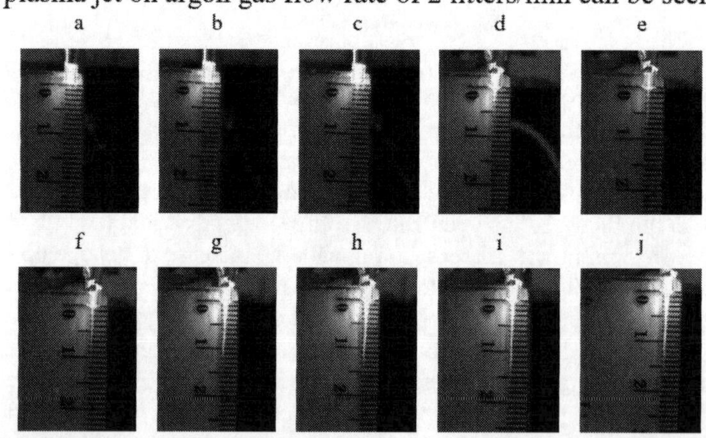

Figure 4. Photographs of plasma jet on Argon gas flow rate of 2 L/min

This flow rate generates the optimum length of plasma (see also Figure 6). This photographs indicates the ten length of the plasma jets. Each length is a length of a jet for a certain value of voltage. So the letter "a" to "j" marking the change in length from the lowest voltage to largest.

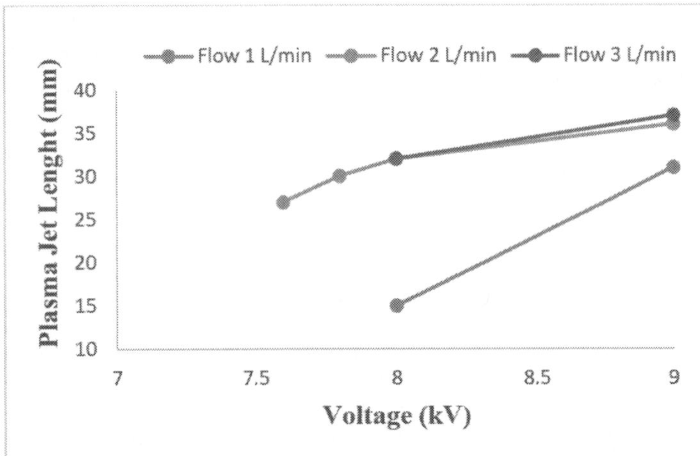

Figure 5. The characteristic of jet length as function of voltage for several flow rate and frequency of power supply at 15 kHz.

Characteristic of Jet Length as Function of Argon Gas Flow Rate (L/minutes)

In this study, the characteristic of jet length as function of Argon gas flow rate (L/min) for voltages variation of 4-10 kV, can be seen in figure 4. This treatment is to determine the length of the plasma jet is based on the increase in argon gas flow rate at a certain voltage.

By depicting the relation between jet length as function of Argon gas flow rate in voltage variation of 4-10 kV obtain the same pattern. The jet length will increase begin the Argon gas flow rate of 1-2 L/min. When the Argon gas flow rate more than 2 L/min, jet length will decrease gradually and when the voltage increase will increase the jet length. At a voltage of 10 kV obtained the longest plasma jets for all Argon gas flow rate. At Argon gas flow rate of 2 L/min obtained the longest plasma jets for all voltages, with a length of 16 mm.

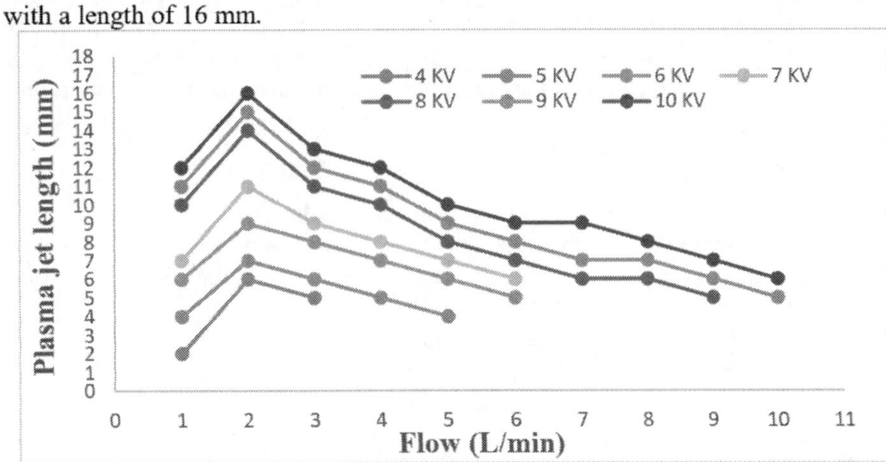

Figure 6. The characteristic of jet length as function of flow rate, for freq. of power 26 kHz.

The typical shape of plasma jet length as a function of argon gas flow rate, the minimum voltage for generating plasma jets is 4 kV at the same argon gas flow rate, the jet length will increase along with the increase of applied voltage and the maximum length of the jet is in voltage of 10 kV at the same Argon gas flow rate. While the minimum jet length of 2 mm occurs at voltage of 4 kV with Argon gas flow rate of 1 L/min, and the maximum jet length of 17 mm occurs at a voltage of 10 kV at Argon gas flow rate of 2 L/min.

Based on observations in this study, it was found that the difference does not only happen on the length of the jet but there are also differences in the plasma jet forms in terms of the diameter of the jet. At lower Argon gas flow rate is obtaining the smaller diameter plasma jet and at the higher gas flow rate is obtaining the larger diameter plasma jet. It occurs due to the higher gas flow rates resulting the higher pressure of gas thereby increasing the ability to push the plasma jet out of the capillary column so that the diameter will be larger and due to the magnitude of this pressure results in reduced length of the plasma jet.

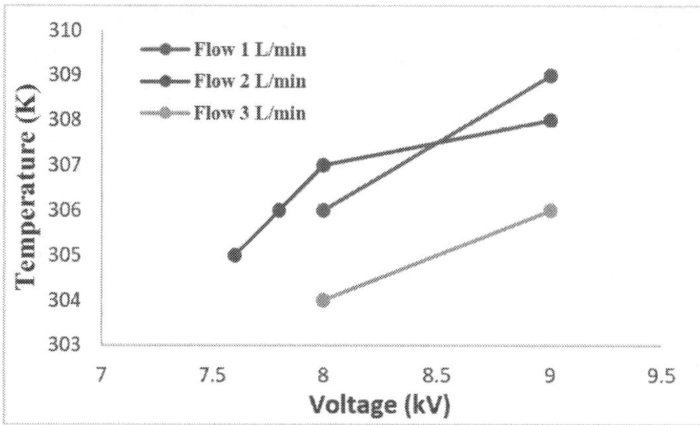

Figure 7. The temperature of plasma jet as function of applied voltage, with frequency of 15 kHz

Figure 7 shows the temperature of the plasma jet as a function of voltage for AC power supply frequency of 15 kHz. In this picture, it can be seen that at a lower voltage the biggest plasma jet temperature of 307 K for an operating voltage of 8 kV. Temperature in figure 7 are closed with rotational temperatures that where found by Hong et al [8]. Overall Figure 7 shows that the larger applied voltage, the greater the temperature of the plasma jet

4. Conclusions

On the characteristics of the current (I) - voltage (V) is obtained that the electric current will increase with increasing of voltage named as "ohmic" region for beginning of discharge. After the plasma jet appears the current tends to volatile saturate. The length of plasma jet increases with increasing applied voltage both for frequencies of voltage signal at 26 kHz and 15 kHz. The influence of argon gas flow rate to plasma jet length has been obtained the optimum value of flow rate at 2 litters/minute. We have found that the temperature of the plasma jet is getting higher with the growing power voltage.

5. References

[1] Harry J E 2010 *Wiley-Vch Verlag GmbH & Co. KGaA*, Weinheim

[2] Nasruddin, Kanae Yukari, Nakajima, Mukai, Komatsu Emi, Setyowati Heni, Rahayu Esti, Nur Muhammad, Ishijima Tatsuo, Enomoto Hiroshi, Uesugi Yoshihiko, Sugama Junko, Toshio, Nakatani 2015 *Plasma Processes and Polymers*, 12, 1128–1138

[3] Nur M 2011 *Diponegoro University Press,* Semarang

[4] Zhang S, Sobota A, van Veldhuizen E M and Bruggeman P J 2015 *J. Phys. D. Appl. Phys.* 48 015203

[5] X. Lu, M. Laroussi and V. Puech 2012 *Plasma Sources Sci. Technol.* vol. 21 pp. 777-788

[6] Nasrudin, Yukari Nakajima, Kanae Mukai, Heni Setyowati Esti Rahayu, Muhammad Nur, Tatsuo Ishijima, Hiroshi Enomoto, Yoshihiko Uesugi, Junko Sugama, Toshio Nakatani 2014 *Clinical Plasma Medicine*, 2, 28-35

[7] Kunhardt E E 2000 *IEEE Trans. Plasma Sci.* 28 189–200

[8] Hong Y, Lu N, Pan J, Li J and Wu Y 2013 *Thin Solid Films* 531 408–14

[9] Höllig A, Schug A, Fahlenkamp A V., Rossaint R, Coburn M, Brucken A, Fries M, Kepp O, Lemaire M, Ma D, Magalon G, Michel P P, Neyrinck A, Pype J, Rex S, Sanders R D, Savage S and Stoppe C 2014 *Int. J. Mol. Sci.* 15 18175–96

[10] Olenici-Craciunescu S-B 2011 *Dortmund University Press,* Dortmund

The prospect of uranium nitride (UN-PuN) fuel for 25-100MWe gas cooled fast reactor long life without refuelling

R. D. Syarifah, Z. Su'ud, K. Basar and D. Irwanto
Nuclear Physics and Biophysics Research Division, Physics Department, Faculty of Mathematics and Natural Science, Bandung Institute of Technology Indonesia, Jalan Ganesha 10 Bandung 40132, INDONESIA

E-mail: syarifah.physics@gmail.com

Abstract. The prospect of uranium nitride (UN-PuN) fuel for 25-100MWe Gas Cooled Fast Reactor has been done. This research use helium coolant which has low neutron moderation, chemical inert and single phase. This study use natural uranium and plutonium. Plutonium taken from spent fuel of LWR (Light Water Reactor). So, it can reduced spent fuel in the world. The calculation use SRAC2006 and JENDL 4.0 for the data libraries. First, we calculate PIJ for fuel pin cell calculation and CITATION for core calculation. The reflector radial-axial width is 50 cm. The variation of fuel fraction is 40% until 65%, cladding 10%, and moderator 25% up to 50%. The variation of the power is 75-300 MWth (25-100 MWe). The calculation of survey parameter has been done. The variation of percentage plutonium is 7% up to 13%. We have optimum k-eff value in percentage of plutonium 11%. The high powers cause k-eff value high too. Second, the core configuration divided by three variation fuel (F1, F2, and F3). F1 is located in the central core, F2 middle core and F3 outer core. The variation percentage Plutonium for fuel F1:F2:F3 = 8%:10%:12%. The increasing power level make the burn up level increase. All case can reach burn up time plus than 20 years. The thermal powers increase cause the peak power density increase. The power 150 MWth, 225 MWth, and 300 MWth have excess reactivity (%Δk/k) less than 2%.

1. Introduction

Nuclear power plant has been characterized by substantial growth not only in numbers but also in plant sizes, with economies of scale favouring the large plants [1]. According to the IAEA (International Atomic Energy Agency) definition an SMPR is an electricity producing nuclear power plant in the capacity range of 100 to 500 MWe (bellow 600MWe) [1,2]. In response to important commercial developments, the energy range of small and medium reactors is now taken as being up to around 700 MW [3]. These guidelines are mainly intended for decision makers in developing countries interested in embarking on a nuclear power programme [2]. The main incentives for these designs have been improvements in safety, reliability and economics [3]. Also proposed for use in a relatively isolated area, however, should satisfy some very important requirements such as easy operation, easy maintenance, easy construction and decommissioning, inherent/passive safety, and proliferation resistance [4].
Since June 1990 there is a team for small nuclear electricity reactor has been performed. It is responsible for assessment and design of a small reactor for electricity and/or sea-water desalination

for small islands and isolated in Indonesia [5,6]. This research the reactor is designed without on-site refuelling, its means it has a capability to operate long life consistent with the plant economics and energy security [7]. The design of small fast reactor has been design with higher fuel volume fraction [8-11]. Beforehand, small and medium GFR with modified CANDLE burn up scheme with uranium natural has been analyze before [12-15]. In this research, In this research The prospect of uranium nitride (UN-PuN) fuel for 25-100MWe Gas Cooled Fast Reactor long life without refuelling has been done.

2. Design Concept and Calculation Methods

Table 1 shows parameter design of fuel pin in the reactor. The fuel use uranium plutonium nitride fuel. Silicon carbide for the cladding is good choice to material cladding because it has high melting point, low chemical activity, no appreciable creep at high temperature, and low neutron absorption cross section.

Table 1. Parameter design of fuel pin in the reactor

Parameter	Specification
Fuel material	UN-PuN
Cladding material	SiC
Coolant material	Helium
Fuel volume fraction	40% - 65%
Cladding volume fraction	10%
Coolant volume fraction	30% - 50%
Pin pitch	1.45 cm

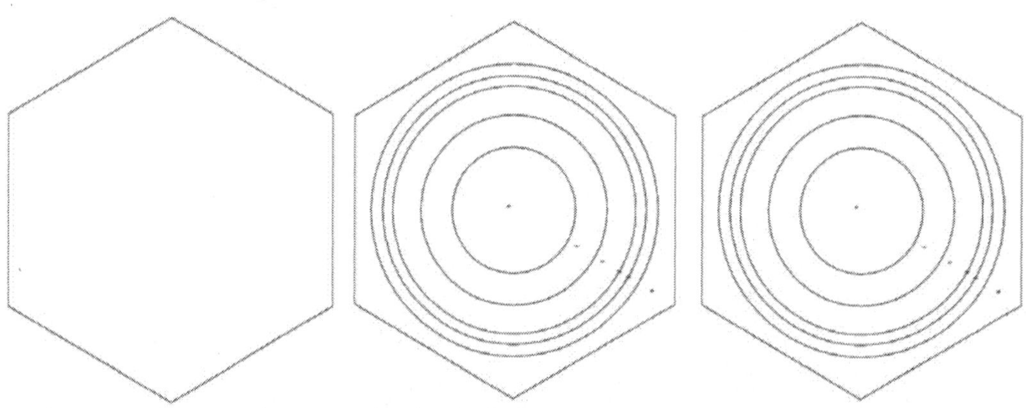

Figure 1. Hexagonal geometries for fuel pin calculation (PIJ) by SRAC2006 results calculation

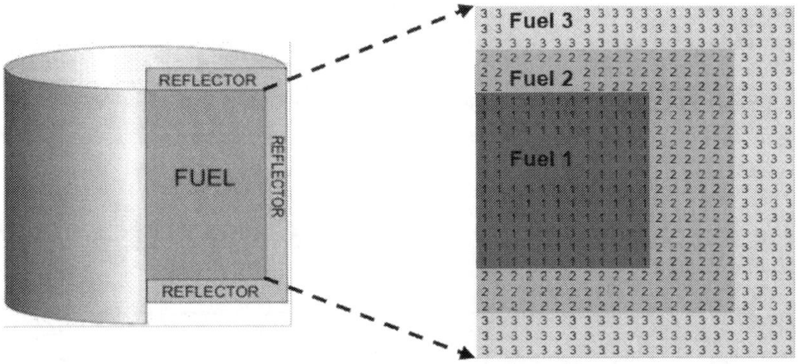

Figure 2. Half heterogeneous core configuration with different width in variation fuel

Fig. 1 show hexagonal geometries for fuel pin calculation (PIJ) by SRAC2006 results calculation. Fig. 2 show half heterogeneous core configuration with different width in variation fuel. There is three variation fuel, F1, F2 and F3 which F1 is in the center, F2 in the middle and F3 in periphery, and after that there is a reflector. Percentage F1:F2:F3 = 8%:10%:12%.

The calculation method use SRAC 2006 and libraries JENDL 4.0. It is designed to permit neutronics calculation for various type reactors. SRAC covers production of effective microscopic and macroscopic group cross section, and static cell and core calculation including burn-up analysis. First, it calculate fuel pin cell PIJ with Collision Probability Method (CPM). In PIJ calculation, we get k-inf, burn-up analysis and more. After that, it continues to calculate core reactor with CITATION, with various core configuration [16].

3. Results and Discussion

Table 2 show the results of parameter optimization UN-PuN fuel in GFR use SRAC calculation. The power of the reactors is varied from 75-300 MWth (25-100 MWe). The core geometry design is set to be pancake geometry, means the diameter width is greater than the height (D>H). The width of reflector is set same for all case i.e. 50 cm. The enrichment variation for F1 : F2 : F3 = 8% : 10% : 12%. The optimum results can be reach if the reactor have effective multiplication factor (k_{eff}) around one ($k_{eff} \geq 1$), means has excess reactivity less than 2% %Δk/k. Multiplication factor is number of fission in one generation divide by number of fission in preceding generation.

$$excess.reactivity(\%) = \frac{(k_{eff} - 1)}{k_{eff}} x100\% \quad (1)$$

All case constrained to have maximum excess reactivity less or around 2% %Δk/k. The burn up level is increases as the power level is increase. For 225MWth the burn up level is less than 150MWth due to the addition of width and height active core. The addition is to find the optimum excess reactivity value (2% %Δk/k).

Table 2. Result of parameter optimization of UN-PuN fuel in GFR

Thermal power (MWt)	Width active core	Height active core	Volume fraction (%)			Burnup level (MWd/ton)	
			Fuel	Cladding	Coolant	Average	Peak
75	175	100	60.5	10	29.5	1.37E+04	2.75E+04
150	175	100	57.5	10	32.5	2.90E+04	5.80E+04
225	195	125	58	10	32	2.64E+04	5.27E+04
300	195	125	55	10	35	3.51E+04	7.03E+04

Table 3 shows the main results or performance UN-PuN fuel in gas cooled fast reactor without refueling. This result was obtained after we calculate the reactor core (CITATION calculation) with all results parameter in Table 2. All case can reach burn up time plus than 20 years. The breeding ratio for 75MWth is largest than the others, due to the thermal power is smallest than the others. When the thermal powers increase cause the peak power density increase. The range of power density is 25 until 114 Watt/cc. For case 2 until 4 has excess reactivity (%Δk/k) less than 2%.

Table 3. Performance of UN-PuN fueled gas cooled fast reactors without refueling

Thermal power (MWt)	Breeding / convertion ratio	Peak power density (W/cc)	%Δk/k
75	1.325	25	2.04
150	1.244	53.5	1.53
225	1.268	77	1.89
300	1.231	114	1.78

Figure 3 show effective multiplication factor (k_{eff}) value of gas cooled fast reactor with various power. The trend of the graph is similar for 150-300MWth. The trend of the graph of 75MWth different because the power is too small, so the number of fission is less than the others cause the graph is decrease until 14 years burn up time after that it increase. The declining graph explains that it is burning and the rising graphs explain breeding. So for 150MWth until 300MWth after the fifth burn up time the reactor is breeding.

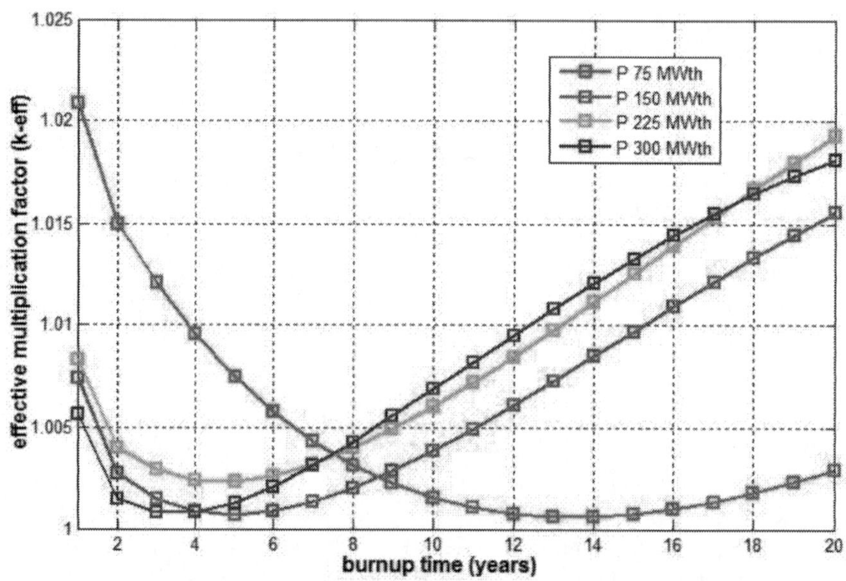

Figure 3. Effective multiplication factor (k-eff) value of gas cooled fast reactor with various power

4. Conclusion

The prospect of uranium nitride (UN-PuN) fuel for 25-100MWe Gas Cooled Fast Reactor has been done. The variation percentage Plutonium for fuel F1:F2:F3 = 8%:10%:12%. The increasing power level make the burn up level increase. All case can reach burn up time plus than 20 years. The thermal powers increase cause the peak power density increase. For all power have similar trend of k-eff value except 75MWth because of the power is too small, but all power experienced two phases of burning and breeding.

5. Acknowledgment

This work was supported by Decentralization Research Grand 2016, DIKTI, Ministry of Research Technology and Higher Education of the Republic of Indonesia which are gratefully acknowledged.

References

[1] IAEA-TECDOC-347 1985 *Small and medium power reactors : Project initiation study phase I* (International Atomic Energy Agency) p 7
[2] IAEA-TECDOC-445 1987 *Small and Medium power reactors 1987* (International Atomic Energy Agency) p 9
[3] IAEA-TECDOC-881 1996 *Design and development status of small and medium reactors system 1995* (International Atomic Energy Agency) p 7
[4] Hiroshi S and Zaki S 1995 *Nuclear Technology* **109** pp 307-313
[5] M Nurdin 1992 Proc. Int. Specialists Mtg. Potential of Small Nuclear Reactors for Future Clean and Safe Energy Sources *Elsevier* pp 145
[6] Zaki S 2006 Strategi pengembangan riset dalam bidang iptek nuklir dalam rangka penyiapan SDM yang berkualitas tinggi *Seminar nasional II SDM Teknologi Nuklir* **ISSN 1978-0176** pp 1-8
[7] IAEA-TECDOC-1536 2007 *Status of Small reactors design without on-site refueling* (International Atomic Energy Agency) p 9
[8] Zaki S 2008 *Progress in Nuclear Energy* **50** pp 276-278
[9] Zaki S 2008 *Progress in Nuclear Energy* **50** pp 157-162
[10] Zaki S Bakrie A and Sedyartomo S 2005 *Progress in nuclear energy* **47** pp. 212-221
[11] Zaki S and Hiroshi S 1993 *Nuclear engineering and design* **140** pp 251-260
[12] Fiber M Zaki S Abdul W Khairul B Menik A and Hiroshi S 2013 *Advanced Material Research* **772** pp 501-506
[13] Menik A M Ali S Ade G A and Zaki S 2010 *Proceeding Seminar dan Workshop Nasional Pendidikan Teknik Elektro (SWNE)* pp 93-98
[14] Menik A Zaki S and Fiber M 2013 *Jurnal Ilmu dasar* **14** pp 11-15
[15] Zaki S and Hiroshi S 2013 *Annals of Nuclear Energy* **54** pp 58-66
[16] Okumura K 2002 *SRACPP version 2002* (Japan Atomic Energy Research Institute) p 1-28

Performance of polymer nano composite membrane electrode assembly using Alginate as a dopant in polymer electrolyte membrane fuel cell

S. Mulijani
Department of Chemistry Bogor Agriculture University,
Darmaga Campus, INDONESIA

E-mail: srimulijani@ipb.ac.id

Abstract. Polymer membrane and composite polymer for membrane electrode assembly (MEAs) are synthesized and studied for usage in direct methanol fuel cell (DMFC). In this study, we prepared 3 type of MEAs, polystyrene (PS), sulfonated polystyrene (SPS) and composite polymer SPS-alginat membrane via catalyst hot pressed method. The performance and properties of prepared MEAs were evaluated and analyzed by impedance spectrometry and scanning electron microscopy (SEM). The result showed that, water up take of MEA composite polymer SPS-alginate was obtained higher than that in SPS and PS. The proton conductivity of MEA-SPS-alginate was also higher than that PS and PSS. SEM characterization revealed that the intimate contact between the carbon catalyst layers (CL) and the membranes, and the uniformly porous structure correlate positively with the MEAs prepared by hot pressed method, exhibiting high performances for DMFC.

1. Introduction

Polymer electrolyte membrane fuel cell (PEMFC) has great potential for movable and stationary power supply application wherein the fuel cell must be capable of operating at high power density with low content of catalyst to reduce its cost. In order to reduce the prime cost of PEMFC, the performance improvement of membrane electrode assembly (MEA) becomes a key point. MEA typically consists of proton exchange membrane (PEM), catalyst layers and gas diffusion layers (GDLs), in which the electrolyte, catalytic particles and reactants have a close contact with each other. The GDLs facilitate transport of reactants and products between fuel cell flow in the field and the ionomer electrolytes-catalysts interface. The state-of the art MEA consists of perfluorosulfonic acid (PFSA) polymer membranes such as Nafion® and Pt/C electrocatalysts. In spite of good physical and chemical stability and high proton conductivity, PFSA suffers from limited operation temperature, high cost and environmental concerns [1-2]. Therefore, significant effort has been made to develop nonfluorinated hydrocarbon polymers based on aromatic plastics, such as sulfonated polystyrene (SPS), sulfonated polysulfone (SPSf), sulfonated poly(ether ether ketone) (SPEEK), sulfonated polybenziimmidazole (SPBI) and sulfonated poly (aryl ether kethone) (SPAEK) as an alternative proton

exchange membranes for PEMFCs. These sulfonated aromatic polymers possess excellent chemical resistance, high thermo-oxidative stability, good mechanical properties and low cost [3-4].
MEA for PEMFCs is generally fabricated by the hot pressing method, wherein a PEM is sandwiched between an anode and a cathode electrode and pressed at a high temperature. However, the MEAs fabricated by the conventional hot pressing method are usually inclined to delamination both at anode and cathode interfaces after long-term operation, leading to degradation in cell performance. Thus, conventional hot-pressing method can only offer physical contact, rather than achieving a true interfacial contact between the PEM and the electrodes during the MEAs fabrication process. Some researchers faced delamination problem in hot-pressing method in case of using liquid methanol as a fuel in a direct methanol fuel cell (DMFC), because swelling of the PEM in methanol solution which can greatly affect the intimate contact between the PEM and electrodes in MEAs. In this context, it is essential to mention some of the recent studies in which effort has been made to improve the interfacial contact in the MEAs for DMFC. Liang *et al.*[5] introduced a simple method for fabricating the MEAs as compared to dry

2. Experimental Methods

Polymer material used in the study was prepared from styrofoam (PS) waste and it's synthesized by sulfonation reaction. Membranes were prepared by dissolving a known amount of sulfonated poly styrene (SPS) in dichloromethane (10%) under constant stirring. Polymer SPS-Na-alginate nanocomposite membranes were prepared by mixing SPS and Na-alginate and dissolving in dichloromethane with ultrsonication process.Three types of proton exchange membranes PS, SPS and SPS-Na-alginate were used as the polymer electrolyte membrane for MEA preparation. Before use, the membranes were pretreated in hydrogen peroxide solution (5 wt.%) at 80℃ for approximately 1 h; after being flushed with deionized water, it was transferred into a 0.5M sulphuric acid solution and boiled at 80℃ for an additional 1 h, then washed with deionized water [8].
The catalyst inks were prepared by dispersing catalysts powder into a mixture of dichloromethane and membrane solution. The catalyst used for anode and cathode layers were Ag/C (20 wt.% Ag/AgCl) and 5% membrane solution. The mixture was ultrasonically dispersed for 30 min before being used. All MEAs used in this study were prepared using hot pressing method at 120-125℃ for approximately 5 min.
Electrochemical impedance was conducted for a single cell by feeding 1.0M methanol at a flow rate of 1mlmin$-^1$ using a high-pressure piston pump and by purging dry oxygen gas into the cathode at a flow rate of 20mlmin$-^1$ at 40°C at a constant dc potential controlled by the potentiostat with applying a small alternating voltage (10mV) and varying the frequency of the alternating voltage from 1×10^4 to $1\times10-^2$ Hz. Scanning electron microscope (SEM) was used to evaluate the cross section of the MEA. The sample with gold sputter coatings was carried out on the desired membrane samples at pressure ranging between 1 and 0.1 Pa. The DMFC performance test was conducted at 40.0C by feeding 1.0M methanol at a flow rate of 1mlmin$-^1$ using a high-pressure piston pump and by feeding dry oxygen gas into the cathode at a flow rate of 20mlmin^{-1} at ambient pressure by fixing the load current, which was controlled with an electric load system.

3. Result and Discussion

The DMFC performance was evaluated through by polarization curve, also known as a voltage–current density (V-I) curve, is commonly used to evaluate fuel cell performance. Figure 1 is a representation of typical polarization compares the performance of the single cells with the membrane PS, SPS, SPS-alginate and MEA-PS, MEA-SPS, MEA-SPS alginate with 1.0 M methanol at 50C. The Figure 1 shows that all of the MEA membranes have better fuel cell performance than that in single membrane. This suggests that the membrane with this type of membrane electrode assembly can improve fuel cell performance. These results also reveal that in the very low current density region, the cell using MEA-SPS exhibited a slightly

lower open circuit voltage (OCV) and performance than the cell with the MEA-PS and MEA-SPS-alginate. This can be attributed to the fact that the rate of methanol cross over for the MEA-SPS is higher than that in the MEA-SPS and MEA-SPS-alginate.
However, in higher current density region particularly above 500 mA/cm2 the cell using MEA-SPS-alginate showed much better performance as compared to the cell with MEA-PS and MEA-SPS. This indicates that the MEA-SPS-alginate has a lower mass transfer resistance than the MEA-PS and MEA-SPS. The lower mass transfer resistance for the MEA-SPS-alginate might be related to the possible formation of hydrophilic pores in the MEA fabrication process. Chaudry and Sengawa [6] have reported increase in hydrophylicity properties and also the conductivity with addition of small inorganic polymer such as monmorilonite, alginate, silica, etc. Earlier reports suggest that sulphonated of polystyrene has ability to increase the hydrophilicity properties of the membrane.

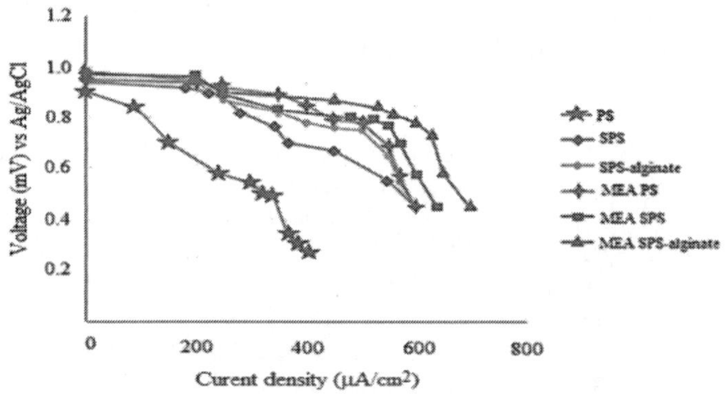

Figure 1. Polarization of comparison of cell performance for membrane

Generally with sulfonated acid membranes, there is no preferential permeation of water or methanol; both decrease with increasing sulfonated content. For PS membranes the total flux of methanol and water increases with increasing temperature and feed methanol concentration. Therefore, sulfonated membranes seems promising and worth investigating for DMFC application. Figure 2 exhibit the result of proton conductivity in different electrode such as carbon, cuprum and iron. In order to provide enough hydration to the membrane and activated the anode catalyst activation of the prepared membranes is necessary. MEA SPS alginate activated membrane appeared to have higher proton conductivity than that the MEA SPS activated membrane. Alginate as a dopants incorporated into the pores of membrane of MEA will affect the high conductivity of membrane. The effect of incorporating alginate as inorganic dopants in the membrane depends on the nature of the dopant in question. If the dopant is not acidic it simply acts as a barrier to both proton and methanol flow. The dopant in this case is non-conducting and acts as a barrier to both vehicular diffusion and Grotthuss diffusion. It obstructs the former by preventing the vehicular transport of hydronium ions, and it also obstructs the latter by disturbing the hydrogen bond network. Apparently this should be particularly effective in blocking ing methanol flow as methanol molecules are much larger than protons, but it also affects proton conductivity adversely. Another effect of higher conductivity of MEA SPS-alginate membrane is due to its contents of sulfonate group in the SPS. The particle has an SO_4^{2-} group attached which attracts protons and continues the hydrogen bond network. This particle would still block vehicular diffusion but will only slightly reduce Grotthuss diffusion, if at all. Depending on the acidity of the particles, it might even increase the proton exchange sites and Grotthuss diffusion and the overall conductivity.

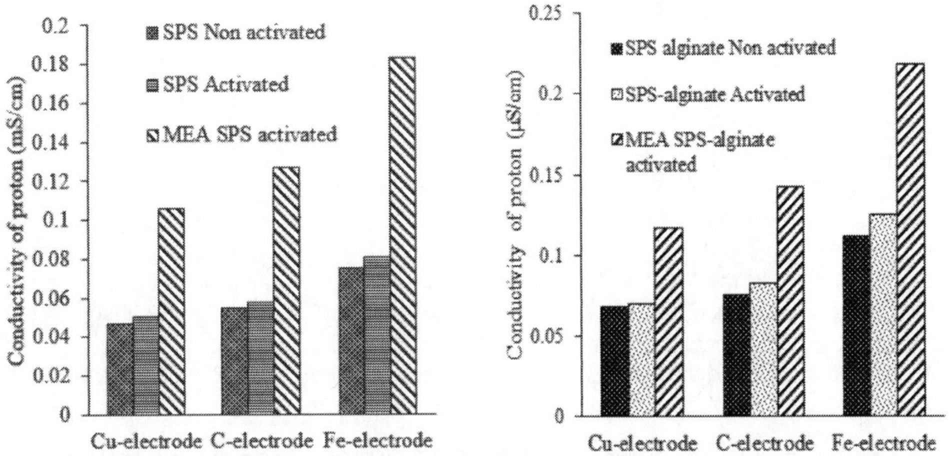

Figure 2. Proton conductivity of SPS and SPS-alginat

Sulfated dopants, such as sulfated zirconia and sulfated silica are especially popular in current research [11] and extensive research has been done with these inorganic particles in PEMFCs. Incorporation of sulfated zirconia particles (ZrO_2/SO_4^{2-}) into Nafion via the sol-gel method has shown increased water uptake and proton conductivity in hydrogen/oxygen fuel cells due to the additional acid sites [12]. However, due to methanol crossover, increased water uptake is undesirable in DMFCs. The acidity of the membrane should be such that overall methanol crossover is reduced while maintaining conductivity. The results revealed that the water up take of MEA SPS (13.68%) was higher compared to SPS (5.15%) and PS (0.69%). The contribution of alginate and sulfonate in the MEA-SPS-alginate membrane would increase proton conductivity and water uptake. While the water up take might increase methanol crossover in the membrane, sulfonate group increases the proton exchange site.

Evaluation of sulfonate group in the sulfonation process of the PS was confirmed qualitatively by FTIR. The process, which occurs exclusively in the phenyl rings should be manifested by the appearance of new vibration characteristic of sulfonic acid groups; by changes in the aromatic C–H out of plane bending region and perhaps by splitting of some bands (e.g. the C–O stretch of the substituted rings), which is due to the reduced symmetry and direct influences of sulfonic acid groups vibration frequencies after the sulfonation process.

Significant differences in FTIR spectra of sulfonated PS were observed compared to pure PS. The peak identified at 1500 cm^{-1} proved the presence of C–C aromatic ring in the PS, which referred to the 2, 4-substitution aromatic ring indicating that only first substitution was involved in this study. As sulfonated PS became water-soluble when the sulfonation degree was approaching 100% the first type of substitution reaction is more favorable. Therefore, the study followed first type substitution. The sulfonation process decreased the intensity of the aromatic C–C vibration and the peak was observed to split to two absorption bands at 1495 and 1475cm^{-1}. These two bands increase with the increasing concentration of sulfonic acid group in the sulfonated PS membranes. Occurrence of a new absorption band at 1028.35 cm^{-1}, 1181.27cm^{-1}, 1275 $cm^{-1}$and 1069 cm^{-1} upon sulfonation indicated presence of the sulfonic acid groups in PS. In summary, the changes observed in the FTIR spectra indicated the presence of $-SO_3H$ groups.

The cross section of membrane morphology was evaluated to support the performance of DMFC. Figure 3 shows the SEM cross-sectional images of membrane PS, SPS, SPS-alginate and MEA SPS-alginate, respectively. It is clear from the SEM images that for the MEASPS-alginate, by hot pressing, the interfacial contact between the membrane and the electrodes proves to be good. However, for the membrane SPS-alginate, which was activated, alginate seems algomerated in the membrane. Dopant of alginate increased both proton conductivity and the strengthen of hot pressing method to reduce the delamination problem that occurred at the interfaces of the MEAs. Liang *et al* reported that delamination problem seems to be more serious both at the anode and cathode interfaces of the conventional MEA; in fact, this could be the main cause for the low performance of the cell with the MEA after being operated for 4 days. The existence of a better interfacial contact at the cathode interface for the MEA SPS-alginate explains why it yielded more stable performance than that of MEA SPS after operating the cell. The prevalence of a good interfacial contact at the cathode interface of the MEA SPS-alginate can be attributed to the effect of alginate as an inorganic dopant into membrane. In addition, the alginate presence in the MEA SPS-alginate also roughens the surface of the membrane, which enhances the contact region of the MEA-SPS-alginate membrane with the electrodes during the fabrication of the MEA. It is pertinent to mention that the solvents methanol/water present in the membrane dispersion play an important role while fabricating the MEA by the hot pressing method.

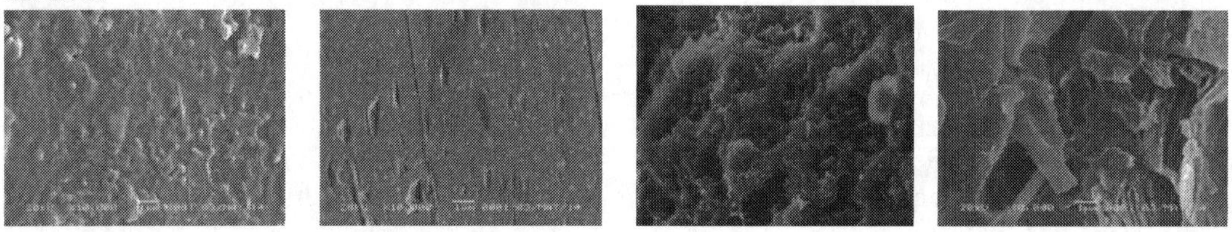

(a) (b) (c) (d)

Figure 3 Cross section morphology of membranes (a) PS, (b) SPS (c) SPS-alginate, dan (d) MEA SPS-alginate

In conclussion, a kind of polymer electrolyte composite inorganic dopant was introduced between the PEM and the electrodes during the MEA fabrication process to achieve a better interfacial contact. The dopant provides a better resistant and stronger binding force between the membrane and the electrodes. The performance of the cell using the MEA-SPS-alginate was found to be more stable. The cross-sectional analysis of the MEAs after the long-term operation revealed that the MEA-SPS-alginate retained a better interfacial contact between the membrane and the electrodes as compared to the MEA-SPS. The proton conductivity results revealed that the MEA-SPS-alginate exhibited a higher value than the MEA-SPS, confirming that the MEA-SPS-alginate had a better interfacial contact between the electrodes and the membranes. The inorganic content needed for the right balance varies. Sulfonate group attached in PS and alginate doped into MEA-SPS-alginate membranes seems to have potential advantages for DMFCs over plain Nafion membranes.

4. References

[1] Woong JA, Yim SD, Choi YW, Sohn YJ, Park SH, Yoon GY, Park GG, Yang TH, Kim KB 2011 *Electrochemica Acta* **56**, 7732-7739.

[2] Kendrick I, Smotkin ES 2014 *International Journal of Hydrogen Energy*. **39**, 2751-2755.

[3] Mulijani S, Dahlan K, and Wulanawati A, 2014 *IJMMM* **2** 36-40.

[4] Lakshmi RTSM, Meier HJ, Schlenstedt K , Choudhary CVV, Varna IK 2005 *J. of Membrane Science*. **26**, 27-35

[5] Liang ZX, Zhao TS and Prabhuram J 2006 *Electrochemica Acta* **51**, 6412-6418.
[6] Chaudhary S, Sengwa RJ 2010 *IJEMS* **18** 147-156.
[7] Cho Y-H *et al* 2012 *International Journal of Hydrogen Energy.* **37**, 5884-5890
[8] Huaneng BJ, Bladergroen P, Sivakumar, Linkov V, Shan J 2012 *Int. J. of Elect. Sci.* **7** 4223-4234.
[9] Wu QW, Zhao TS, Chen R, Yang WW 2010 *Int. J. of Hydrogen Energy* **35**, 10547-10555.
[10] Jalani N 2006 [Report]. - Worcester: Worcester Polytechnic Institute
[11] Pyoungho C, Jalani NH and Ravindra D 2005 *J. of The Elect. Society*, **152** A1548

RR-Interval variance of electrocardiogram for atrial fibrillation detection

N. Nuryani[*1], M. Solikhah[*], A. S. Nugoho[**], A. Afdala[***], E. Anzihory[*]

Eka Anzihory
[*]Physics Department, Sebelas Maret University,
Jl. Ir. SUtami 36A Kentingan Jebres Surakarta 57126, INDONESIA
[**]Information Department, Sebelas Maret University,
Jl. Ir. SUtami 36A Kentingan Jebres Surakarta 57126, INDONESIA
[***]Physics Postgraduate Program, Sebelas Maret University,
Jl. Ir. SUtami 36A Kentingan Jebres Surakarta 57126, INDONESIA

Email: nuryani@mipa.uns.ac.id

Abstract. Atrial fibrillation is a serious heart problem originated from the upper chamber of the heart. The common indication of atrial fibrillation is irregularity of R peak-to-R-peak time interval, which is shortly called RR interval. The irregularity could be represented using variance or spread of RR interval. This article presents a system to detect atrial fibrillation using variances. Using clinical data of patients with atrial fibrillation attack, it is shown that the variance of electrocardiographic RR interval are higher during atrial fibrillation, compared to the normal one. Utilizing a simple detection technique and variances of RR intervals, we find a good performance of atrial fibrillation detection.

1. Introduction

Atrial fibrillation (AF) is a type of heart arrhythmia which is a serious problem of the heart. The problem is originated from the atrial of the heart. The estimation of AF prevalence of AF is estimated to 2.3 million in the United State and 4.5 million in the Europe [1]. A serious risk of AF is stroke [2] which leads to frequent hospitalization and increases mortality. During AF the atrial beats chaotically and irregularly.

Early detection of AF happening in a patient is necessary to conduct appropriate actions which might improve patient care and reduce healthcare costs [3]. These actions are also essentials to prevent the serious AF complications. Therefore, approaches have been proposed for early AF detection. Varied algorithms have been studied to find a sophisticated AF detection system accurately and quickly [4, 5] . Electrocardiographic features are often utilized for AF detection.

An AF detection approach using neural network has been proposed by [6]. The approach detection employs probabilistic neural network and a wavelet feature of electrocardiogram. A different strategy of an AF detection using neural network algorithm is proposed by (Mateo et al., 2013). This algorithm uses radial basis function (RBF) neural network and utilizes electrocardiographic features by cancelling the electrocardiographic QRST complex.

[1] To whom any correspondence should be addressed.

The article explores electrocardiographic features for AF detection. The main explored-feature is variance of RR-interval of electrocardiogram. RR interval is the time interval between two consecutive R peaks of electrocardiogram. RR interval represents heart rate. Heart rate of the heart in patient with atrial fibrillation fluctuates, opposing the normal heart which beats stably. This happen as a result of that the atrial of the heart beats chaotically and irregularly.

2. Method

This article presents a study of variance of electrocardiographic RR interval for atrial fibrillation detection. Firstly, we collect R peaks data of electrocardiogram provided by Physionet [7]. The electrocardiogram is recorded from 23 patients. The recording is 10 hours in duration for each patient with sampling rate of 250 samples per second.

From the R peaks, RR interval is defined. It is the time interval between two consecutive R peaks. RR intervals is then segmented or windowed. The length of segment is 60 RR intervals. A segment is defined as atrial fibrillation or normal by considering the number of atrial fibrillation happening in the segment. If the number of atrial fibrillation is more than 50%, the segment is defined as atrial fibrillation. The other situation is defined as normal. The number of atrial fibrillation refers to the number of RR interval which is defined as atrial fibrillation.

The variance is defined by two definitions. The first definition, variance σ is defined as

$$\sigma = \sqrt{\frac{\Sigma(x-\mu)}{n-1}} \quad (1)$$

The second definition, variance α is defined as the distribution of the data number in defined five ranges, from the lowest to the highest value of RR intervals in a segment. Specifically, we grouped RR intervals –in a segment– into five levels. The variance α is this article is defined as the difference of data number between the levels. For comparison of the detection performance, we use mean value of RR interval.

The detection performance is measured in terms of accuracy. Accuracy is the correctness of both atrial fibrillation and normal events.

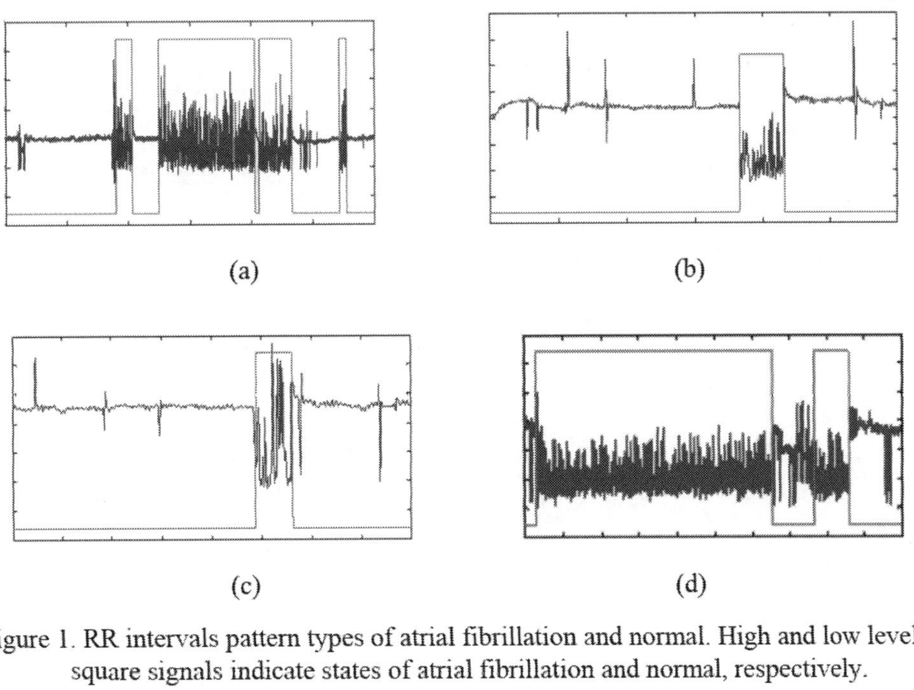

(a) (b)

(c) (d)

Figure 1. RR intervals pattern types of atrial fibrillation and normal. High and low levels of square signals indicate states of atrial fibrillation and normal, respectively.

3. Results

3.1. Description of RR interval during atrial fibrillation

We obtain RR intervals of electrocardiogram of all the patients. The patterns of RR intervals in the state of atrial fibrillation and normal of patients are presented in Figure 1. The line with square forms indicate normal and atrial fibrillation; low and high levels indicate normal and atrial fibrillation respectively. Thus, the most left part of the Figure 1a is normal and the most right part also indicate normal, which both parts are low levels. Figure 1a indicates that the RR intervals rise and decline more significantly during atrial fibrillation than the normal one. In the normal state RR intervals just make a ripple. The similar fact can also been seen in the other RR interval pattern, as presented in Figure 1b and Figure 1c. However, we need to provide attention that, in the normal state RR interval there is a small part in which RR interval also rise to a high magnitude and jump to a very low magnitude; it looks like sparks.

The other the RR interval pattern is a little different, indicated in Figure 1d. The figure shows that in the atrial fibrillation state the RR intervals pattern is similar with the previous figures. In the normal state, the RR intervals fluctuate more steeply, but it is still less steeply than in the atrial fibrillation. In other words, the difference of atrial fibrillation and normal is that RR interval pattern more fluctuates that the normal one. In addition, the fluctuation in the atrial fibrillation state is more steeply.

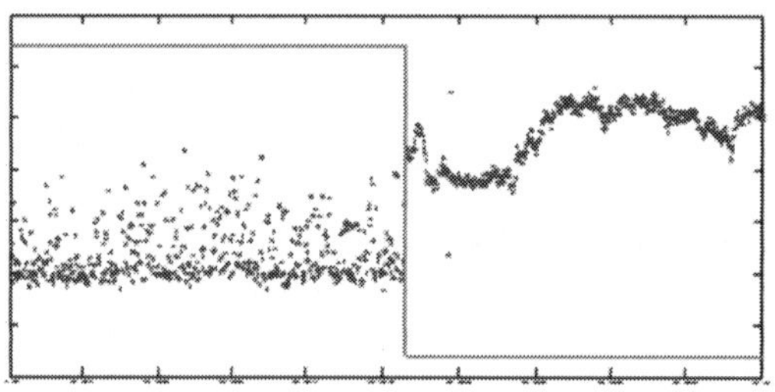

Figure 2. RR intervals pattern types of atrial fibrillation and normal. The

The other RR interval patterns show a different pattern as indicated in Figure 2. The pattern show that the normal RR interval fluctuate with 'fluctuated baseline'. Thus, for the normal state, the difference of this RR interval pattern with the previous one is the fluctuation of the baseline.

Variance of RR interval presented using α and σ are indicated in Table 1. Variance (σ) is higher in atrial fibrillation than that in the normal one, but variance in terms of α is higher in the normal situation than that in the atrial fibrillation. This might be as a result of the difference formulation of variance. In general the two variance indicate different values between atrial fibrillation and normal, which might be used for input or feature of atrial fibrillation detection technique. For addition feature of RR intervals, average of RR interval is presented. The data shows that the average is lower in atrial fibrillation than that in normal one, which is confirmed by previous study [8].

Table 1. Variance of RR interval during atrial fibrillation and normal

	Atrial Fibrillation (mean ± std, ms)	Normal (mean ± std, ms)
Variance (σ)	5136.4 ± 3579.76	1557.68 ± 3160.48
Variance (α)	126.80 ± 37.28	47.38 ± 75.25
Average	646.84 ± 134.88	799.76 ± 156.08

3.2. A simple atrial fibrillation detection using RR interval variance with a threshold

A simple detection for atrial fibrillation using each of the three features are created. The detection method is based on a threshold. In this study the threshold is varied and the accuracy of the different threshold is presented in Figure 3.

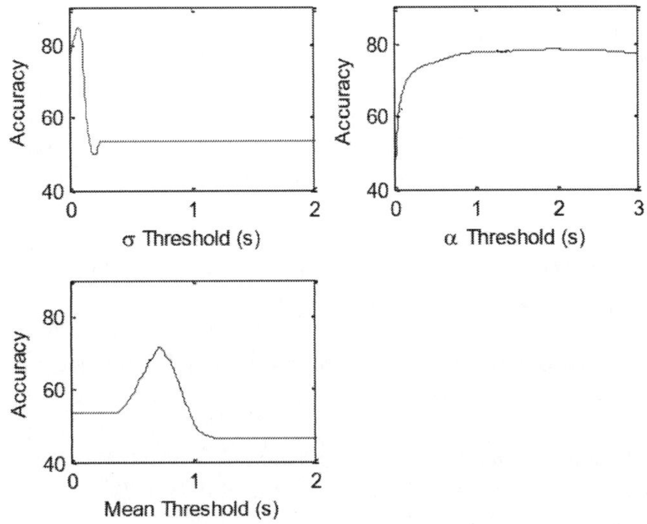

Figure 3. Accuracies of atrial fibrillation detection using different RR interval features –variance α, variance σ and mean– with different thresholds.

Table 1. Maximum accuracies obtained by the detection which use variance σ, variance α and mean of RR interval.

	Maximum Accuracy
Variance (σ)	84.70 %
Variance (α)	78.62 %
Average	71.67 %

In general, different threshold could provide different accuracies and therefore an optimal threshold is necessary to find an optimal accuracy. The maximum accuracies of the three approaches are presented in Table 2. Figure 3a shows the accuracy of the detection using variance σ with different thresholds; the maximum accuracy of this approach is 84.70%. Figure 3b shows the accuracy of the detection using variance α; this approach provides maximum accuracy of 78.62%. Figure 3c indicates that the maximum accuracy obtained by which use mean of RR intervals is 71.67%. Thus, among the three approaches, the detection which utilizes variance σ provide the highest accuracy. Conversely, the lowest is which use the mean. It might imply that variance of RR interval is a better parameter to indicate atrial fibrillation than the mean.

4. Conclusion

A study of atrial fibrillation detection using variance of RR interval is presented. For comparison, the detection which utilize mean of RR interval is compared. Using the variance, the accuracy is higher than which uses the mean. We uses two types of variances: variance α and the proposed variance σ. The proposed variance σ indicates the different number of RR intervals in a certain range with the others.

The proposed variance σ provides higher performance than the variance α. Using σ the atrial fibrillation detection provide 84.70%.

References

[1] M. Lainscak, N. Dagres, G. S. Filippatos, S. D. Anker, and D. T. Kremastinos, "Atrial fibrillation in chronic non-cardiac disease: where do we stand?," *International journal of cardiology,* vol. 128, pp. 311-315, 2008.

[2] J. S. Healey, S. J. Connolly, M. R. Gold, C. W. Israel, I. C. Van Gelder, A. Capucci, *et al.*, "Subclinical atrial fibrillation and the risk of stroke," *New England Journal of Medicine,* vol. 366, pp. 120-129, 2012.

[3] G. D. Clifford, I. Silva, B. Moody, Q. Li, D. Kella, A. Shahin, *et al.*, "The PhysioNet/Computing in Cardiology Challenge 2015: Reducing false arrhythmia alarms in the ICU," in *2015 Computing in Cardiology Conference (CinC)*, 2015, pp. 273-276.

[4] M. Sardana, J. Saczynski, N. Esa, K. Floyd, K. Chon, J. W. Chong, *et al.*, "Performance and usability of a novel smartphone application for atrial fibrillation detection in an ambulatory population referred for cardiac monitoring," *Journal of the American College of Cardiology,* vol. 67, pp. 844-844, 2016.

[5] T. Koivisto, T. Hurnanen, T. Vasankari, T. Kiviniemi, A. Saraste, and J. Airaksinen, "Automatic detection of atrial fibrillation using MEMS accelerometer," in *2015 Computing in Cardiology Conference (CinC)*, 2015, pp. 829-832.

[6] K. Daqrouq, A. Alkhateeb, M. Ajour, and A. Morfeq, "Neural network and wavelet average framing percentage energy for atrial fibrillation classification," *Computer methods and programs in biomedicine,* vol. 113, pp. 919-926, 2014.

[7] A. L. Goldberger, L. A. N. Amaral, L. Glass, J. M. Hausdorff, P. C. Ivanov, R. G. Mark, *et al.*, "PhysioBank, PhysioToolkit, and PhysioNet," *Components of a New Research Resource for Complex Physiologic Signals,* vol. 101, pp. e215-e220, 2000.

[8] J. E. P. Waktare, "Atrial Fibrillation," *Circulation,* vol. 106, pp. 14-16, 2002.

Fiber pad for pressure mapping

H Purwanto, U R Fitriani, A Dwijosutomo, A Marzuki*
Physics Department, Sebelas Maret University
Jl. Ir. Sutami 36 A Surakarta

E-Mail: amarzuki@mipa.uns.ac.id

Abstract. Optical fiber sandwiched pad designed as a pressure mapping sensor has been configured and characterized. Optical fiber sensor was aligned to form a web-like configuration (x-y matrix). Several fibers were positioned to form lines parallel to y-axis while others are in parallel to x-axis. When a mass with a particular surface contour was loaded on the fiber pad, we have shown the dependence of the magnitude of light attenuation on the mass surface contour. Combining these light attenuation results we have successfully constructed a three dimensional contours showing the pressure distribution given by the mass to the fiber pad.

1. Introduction

Optical fiber is a transparent solid conduit used to propagate light from one point to any other point in any path forms. Optical fibers used for optical communication are mainly made of silicate glass. Fibers of this type have low loss but less elastic. In contrast, optical fibers made of polymer are characterized by their high optical loss but show a good elasticity. Because of this reason, these two groups of fiber find their own application area. If delivering light to long distance and hazardous area is our main interest, optical fiber made of glass is the best choice [1-3]. Optical fiber used for ultra-low loss optical communication as well as for most chemical and bio-chemical sensing falls into this group. Optical fiber made of polymer usually called polymer optical fiber (POF) is chosen if elasticity is our main consideration [4-5]. Most mechanical fiber sensors are configured using POF.

Fiber sensor configured to form a kind of fabric has many applications. In medical application, fiber Bragg grating (FBG) has been used to measure plantar strain distribution in human foot [6, 7]. Result of this research can describe the pressure distribution created by foot on the pad at different area of foot such as fore foot, mid foot and hind foot. Similar work intended to measure pressure distribution of foot has also been done. For this purpose, hetero core fiber sensors are embedded in the insole [7]. In this work, fiber sensors are embedded under the heel, thenar, anti-thenar, and big toe. In my paper, fiber sensors were distributed throughout the pad surface. Using this configuration, distribution of force given to the pad can be easily made.

2. Experiment

As shown is Figure 1, several strands of polymer optical fibers are sandwiched in rubber pad. Several strands are aligned parallel to y-axis while several others are parallel to x-axis. Light is launched from one end and series of photo-detectors are positioned at other end. Three small pins are positioned at every point at which optical fiber from y-direction and x-direction meets each other. When a load is positioned in a certain area, all fibers underneath the load will be bent result in light attenuation. By measuring light intensity detected by each fiber from both axis x-axis and y-axis and then adding them to form I_{x-y}, a matrix indicates intensity distribution can be obtained.

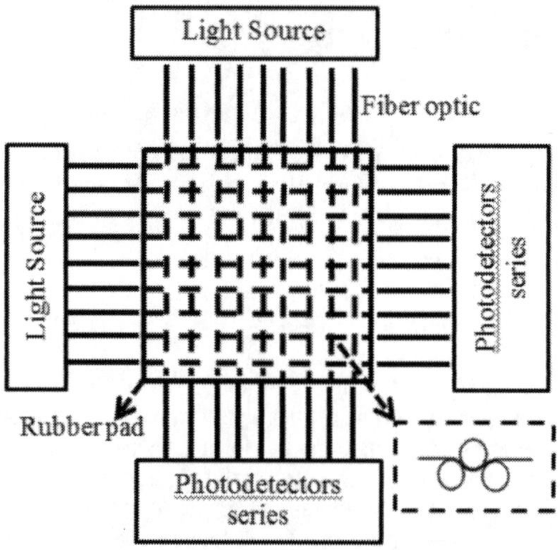

Figure 1. Basic experiment setup

3. Results and Discussion

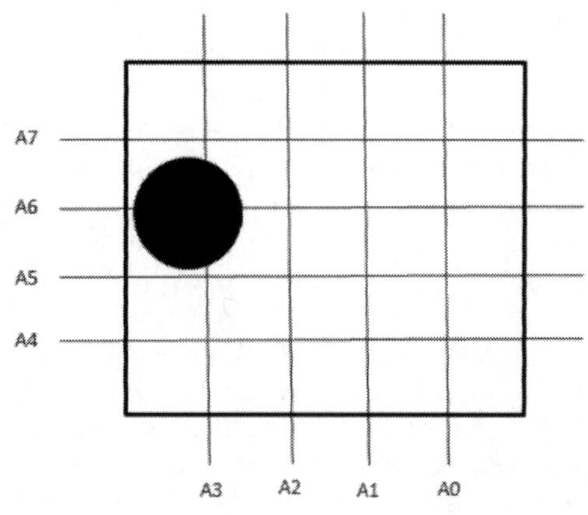

Figure 2. Load position on the fiber pad

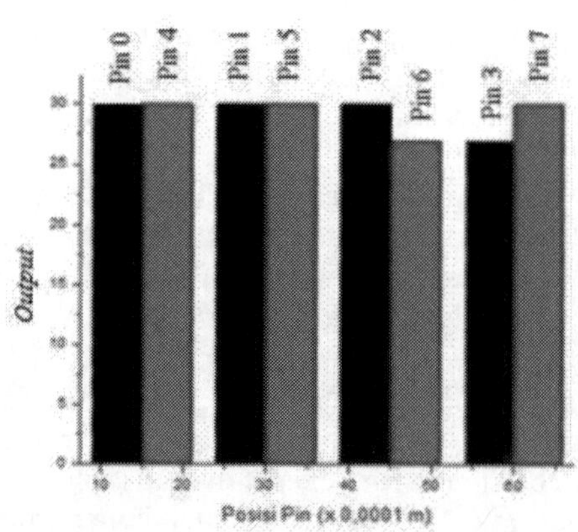

Figure 3 Output power of light detected by photodetector

Figure 2 shows the position of a load placed on the fiber pad. As shown, there are two fibers affected by this load, they are A6 and A3. Because of this load, the upper pin as shown in figure 1 puss the fiber down to some extent that depends on the load mass. Consequently, these two fibers were locally bent and light intensities passing through them decrease (figure 3). Bending loss mechanism was presented by the authors in other paper [4, 8].

		A3	A2	A1	A0
	Loss	3	0	0	0
A7	0	0	0	0	0
A6	3	6	0	0	0
A5	0	0	0	0	0
A4	0	0	0	0	0

Figure 4. Matrix representing pressure distribution experienced by fiber pad

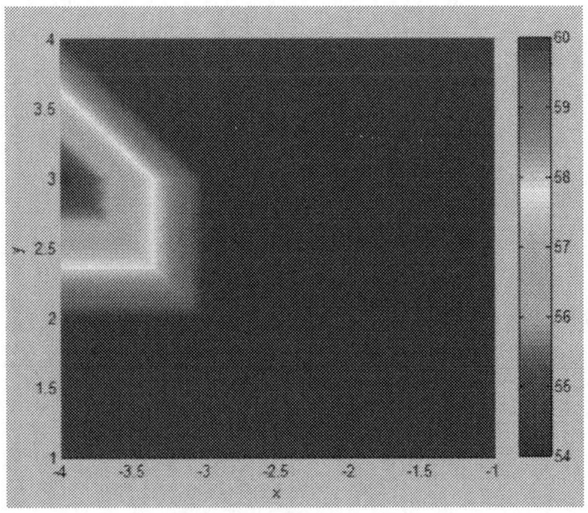

Figure 5. Typical contour plot of pressure mapping showing load position on the fiber pad.

Figure 4 is a matrix constructed based on figure 2 with output power detected by each photodetector taken from figure 3. As shown, A3 and A6 fibers have non-zero output power explaining that they are burdened. Although light attenuation occurs at both fiber does not inform the point source position of load, by adding the output powers of fiber optics between those flowing in y-direction (I_{yi}) to those flowing in x-direction (I_{xi}) we can figure out the position of load (pressure mapping). The intensity of light at each fiber intersection (I_{xi-yi}) can be calculated using equation

$$I_{xi-yi} = I_{xi} + I_{yi} \qquad for\ I_{xi}\ and\ I_{yi} \neq 0 \tag{1}$$

Summation requirement as given in equation (1) is needed to ensure that all area other than the position at which the load is placed have the same color (figure 5).

4. Conclusions

In this paper, we have demonstrated the work of a basic design of a fiber pad used for pressure mapping. The sensor works based on an optical loss due to fiber bending. Basic configuration of fiber sensor consist of two groups of fiber optic strands, one was stranded in x-direction and one in y-direction. We have used matrix to process the light intensity data (in the form of output voltage) recorded from those two groups and finally present the results in the form of contour color plot.

References

[1] Crosby J C, Lucas D, Koshland C P 2013 *Sens Actuators B* **181** 938-942
[2] H.A. Rahman, S.W. Harun, M. Yasin, H. Ahmad 2013 *Optik* **124** (8) 679-681
[3] Buckman A B 1996 *Waste Manage.* **13** (5–7) 517
[4] Marzuki A, Heriyanto M, Setiyadi I D and Koesuma S 2015 *J Phys Conf Ser.* **622** 012059

[5] Ramakrishnan M, Rajan G, Semenova Y and Farrell G 2016 *Sensors* **16** 99
[6] Prasad A S G, Omkar S N, Vikranth H N, Anil V, Chethana K, Asokan S 2014 *Measurement* **47** 789-793.
[7] Otsuka Y, Koyama Y and Watanabe K 2014 *Procedia Eng.* **87** 1465-1468
[8] Yang C and Oyadiji O 2015 *Sens. Actuators A* **222** 67-79

Study of adsorption Ag and Pb in liquid sample using Berea sandstone by commercial laser-induced breakdown spectroscopy (LIBS)

H. Suyanto[1*], N. Wendri[1], U. Agustiningrum[1], M. Manurung[2]
[1]Physics Dept.,
[2]Chemistry Dept., Mathematics and Natural Sciences Faculty, Udayana University
Jl. Kampus Bukit Jimbaran-Denpasar- Bali 80361

*Email : hery6@yahoo.com

Abstract. Qualitative and quantitative analysis of Pb and Ag elements in liquid samples had been done by commercial laser-induced breakdown spectroscopy (LIBS) using adsorption method on a Berea Sandstone. The aim of this study is to identify the thickness of the Berea Sandstone for adsorbing Pb and Ag elements in liquid. The experiment was started with characterizing the Berea Sandstone that contains Si, Na, H, Li, K, Ca, O, N, Be, Ti, Al, Mg and Ba. Some of these elements have ability to adsorb Pb and Ag elements in the liquid. To prove this phenomenon, it is required to look for the experiment parameter optimum conditions such as laser energy, adsorption time and sample temperature. The experiment was conducted by dropping 2 ml standard liquid containing 1000 ppm of Pb and Ag to the Berea Sandstone surface. The result showed that the parameter optimum conditions for analyzing Pb and Ag elements in liquid sample with adsorption method were adsorption delay-time of 15 minutes, laser energy of 120 mJ and sample heating of 80 ^{o}C. The next experiment was focused on the number of adsorption as a function of depth. The data showed that Pb and Ag elements in liquid sample of 2 ml, 1000 ppm were fully adsorbed by the Berea Sandstone until the depth of 0.372 mm and 10.40 mm from the surface, respectively. The data also showed that the limit of detection predicted to about 22.76 ppm.

1. Introduction

Berea Sandstone is a sedimentary rock whose grains are predominantly sand-sized and are composed a lot of quartz (± 93%) held together by silica [1]. The relatively high porosity and permeability of Berea Sandstone makes it a good reservoir rock. Berea Sandstone from Bonbiu village, Gianyar city, Bali Province, Indonesia used in this study has special performance. People in this village usually make the Berea Sandstone a special shape like bowl called Topo used to filter groundwater before drinking or another uses. People usually just see the water after filtering cleaner and used it, but they do not know whether Topo can filter some hazard elements such as Pb and Ag elements as water pollutant or not. Therefore, this work analyzes the ability of Berea Sandstone as a water-filter using laser-induced breakdown spectroscopy (LIBS) through adsorption method.

Laser-induced breakdown spectroscopy (LIBS) has been employed in variety of analytical applications and maintained its constant growth since its introduction in the early eighties [2]. The spectral analysis of the emission lines from the plasma provides the elemental compositions of the target. Therefore, LIBS has many advantages both either qualitative or quantitative analysis with very fast and accurate to gas and solid samples, but not in liquid sample. Liquid sample will splash off when laser beam is focused on liquid sample surface and it does not produce plasma. To anticipate this condition, many researchers worked special methods. One method, analyzed Al^{3+} in solution until 1.5 ppm using droplet method which change liquid phase to gas phase and others, analyzed liquid sample using deposition method which used filter-paper and electrical deposition to aluminum surface [3-7].
Finally, there are two goals of this work, first is to looking for the minimum thickness of the Berea Sandstone to filter or adsorb Pb and Ag elements in liquid samples. Secondly, to determine minimum Pb concentration adsorbed by the Berea Sandstone. All analysis used Laser-Induced Breakdown Spectroscopy (LIBS).

2. Experimental Procedures

2.1. Sample Preparation

The research was conducted in 2 steps that are to determine optimum conditions of parameter (such as delay time reaction, sample temperature, laser energy) and quantitative analysis. For the first step, 2 ml liquid sample containing 1,000 ppm of Pb was dropped onto the surface of Berea Sandstone and time lag treatment of 1, 5, 15, 25, 35 and 35 minutes before laser irradiation of 60 mJ, 80 mJ, 100 mJ and 120 mJ which was chosen in order to find the best adsorption time. The next treatment, by varying the sample heating temperature of 50, 60, 70, 80, 90 ^{0}C after certain time of optimum absorption process in the first treatment.
The second step is quantitative analysis that determine the thickness of the Berea Sandstone for adsorbing Pb and Ag elements in liquid sample and also to determine the minimum concentration (or limit of detection) of Pb element in liquid sample adsorbed by the Berea Sandstone. For this experiment, 3.5 ml liquid sample containing 1,000 ppm of Pb and Ag was dropped onto surface of 0.5 cm and 2.5 cm Berea Sandstone until penetration to another surface. Based on optimized parameter conditions in the first step, the Sandstone was irradiated by laser in the fixed position for four times with accumulation of 4 to get depth profile data [8]. For providing limit of detection data, a 3.8 ml solution containing 50 ppm, 100 ppm, 200 ppm and 400 ppm of Pb were poured to the Berea sandstone surface and analyzed by LIBS.

2.2. Experimental Set-up

The experimental set-up used in this work is presented in figure 1. The commercial LIBS system consists of Neodymium:Yttrium Aluminum Garnet (Nd:YAG) laser (1,064 nm, 7 ns), OOILIBS and AddLIBS software and Ocean Optics HR 2500^{+} spectrometer. For this study, the laser was operated at 10 Hz repetition rate with laser energy of 120 mJ. The laser beam was focused through 10 cm focal length biconvex lens onto the Berea Sandstone surface and the resulted emission radiation was captured by optical multichannel analyzer containing 14,336 CCD pixels in the wavelength region of 200 – 900 nm with resolution of 0.1 nm. All the experiments were conducted in surrounding gas of air at 1 atm pressure.

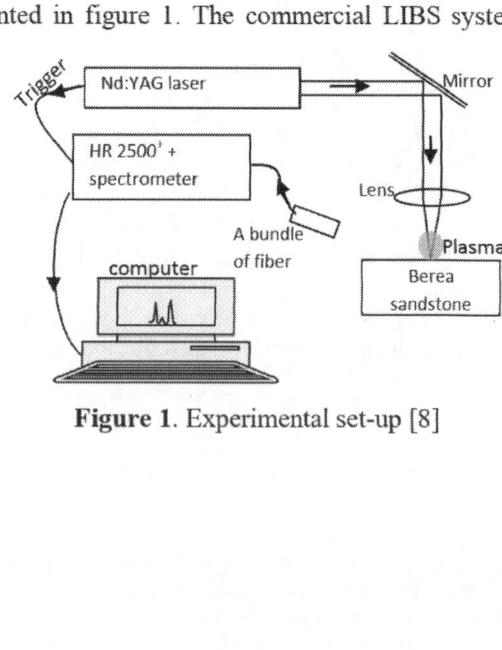

Figure 1. Experimental set-up [8]

3. Results and Discussion

3.1. Determining of experiment optimum conditions

The first step in this work is to identify the optimum conditions of experiment such as delay time adsorption (or reaction) between solution and Berea Sandstone sample, the temperature of sample and laser energy. Prior to irradiation by laser, the Berea Sandstone sample was dropped 2 ml of solution containing 1000 ppm of Pb. The sample, than, analyzed every 10 minutes with focusing laser of 120 mJ on surface of sample which produced plasma and emission intensity of neutral atomic emission Pb I 405.7 nm [9] captured by spectrometer. It is displayed on figure 2.

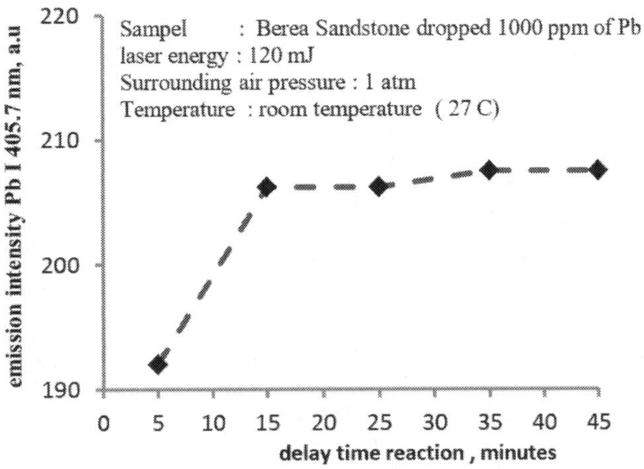

Figure 2. Emission intensity of Pb I 405.7 nm as a function of delay time adsorption (or reaction)

Figure 2 show that, the atomic emission intensity of Pb increases with increasing adsorption time from 5th to 15th minute than remains constant till 45th minute. This rising is predicted as interaction between ionic Pb in the solution with some elements of the Berea Sandstone sample through ionic immobility (adsorption) process whether as ionic exchange, complexion or electrons covalence. Therefore, all analysis data in this experiment was conducted 15 minutes after dropping liquid sample on the Berea Sandstone surface.

To explain ionic immobility process, first, the Berea Sandstone sample was qualitative analyzed using LIBS. It contains of Si, Na, H, K, Ca, O, N, Ti, Al, Mg, Fe, Ba elements and there are no Pb and Ag elements. Based on this data, it can be predicted that ionic immobility process between ionic Pb and the Berea Sandstone sample through ionic exchange process, where element such as Ca will be replaced by ionic Pb as it has similar radius to ionic Pb. Another possibility, the SiO_2 molecule in the sample has free electrons and these electrons will be used together with ionic Pb produced electron covalence process. However, all these ionic immobility processes must be investigated futher.

After obtaining the experiment optimum conditions of delay time absorption, the next step is to know the influence of sample temperature and laser energy to emission intensity of Pb line. After dropping 2 ml solution containing 1000 ppm of Pb, the Berea Sandstone samples were varied its temperature from 50 C to 90 C and were analyzed using varied laser energy from 60 mJ till 140 mJ. Results are exhibited on figure 3.

Figure 3, displays that the neutral atomic emission intensity of Pb I 405.7 nm increases with increasing laser energy till 140 mJ as well as increasing temperature of sample till 80 C. More laser energy applied, Pb atoms propelled with higher speed and adiabatic compression take placed with surrounding air that produce shockwave with higher energy. This energy is absorbed by Pb electrons for exciting to higher level energy, than emit photon of certain wavelength when it experiences transition to ground state. Because of higher energy of shockwave, a lot of Pb atoms excited and as a consequence higher emission intensity produced. However, because of a lot of dust produced when 140 mJ of laser energy focused on surface sample, therefore these all experiment used 120 mJ.

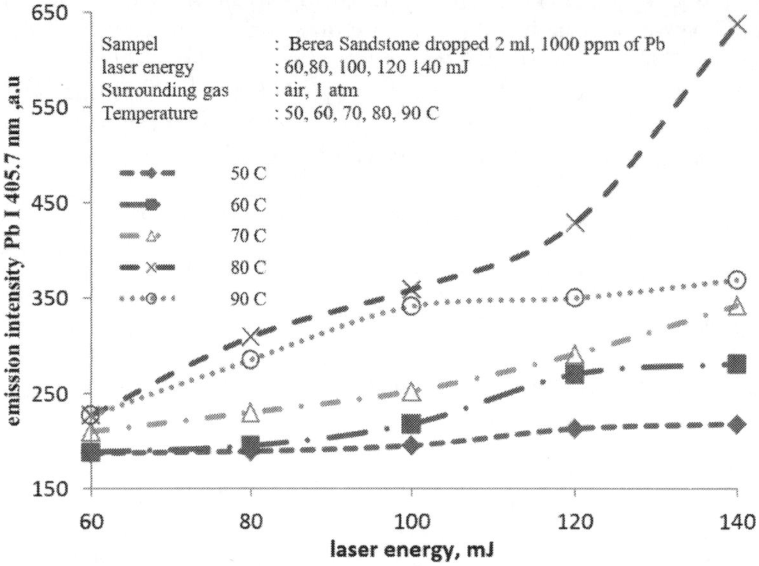

Figure 3. Emission intensity of Pb I 405.7 nm as function of laser energy

In the case of sample temperature, increasing temperature will increase intermolecular bonds between particles and as a result the recoil force rises when laser focused on it. Increasing recoil force, the Pb atoms propelled with higher speed and the same results as increasing laser energy. On the other hand, when the temperature was increased to 90 C, the emission intensity of Pb I 405.7 nm decreased. It might be caused by the break up intermolecular bonds in the sample and as a consequence the recoil force decreased when laser focused on it. Therefore, the experiment optimum parameters used in this work (delay time absorption, sample temperature and laser energy) were 15 minutes, 80 C and 120 mJ, respectively.

3.2 Depth Profile Analysis

Based on the experiment optimum conditions, the next experiment was focused on quantitative analysis concerned on an ability of the Berea Sandstone sample to adsorb ionic Pb element in solution till certain depth on sample called depth profile analysis. The Berea Sandstone sample 0.5 cm thick was dropped 2 ml solution containing 1000 ppm of Pb and waiting till penetrated to another side. Using the experiment optimum conditions, the sample was analyzed every 4 shots until 20 shots laser from surface or front side and then it was analyzed from another side (back side).The results displayed on figure 4. The number of shots stoped on 20 shots to avoid the confinement effect [10].

Figure 4 shows that the emission intensity spectrum of neutral Pb line decreases slowly with increasing the depth or the number of laser shots. It means the Pb element adsorbed or fitered gradualy by sample from the surface. On the other hand, when it was analyzed on the other surface (back side), there was no emission intensity of Pb line. To identify the depth of Pb element filtered by sample, it is better to do linear regression process between the emission intensity as a function of laser shots. Based on data figure 4 obtainned $y = -0.825x + 56.9$. From this equation, the Pb element adsorbed by sample is about 73 laser shots or equevalent to about 0.372 mm from sample surface. This result agrees with figure 4 that there is no signal after 5 mm in depth.

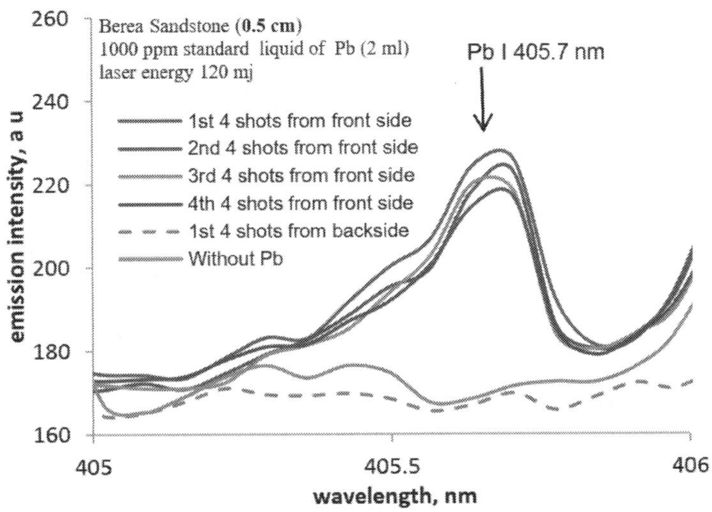

Figure 4. Emission spectrum of Pb in various depth

To know the ability of the Berea Sandstone sample to immobilize another element in the solution, dropped 2 ml solution containing 1000 ppm of Ag and waiting till penetrated to another side of 5 mm thickness. The same procedure as analyzing of Pb, the depth profile was conducted to detect Ag element as a function of the laser shots. The result is displayed on figure 5.

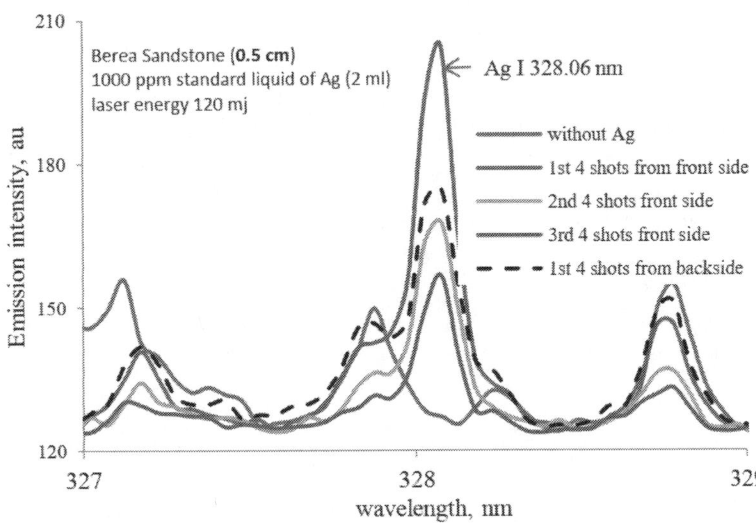

Figure 5. Emission spectra of Ag in various depths on the 0.5 cm Berea Sandstone

Figure 5 exhibits the emission intensity of neutral atom of Ag I 328.06 nm that degrades with the number of laser shots, however, the intensity rises again when it was analyzed from another side (back side). To know the depth of Ag element filtered by Sandstone, it is better to do linear regression process between the emission intensity as a function of laser shots as Pb case. Based on data figure 5, the linear regression equation is $y = -0.0052\ x + 54.592$, that is corelation to the Ag element adsorbed by the Berea Sandstone to about 1.05 cm from sample surface. This result agrees with the figure 5 that there is significant signal after 0.5 cm in depth.

To use the Berea sandstone for filtering the Ag element in liquid sample, the thickness of Berea sandstone is at least 1.05 cm. To prove this statement, the 3.8 ml containing 1000 ppm of Ag element was dropped until it penetrates to the Berea sandstone with thickness of 2.5 cm. The same procedure as before and after analyzing with LIBS, the emission intensity of Ag displayed on figure 6. Figure 6 shows that emission intensity on the back side or 2.5 cm from surface is zero. It means that the Ag element in solution did not penetrate after 2.5 cm from surface.

The immobility phenomenon of Ag element which is adsorbed by Berea sandstone can be described as follow. The ion Ag^+ in solution combines with ammonia molecule in the Berea sandstone to produce a co-ordinate bond $[Ag(NH_3)_2]^+$ which N provides both the electrons for a covalent bond. In this case ion Ag^+ and N act as electron acceptor and electron donor, respectively [11].

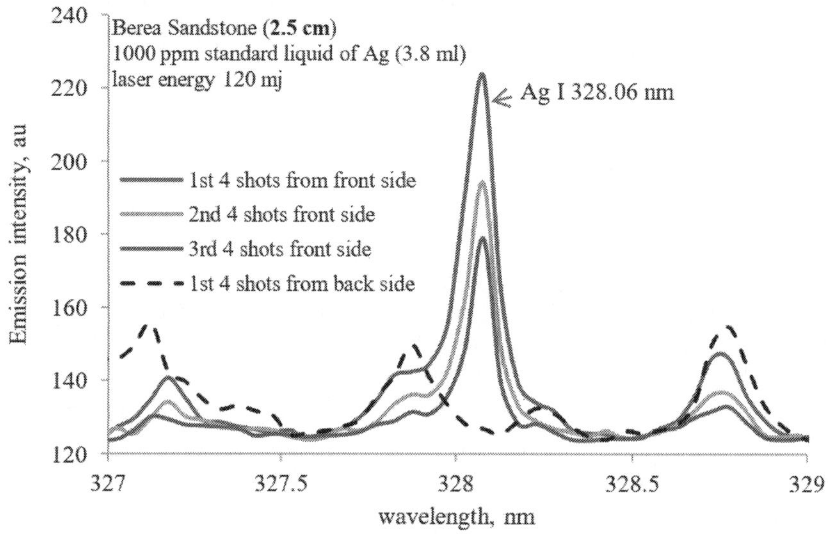

Figure 6. Emission spectra of Ag in various depths on the 2.5 cm Berea Sandstone

3.3. Quantitative Analysis

Quantitative analysis in this study is to know the minimum Pb element in the solution which can be adsorbed by the Berea Sandtone. Therefore, a 2 ml solution containing 50 ppm, 100 ppm, 200 ppm and 400 ppm were poured to the Berea Sandstone surface. After analyzed by LIBS, the emission intensity of Pb of each concentrations were ploted as shown in figure 7.

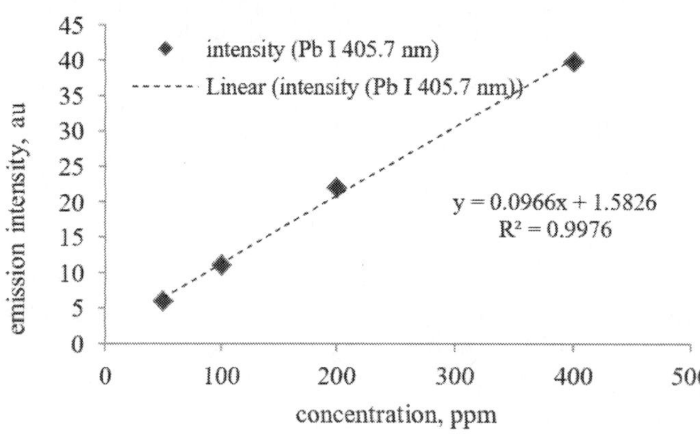

Figure 7. Calibration curves for Pb

Figure 7, states that the emission intensity of Pb decreases linearly from 400 ppm to 50 ppm with a equation y = 0.0966 x + 1.5826. The limit of detection (LOD) of Pb can be calculated by using equation /, where CL, sd, and m are limit of detection (LOD), standard deviation and a slope of calibrtion curve repestively [2,12]. Based on data in figure 7, the LOD of Pb is 22,76 ppm.

In conclusion, the Berea sandstone has ability to adsorb Pb and Ag elements in solution and transform to solid phase. These elements can be detected by LIBS technique with some conditions such as delay time adsorption, the sample temperature and laser energy were 15 minutes, 80 C and 120 mJ respectively. This Berea Sandstone can filter or adsorb 1000 ppm of Pb and Ag elements in solution until the depth of 0.372 mm and 10.49 mm respectively. Besides, the Berea Sandstone also has ability to filter Pb element in solution to about 22.76 ppm

Acknowledgments

Authors are thankful to LPPM-UNUD for financial support to this research through the Unggulan Udayana research grants from Udayana University contract No : 641-22/UN14.2/PNL.01.03.00/2016 on June 15, 2016.

References

[1] Churcher PL, French PR, Shaw JC and Schramm LL 1991 *SPE-21044-MS*
[2] Cremers DA and Radziemski LJ 2006 *Handbook of Laser-Induced Breakdown Spectroscopy* John Wiley & Sons, Ltd
[3] Huang JS, Liu HT and Lin KC 2007 *Analytica chimica acta*, **581** 303-308
[4] Leen Y, Oh SW and Han SH 2012 *Appl. Spectrosc.* **66**(12)1385-1396
[5] Chen Z, Li H, Zhao F and Li R 2008 *J. Anal. At. Spectrom.* **23**(6) 871-875.
[6] Ca´ceres JO, Lo´pez JT, Telle HH and Ure˜na AG 2001 *Spectrochim. Acta Part B* **56**(6) 831-838.
[7] Suyanto H, Manurung M, Sinaga DN 2014 *Jurnal Pendidikan Fisika Indonesia* p-ISSN 1693-1246 **10**(2) 178-185
[8] Suyanto H 2013 *Identifikasi Unsur Utama Penyusun Permukaan Bahan Baja Ringan Dengan Laser-Induced Breakdown Spectroscopy (LIBS)*, konferensi nasional engineering perhotelan IV, Teknik Mesin Univ. Udayana Bali
[9] NIST 2013 *Information system (Electronic structure of atoms)*.http://grotrian.nsu.ru/en. Download June 3
[10] Suyanto H, Kurniawan H, Tjia M and Kagawa K 2002 *Specrtoschim. Acta part:B* **57** 1325-1332
[11] Ryan L and Norris R 2014 *Chemistry Coursebook,* 2ndEdition, Cambridge University Press. P. 51 - 55
[12] Ingle JD and Crouch SR 1988 *Spectrochemical Analysis*, Prentice Hall Inc.

IR spectra of Pb:TZBN glasses

Riyatun, A. L. Sari, H. Purwanto, A. Marzuki*
Physics Department, Sebelas Maret University, Surakarta Indonesia
Jl. Ir. Sutami 36 A Surakarta 57126 Indonesia

*E-mail: amarzuki@mipa.uns.ac.id

Abstract. Tellurite glasses with compositions of $55TeO_2$ - (41- x) ZnO-$2Bi_2O_3$-$2Na_2O$-xPbO where x = 1.0, 1.5, 2.0, and 2.5% mole (Pb:TZBN glasses) were prepared by melting the mixture at 900°C and quenched rapidly. IR spectra were recorded in the range 400 cm^{-1} and 4000 cm^{-1}. Analyses were carried out to study the effect of PbO incorporation into the glasses to the shift the main absorption vibrational spectra. The results were then analysed in term of coordination number of Pb as well as the change in mass reduction.

1. Introduction

In the last view decades, research interest in tailoring an optical devices operating at infrared range has grown very rapidly. Tellurite glasses are of particular interest because they have several advantages, i.e., low glass transition temperature, low melting point, high index of refraction, wide transmissions window, and high rare earth ion solubility [1- 4].

Like all other materials, the properties of glass are affected by their composition and structure (the bonding type between the atoms making up the materials and the way these atoms are packed together). Tellurite glass is a big family of glasses. These glasses are formed by adding many compounds falling into glass intermediates or modifiers. There are many studies intended to investigate structure of various tellurite glasses. From these studies, it is known that there are two types of basic structural unit present in the glass, namely TeO_4 trigonal bipyramids and TeO_3 trigonal pyramid [5]. The proportion of these types depends on the glass composition [6].

This paper presents the infrared spectroscopic analysis on Pb:TZBN glasses with composition $55TeO_2$ - (41- x) ZnO-$2Bi_2O_3$-$2Na_2O$-xPbO where x = 1.0, 1.5, 2.0, 2.5% mole aimed to know the effect of PbO addition on their main absorption peaks and IR edges.

2. Experiment

The glass samples composition we made are 55 TeO_2-(41-x)ZnO-$2Bi_2O_3$-$2Na_2O$-x PbO) here x = 1.0, 1.5, 2.0, 2.5% mole (Pb:TZBN). Appropriate amount of chemical substances were weighed and then poured into a platinum crucible. Melting process was carried out at temperature of 900°C and hold for about 30 minutes in air atmosphere to ensure that the high homogeneity melt was obtained. The melt was then poured into a preheated stainless steel parallel plate and then annealed at temperature of 350°C for 6 hours allowing glass to release its thermal and mechanical stress followed by cooling to room temperature at the cooling rate of 1°C/min. The obtained samples were then cut into 1.5 cm x 1.5 cm and then polished at sides of the glass in a series of fining operation.

The infrared absorption spectra were recorded using Shimadzu Infrared Spectroscopy. For this measurement, samples for about 5 mg was ground and mixed thoroughly with 200 mg KBr. This

Journal of Physics: Conference Series **776** (2016) 012108 doi:10.1088/1742-6596/776/1/012108

mixed powder was then pressed into a pellet. IR absorption spectra measurement of all samples was carried out at wavelength range 400-4000 cm^{-1} at room temperature.

3. Results and Discussion

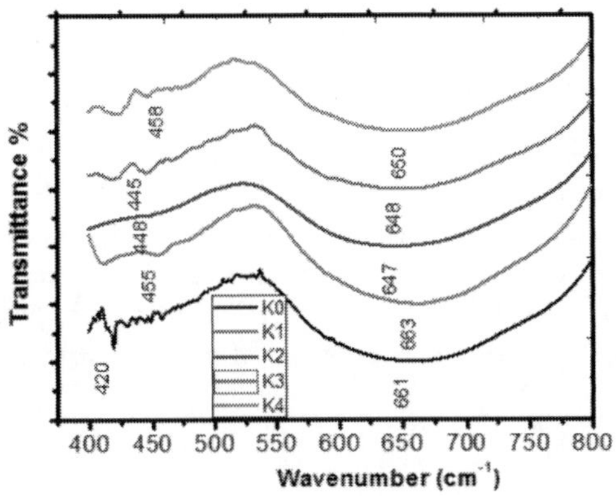

Figure 1. Absorption spectra within IR range of all glass samples recorded at room temperature

Figure 1 shows FTIR spectra of all samples measured at room temperature in the spectral range from 400 cm^{-1} to 4000 cm^{-1}. It can be seen from figure 1 that all samples have similar absorption spectra. For Sample K0, absorption peaks at 419.54 cm^{-1} and 660.65 cm^{-1} are observed. Absorption band at 419,54 cm^{-1} corresponds to vibration frequency due to chemical bonding of Te-O-Te while the absorption band at 660,65 cm^{-1} shows the vibration of Te-O bond in TeO_4 trigonal bipyramidal molecular geometry of TeO_4. Similar peaks are observed in other glass samples. Trend in decreasing wavenumber as the PbO concentration increases can easily be explained in term of molecular vibration

$$\omega = \sqrt{\frac{k}{\mu}} \tag{1}$$

Table 1. Position of the two main IR absorption peaks

Sample	Absorption peak (cm^{-1})	
K0	419,54	660,65
K1	455,22	662,58
K2	447,50	647,15
K3	444,61	648,11
K4	457,15	650,04

Where ω is angular vibration frequency, μ is reduced mass, and k can be related to the stiffness of the chemical bond. By looking at the absorption peaks as shown in figure 1 and table 1, it can be seen that no other bonding modes are observed. Consequently, the values of k of glass samples are relatively the same and give only a small contribution to the ω value change. Similar trends are observed by other researchers. Red shifting of absorption peak in IR spectra which can be related to the increase of PbO content can also be observed in IR edge of all the prepared glass samples as given in table 2.

Table 2. IR edge shifting as function of PBO concentration in all glasses and their transmittance ranges.

Sample	UV- *edge* λ (nm)	IR *edge* λ (nm)	Transmission range $\Delta\lambda$ (nm)
K0	368,02	6750,97	6382,95
K1	370,63	6821,84	6451,21
K2	374,39	6892,72	6518,33
K3	382,14	6963,60	6581,46
K4	389,86	7034,47	6644,61

4. Conclusion

IR spectra of tellurite glasses with compositions of $55TeO_2$ - (41- x) ZnO-$2Bi_2O_3$-$2Na_2O$-xPbO where x = 1.0, 1.5, 2.0, and 2.5% mol (Pb:TZBN glasses) have been studied. In order to find the effect of PbO on their IR spectra, we concentrated only to deal with three distinct IR absorptions by glasses, i.e. absorption around 450 cm^{-1}, 650 cm^{-1} and IR edge. As the PbO concentration increases, we found that the absorption around 450 cm^{-1}, 650 cm^{-1} tend to be red-shifted while the IR edge tend to be blue shifted.

References

[1]. Mohamed N. B., Yahya, A. K., Deni, M. S. M., Mohamed, S. H., Halimah, M. K., and Sidek, H. A. A, 2010 *J. Non-Cryst. Solid*, no.356, pp.1626-1630.
[2]. El-Moneim A A 2002 *Mater. Chem. Phys* **73** *(2–3) 318-322*
[3] Seshadri M, Barbosa L C, Cordeiro C M B, Radha M, Sigoli F A, Ratnakaram Y C 2015 *J. Lumin* ***166*** *8-16*
[4] Marzuki A, Wahyudi, Pramuda A, Nurmalasari I 2015 *AIP Conf. Proc.* **1725**, 020044.
[5] Aida K, Benino Y, Dimitrov V, Komatsu T, Sato R 2000 *J. Am Ceram. Soc.* **835** 1192-1198.
[6] Rosmawati S, Sidek H A A, Zainal A T, Zohir H M 2007 *J. Appl. Sci.* **7** (20) 3051 – 3056.

Multiple Internal Reflections in Acrylic Triangle Waveguide

A. Marzuki*, D. Setiawan, H. Purwanto, M. Yunianto

Physics Department, Sebelas Maret University
Jl. Ir. Sutami 36A Surakarta 57126 Indonesia

E-mail: *amarzuki@mipa.uns.ac.id

Abstract. Theoretical and experimental study of light propagation in acrylic triangle waveguides has been studied theoretically and experimentally aimed to obtain a light cone which can tunnel and bring the light to a certain place. Numerical aperture of light cones which have a length of 7 cm but different cross-section area have been measured using he-Ne laser (λ=632 nm) and the results were compared to those obtained theoretically and no significant different was observed.

1. Introduction

Over decades, multiple internal reflection effect has shown its important role in modern technology. Light propagation in an optical fiber used for optical communication is one of the well-known examples. Using ultra-low loss optical fiber, an optical signal can propagate and maintain its intensity over very long distance. Parallel to the advancement of fiber optic for long distance communication, fiber sensors have become an interesting research topic exploring the unique light propagation in fiber optic under different physical or chemical environment conditions [1-5]. In addition of that, multiple internal reflections have also been used to design light trapping [6-7]. Light is trapped inside the fiber or two reflected surface such that as the light goes deeper inside the trapper more energy will be absorbed.

Tapered optical fiber is one of the fiber types mostly used for fiber sensor [8-10]. The diameter of this fiber can be changed either permanently or temporarily. Gradual change of the fiber diameter can decrease the reflection angle at core-cladding interface results in decreasing light intensity coming out of the fiber.

Similar to the tapered optical fiber, acrylic triangle waveguides presented in this paper is designed to focus the light with no light radiating out from the waveguide sides. Theoretical calculation was made such that for a given maximum incident angle and the base triangle angle, we can find the maximum length of the triangle at which no light coming out from the triangle sides.

2. Experiment

The aim of the research reported in this paper is to find an acrylic triangle configuration that for a certain value of incident angle, the light will be guided through it by multiple internal reflection mechanism. Light is incident at the triangle entrance window and coming out at the other end of it. Between these two windows, it is expected that no light is radiated out. In order to theoretically calculate the triangle numerical aperture of a given triangle configuration that meets the above condition, data of acrylic index of refraction is required. Figure 1 is experimental setup used to measure the index of refraction designed based Brewster angle method. Transverse magnetic (TM) polarized light was incident onto the PMMA sample positioned at the center of a rotational table of a spectrometer. The reflected light intensity was detected by a photo-detector. A curve showing the intensity of reflection angle as function of incident angle was then constructed. From this curve, we can obtain the value of Brewster angle (θ) defined as the incident angle results in minimum reflected light intensity. The PMMA refractive index (n_s) can then be calculated using equation

$$\tan\theta = \frac{n_s}{n_a} \tag{1}$$

where n_a is the air refractive index.

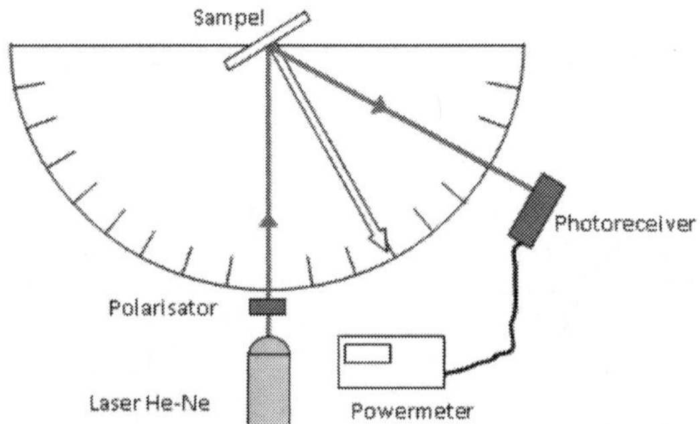

Figure 1. Experimental setup used to measure refractive index.

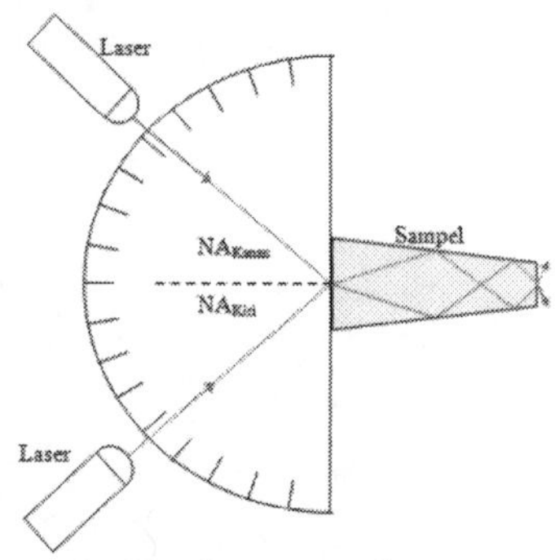

Figure 2. Basic setup for measuring triangle numerical aperture

Figure 2 is basic experimental setup used to measure numerical apperture all acrylic triangle. Samples with the same thicness (0.5 cm) but different length (6.0 cm, 6.5 cm, and 7.0 cm) and different incline angle (83^0, 85^0 and 87^0) were put at the centre of the rotating table. Monochromatic light was incident on the wider window at two fixed points: at the edge of the window and at the axis of the triangle. Laser light was then rotated such that the incident angle observed at the side of triangle wall will get smaller and smaller. Maximum value of the incident angle giving no light radiated out from the triangle edge is defined as the numerical aperture.

3. Results and Discussion

3.1. Refractive index

Reflection coefficient of polarized light in both TE and TM modes are shown in Figure 3. Reflectivity of TM mode polarized light shows a minimum value (Brewter's angle) at 56.33^0. Inserting this value into equation (1) and taking the air index of refraction equals 1, we can find value of the acrylic index of refraction which is 1.500.

3.2. Numerical aperture

Numerical aperture of acrylic triangle waveguide represents an acceptable angle within which light can pass through the waveguide with no light radiate out of from the triangle wall. If light is launched into the triangle from the wider window within this acceptable angle, light is reflected back and forth inside the triangle by multiple internal reflection. No light coming out from wall. Instead, the light will only come out from the small window (Figure 4).

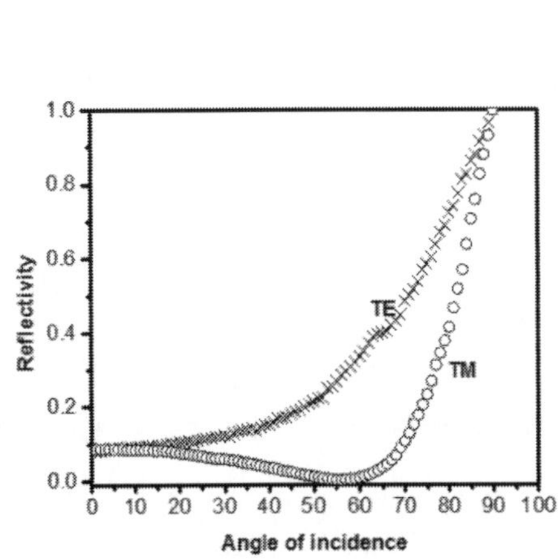

Figure 3. Reflectivity observed for acrylic-air interface

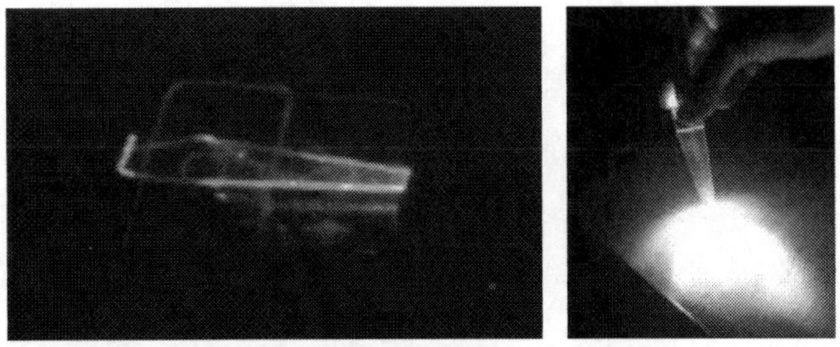

Figure 4. Light propagation in acrylic triangle (a) and acrylic cone (b)

To formulate the relation between the triangle configuration and its corresponding numerical aperture, let's look at figure 5. Light enters the triangle at an angle θ_i and then propagates through the triangle by multiple reflections mechanism. As the point N is shifted to the left to point K and Figure 5 is cut along the S-axis, Figure 5 can then be redrawn to become Figure 6. All points initially at LB wall (Figure 5) are mirrored by the S axis into the AC wall. As the light propagates down to the other end of the triangle, it can be seen that the reflected angles get smaller and smaller. To calculate these reflected angles, let's see Figure 6. From ΔABO, it is found that

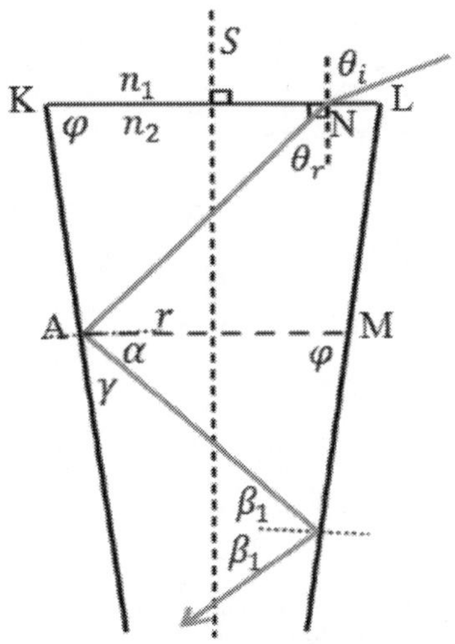

Figure 5. Light propagation in acrylic triangle

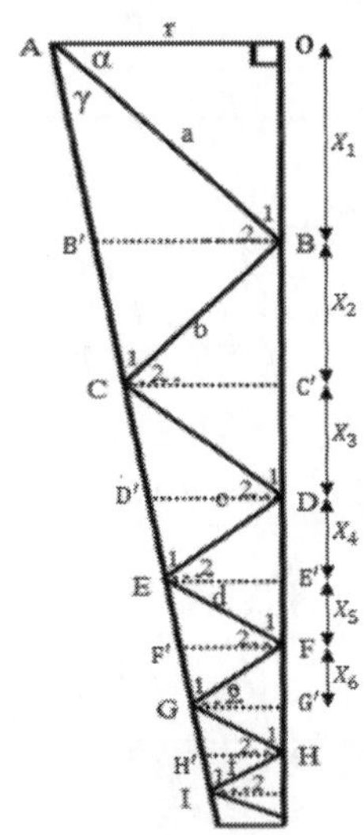

Figure 6. Light propagation after being mirrored by the S axis

$$180° = \angle B_1 + 90° + \alpha \quad (2)$$

or

$$\angle B_1 = 90° - \alpha \quad (3)$$

Since

$$90° = \angle B_1 + \angle B_2 \quad (4)$$

By substituting equation (3) into equation (4), we found

$$\angle B_2 = \alpha \quad (5)$$

Looking back to ΔABC and remembering that the summation of their angles is 180°, it is found that

$$180° = \gamma + (2 \times \angle B_2) + \angle C_1 \quad (6)$$

By substituting equation (5) into equation (6) it found that

$$\angle C_1 = 180° - (2\alpha + \gamma) \quad (7)$$

and

$$\angle C_2 = (2\alpha + \gamma) - 90° \quad (8)$$

By applying similar procedure, other reflected angles can be calculated, they are

$\angle E_2 = (4\alpha + 3\gamma) - 270°$ 2nd reflection

$\angle G_2 = (6\alpha + 5\gamma) - 450°$ 3rd reflection

$\angle I_2 = (8\alpha + 7\gamma) - 540°$ 4th reflection

For ith reflection, its corresponding angle of reflection is

$$\beta_i = 2i\alpha + (2i - 1)\gamma - (2i - 1)90° \quad (9)$$

By taking

$$\gamma = 90° - \varphi + \theta_r$$

$$\alpha = 2\varphi - 90° - \theta_r$$

$$\theta_r = \sin^{-1}\left(\frac{n_1}{n_2}\sin\theta_i\right)$$

Equation (9) can be expressed as

$$\beta_i = (2i + 1)\varphi - \theta_r - 2i.90^0 \quad (10)$$

If φ (incline angle) and n_2 (acrylic refractive index) or θ_r (refraction angle) are known, the ith reflection angle can be calculated. The distance between the two consecutive angle points as shown in Figure 6 are

$$X_1 = r\ \tan\alpha$$

$$X_2 = \frac{r\ \sin\gamma\ \tan\alpha}{\sin(2\alpha + \gamma)}$$

$$X_3 = \frac{r\ \sin\gamma \tan(3\alpha + 2\gamma)}{\sin(2\alpha + \gamma)}$$

$$X_4 = \frac{r \sin\gamma \tan(3\alpha + 2\gamma)}{\sin(4\alpha + 3\gamma)}$$

$$X_5 = \frac{r \sin\gamma \tan(5\alpha + 4\gamma)}{\sin(4\alpha + 3\gamma)}$$

$$X_6 = \frac{r \sin\gamma \tan(5\alpha + 4\gamma)}{\sin(6\alpha + 5\gamma)}$$

in general form

$$X_n = \frac{r \sin\gamma \tan((n-1)\alpha+(n-2)\gamma)}{\sin(n\alpha+(n-1)\gamma)} \quad (11)$$

Taking the triangle height as

$$L = \sum X_n \quad (12)$$

Tabel 1. Numerical apertures comparison between those calculated theoretically and experiment at λ = 532 nm.

Sample	Incline Angle	Triangle height (cm)	Entrance window width (cm)	Numerical Aperture		
				Theoretical calculation	Experiment λ = 632 nm	Percent error
A1	83^0	6.0	2.0	21.69	21.70	0.05
A2	83^0	6.5	2.0	20.79	20.73	0.03
A3	83^0	7.0	2.0	18.18	18.17	0.06
B1	85^0	6.0	2.0	40.27	40.27	0.00
B2	85^0	6.5	2.0	35.18	35.17	0.03
B3	85^0	7.0	2.0	33.81	33.80	0.03
C1	87^0	6.0	2.0	55.20	55.20	0.00
C2	87^0	6.5	2.0	54.84	54.83	0.02
C3	87^0	7.0	2.0	51.44	51.42	0.04

Once all parameters characterize an acrylic triangle are given, its numerical aperture can be calculated. Table 1 shows the numerical aperture values between those calculated based on equation (10) and (12) and those obtained experimentally. These two results show a good agreement. Small percent error might come from mainly uncertainties in preparing sample dimension. The incline angle, triangle height, and entrance window width as shown in Table 1 are the planned dimensions. The actual prepared dimensions are slightly different from those given in Table 1. Samples with the same incline angle show an increase in their numerical aperture with the increase of their heights. In addition, sample numerical aperture increases with increasing their incline angles.

4. Conclusions

This manuscript attempts to find a basic guidance to produce acrylic triangle waveguide theoretically and experimentally. Nine acrylic triangle configurations have been made. It has been shown that the triangle numerical aperture increase with increasing the incline angle and decreases with increasing the triangle length. Results obtained experimentally are in good agreement to those calculated theoretically (percent error less than 0.1 %).

5. References

[1] Tregubov A V, Svetukhin V V, Novikov S G, Berintsev A V, and Prikhodko V V 2016 J. Res. Phys. **6** 131-132.
[2] Ascorbe J, Corres J M, Matias I R, and Arregui F J 2016 Sens. Actuators **B 233** 7-16.
[3] Marzuki A, Heriyanto M, Setiyadi I D, Dedy I, and Koesuma, S 2015 J Phys Conf Ser. **622** 012059.
[4] Moo Y C, Matjafri M Z, Lim H S and Tan C H 2016 Optik **127** (3) 1312-1319.
[5] Weiwei Y, Aiyu Z, Baoshan H and Xinxia C 2008 Sens. Actuators B **130** (1) 21-24.
[6] Ducros C, Szambolics H, Emieux F and Pereira A 2016 Thin Solid Films, **article in press.**
[7] Zhang A and Guo Z 2016 Optik **127** (5) 2861-2865.
[8] Chenari Z, Latifi H, Ghamari S, Hashemi R S and Doroodmand F 2016 Opt. Laser. Technol. **76** 91-95.
[9] Leung A, Shankar P M and Mutharasan R 2008 Sens. Actuators B **131** (2) 640-645.
[10] Verma R K, Sharma A K and Gupta B D 2008 Opt. Commun. **281** (6) 1486-1491.

Journal of Physics: Conference Series **776** (2016) 012110 doi:10.1088/1742-6596/776/1/012110

Fiber optic sensor based on reflectivity configurations to detect heart rate

M. Yunianto[1], A. Marzuki, R. Riyatun, and D. Lestari
Optic and Photonic Research Group, Physics Department, Sebelas Maret University

E-mail: mohtaryunianto@staff.uns.ac.id

Abstract. Research of optical fiber-based heart rate detection sensor has been conducted using the reflection configurationon the thorax motion modified. Optical fiber used in this research was Plastic Optical Fiber (POF) with a diameter of 0.5. Optical fiber system is made with two pieces of fiber, the first fiber is to serve as a transmitter transmitting light from the source to the reflector membrane, the second fiber serves as a receiver. One of the endsfrom the two fibersis pressed and positioned perpendicular of reflector membrane which is placed on the surface of the chest. The sensor works on the principle of intensity changes captured by the receiver fiber when the reflector membrane gets the vibe from the heart. The light source used is in the form of Light Emitting Diode (LED) and Light Dependent Resistor (LDR) as a light sensor. Variations are performed on the reflector membrane diameter. The light intensity received by the detector increases along with the increasing width of the reflector membrane diameter. The results show that this sensor can detect the harmonic peak at a frequency of 1.5 Hz; 7.5 Hz; 10.5 Hz; and 22.5 Hz in a healthy human heart with an average value of Beat Per Minute (BPM) by 78 times, a prototype sensor that is made can work and function properly.

1. Introduction

The human body from a biophysical standpoint is a mechanical, fluid and electrical system.Particularly for the electrical system, organs such as the heart, lungs are viewed as a potential linked to all other organs through the vascular system [1]. Conventionally, the heart activities are observed through an electrical signal characteristic with an electrocardiogram (ECG) [2-5]. Some previous researcheshave studied using microbendingmethods [6], thorax motion [7], as well as interferometry [8] for sensing heart activity-based optical fiber. The advantages of using an optical fiber as a sensor is its small size and light weight [9], and can be connected to a data communication system via an interface device [10]. In this research, the optical fiber used is fiber of plastic optical fiber (POF) type, the prototype that has been created then is used to measure the frequency of human heart pulse, the results are analyzed and compared with the results of thorax motion method.

2. Methods

In this article, the method used to detect the heart rate frequency was a reflection method, in which schematically the series from sensors used in this research are in accordance with figure 1.

Journal of Physics: Conference Series **776** (2016) 012110 doi:10.1088/1742-6596/776/1/012110

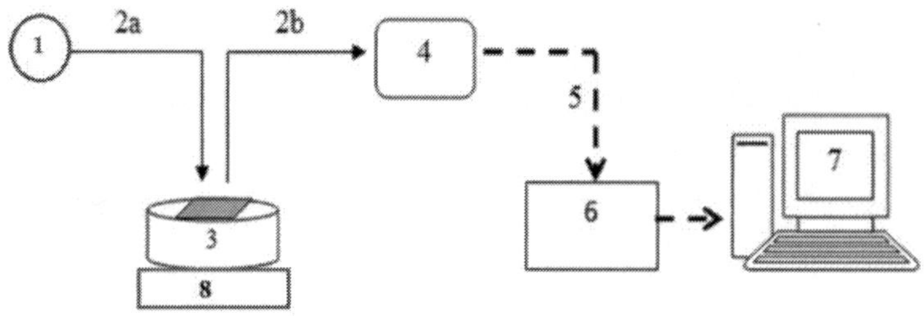

Figure 1. Principle of research work

The working principle of the circuit in Figure 1 was as follow: Fiber transmitter POF type transmitted the light from the source to the reflector membrane, the reflector membrane had been affixed beforehand to the surface of the chest around the middle the heart position in which the position was the greatest heart vibration source, heartbeat subjected to membrane caused the membrane to vibrate. The vibrations affected the light intensity reflected by the reflector which was attached to the membrane. Reflected light from the reflector was captured by fiber optic receiver. Light emitted from the fiber optic receiver was then detected by a detector. Analog data from the detector was converted into digital data and transmitted to the computer via the interface for processing. The membranes used were made of an elastic material in the form of thin rubber, while the selected reflector type was aluminum foil because it could reflect light very well and the thicknesses differences did not give effect to the intensity changes, variation in the diameter membrane was varied, the diameter size was 4 cm, 5 cm and 6 cm.

3. Results and Discussion

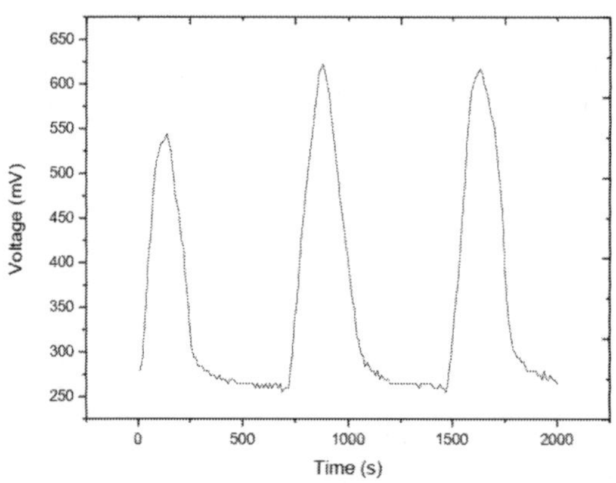

Figure 2. Results of testing the membrane reflector

Figure 2 shows the tests results on the reflector membrane, it is seen that there are significant differences between the voltages captured by a detector which is the value of the reflected light intensity from the reflector membrane when it is given the vibe from the heart rate and when it is not subjected to heart rate vibration. The peaks on the diagram show the light intensity changes periodically, so that it can be concluded that the reflector membrane can work and function properly.

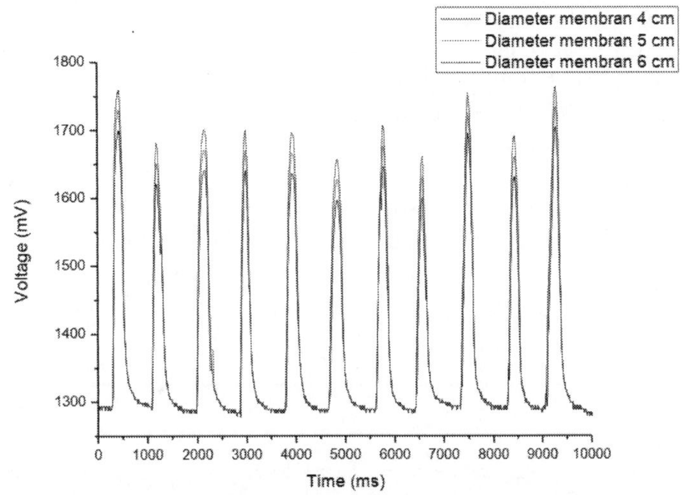

Figure 3. Time function diagram on the output voltage (black: 5 cm, red: 4 cm; and blue: 6 cm)

Figure 3 is a time function diagram on the output voltage detected by the detector, when the reflector membrane is subjected to heart rate vibration, the light intensity coming from the transmitter fiber subjected to reflector and then reflected back into the fiber receiver and is counted as one peak. It is seen in the diagram the one that has the largest voltage diagram is test results diagram using reflector membrane with a diameter of 6 cm. The results obtained for 3 variations done show that the resulting peak is within the same time frame means that the variation of the membrane diameter affects the voltage value but it does not affect the vibration frequency from the heart.
From the data that has been obtained, change on the time function diagram is made into a frequency function diagram. The vibration frequency from the heart rate can be known by wave decomposition process into its constituent waves with operation of Fast Fourier Transform performed by calculating the magnitude of one wave formation(periods) presented in Figure 4.

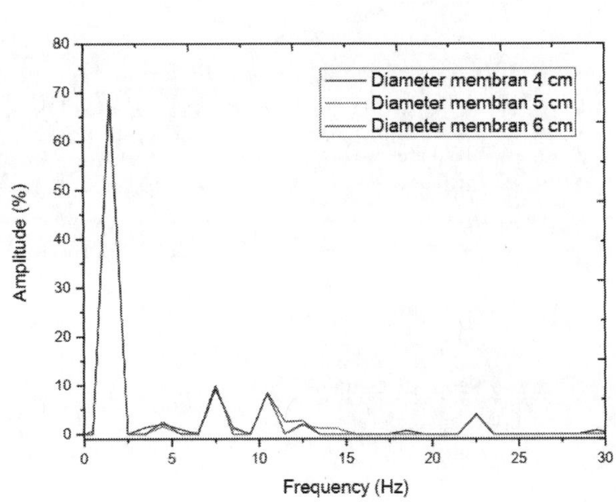

Figure 4. Relationship between frequency and amplitude
(reflector diameter of 4 cm, 5 cm and 6 cm)

Figure 4 shows the relationship between frequency and amplitude; amplitude which is expressed in percent shows the presence percentage of a frequency value in the time function diagram, where the number of overall percentage is 100%. There are four highest frequency values that appear: 1.5 Hz; 7.5 Hz; 10.5 Hz; and 22.5 Hz with a percentage which the value is not much different from each membrane diameter variations as shown in Table 1. This value corresponds and not much different from the research that has been conducted by Allsop et al., [7], where on that research the heart rate of healthy human has been detected at 1.3 Hz; 7.6 Hz; 10.8 Hz and 22 Hz. It also reiterated that the membrane diameter variation does not affect the frequency value that appears in data collection.

Table 1. four highest frequency data appear on variations of the membrane 4 cm diameter, 5 cm, 6 cm

Frekuensi (Hz)	Ø 4 cm (%)	Ø 5 cm (%)	Ø 6 cm (%)
1,5	68,7671	69,6891	69,7429
7,5	9,1781	9,9171	9,9786
10,5	8,2195	8,3886	8,5094
22,5	3,9723	3,98803	3,9934

The next result is the counting of Beat per minute (BPM) from the heart by means of data collection for one minute. Each peaks read on Lab View program shows the number of heartbeats counted and the program calculates the number of BPM and its status after detection is done for 1 minute. In Figure 5 it can be seen that the results of data collection for one minute shows BPM as much as 78 times. This BPM value indicates that the heart rate included in the normal criteria of the range of the value is between 70-80 times per minute. This is according to research conducted by Ali [11].

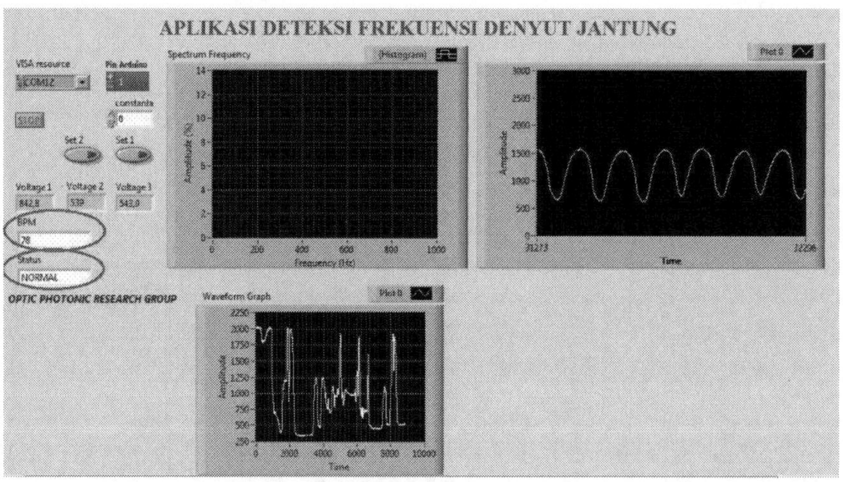

Figure 5. Calculation Results Beat per minute (BPM) with Labview program

4. Conclusion

It has been made the prototype of optical fiber-based heart rate detection sensor. The results show that the sensor prototype made is able to detect the heart rate at 1.5 Hz; 7.5 Hz; 10.5 Hz; and 22.5 Hz, with a Beat Per Minute (BPM) as many as 78 times, so it can be concluded that the sensor prototype made can work and function properly.

5. Reference

[1] Ingebretsen, Richard J 2002 *The Physics of the Human Body.* Companion Manual Physics **3110**.
[2] Jevon P 2009 *ECGs for Nurses: Second Edition.* Wiley-Blackwell
[3] Agung R 2005 *Realisasi Elektrokardiograf Berbasis Komputer Personal Untuk Akusisi Data Isyarat Elektris Jantung*. Jurusan Teknik Elektro Fakultas Teknik Universitas Udayana
[4] Parak J and Havlik J 2011. *ECG signal processing and heart rate frequency detection methods.* In 19th Annual Conf. on Technical Computing **8**
[5] Rizal, Achmad dan Jondri. 2010 *Wireless LAN Electrocardiograph (ECG).* Fakultas Elektro & Komunikasi Institut Teknologi Telkom Bandung
[6] Lau D, Chen Z, Teo, Ju T, Hua NS, Rumpel H, Lian, Y., Yang H, Kei, Pin Lin 2012 *Intensity-Modulated Micro-bend Fiber Optic Sensor for Respiratory Monitoring and Gating During MRI.* Agency for Science, Technology and Research (A*STAR) Biomedical Engineering Programme Grant 102 148 0011.
[7] Allsop T, Lloyd G., Bhamber RS, Hadzievski L, Halliday M, Webb DJ 2014 *23rd International Conference on Optical Fibre Sensors Proc. of SPIE* **9157**, 91574N.
[8] Zazula D, Donlagic D, Sprager S 2012 *Journal of the Laser and Health Academy* **2012**(1)
[9] Lujo I, Klokoc P, Komljenovic T, Bosiljevac M and Sipus Z 2008 *Fiber-Optic Vibration Sensor Based On Multimode Fiber* Radioengineering, 17(2).
[10] Yono, HP and Rohedi, AY 2008 *Jurnal Fisika dan Aplikasinya*, 4 (2), 080204-1-080204-4, Institut Teknologi Surabaya.
[11] Ali N, Rustiadi T, Annas M 2012 *Journal of Physical Education, Sport, Health and Recreation* **1** (4).

Journal of Physics: Conference Series **776** (2016) 012111 doi:10.1088/1742-6596/776/1/012111

Seismic Hazard characterization study using an earthquake source with *Probabilistic Seismic Hazard Analysis* (PSHA) method in the Northern of Sumatra

A. Yahya M, I. R. Palupi, Suharsono
Geophysical Engineering, Universitas Pembangunan Nasional "Veteran" Yogyakarta
Jl. SWK 104 (North Rim) Condongcatur 55283

Email: yahyaasadullah@gmail.com

Abstract Sumatra region is one of the earthquake-prone areas in Indonesia because it is lie on an active tectonic zone. In 2004 there is earthquake with a moment magnitude of 9.2 located on the coast with the distance 160 km in the west of Nanggroe Aceh Darussalam and triggering a tsunami. These events take a lot of casualties and material losses, especially in the Province of Nanggroe Aceh Darussalam and North Sumatra. To minimize the impact of the earthquake disaster, a fundamental assessment of the earthquake *hazard* in the region is needed. Stages of research include the study of literature, collection and processing of seismic data, seismic source characterization and analysis of earthquake *hazard* by probabilistic methods (PSHA) used earthquake catalog from 1907 through 2014. The earthquake *hazard* represented by the value of *Peak Ground Acceleration* (PGA) and *Spectral Acceleration* (SA) in the period of 0.2 and 1 second on bedrock that is presented in the form of a map with a return period of 2475 years and the earthquake *hazard* curves for the city of Medan and Banda Aceh.

1. Introduction

Indonesia has a complex and tectonic conditions because lie on three tectonic plates, including the Indo-Australia plate moves relative to the North, the Eurasian plate is relatively moved to the South and the Pacific plate moves relative to the West. Movements and meetings between the plates forming the collision and causing subduction zones contained in several regions in Indonesia [1]

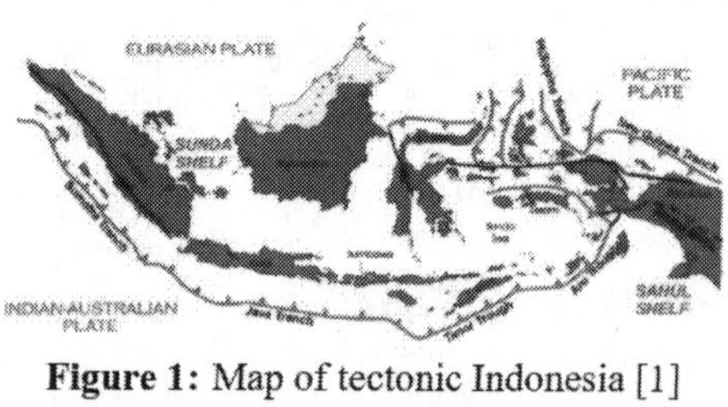

Figure 1: Map of tectonic Indonesia [1]

With this tectonic system, Indonesia be a disaster-prone areas because of potential earthquakes particularly in the northern part of Sumatra. In 2004 there is earthquake with a moment magnitude of 9.2 scale, centered offshore at 160 km in West of Nanggroe Aceh Darussalam and triggered a tsunami [2]. These events take a lot of casualties and material losses, especially in the Province of Nanggroe Aceh Darussalam and North Sumatra. Associated with tectonic conditions in Sumatra that is so active

Journal of Physics: Conference Series **776** (2016) 012111 doi:10.1088/1742-6596/776/1/012111

and causing earthquakes that have a damage in material or fatalities. Therefore it is necessary to carried out an assessment of the potential regional earthquake with earthquake *hazard* analysis which can be used as a basis of reference mitigation and planning earthquake-resistant buildings in the research area.

The purpose of the implementation of this research is to gain an earthquake *hazard* that may occur in the research area based on seismic source characterization by PSHA method. While The purpose of the study is:

1. Make a seismic hazard map *(seismic hazard)* which is described by the value of vibration acceleration in the bedrock at the 2475 year return period earthquake in research area using PSHA method.
2. Make *probabilistic* curve *hazard* in Banda Aceh and Medan in order to determine the contribution of the source of the earthquake *hazard* affecting.

2. Basic Theory

Sumatra subduction zone spread from the northern part of the Sunda Strait to the Andaman Sea. Sumatran subduction zone in the segments affected by the meeting of the Indo-Australian Plate and Plate Micro Burma as part of the Eurasian plate. The Eurasian plate subduction urges under the Indian Ocean to the northwest of Sumatra to the north of Java by varying the speed of movement. The second direction of the plate subduction path forming an acute angle thereby creating a fault zone on the island of Sumatra, which is dominated by horizontal fault.

Sumatra fault line or fault stretches along 1,900 km stretching from Aceh to the Sunda Strait **(Figure 2).** The fracture splitting into two parts of Sumatra island stretching. Sumatra fault is divided into segments amounted to 19 sections with a length of each segment ranges from 60 to 200 kilometers. The segments fault Sumatra is a segment of the Sunda, Segment Semangko, segment Kumering, segments, Manna, segment Musi, segment Ketaun, Segment Dikit, Segment Siulak, segment Sulii, segment Semani, segment Sianok, Segment Barumun, segment Angkola, segment Toru, segment Renun, Tripa segment, segment segment Seulimeum Aceh and [3].

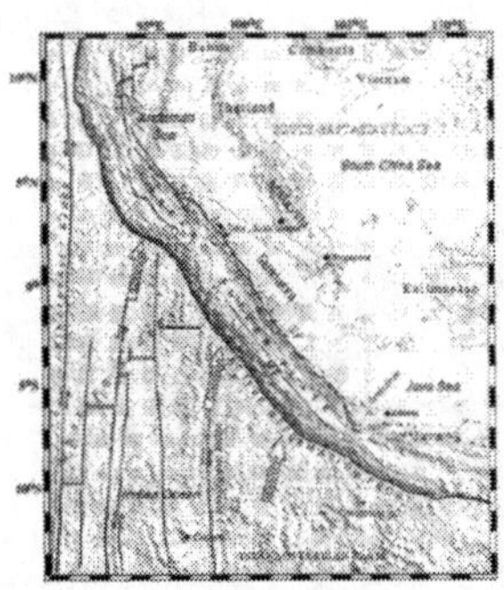

Figure 2 Map tectonic Sumatra Island [3]

3. Method

3.1. Data collection

This study use compile catalog of institutions domestically and internationally, such as the Badan Meteorologi Klimatologi dan Geofisika (BMKG), Engdahl van der hilst and Bulland (EHB) *Bulletin,* and *United States Geological Survey* (USGS) from 1907 to 2014 with a maximum depth 300 km by 700 km radius of the outermost point of the study area. In addition the data used *focal mechanism* of *the International seismilogical Center* (ISC).

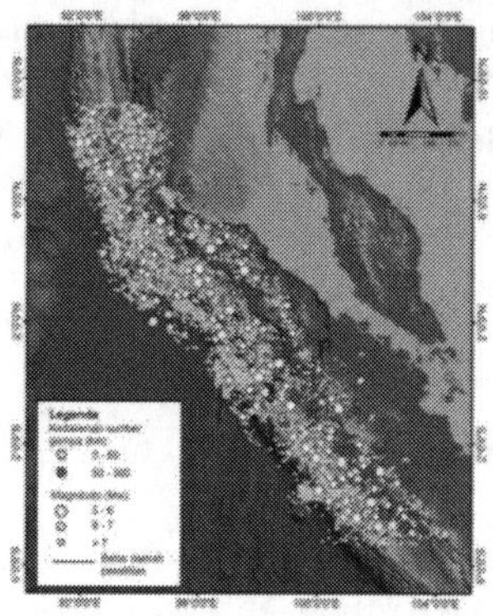

Figure 3. Map of seismicity around Sumatra from compiling catalogs

3.2. Earthquake Data Processing
Magnitude in the catalog that are used are not uniform, then the magnitude is converted to a moment magnitude (Mw) by using empirical correlation by [4]. To calculate the *b-value* and *rate reccurence* mainshock only the data that is used, so the earthquake followup *(foreshock* and *aftershock)* is removed using a program Zmap [5] with the criteria Gardner & Knopoff (1974). Analysis completeness of data [6] is used to remove the data *mainshock* incomplete. *Mainshock* at intervals of 5-6 Mw complete during the interval of the last 51 years, the magnitude of Mw 6-7 complete during the last 51 years and the magnitude of more than 7 Mw complete over 107 year.
3.3. Earthquake Source Modeling
In this study the seismic source model used is the subduction zone and *shallow crustal.* Characterization model is based on the seismotectonic conditions, regional geology and spatial data.
For subduction zone, it is using three-dimensional models. Model subduction zone is divided into two mechanisms: *megathrust* with a depth of 0-50 km and *Benioff* at a depth of 50-300 km. For geometry modeling *megathrust* zones and using catalogs *Benioff* EHB *bulletin* for this catalog has been relocated. Do manufacture *cross section* as a basis for making seismic data to generate profiles hypocenter (Figure 4)

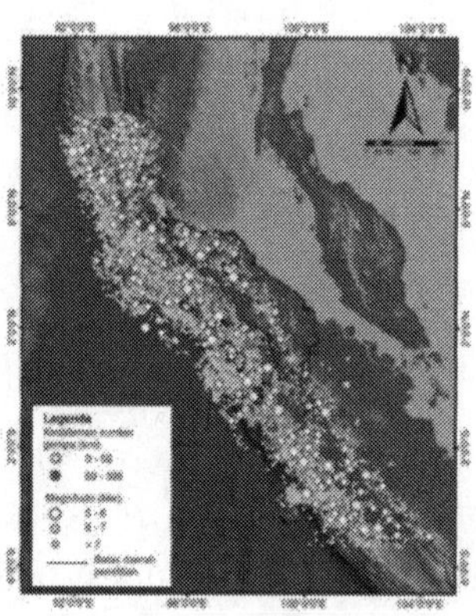

Figure 4. *Cross section* for geometry modeling of subduction

The hypocenter profile then *overlaid* with tomogram created from the data of the P wave velocity deviation *(DVP)* extracted from LLNL-G3Dv3 [7] to generated cross sections as shown in Figure 5. While the interpretation of the geometry of the subduction zone re shown in Figure 6. Referring to the research by [4] subduction zones used in the study were divided into three segments.

For *shallow* crustal *earthquake* source is divided into two: the source of the earthquake source seismic faults that are modeled in 3-dimensions and for a zone that contained the seismic activity but have not yet identified the source accommodated with *shallow* earthquake source *background* is modeled two-dimensional or *area source.*
For fault earthquake source models it is used data from the research results by [4] and then in the re-use map SRTM90m *trace.* Used as many as 14 segments of the fault in this study. Visualize the source of the earthquake fault **is** shown in **Figure 7.**

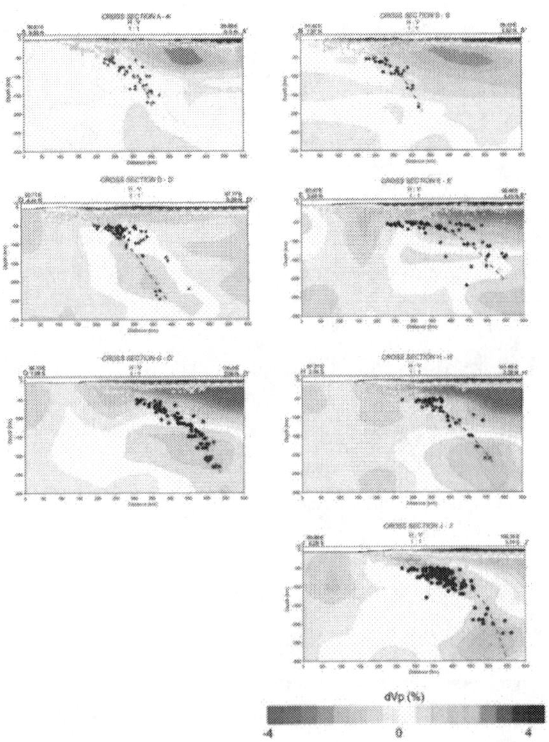

Figure 5 Hypocenter and tomogram Sectional profile extracted from LLNL-Earth3D [7]

Background shallow earthquake data with seismic source area is modeled after *the focal mechanism* of ISC is used as a weighting on the basis of the mechanism of the earthquake that will be used for the analysis of *seismic hazard.* The weighting value given to the mechanism of earthquake dominant in these areas. While the area is not accommodated *focal mechanism,* given the weighting of 0.4 for the *reverse-slip* mechanism, *strike-slip* to 0.4 and 0.2 for *normal-slip* mechanism.

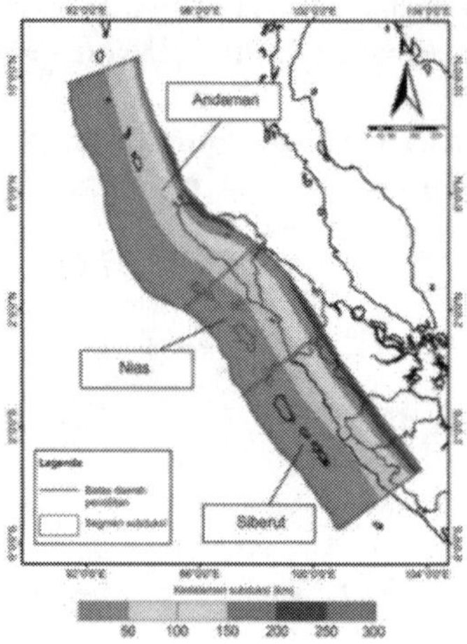

Figure 6. Results of the interpretation of the geometry of subduction

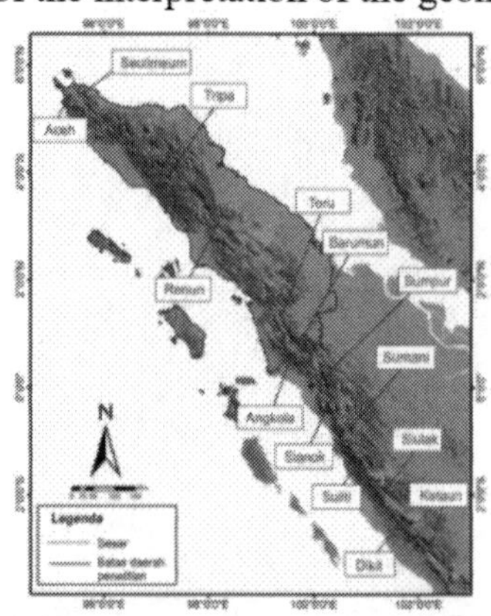

Figure 7. Map of the fault segment that is used in research

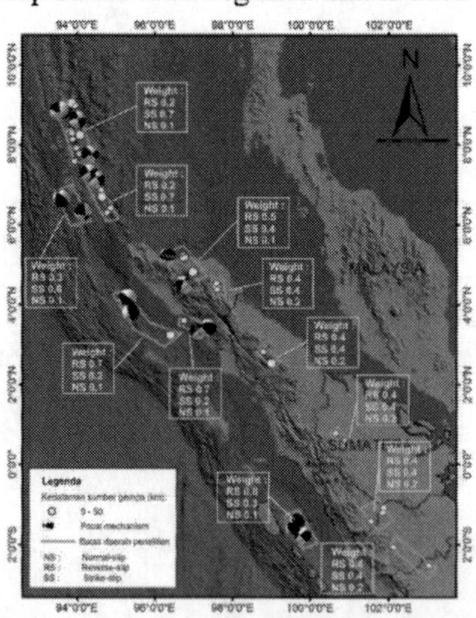

Figure 8. Model *shallow* earthquake source *background*

3.3. b-value and rate Earthquake

Analysis of seismic hazard parameters required *reccurence relationship* that is *b-value* and the *rate* of earthquake. Determination of these parameters obtained from seismic data clustering by source and type of mechanism is then used two methods, they are *Least Square* [8] and *Maximum likelihood* [9]. From both methods obtained the value *of a* and *b-value* which is used to search earthquake *rate* for each source.

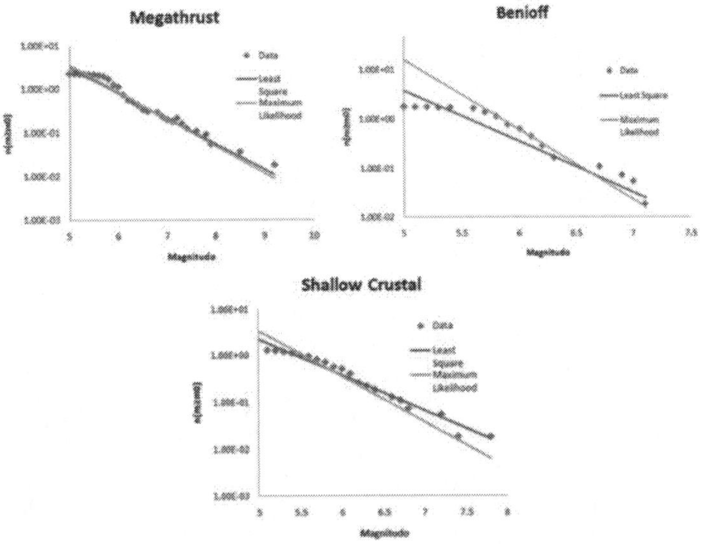

Figure 9. Curve *reccurence relationship* each seismic source

3.4. Maximum magnitude, Slip-rate and Attenuation Function

The maximum magnitude of the earthquake as well as *slip-rate* subduction earthquake sources *megathrust* and fault using the research results of [4]. The magnitude of the earthquake fault is determined by an empirical formula by [10] :

$M_{max} = 5.08 + \log L\ 1{:}16$ (1)

with:

$F_{max:}$ maximum magnitude that can occur $(M)_{w)}$

L : Long fault segment (km)

Table 1 Function Attenuation used

Source quake	**Function attenuation**
megathrust	Youngs et al (1997)
	Zhao et al (2006)
	Atkinson-Boore, *Worldwide* (2003)
Benioff	Atkinson-Boore, *Worldwide* (2003)
	Atkinson-Boore, Cascadia (2003)
	Youngs et al (1997)
Fault & shallow background	Campbell-Bozorgnia, NGA (2008)
	Boore-Atkinson, NGA (2008)
	Chiou -Youngs, NGA (2008)

Because there is no attenuation function is derived specifically for the Indonesian region then used the attenuation function is derived for other regions as well as *worldwide.* Selection is based on the attenuation function similarity of geological and tectonic conditions of the area where the attenuation function is made.

3.5. Logic Tree

The processing using *logic tree* in PSHA to solve uncertainty parameter values that used during the *seismic hazard* analysis. Using *logic tree* provide a more appropriate framework to in this method. So the *logic tree* models applied to four seismic source model that has been created. Examples *logic tree* used in this study looked at Figure 10.

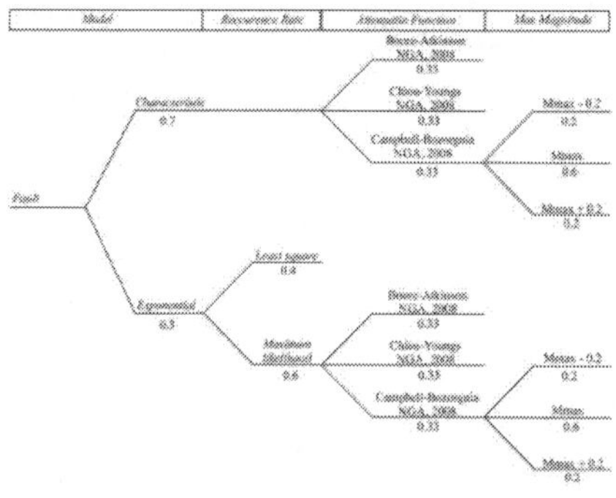

Figure 10. *Logic tree* to the earthquake source fault

3.6. Seismic Hazard Analysis

After obtained the parameters of the previous stage then *seismic hazard* analysis using EZ-Frisk program 7:52 to obtain parameters such as acceleration PGA and SA in the period spectra 0.2 and 1 second. Grid used is 0,1 ° x0,1 ° to a radius of about 50 km of the fault zone and 0,5 ° x0,5 ° outside the zone. The results of seismic acceleration at each grid was mapped using ArcGIS 9.3 program to obtain a map of seismic *hazard* in the northern part of Sumatra at 2475 year return period earthquake. *Hazard* analysis carried out on the layer of bedrock S_B to the value of the S wave velocity *(*$Vs_{30)}$ at 750 m / s.

4. Results and discussion

The results of the analysis of *seismic hazard* in this study is the acceleration of earthquake *hazard* maps for the bedrock in northern Sumatra region covering the provinces of Nanggroe Aceh Darussalam and North Sumatra on condition PGA, SA 0.2 second and 1 second at 2475-year return period earthquake. The visualization can be seen in **Figure 11** to **Figure 13.** As well as *hazard* curves for Banda Aceh and Medan can be seen in **Figure 14** and **Figure 19.**

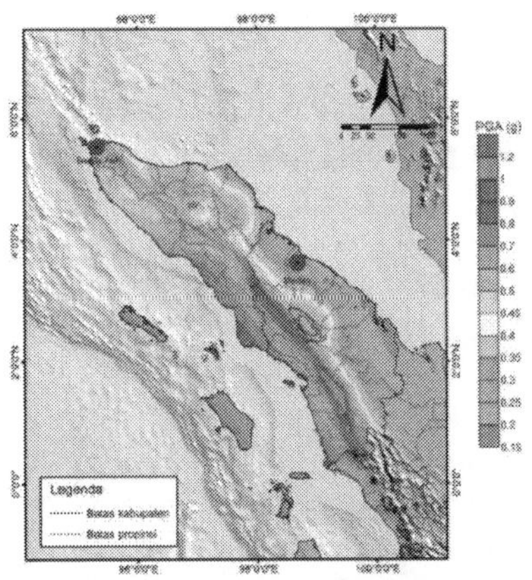

Figure 11. Map of PGA *hazard* conditions in the bedrock over a period of 2475 years of northern Sumatra region

PGA in the bedrock value for the period 2475 earthquake in northern Sumatra region ranged from 0.15 to 1,3g. High acceleration values found on islands and coastal areas in the southwest and around the fault segment. From the pattern contour, *megathrust* and faults are the dominant source of the earthquake in the region.

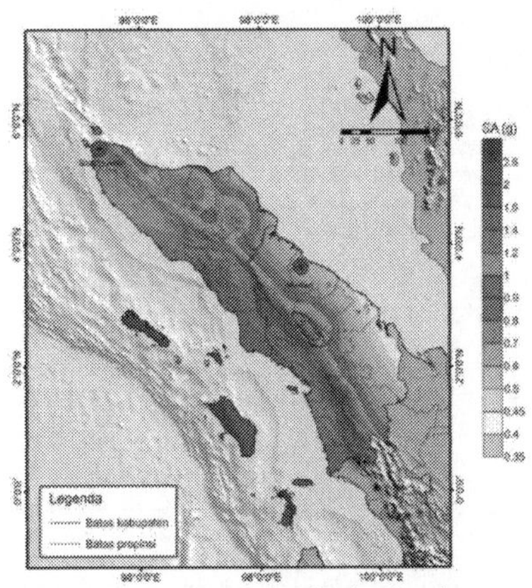

Figure 12. Map of *hazard* conditions SA 0.2 seconds in the bedrock over a period of 2475 years of northern Sumatra region

Value SA 0.2 seconds in bedrock for 2475 year return period earthquake in northern Sumatra region ranged from 0.35 to 2.8 g. Contour pattern looks the same as the map of the PGA, but the value of the acceleration in 0.2 seconds SA is much higher than the PGA.

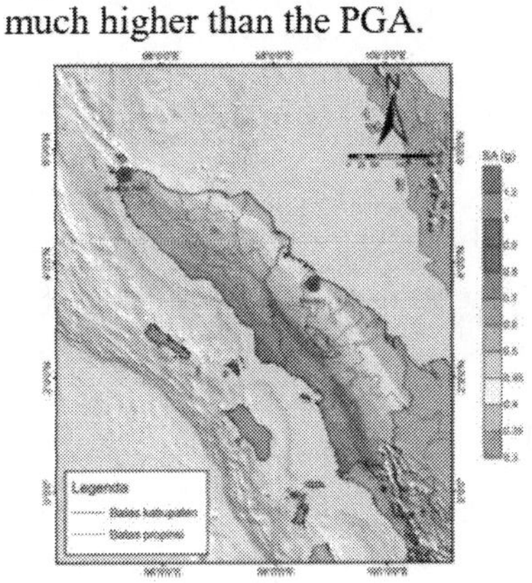

Figure 13. Map of *hazard* conditions SA 1 second in the bedrock over a period of 2475 years of northern Sumatra region.

Value SA 1 second on the bedrock for the 2475 year return period earthquake in northern Sumatra region ranging from 0.3 to 1.3g. In this period of acceleration values decreased significantly compared to the period of 0.2 seconds, especially in the southwest region.

SA 0.2 seconds has an estimated as the highest acceleration of bedrock has relatively low natural vibrating period of less than or equal to 0.2 seconds so the response at this *site* tend to be younger, especially in areas that close to the earthquake source. So the content of the frequency of earthquakes at close distance is still dominant with a high frequency that makes the response of the medium with a low natural frequency or short to a maximum.

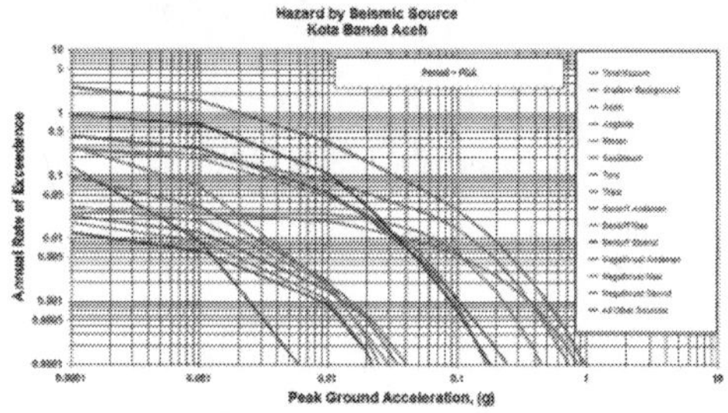

Figure 14. Curve *hazard* conditions in the bedrock PGA Banda Aceh

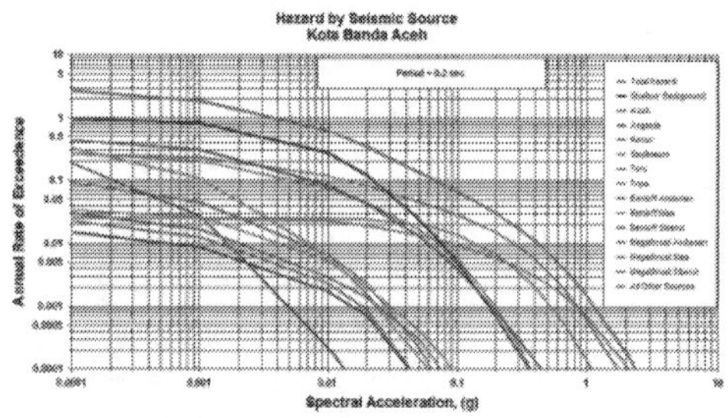

Figure 15. curve SA 0.2 seconds *hazard* conditions in the bedrock Banda Aceh

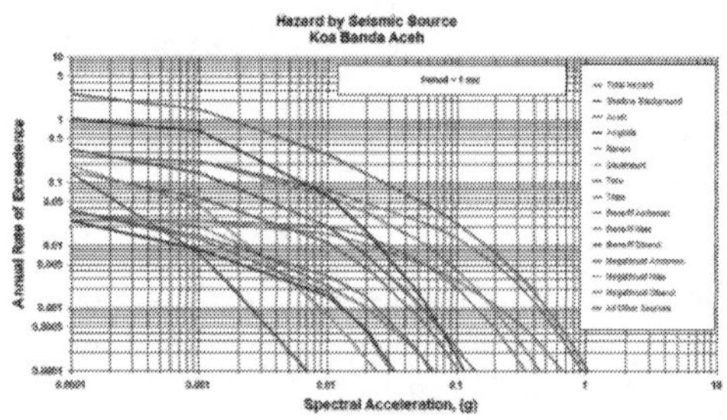

Figure 16. Curve *hazard* conditions SA 1 second on the bedrock of Banda Aceh

Based on the *hazard* curves for the city of Banda Aceh, the dominant source of the earthquake fault segment is *megathrust* in Aceh and Andaman segment. Both of these sources provide significant *hazard* than other seismic sources.

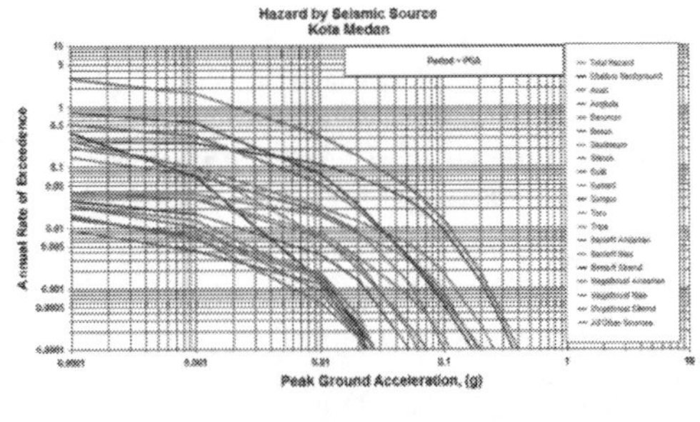

Figure 17 Curve *hazard* conditions in the bedrock PGA Medan

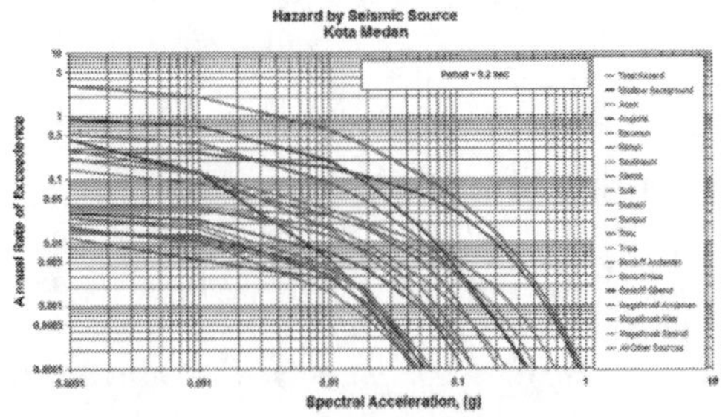

Figure 18 curve SA 0.2 seconds *hazard* conditions in the bedrock Medan

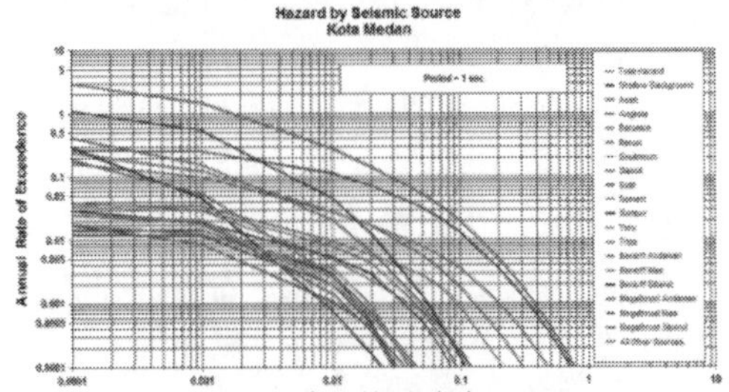

Figure 19 Curve *hazard* conditions SA 1 second on the bedrock of Medan

Based on the *hazard* curves for the city of Medan, the dominant source is *megathrust* earthquake in Nias segment. It is estimated that resources *megathrust* earthquake magnitude and the maximum length of the segment is relatively large so it still feels significant *hazard* at a considerable distance. This is also supported by the value of the *slip-rate* that makes a big earthquake return period to be faster.

5. Conclusion

Based on the results in the form of *hazard* maps by the tremor in the bedrocks, the value of PGA in northern Sumatra ranged from 0.15 to 1,.3g, SA 0.2 seconds value ranges from 0.35 to 2.8 g and SA 1 sec ranging from 0.3 to 1.3g with the dominant source of the earthquake in the region and the *megathrust* fault.

Based on the *hazard* curve to Banda Aceh at 2475 year return period earthquake, equivalent to 0.0004 *annual rate of exceedence,* PGA value is from 0.68g, SA 0.2 seconds of 1.57g and 0.69g SA 1 second of the *megathrust* source. The dominant earthquake in Aceh and Andaman are the fault segments. As for the Banda Aceh and Medan the value of PGA is 0.29g, SA 0.2 seconds of 0.65g and 0.51g SA 1 second of the seismic source *megathrust* dominant segment of Nias.

References

[1] Shah HC dan Boen T 1996 *Seismic Hazazrd Model for Indonesia* Stanford University
[2] Natawidjaja DH 2007 *Gempa dan Tsunami di Sumatera dan Upaya untuk Mengambangkan Lingkungan Hidup yang Aman dari Bencana Alam,* Symposium ITB:Bandung
[3] Sieh K and Natawidjaja DH 2000 *Journal of Geophysical Research* **105(**B12) American Geophysical Union
[4] Irsyam, dkk 2010 *Ringkasan Hasil Studi Tim Revisi Peta Gempa 2010* Bandung
[5] Wiemer S 2001 *Seismological Research Letter* **72**
[6] Stepp JC 1973 *Analysis of the Completeness of the Earthquake Hazard Sample in the Puget Sound Area* NOAA Technical Report ERL 267-ESL 30, Boulder, CO.
[7] Simmons NA, Myers SC, Johannesson G and Matzel E 2012 *J. Geophys. Res.* ***117*** doi:10.1029/2012JB009525
[8] Guttenberg B and Richter C 1944 *Frequency of Earthquake in California* Bulletin of Seismological Society of America, Vol. **34**.
[9] Aki K 1965 *Maximum Likelihood Estimate of b value in the Formula log N = a-bM and Its Confidence Limits.* Bulletin of Earthquake, Res. Ins., Tokyo University
[10] Wells DL and Coppersmith KJ 1994 *New Empirical Relationship Among Magnitude, Rupture Length, Rupture Width, and Surface Displacement,* Bulletin of Seismological Society of America, Vol. **84**

Geological structure analysis in Central Java using travel time tomography technique of S waves

I. R. Palupi, W. Raharjo, S. W. Nurdian, W. S. Giamboro, A. Santoso
Geophysical Engineering, Universitas Pembangunan Nasional "Veteran" Yogyakarta
Jln. SWK 104 Condongcatur, Sleman Yogyakarta

Abstract Java is one of the islands in Indonesia that is prone to the earthquakes, in south of Java, there is the Australian Plate move to the Java island and press with perpendicular direction. This plate movement formed subduction zone and cause earthquakes.
The earthquake is the release of energy due to the sudden movement of the plates. When an earthquake occurs, the energy is released and record by seismometers in the waveform. The first wave recorded is called the P waves (primary) and the next wave is called S waves (secondary). Both of these waves have different characteristics in terms of propagation and direction of movement.
S wave is composed of waves of Rayleigh and Love waves, with each direction of movement of the vertical and horizontal, subsurface imaging by using S wave tomography technique can describe the type of the S wave through the medium. The variation of wave velocity under Central Java (esearch area) is ranging from -10% to 10% at the depth of 20, 30 and 40 km, the velocity decrease with the depth increase. Moho discontinuity is lies in the depth of 32 km under the crust, it is indicates there is strong heterogenity in Moho.

1. Introduction

Earth is composed of several layers. The first layer is a lithosphere, the second is mantle that consisting of the upper and lower mantle and the last is the earth's core. The core layer is hot and heats the mantle that is liquid, causing convection currents. This then leads to convection currents make plate movement which resulted the earthquakes and subduction zones.
Subduction zones are formed because the density is different between the two plates that meet each other. Oceanic crust will infiltrate under the continental crust because the density is higher. The subduction zone's temperature is cooler than the surrounding area, then the subduction zone is lifted the continental crust around it and form a row of volcanoes, such as a series of volcanoes along the western coast of Sumatra and South of Java as the impact of the Australian plate and the Eurasian collission [1]
Widiyantoro (2003) op cit [1] hypothesized that the plates of the ocean in subduction zones have positive anomalies due to tectonic processes and compressed, while the continental plates have a negative anomaly due to decompression.
The purpose of this research is to examine the S wave tomography relationship with the character of the S wave itself of Java.

2. Tectonic Setting of Java

Java tectonic system is still be a hot topic discussed by geologists. According to Hamilton (1979) op cit [2] subduction path of Java, particularly in the Cretaceous is described curved from the Ciletuh (Sukabumi, West Java) continues to the northeast in Meratus (East Kalimantan). Meanwhile, according to another researcher, subduction curved from Ciletuh continue to Karangsambung and turn to Meratus [2].

There are some faults in Java trending east-west direction and parallel to the subduction tertiary pathways, while the structure trending structural pattern with northesast - southwest direction parallel to the Cretaceous subduction path that can be seen in Figure 1 [2]

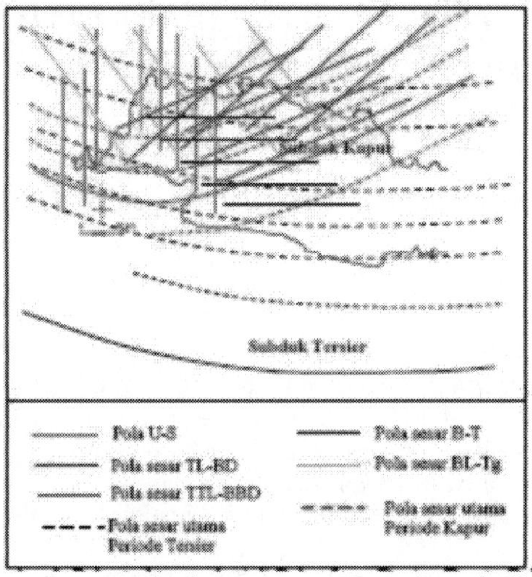

Figure 1. Patterns of Western Java Geologic structure. There is a similar pattern pre-Tertiary rock structure between Ciletuh and Karangsambung [2]

3. Characteristics of S Waves

S wave is a wave that the spreading direction is perpendicular with the propagation. S wave appears after the P wave, and has two directions, namely vertical and horizontal directions. In addition, the S waves can not propagate in the liquid medium.

The negative velocity anomaly (it is means low velocity) for S waves is directed horizontally on a layer of top layer that is associated with Love wave and poitive velocity anomalies (it is means high velocity) on the below layer that associated with Rayleigh waves and the arrival time is slower than the Love wave [1].

Because the S waves can not pass through the liquid medium, then the S wave is namely ScS where waves were only able to pass through the mantle boundary.

4. Methodology

This study uses USGS earthquake data since 1952 to 2012 around Central Java, with 2783 numbers of S wave arrival time data. The data subsequently relocated using the Grid Search to get a better velocity model to get the description of S wave velocity model and it's characteristic in subsurface by tomography technique.

The concept that used in this research is tomography with ray tracing pseudo bending that is based on the principle of Fermat in which the wave propagates through a medium with the fastest travel time (Um and Thurber (1987) op cit [3]).

Wave propagation time can be defined in the following equation:

$$T = \int_{source}^{acceptor} \frac{1}{V} dl \qquad (1)$$

Where T is the travel time, V is the velocity model in the model parameterization and dl is the length of the track. To obtain velocity model inversion time delay process is carried out by dividing the area carefully situations into blocks of a certain size to facilitate the calculation, so that the equation (1) can be written as:

$$t = \sum S\, dl \qquad (2)$$

Where t is the travel time, S is the slowness and dl is the path length for each block [3]
The concept of ray tracing pseudo bending is assume the three points through which the seismic wave, which is denoted by Pk-1, Pk and Pk + 1. Pk is the mid point of Pk + 1 and Pk-1. Then look for a new central point Pk 'criteria latency is smaller than the Pk [4].

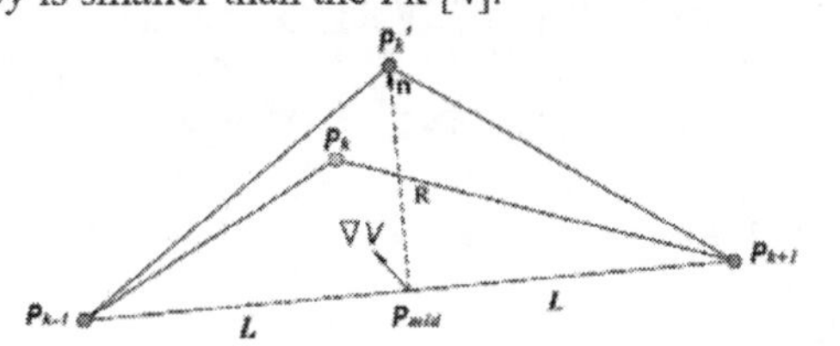

Figure 2. Illustration of three points on the pseudo bending method [4]

5. Discussion

The results of the S wave tomography was made with two cross-section models that cut the Java with Mount Slamet as a central location such as shown in **Figure 3**:

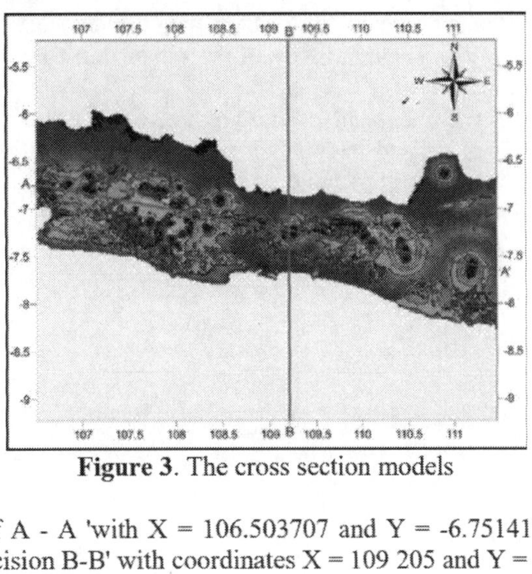

Figure 3. The cross section models

The incision in the form of A - A 'with X = 106.503707 and Y = -6.751412 until X = Y = -7.650037 111.697802 and then the incision B-B' with coordinates X = 109 205 and Y = -5.23 to X = 109 205 and Y = -9.23.

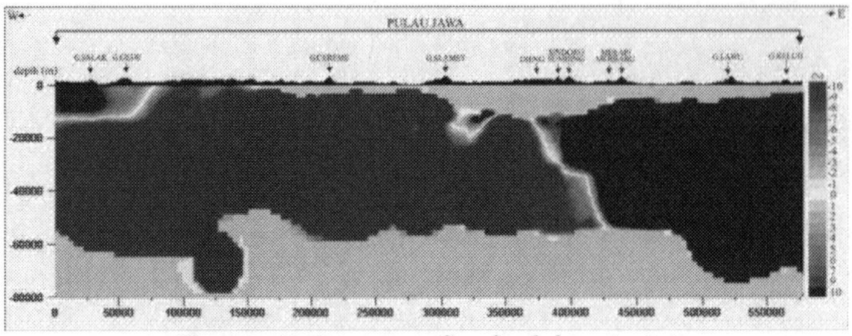

Figure 4. Cross-section incision A-A '

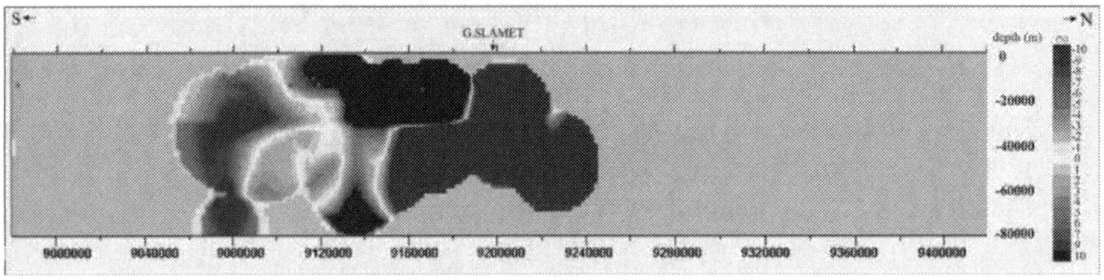

Figure 5. Cross-section incision B-B '

On incision A-A there is a large negative velocity anomalies from West Java to some regions in Central Java precisely on Dieng mountain, then the positive velocity anomaly from Mount Dieng untill East Java. The anomaly arises because the amount of tectonic activity of Java, giving rise to a row of volcanoes. There are two big differences anomalies that indicate differences in the character of the S wave propagation in this region. Referring to the research by [1], in West Java and some region in Central Java was dominated by Love waves associated with negative velocity anomaly while in some areas of Central Java to the East Java, precisely dominated by Rayleigh wave.

In the incision B-B there are positive velocity anomalies on the Ocean plate, exactly in front of subduction zones and negative velocity anomaly on the continental plate that accordance with the hypothesis by Widiyantoro (2003) op cit [1].

Contrast Differences of the velocity anomaly in the Dieng area prove theories about the tectonic structure in Java as proposed by [2] that there is a continuity of the subduction zone of Ciletuh - Karangsambung then turn to Meratus.

In addition, variations of the S wave associated with the danger and exploration zones [5], since the S wave is a shear wave and it is associated with a Poisson's ratio. Distribution of the S wave which is a shear wave at shallow depths of around a few hundred meters which is usually the sediments and bedrock, is a matter that needs to be considered in making infrastructure or buildings.

While tomography models for the Central Java can be seen in Figure 6:

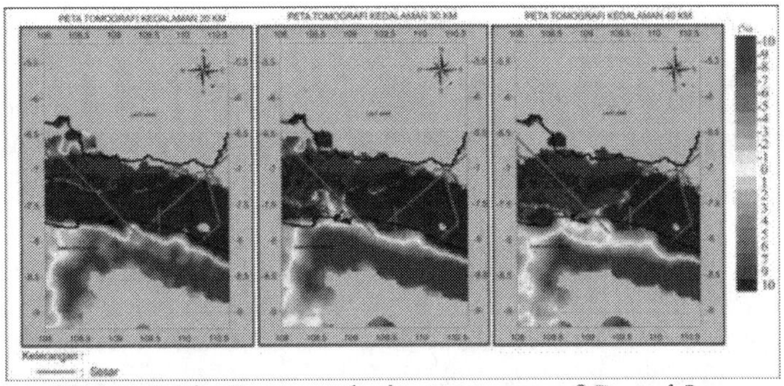

Figure 6. The S wave velocity tomogram of Central Java

In Figure 6 the positive velocity anomaly is bigger with increasng depth. It is accordacne with research by [1]. Beside that there are negative velocity anomalies on the continental shelf. In **Figure 6** there are also differences of positive and negative velocity anomaly at a depth of 40 km which the direction is corresponding with the fault direction. It can be interpreted that there are differences in the geological structure of the area and the dominance of the earthquake that occurred in West Java and some parts of Central Java which is indicated by a large negative velocity anomaly in the area.
Moho depth beneath the continental crust is about 32 km. Figure 6 show tomogram of S wave velocity in the depth of 20, 30 and 40 km. At a depth of 20 km, variation of the S wave velocity anomalies ranging from -10% to 10%, it is show that the area is very heterogeneous, as research by [6] stating that the variation of the S wave velocity anomaly which reached 6% indicated great heterogeneity in the research area, which reached a depth of Moho, exactly at middle and lower crust
Research by [6] also showed the depth of 40 km decrease the velocity and increase in attenuation function. Figure 6 also shows the same thing, at the depth of 30 and 40 km negative velocity anomalies is wider.
Meanwhile, research conducted by Zhao et al (1996); Tape et al (2010) op cit [7] show a very strong heterogeneity in the crust and upper mantle beneath California indicated by the wave velocity description using tomographic techniques. Low wave velocity is usually associated with the fluid. Lin et al (2007); Zhao and Kanamori (1993) and Zhao et al (2005); Lees and Nicholson (1993); Tian et al (2007a) with their own reearchs op it [7] show a correlation between seismic activity and heterogeneity of the crust which refers to the conclusion that the presence of fluid in the earth's crust lies in a weak zone that resulting earthquakes, in this research is the Landers earthquake.
Thus, by examining the earthquake through the wave velocity tomography, a description of the subsurface can be known.

CONCLUSION

The results of the S wave velocity tomography under Java island there are positive and negative velocity anomaly in the area around Dieng-Central Java, where the predominantly negative velocity anomaly is in the western part of Java island. This suggests that in the western part of Java, earthquake occurred more dominant and dominated by Love waves and prove the theories about there is a continuity of the subduction zone in Ciletuh to Karangsambung then turn to Meratus.
At a depth of 20, 30 and 40 km, there is a boundary between the the velocities in the direction of the fault which the low velocity is became wider with the depth increase. Tomographic results also showed that part of the ocean plate on the area in front of the subduction zone has a positive velocity anomaly and on the continental shelf has a negative velocity anomaly.
The variation of the velocity is ranging from -10% to 10%, it is indicates that there is strong heterogenity in the depth of 20, 30 and 40 km, which is the Moho discontinuity.

References

[1] Santosa, BJ. 2008. 2008 *Makara Sains*, Volume **12** No 2 pp 134-145
[2] Haryanto, I, Sunardi E, Sudrajat A, dan Suparka. Hipotesis Mengenai Sejarah Tumbukan Lempeng Zaman Kapur di Indonesia Bagian Barat.
[3] Iskandar, A, Lantu., Aswad S., and Nugraha AD. Tomografi Seismik 3-D pada Lapangan Panas Bumi "X"
[4] Xu, T, Li F., Wu Z, Wu C., Gao E., Zhou B., Zhang Z., and Xu G. 2014. A Sucseccive Three point perturbation method for fast ray tracing in complex 2D and 3D geological models.
[5] Paasche, H., Ohrnberger, M., Werban, U., Fechner, T., Tronicke, J., Krüger, F., Dietrich, P. 2011. GEOTECHNOLOGIEN *Science Report*, Vol. 18, 2011, doi:10.2312/GFZ.gt.18.03
[6] Toksöz, M. Nafi., Robert D. Van der Hilst, Youshun Sun, Chang Li, and Huajian Yao. P 28th Seismic Research Review: Ground-Based Nuclear Explosion Monitoring Technologies
[7] Tong P, Zhao D, Yang D, Yang X, Chen J, and Liu Q. 2014. *Solid Earth*, **5**, pp. 1169–1188

Journal of Physics: Conference Series **776** (2016) 012113 doi:10.1088/1742-6596/776/1/012113

Comparison tomography relocation hypocenter grid search and guided grid search method in Java island

S. W. Nurdian, N. Adu, I. R. Palupi, W. Raharjo

Geophysical Engineering, Universitas Pembangunan Nasional "Veteran" Yogyakarta
Jln. SWK 104 Condongcatur, Sleman Yogyakarta

E-mail: setyonurdian@gmail.com

Abstract. The main data in this research is earthquake data recorded from 1952 to 2012 with 9162 P wave and 2426 events are recorded by 30 stations located around Java island. Relocation hypocenter processed using grid search and guidded grid search method. Then the result of relocation hypocenter become input for tomography pseudo bending inversion process. It can be used to identification the velocity distribution in subsurface. The result of relocation hypocenter by grid search and guided grid search method after tomography process shown in locally and globally. In locally area grid search method result is better than guided grid search according to geological reseach area. But in globally area the result of guided grid search method is better for a broad area because the velocity variation is more diverse than the other one and in accordance with local geological research conditions.

1. Introduction

Indonesia is formed by the interaction of three plates, Eurasia, Indo-Australian and Pacific plate, as a result there are many volcanoes and earthquakes in Indonesia. Indonesia has 129 active volcanoes, or about 15% of volcanoes in the world, one of the island where there are many volcanoes and earthquakes is Java island, it is formed by subduction between the indo-Australia that subducting under the Eurasia, so the research on seismicity of the volcano tectonic earthquakes in Java is needed. One of the method to obtain the seismicity is seismological analysis to determine the location of the hypocenter. But actually the location of hypocenter is often shifted from the actual position because of the velocity model used is less precise. This problem can be solved by relocating the hypocenter using velocity model that accordance with the state of the field, so the hypocenter could represent actual geological conditions.

Hypocenter relocation methods are vary with the advantages and disadvantages of each, in this research used two method that are Grid Search and Grid Guided search. In this research conducted relocation hypocenter of the earthquake on the island of Java, the data recorded from 1952 to 2012 with 9162 P wave and 2426 events recorded by 30 stations. To determine the velocity of dispersion, tomography pseudo bending is carried out centered on the midpoint of the island of Java, Mt Slamet.

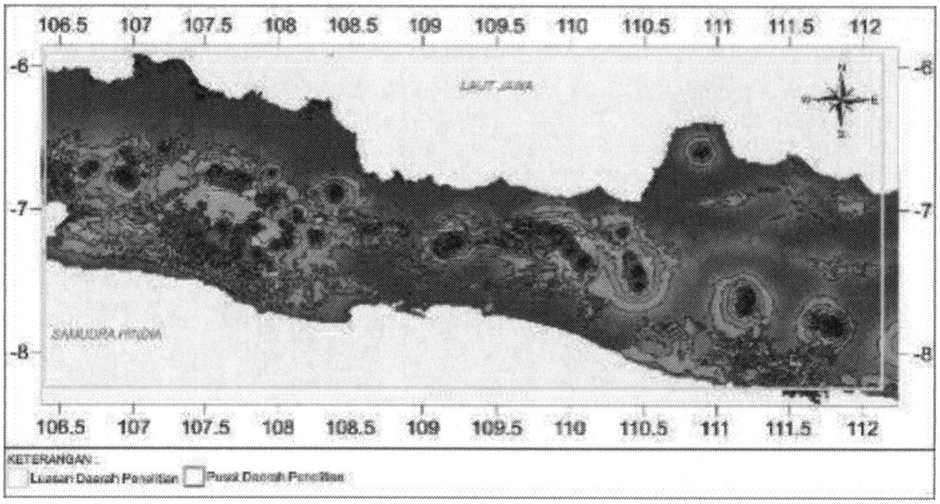

Figure 1 Map Location Research Area

2. Data and Methodology

2.1 Data

Earthquake data recorded since 1952 until 2012 with a 9162 P waves, 2683 S waves, 2426 events were recorded by 30 stations. Tomography process only used the P wave phase.

2.2 Relocation Methods

Grid search relocation method is done by calculating the wave propagation time between the station and the " new hypocenter location" calculated by adding or subtracting a certain distance from the hypocenter positions early and use the 1D velocity table (in this study used AK 135 velocity table) . This method can reduce the duration of computational calculations [1]. If the new hypocenter location has wave propagation time between it and the stations are closer or smaller than the wave propagation time of initial hypocenter location, it is considered as the solution

Guided grid search method is used to relocate the hypocenter because global search technique can describe objective surface function more through by improving efficiency with giving higher probability value on the model that close to the solution than the model that far from the solution (Grandis, 2009 op cit [2]).

Travel time tomography is one way to describe velocity in subsurface. The procedure can describe in three stages : 1. Parameterization of the velocity in the subsurface; 2. Forward modeling by ray tracing from the hypocenter location to the station; 3. Optimization of the parameters obtained in the first stage (Figure 2) [3].
This study uses USGS earthquake data since 1952 to 2012 around Central Java, with 2783 numbers of S wave arrival time data. The data subsequently relocated using the Grid Search to get a better velocity model to get the description of S wave velocity model and it's characteristic in subsurface by tomography technique.

With a simply velocity model that be used, the focus point location can be randomly spread over the subsurface area. The differences between modeled travel times and the travel times of the initial hypocenter is the input for the optimization. If the differences is equal to zero, theoretically an accurate velocity model is obtained [3].
Travel time from hypocenter to the station calculated with the equation (1):

$$T = \int_{raypath} \frac{1}{V}\, ds \qquad (1)$$

where V is the velocity in the model parameterization and ds is the ray segment length. The inversion purpose is to minimize between modeled travel time and real travel time with ray tracing pseudo bending by Um & Thurber (1987) of cit [4].

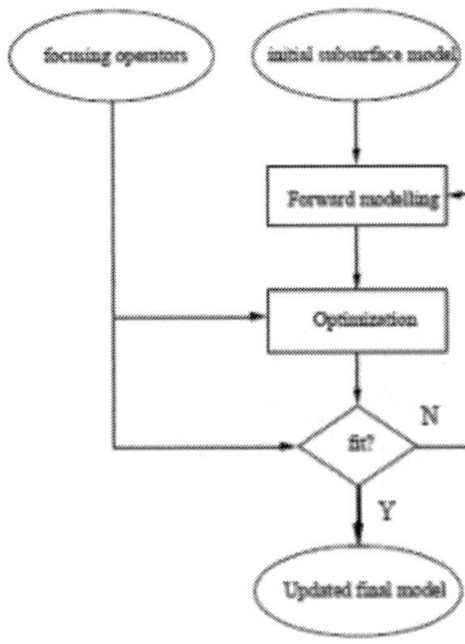

Figure 2. Inversion of focusing operators by means of travel time tomography [3]

Seismic wave propagation length can be calculated by equation (2)

$$P_i = \sum_{j=1}^{N} M_j S_j \qquad (2)$$

with P_i is seismic wave travel time matrix, M_j is seismic wave propagation length in each cell matrix , and S_j is slowness in each cell matrix. The cell that not passed by seismic wave will has zero seismic wave propagation length's value.
The example of the ray tracing in 3D can be seen in **Figure 3 :**

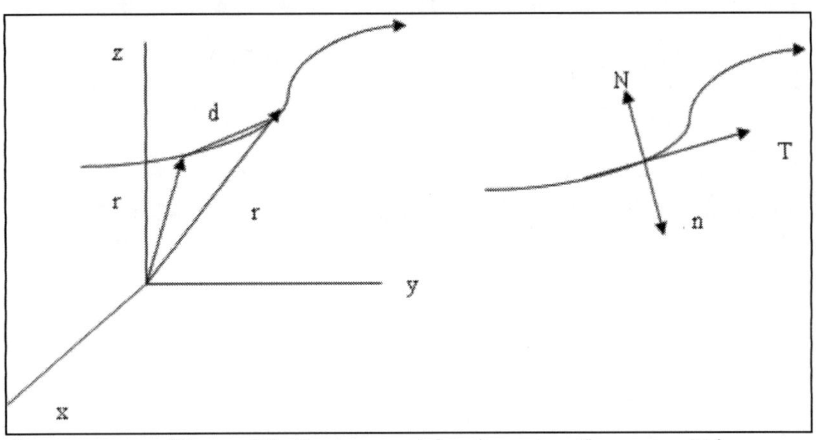

Figure 3. Tracks wave rays. Vector T is the tangential unit vector, the vector N is a normal vector [5]

Normal vector in the direction of the wave always leads to the center of curvature . Both vectors , vector T and N can be expressed in the following equation [5]:

$$\frac{dr}{|dr|} = \frac{dr}{ds} = T \quad (3)$$

$$\frac{d^2r}{ds^2} = N \quad (4)$$

3. Results and Discussion

The Result of this research can be seen in figures 4.

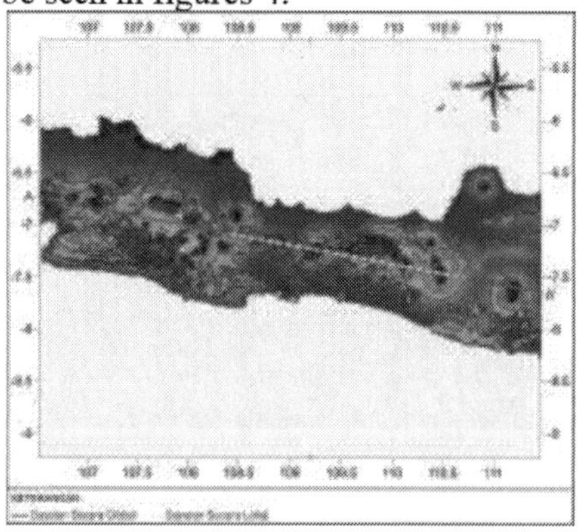

Figure 4. Location Map slicing

To perform a more detailed observations of the study area is observed vertically by making two slicing map in the Slamet mountain. The incision in the form of A - A 'with X = 106.503707 and Y = -6.751412 until X = 111.697802 and Y = -7.650037. These slicing represent the research area and its surroundings. The incision A-A 'cut subduction zones and parallel to the line of volcanoes in Java,

A. The Local (Central Java Area Mt. Slamet)

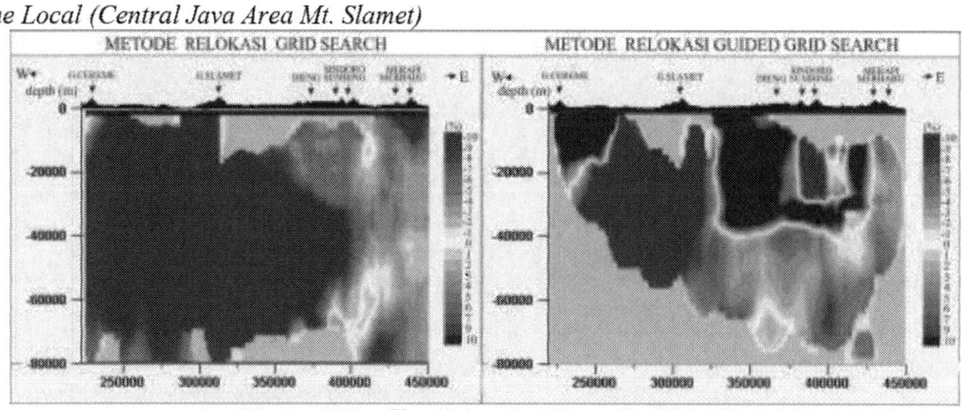

Figure 5. Profile slicing A-A' Locally (Mt. Slamet)

Tomographic inversion results using pseudo bending in slice A-A' describes the state of the subsurface based on the velocity of the P wave,figure 5 shown anomaly with a low velocity at the bottom of each subsurface mount on the relocation of the grid search method. This is caused by found fluid under every mountain pass slice A-A'. However, the value of the P wave velocity in the eastern part of the

Java island was a little high as a state under the surface of the Dieng plateau, Sumbing, Sindoro, Merapi, Merbabu, and Lawu mountain with a standard deviation values ranging from -1% to - 8% compared to western part of the island of Java as Slamet and Cereme mountain with a standard deviation values ranged from -8% to -10 km / s. But there are a differences in guided grid search method, in this method, there is a high velocity anomaly with values standard deviation ranged from 7% to 10% in subsurface Cereme mountain and Dieng plateau , it shows that the area is composed by rocks more compact.

Merapi, Merbabu, and Lawu mountain with a standard deviation values ranging from -1% to - 8% compared to western part of the island of Java as Mount Slamet and mountain Cereme with a standard deviation values ranged from -8% to -10 km / s. But there is a difference in guided grid search method, in this method, there is a high velocity anomaly with values standard deviation ranged from 7% to 10% below the mountain surface cereme and plato Dieng, it shows that the area is composed by rocks more compact.

B. Globally (the entire java island)

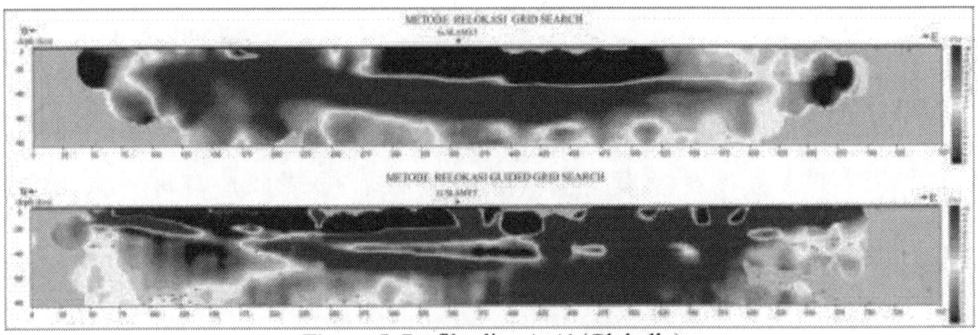

Figure 5. Profile slice A-A' (Globally)

When viewed globally at the cross slice A-A 'grid search method are be found high velocity values at depths of 0-20 km with a standard deviation of 8% -10% rate, while more than 20 km there is a low velocity with value of standard deviation -8% to -10%, the region has a low velocity because it is an area of subduction of the Indo-Australian plate with the Eurasian, so the rocks making up the area in the form of molten rock that melted into magma due to strong impact.

But if we look at the cross slice A-A' guided grid search method there are similarities with the method of grid search pattern, but in this method, the results have a more varied velocity. In the method of guided grid search Low-velocity are values with a pattern of spots on the grid search method, while high-velocity search entirely.

4. Conclusion

The conclusion of this research is result from the comparison of the two methods relocation grid search with the guided search grid as follows:

1. Locally cross-section grid search has a dominant low velocity (-8% to -10%), while the guided grid search method has a low velocity anomaly (-8% to -10%) and high (7% to 10%), geological research area is a cluster of volcanoes, so that the resulting velocity should predominantly low velocity in accordance with the grid search method. This indicated that the relocation of the grid search method is better used in areas that have a narrow area (local).

2. Global cross-section grid search method does not have good velocity variation, while at the cross-guided method varisasi grid search has good velocity result, there is a low velocity anomaly (-8% to -10%) in areas which are volcanic while the grid method search does not show any anomalies. This proves that the guided search grid method is better used for a broad area of research area.

5. References

6.

[1] Andri DN, Ahmad S 2012 **JTM** *XIX* No. 2/2012

[2] Flanagan, M. P., S. C. Myers, and K. D. Koper 2007 *Bull. Seism. Soc. Am.* **97**: 804–825

[3] Indriati RP, Yanti RP, Hidayat W, Santoso A 2015 *Indonesian Conference on Volcanology Mineralogy and Geothermal* Proceeding. ISSN : 2476-9185

[4] Koketsu K and Sekine S 1998 *J. Int.* **132**: 339–46

[5] Nataf HC, and Y. Ricard 2004. 3SMAC *Phys. Earth Planet. Int.* **95**: 101–122

[6] Peter S, Egil H, and Guoqing L 2005 *Bulletin of the Seismological Society of America* **95**(3) pp. 904–915 doi: 10.1785/0120040168

[7] Lippitsch R, Robert S. White, and Soosalu H 2004 *Geophys. J. Int.* **160**: 370-387

[8] Shelly, D. R., G. C. Beroza, and S. Ide 2007 *Nature*, **446**, 305–307

[9] Shelly, D. R., G. C. Beroza, and S. Ide 2007b *Geophys. Geosyst,* 8 Q10014, doi:10.1029/2007GC001640

[10]Um J. and C. H. Thurber 1987 *Soc. Am.* **77**: 972–986

Poisson's ratio analysis (Vp/Vs) on volcanoes and geothermal potential areas in Central Java using tomography travel time method of grid search relocation hypocenter

W. Raharjo[1], I. R. Palupi[1], S. W. Nurdian[1], W. S. Giamboro[1], J. Soesilo[2]
[1]Geophysical Engineering, Universitas Pembangunan Nasional "Veteran" Yogyakarta
[2]Geological Engineering, Universitas Pembangunan Nasional "Veteran" Yogyakarta

Abstract. Poisson's Ratio illustrates the elasticity properties of a rock. The value is affected by the ratio between the value of P and S wave velocity, where the high value ratio associated with partial melting while the low associated with gas saturated rock.
Java which has many volcanoes as a result of the collision between the Australian and Eurasian plates also effects of earthquakes that result the P and S wave. By tomography techniques the distribution of the value of Poisson's ratio can be known. Western Java was dominated by high Poisson's Ratio until Mount Slamet and Dieng in Central Java, while the eastern part of Java is dominated by low Poisson's Ratio. The difference of Poisson's Ratio is located in Central Java that is also supported by the difference characteristic of hot water manifestation in geothermal potential area in the west and east of Central Java Province. Poisson's ratio value is also lower with increasing depth proving that the cold oceanic plate entrance under the continental plate.

1. Introduction

Poisson's Ratio is a change in the geometry of an object depends on the amount of lateral strain and axial strain. The magnitude of Poisson's Ratio also depends on the P waves and S waves velocity [1]. The characteristic of P wave is the direction of propagation is in the same way with the direction of the movement and it is compressed, while the S wave has a characteristic propagation direction perpendicular to the direction of movement and it is the shear wave, so Poisson's Ratio is the ratio between the amount of strain due to compression force against the shear force.
Poisson's Ration can calculated by Wadati diagram, where the arrival time of P wave as the Y axis and the difference between arrival time of S and P wave as the X axis. From the gradient of the linear regression is a Vp/Vs or Poisson's Ratio. Poisson's Ratio value indicates the elasticity properties of rocks at the points (force caused by the earthquake) in the rocks. In regions with geothermal manifestations, Poisson's Ratio value indicates a high degree of saturation of high water, because the presence of water will increase the value of Poisson's Ratio [2].
The existence of high temperature zones beneath the surface gives the variation of the P wave velocity (Vp), S wave velocity (Vs) and the Ratio of both of them (Vp / Vs), which also illustrates the value of Poisson's ratio. In the gas saturated rock, impairment of Vp is greater than Vs so that Vp / Vs or Poisson's Ratio has little value [3] whereas in rocks associated with partial melting, impairment Vs greater than Vp so that the value of Vp / Vs or Poisson's ratio is higher [4].

The purpose of this research is to describe the distribution of Poisson's Ratio value and its relationship with geothermal potential in Central Java as Mount Slamet as the center point using a tomography technique.

A. Geothermal Potential In Central Java

The southern part of Central Java is one of the areas in Indonesia which has a geothermal potential. This is shown by the discovery of geothermal manifestations such as hot water in Cilacap, Kebumen and Wonosobo precisely on the area Panulisan, Cipari, Wadas Malang and Wadas Lintang (Figure 1). Java had a series of volcanic Oligocene-Middle Miocene and Pliocene-Quaternary. Volcanic formation series is related to the subduction of the Hindia-Australian Ocean Plate under Java at the end of Paloegen. Based on the geological map of Kebumen, Majenang, Pangandaran and Banyumas [5] rocks in the area consist of sedimentary rock, intrusion rock, rock as a resulted from the tectonic activities and sediment in the surface ranging from Pre Tertiary to Quaternary.

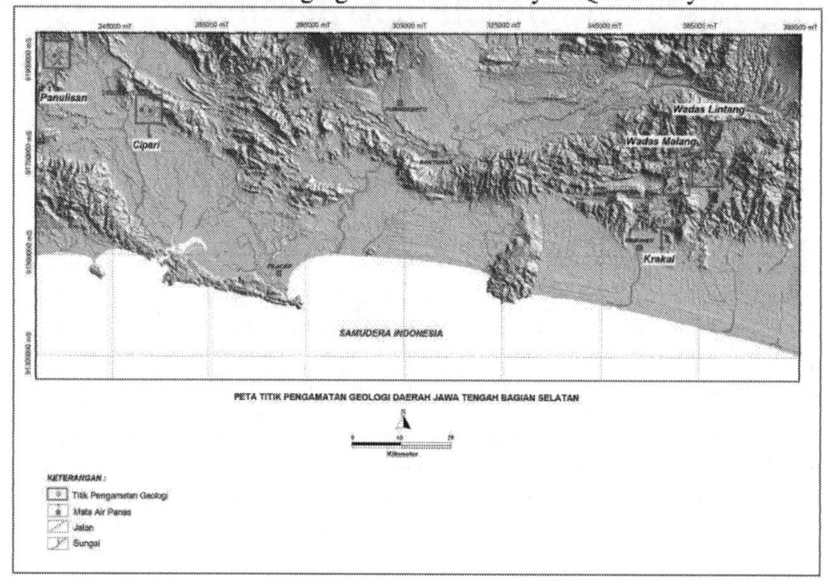

Figure 1. The distribution map of the geothermal manifestation and research area [5]

There are volcanism in the southern part of Central Java-as old Tertiary andesite formation [5]. In Cipari and Krakal, sediment deposition of sandstone and breccia unit are found that has characteristics of volcanic material. At the end of Miocene Pliocene, volcanic activity shifted to the north so that the arc magma is in the north of tertiary magmatism track Oligocene-Middle Miocene period.

Tectonic processes of Java on Plio-Pleistocene led to the appointment of rock deformation at the same area of research in the form of folding structures and faults. The pattern of the old structure back up and cause pattern old structure on younger lithology. Faults formed may contribute to the manifestation of hot water in the area.

In general manifestations of hot water found in Cipari, Panulisan, Wadas Lintang and Wadas Malang have temperature about 36°C-50.8°C. Hot Fluid coming from the hot water in the manifestation area of Cipari, Krakal and Panulisan that have high Cl ion concentration as the chloride water type, while the hot water in the area Wadas Lintang and Wadas Malang is a water type with a predominance of bicarbonate ions HCO_3. Bicarbonate water type is suspected association with hot fluids rising gas containing CO_2 and condenses in the shallow aquifer.

It can be concluded that the establishment of a geothermal system in the Cipari, Krakal, Wadas Lintang and Wadas Malang associated with sedimentary basins. [6] believes that the formation of geothermal systems in sedimentary basins is the result of the circulation of the meteoric water along the fault or fracture zones in areas that have a high heat flow. Meanwhile, there is a breakthrough in the Panulisan form igneous lava dome and Pliocene age. It gives hope about the geothermal manifestation. The existence of tectonic activity in Plio-Pliocene with the fault formation can act as a media release of hot fluid to the surface [5].

2. Methodology

The data used in this research is data earthquake data around Java that are downloaded from the USGS website since 1952 to 2002 with table AK-135 velocity models. The seismic data is relocated using Grid Search to get the value of the travel time calculation of the P and S wave. Then, using an early model of the table AK-135, the relocation of hypocenter inverted using tomographic techniques.

Before the tomographic inversion, performed model parameterize is needed by divided areas into blocks of a given size [7] as shown in Figure 2.

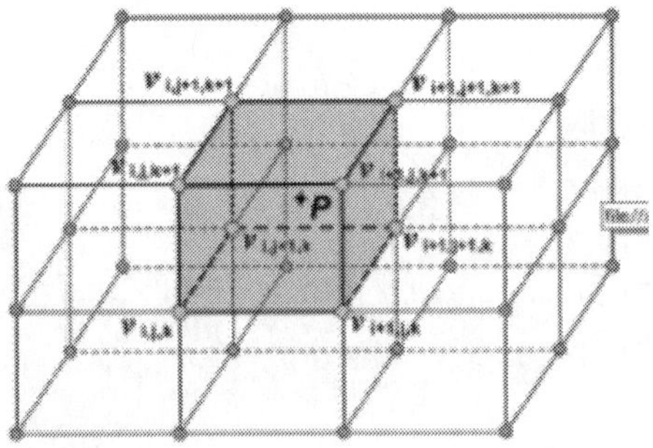

Figure 2. Model Parameterization [7]

In Figure 2 a model parameterization is done by dividing the area into blocks of the same size as the model initial of velocity (v), the point P in Figure 2 is assumed hypocenter point lies in the particular block. By using the Grid Search point P relocated to then be inverted and the resulting tomogram Vp / Vs.

Tomography technique that used can be calculate with equation (1)

$$T = \int_{source}^{receiver} \frac{1}{V} dl \tag{1}$$

with T is the arrival time of P and S waves at the seismic station, V is the velocity model and dl is the length of waves onto the blocks. Source and receiver in equation (1) is the hypocenter and the seismic station [7].

To get a model tomogram, the equation (1) is converted by first changing the equation (1) becomes:

$$T = \sum S_f dL_f \tag{2}$$

With S_f is the value of slowness (1 / V) for the block to f and dL_f is the length of the track on the block to f. Equation (2) is then inverted by first changing the equation (2) becomes:

$$d = Gm \tag{3}$$

With a data d is the arrival time of observation, G is a function forward modeling and m are the parameters of the model that will be sought in this case is V. Equation (3) then inverted is by using the following equation [8]:

$$m = (G^T G)^{-1} G^T d \tag{4}$$

3. Discussion

Sectional slices and tomographic inversion results Vp / Vs shown in Figure 3, 4 and 5 below:

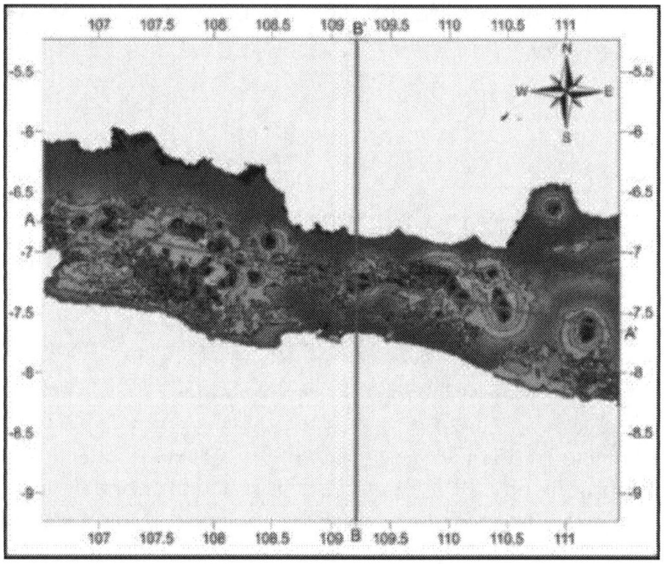

Figure 3. Sectional section of A-A' and B-B'

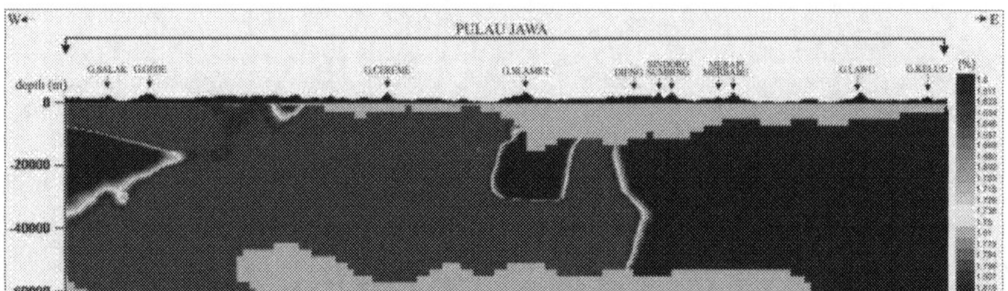

Figure 4. A-A' Tomogram

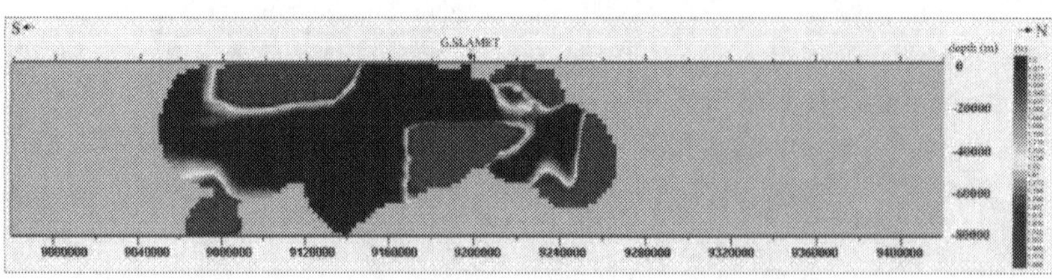

Figure 5. B-B' Tomogram

Figure 4 and 5 show that the high Poisson's Ratio is represented by red and the value of the low by blue color. From Figure 4 it is known that the value high Poisson's Ratio is in West Java until Mount Slamet which are some areas in Central Java, then from Dieng to the east is dominated by the low Poisson's Ratio. This means that in some areas of West Java and Central Java rocks is dominated by partial melting. Whereas in Figure 5 Poisson's ratio is lower dominates under of Mount Slamet associated with partial melting is low, that affect the pattern of the eruption of Mount Slamet is not periodic as Mount Merapi.

Tomogram under Central Java can be seen in Figure 6

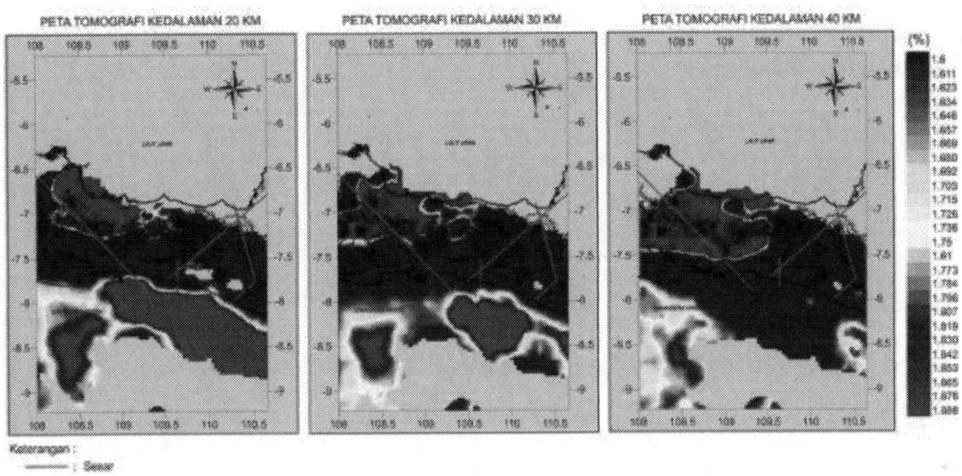

Figure 6. Tomogram under the Java

Figure 6 shows that the high Vp / Vs dominates at a depth of 20 km, and continued to decline up to a depth of 40 km. Lower Vp / Vs is increasing with the depth indicates oceanic plate cooler is cooler and entrance under the continental plate, because the high of Vp / Vs is associated with partial melting that has a high temperature.

4. Conclusion

Variations in the Vp / Vs or Poisson's Ratio can be viewed by using the technique of tomography, where the high Vp/Vs associated with partial melting while conversely associated with gas saturated rock [4]. In this research, it appears that the high Vp / Vs dominates in western of Java until Mount Slamet and Dieng, and to the east is dominated by low Vp / Vs. While the value of Vp / Vs also continued to decline for the increasing depth of proving that the oceanic plate is cooler and entrance under the continental plate. In addition, the value of Vp / Vs linked to areas that would potentially geothermal manifestations, as seen in cross-section tomography there are differences in the value of Vp / Vs in Central Java indicates there are different manifestations characteristic of geothermal areas.

5. References

[1] Sudarmoyo, Herianto, and R. Bobby 2001 *Harsawan Proceeding Simposium Nasional IATMI*
[2] Silitonga T. H., Siahaan E. E., Suroso, 2005 *Proceedings World Geothermal Congress 2005*, Antalya, Turkey, 24-29 April 2005
[3] Bariş, Ş., Nakajima, J., Hasegawa, A., Honkura, Y., Ito, A., Üçer, S. B., 2005 *Earth Planets Space* **57**, 1019-1038.
[4] Takei, Y. 2002 *J. Geophys. Res.*, **107**(B2), 2043, doi:10.1029/ 2001JB000522.
[5] Asikin, S., Handoyo, A., Busono, H., and Gafoer,S 1992 *Geological Research &Development Centre*, Bandung, Indonesia, 24 p.
[6] Anderson, D.N. and J.W. Lund, (editors), 1979 *GRC Special Technical Rept. 7*. Davis, CA
[7] Xu, Thao, Fei Li, Zhenbo Wu, Chenglong Wu, Ergen Gao, Bing Zhou, Zhongjie Zhang, and Guoming Xu *Tectonophysics* ***627*** 72–81
[8] Tarantola A 1987 *Inverse Problem Theory, Methods for Data Fitting and Model Parameter Estimation* (Amsterdam: Elsevier)

Gravity Data Analysis and Modelling for Basin Sedimen of Eastern Java Blocks

L. Khoirunnisa
Luthfia Khoirunnia
Physics Department of Post Graduate Program Gadjah Mada University, Sekip Utara Bulaksumur, Yogyakarta 5528, INDONESIA

E-mail: luthfia.khoirunnisa@mail.ugm.ac.id

Abstract. The study of Eastern Java Basin was conducted by 3D modelling subsurface structure using gravity anomaly. The aims of this research are to describe and 3D modelling basin sedimentary system of Eastern Java Blocks based on gravity anomaly. The modelling construction was performed by inversion technique applying Singular Value Decomposition (SVD) method and Occam optimization. This projection method used equivalent central mass of Dampney with height 5.5 km and error data 1,84 x 10^{-17}. Separation of residual anomaly from the complete Bouguer anomaly on a flat plane was done using the upward continuation. This process uses the principle of low pass filter which passes low frequency. Sedimentary basin appears at a depth of 0.2 km to 1.4 km, is shown by their low anomaly in the area, as well as the visible appearance of basin in 3D modeling shown in figure. The result of inversion with Occam h has an error of 1,2 % and the SVD has an error of 11%. Sedimentary basin was dominant in Probolinggo, partially in Besuki and Lumajang. The formation occurs due to tectonic processes where the tectonic evolution of the material without significant lateral shift is called as the otokton models, and accompanied by the formation of the basin that follows the development of the subduction system, which is semi-concentric pattern. Sediments are dominated by volcanic sediment, the result of sedimentation because of volcanism events and types of volcanic sediments pyroclasts generally occur in a process or event explosive volcanic magma degassing

1. Introduction

Java is one of Indonesian Island part of volcanic arc along Sumatra to Nusa Tenggara [1]. Indonesia has 128 sedimentary basins, base on their age can be divided into Tertiary basin, pre-Tertiary basin, and developing basin since the pre-Tertiary to Tertiary [2]. Eastern Java Basin has a complex structure and stratigraphy of back arc basin of Indonesia. Eastern Java Basin is favorite area in Indonesia for the exploration of petroleum and make this area into a "hot spot" on the exploration [3]. East Java has a thickness of tertiary sedimentary basins around 6 km [4]. Gravity anomaly from 0 mGal to 20 mGal form into sedimentary system from host rock [5]. Gravity method is used to describe the subsurface geological structure by variations in Earth's gravitational field caused by contrast density between rocks. Gravity method can also be used to identify sedimentary basin which is indicated by low Bouguer anomaly. Basin is generally indicated by the appearance of deposits of sediment trapped on it. Surveys of hydrocarbons in Eastern Java are around the edge of the basin, but the survey of deeper basin has not significantly done. The area around Malang has the potential basin for example Probolinggo [6], which indicated by low Bouguer anomaly. Therefore, the gravity method can be used to modelling sedimentary basins in Eastern Java, especially block area of Probolinggo, Besuki, Jember

Journal of Physics: Conference Series **776** (2016) 012115 doi:10.1088/1742-6596/776/1/012115

and Lumajang and to predict that prospects for hydrocarbon exploration and gas. The aims of this research are to describe and 3D modelling basin sedimentary system of Eastern Java Blocks based on gravity anomaly. The research area is described in figure 1.

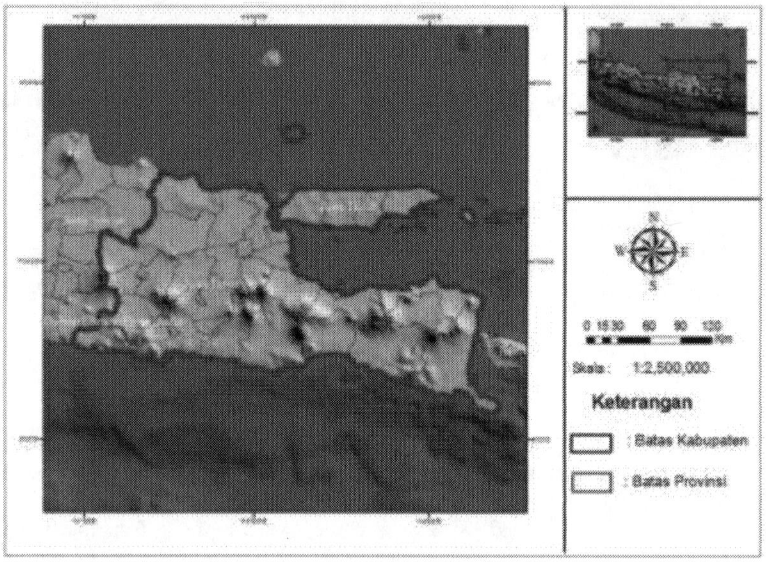

Figure 1 Map of the research area.

2. Gravity Methods

Gravity method is used to describe the subsurface geological structure by variations in Earth's gravitational field caused by contrast density between rocks. Gravity method can also be used to identify sedimentary basin which is indicated by low Bouguer anomaly. Complete Bouguer anomaly data that is still on topography need to project on a flat surface. Its principle is carried variation of heights into the slab mass with the same height to computed upward continuation. This projection method used equivalent central mass of Dampney [7] with height 5.5 km and a data error 1.84×10^{-17}. Separation of residual anomaly from the complete Bouguer anomaly on a flat plane was done using the upward continuation. This process uses the principle of low pass filter which passes low frequency with height 8 km. Modelling basin sedimentary is using residual anomaly. A residual gravity anomaly caused by the presence of objects which are close to the surface. Residual anomalies obtained by subtracting the complete Bouguer anomaly with regional anomalies. Modelling using forward modelling and inverse modelling. Forward Modelling is used to predict the simulated data based on hypothetical subsurface conditions. The simulation data are usually referred to the data or theoretical predictions data. The inversion method uses singular value decomposition (SVD) and occam's principle. Inversion by Singular value decomposition are inversion base, density and heights, inversion by occam's principle are occam density and heights.

3. Results and Discussion

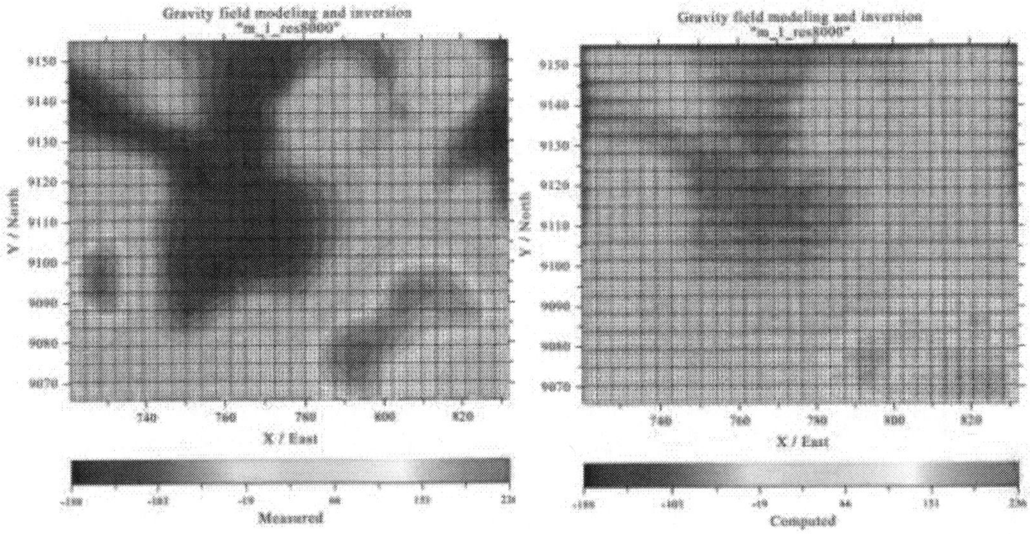

Figure 2 Density optimization of residual anomaly data.

Comparison between measured and computed of density Optimization showed that their pattern was almost same. Density Optimization was meant to get fit data between the model and the density value of measurement, density inversion results of the optimization is shown in Figure 1. Contour map obtained shows an error data of 13%. To obtain a better model with the smallest error density inversion is maximized by using Occam density (Occam d). The results of the optimization process using Occam d shows an error data of 7.6%.

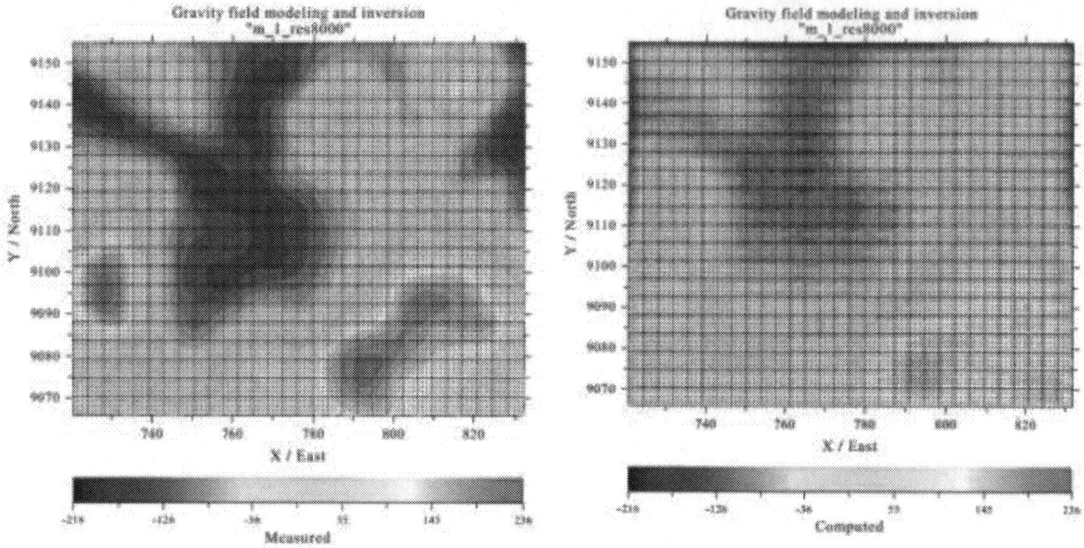

Figure 3 shows the result of occam heights optimization of residual anomaly data. Comparison between measured and computed of occam heights Optimization show that their pattern is almost same. Optimization height of the block (height) is intended to get the size of the minor blocks with a maximum height of each layer, where the height of the block illustrates the depth and thickness of each layer. Contour map obtained shows an error data of 11%. To obtain A better model with the smallest error heights inversion is maximized by using Occam heights (Occam h). The results of the optimization process using Occam h shows in figure 2 with an error data of 1.2%.

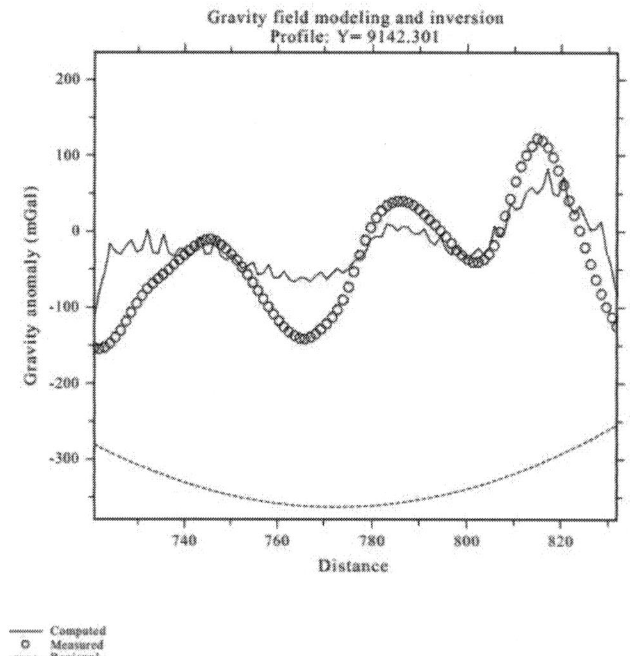

Figure 4 show a cross section at y = 9142.301. Comparison between measured and computed of cross section at profile y = 9142.301 show that their pattern is almost same. It's mean that the proses data of inversion model can be used as model for modelling basin sedimentation.

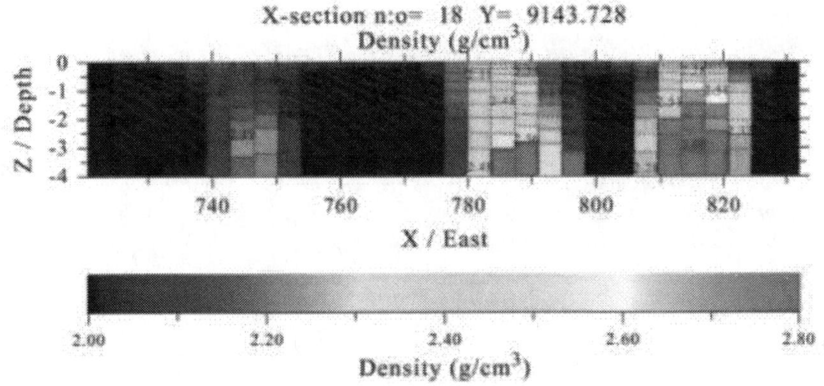

Figure 5 shows a mass with high density ~ 2.8 exactly below Raung's Mountain. We assume that it is intrusion of magma. The position of magma intrusion is located at the longitude 810-820 km. This is make a sense with the information on the data released by the Centre for Geological Research and Development that there is a layer of volcanic breccia which have an insert with andesite lava. The depth of magma intrusion is estimated between 2-8 km. Sedimentary basins shown by their low anomaly that is spread on three areas, at the longitude 720-740 km, 756-776 km and 830-835 km. Sedimentary basin dominant in Probolinggo, partially in Besuki .

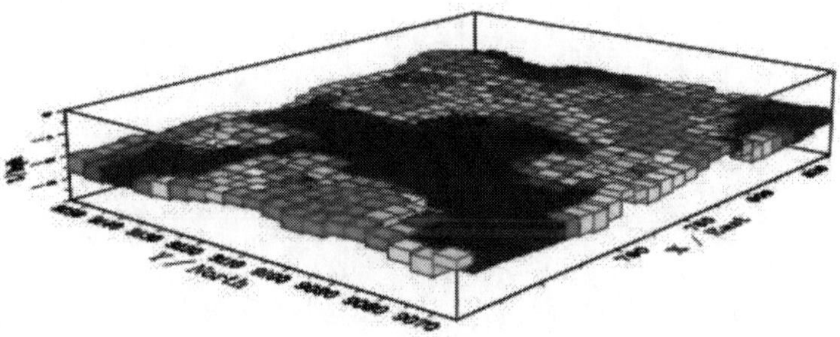

Figure 6 shows 3D modelling sedimentary basin of Eastern Java Blocks with deep at 1.9 km. sedimentary basin is shown by their low anomaly. We modelling basin sedientary system at each layer from 0 km to 3.1 km. The initial model is 30 block mayor toward x axis, 20 toward y axis, and 10 toward z axis at deep model 4 km from surface.

In conclusion, sedimentary basin appear at a depth of 0.2 km to 3.1 km, is shown by their low anomaly. Sedimentary basin is dominant in Probolinggo, partially in Besuki. The formation occurs due to tectonic processes where the tectonic evolution of the material without significant lateral shift is called the otokton models, and accompanied by the formation of the basin that follows the development of the subduction system, which is semi-concentric pattern [8]. Sediments are dominated by volcanic sediment, the result of sedimentation because volcanism events and types of volcanic sediments pyroclasts generally occurs in a process or event explosive volcanic magma degassing [9].

Journal of Physics: Conference Series **776** (2016) 012115 doi:10.1088/1742-6596/776/1/012115

4. References

[1] Hamilton 1979 *Tectonics of the Indonesian Region* USGS Professional Paper 1078
[2] Badan Geologi 2009 *Pusat Survei Geologi, Badan Geologi* Bandung
[3] Satyana A H 2005 *Indonesian Association of Geologists 34th Annual Convention*
[4] Hasan 2008 *Jurnal Berkala Fisika* **11** 37
[5] Panjaitan S 2010 *Buletin Sumber Daya Geologi* **5** 168
[6] Padmawidjaja T 2013 *Jurnal Lingkungan dan Bencana Geologi* **4** 237
[7] Dampney CNG 1969 *Geophysics* **34** 1
[8] Bachri S 2013 *Jurnal Sumber Daya Geologi* 14 19
[9] Einsele, G 1992 *Sedimentary Basin* Springer-Verlag Germany

Calibrating accelerometer sensor on android phone with Accelerograph TDL 303 QS for earthquake online recorder

R. Riantana[1] , D. Darsono[1], A. Triyono[2] and H.B. Azimut[1]
Physics Department of Sebelas Maret University[1],
Geophysics Station of BMKG Yogyakarta, Indonesia[2]

E-mail: rioriantana@yahoo.com

Abstract. Calibration of the android censor was done by placing the device in a mounting at side of accelerograph TDL 303 QS that will be a means of comparison. Leveling of both devices was set same, so that the state of the device can be assumed same anyway. Then applied vibrations in order to have the maximum amplitude value of both censor, so it can be found equality of the coefficient of proportionality both of them. The results on both devices obtain the Peak Ground Acceleration (PGA) as follows, on the x axis (EW) android censor is obtained PGA -2.4478145 gal than at TDL 303 QS obtained PGA -2.5504 gal, the y-axis (NS) on the censor android obtained PGA 3.0066964 gal than at TDL 303 QS obtained PGA 3.2073 gal, the z-axis (UD) on the android censor obtained PGA -14.0702377 gal than at TDL 303 QS obtained PGA -13.2927 gal , A correction value for android accelerometer censor is ± 0.1 gal for the x-axis (EW), ± 0.2 gal for the y-axis (NS), and ± 0.7 gal for the z-axis (UD).

1. Introduction

Vibrations around us may be caused by random sources resulting directly from human activity as well as natural causes, independent from humans (earth-quakes) [1,2]. Vibrations are usually measured on the ground's surface, at specific measuring points, using measuring instruments designed specifically for that purpose, called accelerometers, and the measurements are then, via analog to digital converters and data acquisition cards, registered on a computer disc [3,4].

D'Emilia [5] have done evaluating of aspects affecting measurement of three-axis accelerometers. The result is comparing between two three-axis accelerometers, aiming to highlight the effects of simple solutions for conditioning electronics, interfacing and Analog-Digital Conversion, on the metrological performance of censor itself. A comparison between two solutions for test rig, as for the positioning of the motion with respect to the vertical direction, the motion law and the reference definition [5].

Riantana [6] have done creating Accelerometer Censor Application On Android Mobile as Earthquake Online Recorder using a web server to collect data from the vibration readings in the field by android phone in real time. So that the vibration data in all android phones can be stored by structured and can be visualized using a monitoring graph on web. The system has not been tested and calibrated with standard devices used by BMKG to monitoring the activity of the soil surface vibration. If the system can be tested and calibrated it is estimated the system would save costs up to 99% of the procurement accelerograph TDL.303 QS. So the cost is only 1% of the previous cost. Cost to procurement one

Accelerograph TDL 303 QS is about 25.000 USD, while the cost to procurement this system is just about 250 USD for one point station. Surely this would be very beneficial if the system can be used as an alternative to the existing system [6].

2. Methods

We have calibrated android phone by using Accelerograph TDL 303 QS standart equipment for earthquake meansurement. We did it at Meteorology Climatology and Geophysics Council of Yogyakarta city, Indonesia. Figure.1 shows the flowchart of calibration. Calibration is done by placing both of censor side by side to same level. Then applied vibrations in order to have the maximum amplitude value of both censor, so it can be found equality of the coefficient of proportionality both of them.

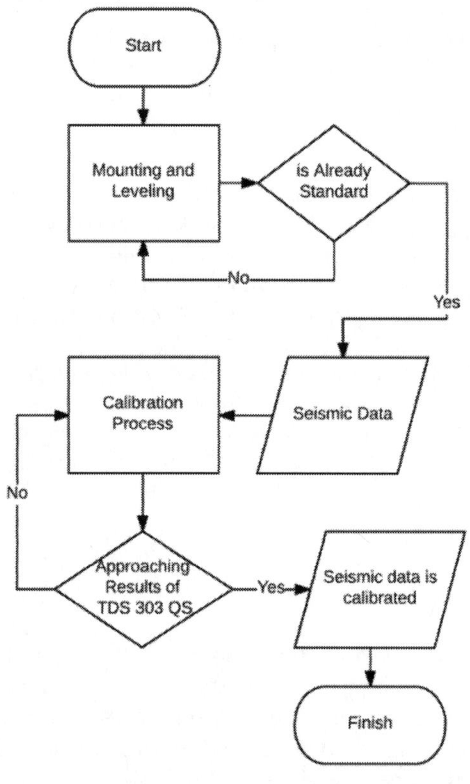

Figure 1. Calibration Flowchart

The data collection is done per session by giving a trigger vibrations from the outside. The obtained data were analyzed the balance on a scale of 0 to chart. The result of recording accelerometer android is already on the balance of 0 to further determine the calibration value and the value of Low and High pass Filter. This calibration value will adjust the value of the accelerometer android and accelerograph TDL 303 QS. The wave function filter is used in order to minimize the wave interference, therefore will reducing noise which influences the results.

3. Results and Discussion

In the data processing, used High and Low pass Filter method on android accelerometer. The wave function filter is to get the value of original wave and minimize noise vibration that can affect the results. Calibration process for accelerometer censor on android phone Xiaomi redmi 2 with Accelerograph TDL.303 QS has done by placing the mounting of android phone beside to

Accelerograph TDL 303 QS with same level on both of devices. Figure.2 shows the results of leveling on android phone.

Figure 2. Leveling On Android Xiaomi Redmi 2

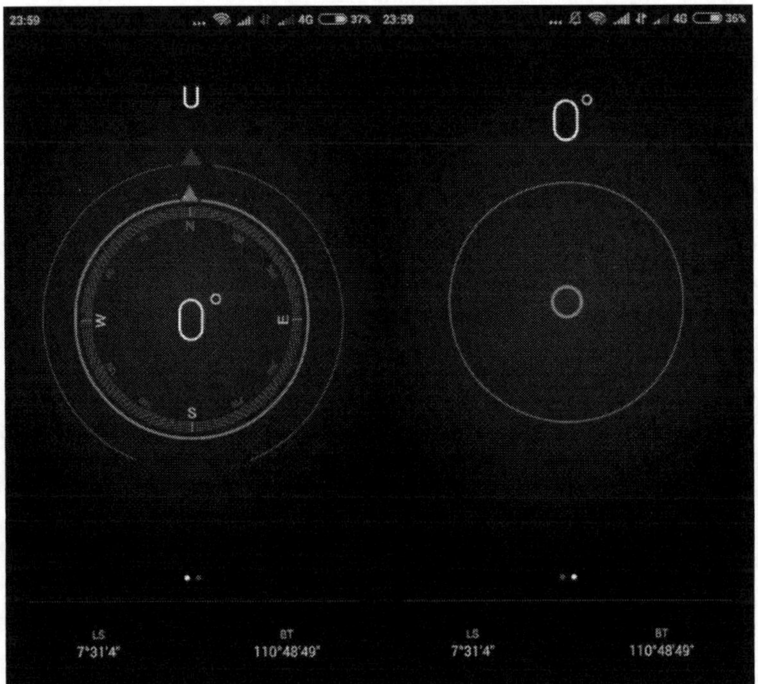

Basically, accelerometer censor on the android phone has units of acceleration m/s^2 and accelerograph has units of acceleration cm/s^2 (gal). In the theory to get the same results, android accelerometer unit is converted into cm/s^2 (gal), the value should be multiplied by 100. But in reality that is not linear. Table 1 shows comparison both of device before the accelerometer on android is calibrated. So we need a lot of data in order to get a lot of variety to processed and equalized furthermore to obtain an equation for each axis of the censor, so that the output obtained by the accelerometer android and accelerograph TDL 303 QS value is approach.

Table 1 Comparison of the maximum amplitude on android phone before calibration with accelerograph TDL 303 QS

Axis	Accelerometer Android (uncalibrated)	Accelerograph TDL 303 QS (gal)
x (EW)	-0.1682934	-2,5504
y (NS)	0.2412843	3,2073
z (UD)	- 0.72002377	-13,2927

After a few takes of data and application the equations to android phone, then obtained the equation for each axis of the accelerometer censor Xiaomi redmi 2 as follows:

$$X = (even\_value(x) - 0.02) \quad (1)$$

$$X = \begin{cases} X < -0.018 \rightarrow X * 13 \\ X > 0.018 \rightarrow X * 13 \end{cases} \quad (2)$$

$$Y = (even\_value(y) - 0.01) \quad (3)$$

$$Y = \begin{cases} Y < -0.018 \rightarrow Y * 13 \\ Y > 0.018 \rightarrow Y * 13 \end{cases} \quad (4)$$

$$Z = (even\_value(z) - 0.687) * 10 \quad (5)$$

The calibration equation above is only for Android devices Xiaomi redmi 2. For other types of Android device need different calibration equations. Figure 3 shows retrieval data test of android graph and TDL 303 QS graph before calibrated. Recording speed on android accelerometer in the server about 50 records per second, whereas the speed record at TDL 303 QS accelerograph is 100 records per second. This makes the results graph vibration formed on the recording accelerometer android is not the same record lenght than accelerograph TDL 303 QS. However, this android accelerometer censor can capture a maximum vibration approaches the maximum vibration from accelerograph TDL 303 QS. Table 2 Shows comparison of maximum amplitude of both devices after calibration. Figure 4 shows the comparison of android graph and TDL 303 QS graph after calibrated. In the last recorded data retrieval for the maximum amplitude of both devices are as follows:

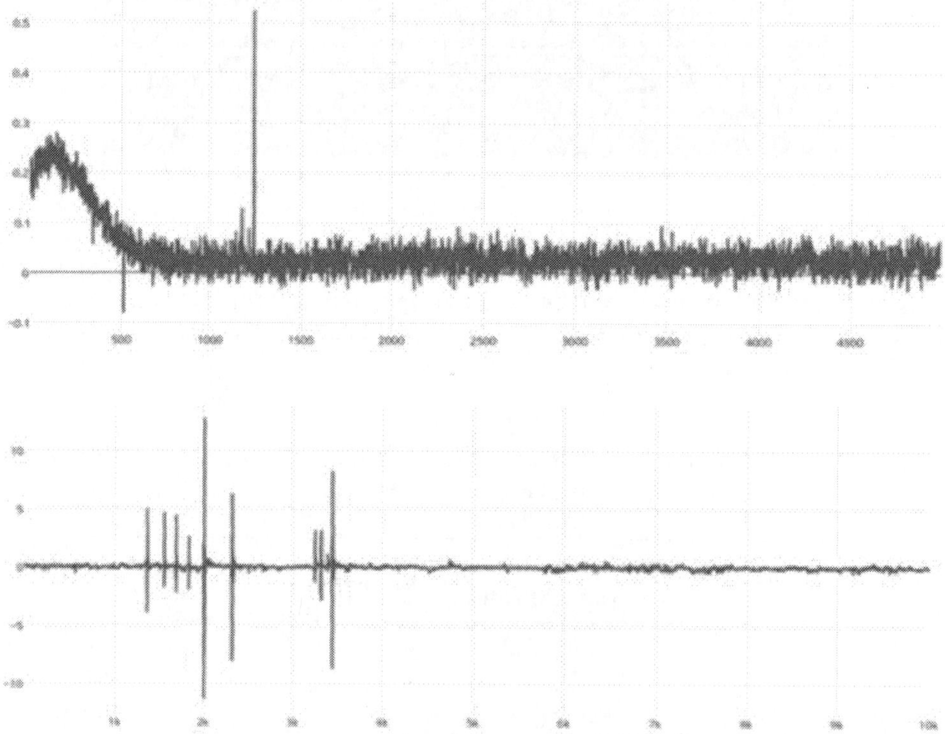

Figure 3 Test retrieval data of X axis on Android Accelerometer graph (top) and TDL 303 QS graph (bottom) before calibrated

Table 2 Comparison of the maximum amplitude on android phone and accelerograph TDL 303 QS after calibration

Axis	Accelerometer Android (gal)	Accelerograph TDL 303 QS (gal)	Correction (gal)
x (EW)	-2,4478145	-2,5504	± 0.1
y (NS)	3,0066964	3,2073	± 0.2
z (UD)	-14,0702377	-13,2927	± 0.7

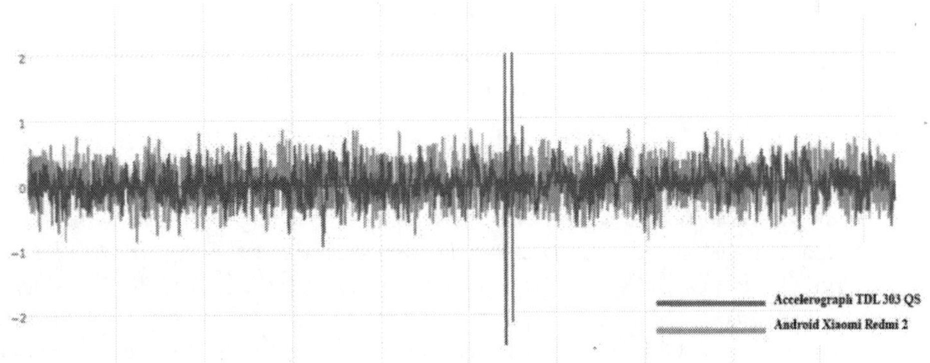

Figure 4 The comparison of X Axis on Android Accelerometer graph and TDL 303 QS graph after calibrated

4. Conclusion

Comparing process of the data, we have a calibration equation :

$$x = even\_value(x) - 0.02 \quad (6)$$
$$x = x * 13 \{x \mid x < -0.018 \text{ or } x > 0.018 \mid x \varepsilon \mathbb{R}\} \quad (7)$$
$$y = even\_value(y) - 0.01 \quad (8)$$
$$y = y * 13 \{y \mid y < -0.018 \text{ or } y > 0.018 \mid y \varepsilon \mathbb{R}\} \quad (9)$$
$$z = ((even\_value(z) - 0.687) * 10 \quad (10)$$

Correction value for android accelerometer censor is ± 0.1 gal for x-axis (EW), ± 0.2 gal for y-axis (NS), and ± 0.7 gal for z-axis (UD).

5. References

[1] Sasaki K, Iinuma H, Kimura N, Ogitsu T, Yamamoto A, Nakayama, Mibe T, Saito N, and Obata T, 2011 *IEEE Trans. Appl. Supercond.* **21** (3) pp. 1748–1751.
[2] Sun XT, Jing XJ, Xu J, and Cheng L, 2014 *IEEE Trans. Industr. Electron.* **10** pp. 5606–6114.
[3] Cattaneo PW, 2002 *Nucl. Instrum. Methods Phys. Res.* Section A **481** (1-3) pp. 632–636.
[4] Al-Naimi K, Villette S, Kondoz A, and Heikkinen AP. 2006 *Vision Image Signal Process* **153** (5) pp. 610–617.
[5] d'Emilia G, Gaspari A, and Natale E. 2016 *Measurement* **77** pp. 95–104.
[6] Riantana R. 2015 *Jurnal Fisika dan Aplikasinya* **11** No 3 pp. 114-119.

Journal of Physics: Conference Series 776 (2016) 012117 doi:10.1088/1742-6596/776/1/012117

Earthquake effect on volcano and the geological structure in central java using tomography travel time method and relocation hypocenter by grid search method

Suharsono, S. W. Nurdian, I. R. Palupi
Geophysical Engineering, Universitas Pembangunan Nasional "Veteran" Yogyakarta Jl. SWK 104 Condongcatur Yogyakarta

Abstract Relocating hypocenter is a way to improve the velocity model of the subsurface. One of the method is Grid Search. To perform the distribution of the velocity in subsurface by tomography method, it is used the result of relocating hypocenter to be a reference for subsurface analysis in volcanic and major structural patterns, such as in Central Java. The main data of this study is the earthquake data recorded from 1952 to 2012 with the P wave number is 9162, the number of events is 2426 were recorded by 30 stations located in the vicinity of Central Java. Grid search method has some advantages they are : it can relocate the hypocenter more accurate because this method is dividing space lattice model into blocks, and each grid block can only be occupied by one point hypocenter. Tomography technique is done by travel time data that has had relocated with inversion pseudo bending method. Grid search relocated method show that the hypocenter's depth is shallower than before and the direction is to the south, the hypocenter distribution is modeled into the subduction zone between the continent of Eurasia with the Indo-Australian with an average angle of 14 °. The tomography results show the low velocity value is contained under volcanoes with value of -8% to -10%, then the pattern of the main fault structure in Central Java can be description by the results of tomography at high velocity that is from 8% to 10% with the direction is northwest and northeast-southwest.

1. Introduction

Java island has a lot of volcanoes and there are many earthquakes , it is caused by the subduction of the Indo-Australian continent with the Eurasian continent to the south of Java. The calculation of earthquake hypocenter position is often shifted from the actual position, this problem needs to be solved by relocating the hypocenter, in this research by Grid Search method. To determine the subsurface conditions it is done based on the travel time tomography method to know the velocity distribution in the subsurface that is useful to analyze the patterns of volcanoes and major geological structures in the province of Central Java.

2. Basic theory

Grid Search Method is a systematic search techniques. This method is one way to obtain a solution inversion nonlinear using a global approach by evaluating the value of an objective function on each model space in a systematic way, so it is necessary to get "a priori" information about minimum and maximum limits interval on each parameter models , Then the interval is discretized in order to obtain a grid that covers the entire space model, where the size of the grid is not always homogeneous.

The advantage of this method is the ease way of systematically evaluating the objective function for each sample on a room model that involves forward modelling, so non-linear problem solution can be resolved without linierization. But sometimes because of space models, this method becomes ineffective because of the many calculation of forward modelling, especially if the problem requires discretizing large scope. However, for a relatively simple matter with little number of parameters and small discretization, this method is efficient enough to be applied [1]

Seismic tomography

Tomography method is one way to describe the condition of the subsurface, which in this research has the purpose to describe the condition of P wave velocity in the subsurface based on some information, they are : velocity models used in reserach area, the P wave travel time from hupocenter to the station and the coordinate location of both of them. Because it uses the travel time data, then it is called delay time technique.

This method uses the travel time (travel time) along trajectory of the hypocenter as a source to the station as a receiver, with the following equation:

$$T = \int_{sumber}^{penerima} \frac{1}{V} dl \qquad (1)$$

Where dl is the length of the segment and V is the velocity of the medium [2]

Wave propagation is calculated for each segment. To simplified the equation discretization technique is used, so that equation (1) can be written in the form :

$$P_i = \sum_{j=1}^{J} M_j S_{ij} \qquad (2)$$

Where S_{ij} is the i wave length that crosses all segments j, M_j is a function of the discrete model on the j segment where J is the total number of segments and P_i is the real data of the travel time. Equation (2) can be solved by using inversion techniques, so that the equation (2) can be rewritten by:

$$d = Gm \qquad (3)$$

Where d is real data, G is a forward modeling function and m is the model parameter to be searched, in this case the P wave velocity [3]

One way to obtain a solution of the parameters models are in the form of a matrix notation, where the solution by reducing the objective function of the model parameters, so the inversion technique to obtain the solution of the equation (3) can be written as follows:

$$m = (G^T G)^{-1} G^T d \qquad (4)$$

$G^T G$ matrix is a square matrix that is sized according to the number of model parameters [1].

3. Data and method

This research use earthquake catalog data that recorded from 1952 to 2012 with a 9162 P waves number, 2783 S waves number, and 2426 event were recorded by 30 stations located around Central Java. Then the process of relocating the hypocenter and tomography performed with the parameters of the AK-135 velocity model.

Table 1. 3D Speed table AK-135

Tabel Kecepatan 3D AK-135		
Depth (Km)	Vp (Km/s)	Vs (Km/s)
0	1.45	0
3	1.45	0
3	1.65	1
3.3	1.65	1
3.3	5.8	3.2
10	5.8	3.2
10	6.8	3.9
18	6.8	3.9
18	8.0355	4.4839
43	8.0379	4.4856
80	8.04	4.48
80	8.045	4.49
120	8.0505	4.5
120	8.0505	4.5
165	8.175	4.509
210	8.3007	4.5184
210	8.3007	4.5184
260	8.4822	4.6094
310	8.665	4.6964

In Grid search method model space is defined as some lattice block, the smaller lattice block will generate more grid blocks to meet the model space. Lattice size of this block is done by trial and error but does not exceed the size of the model space. In this method we are using forward modeling assumptions (forward modeling) where we define the initial model for approaching the actual model. To find out the initial model with the actual model, it can be calculated by the following calculation.

$$E = \sum_{i=1}^{N} \left(t_i^{cal} - t_i^{obs}\right)^2 \qquad (5)$$

To find out the error data obtained, the final result needs to be defined as the root mean square (ERMS) as follows:

$$E_{RMS} = \sqrt{\frac{1}{N}\sum_{i=1}^{N} \left(t_i^{cal} - t_i^{obs}\right)^2} \qquad (6)$$

4. Discussion

The results of the hypocenter relocation using Grid search Method :

The data processing is divided into 10 velocity layers model based on the table of AK135 velocity model, so that the process in the calculation of the relocation of the earthquake is divided into 10 stages based on the depth of each data.

Table 2. Results Relocation Grid search method

Kedalaman (km)	Kecepatan Vp (km/s)	Jumlah data	Grid (km)	Error (%)
0-3	1.45	127	2	3.421524
3.1-10	5.8	374	3	0.776882
10.1-18	6.8	474	5	0.504814
18.1-43	8.0379	2696	6	0.306194
43.1-80	8.045	3185	6	0.432285
80.1-120	8.0505	1411	8	0.424268
120.1-165	8.175	483	8	0.483521
165.1-210	8.3007	243	10	0.303551
210.1-260	8.4822	182	10	0.382312
260.1-310	8.665	167	12.5	0.177332

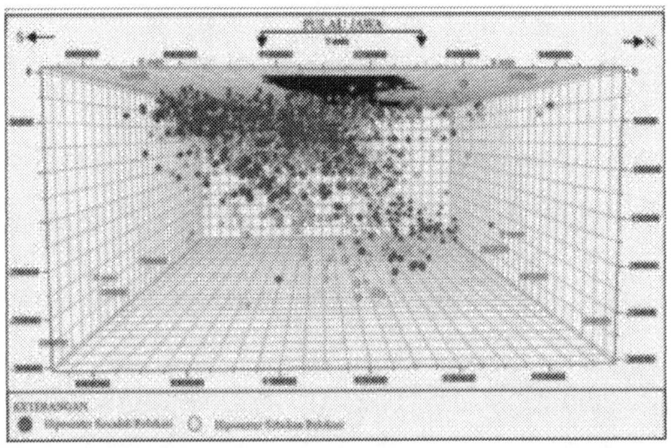

Figure 1. Hypocenter before and after relocating from the east side

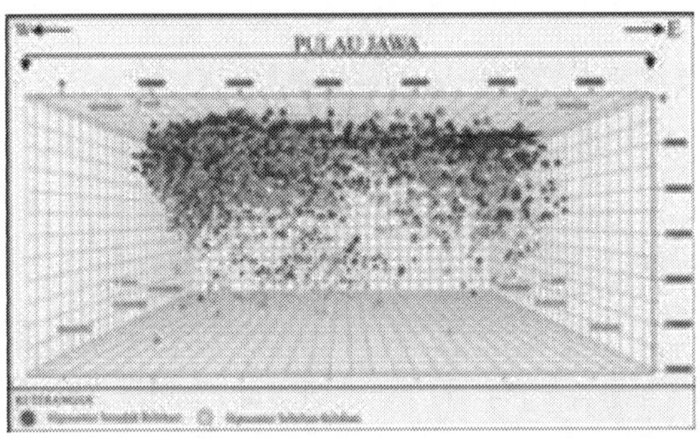

Figure 2. Hypocenter before and after the relocation of the south side

The results of the hypocenter relocation using Grid search method indicates a significant change of position, when viewed from the east side corresponding figure 1 and figure 2, there is a change between the hypocenter before and after relocation. The hypocenter after relocation has a depth shallower than before, a change in position is very varied depths of 2 - 12.5 km in accordance with the grid being used. Then, when viewed from the east side most of the hypocenter after relocation moves towards the south.

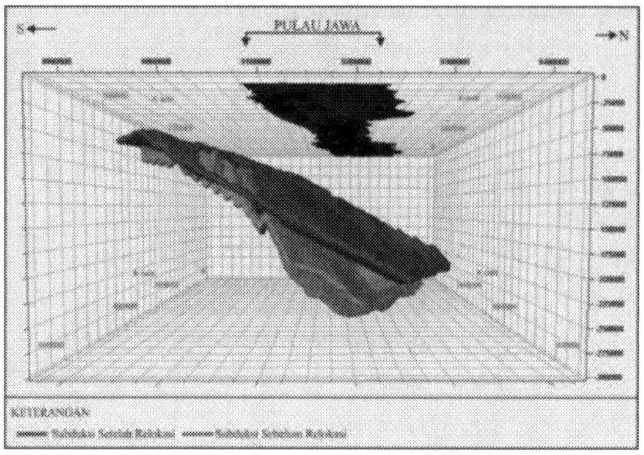

Figure 3. The 3D model Subduction Zone Before and After Relocation Of East Side

The hypocenter distribution can also be interpreted as subduction, because tectonic earthquakes generally occur around the subduction zone. From Figure 3, the trendline of subduction difference before and after relocation, the subduction angle after relocation formed between the Indo-Australian plate with the Eurasian plate at 10.28 ° while the angle before relocation is 14.51 °. It is very useful to determine the age of the subduction. If the subduction angle is high then the plate subduction age older than the smaller angle of subduction. The subduction model after relocation can be seen in Figure 4.

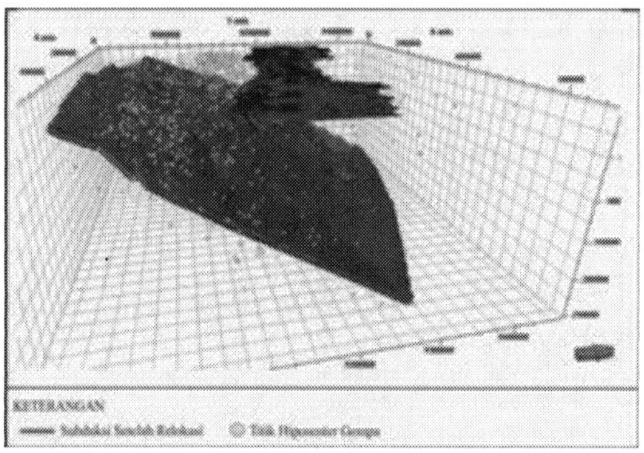

Figure 4. The 3D model Subduction Zone After Relocation

The subduction zone model obtained from the hypocenter distribution, it is describe the location of the collision between the Eurasian continent with Indo-Australian continent, because in this area there are a lot of tectonic earthquake so that trendline hypocenter can be modeled by subduction model such as in Figure 5.

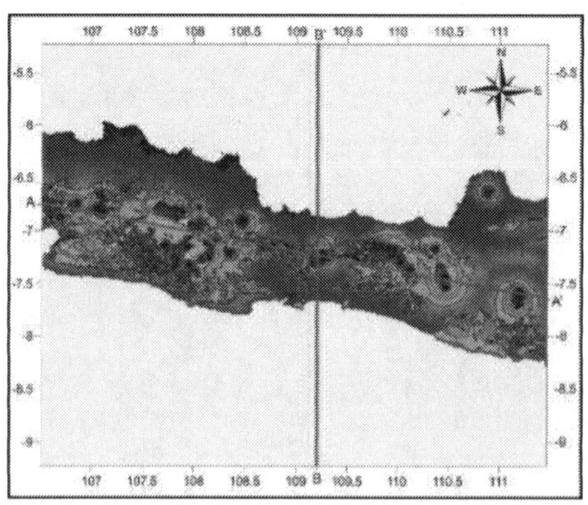

Figure 5. Location incision that cuts the island of Java.

Results of Tomography Pseudobending

To perform a more detailed observations of the research area so vertically observed is needed by make two incisions in Slamet mount for the example. The incision in the form of A - A 'with longitude 106.503707 and latitude -6.751412 until latitude -7.650037 and longitude 111.697802 and then the incision B-B' with coordinates X = 109 205 and Y = -5.23 to X = 109 205 and Y = -9.23 ,. The incision A-A 'cut the mountain range on the island of Java, then the incision B-B' cut Gunung Slamet and perpendicular to mountain range of Java.

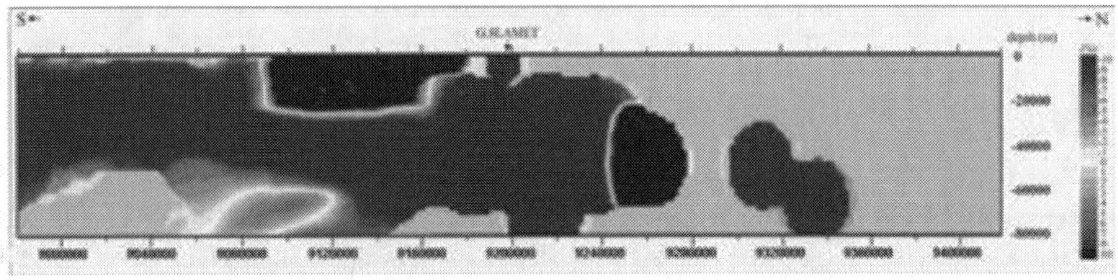

Figure 6. The incision tomography B-B 'wave velocity P.

From the results of incision B-B' (Figure.6) there are anomalies below Slamet mount that has a very low P wave velocity that is -10% than the surrounding areas that has high velocity that with the range from -2% to 10%. If the P wave propagating in the fluid so the velocity will decrease, the P wave that propagates in the bottom of Mount Slamet has a very low speed, it is caused by the fluid form of magma below Mount Slamet, given the mountain of Mount Slamet is active. The depth of the low velocity anomaly below Mount Slamet started 3 km from the summit of Mount Slamet. Then the gray color of the incision showed there was no ray tracing passing through the area.

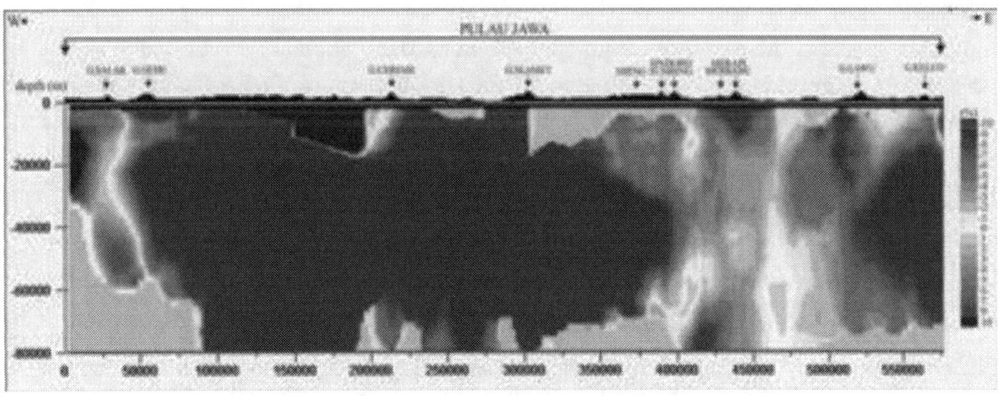

Figure 7. The incision tomography A-A 'wave velocity P.

In Figure 7 there are velocity difference between the eastern and western parts of the incision A-A ', this shows the difference of rock lithology between east and west, in the western part has a low velocity of -8% to -10%, while in the eastern part of the show ie value was -1% to -6%.

It adds to the confidence of previous studies (Reconstruction Tectonics microcontinent South Mountain East Java, [4]) that there microcontinent East Java who infiltrated the Sundaland, which indicated the velocity difference tomographic slice the A-A ', wherein microcontinent East Java has a higher velocity than Sundaland.

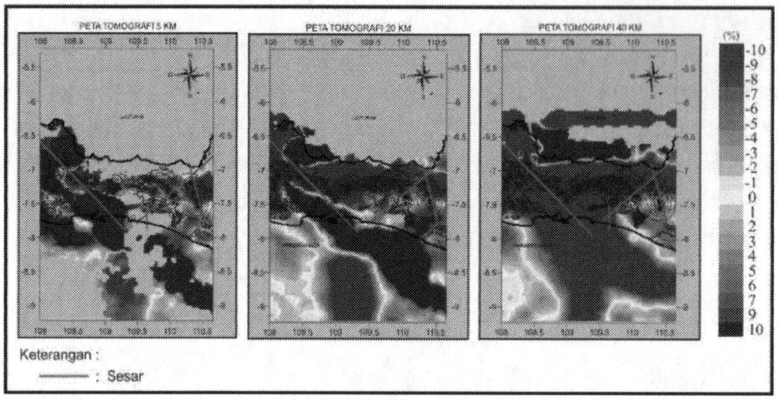

Figure 8. Map of the main structural patterns in Central Java with a tomographic method.

At a depth of 20 km it is clear that in complex fault structures large Cilacap - Pamanukan - Lematang and the complex structure of Kebumen - Muria - Meratus have the high velocity of 8% to 10%, this is because the alteration rocks process altrasi in the area of the fault. Faulting has pores large enough for the fluid that goes into these pores, at a depth of 20 km there is pressure and high temperature to allow for altrasi rocks in the fault complex. While low velocity found on the row of volcanoes and subduction-affected areas for rock bump so strongly that the rock reaches its melting point and becomes magma at the velocity of -8% to -10%.

5. Conclusion

1. Relocation Grid search method result show there are changes in the location of the hypocenter with the dominant direction of motion towards to the south and more shallow than the hypocenter before relocation with the change in direction of motion is 2 km - 12.5 km in accordance with the input grid is used.
2. Tomography result at the depth of 20 km show high velocity anomaly with a value of 8% to 10% contained in a large complex fault structures Cilacap - Pamanukan - Lematang and the complex structure of Kebumen - Muria - Meratus. While the region has a low anomaly with a value of -8% to -10% are mostly found in rows volcano in Central Java, then tomography in A-A 'shows the difference in characteristics between Central Java and East Java based tomographic velocity.

6. References

[1] Grandis, H. 2009. Pengantar Pemodelan Inversi Geofisika. ITB Bandung
[2] Nugraha, AD., Achmad S., Fatkhan. 2011. Jurnal Geofisika Edisi 2011 No.1
[3] Hidayatunnisak, S., Adi S., Muhajjir A.___. Studi Tomografi Seimsik Untuk Menentukan Model Kecepatan Gelombang P Daerah Bali
[4] Hussein, S. 2015, Tectonic Reconstruction microcontinent South Mountain East Java, Proceedings, National Seminar of Earth to-8, Department of Geological Engineering, UGM.

Journal of Physics: Conference Series **776** (2016) 012118 doi:10.1088/1742-6596/776/1/012118

Design of capacitive sensor for water level measurement

A. Qurthobi*, R. F. Iskandar, A. Krisnatal, and Weldzikarvina

Department of Engineering Physics, Telkom University

*email: qurthobi@telkomuniversity.ac.id

Abstract. Capacitive sensor for water level detection has been fabricated. It has, typically, high-impedance sensor, particularly at low frequencies, as clear from the impedance (reactance) expression for a capacitor. Also, capacitive sensor is a non-contacting device in the common usage. In this research, water level sensor based on capacitive principal created using two copper plates with height (h), width (b), and distance (l) between two plates, respectively, 0.040 m, 0.015 m, and 0.010 m. 5 V_{pp} 3 kHz AC signal is used as input signal for the system. Dielectric constant between two plates is proportional to water level. Hence, it can be used to determine water level from electrical characteristic as it inversely proportional to sensor impedance. Linearization, inverting amplifier, and rectifier circuits are used as signal conditioning for the system. Based on conducted experiment, the relationship between water level (x), capacitance (C), and output voltage (V_{dc}) can be expressed as $C(x) = 2.756x + 0.333$ nF and $V_{dc}(x) = 15.755 + 0.316$ V.

1. Introduction

Measurement of liquid level inside a container with various methods has been developed occasionally [1] [2]. Liquid level measurement can be utilized from the characteristic of the liquid itself; such as permittivity, permeability, conductivity, et cetera. One type of sensor which developed for liquid level measurement is capacitive sensor [3] [4] [5] [6].

Capacitive sensor can be categorized as reactive sensor. Hence, it is influenced by its input frequencies. Generally, capacitive sensor has non-contact characteristic. It also needs specific signal conditioning devices [7].

The research on development of capacitive sensor has been done by some researchers. Paczesny, et. al, conducted experiment in the capacitive sensor for liquid level measurement made with ink-jet printing technology [3]. Furthermore, Wei, et. al, conducted a research in implementation and characterization of a femto-Farad capacitive sensor for pico-liter liquid monitoring [4].

In this research, water level measuring device is designed based on capacitive sensor. The sensor is made from two plastic-insulated copper plates and it connected into signal conditioning devices. The change of the water level detected by the sensor will results the change of capacitance value in the sensor and represents as the change of voltage on the output.

2. Sensor Fabrication

The capacitive sensor is formed from two conductive plates (figure 1). Both plates have ability to store electrical charges. The value of capacitor C obtains from equation (1); where l is the distance between two plates, and ε is dielectric constants of the insulator materials between two plates (which it consists of ε_o (permittivity of vacuum) and ε_r (relative permittivity))(equation (2)).

$$C = \varepsilon \frac{A}{l} \quad (1)$$

$$C = \varepsilon_o \varepsilon_r \frac{A}{l} \quad (2)$$

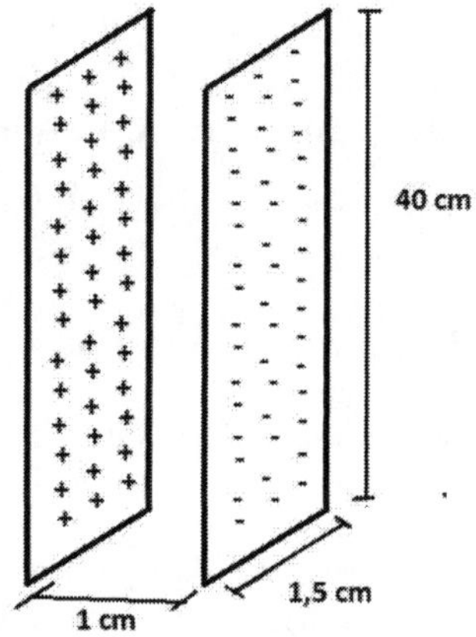

Figure 1. Visualization of charge storage on the plates

From equation (1), the capacitance (C) value is affected by the value of ε, A, and l. Hence, equation (3) and (4), which shows the influence of the variables to sensor's capacitance value, can be derived from equation (1).

$$\ln C = \ln \varepsilon + \ln A - \ln l \quad (3)$$

$$\frac{\Delta C}{C} = \frac{\Delta \varepsilon}{\varepsilon} + \frac{\Delta A}{A} - \frac{\Delta l}{l} \quad (4)$$

$$S = \frac{\Delta C}{\Delta A} = \frac{1}{b} \frac{\Delta C}{\Delta x} = \frac{\varepsilon}{l} \quad (5)$$

Because of the level role as the measuring object, the ideal sensor's sensitivity can be written as equation (5) where l represents level of the liquid and b is width of the sensor. It shows the sensitivity of the sensor depends on water level is a constant if the values of ε and l are also constants.

3. Implementation with Signal Conditioning Devices

The sensor is made from two copper plates which had dimensions, for every plate, are 0.4 m in height (h), 0.015 in width, and 0.001 m in thickness. Both plates separated 0.01 meter from each other and insulated with plastics to avoid direct contact with another substances. It helps to give some room for another substance infiltrated between plates and to reduce, or even remove, the possibility of electrical short-circuit between two plates when ac input is applied. In the experiments, capacitive sensor is used to measure liquid level inside a container with 30 cm of maximum depth. The type of liquid used is water (ε_r = 80.1 in 20° C). Figure 2 shows when the sensor is placed inside of water container. All experiments are conducted on room temperature.

Figure 2. Location of the sensor inside the container

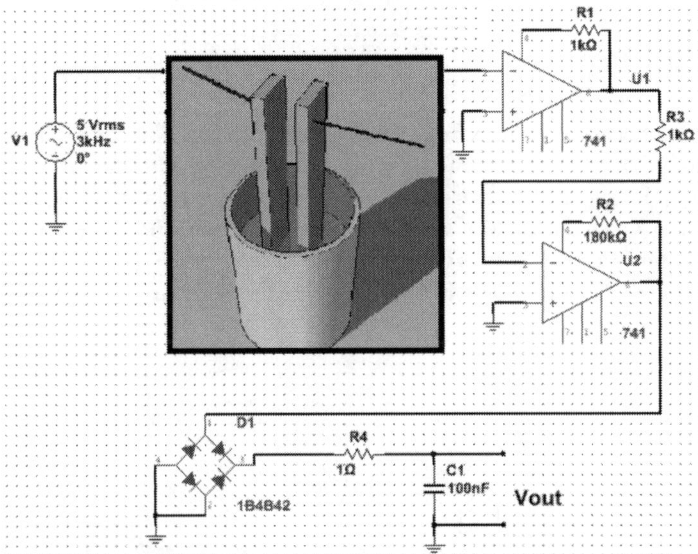

Figure 3. Schematic figure of devices integration

Capacitive sensor is a passive sensor [1, 7]. Therefore, it needs some signal conditioning device(s) to obtain other quantities (voltages or currents) which can be processed on the other conditioning or processing devices. In this research, there are three types of signal conditioning circuits: (1) linearization circuit; (2) Op-Amp inverting amplifier circuit; and (3) rectifier (see figure 3). The sensor is located between power source and linearization circuit. Linearization circuit is a circuit which used to obtain linear correlation between its input and output signal. It used as the feeder of other devices. It consists of resistor, Op-Amp device, and the capacitive sensor. As the ac input applied, the output equation can be expressed as equation (6) where f is the input frequency, R_l is the value of resistor in the circuit, V_{il} represent ac input signal, and V_{ol} is the output of linearization circuit. Assuming the values of V_{il}, R_l, and f are constant, the value of V_{ol} is proportional to C.

$$V_{ol} = 2\pi f C R_l V_{il} \tag{6}$$

Other signal conditioning devices which are used in these experiments are operational amplifier and rectifier. Operational amplifier is operated on the inverting mode since it has linear amplification characteristics (equation (7)) where R_{fa} in the value of feedback resistor, R_{ia} in the value of feedback resistor, and V_{oa} represents the output voltage value of the amplifier. In the other hand, rectifier circuit is used to convert ac signal into dc to approach compatibility with data acquisition or processing devices.

$$\frac{V_{oa}}{V_{ol}} = -\frac{R_{fa}}{R_{ia}} \tag{7}$$

4. Results and Discussion

4.1 Capacitance Value of Sensor

The connection between level (in meter) and capacitance value (in nF) shown by figure 4. The correlation between capacitance value and water level, as informed by the figure, is proportional. The regression correlation equation satisfies equation (8) where x represent water level, and $C(x)$ is the capacitance value of the sensor regarding to water level.

$$C(x) = 2.759x + 0.333 \tag{8}$$

Regarding on conducted experiments, the sensor capacitance has non-zero initial value. This result caused by the non-ideal condition of the conducted experiments. Some part of the sensor which is not submerged into the water also had contribution of the capacitance value. It also creates the non-linear characteristic of capacitance value on the real condition. Figure 4 also shows the change of capacitance value became slower when the sensor submerged deeply into the water. This condition can be expressed, if the sensor impurities are ignored, as equation (9).

$$C(x) = \frac{\varepsilon_o b}{l}\left(\varepsilon_{r_{water}} x + \varepsilon_{r_{air}} (h - x)\right) \tag{9}$$

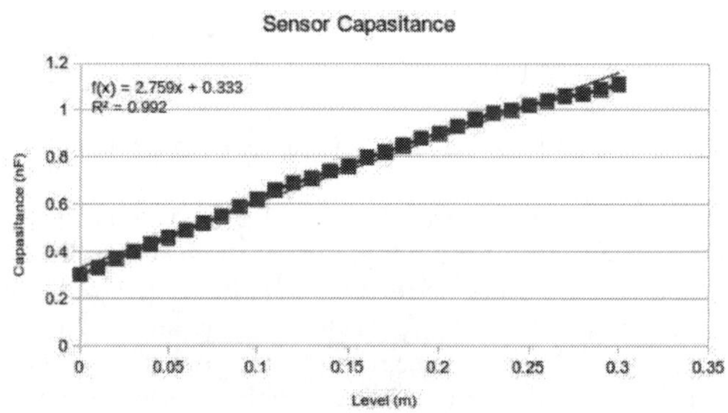

Figure 4 Sensor capacitance value

4.2 Signal Conditioning Output

Figure 5 and 6 shows the connection of the water level and output voltages. In both figures, the output voltage increase almost proportionally with the water level and can be expressed as equation (10) and (11) where $V_{ol}(x)$ and $V_{dc}(x)$ are the output voltages of linearization and rectifier circuit, respectively.

$$V_{ol}(x) = 0.286x + 0.046 \tag{10}$$

$$V_{dc}(x) = 15.755x + 0.316 \tag{11}$$

The correlation between V_{ol} and C is linear and proportional as stated in equation (6). But, the results which displayed on figure 4 and 5 were have a wavy or curved patterns. It occurs because of non-ideal characteristics of op-amp, such as dc imperfection and slew rate. It also caused by the impurities of the sensor itself.
As shown in figure 3, amplification value of the inverting mode op-amp circuit is -180 times. However, the average gain value after the signal pass the rectifier circuit is only 27.639 times. The slew rate, the dc imperfection, the loss in rectifying process, and the loss caused by line impedance are the caused of the problem.

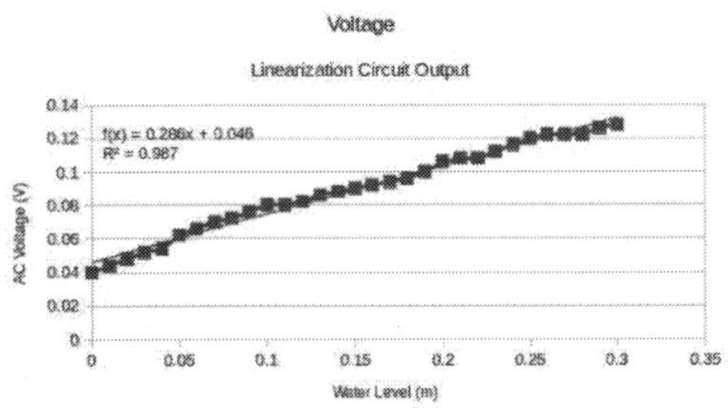

Figure 5. Linearization circuit output voltage

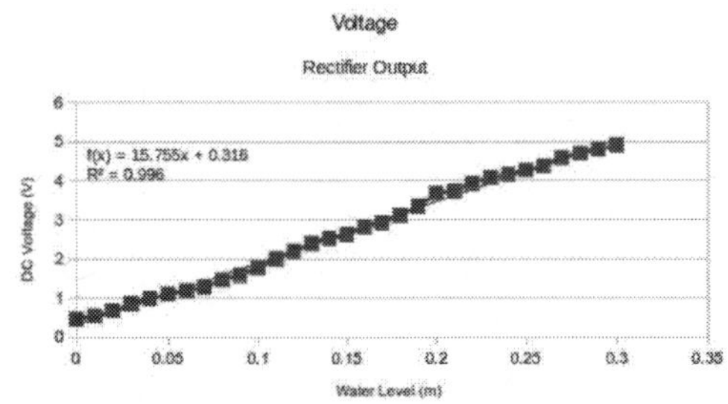

Figure 6. Rectifier Circuit Output Voltage

The effect of non-zero initial condition and non-linear condition can be reduced by either hardware or software approximations. In hardware approximation, adjustment and addtitional filter circuit are added between rectifier and data aquitition device. In the other hand, statistical methods are used in software approximation.

5. Conclusion

A capacitive sensor for water level measurement had been developed and tested. It connected into 5 V_{pp} 3 kHz ac input voltage, and signal conditioning devices(linearization, inverting mode op-amp, and rectifier circuits) to obtain output voltage signal. Depending on the water level, the capacitance of the sensor varies from 0.300 nF (at 0 m) to 1.110 nF (at 0.300 m) and can be approximated by the equation $C(x) = 2.759x + 0.333$. Furthermore, the correlation between water level and dc output voltage can be expressed as $V_{dc}(x) = 15.755x + 0.316$.
The results of the experiments also inform the effect of non-ideal condition. It creates non-constant sensitivity value since the capacitance equation became $C(x) = \frac{\varepsilon_o b}{l}\left(\varepsilon_{r_{water}} x + \varepsilon_{r_{air}} (h - x)\right)$. Moreover, the slew rate of the op-amp at high frequency, the loss in rectifying process, and the loss because of line impedance caused the average gain became under the designed value.

6. References

[1] Doebelin EO 1990 *Measurement Systems: Application and Design*. Singapura: McGraw-Hill
[2] Dunn WC 2006 *Introduction to Instrumentation, Sensors, and Process Control*. Norwood: Artech House
[3] Paczesny D, Tarapata G, Michal M, and Jachowicz R 2015 *Eurosensors* Warsaw pp. 731-735
[4] J. Wei et al. 2009 *Eurosensors*, Delft pp. 120-123
[5] Alfadhl Yahya Khaled, Samsuzana Abd Aziz, and Fakhrul Zaman Rokhani, "Capacitive Sensor Probe to Assess Frying Oil Degradation," *Information Processing in Agrigulture*, pp. 142-148, 2015.

[6] Texas Instrument. (2015, March) Capacitive-Based Liquid Level Sensing Sensor Refecence Design.

[7] Clarence W. de Silva, *Sensors and Actuators: Control System Instrumentation*. Boca Raton, FL: CRC Press, 2007.

Seismic hazard analysis with PSHA method in four cities in Java.

Y. Elistyawati, I. R. Palupi, and Suharsono
Physics Department of *Program Pembangunan Nasional "Veteran"* University,
Jl. SWK 104 (Lingkar Utara) Condongcatur Yogyakarta 55283, INDONESIA

E-mail: Yuppi_elistyawati@yahoo.co.id

Abstract. In this study the tectonic earthquakes was observed through the peak ground acceleration through the PSHA method by dividing the area of the earthquake source. This study applied the earthquake data from 1965 - 2015 that has been analyzed the completeness of the data, location research was the entire Java with stressed in four large cities prone to earthquakes. The results were found to be a hazard map with a return period of 500 years, 2500 years return period, and the hazard curve were four major cities (Jakarta, Bandung, Yogyakarta, and the city of Banyuwangi). Results Java PGA hazard map 500 years had a peak ground acceleration within 0 g $\geq$ 0.5 g, while the return period of 2500 years had a value of 0 to $\geq$ 0.8 g. While, the PGA hazard curves on the city's most influential source of the earthquake was from sources such as fault Cimandiri backgroud, for the city of Bandung earthquake sources that influence the seismic source fault dent background form. In other side, the city of Yogyakarta earthquake hazard curve of the most influential was the source of the earthquake background of the Opak fault, and the most influential hazard curve of Banyuwangi earthquake was the source of Java and Sumba megatruts earthquake.

1. Introduction

The earthquake that happened on July 17, 2006 with a magnitude of 7.7 Mw had occurred around the island of Java and had resulted stunami 2m tall with the type of slow earthquak. Causing substantial losses and the deaths of up to ± 659 inhabitants. Recognizing that an earthquake is one of the natural disasters that can not be predicted and caused the loss of influence on the material nor economy. Based on the research done by the source of the seismic tectonics, the method applied is a method PSHA (Probabilistic seismic hazard analysis) by dividing the seismic source into several sections. Several previous studies had done for the construction of buildings and bridges. One previous study were used as the reference was based on a map of acceleration bedrock of Team Revised 2010 for the whole of Indonesia. This paper is intended to determine the source of the most influential effect in major cities in Java. From the results of this analysis obtained map Java acceleration for a period of 500 years and 2,500 years, as well as the possible hazard curve exceeded 5%.

2. Probabilitic Seisimic Hazard Analysis Method (PSHA)

On the results of the seismic hazard analysis obtained spectral response, maximum acceleration, and time-history. One method applied was the probability or often referred to PSHA developed by Cornell (1968). PSHA method is basically an analysis by performing various scenarios involving frequency

Journal of Physics: Conference Series **776** (2016) 012119 doi:10.1088/1742-6596/776/1/012119

ground motion. The use of PSHA method has the advantage of taking into account factors of uncertainty so that it can estimate the greater the probability for the worst conditions that would occur in a particular area. This method has four stages, namely the identification and characterization of the earthquake, based on the characteristics of the seismicity, Election attenuation function, and a calculation to a combination of factors of uncertainty. The use of earthquake catalog of some site measurements needed to analyze the data completeness. This catalog consists of 4 different magnitude of which body wave magnotide (mb), Surface wave magnitude (ms), Richter local magnitude (ML), and the moment magnitude (MW). The scale is then converted to a magnitude that is MW.

Table 1 Relationship between types magnitude earthquake (Asrurifak, 2010)

Correlation Conversion	Amount of Data (Event)	Range Data	Conformity
$M_w = 0.143\ M_s^2 - 105.1\ M_s + 7.285$	3.173	$4.5 \leq M_s \leq 8.6$	93.90%
$M_w = 0.114\ M_b^2 - 0.556\ M_b + 5.560$	978	$4.9 \leq M_b \leq 8.2$	72.00%
$M_w = 0.787\ M_E + 1.537$	154	$5.2 \leq M_E \leq 7.3$	71.20%
$M_b = 0.125\ M_L^2 - 0.389\ M_L + 3.513$	722	$3.0 \leq M_L \leq 6.2$	56.10%
$M_L = 0.717\ M_D + 1.003$	384	$3.0 \leq M_D \leq 5.8$	29.10%

Converted data were then analyzed the completeness of the data with the first was the stage splitting time of the observation into multiple intervals of time counting backwards [1], then in each interval the seismic data was divided based on those ranges in magnitude[2], then calculated the range of the earthquake and calculate variances in each range of magnitude [3], and then at the last stage graphing the observation interval relationship with the variance magnitude [4].

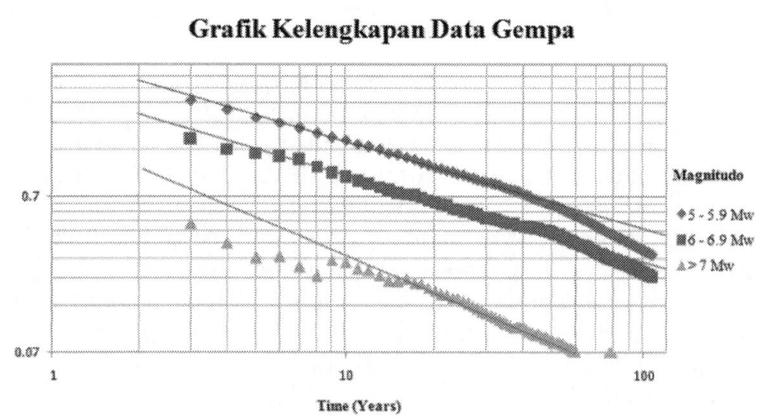

Figure 1 Completeness Model of the data hit the area of Indonesia from 1907 to 2015

In the early stage of the identification and characteristics of the earthquake source was done by dividing the area into sections based on the source of the earthquake. Indonesia, which is between the three major continental plates, the area Brada on active tectonic zone. The existence of tectonic activity that work resulted in the formation of mountains and forming pathways plate meetings, these areas which later became the source of the quake. Java has Java-Sumba subduction zone that stretches from the Sunda Strait to Bali Basin, but (Newcomb & McCann, 1987) divides the segment into two parts, namely Java and Sumba segment.

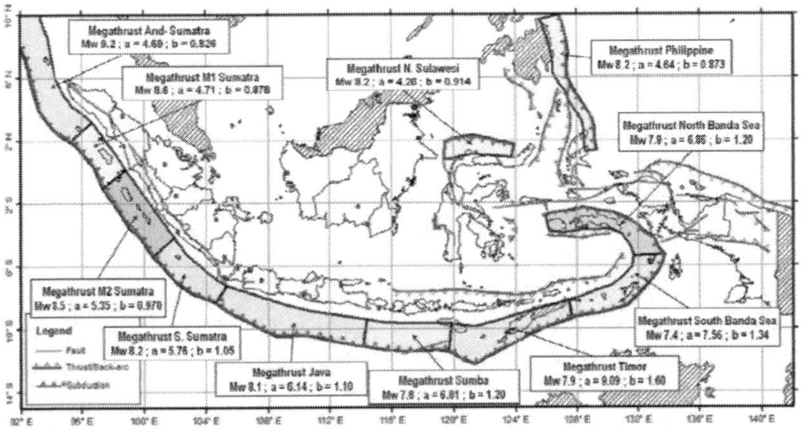

Figure 2 Megatrust segment division of earthquake source in the region of Indonesia (Asrurifak, 2010).

In addition, the development fault in Java Island, including an active fault, so that it influenced the occurrence of the earthquake. In this study the fault that was applied for Java area including fault Cimandiri, Lembang fault, Pati, Lasem, and Opak. In this study seismic source zones were divided into several sections including interface with the depth of 0-50 km zone, intraslab zone at a depth of 50-300 km, the zone at shallow crustal depths of 0-50 km consist of fault and shallow background.

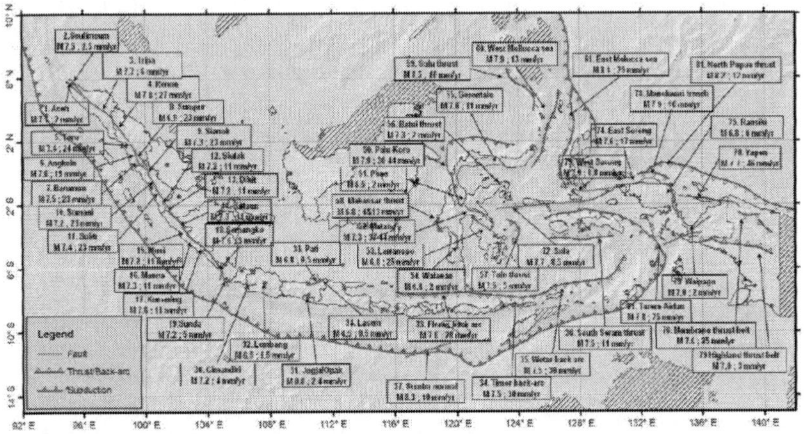

Figure 3 The maximum magnitude and slip-rate of the sources of earthquakes (Asrurifak, 2010).

In the second stage based on the characteristics of the seismicity could be seen through the parameter b-value and earthquake rate by applying the least square method and maximum likelihood. In the area of Java the separation into two parts, a megathrust area and Benioff. Least square method can be expressed by the was conducted formula log N (m) – a - bm with information N (m) was the frequency of earthquakes per time unit, m is the magnitude of the earthquake, a is a / ln10, and b is β / ln10. Applying the Maximum likelihood method because a value can produce -b¬ more stable. Meanwhile, a parameter-b¬ relations and annual rate (v) expressed by the formula v = exp (a-βm). This method can be expressed by b = log e / (-Mi), is the middle value of the data magnitude greater than Mi.

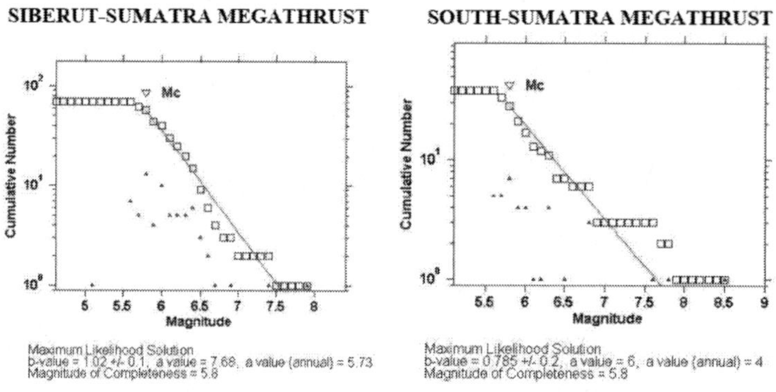

Figure 4 b-value analysis for Siberut segment Sumatra and South Sumatra megathrust implementing the least square method and the maximum likelihood

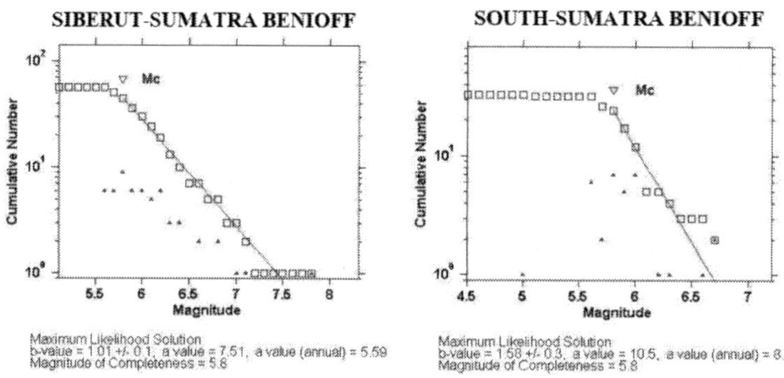

Figure 5 analyzes of the b-value segment of Sumatra and South Sumatra Siberut Benioff implementing the least square method and maximum likelihood

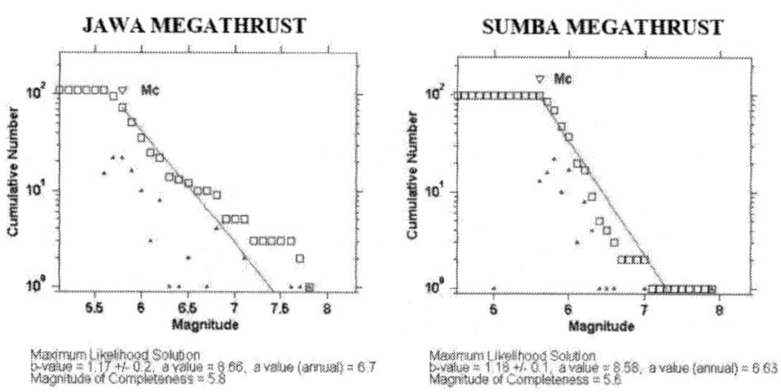

Figure 6 b-value analysis for Java and Sunda megathrust segments implementing the least squares method and maximum likelihood

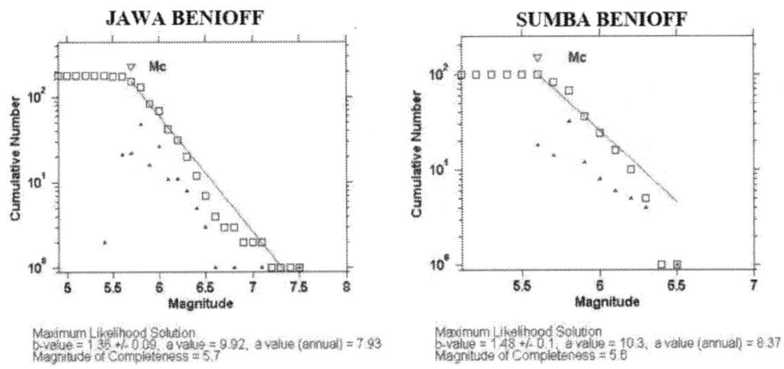

Figure 7 b-value analysis to segment the Javanese and Sundanese Benioff implementing the least square method and maximum likelihood.

Then we get the parameter b-value and the annual rate (v) for megathrust, Benioff, Shallow crustal and deep backgrond in Table 2.

Table 2 Parameter values of b-value and the annual rate

No	Keterangan	b	a	Mc	β	v
1	Siberut Megathrust	0.785	4	5.8	1.81	0.28
2	Southern Sumatra Megathrust	1.02	5.73	5.8	2.35	0.65
3	Jawa Megathrust	1.17	6.7	5.8	2.69	0.82
4	Sumba Megathrust	1.18	6.63	5.6	2.72	1.05
5	Timor Megathrust	1.66	9.3	5.6	3.82	1.01
6	Siberut Benioff	1.58	8.67	5.8	3.64	0.32
7	Southern Sumatra Benioff	1.01	5.59	5.8	2.33	0.54
8	Jawa Benioff	1.36	7.93	5.7	3.13	1.51
9	Sumba Benioff	1.48	8.37	5.6	3.41	1.21
10	Timor Benioff	1.61	9.33	5.8	3.71	0.98
11	Shallow Crustal	1.24	7.58	5.8	2.86	2.44
12	Deep Bacground	1.61	9.37	5.8	3.71	1.08

In the third stage of determining the attenuation function to parts of Indonesia are still using the attenuation of other regions that have almost the same characteristics. This can not be avoided because until now, the data grund-mention for the Indonesian region is still not enough to manufacture atenuation. For it is based on Indonesian Earthquake Map Revision Team 2010 attenuation function Indonesian territory based on those conditions of each seismic source composed as follows:

• Sources of shallow crustal earthquake, the model applied was the model of the earthquake source fault and shallow background with attenuation function (Boore-NGA Atkinson, 2008; Campbell-Bozorgnia NGA, 2008; Chiou-NGA Youngs, 2008).

• Source earthquake subduction interface (megathrust), the model applied was the model of the earthquake source subduction with attenuation function is used (Geomatrix subduction (Youngs et al., 1997); Atkinson-Boore BC rock and global source subduction, 2003; Zhao et al, 2006) ,
• Source earthquake Benioff, the model used is the model of the earthquake source deep background with attenuation function is used (Atkinson and Boore, Cascadia 2003; Youngs et al., 1997; Atkinson and Boore, Wordwide 2003).
In the fourth stage the calculations for uncertainty factors one with a tree logic models. The logic tree approach made it possible to use models by determining the weight factor that described the percentage chance relative of accuracy a model against other models.

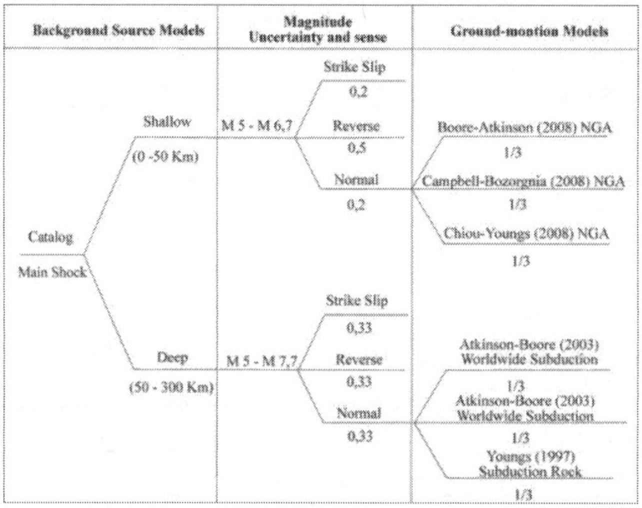

Figure 8 Logic Model tree for earthquake Background

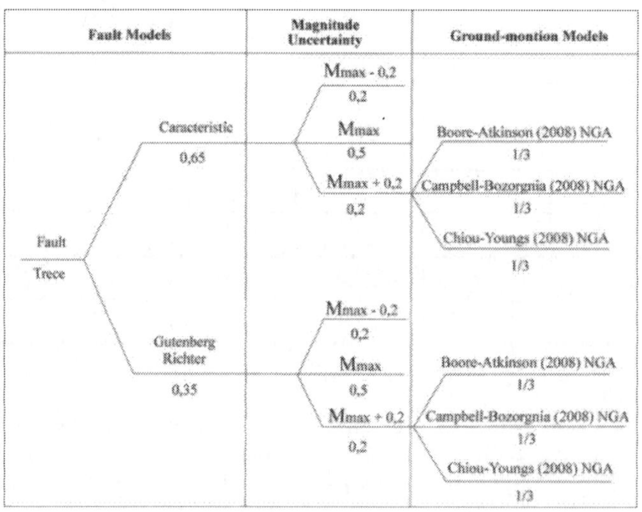

Figure 9 Logic Model Fault tree for earthquake

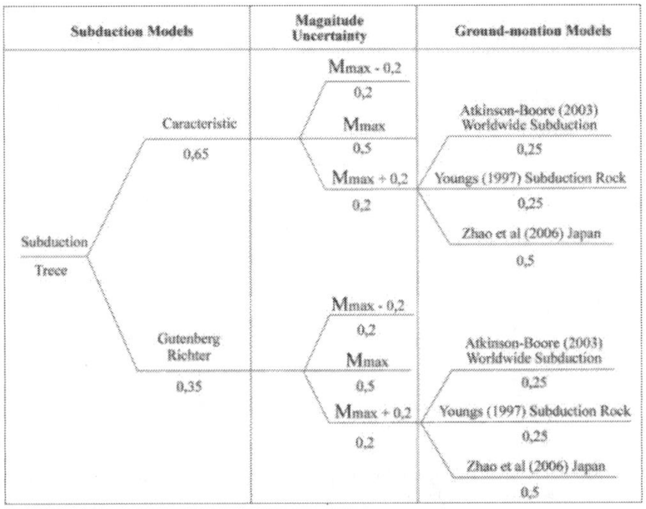

Figure 10 Logic Model tree for subduction earthquakes

3. Results and Discussion

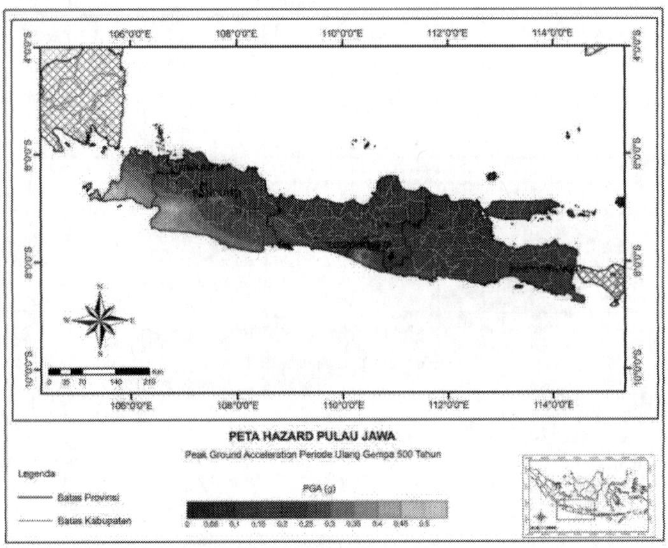

Figure 11 Map of the island of Java with a possible hazard exceeded 5% in a period of 500 years anniversary

Based on the analysis using a map obtained PSHA accelerated renewal of the basis for Java is composed of a map of the acceleration peaks, spectra of 0.2 seconds, and 1 second spectra with the possibility exceeded 5% return period of 500 years and return a period of 2500 years. For peak acceleration (PGA) of bedrock with the possibility of 5% exceeded the return period of 500 years can be seen in figure 10. For this area between dominated by the acceleration of 0 g - 0.2 g with a range of acceleration from 0 g $\geq$ 0.5 g. The value of the acceleration in the bedrock to Jakarta 0.15 g - 0.2 g, to the city of Bandung 0.25 g - 0.4 g, Yogyakarta 0.2 g - 0.4 g and 0.2 g Banyuwangi - 0.25 g.

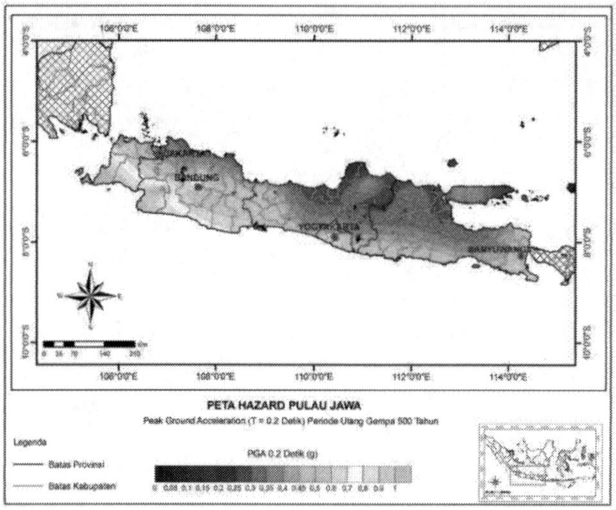

Figure 12 Map of the island of Java hazard spectra of 0.2 seconds with the possibility of 5% exceeded the return period of 500 years

To map the basic spectral acceleration bedrock 0.2 seconds with the possibility of 5% exceeded the return period of 500 years can be seen in figure 11. This Wilayh dominated bedrock acceleration anatara 0.25 g - 0.7 g acceleration range from 0 g ≥ 1 g. Value of acceleration on the bedrock of the city of 0.4 g - 0.5 g to 0.6 g of Bandung - 0.9 g, Yogyakarta 0.35 g - 0.1 g and 0.45 g Banyuwangi - 0.6 g.

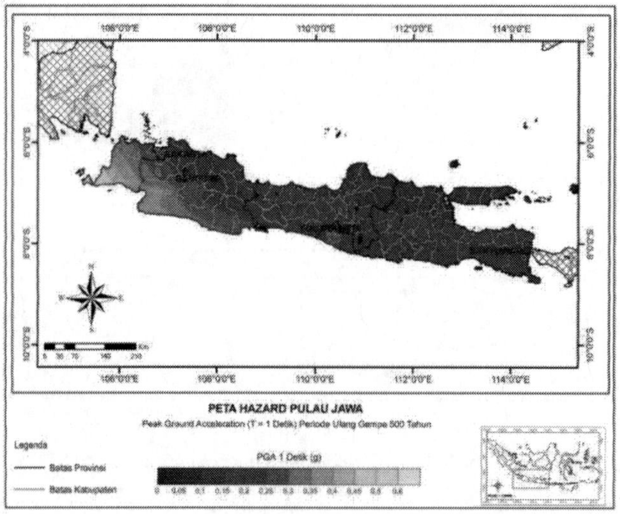

Figure 13 Map of the island of Java hazard spectra of 1 second with the possibility of 5% exceeded the return period of 500 years

To map the acceleration base rock with spectra 1 second possibility is exceeded 5% in the period of 500 years can be seen in figure 12. For this area dominated the bedrock acceleration between 0 g - 0.25 g with a range of acceleration from 0 g ≥ 0.6 g , Value of acceleration on the bedrock of the city of 0.2 g - 0.3 g to 0.2 g Bandung - 0.35 g, the city of Yogyakarta, Bandung almost equal to 0.2 g - 0.35 g, and the city of Banyuwangi 0.15 g - 0.25 g.

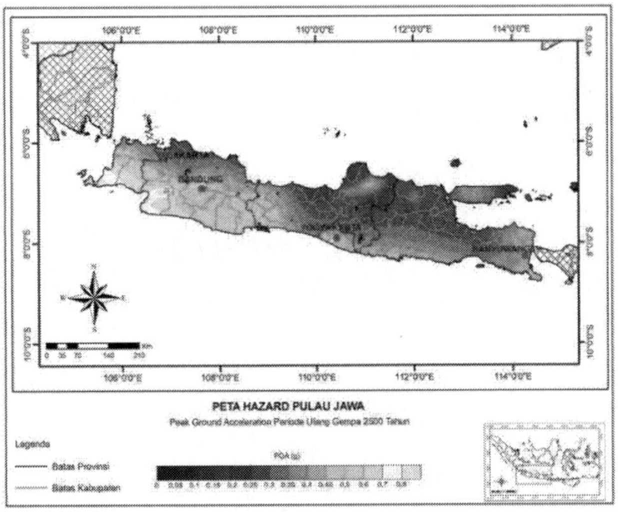

Figure 14 Map of peak acceleration hazard Java with the possibility exceeded 5% in the period of 2500 years.

For peak acceleration (PGA) rock basis with the possibility exceeded 5% in the period 2500 years can be seen in figure 13. For this wilayh dominated bedrock acceleration anatara 0.25 g - 0.6 g with a range of acceleration from 0 g ≥ 0, 8 g. Value of acceleration on the bedrock of the city of 0.3 g - 0.4 g, to the city of Bandung 0.45 g - 0.7 g, Yogyakarta 0,3g - 0.8 g and 0.35 g Banyuwangi city - 0.45 g.

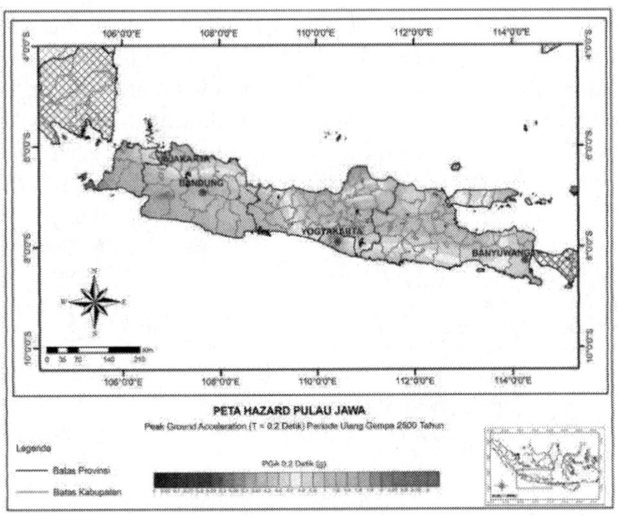

Figure 15 Map of the island of Java hazard spectra of 0.2 seconds with a chance exceeded 5% in the period of 2500 years

To map the basic spectral acceleration rock 0.2 seconds with a chance exceeded 5% in the period 2500 years can be seen in figure 14. This area between dominated bedrock acceleration of 0.45 g - 1.2 g with a range of acceleration from 0 g ≥ 3 g. Value of acceleration on the bedrock of the city of 0.6 g - 0.7 g to 0.9 g of Bandung - 1.2 g, Yogyakarta 0.6 g - 2 g, and the city of Banyuwangi 0.8 g - 1 g.

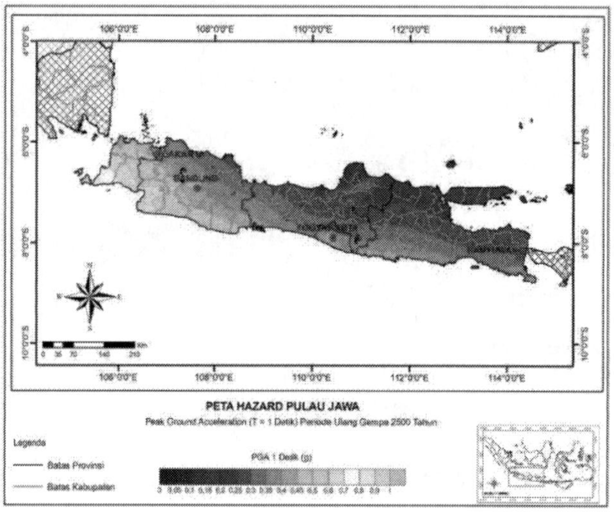

Figure 16 Map of the island of Java hazard spectra of 1 second with a chance exceeded 5% in the period of 2500 years

To map the basic spectral acceleration rock 1 second with a chance exceeded 5% in the period 2500 years can be seen in figure 15. This area rock acceleration between dominated basic 0.25 g - 0.7 g acceleration range from 0 g ≥ 1 g. Value of acceleration on the bedrock of the city of 0.4 g - 0.45 g to 0.45 g Bandung - 0.5 g, Yogyakarta 0.3 g - 0.5 g and 0.3 g Banyuwangi - 0.35 g.
In addition to a map obtained acceleration rock base, also obtained hazard curves for the four major cities of Jakarta, Bandung, Yogyakarta, and the city of Banyuwangi. In the city hazard curve bedrock peak acceleration with the possibility exceeded 5% can be seen in Figure 16 (a). Seen in the city's most influential source of an earthquake originating from the fault Cimandri. In addition there are sources of the fault zone megathrust earthquake on Java, but not very big. Hazard curve acceleration of bedrock in the city of Bandung in Figure 16 (b) of the most visible source of seismic influence derived from fault, fault Lembang. For megathrust fault Cimandiri and Java still have an influence on this town, but not very big.

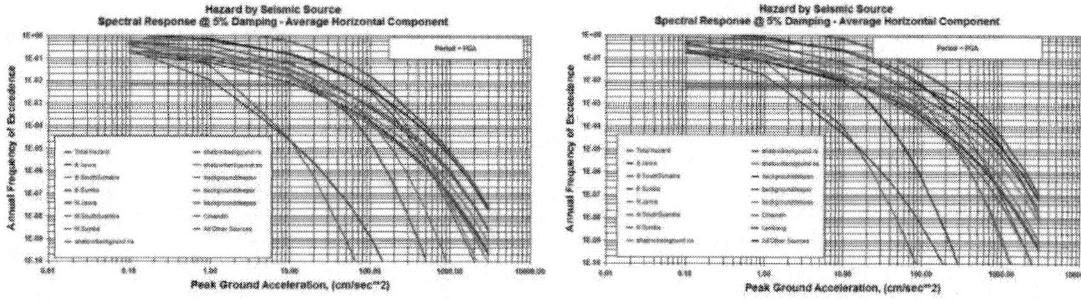

(a) (b)

Figure 17 Curve hazard peak acceleration in Jakarta (a) and the city of Bandung (b) the possibility exceeded 5%

Hazard curve acceleration of bedrock in the city of Yogyakarta in Figure 17 (c) of the most visible source of seismic influence derived from fault, fault Opak. For Lasem fault, fault Pati, megathrust Sumba and Java megathrust still have an influence on this town, but not very big. While the hazard curve acceleration bedrock Banyuwangi shown in figure 17 (d) of the most visible source of seismic influence comes from megathrust Java and Sumba megathrust.

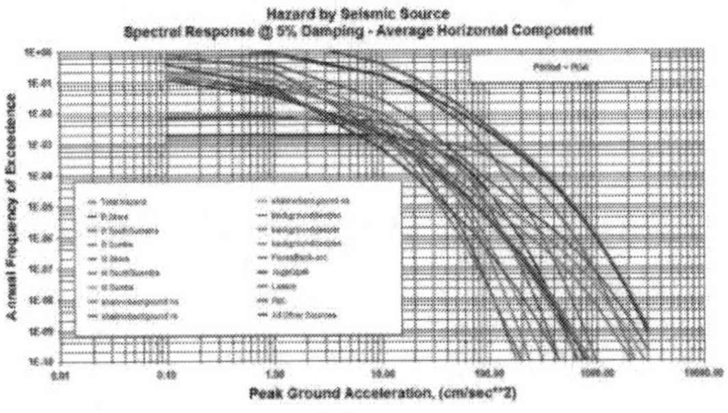

(c)

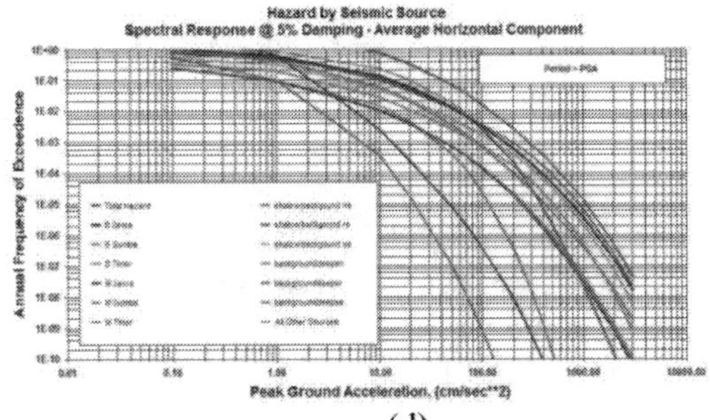

(d)

Figure 18 Curve hazard peak acceleration in Yogyakarta (c) and Banyuwangi (d) the possibility exceeded 5%

Based on the results of the analysis of the PSHA could be concluded that influence as the source of the earthquake the most important in several big cities in Java derived from the fault that developed as in the city of the earthquake on the Cimandiri fault, Bandung sourced from Lembang fault, and the city of Yogyakarta sourced from Opak fault. Yet, do not rule out the influence of the larger zones to Banyuwangi that the most influential source of the earthquake derived from two segments, namely Java and Sumba Megathrust. Value of acceleration on the bedrock of the island of Java was becoming southwest towards higher value.

The value of the acceleration in the higher bedrock towards the southwest visible on the map bedrock acceleration spectra exceeded 0.2 seconds with the possibility of 5% on a 500-year return period and a return period of 2500 years. Value bedrock acceleration in 0.2 seconds maps spectra return period of 500 years for the city of 0.4 g - 0.5 g to 0.6 g of Bandung - 0.9 g, 0.35 g Yogyakarta - 0, 1 g, and the city of Banyuwangi 0.45 g - 0.6 g. Acceleration value bedrock on map 0.2 spectra return period of 2500 years for the city of 0.6 g - 0.7 g to 0.9 g of Bandung - 1.2 g, Yogyakarta 0.6 g - 2 g, and the town of Banyuwangi 0.8 g - 1 g.

4. References

[1] Asrurifak M 2010 *Peta Respon Spektra Indonesia Untuk Perencanaan Bangunan Tahan Gempa Berdasarkan Sumber Gempa 3D Dalam Analisa Probabilitas,* Disertasi Doktor Jurusan Teknisipil, ITB

[2] Aki K 1965 *Maximum Likelihood Estimate of b value in The Formula log N = a-bM and Its confidence Limits* Bull Earthq 43 237240

[3] Irsyam, M., Sengara, W., Aldiamar, F., Widiantoro, S., Triyoso, W., Hilman, D., Kertapati, E., Meilano, I., Asrurifak, M., Ridwan, M., Huhardjono., 2010. *Ringkasan Hasil Studi Tim Revisi Peta Gempa Indonesia 2010,* Kementrian Pekerjaan Umum

[4] Jimmi N, Guntur P, Bambang S, Sri Widiyanto (2014) *Analisis Hazard Gempa Dan Isoseismal Untuk Wilayah Jawa-Bali-NTB,* Paper Jurnal Meteorlogi Dan Geofisika Vol 15 No 1 Tahun 2014

[5] Widodo P, 2012 *Seismologi Teknik & Rekayasa Kegempaan, Pustaka Pelajar,* Yogyakarta

Identifying of ground water level by using geoelectric method in Karanganyar, Central Java, Indonesia

S. Koesuma[1*], **Sulastoro**[2]
[1]Department of Physics. Sebelas Maret University. Surakarta Indonesia
[2]Department of Civil Engineering. Sebelas Maret University. Surakarta Indonesia.

E-mail : sorja@uns.ac.id

Abstract. This study aims to determine ground water level in Karanganyar regency, Central Java Province, Indonesia. Karanganyar regency is located in west flank of Lawu volcano, the third highest volcano in Central Java Province. Karanganyar lays from the top submit of Lawu volcano to down town of city with altitude 3265 m to 88 m. Same as other mountain area, Karanganyar has a lot of ground water potential. We use geoelectric method to finds out how deep of ground water level. The survey locations are distributed surround Karanganyar regency which contain 22 sites, in period survey of 2013 – 2015. Schlumberger configuration is used for acqusition data with lenght of current electrode distance varies from 1 m to 700 m. The result shows that ground water level are located in depth from 50 meter to 150 meter with lithology of tuff and sand. In Munggur and Kedung Jeruk sites, we found two potential aquifers, which are shallow and deep aquifers.

1. Introduction

Karanganyar regency is located in coordinates between 110°40′E to 110°70′E and 7°28′S to 7°46′S, with average height is 511 meters above sea level. This regency is located in eastern part of Central Java province, Indonesia. Most of Karanganyar regency area is mountains with its summit is Lawu volcano, which its height around 3265 meter. The geology of Lawu volcano was first presented in Van Bemmelen [1], further in Sampurno and Samodro [2] has made a geological map named Ponorogo sheet with 1 : 100.000 scale. The Lawu volcano complex can be distinguished as having older and younger parts. The older products are deeply carved by erosion, while the younger parts are characterized by a smoother surface [3]

Similar to other mountains area in southern part of Java island, Karanganyar regency where located with altitude from 88 meter to 3265 meter, has a lot of potential of ground water. Many water companies has explore the ground water in Lawu volcano and then sell it to other city near Karanganyar regency. Some industries also takes ground water for their production process, most of them located in Palur, near Surakarta. And most widely used of ground water are distributed by Government Water Company which is in 2015 has delivered 82.502 m^3 to residents in Karanganyar regency [4]. This research's purpose to determine the ground water level in 22 sites of Karanganyar regency by using geoelectric method.

* To whom any correspondence should be addressed.

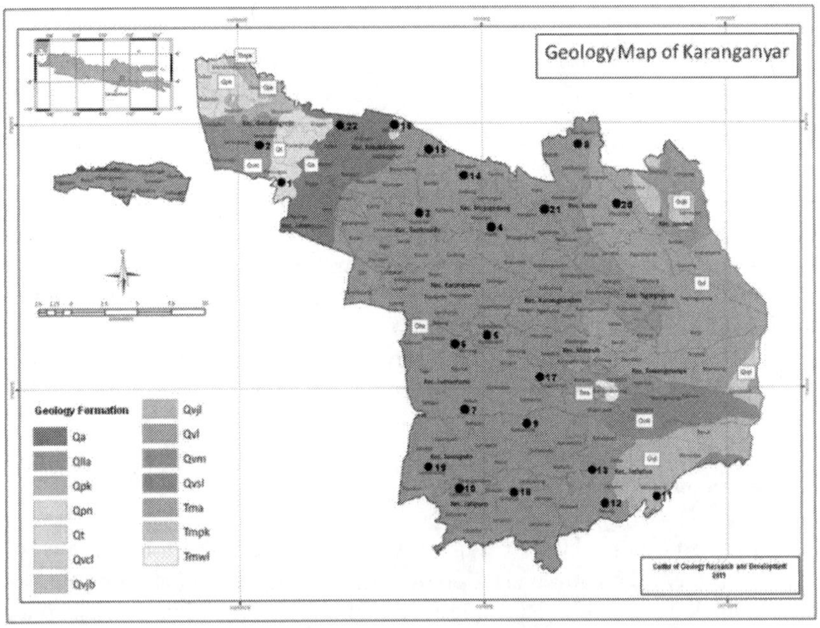

Figure 1. Geology map of Karanganyar regency. Solid circle is location of survey. Inserted map is Java island.

2. Method

The principle of geoelectric method is by injecting an electric current into the earth through two current electrodes, which causing a potential difference. The potential difference that occurs is measured by two potential electrodes. The injected current and potential difference is used to calculate resistivity value for each particular electrode configuration beneath the measuring point. This geoelectric method is more effective and suitable for exploration near surface survey for instance in determining the depth of the basement [5], water aquifer [6], seawater intrusion [7] and exploration of geothermal [8], [9].

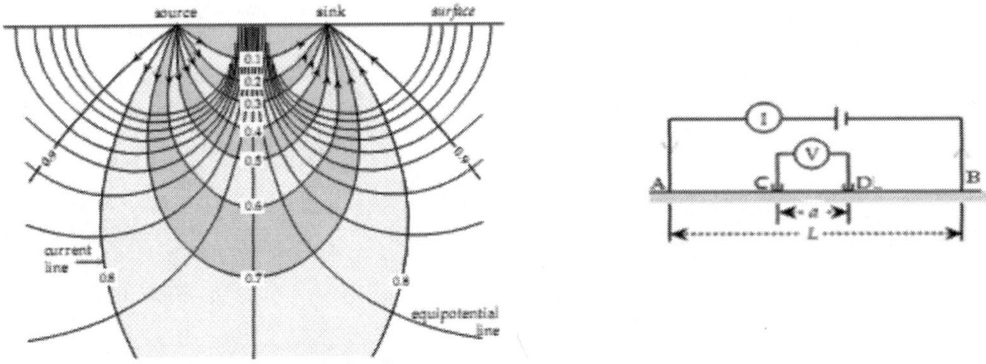

Figure 2. Distribution of current and equipotential line in geoelectric method (left). Schlumberger electrode configuration (right).

$$\rho_a = \frac{\pi}{4}\frac{V}{I}\frac{\left(L^2 - a^2\right)}{a} \tag{1}$$

The Schlumberger array configuration consists of four electrodes. The outer two electrodes are current electrodes as source, and the inner two electrodes are the potential electrodes as receiver. The potential electrodes are installed on the center of the electrode array with a certain separation, typically less than one fourth or one fifth of the spacing between the current electrodes. The current electrodes distances are increased to a greater separation during the survey while the potential electrodes remain in the same position until observed potential becomes too small to measure. In this research, we used Schlumberger electrode configuration with interval current electrodes maximum is 700 meter. Figure 2 shows current–potential distribution in Schlumberger configuration and equation 1 is formula to calculate the resistivity value [10].

3. Result and Discussions

The data processing method uses curve matching method. Principally, this method compare or match between field measurement data and master curve data in order to obtain an errata between field data and model data as small as possible. By using this method, it will be known resistivity values and the depth of each layered rocks.

The following is shown the results of processing data which have been plotted in double log graph. Y axis is resistivity and X axis is distance of current electrodes.

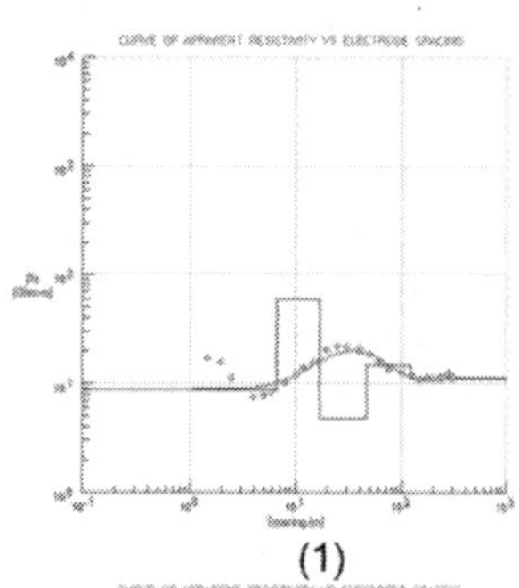
(1)

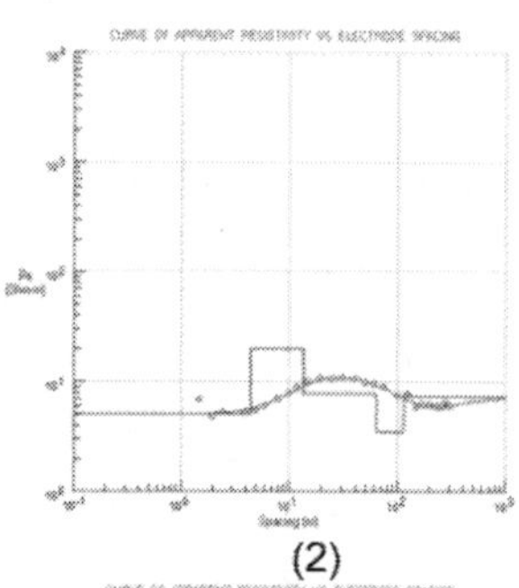
(2)

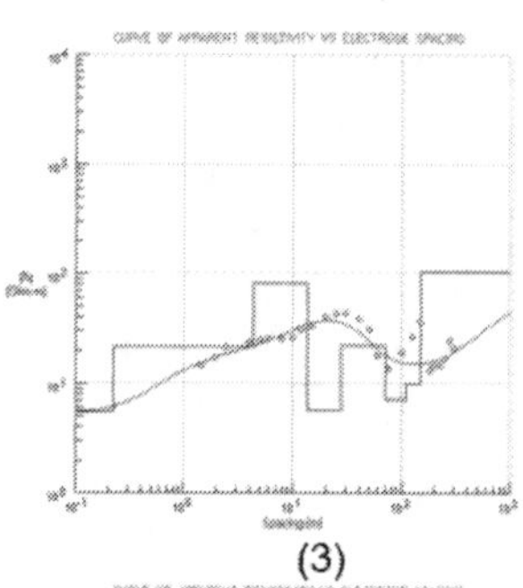
(3)

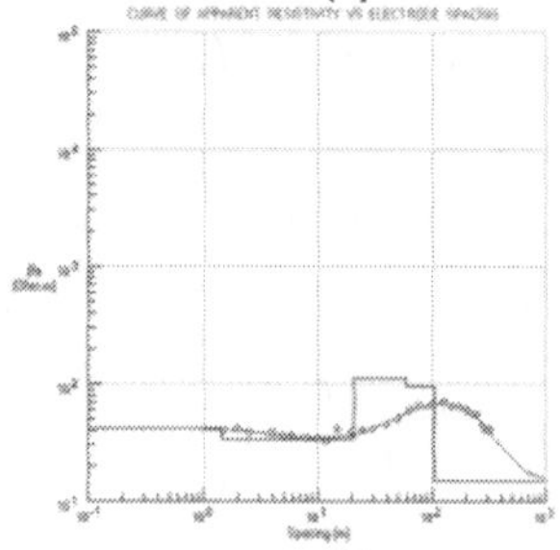
(4)

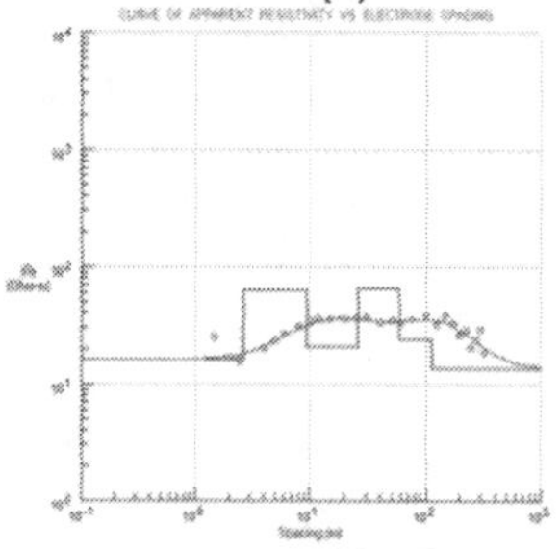
(5)

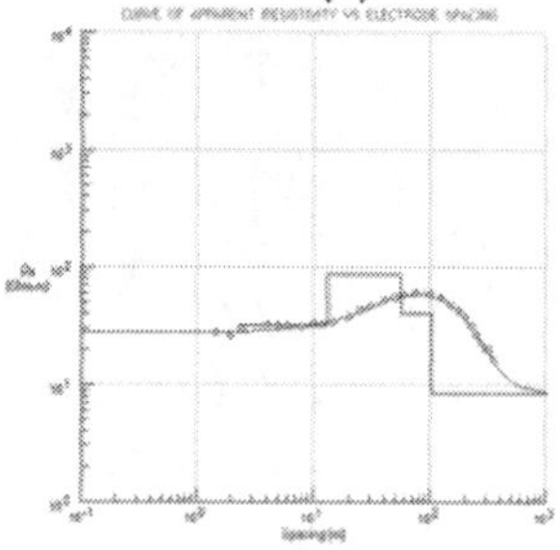
(6)

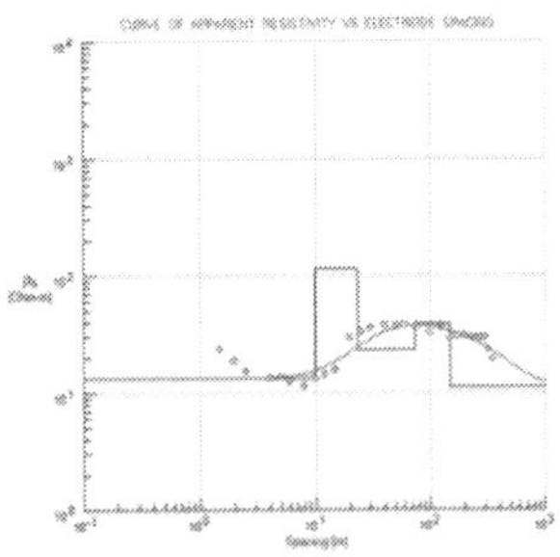
(7)

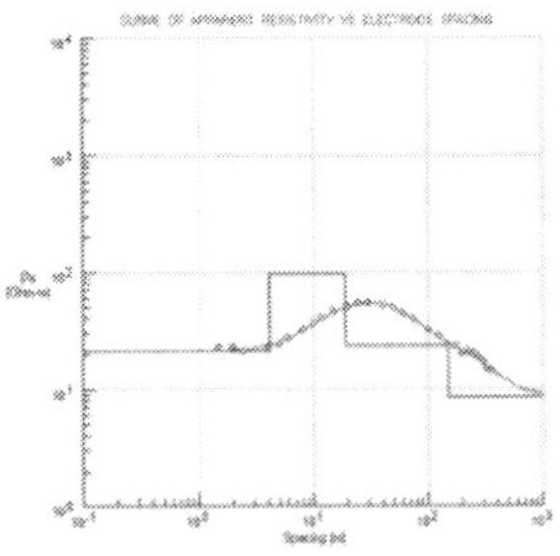
(8)

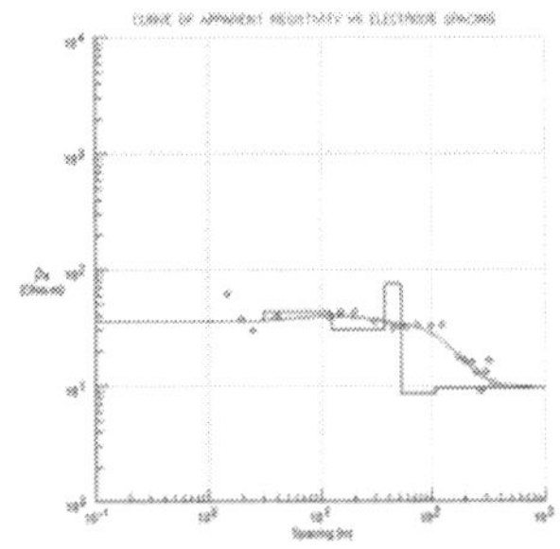
(9)

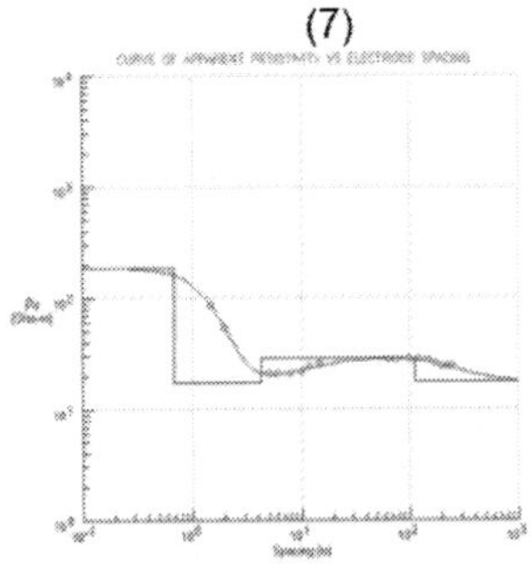
(10)

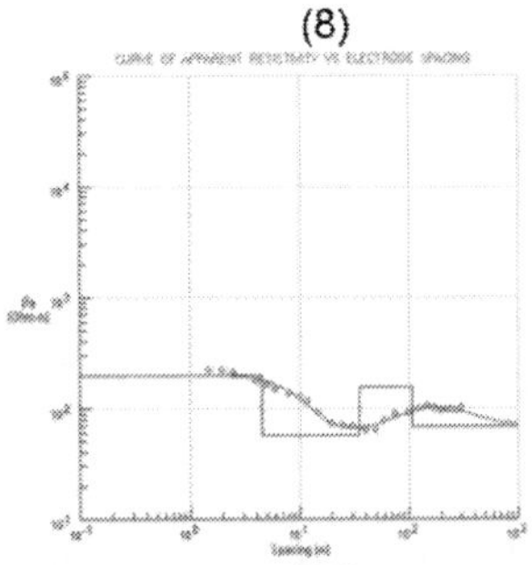
(11)

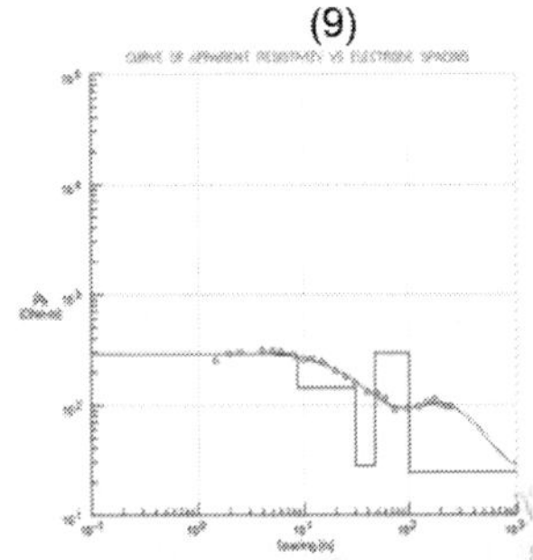
(12)

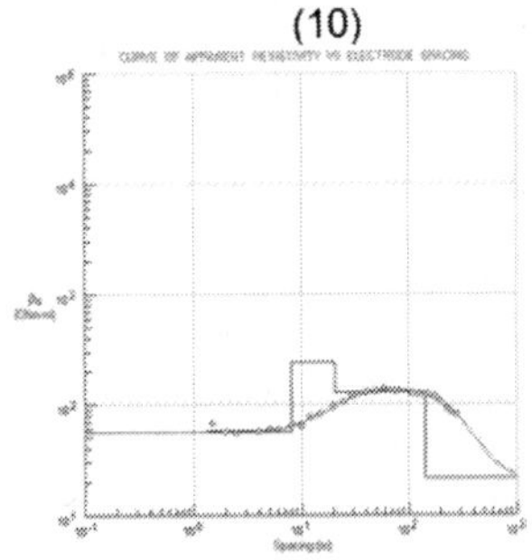
(13)

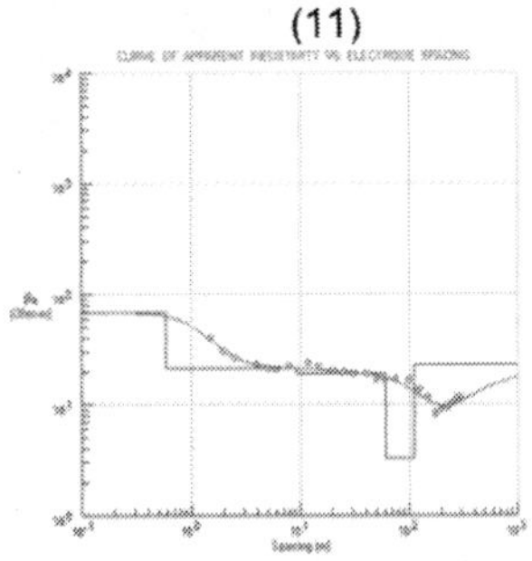
(14)

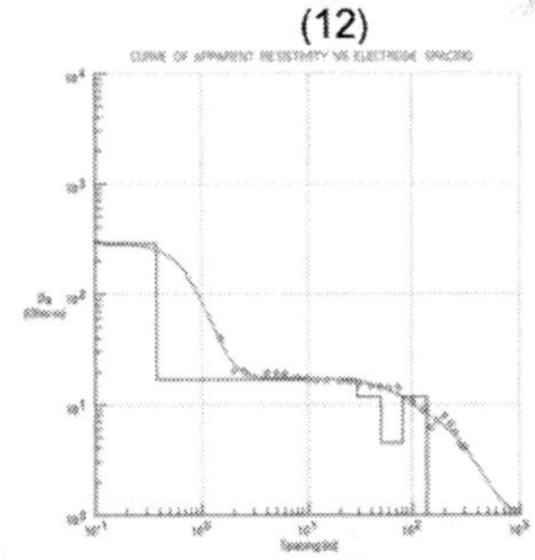
(15)

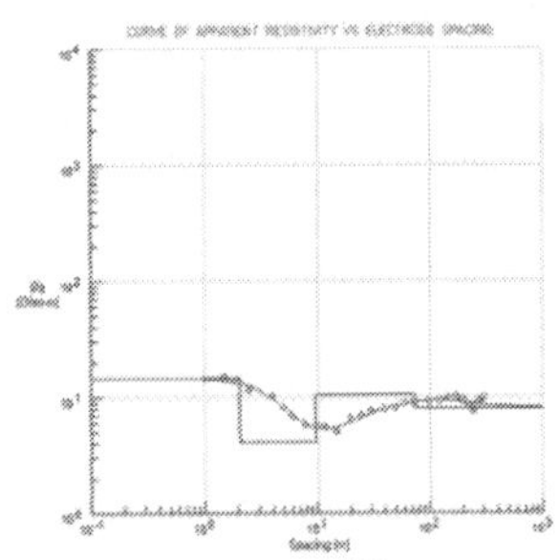
(16)

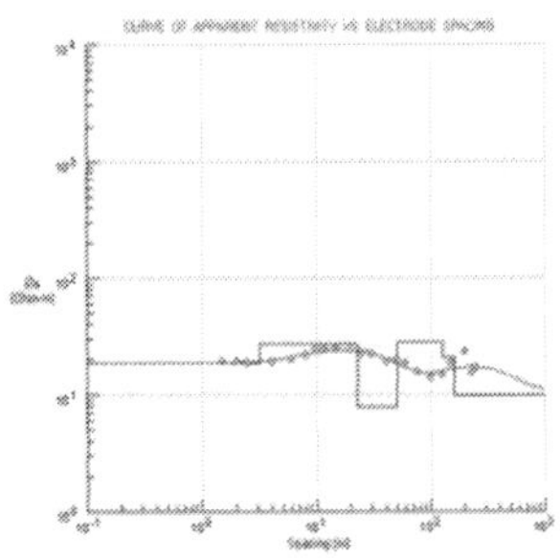
(17)

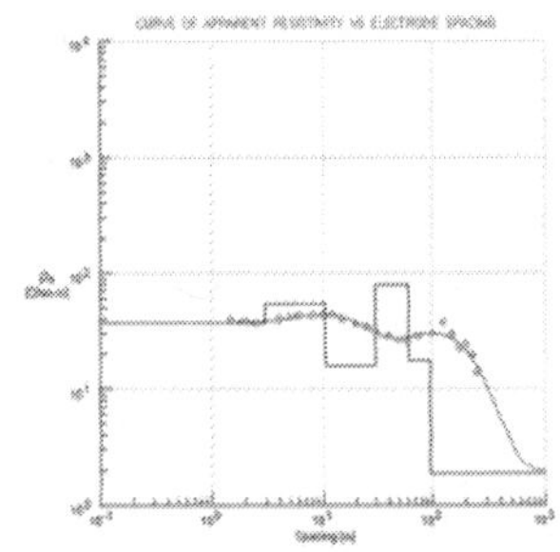
(18)

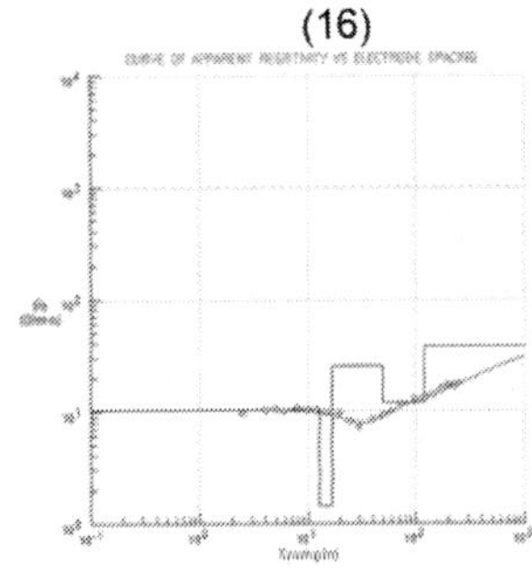
(19)

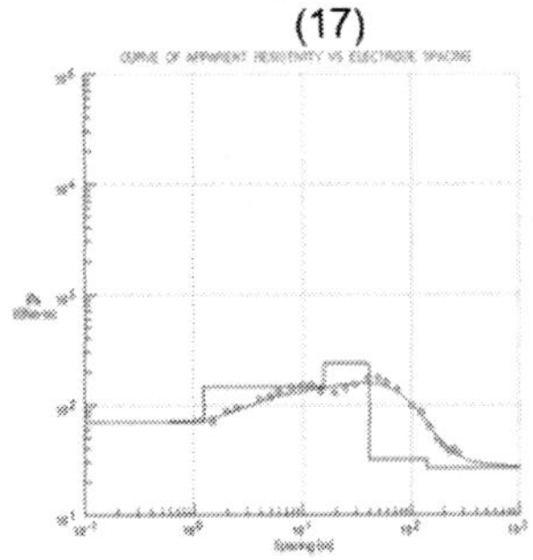
(20)

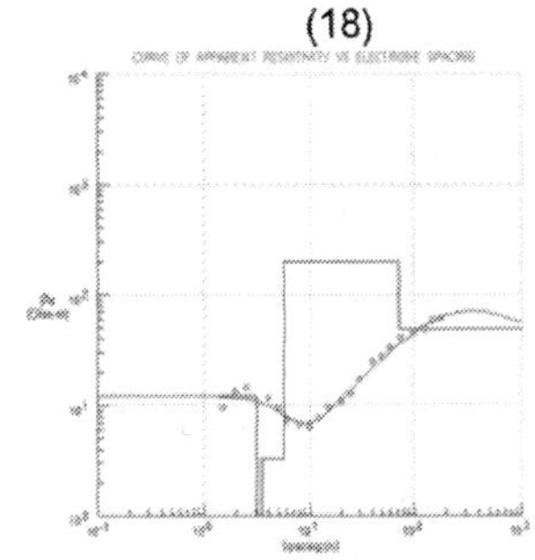
(21)

Journal of Physics: Conference Series **776** (2016) 012120 doi:10.1088/1742-6596/776/1/012120

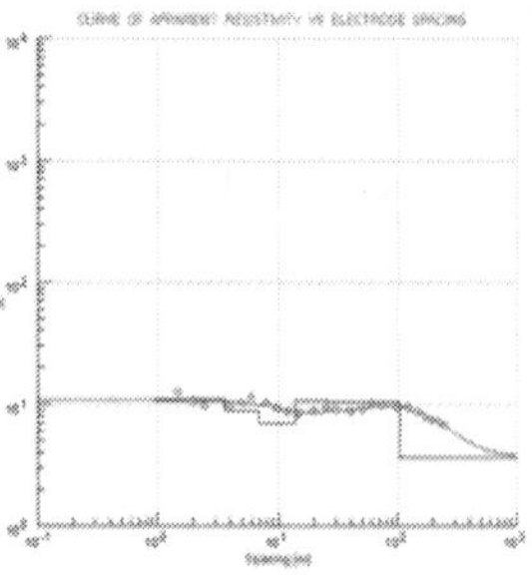

(22)

Figure 3. Plotting the measurement data on a double logarithmic graph. The yellow circle is the measurement data, the red curve is the result of interpolation and the blue line is a layered model and resistivity value

The processing data obtained that variations in the depth of the groundwater (aquifers) were measured at 22 survey sites are varying from 50 meters to 150 meters. This result shows that the groundwater table is not the same in Karanganyar regency. It may also infleunce by water resources and topography. So it can be said that ground water table follows contour of the existing topography. The model of underground layer and the determination of the depth of the aquifer is based on the value of low resistivity in a unit cross section of the model. Aquifer usually contains a lot of water so it has a low resistivity values. Moreover, it can be also analogized with some lithology, for instance clay, sand, or tuff or other kind of rock or mineral [10].

We found two potential aquifers in Munggur and Kedung Jeruk sites, shallow and deep aquifer. In Munggur the potential aquifer found in 60 meter and 111 meter for shallow and deep aquifers respectively. In Kedung Jeruk sites the deep of the aquifer are 50 meter and 135 meter for shallow and deep aquifer. This two sites are in adjacent area with distance around 5 kilometer, where could be from the same water source (see figure 1). The all results of the aquifer depth of 22 locations including its lithology are summarized in table 1 below.

Table 1. Survey location and depth of aquifers

No	Location	Coordinates		Elevation (m)	Depth Aquifer (m)	Lithology
		S	E			
1	Plesungan Gondangrejo	7 ° 31,725'	110 ° 51,131'	139	121	Sand clay
2	Jeruksawit Gondangrejo	7 ° 31,173'	110 ° 51,526'	114	63	Tuff
3	Kalijirak Mojogedang	7 ° 33,431'	110 ° 57,199'	149	70	Sand clay
4	Pojok Mojogedang	7 ° 33,780'	111 ° 00,695'	312	103	Tuff
5	Ngadiluwih Matesih	7 ° 37,994'	110 ° 59,950'	276	58	Grainy tuff
6	Blorong Jumantono	7° 38,786'	110 ° 59,602'	274	101	Tuff
7	Kebak Jumantono	7 ° 40,847'	111 ° 00,169'	323	150	Sand clay
8	Tawangsari Kerjo	7 ° 30,316'	111 ° 04,559'	245	149	Middle tuff
9	Kedawung Jumapolo	7 ° 41,600'	111 ° 02,386'	496	108	Middle tuff
10	Jatimulyo Jatipuro	7 ° 43,927'	110 ° 59,220'	299	112	Sand
11	Wonokeling Jatiyoso	7 ° 43,248'	111 ° 06,936'	963	106	Sand
12	Jatiyoso Jatiyoso	7 ° 44,263'	110 ° 51,526'	645	103	Sand
13	Tlobo Jatiyoso	7 ° 42,995'	111 ° 04,616'	631	143	Sand
14	Munggur Mojogedang	7 ° 31,790'	110 ° 59,473'	236	60 (111)	Sand

15	Kedung Jeruk Mojogedang	7 ° 31,060'	110 ° 58,380'	174	50 (135)	Sand clay
16	Kaliwuluh Kebakkramat	7 ° 30,258'	110 ° 55,976'	96	72	Sand clay
17	Tunggulrejo Jumantono	7 ° 39,450'	111 ° 02,150'	414	128	Sand
18	Jatiharjo Jatipuro	7 ° 43,995'	111 ° 01,333'	393	61	Sand
19	Lemahbang Jumapolo	7 ° 43,090'	110 ° 58,526'	282	50	Sand clay
20	Plosorejo Kerjo	7 ° 33,206'	111 ° 05,315'	510	139	Sand
21	Pendem Mojogedang	7 ° 33,273'	111 ° 01,783'	355	72	Sand
22	Waru Kebakkramat	7 ° 30,758'	110 ° 54,091'	88	140	Tuff

References

[1] Bemmelen RWV 1949 *Government Printing Office, The Hague* 732
[2] Sampurno and Samudro H, 1997 "*Geological Map of The Ponorogo Quadrangle, Jawa.*" Geological Research and Development Centre, Bandung, Indonesia
[3] Hartono U, 1994 "*The Petrology and Geochemistry of the Wilis and Lawu Volcanoes, East Java, Indonesia*," Ph.D Thesis. Univ. Tasmania
[4] BPS, "*Karanganyar Regency in Figures*," p. 345 pp, 2016.
[5] Kayode JS, Adelusi AO, Nawawi MNM, Bawallah M, and Olowolafe TS 2016 *J. African Earth Sci.* **119**, pp. 289–302
[6] Atwia MG, Abu-Heleika MM, and El-Horiny MM 2013 *J. African Earth Sci.* **80** 8–20
[7] Bouderbala A and Remini B 2014 *Acta Geophys.* **62**(6) 1352–1372
[8] Karaman A 2013 *Turkish J. Earth Sci.* **22**(4) 664–670,.
[9] Sircar A, Shah M,Sahajpal S, Vaidya D, Dhale S, and Chaudhary A 2015 *Geotherm. Energy*, **3**(1) p. 22,.
[10] Lowrie W 2007 *Fundamentals of Geophysics*, 2nd edition. Cambridge University Press

Identification of aquifer potential by using resistivity method: A case study in Kedawung and Sambirejo district, Sragen, Central Java, Indonesia

Darsono[*], B. Legowo, S. Koesuma
Geophysics Laboratory, Deparment of Physics, Universitas Sebelas Maret, Indonesia

[*]E-mail : darsono.uns.fisika@gmail.com

Abstract: It has been done geophysics survey by using resistivity method with Schlumberger configuration in Sambirejo and Kedawung subdistrict, Sragen regency, Indonesia. This research aims to identification of aquifer potential in those area. Totally there are 22 site surveys where 11 sites located in Sambirejo and 11 sites located in Kedawung subdistrict. Data collection was performed by using Resistvitymeter OYO McOHM-EL with length of current electrode from 1,5 meter up to 350 meter. Data processing was done by using IP2win software, while cross section was processed using Rockwork software.
The result shows that in Sambirejo subdistrict, the aquifer layer consist of clayey sand and sand, While in Kedawung subdistrict, The aquifer layers consist of clayey sand, sand, gravel sand, gravel and breccias. Identification of groundwater potential zones as good and very good category was show that the sounding point was have a huge of groundwater source for in Sambirejo subdistrict is TS2, TS5, TS7, TS8, and TS9 , while in Kedawung subdistrict is TS13, TS14, T15, TS16,TS17,TS18, TS19, TS20, TS21 dan TS22, respectively.

1. Introduction

Water is a vital requirement for living things, it is very importance for plants, animals and particularly human . Population growth and development in various fields will increase the needs for water. Groundwater is essential for drinking water, household, industrial, irrigation, mining, urban and more. At world level, groundwater was used 50% for drinking water, 20% for irrigation water, 40% for the needs of industry [3]. Water scarcity is problem for human and other living things.
Ground water is water that contained in a layer of soil or rock beneath the subsurface. It is water that moves in the soil contained in the space between grains of soil that seeped into the ground and joined to form soil layers called aquifers. An aquifer is a body of saturated rock through which water can easily move [12]. Aquifers must be permeable, porous and saturated. There are some common rock which is as a good aquifer, for instance sandstone, conglomerate, fractured limestone, unconsolidated sand, gravel and fractured volcanic rocks [1].
Sragen is a regency in Central Java province. The capital is located in Sragen, about 30 km eastern of Surakarta with coordinate $7^0 15' - 7^0 30'$ S and $110^0 45' - 111^0 10'$ E. The total area of Sragen regency is

about 941.55 km^2 and is divided into 20 sub-districts. Within the past decade, every dry season, several villages in Kedawung and Sambirejo district, was always experiencing a shortage of water. The drought that hit in this area not only on agricultural land, but also springs or resident wells. Therefore, the search for ground water resources have been become an important issue to villages was experiencing drought. This research was conducted to identified aquifer potential in Kedawung and Sambirejo district.

2. Method

Electrical Resistivity method is one geophysics method which is used in the investigation of groundwater. Some recent studies about detection ground water using Vertical Electrical Sounding (VES) method [4,5,6,7,8,9]. The VES is a geoelectrical method for measuring vertical change of electrical resistivity. The VES gives detail information on the vertical succession of different conducting zones or formations and their individual thickness and true resistivity below a given point on the earth surface [2].
The principle of VES method is injected electric current into the earth through two electrode currents. The potential difference that occurs is measured through two electrodes potential. The results of measurement of current and potential difference for each particular electrode spacing, can be determined variations of resistivity value below the measuring point (the point of sounding). The following equation is used to calculate the resistivity value [2] :

$$\rho_{semu} = K_{SC}\frac{\Delta V}{I} \quad (1)$$

$$K_{sc} = \pi\frac{(a^2 - b^2)}{2b} \quad (2)$$

ρ_{semu} is apperent resistivity; K_{SC} is geometry factor; ΔV is potential different; I is electrics current; a is distance between center point to current electrode; b is distance center point to potential electrode.

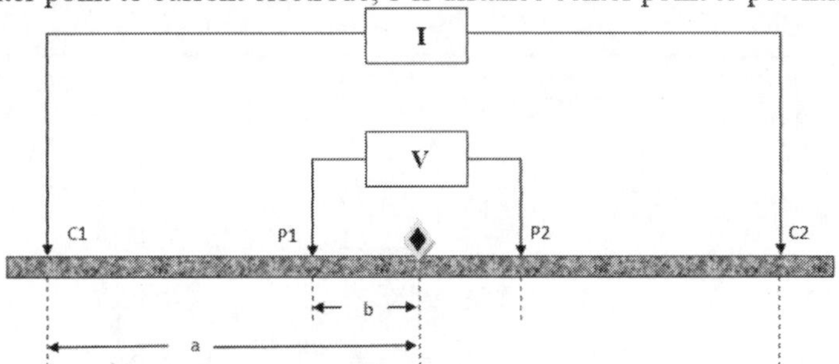

Figure 1. Schlumberger array

C1, C2 and P1, P2 is current electrodes and potential electrodes, recpectively.

3. Geology

Figure 2 shows regional geology map of Sragen regency where consist of six rock formations i.e. Kalibeng formation, Kabuh formation, Notopuro formation, Young Volcanic Deposits of Lawu and Alluvium deposits. The study area is located in Young Volcanic Deposits of Lawu, where composed by volcanic sandstone, volcanic silt-claystone, breccias and lava [11]. The survey location is located at Kedawung

Journal of Physics: Conference Series **776** (2016) 012121 doi:10.1088/1742-6596/776/1/012121

Subdistrict and Sambirejo Subdistrict, Sragen regency, Central Java, Indonesian. it is lies at $7^0 27' - 7^0 32'$ S and $111^0 00' - 111^0 06'$ E.

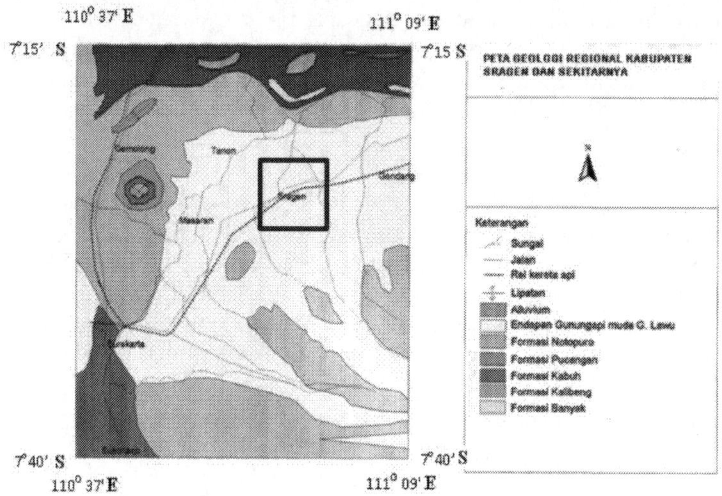

Figure 2. Regional geology map [11]

Data aquisition was done by using Resistvitymeter OYO McOHM-el model 2119C with length of maximum current electrode spacing (AB) is 700 meter. Amount of data as much as 22 points that is 11 sounding points in Sambirejo district and 11 sounding points in Kedawung district. The data obtained from the field surveys were processed to obtain the apparent resistivity.

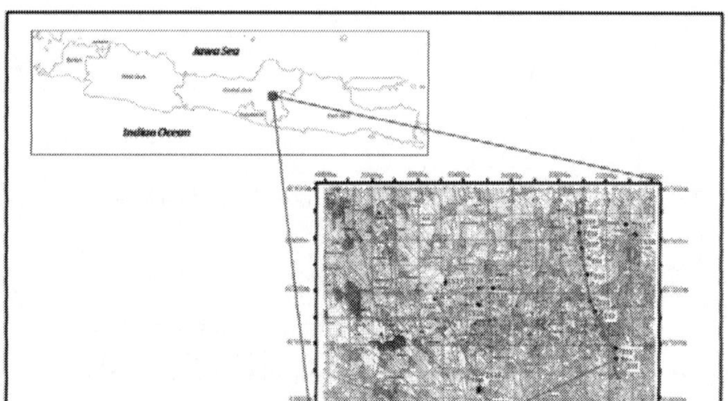

Figure 3 . Studi area. VES locations (black dotted), solid line with letter is cross section line. Inserted map is Java island.

4. Result and Discussions

Processing data is done by using partial curve matching techniques with IPi2win software which the outputs are resistivity values, thickness and depth of layer. Further step is interpretation of result by considering the geology of the study area and list of resistivity of rock, where the research areas located in the young volcanic deposits of Lawu mountain. These area are composed by sand stone, clay rocks and breccias. Lithology of the aquifer layer is detected in the form of clayey sand, sand, gravel sand, gravel, and breccias, while lithology of aquiclude layer is detected in the form of clay, sandy clay, and lava.

Table 1 shows the depth and thickness of aquifers in Sambirejo district. Aquifer layer in the Sambirejo district were found as much as 1 to 4 layers with variation in thickness. The aquifer layer consist of clayey sand and sand, clayey sand layer have thickness batween 9.67 metre to 55.5 metre, while sand layers have thickness batween 17.5 metre to 49.2 metre. Table 2, shows the depth and thickness of aquifers in Kedawung Subdistrict. Aquifer layer in the Kedawung Subdistrict were found as much as 1 to 4 layers with differentiation of thickness. The aquifer layers consist of clayey sand, sand, gravel sand, gravel and breccias. The thickness of clayey sand ranging from 22.3 to 50.4 meter, and sand layer have thickness range of 13.4 to 67.58 meter, gravel sand layer have thickness batween of 28.1 to 44.76 meter, gravel layer have thickness of 6.9 to 49.3 metre, and breccias layer have thickness of 85.9 metre.

Table 1. Position, thickness, depth and groundwater potential zone in Sambirejo district

VES	Location	Depth (m)	Thickness (m)	Lithology	Groundwater potential
TS1	Geblak Musuk	32.2 - 55.3	23.1	Sand	Moderate
		55.3 – 107	39.6	Clayey Sand	
		124-147	24,1	Clayey Sand	
TS2	Sidoharjo Musuk	55,5 – 73,1	17,5	Sand	Very Good
		73,1 – 114	40,7	Clayey Sand	
		114 -154	40,2	Sand	
TS3	Gempol	30,5 – 58,4	27,7	Sand	Moderate
		58,4 – 71,9	13,5	Clayey Sand	
		71,9 – 81,6	9,67	Clayey Sand	
TS4	Gempol	16,8 – 41	24,2	Sand	Moderate
		104 – 128	24,3	Clayey Sand	
TS5	Gempol	34,5 – 74,1	39,6	Sand	Good
		74,1 – 114	39,9	Clayey Sand	
TS6	Blimbing	23,1 – 78,4	55,5	Clayey Sand	Moderate
TS7	Dagangan Blimbing	32,77 – 58,16	25,39	Sand	Very Good
		113,37 – 168,3	54,96	Sand	
TS8	Sidorejo Blimbing	33,74 – 81,54	47,8	Sand	Good
TS9	Sidorejo Blimbing	23.95 – 58,75	34.80	Sand	Very Good
		101.63 – 148,29	46,66	Sand	
TS10	Jatiarum Kliro	71,54 – 110,25	38,71	Sand	Moderate
TS11	Dawung Kliro	36,09 – 43,66	7,47	Sand	Poor
		43,66 – 58,41	14, 77	Sand	

Journal of Physics: Conference Series **776** (2016) 012121 doi:10.1088/1742-6596/776/1/012121

Table 2. Position, thickness, depth and groundwater potential in Kedawung district

VES	Location	Depth (m)	Thickness (m)	Lithology	Groundwater Potential
TS12	Punthuk Mojodadi	43 – 56,4	13,4	Sand	Moderate
		145 – 170	25	Sand	
TS13	Mojodoyong Mojodadi	34,15 – 86,52	52,37	Sand	Very Good
		86,52 – 112,3	25,78	Sand	
		170,9 – 209,3	38,4	Sand	
TS14	Gempol Mojodadi	23,8 – 66,7	42,9	Gravel sand	Good
TS15	Gondang Mojodadi	43,7 – 101	57,3	Sand	Good
TS16	Wungurejo Mojodadi	50,74 – 118,3	67,58	Sand	Very Good
TS17	Wungurejo Mojodadi	34,3 – 63,4	29,1	Gravel	Good
		63,4 – 70,3	6,9	Gravel	
		70,3 – 92,7	22,3	Clayey sand	
		92,7 – 143	50,4	Clayey sand	
TS18	Ngabean Kedawung	46,93 – 98,9	51,97	Sand	Good
TS19	Sukorame Kedawung	20,8 – 52	31,2	Gravel sand	Very good
		82,6 – 168	85,9	Breccia	
TS20	Mojokerto Mojokerto	68,9 – 126	56,6	Sand	Good
TS21	Candirejo Wonokerso	62 – 95,8	33,7	Sand	Very good
		95,8 – 145	49,3	Gravel	
TS22	Nglaban wonokerso	73,61 – 118,4	44,76	Gravel sand	Very Good
		118,4 – 152,9	34,5	Gravel sand	
		152,9 – 181,1	28,21	Gravel sand	

We made a cross section from several VES points, i.e. A-B section, C-D section, C-F section and E-A section. The result of each section is interpreted from result of VES. Based on these cross section, it can provide delineation of the position, thickness of the aquifer layer and layer which is not an aquifer (aquiclude).

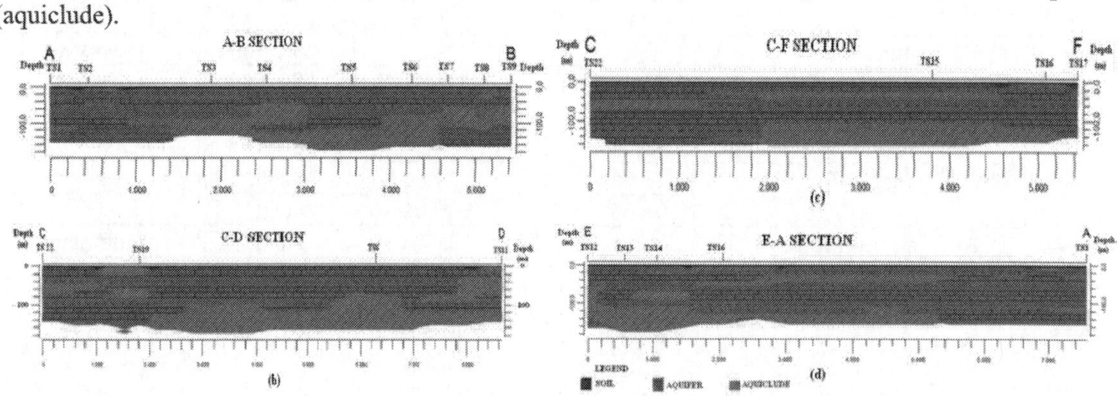

Figure 4. Cross section

5. Conclusions

The result show that in Sambirejo district, the aquifer layer consist of clayey sand and sand, clayey sand layer have thickness batween 9.67 metre to 55.5 metre; while sand layers have thickness batween 17.5 metre to 49.2 metre. In Kedawung district, the aquifer layers consist of clayey sand, sand, gravel sand, gravel and breccias. The thickness of clayey sand is ranging from 22.3 to 50.4 metre, sand layer have thickness range of 13.4 to 67.58 metre, gravel sand layer have thickness batween of 28.1 to 44.76 metre, gravel layer have thickness of 6.9 to 49.3 metre, and breccias layer have thickness of 85.9 metre. The entire study area can be classified as very good, good, moderate and poor for groundwater potential zones. Identification of ground water potential zone as good and very good category was shown in Sambirejo district are TS2, TS5, TS7, TS8, and TS9 , while in Kedawung district are TS13, TS14, T15, TS16, TS17, TS18, TS19, TS20, TS21 and TS22.

6. Acknowledgement

We are grateful to LPPM Universitas Sebelas Maret for providing financial assistance through the MRG grant in this work.

7. References

[1] Freeze RA, Cherry JA, Groundwater, Hempstead H, Prentice Hall International, 1979.
[2] Telford WM, Geldart LP, Sheriff RE, Keys DA, *Applied Geophysics.* Cambridge University Press, London. 1976.
[3] UNESCO, Groundwater resources of the World and their use . In: Zekster IS, Everett LG (eds), IHP-VI Series on Groundwater, 6. 2004.
[4] Srinivasan K, Poongothai S, Chidambaram S, 2013 *European Scientific Journal* **9**(17) ISSN: 1857 – 7881 (Print)e-ISSN 1857- 7431
[5] Anomohanran O, 2011 *International Journal of the Physical Sciences* **6**(33) pp. 7651-7656
[6] Keleko TDA, Tadjou JM, Kamguia J, Tabod TC, Feumoe ANS, Kenfack JV, 2013 *Journal of Water Resource and Protection* **5** (6)
[7] Obianwu VI, Atan OE, Okiwelu AA 2015 *Applied Physics Research* **7**(2) Canadian Center of Science and Education
[8] Ahilan J and Kumar GRS, 2011 *Archives of Applied Science Research* **3**(2) 414-421
[9] Nejad HT, Mumipour M, Kaboli R and Najib OA, 2011 *Journal of Applied Sciences*, **11**3765-3774
[10] Sajeena, S, Hakkim AVM and Kurien EK, 2014 *International Journal of Engineering Inventions*, **3**(6): 17-21
[11] Suharyadi 2005, *Laporan Penyelidikan Geolistrik Untuk Eksplorasi Air Bawah Tanah Di Daerah Kedawung dan Gondang Kab. Sragen Propinsi Jawa Tengah*, Laboratorium Geologi Tata Lingkungan Jur. T.Geologi Fak. Teknik Universitas Gadjah Mada, Yogyakarta
[12] Todd DK, *Groundwater hydrology (third Edition):* John Wiley and Sons, New York, 535 p. 1980.

Developing a playground as catchment area in effort to maintaining groundwater in Jaten village of Karanganyar district of Central Java, Indonesia

Budi Legowo[1], Darsono[1], Daru Wahyuningsih[2]
1. Physics Department, Universitas Sebelas Maret,
2. Physics Education Department, Universitas Sebelas Maret ,

E-mail: pakbeel@staff.uns.ac.id

Abstract. Changes in land use for housing indirectly disturb the hydrology balance of the area. Groundwater conservation efforts can be done by keeping the function the catchment area. One of the housing developer's obligations is providing open spaces (the playground) to play or activity of the residents. Playground in *Bumi Graha Indah* Housing, Jaten village, Karanganyar district, Central Java, Indonesia has a fundamental issue, that is, in the rainy season the water is difficult to seep due landfill process are not well planned. It causes the playground become in muddy conditions with tall grass, so that reduces the function as a playground and or activity the residents. In the dry season, the soil dry of landfill caused dust scattering and disrupt the activities of people around the playground. Lack of water resources lead watering process for solving the problem of dust during the dry season was considered ineffective. Structuring drainage combined with modified recharge wells can be used to catch water runoff housing. This modification of water catchment areas can make playground dry quickly after rain so the activities of people are not bothered when utilizing the open space provided. Surface runoff water absorbed in open aquifer so that the hydrological balance always be maintained. Adequacy groundwater in the area playground can be used to sprinkler dust and backup needs clean water residents by creating wells and reservoir stocks.

1.Introduction

Land used changes for housing led to a reduction of open space, it will indirectly disrupt the hydrological balance of a region (Sulistio, 2012). Groundwater conservation efforts can be done by keeping the function of the catchment area (Iriani, Gunawan, & Besperi, 2013). One of the obligations of housing developers in Karanganyar is to provide an open space (playground) to play or move residents (LG Karanganyar, 2012). Changes of land use of Bumi Graha Indah Housing in Karangayar, one of which is useful as a catchment area have caused a decrease in groundwater level static (mas) and the poor quality of groundwater. Efforts should be made to harness science and technology for society as a way to preserve local groundwater and hydrological balance.

As the rainy season playground conditions perpetually flooded and lead to side effects were less pleasant as high grass and muddy. In addition to the function of absorbing less maskimal while doing backfill, drainage system one also further aggravate the situation. It took 2-3 days in hot weather for a pending dry playground and reusable. Different conditions during the dry season, the soil dry landfill causes dust so easily carried by the wind, causing air pollution and the environment. Watering needs to be done to anticipate this. Limitations of water resources led to the watering process can not be done with the maximum.

Journal of Physics: Conference Series **776** (2016) 012122 doi:10.1088/1742-6596/776/1/012122

Groundwater conservation efforts can be maintained by maximizing the playground provided by the developer to place the residents move as a catchment area by applying a modified recharge wells (Iriani, Gunawan, & Besperi, 2013). By doing some adaption recharge wells accordance with the requirements as a playground, expected water runoff can quickly seep into the soil and into groundwater resources stocks so as to ensure regional hydrological balance.

2. Methodology

The technology will be transferred to partners as catchment wells is modified by utilizing the playground as catchment area as groundwater conservation efforts in the region XI RW Bumi Graha Indah Housing Jaten Karanganyar, as shown schematically in Figure 1. The cross sectional modified catchment wells is expected to hold water playground surface runoff so that it becomes easy to dry and can be used effectively by citizens. Air infiltration can maintain the balance hydrology, so as to ensure the availability of groundwater areas devotion. Adequacy of groundwater can be used also as a backup water supply when the main water source (PDAM Karanganyar) experience problems.

In general, infiltration wells hold water from households (gutters), the water will be wasted through runoff drain water when the volume exceeds the capacity of the wells (Iriani, Gunawan, & Besperi, 2013). Further modifications is done by creating a roving cast concrete that serves as a funnel to direct the water to the water catchment wells.

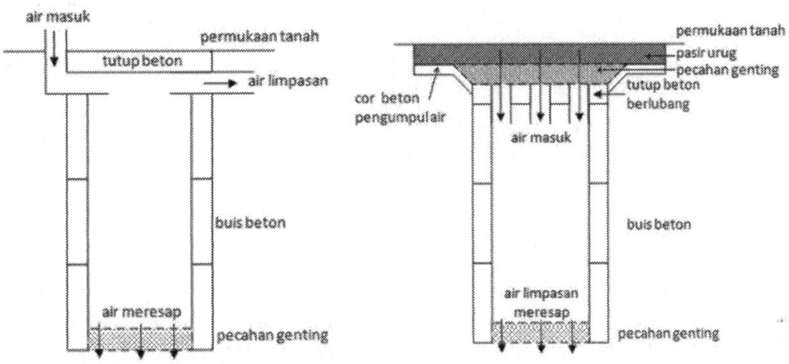

(a) (b)

Figure 1. Modification playground as well catchment area
(a) regular recharge wells, (b) modifications absorption wells

Water will be inserted through the hollow concrete lid. The stored water is rainwater on playground and other runoff toward the field. Not given channel runoff with anticipation as well has a maximum volume, the surface of stagnant water will seep along the water catchment bottom of the well.

In the case of RT 05 RW XI Bumi Graha Indah Housing Jaten Karanganyar modifications were also made on the surface of the playground, that is by doing dredging old land and hoarding re-use landfill plated. The first layer is broken tiles that make water easily seep into the well. The second layer is sand landfill that serves to provide comfort and safety on the use of playground time is used by citizens to conduct the activity. In anticipation of the excess volume of water runoff will be made of modified catchment wells installed sebagnyak three parallel, in the sense of absorption wells are not interconnected but but has a function to support each other as a funnel connected using concrete cast around as shown in Figure 2a. The existence of control wells in Figure 2b has two functions. The first function is to perform palyground watering during the dry season and as a groundwater source backup when the main water source of taps Karanganyar having problems. The second function is as a static groundwater level control primarily in RW XI. Planned regular basis every 6 months will be measured static groundwater table directly (in situ) and or using identification techniques geophysical, geoelectric resistivity involving Student of Geophysical Group of Geophysics Laboratory in Mathematics and Natural Science Faculty of Sebelas Maret University. In particular, the fulfillment of infrastructure control wells will be funded nongovernmental through the filing of the village development program of Jaten District of Karanganyar

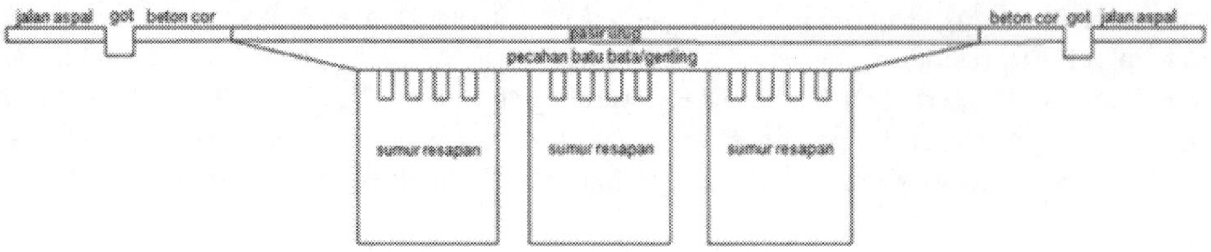

(a)

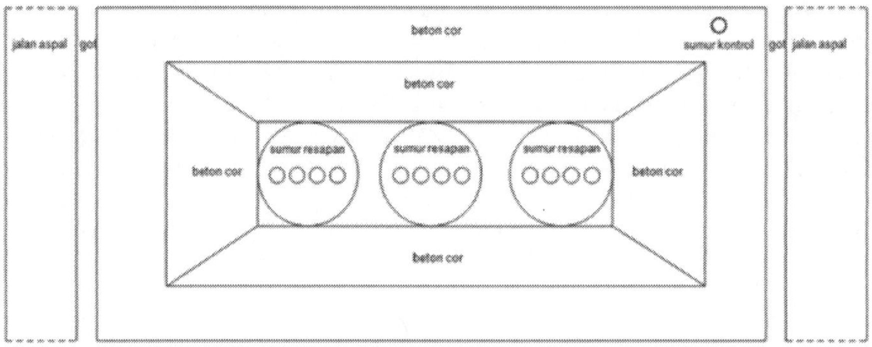

(b)

Figure 2. Modification playground as catchment area; (a) Side, (b) Top

3. Results and Discussion

Utilization step of science and technology for communities in RT 05 RW XI Bumi Graha Indah Housing in Jaten Karanganyar is expected to result in the form of an indication of geographic information ketinggaan static groundwater level pengabidan area before and after the engineering playground as a catchment area on a regular basis every 6 months. Catchnment engineered as a playground area and periodic analysis of the geographic indication would be delivered in a national seminar related to the main topic of geophysics groundwater conservation.

4. References

[1] Iriani, K., Gunawan, A., & Besperi. (2013). Perencanaan Sumur Resapan Air Hujan untuk Konservasi Air Tanah di Daerah Permukiman (Studi Kasus di Perumahan RT. II, III dan IV Perumnas Lingkar Bengkulu). *Jurnal Inersia* , 9-21.

[2] Sulistio, H. (2012). Konservasi Daerah Tangkapan Air Cipta Graha. *Media Sains* , 36-41.

[3] Pemda Karanganyar. (2012). *Peraturan Daerah Kabupaten Karanganyar No. 9 tahun 2012 tentang Pengelolaan Ruang Terbuka Hijau.* Karanganyar: Pemda Karanganyar.

[4] Legowo, B., Darsono, & Wahyuningsih, D. (2015). *Pelestarian Airtanah Wilayah RW XI Desa Jaten Kecamatan Jaten Kabupaten Karanganyar melalui Pengembangan Playground sebagai Cacthment Area.* LPPM UNS: Laporan P2M IbM.

Risk analysis of landslide disaster in Ponorogo, East Java, Indonesia

S Koesuma[1,3*], A P Saido[2], Y Fukuda[3]

[1]Department of Physics. Universitas Sebelas Maret. Surakarta Indonesia
[2]Department of Civil Engineering. Universitas Sebelas Maret. Surakarta Indonesia.
[3]Graduate School of Sciences, Kyoto University, Kyoto, Japan
[*]E-mail : sorja@uns.ac.id

Abstract Ponorogo is one of regency in South-West of East Java Province, Indonesia, where located in subduction zone between Eurasia and Australia plate tectonics. It has a lot of mountain area which is disaster-prone area for landslide. We have collected landslide data in 305 villages in Ponorogo and make it to be Hazards Index. Then we also calculate Vulnerability Index, Economic Loss index, Environmental Damage Index and Capacity Index. The risk analysis map is composed of three components H (Hazards), V (Vulnerability, Economic Loss index, Environmental Damage Index) and C (Capacity Index). The method is based on regulations of National Disaster Management Authority (BNPB) number 02/2012 and number 03/2012. It has three classes of risk index, i.e. Low, Medium and High. Ponorogo city has a medium landslide risk index.

1. Introduction

Ponorogo is a regency in East Java Province, which is located between 111 ° 17 ' – 111° 52' E and 7°49'N – 8°20'S with an area of 1371.78 km^2. Ponorogo located in southern part of Java island which is a subduction zones between the Australian plate and the Eurasian plate. The impact of subduction process is the uplifting and folding of geological strata forming such varied geomorphology sloping plains, hills and mountains. Such geological conditions makes Ponorogo have the potential threat of landslide disaster.

There is a study concerning landslide in Ponorogo, Basofi et. al [1] has a classification of landslide by using Natural Breaks and Head/tail Breaks methods, and then compare the both of result to get the landslide hazard map. The hazard map result is slightly different with our results, especially in the southern part of Ponorogo regency.

Some researchers has clasified the landslide into several types. Varnes [2] has clasified the landslide into 29 types which are divided in five major components, i.e. (i) relating to movement, (ii) relating to material, (iii) relating to geologic, geopmorphic, geographic, or climate setting, (iv) relating to size or geometry and (v) relating to age or state of activity. Hungr [3] accomplished the landslide types into 32 types. It is not added some new type of landslide but because of its ambiguities in different point of view, especially where a landslide type is graded from other, without a clear-cut boundary. The USGS special report [4] shows a good ilustration of some landslide types and its caused. This paper will discuss the risk analysis of landslide in Ponorogo regency based on hazard, vulnerability and capacity parameters. Landslide hazard map in Ponorogo regency is shown in Figure 1.

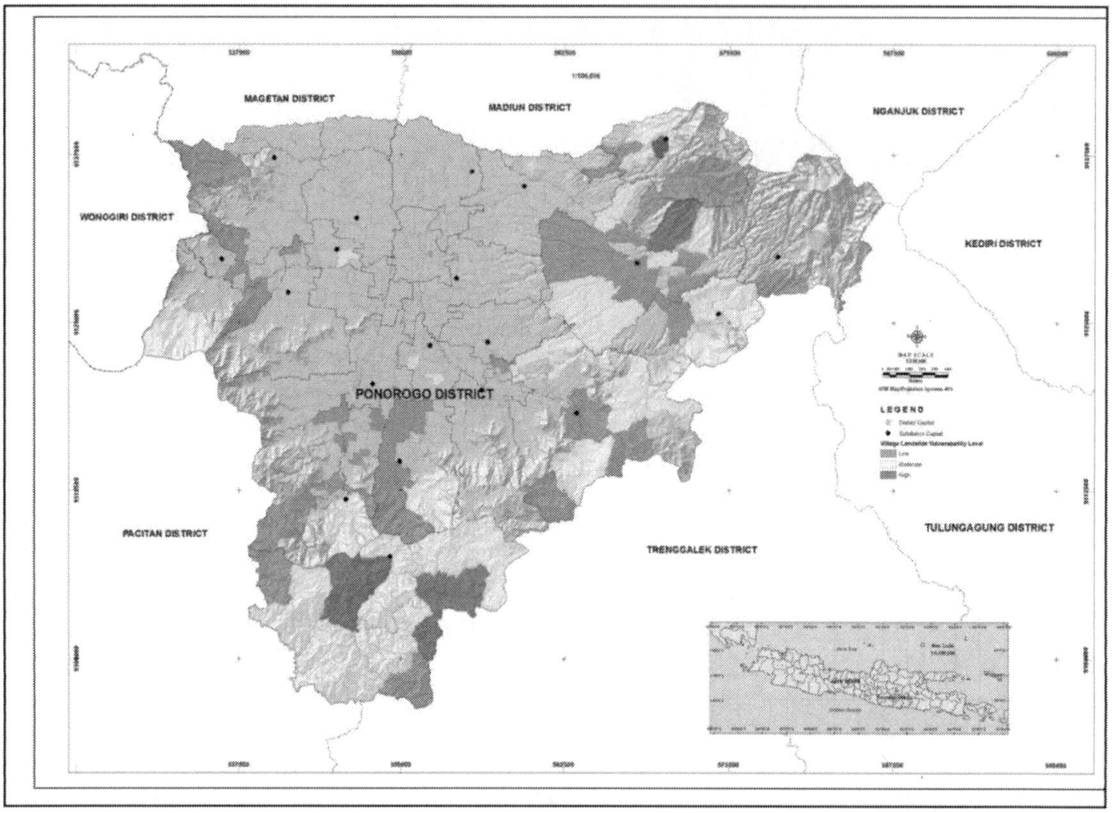

Figure 1. Landslide Hazard Map. Inserted map in right bottom is Java island.

2. Method

The risk analysis map is composed of three components H (Hazards), V (Vulnerability, Economic Loss index, Environmental Damage Index) and C (Capacity Index). We used the method proposed in the regulations of National Disaster Management Authority (BNPB) number 02/2012 [5] and number 03/2012 [6], in formula form:

$$R = H\,x\,\frac{V}{C} \quad (1)$$

The H hazards map shown in figure 1 is based on event of landslide which happened in an area. The map is composed in village-based data, so we collected the landslide data in village in the last 3 years. Most of landslide event are in southern part and western part of Ponorogo. These area are hills, mountains and foot of mountain. The V component is composed by three parameters which are vulnerability of residents againts landslide disaster (figure 2.a), economic loss caused by disaster and environmental damage caused by landslide (figure 2.b.). Those three parameters are summed and then mapped together with other parameters.

Stakeholder Capacity Index in disaster management is a hint of a capability of being owned by the region in the implementation of disaster management or a regional capability and communities to take action to reduce the threat and potential catastrophic losses as a result of a structured, planned and integrated. We make a questionnaire to calculate Stakeholder Capacity Index based on BNPB regulation number 3/2012 [6], and we found that the capacity index is 55 point (Medium index), and then mapped as shown in figure 2.c.

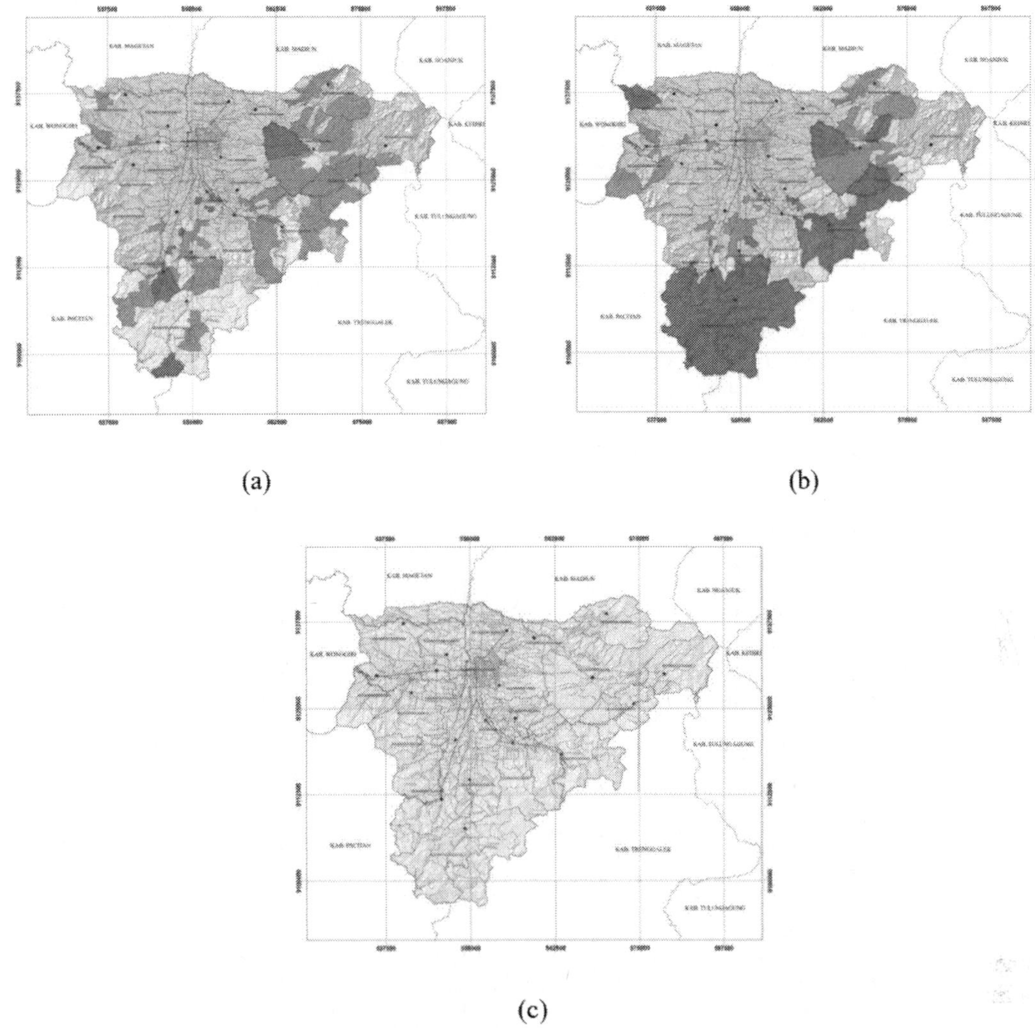

Figure 2. Map of (a) Vulnerability, (b) Economic Loss – Enviromental Damage, (c) Capacity

3. Discussions

Kohler et.al [7] stated that Risk analysis consists of hazard analysis and vulnerability analysis, together with analysis of protec tive capabilities. Some authors treat the analysis of the protective capabilities of the local population (coping strategies) as part of vulnerability analysis, others as a third component of risk analysis, others see it as an additional chapter, and as such a component of risk assessment and not risk analysis. Otherwise, Bell and Glade [8] did a quantitative risk analysis for landslide, and divided many parameters which is affect the risk analysis. For instance in consequence analysis, there are element risk, vulnerability, probability of spatial impact, probability of temporal impact and probability of seasonal impact. Other researcher Guzzeti et.al [9] propose a new tool to inventory landslide map, which is consist three aspects (i) analysis of surface morphology, chiefly exploiting very-high resolution digital elevation models (DEMs), (ii) interpretation and analysis of satellite images, including panchromatic, multispectral and synthetic aperture radar (SAR) images, and (iii) the use of new tools to facilitate field mapping.

The landslide risk map that have been made is based on village area, actually the data in each village shows some variations. There are ome villages with low, medium and high value respectively, but after an average of all value, the risk index is become medium value. There are 79 village which have landslide hazard, we resume the component in the following table. Based on the calculation on risk

Journal of Physics: Conference Series **776** (2016) 012123 doi:10.1088/1742-6596/776/1/012123

analysis in formula (1), we found that Ponorogo city have a medium landslide index with average point 9.5 [5].

Table 1. Village with landslide disaster risk

No	Sub	Village	Hazards	Vulnerability	Economic	Environment
1	Ngrayun	Desa Baosan	10	0,8	1,8	0,3
2		Desa Wonodadi	5	0,8	1,8	0,3
3		Desa Sendang	10	0,8	1,8	0,3
4		Desa Mrayan	10	0,8	1,8	0,3
5		Desa Binade	5	0,4	1,8	0,3
6		Desa Baosan Lor	15	0,8	1,8	0,3
7		Desa Ngrayun	10	0,8	1,8	0,3
8		Desa Temon	10	0,4	1,8	0,3
9		Desa Selur	15	0,8	1,8	0,3
10		Desa Cepoko	10	0,8	1,8	0,3
11		Desa Gedangan	10	1,2	1,8	0,3
12	Slahung	Desa Tugurejo	5	0,8	1,8	0,3
13		Desa Senepo	5	0,4	1,8	0,3
14		Desa Slahung	10	1,2	1,8	0,3
15		Desa Broto	5	0,4	1,8	0,2
16		Desa Plancungan	5	0,8	1,8	0,1
17		Desa Galak	5	1,2	1,8	0,1
18		Desa Crabak	5	1,2	1,8	0,1
19		Desa Mojopitu	5	0,8	1,8	0,1
20		Desa Gundik	5	0,8	1,8	0,1
21		Desa Janti	5	0,4	1,8	0,1
22	Bungkal	Desa Pelem	5	0,4	1,8	0,3
23		Desa Koripan	5	0,8	1,8	0,3
24		Desa Bekare	5	0,8	1,8	0,2
25		Desa Nambak	5	0,4	1,8	0,1
26		Desa Kalisat	5	0,4	1,8	0,1
27		Desa Munggu	10	0,4	1,8	0,3
28		Desa Ketonggo	5	0,4	1,8	0,1
29		Desa Kunti	5	0,4	1,8	0,1
30		Desa Bancar	10	0,4	1,8	0,1
31		Desa Padas	5	0,4	1,8	0,3
32		Desa Bungu	5	0,8	1,8	0,1
33		Desa Bedi Wetan	5	0,4	1,8	0,1
34		Desa Bedi Kulon	5	0,8	1,8	0,1
35	Sambit	Desa Kemuning	5	0,4	1,8	0,3
36	Sawoo	Desa Tumpuk	5	0,4	1,8	0,2
37		Desa Pangkal	5	0,8	1,8	0,3
38		Desa Tumpak	5	0,4	1,8	0,3
39		Desa Sriti	10	0,4	1,8	0,2
40		Desa Temon	10	0,4	1,8	0,3
41		Desa Sawoo	5	0,8	1,8	0,3
42		Desa Prayungan	10	0,4	1,8	0,3
43		Desa Tugurejo	10	0,4	1,8	0,2
44		Desa Grogol	10	0,4	1,8	0,3
45		Desa Ketro	10	0,4	1,8	0,1
46		Desa Bondrang	10	0,4	1,8	0,2
47		Desa Ngindeng	10	0,4	1,8	0,2
48	Sokoo	Desa Ngadirojo	10	0,4	1,8	0,3
49		Desa Klepu	10	0,4	1,8	0,3
50		Desa Suru	5	0,4	1,8	0,3
51		Desa Sooko	10	0,4	1,8	0,1
52		Desa Jurug	10	0,4	1,8	0,2
53	Pudak	Desa Banjarjo	5	0,4	1,8	0,2
54	Pulung	Desa	10	0,4	1,8	0,1
55		Desa Bedrug	5	0,4	1,8	0,1
56		Desa Singgahan	5	0,4	1,8	0,1

57		Desa Patik	10	0,4	1,8	0,1
58		Desa Pulung	5	0,8	1,8	0,1
59		Desa Pulung	5	0,4	1,8	0,1
60		Desa Sidoharjo	5	1,2	1,8	0,3
61		Desa Plunturan	5	0,4	1,8	0,1
62		Desa Pomahan	10	0,4	1,8	0,1
63		Desa Serag	5	0,4	1,8	0,2
64		Desa Munggung	15	0,4	1,8	0,3
65		Desa Bekiring	10	0,4	1,8	0,2
66	Jetis	Desa	10	1,2	1,8	0,1
67		Desa Jetis	10	1,2	1,8	0,1
68	Kauman	Desa Ploso Jenar	10	0,8	1,8	0,1
69	Jambon	Desa Srandil	5	0,4	1,8	0,1
70	Badegan	Desa Dayakan	10	0,8	1,8	0,1
71		Desa Karang Joho	5	0,8	1,8	0,1
72		Desa Tanjung	5	0,8	1,8	0,1
73	Sampung	Desa Gelang	5	0,4	1,8	0,1
74		Desa Pohijo	5	0,8	1,8	0,3
75		Desa Jenangan	5	0,4	1,8	0,3
76	Ngebel	Desa Ngrogung	5	0,4	1,8	0,1
77		Desa Sahang	10	0,4	1,8	0,1
78		Desa Talun	5	0,4	1,8	0,1
79		Desa Ngebel	10	0,4	1,8	0,1
Total			**586**	**46.8**	**142.2**	**15.4**
Average			**7.4**	**0.59**	**1.8**	**0.19**
Risk Index			**7.4 x (0.59 +1.8+0.19)/2**			**9.5**
					Medium	

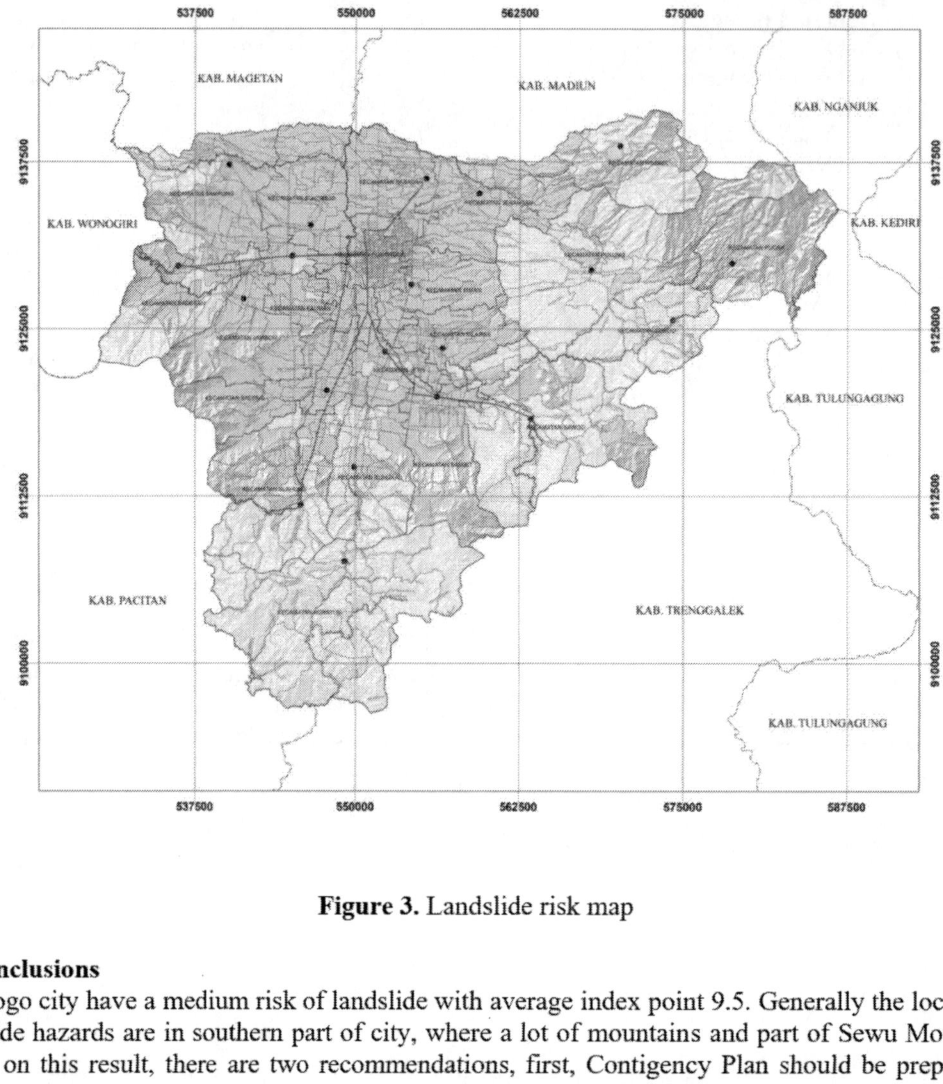

Figure 3. Landslide risk map

4. Conclusions

Ponorogo city have a medium risk of landslide with average index point 9.5. Generally the location of landslide hazards are in southern part of city, where a lot of mountains and part of Sewu Mountains. Based on this result, there are two recommendations, first, Contigency Plan should be prepared by stakeholder, business sector and communities. Secondly, Disaster Mitigation Plan on village level should be made in order to more focus on landslide disaster prevention.

5. References

[1] Basofi A, Fariza A, Ahsan AS, and Kamal IM, 2015 *Int. Conf. Sci. Inf. Technol.*, pp. 337–342
[2] Varnes DJ 1978, *Transp. Res. Board Spec. Rep.*, no. 176, pp. 11 – 33
[3] Hungr O, Leroueil S, and Picarelli L, 2014 *Landslides*, vol. 11, no. 2, pp. 167–194
[4] USGS 2004 *Highw. Res. Board Spec. Rep.*, no. July, pp. 1–4
[5] BNPB, 2012 *"Regulation of the Head of National Disaster Management Agency of the Republic of Indonesia Number 2 Year 2012 on General Guidance on Study of Disaster Risk,"* p. 62
[6] BNPB, 2012 *"Regulation of the Head of National Disaster Management Agency of the Republic of Indonesia Number 3 Year 2012 on Regional Capacity Assessment Guide in Disaster Relief,"* p. 42
[7] Kohler A, Jülich S, and Bloemertz L, 2004 *Risk Manag.*, pp. 18–67 [8] Bell R and Glade T, *Nat. Hazards Earth Syst. Sci.*, vol. 4, no. 1, pp. 117–131
[9] Guzzetti F, Mondini AC, Cardinali M, Fiorucci F, Santangelo M, and Chang KT, 2012 *Earth-Science Rev.*, **112**, no. 1–2, pp. 42–66